Human–Computer Interaction:

Ergonomics and User Interfaces

Volume 1

HUMAN FACTORS AND ERGONOMICS

Gavriel Salvendy, Series Editor

Bullinger, H.-J., and Ziegler, J. (Eds.) : Human–Computer Interaction: Ergonomics and User Interfaces, *Volume 1 of the Proceedings of the 8th International Conference on Human–Computer Interaction*

Bullinger, H.-J., and Ziegler, J. (Eds.) : Human–Computer Interaction: Communication, Cooperation, and Application Design, *Volume 2 of the Proceedings of the 8th International Conference on Human–Computer Interaction*

Human–Computer Interaction:

Ergonomics and User Interfaces

Proceedings of HCI International '99 (the 8th International Conference on Human–Computer Interaction), Munich, Germany, August 22–26, 1999
Volume 1

Edited by

Hans-Jörg Bullinger
Fraunhofer IAO

Jürgen Ziegler
Fraunhofer IAO

1999 LAWRENCE ERLBAUM ASSOCIATES, PUBLISHERS
Mahwah, New Jersey London

Copyright © 1999 by Lawrence Erlbaum Associates, Inc.
All rights reserved. No part of this book may be reproduced in any form, by photostat, microfilm, retrieval system, or any other means, without prior written permission of the publisher.

Lawrence Erlbaum Associates, Inc., Publishers
10 Industrial Avenue
Mahwah, NJ 07430

Library of Congress Cataloging-in-Publication Data

Human–Computer Interaction : Ergonomics and user interfaces, Volume 1 / edited by Hans-Jörg Bullinger and Jürgen Ziegler.
p. cm.
Includes bibliographical references and index.
ISBN 0-8058-3391-9 (cloth : V.1 : alk. paper),
ISBN 0-8058-3392-7 (cloth : V.2 : alk. paper).

1999

Books published by Lawrence Erlbaum Associates are printed on acid-free paper, and their bindings are chosen for strength and durability.

Printed in Great Britain

Content

Part 2
User Interfaces and Interactions

Part 3
User Interface Design

Part 5
Design and Development

Preface

HCI International '99 is the 8th conference in a series of major events related to research, development, and practice in human-computer interaction. HCI International '99 is held in Munich, Germany, August 22-27, 1999. The conference provides a major international forum for exchanging and discussing latest results and developments related to analyzing, designing, developing, applying, and evaluating information and communication technologies for work, leisure and personal growth. Under the general theme of Creating New Relationships, new links and synergies are explored between information technologies and their users, among people working together, and in the context of the rapidly evolving global information society. Four major areas are in the focus of the program:

- Ergonomics and Health Aspects of Working with Computers
- Human-Computer Interfaces
- Organizational Aspects of Information and Communication Technologies
- Communication and Interaction in Information Networks.

The two-volume conference proceedings contain 540 papers from 30 different countries. The papers included in the program were selected on the basis of reviews done by the program committee members and by leading researchers organizing special sessions centered around topics of particular interest. The first volume "Human-Computer Interaction – Ergonomics and User Interfaces" begins with a set of papers related to health and ergonomics issues of the physical work environment. The major part of Vol. 1 is dedicated to contributions addressing novel concepts, techniques, and studies for user interfaces and interaction techniques. A number of papers in this volume present research in usability engineering as well as methods and processes for the design and development of user interfaces.

Volume 2 "Human-Computer Interaction – Information, Communication, and Cooperation" puts the main emphasis on the informative and communicative aspects of computer use. A larger number of contributions is concerned with computer-supported cooperation using a wide variety of different techniques. In keeping with the increased focus of HCI International '99 on internet issues and aspects of the global information society, many papers in this volume are centered around information and communication networks and their implications for work, learning, and every-day activities. Due to the growing number and diversity of groups utilizing modern information technologies, issues of accessibility and design for all are becoming more and more pertinent. A range of papers in this volume address these issues and provide latest research and development results.

The organisation of HCI International '99 would not have been possible without the diligent efforts of a large number of volunteers and the staff members in the conference office. We thank particularly the international program committee, which is listed after this preface. Special thanks go to the organizers of the Symposium on Human Interface (Japan), which is held jointly with the conference, as well as the cooperating scientific societies supporting the conference. We are also most grateful to our corporate conference sponsors: DaimlerChrysler Services (debis), Fraunhofer-Gesellschaft, Siemens, Oracle, and SAP.

Invaluable support for the submission and review process and for preparing the proceedings on hand has been provided by the conference office team Anke Ahrend, Marion Kostendt and Katja Roglin, and the organizing committee members Rolf Ilg and Paulus Vossen. Anne Duffy and Art Lizza from Lawrence Erlbaum and Assiocates were most supportive throughout the preparation of the publication.

Our final thanks goes to all HCI International '99 authors who have enabled us to set up a high-quality, diverse, and exciting conference program.

Hans-Jörg Bullinger (General Conference Chair)
Jürgen Ziegler (Program Chair)

August 1999

Acknowledgements

General Conference Chair
Hans-Jörg Bullinger
Fraunhofer IAO and University of Stuttgart, IAT, Germany

Program Chair
Jürgen Ziegler, Fraunhofer IAO, Germany

Organizational Chair
Rolf Ilg, University of Stuttgart, IAT, Germany

Organizing Committee
Anke Ahrend
Marion Kostendt
Maren Rehpenning
Katja Roglin
Paul Hubert Vossen
(all Fraunhofer IAO, Germany)

Student Volunteer Chair
Michael Koch, TU Munich, Germany

Conference Advisory Board
Peter Gorny, Germany
Satoshi Goto, Japan
Bengt Knave, Sweden
Richard Koubek, USA
Holger Luczak, Germany
Takao Ohkubo, Japan
Ralf Reichwald, Germany
Susumu Saito, Japan
Gavriel Salvendy, USA
Ben Shneiderman, USA
Constantin Skarpelis, Germany
Michael J. Smith, USA
Jean-Claude Sperandio, France
Hiroshi Tamura, Japan
Thomas J. Triggs, USA
Constantine Stephanidis, Greece

Program Committee
Arne Aaras, Norway
Julio Abascal, Spain
Beth Adelson, USA
Yuichiro Anzai, Japan
Udo Arend, Germany
Albert Arnold, Netherlands
Elena Averboukh, Germany
Sebastiano Bagnara, Italy
Evangelos Bekiaris, Greece
Massimo Bergamasco, Italy
Eric Bergman, USA
Nigel Bevan, UK
Martin Böcker, Germany
Kenneth R. Boff, USA
George Boggs, USA
Valentina M. Bondarovskaia, Ukraine
Gunilla Bradley, Sweden
John Carroll, USA
Frue Cheng, Taiwan
Yee-Yin Choong, USA
Gilbert Cockton, UK
Barbara G. F. Cohen, USA
Martha E. Crosby, USA
Ulrich Eisenecker, Germany
Klaus-Peter Fähnrich, Germany
Daniel Felix, Switzerland
Shin 'Ichi Fukuzumi, Japan
Kazuo Furuta, Japan
Ephraim P. Glinert, USA
Tamara A. Grinchenko, Russia
Vincent Grosjean, France
Seppo Haataja, Finland
Martin Helander, Sweden
Michitaka Hirose, Japan
Erik Hollnagel, Norway
Egon Hörbst, Germany
Pekka Huuhtanen, Finland
Sheue-Ling Hwang, Taiwan
Ilias Iakovidis, Belgium
Julie A. Jacko, USA

Bente Jensen, Denmark
Gerd Johannson, Sweden
Henry S. R. Kao, Hongkong
Waldemar Karwowski, USA
Christoph Kasten, Germany
Halimahtun Mohd Khalid, Malaysia
Peter Kern, Germany
Kinshuk, Germany
Christophe Kolski, Finland
Ilona Kopp, Germany
Danuta Koradecka, Poland
Heidi Kroemker, Germany
Helmut Krueger, Switzerland
Yasufumi Kume, Japan
Masaaki Kurosu, Japan
Thomas Laeubli, Switzerland
Ron Laughery, USA
Mark R. Lehto, USA
Kee Yong Lim, Singapore
Soo-Yee Lim, USA
Yu-Chin Lin, China
Holger Luczak, Germany
Joachim Machate, Germany
Christoph Maggioni, Germany
Jim Maida, USA
Ann Majchrzak, USA
Marilyn Mantei-Tremaine, USA
Aaron Marcus, USA
Hans Marmolin, Sweden
Ian McClelland, Netherlands
Michael McNeese, USA
Hiroyuki Miki, Japan
David L. Morrison, Australia
Masaki Nakagawa, Japan
Jakob Nielsen, USA
Joachim Niemeier, Germany
Alexander Nikov, Bulgaria
Shogo Nishida, Japan
Lucas Noldus, Netherlands
Celestine Ntuen, USA
Ahmet Fari Oezok, Turkey
Katsuhiko Ogawa, Japan
Leszek Pacholski, Poland
Barbara Paech, Germany
Thomas Plocher, Japan
Peter G. Polson, USA
Girish Prabhu, USA
Christian Rathke, Germany
Patrick Rau, USA
Matthias Rauterberg, Netherlands
Goonetilleke Ravindra, Hongkong
Fieny Reimann-Pijls, Belgium
Manfred Rentzsch, Germany
Dominique L. Scapin, France
Hans Schedl, Germany
Lawrence Schleifer, USA
Matthias Schneider-Hufschmidt, Germany
Stephen Scrivener, UK
Pentti K. Seppala, Finland
Richard So, China
Kay M. Stanney, USA
Christian Stary, Austria
Constantine Stephanidis, Greece
Alistair G. Sutcliffe, UK
Vappu Taipale, Finland
Michael Tauber, Germany
Gunnar Teege, Germany
David Thompson, USA
Manfred Tscheligi, Austria
Masato Ujigawa, Japan
Roelof van den Berg, Netherlands
Jean Vanderdonckt, Belgium
Matti Vartiainen, Finland
Seppo Vaeyrynen, Finland
Valery F. Venda, Canada
Kim Vicente, Canada
Paul Hubert Vossen, Germany
Tomio Watanabe, Japan
Harald Weber, Germany
Gunnella Westlander, Sweden
Nong Ye, USA
Wenli Zhu, USA
Bernhard Zimolong, Germany

PART 1

HEALTH AND ERGONONOMICS

Age and Implementation of new Information Technology in Banking Tasks

Huuhtanen P., Ristimäki T., Leino T.
Finnish Institute of Occupational Health, Topeliuksenkatu 41 a A
FIN-00250 Helsinki, Finland
Email: pekka.huuhtanen@occuphealth.fi

1 Introduction and Objectives

The financial service sector has undergone continuous changes in information technology. Electronic transactions and increased self-service for the customers, e.g. via internet, has altered the work organisation and division of tasks. Simultaneously, the labour force has been reduced dramatically in bank organisations in Finland. The aim of this study is to analyse the relationship between age and the evaluation of changes during an implementation process of new data systems.

A multilevel process model of change was adopted; it combines technological, organizational and psychological change processes (Huuhtanen 1997). It is hypothesized that the evaluations of change vary according to the phase of change and by the previous experiences of changes at work. In addition, employees in different age groups have different educational background and work experience. It is assumed that these factors often explain evaluations of technological changes more than does the chronological age as such.

2 Material and Method

The results of this paper are based on the first phase of a longitudinal study on technological changes in financial firms in 1997-99 in Finland. The study group consisted of customer service employees in two Call Centers of a nationwide bank in Finland in 1998. The first center was started in 1992, the other in 1998. The mean age of the subjects (n=98) was 43 years (sd 7 years). The response rate was 82%.

Survey data were collected on the impact of new data systems on work content and mental demands, on different aspects of the implementation process, training and user support, and on the usabilitity of the new systems. Furthermore, questions were asked about experienced stress, strain symptoms and ergonomic issues related to the physical work environment and computer work station. Changes in work characteristics during the previous six months were asked by giving the alternatives: (1) decreased a great deal, (2) decreased somewhat, (3) remained the same, (4) increased somewhat, or (5) increased a great deal.

Usability of the computer applications was measured by 26 usability items developed by McNeive and Ryan (McNeive and Ryan 1997). Age was analyzed by three age groups: under 35 years, 35-44 years, and over 45 years.

3 Results

90% of the subjects worked with computers more than four hours daily, the youngest group (under 35 years of age) somewhat less (83%) than the others. 52% of the oldest group but only 8% of the youngest group typically worked continuously for more than two hours at the terminal. The oldest group was, correspondingly, less satisfied with the amount of breaks at work. They also felt more often than the others that the impact of technological changes on their work was significant (89% vs 62% among the youngest group).

Evaluation of the feeling of being productive at work, the opportunities to use one's abilities, level of interesting work, and job appreciation as a result of new technology was most positive in the oldest age group. In contrast to these positive evaluations, the older workers felt more often than the others that the work pace, things to be remembered, rules to be taken into consideration, difficulty of tasks, and monitoring of their work had increased.

Regarding problems with system usability, the oldest group had experienced more problems than the youngest group with the following issues: working out how to use the system, understanding how the information on the screen relates to what the employee is doing, finding the information wanted, information difficult to read, too many colours, an inflexible help facility, losing track of where you are in the system, system response times are too quick, information does not stay on the screen long enough, working out how to correct errors, and having to spend too much time correcting errors.

In general, the results are in line with many previous studies as regards mainly the positive impact of new technology on the content of work. In the Call Centers under study, no major differences were found between the age groups regarding the evaluation of the implementation process. When evaluating the support and training given by the supervisors and data experts during the change, the oldest employee group was in general more satisfied than the youngest group. However, age did not correlate with job satisfaction or with experienced stress.

4 Conclusions

A follow-up survey with interviews will be conducted in 1999-2000. Based on it, more elaborate hypotheses regarding the relationship between age and new tools will be tested. Preliminary findings suggest that the risk of information overflow is high especially among older banking employees. More emphasis should be laid on the ergonomics of user interfaces and on the work organisation, e.g. flexibility of working hours.

Previous studies have revealed that time is an important factor when individual responses to stress factors and changes at work are evaluated. Important questions to be analyzed are, e.g., how permanent are the differences between the age groups regarding the usability evaluations, and do these evaluations change along with the increased competence in new tasks and software application. In addition, only a longitudinal study design offers possibilities to analyze how employees in different age categories cope with continuous changes under the increased productivity demands.

5 References

Huuhtanen, P.: Towards a multi-level model in longitudinal studies on computerization in offices. Int J Human-Comp Interaction 1997; 9 (4), 383-405.

McNeive, A. and Ryan, G.(1997): Joint AIB/IBOA case study on the introduction of phase 1 of the new branch banking system (NBB), Final Report, University College, Dublin.

Implementation of Information Technology in Call Centers

Ristimäki T., Leino T., Huuhtanen P.
Finnish Institute of Occupational Health

1 Introduction and Objectives

The implementation of information technology (IT) is a process in which organizational, social and technological aspects interact. It is widely recognized that successful implementation of information technology must take into account not only the technical aspects, but also the organizational and social factors involved in the implementation (Zauchner et al. 1997, Eason 1996). For example, it has been found that participation in the implementation process has positive effects on strain and job satisfaction (Zauchner et al. 1997). Other organizational and social factors are, for example, the practices of organizational planning, the management of change, leadership, interaction, communication, training and IT support.

In this paper the implementation of information technology is analyzed in two call centers of a bank, which provide a variety of banking services by telephone to customers. It is important to emphasize that call center work is extremely intensive, using the latest information and telecommunication technologies. In addition, the number of call centers is continuously rising. (Richardson and Marshall 1996).

This study aims to analyze some of the organizational and social factors during the implementation process of a new information system (IS) in call center work.

2 Material and Methods

This paper is part of a longitudinal study "Mastery of technological changes in financial firms in the information society". The study is being carried out in 1997 - 2000 in three units of a nationwide bank (n=169) and five units of a nationwide insurance company (n=97) in Finland. All units use and implement the latest IT. This paper's study group consists of the employees of two call centers, most of them customer service employees of the studied bank (n=119).

Qualitative as well as quantitative data were collected. The methods used were a questionnaire survey (response rate 82%) and semi-structured interviews (30 one-hour interviews with the employees, superiors and people involved in the implementation project). The questionnaire and interviews concentrated on the changes in working conditions and the implementation of new information systems during the past half year. The study included topics such as the description of work and the implementation process, comparison of the new and old information systems and work tasks, management style, organizational climate, participation in the change, communication, training and IT support. Questions were also asked about health, job satisfaction, experienced stress and strain symptoms.

3 Results

The interviewed customer service employees could be grouped into three categories depending on their attitude to the implementation of new IT, and the matters they emphasized: 1) Most of the interviewed employees emphasized technical difficulties. According to their experience, technical difficulties after the implementation constitute the biggest disadvantage in the implementation process. 2) The second largest group voiced their doubts about their ability to learn the new information system and about its usability. Contrary to the fear of possible technical difficulties, most doubts about the ability to learn the new system or its usability proved to be wrong after the implementation. 3) The third group consisted of a couple of the interviewed employees who expressed their enthusiasm about the new IT and its capacity to make working easier. These people were eager to implement the new IT. They didn't want to call attention to the technical difficulties, because in the long run new IT would make working conditions better.

3.1 Technical Difficulties and Psychological Well-being

All the interviewed employees agreed that during the past half year there had been lots of technical difficulties related to the implementation of the new IT. This had affected their work. The interviewed employees felt that they could deal with the negative feelings that the technical difficulties caused as such. But the way in which the technical difficulties affected the actual customer service made the situation burdensome. Continuously telling the customers that the computer applications didn't work, explaining why they didn't work, and facing dissatisfied customers was stressful. In addition, the described situations caused a role conflict to the customer service employees, as they had to suppress their own feelings in order to serve the customers in the way their employer expects. According to the interviews the technical difficulties encountered in the recently implemented information systems *tire, annoy, irritate, make one angry, provoke, overstrain, vex, make one loose one's temper, are stressful, troublesome, burdensome, unpleasant, and really embarrassing.* As shown by the qualitative data, the technical difficulties in IT affect the employees' psychological well-being.

Some of the interviewed employees pointed out that at least some of the technical difficulties were due to the fact that the IT is designed and tested by people who know too little about the end-user's work. They felt that more thoroughly planning, testing and piloting of the program could prevent this. The system designers' should acquaint themselves better with the end-user's every-day work in order to understand all aspects of it. It was also suggested that the co-operation between the end-users and system designers should be more efficient, i.e. the participation of the end-users in the implementation process could prevent many of the technical difficulties faced by the end-users today.

Both the qualitative and the quantitative data show that only a few of the call center workers were able to participate in the planning. In the questionnaire, 14% answered that they were able to participate in the planning of the implementation process, and 9% said that they were able to participate in the planning of a new IS. In addition, most of the people who were involved felt that their participation was insufficient. 10% of those who participated in the planning of the implementation process, and 7% of those who were involved in the planning of a new IS, felt that their participation was insufficient.

3.2 Communication, Training and IT Support

Other organizational or social factors examined were the sufficiency of communication, training and IT support. As regards communication, the employees felt that they had received enough information, especially about the bank's new strategy and vision. The problem was one of information overflow, rather than the lack of information. Many employees were unhappy about the quality of the communication, especially it's comprehensibility. In the questionnaire, only 39% agreed with the statement 'It has been easy to grasp the most important messages from the communication' and 53% agreed with the statement 'The communication has been easy to understand'.

Most employees felt that they had received a sufficient amount of training (69%). In the qualitative data, the ones who were unhappy with the amount of training pointed out that this led to a situation where the learning process of the new IS took place only after the implementation during customer serving situations.

Most employees (71%) felt that they had received a sufficient amount of support in the use of a new IS. IT support was used widely: the employees sought support from the computer, manual instructions, their co-workers, the unit's experts, the company's help desk, and their superiors. In the qualitative data especially the help from co-workers was experienced to improve psychological well-being and job satisfaction.

In the quantitative data, the correlations of communication, training and IT support with job satisfaction and psychological well-being were examined. Psychological well-being was measured with nine questions concerning, e.g. anxiety, concentration and self-confidence (Elo et al. 1992). Cronbach's alpha of the sum scale was 0.93. Also a sum scale was made of six measures of communication. Cronbach's alpha reliability coefficient of the sum scale was 0.88. Fluent communication correlated positively with job satisfaction ($p<0.001$) and psychological well-being ($p<0.001$). Similarly, the relationships between a sufficient amount of training and support in the use of the information system and job satisfaction were statistically significant ($p<0.01$). A sufficient amount of training correlated statistically significantly also with psychological well-being ($p <0.001$).

4 Conclusions

The results confirm the relationship between technical functioning and management of change, and psychological well-being in the implementation process of new IT. Technical difficulties in IT affect the employees' psychological well-being. From the end-users' point of view some of these technical difficulties could be prevented if the system designers were better acquainted with the end-user's work, and the co-operation between the end-users and system designers could be improved during the implementation process.

Similarly, efficient communication, a sufficient amount of training and support in the use of the information system correlated positively with job satisfaction and psychological well-being. It is therefore concluded that more attention should be paid to the organizational and social factors involved in the implementation process of a new information system.

5 References

Eason K. (1996). Implementation of information technology in working life. In Rantanen J. & co (Eds.): *Work in the Information Society. People and work.*

Research reports 8., (Proc. of the International Symposium on Work in the Information Society, Helsinki, Finland, May 20-22, 1996), pp. 71-78. Helsinki: Finnish Institute of Occupational Health.

Elo A-L., Leppänen A., Lindström K., Ropponen T. (1992). *OSQ. Occupational Stress Questionnaire: User's Instructions.* Helsinki: Institute of Occupational Health.

Zauchner S., Korunka C., Vitouch O. & Weiss A. (1997). The "Second Vienna Implementation Study:" I. Contextual Factors Modifying the Effects of Continuous Implementation of Information Technology. In Salvendy, G. Smith, M. & Koubek, R. (Eds.): *Design of Computing Systems: Cognitive Considerations (Vol.1), Proc. 7th Int. Conference on Human-Computer Interaction* (HCI International '97, San Francisco, USA, August 24-29, 1997), pp. 383-386. Amsterdam: Elsevier.

Richardson R. and Marshall J. N. (1996). The growth of telephone call centres in peripheral areas of Britain: evidence from Tyne and Wear. *Area*, 28 (3), 308-317.

Computer Usability Problems and Psychological Well-being

Leino T., Ristimäki T., Huuhtanen P.
Finnish Institute of Occupational Health

1 Introduction

In this paper, the key psychological question concerning new technology is the good usability of computer applications. The key question arises from the fact that, although we nowadays often use computers in our work, the work is always work with the mind or by the mind (Hollnagel 1997). This means that computer applications should be optimal from the user's point of view, so that the user can carry out the required tasks successfully and without difficulty. (Mitchell & al. 1997, Izso & Zijstra 1997).

Most of the empirical studies on psychological well-being at work have shown that control over one's work and social support improve psychological well-being (Karasek 1979). Previous studies have nevertheless shown that in work environments where information technology and computers are important, the good usability of computer applications can improve the worker's psychological well-being and job satisfaction (Harrison & al. 1994, Izso & Zijstra 1997).

In Call Center work the customer service employees use the latest information and telecommunication technology. They provide a variety of sales, marketing and information services remotely by telephone connected to computer applications (Richardson and Marshall 1996). There is no face-to-face contact with the customer. The employee is able to discuss, on the phone the best possibilities to solve the customer's problems. The action that the customer service employee then takes, depends on the employee's understanding of the customer's problem and possibilities to find a solution by using the application quickly for finding all the information needed.

2 Material and Methods

The study is part of a longitudinal study "Mastery of technological changes in financial firms in the information society", which is being carried out in 1997-2000 in two Call Centers of a nationwide bank in Finland. The study group consists of the employees in two Call Center's, most of them customer service employees (n=119). The basic principles of work organization, model of customer service, job contents and computer applications were the same in both units. The first center was started in 1992, the other in 1998. Since 1992 computer applications have been used as sources of information, but in 1998 still newer computer applications were taken into use.

The aim of the study was to analyse the usability of one of the latest computer applications and to study the role played by usability problems in explaining the psychological well-being of customer service employees in Call Centers.

Both qualitative and quantitative data were collected. The methods used were a questionnaire and interviews.

The employees were asked to assess the latest and also most widely used application, "the contact call to the customer", which had been taken into use about a month before the survey. The usability was measured with an inventory of 25 usability items developed by McNeive & Ryan (McNeive & Ryan 1997). In the interviews, the usability was charted by asking the employees how the computer application helped to carry out the required task in the customer service situation.

3 Results

The overall response rate was 82% and the mean age of the subjects was 43 (sd 7) years. The factor analysis of the inventory of usability yielded a 4-factor orthogonal solution. On the basis of the 4 items of the factor analysis, 4 variables of usability were constructed and named "understanding", "accessibility", "errors" and "ergonomics". The share of the explained variance in multiple linear regression (r2) was 20%.

Psychological well-being was measured with nine questions concerning, e.g. anxiety, concentration and self confidence (Elo et al. 1992), work control with 4 and social support with 4 questions, which all have been validated in the studies at the Finnish Institute of Occupational Health. Cronbach's alpha of the summed psychological well-being was 0.92, social support 0.79, work control 0.77, and the measures of usability ranged from 0.82 to 0.85.

Multiple linear regression analyses were computed with psychological well-being as a dependent variable. The independent variables of the regression model were age, work control, social support and "understanding". The most powerful variable explaining psychological well-being was "understanding" ($p<0.01$)

The following descriptions of usability came up in the interviews of 24 customer service employees:

"We have to use many different applications during the same customer service situation, and it makes the work mentally heavy for me."

"I think the usability problems stem from the fact that this is a big company with many different computer applications, and some of them date back to the year X. Then they have made some additions at intervals of say two or three years, and now again some additions. So after many different additions the whole program turns out to be difficult to understand...you really have to struggle to understand how to use the applications."

"The problem is that the applications are so different and the commands are different. When working with one application, you use F11, put Finnish marks without any zeroes, but then add zeroes to Finnish pennies. Some applications insist that if you have zero pennies, you have to put Finnish marks with a comma, but no zeroes. Some applications say that you can end it with the command F8, some say you can end it with the command "clear". But if you remember it wrong and end some application with command "clear" you can accidentally destroy all you have written before this command".

"There is so much of that old stuff in the applications, and then you have to search from both the old applications and the new ones. That's why it clogs up the work.

4 Conclusions

Most empirical studies on psychological well-being at work have shown that work control and social support explain psychological well-being. In this study, where the target group consisted of employees working solely with telephone and computer applications, the most powerful variable explaining psychological well-being was the item "understanding". This item of the sum scale of usability refers to the psychological, cognitive understanding of the program of the computer application.

Also the fact that the employees have to use different commands for the same operation in different applications makes them perceive the whole application as confusing and mentally overloaded. The conclusion that can be drawn is that usability is directly related to psychological well-being among the Call Center workers. This result of the regression analyses and the conclusion from the interviews guide the researchers to concentrate more on the study of mental models and their correspondence to computer applications.

5 References

Elo A-L., Leppänen A., Lindström K., Ropponen T. (1992). *OSQ. Occupational Stress Questionnaire: User's Instructions.* Helsinki: Institute of Occupational Health.

Harrison M. C., Henneman R.L. and Blatt L.A. (1994). Design of Human Factors Cost-Justification Tool. Pub. Bias, R.G. ja Mayhew, D.J. (ed.) Cost-Justifying Usability. Academic Press.

Hollnagel E. (1997) Cognitive ergonomics or the mind at work, Proceedings of the 13th Triennal Congress of the International Ergonomics Association, Tampere 1997.

Mc Neive A. and Ryan G.M. (1997) Joint AIB/IBOA case study on the introduction of phase 1 of the new branch banking system (NBBS), Final report, University College, Dublin.

Mitchell C.M., Morris, J.G. and Ockerman J.J. (1997) Recognition-Primed Decision Making as a technique to support reuse in software design, Pub. Zsambok C.E. and Klein G. (ed.) Naturalistic Decision Making, Lawrence Erlbaum Associates.

Richardson R. and Marshall J.N. (1996). The Growth of telephone call centres in peripheral areas of Britain: evidence from Tyne and Wear. Area, 28 (3), 308-317.

Izso, L. & Zistra, F. (1997) Efficiency in work: an approach to interface evaluation and -design. Proceedings of symposium, Symposium No. 23, Verona 2-5 April 1997.

Electronic Document Handling – A Longitudinal Study on the Effects on Physical and Psychosocial Work Environment

Carl Åborg (1, 2)
1. Futura, Statshälsan Research and Development Occupational Health, Uppsala, Sweden
2. Department of Human-Computer Interaction, Uppsala University, Uppsala, Sweden

1 Introduction

Information technology is rapidly transforming working life. The use of computer networks are growing fast and steadily, and is dramatically changing the methods for information distribution. During the 80:s new techniques for document handling, where paper-based information is replaced by electronic information, have been introduced at many workplaces. The information shall be available when and where you need it, the speed, quality and effectiveness of document handling shall increase. This technique has several different names, e.g. Electronic Document Handling (EDH). We have studied the type of EDH where the information is scanned from paper documents into the computer system, is stored on computer media, presented on a computer screen and available for use within the computer system, often as part of a "work flow" and "case-handling" system. This technology has a great impact on working conditions, physical and psychosocial work environment and thereby on health and well being of individuals. The use of computers and visual display units (VDU:s) at work has in numerous studies shown to produce increased mental and physical work load and increased risk of a number of somatic and mental health symptoms, especially eye strain, neck/shoulder problems and psychosomatic symptoms (Grieco et al 1995, Aronsson, Åborg, Örelius 1988, Aronsson, Dallner, Åborg 1994).

One clear result of these earlier studies is that personnel working more than 6 hours per day at a visual display unit (VDU) show more symptoms of ill health than others. (Aronsson, Åborg, Örelius 1988, Aronsson, Dallner, Åborg 1994). Computer systems to be used by skilled professionals are often poorly designed and not very usable. This leads to inefficient use and to a variety of cognitive problems, such as confusion, lack of overview and memory overload (Nielsen 1993). From the extensively used theoretical model for measuring stress at work called the "demand-control model" (Karasek, Theorell 1990), we can predict that this situation can produce stress reactions and stress related diseases.

Most studies of VDU use and health effects are cross-sectional, comparing different groups of users. To understand the relations between different work environment factors, work content, use of specific techniques/computer systems and user reactions they have to be studied over time, using a longitudinal design.

2 Objectives

This project was designed to study the effects on physical and psychosocial work environment, and on self-reported health and well being, by introducing an electronic document handling (EDH)-system at a number of Swedish work places.

3 Design and Methods

The study design was longitudinal, data was collected with several different methods on three occasions; before, 6 months after and 18 months after the introduction of the EDH-system.

Four different workplaces at two Swedish State authorities and one office at a private company took part in the study.

Methods used were **interviews, observation interviews, questionnaires, video recordings, technical measurements and expert observation and examination.**

The **interviews** were semi-structured and concerned work content, work load, personal control, influence and participation in decision making, peer- and supervisory relations, social support, contact and collaboration. In the interviews after the introduction of EDH the subjects were asked to express their view on the work with that system and on changes in the work situation caused by the EDH-system. Each interview lasted for about one and a half-hour, took place in the subjects workroom and was conducted by an experienced psychologist. 16 persons were interviewed before the introduction of the EDH-system and 6 persons after.

The usability of the EDH-system was evaluated by an **observation interview**, according to the recently developed ADA-method (Åborg, Sandblad 1996).

An experienced evaluator observed and interviewed the users during his or her ordinary work with the system. If necessary the interview was completed after the observation period. The interview was based on an interview guide, containing a list of usability aspects, such as response times, error controls, disposition of screen area, feed-back functions and input functions. A total number of 8 persons, at 3 different workplaces, were studied for the usability evaluation.

The **questionnaires** concerned job content, physical and psychosocial work environment and mental and somatic health symptoms, and was based on standardized, well-tested questionnaires, (Andersson, Åborg 1992). Before the EDH introduction 60 persons were asked to fill in the questionnaires, and 56 of them did so. 6 months after the introduction the questionnaires was distributed to 38 users of the system and it was answered by 37 of them. The third time, 18 months after the introduction, 22 of 29 subjects completed the questionnaires.

4 Results

The interviews, questionnaires, video recordings, and expert observations showed increased workload, more repetitive and monotonous tasks, more constrained, static work postures and less task variability after the introduction of EDH. The questionnaires showed an increase in psychological and psychosomatic complaints and in eye complaints, and a constant, very high frequency of musculoskeletal symptoms, especially from the right side (the right shoulder, arm and hand).

The interviews showed that a majority of the subjects believed the EDH-system to have increased the effectivity of their work, but in the same time they had experienced increased time pressure. A majority also had experienced what can be called "technostress", stress reactions caused by a combination of information overload and poor ergonomics (badly working machines and badly designed software, etc.). Work collaboration and contacts with peers and supervisors had decreased, but were still on an acceptable level.

The usability examination showed a number of problems in the human-computer interaction. The most frequent problems were related to lack of overview and lack of consistence. The interface design led to extensive use of computer mouse.

5 Discussion

This study leads to the conclusion, that the introduction of EDH-systems can result in increased risks of work-related musculoskeletal disorders, (especially "mouse-arm syndrome"), eyestrain and stress-related mental and somatic symptoms.

The effectiveness of work can improve in the short run, but in order not to risk the health of the users an ergonomic strategy for the design of work organization, computer systems, job tasks and work-stations is essential.

6 References

Andersson, P., Åborg, C. (1992). *Bildskärmsformulären. Referensdata insamlat åren 1988-1991* (VDU questionnaires. Reference data 1988-1991) Futura, Statshälsan, Karlskrona, Sweden (in Swedish)

Aronsson, G., Dallner, M. & Åborg, C. (1994). Winners and Losers From Computerization: A Study of the Psychosocial Work Conditions and Health of Swedish State Employees, *International Journal of Human-Computer Interaction*, Vol 6 No 1 pp 17-37.

Aronsson, G., Åborg, C., Örelius, M. (1988). *Datoriseringens vinnare och förlorare*, (Winners and losers from computerization), *Arbete och Hälsa* 1988: 27 National Institute of Occupational Health, Solna, Sweden (in Swedish, summary in English)

Grieco, A., et al, (1995). *Work with display units '94*. Elsevier Science Publishers, Amsterdam.

Karasek, R., Theorell, T. (1990). *Healthy work: Stress, productivity and reconstruction of working life.* Basic books, New York.

Nielsen, J. (1993). *Usability Engineering*. Academic Press, Inc., San Diego.

Åborg, C., Sandblad, B. (1996). *A method for evaluation of cognitive usability in human-computer interfaces.* 25th International Congress on Occupational Health, Book of Abstracts, Stockholm.

Ergonomics in VDU work: A randomized controlled trial

Ketola R, Häkkänen M, Toivonen R, Takala E-P, Viikari-Juntura E
Finnish Institute of Occupational Health, Helsinki

1 Introduction

Computers facilitate and make work more efficient in many professions. In Finland in 1996, 57% of the women and 55% of the men used a video display unit (VDU) in their work (Statistics Finland 1996). The use of graphics software and non-keyboard input devices, e.g. mouse, has increased rapidly causing new demands for the design of office work places. Poorly designed VDU work has been associated with a variety of physical and psychosocial problems. Redesign and improvements in ergonomics have generally been recommended as a solution for musculoskeletal disorders in VDU work. Scientific evidence on the effects of ergonomics is, however, scanty.

We carried out a randomised controlled trial on the effects of ergonomic changes in VDU work. The study includes three different interventions and investigates their effects on neck, shoulder and upper extremity symptoms.

2 Study design

The study was carried out in three administrational centres of a medium-sized city in Finland. Employees working in the office at least 4 hours per week (n=515) were asked to fill in a questionnaire in January 1998. Organisational and environmental factors of VDU work, time working with input devices and various types of software, general health, psychosocial stress, and musculoskeletal symptoms were inquired. A total of 410 employees (80%) replied to the questionnaire.

The questionnaire survey showed that employees had had VDU-work approximately 40% of their working time during the last month. Word processing, e-mail, CAD, and statistical programs were most commonly used. Keyboard was used on the average 66%, and mouse 28% of VDU-working time. Ability to work was generally rated as high: 8 for the men and 9 for the women (scale 1-10).

The study population for the interventions was selected on the basis of musculoskeletal symptoms; i.e. symptoms in the neck- shoulder- upper limb region in at least one and at most eight of a total of anatomical areas. In addition the subjects should use the mouse for more than 10% the time working with VDU. These criteria were met by 120 workers. They were stratified by the administrational centre and randomised into three intervention groups.

For the first group ergonomic changes were designed and implemented by two ergonomists using a participatory approach. The second group attended a lecture of ergonomics in VDU work. They also received a VDU work checklist and were encouraged to make changes in their workplaces. To the third group, an article was given about musculoskeletal symptoms in office work. The interventions were performed in March – June 1998.

Before the interventions baseline data were collected by two blinded observers. The data included measurements of work place layout, placement of keyboard, screen, mouse, and all other equipment, e.g. supports. Moreover, anthropometry, viewing angles and distances were measured, and the posture and patterns of supporting the arms were observed (Kylmäaho et al). The employees filled in a diary of musculoskeletal discomfort for a two-week period before the intervention. To assess the exposure we used a computer program counting the number of typed keys and the number of performed mouse clicks for the same period (Toivonen et al.).

The follow-up measurements were carried out after 2 months and 12 months after the intervention.

In the analysis musculoskeletal symptoms and local discomfort were compared between the groups. Exposure to VDU work will be handled as a covariate. The costs of the different interventions will also be assessed.

3 Results

Table 1. Prevalence of severe pain (pain > 14 days) during preceding 30 days at 12 months follow-up

Anatomical area	**Expert assisted ergonomics n = 31**	**Education n = 32**	**Control n = 31**	**p value**
Neck radiating	**6.5**	**3.2**	**14.3**	**ns**
Right shoulder	**6.5**	**0**	**10.3**	**ns (0.09)**
Left shoulder	**3.2**	**3.2**	**0**	**ns**
Right elbow	**0**	**0**	**6.9**	**ns (0.09)**
Left elbow	**3.2**	**0**	**3.5**	**ns**
Right wrist	**3.2**	**3.1**	**13.8**	**ns**
Left wrist	**0**	**0**	**3.5**	**ns**
Right fingers	**3.2**	**0**	**6.9**	**ns**
Left fingers	**3.3**	**0**	**3.7**	**ns**

No statistically significant differences were seen between the groups one year after the intervention. The group with expert assisted ergonomic changes and the group with education had a tendency for less severe pain in the right shoulder and right elbow (Table 1). An analysis of severe pain before and after the intervention showed that the prevalence tended to reduce in the group with expert assistant ergonomic changes and group with education, whereas the prevalence in the control group was stable.

No statistically significant differences were seen in local discomfort scores.

4 Discussion

We saw a tendency for less severe pain in the right shoulder and right elbow in the two groups with more intensive intervention. The prevalence of severe pain was low in many anatomical areas and the circumstances were therefore less feasible for major effects.

We will do further analyses of the effects of the interventions via logbook data.

5 References

Kylmäaho E, Rauas S, Ketola R, Viikari-Juntura E: Supporting the forearm and wrist during mouse and keyboard work. A field study. *In Abstract book of HCI International '99, 8th International Conference on Human –Computer Interaction*, August 22-27, 1999, Munich, Germany.

Toivonen R, Takala E-P: Assessment of dose-response relationship in VDU work. *In Abstract book of HCI International '99, 8th International Conference on Human –Computer Interaction,* August 22-27, 1999, Munich, Germany.

Supporting the forearm and wrist during mouse and keyboard work. A field study

Kylmäaho E, Rauas S, Ketola R, Viikari-Juntura E.
Finnish Institute of Occupational Health, Helsinki, Finland

1 Introduction

A high number of people use the computer daily. Many studies have found relationships between video display unit (VDU) work and musculoskeletal problems. Especially disorders of the upper extremities have been related to a long duration of VDU work per day (Punnett and Bergqvist 1997). Some studies suggest that supporting the forearm decreases the risk of musculoskeletal symptoms in mouse and keyboard work (Aarås et al. 1997, Karlqvist et al. 1998). Supporting the entire forearm reduces static trapezius load during mouse and keyboard work (Aarås et al. 1995), but supporting only the wrist during typing may increase the trapezius load (Bendix and Jessen 1986). It has been commonly recommended to support the wrist and forearm during computer work. It is not clear, however, whether the location and extent of the wrist-forearm area that is supported, the degree of supporting, the type of support used or the continuity of supporting has any effect on musculoskeletal symptoms.

Our aim was to describe in a semiquantitative way how VDT workers support their forearms and wrists during mouse and keyboard work. These data formed a part of exposure information against neck and upper arm disorders in an intervention study (Ketola et al. 1999).

2 Materials and methods

The study population consisted of 109 office workers; 64 women and 45 men. Their mean age was 47.9 ±7.9 years. Two experienced physiotherapists interviewed the subjects and observed and made videorecordings of them in their regular computer work for about 15 minutes. Dimensions of the workstation were also measured and the equipment used was recorded.

The observers used a diagram of hands and forearms, in which they marked the contact area of the forearm and wrist (Diagram 1.). They also observed the degree of supporting (not at all, lightly, with full weight), the continuity of supporting (not at all, occasionally, all the time) and the type of supporting. In the analysis the contact area of the forearm was estimated as proportional area of the potential maximum forearm support area. Contact in the wrist and hand was recorded as a dichotomous variable: support or no support. These were done separately for the right and left arms and during mouse and keyboard work. Fingers were not included in the observation. According to a survey carried out prior our measurements, the subjects used computers on the average 46% of their work day. Word processing and CAD programs were most often used.

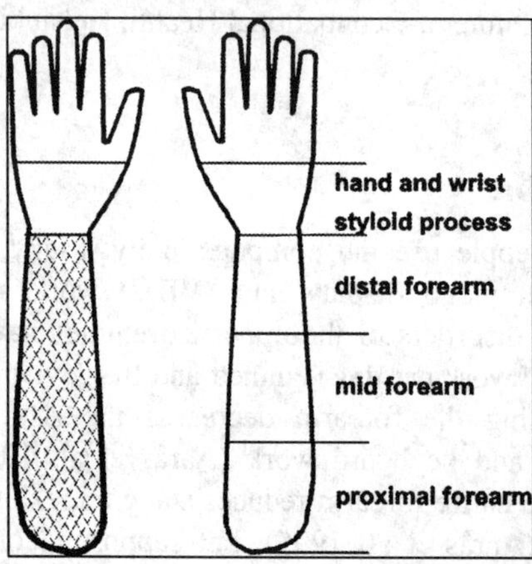

Diagram 1. The diagram for location and contact area of support. On the left side the potential maximum forearm contact area is shaded.

3 Results

The mouse was generally used by the right hand, 10% of the 109 workers used it by the left hand. Two subjects did not have mouse. 39% of the subjects used a wrist pad in mouse and keyboard work, 28% used a wrist pad in either keyboard or mouse work and 33% had no wrist pad. 70% had adjustable chairs, 61% had armrests and 75% adjustable desks. Half of the subjects utilized armrests of the chair during work, others used pads or rested their arms on tabletop. Three workers had movable armsupports.

Table 1. shows the percentage of the subjects supporting the wrist and forearm during keyboard and mouse work. During keyboard work the subjects supported the wrist and forearm frequently, but they used a smaller forearm area and supported more occasionally compared to mouse work.

During keyboard work 5% of the workers supported all forearm parts of the right side. At least the distal forearm was supported by 28% of the subjects and proximal part by 31%. On the left side the same values were 9%, 34% and 42%, respectively. During mouse work 19% of the subjects supported all forearm parts in the mouse-hand. At least the distal forearm was supported by 82% of the workers and proximal part by 32%. On the other arm the same values were 15%, 43% and 51%, respectively.

Table 1. The percentage of subjects supporting the wrist and forearm during keyboard and mouse work.

	Keyboard work n=100		Mouse work n=96	
	Right arm	Left arm	Mouse arm	Other arm
Support wrist (%)	72	78	97	92
➤ Support wrist all the time lightly or with full weight (%)	13	16	94	51
Support forearm (%)	50	65	90	71
➤ Support forearm all the time lightly or with full weight (%)	7	16	39	38
Contact area of the forearm more than 50% (%)	4	7	16	14

Layout of the workstation seemed to affect the way of supporting. Narrow keyboardtables, chairs without (or unadjustable) armrests and unadjustable desks resulted in less frequent supporting and also smaller area of the forearm supporting in both mouse and keyboard work.

4 Conclusions

The results show that supporting the wrist and forearm is more common during mouse than keyboard work. During mouse work more subjects supported the wrist and forearm all the time compared to keyboard work. They also tended to support more often the entire forearm and use larger areas of support with mouse work. The layout of the workstation seems to affect the way of supporting. Narrow keyboardtables, chairs without (or unadjustable) armrests and unadjustable desks resulted in less frequent supporting and also smaller area of the forearm supporting in both mouse and keyboard work.

5 References

Aarås, A., Fostervold K.I., Thoresen, M., & Larsen, S. (1995). Postural load at VDU work. In M. Kumashiro (Ed.): *Proceedings: The paths to productive aging*, pp. 151-156.

Aarås A., Fostervold K.I., Thoresen M. & Larsen S. (1997). Postural load during VDU work: a comparison between various work postures. *Ergonomics*, 40(11), 1255-1268.

Bendix T. & Jessen F. (1986). Wrist support during typing – a controlled, electromyographic study. *Applied Ergonomics*, 17(3), 162-168.

Karlqvist L.K., Bernmark E., Ekenvall L., Hagberg M., Isaksson A. & Rostö T. (1998). Computer mouse position as a determinant of posture, muscular load and perceived exertion. *Scandinavian Journal of Work, Environment & Health*, 24(1), 62-73.

Ketola R., Häkkänen M., Toivonen R., Takala E-P. & Viikari-Juntura E. (1999). Ergonomics in VDU work: A randomised controlled trial. *In Abstract book of HCI International '99, 8th International Conference on Human-Computer Interaction*, August 22-27, Munich, Germany.

Punnett L. & Bergqvist U. (1997). Visual display unit work and upper extremity musculoskeletal disorders. A review of epidemiological findings. *Arbete och Hälsa*, 1-161.

Assessment of dose-response relationship between VDU work and discomfort

Risto Toivonen, Esa-Pekka Takala
Finnish Institute of Occupational Health, Helsinki

1 Introduction

There is a general agreement of the relationship between visual display unit (VDU) work and a high prevalence of musculoskeletal symptoms. Individual, psychosocial, work organisational and biomechanical factors have all been associated with problems in the upper body and extremities (Punnett and Bergqvist 1997). However, exact epidemiological data about the relationship between physical workload and musculoskeletal symptoms is missing.

The aim of this paper is to introduce a method to be used in the assessment of dose-response relationship between VDU work and discomfort.

2 Hypothesis

Feeling of comfort/discomfort in a body part can be used as an indication of localised muscle fatigue during work. Uncomfortable feeling may reflect metabolic changes in a muscle. These changes will disappear if muscle is allowed to rest. However, if not enough rest is allowed, more advanced changes in a muscle may develop and manifest as musculoskeletal symptoms. This hypothetic situation is presented in figure 1. Symptoms are assumed to cumulate during the workweek. During the weekend there is a long enough pause for muscles to take a sufficient rest.

We can further set a hypothesis that the amount of daily workload has an effect on the cumulation of discomfort.

To assess dose-response relationship in VDU work, we should, according to our hypothesis, see a relationship between computer usage and feeling of discomfort in different body parts. We have used the following methodology for the assessment of these items.

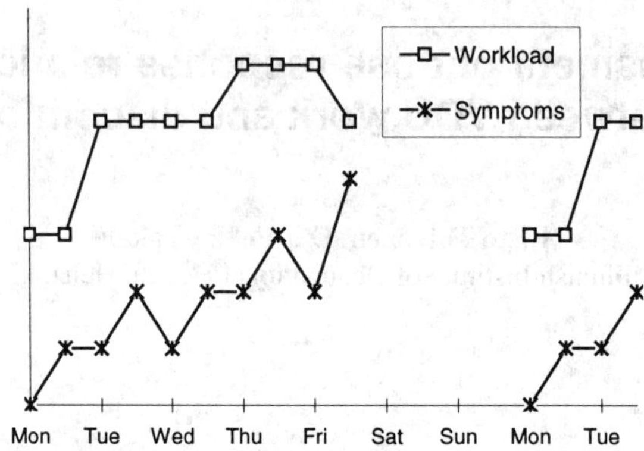

Figure 1. A hypothetic model of dose (workload) and response (musculoskeletal symptoms)

3 Diary of discomfort

VDU workers were asked to fill in a diary of discomfort three times per workday: at the start of the day, at noon, and at the end of the workday.

The diary had 19 questions about the feelings of different anatomical areas. The answer was given on a scale of five levels ranging from (5) "feel good" to (1) "feel very uncomfortable". An anatomical map indicating each body part in question was included in order to help responding. In the morning and at the end of the day there was an additional question about the overall feeling of the moment.

We have successfully used this diary in an intervention study (Ketola et al. 1999) among 109 VDU workers. They filled in the diary for three weeks before and for two weeks after the intervention.

4 Monitoring of Computer Usage

During the period when the worker filled in the diary, a special program (WorkPace™, Niche Software Limited, New Zealand) monitored continuously keyboard and mouse usage. Data were saved into a daily file containing exact history of Windows™ keyboard and mouse events with the accuracy of 10 ms. The data will be processed into a set of parameters that characterise the intensity of usage for the morning and afternoon hours. We have used the following parameters:

- Duration of computer usage (in minutes)

  Pause in keyboard usage longer than 2 minutes was considered as a break, and was not included in the calculation of duration.

- Duration of usage of mouse (in minutes) in a day

  Pause in mouse usage longer than 30 seconds was considered as a break, and was not included in the calculation of duration.

- Number of typed keys

- Number of clicks of mouse performed

5 Discussion

With this method it should be possible to evaluate the relationship between computer usage (keystroke rate, mouse clicks, duration of computer usage) and musculoskeletal discomfort.

Figure 2 presents the relationship between the feeling of discomfort in the right shoulder and cumulative usage of computer based on the data collected in our study. In the figure the means (std err.) of 61 female office workers' feelings during the five week period are presented. As can be seen, computer usage cumulates almost linearly toward the end of the week. However, the means of feeling seem to reach their level of maximal discomfort well before the Friday.

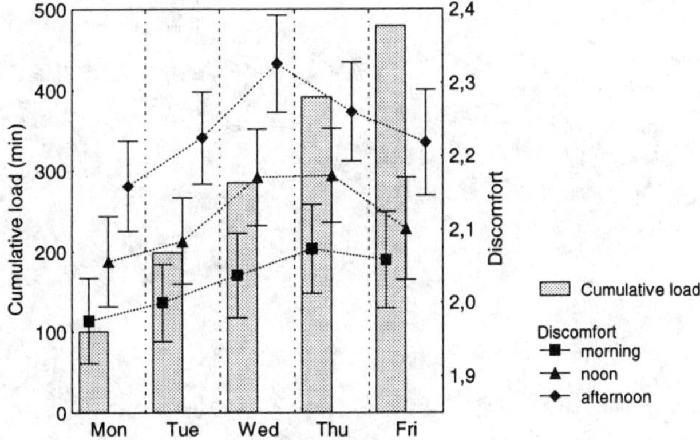

Figure 2. Relationship between discomfort in the right shoulder and cumulative workload, means and standard errors. Pooled data of five weeks of 61 women.

In VDU work there are also other factors than keyboard and mouse usage acting as exposures. Work posture, for example, affects the overall work load. In the analysis of dose-response relationship, these additional factors should be taken into account as modifiers of exposure.

It is also apparent, that combining the data of computer usage into a few parameters per half a day does not give the best possible conception of a dose. In VDU work there are some faster and slower phases, pauses etc. during a workday. This variation of dose in time (Mathiassen and Winkel 1991) should be considered thoroughly in the analysis. Our methods in collecting data will allow analysis of these time-related aspects of doses and responses.

6 References

Ketola, R., Häkkänen, M., Toivonen, R., Takala, E.-P.& Viikari-Juntura E. (1999). Ergonomics in VDU work: A randomized controlled trial. In *Proc. 8th Int. Conference on Human-Computer-Interaction* (HCI International '99, Munich, Germany, August 22-27, 1999)

Mathiassen, S. E. & Winkel, J. (1991). Quantifying variation in physical load using exposure-vs-time data. *Ergonomics,* 34 (12), 1455-1468.

Punnett, L. & Bergqvist, U. (1997). Visual display unit work and upper extremity musculoskeletal disorders, a review of epidemiological findings. *Arbete och Hälsa 1997:16.*

A Pragmatic and Inclusive Approach to Assess Health and Safety Aspects at VDU Workplaces

Harald WEBER
Institute of Technology and Work, University of Kaiserslautern, Germany

1 Introduction

European legislation forces employers to assess health and safety conditions at work (Council Directive 89/391/EEC; European Commission 1989). An additional specification regarding work with visual display units (VDU) was released 1990 (Council Directive 90/270/EEC; European Commission 1990). All member states meanwhile have transfered these directives into national law. Nevertheless, in none of the states assessment tools are known that take into account adaptations at workplaces for employees with disabilities. To fill this lack an accessible assessment tool was developed that guides employees through an assessment, evaluation and documentation procedure.

2 Methodology

The development of this assessment tool based on a VDU workplace model, which is an adaptation of a general work system model (Luczak, H. 1998). This VDU workplace model identifies the relevant components of a workplace system (e.g., hardware, social context, tasks) and limits the analysis focus to the micro level. Literature surveys referring to these components indicated a lack of safety and health information regarding input and output devices for people with disabilities. To collect this information, surveys were launched to gather data. The target groups for these surveys were: a) technical counselors at job centres specialized on disability specific workplace adaptations (n = 99), b) experts at trade unions and at safety and health authorities (n = 97), and c) manufacturers of input and output devices (n = 53).

The collected data were used both to value the 'significance' of several criteria of devices and to build a set of questions for the analysis tool. The high-level structure of the tool was taken from the according law (in Germany: *ArbSchG*,

BildscharbV). The sub-structure used the VDU workplace model for refinement. Finally, the low-level contents (= relevant questions) were derived from the results of the surveys. Respond rates (a: 58 %, b: 35 %, c: 36 %) indicated a strong interest in the topic.

The complete set of questions after this procedure had a size of approx. 230. However, only an average of 120 questions were relevant for a specific workplace, so an algorithm was needed to select the appropriate questions for a given workplace. Flow charts for each workplace component were created to decide about disability specific question alternatives.

Additionally, next to this assessment module, an evaluation and documentation module was created. Furthermore, a qualification module to raise the users' health and safety awareness was developed and tested.

3 Tool Description

The tool was implemented as software to fulfil the criteria of universal design for the expected target groups. To support older computer platforms a solution was created with low resource requirements. Nevertheless, economic constraints limited the product to Windows-based operating systems (3.*, 95, NT, 98) and emulations on other platforms. System resources needed are: 4 MB RAM, 256 colors display, 18 MB hard disk space, CD-ROM drive. For blind users a sound card, loudspeaker and/or headphones are required.

The software (called GEA) supports different kinds of input and output strategies, i.e., stick operation (e.g., headstick, one finger input), keyboard- or mouse-only operation, speech control and natural speech output. It was tested and improved in co-operation with employees who require the mentioned input/output strategies. The CD-ROM also contains handling instructions for blind users (as an electronic equivalent to the written manual), information about additional assitive software products to enhance accessibility, and an easy to use and accessible offline browser to read the analysis results.

The software is divided up into three main parts. The first part explains the handling to the user (introduction). The second part allows the analysis of the workplace and generates the result and documentation, the third part introduces the 'most important' risks at VDU workplaces to the users (qualification module). In average 30 - 45 minutes are needed for the introduction and analysis part, another 15 minutes for the qualification module.

Due to potential restricting company or trade regulations the software can easily be removed from a system without any change of relevant system files. Additionally, anonymous use is supported (privacy, data protection) as well as the use of selected modules instead of being forced to answer the whole set of questions. Finally, to avoid stigmatization the tool assesses workplace conditions for people with *and* without disabilities (universal design approach).

4 Evaluation

Experiences were collected during the development as well as through final testing. During development the main focus was set on those aspects being most important from the user's point of view: *comprehensibility*, *completeness* and *accessibility*. Comprehensibility was tested through detailed observations during analyses, where users were instructed to verbalize their thoughts immediately. Completeness and accessibility were tested through interviews at the end of each analysis. The results were used for continuous improvements in the design.

Final testing of the tool was carried out in the field. The main test criteria were: *objectivity*, *reliability*, *validity* and *utility*. Objectivity was not measured, but both, creation as a software tool (deterministic behavior) and instructions for use in a recommended environment (no stress, no interruptions, etc.) contribute to optimize the tool's objectivity.

Reliability was measured twice, once during design to get hints for improvements (n = 15; aged 27-40, av. 30; 2 female, 13 male) and again at the end to control the results of this process (n = 11; aged 26-36, av. 30; 4 female, 7 male). Unfortunately, the tool's dynamic structure creates approx. $2.1*10^{11}$ different paths through its set of questions, i.e., a complete test of all variations is impossible. A heuristic approach was proposed to cluster 'similar' questions and to estimate the overall (test-retest) reliability. Whereas in the first reliability test 24 of 43 clusters appeared not to be valid, the second measure after reworking the set of questions resulted in 6 non-valid clusters (mainly software ergonomic aspects). The estimation of the tool's overall reliability resulted in $r_{test\text{-}retest} = 0{,}814$, which is suffient to proceed.

Obviously, due to the limited possibility to calculate the reliability it is also impossible to estimate the validity of the overall tool. The lack of a tool to identify safety and health aspects from the users point of view at workplaces equipped for employees with disabilities makes a direct comparison for validity calculation impossible and requires a different arrangement. Here, a comparison between the users' answers to the clusters (identified during reliability measurement) and the answers from an expert in the field of safety and health was used to get hints about each single cluster correlation. It is important to emphasize, that this approach compares subjective user data with objective expert data, i.e., the result is not the validity of the tool, but an indicator for how close the data are. A Multitrait-Multimethod approach (Campbell & Fiske 1959) was chosen to explore the closeness of data in a 86 x 86 matrix. Due to experimental constraints (see above) just 28 correlations (from 43) could be calculated. 15 correlations were suffiently high to indicate strong correlations, 13 were not. Both results can be used from experts to decide, which data can be taken from the user's analyses and which data need further examination and communication with the users (see Fig. 1).

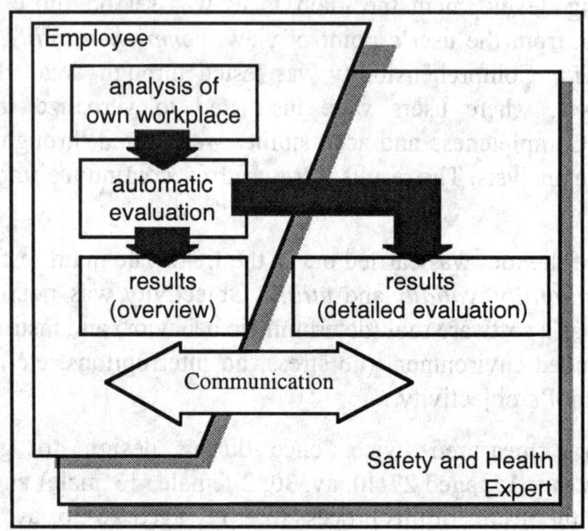

Figure 1: Recommended use of the analysis tool

Finally, users were asked to assess different aspects of utility according to a scale from 1 to 6 (1=very good, 2=good, 3=satisfactory, 4=sufficient, 5=unsatisfactory, 6=not acceptable), the number of participating users is mentioned in braces. The time needed to use the tool was valued with 1.9 (n=14), the manual design with 1.8 (n=13), manual comprehensibility with 1.6 (n=13), manual completeness with 1.9 (n=12), software design with 1.7 (n=15), software completeness with 2.0 (n=13), number of questions with 2.1 (n=14) and the personal benefit with 2.5 (n=14). Regarding the ease of use the usability was valued with 1.9 (n=15), software clarity with 2.1 (n=15) and software comprehensibility with 2.5 (n=15).

5 Discussion

Despite it was strived for a complete analysis of a workplace (according to the basic workplace model) insufficient reliability coefficients led to the exclusion of software ergonomic aspects from the set of questions. This result corresponds to the scientific literature regarding the conclusion, that single users cannot sufficiently assess the ergonomic quality of a software product (e.g., Prümper 1997). Economic constraints make it necessary to strive for a compromise between scientific conciseness and usability criteria during workplace analyses. The compromise developed in this project neither allows to completely replace an expert's analysis by an employee's analysis nor does it measure objective data for further processing. Nevertheless, the software can be used for a division of labor between employees and experts with a benefit for both parties. Employees will be introduced with the most important safety and health spots at their workplaces. This awareness achieved will be of additional benefit when workplaces are changing or employees additionally work at home (without safety inspections). Experts find a means of communication with the employees when discussing the results of an analysis. While the employees perception of safety and health aspects might be different to those of an expert, the difference also contains worthful information for the expert.

6 References

Campbell, D. T. & Fiske, D. W. (1959). Convergent and discriminant validation by the multitrait-multimethod matrix. *Psychological Bulletin*, Vol. 56. pp. 81 - 105

European Commission (1989). *Council Directive 89/391/EEC of 12 June 1989 on the introduction of measures to encourage improvements in the safety and health of workers at work.* (OJ L 183, 29.6.1989).

European Commission (1990). *Council Directive 90/270/EEC of 29 May 1990 on the minimum safety and health requirements for work with display screen equipment* (Fifth individual Directive within the meaning of Article 16(1) of Directive 89/391/EEC). (OJ L 156, 21.6.1990)

Luczak, H. (1998). *Arbeitswissenschaft*. 2nd Edition. Berlin, Heidelberg, New York.

Prümper, J. (1997). Der Benutzungsfragebogen ISONORM 9241/10: Ergebnisse zur Reliabilität und Validität. In: Liskowsky, R., Velichkovsky, B. M. & Wünschmann, W. (Eds.). *Software-Ergonomie '97, Usability Engineering: Integration von Mensch-Computer-Interaktion und Software-Entwicklung*. Stuttgart. pp. 253 - 262

The Comparison of Preferred Settings between PC and CAD Workstation

Wen-Hsin Hsu[1] and Mao-Jiun Wang[2]
[1]Center for Industrial Safety and Health Technology, Industrial Technology and Research Institute, Hsinchu, Taiwan, R.O.C.
[2]Department of Industrial Engineering and Engineering Management, National Tsing Hua University, Hsinchu, Taiwan, R.O.C.

1 Introduction

The wishful thinking of the traditional VDT guides that average person sits upright with right angles at elbows, hips and knees is in dispute. Many studies have shown that the VDT users preferred to sit in a backward leaning posture with a longer viewing distance and a larger elbow angle and that the physical discomfort was reduced with preferred settings (Grandjean *et al.*, 1983; Ong, 1988).

Among factors that can possibly affect VDT workstation preferred settings, there are two factors deserving more attention, i.e. the different VDT workstation type and the preference difference in different ethnic groups. However, there is very little literature exploring these effects. Van der Heiden and Krueger (1984) reported a preferred forward monitor tilt setting (top of the screen toward the operator) at CAD workstations which was quite different from the preferred backward monitor tilt setting at PC workstations. As for the preference difference in different ethnic groups, though it can be concluded from previous studies that the preferred settings of VDT workstations are not greatly influenced by anthropometric factors (Grandjean 1984). However, Ong (1988) still reported that the Singapore VDT users preferred a lower sitting and working height compared with European VDT users which was possibly due to differences in anthropometric dimensions.

Therefore, the present study was aimed to obtain the preferred settings of Taiwanese VDT users for both PC and CAD workstations, to see if there existed some differences between them and to compare them with previously reported data. Thus, these data will be useful for the manufacturers of computer workstation furniture and so be beneficial to VDT users.

2 Methods

2.1 Subjects and Their Jobs

A total of thirty experienced VDT users from a semiconductor company participated in the laboratory study. None of them ever received training on ergonomics VDT design guides. All subjects were full-time VDT users with at least five month's job experience.

2.2 The Experimental Workstation and Experiment Procedure

An adjustable VDT workstation was designed based on a recently constructed anthropometric database for Taiwanese workers (Wang *et al.*, 1999). The working height and screen height can be adjusted respectively by turning the electrical switches. Screen distance can be adjusted by moving the mechanical sliding track. The height-adjustable chair was equipped with armrest adjustable in height and backrest adjustable both in height and inclination. Two types of VDTs were used: a 14" color monitor for PC workstation and a 17" color monitor for CAD workstation.

Each subject was seated and asked to adjust the workstation to his/her preference. All the adjustable components were reset at their lowest or erect positions after each measurement. Mark balls were adhered to the specified body positions before the experiment to enhance data measurement accuracy. For PC workstation, an article of about 300 words was assigned for typing. For CAD workstation, a simple IC design pattern was assigned for drawing. The experiment task usually took about one hour

3 Results and Discussion

3.1 Comparison of Preferred Settings between PC and CAD Workstation

Descriptive statistics and the mean differences of preferred settings between PC and CAD workstation are provided in table 1. A preferred backward leaning posture with upper arm slightly elevated and with a downward gaze angle were observed for most of the subjects on both workstations. The results show that there were differences between the actual preferred posture and the design standard posture suggested in general VDT workstation design guides, such as the ISO 9241. The preferred settings which are proportional to the actual dimensions for PC and CAD workstation were illustrated in figures 1. Neither the preferred working desk height nor the preferred seat height showed significant difference between PC and CAD workstation. But the differences on preferred monitor screen center and top height were significant. Perhaps, these differences were mainly caused by the difference in monitor size.

Table 1. Descriptive statistics, mean differences in preferred settings and independent samples t-test for equal of means between PC and CAD workstations

Variables/M ± SD	PC	CAD	Mean difference	
Eye ht. (mm)	1128.6±42.7	1104.9±35.6	23.7	
Monitor ht (mm) (screen top)	1036.0±33.7	1113.8±36.2	-77.9	***
Monitor ht (mm) (screen center)	936.0±33.7	983.8±36.2	-47.9	***
Monitor ht (mm) (monitor stand)	752.0±33.7	738.8±36.2	-13.1	
Eye-monitor screen top ht. diff. (mm)	92.6±44.2	-8.93±46.0	101.5	***
Neck angle (deg)	11.9±8.07	3.4±5.9	8.5	**
Monitor tilt (deg)	8.5±3.7	3.5±3.9	4.9	***
Gaze angle to monitor center (deg)	18.5±6.8	12.5±6.9	5.9	*
Eye to monitor dist. (mm)	578.0±86.1	609.9±84.9	-31.9	
Working desk ht	670.0±31.1	665.6±31.5	4.5	
Keyboard ht (home row) (mm)	700.0±31.1	695.6±31.5	4.5	
Elbow ht (mm)	712.1±41.7	715.5±33.8	-3.4	
Elbow ht - keyboard ht. diff. (mm)	12.0±22.3	19.87±21.5	-7.9	
Seat ht (mm)	433.9±8.0	432.8±8.4	1.1	
Upper arm flexion angle (mm)	31.7±7.8	32.33±14.4	-0.7	
Elbow angle (deg)	100.3±8.8	115.9±12.4	-15.6	***
Trunk inclination angle (deg)	102.2±5.8	107.7±10.4	-5.5	
Back rest angle (deg)	108.3±4.8	108.0±3.1	0.7	
Monitor screen to desk edge dist. (mm)	364.1±65.7	415.4±49.2	-51.3	*

Equal variances not assumed according to Levene's test. *p<0.05 **p<0.01 ***p<0.001

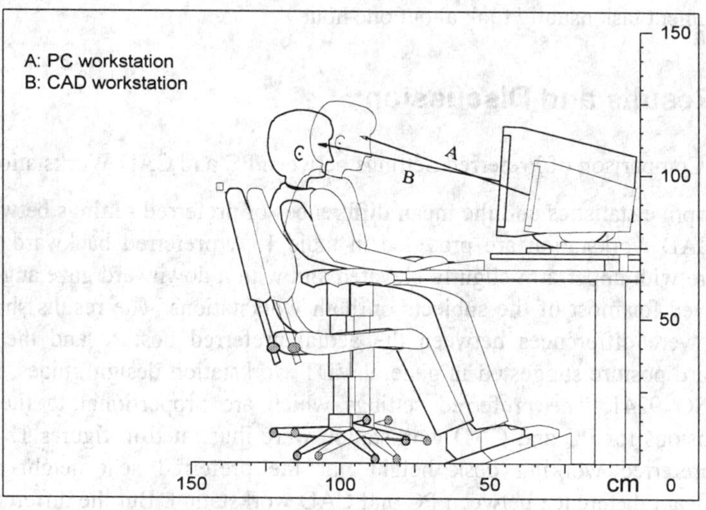

Figure 1. Comparing preferred settings between PC and CAD workstations

The postulate was supported by the insignificant difference in the base stand of the monitor (1.3cm). The above findings suggest that the adjustable dimensions in working desk height, seat height and monitor base stand height for the design of PC and CAD workstations should not be differed. However, the CAD users did prefer a greater monitor screen to desk edge distance than that of PC users. It suggests that the depth of the desk should be greater for CAD workstation due to the combined effects of greater preferred monitor screen to desk edge distance and the usually greater monitor depth.

3.2 Preferred Settings for Different Ethnic Groups

Table 2 shows the comparison of preferred settings among different ethnic groups. Two major trends can be observed from the results. First of all, the preferred monitor center height, keyboard height and seat height for both PC and CAD workstations in this study were found to be significantly lower than those reported in references (a) and (b). Ong (1988) reported a similar finding and attributed the result to the differences in anthropometric dimensions. However, as observed in table 2, the mean subjects stature for PC workstation in this study was even higher than that reported in reference (a). It seems that there existed some additional factors other than the anthropometric differences, which resulted in the above phenomenon. One possible explanation is that the operators have accustomed themselves to the lower settings of VDT workstation at their workplace. The postulation was supported by further examination of the settings of the subjects' field workstation which showed a significantly lower mean keyboard height (66 cm) and seat height (41 cm).

Table 2. Preferred settings compared with previous research data

Reference Mean/Range	a PC (Switzerland)	b CAD (Switzerland)	c. PC (Singapore)	d PC (Taiwan)	e CAD (Taiwan)
Stature (cm)	168	–	157	169 (150-183)	163 (152-178)
Screen ht. (cm)	103 (92–116)	113(107–115)	105 (99–113)	93 (89–99)	98.4 (95–107)
Keyboard ht. (cm)	79 (71–87)	73 (70–80)	74.3 (57–92)	70 (66–77)	69.5 (65–76)
Seat ht. (cm)	48 (43–57)	54 (50–57)	46 (41–54)	43 (42–46)	43 (42–46)
Eye to scr. dist. (cm)	76 (61–93)	70 (59–78)	56 (42–67)	58 (43–76)	61 (46–74)
Monitor tilt (deg)	4 (2 – 13)	-7.7 (-15–1)	18.9 (10– 20)	8.5 (0–15)	3.5 (-2–14)
Gaze ang. (deg)	9 (-2–26)	–	–	18.5 (8–33)	12.5 (2–24)
Neck ang. (deg)	51† (34–65)	–	–	12 ‡(-2–27)	3.4‡ (-13– 8)
Trunk incl. ang.(deg)	104 (91–120)	–	104 (82–124)	102 (89–109)	108 (89–124)
Upper arm ang. (deg)	23 (1–50)	–	33.4 (0–90)	31.7 (13–46)	32.2 (4–65)
Elbow ang. (deg)	99 (75–125)	–	94 (63–151)	100 (82–122)	116 (92–137)

a. Grandjean *et al.* (1983) b. Van der Heiden *et al.* (1984) c. Ong. *et al.* (1988) d. Current study: PC e. Current study: CAD † Angle formed by C7 and ear hole referenced against the vertical plane ‡ the angle of Reid's line with reference to the horizontal plane plus 15 °

Secondly, the preferred eye to monitor distance (about 58 cm) for PC workstation was found to be similar to that of reported in reference (c). But the distance is relatively shorter than those reported in references (a) and (b) for PC and CAD workstations.

4 Conclusions and Recommendationds

The findings of this study reveal that the Taiwanese VDT users preferred to sit in a pronounced backward leaning posture with upper arm slightly elevated, and elbow angle greater than 90° for both PC and CAD workstations. But there are some differences between the posture recommended by general VDT workstation design guideline and the actual preferred posture. As for the comparison of preferred settings between PC and CAD workstation, the main differences occurred at the head-neck posture and gaze angle which were mainly caused by the difference in monitor size. A more erect but still backward monitor tilt setting was preferred for CAD workstation. Since there is no significant difference in the working height, seat height and monitor base stand height between PC and CAD workstation, similar adjustable dimensions should be specified for designing the two types of VDT workstations.

The preferred posture angles of upper body showed some consistency with previous studies based on western data. But the preferred monitor center height, keyboard height and seat height were significantly lower than the previous reports based on western data. It seems that there existed some additional factors other than the anthropometric differences such as the influence of the subjects' field VDT workstation settings which might attribute to the above phenomenon.

5 References

Grandjean, E., Hünting, W. and Pidermann, M. (1983). VDT workstation design: preferred settings and their effects, *Human Factors*, 25, 161–175.

Ong, C. N., Koh, D., Phoon, W. O. and Low, A. (1988). Anthropometrics and display station preferences of VDU operators, *Ergonomics*, 31, 337–347.

Van der Heiden, G. and Krueger, H. (1984). Evaluation of ergonomic features of the computer vision instaview graphics terminal. In E. Grandjean and E. Vigliani, eds., *Ergonomic Aspects of Visual Terminals*. Taylor and Francis, London.

Wang, E. M. Y., Wang, M. J., Yeh, W. Y. Shih, Y. C. and Lin, Y. C. (1999). Development of anthropometric work environment for Taiwanese workers, International Journal of Industrial *Ergonomics*, 23, 3–8.

Visual and Lighting Conditions for VDU Workers

Hans-Henrik Bjørset[1], Arne Aarås[2] and Gunnar Horgen[3]
[1] K.O. Thornesveg 11, N-7033 Trondheim, Norway
[2] Alcatel STK AS. P.O. Box 60 Økern, N-0508 Oslo, Norway
[3] Buskerud College, Department of Optometry, P.O. Box 235, N-3601 Kongsberg, Norway

1 Introduction

A total renovation in two buildings at Alcatel Telecom Norway, Oslo, was carried out in 1991-92. During this process, unfortunately non-ergonomically luminaires were mounted. After a short periode of time the VDU operators reported visual problems in terms of bad lighting conditions, serious glare problems, eye fatigue, blurred vision and increased sensitivity to light. Headache was also reported. It was therefore decided to carry out a study of these problems and in particularly look for a relationship between the visual discomfort and the lighting conditions. First, the visual and lighting conditions were examined in a laboratory study (Bjørset and Aarås 1996). At the same period of time another laboratory study was performed, showing that there was a significant lower static trapezius load for sitting position with support of the forearms on the table top compared with sitting and standing without such support. These results were valid for data entry work when using keyboard and mouse (Aarås & al 1997). Therefore, multipisciplinary team was established to carry out a longitudinal epidemiological study, where the effect of introducing ergonomically luminaires and workplaces should be evaluated. In addition eye examination should be performed and optometric corrections given if needed. The three inverentions were implemented in a serial way.

2 Aims of the Study

Will improved lighting conditions reduce visual discomfort for the VDU operators?

3 Material and Methods

3.1 Design of the Study

The study is a prospective parallel group design with three groups of male VDU workers, performing software engineering. Approximately 50 subjects participated in each group. Two groups (T and S) got new improved lighting system, while the third group (C) acted as a control group which continued with the initial lighting system. After three years, also group C got the lighting intervention.

3.2 Visual Discomfort and Pain in the Musculoskeletal System

Pain intensity and duration was assessed on a 10 cm Visual Analog Scale (VAS) for the last month and the last six months periods before and after interventions. The lighting condition and glare problems were reported on VAS according to defined criteria.

3.3 Psychosocial Factors

The psychosocial questionnaire deals with: The variation of the VDU work, job control, possibility to make contact that you feel you need. In additon, self realization in tems of learning, increased skills and utilization of own capability (Westlander 1987).

4 The Lighting Intervention

The initial lighting, which applied simple luminaires with non-effective louvres and downward lighting distribution only, gave very poor lighting conditions regarding illuminances, luminances and luminance distributions and caused too much glare. This was mostly due to the high luminance of the fluorecent tubes which were directly seen from many positions in the room. The new lighting system with localized lighting applies suspended luminaires with a light distribution preferably about 25 % diffused upwards and 75 % downwards, through an effective semidiffuse reflector-louvre system. A recommendable positioning would be with one luminaire, 2x36 W fluorescent tubes, at each side of the VDU workplace (Bjørset 1987). A number of other parameters was considered and is reviewed in (Bjørset 1997).

5 Results

Both the illuminance and the luminance increased to satisfactory levels for the relevant areas: The average maintained illuminance from below 300 lux to more than 600 lux and the average luminance levels of the ceiling and the walls from around 30 cd/m^2 to more than 80 cd/m^2. The operators of the T and S groups reported highly significant improvements of the lighting and viusal conditions (p=0,0001), see figure 1, and highly signficant reduction of glare problems (p=0,0001), see figure 2, after the lighting intervention.

The control group (C) assessed a small significant worsened visual condition after the lighting intervention (p=0,008), while no significant change was reported on the glare (p=0,30) (Bjørset and Aarås 1996). At followup, after optometry, no significant changes were reported in any of the groups regarding lighting and visual conditions and glare by comparing with the values after the lighting intervention. After 3,5 years, also group C, which had got lighting intervention, reported significant improvement of the lighting and visual conditons. Now, there is no longer significant differences between the three groups.

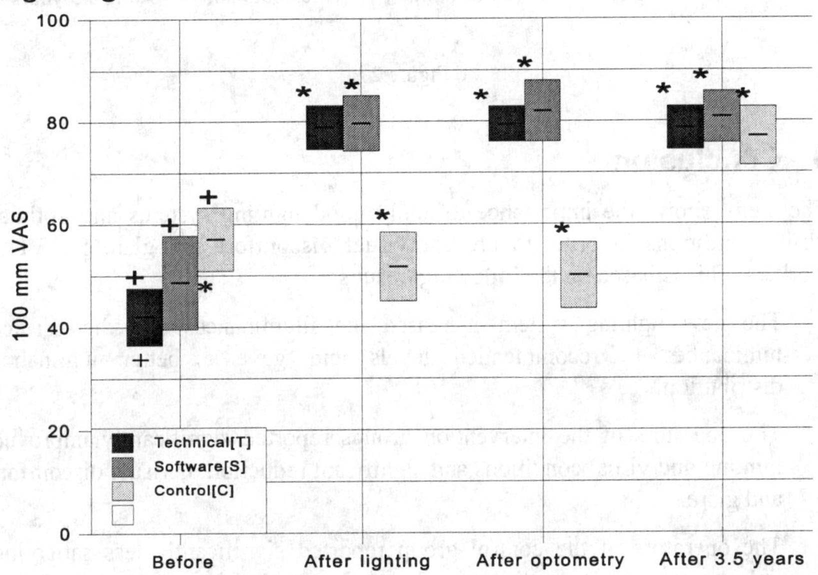

Figure 1

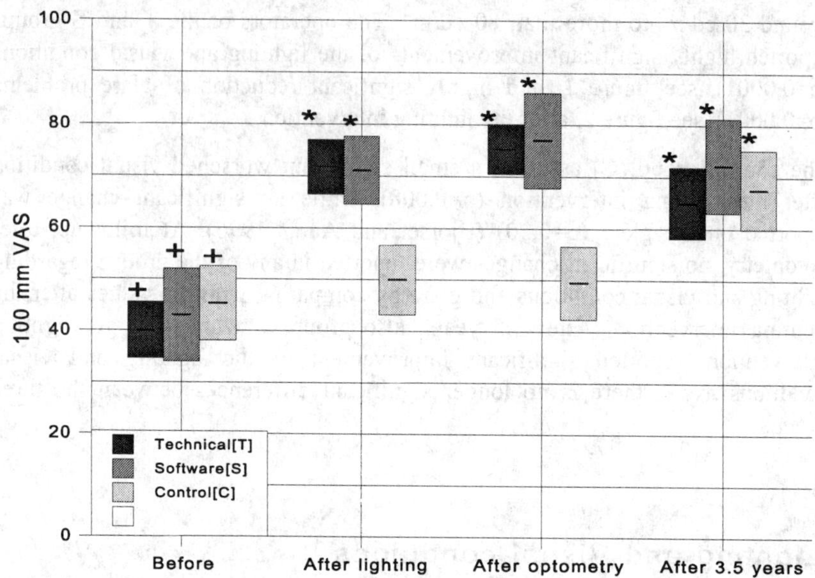

Figure 2

6 Conclusion

The study shows the importance to apply good lighting systems and optimal visual conditions in order to prevent visual discomfort and glare for VDU workers. This is based on the following results:

- The new lighting system increased the illuminances and the surface luminances to recommended levels and gave a better luminance distribution.
- The operators of the intervention groups reported significantly improved lighting and visual conditions and significant reduction of visual discomfort and glare.
- The operators of the control group reported significantly less satisfying lighting and visual conditions and significant more visual discomfort and glare compared with the intervention groups. However, when this group got the lighting intervention, it also reported significant improved lighting and visual conditons and significant reduction of visual discomfort and glare. The three groups are no longer significant different.

7 References

Bjørset, H-H. (1987). Lighting for visual display unit workplaces. Work With Display Units 86. Proceedings part II, 683-687, Elsevier Science Publishers B.V. (North Holland).

Bjørset, H-H. and Aarås, A. (1996). Visual and Lighting Conditions. In: Advances in Applied Ergonomics. Eds: A Øsak and G.Salvendy. USA Publishing, 215-220.

Bjørset, H-H. (1997). Visual conditions for VDU workplaces. In: The Workplace, Volume 2: Major Industries & Occupations, 191-214. Eds. D. Brune, G. Gerhardsson, G.W. Crockford and D. Norback, International Occupational Safety and Health Information Centre, Geneva, International Labor Office, Geneva and Scandinavian Science Publisher AS, Oslo.

Aarås, A., Fostervold, K. I., Ro, O., Thoresen, M. and Larsen, S. (1997). Postural load during VDU work: a comparison between various work postures. Ergonomics, Vol. 40, no. 11, 1255-1268.

Westlander, G. (1987). How identify organizational factors crucial of VDU-health? A context-oriented method approach. In: Knave, B. and Wideback, P. G. (Eds) Selcted Papers Presented at the Conference on Work with display units, Stockholm Sweden. North Holland, Amsterdam, 816-821.

OPTOMETRIC EXAMINATION AND CORRECTION OF VDU WORKERS

Gunnar HORGEN[1], Arne AARÅS[2].
[1] Kongsberg College, Department of Optometry – P.O.Box 251
N- 3603 Kongsberg, Norway
[2]Alcatel/STK. P.O. Box 60 Økern, N- 0508 Oslo, Norway.

1 Introduction

In 1991 – 1992 a total renovation of two buildings at Alcatel/Telecom Norway, Oslo, was carried out. Unfortunately non-ergonomic luminaries were mounted, and complaints of bad lighting, poor visual conditions, eye fatigue and blurred vision were reported. It was therefore decided to carry out a study of these problems, and in this study we look particular for problems related to the eye problems and optometric corrections. First the lighting and visual conditions were examined in a laboratory study (Bjørset and Aarås 1996). At the same time another laboratory study showed that there was significantly lower static trapezius load for sitting position with support of the forearms on the table top compared with sitting and standing without such support. These results were valid for data entry work when using keyboard and mouse (Aarås et al. 1997). A multidisciplinary team was established, and the effect of introducing ergonomically luminaries and workplaces and optometric corrections was studied. The three interventions were done in a serial way, and in this study the effect of optometric intervention is evaluated.

2 Aim of the study

Will optometric corrections reduce visual discomfort and headache among VDU-operators.

3 Material and methods

All clients were given a complete optometric examination, and the corrections given were based on this examination. The examinations focused on optimum visual acuity and comfort for the VDU-worker. This implies that the corrections were modified in order to fit the physical dimensions of the actual workplace. Binocular problems was evaluated and corrected according to both the amount of fixation disparity measured, and the phoria measurements (Borish 1975 and Sheedy 1977, 1995)

3.1 Criteria for correction

Criteria for prescribing an optometric correction are not easily given, and some controversy exists around this topic. The criteria were based on clinical experience and earlier studies (Metling 1992, Horgen et al 1989, 1995, Borish 1975)

3.2 Spectacle lenses

The corrections are meant to be worn while working on a VDU, and single vision lenses are the lenses of choice (Horgen 1989,1995). The lenses recommended were white, organic lenses with antireflection coating. Antireflection coating improves the quality of the lenses by minimizing the reflections in the lenses, although there is no scientific evidence that this improves the performance of the wearer (Sheedy 1995). However, clinical experience strongly support the fact that users prefer antireflection coated lenses. Organic lenses were chosen because of lighter weight.

3.3 Study design.

The study is a prospective parallel group design with three groups. Each group consists of approximately 50 subjects. One group, designated "T" for technical group, one designated "S" for software group and one designated "C" for control group. The T and S groups got new luminaries, arm support and finally optometric examination and corrections if needed.

Intensity and frequency of visual discomfort were evaluated by using a 10 cm. visual analogue scale (VAS). Three points on the scale indicated "none", "some" and "very much" discomfort. Location of the different problems was also recorded. The evaluation was for the last month and for the last six months before and after the intervention, and after 3,5 years.

Severity and frequency of headache for the last month and the last 6 months, were evaluated on the same type of VA-scales as visual problems.

4 Results

Visual discomfort last month shows a significant reduction in both intervention groups after lighting and ergonomic intervention (p=0.03), but no reduction in

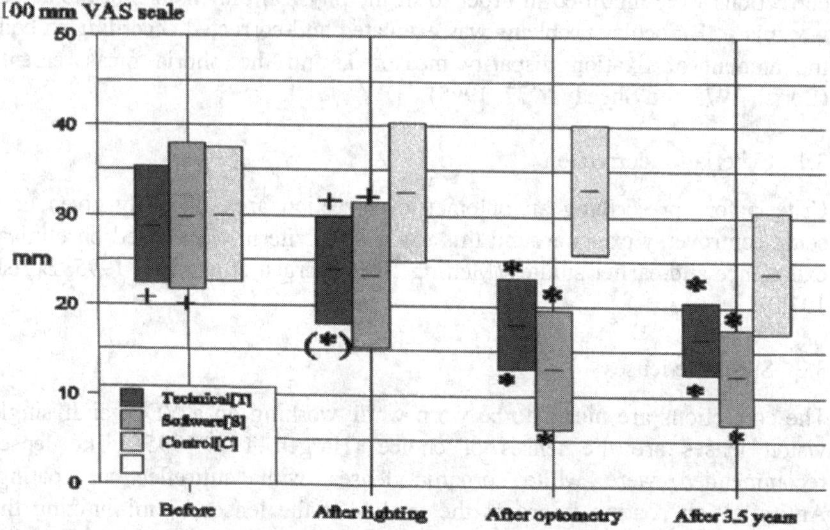

Fig: 1: Visual discomfort last 6 months. (A star indicates significance at 0.05 level. A star in brackets indicates a tendency (0.05<p<0.09)).

the control group. After optometric intervention, there is a tendency towards reduction in visual problems only in the technical group (p=0.07). After 3,5 years, there are significant differences between the intervention groups and the control group. (p=0.0001), and there is also a difference in the rate of reduction (p=0.0004). Looking at visual discomfort last 6 months (fig.1) there is a tendency towards reduction in the technical group after lighting and ergonomic intervention (p=0.008), and significant reductions after optometric intervention in both intervention groups (p=0.003 - 0.004). At the end of the study period, there are significant differences between intervention groups and the control group (p=0.0002), and also in the rate of reduction (p=0.01).

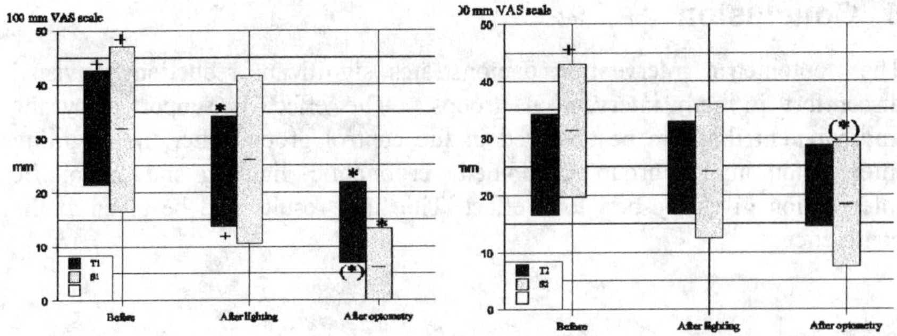

Fig. 2 (left) showing visual discomfort last 6 months for the group who got new corrections. Fig. 3 (right) shows visual discomfort last 6 months for the group who did not get new corrections.

The two intervention groups were divided into two subgroups, one who have got new corrections (T1, S1), and the other who did not need new corrections (T2, S2). There is significant reduction of visual problems after lighting and ergonomic intervention (p=0.04) and tendency towards improvement after optometric intervention (p=0.07) in T1. In the software subgroup (S1) there is no change after lighting and ergonomic intervention, but tendency towards improvement after optometric intervention (p=0.08). At the end of the study period there are significant improvements in both T1 and S1 groups.

In the groups T2 and S2, there were no significant changes.

The T1 group started up with more problems than the T2, they got a reduction down to the same level after lighting intervention, and continued to improve after having new corrections. T1 ended up with a tendency towards having less problems than the T2 group. In the software group, the tendencies are not as clear, but S1 shows slightly more reduction of problems compared to S2, however not significant (p= 0,20).

One implication might be that uncorrected/wrongly corrected ametropes are more sensible to the lighting conditions than emmetropes or corrected ametropes.

After 3.5 years, where an intervention in about half of the control group is done, we can observe a tendency towards a reduction in visual problems. The total results after full intervention in the control group will be given at the conference.

Headache last month shows a significant reduction in the technical group after lighting and ergonomic intervention (p=0.02). In the software group the reduction is after optometric intervention (p=0.03). After 3,5 years, there is no difference between the groups.

5 Conclusion

The optometric intervention demonstrates significant reduction of visual discomfort in both intervention groups. The effect is supported by the improvement that can be observed in the control group after the lighting intervention in this group. Combined ergonomic, lighting and optometric intervention gives the best total effect. The final results will be given at the conference.

6 References

Bjørset, H-H. and Aarås, A., (1996). *Visual and Lighting Conditions.* In: Advances in Applied Ergonomics. Eds: A Øsak and G. Salvendy. USA Publishing, 215-220

Borish, I.M., 1975: *Clinical Refraction.* Proff. Press. Chicago - USA.

Metling, D. 1992: *Standards and methods for visual examination of VDU users - a critical revue.* WWDU - 1992. B2, pp. b-7

Sheedy, J.E., 1977: *Phoria, Vergence and Fixation Disparity in Oculomotor Problems.* American Journal of Optometry & Physiological Optics Vol 54, No. 7 p. 474-478

Sheedy, J.E. 1995; *Vison at computer Displays.* Vision Analysis. 136 Hilcroft Way. Walnut Creek, C.A. 94596

Aarås, A., Fostervold, K.I., Ro, O., Thoresen, M., and Larsen, S., (1997). *Postural load during VDU work: A comparison between various work postures.* Ergonomics Vol 40, No 11, 1255-1268

Horgen, G., Aarås, A., Fagerthun, H.E., Larsen, S.E., 1989: *The work posture and postural load of the neck/shoulder muscles when correcting presbyopia with different types of multifocal lenses on vdu-workers.* Proccedings of the third international conference on human/computer interaction, Boston Massachusetts 18-22, vol.1. Elsevier.

Horgen, G., Aarås, A. 1995: *Is there a reduction in trapezius load when wearing progressive lenses over a three months period.* Applied Ergonomics 1995, no 3. pp 165-171

Can a more neutral position and support of the forearms at the table top reduce pain for VDU operators. Laboratory and field studies.

Arne Aarås[1], Ola Ro[2] and Gunnar Horgen[3]

1. Alcatel STK A/S, P.O.Box 60 Økern, N-0508 Oslo, Norway,
2. Premed. P.O.Box 275, Økern, N-0511 Oslo, Norway.
3. Buskerud College, Department of Optometry, P.O. Box 235, N-3601 Kongsberg, Norway.

1 INTRODUCTION

In order to evaluate the health effects of both supporting the forearms and keep it in a more neutral position when doing VDU (Visual Display Unit) work, laboratory and field studies were carried out.

2 LABORATORY STUDIES

2.1 Materials and Methods

A laboratory study with 20 experienced VDU workers was performed. The aim of the study was to compare the postural load during data entry work when sitting with and without possibility to support the forearms on the table top by using electromyography (EMG). Figure 1. The upper static trapezius load when using keyboard was significantly less when sitting with supported forearms on the table top (0.8 % MVC (Maximum Voluntary Contraction)) compared with sitting (3.6 % MVC) without support, as group mean values respectively. The static load when using mouse was significantly less with the forearm supported, (0.1 % MVC vs. 1.2 % MVC) without support, as group mean values (Aarås et al 1997).

In another laboratory study, 13 experienced VDU workers participated. The aim was to compare the muscle load of the extensors of the foream when using the forearm in a more neutral position compared with a pronated one during mouse work. The load on the extensor digitorum communis, extensor carpi ulnaris and musculus trapezius was measured by EMG. The muscle load of the forearm was significantly less when using the forearm in a more neutral position compared with a pronated one. This was true for extensor digitorum communis regarding the static (p=0.0005) and median (p=0.001) values. The same clear tendency was also found regarding the static muscle load from the extensor carpi ulnaris (p=0.06).

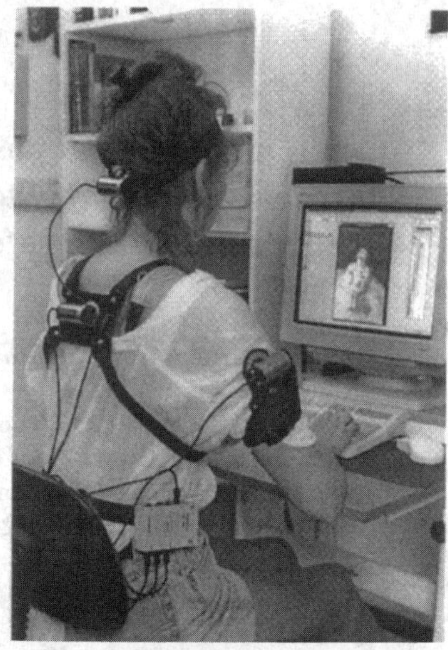

Figure 1. The physiometer records EMG on 4 channels and postural load on 6 channels. Inclinometers are attached to the upper arm, head and back.

3 FIELD STUDIES

3.1 Materials and Methods

The field studies were to evaluate if the results found in the laboratory studies were valid in real work situation. Secondly, to find out if a reduction in postural load leads to a reduction in the incidence of musculoskeletal illness. The first field study has a prospective parallel group design with two intervention groups and one control group, each consisting of approximately 50 subjects. Only experienced male VDU workers were included in the study. The main task in each group was software engineering. Before the intervention, there was no possibility for the operator to support the forearms on the table top. After intervention, the important ergonomic feature was that the table top had an increased depth compared to the old workplaces. This allowed the operator to support the whole forearms and hands when operating keyboard and "mouse" (Aarås et al 1998). In addition, the project aimed at evaluating the effect on visual discomfort by introducing ergonomically luminaires and optometric corrections if needed. The three interventions were implemented in a serial way (Aarås et al. 1998).

3.2 Results

Supporting of the forearms on the table top did reduce the static trapezius load in real work situations when the operators used both keyboard and "mouse" as input devices. Operators from one of the intervention group (the T group) recorded a reduction in static load as a mean group value from 1.5 % MVC to 0.2 % MVC, p=0.002. Similar data for the other intervention group (S group) was 1.5 % MVC vs. 0.3 % MVC, p=0.003. Figure 2. The static trapezius load was not significant different in the control group (C group) (1.1% MVC) compared with the intervention groups before intervention (p=0.3). However, the T and S groups had significantly less static load after intervention compared with the C group (T group, p=0.002, S group, p=0.003).

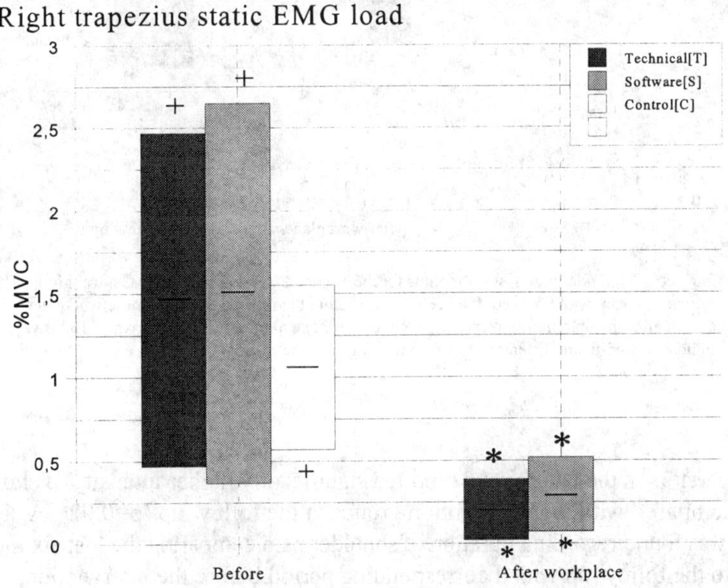

Figure 2. Right trapezius static EMG load for the Technical, Software and the Control group. The values are given as mean, with 95 % confidence interval, before and after the workplace intervention. A * indicate a significant difference when comparing with the column containing a +.

The average intensity of shoulder pain the last six months before follow-up (two and a half years after the intervention) compared with six months prior intervention showed a significant reduction in the S group (p=0.02) and a clear tendency to reduction in the T group (p=0.08), while no significant changes were found in the

C group. Figure 3.

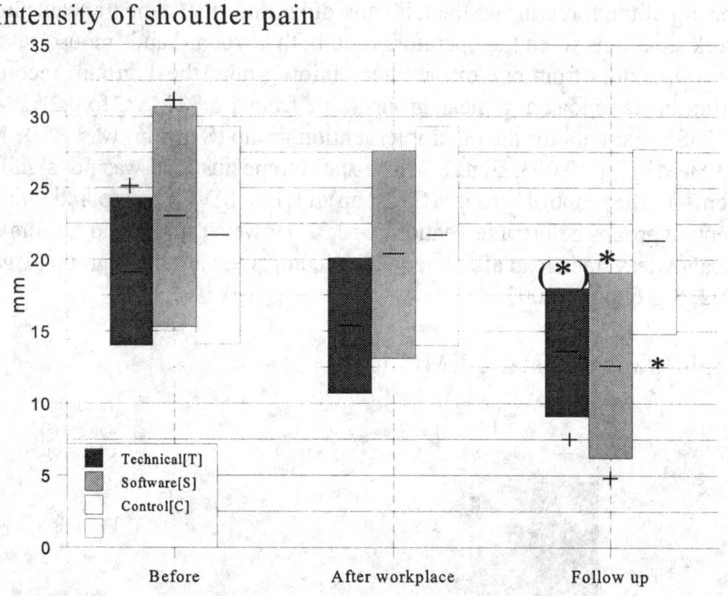

Figure 3. Intensity of shoulder pain last six months for the Technical, Software and the Control group. The values are given as mean, with 95 % confidence interval, before the interventions, after the workplace intervention and after the optometric intervention. On the ordinate, 0 indicates no pain while 100 is very severe pain. A * indicate a significant difference when comparing with the column containing a +. A (*) is very close to a significant difference.

The operators in the latter group reported significantly higher intensity of shoulder pain compared with the intervention groups at the follow-up (p=0.02). A similar result was found regarding duration of shoulder pain comparing the last six months prior to the follow-up with a corresponding period before the intervention.

In the second study, 70 experienced VDU workers with pain in the forearm and shoulder were divided randomly in two groups. The aim of the study was to evaluate pain development when using a mouse with a more neutral position of the forearm and wrist compared with pain development when using a traditional mouse with a more pronated forearm. After six months, a significant reduction was reported regarding pain intensity and frequency of wrist/hand, forearm, shoulder and neck (p<0.009). The control group using the traditonal mouse reported only small changes in the pain level (p>0.24) (Aarås et al. 1999).

4 CONCLUSION

Giving VDU operators the possibility to support of the forearms on the table top in front of the them and operating the mouse with a more neutral position of the forearm, reduced significantly the load on the extensor digitorum communis, extensor carpi ulnaris and musculus trapezius with a corresponding reduction of pain in the upper part of the body.The static load seem to be predictor for the development of musculoskeletal illness in the upper part of the body.

5 REFERENCES

Aarås, A., Fostervold, K. I., Ro, O., Thoresen, M. and Larsen, S. (1997). Postural load during VDU work: a comparison between various work postures. Ergonomics, Vol. 40, No. 11, 1255-1268.

Aarås, A., Horgen, G., Bjørset, H-H., Ro, O. and Thoresen, M. (1998). Musculoskeletal, visual and psychosocial stress in VDU operators before and after multidisciplinary ergonomic interventions. Applied Ergonomics Vol. 29, No 5, 335-354.

Aarås, A., Ro, O. and Thoresen, M. (1999). Can a more Neutral Position of the Forearm When Operating a Computer Mouse reduce the Pain Level for VDU Operators? International Journal Human-Computer Interaction. In print. Vol 11, (2).

Bjørset, H-H. and Aarås, A. (1996). Visual and Lighting conditions. RF: Ahmet F. Øzok and Gavriel Salvendy. Proceedings of the 1st International conference on Applied Ergonomics (ICA`96), Istanbul, May 21-24, USA Publishing, 215-220.

Horgen, G and Aarås, A. (1996). Optometric examination and corrections of VDU. Ed: A hmet F. Øzok and Gaveriel Salvendy. Proceedings of the 1st International conference on Applied Ergonomics (ICA`96), Istanbul, May 21-24, USA Publishing, 227-230.

Variation of sitting posture in work with VDUs. The effect of downward gaze.

Knut Inge Fostervold and Ivar Lie
Department of Psychology, University of Oslo. P.O.box 1094 Blindern
N-0317 Oslo, NORWAY. e-mail: k.i.fostervold@psykologi.uio.no

1 Introduction

Over the last fifteen years information technology and the use of visual display units (VDU) have become a predominant device in the industrialised world. Due to its rapid implementation very little was known about possible health risks ensuing from prolonged interactive VDU work. This fact led to immense interest among scientists and health professionals, when reports about adverse health reactions started to appear (Dainoff and Happ 1981, Grandjean 1984).

Initially, eye problems and musculoskeletal problems were the main focus of interest. At that time, ergonomists were fairly acquainted with health problems caused by sedentary and repetitive work. Since most VDU-operations are sedentary, necessitating repetitive body movements and constrained postures, earlier data, especially from typists, were used as a frame of reference. Visual and oculomotor consequences of sustained excessive near-work have been research topics in visual perception and optometry for almost 50 years (Brozek et al. 1950, Tyrrell and Leibowitz 1991). Since typical VDU-work tasks are dominated by intensive visual work at short range, renewed interests in these issues were a logical consequence. Other health problems discussed in the literature are psychosocial factors, concentration problems, symptoms of skin irritation, headache, dizziness and general fatigue (Grandjean 1984, Bergqvist et al. 1992). However, in spite of considerable scientific and ergonomic effort many VDU operators still experience physiological problems related to their work.

Consequently, several guidelines have been developed recommending preventive and remedial actions for the VDU workplace. A recommendation that has achieved widespread acceptance is the high monitor placement in which the VDU-screen is positioned slightly below eye level. In this position, the midpoint of the screen requires a gaze angle approximately 10°- 20° below the horizontal line. Recently, this recommendation has been challenged. Today, a growing number of researchers and ergonomists advocate VDU positions that require downward gaze inclinations. This recommendation is based on research indicating reduced visual fatigue and musculosceletal strain in subjects working with the midpoint of the VDU 30°- 45° below the horizontal line (Ankrum and Nemeth 1995, Jaschinski et al. 1995, Lie and Fostervold 1995, Lie et al. 1997, Burgess-Limerick et al. 1998).

Empirical studies investigating the relationship between monitor placement and sitting posture are relatively scarce. However, in some instances laboratory research has shown increased head and trunk flexion as a function of downward gaze (Villanueva et al. 1996, Bauer and Wittig 1998, Burgess-Limerick et al. 1998, Turville et al. 1998). Others have reported no difference in head flexion when comparing gaze angles of 15° and 30° (Aarås et al. 1997). The objective in this study was to investigate the relationship between sitting posture and monitor placement in an ordinary office environment.

2 Methods

The study was an integrated part of a large field study in Norway. A sample of 150 subjects participated. The sample consisted of 111 women and 39 men. The mean age was 40,9 years, with a standard deviation of 9,7 years. The Subjects were divided randomly into a Downward gaze group and a Standard gaze group.

A new computer desktop was designed for the downward gaze group. Sitting posture was recorded, by measuring body movements of the head, right upper arm and back, continuously during 45 minutes of normal labour. Three dual axis inclinometers (Physiometer PHY-400) were used for these measurements. The postural angles were determined as deviations from an upright standing position with neutral head and trunk. Shoulders and arms were hanging down while the subject fixated a red mark at eye level. Flexion/extension and abduction/adduction were recorded for the right upper arm while flexion/extension and sideways flexion was recorded for the head and the back. The measurements were recorded twice, prior to the implementation of the new desktop and one-year later.

The gaze inclination was defined as the intermediate angle lying between a line drawn from the eye to the midpoint of the VDU and a horizontal line drawn at eye level, provided the head-erect sitting posture. Desktop and monitor placement in the standard gaze group was optimised according to a gaze inclination of 15°. The downward gaze group was originally planned with a gaze inclination of 40°. However, due to glare problems caused by inferior luminaries the gaze angle was reduced to about 30° below the horizontal line.

3 Results

The results revealed that only minor pre-test differences was present between the two conditions. Posture angles recorded one year later revealed only minor changes in the standard gaze group compared with the pre-test measures. All changes was within 2° from the pre-test, except for shoulder abduction were an significant increase of 4,06° was observed (t= -2,33, df=60, p=.023).

The posture angles recorded one year later for the downward gaze group revealed significant increases of 4,92° for median head flexion (t= -3,58, df=64, p= .001), 6,11° for head flexion recorded when subjects were looking at the upper most line of the VDU (t= -3,53, df=64, p=.001) and 5,95° for head flexion recorded when subjects were looking at the lowest line of the VDU (t= -3,45, df=64, p=.001). A significant increase of 4,75° was also found for back flexion (t=-3,45, df=64, p=.001). For shoulder abduction the increase of 4,38° was almost significant (t=-1,95, df=64 p=.056). Head sideways, back sideways and shoulder flexion showed only small insignificant increases. Postural angles recorded at the pre-post-test are shown for both groups in table 1.

Table 1. Pre-post-test means of median postural angles.
Downward and standard gaze inclination.

Body part and type of measure.	*Down. gaze Pre-test*	*Down. gaze Post-test*	*Stand. gaze Pre-test*	*Stand. gaze Post-test*
Head flexion	16,74°	21,66°	16,05°	16,11°
Head sideways	1,64°	3,51°	1,63°	1,22°
Head flex., VDU's upper line	1,14°	7,25°	0,29°	0,61°
Head flex., VDU's lowest line	6,47°	12,42°	6,80°	7,19°
Back flexion	10,19°	14,94°	8,43°	10,13°
Back sideways	-0,08°	1,23°	-0,17°	-0,94°
Shoulder flexion	4,07°	4,88°	2,74°	2,58°
Shoulder abduction	17,89°	22,27°	19,64°	23,70°

Between group comparisons revealed significant differences for median head flexion ($F_{(1,124)}$ = 6,74, p= .011), head flexion recorded when subjects were looking at the upper most line of the VDU ($F_{(1,124)}$ = 6,53, p= .012) and head flexion recorded when subjects were looking at the lowest line of the VDU ($F_{(1,124)}$ = 6,08, p= .015). The analysis for back flexion showed a between group difference that was nearly significant ($F_{(1,124)}$ = 3,47, p= .065).

Postural shifts per minute were recorded in relation to some of the postural angles. The analysis did not reveal any significant differences between the two conditions at the post-test. One-year later, the results for head flexion upper and lowest line of the VDU, revealed small increases in postural shifts per minute in both groups. For shoulder flexion and abduction, a decrease was present.

However, in the standard gaze group only the decrease in postural shifts per minute found for median shoulder abduction was significant (t=2,45, df=60, p= .017). In the downward gaze group the decrease found for both median shoulder flexion (t=3,21, df=62, p= .002) and median shoulder abduction was significant (t=2,26, df=64, p= .027). Postural shifts per minute recorded at the pre-post-test are shown for both groups in table 2.

Table 2. Pre-post-test means of postural shifts per minute.
Downward and standard gaze inclination.

Body part and type of measure.	*Down. gaze Pre-test*	*Down. gaze Post-test*	*Stand. gaze Pre-test*	*Stand. gaze Post-test*
Head flex., VDU's upper line	3,48	4,10	3,66	3,89
Head flex., VDU's lowest line	6,12	6,71	6,48	6,76
Shoulder flexion median	12,93	10,42	11,35	10,34
Shoulder abduction median	16,13	13,95	15,15	12,70

Between group comparisons revealed no significant differences. However, a tendency was present for postural shifts at median shoulder flexion ($F_{(1,122)}$ = 3,59, p= .061).

4 Discussion

In this study, the objective was to investigate the effect of lowering the midpoint of the VDU from 15° to 30° below the horizontal line on sitting posture. The results obtained show an increase in head and back flexion in the downward gaze group compared to the standard gaze group. The difference was 5,5° for median head flexion and 4,8° for median back flexion. The difference in head flexion related to the midpoint of the VDU was 6,2°. Thus, alterations in postural angles accounted for 11° of the altered gaze angle, leaving about 4° to alterations in eye position. Our finds correspond with postures reported by Burgess-Limerick et al. (1998). However, others have reported larger head tilts in conjunction with lowered monitor placement (Villanueva et al. 1996, Bauer and Wittig 1998, Turville et al. 1998). Thus, the results indicate that the downward gaze inclination was compensated for by both increased head flexion and increased back flexion. In this context, it is intriguing to note that most people engaged in work at close range under natural conditions prefer a similar, forward leaning sitting posture.

Increased shoulder abduction was observed in both groups. Most likely, this increase was caused by the underarm support provided by the new and the adjusted desktop used by the downward and standard gaze group, respectively. The causal relationship for the observed decrease in postural shifts per minute is probably similar. Aarås et al. (1997) have previously shown reduced trapezius activity in subjects working with full underarm support.

To conclude, the downward gaze inclination represents not only a shift of monitor placement but also has implications for the preferred sitting posture. Prevailing guidelines have regarded the erect sitting posture as preferable. However, the scientific basis for the contention of the superiority of an erect sitting posture in VDU work is questionable. Moreover, even if we accept that the erect sitting posture is preferable when working with high monitor placements this is not necessarily the case when working with a downward gaze inclination.

5 References

Aarås, A., Fostervold, K. I., Ro, O. Thoresen, M. & Larsen, S. (1997). Postural load during VDU work. A comparison between various work postures. *Ergonomics*, 40(11), 1255-1268.

Ankrum, D. R. & Nemeth, K. J. (1995). Posture, Comfort, and Monitor Placement. *Ergonomics in design*, April, 7-9.

Bergqvist, U., Knave, B., Voss, M. and Wibom, R. (1992). A longitudinal Study of VDT Work and Health, *International Journal of Human-Computer Interaction*, 4(2), 197-219.

Bauer, W. and Wittig, T. (1998). Influence of screen and copy holder positions on head posture, muscle activity and user judgement. *Applied Ergonomics*, 29(3), 185-192.

Brozek, K., Simons, E. and Keys, A. (1950). Changes in performance and in ocular functions resulting from strenuous visual inspection. *American Journal of Psychology*, 63, 51-56.

Burgess-Limerick, R. Plooy, A. and Ankrum, D. R. (1998). The effect of imposed and self-selected computer monitor height on posture and gaze angle. *Clinical Biomechanics*, 13, 584 – 592.

Dainoff, M. J. & Happ, A. (1981). Visual Fatigue and occupational Stress in VDT Operators. *Human Factors*, 23(4), 421-438.

Grandjean, E. (1984). *Ergonomics and Health in modern Offices*. London; Taylor & Francis.

Jaschinski, W., Koitcheva, V. and Heuer, H. (1995). Accommodation and vergence during inclined gaze. *Perception*, 24, supplement A 70-71.

Lie, I. & Fostervold, K. I. (1995). VDT - Work With Different Gaze Inclination. In Grieco, A. Molteni, G. Piccoli B. and Occhipinti E. (Eds): *Work with Display Units'94*, pp 137-142. Amsterdam: Elsevier Science.

Lie, I. Fostervold, K. I. Aarås A & Larsen S. 1997, Gaze Inclination and Health in VDU-operators. In Seppälä, P. Luopajärvi, T. Nygård, C. H. & Mattila, M. (Eds): *From Experience to Innovation, IEA'97 (vol. 5)*, pp 50-53. Helsinki: Finnish Institute of Occupational Health.

Turville, K. L., Psihogios, J. P., Ulmer, T. R. and Mirka, G. A. (1998). The effects of video display terminal height on the operator: a comparison of the 15° and 40° recommendations. *Applied Ergonomics* , 29(4), 239-246.

Tyrrell, R. A. & Leibowitz, H. W. (1991). The Relation of Vergence Effort to Reports of visual Fatigue Following Prolonged Near Work. *Human Factors*, 32(3), 341-357.

Villanueva, M. B. G., Sotoyama, M. Jonai, H., Takeuchi, Y. and Saito, S. (1996). Adjustment of posture and viewing parameters of the eye to changes in the screen height of the visual display terminal. Ergonomics, 39(7), 933-945.

Health Improvements Among VDU Workers After Reduction of the Airborne Dust in the Office. Three Double-blind Intervention Studies

Knut Skyberg, Knut Skulberg and Wijnand Eduard (1)
Arnt I. Vistnes (2) Finn Levy and Per Djupesland (3)
1. National Institute of Occupational Health,
PO BOX 8149 Dep, N-0033 Oslo, Norway
2. Institute of Physics, University of Oslo
3. Ullevål University Hospital, Oslo

1 Background of the Programm

VDU users frequently report skin and mucous membrane problems. These symptoms may at least in part be due to physical and chemical exposures in the office environment (Skov et al, 1990). Dust particles, static electric (Lindén and Rolfsen, 1981) and electromagnetic fields (Sandström et al, 1995) around the VDU have been suggested as exposures that might be responsible for the complaints reported. A case-control study of VDU workers indicated that the intensity of cleaning influenced the risk of perceiving skin symptoms (Sundell et al, 1994). Mucous membrane and general symptoms have been associated with dust exposure in other studies on indoor air problems (Raw et al, 1993) (Gyntelberg et al, 1994). Acoustic rhinometry has previously been used as a method in a cross-sectional study of the indoor air in schools (Wålinder et al, 1997).

Previously one blinded intervention study succeeded in reducing symptoms by increasing the air ionization (Wyon, 1992). Another experimental study showed a slight increase in symptoms when the mechanical ventilation was reduced Jaakola et al, 1991). The importance of total volatile organic compounds (TVOC) has been evaluated by a Nordic expert group. They concluded that some VOCs most likely may cause health complaints in office environments. However, TVOC as a risk index for health and comfort in buildings is not established on a scientific basis (Andersson et al, 1997).

To further elucidate the possibilities of preventing skin and mucous membrane symptoms among VDU office workers, we initiated an intervention program, where exposure factors and subjective as well as objective health indicators were measured. In the study of skin irritation colorometry was used, while in the study of mucosal effects acoustic rhinometry was used. A double blind intervention study design was chosen. So far two independent interventions have been completed.

2 First Intervention - Skin Symptoms and Static Electric Fields

In the first intervention we wanted to test the hypothesis that a reduction of static electric fields around the VDU may reduce facial skin symptoms of VDU users.

Four companies participating in a screening questionnaire study were chosen for the first intervention. Among those employees reporting facial skin complaints, 120 employees were selected. They were assigned to an intervention or a control group, so that the two groups had a similar distribution of age, gender and allergy. The intervention group got antistatic treatment of all sides of the VDU, and an antistatic mat was placed on the floor. Both the VDU surface and the antistatic mat were connected to ground. The control group got treatment with a similar appearance but with no effect on the electrostatic field. Before intervention, after two weeks and after four weeks exposure and health indicators were measured.

In the intervention group the static electric fields measured in front of the VDU were reduced as compared to placebo. A similar reduction in the intervention group was found in the localized static electric fields from the top and sides of the VDU, called «hot spots». The reduction of the skin symptom index was largest in the intervention group and there was a statistically significant difference between the intervention and placebo groups ($P<0.05$). Multiple linear regression analysis showed that only in offices with a high dust level a reduction of static electric fields was associated with a fall in skin symptoms.

3 Second Intervention - Mucous Membrane Symptoms and Cleaning

In the second intervention we wanted to study if an intensified cleaning of the office might decrease the concentration of airborne dust in the office, followed by a reduction in mucous membrane symptoms, and reduced nasal airways congestion.

Participants for the second intervention were recruited from one of the companies taking part in the first intervention. The 120 non-smoking employees reporting most mucous membrane complaints were selected. They were assigned to an intervention or a control group, so that the two groups had a similar distribution of mucous membrane symptoms, age, gender and allergy. In the intervention group the offices were thoroughly cleaned, including removal of mineral fibre mats from the ceiling. The placebo group received only a quick, superficial cleaning of the office. Dust concentration and health indicators were measured before and after intervention or placebo treatment.

The dust level was reduced near the VDU after the intervention, and the reduction was statistically significant for offices with a high dust concentration before cleaning. A mucous membrane symptoms index score was reduced in the intervention group, compared to placebo ($p<0.05$). Among those participants with a narrow nasal airway, there was a larger increase in nasal volume 22-52 mm from the nasal orifice in the intervention group as compared to placebo. This effect was most pronounced among persons younger than 40 years and among participants with a history of rhinitis or asthma.

4 Third Intervention - Mucous Membrane Symptoms and Airborne Dust Filtration

The third intervention investigates health improvements after removal of airborne dust using electrostatic filter units. Both upper and lower airways symptoms and function has been investigated. The project was implemented in early 1999. Results will be presented at the conference.

5 Conclusions

Two intervention studies on the physical environment of VDU users have been performed. In the first study we found that a reduction of the static electric field around the VDU prevented skin symptoms. In the second study an intensive cleaning of the office reduced the airborne dust, and upper airways symptoms and nasal congestion were reduced. These interventions indicate that dust and static electric fields should be given attention when the work environment for VDU users is evaluated. A third intervention testing the effect of air cleaning is carried out in 1999.

6 References

Andersson L, Bakke JV, Bjørseth O, Bornehag C-G, Clausen G, Hongslo JK et al. (1997). TVOC and health in non-industrial indoor environments. Report from a Nordic Scientific Consensus Meeting at Långholmen in Stockholm, 1996.

Gyntelberg F, Suadicani P, Nielsen JW, Skov P, Valbjørn O, Nielsen PA et al. (1994). Dust and the sick building syndrome. *Indoor Air* 4, 223-238.

Jaakkola JJK, Heinonen OP, Seppänen O. (1991). Mechanical ventilation in office buildings and the sick building syndrome. An experimental and epidemiological study. *Indoor Air*, 2, 111-121.

Lindén V, Rolfsen S. (1981). Video computer terminals and occupational dermatitis. *Scand J Work Environ Health*, 7, 62-63.

Raw GJ, Roys MS, Whitehead C. (1993). Sick building syndrome: cleanliness is next to healthiness. *Indoor Air*, 3, 327-345.

Sandström M, Hansson Mild K, Stenberg B, Wall S. (1995). Skin symptoms among VDT workers and electromagnetic fields - a case referent study. *Indoor Air*, 5, 29-37.

Skov P, Valbjørn O, Pedersen BV. (1990). Influence of indoor climate on the sick building syndrome in an office environment. *Scand J Work Environ Health*, 16, 363-371.

Sundell J, Lindvall T, Stenberg B, Wall S. (1994). Sick building syndrome (SBS) in office workers and facial skin symptoms among VDT-workers in relation to building and room characteristics: two case-referent studies. *Indoor Air* 1994; suppl. 2.

Wyon D. Sick buildings and the experimental approach. (1992). *Envir Technology*, 13, 313-322.

Wålinder R, Norbäck D, Wieslander G, Smedje G, Erwall C. (1997). Nasal mucosal swelling in relation to low air exchange rate in schools. *Indoor Air*, 7, 198-205.

Indoor Foliage Plants Reduce Mucous Membrane and Neuropsychological Discomfort Among Office Workers

Kaj Bo Veiersted [1], Tove Fjeld [2]
[1] National Institute of Occupational Health, P.O. Box 8149 Dep, N-0033 Oslo, Norway
[2] Department of Horticulture and Crop Sciences, P.O. Box 5022, Agricultural University of Norway, N-1430 Aas, Norway

1 Introduction

The indoor climate in office buildings has been changing through the last decades due to the introduction of new building materials, ventilation systems and increased emphasis on energy saving. During the same period it seems that health and discomfort problems are increasing among office workers (Skov et al 1990). The objective of the present study was to assess the effect of foliage plants in the office on health and discomfort symptoms among office personnel. The results are previously published (Fjeld et al 1998).

2 Subjects and Methods

A cross-over study with randomised period order was conducted among 51 healthy subjects (27 males and 24 women); one period with plants in the office, and one period without. The plants used for the intervention were: Four *Aglaonema commutatum,* two *Dracaena deremensis,* four *Epipremnum aureum* and three *Philodendron scandens* on the window bench. In the back corner of the office, close to the entrance, one terracotta container was placed with one 1.75 meter high *Dracaena fragrans* and four *Epipremnum aureum*

The subjects completed a questionnaire every second week regarding 12 different health symptoms during two spring periods of 3 months in 1995 and 1996. The questionnaire included the following 12 symptoms: (1) fatigue, (2) feeling heavy-headed, (3) headache, (4) nausea/drowsiness, (5) concentration problems, (6) itching, burning, irritation of the eyes, (7) irritated, stuffy or «running» nose, (8) hoarse, dry throat, (9) cough, (10) dry or flushed facial skin, (11) scaling/itching scalp or ears, (12) hands with dry, itching, or red skin. The questions were modified after (Anderson *et al.* 1993). Each symptom could be given one of the following scores: 0 (no problems), 1 (small problems), 2 (moderate problems) or 3 (severe problems). The scores given should reflect problems *the same day* as the questionnaire was filled in.

Since most of the subjects did not complete all the questionnaires, (due to travelling, illness etc.), a mean score was calculated for each person for every single symptom in the two periods; spring 1995, and spring 1996. The statistical analyses are based on these mean scores, together with the mean sum score (summarised for all 12 symptoms). A two-sided Wilcoxon signed-rank test was used to decide if a mean difference between the periods with and without plants was statistically significant. Significance level was 5%.

If the effect of an intervention depends heavily on whether it is given in the first or in the second period (carry-over effect), this will destroy the cross-over design. In the present study no carry-over effects were seen in any of the symptom scores.

3 Results

It was found that score sum of symptoms was 23 % lower during the period when subjects were exposed to plants in their offices compared to the control period (Table I). Complaints regarding cough and fatigue were reduced with 37% and 30 % respectively, if the offices contained plants. The self reported level of dry/hoarse throat and dry/itching facial skin each decreased approximately 23 % after plant-intervention. A significant reduction was obtained in neuropsychological symptoms and mucous membrane symptoms, while skin symptoms seemed to be unaffected by the plant-intervention.

Table 1. Effects of plant intervention on health and discomfort symptoms.

Period :	without plants	with plants	Difference
Mean score	7.1	5.6	1.5
Standard deviation	4.7	3.3	3.1
Min-max	0.0-18.6	0.6-13.0	-4.8-11.4
p value			0.002

4 Discussion

This study suggests that foliage plants in the office environment reduce health and discomfort symptoms, especially neuropsychological and mucous membrane symptoms. The findings may have three main explanations 1) improvement of air-quality by the plants, 2) an increase in general well-being due to the perception of foliage plants, or 3) a «pleasing-effect» of increased attention.

Previous studies have shown that commonly used species of indoor foliage plants may reduce the content of air contaminants (Wolverton et al 1985). The plants may also increase the air humidity with up to 15% (Pearson 1994). An increase in humidity and a decrease in CO_2 due to improved ventilation system have recently been shown to reduce complaints of «sick building syndrome», a reduction maintained up to 3 years (Bourbeau et al 1997). However, it appears from earlier measurements in a sample of the used offices that the air-quality and the ventilation capacity in our setting are better than the minimum requirements. It is likely that the micro-climate around the plants will have a somewhat higher level of humidity. This might affect the subjects perception of the air quality, especially since most of the plants were placed close to the subjects. Hence, the effect of plants on the local air quality, especially the humidity, may be one explanation of our results.

The reduction in health and discomfort complaints during the period with plants may also be explained by increased subjective well-being. Several studies have shown that passive experiences with environments or view-settings with vegetation tend to have positive influence on well-being. Some data (Ulrich & Parsons 1992)indicates that the benefit of viewing vegetation goes far beyond aesthetic, and includes not only psychological effects, but also measurable physiological effects (Ulrich 1981). Relief from stress may be accomplished faster and more completely if the setting is dominated by vegetation than if it is an urban one with little or no vegetation (Ulrich et 1991, Ulrich 1984). Scientific investigations regarding effects of indoor plants and flowers on emotional states and stress recovery, is lacking. The results of the present study suggests, however, that similar favourable responses may be obtained by indoor plants. This second interpretation may the also explain our findings.

The third explanation of our findings was that they are an effect of increased attention, a pleasing effect. An attempt to minimise this effect was performed by giving the control group the opportunity to have a nature poster on the office wall. One third of the controls accepted this offer. Furthermore, the service of the plants were performed after ordinary working hours to reduce contact with the intervention group and all participants were contacted regularly for an informal conversation. Another argument against a strong pleasing effect in this study, is that not all variables investigated showed a significant reduction in complaint level. In the main only those that *a priori* were assessed to be related to cleaner and more humid air were affected. Thus, we consider an effect of increased attention paid to the subjects to be of minor importance in the present study.

In conclusion, the present study indicate that plants in offices may reduce mucous membrane and neuropsychological discomfort symptoms markedly. Improved air-quality and/or a perceived nicer office environment may be the main explanations to these findings.

5 References

Skov P, Valbjørn O, Pedersen BV: Influence of indoor climate on the sick building syndrome in an office environment. *Scand J Work Environ Health* 1990;16:363-371.

Fjeld T, Veiersted B, Sandvik L, Riise G, Levy F: The effect of indoor foliage plants on health and discomfort symptoms among office workers. *Indoor Built Environ* 1998;7:204-209.

Anderson K, Fagerlund I, Stridh G, Larsson B: *The MM-questionnaires*. A tool when solving indoor climate problems. Institutt for Miljømedisin, Örebro Hospital. 1993.

Wolverton BC, McDonald RC, Mesick HH: Foliage plants for indoor removal of the primary combustion gases carbon monoxide and nitrogen dioxide. *Jour. of the Mississippi Acad. of Sci.* 1985; 30: 1-8.

Pearson S: Easy breathing in the office. *Horticulture Week*, Nov. 17th, 1994:24-25, 27.

Bourbeau J, Brisson B, Allaire S: Prevalence of the sick building syndrome symptoms in office workers before and after six months and three years after being exposed to a building with an improved ventilation system. *Occup Environ Med* 1997;54:49-53.

Ulrich RS, Parsons R: Influences of passive experiences with plants on individual well-being and health, in: Relf D (ed): *Human benefits of plants: Well-being and social development*. Portland, Oregon, 1992: 93-105.

Ulrich RS: Natural versus urban scenes. Some psychophysiological effects. *Environment and Behavior* 1981; 13(5):523-556.

Ulrich RS, Simons RF, Losito BD, Fiorito E, Miles MA, Zelson M: Stress recovery during exposure to natural and urban environments. *J Environmental Psychology* 1991; 11: 201-230.

Ulrich RS. View through a window may influence recovery from surgery. *Science* 1984;224:420-421.

Difference in the ANS activity between gaze angles while seated

Kaoru Suzuki[a] and Dennis R. Ankrum[b]
[a]College of Engineering, Hosei University, 3-7-2 Kajinocho, Kogani-city, Tokyo 184-8584 Japan
E-mail: suzuki@ergo.is.hosei.ac.jp
[b]Human Factors Research, Nova Solutions, 10007 San Luis Trail, Austin TX 78733-1253 USA
E-mail: Ankrum@aol.com

1 Introduction

The autonomic nervous system (ANS) regulates most bodily functions. While it has been established that ANS activity is affected by both physical and mental activities, the reverse might also be true.

For seated VDT operators, the lower gaze angle has several advantages. The advantages might not be limited to physical ones, such as reduced convergence effort and a reduced risk of dry eye syndrome. For example, the activity of the sympathetic and parasympathetic nervous systems, components of the ANS, are related to mental strain. If a lower gaze angle stabilizes its activity, that might have a desirable effect.

An experiment was conducted on seated subjects performing a VDT task. Both monitor placement and task difficulty were varied to compare ANS activity between gaze angles and task difficulty, and to examine interactions between gaze angle and task difficulty.

2 Evaluation of the ANS activity

ANS activity can be estimated from several physiological measures. In many cases, the measures and evaluations calculated from them vary synchronously. However, in some cases, they vary asynchronously. Hence, to evaluate ANS activity, it is important to observe multiple physiological measures.

In this study, the electrocardiogram (ECG) signal, the pulse signal observed at a lobe, and the continuously-measured systolic blood pressure (BP) were recorded. The instantaneous heart rate (HR) was calculated from inter-beat intervals corresponding to R-wave intervals represented on the ECG (Fig. 1). The heart rate variability (HRV) index (Suzuki and Hayashi 1993, not variance) was also calculated from the intervals. The time lag, or latency, between a beat and a corresponding pulse (R-P, Nagashima *et al.* 1994) was measured (Fig. 1).

Generally, both systolic BP and instantaneous HR increase with increased sympathetic nervous system activity, while the R-P latency and HRV index decrease. The HRV index rises when a high frequency component of the heart rate variability becomes dominant. A remarkable rise of the index often implies that the sympathetic nervous system's activity is reduced. Peripheral blood flow tends to be reduced with a rise in the sympathetic nervous system's activity, because of the tightening of peripheral blood vessels. Consequently, the R-P latency would be expected to shorten with an increase in sympathetic nervous system activity.

In general, the sympathetic nervous system stimulates activity, while the parasympathetic nervous system governs one's vegetative or refreshing activity. When mental strain, or a part of the mental workload, is moderate, the sympathetic nervous system's activity rises with the strain, generally increasing performance. When mental strain is excessive, sympathetic nervous system activity is suppressed. The parasympathetic nervous system's activity becomes relatively dominant, in many cases reducing performance. Consequently, the analysis is requested to detect and evaluate both of the phases.

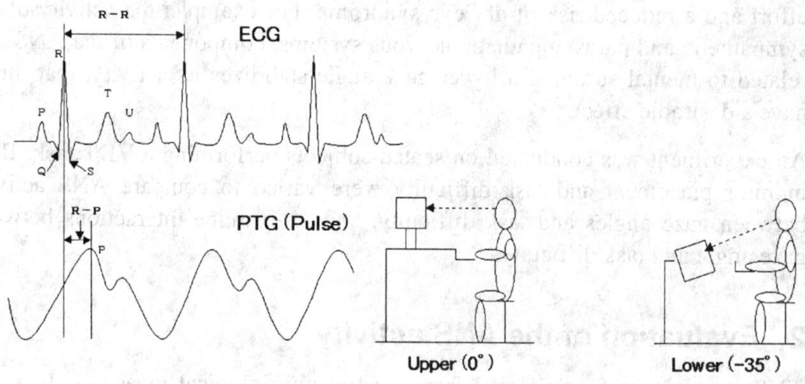

Fig. 1 The R-R interval and the R-P latency. **Fig. 2** The display placements.

3 Experiment

An experiment employing a mental arithmetic task with a time limit was conducted on 12 male subjects (age: 19-24). The time limit was determined according to each subject's mental arithmetic performance measured without the time limit. The task forced the subject to add two one-digit numbers on a display and to enter the least significant digit of the answer on a keypad. Each set consisted of 15 successive correct answers, with the count reset when the time limit was exceeded or an incorrect response was entered. The successful completion of 20 sets terminated the task. There were two monitor placements (Fig. 2), and two levels of time limits (shorter/longer). Each subject performed the task under four experimental conditions. Task performance, i.e. time required to answer and correct/incorrect, was recorded. The NASA task load index (NASA-TLX, Haga 1994, Japanese version) was employed to evaluate the subjective workload.

4 Results and Discussions

Fig. 3 and Fig. 4 show mean values and (Fig. 3 only) 95% confidence intervals. As shown in these figures, strain perceived by the subjects was higher when the time limit was shortened. The lower monitor placements showed a desirable tendency to decrease the error rate and the mental workload perceived by the subjects. However, the differences were not statistically significant.

Fig. 5 and Fig. 6 show mean values and 95% confidence intervals of the physiological evaluations. The data obtained in the first trials were discarded, as subjects tended to show singular values, most likely a result of inexperience with the task. Each of the evaluations was sampled at the end of a set, and then the difference between the sample and the subject's resting average was calculated. The systolic BP (Fig. 5), the R-P latency, and the HRV index varied similarly. The difference in each of these evaluations between the display placements was significant ($p<0.01$). In many cases, increases in systolic BP, decreases in R-P latency, or decreases in the HRV index indicates activation of the sympathetic nervous system. One can therefore assume that the lower monitor placement tends to suppress rising of the sympathetic nervous system's activity, which is caused by the moderate strain. This assumption is coincident with the desirable tendencies shown by the lower monitor placement regarding error rate and perceived workload. However, the shorter time limit in the low monitor condition showed evidence of a lower rising of the sympathetic nervous system's activity. A possible interpretation, coincident with the error rate and the perceived workload, but is not verified, is that since subjects were more composed in the lower monitor placement, resulting in possible boredom or irritation at the longer time limit.

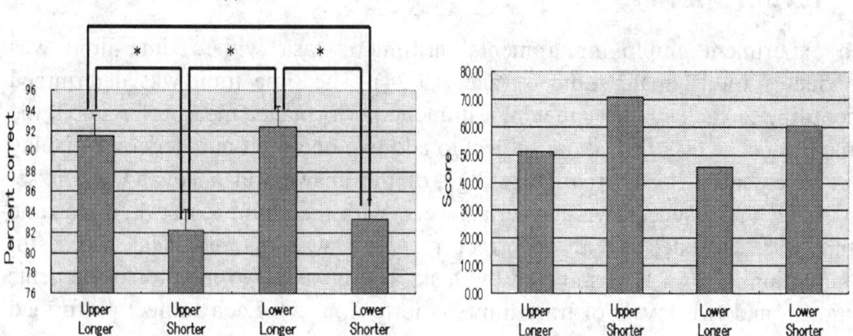

Fig. 3 Percentage of the correct answers. **Fig. 4** The subjective evaluations.

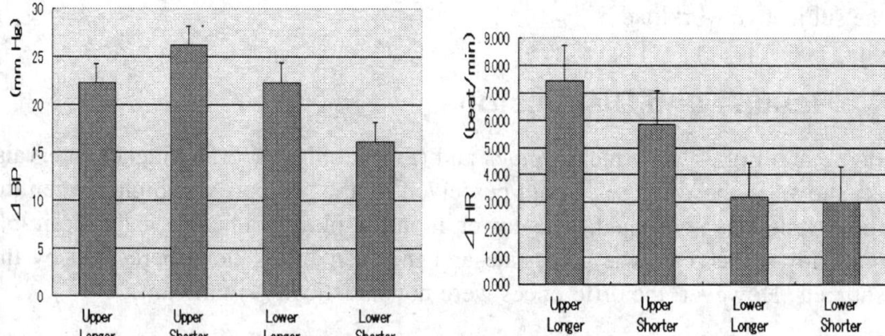

Fig. 5 The increases in the systole BP. **Fig. 6** The increases in the instantaneous HR.

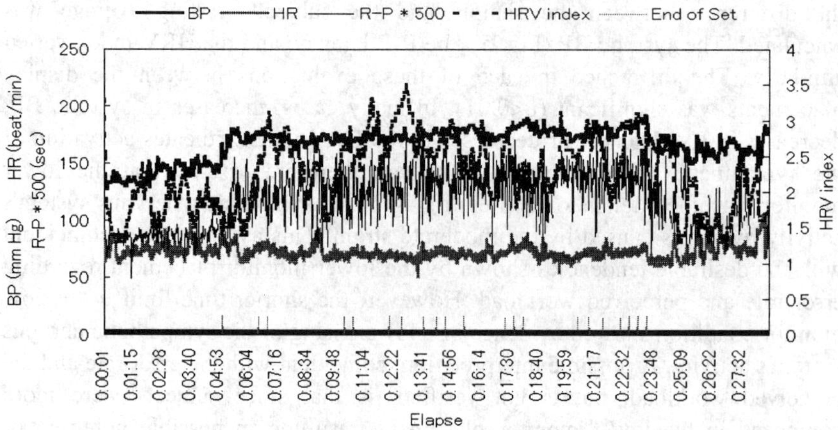

Fig. 7 Changes in the physiological evaluations.

Fig. 7 shows a rare example of changes in the instantaneous HR, systolic BP, HRV index, and R-P latency. With the start of the successive task execution, the R-P latency tended to shorten, and the HRV index tended to increase. Conversely, the systolic BP tended to increase, and the instantaneous HR tended to remain stable, possibly because the sympathetic nervous system's activity was partially suppressed by the excessive mental strain. This is a case, in which the physiological measures and evaluations vary asynchronously with each other. This kind of asynchronous behaviour can cause the instantaneous HR (Fig. 6) to vary dissimilarly, in comparison with the systolic BP, R-P latency, and HRV index.

5 Conclusions

The study examined the effects of gaze angle on ANS activity during seated VDT operations. A non-statistically significant trend was found towards both reduced error rate and perceived mental workload with a low gaze angle. The effect on B-P and instantaneous HR were complicated by the interaction between the gaze angle and the task difficulty and difficult to interpret. The results showed the importance of observing several physiological measures to evaluate its activity.

6 References

Suzuki, K. & Hayashi, Y. (1993). On a simple and effective method to analyze heart rate variability. In Smith, M. & Salvendy, G. (Eds.): *Human Computer Interactions: Applications and Case Studies (Vol. 1), Proc. 5th Int. Conference on Human-Computer Interaction* (HCI International '93, Orlando, USA, August 8-13, 1993), pp. 914-919. Amsterdam: Elsevier.

Nagasgima, K., Hoshiai, K., Aiba, T. *et al.* (1994). Experimental study on an evaluation of autonomic function using digital plethysmography. *Bulletin of Tokai University, School of High Technology for Human Welfare*, 4, 215-221. (in Japanese)

Haga, S. (1994). The Japanese version of NASA task load index. *RTRI Report*, 8(1), 15-20. (in Japanese)

Study of Adaptive Input System for Telecommunication Terminal

-Evaluation of Discriminant which Controls Computer key-input-

† **Seiji Kitakaze** ‡ **Yoshinori Arifuku**, **Tatsuro Imaoka** § **Shin'ichi Fukuzumi**
† 1st C&C Systems Operations Unit, NEC Corporation
‡ Koisikawa Research Center, Telecommunications Advancement Organization of Japan
§ Human Media Research Laboratories, NEC Corporation

1. Objective

For the swift realization of a modern, information-centered society, an approach that will facilitate elderly and physically handicapped people using public infrastructure, through the support of conversation aids, transmission of ideas, education and recruitment, is required as an urgent task.

The writers have accumulated particular data on the characteristics of movement of physically handicapped subjects that have difficulty inputting data into a computer, for each individual according to the content of the disability. Through registration, we have studied a system that will allow the subjects to input data freely, regardless of the disability.

2. Significance

In this paper, we look at the time-related factors that control keyboard input, and analyze the relationship with the disability and consider an input system appropriate for the users.

At first, we confirmed four parameters, including "keyboard input time", as being valid as time-related parameters affecting keyboard input.
Next, through the analysis of actual test data, we adjusted the parameters according to the content of the disability so that input at an almost normal level was possible.
Furthermore, through analysis of the dispersion of experimental data, we could statistically process the data and propose a discrimination formula. Following the calculation, a method of discriminating for each user could be set, and through re-input tests, it was confirmed that the error rate for each user were reduced and the validity of the formula could be confirmed.

3. Method of technical approach

3.1 Setting of time parameter that controls computer key input

By analyzing one sequence of action from pressing down a key to raising the finger and releasing it from the key after input, we determined parameters that control key input.

In the Fifure1, the time interval from when a key is first pressed to when the key data is registered as input, is **'key input time:T1'**.
The time interval from when the key data is registered to when the auto repeat function starts to operate, causing the data to be input repeatedly, is **'key repeat start time:T2'**.
The time interval from when the key data is again registered to be repeated to when the key is more continuously pressed and so the auto repeat function operates, causing the data to be input repeatedly once more, is **'key repeat interval:T3'**.
The time interval from when one key is released to when the next key can be pressed again, is **'key input invalid time:T4'**.

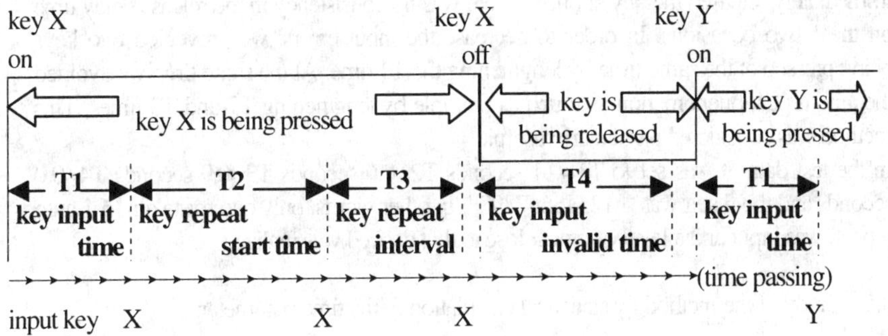

Figure 1 The transition of key input and its relationship to evaluation parameters

3.2 Study to realize normal input by adjusting of parameters for each type of disability

By making key input time T1 a long value, even if a key is mistakenly pressed, input will not occur immediately. This is particularly useful for users who often mistake to key pointing, due to tension or tremble in the hand and fingers.
By making T2/T3 a long value, even if the user is slow to release their fingers, key repeat will not happen immediately. This is particularly useful for users who, due to tension or tremble in the hand and fingers, being unable to release their fingers immediately after pressing a key, cause an unexpected auto-repeat to occur.
By making T4 a long value, if a key is mistakenly touched immediately after releasing a key, input will not occur. This is particularly useful when, due to tension or tremble in the hand and fingers, key release is unstable and there is a danger of touching a key other than expected.

3.3 Test system design

Based on the thinking outlined above, the test system was designed, which equipped the four time parameters, containing a wide variety of keyboard sizes and keyboard covers. Test data was analyzed based on representative types of disability. The test start time parameters T1~T4 were set with the standard values for able-bodied participants.

3.4 The trial of the parameter adjustment to bring about desirable input

From the test data gained from wrong input, we attempted to adjust the parameters to bring about desirable input. At first, using the set parameters, for the test data 「1234567890」, the test subjects (suffering from cerebral palsy) response was 「123345*66666666666*7*8*89*99*0」. For the two types of mistaken input (5*6* , and 7*8*) and (***66*··*66*** and ***99***) ,in the first case two keys alongside each other were pressed by the trembling at the same time and in the second case auto-repeat occurred due to failure to release the finger quickly enough.

This test subject pressed two keys at the same time on two occasions, and realizing immediately released the key at once. There was no consistency in the release delay time on these two occasions. In order to decrease the input errors, we prevented two keys being pressed at the same time by lengthening the T1 time. At the same time we avoided the auto-repeat phenomenon as much as possible by lengthening T2 and T3 times, This method was considered particularly useful.
In the test data, it was set to T1=0.1 seconds T2=0.6 seconds T3=0.9 seconds T4=0.0 seconds and the result was 「12345*6*67890」. In other words, only one mistaken (***6*** : auto repeat) was input and a level of input close to that desired was realized.

3.5 Study of the method by statistical calculation of the time parameter

Based on the hypothesis of regular distribution for time parameter measurement data, We set the allowable range for regular input data and the mistaken input data rejection rate and studied a formula that could calculate the time parameters statistically.

A basic formula. The distribution of test data for key input time T1 (regular distribution, average M, standard deviation σ) to smaller than $M - 2 \cdot \sigma$ is about 2%, and to smaller than $M - \sigma$ is about 16%. As a result, if we set $T1 = M - n \cdot \sigma$ (n: decided from maximum range of reject rate), input data larger for T1 is registered as input normally.

A corrected processing considering mistaken input. If, while holding down a key, a mistake is realized and the key is hastily released, the distribution of data for mistaken input time T1` (regular distribution, average N, standard deviation $\rho$) to larger than $N + 2 \cdot \rho$ is about 2%, and to larger than $N + \rho$ is about 16%, if as in the previous item $T1` = N + n \cdot \rho$ (n: decided from maximum range of reject value), input data smaller than T1` is normally rejected.

Based on the above, if T1` <T1, T1` is adopted. Where T1'>T1, as the center between the two value, T1" = 1/2 (T1+T1`) is adopted.

4. Result

4.1 The improvement of key input by setting discrimination formula

We confirmed that the distribution of measured data is nearly close to regular distribution, and normal input is more generally right than mistaken input (Figure 2 Reference).

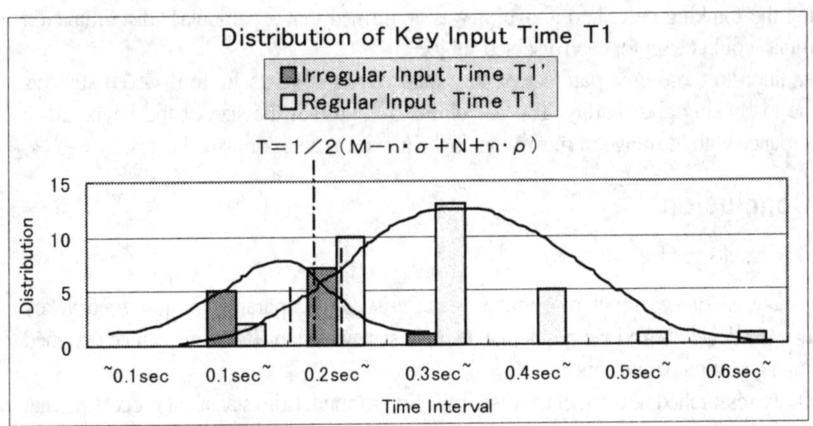

Figure 2 Test data for testee from cerebral palsy

And the discrimination formula was set using the calculation formula and evaluation and comparison were carried out before and after setting discrimination formula.

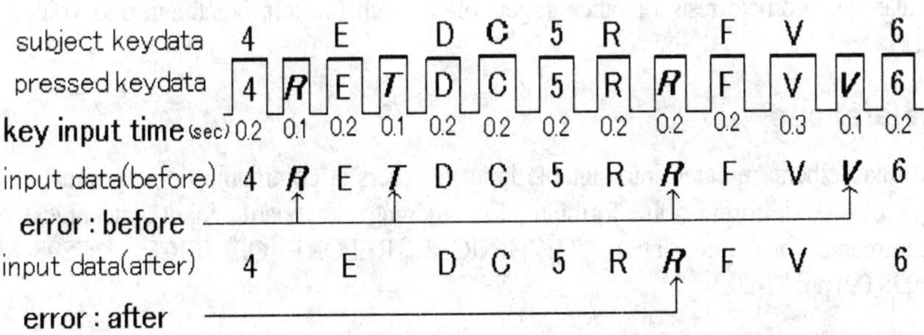

Figure 3 The improvement in key input by setting discriminant-comparison of before／after-

With the example of the cerebral palsy, key pointing is unstable due to the tension or tremble, then often touched next key, i.e. before setting the discrimination formula (： T1 is standard = 0.1sec), many mistaken input is occurred as shown in the Figure 3(**error** ：

before). Mistaken input due to instability in pointing is as follows 4 cases： ***R,T,R,V***.

Toward this, the results of re-input using the discrimination formula (T1 = 0.2sec) showed that undesired data input caused by "pointing instability due to the tension or tremble " had dropped 25% as shown in Figure 3 (**error：after**).

4.2 Decision of the elements of Discriminant

Using the thinking described above, it was confirmed that a particular "discrimination formula" could be set for each disabled subject.
In addition to these time parameters, keyboard covers were useful to disabled subjects prone to shaking and rigidity. The usefulness of adjusting the size of the keyboard in accordance with the range of movement in the hand was also confirmed in tests.

5. Conclusion

5.1 The results so far

We analyzed of key input movement characteristics, and parameters that control key input were decided. And we check that an input system suitable for users can be obtained through adjusting parameters.
Next, we researched how to set the discrimination formula from statistical processing that takes into account a wide range of input data, and we confirmed the improvement in key input.
More, we decided all the factors for the discrimination formula.

5.2 Future subjects

We consider to evaluation of an appropriate input system based on the set discrimination formula, and promote tests of other targets of research (switch, breathing, and voice input).

References

1. Arifuku, others: research into suitable input for users at communications terminals-study of a discrimination formula for judging keyboard input-, technical telecommunications information, TECHINICAL REPORT OF IEICE, HCS98-29(1998-09) pp. 73-80

2. Arifuku, others: research into suitable input for users at communications terminals-evaluation of particular discrimination formula for each user for key input (1)-, 13th rehabilitation engineering conference paper (1998.8), pp.39-44

3. Yasuko Miyazaki, Kazuhiro Fujita (1995) Study of the key input confirmation time for cerebral palsy sufferers, 10th rehabilitation engineering conference paper, pp245-248, 1995

Study on Stress Management

-Relationship among Body Sway, Accommodation, and Mental Fatigue by VDT Work-

Shin'ichi FUKUZUMI* and Toshimasa YAMAZAKI**
*Human Media Research Laboratories, NEC Corporation
** Fundamental Research Laboratories, NEC Corporation

1 Introduction

Stress management is the management of office work to reduce workers' stress levels by evaluating their mental and physical stress levels and taking measures against stress (Phillips 1988). Stress management for VDT work has to be studied as a problem of serious stress.

Primarily, when measuring workers' mental and physical fatigue or stress levels caused by VDT work, changes in physiological functions before and after the work are measured as an index of fatigue or stress (Yamazaki, et al 1990). However, the changes in these functions differ depending on whether the worker is tired or not, and also differ depending on the condition of the worker (Fukuzumi, et al 1987). For this reason, in order to evaluate their stress levels, it is necessary to measure not only the changes in their physiological functions and subjective symptoms before and after the work but also the change in them during a few weeks.

In this study, eye accommodation and body sway were chosen as the physiological functions for evaluating stress levels, because eye accommodation is known to fail due to visual work or mental loads (Östberg 1980) and body sway is known to change by mental and physical fatigue (Oman et al. 1989).

2 Experiments

2.1 Environment of the experiments

During the VDT work for the experiments, the environment was adjusted as follows; illuminance of the screen was 500 Lux, text luminance was 35 cd/m^2, background luminance was 10 cd/m^2 and viewing distance was about 60 cm (ISO 1992).

2.2 Physiological measurements

As indexes of the physiological functions, diopter and accommodation time were used for eye accommodation and moving area was used for body sway.

For eye accommodation, accommodo-meter (Nidek, AA-2000) was used. As an index for eye accommodation, the ratio between pursuit distance to a target (Ar: diopter) and the target moving distance (As: diopter) was used. Figure 1 shows the relation between the stimulus and the response.

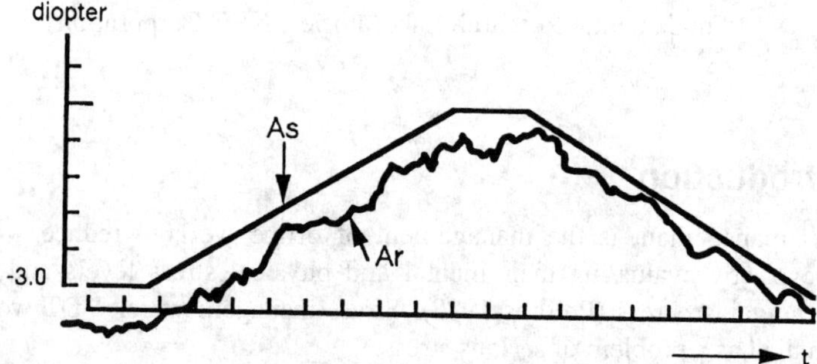

Figure 1: The relationship between the target stimulus and eye accommodation response.

For body sway, the equilibrium function meter (NEC San-ei, 1G06) was used. Moving area was measured for 50 sec. During the measurement, Romberg's posture was used.

2.3 Participants

One 28-year-old male subject and one 20-year-old female subject participated in this experiment.

2.4 Procedure

The relationships between subjective symptoms and physiological functions in the VDT workers' mental stress levels were considered. For two weeks, the subjects carried out VDT work for about one hour every day. Their subjective symptoms for mental fatigue were collected and their body sway and eye accommodation were measured before and after the VDT work. A questionnaire consisting of 30 items, such as "eye fatigue", "headache", "back pain", "irritating", and so on, was used for the subjective symptoms,.

3 Results

3.1 Questionnaire results

A principal component analysis was applied to the results of the questionnaire. From this, "vitality", "head", "eye", and "skeletal muscle" were extracted as factors of fatigue. For the physiological functions, it was found that when the visual fatigue of the subjects increased, the body sway area increased and the characteristics of eye accommodation for target chase failed.

3.2 Physiological measurements results

Figure 2 shows the results of eye accommodation and body sway during the two weeks.

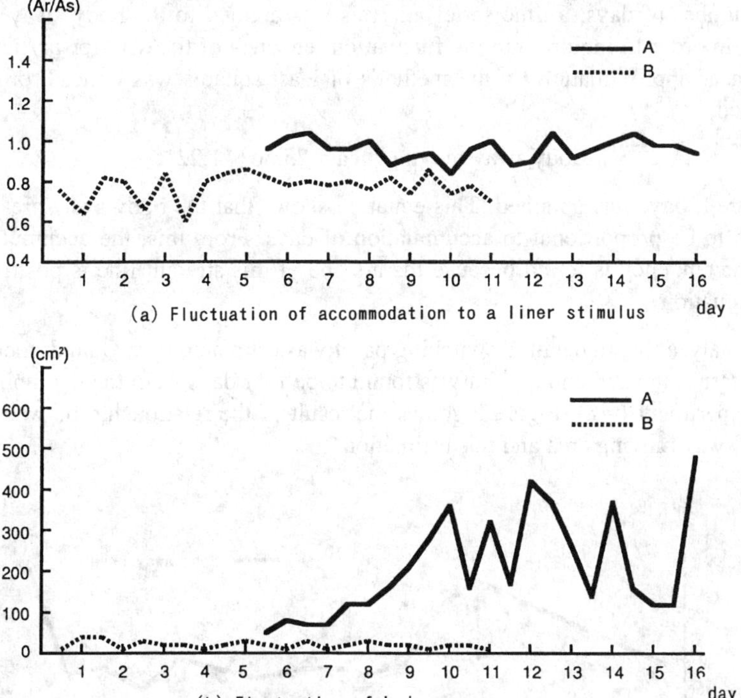

Figure 2: the results of eye accommodation and body sway during the two weeks.

(a) accommodation to a Liner stimulus, (b) body sway area

From this, the function of eye accommodation has not changed significantly for each participant. However, for participant A, the function of body sway failed in the latter half of the week.

4 Discussion

To clarify the relationship between the physiological functions and the subjective symptoms, a principal component analysis was applied to the data. The accumulation contribution rate for these three components was 70.3%

From this, the first component showed whole body fatigue related to body sway, the second showed asthenopia, and the third and fourth components showed eye accommodation.

To clarify the relationship between the change in the body sway function and the number of days, a time series analysis was applied to the body sway data. First, in order to inquire into the fluctuation tendency of the data for participant A, linear approximation by the methods of least squares was done. From this method,

$$\text{Body sway moving area} = 75.56 + 11.22 * t$$

where t: day, was obtained. This equation shows that the body sway area was found to be proportional to accumulation of days. From this, the accumulative fatigue tendency is found because the a slope of this straight line is positive in this equation.

To analyze this in detail, a switching point was estimated by a Quandt method. As a result, the switching point was found to be four days from the beginning of the experiment (t=8). Figure 3 shows the result of the relationship between the body sway moving area and this estimation.

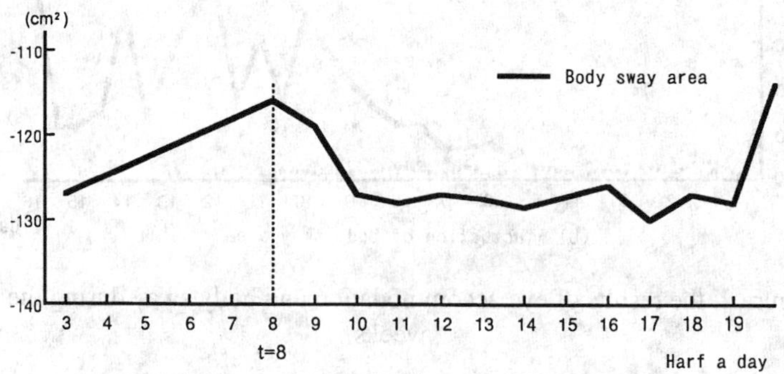

Figure 3: Result of a time series analysis (participant A, body sway area).

This figure indicates that after four days from the beginning of the experiment, the change in the body sway moving area is large.

5 Conclusion

From the obtained data and analysis, the body sway area was found to be related to asthenopia and accumulation fatigue, and also found to be a possible index of stress.

To evaluate workers' fatigue due to VDT work, it is important that workers' physiological conditions are considered.

6 References

Phillips, D. A. (1988). *The Social Effects of Computer Communications in Organizations, Computer Communication Technologies for 90's*. J. Raviv (Editor). North-Holland: Elsevier. 219-223.

Yamazaki, T., Kamijo, K. & Fukuzumi, S. (1990). Quantitative Evaluation of Visual Fatigue Encountered in Viewing Stereoscopic 3D Display –Near Point Distance and Visual Evoked Potential Study-. *Proceedings of SID*, 31 (3), 245-247.

Fukuzumi, S. & Honda, K. (1987). Physical Condition and Visual Accommodation of Software Engineers. *Proceedings of the third Symposium on Human Interface*. 213-216 (In Japanese).

Östberg. O. (1980). Accommodation and visual fatigue in display work, In Grandjean and Vigliani, E. (Eds.). *Ergonomics Aspects of Visual Display Terminals*. Taylor & Francis. 41-52.

Oman, P. W., Gomes, C. S., Rains, K & Morandi, M. (1989). Posture and VDU operator satisfaction. *SIGCHI Bulletin*, 20 (3), 52-57.

ISO (1992). ISO 9241-3: *Ergonomic requirements for office work with visual display terminals –Part 3, Visual display requirements*.

Influences of Ergo-meter Work Load Stimulated Fluctuations in CFF Values of Circadian Rhythm on VDU Work Load Related Central Nerve Fatigue

Masahau Takeda and Satoru Kishino
Department of Industrial Engineering Musashi Institute of Technology
1-28-1 Tamazutsumi Setagaya-Ku, Tokyo, Japan 158-8557

1. Objectives of this study

VDT activities have penetrated into all aspects of society to present a problem in modern society by causing such health hazards as central nerve fatigue and dysfunctions of shoulders, jaws and arms. One such problem is fatigue caused by overall lack of exercise due to the static nature of VDT activities. One suggested countermeasure to combat this problem is to separate VDT activities into 60-minute intervals with a 50-minute working time and a 10-minute recess. Seated operator produces blood circulation failure through out the body and as a result, activity of the central nervous systems deteriorates. Of break spend it, and pointer of a person is not instructed for VDT operator. The VDT operator has not been instructed on what to do during the rest period. So we drew up a hypothesis that lightening of the central nervous system fatigue could be achieved by a light overall body exercise, to be completed during the rest period. The objective of this study is to promote activation during the recess period by implementing ergo-meter work loading in order to relieve poor localized blood circulation and stimulate the monotonousness of VDT activities. Experiments were conducted under the hypothesis that central nerve fatigue may thus be reduced. Since the author et al have found in previous studies that monitoring fluctuations in CFF values was an easy and effective method for measuring fatigue, it was used in this experiment as an index. The experiment was planned taking into account special considerations for the influences that CFF values have on circadian rhythm.

2. Experiment methods

2. 1 Subjects and equipment

Ten healthy young male subjects aged between 21 to 23 were selected and divided into 2 groups where Group A was assigned to carryout only VDT activities and Group B was subjected to both VDT activities as well as ergo-meter work loads. The equipment used in the experiment included CFF measurement apparatus, an ergo-meter and a tank-plant- simulator.

2. 2 Experiment 1

Experiment 1 CFF measurements were taken over a two month period in order to ascertain the circadian rhythm of the ten subjects.

2. 3 Experiment 2

There were significant individual differences with regard to the loading of ergo-meter exercises completed during the rest period. Therefore, I tested ergo- meter motion for three minutes so that the loading could be adjusted to take into account the stamina of each individual. This could then be set. For this experiment the loading was set between 40 and 75 watts. This physical loads, every 110 times becomes not more than minute maximum heart rate with each subjects of five motion groups. Experiment 2 used scenario 1 or2 of tank - plant - simulator on VDT as work load. A subject did it with work load to let operate four control- bar with cursor up and down. The subject curled four control –bars (lifting and lowering motion) The activity performed had target water temperatures and amounts and these were permitted to work against each other to see how close they could get to their targets. A valve was used to control both water amount and heat. We allowed the subject to complete one cycle in three minutes and then measured the CFF. One cycle consisted of ten repetitions followed by ten minutes rest. The methods of resting between groups differed with Group A sitting passively and Group B completing 3 minutes of ergometric exercise, after which they were permitted to sit and rest for 7 minutes. In experiment 2 the circadian rhythm tended to both ascend and descend and in both cases the time was recorded at four hours. For Group A, there were two types of nuclear power tank plant simulation controlling activities, Scenarios 1 and 2, were randomly assigned for 3 minutes then a survey of subjective symptoms of fatigue and measurement of CFF values were repeated for approximately 50 minutes followed by a 10-minute recess at seated position. This routine was defined as a singular unit of experiment and 4 units were conducted at the rise of circadian rhythm and 4 additional units were conducted on a separate day when the circadian rhythm declined. Group B was made subject to 3 minutes of ergo-meter work load at an average of 50W (adjusted to weight) to maintain a maximum heart rate of 110 during Group A's seated recess period. Other conditions were the same for Group A and B. In addition, the amplitude of fluctuations in the center temperatures and water volume of the designated tanks for Scenarios1 and 2 was selected as an index for workability.

3. Results

3. 1 Results 1

The results of experiment 1 are shown in table 1. The figures in table 1 are estimated as the optimum cosine curve. The following itemization is a one-line description of the main results from experiment1.

- There are individual differences for each subject with respect to the circadian rhythm in CFF.
- The Circadian rhythm in an average of 24 hours, ranged between 21.7 hours at the lower end 28 hours at the upper end.
- Using this effect, we divided ten subjects into 2 so that homogeneity could be achieved between groups.

Tab.1 Parameter for optimam cosine curve

	Data number	A period	Mesor	Amplitude	Vertex phase	optimam cosine curve
subject A	56	26.1	35.3	0.5	12:14	Y=35.3+0.5×cos(12.9×t-183.60)
B	80	28.0	46.0	1.1	9.55	Y=46.0+1.1×cos(12.9×t-148.75)
C	56	23.8	36.2	1.8	14:55	Y=36.2+1.8×cos(12.9×t-223.72)
D	51	28.0	25.4	0.9	18:14	Y=25.4+0.9×cos(12.9×t-273.44)
E	46	23.7	32.7	0.7	19:20	Y=32.7+0.7×cos(12.9×t-290.11)
F	66	27.7	36.8	1.0	14:50	Y=36.8+1.0×cos(12.9×t-222.43)
G	42	28.0	33.4	0.8	13:53	Y=33.4+0.8×cos(12.9×t-208.21)
H	70	21.7	33.8	1.2	17:07	Y=33.8+1.2×cos(12.9×t-256.81)
I	85	27.0	32.4	1.1	13:14	Y=32.4+1.7×cos(12.9×t-198.48)
J	57	24.9	30.6	1.2	15:26	Y=30.6+1.2×cos(12.9×t-231.59)

3. 2 Results 2

Figure 1-Figure 4 shows the results for experiment 2.
Vertical line represents the rate of change of CFF , and measurement for the number of times is represented by the quadratic axis. In addition, the cut in the diagrammatic chart shows resting spells.
The following itemization is one-line description of results of experiment 1.

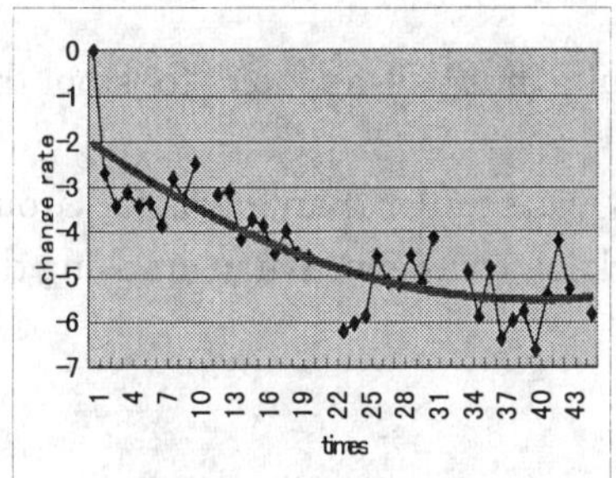

Fig 1 Average change rate for CFF of Group with out motion (When Circadian rhythm rises)

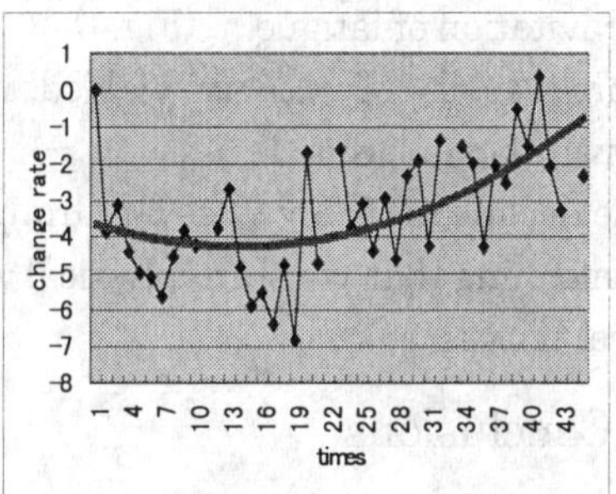

Fig 2 Average change rate for CFF of Group with motion (When Circadian rhythm rises)

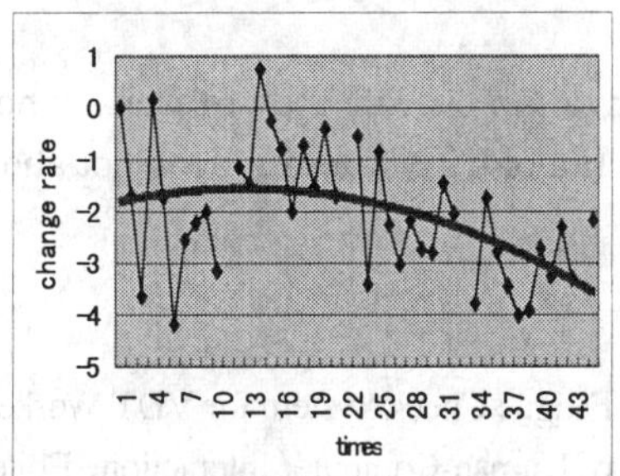

Fig. 3 Average change rate for CFF of Group without motion (When Circadian rhythm drops)

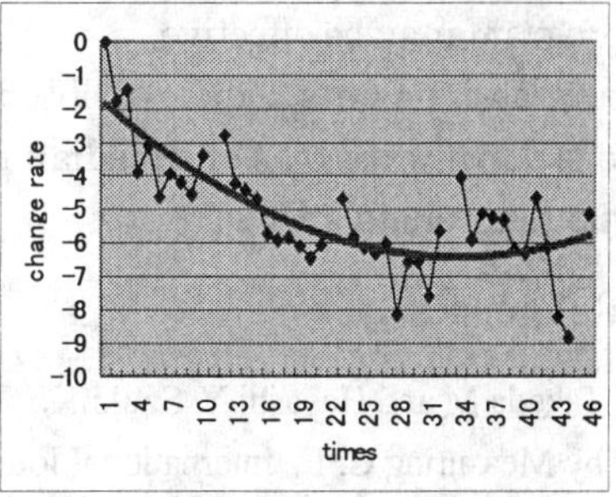

Fig. 4 Average change rate for CFF of Group with motion (When Circadian rhythm drops)

- When Circadian rhythm of CFF rises, CFF of group without motion is gentle and shows a downward bias. (Fig. 1)
- Recovery of fatigue by three spells of rest is not found (Fig. 1)
- When the Circadian rhythm of CFF rises, CFF of the motion group momentarily shows signs of fatigue but then is seen to assume a climbing trend afterwards. (Fig.2)
- Effect of motion by ergo-meter where there are three spells of rest can be seen noticeably. (Fig2)
- When Circadian rhythm of CFF drops, the CFF of the static group is gentle and shows a downward bias. (Fig.3)
- The effect of having three rest periods shows a ratchet effect with regard to fatigue. (Fig.3)
- When Circadian rhythm of CFF dropped, CFF of the motion group shows

a gravitation of fatigue. (Fig.4)

・Three spells of activity yielded results but wasn't able to stop the downward bias (Fig.4)

・The decline in CFF values for Group B was smaller than it was for Group A, confirming that constant physical work load was effective in arousing the central nerves.

4. Conclusions

Due to recent increases in monitoring tasks in VDT activities, there are concerns that lack of activity induces reduced level of central nerve arousal, thereby hampering workability during emergencies. This study has proved that the incorporation of a proper amount of overall physical activity into the program may be effective.

Future assignments could possibly include an overall body exercise when there is down ward bias of circadian rhythm of CFF where it does not show fatigue Lightening effects.

5 References

1. Takeda.M and Hayashi.Y, Szuki.K, The Fatigue Check System for VDT Workers by Measuring CFF, International Journal of Human-Computer Interaction, Edited by M.J.Smith and G.Salvendy, ELSEVIER, Vol.19, pp.618-623(1991-8).
2. Takeda.M and Hayashi.Y, Szuki.K, Analysis of Uric Properties for Stress Caused by VDT Work and Relationship among CFF, HRV and VEP, International Journal of Human-Computer Interaction, Edited by M.J.Smith and G.Salvendy, ELSEVIER, Vol.19, pp.903-908(1991-8).
3. Takeda.M and Hayashi.Y, Suzuki.K, CFF Values for Stress Caused by VDT Work and Relationship among Analysis of Uric Properties Symbiosis of Human and Artifact, Human and Social Aspects of Human－Computer Interaction, Edited by Y.Anzai & H.Mori, Vol.20, pp.785-790(1995-7).
4. Takeda.M and Hayashi.Y, Does Circadian Rhythm of VDT Operator Cause Fluctuation of CFF Value that to Mask Fatigue Variations During Work Load？, International Journal of Human-Computer Interaction, Design of Computing Systems, Edited by G.Salvendy, M.J.Smith & R.J.Koubek, ELSEVIER, pp.563-566(1997-8).

Performance and Muscle Load among Young and Elderly

Laursen B., Jensen B.R., Ratkevicius A.*
Department of Physiology, National Institute of Occupational Health, Copenhagen, Denmark

*NMR-center, Panum Institute, University of Copenhagen, Copenhagen, Denmark

1 Introduction

The share of elderly employees is rising. Age-related changes in physical capacity (Ilmarinen et al.; 1991) and muscle function (Kinoshita and Francis, 1996) may influence the workload when the work demands remain unchanged. This might increase the risk of developing work-related symptoms and pain in the musculoskeletal system. Some of the age-related physiological changes could be important in computer work, especially those concerning the ability to perform precision work. The aim of the present study was to test for age-related differences in working speed and muscle activation patterns when performing computer work.

2 Methods

Eight young (25 years, range 22-28 years) and nine elderly (63 years, range 56-70 years) experienced female computer users participated in the study. The subjects performed computer work at a height-adjustable desk with forearms supported. During a 4-hour period, the subjects performed a computer mouse test, consisting of 11 tasks, five times. These tests were based on an ISO standard (ISO 9241, 1994). Six tasks consisted of clicking multidirectional targets of different width (8, 16, 32 pixels, corresponding to 2.5, 5, and 10 mm on the screen and approximately 0.3, 0.6, and 1.2 mm for the mouse) in a self-determined speed and a predefined speed (40 min^{-1}). A circle was traced in a predefined speed (1 rev / 8 s). The remaining tasks were horizontal and vertical (relative to the computer screen) clicking a 16-pixel wide target, double clicking (40 min^{-1}), and dragging (20 min^{-1}). For each task, the error rate (failing/attempted operations) and working speed (successful operations$\cdot min^{-1}$)

were calculated. Electromyography (EMG) was recorded from three forearm muscles: the wrist extensor muscle group (WE), the wrist flexor muscle group (WF), and the finger extensors (FE). EMG was normalised to the maximum EMG during maximal contractions. For each task the static EMG level (exceeded in 90% of the time), the mean level, and the peak level (exceeded in 10% of the time) were calculated.

3 Results

Results from the experiments are presented in tables 1 and 2.

Table 1. Error rate and working speed for the young and elderly.

	Task, target width (pixels)	Young/elderly, error rate	Young/elderly, working speed
Self-determined speed	clicking, 32	6.0 / 7.4 %	68.4 / 55.9 min^{-1}
	clicking, 16	6.8 / 8.7 %	55.5 / 46.9 min^{-1}
	clicking, 32	11.4 / 15.8 %	43.5 / 37.9 min^{-1}
Predefined speed	clicking, 8	1.4 / 4.2 %	Predefined 40 min^{-1}
	clicking, 16	2.5 / 7.8 %	
	clicking, 32	7.9 / 16.5 %	
	Horizontal clicking, 16	2.4 / 5.3 %	
	vertical clicking, 16	2.8 / 8.4 %	
	Tracing	0.235/ 0.263 min^{-1}	Predefined
	Double clicking, 16	11.8 / 35.1 %	Predef. 40 min^{-1}
	Dragging, 16 (20 min^{-1})	3.6 / 8.3 %	Predef. 20 min^{-1}

For the self-determined speed trials, the working speed decreased significantly when the precision demand increased ($p<0.05$). However, the elderly group performed 13-18% slower ($p<0.05$). The working speed of the elderly corresponded to the young group, if the target size for the elderly was increased by a factor of two.

The error rate increased with increasing precision demand in the self-determined speed tasks with no significant difference in error rate between the elderly and the young group. For the predefined speed trials, the elderly group made more errors, compared to the young group ($p<0.05$).

Table 2. EMG levels for the young and elderly group (median of subjects). The 11 tasks are the same as above.

Mean EMG %EMGmax		Young			Elderly		
		WE	WF	FE	WE	WF	FE
Self-determined speed	Clicking, 32	4.3	3.6	8.4	6.8	6.1	10.4
	Clicking, 16	4.2	3.2	8.7	6.6	5.9	10.3
	Clicking, 8	4.3	3.5	7.9	6.7	6.2	10.4
Predefined speed	Clicking, 32	3.9	1.8	5.8	6.2	5.8	8.7
	Clicking, 16	4.0	1.8	6.7	6.6	5.9	8.3
	Clicking, 8	4.0	2.0	6.7	6.3	6.3	9.6
	Horizontal click, 16	3.8	1.7	5.9	5.8	5.0	9.1
	Vertical click, 16	3.7	1.7	5.4	6.1	5.1	7.4
	Tracing	3.3	1.7	5.1	4.9	4.4	7.3
	Double clicking, 16	4.5	3.7	7.1	7.5	7.5	10.5
	Dragging, 16	3.6	2.3	5.9	6.6	7.3	9.3

The normalised EMG was higher in the elderly group, compared to the young group. For the wrist extensors, the elderly group had higher EMG than the younger group ($p<0.05$). For the wrist flexors and finger extensors, there was a tendency towards a higher EMG level for the elderly. The static and peak EMG showed the same features as the mean EMG.

4 Discussion

In general, the extensor muscles had the same or higher EMG compared to the flexors. This is probably because the extensor muscles are activated most of the time, in order to facilitate movements easier and to avoid clicking the mouse.

One explanation for the higher relative EMG level for the elderly may be more co-activation of antagonist muscles, which has been found in another laboratory study (Spiegel et al 1996). The muscles responded significantly different to the different mouse tasks, although the differences were minor. The task causing the highest muscle activity was double clicking for the wrist flexors and extensors, and the self-selected speed clicking tasks for the finger extensors. Increased precision demands cause an increased muscle activity when the working speed was fixed in agreement with what has been found previously (Laursen et al.; 1998). When the speed was self-selected, the precision demand had no effect on muscle activity.

In conclusion, the elderly performed the mouse work slower than the young group but with the same error rate, when the working speed was self-selected. Double clicking caused many errors for the elderly group. The relative muscle load was higher for the elderly group. There were small but significant differences in muscle load between the different mouse tasks.

5 References

Ilmarinen J., Vouhevaara V., Korhonen O., Nygård C.H., Hakola T., & Suvanto S. (1991). Changes in maximal cardiorespiratory capacity among aging municipal employees. *Scand J Work Environ Health,* 17 (suppl1), 99-109.

ISO 9241 (1994). Visual display terminals (VDTs) used for office tasks - Ergonomic requirements. Part 9: Non-keyboard input device requirements (committee draft).

Kinoshita H. & Francis P.R. (1996). A comparison of prehension force control in young and elderly individuals. *Eur J Appl Physiol,* 74, 450-460.

Laursen B., Jensen B.R., & Sjøgaard G. (1998). Effect of speed and precision demands on human shoulder muscle electromyography during a repetitive task. *Eur J Appl Physiol,* 78, 544-548.

Spiegel K.M., Stratton J., Burke J.R., Glendinning D.S., & Enoka R.M. (1996). The influence of age on the assessment of motor unit activation in a human hand muscle. *Exp Physiol,* 81, 805-819.

Forearm Muscular Fatigue during 4 Hours of Intensive Computer Mouse Work – Relation to Age.

Jensen B.R., Laursen, B. and Ratkevicius, A. *
Department of Physiology, National Institute of Occupational Health, Copenhagen, Denmark.
* NMR-Center, Panum Institute, University of Copenhagen, Denmark

1 Introduction

Muscles undergo a whole variety of morphological and neurological changes with age. A reduction in muscle cross-sectional area, a decrease in total number of motor units, an increase in average motor unit size and a general slowing of skeletal muscles have been reported in aging humans (e.g. Grimby and Saltin 1983; Davis, Thomas and White 1986). The precision of force exertions also appears to decrease with age (e.g. Cole 1991; Galganski, Fuglevand and Enoka 1993). However, it remains unclear if these age-related changes in neuromuscular function affect the performance of computer work. Reports on age related changes in muscle fatigue resistance are contradictory (e.g. Davis, Thomas and White 1986; Klein, Cunningham, Paterson and Taylor 1988). Thus, it is even less clear if aging contributes to muscle fatigue during intensive computer mouse work performed for prolonged periods of time.

The aim was to investigate to which extent 4 hours of intensive computer mouse work elicited electromyographical signs of muscle fatigue in young and elderly subjects. Furthermore, to study if development of muscle fatigue is related to work intensity in terms of number of mouse clicks per unit of time.

2 Method

Eight young (25 years, range 22-28 years) and nine elderly (63 years, range 56-70 years) experienced female computer users participated in the study. They performed four 50-min sessions of intensive computer mouse work. The sessions were separated by 10-min breaks. The subjects performed a combination of computer mouse tests and electronic painting. The computer mouse test was performed in the beginning of the first session and at the end of each of the four sessions. Each computer mouse test consisted of 11 tasks performed for 100 s each. The 11 tasks included one-directional and multi-directional clicking, double clicking, tracing, dragging, different precision demands and combinations of predefined and self-determined speed. The computer work was performed with self-determined speed for 2/3 of the time and with a predefined speed for 1/3 of the time. During the periods with self-determined speed the subjects were encouraged to perform the work fast and correct. The workstation was carefully adjusted to each subject, the forearms were supported by the table and a mouse pad for the computer mouse was placed at forearm distance and to the right side of the keyboard (keyboard was not used in the present study).

Forearm muscle electrical (EMG) activity was measured with surface electrodes (Neuroline 70002-k, Medicotest) at the wrist extensor muscles (WE), wrist flexor muscles (WF), and the finger extensor muscles (FE), low-pass filtered (460 Hz), and sampled on a computer with a frequency of 1024 Hz. Mean power frequency and root-mean-square amplitude were calculated for each muscle group. Root-mean-square amplitudes were expressed relative to maximum EMG values (1 sec) obtained during maximum contractions (%EMGmax). Maximum EMG values was measured during handgrip contractions, wrist extensions and finger extensions. Average EMG values of root-mean-square amplitudes and mean power frequencies for computer mouse test 1 and 5 were calculated. Work intensity in terms of number of mouse clicks per time was recorded at the computer for the entire working period.

3 Results

For the total group, including young and elderly subjects, mean power frequency decreased significantly from computer mouse test 1 to computer mouse test 5. For each of the two groups a significant decrease was found for two of the forearm muscles while for the third muscle a tendency to a decrease was seen (table 1). Median values of the root mean square amplitudes during computer mouse test 1 were 4 %EMGmax (WE), 3 %EMGmax (WF) and 7 %EMGmax (FE) for the young group and 7 %EMGmax (WE), 6 %EMGmax (WF) and 10 %EMGmax (FE) for the elderly group of subjects. The median root mean square amplitudes remained at the same level throughout the 4 hours working period.

The total number of mouse clicks during the four-hour working period was 7500 (range 6940-9090) clicks for the young group and 6360 (range 5210-7500) clicks for the elderly group. The number of mouse clicks was different between the two groups.

Table 1: Mean power frequency measured in session 1 and session 4. WE - wrist extensor, WF - wrist flexor, FE - finger extensor.

	WE		WF		FE	
	Young	Elderly	Young	Elderly	Young	Elderly
Test 1	104 Hz	112 Hz	119 Hz	121 Hz	136 Hz	126 Hz
Test 5	96 Hz	104 Hz	116 Hz	113 Hz	125 Hz	121 Hz

The change in mean power frequency from computer test 1 to computer test 5 was not significantly related to the work intensity measured as the total number of clicks for the 4 hours working period. No relation between relative change in mean power frequency and work intensity was found for any of the 3 muscles neither for the young or the elderly group of subjects.

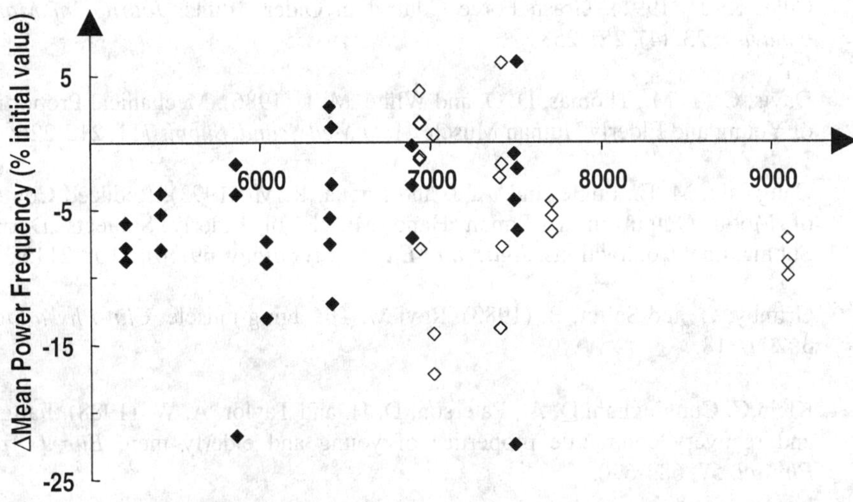

Figure 1: Change in mean power frequency relative to total number of mouse clicks during a four hour working period. N=17. Open symbols: 8 young subjects. Closed symbols: 9 elderly subjects.

4 Conclusion

Forearm muscle fatigue, measured as a decrease in mean power frequency could be documented during four hours of intensive computer mouse work.

Electromyographical signs of muscle fatigue were found for forearm extensor as well as forearm flexor muscles.

No difference regarding relative change in mean power frequency was found between the young and the elderly group.

Finally, the total number of mouse clicks per unit of time is not related to the degree of forearm muscle fatigue and can therefore not be used as a predictor for development of electromyographical signs of fatigue during prolonged computer work.

5 References

Cole, K. J. (1991). Grasp Force Control in Older Adults. *Journal of Motor behavior.* 23, (4), 251-258.

Davis, C. T. M., Thomas, D. O. and White, M. J. (1986). Mechanical Proporties of Young and Elderly Human Muscle. *Acta Med Scand, Suppl.* 711, 219-226.

Gangalski, M. E., Fuglevand, A. J. and Enoka, R. M. (1993). Reduced Control of Motor Output in a Human Hand Muscle of Elderly Subjects During Submaximal Contractions. *Journal of Electrophysiology.* 69, (6), 2108-2115.

Grimby, G. and Saltin, B. (1983). Review. The aging muscle. *Clin Physiology,* 3, 209-218.

Klein,C. Cunningham,D. A., Paterson, D. H. and Taylor, A. W. (1988). Fatigue and recovery contractile properties of young and elderly men. *Eur J Appl Physiol.* 57, 684-690.

Mental Workload in IT Work: Combining Cognitive and Physiological Perspectives

Anker Helms Jørgensen[1], Anne Helene Garde[2], Bjarne Laursen[2], and Bente Rona Jensen[2]
[1] Institute of Psychology, Copenhagen University
Njalsgade 88, DK-2300 Copenhagen S, Denmark
anker@axp.psl.ku.dk
[2] National Institute of Occupational Health
Lersø Parkalle 105, DK-2100 Copenhagen Ø, Denmark
{ahg,bl,brj}@ami.dk

1 Introduction

With the introduction of graphical interfaces to computer systems, the mouse also appeared. Mastering a mouse is a complex skill, which involves a range of perceptual, cognitive and motor processes. Automated cognitive processes play a major role, as these can be carried out with virtually no attention and as they are fundamental to the development of highly skilled performance.

IT-work including use of mouse seems to be a cause of musculoskeletal disorders, but very little research has been done to explore causes and remedies (Jensen 1997). Our research aims primarily at identifying essential cognitive *and* physiological factors and their interplay in IT-work - in particular in mouse work with emphasis on the role of *automaticity* - and secondarily at identifying their potential role in developing musculoskeletal disorders. We have initially focussed on the concept of *mental workload* (Gopher and Donchin 1986) as it involves both cognitive and physiological components.

In this paper we first outline the major features of the concept of mental workload. Next we concretize it and illustrate it by describing four empirical evaluations of a range of measures of mental workload. Finally, we discuss some general features of the concept.

2 The Concept of Mental Workload

The concept of mental workload was developed in the American Human Factors tradition. The overall aim was to develop valid and reliable measures of mental workload for designing safe man-machine systems, e.g., in aviation. The concept has an intuitive appeal, but it is multi-facetted and hard to define. The concept is often construed in the context of multiple-resource theories of cognitive resources and attention. For authoritative reviews of the concept see (Gopher and Donchin 1986; O'Donnell and Eggemeier 1986).

There are four principal types of measures of mental workload:

1. Primary task measures: these directly relate to the task being undertaken by the subjects, e.g., performance in IT-work.
2. Secondary task measures: these attempt to tap the current cognitive resources by the dual-task approach where subjects – besides the (major) primary task – also carry out a (minor) secondary task, e.g., to estimate elapsed time.
3. Physiological measures: these tap physiological responses to mental workload, e.g., pupil diameter, heart rate, and respiration rate.
4. Subjective measures: these tap subjects' experience of the mental workload (e.g., stress, effort and difficulty), frequently administered as questionnaires after the task is completed.

Two of the most salient features of techniques for measuring mental workload are *sensitivity* (the capability to discriminate significant variations in the mental workload) and *diagnosticity* (the capability to discriminate the mental workload imposed on different resources).

3 Measures of Mental Workload

Walter W. Wierwille and his colleagues conducted four interesting and relevant evaluative studies of measures of mental workload. The studies were conducted in a flight simulator and employed the same research design and procedures. Thus, they can be compared and are therefore ideal for illustrating applications of the concept. Indeed, three of the studies were briefly compared by O'Donnell and Eggemeier (1986).

In each study between 14 and 20 measures of mental workload were evaluated by identifying their sensitivity to three levels of workload (low, medium, high), which were imposed as primary tasks on the subjects. In total 22 measures were evaluated. The primary tasks in the four studies were a psychomotor task: flying the aircraft manually (Wierwille and Connor 1983), a communication task: vocal dialogue with air traffic controllers (Casali and Wierwille 1983), a perceptual task: visual monitoring of displays (Casali and Wierwille 1984), and a central processing task: calculation of bearing, drift, etc. (Wierwille et al. 1985). Table 1 below provides an overview of the studies. We excluded primary task measures as they do not compare across studies. Of the remaining measures we only included those that were found to discriminate significantly between levels of mental workload.

Table 1
Overview of measures employed in the four evaluation studies which significantly discriminated levels of mental workload.

Measure of mental workload	*Primary task* Psycho-motor	Commu-nication	Per-ceptual	Central processing
Secondary task				
Time estimation SD	•	√√	√	√√
Tapping rhythm		•	√√	•
Physiological				
Heart rate mean	√	•	•	•
Respiration rate	•	•	√√	•
Eye pupil diameter	•	√√	•	•
Eye blink frequency	•	•	•	√√
Eye fixation fraction				√√
Subjective				
Cooper-Harper scale	√√√	√√	√√	√√
WCI/TE scale	√√√		√√	√√
Multi-descriptor scale		√	√√	•

Legend: √ The measure was evaluated and did discriminate.
• The measure was evaluated but did not discriminate.
blank The measure was not evaluated.

Explanations of measures
SD Standard Deviation
WCI/TE Workload-Compensation-Interference/Technical Effectiveness

The number of tickmarks in each cell denotes the number of significant discriminations where three is maximum (low-medium, medium-high, low-high).

4 Discussion

The following discussion is based on table 1 above and (Gopher and Donchin 1986; O'Donnell and Eggemeier 1986; and Wierwille and Eggemeier 1993). A striking feature of the table is the massive sensitivity of the subjective measures to discriminate significant variations in the mental workload as compared to the other measures. Indeed the subjective measures seem to capture a valid, general aspect of mental workload. The physiological measures on the other hand appear to have high diagnosticity, i.e. they discriminate mental workload imposed on different resources. There are, however, many more facets to it:

The *subjective measures* excel in face validity as they depend directly on the subjects' actual experience. An advantage is that they are easy to obtain. However, the subjective measures shown in the table do not identify particular constituents of the mental workload (low diagnosticity), be they cognitive or physiological.

Secondary task measures are relatively easy to obtain. In contrast to the subjective measures, the mental workload can be measured concurrently with the primary task. A drawback is their potential intrusion in the primary task. Selecting and interpreting a secondary task measure can be difficult, as the interplay between primary task resources and secondary task resources is subtle.

The *physiological measures* appear to have high diagnosticity. As seen in the table, heart rate mean discriminates between levels of a psychomotor task, but not any of the other tasks. However, physiological measures are in general sensitive to combined mental workload *and* motor demands. It is therefore important to keep the physical workload constant, if physiological measures are to reflect only the mental component. An advantage of physiological measures is that they provide for a continuous record of the workload, which allows for identification of local variations.

5 Conclusion

All types of measures (subjective, secondary task and physiological) appear to reflect mental workload. They all have advantages and drawbacks. It is therefore recommended to include all three types of measures for evaluation of mental workload in IT-work – and of course also the primary task measures that have been excluded from the present description as they do not compare across tasks.

6 References

Casali, J. G. & Wierwille, W. W. (1983): A comparison of rating scale, secondary-task, physiological, and primary-task workload estimation techniques in a simulated flight task emphasizing communications load. *Human Factors*, 25, 623-641.

Casali, J. G. & Wierwille, W. W. (1984): On the measurement of pilot perceptual workload: A comparison of assessment techniques addressing sensivity and intrusion issues. *Ergonomics*, 27, 1033-1050.

Gopher, D. & Donchin, E. (1986): Workload: An exploration of the concept. In Boff, K., Kaufman, L. & Thomas, J. (eds): *Handbook of Perception and Performance (vol. II)*. New York: Wiley, ch. 41.

Jensen, B. R. (1997): Review on risk factors during VDU-work including work with input devices. Copenhagen: National Institute of Occupational Health.

O'Donnell, R. D. & Eggemeier, F. T. (1986): Workload assessment methodology. In Boff, K., Kaufman, L. & Thomas, J. (eds): *Handbook of Perception and Performance (vol. II)*. New York: Wiley, ch. 42.

Wierwille, W. W. & Connor, S. A. (1983): Evaluation of 20 workload measures using a psychomotor task in a moving-base aircraft simulator. *Human Factors*, 25, 1-16.

Wierwille, W. W. & Eggemeier, F. T. (1993): Recommendations for mental workload measurement in a test and evaluation environment. *Human Factors*, 35, 263-281.

Wierwille, W. W., Rahimi, M. & Casali, J. G. (1985): Evaluation of 16 measures of mental workload using a simulated flight task emphasizing mediational activity. *Human Factors*, 27, 489-502.

Investigation of the Use of Non-Keyboard Input Devices (NKID)

Buckle, P.[1], Haslam, R. A.[2] Woods, V.[1] and Hastings, S.[2]
1. University of Surrey; 2. Loughborough University

1 Introduction

Input devices to computers have evolved rapidly in recent years. Whilst the original introduction of computers to offices provoked concern with health risks associated with prolonged, high frequency keyboard use, there is now increasing concern over the use of non-keyboard input devices (NKID). The widespread development of graphical computer interfaces has led to a proliferation of devices that enable the user to move the cursor around the screen. The devices developed include the so-called mouse, tracker balls, joysticks, and increasingly the use of touch screens. The use of NKID is now spreading out of the office and increasing in industrial and manufacturing applications. There appears to be a surprisingly limited literature pertaining to these tools, given their widespread use in offices and elsewhere.

A small number of studies have considered potential health and other risks associated with their use. Thus for example Fogleman and Brogmus (1995) reported that mouse related worker compensation claims were increasing. Hagberg (1995), Karlqvist et al (1994) and Harvey and Peper (1997) have all examined the physiological implications of mouse use and have reported higher than would be expected levels of discomfort in a number of body parts. Substituting other NKID for the mouse may improve some aspects of physical loading, whilst increasingly placing stress elsewhere e.g. the likely increase in physical load on the shoulder girdle involved in touch screen applications.

The study reported in this paper is the first in a series that has been designed to investigate the extent of use and problems associated with NKID. Whilst the full study involves an epidemiologic, laboratory and participatory study design, the focus of this paper is a survey of a number of organizations in order to ascertain the scope and scale of the problems. The study is being undertaken at the universities of Loughborough and Surrey in England. The research is funded by the Health and Safety Executive.

2 Aims of the Complete Study

The aims of the complete study are as outlined here, although the stage reported in this paper has addressed only limited aspects of these.

1. determine the extent to which different NKID are being used
2. document patterns of NKID use, ie tasks, periods and proportion of time for which they are used
3. describe the range of workstation configurations occurring in practice, especially with regard to NKID placement and use of aids such as mouse mats and arm rests
4. identify the extent of individual variation among users of NKID in their manner of device operation
5. measure prevalence of musculoskeletal and other complaints associated with NKID use
6. examine differences in user behaviour when operating NKID that may have resulted from a prior musculoskeletal problem
7. identify desirable and undesirable aspects of NKID in respect of users comfort and health
8. comment on, and provide guidance where feasible, on the suitability of NKID in terms of their suitability for different generic tasks (eg wordprocessing, CAD)
9. collate and validate best practice advice concerning use of NKID

3 Organizations Studied

The use of NKID varies according to the type of organization and the application concerned. Little has been systematically recorded with respect to how and where such devices are being used. One hundred organizations have been surveyed. They have been selected to represent a wide range of industrial and service sectors. Data have been collected from the information technology groups, occupational health groups and human resource mangers. They have been asked to provide information relating to variables that may be considered as both potential causes or problems, as well as health outcomes (see table1.)

The types of NKID, their application and the extent of their use has been considered. It is believed that this is the first survey of its kind.

Table 1 Variables considered

Exposure Data	**Outcome Issues**	**General Information**
Applications used	Ease of use	Type/function of organization
NKID devices	User complaints	Total employees
Duration of use	Maintenance concerns	Number of desktop computer users
Frequency of use	Musculoskeletal problems	Number of laptop computer users
	Other health issues	
	Sickness/absence data	

4 Additional Studies

4.1 End Users

The ergonomics participatory approach demonstrates the importance of including the end users in any evaluation of work systems. Thus this project will also examine users, their workstations and work systems. Interviews will be held with subjects to explore patterns of NKID use and factors affecting this. This will be important for subsequent user survey questionnaire design. Respondents to the questionnaire will be asked to indicate the amount of time engaged in different computing tasks such as word processing input, word processing editing, data entry, programming, computer aided design etc. From these data and from observations of workers at the workplace the researchers will estimate exposure levels. These will be validated through direct observation using video recording.

4.2 Other Aspects of the Study Design

The results from these stages will be used to develop an ergonomics check-list for NKID which will be used in later phases of the project. Subsequently a cross-sectional symptom survey and exposure assessment will be undertaken on a sample of users and laboratory investigation of some NKID (those considered a priority following the earlier stages of the study) will also be undertaken.

5 Conclusions

This study will provide a basis from which priorities for ergonomic intervention, including design, can be set. The changing patterns of NKID use suggest that these issues will become of increasing concern to ergonomists in the near future. It is believed that only through a systems approach involving participatory ergonomics can the problem be fully evaluated and solutions found.

6 References

Fogleman M and Brogmus G, 1995, Computer mouse use and cumulative trauma disorders of the upper extremities, *Ergonomics*, 38, 2465-2475.

Hagberg M, 1995, The "mouse-arm syndrome" - concurrence of musculoskeletal symptoms and possible pathogenesis among VDU operators. In: *Work with Display Units 94*, edited by Grieco A, Molteni G, Piccoli B and Occhipinti E (Elsevier: Amsterdam), 381-385.

Harvey, R and Peper, E, 1997, surface electromyography and mouse use position, *Ergonomics*, 40, 781-789

Karlqvist L, Hagberg M and Selin K, 1994, Variation in upper limb posture and movement during word processing with and without mouse use, *Ergonomics*, 37, 1261-1267.

Fatigue and Muscle Activity in Fast Repetitive Finger Movements

Thomas Läubli, Michael Schnoz, Joseph Weiss and Helmut Krueger
Institute of Hygiene and Applied Physiology, Federal Institute of Technology, Zurich, Switzerland

1 Aim

Although repetitive finger movements are known risk factors for the development of work-related musculoskeletal disorders, the underlying mechanism that causes discomfort and pain is not well understood (Kuorinka *et al*, 1995, Läubli and Krueger, 1992, Weiss et al. 1998). Maeda (1977) postulated a model on the development of work-related neck-shoulder pain. It stressed the importance of the combined strain resulting from many workload factors such as repetitive finger movements, long working hours, psychological stress, room climate, and others. However, the mechanism in the development of muscle pain at low-intensity work is still not established. Actual guidelines are based on practical experience, performance data and subjective symptoms, and are not deduced from physiological knowledge. A prevailing hypothesis postulates that specific motor units are active continuously during low-level contraction (Seidel and Bräuer, 1988). Long lasting activity of single muscle fibers may overload their metabolic capacity and finally trigger nociceptive nerves (Hägg, 1991).

To approach the pathophysiology of work-related musculoskeletal disorders, it is of interest to investigate whether fast repetitive finger movements induce constant activity of neck and shoulder muscles. We therefore examined the muscle activity during a fast and fatiguing tapping task.

2 Methods

2.1 Subjects

9 right-handed subjects participated in the study (7 males and 2 females aged between 24 and 39 years). All test subjects were healthy and did not suffer from musculoskeletal disorders.

2.2 Posture

The subjects sat at a table on an ergonomic office chair with a high back support at upright position. The heights of table and chair were adjusted such that the knees were at right angles and the thighs parallel to the floor. The working and non-working hands and arms were kept in a symmetrical and relaxed position, the upper arms hanging down vertically, the forearms parallel to the table. The right forearm laid on a wooden board with a built-in 8-key input device in its center. The left forearm laid on a similar rest without an input device. The finger tip of the index finger was positioned on one of the rigid keys. Such a posture does not require activation of the trapeze muscle.

2.3 The Tapping Task

The subjects were asked to tap at a speed of 5 Hz as long as possible (but not longer than 480s). The tapping rates were paced by an acoustic metronome. The subjects tapped with the index finger of the right hand on the rigid key.

2.4 Measurements

EMG activities of the following muscles of the right body side were registered: m. extensor digitorum communis (referred to as "extensor" in the following), m. flexor digitorum superficialis ("flexor"), m. biceps brachii ("biceps"), m. triceps brachii ("triceps"), m. trapezius pars descendens ("trapezius p. descendens"), m. trapezius pars transversa ("trapezius p. transversa").

Electrodes: Surface EMG was measured by pairs of round Ag/AgCl surface electrodes with a diameter of 5 mm (center-to-center distance of an electrode pair 20 mm). The reference electrode was placed laterally on the right elbow.

EMG-amplifiers and registration: Two amplifiers were in use to process the six EMG signals: i) a 4-channel amplifier with guard-drive, anti-aliasing low-pass filter (-100dB/oct, switched capacitor technique, set at 300Hz), and 50Hz notch filter ("amplifier 1"); ii) a 16-channel amplifier with low-pass filter (-3dB/oct, set at 490 Hz), high-pass filter (set at 5 Hz), and no notch filter ("amplifier 2"). The high-quality amplifier was used for the EMG-processing of extensor, flexor, trapezius p. descendens, and trapezius p. transversa. Amplifier 2 was used for the biceps and triceps. The data was acquired using a *LabView 4.01* program (National Instruments Inc. 1995) on a PC with Windows-NT. The sampling rate was 980Hz.

Applied force: For the tapping we used a key on a 8-key rigid keyboard. The actual force exerted during key tapping was measured by a strain gauge and registered with the same *LabView* program.

Acoustic clicks: The acoustic clicks (metronome) were produced by the *LabView* program and recorded together with the EMG and applied force data.

2.5 MVC/Relaxation

Prior to the experiments, the subjects were instructed to apply maximum voluntary contraction (MVC) during 5 seconds with each of the 6 muscles in sequence. The signals were used for the standardization of the EMG. Next, the subjects were instructed to relax completely, and the rest EMG potentials were registered. Both MVC and rest measurements were repeated at the end of the experimentation session.

2.6 Analysis

RMS and MVC: The measured signal of each muscle was filtered, rms-averaged with a 50ms moving window, and standardized to parts of MVC-values.

Muscular strain: Percentile calculations, based on 10s intervals, were used as indicators of the muscular strain: The 5th percentile for tonic (static) activity, the 95th percentile for phasic (dynamic) contraction, and the 50th percentile (median) as an indicator of average activity.

3 Results

3. 1 M. Extensor Digitorum (lifts finger)

In the extensor, the phasic activity (95th percentile of MVC) lay between twelve and seventy percents of MVC. The static component (5th percentile) ranged between two and ten percents of MVC. Subjects that only could tap for a short period (< 150 s) tended to have a more pronounced increase of both the phasic and dynamic muscle activity ($p < 0.05$) than subjects that could tap for more than 200 seconds (fig. 1, above).

3.2 M. Trapezius Pars Descendens (helps to lift shoulder)

The phasic activity of the trapeze muscle (95th percentile) ranged between one and thirty percents of MVC. The static components (5th percentile) lay between less than one and ten percents. Tapping endurance time was not correlated whether with the level nor with an increase of trapeze muscle activity. The trapeze muscle activity tended to be higher in subjects with higher activity of the extensor muscle (fig. 1, below).

4 Discussion

After a few minutes, fast tapping caused discomfort in most subjects. A shorter tapping endurance time was correlated with a stronger increase of both the static

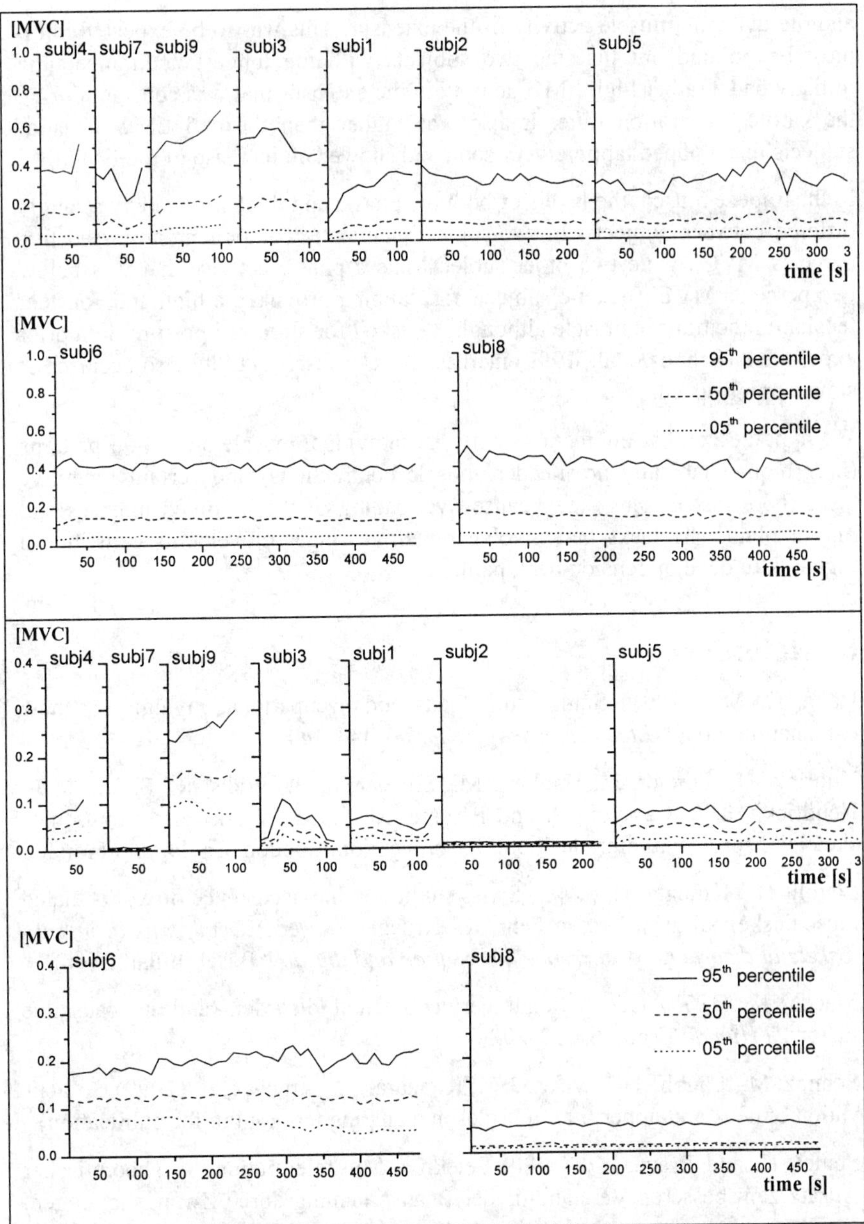

Fig 1 5th, 50th, and 95th percentiles of EMG-activity (10s intervals). All subjects, in the order of individual ability to continue the 5Hz tapping task. Above: M. extensor digitorum. Below: M. trapezius pars descendens.

and the dynamic muscle activity of the extensor. This was to be expected but it must be pointed out that the two subjects with the longest endurance time initially had a rather high EMG activity in the extensor that was constant during the whole registration time. It also was higher than the end EMG level of subjects that stopped tapping very soon and showed an increase of the EMG.

In the trapeze muscle the levels of both the phasic and dynamic activity strongly differed between subjects. In two subjects the static component surpassed five percents MVC, while two other subjects had a phasic activity that was below two percents MVC. In some subjects fast tapping provoked a high and constant tension in the trapeze muscle although we asked for a relaxed posture. In a prior experiment (Schnoz et al., 1999) increased neck muscle activity also occurred at slow tapping speed.

We hypothesize that interactions between individual muscle activation patterns (here biomechanically not needed muscle contractions) and certain repetitive tasks (here fast tapping) may cause overloading of the involved muscles and may explain why tasks that involve repetitive finger movements constitute a high risk to develop tension neck pain.

5 References

Hägg, G. M. (1991) Static work loads and occupational myalgia - a new explanation model. *Electromyograph Kinesio*, 141-144

Kuorinka, I., Forcier, L., Hagberg M., Silverstein, B., Wells, R., Smith, M.J., Hendrick, H.W., Carayon, P. and Pérusse, M. (1995) Work related musculoskeletal disorders: a reference book for prevention. London: Taylor and Francis.

Läubli, Th., Krueger, H. (1992) Swiss studies on the occurrence of work-related musculoskeletal disorders, in Fehr, K., Krueger, H. (eds) *Occupational musculoskeletal disorders: occurrence, prevention and therapy*. Basel: Eular, 25-42

Maeda, K. (1977) Occupational cervicobrachial disorders and its causative factors. *J Human Ergol*, 6, 193-202

Schnoz, M., Läubli, Th., Weiss, J.A., Kitahara, T., Krueger, H. (1999) Co-activation of neck and upper limb muscles in a tapping task (subm for publication)

Seidel, H. and Bräuer, D. (1988) Selektive Muskelermüdung – Theoretischer Ansatz zum besseren Verständnis der Beanspruchung durch Zwangshaltungen; *Kurzfassung der Beiträge des siebten Kongresses der Gesellschaft für Arbeitshygiene und Arbeitsschutz der DDR* (in German). Dresden: 185

Weiss, J.A., Krueger, H. and Läubli, Th. (1998) Repetitive finger movements and antagonistic muscle activity. *Ergonomics* (accepted for publication).

QUALITY OF INFORMATION IN INTERNET AS AN INFORMATION SECURITY BASIS

Tamara Grinchenko
Institute of Applied Informatics, Kyiv, Ukraine
Chervonoarmiyska St. 23b 252004
Tel.: 228-57-62,
E-mail: grin@iprinet.kiev.ua

1 Introduction

«In the beginning was the Word»...that is some information. There is no question that information has huge effect. It can promote development of person and community, it can kill as well. «Oh, evil tongues, they are more dangerous then pistol». Today, in Internet days, these words of Russian literature classic (A. S. Griboedov) are very actual. What is the cause of appearing such huge quantity of "evil" information (pornography, especially children one, terrorist's manual, aggressive games, racism propaganda and so on) in Internet?

The notions of «well» and «evil» were stated in high antiquity. We think that there were a lot of evil-bearing people in all the times. They alike other people wanted manifest themselves through the word and wanted to be heard or to be read.

When authoring was destiny of selected people (because of restriction in publishing, illiteracy, mass-media control) the last ones were limited oneself with sounding the words or picturing them on the mountains, walls, desks, fences. Now, in the epoch of Internet, everyone having access to it, may manifest himself. So Internet is so manifold.

What are the methods to overcome the negative sides of Internet?

Today there are some methods of information filtration of Internet by filter programs. These programs are divided into corporative ones and for home use. The first ones carrying out control over efficiency of Internet use by office collaborators. The second programs intend to protect morals of minors.

In Ukraine, basing on doctrine of Internet's danger,there are many publications regarding restriction of its use, and control over information. That brings the question of creating some «Ministry of truth» (J.Orwell, «1984»). The obligation of this Ministry was to correct the information according to political order.

Now, Ukraine has declared its tendencies of integration into European community. Violation of constitutional rights providing free access to information, can slow down this integration process. So today the quality of Internet content is of grate importance.

2 Goal

In this paper we propose some approach to the problem of improving the quality of information available over the Internet.

3 Method

As far as the world history shows, «evil» is irresistible. So the only way to reduce the «evil» impact on life it is to create a lot of«well». In this creation two aspects are important. The first aspect is connected with augmenting the positive information available over the Internet. In addition this aspect in the grate extent dependents on the second aspect - moral principles of the Internet creator, his beliefs and the feeling of obligation to society.

So, we want to propose the formation (it is desirable to declare the competition) of some «Codex of Honour» of member of International information common wealth, something like to «10 commandments». The first commandment, which we propose could sound like:» do not make harm». We think about ergonomics of soul.

In our Internet activity we strive to became the source of the quality information.

For some concrete definition of the «quality information» notion we were following the principals of Program of creating the national informatization infrastructure of the USA, proposed by President B. Klinton and Vice President Al. Gore.

In the base of the program the following principles are taken:

- information is the most important factor of the government administration;
- information is the most important resource, which requires saving, development and defence;
- information is the most important commodity, to which the users must have access.

The Program serves the purposes:

using the opportunities of the sphere of informatozation for the development of the enlightenment, science, culture, health protection, preparation of the rising generation for the life in XX1 century, building the more open democratic society.

One of the very important ways to achieve these purpose is integrating knowledge of information security into the curriculum. Analogous approach is proposed in as concern as computer security.

The technical aspects of security are closely related to computer science and engineering. Many of the goals and concepts are similar. (Cynthia E. Irvine, Shiu-Kai Chin, Deborah Frincke, 1998).

Regarding information security it is necessarily to take into consideration ethics aspects of this problem.

4 Results

We have put the described approaches into the base of creating the system of sites, representing Ukraine in Internet.This system was developed by hypertext department of Institute of applied informatics and is a part of Ukrainian National program «The open World» («Developing and creating information and telecommunication infrastructure of the state for integration of Ukraine into the global information common wealth»).

We present next sites on Ukraine: (Tamara A. Grinchenko, Anatoliy A. Stogny, Valentina Bondarovskaya, 1998).

«Welcome to Ukraine» (http://www.ukraine.online.com.ua/) (Figure 1.)

«Welcome to Kyiv» (http://www.kyiv.com.ua/) (Figure 2.)

«Lost churches of Kyiv'' (http://www.iprinet.kiev.ua/oldkiev)

«Welcome to Ukraine» - magazine (http://www.wumag.kiev.ua/) (Figure 3.)

These sites form information image of Ukraine as a big European country, which is interesting for clobal common wealth.

The site «Welcome to Ukraine» contains analytical articles on main information aspects: state, geographical position, history, population, culture, science, economy, education, lifestyle etc.

The site «Welcome to Kyiv», contains information on the capital of Ukraine, its meria, history, science, culture, industry, useful information for Kyiv's guest, walking on Kyiv streets etc.

Kyiv is the city of ancient culture with a lot of historical landmarks. Unfortunately many of them had been destroyed in the XX century.

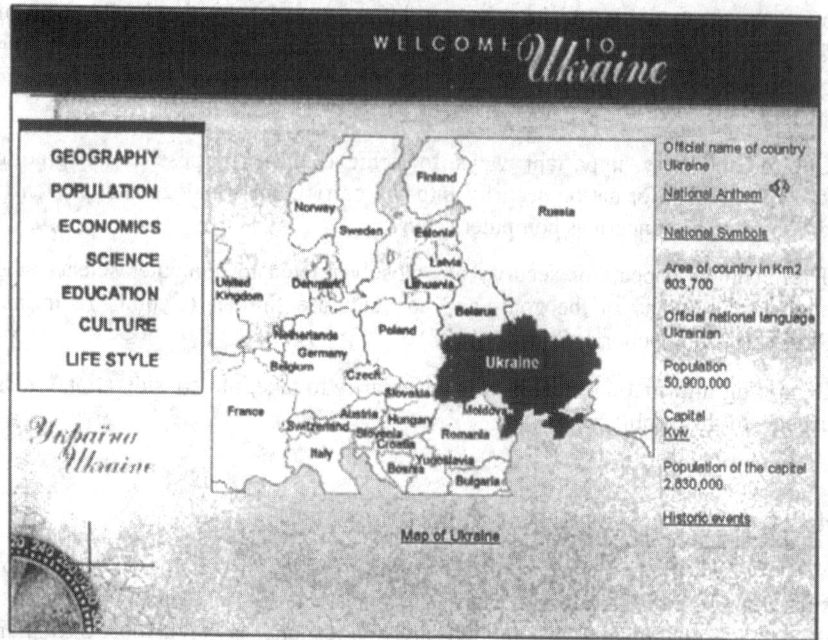

Figure1.

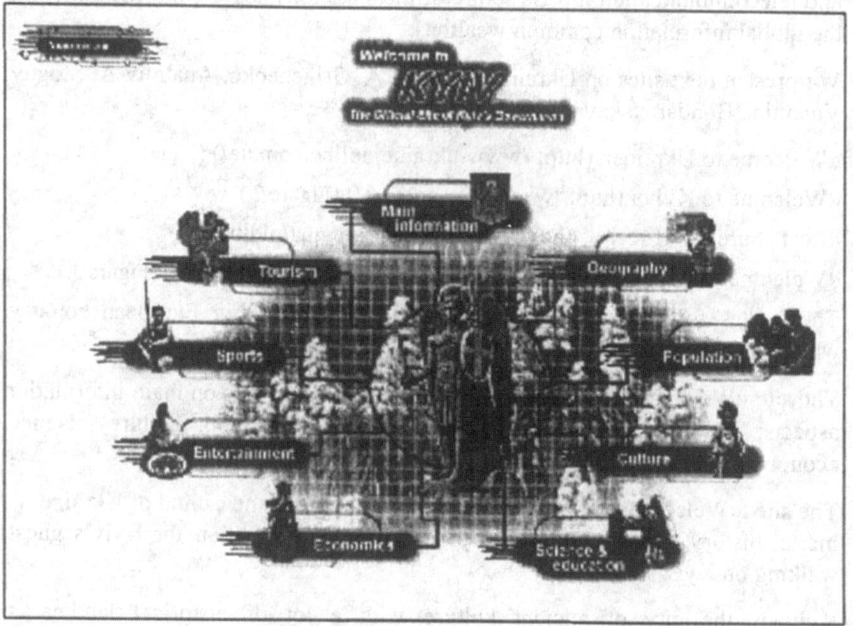

Figure 2.

The site«Lost churches of Kyiv'' devots to computer reproduction of the destroyed churches.

The site «Welcome to Ukraine» - magazine represents our country abroad, givs information regarding culture and history legacy, economic possibilities and presents key firms also.

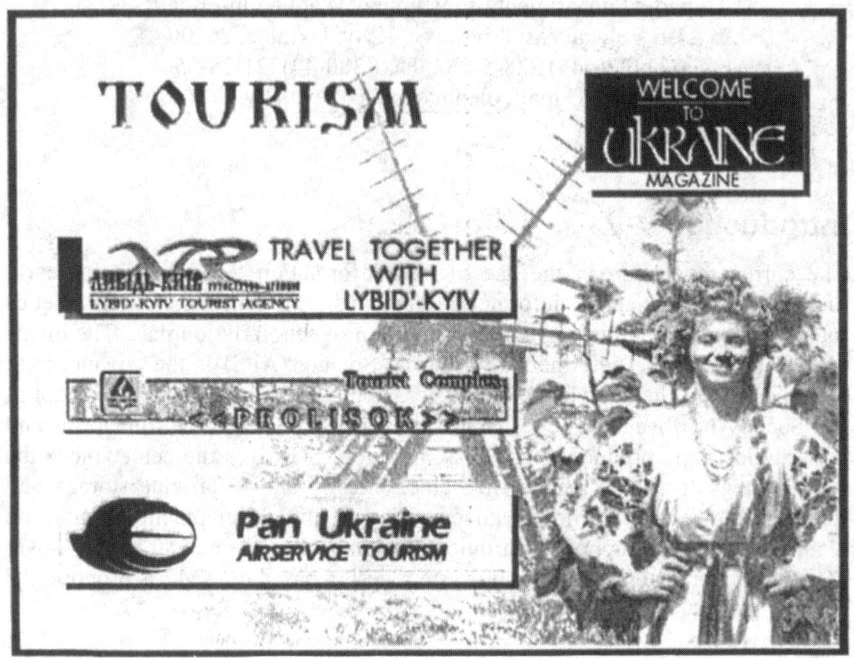

Figure 3.

5 Conclusion

This system of sites on Ukraine enriches Internet with a new quality information.

6 Reference

Cynthia E. Irvine, Shiu-Kai Chin, Deborah Frincke. (1998). Integrating Security into the Curriculun, p. 25-30. Journal «Computer» Vol.3, №12.

Tamara A. Grinchenko, Anatoliy A. Stogny, Valentina Bondarovskaya. (1998). Integration of Ukraine into information world space: cross-cultural psychological relationships. XIVth Congress of International Association for cross-cultural psychology. August 1998. Bellingham, Washington, USA.

Authoring Tools for Internet Publishing

Michael V. Olenin, Ph.D., Tamara A. Grinchenko, Dr.
Hypertext department of Institute of Applied Informatics
23-b Velyka Vasylkivska St, Kyiv, Ukraine, 252004
Tel:(380-44) 228-57-62, Fax:(380-44)221-28-35
E-mail: olenin@hytec.kiev.ua

Introduction

The current explosion in the use of new information technologies for early dissemination of scientific information dovetails with many scientific Societies interest electronic publishing and consumption of scholarly journals. The often-stated objective of The American Physical Society (APS) is the advancement and diffusion of the knowledge of physics. In the last few years, a major goal of the Society has been to move its scholarly publications away from print and toward electronic production and dissemination. SGML is the center piece the APS strategy to accept manuscripts electronically and to provide storage and delivery choices. The work is concerned with the development of mark-up technology tools to support electronic publishing of mathematical journals over Internet. A framework of authoring system design based on XML is proposed.

Objectives

One of the hardest problems is to write the authoring tools for an SGML/XML system. A good tool has to provide a natural interface for authors; most of them do not wont to know the principles of markup languages. It may also have to enforce strict and complex rules, possibly after every keystroke. Many current authoring tools are therefore tailored to a limited number of specific applications.

We have experience in developing and implementation of the first in Ukraine hypertext system HYPSY with some intellectual features such as context searching, composition language for generating new hypertexts, transaction management for collaborative authoring. We have also implemented visual SGML-HTML editor and browser for mathematical documents marked up using standard specification for patent abstracts. Now our interests focused on exploring authoring tools for electronic publishing and design principles of intellectual hypertext systems.

The main goal is R&D methods and authoring tools for Internet publishing based on eXtensible Mark-up Language (XML) for mathematical application.

Methods

The pilot implemented authoring system contains such components as visual XML-editor of mathematical documents such as papers and journals (collections of papers with corresponding layout for printing and browsing over Internet), browser of XML-documents, transaction management system to support document version control in distributed environments.

1. Entity Definition

Before creating an XML document, we need to define the elements (fields) and attributes (field descriptors) it will use, as well as whether or not it will reference or incorporate a Document Type Definition (DTD). The common approach to design XML document editor is to make visible object model of document on screen and provide user with opportunities to manipulate with contents of these objects. The document object model is corresponding to the XML marked up structure (DTD) for specific class of documents.

2. Document Creation

With authoring tools such as Vervet's XML Pro an XML document can be defined by dragging elements into the free structure on the left and defining each element's attributes in the box on the upper-right-hand side of the screen.

3. XML describes information, not its layout. This lets you separate presentation from content. With the use of an XSL or CSS stylesheet or server-based scripting, we can take the same XML document and present it in the most appropriate format, be it in a database record, a published article, or a series of abstracts.

Principles of a construction of the editor

In present work we attempt to provide user with both mark-up mechanisms of describing as document structure and content (XML) and it's style-sheets in the form how it is required for mathematical papers. User can manipulate with visual document objects on the screen and see the document image just how it will be printing. It is common WYSIWYG way implemented in many document editors but without XML. We propose an approach that allows hiding complex mark up mechanism from users. That was required of exploring and embedding more complicated components with intellectual features, such as automatically recognizing relationships between XML model document objects and their visual stylesheet representation images on the screen during authoring process.

This work is focused on the topic of construction and implementation of Structured Document Editor (SDE).

The version of mark-up language used in SDE is mathematical subset derived from version 3.4 of the WIPO (World Intellectual Property Organization) standard ST.32, extended by hyperlink definitions of XML.

Objects of the editor are: the document, section of the document, page of the document, paragraph, figure, table, group of the formulas, mathematical

formula, chemical formula etc. Appropriate tags of the markup language describe each of these objects in the document. The markup is determined as the text, which is added to contents of the document and describes a structure and other attributes of the document, irrespective of processing, which can be executed above this document. The markup includes the descriptions of types of documents, reference and descriptors (tags). With reading or creation of the XML-document by the editor the analysis of markup of the document happens. Outcome of that is the internal representation of an electronic model of the document. Each object of a model can be described by the properties (coordinate, height and width with visualization, various types of selection, type of the font for the texts, type of the formula etc). Above objects of this model in the editor the methods of their processing similar to the modern editors (copying, moving, resizing, substitution etc) are realized.).
The program of the editor consists of a number of units, basic of which are:

- The parser of the source of the document (for a new document a source code is empty); the parser produces the analysis of the document and forms objects of its electronic image;
- The viewer of the document;
- The interactive unit of editing of visual objects of the document;
- The unit of the output of the document (on printing).

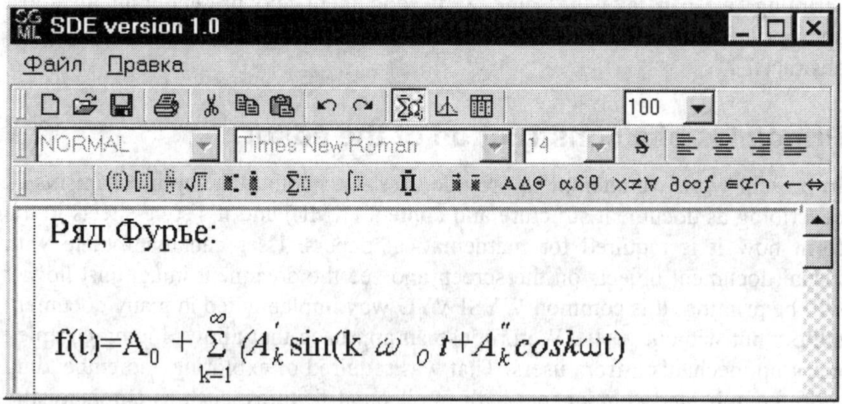

Fig. 1. The work window of the editor.

The viewer of the document is made out also as the separate application and can independent be used for review of the document. The hyperlinks allows using the viewer as browser for navigation on a structure of the document. The work window of the editor is represented in a Fig. 1.

The mark-up format of mathematical formula presented in Fig.1 is

<DF ALIGN= “CENTER”> f(t) = A _{</I> 0} <I> + <SUM> <FROM> </I> k = 1 <TO> &165; <OF> (A <SP> <I> ' </SP> <SB> k </SB> </I> sin (k <I> &wgr; <SB> 0 </SB> t + A <SP> " </SP> <SB> k </SB> cosk </I> &wgr; t) </SUM></DF>

For the tool environment of development of the editor SDE was the programming system DELPHI. The program of the editor can be executed in the Windows'95 environment and requires minimum size of the main memory 16-MB. The program with the library occupies 2Mb of memory on the hard disk.

Results

The main results are:

- problem solutions on architecture of authoring system for mathematics;
- XML editor with creating layout features to publish documents both in electronic and common hard copy ways.

Conclusions

We assume that XML authoring tools with WYSIWYG features is very useful for scientist to present his scholarship works over Internet and receive it hard copy forms. Received results can be useful for incoming Ukraine to world information exchange via Internet.

Psychophysiological effect of the information technologies on users and a psychophysiological approach for its detection

Alexander Burov

ELAN Ltd., Centre of Ecology, Ergonomics and Safety, Kyiv, Ukraine

1 Backgrounds

The human's occupational health, fitness for work and work quality depend on both external (beyond his control) and internal (professional knowledges and skills, psychophysiological opportunities) factors.

First of all, it relates to a work with information technologies which could be used to effect on human's motivation, security, effectiveness, health as in general, as during the particular time. Such influence of the information technology on a human can have a cumulative result if it is used regularly.

The purpose is a development of a method and a system for psychophysiological detection of changes of a human psychophysiological state when working with information technologies.

2 Factors effecting on the user of information technologies

The development and perfection of information technologies has resulted in that problems connected with labour activity become problems of every day life because borders between work in industrial conditions and at home has practically disappeared, if the question is a work with computer of manager in office, operator on production, programmer, student or ordinary computer user. In any of these cases the human interacts during work with production means (information technologies), industrial environment (on the workplace), subjects of work (information streams, data bases).

Thus, as a rule, hir or she is subjected to effect of large number of factors, various on the nature, forms of display, nature of action. In some conditions the affect of these factors can render the adverse influence on human health (to result in trauma, illness, decrease of serviceability) or on safety and efficiency of work carried out.

A high neuro-emotional pressure, low or moderate physical workload are a characteristic feature of user (operator) work with information technologies.

Though they are developed the normative documents (standards, methodical recommendations etc.) which regulated the interaction of human and technical (technological) means of work and are intended for good safety and efficiency of human work they do not decide finally the problem of work safety because basically they are aimed to the valuation of constant conditions of activity (operation) and in majority of cases do not take into account dynamic character of these processes. At the same time, human opportunity to regulate the work process by himself with the purpose to increase a comfort of activity, as a rule, are rather limited.

As a whole, fitness for work and reliability of operator depend on 4 groups of factors: professional knowledge and skills; psychophysiological professionally important qualities; functional condition of person and factors of environment. These factors form the target parameters of operators, affecting work efficiency as by value, as by time.

It is necessary to note, that no measures on perfection of software do not allow to exclude human errors. It is stipulated not so by quality of HCI programming, as by insufficientness of knowledge about psychophysiological opportunities of human who is included in information space, is opened for external effects and is not protected from direct change of the information model of world. The work in opened information networks (Internet) or advanced programme systems (WINDOWS etc.) is convenient means of "information" programming of human, effect on his or her tastes, desire and actions. His or her psychophysiological condition in combination with factors of industrial environment on workplace has the considerable significance for the protectiveness.

In practice, at the work design, as a rule, they are not taken into account such characteristic of topology of working premise, as its space arrangement (in addition to volume and form), as well as, natural fluctuations of human's psychophysiological parameters. At the same time the experience of practical use of registration technique for complex of parameters of geodynamic processes shows that at valuation of operatots workplace it is necessary:

- to take into account geodynamic and other physical factors when choosing a workplace location of user with higher status of safety;
- to carry out the monitoring of human psychophysiological characteristics (for all stages of his or her professional biography since the moment of coming into trade up to leave on age or owing to illness) possessing the individual oscillatory nature in various phases of life and on more short intervals of time (year, month, day,...);
- carrying out of serviceability check to execute taking into account of real geophysical anomalies for valuation of degree of stability and of adaptiveness of organism to various psychophysiological and emotional stresses (recent researches according to the program "Sun-climat-human" have shown that serviceability of healthy people on days of geophysical anomaly was increased by 4-5 %, speed of reaction - by 3-5 % with increasing of the errors numbers - by 5-9 %).

3 Method

On the basis of laboratory system for psychophysiological researches they were developed the systems of valuation of a human's psychophysiological state, wich ensure an indirect valuation of physiological "cost" of human performance by results of specially developed complex of psychophysiological indexes.

4 System of monitoring

The system of individual monitoring of human mental fitness for work (SIM) is used for objective individual valuation of functional state of fitness for of user of information technologies in beginning of workday.

The valuation of functional state is intended, first of all, for use by users with the purpose of corrections of own functional state, revealing of factors affecting mental fitness for work. The system is oriented on maintenance of high professional fitness for work and long preservation of user health at the expense of optimization of work stress. The achievement of these two purposes simultaneously is stipulated by important physiological feature of mental fitness for work - it reaches maximal significances at moderate pressure of physiological systems, in relatively quiet state. Too high pressure during the operating time is not only adversely for health, but also results in decrease of efficiency of human work. When levels of pressure (overactivity, increased emotionality of situation perception) are too high, as well as when levels of mobilization (slackness, sleepiness, overwork) are too low, the mental fitness for work is reduced, the probability of faulty actions and of non-optimal decisions is increased. The optimization of pressure in work activity permits to improve as the professional efficiency of work, as the health.

The professional qualification of user is acquired during many years and does not vary considerably from day to day. At the same time, ability to realize the qualification largely depends on the functional state of human. The human physiological state and his or her mental fitness for work are not always identical and can vary as under influence of internal reasons (illness state, biorythm, fatigue), as external circumstances (quality of rest, home and service mutual relations, trouble, time of shift etc.). Subjective, internal valuation of own fitness for work is poorly developed at human, and it is frequently far from validit. Usually a human notices only the considerable deviations of the functional state and mental fitness for work from usual, requiring medical interference.

The main task of the SIM is the granting to user of objective valuation of his or her functional state to make a decision on basis of this valuation to correct operatively his or her state in direction of optimum characteristics or hereinafter to avoid or to prevent the action of factors which reduce his or her fitness for work.

The efficiency of monitoring realized as computer systems of daily check is checked on more than 90 users and permits to lower the risk of faulty actions, as well as to optimize the working pressure of user.

Psychological Aspects of the Organization of the Distance Learning

Serge S. Azarov, Olga V. Makyeyev
Institute of Applied Informatics

Distance learning absorbs in itself almost all modern forms and methods of the tradition learning process, but "distance" or remoteness bear in a context of learning process completely new accents "without the contact". To recognise and to take into account these accents in conditions of the future "Global information society". It is a very important problem, because all modern new methods of the organisation of the learning process are directed on the permanent decrease of a degree of the direct participation of the teacher in learning process and transference of accents on the self-education. We see two ways of the decrease of manifestations "without the contact", namely:

- Increase of the efficiency of information interfaces using effects of "attachment" and "full immersing", which are characteristic for systems of a virtual reality;
- Organisation of the active feedback and modification of the concept of the control of learning process results as subsystems of diagnostics and testing of a degree of the perception of an investigated material.

This article purpose is the holistic view of the psychological aspects of the distance learning, which is based on the theory of the distance learning.

1 Basis Positions of the Theory of the Distance Learning

According to traditional definition the binary teaching – learning process is active interaction of a pupil with the teacher, in which the concrete conscious purpose is reached. This purpose was formed as a result of concrete needs: mastering by the certain knowledge and methods of their application.

But we shall mark, so far as the development of the learner happens during of own activity, in so far as the basis of the binary process is necessary considered the "learning", instead of the "teaching". Let's pay attention to that fact, the learning process is characterised a large degree of mobility and variability, as it leans on psychology of mastering of the material by the learner, though the learning process is realised in the certain sequence and the logic of educational process is closely connection with logic of an educational subject.

Proceeding from above-stated, we have selected (that is agreed with traditional paradigm of teaching) that constant components of the learning process: apperception, perception and control of the learning process results.

All these components function as the whole, that is learning process. When model of this process has the view of the semantic net, which connect the apperception, the perception and the control of the learning process (fig.1).

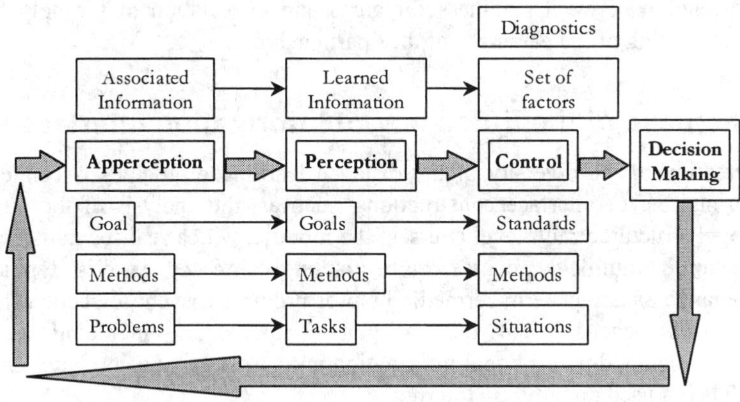

Fig. 1. The model of the learning process

Such model of the learning process can be consider as a basis for forming of a didactic expediency of the computer application for the maintenance of learning cognitive and educational activities, it determines of the directions and features of using of the computers for a learning process, as well. The hierarchy of model allows us to analyse connections between components of a learning process.

The model of the learning process gives a possibility to study of the ways of a development and perfecting of forms, methods and components of a learning process. The constructed model of the learning process enables for a research of metamorphoses of the paradigm of the learning process, while forming paradigms of a training and distance learning and the modification components of the learning process (contents, feedback and others).

Basing on the model of the learning process, we have formulated the list of the phenomenon of the distance learning and proposed the methods of the solution of the psychological problems, which are connected with peculiarities of this learning.

The essence of a phenomenon of the distance interaction is the accent "without the contact". It predetermines main requests to the organisation of an educational material and learning process. From the psychological party, if the teacher is absent, then the success of the learning process is possible for want when the introduction of the effect "attachment", which stimulates interest to a prolongation of learning. We imply an adequate replacement or imitation of direct interactions by computer methods. The methodology of the distance learning requires unordinary solutions, which are directed on the maintenance of the correlation between readiness for gains and expectation of the help from enclosing (using of the paradigm of the "partnership").

2 Increase of the Efficiency of Information Interface

The creation of the diversity of the distance learning prototypes is connected with a number of interface constructions, such as: the analysis of the pupil's typology; structured multilevel courses; the module, which realises transition at a structured multilevel course correspondingly of the pupil's typology; maintenance an adequate hypermedia help or prompts (psycho-didactic effect); the "Virtual teacher" and the "Virtual competitor "; including a new apperceptional and perceptional information in a structure of a computer course (which is not used directly) by Internet.

3 Diminution of the Formality of a Course

Some constructions make a system of the distance learning less formal, it is necessary to enter into it a structure a number of constructions. These are: to organise a course as a form of the hypertext or hypermedia structure, which ensures the transition to previous levels of the submission of knowledge or "repetitions" in case of the difficult perception; to enter a system of the help, but it not presents a correct solution (directing, as used in Software), and stimulating and raising motivation (as the award for activity or for demonstrating of the certain knowledge).

4 Personalising

For the introduction of modes of personalising the system of the distance learning is necessary to supply by subsystem of the recognition of the pupil's typology, which would install the correspondence between the pupil and the learning process model. The efficiency of using of the factors "Virtual teacher" and "Virtual competitor" (the mode "attachment") is depended on perception of the "properties" (that is, individuality) by the user. Thus, the designing of the indicated elements should take into account a type of the person and psychological features of the pupil.

To improve the perception of a material is possible by an individual selection of an appropriate *verbal level* and granting *a possibility of the choice* of a model learning (as a matter of fact of methods and forms) to the pupil (we shall mark, the structure of a course should be saved).

5 Control in the Learning Process

The efficiency of the learning process is stipulated by activity and systematic character of the feedback (diagnostics of results of the learning process).

The problem of personalising of the diagnostics process is decided by creating of a system of the standards (basis of the measurement standards) and accordingly situations (test dates). It depends on a model of the learning process, learning purposes and type of the person.

To achieve interest during testing is possible, if to develop a testing system in the correspondence with principles: **principle of the intellectuality; principle of the friendship interface; principle of co-operation; principle of three problems** (it directs on a solution of three problems of the didactics process: tutoring, control and diagnostics of knowledge and skills); **principle of constructing; principle of the entertainment** (use of game moments during testing).

6 Introduction of Elements of the Direct Dialogue

The learning and tutoring includes social oriented function: the pupils should produce skills of the contact with the people of an equal level of knowledge and with the authoritative experts. The realisation of this function is possible with the help of Internet-conferences, granting of channels of links for direct contact between the pupils and with the teacher, creation of a system "hot line".

Thus, in the article possibilities of updating of the methodology of the distance learning are considered, such as: the introduction of the mode "attachment", personalising and other, which took into account the phenomena of the distance learning and were directed on the increase of the efficiency of the learning process. The dates of modes are directed on a diminution of the effect "contractions of a set of choices of methods ".

7 References

Ackoff R. L., Emery F. E. (1972). *On Purposeful Systems.* Chicago and New York.

George F. H. (1977). *The foundations of cybernetics.* London.

De Volder M. L., Lens W. (1982). Academic achievement and future time perspective as a cognitive-motivational concept. *Journal Person. Soc. Psychology.* V. 42. 3. pp. 566-571.

Kulhavy R. W., Andre T. (1972). Conditions under which feedback facilitates learning from programmed lesson. *Journal of Educational Psychology.* V. 63

Maslow A. H. (1995). Deficiency motivation and growth motivation. *Nebraska symposium on motivation* / Ed. By R. M. Jones. Lincoln.

Mowrer O.H. (1989). Stimulus-Response Analysis of Anxiety and its Role in Reinforcing Agent. *Psychological Review.* V.66.

Pask A. G. S. (1975).*Cybernetics of Human Learning and Performance. Guide to Theory and Research.* London.

Wartofsky M. W. (1979). *Models, Representation and the Scientific Understanding.* Dordrecht, Holland.

PSYCHOLOGICAL, ETHIC AND SOCIAL PECULIARITIES OF USING NEW INFORMATION TECHNOLOGIES

Valentina M. Bondarovskaia, Anatolyi A. Stognyi
Institute of Applied Informatics
23b, Velyka Vasylkivska str.,
252004, Kyiv, Ukraine
bond@iprinet.kiev.ua

1 Introduction

The objective of our investigation was to develop principle approaches to designing of the system for exposing the factors and means of "hidden" influence of information technology virtual world to individual, group and mass consciousness, to analyzing of psychological, ethic and social consequences of the information technological culture absence nowadays, and to working out the methods of support information psychological security of all categories computer users.

Experience of analyzing modern researches in the field of psychological and ergonomics aspects of the information and communication technologies(ICT) using shows that during the last ten years the intrusion of ICT upon the external and internal world of person have been realized (.Rantanen,1996; Weinzenbaum,1996; .Baron and .Richardson, 1994; Проблемы…,1996).

We supposed that the typical factor when committing a breach of information-psychological security is incorrect setting of tasks related with informatization and automatization in organizational management, in particular, underestimation of human aspect and individual peculiarities, systems of individual conceptions which solving the problems. In the other hand, individual's information-psychological security at all levels is under a threat of new information technology. We supposed that it is resulted in breaches of a person's orientation, structure of a person's values, goals and motives, formation of his/her aggressive behavior.

A special situation turned out in New Independent Countries of Former Soviet Union. In soviet time there was formed a special type of person, small part of big social machine(I.Stalyn). In the area of personal responsibility, the "soviet" citizen was culturally trained in learned helplessness and dependency. The state oversaw the planning, organizing and implementing of all facets of each individual's life. The situation created a patronymic system that functioned like a family. Consequently, people believed that it was the responsibility of other people to decide the problems of each individual and family, creating a lack of responsibility in people of their own welfare and a lack of living with freedom.

Ethic norms that were developed in soviet psychological and social space are related with lie on all level of human life, compromise between good and evil("the end justifies the means")(Lefebvre, 1992) and lack of respect to private property and private life.

Consequently the postsoviet society and postsoviet person in the world of new information technologies entered with their problems, specific psychological models of behavior.

2 Methods

We carried out the cross-cultural psycho-ergonomics analysis of human-computer interfaces. The main directions of our analysis were: cognitive, emotional, ethic, pattern proper(the system of values and attitudes) peculiarities of human being in information space. On the cognitive level of the analysis we considered the peculiarities of person's perceptive activities in information space in different cultures, the color coding strategies on the different interfaces, the ergonomicity of the images in typical contemporary Internet interfaces and possible psychological and social consequences of human (in particular teenagers)being in aggressive or with low level of ergonomicity information technology space.

On the emotional level of the analysis we paid attention the sources of the formation of computer users aggressive behavior: the content of computer games and some Web pages, dynamic ergonomics peculiarities of interfaces, etc.

On the ethic level we considered the intervention of the violence to information technology space(aggressive advertising, violent games and information, etc.) and ethic systems of totalitarian and post-totalitarian societies.

We developed a questionnaire for students(16-20 years). We asked young people about the influence of ICT on their relationships with family, friends, their state of health, how the "life in Internet" reflect on their safety feeling, formation of aggressive behavior(mine and my friends). We interrogated 180 students of Kiyo high schools and 120 national Technological University.

3 Results

Human being in the virtual information technology world relates with changing of living surrounding, work conditions, communication systems, person's selfexpressing and self-esteem. There are some direct psychological, ethic, social consequences of human being in ICT virtual world. This phenomenon has found a special reflection in East Europe and Former Soviet Union because elemental intrusion of ICT upon "other" unpreparing surrounding of posttotalitarian person has brought deep contradictions on his/her internal world:

- person that was inclined to the totalitarian social space had the double moral, unpretentiousness, was declining the responsibility about own life, was ready to work in uncomfortable conditions. postcommunist person was not aquatinted with the legislation system. The law for him/her was substitute by terms of honesty and decency;

- there is a special reflection of free democratic liberal structures in virtual ICT world on the posttotalitarian internal person's world and his/her behavior psychological models. There is an ethic conflict between the person's feeling "All is possible in ICT space" and common ethic norms(the respect to private property, private life, for example);

- specific influences ICT on consciousness of teenagers on posttotalitarian societies: the substitution of real difficult life in the society at transition time to passing away the time near computer, the taking upon themselves even decisions that can have the influence to the safety of nations, military and bank systems.

There are some common consequences of human being in ICT space:

- a break between the structure of person's activity and communication in midst of the ICT and in the midst of his/her ordinary life;

- the influence of violent content of computer games and some Web sites to the human behavior, especially to psychological behavior models of teenagers. The direct psychological relations between virtuality of artificial persons death in computer games and aggressive teenagers behavior in the real life;

- the influence of ergonomics culture absence in creating of human-computer interfaces in Web sites to emotional status of these Websites users;

-the formation of aggressive behavior of different age people under the influence of aggressive interfaces negative and violent information. There are a lot of family conflicts that are the direct consequences of ICT life of one or more members of family. There is some gender contradiction on the family level: we men are ICT users, we are higher that you, women and girls, kitchen people.

Nowadays it is important to create and to share the technology of individual and group psychological defense, ergonomics, psychological and ethic support of the new virtual information space safety for users.

5 References

1. Rantanen J.(1996). Human-technology interface: Problems and challenges. In "Work in the information Society. 20.-22.5.1996.-Finish institute of Occupational health.-Helsinki.

2. Weizenbaum J.(1996). Effect of television - a critical opinion.Ibid.

3.Baron R.A. and .RichardsonD.R.(1994). Human Aggression.-Plenium Press.-N.Y.and London.- 1994.

4.Проблемы информационно-психологической безопасности(1996)(Ed. by A. Brushlinskyi and V.Lepskyi).-Moscow.

5. Lefebvre V.A.(1992)A Comparative Analysis of Western and Soviet Ethical Systems. D.Redial Publishing Company, Boston/London/Holland.

The Comfort Evaluation of VR in Vibration and Slopes Environment

Chien-Hsu Chen, Ming-Chyuan Lin, Dao-Yuan Chen, and Hsin-Tien Wu
Department of Industrial Design, National Cheng-Kung University, Tainan, Taiwan, Republic of China

1 Introduction

Comfort and discomfort in ergonomic studies are major concerns of design the chair or seat. Several researchers have done such evaluation in those areas. Riding bicycle could be defined as a kind of "sitting", however, bicycle riding is very rare in comfort and discomfort researches.

In common parlance, comfort is a kind of concept and can refer to the feelings of either comfort or discomfort. For different riding environment such as mountain climbing or bad road, comfort would be the major factor to influence riding and control efficiency. In the off-road conditions, slope and vibration were discussed quiet often. In our research, we investigated these two factors in bicycle riding experiment and applied virtual reality (VR) techniques to make the experiment more real.

There are some literatures related the comfort problems of head-mounted display (HMD) in virtual reality environment. However, the primary objectives of our research were to (1) investigate the influences caused motion sickness and headache under the different slopes and vibration conditions with having HMD, and (2) investigate the physiological measurements such as the pressures of seat, riding speed, and the vibration of hand during bicycle riding experiment.

2 Experiments

The experiment involved data collections under different slopes and vibrations with having HMD. We tried to evaluate the effects of different slopes and vibrations according the subjects' physiological measurements. We also defined general comfort rating (Table 1) and evaluation check list as indexes of subjects' responses after experiments. The detail of experiments will discuss as followings:

2.1 Subjects

Ten younger volunteers (2 women and 8 men) were involved in this study. Their ages ranged from 22 to 28 years, with a mean age of 24 years. Subjects must wear tights to make sure hip completely touch the pressure sensors on the seat of bicycle.

2.2 Equipment

The whole system is showed in figure 1 included a roller platform for simulating the riding of bicycle. We adopted the roller platform by stick on rubber pads to simulate vibrations. We also added a 400 pulse/rev rotary encoder to collect the riding speed. The signal of rotary encoder was decoded by decoding circuit and connected to DI/DO interface card. There were two synchronous hydraulic jacks to rise the whole platform to simulate the slope conditions.

The bicycle is the Giant Co. LTD's product with front and rear full suspension and mounted with three pressure sensors at seat. The pressure sensor's model is LM-A super micro load-cell and the maximum load is 100kg/cm^2. The pressure data was enlarged through load-cell amplifier and processed by A/D transfer interface card. There was a data glove to detect the vibration of hand. The VR scene was produced by camera recorded from real riding environment and controlled by self-development program.

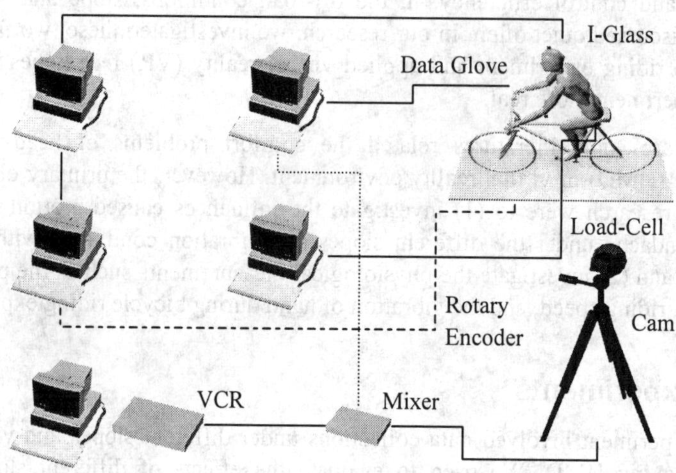

Figure 1. The diagram of the whole system.

1	I feel completely relaxed.
2	I feel quite comfortable.
3	I feel barely comfortable.
4	I feel uncomfortable.
5	I feel just fine.
6	I feel restless and fidgety.
7	I feel cramped.
8	I feel numb.
9	I feel unbearable pain.

Table 1. The general comfort rating (Shackel, Chidsey, & Shipley 1969).

2.3 Procedure

The first step of experiment is settled down the equipment and explained the sequence of experiment to subjects. Subjects have three minutes of warm up exercise to be practiced with the equipment. When warm up is done, subjects started to proceed the experiment under different slopes and vibrations. The whole procedure was supervised and recorded by camera. The data involved pressures of seat, riding speed, and the vibration of hand was collected by computers. Finally, we asked the subjects to write down the comfort rating scale and interviewed with subjects by using the evaluation check list to collect information of comfort at the end of experiment.

3 Result and Discussion

The aim of this research is to evaluate the affection of comfort in VR under different vibration and slope conditions. And the physiological measurements were recorded and observed.

From the evaluation check list, we could find out the reasons of motion sickness and headache are the fatigue of eyes, having muscles stiff and the fear feelings to riding platform. However, the balance of riding is not a significant effect caused motion sickness and headache. But the slope will influence the balance and cause subjects to have muscles stiff. However, on the same slope condition, increase the vibration will reduce the vibration of hand. Under the vibration condition, also increase the slope will reduce the vibration of hand.

For three pressure sensors at seat, we find out the pressure of rear part is not expected to have a significant increase during up-hill situation. The primary reason is the body inclined to the front in order to balance the gravity under up-hill situation.

4 Conclusion

Virtual reality is a modern tools for simulation the real world. If the confusion and perplexity more than we can accept, VR will lose its initial standpoint. This research proposed an alternative and possible direction for VR using in bicycle simulation.

The way people using VR is supposed to be more common and diversified. However, vibration and slope conditions will challenge VR to be more practical using. We hope this study will find the key to prove VR in ergonomics design. Further development is needed for comparing human motion from the real world to the virtual reality environment, for analyzing the ergonomics, and for developing the optimal platform to simulate bicycle riding.

5 References

Shackel, B., Chidsey, K. D., & Shipley, P. (1969). The assessment of chair comfort. Ergonomics, 12, 269-301.

The Maintenance of Habituation to Virtual Simulation Sickness

Peter A. Howarth and Kevin J. Hill
Visual Ergonomics Research Group [1]
Department of Human Sciences, Loughborough University,
Leicestershire, LE11 3TU, England.

1 Introduction

The increasing use of Head-Mounted Displays (HMDs) for the viewing of virtual environments has been accompanied by a heightened awareness that, for some people, immersion in this way produces symptoms akin to those of motion sickness. However, this 'virtual simulation sickness', or 'cybersickness', must have a different genesis from that of motion sickness, as closing the eyes prevents the former but not the latter. Although it is known that most people can adapt (or 'habituate' if the stimulus is one that is repeated) to real motion (Griffin, 1990), the characteristics of habituation to the *appearance* of motion are not yet known.

In order to examine this habituation, we extended the work of Ramsey (1996) by examining two groups of participants who had different amounts of visual stimulation (Hill and Howarth, 1999). Of the 26 participants who started that experiment, 19 completed the full five days, during which time they were exposed to a virtual environment for twenty minutes (unless their symptoms became too severe) on a daily basis. Eleven of these participants were also exposed for an extra fifteen minutes daily, to provide additional visual stimulation. All 19 participants showed habituation over the week, with those who had been given the extra exposure showing the greater effect. We have now re-examined 10 of these participants (those available) after a time lapse of six months. During this time none of the participants experienced any immersion in a virtual environment, and we report here that the level of habituation previously seen has not been maintained.

[1] http://www.lboro.ac.uk/departments/hu/groups/viserg/viserg1.htm

2 Methods

2.1 Participants

All ten participants reported that they were healthy, were not suffering from any vestibular dysfunction and were not taking any medication during the experiment. All were volunteers, and most were staff or postgraduate students of Loughborough University. They were aged between 23 and 50 yrs. (mean 32.8, s.d. 9.0); two were female, the remainder were male. Participants 1-4 had undertaken the 20 minute immersion in the previous experiment, and the other six had experienced the extra exposure.

2.2 Equipment

A Virtuality Dynovisor Head Mounted Display was used to present the visual stimulus. This HMD does not track the head position, and thus the system can be considered to be a „personal viewing system." The software used was a CD game called Wipeout, which was run on a Sony PlayStation®. The game is one where competitors race a hovercraft around a futuristic circuit, and as the hovercraft can move about a number of axes, the game provides a particularly powerful nauseogenic stimulus.

In the previous experiment, symptoms were assessed by asking participants to rate their feelings of nausea at one minute intervals during immersion. This rating was on a four-point scale, as used by Regan (1995), where 1= no nausea, 2 = some symptoms, no nausea, 3= mild nausea, 4= moderate nausea (at which point the trial was stopped, for ethical reasons). In the current experiment the trial was stopped as soon as a *change* in rating was reported - although participants had not been informed beforehand that this would occur.

3 Results

The table shows the time taken for each subject to first report an increase on the rating scale, on three measurement occasions. These occasions were the first and last days of their daily immersion in the previous experiment, and six months after their last immersion.

There was appreciable habituation between day one and day five, as shown by the increased time taken to report a change in symptom (sign test, $p < 0.01$). The habituation does not appear to have been fully maintained during the time that participants did not play the game, as the times recorded after the six month break are significantly shorter than those seen on the final day of the week's exposure (sign test, $p < 0.01$). However, most of the times seen after the six month break are slightly longer than those reported on the participants' first exposure to the stimulus, suggesting that some habituation has remained.

Table 1. The time elapsed (minutes) before the first report of a symptom increase.

Subject	Day 1	Day 5	6 months
1	5	12	2
2	2	11	2
3	1	1	1
4	1	6	4
5	14	*	14
6	1	6	3
7	6	*	7
8	4	*	5
9	3	10	6
10	1	6	3

(* No increase occurred during the 20 minute immersion on these occasions).

4 Conclusions

The habituation that occurred in our previous experiment between days one and five can be clearly seen in the results shown in table 1. Three of the ten participants failed to show *any* symptom change after immersion by the end of the week's habituation, and all but one of the other participants took appreciably longer to first report a change. After an extended period without exposure, however, the situation had changed dramatically, and *all* of the participants showed some symptom increase during the trial - half of them reporting a change within the first three minutes.

Although it is clear that the habituation was reduced during the intervening six months that the subjects were not exposed to virtual environments, it does not appear to have been entirely eliminated. Six participants took longer to report a symptom change during this final exposure than they had on the first exposure, whilst only one took less time to report a change.

At first sight, our finding of a reduction in habituation is puzzling in the light of the results of Regan (1995), who reported a consistent decrease over time in the number of participants reporting symptoms. However, the difference can be explained by the different experimental conditions employed. Regan's participants were tested on four occasions, and there were breaks of (i) four months between the first and second immersions, (ii) four months between the second and third immersions, (iii) one week between the third and fourth immersions. During the week's exposure of our previous experiment, our participants were immersed every day and habituation was manifest as a consistent daily decrease in reported nausea (Hill and Howarth, 1999). However, the habituation that Regan's participants experienced is likely to have been weak because of the long time intervals between immersions in her experiment. As a consequence, only a very small proportion of participants would be expected to show any change during her study - which is the picture seen - and if there was little habituation in the first place, there would be little scope for dehabituation.

The pattern of habituation to the appearance of motion, and the subsequent loss of habituation, is consistent with the human response to real motion. Griffin (1990) reports that some experienced boat owners are sick on their initial voyages of every season, after only a few months ashore. Our finding of a reduction in habituation after extended non-exposure to the stimulus is entirely consistent with this report, and even though the physical stimulus in the two cases differs it appears that each can produce habituation which is not fully maintained over a period of a number of months.

5 References

Hill, K.J. & Howarth, P.A. (1999). Habituation to side effects of immersion in a virtual environment *(in preparation)*

Ramsey, A. (1996). Incidence and adaptation to VR induced symptoms and effects due to changes in passive display speed and repeated immersion in a VE. *VIRART Report 96/129; Nottingham University, England*

Regan, E.C. (1995). Some evidence of adaptation to immersion in virtual reality *Displays,* 16(3), 135-139.

Griffin, M.J. (1990). Motion Sickness. In *Handbook of Human Vibration.* Academic Press Limited, London.

Virtual Reality Induced Symptoms and Effects (VRISE) in Four Different Virtual Reality Display Conditions

Amanda Ramsey, Sarah Nichols & Sue Cobb[1]
Virtual Reality Applications Research Team (VIRART)
Division of Manufacturing Engineering and Operations Management
University of Nottingham
University Park
Nottingham
NG7 2RD
Tel: +44 (0)115 951 4040
Fax: +44 (0)115 951 4000

1 Introduction

In the early 1990s there was a rapid increase in the development of commercial Virtual Reality (VR) systems and expectations of widespread application of the technology in industrial, public and domestic environments. At that time the interest was mostly in systems using Head Mounted Displays (HMDs) and datagloves for personal viewing and interaction with a Virtual Environment (VE). A previous research project conducted at VIRART examined the general Health and Safety Implications of Virtual Reality (Nichols et al., 1997; Cobb et al., 1999). In recent years, the increase in power and capacity, and reduction in cost of personal computers, and advances in projection display technology has resulted in increased interest in the potential of desktop and projection based VR. Yet, to date there have been no comparisons of the effects of viewing the same environment on different display mediums.

This paper presents a study in which Virtual Reality Induced Symptoms and Effects (VRISE) experienced by 149 participants were compared when they

[1] Epzsvc@epn1.maneng.nottingham.ac.uk

viewed the same VE on four different displays under normal viewing conditions: HMD, Desktop, Reality Theatre and Projection Screen. By normal viewing conditions we mean that the subject has control of their movement and interaction within the environment and this is normal for HMD and PC displays. In conditions of no user control (Reality Theatre and Projection Screen), movements and interactions were completed by an expert controller whilst the subject passively watched. The specific impact of user control and the influence of lighting conditions within the environment on symptom reportage will be reported elsewhere.

2 Experimental Method

2.1 Subjects

149 participants (77 males and 72 females, with a mean age of 24 years 5 months) were drawn from Student and Staff populations from the University of Nottingham and The Queens Medical Centre. All were healthy volunteers with normal vision and no history of migraine, epilepsy, back or neck pain or any other serious injury or illness. All participants were asked not to drive for at least half an hour following the trial.

2.2 Apparatus

The same Virtual Factory environment was viewed on all display mediums for thirty minutes. The environment consisted of a basic factory with an entrance, office, canteen, machine bay and fume cupboards, participants explored the environment and had to rectify a number of Health and Safety hazards. The display mediums were selected as to represent a range of commercially available display types. An outline of the trial and some technical specifications of the displays can be found in Table 1. For all four viewing conditions the frame rate ranged from 15-25 Frames per second (FPS), depending on the content of the VE at any one moment.

The procedures for all the experiments were the same. The Simulator Sickness Questionnaire (SSQ) (Kennedy et al, 1993) was administered to all participants pre and post immersion. Immersion in the virtual environment was for a period of 30 minutes although participants were informed that they could leave the VE at any time. Participants were held for at least thirty minutes after exposure or until all symptoms has gone.

Table 1: Trial outline and technical specifications of display medium

Display	Participants	Technical Specification
HMD	97 (49M, 48F)	Virtual Research V8, HMD resolution 1920*480 per eye, display resolution 640*480, Field of View (FOV): 60° diagonal; Weight: 0.82 kg.
Desktop	19 (9M, 10F)	Viewed on 17" monitor and produced by a Pentium 166 PC.
Reality Theatre	17 (8 M, 9F)	Viewed on a concave screen 7.5m horizontally by 2.5m vertically, FOV 150°*43°. The resolution was 1024*3556.
Projection Screen	16 (11M, 5F)	A Plus UP-1100 data projector with a light output of 1000 ansilumens was used. Viewed on a 1220mm x1200mm screen, 1725mm diagonal, 750mm off the floor.

3 Results

3.1 Symptom reports

Table 2 illustrates the pre, post and pre-post change scores for all participants. One way ANOVAs were performed on both the change in symptoms from pre to post immersion and the post immersion symptom scores. A square root (k-x) transformation was applied to the data to ensure that the assumptions required for an ANOVA to be performed were met. The results of these were as follows. The non-transformed data is presented in Table 2.

For the post immersion symptom scores there were no significant differences between symptoms reported on the different display symptoms. However, when pre immersion symptom scores were taken into account, a significant main effect was found for nausea ($F = 4.202$; $df = 3, 143$; $p<0.01$) and disorientation symptoms ($F= 2.918$; $df=3, 143$; $p<0.05$). Post-hoc t-tests were performed on all three symptom subscales and total symptom scores to examine the causes of these changes.

A significant difference was found between the pre-post change in nausea symptoms reported by participants in the HMD and Desktop conditions ($t=2.332$; $df=114$; $p<0.05$). This result indicates that participants in the HMD condition experienced a higher increase in than participants in the desktop condition. A significant difference was also found between symptom changes

Table 2: SSQ Results from SSQ administration

SSQ subscale	SSQ mean score (SD)			
	HMD	**Desktop**	**Projection**	**Reality Theatre**
Pre immersion				
Nausea	5.84 (8.94)	11.05 (11.14)	11.58 (14.09)	9.54 (12.62)
Oculomotor	9.36 (12.21)	9.18 (9.65)	11.91 (12.86)	16.50 (12.64)
Disorientation	2.41 (7.21)	2.20 (5.21)	3.98 (8.51)	7.37 (9.99)
Total	9.24 (10.65)	9.84 (8.66)	12.29 (13.24)	15 62 (11.32)
Post immersion[2]				
Nausea	17.11** (19.81)	14.06 (17.23)	17.89 (18.39)	10.10 (11.91)
Oculomotor	21.18** (19.60)	15.96* (14.49)	24.16* (17.77)	19.17 (18.00)
Disorientation	20.95** (22.56)	12.45* (19.07)	14.79* (16.45)	17.02* (20.61)
Total	19.55** (17.58)	15.35* (13.25)	26.18* (21.59)	22.88* (16.88)
Pre-post change				
Nausea	11.21 (17.47)	3.01 (17.43)	4.09 (9.70)	0.56 (8.58)
Oculomotor	11.71 (13.93)	6.78 (10.69)	8.66 (11.46)	2.68 (14.66)
Disorientation	18.51 (21.33)	10.26 (17.88)	8.95 (12.93)	9.83 (18.26)
Total	10.26 (13.42)	5.51 (11.04)	10.15 (9.67)	7.26 (13.90)

reported in the HMD and Reality Theatre conditions, where participants in the HMD condition experienced a higher increase in symptoms than participants in the Reality Theatre condition (t=2.753; df=112; p<0.01). A significant difference was found between the oculomotor symptoms experienced by participants in the HMD and Reality Theatre, where participants in the HMD experienced higher level of oculomotor symptoms than in the Reality Theatre (t=2.468; df=111; p<0.02). A significant difference was found between the disorientation symptoms experienced by participants in the HMD and Reality Theatre, where participants in the HMD condition experienced a greater increase in symptoms than those in the Reality Theatre (t=2.258; df=112; p<0.03).

---

[2] * = t-test indicates significant increase in SSQ score pre-post immersion p<0.05, ** = p<0.01.

4 Discussion

This study identified differences in VRISE experienced under alternative VE display conditions. Specifically significant differences were found between HMD and desktop (for change in nausea symptoms) and HMD and Reality Theatre (for nausea, oculomotor and disorientation). In both cases, participants experienced higher increases in symptoms as a result of participation in the HMD condition.

Disorientation and Total SSQ scores significantly increased post immersion for all display types. Oculomotor scores significantly increased for HMD, Desktop and Projection screen, and Nausea increased significantly only in the HMD condition. These results, and examination of the symptom profiles, suggest that different display mediums provoke different types and levels of sickness symptoms. However, display type is only one variable involved in the multifactorial production of VRISE. In particular, close examination of the influence of VE design and the impact of individual participant characteristics is required, and is being completed as part of on-going research at VIRART.

5 References

Cobb, S.V.G, Nichols, S.C., Ramsey, A.D. & Wilson, J.R. (1999) Virtual Reality-Induced Symptoms and Effects (VRISE). *Presence: Teleoperators and Virtual Environments* **8**(2), 169-186.

Nichols, S., Cobb, S. & Wilson, J.R. (1997) Health and Safety Implications of Virtual Environments: Measurement Issues. *Presence: Teleoperators and Virtual Environments.* **6**(6), 667-675.

Kennedy, R.S., Lane, N.E., Berbaum, K.S. & Lilienthal, M.G. (1993) Simulator Sickness Questionnaire: An enhanced method for quantifying simulator sickness. *The International Journal of Aviation Psychology*, **3**(3), 203-220.

6 Acknowledgements

The work presented in this paper was funded by UK Health and Safety Executive grant 3812/R67.130. The authors also wish to thank Tim Whitehouse of The Centre for Industrial and Medical Informatics for his cooperation in this experiment.

An Investigation into the Predictive Modeling of VE Sickness

Eugenia M. Kolasinski, Ph.D. and Richard D. Gilson, Ph.D.
University of Central Florida
Orlando, FL USA

1 Introduction

Research has documented that, as with other simulated environments, simulator sickness can also occur in conjunction with exposure to Virtual Environment (VE) systems. Sickness associated with VE exposure ranging from queasiness to nausea and vomiting may both discomfort and discourage users. If it could be predicted who will experience sickness in VE systems, then it may be possible to identify at-risk individuals, warn them, and, perhaps, even train them in some way to reduce their risk. This research investigated the prediction of sickness in a VE.

Reschke (1990) investigated the prediction of space motion sickness and met with some success using both linear discriminant analysis and logistic modeling techniques on reported cases of sickness in weightlessness environments. This suggests that models might be successfully developed for predicting simulator and, specifically, VE sickness as well.

Kolasinski (1995) reviewed motion and simulator sickness literature and identified 40 factors that may be associated with sickness occurring in VEs. The factors fell into three global categories: simulator-related, task-related, and individual-related. As an exploratory investigation into the modeling of sickness, this research focused on individual-related factors only.

2 Method

2.1 Experimental Design

The goal of this research was to determine if sickness could be modeled on characteristics of an individual. Linear regression techniques were selected because of their wide general applicability and the exploratory nature of the study. A pre-experiment power analysis (Kolasinski 1996) revealed that a sample size of 40 would be adequate for the regression.

2.2 Independent Variables

The four independent variables were four individual-related characteristics: age, gender, mental rotation ability, and pre-exposure postural stability. Educational Testing Service's Cube Comparison Test (CCT) provided the measure of mental rotation ability. Possible scores on the CCT range from -42 to +42, with higher values on the CCT indicating better mental rotation ability. Postural stability was assessed using the video-based posture test equipment detailed in Kennedy and Stanney (1996) and using a measure - the Prototype value - proposed therein. As per the recommendations of Kennedy and Stanney, the stance employed was the Tandem Romberg (i.e., heel-to-toe) with arms folded across chest and eyes closed. Two 30-second trials of the postural test were administered prior to VE exposure. The trial ended either at 30 seconds or when the participant could no longer maintain the position. The trial time was recorded. The Prototype value is a refinement of the time measure for individuals who are able to perform the test well. Lower values of the Prototype reflect better postural stability. There were two Prototype values, one for each of the two pre-exposure postural stability trials. As per the recommendations of Kennedy and Stanney, the mean of these two values was computed to provide the measure of postural stability used for analysis.

2.3 Dependent Variables

Scores obtained from the Simulator Sickness Questionnaire (SSQ) provided the measures of sickness. The SSQ (Kennedy, Lane, Berbaum, and Lilienthal 1993) is a 16-item symptom checklist. Each symptom is rated by the individual as either "none," "slight," "moderate," or "severe." The ratings from the 16 symptoms are used to compute four scores: scores for Nausea, Oculomotor Discomfort, and Disorientation subscales, as well as an overall Total Severity score. For each of these measures, higher scores indicate more sickness. The regression analysis was conducted for the Total Severity scores only. To correct for skew in the scores, the natural logarithm transformation was applied (a value of 1 was added to each score before the logarithm was taken so that all natural logarithm values were defined).

2.4 Apparatus

The VE was produced by a PC-based VE system. The PC used was a Compaq Presario CDS 972, 75 MHz Pentium. The Head-Mounted Display (HMD) used was the i*glasses!™ manufactured by Virtual i*O™ (now being manufactured by Virtual Research Systems, Inc.). The stimulus was the computer game "Ascent," produced by Gravity, Inc. for Virtual i*O™. This game of rock-jumping through virtual canyon walls came bundled with the i*glasses!™. Direction of movement in the game is controlled by a tracker on the back of the HMD. Selection of the position to jump to is made with a left button click on a standard mouse. Ascent can be played in either stereo- or monoscopic mode and stereoscopic mode was used for all research participants. Ascent consists of 10 levels. The level reached by the participants ranged from 1 to 5, with a median and mode of 3, reached by over half of the participants ($f = 22$).

2.5 Participants

Research participants were recruited from undergraduate psychology and industrial engineering courses at the University of Central Florida (UCF). Course credit was offered for participation. Participants gave full informed consent and the experimental procedure was approved by the Research Subjects Committee of the UCF Department of Psychology. Data from 40 participants were analyzed. All participants were at least 18 years of age, did not have a personal history of epilepsy, were not color blind, and were not under the influence of alcohol or other drugs. To minimize adaptation effects from other simulation environments, prospective participants were not allowed to participate in the research if they had been in any type of VE system within 30 days prior to participating in the experiment. To minimize the effects of any existing illness, prospective participants were asked to arrive for the experiment in their usual state of good fitness. Data from participants suffering from colds or flu, taking certain medications, or who were highly symptomatic prior to exposure were not included in the analysis.

2.6 Procedure

Prior to immersion, participants were evaluated for their eligibility to participate in the research through the use of a pre-exposure SSQ and a color blindness test. The CCT and the pre-exposure postural stability test were then administered. Participants were immersed in the VE and were asked to play Ascent for 20 minutes. Only one participant stopped play before 20 minutes and a discussion of this participant appears elsewhere (Kolasinski and Gilson 1998). After exiting the VE, the post-exposure SSQ was administered.

3 Results

3.1 Summary Statistics

The age of research participants ranged from 19 to 46 years with a mean of 22.7 years (*SD* = 4.73) and a median of 22 years. Of the 40 participants, there were equal numbers of males and females. Scores on the CCT ranged from 2 to 38 with a mean of 18.4 (*SD* = 9.46) and a median of 19.5. The mean Prototype value ranged from 1 to 8.5 with a mean of 3.71 (*SD* = 1.93) and a median of 3. Total Severity scores ranged from 0.00 to 138.38 with a mean of 21.22 (*SD* = 26.81) and a median of 13.09. Summary statistics for the subscale scores, which are not a part of this analysis, are reported elsewhere (Kolasinski and Gilson 1998).

3.2 Regression Analysis

The search for a model involved examination of scatter plots and Pearson correlations (to examine relationships among the variables) and sequential variable selection procedures and best subsets regression (to identify candidate models). Once a candidate model was selected, fit of this model was assessed by verifying the assumptions underlying linear regression and analyzing residuals to assess high-influence points. For prediction of the transformed Total Severity score, it was concluded that the best linear model was

$$\underline{\text{lntotal}} = 2.3993 - 0.1190\,\underline{\text{age}} + 0.0198\,\underline{\text{genmra}} + 0.0064\,\underline{\text{agemra}} + 0.0269\,\underline{\text{agepro}} - 0.0313\,\underline{\text{mrapro}}$$

where lntotal represents the natural logarithm of one plus the Total Severity score; age represents an individual's age; genmra represents the product of gender (coded -1 for males and +1 for females) and CCT score (denoted as mra); agemra represents the product of age and mra; agepro represents the product of age and the Prototype value (denoted as prepro); and mrapro represents the product of mra and prepro. This model is significant (F = 3.94, p = 0.006) and, based on the R^2 value, explains 36.7% of the variance in the transformed Total Severity score.

4 Discussion

This research was an exploratory investigation into the modeling of VE sickness on characteristics of an individual using linear regression techniques. For sickness measured as a function of the transformed Total Severity score (referred to as simply "sickness"), a model was developed that explains approximately 37% of the variance in sickness and fits the data well based on statistical regression diagnostics.

The model developed in this study portrays a complicated relationship between sickness and age, gender, mental rotation ability, and pre-exposure postural stability. The four characteristics all interact with each other in their effect on predicted sickness and there is no clear relationship between sickness and any one of these four variables. The utility of the model developed lies not in its use for exact prediction but, rather, in the fact that it could be developed - thus indicating that sickness can be successfully modeled (in a statistic sense) on characteristics of an individual.

5 References

Kennedy, R.S., Lane, N.E., Berbaum, K.S., & Lilienthal, M.G. (1993). A simulator sickness questionnaire (SSQ): A new method for quantifying simulator sickness. *International Journal of Aviation Psychology, 3*(3) 203-220.

Kennedy, R.S. & Stanney, K.M. (1996). Postural instability induced by virtual reality exposure: Development of a certification protocol. *International Journal of Human Computer Interaction, 8*(1), 25-47.

Kolasinski, E.M. (1995). *Simulator sickness in virtual environments* (ARI Technical Report 1027). Alexandria, VA: U.S. Army Research Institute for the Behavioral and Social Sciences. Available on the World Wide Web at http://www.cyberedge.com/4a7.html

Kolasinski, E.M., (1996). *Prediction of simulator sickness in a virtual environment* (Doctoral dissertation, University of Central Florida, Orlando, FL, 1996). *Dissertation Abstracts International, 57*(03). (University Microfilms No. 96-21485). Available on the World Wide Web at http://www.hitl.washington.edu/projects/knowledge_base/virtual-worlds/kolasinski/

Kolasinski, E.M. & Gilson, R.D. (1998). Simulator sickness and related findings in a virtual environment. *Proceedings of the 42nd Annual Meeting of the Human Factors and Ergonomics Society*, vol. 2, 1511-1515.

Reschke, M.F. (1990). Statistical prediction of space motion sickness. In G.H. Crampton (Ed.), *Motion and space sickness* (Ch. 14). Boca Raton, FL: CRC Press.

The Search For A Cybersickness Dose Value

Richard H.Y. So
Human Performance and Virtual Reality Laboratory
Department of Industrial Engineering and Engineering Management
Hong Kong University of Science and Technology
Clear Water Bay, Kowloon,
Hong Kong SAR
rhyso@ust.hk

1 Introduction

Cybersickness refers to the sickness phenomenon associated with the use of virtual reality systems. When a user views a wide field-of-view moving scene inside a virtual environment, an illusion of self-motion (vection) in the opposite direction can occur and the experience can be nauseogenic. This paper presents the initial development of a Cybersickness Dose Value (CSDV) and explores how this CSDV can be used to predict the severity of cybersickness.

2 Problems of Cybersickness

A virtual reality (VR) system enables a user to interact with a computer-generated 'virtual' environment (Furness and Barfield, 1995) and is useful in operator training applications (e.g., driving simulation: Bayarri *et al.*, 1996; flight simulation: Haas, 1984; equipment simulation: Lin *et al.*, 1996). However, symptoms of motion sickness (e.g., nausea and headache) have been reported among users who navigate through a virtual simulation for 20 minutes or longer (e.g., passive navigation: So, 1994; active navigation: Regan, 1995; Finch and Howarth, 1996; Stanney and Kennedy, 1997, 1998; Kolasinski and Gilson, 1998; Draper, 1998). Some studies reported sickness occurrence even after 10 minutes of exposure (e.g., Wilson *et al.*, 1997). This type of sickness has been referred to as 'cybersickness' (McCarley and Sharkey, 1992) and its occurrence has hindered the widespread applications of VR simulators. A review of literature indicates that cybersickness has been the subject of many studies (see Table 1). The results of these studies have provided a body of knowledge, which can be used to predict levels of cybersickness. The rest of the paper will explore the possibility of formulating a Cybersickness Dose Value (CSDV) for predicting the levels of cybersickness.

Table 1 Summary of experimental studies of cybersickness (excluding studies of simulator sickness in a non-VR environment, adapted from Lo and So, 1998).

Variables investigated	**References**
Duration of exposure (including with / without exposure)	Bliss *et al.* (data from Kennedy and Stanney, 1997); Lampton *et al.* (1994a,b); Lo and So (1998); Kennedy *et al.* (1995); Kolasinki and Gilson (1998); Regan and Price (1993a,b,c); Regan (1995); Salzman *et al.* (1995); So (1994); Stanney and Kennedy (1997, 1998); Wilson *et al.* (1997)
Types of VR displays	Costello and Howarth (1996); Howarth and Costello (1997); Lampton *et al.* (1994); Wilson *et al.* (1997)
Display's field-of-view	DiZio and Lackner (1997)
Display lags	DiZio and Lackner (1997); So (1994)
Stereoscopic presentation	Ehrlich (1997)
Method of navigation	Finch and Howarth (1996); Rich and Braun (1996);
Age, gender, posture stability	Ehrlich (1997); Ehrlich *et al.* (1998); Kolasinski (1996); Regan & Price (1993b,c); Rich & Braun (1996)
Sitting Vs Standing	Regan and Price (1993b)
Amount of head movement interactions	Regan (1995); So (1994)
Drug treatment	Regan (1995)
Axes of scene movement (rotational axes)	Lo and So (1998); So and Lo (1998a)
Types of scene	Draper, 1998; Kennedy *et al.*, 1994; Kennedy *et al.*, 1996; Slater *et al* (1996); So and Lo (1998b,c)

3 Previous studies of Motion Sickness Dose Value

Predicting levels of motion sickness with a dose value is not new. The development of Motion (sea) Sickness Dose Value (MSDV) for predicting sea sickness can presently be used to predict the percentage of sea passengers who will vomit for a given voyage (British Standard 6841). A review of literature indicates that at the center of the MSDV, there is a basic unit, acceleration, which quantifies the physical motion that is responsible for symptoms of motion sickness (Griffin, 1990, British Standard 6841). The MSDV is essentially a time integral of this basic unit (i.e., frequency weighted vertical ship acceleration):

$$MSDV = \left[\int a^n(t)\,dt\right]^{1/n}$$

where a is the frequency-weighted vertical ship acceleration (ms^{-2})
n is 2 or 4

Griffin (1990) reported that for n = 2 or 4, a good correlation has been found between the values of the MSDV and the percentage of vomit incidence among sea passengers.

4 A proposed Cybersickness Dose Value (CSDV)

A literature review on studies of cybersickness indicates that the levels of cybersickness increase with increasing duration of exposure (see row1, Table 1). A Cybersickness Dose Value (CSDV) having a similar structures as the MSDV is, therefore, proposed:

$$CSDV = \int_0^T [basic\ unit]\,dt$$

Where T is the duration of a VR simulation

As seen from the proposed formula, a basic unit, which quantifies the dominant factors responsible for the generation of cybersickness, has to be identified. According to the sensory conflict theory (Reason and Brand, 1975), cybersickness can be classified as a type of vection-induced motion sickness (Hettinger and Ricco, 1992). This implies that visual scene movements during a Virtual Reality (VR) simulation will introduce illusion of self-motion (i.e., vection). In the absence of appropriate physical motion, the vection will give rise to symptoms of cybersickness. Following this theory, the dominant factor responsible for cybersickness will be visual scene movement. So and Lo (1998b) developed a metric to quantify the visual scene movement during a VR simulation. This metric is called 'Spatial Velocity (SV)'.

Experimental results have shown that both the rated levels of nausea and the total sickness severity scores measured using the Simulator Sickness Questionnaire developed by Kennedy *et al.* (1993) increased linearly with the 'spatial velocity' measurements (Figure 1). Details concerning the 'Spatial Velocity (SV)' metric can be found in So and Lo (1998b). Currently, experiments are being conducted to study the effects of frequency of scene movement on the level of cybersickness. Assuming a frequency weighting can be developed for the SV metric, the proposed CSDV formula can be modified as follows:

$$CSDV = \int_0^T [frequency\ weighted\ SV]\, dt$$

Inspection of Table 1 shows that the effects of task-related, display-related, and subject-related parameters on cybersickness have been studied. In particular, the effects of field-of-view (DiZio and Lackner, 1997; Draper, 1998); stereoscopic presentation (Ehrlich, 1997); lags (So, 1994; DiZio and Lackner, 1997); gender (Kolasinski, 1996; Rich and Braun, 1996); and methods of navigation (Rich and Braun, 1996) have been examined. In order to account for these effects, the proposed CSDV can be modified as follows:

$$CSDV = [D]x[T]x[S]x \int_0^T SV_w\, dt$$

where [*D*] is the display-related scaling factor
[*T*] is the task-related scaling factor
[*S*] is the subject-related scaling factor
SV_w is the frequency weighted Spatial Velocity metric

As the proposed CSDV is intended to predict the 'average' level of cybersickness associated with a particular VR simulation, the subject-related scaling factor would be derived from the average data of a population rather than individuals. On the other hand, both the display and task-related scaling factors can be specified to a particular VR simulation program and apparatus.

5 Final Remarks

Based on the existing knowledge concerning cybersickness, an initial form of a Cybersickness Dose Value (CSDV) is proposed to predict the levels of cybersickness. This CSDV is essentially a time integral of a unit called 'Spatial Velocity, SV' developed to quantify the visual scene movement during a Virtual Reality simulation. With the proposed CSDV, the author intends to stimulate further research towards the modelling and prediction of the occurrence and severity levels of cybersickness.

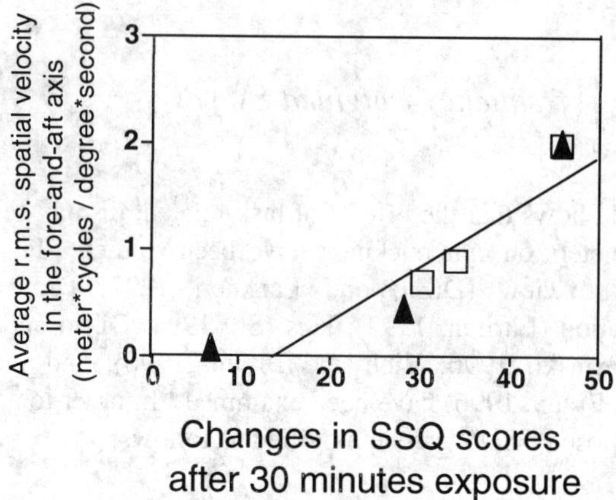

Figure 1 Levels of SSQ Total Sickness Severity scores with different levels of Spatial Velocity (SV). (▲, ❑: data from two experiments) adapted from Lo and So, 1999).

6 References

[Due to the page limit, references list is not shown. A complete list of references can be obtained from the author]

Keyboard Tactile Feedback and its Effect on Keying Force Applied

David A. Thompson
Professor of Industrial Engineering, Emeritus
Stanford University[1]

1 Introduction

Keyboards are the primary communication medium for entering information into computers, and as such are extremely important in the information age. However, they have also become the primary source of physical injury to computer users. Other sources outline the nature and extent of these keyboard injuries (Bammer and Blignault 1987; Maeda et al. 1982; Rosignol et al. 1987; Guggenbuhl and Krueger 1990, Thompson 1995, and Matias and Salvendy 1998); and they will not be repeated here.

Force has been discussed as one of the primary causes of musculoskeletal injuries to the upper extremities of the sort experienced by computer keyboarders (Putz-Anderson 1998). Excessive keying force has also been implicated in operator injuries by Rose, 1991' Loricchio, 1992, Hargreaves, 1992, Feuerstein et al. 1997 and Radwin, 1999.

The present investigation has evaluated one aspect of the effect of keyboard design on the forces that keyboarders apply in the normal performance of their work. Clearly, the amount of force required per key should be minimized, because of the high repetition rate of the work. Text and data entry operators typically enter 12,000 to 15,000 keystrokes per hour (84,000 to 105,000 keystrokes per 7-hour day.

This study focuses on one of the keyboard design characteristics that directly influences the "feel" of the keyboard, tactile feedback, defined by a force-

[1] Contact address: Portola Associates, 920 Incline Way, Suite F, Incline Village, NV 89451

displacement curve of the key action while being depressed. It has been discussed in the computer ergonomics literature for many years (Cakir, et al., 1979; Greenstein and Arnaut, 1987, and Potosnak, 1988) and recommended in the ANSI Standard for VDT Workstations (ANSI/HFS100-1988) The present study tests the propensity of keyboarders to apply more force in the absence of tactile feedback.

Bruner and Richardson (1984) tested the phenomena in 1984, comparing keyboards with a linear-spring (no tactile feedback), a snap-spring, and an elastomer key mechanism (both with differing forms of tactile feedback) with each other. They found a modest performance difference favoring the elastomer keyboard. They did not evaluate key forces and operator workload, however.

The present study compares a keyboard with a linear-spring, a collapsing (snap) spring, and a rubber dome key mechanism to assess the actual forces that keyboarders apply as a function of the minimum force required to depress the key. The effect of tactile feedback is to provide an "early warning" stimulus to the fingertip as the key is being depressed, in advance of the key striking bottom, allowing the finger to anticipate the key bottoming and begin to withdraw prior to actually striking bottom. This may reduce the force exerted by the fingers, reducing fatigue, contact shock when bottoming, and possible pain and injury. The experimental hypothesis being tested is that the effect of tactile feedback is to reduce the amount of force applied to the keys.

2 Methodology

Nine experienced VDT keyboarders were recruited from a local temporary help agency to type on each of three keyboards, using a latin-square rotation among the keyboards to eliminate the sequence effect. They each typed "standard" office text during a 5-minute warm-up period and a 30-minute test period on each keyboard. The keyboards were supported on a digital force plate to measure the keystroke forces directly and record them with appropriate software. The force plate, digitizer, and software were purchased from Mecmesin Co. In England. Ten 30-second samples were taken systematically during each test run. The study was described to the subjects as a general keyboard comparison, but the differences between the keyboards were not discussed. They were instructed to type as they normally would were this keyboard provided to them in a typical work situation.

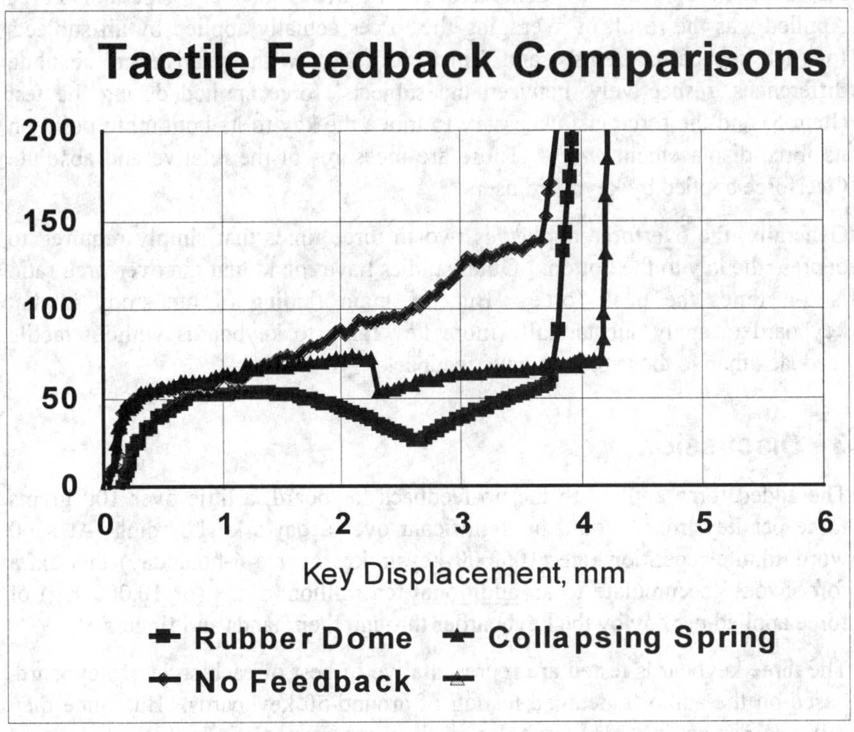

The tactile feedback characteristics of the three keyboards tested are described in the following **Figure 1.**

The results of the study are shown in **Figure 2.** The data samples taken from each keyboard (collapsing spring, rubber dome, and no feedback) were averaged, and are shown for each of the critical points on the force-displacement curve in grams force (100 grams force equals 1 Newton).

	Characteristic	Spring	Rubber Dome	None
1	Key: Nose	53		50
2	Key: Peak	77	55	
3	Key: Inflection	24	25	
4	Key: Bottom	73	60	135
5	**Operator force applied**	**164**	**178**	**282**
6	Overforce Ratio, 5/4	2.2	3.0	2.1
7	Overforce Amount, 5-4	91	118	147

Characteristics 1-4 are specific to the design of each keyboard, and are independent of the user. They were measured separately when the force-displacement curves were generated for **Figure 1** above. Operator Force Applied was the result of averaging the forces actually applied by all subjects for each keyboard. Items 6 and 7 in **Figure 2** show the relative and absolute differences, respectively, between the subjects' force applied during the test (Item 5) and the force just necessary to move the key to its bottoming point on its force-displacement curve. These are measures of the relative and absolute Overforce applied by keyboard users.

Generally, the overforce applied is two to three times that simply required to depress the key to the bottom. Other studies have found that the overforce ratio is 3-4 times the peak force. But the main finding of the study is that keyboarders apply substantially more key force to keyboards without tactile feedback than to those that do have feedback.

3 Discussion

The added force applied to the no-feedback keyboard, a little over 100 grams force per keystroke, would be significant over a day's keyboarding. At a 60 word/minute repetition rate (105,000 keystrokes over a 6-hour day) this extra force would accumulate to an additional ten million grams (or 10,000 Kg) of force applied per day by the keyboarder through their hands and fingers.

The three keyboards tested are representative of best of each style of keyboard, based on the author's detailed testing of around 50 keyboards. But since they differ somewhat in their keyset design, it is necessary to do both a relative and an absolute analysis. And, in this regard, they are similar in their proportion of overforce but markedly different in the absolute amount of overforce. This might, however, be an effect of the rotation between keyboards, so that the keying behavior based on the feel of each keyboard may have carried over to the other keyboards. Were each keyboard used exclusively for a week or so, the keying forces unique to each keyboard may represent greater differences. In any event additional testing is required to explore this area.

4 References

ANSI/HFS 100-1988, *American National Standard for Human Factors Engineering of Visual Display Terminal Workstations,* Human Factors Society, 1988.

Bammer, G. & Blighault, I. A Review of Research on Repetitive Strain Injuries, in Buckle, P. (Ed): *Musculoskeletal Disorders at Work, Proceedings of a Conference at the University of Surry,* Guildford, April 13-15, 1987.

Bruner, H. & Richardson, R., Effects of Keyboard Design and Typing Skill on User Keyboard Preferences and Throughput Performance, *Proceedings of the Human Factors Society 28th Annual Meeting*, 1984, p. 267.

Cakir, A., Hart, D., & Stewart, T., *Visual Display Terminals,* J. Wiley & Sons, 1979

Greenstein, J. and Arnaut, L, Human Factors Aspects of Manual Computer Input Devices, Ch. 11.4, in Salvendy, G., ed., *Handbook of Human Factors*, J. Wiley & Sons, New York, 1987.

Guggenbuhl, U. & Krueger, H., Musculoskeletal Strain Resulting from Keyboard Use, in Berlinguet, L. & Berthelette, D. (Eds.) *Work With Display Units 89*, Elsevier Science Publishers BV, (North Holland), 1990.

Hargreaves, W., Rempel, D., Halpern, N., Markison, R., Kroemer, K., & Litewka, J., Toward a More Humane Keyboard, in Bauersfeld, P. Bennett, J. & Lynch, G., (Eds.), CHI 92 Conference Proceedings, May 3-7, 1992, Monterey, CA, p. 365

Loricchio, D., Key Force and Typing Performance, *Proceedings of the Human Factors Society 36th Annual Meeting*, 1992, 281-282.

Maera, K,, Horiuguchi & Hosokawa, M., History of the Studies on Occupational Cervicobrachial Disorder in Japan and Remaining Problems, *J. Human Ergology,* 11:17-29, 1982.

Matias, A. & Salvendy, G., Carpal Tunnel Syndrome Causation among VDT Operators, *Occupational Ergonomics*, 1(1), 1988, 55-66.

Potosnak, K., Keys and Keyboards, Ch. 21, in Helander, M. (Ed.), *Handbook of Human-Computer Interaction,* Elsevier Science Publishers BV, (North Holland), 1990.

Putz-Anderson, V., *Cumulative trauma disorders: A manual for Musculoskeletal diseases of the upper limbs*, Taylor & Francis, 1988

Radwin, R. & Ruffalo, B., Computer key switch force-displacement characteristics and short-term effects on localized fatigue, *Ergonomics*, 42:1, 160-170

Rose, M., Keyboard operating posture and actuation force: Implications for muscle overuse, *Applied Ergonomics,* 22:3, 198-203, 1991.

Rossignol, A. et al., Video Display Terminal Use and Reported Health Symptoms Among Massachusetts Clerical Workers, *JOM*, 29:2, February, 1987, 112-118.

Thompson, D., Analysis of VDT Keyboard Tactile Feedback, in Bitner, A., & Champney, P. (Eds), *Advances in Industrial Ergonomics and Safety VII*, Taylor & Francis, 1995, 613-618

An Alternative Keyboard for Typists with Carpal Tunnel Syndrome

Peter J. McAlindon
McAlindon Enterprises, Inc.

Florian G. Jentsch
University of Central Florida

1 Introduction

The de-facto standard QWERTY keyboard was developed over a hundred years ago and is one of the few visual display terminal (VDT) workstation components that has not made the same technological strides as the rest of the workstation components. It is this lack of advancement that has caused the QWERTY keyboard to become suspect in causing repetitive strain injuries (RSIs) such as carpal tunnel syndrome (CTS). Keyboard and mouse operations require unnatural physical positioning of the arms, hands, and fingers; in typical operations elbows are flexed and wrists are ulnarly deviated, pronated, and extended (Duncan & Ferguson, 1974). Such positions put operators at risk of developing cumulative trauma disorders (CTDs). CTDs are caused from continuous repetitive motions of the hand, wrist, and arm. In extreme cases, these compromising positions can cause severe wrist trauma such as CTS as well as muscle strains in the shoulders, neck, and arms of the typist.

The purpose of this study was to investigate the ergonomic, biomechanical, and typing performances of a newly designed alphanumeric keyboard called the Alphanumeric Input Device for those with Carpal Tunnel Syndrome (AID-CTS keyboard) – one of the first ergonomically designed keyboards that eliminates finger movement and drastically reduces wrist movement while maximizing typing comfort.

2 Method

2.1 Participants

Forty-four touch typists (40 females and 4 males) between the ages of 18 and 55 (mean = 28, s.d. = 9.9) fulfilled the initial requirements to participate in the research. Pre-experimental typing performances ranged from 32 to 88 GWPM (mean = 49.9, s.d. = 12). An orthopedic surgeon determined the degree of disability for each participant. Based on this classification, the participants were assigned to cohort groups with similar disabilities. Within each disability cohort group, participants were then separated into performance groups. They were then randomly assigned to the six experimental groups: Two groups used the QWERTY keyboard, two used the ergonomic split keyboard, and two used the new AID-CTS keyboard.

2.2 Keyboards

QWERTY keyboard. The de-facto alphanumeric input device–the QWERTY keyboard (see Current, 1954, for a complete history of the QWERTY keyboard) has a layout that consists of four parallel rows of keys that in sum comprise the 26 letters of the alphabet, 10 numeric keys, and several other specific symbol or function keys.

Key contoured, split keyboard. A number of efforts have been made to improve the ergonomic/biomechanical characteristics of the QWERTY keyboard, while maintaining its basic layout. The basic reason for splitting the keyboard is to eliminate ulnar wrist deviation, a suspect static position in the development of CTS. A number of these keyboards also allow the sections to be tilted outward to pronate the hand. The keyboard used in this study has a sculpted keying surface, separated alphanumeric keypads, thumb keypads, and closely placed function keys. This keyboard uses the QWERTY key layout.

New AID-CTS keyboard. A new type of alphanumeric input based on the chording concept was designed to make typing less physically traumatic, increase typing efficiency, and facilitate typing task learning. The AID-CTS keyboard, as depicted in Figure 1, is an alphanumeric input system that uses a pair of devices each comprised of an inverted dome upon which the hands rest. Each dome is flexibly coupled to a base. The design alleviates many of the problems of key spacing, key size, and key force that are part of every traditional QWERTY type keyboard. The dome design was chosen because it closely approximates the at rest posture of the hand, which reduces static muscle fatigue and increases long-term comfort.

The AID-CTS keyboard is an extremely flexible typing device and was developed to accommodate the user's needs. In fact, AID-CTS keyboard users are expected to be the ones that a) have an upper extremity disability, b) suffer from CTS, or c) are worried about CTS risk as it relates to typing and are willing to consider a keyboard alternative. Different attachments can be used in place of the dome (e.g., ball or flat board). Other features of the AID-CTS keyboard include adjustable dome movement force and displacement, adjustable tilt and height, and complete self-containment for use in underwater or hostile environments. In addition, the AID-CTS keyboard is a perfect candidate for miniaturization and can be used by one or both hands.

As a chordal device, the AID-CTS keyboard typing methodology entails creating a keystroke via a combination of positions of the two domes. For example, referring to Figure 2, moving the selector dome to the "dark gray" position enables access to the "dark gray" concentric circle of the character dome (here shown to contain the letters i, o, p, l, m, n, h, and u). Moving the selector dome to the "light gray" position would enable the character dome to access r, t, f, d, c, s, a, and e. Once a position on the selector dome is selected, the characters on the character dome can be typed by moving the character dome into the direction of the character the user wishes to type. The lateral movements of each dome are the same for all characters (i.e., the characters on the outer character rings require the same lateral displacement as those on the inner ones).

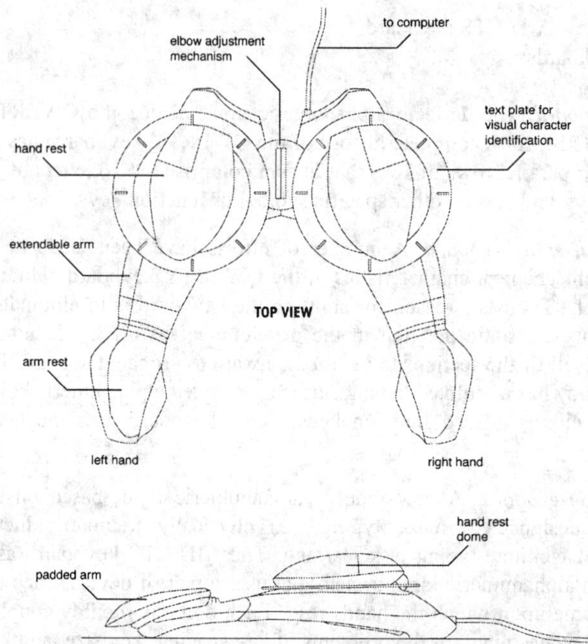

Figure 1: The AID CTS keyboard for use by individuals with upper extremity disability (Model shown for illustration purposes only. Experimental models were simplified prototypes).

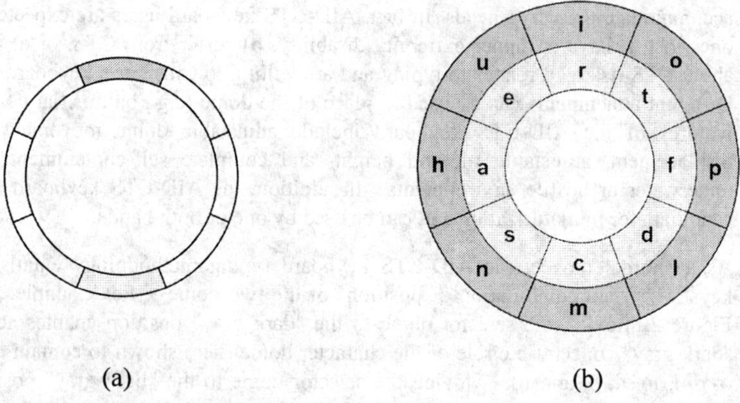

Figure 2. Top view of (a) Selector AID-CTS keyboard (b) Character AID-CTS keyboard: An example key chording activation scheme. (Note: six more selector positions on the Selector AID-CTS keyboard exist and six more concentric circles need to be added to the Character AID-CTS keyboard to allow for 64 'keys'.)

The AID-CTS keyboard design was kept as close to QWERTY specification to control for statistical confounding. In addition to force and displacement considerations, character size and font was the same for each keyboard for easy visual 'key' activation reference.

3 Results

AID-CTS keyboard users' absolute flexion/extension wrist acceleration movements, when compared to QWERTY movements, were reduced by an average of 70% whereas absolute acceleration in the ulnar/radial plane were reduced by an average of 43%.

In regard to typing performance, results indicate that users of the AID-CTS keyboard typed an average of 30% of their regular QWERTY keyboard speed in as little as three hours. Split keyboard typists typed an average of 75% of their regular QWERTY keyboard speed.

Mental workload measures were significantly higher for the AID-CTS keyboard group for the first few sessions of testing due to adaptation to the AID-CTS keyboard method of typing. Otherwise, they were not significantly different between the groups.

In terms of subjective comfort in using the keyboards, the AID-CTS keyboard was found to have less wrist and finger fatigue when compared to the split keyboard and QWERTY keyboards. There were no significant differences between the split keyboard and the QWERTY keyboards with respect to fatigue.

Character error analysis revealed that the eight positions of dome movement were for the most part proportionally balanced. This finding indicates that of the eight positions, no one position was more difficult to actuate than any other position.

3.1 Acquisition/learning curves

The main effect of session was found to be statistically significant, $F(8,54) = 3.91$, $p < 0.0001$. Gaining proficiency was much more pronounced in the split keyboard and AID CTS groups, however. Figure 3 indicates a linear performance improvement for the AID-CTS typists with no apparent asymptote. This suggests that AID-CTS typists had not reached a performance plateau after nine sessions (or 3 hours of using the device). The split keyboard typists showed signs of leveling off at session 10 or after approximately two hours of using the device. The "learning curve" for QWERTY group was substantially flatter, with only slight improvements over the same 3-hour period.

4 Discussion

The objective of this study was to determine how the AID-CTS keyboard compared to the traditional QWERTY keyboard and its "ergonomic" derivative, the split keyboard. The results of this study support the notion that the AID-CTS keyboard has the potential to become an effective alternative device. The results indicated a clear ergonomic advantage of the AID-CTS device when compared to keyboards using the traditional QWERTY key layout. In addition to completely eliminating finger movements, the AID-CTS device also led to a marked reduction in wrist motions. In the planes of flexion/extension and ulnar/radial motion, significantly less movement was recorded when using the AID-CTS device than when using traditional keyboards. Further, subjective perceptions of pain and discomfort in the hand/wrist/arm system were significantly lower for participants who used the AID-CTS devices than for those who used keyboards based on the QWERTY layout. In fact, two participants (or about 12% of the group) who used the traditional QWERTY keyboards had to be excused from the study prematurely because their condition did no longer allow them to continue their participation. In contrast, no participant from the group using the AID-CTS devices showed similar deterioration that required termination of their participation. This suggests that while the use of the AID-CTS keyboard comes (at lease initially) at a performance price, if does offer promise for those who can not use QWERTY keyboards or must avoid wrist and finger movements.

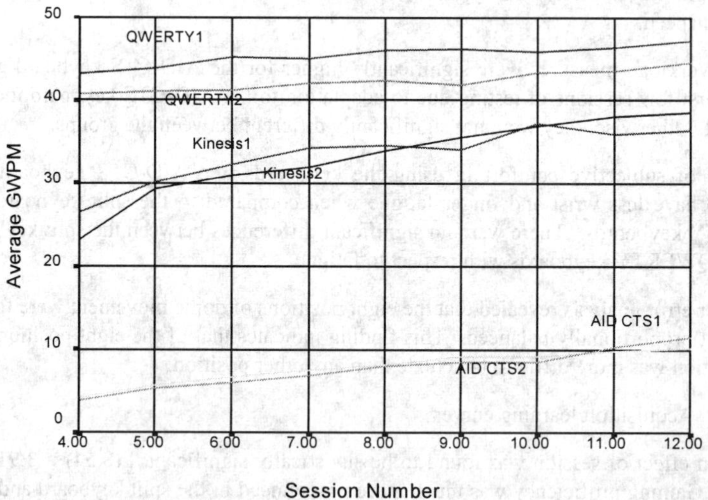

Figure 3. Average gross words per minute as a function of keyboard and session.

5 References

Current, R. N. (1954). *The typewriter and the men who made it.* University of Illinois Press.

Duncan, J., and Ferguson, D. (1974). Keyboard operating posture and symptoms in operating. *Ergonomics*, 17(5), 651-662.

This material is based upon work supported by the National Science Foundation under award number: DMI-9661259. Any opinions, findings, and conclusions or recommendations expressed in this publication are those of the authors and do not necessarily reflect the views of the National Science Foundation.

Effect of Menstrual Cycle on Monotonous Works demand High Awaking Conditions

Keiko Kasamatsu*, Michiko Anse*, Mariko Funada, Kyoko Idogawa***, Satoki P. Ninomija***
*Aoyama Gakuin University, Tokyo, JAPAN, ** Hakuoh University, Tochigi, JAPAN, *** Seitoku University, Chiba, JAPAN
E-mail: ninomija@ise.aoyama.ac.jp , kasamatsu@ise.aoyama.ac.jp

1 Introduction

In June of 1997, the Equal Opportunities Law for Employment for both sexes was reformed and the Protection of women in the Labor Standards Act was abolished in Japan. Therefore the proper working fields for female have been steadily increasing. Female have menstrual cycle, which is the innate female rhythm. We know that menstrual cycle has various influence on her body both physiologically and psychologically[1)]. We'd like to make out the data for the female workers as labor resources are taken into employment on proper environment and contribute to the improvement of productivity. In this research, we've done the two kinds of mental summation of random two-digit numbers as the monotonous work on the three phases, Menstruation, Follicular Phase, and Premenstruation of menstrual cycle. Also, we investigated by the Questionnaire and the Y-G (Yatabe-Guilford) examination, i.e., character examination. We've tried to group the subjects in addition the experiment results.

2 Methods

2.1 Menstrual Cycle

We've investigated the following three phases whose fluctuations are large and, therefore important by the previous research, i.e., Menstruation , Follicular phase, and Premenstruation (Fig 1). Menstruation : 5 days after the first date of menstrual bleeding. Follicular phase : About 7 days after the arrest of bleeding. Premenstruation : About 5 days before Menstruation.

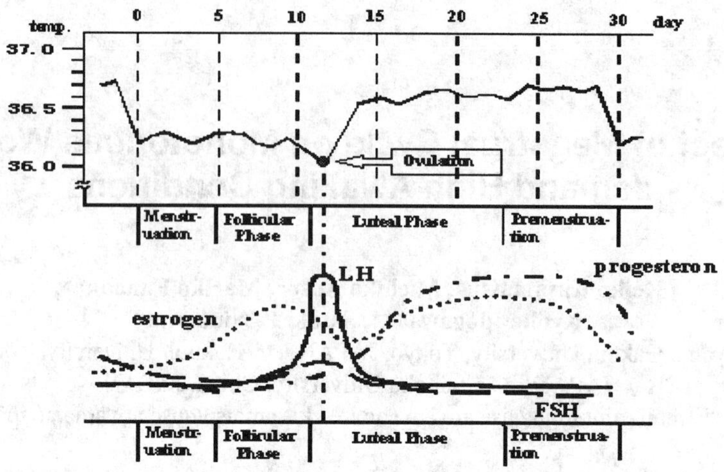

Fig. 1 Menstrual Cycle

2.2 Subjects

There were 7 female university students in this experiment. All of them have done preliminary experiment and understood about the details of the task well. We have investigated their physical and psychological condition before the experiment start.

2.3 Experimental Methods

We've represented the sets of the random 2 digit numbers in the central part of the display of personal computer. Then, we've made the subjects add the 2 digit numbers (e.g."23+45") in their heads and input the answer using the ten-keys. We excluded the continuation of the some 2 digit numbers such as "11", "22" or "33". We've done the two kinds of experiments by the difference of method that represents the questions. In Experiment 1, the question was presented until the subject input the right answer. In Experiment 2, we've set the represent time 3 seconds as the task required high concentration and memory. The subject finished the experiment if she has forget the her task and been not able to input the right answer. The time of experiment 1 was 10 minutes, and experiment 2 was maximum10 minutes. But the subject G couldn't do experiment 2 on all three phases. We measured the all experimental time, all processing numbers as task amount, the time until inputting the first response, the time until inputting the right answer, error numbers and appearance time of grouped waves.

2.4 Questionnaire and Y-G Examination

We've referred to the Menstrual Distress Questionnaire of Moos[2)] and made out our Questionnaire. We selected 46 symptoms that predicted symptoms concerning the menstrual cycle. The questionnaire is optional system with SD(Semantic Differential) method and a pair of each item has 7 scales. Each subject filled in the questionnaire before the experiment started about three phases. We analyzed using the factor analysis to investigated the effect on menstrual cycle about the subjects. We've grouped items by factor loading indicated the degree that the items depend on each factor. We've used the Y-G (Yatabe-Guilford) character examination as character examination and done on the day except experiment day.

3 Results

3.1 Experimental Results

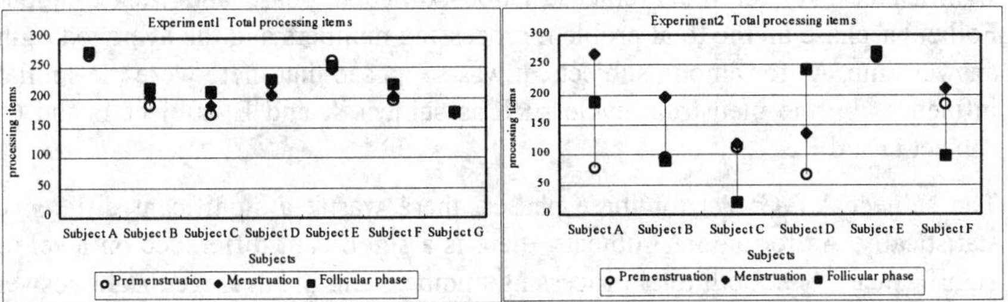

Fig. 2 Total processing items of each subject
Left: Experiment 1, Right: Experiment 2

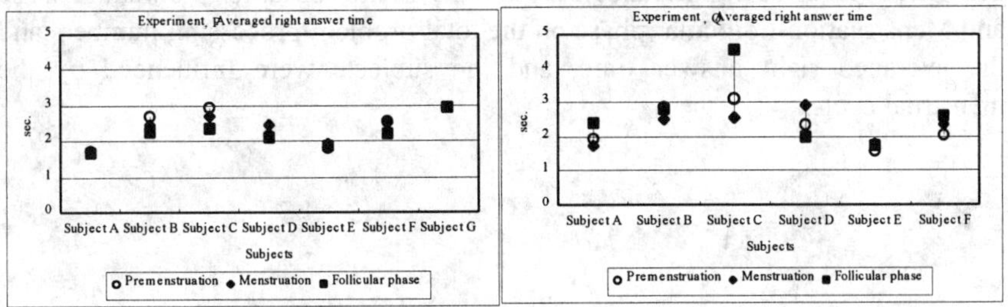

Fig. 3 Averaged right answer time of each subject
Left: Experiment 1, Right: Experiment 2

Judging from the experiment data, in the experiment 1, as for all subjects there was a little differences during phases about all processing items and averaged right time. In the experiment 2, as for the other subjects except subject E there were differences during phases because of stimulus that the question disappeared. As for the subject C and D, there were differences in averaged right time (Fig 2, 3). The subjects were needed concentration for the question disappeared. So there was the individual difference. We examined if there were differences on these experimental data statistically. Therefore, the analysis of variance and the test of difference with averaged value were used in all processing items and averaged right time. However, the experimental time wasn't fixed in the experiment 2. Only experiment 1 was statistically given to test this time. As a result of analysis of variance, it is not easy to be influenced for subject A, E and G by the menstrual cycle. Then, the group of all subjects and the group of subject A,E,G and the group of subject B,C,D,F were analyzed separately in the test of the difference by the average value.

All subjects: As for three phases, there was a significant difference (level of significance 5%) for the Premenstruation-Follicular phase and Menstruation-Follicular phase on the total problem processing numbers and the averaged right answer time. As for among subjects, it was supposed that there were the similar influences by the menstrual cycle such as subject A and E, subject B and C, subject D and F .

The subject A,E,G: As for three phases, there wasn't a significant difference statistically. As for among subjects, there is a significant difference on level of significance 1% at both total processing numbers and the averaged right answer time. Then, it suggested that they weren't influenced easily by the menstrual cycle though there was an individual variation.

The subject B,C,D,F: As for among three phases, there was a significant difference (level of significance 5%) for the Premenstruation-Follicular phase and Menstruation-Follicular phase on the total problem processing numbers and the averaged right answer time, and the subjects were influenced by the menstrual cycle.

3.2 Analysis Results of Questionnaire and Y-G Examination

The factor analysis was used in the questionnaire of 46 symptoms. The symptoms of three factors were arranged in order with the large factor loading. Three factors were interpreted at each phase. As for premenstruation, the factor loading such as "Feel tired", "Depression" is large. The symptoms which appears as the first factor are concerned with autonomic reactions such as "Hot flash", "Dizziness Anemia". As for the second factor, physical change such as “Cold sweat” and “Ringing in the ears” is developed that is, there are many symptoms concerning to Pain. As for the third factor, there are many symptoms concerning to psychological change as such “No drowse” and “Feeling of unhappiness”. As for menstruation, the symptoms of the first factor are feelings, efficiency and behavioral change such “Tension” and “Loneliness”. The symptoms of the second and third factor are similar to physical change of premenstruation. There are symptoms such as “Vacantly” and ”Chest pains” in the second factor, and such as “Merry mood” and “Tearful” in the third factor. As for follicular phase, there is physical change such as ”Chest pains” and “Heartbeat short of breath” in the first factor. There is behavioral change such as ”Loneliness” and “Depression” in the second factor. As for the third factor, there is psychological change such as ”Forgetfulness” and “Mood stability”. Next, the Table 1 is the results of Y-G character examination and the change of symptoms in the questionnaire on the each subject. As a result, we supposed that there was relationship between Y-G character examination and the questionnaire.

Table 1. Y-G examination and symptoms of Questionnaire by each subject

Y-G	Subject	Questionnaire
B group	Subject E	Many symptoms of psychological change. Especially menstruation.
	Subject F	Many symptoms of psychological change. The symptom of premenstruation and menstruation are similar.
C group	Subject B	Many symptoms of psychological change. The symptom of premenstruation and menstruation are similar.
D group	Subject A	Many symptoms of physical change. The symptom of premenstruation and menstruation are similar.
	Subject C	Same answer tendency in all phase.
	Subject D	Many symptoms of physical change. The symptom of premenstruation and menstruation are similar.

4 Discussion

In the experiment 2, the work amount of almost subjects diminished and there were a lot of fluctuations in the work amount by the each phase because of stimulation that producing question disappeared. This experiment needs high awaking conditions, therefore we supposed that it was necessary for subjects the more concentration than the experiment 1. Judging from the examination of difference with averaged value, we supposed that it was possible to group the subjects into the type were easy to influence by the menstrual cycle such as subject B, C, D, F or were not such as subject A, E, G. As a result of questionnaire, there were negative feelings and behavior in the menstruation. As for the questionnaire and character examination in the each subject, we supposed that it was possible to group into the subjects complain of symptoms in psychological change or in physical change by means of character type.

5 Conclusions

As for the classification of the subjects in the experiment results, it was possible to classify into a type from which task performance was influenced easily by the menstrual cycle and a type not so, and type influenced easily by stimulation which is that problem disappears and type not so. As a results of the factor analysis, there were the differences of symptoms by each phase. We supposed that there was the relationship between the results of Y-G character examination and the answer of the questionnaire. We couldn't investigate about individual differences enough. However, in future it is necessary to divide into individual classes and to investigate the various individual characteristic and the differences of the effect by the menstrual cycle.

6 References

Keiko Kasamatsu et al., Research concerning influence on work efficiency of analog and digital VDT work caused by menstrual cycle, *Proceedings of WWDU'97 Tokyo*, p47-48, 1997

Rudolf H. Moos, Ph.D., The Development of Menstrual Distress Questionnaire, *PSYCHOSOMATIC MEDICINE*, Vol.XXX, No.6,1968

IT-quality and employees ability to cope, and how age affects subjective IT-quality

Linn Anette Solberg
Center for Usability Studies, SINTEF Telecom and Informatics

1 Introduction

Implementation of new a IT-system can make the employees in companies stressed (Turnage, Cornelius and Sansing 1994). Stress can lead to several physiological, psychological and behavioural consequences for the employees (Eisenberg and Goodall 1993; Smither 1994; Zaleznik, Vries and Howard 1997). These problems can again have serious consequences for the companies. Studies on how to make the IT-changes easier are therefore important for both employees and companies. In this article we have focused on IT-quality[1] and the effect it has on the employees' ability to cope.

Elderly employees seem to have bigger problems with IT-changes than younger employees do (Stammers et al. 1991; Lindström et al. 1989). We will look at how the differences can be explained by IT-quality.

2 Methods

The results in this article have come from a survey and focus groups.

2.1 Survey

A questionnaire was worked out on the basis of a pre-project. Data from 649 individuals (34% women and 66% men, overall age 41 years) was tested. The questionnaire contained questions regarding:

---

[1] With technology quality do we here mean the usability to the technology. A product with good usability is: easy to learn the use of, easy to remember the use of, does solve tasks on an efficient way, is secure in use and the user likes the product.

- *IT-quality*; To measure the quality of the IT-systems an international survey SUMI (Software Usabilty Measurement Incentory) was used. This survey gives a measurement of the quality of a data product, based on the users experiences. SUMI contains one global scale and eight sub-scales: efficiency, affect, helpfulness, control and learnability.
- *Technostress*; any effects on attitude, thoughts, behaviour or the body's physiology that are caused by either direct or indirect by technology (Weil and Rosen 1997).
- *Negative consequences of IT-change*; e.g. feeling of being watched and controlled, less contact with co-workers, and being unsure about what one knows.
- *Positive consequences of IT-change*; e.g. easier to organise the work, and being more productive.

2.2 Focus groups

Nine companies were visited - all of them had implemented new data technology. 74 employees over 45 years old (45% women and 55% men, overall age 54 years) participated in thirteen different focus groups. Issues about IT-changes were discussed.

3 Discussion of results

3.1 IT-quality

There was a connection between IT-quality and technostress (r=-0.30**[2]). This connection was negative; this means that the poorer the quality of an IT-system the more technostress the employees experienced. It was the affect and control part of IT-quality that effected the employees' technostress the most. Poor IT-quality can result in less efficiency, which again can lead to frustration, stress and increased workload for the employees. Poor learnability can also result in increased workload because the employees spend too use much time learning the system. They may feel that they are wasting their time on learning the IT-system instead of doing their work. A system that is difficult to learn will also give less help to the employees when doing their work.

There was a connection between IT-quality and how the employees experienced the consequences of the IT-change. The better quality of the system that was introduced, the fewer negative (r=-0.29**) and more positive (r=0.21**) consequences there were.

---

[2] * = p<0.05, ** = p,0.01.

As the results indicate the quality of the IT-system is important for how the employees cope with the IT-change. Involvement of the employees does improve the IT-quality (r=-0.23**). This is because the employees know more of what is going on and therefore their expectations are more realistic. The IT-designers also get more knowledge about what the employees need and expect, and what sort of tasks they will use the system for. It is also important that the system is being thoroughly tested before it is put into use.

The IT-quality also has a positive connection depending on the quality of training (r=0.38**). It is especially the IT-product's helpfulness (r=0.36**) and learnability (r=0.38**) that affect the quality of training the most. One explanation of the result can be that it is easier to arrange good training if the IT-system is good – so the training really becomes better. The training can also be perceived as good because it is easier to learn an IT-system of good quality. The employees can therefore give the training credit for something that really is caused by a good IT-system. The third reason can be that companies that have a poor IT-system, have the habit of not involving the users in the development and introduction of the IT-system. The users will not be asked for feedback on the training, therefore, both the IT-system and the training will be less adjusted to the users needs, and consequently the users perceive them as poor.

3.2 IT-quality and age

The IT-quality was not evaluated equally among all the employees. Age had a significant effect on the subjective IT-quality (F=8.26**). In all parts of IT-quality, except for 'affect', age had significant effect; efficiency (F=12.44**), helpfulness (F=5.73**), control (F=10.20**) and learnability (F=25.50**). This means that the older employees evaluate the system lower.

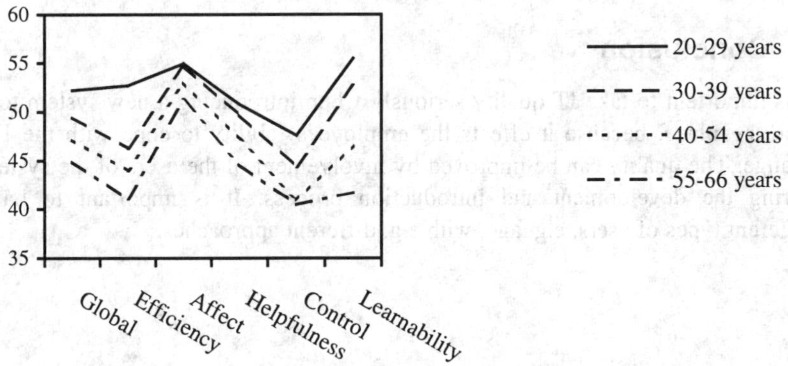

Figure 1. IT-quality and age

Through the focus groups several reasons for the effect of age on IT-quality become visible.

One reason for the difference can be that elderly employees can have other criteria for quality than younger ones, such as experience with other ways of doing the same work. The IT-system will then be compared with what they have been doing before. Several also mention that the experience can help the elderly employees to put the technology in its "right place" – as a tool. They said that it is not the age that creates the differences but the experiences: "Experience makes us more selective of what we "let in", and we become more conscious regarding what we do". They are therefore more critical than younger employees.

Elderly employees are also often more used to doing a task in a special way, and are therefore more dependent on the IT-system fitting the task. Young employees have grown up with IT, and elderly have not. This difference can have affect on their evaluation of an IT-system's quality.

Several also felt that elderly employees more often than younger ones feel forced to begin using IT, and that there is more resistance among elderly employees and therefore they are more negative to the IT-system.

Some did believe that elderly employees are a bit slower than younger ones when learning new subjects. However, others felt this is a myth and self-fulfilling prophecy, and that in fact elderly employees learn as fast as younger ones. The results from scientific research do also show different findings according to learning and age. Some find that elderly employees need specially designed training (Wallensteiner 1985; Czaja, Hammond, Blascovich and Swede 1989). Other studies show little difference between the age groups (Stammers et al. 1991; Staufer 1992).

4 Conclusion

It is important to take IT-quality seriously when introducing a new system to a working place, because it effects the employees' ability to cope with the IT-change. The quality can be improved by involvement of the users of the system during the development and introduction process. It is important to have different types of users, e.g. age, with e.g. different approaches.

5 References

Andersen, R. E., Crespo, C. J., Bartlett, S. J., Cheskin, L. J. & Pratt, M. (1998). Relationship of physical activity and television watching with body weight and level of fatness among children. *Journal of the American Medical Association*, 279, 938–942.

Czaja, S., Hammond, K., Blascovich, J. J. & Swede, H. (1989). Age related differences in learning to use a text-editing system. *Behavior and Information Technology*, 8, 4, 309-319.

Eisenberg, E. L. & Goodall, H. L. (1993). *Organizational communication – balancing creativity and constraint.* New York: St. Martin's Press.

Lindström, K., Pakkala, K. & Torstila, I. (1989). Coping with technological change in banking and insurance. In Smith, M. J. & Salvendy, G. (Eds.): *Work with computers: Organizational management, stress and health aspects*, pp. 256-263. Amsterdam: Elsevier Science Publishers B. V.

Smither, R. D. (1994). *The psychology of work and human performance.* New York: HarperCollins.

Stammers, R. B., Wong, C. S. K., Davies, D. R., Glendon, A. I., Taylor, R. G. & Matthews, G. (1991). Office Technology Skills and the Older Worker. Designing for Everyones. In Queinnec, Y. & Daniellou, F. (Eds.): *Proceedings of the 11th Congress of the International Ergonomics Association (Vol. 2)*, pp. 1583-1585. Amsterdam: Elsevier.

Staufer, M. (1992). Technological change and the older employee: Implications for introduction and training. *Behaviour & Information Technology,* 11, 1, 46-52.

Turnage, J., Cornelius, T. & Sansing, L. (1994). A comparative study of technology stress in the workplace. i Bradley, G. E. & Hendrick, H. W. (Eds.): *Human factors in organizational design and management – IV*, pp. 739-744. Amsterdam: Elsevier Science Publishers B.V.

Wallensteiner, Ul (1985). Human factors and the challenge of aging. *Proceedings of the Human Factors Association of Canada, 18th Annual Meeting, Hull, Quebec, 27-28 September*, pp. 81-84. HFAC, Rexdale, Ontario.

Weil, M. M. & Rosen L. D. (1997). *Technostress: coping with technology.* John Wiley & Sons.

Zaleznik, A., Vries, M. F. R. K. & Howard, J. (1977). Stress reactions in organizations: syndromes, causes and consequences. *Behavioral science*, 22 (3), 151–162.

Factors Associated with the Experience of Stress and Well-Being in Computer-Based Office Work

Pentti Seppala
Finnish Institute of Occupational Health, Topeliuksenkatu 41 a A, FIN-00250 Helsinki, Finland

1 Introduction

Many studies on VDT work have shown that physical and mental well-being in computer-based office work is related to various characteristics of the work and the work environment, as well as of the individual using the computer. The majority of the studies in the '80s indicated that the complaints of discomfort in the eyes and in the musculoskeletal system, and the experience of stress, were associated with working with computers. The etiology of the problems is not clear, however. Especially studies in the '90s have shown the importance of psychosocial factors and the need for a systems approach for explaining the multifactorial relationships (Smith and Cohen 1997). The balance model introduced by Smith and Sainfort (1989) provides a useful framework for examining the complex interaction of using computerized technology in an organization. The *individual* with his/her physical characteristics, perceptions and behavior is at the center of this model. The individual has access to available *technologies* when performing his/her specific work tasks. The *task* requirements also affect the skills and knowledge needed. The tasks are carried out in a work setting that comprises the *physical and social environment.* There is also an *organizational structure* that defines the nature and level of individual involvement, interaction and control. This framework of the balance model was utilized for examining stress and well-being of office workers in a large municipal organization. The aim was to analyze the extent to which perceived stress, disorders of the upper limbs, and visual strain were associated with characteristics of the work and of the organization, introduction of technological change, problems with the workstation, and experience in using computers.

2 Material and Method

A questionnaire on the use of information technology and stress at work was administered in three organizational units - town office, real estate office, and town planning office - of a large municipality. 380 employees from several professional groups, e.g. architects, engineers, lawyers, clerical workers, draftspersons, etc., answered the questionnaire. The age of the respondents ranged from 21 to 62 years (mean 45.2 and standard deviation 8.7). 61% of the respondents were women and 39% were men. The length of work experience varied from 1 to 42 years (mean 11.8 and standard deviation 8.3).

The questionnaire comprised questions from the following areas: 1) gender, age, education, 2) job title and tasks, 3) use of a computer in performing one's tasks, 4) application softwares used, 5) ease of learning and usability of softwares, 6) introduction of technological change, 7) experienced skills in using application softwares, 8) stressors related to the organization and the use of a computer, 9) problems with the workstation and the work environment, 9) perceived stress, problems with mental well-being, vision and the musculoskeletal system.

Composite scales describing different dimensions of human-work-computer interaction were formed, based on factor analyses of the data. The associations between experienced stress and strain and the characteristics of the work and technology were analyzed by means of correlation and regression analyses.

3 Results

The statistically significant differences in the experiences of job characteristics and perception of stress and various symptoms between different categories of age, gender, vocational education and occupation are presented in Table 1.

G*ender* was related to the concern about changes in the organization and security of employment (women were concerned more often than men), task variety (men had more variable tasks), receiving information on technological changes (men were better informed), problems of the work station and amount of daily use of computers (women used more and experienced problems more often). Gender was also connected to perceived stress and symptoms, so that women, more often than men, reported all the stress as well as other symptoms inquired.

The responses of the different *age* groups differed on the scales regarding changes in tasks and responsibilities (those aged 41-45 years perceived changes more often than those aged over 51 years), lack of competence and difficulties in using computers (those aged over 55 years experienced difficulties more often than the younger persons), influence of information technology on job content (positive influence for age group -30 years) and occurrence of skin problems (age group -30 years had more problems).

The level of *vocational education* was associated with stress related to the amount of work and haste (those with a higher education experienced more often), concern about changes in the organization and job security (those with less education were concerned more often), changes in tasks and responsibilities (more changes for those with lower or middle-level education) and task variety (more changes for those with a higher education). The different educational categories differed in the daily use of computers (less educated persons used more). Feelings of lack of competence and difficulties in using computers differed in the educational categories (less educated clerical workers experienced difficulties less often than the others).

Table 1. Statistically significant differences of composite scale values between different categories of gender, age, vocational education and occupational groups. T-tests (gender) and F-tests.

Composite scales	Gender	Age	Vocational education	Occup-ation
I. Work and organization				
1. Quantity of work and haste			.01	.01
2. Changes in organization	.02		.01	.02
3. Changes in tasks and responsibilities		.001	.02	.01
4. Demands and difficulty of work				
5. Task variety	.01		.01	.0001
II. Information technology				
1. Information on changes	.001			.01
2. Implementation and training			.05	.02
3. Problems of workstation	.05			
4. Environmental problems				
5. Amount of daily use of computers	.0001		.0001	.0001
III. Experiences in using computers				
1. Competence, difficulties		.01	.01	
2. Stress sources in using computers	.01			.02
3. Influece of computers on job content		.01	.05	
IV. Stress and symptoms				
1. Mental stress, anxiety	.05			
2. Skin symptoms	.001	.05		
3. Problems in upper extremities	.0001		(.06)	.05
4. Visual strain	.0001			
5. Problems with vision	.05			(.07)

The experience of problems in using computers differed also in the different *occupational groups* (clerical workers had less problems than the others). Persons in administrative management and specialists had experienced stress related to the amount of work and task variety more often than the others. The persons in administrative management were also better informed about changes than the other groups. Regarding perceived stress and various symptoms, the occupational categories differed only on the scale of disorders in the upper extremities. Draftspersons and clerical workers had experienced these problems most often.

The *correlation analysis* showed that the features of the process of introducing information technology (opportunities to participate in planning, adequacy of training, etc.), stress when using computers, characteristics of work organization and collaboration, need for additional training, as well as problems with the work station, correlated significantly with the experience of overall stress, fatigue, headache, tension and anxiety. The experience of stress correlated most strongly with a rapid work tempo, amount of work, overlapping tasks, and various sources of stress when working with computers (insufficient skills, failures in software and hardware, difficulties in getting help, etc.). These same factors correlated also with the disorders of the upper extremities. Also visual discomfort correlated with the above-mentioned sources of stress when working with computers.

Table 2. Factors explaining the experience of mental stress and anxiety

Job characteristics	Regression coefficient	R-square	F	df	p<
		41 %	50.03	4, 292	.0001
Intercept	0.34				.001
Workload, haste	0.18				.0001
Management, atmosphere	0.14				.0001
Work demands, competence	0.16				.0001
Defects of workstation	0.16				.0002

Table 3. Factors explaining the experience of disorders of the upper extremities

Job characteristics	Regression coefficient	R-square	F	df	p<
		23 %	16.26	5, 271	.0001
Intercept	0.28				.06
Workload, haste	0.14				.0001
Management, atmosphere	0.10				.01
Work environment	0.11				.005
Defects of work station	0.12				.02
Hours of computer use	0.07				.005

According to *regression analyses* (Table 1 and 2), the composite score of mental stress and anxiety was best explained by the composite scores of 1) the amount of work and haste, 2) management, collaboration and work atmosphere, 3) the demands and difficulty of work and 4) defects of the workstation.

Disorders of the upper extremities were best explained by 1) the amount of work and haste, 2) management, collaboration and work atmosphere, 3) problems of the work environment, 4) defects of the workstation and 5) hours of daily use of a computer.

4 Conclusions

Stress and mental and physical well-being in computer-based office work seem to be connected in a complex way to various characteristics of an organization, to technology, jobs and the employees. Features of the organization and work tasks are important mediating factors in the experience of stress in work with computers. Information technology and computers are only one element which is associated with stress in various ways in different organizational settings. The results of the current study are in accordance with the previous findings reviewed by Smith and Cohen (1997). The stress experienced by those with a higher education level and in administrative and expert jobs is related to time pressure, amount of work and enlarged responsibilities. Those who work more with various computing and design tasks are stressed by failures in equipment and software, and by insufficient information when new technologies or changes are introduced, as well as problems with the workstation and the environment. Problems in the division of labor, job design and roles correlated with perceived lowered mental and physical well-being. This implies that dissatisfaction with various organizational and technological factors at work, and problems of well-being are interrelated and cumulated in certain groups of employees.

Of the individual factors, age was not linearly associated with most of the problems. Middle-aged employees were most stressed about the changes in their tasks and responsibilities, and the oldest age group was concerned about their competence in using computers. Gender correlated significantly with the experienced stress and well-being. It must be noted, however, that gender and task characteristics were closely connected in this material, so that women typically performed those tasks which have also many risk factors. When interpreting the findings, one must also note that rapid changes where going on in technology and municipal services at the time of the study.

5 References

Smith, M.J. & Cohen, W.J. (1997). Design of computer terminal workstations. In Salvendy, G. (Ed.): Handbook of human factors and ergonomics. Second edition. New York: John Wiley & Sons 1997.

Smith, M.J. & Sainfort, P. C. (1989). A balance theory of job design for stress reduction. International Journal of Industrial Ergonomics, 4 , 67-79.

Visual Strain and Eye Disorders in Display Uses

Valentina Kovalenko, M.D., Ph.D.
P.O. Box 9141, 310058 Kharkov, Ukraine
Phone: 0572 - 373-428. Fax: 0572 - 791-313

It is known, that work with display units for a long time leads to definite disorders of the functional body state and health in users, among them visual discomfort is situated at the first place (WHO, 1987). Additional information concerning this problem was obtained for the last years. Some alterations in visual organ itself were revealed, in particular in our special experimental research (V.V.Kovalenko, 1994).

The aim of this investigations was detailed examination of visual capacity in designers, who used computers in their work for 4 - 5 hours per day, in order to detect profound mechanisms of visual disorders and found the ways for softening of negative influence to human visual system of work with display units.

Our ergophthalmological approach has been used in this elaboration. It consisted in detailed questionnairing, careful ophthalmological examination with functional tests and physiological visual function study during the working activity. The latter included the evaluation of accommodation (the nearest point of clear vision, NPCV; the time of maximum accommodation, TMA), convergence (the nearest point of convergence, NPC) and retinal function (the time of the following colour contrast, TFCC); general reaction of the eye to work carried out was measured with wink movements frequency, WMF.

139 designers from 21 to 50 years old have been examined with questionnairing and ophthalmological methods. 60 persons of them were subjected to physiological measuring of visual functions - four times during the working day.

According to the data obtained the average frequency of visual fatigue was 69,8% without clear differences connected with age. The astenopic complaints were: headache, dizziness, head heaviness, nausea and even vomiting, a feeling of eye fatigue, conjunctival hyperimia, lacrimation, eyeball pain and heaviness, burning below the eyelids, pain in the depth of the orbital cavity and also eyesight difficulty, dark or colour spots before eyes, irritation from the light.

6,2% of specialists examined suffered from diseases of eye front section namely chronic conjunctivitis and blepharitis. At the same time 48,6% of them had stagnative phenomenon in this area without the signs of inflammation.

Vessels disorders of eyegrounds have been revealed in 41,8%: the dilatation of retinal viens, the optic nerve disk hyperemia, the disk border vague and others. These alterations were connected with vessels dystonia and musculosceletal disorders in many cases. Such changes of eyegrounds were most frequently in the persons, who suffered from musculosceletal disorders in the neck area and shoulder zone.

Visual activity for long distance was disordered in 59,7% of designers examined independently of the age. The frequency of myopia was 38,5%. Weakness of accommodation has been found in 51,4%, more frequently when myopia. Age-related changes of accommodative function were detected in specialists with normal vision when asthenopia only. Anizoaccommodation as very delicate sign was found in 18,5%, more frequently in persons with normal vision when asthenopia.

The results of physiological measuring of visual function parameters during the working day have shown some peculiarities. General reaction of eye to working activity was sufficiently visible (change of WMF from 10,57±0,54 to 11,58±0,57 winks) on the background of initial exceeding of parameter when comparing with mean statistical value. Displacements of both parameters of accommodative function were distinct and considerable, in particular NPCV (from 8,16±0,41 to 9,79±0,48 cm). Retinal function changed clearly too (TFCC has decreased from 10,8±0,51 to 9,13±0,49 s). Oculomotor muscles function only remained without clear changes under influence of work (NPC values 7,15±0,40 and 7,51±0,43 cm accordingly correspond to faint tendency towards an increase of the parameter).

All the above changes of physiological parameters of visual system are evidence of visual fatigue and weakness of accommodation and retinal function.

The data obtained when examining of the group of designers, who used display units, have shown essential alterations in visual organ in combination with manifest visual fatigue. Analysis of asthenopic complaints testified about availability of visual fatigue of mainly mixed forms with expressed retinal component. This conclusion is confirmed with physiological measurements results, which reflect distinct reaction to work carried out of accommodative apparatus, retinal function and the whole eye.

It is perfectly obvious those numerous alterations in visual organ, which have been found when ophthalmological examination, were results of working activity with using display units. When comparing these data with results of our previous investigation, in which operators of computing centers were examined (V.V.Kovalenko, 1997), we have to note the resemblance of changes revealed. However these disorders were found more expressed in designers, apparently in consequence of specific matter of visual tasks.

Mechanism for forming of functional and structural alterations in human visual system obviously consists in accumulation of daily small deformation with following their transformation to neuromuscular disorders transient at first and after that steadfast. Hence the strategy of preventive and rehabilitation measures follows. Its matter is contained in making inadmissible the transition from transient form of disorders to stable one. Special model of programme for such purpose has been elaborated and tested.

References

V.V.Kovalenko. Visual organ state in operators when working with display units. Medical Occupation, 1994, N9-12, p.165-167.

Valentina Kovalenko. Ergopthalmological characteristics of VDT-operators. HCI Internationa'97. Poster Sessions: Abridged Proceedings. San Francisco, California, 1997, p.107.

WHO. Offset Publication N99. Visual display terminals and worker's health. World Health Organization. Geneva, 1987.

VDT Use and its Relation to Job Stress in Various Sectors of Industry in Finland: Qualitative Differences

Kari Lindström and Vesa Hottinen
Finnish Institute of Occupational Health, Department of Psychology

1 Introduction

The use of VDTs and personal computers is common in most occupations and in nearly all branches of industry. The context of VDT use, however, differs quite a lot depending on the tasks, occupations and the sector of industry (e.g. Lindström et al. 1997). The wide variety of tasks in which VDTs and various softwares are used complicates the monitoring of their possible stress and health effects. Earlier studies have also shown that it is important to consider the temporal issues in Human-Computer Interaction when studying VDT work and its relations to job and organizational factors and well-being (Carayon 1997). Lack of control and lack of social support have been chronic stressors affecting psychological distress among VDT workers. Lack of control (Carayon et al. 1995) and high physical work load (Lindström et al. 1997) have related to musculoskeletal discomfort. But all these effects are usually based on multifactorial relationships. Eye discomfort has been explained inconsistently by various factors when the same VDT -users have been followed (Lindström et al. 1997). Thus, the relationships between VDT use, work content and organizational factors, as well as various stress effects, most probably differ according to the temporal as well as individual and task-related factors.

The aim of this paper was to describe VDT-related work at the national level, and to investigate how VDT work, its amount and the difficulties associated with software use were related to job and organizational factors and well-being in such a heterogeneous sample.

2 Material and Methods

A national survey on work and health, carried out in Finland in 1997, included also questions on VDT use and job stressors and well-being (Kauppinen et al. 1997). A computer-aided telephone interview was focused on 3202 individuals aged 24 to 64 years; this group formed a random sample of the Finnish working aged population. Of those interviewed, 2156 were at work at the time of the interview and 54% of them used some kind of VDTs or microcomputers at work. The present analysis was conducted among those using VDTs (n=1167). In this national level data we analyzed the prevalence and the amount of VDT use, and the difficulties associated with software use according to sex, age and different branches of industry, and also how the VDT use and the software problems correlated to job and organizational factors and some subjective symptoms.

Table 1. Use of VDTs or microcomputers in Finland by gender in different branches

	Total		Men	Women
Branch	n	%	%	%
Agriculture, forestry	138	24	24	24
Industry (metal, forest, other)	414	57	57	57
Energy production and water supply	30	67	67	67
Construction	90	20	17	67
Wholesale and retail	223	64	66	63
Hotels and restaurants	51	25	42	21
Traffic and communication	163	41	33	60
Banking and insurance	63	98	95	100
Private service	209	73	77	67
Public administration	104	82	83	81
Education and training	190	62	72	57
Health care and social welfare	330	46	62	45
Other public services	141	50	51	49

Because the number of employers in some branches was quite small and the type of work differed greatly between the branches, most of the branches were combined in the analysis into three joint groups. These were:

- Production (including metal manufacturing, the paper industry, and other production)
- The private service sector (including retail and wholesale trade, hotels and restaurants, banking and insurance)
- Public services (health care and social welfare, training and education, other public services)

The women and men in these groups were analyzed separately. Some branches were excluded because they were difficult to include in any of the major groups.

3 Results

3.1 Amount of VDT Use and Problems in Software Use and in Learning

The proportion of VDT or microcomputer use (Table 1) was relatively low in agriculture and forestry (24%), in the construction industry (20%), in restaurants and hotels (25%) and in health care and social welfare (46%). In the banking and insurance sectors most employees used VDTs (98%). The frequencies in the amount of daily use of a VDT over 4 hours daily was 21% in the total sample, and it was highest in banking and insurance (78%), in public administration (41%) and in services for private business (35%). The amount of VDT work differed also according to the workers' age, sex and education. Those respondents who did not use computers in their work were most often over 54 years of age. Daily VDT use exceeded 4 hours for 24% of the women and 18% of the men.

Difficulties to use and learn the applied softwares were reported by 14% of the respondents; 19% of the men and 10% of the women. Especially among men the difficulties increased clearly with age; 38% of the men over 55 years reported difficulties. Sectorially, most difficulties in coping with software were encountered in agriculture and forestry 24%, in energy production and water supply 29% and in public administration 22%. Least problems were reported in banking and insurance, only by 2%.

3.2 Relations to Job and Organizational Factors and Well-being

The relations of VDT use to job stressors and well-being were analyzed first separately among men and women (Table 2) and within various sectors of industry.

A high amount of daily VDT use among men related in general to a higher educational level, time pressure, high mental work load and low control, as well as to more strain symptoms. In production and in the private service sector, however, the better learning opportunities at work were related to a greater amount of VDT use. The good social support from coworkers related to a greater amount of daily VDT use in production and in private services, but this corelation was negative in the public service sector. In industry, the strain symptoms were not related to the length of VDT use.

Among women, only a higher prevalence of eye discomfort related to the high amount of daily VDT use, especially in the public service sector.

However, the length of daily VDT use was related, among women, to high mental work load in production and private services, but not in the public service sector.

Learning opportunities among women in industry were also related to long daily use of VDT. Also job satisfaction related among women differently in the different sectors to the length of daily VDT use. In industry, it improved with more VDT work, whereas in private and public services it decreased.

Table 2. Correlations (Pearson's r) of the amount of daily VDT use and software problems to job and organizational characteristics and well-being (n = 1168) ($r \geq 0.08$, $p < 0.01$)

	Amount of VDT use			Problems in software use		
	Men	Women	All	Men	Women	All
Age	-0.06	-0.02	-0.02	0.27	0.16	0.20
Educational level	0.29	0.09	0.18	-0.12	-0.07	-0.11
Time pressure	0.23	0.04	0.13	0.04	0.04	0.04
Mental work load	0.12	-0.05	0.02	0.16	0.06	0.11
Job control	-0.12	-0.03	-0.04	-0.12	-0.11	-0.09
Use of skills	0.03	-0.06	-0.02	-0.01	-0.06	0.00
Learning opportunities	0.11	0.05	0.08	0.05	0.03	0.05
Coworker support	-0.01	-0.00	0.02	-0.12	-0.12	-0.14
Supervisory support	-0.03	-0.00	0.02	-0.19	-0.13	-0.14
Job satisfaction	-0.08	-0.06	-0.07	-0.14	-0.10	-0.11
Strain symptoms	0.16	0.06	0.11	0.11	0.22	0.17
Eys discomfort	0.04	0.12	0.11	0.04	0.05	0.02
Neck & shoulder pain	-0.02	-0.02	0.01	0.06	0.03	0.01
Wrist & finger pain	-0.05	0.02	0.00	0.10	0.03	0.06

The difficulties associated with software use and learning were related both among men and women to higher age, low job control, and low social support from coworkers and supervisors (Table 2). Sectorially, however, in the private service sector, coworker support among women was not related to software problems. High mental work load associated with problems in software use in all other sectors, except for women in the public service sector.

Both among men and women, in general, the software problems were related to more strain symptoms and lower job satisfaction. However, the association between software problems and strain was not found among men in private services, whereas frequent problems with software related to low job satisfaction among women only in production.

In addition to these relationships, the men who reported problems in software use generally perceived higher mental work load and had a lower educational level.

4 Discussion

In conclusion, the percentage of employees using VDTs varied greatly among the different branches of industry and was clearly lower in the older age groups. Also the daily length of VDT work differed among the branches. Difficulties in software use and in learning were reported by 14% of those using VDTs. The difficulties were more common among men than women, and especially among older workers and those with a lower basic education.

Interestingly, the longer daily VDT use was associated more clearly with job stressors and strain symptoms among men than women. Women suffered mainly from eye discomfort. Regarding job stressors and well-being in VDT work, sectorially the situation was best in production as compared to public service work. Low job control has also been frequently found to lead to psychological distress among VDT workers (Carayon et al. 1995, Lindström et al. 1997).

The difficulties in software use were associated with older age and low job control and low social support, and also to elevated strain and lower job satisfaction. Sectorially, the situation of women was best in public services. The software problems were, however, mainly related to individual background factors and to the organizational practices. The lack of social support has been found also earlier to be a risk factor for VDT-related strain (Carayon et al. 1995), and especially supervisory support is important when implementing new VDTs and software.

The qualitative differences in VDT-related job stressors and strain were clear between the genders and various branches. The national surveillance of VDT use and its consequences on work stressors and well-being seemed to have sigificance in seeking out the high risk sectors and individuals in need of preventive actions, such as training and job redesign.

5 References

Carayon, P. (1997). Temporal issues of quality of working life and stress in human-computer interaction. *International Journal of Human-Computer Interaction*, 9 (4), 325-342.

Carayon, P., Yang, C., & Lim, S.-Y. (1995). Examining the relationship between job design and worker strain over time in a sample of office workers. *Ergonomics*, 38, 1199-1211.

Kauppinen, T. et al. (eds). (1997). *Work and health in Finland in 1997.* Helsinki: Finnish Institute of Occupational Health. (In Finnish with English Summary.)

Lindström, K., Leino, T., Seitsamo, J. & Torstila, I. (1997). A longitudinal study of work characteristics and health complaints among insurance employees in VDT work. *Int. Journal of Human-Computer Interaction*, 9 (4), 343-368.

The Physiological Measurement of User Comfort Levels: An Evaluation Experiment for Comparing Three Types of CRTs

Sufang Chen[1], Yoshio Machi[1], Xiangshi Ren[2] and HunSoo Kim[3]
[1]Department of Electronics Engineering,
[2]Department of Information and Communication Engineering,
Tokyo Denki University
2-2 Kanda-Nishikicho, Chiyoda-ku, Tokyo 101-8457, Japan
Phone: +81-3-5280-3360, Fax: +81-3-528-3567
Email: s_chen@dserve5-5.d.dendai.ac.jp, ren@c.dendai.ac.jp
[3]Technology Center, Samsung Display Devices Co.Ltd
575, Shin-Dong, Pladal-Gu, Suwon City, Kyungki-Do, Korea, 442-391
Tel: +82-331-210-7963, Fax: +82-331-210-7799
Email: khs@sdd.samsung.co.kr

1 Introduction

One of the goals of human computer interface research is to make computers comfortable for users. Recovery time, long term usage and psychological attitude are strongly affected by user comfort levels and they have an important effect on productivity (Tamura, 1998). Extensive literature exists which reports subjective preferences. However, subjective user satisfaction does not distinguish between psychological factors and measurable physiological effects and responses. The use of technology to measure physiological responses promises a more accurate evaluation of user comfort levels and psychological responses. One important index is the human brain's alpha waves. While much research has focused on the heart rate (HR) and the heart rate variations (HRV) of computer users (Ohsuga, 1993), not much research has focused on alpha brain wave responses. One reason is that beta waves are shown when users are doing tasks while alpha waves are shown when users finish their tasks and are relaxing their minds (Andreassi, 1995). We combined brain wave and heart rate monitoring with a galvanic skin response monitor to objectively evaluate computer user comfort levels.

In this study we developed an evaluation method which allows us to measure the comfort levels of computer users according to their brain wave responses. We then applied this method in an experiment which evaluated three different types of Cathode Ray Tube (CRT) displays. Other research which evaluated CRTs has, for example, used subjective evaluations in response to changes in CRT chrominance levels, however, not these did not directly measure the users comfort levels.

In this experiment we monitor the user's brain waves both during and immediately after the user completes each task i.e. during a *break* after interaction with the computer. Thus, we can observe alpha waves clearly during the break. We consider that this "break monitoring" is an important facet of human computer interaction because the after effects of computer use are important to computer users e.g. for recovery time, long term use, psychological attitude to the technology in general etc. Concretely, we used this idea to evaluate three displays, which included a normal type CRT display commonly available in the market place, and also newly developed bio-television monitors which emit the far infrared ray. The far infrared ray is widely used in various field, especially medical field (Takajima, 1996; Takeuchi et al., 1996).

2 Method

2.1 Participants

Twenty-two Chinese subjects (15 male, 7 female) were tested in the experiment. They were from 24 - 40 years old and had normal vision. All subjects were regular users of computers with CRT displays and mouse-type pointing devices.

2.2 Apparatus

The hardware used in this experiment was: a Brain Wave monitor (SYNAFIT 5000, NEC SANEI Corp.), a Galvanic Skin Response sensor (GSR MP100, BIOPAC system Crop.), an electrocardiogram (Bio-view G (2G52), NEC SANEI Crop.) and a personal computer (Power Macintosh 6300/120). The software was: Topography (SYNAFIT 5000, NEC SANEI Corp.), an electroencephalogram (EEG) analytic program used for bit map reading.

Three types of displays were used in the experiment:

- The A type is a normal display which is commonly sold in the market.
- The B type is a new bio-television monitor which increases the radiation of the far infrared rays (wavelength is 5 μ m-15 μ m, radiation temperature is 36°C which is equal to temperature of electromagnetism wave) from the

cathode ray tube (CRT). In addition the tube surface is coated with aluminium trioxide (Al_2O_3) which strongly radiates the infrared rays to the side of the tube.

- The C type is one whose radiation of the far infrared rays is radiated intermittently, with a fan installed in the display. The surface of this tube is also coated aluminium trioxide (Al_2O_3).

2.3 Design and Procedure

Two types of simple tasks were designed: (i) typing an article with a word processor, and (ii) drawing a picture with a drawing software.

Before beginning the test session the experiment was explained to the subjects.

Each subject was asked to follow these steps using each type of display.

(1) Close your eyes and keep your mind quiet in a relaxed state for three minutes.

(2) Do the first task continuously for seven minutes.

(3) Close your eyes again and keep your mind quiet in a relaxed state for five minutes.

(4) Do the second task continuously for seven minutes.

(5) Close your eyes and keep your mind quiet in a relaxed state for three minutes.

The order for the two tasks and the three displays was varied for the different subjects. The difference between the A, B, and C displays was not explained to the subjects.

The data for each subject were recorded automatically as follows:

(a) The alpha 1 brain waves were recorded in steps (1), (3), and (5) above.

(b) The ratio of lower frequency (LF) to high frequency (HF) heart rates was recorded in all steps (1)-(5).

(c) A voltage value for GSR was recorded in all steps (1)-(5).

3 Results and Discussion

Statistical analyses were performed using a pair t test.

On the basis of brain wave and heart rate monitoring the results show that the C display was the most comfortable as follows: (i) significant increases in alpha 1 brain waves ($p < 0.01$); (ii) changes in the value of range of LF to HF heart rates i.e. the autonomous nerve index ($p < 0.05$).

On the basis of GSR (Max-Min) ($p < 0.05$) it was found that display B and display C were almost the same in the depth of relaxation achieved, both being more effective than display A. However no significant difference between B and C was found based on GSR values. One reason may be that display C was not given enough scope for the special functions for which it was designed. Other factors might influence the GSR results. For example, the brain wave monitoring hat might have been too tight or the unfamiliar monitoring equipment may have affected GSR levels.

4 Conclusion

First we proposed a unique strategy, "break monitoring", for physiological measurements. The experiment demonstrates that displays C and B (especially C) were more comfortable than display A. These displays were intermittently radiating the far infrared rays from the cathode ray tube (CRT). The tube surfaces of the C and B displays are coated with aluminium trioxide (Al_2O_3) which strongly radiates the far infrared rays to the side of the tube. Based on these results we can say that these displays reduce problems such as tiredness and stress compared to the normal displays of the type A used in this experiment.

We believe that our proposal and the results can impact research on user interface comfort levels.

5 References

Tamura, H. (1998). *Human interface*. Omen Publish. 61-65.

Ohsuga, M (1993). To evaluate the workload with index of autonomous nerve system. *Proc. of the Society of Instrument and Control Engineers*. 29-8, 979-986.

Andreassi, J. L. (1995). *Psychophysiology: Human Behavior & Physiological Response (The Third Edition)*. Lawrence Erlbaum Associates, Publishers Hove, UK, 21-25, 166-171.

Takajima, H. (1996). *Easy to learn far-Infrared ray engineering*. Kogyo Chosakai Publishing Co., Ltd..

Takeuchi .T., Takeuchi . A., Yokoyama. M. (1996). Clinical experiences of far-Infrared whole body hyperthermia by the use of RHD2002, Proc. of the 7th International Congress on Hyperthermic Oncology, 272-274.

Coping with Psychophysiological Stress during Multi-Tasking in Human-Computer Interaction

Wolfram Boucsein and Florian Schaefer
Physiological Psychology, University of Wuppertal, Germany

1 Introduction

Inadequate temporal structures in human-computer interaction constitute a major source of psychophysiological stress-strain processes (Boucsein 1999). This has been repeatedly demonstrated by our group while investigating the adverse effects of system response times for more than a decade. The term system response time refers to temporal delays caused by both hardware and software features of the computer system. Prolonged system response times may not only reduce work efficiency, they may also impose severe stress on the user, as shown in physiological and subjective measures (Kuhmann 1989). Although computers have become incredibly fast, temporal delays in human-computer interaction continue to be a problem. Increases in hardware speed are often jeopardized by bulky software packages, time-consuming input and output devices, extensive network functions, and huge data banks to which many users may have access simultaneously.

An additional resource for coping with stress-strain processes induced by system response times may be using the multi-tasking capabilities of modern computer systems. Instead of merely waiting for the computer to respond, the user may decide to work on several processes simultaneously. However, if system response times do not exceed a certain duration, the benefit from switching to another task instead of waiting may be more than outweighed by the additional mental load that results from scheduling. The term scheduling refers to the need for organizing the work flow of different tasks running in parallel. Maintaining an optimal scheduling is an additional mental challenge for the user and thus becomes another source of stress. Coping with this type of stress can be facilitated by providing appropriate feedback on the temporal aspects of processes running in the background while a main task is performed. Therefore, process indicators play an important role in multi-tasking systems.

The aim of the two studies to be reported here was to establish the relationship between the duration of system response times (i.e., the duration of processes running in the background), the type of feedback given by process indicators, and the stress-reducing properties of multi-tasking. We expected the stress-reducing features of multi-tasking to become more prominent with increasing background process duration. Furthermore, we expected the stress imposed by the need for scheduling to be reduced if the temporal flow of processes running in the background is visualized by means of process indicators.

Two types of process indicators were applied, a static and a dynamic-relative indicator. In order to create a realistic environment for multi-tasking, a simulated CAD task was programmed permitting two different background processes. The task consisted of a series of three pages with 12 screws, each of which was to be calculated and inserted. Users were asked to get a screw symbol with the mouse from a graphic board, insert it at the predetermined position and measure the required length by using two mouse clicks. Background process 1 was started by entering the numbers for thickness of material and required strength for the screw in a separate window. The results of this process were the minimal length and diameter required for the screw. Those numbers were entered in another window for background process 2 that resulted in the norm length and diameter of the screw fitting for the particular purpose. Finally, the numbers for length and diameter were entered in a window in order to generate the selected screw, and the screw was draught to its place with the mouse. (For more details, see Boucsein et al. 1998.)

2 Background Process Duration

A first study with 18 male engineering students was performed to determine the minimal background process duration for a successful multi-tasking. Three different values for the process duration (10, 20 and 40 s) were applied in counterbalanced order. Electrodermal activity, heart rate, electromyogram recorded from the neck, blood pressure, respiration rate, and electrooculogram were continuously recorded, together with different working strategy parameters such as mouse movements, processes in use and activity breaks. Subjective ratings were taken after each CAD page. Covariance analyses were applied after eliminating general habituation effects over trials.

Psychophysiological parameters did not yield significant effects, presumably because of the different work strategies used during the three different levels of processing times as could be shown by strategy parameters. Besides causing longer working periods, prolongation of background process duration times resulted in less arousal and more emotional strain. The highest error rate and the strongest self-rated bodily pain symptoms occurred with processing times of

20 s. Our conclusion was that the 20 s background process duration might have the most benefit from introducing process indicators.

3 Type of Background Process Indicator

As a consequence, the background process duration of 20 s was used in a second study with 42 male engineering students. Three different experimental conditions were applied in counterbalanced order: A process indicator (an hour-glass showing up until the background process was done), a dynamic-relative process indicator (a horizontal bar filling from left to right proportional to percent-done of the background process), and no process indicator as a control condition. We expected the dynamic indicator to further reduce mental strain, but possibly increase emotional strain because of its property to push towards the end of the task in progress. Psychophysiological, subjective, and performance measures as well as the statistical evaluation were the same as in the first study.

The presence of process indicators accelerated working speed and cardiorespiratory activity, regardless of the indicator type. The rate of utilization of the multi-tasking features increased significantly with the presence of process indicators: In only 53 % of the time there was no background process active as compared to 56 % under the no-process-indicator condition. This rate of utilization difference was better reflected in subprocess 1 than in subprocess 2 where the significance was only marginal. However, only in less than 2 % of the time both subprocesses were simultaneously activated. The only physiological measures affected by the introduction of process indicators were the systolic blood pressure and the respiration rate, both increasing significantly. In the subjective domain, anxiety/depression decreased slightly but significant. On the other hand, there was a marginal significant decrease in subjective well being with increasing complexity of the process indicators. The psychophysiological changes observed can be easily explained with the increase in performance.

4 Discussion

In general, there were not many differences between physiological measures under the different experimental conditions in both studies. Instead, we observed marked differences in various measures of working strategy. Our interpretation is that our subjects had so many degrees of freedom in the CAD task that they were able to compensate for both kinds of possible stress inducing factors (mental and emotional) by changing their working strategies accordingly. In our former work on system response times, we used highly determined task sequences with almost no degrees of freedom. Therefore, the present results do not directly compare since the CAD task used allowed for multiple degrees of freedom. Those may be used as an additional resource for preventing stress caused by adverse factors in the work flow such as forced waiting periods. The introduction of process indicators increased working speed and accordingly cardiorespiratory activity. However, there is no increase in neck muscle tension and no stress relevant change in electrodermal or subjective measures. Instead, subjective anxiety and depression shows a small though significant decrease.

5 Conclusion

Our general conclusion is that using multi-tasking computer systems may increase the amount of work and enhance performance but does not seem to increase psychophysiological stress. Moreover, the kind of process indicator used does not really matter, although the presence of such indicators shows superiority over the absence of process indicators. It is concluded that multi-tasking may facilitate action regulation as an additional resource for coping with psychophysiological stress during human-computer interaction.

References

Boucsein, W. (1999). The use of psychophysiology for evaluating stress-strain processes in human computer interaction. In: R. W. Backs, & W. Boucsein (Eds.). *Engineering Psychophysiology*. Mahwah, N. J.: Lawrence Erlbaum (in print).

Boucsein, W., Figge, B. R., Göbel, M., Luczak, H., & Schaefer, F. (1998). Beanspruchungskompensation beim Multi-Tasking während der Bearbeitung einer CAD-Simulation. *Zeitschrift für Arbeitswissenschaft, 52 (24 NF)*, 221-230.

Kuhmann, W. (1989). Experimental investigation of stress-inducing properties of system response times. *Ergonomics*, *32*, 271-280.

Acknowledgment

This research was performed in co-operation with Holger Luczak and Matthias Goebel, Technical University of Aachen, Germany, and supported by the German Research Council (Deutsche Forschungesgemeinschaft), Grants Bo554/15-1 and Lu373/11-1.

How to determine optimal system response times in interactive computer tasks

Olaf Kohlisch
Physiological Psychology, University of Wuppertal, Germany

Introduction

For each computer task and sub-task, a system response time (SRT) of optimal duration can be determined. SRTs being longer or shorter than a once identified optimal duration have been demonstrated to have a negative impact on the users' work motivation and related attitudes (Thum, Boucsein, Kuhmann and Ray 1995), on error rates, task completion times, and the general level of physiological arousal (Kohlisch and Kuhmann 1997), and on electrocortical correlates of cognitive processes during computer work (Schaefer and Kohlisch 1995). Particularly the guideline 'the faster the better' does not seem to be appropriate for setting SRTs in interactive computer tasks. However, software developers usually ignore the negative impact of sub-optimal SRTs concerning the computer user, although all modern computer systems meet the technical requirements for optimizing SRTs. One reason may be, that optimal SRTs for particular tasks cannot be inferred using a rule of thumb. Therefore, it was asked in the present experiment whether optimal SRTs can be determined by psychophysical methods.

Method

A comparison was made among four psychophysical methods for determining optimal SRTs (see Luce and Krumhansl 1988): (1) The method of adjustment, that is, the subject terminates each SRT when having the desired duration. (2) The up-method of thresholds and (3) the down-method of thresholds. Here, the experimenter varies SRT duration upward or downward and the subject signals if it meets the desired duration. (4) A procedure using graphic representations of SRTs being adjusted by the subject. This method can be regarded as a variant of cross-modal matching.

Forty-eight subjects were randomly assigned to one of these methods. They performed six trial blocks of 25 min duration each, solving a simulated travel-booking problem. During each trial block, they had to serve virtual customers. All information referring to a customer was always presented on the computer screen. Serving a customer comprised several sub-tasks. For instance, subjects had to inspect a passenger list for a control number, identify the positions of the preferred seats on a plan, and to make a decision under time pressure. Four of the necessary operations for solving each booking-task were assumed to cause SRTs.

Subjects were told that the aim of the experiment was to optimize SRTs for a task to be implemented on a terminal, making use of a multitasking operating system such as Windows or Unix, and connected to an international computer network. It was supposed that the transfer of data within the network would cause a total SRT of 13 s per booking. It was further assumed that a programmer would be able to distribute the total SRT of 13 s across the four 'gaps' between sub-tasks in any subdivision by using the multitasking-ability of the system. Subjects' judgments should help the programmer identifying the optimal SRTs.

In trial block one and two, subjects learned to solve the booking problem with respect to certain performance criteria. During that phase, experimental conditions were similar for all subjects. Afterwards, subjects were assigned to one of the four psychophysical methods. They learned how to judge optimal SRTs using the respective method in trial block three and four. While performing the task, they assigned a certain portion of the total SRT of 13 s to each of the four 'gaps' between sub-tasks. Under each condition, subjects could earn a monetary reward, if the sum of partial SRTs was close to 13 s. All subjects earned the reward.

Under the method of adjustment, all SRTs controlled by the simulation software had an infinite duration. Subjects terminated SRTs through a mouseclick on a specific button. Mouseclick latencies were recorded as preferred SRTs. After serving a customer, these subjects always received feedback about the deviation from a total SRT of 13 s.

Under the up-method and the down-method, all SRTs were either prolonged or shortened by the simulation software after serving a customer. Subjects could stop the systematic change in duration of any SRT after each booking. For this purpose, a specific screenpage was displayed where SRTs could be determined through a mouseclick on an appropriate button if having the preferred duration. If all SRTs were set, those times were recorded and subjects were informed about the deviation from a total SRT of 13 s. Then the simulation software reset all SRTs to certain initial values, and the procedure restarted.

Under the graphic procedure, a particular screenpage was always presented after serving three customers. Subjects could change the length of four bars representing the duration of the different SRTs by using the mouse. The page could not be left until the sum of SRTs was 13 s. Resulting times were recorded.

In trial five, all subjects made their final estimation of optimal SRTs according to the method to which they were assigned. Afterwards, each of the four samples of SRT recordings per booking task and subject were averaged. If the sum of average SRTs was not 13 s, these were adjusted by linear interpolation. The results were considered as the individual optimized SRTs. During the last trial, all subjects performed the task under constant individual optimized SRTs according to their subdivision from trial five. Subjective ratings of well-being and mental workload were collected after each trial block. Physiological measures were made during a rest period and during the sixth trial. They involved nonspecific skin conductance responses, instantaneous heart rate, and power spectrum of heart rate.

Results and discussion

The fifth trial yielded three statistically different subdivisions of total SRT. Subjects who had been assigned to the method of adjustment or the up-method produced two of those distinct subdivisions of total SRT. Subjects using the graphic procedure and the down-method assigned similar portions of total SRT to corresponding sub-tasks. During the sixth trial, the latter two groups outperformed subjects who had optimized their SRTs by using the method of adjustment or the up-method (see table 1). Consequently, when considering profits from correctly served customers and costs from wrong and abortive bookings, SRT subdivisions resulting from the graphic procedure and the down-method are recommendable. The graphic procedure has a further advantage. It is self-evident. In comparison with the other methods, subjects will learn it easiest.

Table 1: Performance by psychophysical method applied to optimize SRTs (Means).

	method of adjustments	up-method of thresholds	down-method of thresholds	graphic procedure
customers served	43.9	46.5	47.9	50.0
wrong bookings	12.4	7.0	5.1	5.2

It can be inferred that all persons optimized subjective costs on expense of their performance. Such costs are related with self-perception and effort. As an average over all subjects, subjective ratings indicated a high level of well-being and a low level of mental workload during the sixth trial. At the same time, psychophysiological variables indicated a moderate level of arousal. Differences between groups were marginal. These results reflect the successful coping with deficits in temporal aspects of work concerning SRT subdivisions. Regardless whether such deficits were present or not, all subjects expended a similar amount of effort in performing the task, and they all felt comfortable while working.

However, similar efforts resulted in different levels of performance. In comparison with subjects working under an optimal subdivision of total SRT, subjects working under a sub-optimal subdivision served relatively few customers, and they made no particular efforts as to avoid errors. But this way of dealing with the task enabled them to keep subjective costs low.

After all, it can be concluded that poor performance during computer work can be prevented by optimizing SRT duration. Optimal SRTs will be obtained when using graphic representations of SRTs to be adjusted by the subject. This procedure is very simple to learn. Tools for graphic subdivision of total SRT can be easily programmed with each modern computer. They should be applied as standard tools for determining the optimal SRTs whenever developing interactive computer tasks. For achieving a high user satisfaction and a high efficiency of computer work, SRTs should always be optimized.

References

Kohlisch, O., & Kuhmann, W. (1997). System response time and readiness for task execution - the optimum duration of inter-task delays. *Ergonomics*, 40, 265-280.

Luce, R. D. & Krumhansl, C. L. (1988). Measurement, Scaling and Psychphysics. In Atkinson R. C., Herrnstein R. J., Lindzey G. & Luce R. D. (Eds.), *Stevens' Handbook of Experimental Psychology. Vol. 1: Perception and Motivation*, pp. 3-74. Chichester: Wiley.

Schaefer, F., & Kohlisch, O. (1995). The effect of anticipatory mismatch in work flow on task performance and event related brain potentials. In Grieco A., Molteni G., Piccoli B., & Occhipinti E. (Eds.), *Work with display units 94*, pp. 241-245. Amsterdam: Elsevier.

Thum, M., Boucsein, W., Kuhmann, W., & Ray, W. J. (1995). Standardized task strain and system response times in human-computer interaction. *Ergonomics*, 38, 1342-1351.

Author's present address: *IfADo* Institut fuer Arbeitsphysiologie, Ardeystrasse 67, D-44139 Dortmund, Germany. *Email*: kohlisch@arb-phys.uni-dortmund.de

Office Ergonomic Interventions

Michelle M. Robertson
Liberty Mutual Research Center
Hopkinton, MA 01748

1 Introduction

Office ergonomics intervention programs serve as a powerful means of enhancing organizational effectiveness. When a successful office ergonomic intervention program is implemented, the result is an increased ability for the worker to change his/her work environment. Controlling the work environment leads to enhanced organizational effectiveness in several ways. Ergonomic interventions including work system design changes can result in lower worker's compensation rates, loss work days, and reduced injury and illness rates. Intervention programs that include workplace design, job design, work organization and training programs can influence how workers use and interact with the work environment and organization. High quality ergonomic intervention programs incorporate a "participatory" approach, in which end users, managers and others are involved. It is this participatory aspect, along with ergonomic intervention programs, that forms the basis for creating an improvement organizational and work environment and continual change in the company. In this paper, we discuss two case studies that illustrate workplace design, the concepts of learning, participation, and subsequent change, due to ergonomics intervention programs. In each case study we illustrate how an ergonomic intervention program positively affects individual interaction with the workplace; changes that occurred to the workplace; and how the organization demonstrated the impact of the training on organizational performance measures.

2 Importance of Ergonomic Intervention Programs

A successful office ergonomic intervention program, should be incorporated into the overall organizational strategic plan for health and stress reduction programs. The linking of the ergonomic program objectives with the organizational policies and health prevention goals establishes the importance and commitment aimed at reducing negative health effects, e.g., work related musculoskeletal disorders (WMSDs). Rarely are interventions evaluated for their effectiveness and even more so for their cost-effectiveness (Silverstein, 1987; Kilbom, 1988). It is important to establish an evaluation plan to based on organization performance measures. Such an evaluation plan will be described further in this paper.

3 Phases of an Ergonomic Intervention Evaluation Model

The following intervention evaluation model is modified from a training evaluation model (Robertson & Robinson, 1995). When the workspace design and training objectives are identified, specific evaluation assessments of each intervention effectiveness must be defined and established. The model described below includes ergonomic intervention programs that involve office workspace redesign, office ergonomic training, and ergonomic workstation evaluation..

3.1 Four Level Evaluation Model

The four level evaluation model includes: Level I: User Reaction to the workspace design and to the ergonomic training program; Level II: Knowledge gained from ergonomic training program; Level III: Behavior changes relevant to office ergonomic practices; and Level IV: Organizational performance results. For evaluating an ergonomic intervention program that includes workspace redesign and training, we suggest incorporating some of the following measurements at each of the evaluation levels. For Level I measurements may include a post-training questionnaire asking the usefulness and value of the training. Also, a workplace satisfaction survey could be administer to measure the user's perceived satisfaction of the work environment in supporting their needs. For Level II: various measurements may be taken *prior* to the ergonomic interventions, such as an observational and objective analysis of posture and work habits, musculoskeletal discomfort and symptoms surveys, and a pre-knowledge VDT ergonomic training test. A *post-ergonomic intervention* discomfort and symptom survey; work environment satisfaction survey, which are identical to the pre-intervention survey measures, and a *post-training* ergonomic knowledge test should be given. Changes in gain scores may reflect the immediate effect of the intervention. For Level III: various measurements may be taken *after* the ergonomic interventions in order to evaluate the behavior changes of the users and the transfer of the ergonomic knowledge to the workplace. These measures may include an observational and objective ergonomic analysis, interviews of what the user changed in their workplace as a result of the training, and open ended survey questions. These questions ask the user how they are going to *use* the training when they return to work. For Level IV, various organizational and safety performance measurements may be taken related to the results of intensive VDT work, such as reported CTDs and WMSDs, time off work, worker's compensation rates, and health and stress related costs. Pre and post training organizational health and safety performance measurements should be benchmark and tracked over time as well as tested for significant changes overtime (e.g., time series analysis). Tracking these pre and post organizational and safety performance measures, as well as other training costs, determines the basic variables for calculating a Return on Investment of the ergonomic intervention programs.

4 Ergonomic Intervention Programs

Two computer intensive technology companies designed, developed, implemented and evaluated their office ergonomic intervention programs. These programs includes workspace redesign, ergonomic training and workstation evaluation. The goals of these intervention programs were to; 1. Reduce adverse health effects from VDT work; 2. Impart knowledge about how to effectively use their new ergonomically designed VDT workstations; and 3. Demonstrate a positive effect on performance. The content of these office ergonomic training programs included: definition of ergonomics, basic physiology of the upper extremities, causes of discomforts and injuries, ergonomic principles regarding workstation layout, techniques on how to adjust and use their workstations properly, recommendations for analyzing the employees workstation, procedures to follow when they feel uncomfortable (management's policies on who to contact), and relaxation and exercise techniques to relieve VDT stress. For company 1, a three tiered training program was designed for senior managers, supervisors and employees (Robertson & Robinson, 1995). Each of these training programs included specific content areas that were related to their job responsibilities including how they were expected to respond when employees approached them with ergonomic problems. This allows the opportunity to support interaction and participation between employees and managers to support the changes in the work environment. For company 2, a similar approach was taken however there was more focus on how to allow for a high level of user control as the ergonomic workstations were designed with the intent to allow for high flexibility and mobility (Robertson & Robinson, 1995). Managers involved in this training program were faced with different change issues as employees were applying their ergonomic knowledge in a more systematic manner by re-arranging their workstations and work environment components. In both companies, ergonomically designed workstations were provided with a moderate to high level of workspace adjustability and flexibility allowing for a high level of user control. Furthermore, onsite individual ergonomic evaluation of the user workstations were conducted when an user expressed a need to have some ergonomic assistance. These ergonomic evaluation were performed by trained ergonomists and they involve an objective evaluation of the workstation, review of the training materials, and discussion of any work related musculoskeletal discomforts.

4.1 Office Ergonomic Intervention Program Results

Overall, positive effects of the office ergonomic intervention programs for both companies were demonstrated. For company 1, all four levels of the evaluation model were applied and clearly documented. Significant results of the ergonomic program was accomplished for all 2500 individuals included in the program. Company 2 demonstrated positive results for three of the evaluation levels and results of the fourth level showed positive results in that there was a significant decrease in self-reported musculoskeletal discomforts for the twenty individuals involved in the program. For Level I evaluation, the workers from both companies rated the training highly favorable, useful, informative, and relevant to the workstation design. For Level II evaluation, there were significant changes in the amount of knowledge gained by the worker's concerning VDT ergonomics as measured by self-reporting knowledge gained scores for company 1 and 2.

For Level III evaluation, significant positive behavioral changes of the workers as measured by self-reported behavioral changes and observed changes were found and reported for both companies. For company 1, over 80% of the employees reported that they had applied the ergonomic knowledge to their jobs, and included the correct placement of the VDT screen, the position of their wrists at the workstation, and their sitting posture. Follow-up observations and interviews by the corporate ergonomist confirmed these self-reported behavioral changes. For company 2, all participants exhibited a high level of user control as they continually adjusted their workstations to meet various job demands. This involved arranging the workstation for sitting and standing postures. In addition, a significant decrease in overall discomfort was found in this group. Follow-up interviews and observations by the ergonomist supported these results, as over 80% of the trainees said that they were able to apply many of the principles taught in class to their workplace. Of the changes to the workplace reported, most were adjustments to the chair, placement of the monitor, workstations configuration and layout, and height adjustments of the keyboard and working surfaces. Many of the participants reported that the awareness developed from training led to changes in posture and an increase in the number of breaks for exercises or movement. In company 1, there was some ability to change the workstation configuration within some defined constraints which provided some means of user control at the individual level. For company 2, the workstations were highly mobile and groups of workers were able to change not only the configuration of their individual workstation, but the overall configuration of the group's work area. One manager was very supportive and encouraged his workgroup to change their workstations configurations within certain boundaries to support their varying job processes.

For Level IV evaluation, a Return on Investment analysis for the office ergonomic intervention for company 1 revealed that the program resulted in a positive payback for the workers and their companies. A significant decrease in work related musculoskeletal disorders and overall loss work days for company 1 resulted. For company 2, as the user's knowledge of control of their workstation increased reports of stress decreased and a reduction in musculoskeletal discomforts was reported.

Other successful components of the intervention programs were the commitment by top management to the ergonomic program itself, active involvement of the employees, positive response by management to employee's requests regarding VDT workstation redesign or reconfiguration, and continuous support of management in applying the VDT ergonomic principles to the work environment.

5 Conclusion

In these two studies, overall positive effects of the office ergonomic intervention programs were demonstrated for each of the program evaluation levels. These studies also support the contention that the combination of user control, participation and training are important--- since they can provide the worker with a high degree of control through an increased knowledge of office ergonomics and the ability to effectively apply ergonomic principles to their office work environment. With an increase in office ergonomic knowledge as well as implementation of an evaluation process the value of ergonomic interventions can be demonstrated. A well designed ergonomic program, coupled with an ergonomically designed work environment provides the foundation for creating a responsive environment for the employee and manager. As these individuals interact with one another, applying and seeking ergonomic solutions, a sense of participation is created forming the basis of a positive change management process. These two case studies show that a work system approach decreases health and stress effects as well as a reduction in ergonomic problems associated with WMSDs. Ergonomic workstation design, workstation evaluations, and training can be successful intervention strategies at both the individual and organizational level in preventing WMSDs, related VDT job stress and enhancing organizational effectiveness.

6 Selected References

Robertson, M.M. & Robinson, M. (1995). Enhancing user control of VDT work environments: Training as the vehicle. *Proceedings of the 39th Annual Meeting of the Human Factors and Ergonomics Society*,(pp. 417-421),Santa Monica, CA.

Silverstein, B.A., (1987), Evaluation of interventions for control of cumulative trauma disorders, in ACGIH (Ed.) *Ergonomic Interventions to Prevent Musculoskeletal Injuries in Industry*, pp. 87-99, Chelsea, MI: Lewis Publishers. (Industrial Hygiene Science Series, 2).

PART 2

USER INTERFACES AND INTERACTIONS

Extension of the IdeaBoard for Educational Applications

Takeshi Sakurada, Hirokazu Bandoh and Masaki Nakagawa
Dept. of Computer Science, Tokyo Univ. of Agri. & Tech.
Naka-cho 2-24-16, Koganei, Tokyo, 184-8588, Japan
phone: +81-423-88-7144, fax: +81-423-87-4604
e-mail: takes@hands.ei.tuat.ac.jp

1 Introduction

Since computer literacy education has commenced in schools, the number of personal computers installed at each school is increasing. However, the way in which students use the computer seems to have some intrinsic problems. The students get absorbed in their own matter and do not concentrate on what their teacher is saying. Moreover, it is difficult for the teacher to see the students' faces because they are hidden by displays. Although information processing is available at hand, a PC for each student may destroy the interactive and collaborative relation between the teacher and the students which has been inherent in a class room with a blackboard where the teacher can focus the students' attention on her/his writing on the blackboard.

Although, exercises and self-study with the computer are indispensable for the students to learn for themselves at their own pace, the necessity of combining the information processing power of the computer with the potential advantages of the blackboard is not lessened.

We have made an interactive electronic whiteboard system named "IdeaBoard" (interactive, dynamic, electronic assistant board) similar to Liveboard (Elord et al. 1992), and developed new human interaction techniques suitable for a large surface with an electronic marker as well as some educational applications which exploit the system (Nakagawa et al. 1997).

This paper presents two extensions to the IdeaBoard. One extension is to separate the control information from the contents of the board. The second extension is to connect multiple boards.

2 Computerization of a Whiteboard

2.1 Separation of Control Information

One of the design guidelines of the IdeaBoard was the consistency with the desktop GUI. This makes computer users feel comfortable to use the IdeaBoard. Moreover, it is advantageous to show computer operations for computer literacy education. On the other hand, the design policy that seeks to hide the computer operations from the users is the other choice which may be effective for teaching general subjects like languages, social studies and so on. Therefore, we try to computerize the blackboard rather than extending the desktop GUI.

So far, both of the contents and the control information such as icons and menus have been displayed on the IdeaBoard as on the desktop monitor. The control information includes not only the above-mentioned items that control the system, but also scenarios of lectures that are used by the teacher. It can be displayed over the contents (Figure 1) or side by side with the contents (Figure 2). In either case, the control information occupies some area and decreases the display area for the contents. It may also hinder the students from focusing on the contents. Moreover, the scenarios of lectures that had better be hidden from the students appear on the board.

To show control information to the teacher while reserving the largest area for the contents, each type of information should be displayed on different screens rather than on the same screen. Therefore, we designed a system as shown in Figure 3, which separates the display for contents from the display for control information. On the control information screen, status of the system as well as menus and buttons for controlling the system are displayed. It may also display the teachers' scenarios of lectures. By separating the two types of information, the control information has been removed from the IdeaBoard. Thus, all the area on the IdeaBoard is reserved for the contents and the students' attention can be focused on the contents without being bothered by the computer operations.

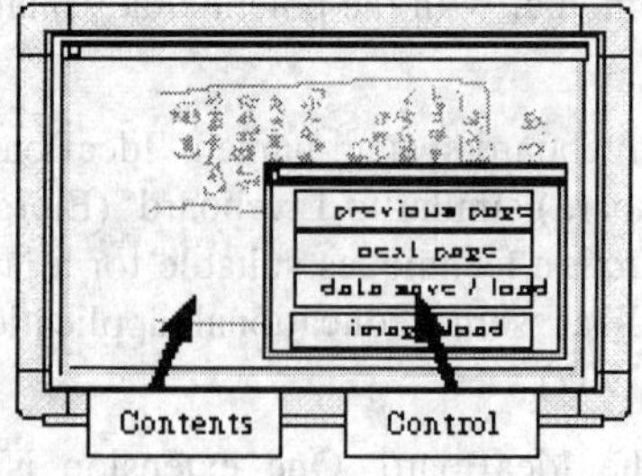

Figure1. Control information overlaid on contents.

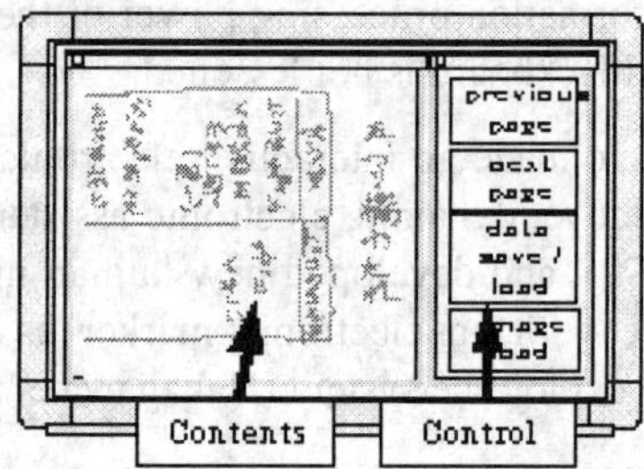

Figure2. Contents and control information displayed side by side.

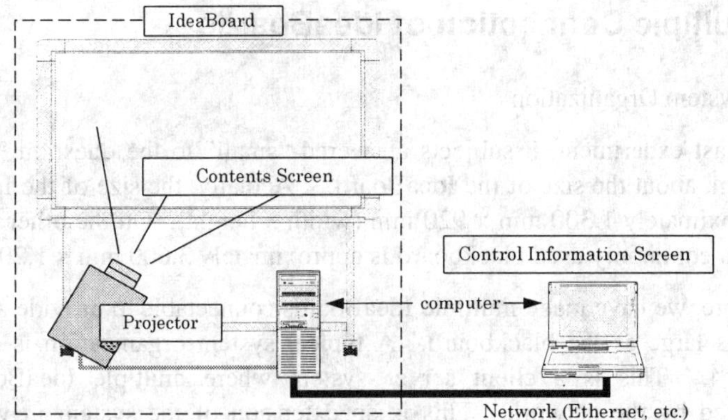

Figure 3. Separating a display for contents from a display for control information.

2.2 Preliminary Evaluation of Separating Control Information from Contents

To find whether complete separation of control information is effective or not, a preliminary evaluation has been done by preparing two types of systems: a system ***A*** in which everything is operated on the IdeaBoard without using the control information screen, and a system ***B*** in which the control information is completely separated from the IdeaBoard. The system ***B*** only accepts writing on the board with an electronic marker. Controls such as loading of pre-stored handwritten data, and storing of data must be done on the control information screen. We interviewed ten students (subjects) after they used the above two systems. A summary of their opinions is shown in Table 1.

Table 1. Subjective opinions on the two types of systems.

Systems	No. of subjects who prefer the system ***A***	No. of subjects who prefer the system ***B***
Easy to operate.	6	4
Easy to read contents.	4	6
Can concentrate on contents.	2	8

From Table 1, we see that the students can better concentrate on the contents with the system ***B*** rather than with the system ***A***. On the other hand, to operate the system ***B*** people have to move to the control information display every time some operation needs to be performed and this makes the system uneasy to use. Moreover, the audience do not understand what has happened when the contents are changed suddenly because they do not know what the teacher has done on the control information display.

From the above, we can infer that simple separation of the control information is not necessarily appropriate. When the contents are changed, system controls had better be displayed on the contents screen while when the contents are not changed drastically, they may be issued on the control information screen.

3 Multiple Connection of IdeaBoards

3.1 System Organization

In the last experiment, 8 subjects answered "small" to the question "What do you think about the size of the IdeaBoard?" Actually, the size of the IdeaBoard is approximately 1,300 mm x 920 mm (width x height). On the other hand, the size of a commonly used blackboard is approximately 3,600 mm x 1,200 mm.

Therefore, we have made multiple IdeaBoards connectable to provide a working space as large as the blackboard. A typical system organization is shown in Figure 4. This is a client server system where multiple IdeaBoards are connected by the network. This is an extension of the system in which the control information is separated and control information can be displayed on the server.

3.2 Transfer of the Displayed Contents on the Whiteboard

The written contents on each IdeaBoard are transferred to the next IdeaBoard by this function. On a conventional blackboard, when we write something up to the end of the blackboard, we usually erase material at the other end and then start writing again. This way of writing is inconvenient because we cannot write a series of contents continuously. On the connected IdeaBoards, therefore, a function called "transfer of the displayed contents on the IdeaBoard" is provided so that the contents can be continuously written by shifting one frame of contents to the next IdeaBoard. As shown in Figure 5, after the contents C_3 are written on the IdeaBoard I_3, the contents C_2 are moved to the IdeaBoard I_1, then the contents C_3 are moved to the whiteboard I_2 and the display area of the IdeaBoard I_3 is cleared. Then, new contents C_4 can be written next to C_3. Of course, C_1 can be reloaded when necessary.

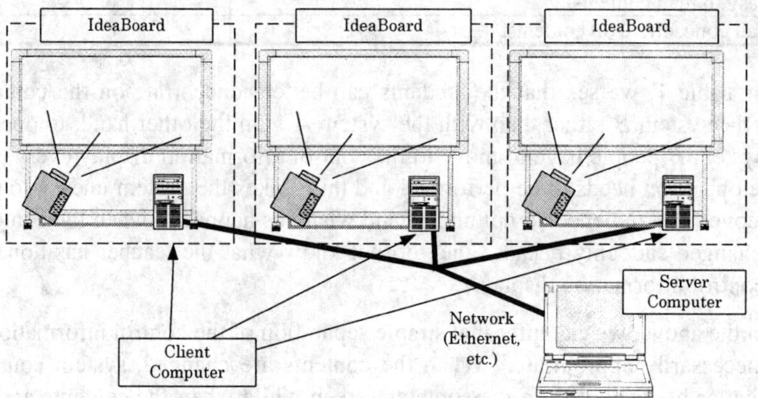

Figure 4. System Organization.

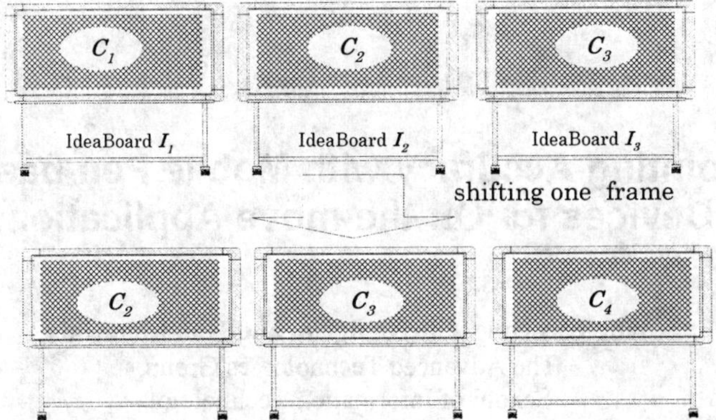

Figure 5. Transfer of the displayed contents on the IdeaBoards.

3.3 Use for Remote Classrooms

The system can be used not only for one classroom by placing multiple IdeaBoards, but also for remote classrooms where IdeaBoard(s) at each site displays synchronized contents with the other IdeaBoard(s).

4 Conclusion

This paper has presented two extensions to our interactive electronic whiteboard system named IdeaBoard. One extension is to separate the control information from the contents of the board so that students are not bothered by the control information. Preliminary evaluation has shown that hiding control enables the audience to focus their attention on the contents, although it may prevent them from predicting what is going to happen. The second extension is to connect multiple electronic whiteboards. This allows us to realize a computerized surface as large as real blackboards. Moreover, it may provide collaborative usage between remote sites. More rigorous evaluation remains to be done.

5 References

Elord, S., Bruce, R., Gold, R., Goldberg, D., Halasz, F., Janssen, W., Lee, D., McCall, K., Pedersen, E., Pier, K., Tang, J. and Welch, B. (1992) Liveboard: a large interactive display supporting group meetings, presentations and remote collaboration. *Proc. CHI'92*, 599-607.

Nakagawa, M., Oguni, T. and Yoshino, T. (1997). Human interface and applications on IdeaBoard, *Proc. IFIP TC13 Int'l Conf. on Human-Computer Interaction*, Sydney, pp.501-508.

Pointing Accuracy with Mobile Pen-based Devices for On-the-move Applications

Ying K. Leung, Chris Pilgrim and Kon Mouzakis,
The Advanced Technologies Group,
School of Information Technology,
Swinburne University of Technology,
P.O. Box 218, Victoria 3122,
Australia

1 Introduction

The popularity of mobile computing devices is mainly attributed to two important features: portability and connectivity. These features bring about many new applications, enabling the user to perform useful work on the move. Pen-based devices in particular are gaining wide acceptance because of their natural physical interface and the ability for the user to input data without a keyboard.

Despite the diverse application domain of pen-based devices, there are only a limited number of selection strategies available for user interaction (Ren and Moriya, 1997). The use of the pen (or the pointing device) to select an item on a menu or a check box is the most common action taken by the user with these mobile devices. The scarcity of screen space of mobile devices is invariably one of the most constraining factors confronting the interface designer, as the designer often has to compromise between visibility of interface controls and the number of controls that can be displayed. For applications involving the user interacting with the device on-the-move, the lack of screen space places another level of complexity in the user interface design as pointing accuracy invariably diminishes.

Whilst Fitts' law (Fitts 1954) provides a useful basis for estimating user performance in pointing tasks, more specific design guidelines are needed to deal with situations involving the user on the move. Given the inherent constraint of limited available screen space, design guidelines on the physical size of interface controls and their placement would be very useful. This paper describes an experimental study of pointing accuracy with pen-based device for on-the-move applications. The results of this experiment provide practical guidelines to the user interface designer regarding the placement and size of the interface controls for such applications.

2 Experiment

2.1 Subjects

Twenty-five undergraduate and postgraduate students and staff members from Swinburne University, both males and females, participated in this experiment. The age of the subjects ranged from 18 years to 45 years.

2.2 Procedure

A motorised treadmill running at 4 feet per seconds and an Apple Newton 120 PDA device were used for the experiment. A program was written to generate a series of targets for user interaction, and information on the points of interaction and reaction times were captured. Before the data was collected on the treadmill, the subject calibrated the Apple Newton with the pointing device. The subject then used the treadmill for about half a minute for acclimatisation, before the thirty targets were presented one after the other on the Newton PDA.

The Apple Newton was held in portrait mode for interaction. Each target was a vertical cross of 2 pixels thick and 18 pixels in length and width, presented randomly over a blank screen. The Apple Newton has an effective screen area of 240 pixels by 320 pixels in portrait mode, with a screen resolution of about 85 dots per inch. The experiment took about 5 minutes to complete. The subject was asked to point to the target as accurately and quickly as possible. Subjects' performance data in terms of pointing accuracy and response time were recorded by the Apple Newton.

	Error in X direction (pixels)	Error in Y direction (pixels)	Response Time (seconds)
Mean	4.386	4.501	1.205
Standard Dev	3.684	3.857	2.228

Table 1 Summary of results collected from 25 subjects

3 Results and Discussion

The results of the 25 subjects collected in the experiment are summarised in Table 1 above as follows:

Using a two-tailed t test (α=0.05), it can be shown that the difference in subjects' accuracy in the X and Y dimensions is not significant (t=-0.401, df=24).

To examine the effect the different parts of the display has on user performance, the screen was arbitrarily divided into two regions, the centre region and the outer region. The centre region, which covers about 30% of the entire display area, has an area of 150 pixels and 150 pixels in centre of the display. More specifically, we compared user performance in radial error, response time and the number of errors made, for targets in each region. An error was registered when the radial distance was more than 10 pixels away from the centre of the target. The results are summarised in Table 2 as follows:

Centre Region	Radial Error[1] (pixels)	Response Time (seconds)	No. of Errors (out of 30 trials)
Mean	7.049	1.220	1.64
Standard Dev	5.768	.378	3.277

(2a)

Outer Region	Radial Error (pixels)	Response Time (seconds)	No. of Errors (out of 30 trials)
Mean	6.894	1.189	2.28
Standard Dev	7.430	0.155	3.703

(2b)

Table 2 Summary of results for (a) the centre region and (b) the outer region

Using a two tailed test (α=0.05), the difference in subject performance in terms of radial error (t=0.132, df = 24) and response time (t=0.424, df = 24) is not significant between the centre and outer regions. However, there is significant difference in the number of errors made by the subjects (t=-2.691, df = 24, $p<0.05$).

[1] Distance from the centre of target

The results of this experiment showed that whilst the subject can operate the pen accurately in both X and Y dimensions, the placement of the interface controls should be given careful consideration. Interface controls should not be placed on the periphery of the screen; in this study, subjects tended to sacrifice accuracy for speed when pointing to targets in the peripheral area of the screen.

In this experiment, the mean overall error rate was just under 2 per 30 interactions for a target size of 18 x 18 pixels (about 0.21" x 0.21") at a walking pace of 4 feet/second with a mean response time of 1.205 seconds. Whilst the resolution of the display screen varies from system to system, it should be pointed out that it is the physical size of the interface control that is important. The user needs to be able to see the target to point at it accurately. The results from this experiment can be used as a basis for design tradeoffs: response time and accuracy may be compromised by walking speed and/or target size.

Fitts (1954) defined an Index of Movement Difficulty (ID) and the Movement Time (MT) as:

$$ID(bits) = \log\left(\frac{2A}{W}\right) \qquad and \qquad MT(\text{sec}) = k_m + C_m.ID$$

where A is the intended movement, W is the width of the target, k_m is a delay constant that depends upon the body member, and C_m is a measure of information handling capacity in seconds/bit. For hand movements, C_m has a typical value of 0.177 seconds/bit. In the context of pen-based devices for on-the-move applications, it is envisaged that the k_m term will be more dominant because of the additional time required for the user to move the pen to the target while walking. Consequently, the C_m term will be relatively less dominant, given the pen movement is typically small; in this experiment the maximum pen movement was about 9 cms.

To develop a preliminary refined model of Fitts' law for on-the-move applications, the mean values for ID have been calculated for each of the 30 targets. W is fixed at 18 pixels while A has been estimated to be the radial distance between the previous target and the current target. The vertical distance involved in moving the pen from one target to the next has been estimated to be negligible.

Only the mean values of ID and MT for on-target interactions have been included in the calculation. Figure 1 shows a scattered plot of the MT versus ID. Using a linear regression analysis, k_m and C_m are determined to be 4.95 and –1.50 respectively, by the method of least squares.

The negative value of C_m suggests that the longer the hand movement the quicker the response time. It appears that Fitts' law may not readily be applied to mobile devices for on-the-move applications. This may be because for a long

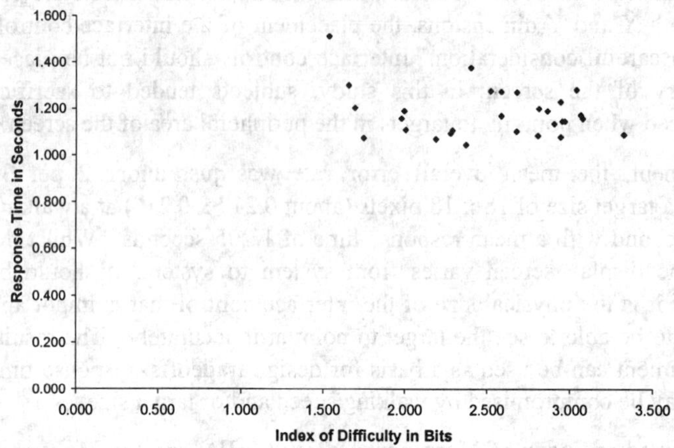

Figure 1. A scatter plot of response time versus index of difficulty.

distance target the movement strategy is to move quickly first and then adjust direction and speed, while for a short distance target, the strategy involves slow movement and direction correction simultaneously. Furthermore, there are two distinctive differences between Fitts' law and the way the user interacts with the mobile device. First, Fitts' law assumes the user interacts with a fixed target, while in this experiment, the mobile device is invariably affected by the movement of the subject. Second, the subject may move the hand-held device and the pen to coordinate the targeting action.

This experiment has provided the user interface designer with useful information about the placement and size of interface controls for pen-based interfaces. Further experimental work on walking speed and target size is required to derive a more comprehensive model of how users interact with these devices for on-the-move applications. Fitts' law can then be modified and applied to a broader domain.

4 References

Fitts, P.M. (1954), The information capacity of the human motor system in controlling amplitude of movement, Journal of Experimental Psychology, Vol 47, No. 6, pp381-391.

Ren, X. & Moriya, S. (1997), Selection strategies for small targets and the smallest maximum target size on pen-based systems, IEICE Transactions on Information and Systems, Vol E00-D No 7, 1-6.

OCR-Oriented Automatic Document Orientation Correcting Technique for Mobile Image Scanners

Kenichiro Sakai, Tsugio Noda
Fujitsu Laboratories, Ltd.
E-mail: ksakai@flab.fujitsu.co.jp, noda@flab.fujitsu.co.jp

1 Introduction

The progressive miniaturization of notebook personal computers and the spread of communication tools such as portable phones, e-mail, and the Internet have accelerated the mobile use of notebook personal computers. This enables us to gather digital image data and transmit it to another person or appliances in the mobile environment.

The use of the digital still camera has been increasingly widespread all over the world. It is suitable for taking a picture of a 3D object, e.g. landscape, but is unsuitable for capturing paper documents because of the lack of scanning resolution. Furthermore, the demands for clipping and collecting document information outside the home and office have been dramatically increasing in recent years.

2 Background

To meet the increased demands for clipping and collecting articles from any type of paper document such as newspapers, magazines, and business papers even when we are away from home or the office, we developed a high-performance mobile image scanner, called “Pen Scanner”, which is extremely small (183(W) x 17.6(D) x 14.7(H) mm) and lightweight (80g), and scans documents with high resolution (400dpi) and at high speed (less than 3sec/A6).

In this development, we aimed to achieve orientation-less easy scanning, which enables users to scan a document in any direction they like without any instructions to the computer. We would like to introduce an improved technique, focused upon the user interface of the mobile image scanner.

3 Subject

Because the mobile image scanner is very small and lightweight as mentioned above, the user can scan any document in any direction, e.g. from top to bottom or left to right, according to their liking or the kind of documents to be scanned. This is very convenient for the user, but this causes one serious problem.

The problem is that the scanned image may be rotated 90, 180 or 270 degrees and/or mirrored horizontally according to the scanning directions. There are eight patterns of scanned image according to the combination of rotation and mirroring even if the same part of the document is scanned. Therefore, in conventional handheld image scanners, users have to instruct the application program in what direction they are going to manipulate the scanner before scanning or they have to rotate and/or transpose the scanned image after performing every scanning operation. These operations may irritate users.

To solve this operational problem and to make the scanning operation easier, we developed a new algorithm to detect the orientation of the document image, and implemented an auto-correction technique.

4 Algorithm

As shown in Figure.1, an input character image (a) extracted from a scanned document image is rotated by 90, 180, and 270 degrees (b through d), and the input character image and above three rotated images are mirrored horizontally (e through h). OCR processing is executed for those eight images to obtain the degree of recognition certainty for each image. The orientation of the character image with the highest degree of recognition certainty is taken as the correct orientation (e).

Then the correct document image is obtained by converting the scanned document image to this detected orientation. For example, in Figure 1, detected

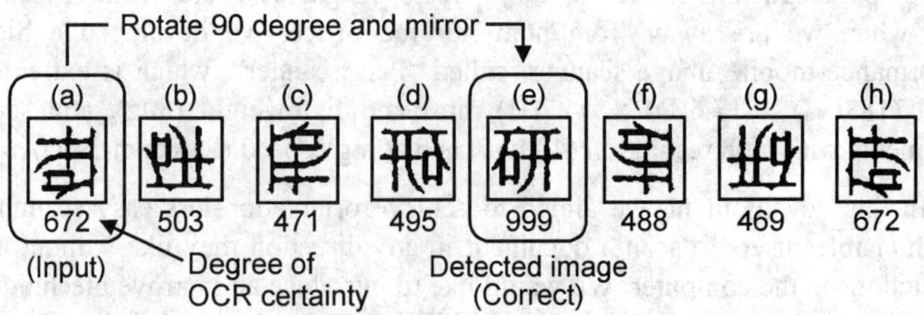

Fig.1 Principle of detecting correct character image orientation.

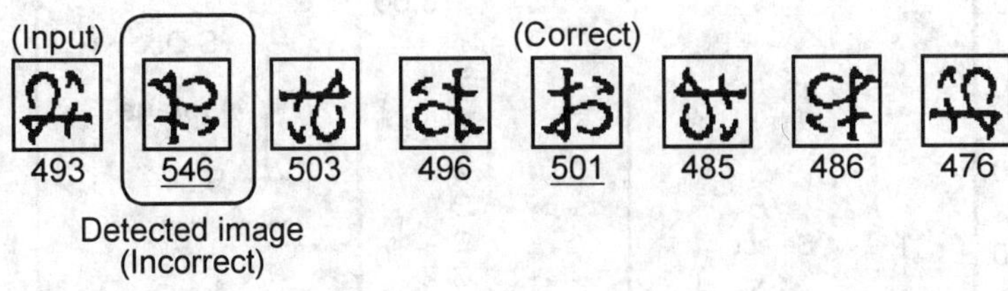

Fig.2 Example of a character image in which OCR certainty is low.

image (e), which image is rotated 90 degrees and mirrored horizontally, and the input image (a) are the same. So the correct document image is obtained by rotating 270 degrees and mirroring the scanned document image.

While we were verifying above algorithm, we found two cases of orientation detection failure caused by some kind of character images. So we developed the character image selecting method in order to improve the accuracy of detecting orientation.

The first failing case is caused by a character image in which OCR certainty is low. For example, as shown in Figure 2, OCR certainty has the highest value for the wrong orientation and orientation detection fails. We investigated the average degree of OCR certainty of each kind of character such as Kanji (i.e. Chinese character), Hiragana and Katakana in Japanese, Alphabet and Symbol. As a result of this investigation, as shown in Table 1, OCR certainty of Kanji is higher than any other kind of character. By using Kanji character images for detecting the orientation of the character image, increased accuracy of orientation detection will be expected.

However, whether the kind of character image is Kanji or not isn't clear before OCR processing is executed. So we contrived a process that estimates the likelihood that the extracted character is Kanji or not. In this process, "percentage of black pixels" and "the ratio of long side to short side" of the character image are compared with predetermined thresholds shown in Figure 3. The character image with percentage of black pixel over 20% and the ratio of long side to short side below 1.5 is selected as a Kanji character image. This process improves the accuracy of orientation detection from 83% to 97%.

Table.1 The degree of OCR certainty.

Kind of character	OCR certainty
Kanji	946
Hiragana	874
Katakana	885
Alphabetic	854
Symbol	735

(Average)

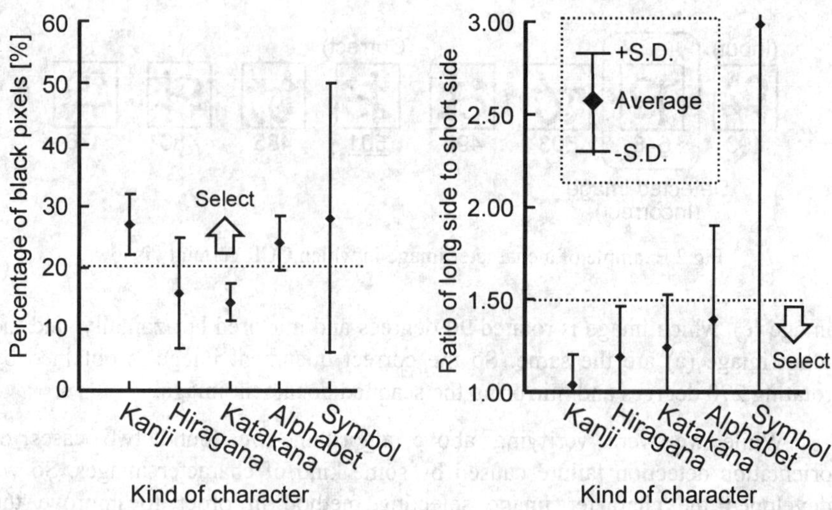

Fig.3 Characteristics of character images.

The second failing case is caused by a character image which shape is symmetrical as shown in Figure 4. There are many symmetrical Kanji characters the same as Alphabetic characters such as 'A', 'D', 'Z'. The OCR processing recognized the symmetrical scanned character image and its mirrored image as the same character, and the degree of OCR certainty of the mirrored image is almost the same value as that of the scanned character image. As a result, orientation/mirroring detection may fail.

To avoid this problem, we provided a checking process to be used after OCR processing has been done. In this step, the eight degrees of OCR certainty of all rotated and mirrored character images are compared with a predetermined threshold, and if there are two or more degrees over the threshold, this character image is rejected, not to be used for orientation detection. This process improves the accuracy of orientation detection from 97% to 99%.

Fig.4 Examples of a character image in which shape is symmetrical.

By applying the above processes, two characters that have been detected as the same orientation are selected, and the orientation of the scanned document image is determined to process the same orientation as these two character images. As a result of this, the accuracy of orientation detection is improved 100%.

5 Conclusion

Applying these algorithms to sample documents chosen from magazines, newspaper and other ordinary paper documents, correction of document orientation achieves 100% accuracy. This technique can be applied to any document image, except for picture or graphical images which don't include any characters in the image. The correction processing takes about 0.4 seconds on an ordinary personal computer with an MMX Pentium-233 MHz CPU. It is short enough for practical use.

This technique dramatically reduces user operations on the computer before and after scanning, and makes scanning operation easier, so that user can manipulate the scanner in any direction and always obtain the correct document image.

6 References

Kamada, H, Fujimoto, K. & Akimoto, H. (1997). High-speed, High-accuracy Document Recognition Method using Compressed Character Features and Original Character Features. *Technical Repot of IEICE*, PRMU96-205, 97-104.

Fujimoto, K. & Kamada, H. (1996). Fast and Precise Character Recognition by Estimating Recognition Probability. *Proceedings of the 1996 Engineering Science Society Conference of IEICE*, D-361, 364.

An Easy Internet Browsing System with Sophisticated Human Interface Using 3-D Graphics

Kimiyoshi Machii, Shigeru Matsuo, Yoshitaka Atarashi, Satoko Kojima*, Masahiro Yaezawa*, Kazuyoshi Koga, Yasuhiro Nakatsuka and Koyo Katsura
Hitachi Research Laboratory, *Design Center, Hitachi Ltd., Japan.

1 Introduction

Recently, the number of internet users is increasing. Messages are sent by e-mail and information is gotten from the WWW (World Wide Web) by HTTP (HyperText Transfer Protocol). Although most users do these tasks with a personal computer (PC) on their desks, the desire for "non-desktop" computers is also increasing. With "non-desktop" computers, people will be able to access the internet while sitting in their cars or walking around town. These situations will be realized soon by high-speed communication technologies.

On the other hand, several electronic products for accessing the internet have been developed. Some of them are special for the internet, others have embedded add-on functions to access the internet. These are home game machines, mobile telephones, car navigation systems and so on. We focused on a "non-desktop" internet browsing system as a successor to car navigation systems.

As the market grows, functions of car navigation systems are improving. Since 1997, network services for cars have become available. Using the services, drivers can get area information about restaurants, parking areas, amusement parks and so on. However, these services have been supplementary to the car navigation function. We think it is now time for such services to become a main function of car information systems. A network-based system does not need any external media such as CDs or DVD-ROMs whose contents are soon out of date. Besides, information got from the network is not the only thing provided, maps and/or routes to the target place are supplied.

Thus, we developed an internet browsing system for cars. It consists of terminals and a server for the information services. The terminal has a compact sized body and graphical user interface function using 3-D graphics technology. This paper describes the concepts and implementation of the system.

2 Prototype of the Internet Browsing System

2.1 System Configuration

Figure 1 shows the system configuration. The information sites are on the internet. They can be linked to the terminal through a provider site and a cellular phone. The terminal is set on a dashboard and it is connected to the cellular phone. The power supply and display for the terminal are already installed in the car. The user controls the system with a wireless device.

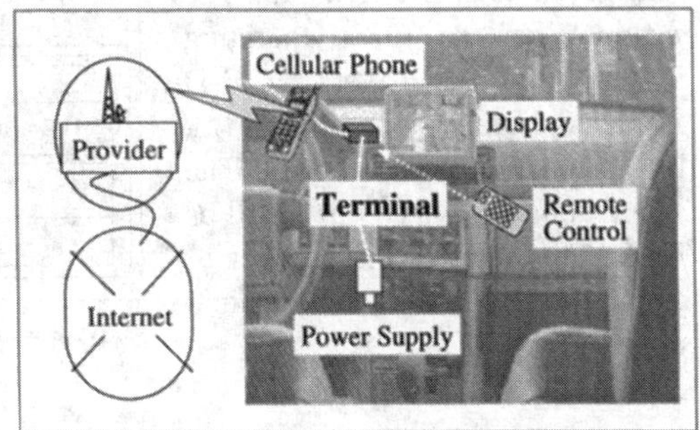

Figure 1 : System configuration

2.2 Simple and Small Sized Terminal

The internet terminal in a car should be compact so it can be set on a small area such as a dashboard. It also should not cost much. However, existing car navigation systems with networking capability are large and expensive. To solve the problems, we developed a low-cost terminal using a graphics processor "Q2i" (Quick 2-D Renderer Improved) (Matsuo, et al. 1997).

Figure 2 shows the hardware configuration and photograph of the terminal board. The terminal consists of a CPU "SH3" (Super H RISC Engine 3), a graphics processor "Q2i", a 2MB flash memory, a 512KB ROM, an IrDA port and 2 serial ports. They are placed on a card-sized board, with about 10-centimeter width and length.

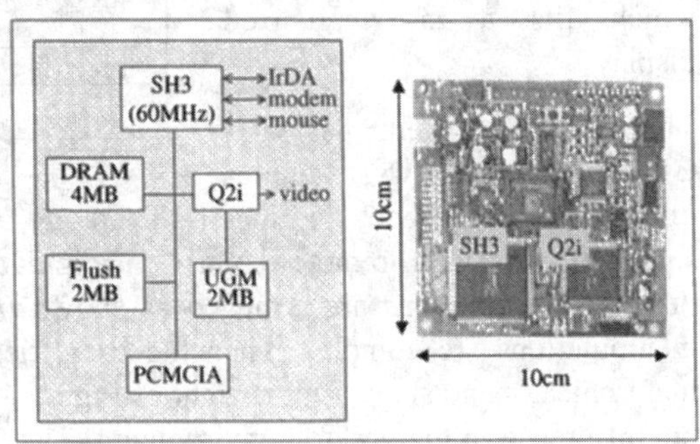

Figure 2 : Hardware configuration

We could realize such a small size for the following reason. Existing systems need three kinds of memories for graphics processing. They are used for storing commands, target data, and display data. As Q2i has a UGM (Unified Graphics Memory) structure that can integrate these memories into one DRAM, the area of the graphics portion of the system can be reduced to about a 2-centimeter width and length that is appropriate for the compact system.

2.3 Software Configuration

Figure 3 shows the software configuration of the terminal. The terminal has an OS (Operating System), device drivers, a 3-D graphics library (Iimura, et al. 1996) and a WWW browser. The 3-D graphics library is as small as 70kB. This reduces the number of memories and makes the terminal small.

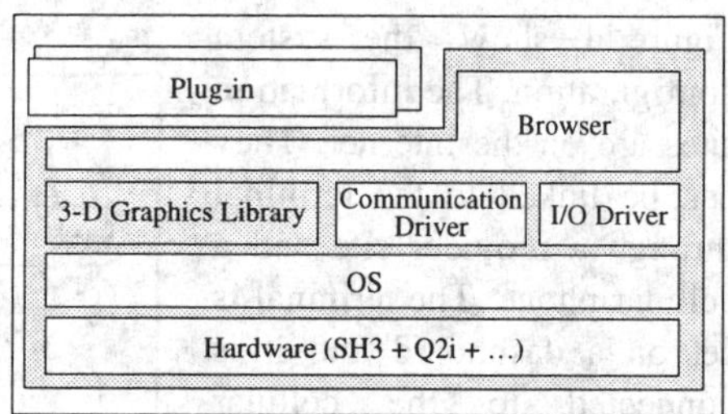

Figure 3 : Software configuration

3 Menu Selection Using 3-D Graphical User Interface

3.1 Needs of the Intuitive Graphical User Interface

In the car cabinet, people can not use the large-sized display often used in a desktop environment. Current popular displays for a car are no more than 320x240 dots. So the system has to supply graphical user interface (GUI) which fits to a small-sized display.

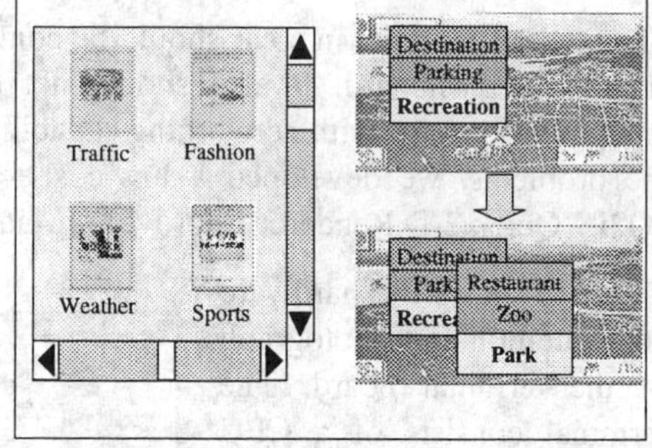

Figure 4 : Examples of 2-D GUI

The existing system has 2-D GUI. Figure 4 shows examples of 2-D GUI using icons and menus. There are two problems when applied to a small display with remote control manipulation. First, few icons can be displayed when icons are arranged as shown in the left of Figure 4. Second, manipulations are complex when the user tries to select an object from the hierarchical menu shown in the right of Figure 4. These problems are caused by the difference in the desktop environment. The 2-D icon and hierarchical menu selection better fit the large display with mouse manipulation.

To solve the above problems, we developed a menu selection scheme using a 3-D GUI.

3.2 Model

Figure 5 shows the model of the 3-D menu. The model consists of a circular table with icons on it. The table is divided into eight sectors which each have a category assignment. Icons represent the target contents and they are arranged on the appropriate sector in the order of importance.

At first, the user's viewpoint is above the center of the table. The user can select a category by turning the table. Next, the user walks into the sector to find the icon that he is interested in.

With the 3-D GUI, the system can put a lot of icons in a display and icon selection from hierarchical menu selections becomes easy. Users can look at almost the whole menu at once. So, the 3-D GUI is suited to remote control.

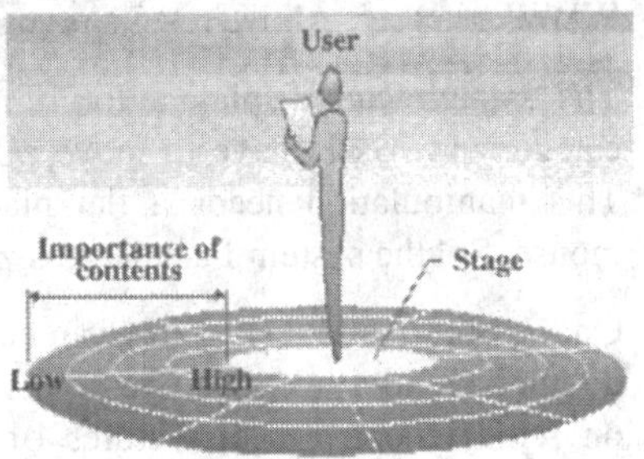

Figure 5 : Model of 3-D menu

3.3 Implementation

Figure 6 shows an example view displayed by the 3-D GUI. It shows bookmarks of WWW pages. Icons nearest to the center of the table represent the category of the sector. Each icon in the sector represents a thumbnail sketch of the web page. Icons less important are arranged farther from the center of the table, so are displayed small. The more important the icons are, the nearer they are arranged and the bigger they are displayed.

If the user moves the switch to the right (left), the table turns to the left (right) and the next category is displayed in front. Just these manipulations allow the user to select a category. The user moves the switch back and forth to select an icon. These movements between the views are animated for intuitive representation.

Figure 6 : Menu selection using 3-D GUI

We measured the performance when displaying the 3-D menu. We could get 10.4 frame/s on a 320x240 display which is practical as the GUI in a car.

3.4 Related Work

Robertson (Robertson, et al. 1998) developed the "Data Mountain" system. The system is a document management system using 3-D graphics technology. The system allows users to place icons representing the document on arbitrary positions in a 3-D desktop virtual environment. Users can access the document selecting the icon.

They evaluated the system comparing an existing WWW browser. Using "Data Mountain", the retrieval speed was higher, incorrect retrievals were fewer, and the failure retrievals in two minutes were less than when using an existing WWW browser. Above results revealed effects of 3-D GUI.

The system can display a lot of icons. But icons are overlapped and not categorized. Users have to move a mouse in every direction to select an icon. This manipulation needs a flat place on which the user can manipulate the mouse. So, the system fits a large-sized display using a mouse on a desktop.

On the other hand, using our system, icons move to the front of a display. Users do not have to move their eye points and hands too much when moving a switch on remote control. So, features of our system fit a small-sized display and remote control.

4 Conclusion

This paper described the internet browsing system using 3-D GUI. We have focused focusing on internet services for cars for which we developed the terminal. We realized a 10-square-centimeter size board that is compact enough for cars. We proposed the 3-D GUI and implemented it on the board. It works at a practical speed.

5 References

Iimura, I., Koga, K., Katsura, K., Abe, K. & Kikuchi, A. (1996). A Basic Graphics Library Fitting for Consumer Information Systems (in Japanese), *Information Processing Society Japan, SIG Graphics & CAD*, pp.7-12.

Matsuo, S., Koga, K., Shimomura, T., Katsura, K., Nakatsuka, Y. & Yamagishi, K. (1997), Development of Graphics Processor Q2 for Consumer Information Systems (in Japanese), *The Institute of Electronics, Information and Communication Engineers, National Convention, 1997 Spring*, p.377.

Robertson, G., Czerwinski, M., Larson, K., Robbins, D., Thiel, D. & Dantzich, M. (1998), Data Mountain : Using Spatial Memory for Document Management, *UIST 98*, pp. 153-162.

User Adaptation of the Pen-based User Interface by Reinforcement Learning

Shun'ichi TANO, Mitsuru TSUKIYAMA

Graduate School of Information Systems,
University of Electro-Communications
1-5-1 Chofugaoka, Chofu-shi, Tokyo 182-8585, JAPAN
Tel:+81-424-43-5601, Fax: +81-424-43-5681
tano@is.uec.ac.jp

1. Introduction

The interface based on the pen and the paper is so natural that most of the people can easily use it. We can use it without any perceptual load. Therefore the usage of the combination, i.e. the pen and the paper, does not interrupt our thinking process (Wilcox 1997, Tano 1997). However the current pen-based user interface is far from it. For example, concerning the handwriting, we are usually forced to write in the prefixed area or to write each character separately so that the system can distinguish each character correctly. Concerning the pen gesture, we are usually forced to be familiar with the implemented gesture commands.

Our goal is to realize the natural pen-based user interface that does not disturb the thinking process by adapting itself to the user preference (Tsukiyama and Tano 1998).

2. Problems and Goal

We have extracted two serious problems from many drawbacks, which often cause the interruption of the human's thinking process.

First one is the weakness in grouping of the handwriting strokes. It is often the case that the system fails to group the handwriting strokes correctly and eventually can not recognize it. It is well known that the recognition rate of Japanese characters is quite high if the grouping is completed adequately. If the capability of the grouping is improved, the quite high recognition rate will be achieved.

Second one is the low adaptability in the gesture commands. Usually the system forces the users to use the prefixed command sets. Even in the sophisticated systems, they force the users to evoke the special application to register the individual gestures.

Our goal is to realize the pen-based system (1) in which the user can write the characters and use the pen gesture without any restriction on the location, size and time, (2) which has no mode to distinguish between the handwriting and the pen gesture, and (3) which allows the users to use their own writing style and gesture.

3. Approach

To realize the natural user interface, we think that the passiveness of the user interface is the most essential feature. The user interface should silently watch the users' behavior over their shoulders and adapt itself to their preference and manner. So we have to design the new architecture, which resolve the two problems mentioned above.

Fig.1 shows the relation between the user and the pen-based system. The user inputs the handwriting and the gesture. Then the system reacts to it.

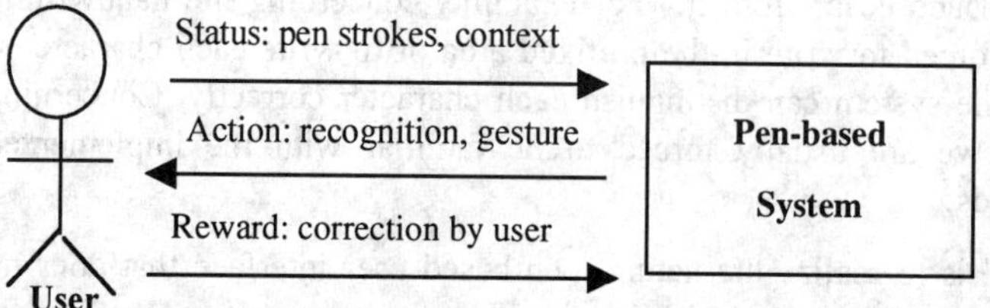

Fig.1 System model by reinforcement learning

By the terminology of the reinforcement learning, the scheme can be described as follows. The pen-based system gets the status data that consist of the pen strokes and the contextual data. The pen strokes come from the handwriting and the contextual data can be deduced from the other drawing and handwriting on the paper. The system decides the appropriate action and executes it. The users judge the system's action and respond it. Sometimes they are satisfied with it and keep writing. Sometimes the action is wrong, and then they begin to correct it irritably. The system can adapt itself by taking the response given by the users as the reward or the penalty. We found out that two problems could be uniformly solved by the reinforcement learning architecture.

4. Learning Algorithm

We used the Q-learning algorithm (Watkins 1992) as the reinforcement learning. Our system is realized by the two-staged Q-learning architecture shown in fig.2.

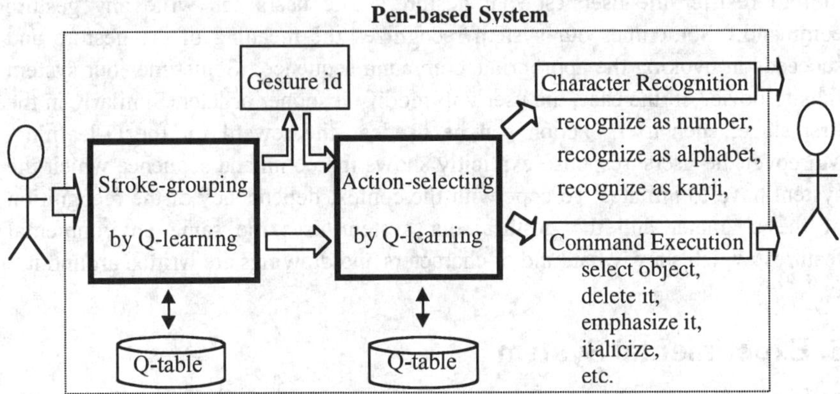

Fig.2 Two layered system architecture

The user inputs the handwriting, then the first Q-learning block groups the set of strokes based on the Q-table. The grouped strokes are sent to the gesture id block to assign it the tentative gesture id on the assumption that it is a gesture. The second Q-learning block decides the appropriate action. The action can be divided into two categories, i.e. the character recognition actions and the gesture commands. The examples of each command are listed in the fig. 2.

The result of the action appears as the change in the display to the users. According to the user response, the system adapts itself by the two-staged Q-learning.

In the first Q-learning stage, our system learns how to group the handwriting strokes. In other words, it finds out the correct separable point in the handwriting strokes and cuts the sequence of strokes at the point to make up one character.

To learn how to group and classify, our system uses (i) the stroke features which are calculated from the un-grouped input strokes, such as the ratio of the length and breadth, (ii) the environmental features which describe what kind of characters and drawings are already written around the current input strokes, and (iii) the users' response to the system's behavior. When the system's behavior is wrong, the user will modify it sooner or later. Such users' action can be used as the reward for the Q-learning. Although the user sometimes responds with it immediately, sometimes responds with it later, Q-learning can handle the time-delayed response.

In the second Q-learning stage, our system learns how to estimates the input as a number, an alphabet, a Kana character, a Kanji character or a pen-gesture and the user preference in the gesture commands. Our system does not have any tool to resister the users' special gesture. The users can write any gesture command. Sometimes our system recognizes the meaning of the gesture and succeeds in evoking the appropriate command sequence. Sometimes our system fails to do so. In the case, the user will modify it sooner or later. Similarly in the first stage, such users' action can be used as the reward for the Q-learning. Moreover the users' response explicitly shows the command sequence which the system have to imitate. To cope with the context dependency of the recognition of the character and the gesture, our system uses the same environmental features, which show what kind of characters and drawings are written around it.

5. Experimental System

We have developted the experimental system to confirm the feasibility of the learning algorithm, evaluate the performance of the recogniton and validate the adaptability of the user interface scheme. Fig.3 is the snapshot of our system.

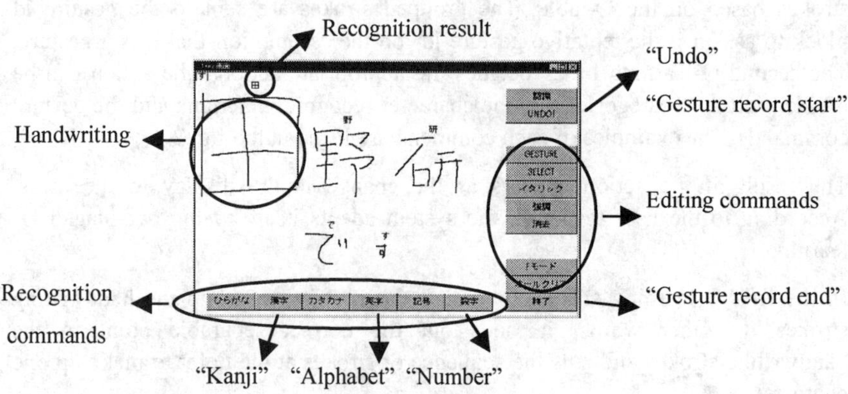

Fig.3 User interface

The users can input the handwriting anywhere in any size at any speed. The system groups the set of strokes, decides the appropriate action, and displays the result of the action. The recognition result is initially displayed in the small size as shown in fig.3 and, in a short time, substitutes the handwriting at the same size. The gesture commands are promptly executed. In case of the wrong actions, the users correct it by the "Undo" botton, the editing command bottons and the recognition command bottons. Taking the users' response as the reward/penalty, the system adapt itself.

6. Evaluation

The feasibility and the effectiveness were successfully exemplified by the experimental system. We gathered the stroke data which consist of the nearly thousand of characters by a dozen of persons.

Concerning the grouping, surprisingly, the recognition rate 71.5% was achieved only by the shape features (shown in the upper line in fig.4). The lower line shows the recognition rate in case of the data set different from the learning data set.

Note that the handwriting recognition algorithm does not help the grouping and the classification. We think that the tight connection to the handwriting recognition engine will greatly improve the accuracy. Concerning the gesture adaptation, our system could follow the user preference in the real field.

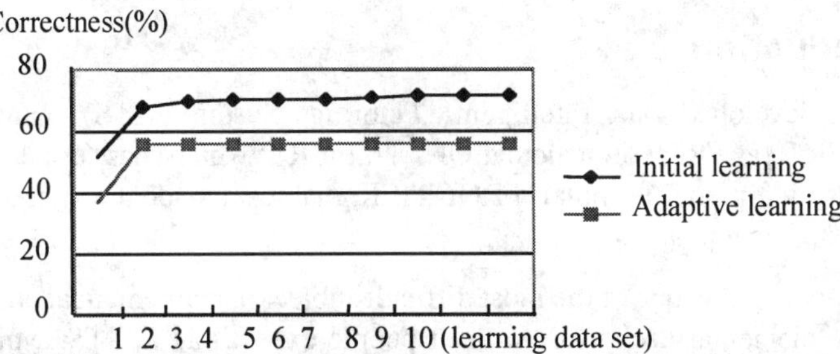

Fig.4 Initial learning and adaptive learning

References

Wilcox, L., Schilit, B., Sawhney, N. (1997). DYNOMITE: A Dynamically Organized Ink And Audio Note Book, *CHI97*, pp186-193

Tano, S., et. al. (1997). Design Concept Based on Real-Virtual-Intelligent User Interface and Its Software Architecture, *HCI-97,* pp.901-904

Tsukiyama, M. and Tano, S. (1998). The design of the pen-based user interface by reinforcement learning, *IEICE NCL97-57*, pp. 17-24 (in Japanese)

Watkins, C. and Dayan, P. (1992). Technical note : Q-Learning, *Machine Learning*, Vol.8, No.3, pp279-292

FRONTIER: An Application of A Pen Based Interface to An ITS For Guiding Fraction Calculation

Kenzi Watanabe[(1)], Teisuke Yamaguchi[(1)], Hiroyuki Egashira[(1)], Joe Ito[(1)], Yasuhisa Okazaki[(1)], Hiroki Kondo[(1)], Masayoshi Okamoto[(2)] and Takatoshi Yoshikawa[(2)]
(1) Saga University, Japan. (2) SANYO Electric Co., Ltd.,Japan.
(1) 1, Honjyo, Saga 840-8510, Japan.
TEL: +81-952-28-8828 FAX: +81-952-28-8650
E-mail: watanabe@is.saga-u.ac.jp

1 Introduction

We have developed an Intelligent Tutoring System (ITS) named as "FRONTIER" (a FRaction calculatiON TraInER), which has a pen based interface using a pen X terminal. FRONTIER enables a student writes fraction expressions using a pen.

Fraction calculation is one of the most difficult subject in mathematical education in Japanese elementary schools. A lot of people expect that an ITS, which has flexible and individualized tutoring facilities, for fraction calculation will be developed.

We have developed an ITS for guiding fraction. Distinctive feature of the system is that the system can recognize student's context of solving processes and guide them in accordance with their context (Watanabe, Okazaki, Tadaki and Kondo 1994).

As using this system, a student inputs not only a final answer but also its solving processes. Since he/she should input many fraction expressions, we develop a pen based interface to enable he/she can input and correct fraction expressions easily, naturally and quickly.

A purpose of our research is to develop an ITS for practical use, so that we must achieve that every function of the system, especially response of the interface and a performance of handwriting character recognition, attains a practical level. The interface and concurrent handwriting character recognition contribute to push up the performance of them to a practical level.

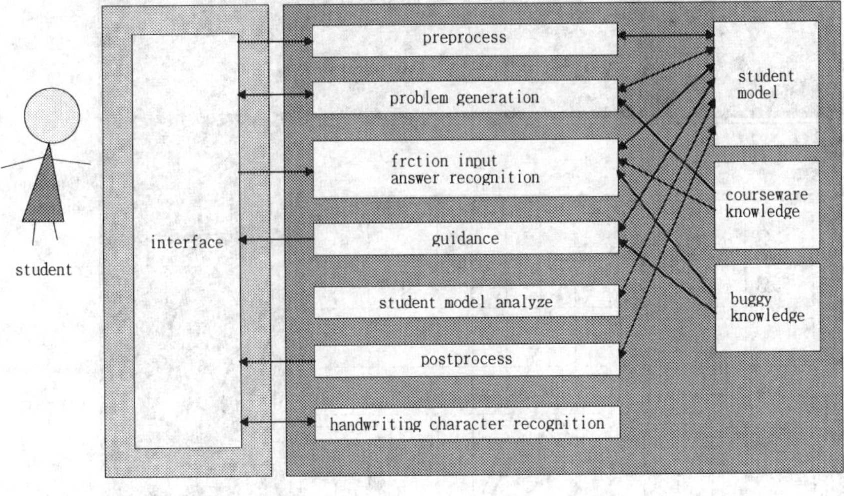

Figure1: The architecture of the system.

2 Outline of The System

2.1 System Configuration

Figure 1 shows the architecture of the system. FRONTIER is implemented on two computers connected with LAN, a pen X terminal (SANYO PXT-M001) and UNIX workstation (Sun SPARC station 5). The pen X terminal plays a role as an interface. The UNIX workstation has intelligent modules including handwriting character recognition, problem generation, answer recognition, courseware knowledge, buggy knowledge and student model. Every module except the interface, which is written by C, is written by Prolog.

2.2 Progress of Lessons

FRONTIER is a drill type ITS. At first, the system creates and presents a problem according to a selected subject by a student. He/She inputs not only a final answer but also solving processes. After it, the system recognizes student's answer with solving processes using courseware knowledge. If there is a mistake in the answer, the system identifies an error origin by using buggy knowledge and presents a guidance message according to the origin.

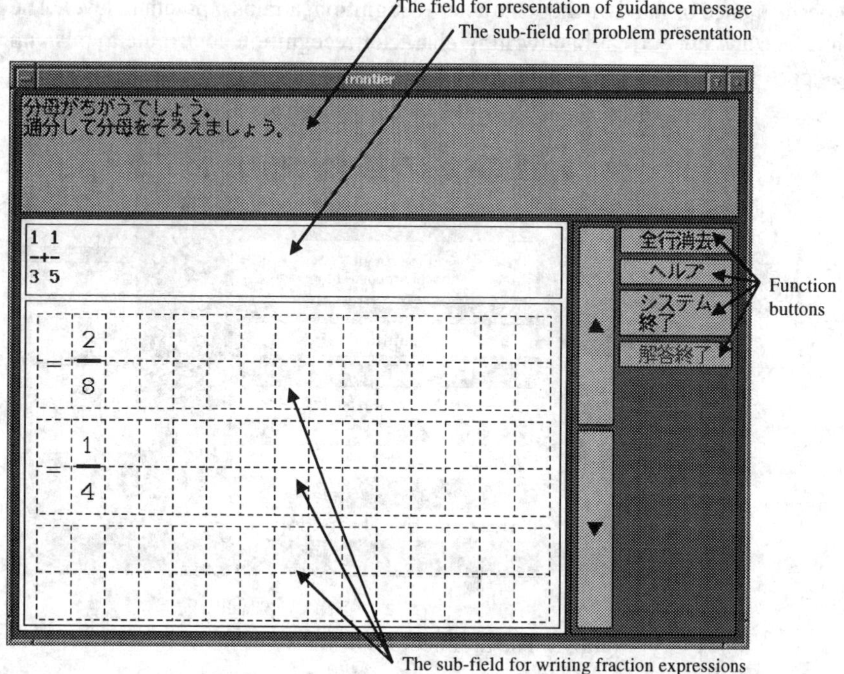

Figure2: The interface.

3 The Interface of The System

3.1 Window Configuration

Figure 2 shows the interface. The interface window has three fields, "the field for fraction input,"' "the field for presentation of guidance message" and "the field for function buttons." The field for fraction input is divided into two sub-fields. One is a sub-field for problem presentation, the other is for writing fraction expressions.

The sub-field for fraction writing is combination of two rows of squares and middle line. This form is reflected fraction form which is two dimensional arrays. The upper row is for numerators and the lower row is for denominators. The middle line is for center lines, some mathematical symbols ('+,' '-' and '=') and integer parts of a mixed fractions.

3.2 Input Operations

A student writes characters into the squares and onto the middle line in the sub-field for fraction input. If the system recognizes the character successfully, the recognized character is shown at the same place. If it fails, the system erases the handwriting. So he/she has to write character again. He/She wants to write a center line, he/she should write a long line at the middle line of the sub-filed. Because the system recognizes a short line as '-'.

FRONTIER also has two functions for fraction correction. One is "overwriting function"' which replaces a character, the other is "gesture functions." The gestures are special handwriting symbols bounding functions which are "erasing one character"' and "erasing one fraction expression."'

4 Roles of the Sub-field for Fraction Writing

The sub-field for fraction writing plays three important roles. The system separates handwriting characters to isolated character at first in the recognition processes, then the system recognizes each one. This system can use the square in the sub-field as a criterion at the separation. This is the first role.

The system translates fraction expressions to terms of Prolog. Since this sub-field reflects the form of fractions, it plays a role as a criterion for the translation. So that, the system can recognize the form of fractions. This is the second role.

The last is a role as a student's help for writing characters. The sub-field shows a student a position to write in. It guides correct form of fractions implicitly.

Although the sub-field may restrict free writing, the roles mentioned above contributes to push up the system to a practical level.

5 The Concurrent Handwriting Character Recognition

There are three processes, which are timer process, display process and analyzing process, work together for recognition of one handwriting character.

The timer process watches the end of writing and informs it to the other processes. The display process displays handwritings. The analyzing process works following steps. The process cuts an area of the handwriting at first. Then the process classifies the area into three categories which are a character, a center line and a gesture. If it is a character or a center line, the process tries to recognize it and shows the recognized character. If it is a gesture, it executes a function.

The recognition method of handwriting character is "the online character recognition method" (Okamoto, Yamamoto, Sawada, and Yamamoto 1996).

Every recognition process is executed as concurrent jobs. Since a student is not forced to wait till the recognition is over, he/she can write other characters continuously. It enables quick response of the interface.

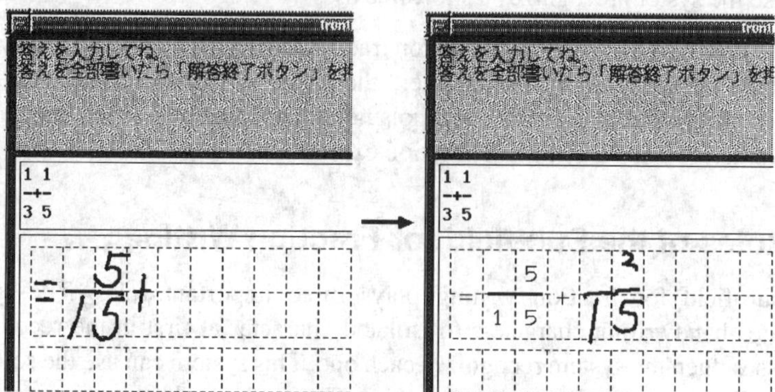

Figure 4: Progress of fraction writing and its concurrent recognition.

6 Evaluation

We evaluated elapsed time for recognizing a character and successful rate of recognition. The average elapsed time for 65 characters is 20.1 (ms). The successful rate for 104 characters is 91.3%. Those results show the system has enough practical performance.

7 Conclusion

We have developed a pen based interface using a pen X terminal for an ITS for guiding fraction calculation named as FRONTIER. We tried to develop a practical ITS. The results of the evaluation show the interface is practical enough. FRONTIER is a good example to apply a pen based interface to an educational systems.

References

Watanabe, K., Okazaki, Y., Tadaki, S. and Kondo, H., (1994). An Implementation of an ITS for Guiding Fraction Calculation. *The Transactions of the Institute of Electronics, Information and Communication Engineers*, Vol. J77-A, No. 3, 518 - 529 (in Japanese).

Okamoto, M., Yamamoto, H., Sawada, K. and Yamamoto, K., (1996). On-Line Handwriting Character String Separation Method Using Network Expression. *Proc. of 13th International Conference on Pattern Recognition*, Vol 4, Track D, 422-425.

A Paper Based User Interface for Editing Documents

Yoji Maeda and Masaki Nakagawa*
Mitsubishi Electric Corporation
325 Kamimachiya, Kamakura, Kanagawa, 247-8530, Japan
e-mail : maeda@tokyo.jst.go.jp

*Dept. of Computer, Information and Communication Sciences
Tokyo Univ. of Agri. &Tech.
Naka-cho 2-24-16, Koganei, Tokyo, 184-8588, Japan
phone: +81-423-88-7144, fax: +81-423-87-4604
e-mail : nakagawa@cc.tuat.ac.jp

1 Introduction

Generally speaking, document preparation has become easier due to the development of word processing. Nevertheless, some people dislike typing a keyboard and feel pains to operate PCs.

On the other hand, document OCRs have been widely used to utilize existing documents to prepare new documents. However, their recognition performance is still unsatisfactory. Especially, the Japanese documents are difficult to read because the mixture of Kanji, phonetic characters, numerals and even alphabets makes their segmentation and recognition very hard. Therefore, a user has to correct recognition errors using word processing software. Then, the user needs to learn how to type and how to use the word processor.

We are developing a system that recognizes correction marks written on a printout of a document and apply them to the document automatically, thus allows a user to edit documents without using a keyboard.

So far, some researchers have reported systems for correcting a document without using a keyboard. One of the earliest works was done by Suenaga (1980). His system read a document with handwritten correction marks directing deletions and permutations of characters and performed designated corrections. However, the corrections were made on the document image, so that they were not reflected to the document codes. Moreover, correction marks must be written without touching characters in a document in order to make their detection easy.

Bunke, Gonin, and Mori (1997) have published a system that reads correction marks and reflects them to the codes of a document. Their system recognizes correction marks written in colored ink on a document and performs corrections to the document codes.

Our system does not restrict the correction marks to be written in colored ink. We have developed a new method which can detect correction marks robustly, even when they are written in the black ink on characters on a document, so that the users don't have to pay attention to the way of writing correction marks. They can direct correction of a document with the same manner as having carried out from the time of childhood, that is, using paper and pencil and the system performs the correction with their help. Moreover, the system allows the users to select the right character among candidates or redirecting character segmentation rather than rewriting all the wrong characters by utilizing the candidates of character recognition and segmentation.

The advantage of our system is that the users need not use a keyboard and word processor. They can do it in the same way as they do with paper and pencils.

This method can be applied to the documents prepared by other methods such as those typed by someone or prepared by the voice recognition (their performance is ever lower than OCR).

2 Document Correction System with a Paper Based UI

Office workers who don't use word processors by themselves often prepare a draft in the following ways:

- Writing a draft wholly by hand.
- Using an existing document and designating some modifications by handwriting.
- Using a portion of an existing document and combining it with a written draft by hand.

Then, a secretary prepares a document using some word processor. A person who made the draft reviews and polishes the document several times until the final version is prepared by the secretary.

We are developing a system to realize this document preparation cycle by replacing the secretary by a document processing server as shown in Fig. 1. Although highly matured document recognition technology is needed to realize this system, the human interface is also as important as the recognition engine.

Especially important is the HI to correct recognition results with pen in the same way that people indicate errors in a document. Our system reads a document with or without correction marks, interprets them and revises the document like a secretary.

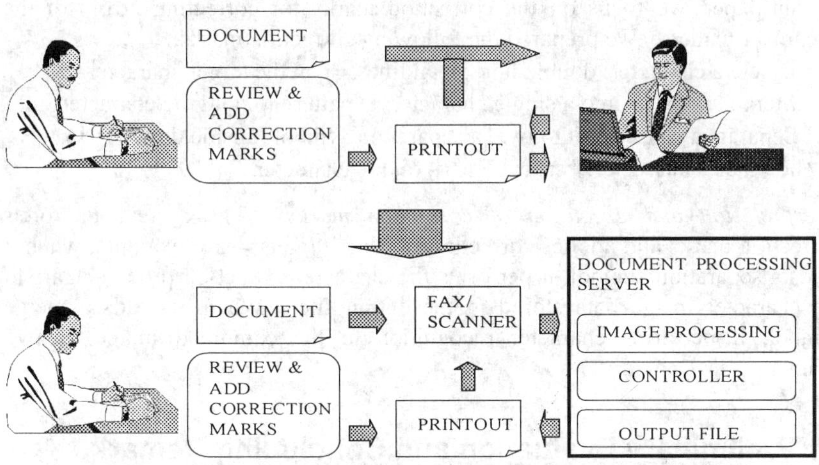

Fig.1 System Organization.

2.1 Document Processing Server

Image processing. First, an initial document without correction marks is digitized and inputted by a scanner and character segmentation and recognition are executed on the digitized image. Thereafter, a printout with correction marks is digitized and inputted by a scanner and correction marks are extracted, recognized and executed on the digitized image.

Output file. A recognition processing result is memorized with processing data such as the information on character image position, character-string position, character recognition candidates, character images etc.

2.2 Process

The process has three steps: document recognition, correction mark recognition and their interpretation and execution.

Document recognition. This step recognizes a document image and outputs the result with gray color.

Correction mark recognition. The user writes correction marks on the printout paper. The system separates them from the original document image by dropping gray color images (This can be easily done by deleting isolated small dots). Then, the system extracts and classifies correction marks into the following three classes:

- Correcting errors of the document reader.
- Correcting characters.
- Correcting document layout.

In this paper, we focus on the correction marks for correcting errors of the document reader. We prepared the following correction marks:

- Delete a character: double horizontal lines(=) on the target character.
- Merge two parts into a single character: U pattern to hold the character.
- Separate a character into two: a separation symbol (upside down of T).
- Replace a character: vertical line (|) on the character.

Interpretation and execution of correction marks. The system interprets correction marks and applies corrections to their targets. For example, when it reads a separation symbol, it performs the character extraction process again to the character image data stored in the output file. When, it reads a merge symbol, it performs character recognition to the combined image of two characters designated.

3 Preliminary Evaluation and Concluding Remark

We made a prototype system on a PC and asked several people to try the system. Some of them were surprised by the result since their pencil corrections were reflected to the result. Others said that they prefer editing on PC. At the moment, a set of correction marks to be recognized is too small. Moreover, recognition of correction marks sometimes fails. After solving simple problems revealed in the prototype system, we will evaluate this method more precisely. By this initial prototype, however, we have confirmed that the people who do not want to use a keyboard or PC can accept the method.

4 References

Suenaga,Y. (1980). A facsimile based text editor. *The Transaction of the Institute of Electronics, Information and Communication Engineers*, J63-D, 12. 1072-1079 (in Japanese).

Bunke, H., Gonin, R. & Mori, D.(1997). A tool for versatile and user-friendly document correction. *Proc.4th ICDAR'97,(Vol.1)* Aug.1997, 433-438.

Interface Using Video Captured Images

Yu SHIBUYA and Hiroshi TAMURA
Kyoto Institute of Technology

1 Introduction

Since several years ago, multimedia has been expected to make an effective human-computer interaction. Until now, multimedia might be mainly used for "output", that is information flow from machine to user. There is another direction of information flow, that is "input" from user to machine. Traditional type of input method is keyboard, mouse, touch screen, and so on. Few years ago, we introduced an another interaction method named "Action Interface" (Shibuya 1998), a kind of gesture interfaces. In the Action Interface, user's body image is captured and his action is detected by the system. Depends on his action, the system makes suitable reactions to the user.

There are several commercial or experimental uses of human action for human-computer interface. In the MANDALA© VR system (Vivid), a player's body image is captured, then combined with computer graphics, and finally displayed on the monitor in front of the player. The player can make a tune with virtual musical instruments displayed on the monitor. Trans Plant (Christa, 1995) is an artistic use of combined users' live body image and computer graphics. Depending on the players' movement, artificial plants are growing up. Reality Fusion provides an interface that enables users to interact with a computer simply by using hand or body motion (Reality Fusion, 1997).

Now a day, it is easy and not so expensive to combine a video camera with PC. The video camera has a possibility to be an another input method in human-computer interaction. Multimedia should be used for "input". With thinking about these situations, and applying the method of Action Interface, an another type of interface is introduced in this paper. This interface is called "Variable Sheet interface".

Variable Sheet is a sheet on which some pictures are drawn. With changing the pictures, user can make various interactions with the system. For example, if a piano keyboard or a drum is drawn on the sheet, a user can play piano or drum by touching the instrument drawn on the sheet.

2 System Configuration

The system configuration of Variable Sheet interface is shown in Figure 1. The system consists of video camera, PC, and the Variable Sheet. The PC can identify the sheet with checking a color code drawn on it. Depends on the color code, the PC will locate the sensing points on the prefixed position of the sheet. User's hand image is captured through the video camera and processed by the PC. If a user overlaps his hand on one of the sensing points, then the PC understands that the user touches the point.

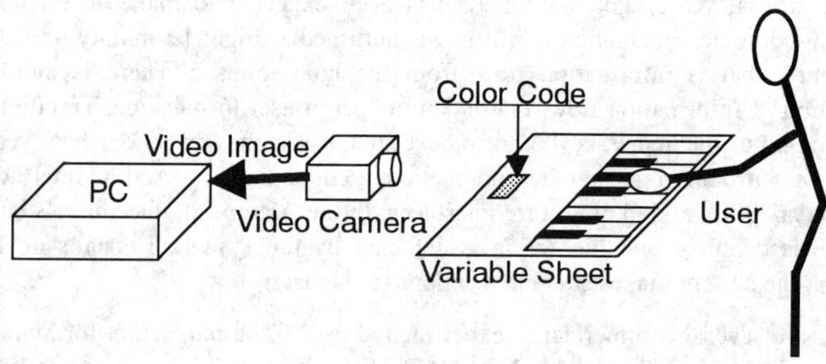

Figure 1. System Configuration

In order to detect the overlapping of the hand, similar method with Action Interface is used in this Variable Sheet interface. The system checks the color changing in the desired region of the captured video image. If the color of it changes to the user's skin color, then the system decides that there is his hand. In order to fasten the sensing, the sensing region is restricted to the surrounding region of the sensing point and mosaic image of sensing region is made before checking. Furthermore, for the stable sensing, HSL (Hue, Saturation, and Lightness) color space is used to express color instead of RGB (Red, Green, and Blue).

One of the most significant points of system configuration is to select the suitable camera position. As shown in Figure 1, the location of video camera should be behind the sheet and a few cm higher than the sheet. The camera angle against to the sheet and the zoom level of camera is also important. The camera view should cover the whole area of the Variable Sheet and the area should be located lower part of camera view. After clearing these conditions, a camera view of this system should be like Figure 2.

Figure 2. Camera View of the System

In figure 2, a piano-like keyboard is drawn on the sheet. On each key, there is a sensing region and if the user's hand overlaps one of these sensing regions, then the system recognize that the user hit the key, and make sound to the user.

It is possible to overlap a hand on several sensing regions at once but it is not suitable to recognize the every sensed region are hit by the user. Usually, users want to touch only one of them. In order to avoid such miss operation, the system checks the each sensing region in the fixed order. The system will scan the sensing regions from lower right corner to upper left corner of the camera view. Even if the system detects the user's hand, the system will not make response immediately. The system will scan again and if there are no other sensed regions which are lower scanning order, then the system make response. This scanning doesn't take so long time that the user can make interaction smoothly.

3 Consideration

Variable Sheet is easy to configure and adaptable for several situation. Similar kind of interaction might be possible using touch screen. However, comparing with touch screen, Variable Sheet has several advantages. It is easier to configure the system of Variable Sheet than that of touch screen. Furthermore, because there is no restriction about the material of the sheet, it is possible to change the touch feeling with changing the material. It is very difficult to do such thing with touch screen.

Some experimental system of Variable Sheet have been realized and used. Currently, three kinds of pictures are drawn on the various sheets. They are piano-like keyboard, drums, and animals. Users can play a simple music with keyboard and make rhythm with drums. Furthermore, if users touch an animal on the sheet, there is a cry of the animal.

In order to evaluate the response time of the system, maximum number of sensing per minute is examined. At this moment, an old PC is used in the system. However, the system can sense about 120 times per minute without error. This response time will be improved with using a more modern PC.

Depends on the subjective evaluation of the users, the Variable Sheet is fun to use and most of them were interested in the system. However, the accuracy of the system is not so high, so some users recommended that the touch screen was better than Variable Sheet for the task which need accuracy. Anyway, from these experiments, Variable Sheet is available to use for entertainment purpose. Furthermore, if the requested accuracy is not so high, the Variable Sheet should be an attractive interface.

4 Conclusion

In this paper, a new kind of interface, named Variable Sheet interface, is introduced. Variable Sheet interface has so much flexibility that it is easy to make a new interaction with changing the pictures drawn on the sheet. Another application of Variable Sheet will be introduced and objective evaluations will be done in near future.

References

Christa, (1995). http://www.mic.atr.co.jp/~christa/link6.html

Reality Fusion, (1997). http://www.realityfusion.com/

Shibuya, Yu, Takahashi, Kazuma & Tamura, Hiroshi (1998). Introduction of Action Interface and Its Experimental Uses for Device Control, Performing Art, and Menu Selection, Proc. the 7th IFAC Symposium on Man-Machine Systems, pp. 71-76.

Vivid. http://www.vividgroup.com/

Design of Interface for Operational Support of an Experimental Accelerator with Adaptability to User Preference and Skill Level

Makoto Takahashi, Yoshinori Kuramochi, Shigeo Matsuyama, Masanori Fujisawa and Masaharu Kitamura
Graduate School of Engineering, Tohoku University, Japan

Introduction

The interface design of complex engineering system has been considered as the important issues for effective and reliable operation of the system. More attention is being paid to interface design for newly developed systems compared with the ones developed a few decades ago, when design of hardware system had been the primary issues. Our target system, twenty years old experimental accelerator has gone through incremental hardware modification to maintain its functionality. As it is a system primarily for experimental purpose, little attention has been paid to interface design so far. As this system is supposed to be operated by experimenters themselves with varying level of expertise, the modification of interface design has been necessary to realize more effective and safety operation. The purpose of the present study is to develop a modified interface design for the accelerator system through the detailed analysis of operators' behavior with the method of cognitive task analysis. The concept of adaptable interface design, which is revised concept of our preference-based interface design (Takahashi.,et.al.,1998), has been adopted as the basic principle of this modification scheme.

Target System

The accelerator experimental facility consists of two major components; (1) the accelerator itself and (2) the beam transport system. The ionized particles generated and accelerated in the accelerator are then transported along with the beam line. Six target positions with different angles are prepared for different experimental purposes. The orientation and convergence of the beam are controlled by the set of magnets installed along with the beam line. In the present study, the beam transport system (BTS), that requires more complex operation compared with accelerator, has been taken as the main object for interface modification.

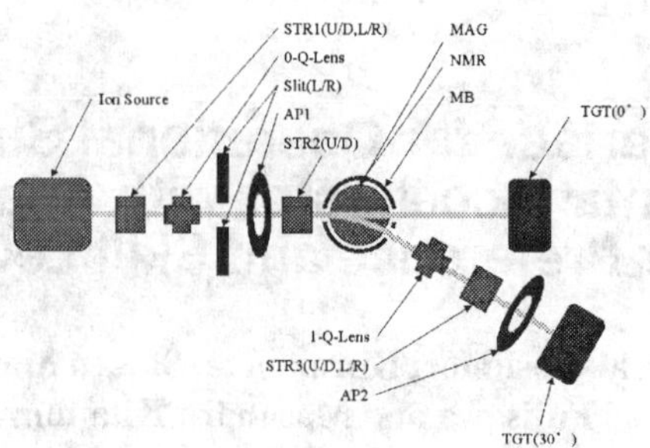

Fig.1 Target System:Beam Transport System of Experimental Accelarator

The configuration of the BTS is shown in Fig.1. The difference in experimental conditions makes BTS operation more complex and difficult for operator with insufficient expertise. Although the behavior of the ion beam is governed by the physical laws, it is not straightforward to guide the beam to the desired target position by controlling several parameters simultaneously. Fully automatic operation is almost impossible because of the inadequacy of the system model and inevitable influence of varying hardware conditions. The goal of the present study has been set to provide more effective support for the operators with various level of expertise through modifying the interface design.

Cognitive Task Analysis

As for the first step for interface modification, the detailed cognitive task analyses has been performed in order to clarify how the operators actually behave during the BTS control. As our purpose is to develop interface for operators with different levels of expertise, the behavior of typical two operators, one expert and one novice, has comparatively analyzed to examine the differences in the operational procedures and in the way of controlling parameters. The video records and interviews were major source of information for task analysis. The history of controlled parameters and focused parameters were investigated along with the protocols and interviews.

The results of the cognitive task analysis revealed that the BTS control consists of three phases shown below;

Phase 1: initial beam conditioning.
Phase 2: beam bending for desired direction
Phase 3: final beam shaping.

The task flow of the expert operator has been represented in the form of task flow diagram(TFD). The example of the TFD for Phase 3 is shown in Fig.2. The goal of Phase 3 is to maximize the target beam current (TGT in Fig.2), while minimizing the leak current(AP1,AP2,Slit) by controlling the eight control parameters. The TFDs are mainly based on the cognitive task analysis of expert opera-

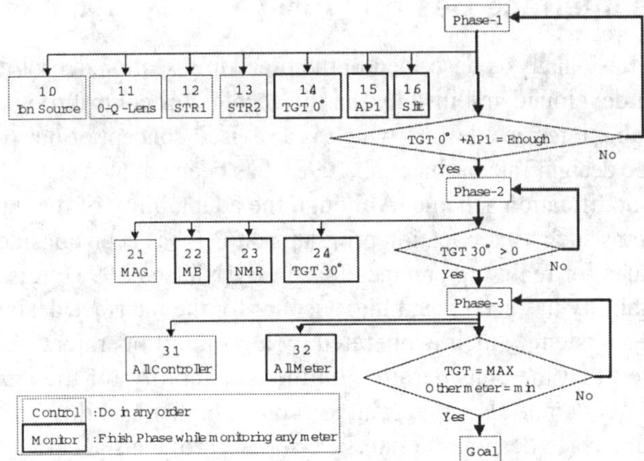

Fig.2 Task Flow Diagram of Phase3

tor and the distinct differences in the pattern of operational sequences have been observed between the novice and the expert operators. The novice operator tended to focus attention within the narrow range, while the expert periodically examined related parameters more extensively and systematically. The TFDs have been utilized to clarify the basic user requirements for the modified interface.

The importance of utilizing log data to set initial parameter conditions has also been revealed by task analysis. The operation always starts with the searching in previous log data (in handwritten format) with similar experimental conditions. The operators use these parameter values as the initial parameter settings. As the log data has been recorded on paper files, these procedures (searching and setting) have been found to be quite time consuming task.

The need for support for multivariate optimization problem has been pointed out during the interview with operators. The procedure of operators to achieve the optimal beam condition (beam current intensity, stability, focussing, etc.) can be considered as the multivariate optimization problem. In the recorded history of the parameter adjustment of the expert operator, the try-and-error type optimization process has been observed and these processes have been found to be most time consuming. The support for this optimization process is eagerly requested by the operators during the interviews.

The cognitive task analysis described so far has revealed the user requirements for specific task phase and these requirements are then utilized as the guideline for developing prototype interface design.

Prototype Interface Design

Based on the task analysis described in the preceding section, a prototype interface has been developed and installed in the accelerator control room. The concept of adaptable interface design, which is a revised concept of our preference-based interface design(Takahashi.,et.al.,1998), has been adopted as the basic principle of this modification scheme. Although the adaptability of the interface for ordinary systems such as application programs on PC, has been considered as the important issues for realizing enhanced usability(Nielsen,1993), it is quite rare that the adaptability has been taken into account for the interface design of large scale, complex system, which is operated by a group of operators. Authors believe, however, that the consideration on the adaptability for the interfaces of huge, complex system is also important to achieve higher usability. In our notion of adaptable interface design, the interface should provide selectable options of form and contents, among which operator can select desirable ones by themselves. In order to avoid negative effects of introducing selectable options, the following guideline has been applied;

a. The extent of selections varies according to the level of expertise.
b. The unused option should be eliminated eventually.
c. The operators are given enough time in advance to know well about the contents of selectable options.

The example image of prototype interface is shown in Fig.3. The prototype interface has been developed by using LabVIEW (National Instruments Co. Ltd.). Two CRT displays provide visual information to operators. At this stage of development, the proposed interface is not supposed to replace existing hard-wired interface. Instead, it provides additional way of monitoring the state of the BTS.

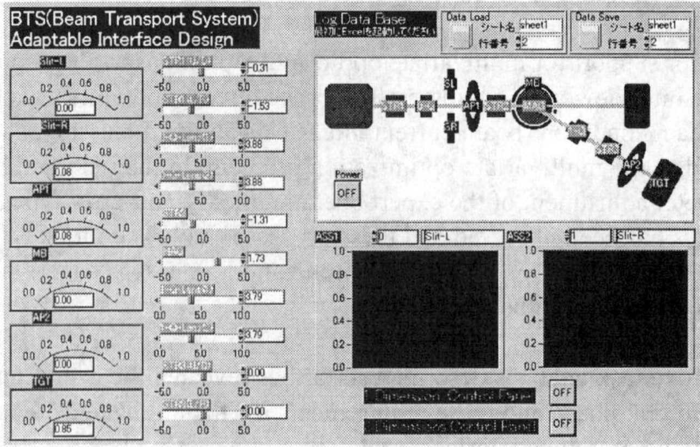

Fig.3 Example Image of Prototype Interface

The proposed interface consists of two main windows; one is for overview of the BTS and the other is for the control and monitoring of subcomponents.

The proposed interface has been made adaptable in following ways. One is that the display format can be selected based on the operators preferences. The other is that the support functions both for novice and expert are available upon request. For example, the initial log data search and setting, which has been found to be important and time consuming task in the task analysis, is supported both for novices and experts. The support function for multivariate optimization procedure is provided mainly for novice operators. This function has two modes; one is single parameter mode and the other is two parameters mode. In single parameter mode, the operator select one specific control parameter and the support display provides the sensitivities to the resultant parameters. In two parameters mode, the operator selects one pair of parameters and the support display also provides the estimated value of the target in two-dimensional plane. These support functions, which has been designed based on the nominal behavior of expert operator, are offered mainly for novice operators.

The prototype interface has been evaluated by the expert operators and the review results were quite affirmative. Although the proposed system requires further feedback from the actual operators, the authors believe that the basic validity and effectiveness have been confirmed as far as the BTS is concern.

Conclusion

The method of cognitive task analysis has been successfully applied to the process of interface modification of the experimental accelerator system. The prototype interface has been developed, which aims to support extensive level of expertise. Although this project is still going on aiming at realizing fully computerized interface, the prototype system developed and actually installed at the accelerator facility has shown the basic validity of the present approach.

Acknowledgments

This work was supported by Grant-in-aid for Scientific Research (A)(2)-09308012

References

Takahashi,M.,et.al.(1998). Preference-based MMI for Complex Task Environments, Proc. of the 7the IFAC/IFIP/IFORS/IEA Symposium of Analysis, Design and Evaluation of Man-Machine Systems, Kyoto Japan,503-508.

Nielsen,J.(1993). Usability Engineering. Academic Press.

Improving the Usability of User Interfaces by Supporting the Anticipation Feedback Loop

Marcus Plach[1] and Dieter Wallach[2]
[1]Saarland University, [2]University of Basel

1 Introduction

Designing a user interface is a demanding cognitive task. In order to develop effective methods and software-based tools which could support and assist designers, a better understanding of the mental activities involved in design is needed (Carroll 1996a). In investigating cognitive processes in design, researchers have traditionally focussed on the perspective of the individual mind. However, essential constraints of design problems emerge from the need to share and anticipate users' knowledge. In particular, during design sessions, the designer has to take the perspective of the user. That is, the designer has to anticipate the user's (sub)goals, assumptions, and understanding of the information presented at any point in the interaction sequence. Whereas there exist well known formal methods for user modeling in the human-computer interaction literature, e.g. GOMS (Card, Moran and Newell 1983) or TAG (Payne and Green 1986), we are concerned with the informal and heuristic way designers deal with this aspect of interface design. For this heuristic mode of mental activity we adopt the term *anticipation-feedback loop* (AFL) which was originally introduced in the context of intelligent user interfaces by Wahlster (1991).

In usability engineering the users' perspective formally comes into play in the form of design or prototype evaluation methods. Besides empirical usability testing, *discount usability methods* have been proposed to evaluate the usability of (prototype) interfaces (see Nielsen and Mack 1994). In general, two different classes of such methods can be distinguished. On the one hand there are guideline-oriented methods like *heuristic evaluation* (Molich and Nielsen 1990). On the other hand, there exist methods like *cognitive walkthrough* which represent a model-based evaluation technique (e.g. Wharton, Rieman, Lewis, and Polson 1994).

The important point is that, opposed to guideline-oriented methods, with the model-based method users' cognition must be explicitly taken into account. Along the same line, other authors have stressed that the explicit posing of questions concerning concrete usage scenarios of the to be designed artifact is an effective method for guiding design (Carroll 1996b). In this study, it is hypothesized that a supporting framework derived from the cognitive walkthrough method leads to more thorough perspective taking in the AFL than guideline-oriented methods. Moreover, we expect such a method to help designers to give constraints imposed by users' goals and assumptions a higher priority. To test this hypothesis, an experimental study was conducted.

2 Methods

The task for the subjects was to design the user interface for a numerical pager (a small mobile telecommunication device that can only receive messages). Subjects were given a design brief consisting of the following instruction package: rough functional requirements of the device, technical constraints, and think aloud instructions adopted from Ericsson and Simon (1993).

Depending on random assignment to one of two conditions, subjects received either a checklist of nine standard context-independent usability guidelines, e.g. 'provide explicit feedback' (condition 1) or the rationale of the cognitive walkthrough method (condition 2). Subjects were advanced students of computer or information science who had successfully designed user interfaces as partial requirement of a human-computer interaction class. Design sessions were videotaped, and the verbal reports were transcribed.

3 Analysis

In order to measure potential effects of different supporting methods on design cognition in the AFL, two category systems were developed and used to code verbalizations according to the theoretical questions of the present study: (1) direction of constraining and (2) depth of perspective taking. For reasons of space restrictions only the first category system is described in more detail.

In general, constraints imposed by assumed user goals and plans could be viewed as any other technical constraint. But, because these constraints are wanted per se in user interface design, they should be given the highest priority. In contrast, functional constraints including usability guidelines like 'be consistent' are not valued for their own sake, that is, they are determined by the overall system configuration (Smith and Browne 1993). Moreover, the order in which different constraints are taken into account by the problem solver has a marked effect on possible solutions because 'later decisions are constrained by earlier decisions in that they are taken within the context of an existing partial solution, and each solution further limits the range of possible alternatives' (Logan 1989, p. 189).

As has been stated earlier, a users' cognition is central in the cognitive walkthrough approach. For subjects who were given this method as a conceptual framework, we therefore expect to find more verbalizations indicating that user goals lead to functional specifications and that these are evaluated with respect to a certain user cognition. Such a scheme would reflect a top-down problem solving strategy. In contrast, subjects in the guideline condition of the study were expected to be more oriented towards functional concepts. That is, we expected to find more verbalizations where functional specifications lead to detail solutions and solutions which are evaluated from a functional view point.

To examine this hypothesis, two independent raters coded segments of verbal protocols according to four categories: (1) anticipated user cognition leading to functional specification, (2) functional specification leading to evaluation with respect to a user cognition, (3) functional specification leading to detail solution, and (4) solution leading to evaluation with respect to a functional concept. The following annotated sample shows examples of segments coded according to this scheme:

1. 'Then the user wants to read a new message [`user cognition`], hm, it would be best, if this could be done somehow automatically [`functional specification`]'.
2. 'Do I need other types of signaling messages [`functional specification`], the user might be in a situation when he doesn't want an annoying loud beep [`user cognition`]'.
3. 'We should build something like a memory full thing [`function`]; we could do this with an unexpected behavior, like messages blinking, instead of permanently showing them [`solution`]'.
4. 'With 'back' we get there [`solution`], yes, we should keep that consistent [`functional specification`]'.

4 Results

The ratio of the number of concordantly coded segments to the number of all coded segments is 0.87, indicating a satisfying degree of inter-rater reliability. The following table shows the results of the protocol analysis for the eight subjects according to the category system "direction of constraining". As can be seen in Table 1, the effect of the different methods used to guide the design for usability on the cognitive processes of designers is quite remarkable. The numbers given in the table represent proportions relative to the total number of segments coded for each subject.

Table 1. Direction of constraining

Direction	*Cog.Walkthrough*	*Guidelines*
u.cog->func	26%	11%
func->u.cog	14%	6%
func->solution	20%	34%
solution->func	40%	49%

Subjects in the cognitive walkthrough condition clearly verbalized more statements in which an anticipated user cognition seems to drive design problem solving. That is, we find more instances of an anticipated user cognition leading to functional specifications (u.cog→func) and more functional specifications evaluated with respect to the constraint of an anticipated user cognition (func→u.cog). On the other hand, and also consistent with the hypotheses of the present study, subjects who were given guidelines appear to be more concerned with the functional level of the problem. We find a higher percentage of functional concepts leading to detail solutions (func→solution) and solutions that were evaluated in the light of a functional concept (solution→func).

5 Discussion

The results of this study suggest that different methods that help to focus on usability have an notable impact on problem solving processes. The results for the second category system used to measure 'depth of perspective taking' also supports this view (not reported here). Subjects in the cognitive walkthrough condition were clearly much more concerned with constraints implied by the anticipation of user goals and assumptions than subjects who were supported by context-independent usability guidelines. This finding is an indicator for a more pronounced top-down problem solving strategy which is generally considered superior to a bottom-up strategy. Guidelines apparently lead to focussing on the functional level of the problem structure and a bottom up oriented strategy, resulting in mental activities that are more remote from user cognition. In sum, the results found in this study support the view that methods that help the designer to anticipate concrete user cognition in the AFL, in fact, *strengthen user-orientation*. Software-based design supporting tools which embody methodologies like cognitive walkthrough might therefore provide surplus benefits beyond those obtained by purely guideline-oriented tools. Further empirical investigations will be needed to pin down in which way the differences found for the different methods influence specific usability aspects of the actual interface.

6 References

Card, S.K., Moran, T.P., & Newell, A. (1983). *The psychology of human-computer interaction*. Hillsdale: LEA.

Carroll, J.M. (1996a). Human-computer interaction: psychology as a science of design. *Annual Review of Psychology*, *48*, 61-83.

Carroll, J.M. (1996b). Becoming social: expanding scenario-based approaches in HCI. *Behaviour & Information Technology, 15*, 266-275.

Ericsson, K.A. & Simon, H.A. (1993). *Verbal reports as data (rev. ed.).* Cambridge: MIT Press.

Logan, B.S. (1989). Conceptualizing design knowledge. *Design Studies*, *10*, 188-195.

Nielsen, J. & Mack, R.L. (1994).*Usability inspection methods*. New York: Wiley.

Molich, R. & Nielsen, J. (1990). Improving a human-computer dialogue. *Communications of the ACM*, *33(3),* 338-348.

Payne, S. J. & Green, T.R.G. (1986). Task-action grammars: A model of the mental representation of task languages. *Human-Computer Interaction*, *2*, 93-133.

Smith, G.F. & Browne, G.J. (1993). Conceptual foundations of design problem solving. *IEEE Transactions on Systems, Man, and Cybernetics, 23*, 1209-1219.

Wahlster, W. (1991). User and discourse models for multimodal communication. In J.W. Sullivan & S.W. Tyler, *Intelligent User Interfaces*, (pp. 45-67). Reading: Addison-Wesley.

Wharton, C., Rieman, J., Lewis, C., & Polson, P. (1994). The cognitive walkthrough method: a practitioner's guide. In J. Nielsen & R.L. Mack (eds.), *Usability Inspection Methods*. New York: Wiley.

USER MODELLING IN A GUI

Maria Virvou
Department of Computer Science, University of Piraeus,
80, Karaoli and Dimitriou St., Piraeus 185 34, Greece
mvirvou@unipi.gr

Anna Stavrianou
Department of Computer Science, University of Glascow,
Glasgow G12 8QQ, Glasgow, UK
stavriaa@dcs.gla.ac.uk

1 Overview

This paper describes a graphical user interface that reasons about user's actions. This graphical user interface is called "Intelligent File Manipulator" and is used for a file manipulation program. Intelligent File Manipulator executes users' actions in a similar way as the Windows 95 Explorer. However, it also reasons about each user action so as to find out whether the user really meant the action that s/he issued. The main aim of Intelligent File Manipulator is to provide spontaneous assistance and advice to users who have made an error with respect to their hypothesised intentions. As Cerri and Loia (1997) point out if error diagnosis is to be performed then a user modelling component should be incorporated into the system. Therefore, Intelligent File Manipulator has a user modelling component that keeps track of intentions and possible confusions of each individual user.

Intelligent File Manipulator has originated from another system, called RESCUER, that was meant to provide assistance to users of command-language interfaces such as the interface of UNIX (Virvou, 1998). The requirements analysis of RESCUER was based on an empirical analysis that involved real UNIX users (Virvou et al., 1999). This analysis revealed that users often issued commands that they did not really mean. These lead them to problematic situations with respect to their original intentions. Similar problems may occur in graphical user interfaces despite the fact that they are user friendlier than command-language interfaces. As McGraw (1994) points out, graphical user interfaces may prove difficult to traverse and use if they are poorly designed. Indeed, users may accidentally make mistakes by clicking on the wrong command or selecting the wrong files etc. However, there may be ambiguities about errors since there may be different explanations of observed incorrect user's actions, as (Mitrovic et al., 1996) point out. Intelligent File Manipulator always attempts to resolve ambiguities about errors by combining two different reasoning mechanisms: a simulator of human error generations and a limited plan recognition mechanism.

2 Reasoning about user's actions

The user modelling component of Intelligent File Manipulator generates hypotheses about user actions that may not have been in accordance with the user's hypothesised intentions. Hypotheses are generated based on two reasoning mechanisms as already mentioned. One reasoning mechanism is the simulator of human error generations. This is based on a cognitive theory that attempts to formalise the reasoning that people use. This theory is called Human Plausible Reasoning (Collins & Michalski, 1989). Intelligent File Manipulator uses an adaptation of this theory to simulate possible errors of a user. For example, if a user has clicked on a certain file having a certain name s/he could be considered to have intended to click on a different file having a similar name.

A second reasoning mechanism is a limited plan recognizer which is used by the system to improve control. Intelligent File Manipulator evaluates the user's actions in terms of their relevance to his/her hypothesised goals. As a result of that every user action is categorised in one of four categories:

Expected: In this case the action is expected by the system in terms of the user's hypothesised goals.

Neutral: In this case the action is neither expected nor contradictory to the user's hypothesised goals.

Suspect: In this case the action contradicts the system's hypotheses about the user's goals.

Erroneous: In this case the action is wrong with respect to the user interface formalities and would normally produce an error message. One important assumption about users is that they do not intend to produce an error message, therefore actions like this are considered unintended.

The categorisation of user actions in the above categories is done based on what we call "instabilities". Instabilities are added and/or deleted from a list as a result of user actions. For example, the creation of an empty folder adds an instability to the file store because the system would expect a subsequent user action by which the folder would acquire a content or be deleted.

Instabilities are deleted when an action results in a file store state that should not contain them. For example, an instability associated with the existence of an empty folder is deleted if this folder is removed or if it acquires some content. In this sense the deletion of an instability represents the continuation of a user plan that started earlier. An action is considered expected if it deletes at least one of the existing instabilities of the file store state. It is considered neutral if it neither adds nor deletes instabilities and suspect if it only adds instabilities although there are already other instabilities that have not been deleted. However, Intelligent File Manipulator uses the categorisation of user actions as a way of acquiring some idea about which action may need more attention or not. By no means does it intervene based only on the categorisation of commands.

3 Algorithmic approach

The algorithmic approach of Intelligent File Manipulator can be summarised in the following steps:

1. The user issues an action.
2. The system reasons about the action so as to categorise it in one of the four categories.
3. If the action is categorised as "expected" or "neutral" it is executed.
4. If the action is categorised as "suspect" or "erroneous" then it is transformed based on an adaptation implementation of Human Plausible Reasoning. The transformation of the given action is done so that a similar alternative can be found which would not be suspect or erroneous.
5. The system reasons about every alternative action so that it can categorise it in one of the four categories in a similar way as in step 2.
6. If an alternative action is categorised as "neutral" or "expected" it is suggested to the user. Expected actions have priority over neutral ones.
7. If an alternative action is categorised as "suspect" or "erroneous" then it is ignored and is not suggested to the user.
8. If no better alternative can be found then the user's action is executed without the user realising that the system was alerted.
9. After the execution of the action instabilities are deleted or added accordingly.

An example of a user interaction with Intelligent File Manipulator is the following:

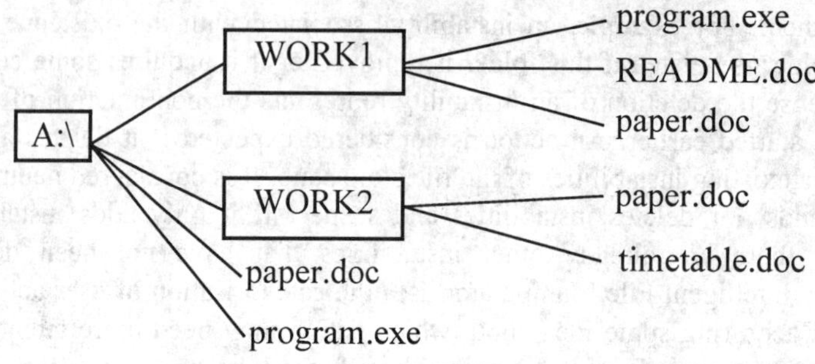

Figure 1: Initial state of the file store

Initially the user had two non-empty folders "WORK1" and "WORK2" and two files "paper.doc" and "program.exe" in her/his root directory (Figure 1).

The user created a new folder "WORK3" where s/he intended to move the files "paper.doc" and "program.exe" from the root directory.

The files were successfully 'cut' but they were accidentally 'pasted' into the old folder "WORK1" instead of the new folder "WORK3". However, in the folder "WORK1" there were already files having these names, which would run the risk of being accidentally replaced by other files.

In this case, a standard explorer would produce a message asking the user to confirm that s/he would want the files to be replaced. Intelligent File Manipulator would produce a more informative message asking the user whether s/he wanted to replace the existing files of "WORK1" or perhaps s/he really intended to paste these files into "WORK3" which was new and empty.

The reason why Intelligent File Manipulator would select "WORK3" to suggest it to the user is based on the fact that "WORK3" is empty and also its name is very similar to "WORK1" which was selected by the user.

4 Conclusions and future plans

Intelligent File Manipulator reasons about user's actions in a way that it can produce helpful and informative messages to possible user mistakes. The requirements analysis of Intelligent File Manipulator has been based on empirical data collected from real user interactions with operating systems which were not restricted to graphical user interfaces. At the present phase

Intelligent File Manipulator has not yet been evaluated by real users. Evaluation has been planned to be done in the near future. Users will be asked to perform certain tasks using Intelligent File Manipulator and then they will be asked to answer a questionnaire concerning the degree of help they thought they received. In addition a different group of users will be asked to perform certain tasks using a standard GUI such as Windows 95. A human expert will be watching over their shoulder. Then the remarks of the human expert will be compared to the reasoning that Intelligent File Manipulator would have produced.

5 References

Cerri, S. A., & Loia, V. (1997). A Concurrent, Distributed Architecture for Diagnostic Reasoning. *User Modeling and User-Adapted Interaction*, Vol. 7, No. 2, pp. 69-105.

Collins, A., Michalski R. (1989). The Logic of Plausible Reasoning: A core Theory. *Cognitive Science* 13, pp 1-49.

Mitrovic, A., Djordjevic-Kajan, S., Stoimenov, L. (1996). INSTRUCT: Modeling students by asking questions. *User Modeling and User Adapted Interaction*, Vol. 6, No. 4, pp. 273-302.

McGraw, K. L. (1994). Performance Support Systems: Integrating AI, Hypermedia and CBT to Enhance User Performance. *Journal Of Artificial Intelligence in Education*, Vol. 5, No. 1, pp. 3-26.

Virvou, M. (1998). RESCUER: Intelligent Help for Plausible Errors. *Proceedings of ED-MEDIA, ED-TELECOM 98, World Conferences on Education Multimedia and Educational Telecommunications*. Vol. 2, pp. 1471-1476.

Virvou, M., Jones, J. and Millington, M. (1999). Virtues and Problems of an active help system for UNIX. *Intelligent Systems for UNIX*, Hegner, S., Mc Kevitt, P., Norvig, P. and Wilensky, R. (Eds.), Dordrecht, The Netherlands: Kluwer Academic Publishers

Using a Painting Metaphor to Rate Large Numbers of Objects

Patrick Baudisch
Integrated Publication and Information Systems Institute (IPSI)
German National Research Center for Information Technology (GMD)
64293 Darmstadt, Germany
+49-6151-869-854
baudisch@gmd.de

1 The problem

When retrieving information from databases or search engines, or when configuring user profiles of information filtering systems, users have to describe what objects to retrieve. While some systems require users to describe their needs using textual input, other systems simplify the users' task by proposing a set of items, so that the user's task is reduced to picking or rating these items (e.g. relevance feedback (Haines 93)). This task is generally much simpler than writing queries from scratch. In interfaces of this type, users have to provide information of the type "does this item represent my information interest", "do I like this item" or "how much do I like this item". Similar problems are encountered in utility theory, when assessing the user's value functions (Keeney and Raiffa 76). As an example, Figure 1 shows such a selection user interface. It allows users the selection of TV channels, e.g. to configure their user profile for a TV recommender system. The interface contains about sixty toggle switches.

Assessing a large number of items can be time consuming. How can interfaces handle hundreds of such selection or assessment tasks in an efficient way? Sometimes it is possible to aggregate objects or to provide good defaults, so that only a few objects have to be manipulated in the first place. But what to do if there are no good defaults or too many of them? The set of TV channels that users can receive depends not only on the carrier, such as cable or satellite, but also varies widely depending on the local provider and subscriptions to pay-TV. If the interface provided extra buttons for every useful default configuration, the number of these buttons could easily exceed the number of actual toggles in the interface.

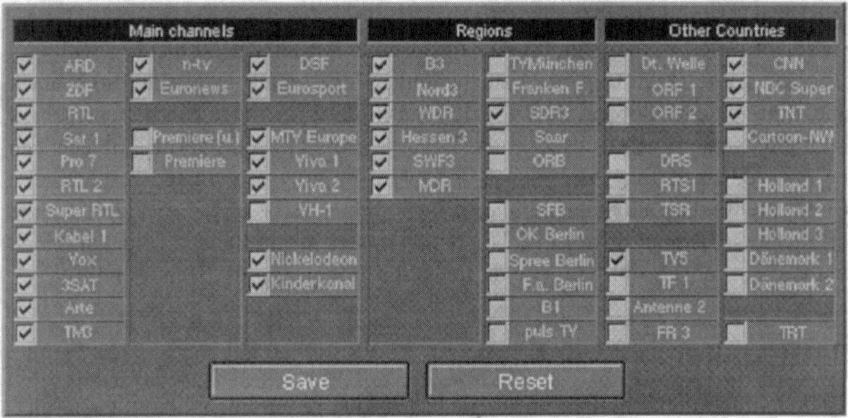

Figure 1: Dialog allowing users to input their personal TV channel profile. Each TV channel name is associated with a toggle switch.

2 Multiple select and painting

How are large numbers of interface objects handled in other application areas? One common technique is multiple select. First, users select a subset of items, e.g. cells in spreadsheet programs, icons in desktop GUIs, or pixels in paint programs. Usually this can be done with a mouse drag operation or using multiple mouse clicks while keeping a qualifier key depressed. Then, users select a method to be applied, e.g. clear the selected cells, move the selected icons, or set the selected pixels to a specific color. This order (select items first, then select method) is called *noun-verb* order (Smith 82). The noun-verb order allows restricting the list of available methods to those methods that are applicable to the items within the current selection (e.g. "empty trashcan" only to non-empty trashcans).

Paint programs offer more possibilities. Since painting programs deal with a single object type only, i.e. pixels, the noun-verb application order is not imperative. Therefore, paint programs provide both the noun-verb order as described above, and the verb-noun order, called painting. In the latter case a function is chosen, e.g. a painting tool such as pen or an airbrush, and then applied continually to all subsequently selected pixels.

The noun-verb order is preferable if several methods are to be applied to the same, possibly complex, selection. Otherwise painting has two advantages over selection. First, if the same painting tool is used several times in a row, then the tool has to be chosen only the first time, which saves interactions in all subsequent paint actions. Second, since the manipulation of painted items takes place immediately, painting gives better visual feedback.

We propose to apply these interaction techniques to the problem of object assessment stated at the beginning of this article. An interface object providing a Boolean value range (e.g. a toggle switch) is functionally equivalent to a black and white pixel; an object with a wider, but finite value range can be represented by a gray scale pixel. Consequently, larger layouts of such interface objects can be handled as if they were black-and-white or grayscale image respectively. Consequently, all advantages of pixel painting can be made available to the efficient assessment of objects. The resulting sets of interface objects combined with one or more painting methods are called "toggle maps" (Baudisch 98).

3 Requirements and layout

To maximize the benefit from toggle maps, two requirements have to be met. 1) Individual items should bear no or only small labels and should not require much time for decision making. Otherwise, users might prefer to release the mouse button and click switches individually. 2) It must be possible to manipulate several toggle switches per mouse drag. Otherwise there is no speed-up. This requires two things. First, there must be a need to manipulate a significant number of switches during individual sessions. Second, a significant frequency of co-occurrence between toggles has to exist and to be reflected by the layout (see below).

The goal of toggle map layout is to maximize usage speed by optimizing the factors stated above. To minimize recognition time (Requirement 1), layout should group buttons according to subjective similarity, so that recognizing one button already provides information about the neighboring buttons. To maximize the scope of paint operations (Requirement 2), layout should group buttons that are frequently manipulated together. Common techniques to accomplish this are non-metric multidimensional scaling and disjoint cluster analysis, e.g. using bottom-up clustering (Lindsey 77). Due to space limitations, we cannot discuss toggle map layout in detail here. However, the requirements toggle map layout tries to meet are very similar to the layout requirements of spatial menus (e.g. Norman 91, p. 261-280), so that many concepts can be transferred. In general, layouts based on frequency of co-occurrence usually offer good results (McDonald 86). Nonetheless, toggle map layout is more demanding than menu layout. Toggle map layout not only has to assert relative proximity of similar items but also direct adjacency, optimized for the respective paining tools.

4 Assessing objects using toggle maps

Using the techniques described above, the TV channel interface shown in Figure 1 can be handled as a toggle map. In this example, a tool which paints filled rectangles proved to be particularly efficient. Experiments proved that "paintable" interfaces such as the one shown in Figure 1 provide significant speedups over click-only interfaces (Baudisch 98).

Alternatively to creating layouts using multidimensional scaling, it is possible to apply the toggle map concept to domains whose a regular internal structure practically imply a layout. In (Baudisch 98) an efficient timer interface was presented that used a tabular layout with days forming rows and hours forming columns. Layouts based on such strong internal structure have the advantage of meeting the requirements especially well, because both subjective similarity and frequency of co-occurrence between neighboring objects are unusually high.

Figure 2 gives another example of such a domain with a strong internal structure. The shown toggle map allows users to express their preferences concerning computer monitors, e.g. to retrieve the corresponding classified ads from a database. In the shown state, the user is looking for a rather large screen with a high resolution, e.g. for CAD applications. The interface contains buttons only for those feature combinations that correspond to available objects. Buttons can take a discrete range of grayscales to represent different grades of like and dislike. They can be painted very efficiently, e.g. using an airbrush.

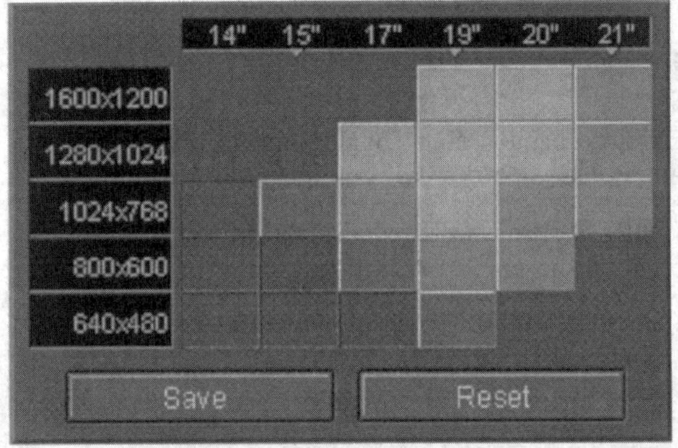

Figure 2: Toggle map allowing users to enter their preferences about computer monitors.

If the internal structure of the data objects is more than two-dimensional, it is possible to adapt the implied toggle map layout within bounds. Figure 3 shows two examples that are derived from the example shown in Figure 2 by successively introducing two more dimensions. To keep the original 2D painting interaction applicable, we used so-called *explosion displays* that solve the problem of occlusion. Layers of toggles are spread, the resulting gaps provide access to buttons that would otherwise be occluded. Using this technique, toggles at any depth can still be accessed directly with a 2D input device, such as a mouse. To provide maximum efficiency, additional n-dimension tools can be introduced. Figure 3a sketches the usage of a tool that allows painting filled (hyper) cubes by dragging the mouse from one vertex to the diametrically opposed one.

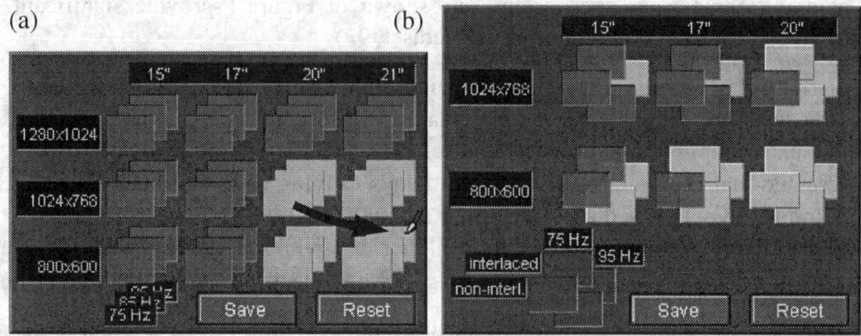

Figure 3: A three (a) and a four-dimensional (b) toggle map.

5 Conclusions and Future work

In this article, we discussed how painting interactions can be applied to the selection and assessment of objects, especially objects from domains having a regular internal structure facilitating layout. Future work will include experiments with computer generated layouts and n-dimensional painting tools.

6 References

Baudisch, P. (1998) Don't click, paint! Using Toggle Maps to Manipulate Sets of Toggle Switches. In *Proceedings of UIST '98,* San Francisco, pp. 65-66.

Haines, D. and Croft, W. (1993). Relevance Feedback and Inference Networks. In *Proceedings of SIGIR '93*. Pittsburgh, PA: ACM Press. pp. 2-11.

Keeney R.L., and Raiffa, H.R. (1976). *Decision with Multiple Objectives: Preferences and Value Tradeoffs.* John Wiley and Sons.

Lindsey, P.H., and Norman, D.A. (1977) *Human information processing*, New York: Academic Press.

McDonald, J.E., Dayton, T., McDonald, D.R. (1986). *Adapting menu layout to tasks* (MCCS-86-78). Las Cruces, NM: Memoranda in Computer and Cognitive Science, Computer Research Laboratory, New Mexico, State University.

Norman, K. L. (1991). *The Psychology of Menu Selection: Designing cognitive control at the human/computer interface,* Norwood, New Jersey, Ablex.

Shneiderman, B. (1998). *Designing the User Interface: Strategies for effective human-computer interaction*, 3rd edition, Reading MA: Addison-Wesley.

Smith, D.C. Irby, C. Kimball, R. Verplank, W., and Harslem, E. (1982). Designing the Star User Interface, *Byte* 7(4):242-282. McGraw-Hill, Inc.

IconStickers: Converting Computer Icons into Real Paper Icons

Itiro Siio[a], Yoshiaki Mima[b]
[a] Faculty of Engineering, Tamagawa University, siio@eng.tamagawa.ac.jp
[b] IBM Research, Tokyo Research Laboratory, mima@acm.org

1 Introduction

Many users of computers based on the desktop metaphor interface, prefer to put icons on the desktop (outside folders), or in upper-hierarchy folders . The real estate on the computer screen is therefore very quickly occupied by icons (Barreau and Nardi 1995) (Nardi and Barreau 1997).

Why not use the real world as an extension of the computer screen? This would be a natural solution, since people think of the screen as an extension of the real world. This paper describes the concept, prototype, and applications of IconStickers, which are paper icons linked to original icons in a computer.

2 IconStickers

We propose a method for taking icons off the computer screen in the form of paper icons, called IconStickers, so that they can be handled in the real world.

Figure 1: Example of an IconSticker. A bitmap image, name, and bar code are printed on a 28 x 89 mm sticker.

An IconSticker is a piece of paper on which an icon image, icon name, and machine-readable information (bar code) are printed. It is a real-world representation of a computer icon. Figure 1 shows an example of an IconSticker. It is printed by dragging-and-dropping a computer icon over an "Exit" icon, as if the user were taking it off the computer screen (Figure 3). When an IconSticker is scanned by a bar-code reader (Figure 4), the original computer icon is accessed, and is opened on the computer screen. Since IconStickers are printed on sticker labels, a user can easily arrange them in the real world, as shown in Figure 2.

3 Implementation

Figure 2 shows an overview of the IconSticker environment. A label printer and a bar-code reader are connected to a personal computer.

The IconSticker manager program is coded in C++ on Mac OS. The program has two functions. One is to generate an IconSticker image and print it when a Mac OS icon is dragged-and-dropped onto the icon of the program. The other is to "open" the original Mac OS icon on the computer screen when an IconSticker is scanned with a bar-code reader. The following sections give details of these functions.

The manager program icon is normally placed in the lower left corner of the display, as an "Exit to the real world" icon. When a user drags-and-drops an icon of a document or an application, the label printer by the display prints an IconSticker corresponding to the icon. This makes it appear as if the user were taking the icon off the computer screen (Figure 3).

Figure 2: IconStickers make it possible to arrange icons outside the computer screen.

In reality, not the icon but an "alias" of it is taken off the screen. The alias is a Mac OS icon that has a pointer to the original icon, and operations on the alias are automatically applied to the original icon (Apple Computer, Inc. 1995). Although aliases are similar to "short cuts" in Windows or "symbolic links" in the Unix file system, they have superior ability to track the original icons when their names and locations are changed.

The prototype system creates an alias for each dragged-and-dropped icon, and stores it in a predefined directory. In a network configuration, aliases are stored in a predefined file server. Each alias is identified by its name consists of 8 alphanumeric characters, indicating the time stamp when it is created.

As shown in Figure 1, an IconSticker has an icon image, an icon name, and machine-readable symbols to identify the dragged-and-dropped computer icon. The name of the alias is coded in the machine-readable symbols. In the current prototype, bar code is used for the machine-readable symbols.

The second function of the IconSticker manager program is to open or invoke the original computer icon when the corresponding IconSticker is scanned with a bar-code reader, in the same way as when a computer icon is double-clicked. When the bar-code reader detects an encoded name in an IconSticker, the program calls OS functions to open the alias icon with the same name.

Since an alias is actually created when an IconSticker is printed, the result of scanning an IconSticker is exactly the same as that of double-clicking the alias icon. An experienced user familiar with alias manipulations could easily understand the function of IconStickers.

Combining a mouse and bar-code reader has the merit of allowing users to handle computer icons and IconStickers without changing devices. It also helps to provide a mental model that enables users to regard icons and IconStickers as the same thing. In the current prototype, a pen-mouse and bar-code wand are joined together

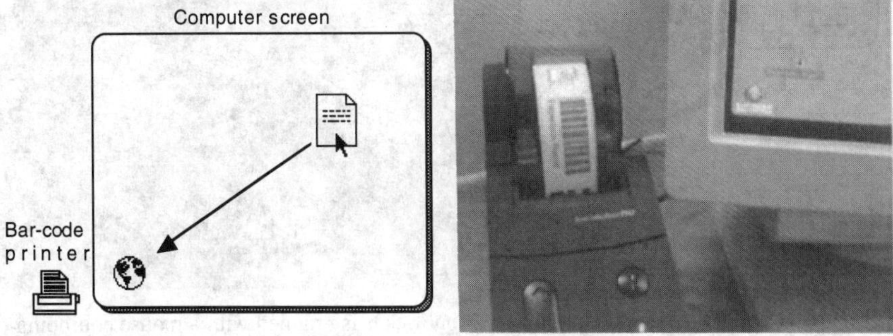

Figure 3: By dragging-and-dropping an icon onto the "Exit to real world" icon, a corresponding IconSticker is printed by a label printer next to the display.

to provide seamless operations (Figure 4).

4 Applications

IconStickers can be attached to items in a real working space, as shown in Figure 2. As a result, the variation in the layout of icons is broader than the computer display. They can be pasted into a scrapbook by people who want to manage and classify computer icons in the real world. IconStickers for frequently used applications and documents can be attached to places where users can easily access them, such as the frame of a computer display or keyboard.

Adding links to computer objects augments the capabilities of everyday items. For example, attaching an IconSticker to associates;

1, A hard copy of a document with the original soft copy.

2, A telephone with a telephone database.

3, A display or a keyboard with a control panel software.

4, A printer with a queue monitoring program.

5, A product or its manual with a URL containing the information about the product.

Figure 5, shows some examples of the above uses of IconStickers.

IconStickers allow the use of paper-based information distribution, sharing, and storing methods. IconStickers can be attached to real bulletin boards or postcards. They can even be copied by using a normal copier for paper documents. With the current prototype, it is possible to create a document-sharing system by storing aliases and original documents on a file server.

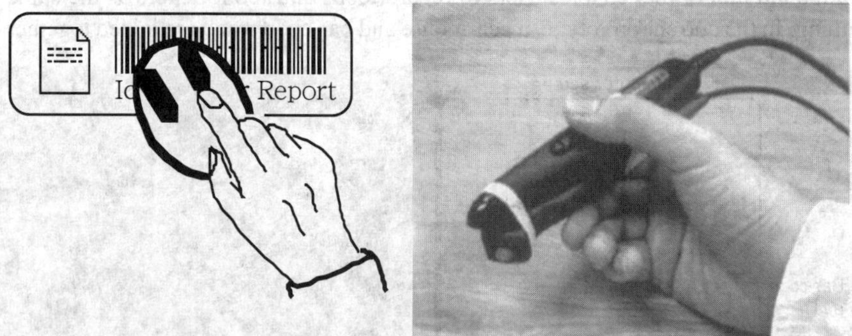

Figure 4: When the bar code part of the IconSticker is scanned with a mouse equipping with a bar-code reader, the original computer icon is opened on the display (left). A combination of a mechanical pen-mouse and a bar-code wand is used in the current prototype (right).

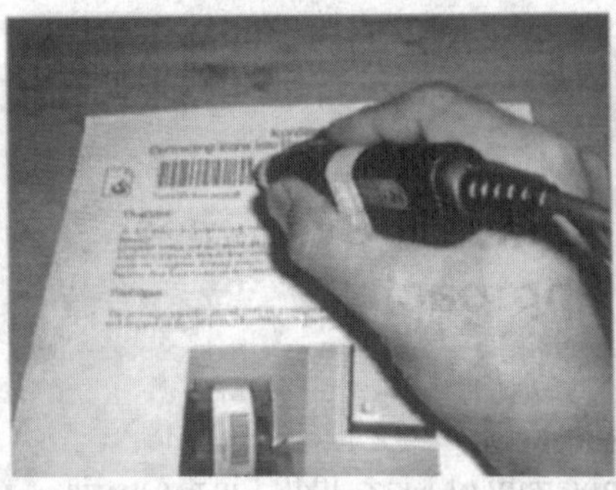

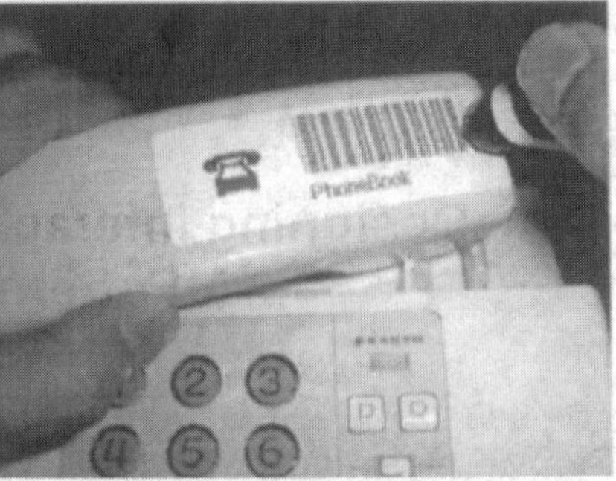

Figure 5: Examples of IconSticker applications. It can be used to find the original soft copy (left), or to access a telephone database (right).

5 Related work

InfoBinder (Siio 1995) is a small box-shaped device used with projector that projects a window system and its desktop into a real desktop. In this system, when the user closes a window, an icon is projected onto the InfoBinder. This experimental system is also an attempt to "materialize" icons. Conceptually, IconSticker is an extension of InfoBinder.

The MetaDesk (Ishii and Ullmer 1997) system is a system has Phicons (physical icons). Phicons represent, for examples, buildings, and can translate, rotate, and scale a map projected onto a table. Like IconSticker, it uses physically operable icons, but the shapes, objectives, and functions of these icons are different. IconStickers are tightly coupled with icons according to the desktop metaphor, and have logical links to the original iconic system.

6 References

Apple Computer, Inc. (1995). *Inside Macintosh CD-ROM*, Addison-Wesley, ISBN0201946114.

Barreau, D. K. and Nardi. B. (1995). Finding and Reminding: File Organization from the Desktop, *ACM SIGCHI Bulletin*, Vol.27, No.3, pp.39-43, July.

Ishii, H. and Ullmer, B. (1997). Tangible Bits: Towards Seamless Interfaces between People, Bits and Atoms, *Proceedings of CHI '97*, pp. 234-241, ACM Press, March.

Nardi, B. and Barreau, D. K. (1997). "Finding and Reminding" Revisited: Appropriate Metaphors for File Organization at the Desktop, *ACM SIGCHI Bulletin*, Vol.29, No.1, January.

Siio, I. (1995). InfoBinder: A Pointing Device for Virtual Desktop System, *Proceedings of HCI International '95*, pp. 261-264, Elsevier Science, July.

Designing Interactive Music Systems: A Pattern Approach

Jan O. Borchers
Telecooperation Research Group, University of Linz, 4040 Linz, Austria
http://www.tk.uni-linz.ac.at/~jan/

1 Introduction

Music is a structurally very complex type of multimedia information. Today, computer systems easily play back music in high quality, but an ideal system should offer many more, media-appropriate ways to interact with musical data. Users should be able to hum a melody to locate the tune they want, then conduct it, or play to it with adjustable computer support, alone or with others via the Internet. Such interactivity would encourage musical creativity, and let users learn about musical concepts by playing with them. This paper shows how we used a pattern language approach for software design, interaction design, and the application domain "music" to build a system with these qualities.

2 The Pattern Approach

Designing such a system requires expertise from software engineering, user interface design, and music to be brought together. We used *pattern languages* to capture and describe this knowledge for all three domains in a uniform way. Pattern languages were originally developed to capture proven solutions to common design problems in urban architecture (Alexander et al. 1977, 1979). Each pattern captures one such "guideline"; the pattern language connects them into a structure, helping the designer to create an overall design of high quality.

The idea has been adapted successfully by software engineering (Coplien & Schmidt 1995), and is currently beginning to find its way into human-computer interaction (HCI) design (Bayle et al. 1997, Tidwell 1998).

2.1 A Formal Pattern Language Definition

To define the components of a pattern language regardless of the problem domain it addresses, we introduce a formal notation.

A *pattern language* is a directed acyclic graph. Each node represents a *pattern*. There is a directed edge from pattern p_1 to p_2 if p_1 *recruits* p_2 to complete its solution. Edges pointing *away* from a pattern are its *consequences*, showing what lower-level patterns need to be applied next. Edges pointing *to* a pattern are its *context*, the situations in which it can be applied. This relationship establishes a *hierarchy* within the pattern language. It leads the designer from patterns addressing large-scale design issues, to patterns about small details.

Each *pattern* is a set $p=\{n, f_1...f_i, s, e_1...e_j\}$ of a *name* n, *forces* $f_1...f_i$, the *solution* s, and *examples* $e_1...e_j$. It describes a commonly encountered design problem, and suggests a solution that has proven useful in this situation.

While this formal notation helps to clarify the structure of patterns and pattern languages, patterns are actually written texts, to make them easy to read and understand even for people from other professions. Each part of a pattern, and its connections to other patterns, are usually presented as several paragraphs in the pattern description (Alexander et al. 1977).

The *name* of a pattern helps to reference it easily, communicate its central idea quickly, and build a vocabulary. The *forces* are aspects of the design context that need to be optimised. They usually come in pairs that contradict each other. The *solution* describes a proven way to balance these forces optimally for the given context. The *examples* show that the solution has been applied successfully in existing designs. For a more detailed description of the concept of pattern languages, see (Alexander et al. 1977, Alexander 1979).

2.2 Software, HCI, and Musical Design Pattern Languages

A key idea of this paper is to use the pattern approach not only in software or interaction design in isolation, but to structure the problem domain, in this case "music", into a pattern language as well. The rationale behind this approach is that design patterns can be defined for any discipline where some creative design work takes place. In music, the "designer" is a composer or player. Patterns can describe important aspects of how a composition is crafted, or how choices between alternative chord sequences, voices, or single notes are made.

As in the other disciplines, it is obvious that the knowledge captured in a pattern language cannot represent the intuition and creativity of a design expert. But it *is* possible to communicate basic principles of making good designs, or good music, using the pattern approach. Casting the expertise of all disciplines involved into patterns (Fig. 1, only selected patterns shown) has two advantages:

For the development team: By studying each other's pattern languages, it becomes easier for the different professions in the interdisciplinary development team to understand each other's design principles, paradigms, and professional values. The common vocabulary of design patterns from the various professions facilitates communication within the design team.

For the user: Architecture and software engineering have shown that pattern languages are a very suitable format for communicating expertise. An interactive system such as the exhibit we had in mind can use patterns to didactically structure its presentation of musical concepts. These musical patterns can be embodied in user interface objects and relationships that users can see and interact with. The musical concepts become easier to understand.

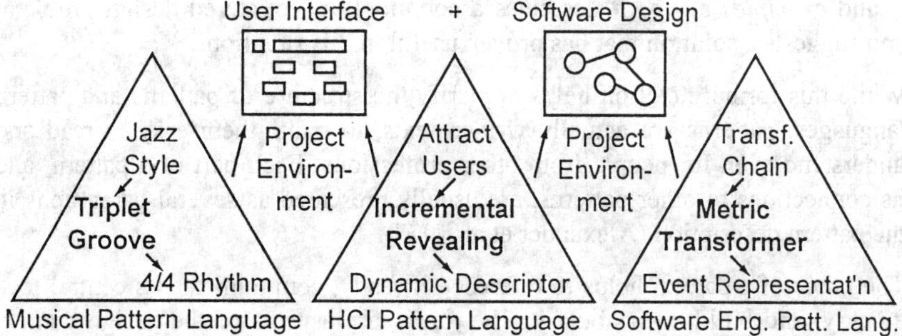

Fig. 1: Design pattern languages for Music, HCI, and software engineering are used in a project environment "interactive exhibit" to create a user interface and software design.

3 Pattern Examples

To demonstrate the use of patterns, we will give examples from each domain involved. They have been shortened, and references to other patterns (in italics) left unresolved, to fit the format of this paper.

3.1 Musical Design Pattern: *Triplet Groove*

Context: Playing music in the *Jazz Style.*

Forces: Players need to create a swinging feeling that the straight rhythm from other musical styles does not convey. **But:** Sheet music cannot include all rhythmic variances; it would become unreadable.

Solution: Where the score contains an evenly spaced pattern of eighth notes, shift every second eighth note backwards in time by about one third of its length, shorten it accordingly, and make the preceding eighth note one third longer. The length ratio changes from 1:1 to 2/3:1/3. Two straight eighth notes become a triplet quarter and eighth note. The result is a "laid-back groove".

Examples: Any recorded Jazz piece features this rhythmic shift. The actual shift percentage varies widely: usually, the faster a piece, the less shifting takes place.

Consequences: This pattern uses an underlying straight beat like *4/4 Rhythm.*

3.2 Interaction Design Pattern: *Incremental Revealing*

Context: Decide how to unfold contents and features of an interactive system so that it conveys a *Simple Impression* to *Attract Users,* and *Engages Users.*

Forces: A simple impression is important to make a system look non-intimidating and inviting, especially for novices. **But:** To keep users engaged, the system needs to convey its depth of features and contents as well.

Solution: Initially, only present a concise and simple overview of the system's functionality. When the user actively shows interest in a certain part of this overview, offer additional information about it, revealing in successive stages what lies behind the initial presentation.

Examples: Desktop GUIs hide menu entries in menu bars until the user selects a menu. *WorldBeat* has a simple main selection screen with only names and icons for composing, conducting, etc.; when the cursor is over an item, a short text explains it; when it is selected, the system switches to the new page.

Consequences: Incremental revealing is easier when the contents have a *Flat & Narrow Tree* structure. To show what lies behind a user interface object, use *Dynamic Descriptors* (as in Mac OS Balloon Help, or Windows ToolTips).

3.3 Software Design Pattern: *Metric Transformer*

This pattern addresses the problem that, in a *Transformer Chain* of modules that modify an incoming stream of music in *Event Representation,* one module needs to modify the timing, for example, to implement a component that allows the user to apply *Triplet Groove* to music that is being played. This pattern has been described in detail in (Borchers & Mühlhäuser 1998).

4 The WorldBeat Exhibit

WorldBeat is an interactive music exhibit that the author designed for the *Ars Electronica Center (AEC),* a technology "museum of the future" in Linz, Austria. *WorldBeat* offers visitors new ways of interacting with music. Using just a pair of infrared batons, they navigate through the system, play virtual instruments like drums or a guitar, conduct a computer orchestra playing a classical piece, and improvise to a Blues band with computer support. Users can locate tunes by humming their melody, and try to recognize instruments by their sound. Details can be found in (Borchers 1997).

Many design aspects of *WorldBeat* are applications of the musical, HCI, and software engineering patterns, as described in the examples. The success of *WorldBeat* indicates the validity of our approach: AEC visitors rated it among the top three of the center's exhibits, and the system received the *1998 Multimedia Transfer Award* among over 150 international contestants.

We are currently designing a new exhibit to let people compose in the twelve-tone style of Arnold Schönberg. We have successfully cast many aspects of his composition theory into a pattern language. Also, many HCI and software engineering patterns we used in WorldBeat carried over very well to this new project, and helped to communicate experience to new design team members.

5 Further Research

Our pattern languages have to be refined, extended, and validated further. The approach needs to be applied to different application domains, and our formal notation suggests computer support for working with pattern languages. The author will use the pattern format to teach HCI design to students, to see how novices benefit from this approach.

6 References

Alexander, C., Ishikawa, S., Silverstein, M., Jacobson, M., Fiksdahl-King, I. & Angel, S. (1977). *A Pattern Language: Towns, Buildings, Construction.* New York: Oxford University Press.

Alexander, C. (1979). *The Timeless Way of Building.* Oxford University Press.

Bayle, E., Bellamy, R., Casaday, G., Erickson, T., Fincher, S., Grinter, B., Gross, B., Lehder, D., Marmolin, H., Moore, B., Potts, C., Skousen, G. & Thomas, J. (1998). "Putting it all together: Towards a pattern language for interaction design". *SIGCHI Bull.* **30**(1), 17–23. New York: ACM.

Borchers, J. (1997). WorldBeat: Designing a Baton-Based Interface for an Interactive Music Exhibit. *Proc. CHI'97* (Atlanta, GA), 131–138. New York: ACM. See also the video proceedings of that conference.

Borchers, J. & Mühlhäuser, M. (1998). Design Patterns for Interactive Musical Systems. *IEEE MultiMedia* **5**(3), 36–46. Los Alamitos: IEEE Computer Society.

Coplien, J. & Schmidt, D. (1995). *Pattern Languages of Program Design.* Software Patterns Series. Reading, MA: Addison-Wesley.

Tidwell, J. (1998). Interaction Design Patterns. *PLoP'98 Conf. on Pattern Languages of Programming* (Monticello, IL). Presentation; extended version at http://www.mit.edu/~jtidwell/interaction_patterns.html.

Dextrous Interaction with Computer Animations using Vision-Based Gesture Interface

Jakub Segen and Senthil Kumar
Bell Laboratories, USA
segen@lucent.com, senthil@bell-labs.com

1 Introduction

It has been only six years since the publication of Neal Stephenson's "Snow Crash" [8] and much of the technology predicted in this techno fiction already exists. Realistic 3D scenery and avatars with human-like motion can be animated on a PC, along with continuous speech recognition, speech synthesis, and 3D sound. This could create the medium for the next century, for interactive entertainment, education, business and social gatherings on the web. However, one part is still missing – a simple and natural interface to 3D animations. Conventional input devices such as a mouse or joystick control only two parameters, and are counter-intuitive in 3D interaction. New devices such as sensing gloves or body suites are too cumbersome for the general use.

A technology that promises a natural and unconstrained spatial interface is a direct gesture input, where gestures and control values are extracted from hand images acquired by video cameras ([1, 3, 4]. Such optical interfaces promise multiple degrees of freedom for control, high precision, speed, and ultimately very low cost.

Research on video-based gesture recognition in Bell Labs started ten year ago. Our first real-time gesture recognizer was operating 1n 1990 at four frames/sec, using a system of thirty processors [5]. A recently developed practical gesture interface system, GestureVR [6] runs at the rate of 60Hz on a PC. It recognizes gestures reliably, computing up to twenty simultaneous parameters from two hands. This paper describes several variants of GestureVR, and shows how this system can be used as an intuitive input to spatial applications.

2 GestureVR

The basic component of GestureVR is a planar gesture recognition unit. It acquires hand images from a camera, recognizes gestures and estimates pose parameters. Two static and one dynamic gestures are recognized, shown in Figure 1. The *Point* gesture has an extended index finger, while the thumb can be extended or folded. In the *Reach* gesture all fingers extended and separated. *Click* is a dynamic gesture, formed by a rapid bending of index finger. All other gestures are called *Ground*.

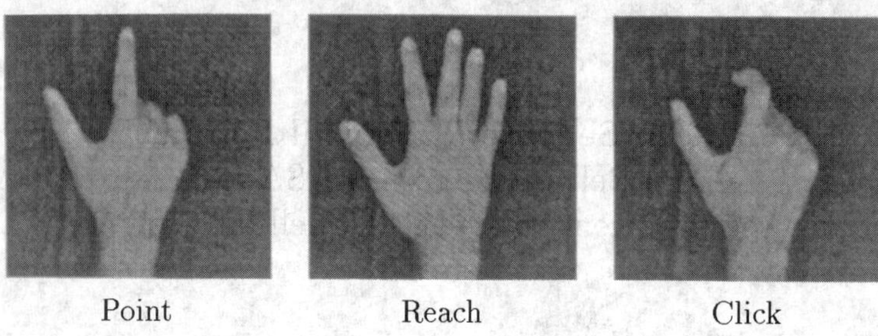

Point Reach Click

Figure 1: Gesture Types

For the *Point* gesture the planar recognition unit computes the planar *pose*, that is the position and orientation in the image plane, for the index finger and the thumb (if it is extended). The simplest version of the GestureVR is the planar recognizer, that computes the pose of the index finger. We refer to it as the 3-DOF (three degrees-of-freedom) system. A recent extension of the 3-DOF system is s single camera 4-DOF system, which computes a fourth parameter - the height of the tip of the index finger, [2].

The planar recognizer is very stable, precise, and fast. It has been tested with users of different age, gender, skin complexion and hand size. The results show very low recognition errors (less than 1/500) on static gestures, while the dynamic gesture ("Click") requires a brief user training to reach an acceptable error level. The angular jitter of the pose parameters is less than 1/2 degree. The system operates at rate of 60Hz, which is limited only by the field rate of the NTSC cameras.

2.1 Multi-camera Systems

To compute control parameters for 3D applications planar gesture recognizers are combined into stereo systems, which can compute a five parameter pose for the thumb and the index finger. The pose of a finger consists of the components: $(X, Y, Z, \alpha, \epsilon)$, where, (X, Y, Z) denotes the position of the finger tip and (α, ϵ) denotes the azimuth and elevation angles of the finger's axis. The basic 3D version of GestureVR is a 5-DOF system, that delivers a five parameter pose of the index finger. Its extensions compute pose parameters for the index finger and the thumb, from one or both hands. A 2x5-DOF system computes five parameter pose of index finger from each hand, for a total of ten parameters. A 10-DOF system delivers the five parameter pose from the index finger and the thumb of one hand, and a

2x10-DOF system computes poses of these two fingers for both hands for a total of twenty control parameters. The reason for using these separate versions of GestureVR instead of just one maximal 2x10-DOF system is the trade-off between the number of degrees of freedom and some of the performance parameters. For example, the 5-DOF and 2x5-DOF systems, which do not rely on the thumb parameters, operate over a greater range of hand orientations than the systems using the thumb. The two-hand systems have somewhat lower recognition rate than single hand systems, since the gesture interpreter must identify left and right hand and resolve or flag occlusions.

In a stereo system that uses two cameras the range of the elevation angles is about 80 degrees centered at the horizontal plane. This range can be increased to about 150 degrees using four cameras. Our implementation of the four camera system uses only two cameras at a time, selecting at run time the best pair of cameras to acquire each video field. Such a system has nearly all of the advantages of an implementation that would simultaneously process all four video streams, but it allows us to keep the cost of the video acquisition hardware and image processing at about the same level as in a two camera system.

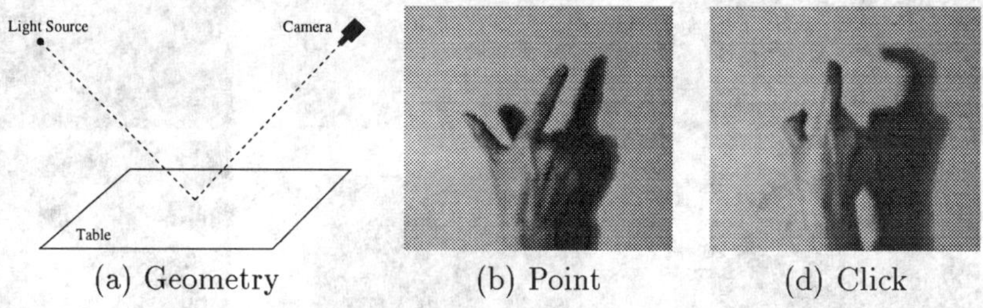

(a) Geometry (b) Point (d) Click

Figure 2: Shadow Gestures

2.2 Shadow Gestures

One can avoid the cost and complexity of using multiple cameras utilizing a shadow of the hand in lieu of the second image [7]. A shadow based stereo system uses a single camera and a point light source, Figure 2. It performs similarly to the two-camera 5-DOF and 10-DOF systems, as long as the hand remains above a certain height (about 10 cm), so that the hand and shadow images can be separated. Despite this constraint, and the need to control the ambient illumination, this system can be very useful as a simple and inexpensive 3D interface.

3 Applications

GestureVR provides an intuitive input to spatial applications, that require multi-dimensional control. The type of the recognized gesture invokes a command, while the pose components are translated into spatial arguments.

In the **3D Fly-thru** the user navigates above a landscape, using the 5-DOF system. Figure 3 (a) shows diving into the Yosemite Valley. Flight velocity is controlled by the position of the pointing finger along for-and-aft axis (Y value). Moving the hand forward increases the speed and bringing it towards the user decreases it. Placing the hand aft of the origin results in a backward motion. Moving up the hand increases the elevation. The azimuth angle controls the turning speed, that is pointing directly ahead makes a straight flight, a slight left turn of the hand starts a left turn, increasing the hand angle makes turning faster, etc. Similarly, the pitch angle is controlled by the elevation angle of the hand. The motion can be stopped using the "Reach" gesture.

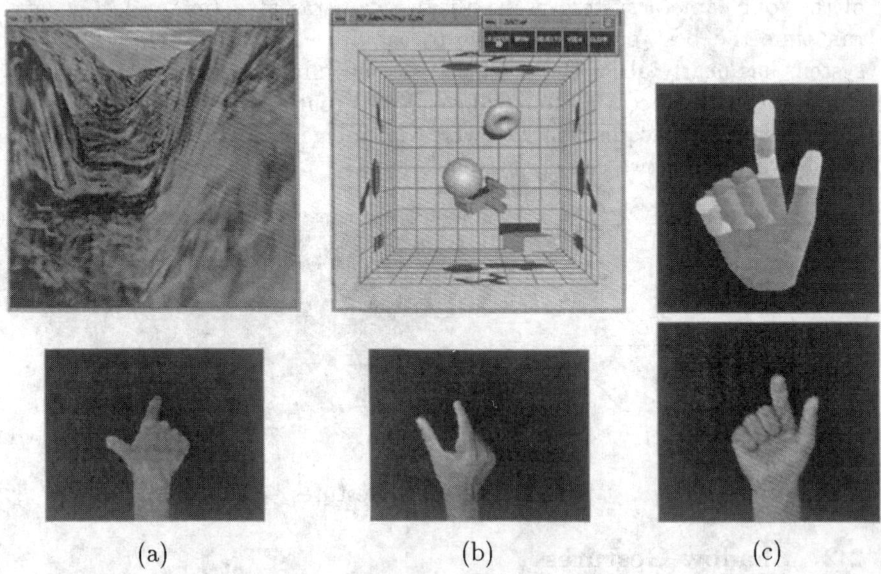

(a) (b) (c)

Figure 3: (a) Fly-thru. (b) Scene composer. (c) Hand model control

The **Scene Composer** (Fig. 3 (b)) uses the 10-DOF system to assemble 3D scenes. The hand controls a robot gripper, that can grasp and move objects. Other modes, invoked with the "Reach" gesture, enable rotating the scene by twisting the hand, or drawing a surface with a sweep of the hand.

The **Articulated Hand Control** (Fig. 3 (c)) enables a partial control of an articulated 3D model of a hand. The 10-DOF system with a bit of inverse kinematics is used to control the index finger and the thumb.

The **Doom** interface (Fig. 4) uses the 3-DOF system to play the game.

The mapping of controls is similar to that in the Fly-thru, but restricted to a plane. The "Reach" gesture is used to invoke actions such as opening doors, and "Click" fires the gun.

Move Open Door Fire

Figure 4: Playing Doom

References

[1] M. W. Krueger. *Artificial Reality II.* Addison-Wesley, 1991.

[2] S. Kumar and J. Segen. Gesture based 3D Man-Machine Interaction using a Single Camera. In *Proc. IEEE Intern. Conf. on Multimedia Comp. and Systems, ICMCS'99, Florence, Italy.* 1999.

[3] C. Maggioni. GestureComputer - New Ways of Operating a Computer. In *Proc. International Conference on Automatic Face and Gesture Recognition*, pages 166–171. June 1995.

[4] J. Segen. Controlling Computers with Gloveless Gestures. In *Proceedings of Virtual Reality Systems.* 1993.

[5] J. Segen and K. Dana. Parallel Symbolic Recognition of Deformable Shapes. In H. Burkhardt, Y. Nuevo, and J. C. Simon, editor, *From Pixels to Features II.* North-Holland, 1991.

[6] J. Segen and S. Kumar. GestureVR: Vision-Based 3D Hand Interface for Spatial Interaction. In *Proc. ACM Multimedia Conference.* 1998.

[7] J. Segen and S. Kumar. Shadow Gestures: 3D Hand Pose Estimation using a Single Camera. In *Proc. CVPR, Fort Collins, CO.* 1999.

[8] N. Stephenson. *Snow Crash.* Bantam Spectra, 1993.

Perceptual Intelligence: learning gestures and words for individualized, adaptive interfaces

Alex Pentland, Deb Roy, Chris Wren
M.I.T. Media Laboratory
E15-387, 20 Ames St., Cambridge MA 02139
http://www.media.mit.edu/~pentland

1 Introduction

In the language of cognitive science, perceptual intelligence is the ability to solve the frame problem: it is being able to classify the current situation, so that you know what variables are important, and thus can take appropriate action. Once a computer has the perceptual intelligence to know who, what, when, where, and why, then statistical learning methods are usually sufficient for the computer to determine what aspects of the situation are significant, and to react appropriately.

A key idea in building interfaces based on the idea of perceptual intelligence is that they be adaptive both to overall situation and to the individual user. Thus the interface must learn user behaviors, and how they vary as a function of the situation. For instance, we have built systems that learn user's driving behavior, thus allowing the automobile to *anticipate* the driver's actions (Pentland and Liu 1999), systems that learns typical pedestrian behaviors, allowing it to detect unusual events (Oliver et al 1998).

Most recently we have built audiovisual systems that learn word meanings from natural audio and visual input (Roy and Pentland 1998). A significant problem in designing effective interfaces is the difficulty in anticipating a person's word choice and associated intent. Our system addresses this problem learning the vocabulary of each user together with its visual grounding.

2 Smart Rooms

Our first smart room was developed in 1991; now there are smart rooms in Japan, England, and several places in the U.S., and can be linked together by ISDN lines In this section I will describe some of the perceptual capabilities of my smart room project, and provide a few illustrations of how these capabilities can be combined into interesting applications.

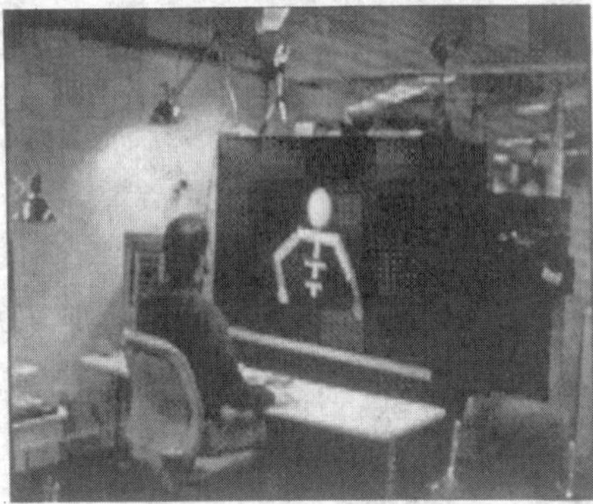

The system uses 2-D observations to drive a 3-D dynamic model of the skeleton. This dynamic model uses a control law that chooses 'typical' behaviors when it is necessary to choose among multiple possible legal trajectories It is real-time, self-calibrating, and has been successfully integrated into application ranging from physical rehabilitation to a computer-enhanced dance space.

2.1 Real-Time Person Tracking using a Dynamic 3-D Model

The human body is a complex dynamic system, whose visual features are time-varying, noisy signals. Accurately tracking the state of such a system requires use of a recursive estimation framework. The elements of the framework are the observation model relating noisy low-level features to the higher-level skeletal model and *vice versa*, and the dynamic skeletal model itself. This extended Kalman filter framework reconciles the 2-D tracking process with high-level 3-D models, thus stabilizing the 2-D tracking by directly coupling an articulated dynamic model with raw pixel measurements. Some of the demonstrated benefits of this added stability include increase in 3-D tracking accuracy, insensitivity to temporary occlusion, and the ability to handle multiple people.

The system runs on an SGI O2 machine at 30 Hz, and has performed reliably on hundreds of people in many different physical locations, including exhibitions, conferences, and offices in several research labs. The 'jitter' or noise observed experimentally is 0.9 cm for 3-D translation and 0.6 degrees for 3-D rotation when operating in a desk-sized environment (Wren and Pentland 1997). We have used the recovered 3-D body geometry for several different gesture recognition tasks, including a real-time American Sign Language reader Starner, Weaver and Pentland 1998), and a system that recognizes T'ai Chi gestures, and trains the user to perform them correctly (Becker and Pentland 1996). Typically these systems have a gesture vocabularies of 25 to 65 gestures, and recognition accuracies above 95%

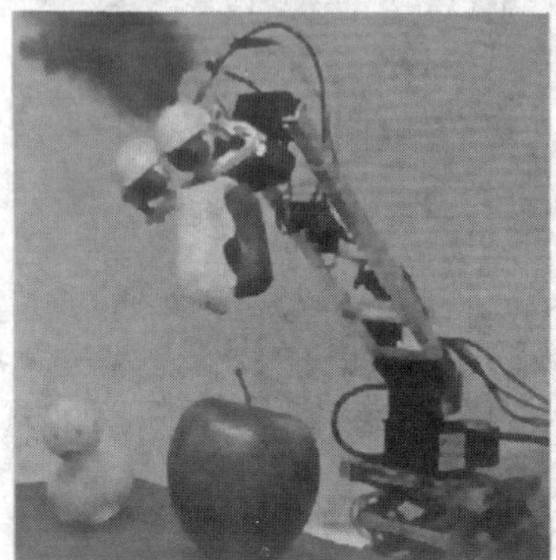

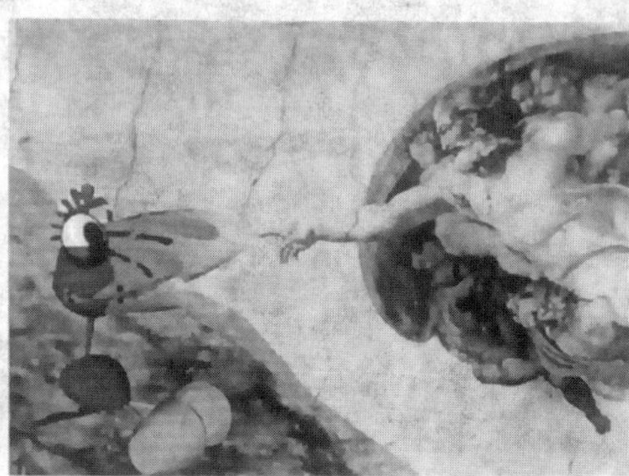

Toco the Toucan in both his robotic and computer graphics form. This demo of word and gesture learning for human-machine interactions was called ``the best demo at SIGGRAPH '96'' by the Los Angeles Times.

2.2 Audio-visual Interpretation

Traditional HCI interfaces have hard-wired assumptions about how a person will communicate. In a typical speech recognition application the system has some preset vocabulary and (possibly statistical) grammar. For proper operation the user must restrict what she says to words and vocabulary built into the system. However, studies have shown that in practice it is difficult to predict how different user will use available input modalities to express their intents. For example Furnas *et al* did a series of experiments to see how people would assign keywords for operations in a mock interface. They conclude that: "There is no one good access term for most objects. The idea of an "obvious", "self-evident,' or "natural" term is a myth! ... Even the best possible name is not very useful...Any keyword system capable of providing a high hit rate for unfamiliar users must let them use words of their won choice for objects." Our conclusion is that to make effective interfaces there need to be adaptive mechanisms which can learn how individuals use modalities to communicate. We have therefore built a trainable interfaces which let the user teach the interface which words and gestures they want to use and what the words and gestures mean.

Our current work focuses on a system which learns words from natural interactions; users teach the system words by simply pointing to objects and naming them. This work demonstrates an interface which learns words and their domain-limited semantics through natural multimodal interactions with people. The interface is embodied as an animated character named Toco the Toucan. Toco is able to learn semantic associations (between words and attribute vectors) using gestural input from the user. Gesture input enables the user to naturally specify which object to attend to during word learning. For additional information see Roy and Pentland 1998 .

The author, wearing a variety of new technologies. The glasses (built by Microoptical of Boston) contain a computer display that is nearly invisible to others. The jacket has a keyboard that is literally embroidered into the cloth. The lapel has a `context sensor' that can classify the user' surroundings, the computer is too small to be visible.

3 Smart Clothes

So far I have presented matters mostly from the smart room perspective, where the cameras and microphones are passively watching people move around. However when we build the computers, cameras, microphones and other sensors into a person's clothes, the computer's view moves from a passive third person to an active first-person vantage point. This means that smart clothes can be more intimately and actively involved in the user's activities, making them potentially an intelligent Personal (Digital) Assistant (Starner et al 1997, Mann 1997, Pentland 1998).

The mobility and physical closeness of a wearable computer makes it an attractive platform for perceptual computing. A camera mounted in a baseball cap can observe the user's hands and feet. This allows observation of gestures and body motion in natural, everyday contexts. In addition, the camera acts as an interface for the computer. For example, the user's fingertip becomes the system's mouse pointer through tracking the color of the user's hand. Similarly, hand tracking can be used for recognizing American Sign Language.

4 Conclusion

It is now possible to track people's motion, identify them by facial appearance, and recognize their actions in real time using only modest computational resources. By using this perceptual information we have been able to build smart rooms and smart clothes that have the potential to recognize people, understand their speech, allow them to control computer displays without wires or keyboards, communicate by sign language, and warn them they are about to make a mistake. We imagine eventually building a world where the distinction between inanimate and animate objects begins to blur, and the objects that surround us become more like helpful assistants or playful pets than insensible tools.

Papers are available at our web site, http://www.media.mit.edu/vismod

Becker, D., and Pentland, A., (1996) "Using A Virtual Environment to Teach Cancer Patients T'ai Chi, Relaxation and Self-Imagery," M.I.T. Media Laboratory Perceptual Computing Technical Report No. 390.

Mann, S. "Smart Clothing: The Wearable Computer and WearCam", Personal Technologies 1(1), 1997

Oliver, N., Rosario, B., Pentland, A., (1998) Statistical Modeling of Human Interactions, NIPS 98, Denver CO, Dec 4-6

Pentland, A. (1996) Smart Rooms, Smart Clothes Scientific American, Vol. 274, No. 4, pp. 68-76, April 1996.

Pentland, A.,"Wearable Intelligence" Scientific American Presents, Vol 9, No. 4, Dec. 1998

Pentland, A., and Liu, A., (1999) Modeling and Prediction of Human Behavior, Neural Computation, 11, 229-242, Jan 1999

Roy, D., and Pentland, A., (1998) "An Adaptive Interface: Learning worlds and their audio-visual grounding" ICSLP98, Sydney, Austrialia, Dec. 1998

Starner, T., Weaver, J., and Pentland, A., (1998) Real-Time American Sign Language Recognition from Video Using Hidden Markov Models, IEEE Transactions on Pattern Analysis and Machine Vision, Dec. 1998.

Starner, T., Mann, S., Rhodes, B., Levine, J., Healey, J., Kirsch, D., Picard, R., and Pentland, A., (1996) "Visual Augmented Reality Through Wearable Computing," Presence, (M.I.T. Press), Vol 5, No 2, pp. 163-172.

Wren, C., and Pentland, A., (1998) "Dynamic Modeling of Human Motion,' IEEE Face and Gesture Conf., Nara, Japan., also M.I.T. Media Laboratory Perceptual Computing Technical Report No. 415

Let Your Hands Talk - Gestures as a Means of Interaction

Joachim Machate[#] and Sven Schröter*
Interactive Products, Fraunhofer IAO, Stuttgart, Germany
* University Dortmund, Dortmund, Germany

1 Introduction

Since computers exist, a crucial wish of people is to be able to communicate with these machines in a natural way. Science fiction films and literature provide lots of scenarios which lead to the impression that at one certain day people will be able to communicate with a computer or robot just like with any other human (Adams 1979, Negroponte 1995, Stork 1997). Of course, nowadays the situation is different. After breaking the barriers of command-line interaction and the need for remembering not very meaningful codes, we learned about new concepts of interaction called direct manipulation. Graphical user interfaces and pointing devices became state of the art regarding human-computer interaction. But still an intermediate medium was needed in order to establish the power of direct manipulation concepts, i.e., a mouse or some other kind of pointing device. The use of a touch screen brought in a higher degree of directness, but shortened the possibilities for direct manipulation. Especially dragging and dropping requires a certain degree of training before being successfully applied with a touch screen. Nevertheless, just touching some virtual button surely provides a greater degree of naturalness. No need to remember any codes, no need to learn how to operate a mouse, just touch. In the late sixties, Myron W. Krueger started to work on video-based human computer interaction. Several projects like *Videoplace* or *Painting the town*, that are based on the calculation of the human body's silhouette, realized his vision of an intuitive and non-immersive input modality (Krueger 1991). At the end of the nineties, speech recognition and machine translation seem to have made their way also into the homes of the residential user. Speech products became affordable at a price which was absolutely unrealistic some years ago. If we consider human-human interaction, speech provides only one communication channel. Others comprise facial expression, postures, glance, and gestures. Comparing these natural communication channels with the way we are used to interact with computers, we can see immediately that current human-computer interfaces are far off from being natural in a human-human communication way. In the following, we will focus on what input by gestures can contribute to human-computer interaction.

2 Home Environment Control

Using hand gestures as an input means for interactive systems (Kurtenbach & Hulteen 1990) is seen as a potential aid for people with special needs because it is easy to learn and highly adaptive (Brinkmann, et al. 1995). Home environment control by means of gestures provides one input modality amongst others for the interaction concept of the HOME-AOM project (Bekiaris, et al. 1997) sponsored by the European Commission (TAP-DE 3003). In order to figure out which kind of gestures are suited best for elderly and disabled people controlling a smart home environment, a series of workshops was conducted in order to derive an initial gesture set which was later condensed into a minimal set derived from results of Wizard of Oz experiments (Nielsen & Mack 1994). The workshops were performed with users, nurses and ergo-therapists in order to bring about hand or finger gestures that can be performed by the target group without causing any pain in the hand wrist or fingers. Starting from an initial set of 84 gestures which could be potentially recognized by the ZYKLOP system (Stark & Kohler 1995), the set finally comprised only 10 hand gestures which were thought as being potential candidates for a final gesture set. Of course, the philosophy of the HOME-AOM project is to provide a set of recommended gestures to the user, but finally leave it up to the user to change this set to any set which is more convenient in respect to the personal requirements and capabilities. The WOz experiments were run with the same test design in two European countries: Germany (12 elderly users) and Greece (12 people with special needs). In the beginning of the experiment, participants were allowed to choose whatever they thought a gesture constitutes. So, even postures were used in order to accomplish the tasks provided during the experiment. This phase was primarily used to get users acquainted to this unusual input modality. Users mostly started using either metaphoric gestures or postures, e.g. shivering in order to increase temperature, they tried to transform real world gestures into the virtual setting, e.g. grasping a virtual door handle and pushing it downwards in order to get a door opened, or they acted as if pressing a button on a virtual remote control. Fingers were sometimes used in order to represent a position of a slider when adjusting a scale. In a later stage, a selection of gestures chosen from the 10 gestures set were used in order to control a selection frame on a graphical user interface (GUI).

2.1 The ZYKLOP System

The gesture recognition system ZYKLOP consists of a usual PC with a 130 MHz Pentium processor. Video signals are captured by an analog CCD color video camera and processed by a framegrabber card. In order to use gestures as an input modality in addition to the keyboard or the mouse, a camera is mounted to a tripod next to the computer monitor and is directed towards the table surface. The basic algorithms building the recognition kernel of ZYKLOP are classical image processing techniques that are part of an image processing pipeline. The main parts are segmentation, feature extraction and classification (Kohler 1999). The system is able to process the images in real time (>15 frames/second). With each processed frame the users' hand is classified into different classes of gestures that can be taught to the system by showing one example per gesture. One command per gesture class can be assigned by the user in order to control the particular application. Besides control of a special GUI like in the HOME-AOM project, other application areas comprise substitution of the mouse, 3D visualization software, 3D CAD, or presentation systems.

3 Work in Collaborative Rooms

Another potential candidate for gesture based input is the use of so-called interactive walls. With such a device being at hand, the question arises which are the best modalities when working with this wall in co-operative teams. E.g., imagine standing directly in front of such a large wall, you will not perceive everything what is shown on it. In order to get the complete picture, a certain degree of distance is required. Furthermore, consider that such a screen will surely not be used in order to display only one thing or application, but rather is a prime candidate for installing a split screen. You may end up with one part of the screen showing some meeting participant being at a remote place, but video-conferencing with the team currently being in this room, another part showing some product design currently being under discussion, and probably the third part showing the latest business and market analysis data. So, if the users are not standing directly in front of the interactive wall, how can they interact with it? One solution could be to use some kind of infra-red or laser beam based pointing device to control the elements of a graphical user interface. A solution, which is realized by video based techniques, has been shown by Kirstein (Kirstein 1998). In this work, the user can point to a graphical element with a laser pointer, and rest a certain amount of time on this element, or perform a simple gesture, e.g. drawing a circle or a zigzag line in order to activate the element. But wouldn't it be very natural, if meeting participants were able to just point to some objects or parts of objects displayed on the screen saying something like "zoom to 200%", "show this in a cake diagram", "color this one blue", etc. By now, this vision of artificial reality (Krueger 1991) - the interaction with a world that is separated from the users world - has been realized by a few projects either with video based techniques (Hoch 1997) or with tracking devices like the data glove (Latoschik 1998, LaViola 1999). In all of these projects the user stands in front of a large scale back projection display. Hoch uses deictic gestures in combination with simple speech commands like „put this there“ to build a tool for scenery design. The features that are calculated from video images are projected onto a simple human body model to get the 3D pointing direction. LaViola presents a tool for scientific visualization that can be controlled by 3D semantic and deictic gestures and simple speech commands.

4 Towards the responsive environment

From the results of his *GlowFlow* project, Myron Krueger concludes: „So that it can respond intelligently, the computer should perceive as much as possible about the participants' behavior“ (Krueger 1991). Gesture as a means of non-verbal communication consists of more than a hand posture and a hand movement. Cuxac explained that a sentence in French sign language can have different meanings depending on the direction and the place relative to the upper body where the gesture is performed (Cuxac 1999). Additionally, the term gesture bears itself different kinds of human motions e.g. gaze, miming, movements of the head or limbs, movements of the whole person. To build a real humanlike communication between humans and computers it is necessary to create so-called responsive environments that are aware of the users mood and react or act intelligently. First attempts to incorporate the user's movements in a responsive environment were undertaken in the *PsychicSpace* project (Krueger 1991) and in the German *Dialogue with Knowbotic South* project (Knowbotic Research 1997). In both projects, the position of the user is tracked either by a tracking device or with a video camera to navigate in a virtual data space. The data space is presented by visual and audio feedback and changes with each body movement.

What has been illustrated, is that gesture based input offer new and challenging ways of interaction. If combined with speech, they open a new dimension towards the naturalness of human-computer interaction.

5 References

Adams, D. (1979). *The Hitchhiker's Guide to the Galaxy*. Ballantine Books, November 1995: reissued edition.

Bekiaris, E., Machate, J. & Burmester, M. (1997). Towards an intelligent multimodal and multimedia user interface providing a new dimension of natural HMI in the teleoperation of all home appliances by E&D users. *La Lettre de l'*

IA, Proceedings of Interfaces 97 (Interfaces 97, Montpellier, France, May 28-30, 1997), 226-229, EC2 & Développement.
URL: http://www.swt.iao.fhg.de/home

Brinkmann, H., Boland, F.M. & Wright, J. (1995). A Gesture Input Interface (G.I.I.) Implemented Using Neural Networks. In I.P. Porrero, & R.P. de la Bellacasa (Eds.) *The European Context for Assistive Technology, Proceedings of the 2nd TIDE Congress* (Paris, April 95), 436-439, Amsterdam: IOS Press.

Cuxac, C. (1999). French Sign Language: Proposition of a Structural Explanation by Iconicity. Presented at *The 3rd Gesture Workshop ´99*, Gif-sur-Yvette, France, March 1999.

Hoch, M. (1997). Human Body Tracking for the Intuitive Interface. In D. Paulus und Th. Wagner (Eds.) *Tagungsband zum 3. Workshop Farbbildverarbeitung*, Erlangen, September 1997, IRB-Verlag Stuttgart.

Kirstein, C. (1998). Interaction with a Projection Screen Using a Camera-Tracked Laser Pointer. *Proceedings of the International Conference on Multimedia Modeling (MMM ´98)*, IEEE Cmputer Society Press.

Knowbotic Research (1997). *Dialogue with the Knowbotic South*, as part of the exhibition INTERACT, Wilhelm Lehmbruck Museum, Duisburg, April 27. - June 15. 1997, URL: http://www.khm.de/people/krcf/

Kohler M. (1999). *New Contributions to Vision-Based Human-Computer Interaction in Local and Global Environments*, PhD thesis, University of Dortmund.

Krueger, M. W. (1991). *Artificial Reality II*, Addison-Wesley.

Kurtenbach, G., & Hulteen, E.A. (1990). Gestures in Human-Computer Communication. In B. Laurel (Ed.): *The Art of Human Computer Interface Design*, 309-318, Addison-Wesley.

Latoschik, M., Fröhlich M., Jung, B. & Wachsmuth I. (1998). Utilize Speech and Gesture to realize Natural Interaction in a Virtual Environment. *IECON´98 - Proceedings of the 24th Annual Conference of the IEEE Industrial Electronics Society, Vol. 4*, IEEE, 2028-2033.

LaViola Jr., J. J. (1999). A Multimodal Interface Framework For Using Hand Gestures and Speech in Virtual Environment Applications. Presented at *The 3rd Gesture Workshop ´99*, Gif-sur-Yvette, France, March 1999.

Negroponte, N. (1995). *Being Digital*. New York: Alfred A. Knopf.

Nielsen, J. & Mack, R.L. (1994). *Usability Inspection Methods*. John Wiley.

Stark, M., & Kohler, M. (1995). ZYKLOP - ein System für den gestenbasierten Dialog mit Anwendungsprogrammen, In Dieter W. Fellner (ed.) *Modelling - virtual worlds - distributed graphics*, International Workshop MVD, Bad Honnef, Bonn, November 1995, Dieter W. Fellner, ISBN 3-929037-98-X.

Stork, David G. (Ed.) (1997). *HAL's Legacy: 2001's Computer as Dream and Reality*. Cambridge, Massachusetts: MIT Press.

THE ARGUS-ARCHITECTURE FOR GLOBAL COMPUTER VISION-BASED INTERACTION AND ITS APPLICATION IN DOMESTIC ENVIRONMENTS

Markus Kohler, Sven Schröter, Heinrich Müller
Informatik VII (Computer Graphics)
University of Dortmund, D-44221 Dortmund, Germany

1 Scenario of interaction and system architecture

The idea of computer vision-based interaction is that a computer observes the user with video-cameras, and interprets his actions using techniques of image recognition. In the past few years, vision-based interaction has found significant research interest, and about 40 projects and systems have come to our knowledge, which emphasize different applications and technical solutions (Kohler 1998). First devices based on this technology did appear commercially (Siemens 1998).

In the very general scenario of this contribution, several box-shaped sensitive regions are defined in a room. The user may select one of those regions by pointing towards it, in order to interact with a process attached to the region by hand gesture input and visual and auditive feedback. If the dialog is terminated, a selected region may also be deactivated. A concrete application which is the background of our development of this generic concept is remote control of home devices linked together by a domestic network. The idea in particular is to provide elderly and disabled users with a natural and intuitive means of controlling home appliances. In that application, the sensitive regions correspond to the different home devices which can be selected and manipulated by pointing at them (figure 1).

In order to implement this scenario, we have developed the ARGUS architecture outlined in figure 2. The system is roughly split in two parts, *image processing* depicted by the large block in the lower part of the figure, and *global environment control* represented by the smaller block at the upper part of the picture. The image processing part itself is subdivided in *low level image processing* and *high level image processing*. The following sections give a closer look at them. More details beyond those presented in this contribution, and references to literature concerning the applied methods can be found elsewhere (Kohler 1999).

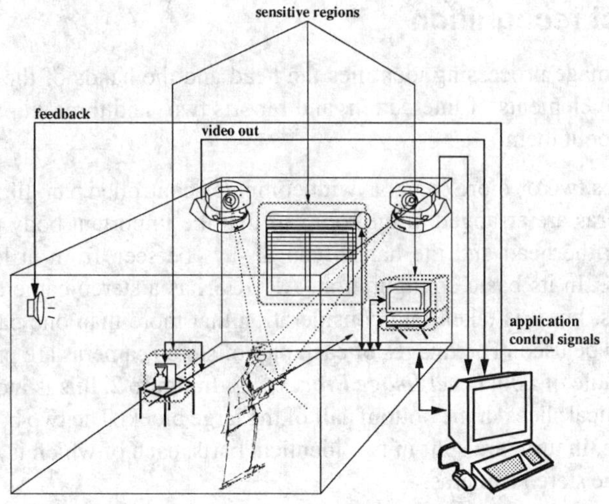

Figure 1: The ARGUS scenario with sensitive regions assigned to home devices.

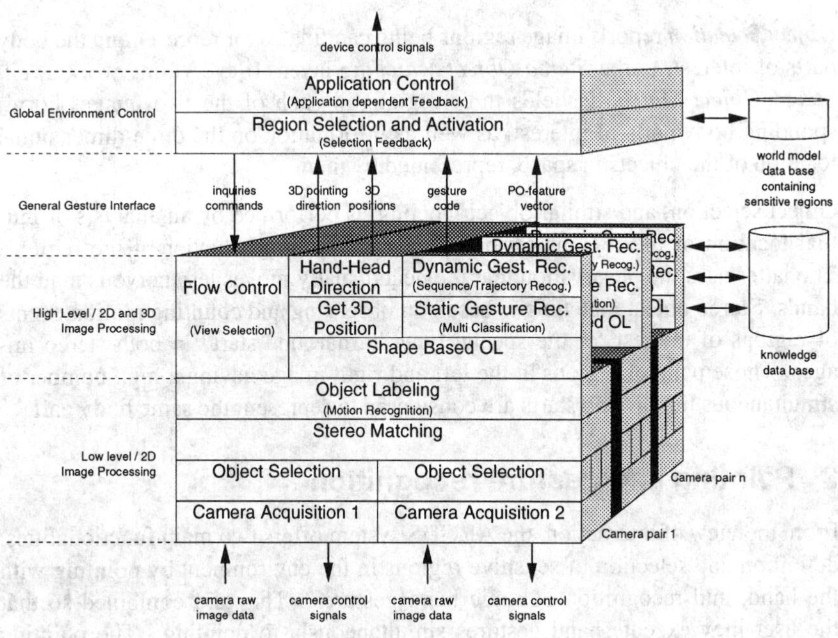

Figure 2: The ARGUS system architecture.

2 Object recognition

Low level image processing identifies the head and the hands of the user which are the basic elements of interaction, and reports two- and three-dimensional information about them.

ARGUS uses two or more cameras with computer-controlled pan, tilt, and zoom. These cameras are arranged in the room so that the important body parts of the user, that is the head and the hands, can always be seen from at least two of the cameras. In its basic configuration, ARGUS has a stereo camera system of two cameras, but it is taken into consideration that more than one pair of stereo cameras can be used. The images of each pair of stereo cameras are processed by an own module of *Low Level Image Processing*. In figure 2, this is indicated by a stack of vertical slices in the bottom half of the large block. The two bottom levels of each slice, in turn, are split in two identical parts, each of which is responsible for one of the stereo cameras.

The *Camera Acquisition* layer of each of those parts performs *object tracking* and *camera control*. Object tracking uses Kalman filtering. The strategy of camera movement and zooming takes into consideration the requirements of motion-based segmentation based on dynamic difference images.

Object Selection reports image regions being candidates for representing the body parts of interest, to the *Stereo Object Matching* layer. If everything works well, *Stereo Object Matching* yields three regions in each of the two images corresponding body parts of interest, as well as information on the three-dimensional location of the objects in space, represented by them.

Object selection, and similar object labeling, is performed by an analysis of mutual location of the regions of interest and of motion behavior of the regions. The latter uses the observation that the head usually moves less nervous than the hands. Stereo object matching is based on observing and counting motion events of regions of interest, of the special type "turnaround start" in both stereo images. Those pairs of regions in the left and right image having a high number of simultaneous turnaround starts are considered to represent the same body part.

3 Pointing and gesture recognition

From the view of interaction, the ARGUS system offers two main functionalities: definition and selection of sensitive regions in the environment by pointing with the hand, and recognition of static hand gestures. They are combined so that the user may execute hand gestures simultaneously to pointing. The pointing direction and gesture recognition are provided by the high level image processing part of ARGUS, depicted as top half of the of the big block in the figure. This module in turn has three parts, *3D-pointing*, *gesture recognition* and *flow control*.

The *3D-pointing* part consists of the two layers *Get 3D Position* and *Hand-Head-Direction* which use information on the location of the objects, obtained from the layer *Shape Based Object Location*, and the 3D-match information and object labels of low-level image processing in order to determine the pointing direction. The pointing direction is defined by the line connecting two reconstructed spatial points representing the location of the pointing hand and the head, respectively. This direction is additionally corrected by an angle of experience.

In the *gesture recognition* part defined by the right-most two-layer block above the *Shape Based Object Location* layer, gesture recognition is performed in the most suitable image obtained by the stereo cameras. Static and dynamic gestures are distinguished which are treated in separate layers *Static Gesture Recognition* and *Dynamic Gesture Recognition*. Static hand gesture recognition consists in first separating the hand region from the forearm region, and then applying a set of shape features to the hand region, in particular Fourier descriptors, to classify the gesture. Dynamic gesture recognition is based on a hidden-Markov model like approach to classification of hand trajectories.

The third part, *Flow Control*, receives the inquiries from global environment control and coordinates the calls of gesture recognition and the 3D-pointing part. An important function of *Flow Control* is to select that camera which allows best recognition of the hand posture. Selection is based on the location of the cameras with respect to the current hand-head direction.

4 Region selection

The results of image processing are reported to *Global Environment Control* through a *general gesture interface* which separates both parts (figure 2). Global environment control is performed in two layers. The *Region Selection and Activation* layer initializes and activates, respectively, the sensitive regions in a regions data base, based on the pointing direction and 3D-user position supplied by the general gesture interface. It also generates the (vocal) feedback during the selection process.

In order to cope with the relatively imprecise pointing direction, the regions are taught to the system in the system-guided selection process in that the user points to the corners of the desired sensitive regions. From this information, a location of the corners is calculated an stored in the regions database. The advantage of this approach is that the regions are adapted to the pointing ability of the user.

5 Device control

If a sensitive region was activated, the *Application Control Layer* becomes active. *Application Control* receives the identifier for the activated region from the general gesture interface. Based on that, it reacts to the gestures, generates the application-dependent feedback, if necessary, and controls the devices assigned to the activated region. The devices have interfaces through which they receive the control signals from the computer.

A problem of interaction with several complex devices is that the user has to remember the associated gestures and their functionality. An investigation shows that only about 24 static gestures are easy to perform, and half of them are very easy. In ARGUS, a reduction of the number of necessary gestures is achieved by grouping the device control operations into device independent classes of similar or equal functionality according to qualitative criteria. The loudness of a tuner or a TV-set, for example, has equal functionality and should be controlled by the same gesture. The "next/previous title" function of the CD player is similar to the "next/previous program" function of the TV-set and is controlled by the same gesture. This reduces the overall number of gestures and makes the dialog more intuitive and easier to remember. Additional functionality is achieved by gesture sequences.

6 Development state

Much of the ARGUS system is still of conceptual nature. A prototype implementation exists which demonstrates that selection by pointing, in combination with simple gestures, can basically be performed. On a Pentium PC, 133 MHz, 16 MB, a rate of 5 to 15 processed frames per second could be achieved. It turned out that rates below 10 frames per second are too low for a natural and comfortable dialog. Experiments showed that the angular resolution of pointing is about 1.15°, that is two points that have an angular distance of at least this amount relative to the user can be distinguished.

7 References

Kohler, M. & Schröter, S. (1998). Hand gesture recognition by computer vision (in German). In Dassow, J. & Kruse, R. (Eds.): *Informatik '98* pp. 201–212. Heidelberg: Springer (also Research Report No. 693 (in English), Department of Computer Science, University of Dortmund, Germany, and http://ls7-www.cs.uni-dortmund.de/research/gesture/)

Kohler, M. (1999). *New Contributions to Visio-Based Human-Computer Interaction in Local and Global Environments*, Ph.D. thesis, Department of Computer Science, University of Dortmund, Germany

Siemens AG (1997). SIVIT – Siemens Virtual Touchscreen. Siemens AG, Abt. ZT IK 5, Munich, Germany

Virtual Touchscreen – a novel User Interface made of Light – Principles, Metaphors and Experiences

Christoph Maggioni and Hans Röttger
Siemens AG, Corporate Technology, Dep. ZT IK 5, Munich, Germany, e-mail: Christoph.Maggioni@mchp.siemens.de and Hans.Roettger@mchp.siemens.de

1 Introduction

Today's interfaces to technical systems commonly rely on keyboards, monitors, mice, switches or buttons that are mostly difficult to understand, to learn and to operate. We will describe a new "Come as you are"-style interface (Krueger 1991), called VirtualTouchscreen, for contactless and natural man-machine interaction. The video-based system is able to detect human hand gestures and movements in real-time and uses them to control a projected graphical user interface (GUI).

For the input channel we developed a vision-based real-time gesture recognition system. It runs on state-of-the-art personal computers without the need for special hardware and requires only a moderate amount of the available computational power thus forming an affordable novel input device. On the output channel the VirtualTouchscreen uses a video beamer to project information on a wide range of arbitrary shaped and sized surfaces. The projected GUI may be designed to be a standard window-based display or may be tailored to the specific needs of the system to be controlled. Through this approach, any input hardware like keyboard, mouse or traditional touch screen can be replaced. Hand gestures within the projected GUI control the desired application.

First approaches for using a computer vision system as an input device for computer applications include Videodesk (Krueger 1991) and DigitalDesk (Wellner 1991). Further research explored 3D hand interaction (Rehg & Kanade 1994; Segen, & Kumar 1998) and even full body interaction as in the MIT Pfinder project (Wren, Azarbayejani, Darrel & Pentland 1997).

A comprehensive overview on further video-based gesture recognition systems can be found in (Huang 1995, Kohler & Schröter 1998).

The Virtual-Touchscreen is one result of the GestureComputer project (Maggioni 1993-1998) sold by Siemens since 1998 under the brand name SIVIT (**Si**emens **Vi**rtual **T**ouchscreen, SIVIT 1999) and being one of the first gesture controlled systems on the market.

2 System Principles

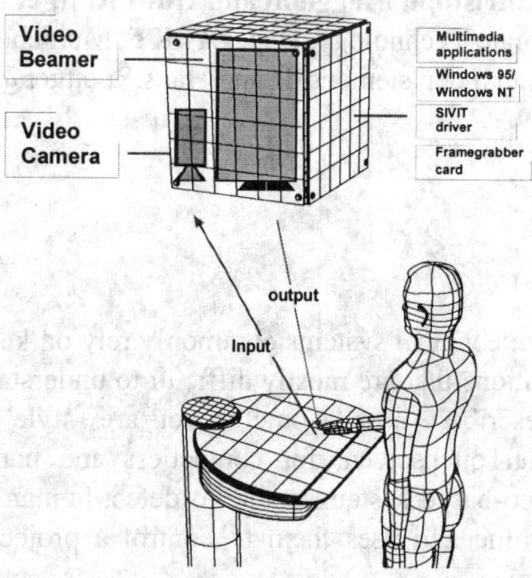

Fig. 1: System setup

Fig. 1 shows the VirtualTouchscreen setup that involves a classical feedback loop. Images of the users hand are acquired by a CCD video camera. To distinguish real world objects and the hand from the projected image we use infrared lighting and a set of infrared filters on the projector and the camera. Using image-processing techniques the hand is detected and its position, orientation and configuration is computed. Gestures are recognized and the interaction of the user's hand with the application is determined (see Maggioni 1998). When the application reacts, the results are rendered and displayed through the video beamer on the interaction surface. When moving the hand the user sees a cursor at the corresponding position in the projected GUI. Since timing is a crucial issue in feedback loops our system updates with the camera video rate of 25 or 30 Hz with a lag shorter than 60 milliseconds. We have chosen standard OS events (mouse_move...) making it possible to control standard applications by gestures.

3 Gestures and Metaphors

The image and gesture processing system is able to detect several different static (pointing, thumb right, thumb left, all fingers…) and dynamic (finger rest, circle…) gestures. To be compatible with the traditional touch-screen technology and to minimize learning effort on the user side we have chosen the pointing gesture (Fig. 2a) to move the GUI cursor. Triggering an event (clicking) is accomplished whenever the pointing gesture is seen to be in a constant position for some time (~ 0.2 sec.). This gesture is automatically performed whenever users try to press a virtual button.

The problem with this mechanism is that sometimes events are triggered without intention. A good solution is to substitute the gesture "resting pointing finger" by other dynamic gestures – for example "small clockwise circle with pointing finger" (Fig. 2b). Another solution is to construct widget sets that mimic switches and controls of the "real physical" world. The state of these widgets is not changed by pressing but by moving the finger over them into a defined direction (Fig. 2c). A good example is an old style power switch.

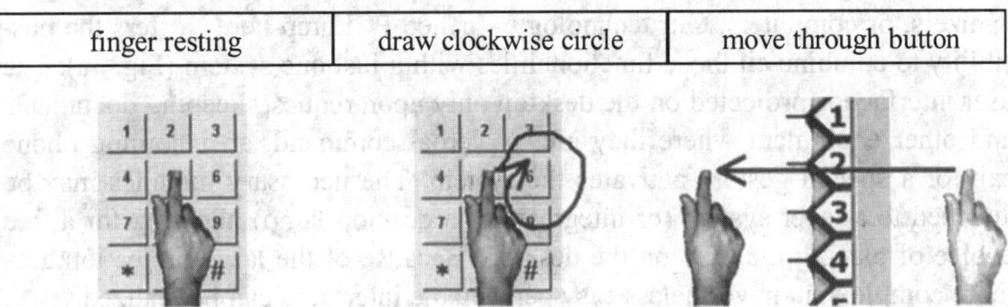

Fig. 2 a/b/c: pointing gestures for event generation

4 Experiences

Since 1998 a few hundred systems have been sold. Our customers – mostly system integrators and multi-media companies - recognized that users like the Virtual Touchscreen and perform well when using it. Without explaining how to trigger a button users mostly choose the pointing gesture and try to use pressure on the projection surface thus successfully controlling the system. Later on they recognize how the system really works and stop touching the projection surface. People very much like direct interaction, for example changing the viewpoint into a 3D scene by moving the finger. So far we do not have enough results to rate the interaction metaphors using dynamic gestures but we started studies on that matter.

The Virtual Touchscreen has proven well as a natural and easy to use human computer interface with the additional benefit of a huge, arbitrary shaped and bright display that is insensitive to vandalism and allows for attractive new designs.

The reliability of our first gesture recognition version was somehow sensitive to changes in lighting conditions or to strong ambient light sources. By using a special projection surface, a new infrared light source and other methods we where able to improve the signal to noise ratio by three orders of magnitude resulting in a system that is insensitive, even to the brightest sunlight.

Integrating these improvements into the second product generation we were able to demonstrate a system with 3,5 meters distance between beamer/camera and projection surface at CeBIT 1999. Now - for the first time - a really invisible device with an interface "made out of light" can be realized.

5 Future Directions

Often desktops are crowded with electronic equipment such as telephones, organizers, or computers. Our technology – called PCLamp here - offers the possibility to combine all those functionalities within just one system (Fig. 3a). The user interface is projected on the desktop only upon request, keeping documents and other equipment where they are. A verbal command, an incoming phone call, or a special gesture activates the system. The necessary modules may be attached to a lever system (or integrated in a desktop lamp) allowing for a free choice of placing the GUI on the desktop. Because of the large projection area more complex but nevertheless easy perceptable interfaces can be realized.

Fig. 3 a/b: Future directions – PCLamp and Virtual Payphone

The same approach will be used in advanced payphones to ease public access to large information sources and to enable additional services like faxing and sending images of real world objects (Fig. 3b).

6 Acknowledgements

We thank all contributors to the GestureComputer project: Subutai Ahmad, Jürgen Finger, Daniel Goryn, Thomas Johannsen, Bernd Kämmerer, Holger Kattner, Uwe Kusch, Christoph Pomper, Rolf Schuster, Sven Schröter and Brigitte Wirtz.

7 References

Huang, T.S. & Pavlovic, V.I. (1995). Hand Gesture Modelling, Analysis and Synthesis, *Proc. International Workshop on Automatic Face- and Gesture-Recognition, Zurich, June 1995*, pp. 73-79

Kohler, M. & Schröter, S. (1998). A Survey of Video-based Gesture Recognition – Stereo and Mono Systems. *Research Report No. 693/1998*, Fachbereich Informatik, Universität Dortmund, 44221 Dortmund, Germany.

Krueger, M.W. (1991). *Artificial Reality II*, Addison-Wesley.

Maggioni, Ch. (1993). A Novel Device for using the Hand as a Human Computer Interface. In Alty, J.L. (Ed.): *People and Computers VIII, Proc. HCI '93*, September 1993, pp. 191-203. Loughborough GB.

Maggioni, Ch. (1995). GestureComputer: New Ways of operating a Computer. *Proc. International Workshop on Automatic Face- and Gesture-Recognition*, Zurich, June 1995, pp.166-171

Maggioni, Ch. & Kämmerer, B. (1998). GestureComputer – History, Design and Applications. In Cipolla, R. & Pentland, A. (Eds.): *Computer Vision for Human-Machine Interaction*, Cambridge Univ. Press.

Rehg, J.M. & Kanade, T. (1994). Visual tracking of high dof articulated structures: an application to human hand tracking. In J. Eklundh (Ed*), Proc. of Third European Conf. on Computer Vision*, vol. 2, Stockholm, Sweden, May 1994, pp. 35-46. Springer-Verlag.

Segen, J. & Kumar, S. (1998). Human-Computer Interaction using Gesture Recognition and 3D Hand Tracking. In Proc. ICIP, Chicago, 1998, pp. 188-192.

SIVIT (1999). SIVIT Product Information at
http://www.atd.siemens.de/td_electronic/produkte/sivit/sivit.htm

Wellner, P. (1991). The DigitalDesk Calculator: Tangible Manipulation on a Desk Top Display. In *Proceedings of the ACM Symposium on User Interface Software and Technology (UIST '91)*, November 1991, Hilton Head, USA.

Wren, C., Azarbayejani, A., Darrel, T. & Pentland, A. (1997). Pfinder: Real-Time Tracking of the Human Body. *IEEE Trans. on PAMI*, 19 (7), 780-785.

Use of Fuzzy Logic in an Adaptive Interface meant for Teleoperation

Fabrice Vienne[1][2], Anne-Marie Jolly-Desodt[2], Daniel Jolly[2]
[1] Ecole des Hautes Etudes Industrielles, Groupe de recherche E.R.A.S.M., 13 rue de Toul 59046 Lille Cedex, France
E-mail: fabrice.vienne@hei.fr
[2] Laboratoire d'Automatique I^{3}D, Bat P2, Université des Sciences et Technologies de Lille - 59655 Villeneuve d'Ascq Cedex, France
E-mail: {Anne-Marie.Desodt, Daniel.Jolly}@univ-lille1.fr

1 Introduction

A teleoperation system is mainly composed of a master robot arm that can be driven by a human operator, of a slave robot arm carrying out the task to be performed, and of an interface universe linking the master and the slave surroundings. This kind of system is used for the implementation of remote tasks or to carry out missions in a hostile universe. In teleoperated devices, the human operator plays a main role in the command loop for which he or she is at the same time the supervisor of the mission and the driver of the master set, able to intervene at any moment in case of a problem in the slave surrounding. So he must be in symbiosis with the system that allows him to stay in contact with the telerobot: for this purpose the operator has to use an interface between the device he supervises and himself (Wawak, Jolly-Desodt and Jolly 1997).

This interface must present relevant information to the human operator at each moment. The problem in teleoperation systems is that data used to manage the device come from a great number of different sensors placed in various places of the teleoperation site, the complexity of such devices increasing continually.

The interface cannot present all this data to the operator at one time because he or she would be unable to cope with so much information.

Due to the lack of reliability of certain sensors at some times, the system has to filter the data before displaying on them interface. Data fusion techniques can manage both the aggregation and filtering functions.

2 Global presentation of the adaptative interface called Telemaque 2

The architecture of the adaptive interface uses five generic agents related to the mission, the help, the display, the operation and the planning (figure 1).

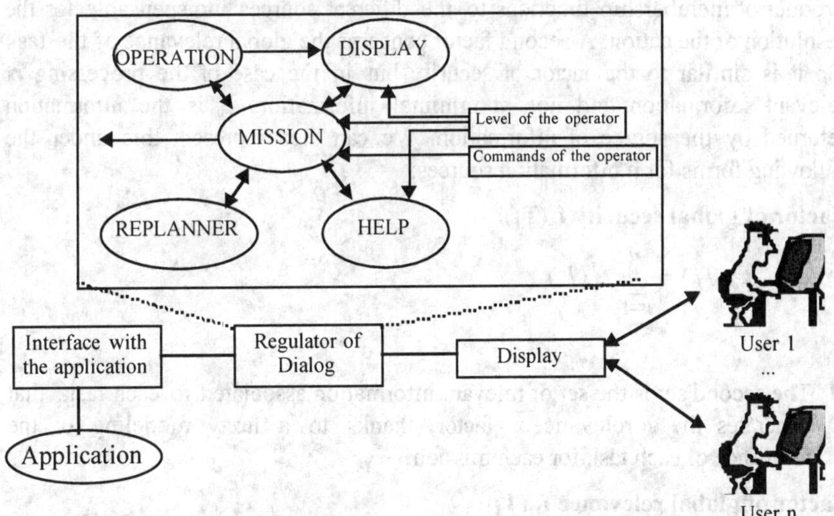

Figure 1. Telemaque 2

The adaptivity of the interface realized allows it to adjust itself to the mission as well as to the human operator. The adaptivity can be symbolized by an automatic modification of windows attributes, but also by the modification of the display of data on the screen placed in front of the operator managing the system (Kolski 1997; Browne, Totterdell and Norman 1990; Patterson 1997; Dimitrova et al. 1998). This adaptive interface is called Telemaque 2. It uses techniques derived from fuzzy sets both to model human operator and to fuse imperfect data generated by the sensors.

3 Concepts of adaptivity

The concept of adaptivity with respect to the mission uses concepts such as relevance and security functions. It is possible, for instance by a simulation of the mission, to determine the ranges of values to which data given by the sensors should belong.

The relevance and security functions are obtained through the comparison of data generated during the mission to these ranges modeled under the form of trapezoidal fuzzy sets (Gottwald 1993):

- Some information is really necessary for the realization of the mission without danger, it can be considered as a minimal set of information; that is to say that without it some serious security problems may occur. The security factor is derived from these considerations.

A factor of global security of task T_i has been created: it corresponds to the product of membership functions to the different sources indispensable for the resolution of the action. A second factor concerns the global relevance of the task T_i: it is similar to the factor of security but in the case of the processing of relevant information and not of minimal information. I_i is the information returned by the source of information. We can then represent this under the following forms for n information sources:

Factor of global security $f_S(T_j)$:

$$f_S(Tj) = \underset{i=1}{\overset{n}{\subseteq}} f_S^i(I_i)$$

- The second set is the set of relevant information associated to each task, that generates the « relevance » factor, thanks to a fuzzy modeling of the relevance of each task for each mission.

Factor of global relevance $f_P(T_j)$:

$$f_P(T_j) = \frac{\sum_{i=1}^{n} m_{ij} \times f_p^i(I_i)}{\sum_{i=1}^{n} m_{ij}}$$

In both cases, processes of aggregation of fuzzy sets under their conjunctive or disjunctive form will make it possible to generate, throughout the mission, the security and relevance factors filtering the information displayed on the interface (Vienne et al. 1996).

4 User modeling

Fuzzy logic is also used for user (i.e. the human operator) modeling with the possibility to manage several users (Nonoka and Da-te 1998). We do not want to use a complete model of the user, because it is more convenient for the interface design to focus on the continuous evolution of this operator. These evolution factors can be divided into four classes: we can first consider the knowledge that the operator has about the system and the mission. The physical condition also has a role to play; it depends on the age of the operator, on the time of day and the time elapsed from the beginning of the mission. The mental workload interferes on the modeling too, then we get specific criteria such as the periodicity in the use of the system (γ) and its ability to evolve (δ). These last two factors γ and δ will modify the user modeling at all times.

In fact, each operator evolves in his or her own way; so the membership function used to model the operator must be modified by δ which represents the evolution factor of this operator with respect to the evolution of a standard operator. This factor depends on the difficulties of the operator to improve his or her performances, but also on his or her mistakes. The value δ is also modified by the erroneous behavior of the operator. Its expressions are delay, repetition, omission (the task was not realized), bad realization (it is a required task that ends in an incorrect result or has been realized at an inappropriate moment) and the inopportune realization (an unexpected task was carried out). So the mistakes taken into account are of two types: the first kind is linked to the success of the mission, the second depends on the correct use of Telemaque 2.

The criteria of periodicity γ will decrease if the human operator does not use the interface very often, so its progression will be slower than that of the standard operator; this will happen if the period between two missions is too long.

Fuzzy modeling of the human operator in human-machine systems is relatively convenient because of the subjective aspect of the information involved. We have also used this concept for the task allocation system of the teleoperation device (F. Wawak and all 1996). This technique doesn't take much computation time and allows a dynamic representation of information which is easily understandable. In Telemaque 2, the system computes at each moment the evaluation of the human operator; this evaluation generates the evolution of the display.

5 Conclusion

The advantages of this method for management of an adaptive interface are (1) the facility given by fuzzy sets to model subjective information and (2) the possibility to adapt it to many industrial areas where information on human operator can and must be used (in the context of teleoperation).

This method is not very costly in calculation time and can therefore execute on a majority of systems. Future works will develop the characterization of the standard operator. Indeed, it is the central point of the modeling of the operator.

6 References

Browne, D., Totterdell, P. & Norman, M. (1990). *Adaptive User interfaces*, Academic Press.

Dimitrova, M., Johannsen, G., Nour Eldin, H.A., Zaprinov, J. & Hubert, M. (1998). Adaptativity in human computer interface systems identifying user profiles for personalised support. In *Proceedings IFAC International Congress on Man-Machine Systems*, Kyoto, Japan, September 16-18.

Gottwald, S. (1993). *Fuzzy sets and fuzzy logic.* Wiesbaden: Teknea.

Kolski, C. (1997). *Interface Homme-Machine application aux systèmes industriels*, Paris: HERMES Editions.

Nonoka, H. & Da-Te, T. (1998). Adaptative human interface based on fuzzy behavior model. In *Proceedings EUFIT'98 International Congress*, Aachen, pp. 1707-1710.

Patterson, P. (1997). A New perspective on HCI Education, research and practice for user interface design. In *Proceedings Interfaces'97 Congress*, Montpellier, France.

Vienne, F., Zunino, G., Jolly-Desodt, A.M. & Jolly, D. (1996). Application of Fuzzy Logic to the Fusion of Information in the framework of man/machine interface. In *Proceedings IEEE SMC International Congress*, Lille, France, 9-12 july.

Wawak, F., Desodt A.M. & Jolly, D. (1996). Modelisation of the criteria coming from the human operator for a fuzzy decision support system. In *Proceedings EUFIT'96 International Congress*, Aix la Chapelle, pp. 1349-1352.

Wawak, F., Jolly-Desodt, A.M. & Jolly, D. (1997). Man Machine Interface based on the fuzzy concepts. In *Proceedings IFAC Conference*, Gand, Belgium, april.

Specifications of Human-Machine Interfaces for Helping Cooperation in Human Organizations

Emmanuel Adam *
SOLVAY Research and Technology, SOLVAY S.A.,
310 rue de Ramsbeek, 1120 Brussels, Belgium
E-mail: emmanuel.adam@solvay.com

1 Introduction

Introduction of computers in workspaces has involved new habits. Cooperative work has turned little by little into computer supported coordinated work (people isolating themselves and communicating only when the task is completed). Thanks to research led on CSCW (Computer Supported Cooperative Work), many local solutions to cooperation problems have appeared and appear yet. This paper concerns the study of a user driven modeling method, applied to a human organization in a workflow context. This method allows specification of human-machine interfaces making cooperation easier in the global system.

2 Social Context

This research have been carried out in an industrial context, more precisely in the patent department of SOLVAY enterprise in Brussels. This department, composed of about sixty persons, has known a digital data growth. This growth has, in one hand, accelerated procedure working, but on the other hand, has instigated new problems of cooperation, particularly of coordination. So, before the setting-up of a tool supporting cooperative work, processes of the department have to be optimized. With this aim in view, we have exploited a reference organizational model: holonic systems. Holonic systems have been proposed by A. Koestler (Koestler 1969), to describe biological and social systems. Each part of a holonic system, called holarchy, is considered both as a

* PhD student (LAMIH, university of Valenciennes) whose thesis is jointly sponsored by the Nord-Pas de Calais Region (France) and SOLVAY S.A.

whole and as a part of a whole. A part, called holon, is: stable, to react against severe perturbations; autonomous, to create its own plans to reach its objectives; cooperative, to develop mutually plans in order to reach the system goal. A holon in a layer is responsible of holons in a lower layer. The more a holon is on the top of the holarchy, the more it has to take strategic decisions, and the more a holon is at the bottom of a holarchy, the more it has to react rapidly. More, according with Koestler, in a holonic system, communications have to follow the holarchy. This concept is currently applied in various domains (such as manufacturing systems, multi-agent systems, robotics, cognitive sciences). In our case, we think that holonic concepts of autonomy, cooperation, responsibility and communication can be very useful to optimize new human organizations in administrative contexts. In our application field, in order to propose organizational solutions, the modeling of the cooperative work has to allow to show the distance between the current organization and a holonic one. As the total department analysis and modeling should not been feasible in an acceptable time, only the procedures that have been judged important or critical by the head of the department have been modeled.

3 Choice of a Method to Model Complex Administrative Systems

Our case study is a patent department which works on base of data flow between a few ten of people regrouped by roles having different levels of responsibility. It can be considered as a complex administrative system. So we have looked for a method able to model this kind of system, with a aim of designing tools supporting cooperative work. A framework of comparison has been designed to confront software engineering methods in the hypothesis of an application to complex administrative systems (Adam and Kolski 1999). This one, based on (Pascot and Bernardas 1993), is composed of 76 criteria merged in 5 dimensions related to: the used formalisms; the type of organization and environment tackled; the underlying methodology; the aimed tools; the study of the communications and relations between the actors. Any of the eight compared methods (MERISE, OMT, UML, 3AR, SADT, OSSAD, CISAD, MKSM), which are usually used in the domain of the software engineering, answers totally to our criteria; that is to say no method allows to model a complex administrative system in order to specify a system helping cooperation, but some of them bring parts of solutions.

So, a modeling method has been set-up, on the basis of the more pertinent parts of these eight methods (Adam, Kolski and Vergison 1998). For the system modeling, the selection criteria are: the clarity (primordial for the confrontation of the models with actors), the representation of the communicated objects (the documents), of the data flows, of the cooperation between actors and the representation of responsibility levels. Our method uses four models: the data model (by data model of OMT method), the data flows (by adapted actigrams of SADT method, which allow to represent responsibility levels), the data processing (by the processing model of CISAD method), and the dynamics (by parameterized Petri nets). Each of these models brings a complementary view on the modeled system. But the more interesting for the specification of human-machine interfaces is the dynamic model. The dynamic model is very important in a complex system where human factors play an important role. The parameterized Petri nets allow to represent loops, quasi-parallelism and to take into account, in a generic and clear way, work interrupts and activity changes (Adam, Mandiau and Vergison 1998) (Fig. 1).

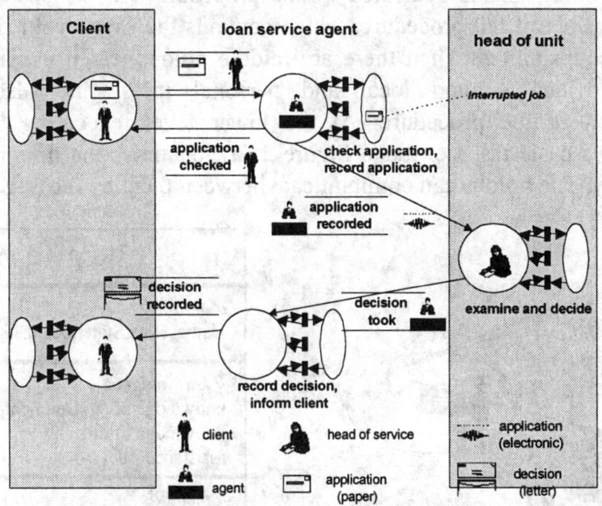

Figure 1. Example of the dynamic model (extract).

In order to better understand how the system is working, and to "virtually" test envisaged organizational solutions, a simulator has been developed. This one contains a motor inferring rules issued from the different levels of the Petri Net. In fact, the net contains three levels of rules : the data flows general rules (associated to transitions), the local rules (associated to places, they correspond with protocols of agreement between office people), and the personal rules (associated to tokens representing actors). These rules are also used by our system to guide actors toward more cooperation and more stability. But before to set-up a computer system, the modeled procedures have to be optimized, if necessary. Organizational optimization of the procedures are made by their actors, on the basis of the different models, before to be really changed. That is why these models have to be very clear to not be misinterpreted.

4 Multi-Holon Systems

Each actor in the organization has to be assisted by a system which helps him or her both to reinforce his or her stability and to make the cooperation with others actors easier. A multi-agent distributed system, designed in a holonic way, fulfils these needs. In order to meet holonic properties, a holonic multi-agent system (MAS) has to be composed of agents more and more cognitive at the top of the system, and of agents more and more reactive at the bottom. Moreover, a holon has knowledge of the others, and, although it has a relative autonomy, some of its actions are fixed by the global system. That is why a holonic MAS can be considered as a multi-holon society (Mandiau et al. 1999). It is composed of three levels, corresponding to the three levels of rules appeared during the simulation phases, and is dedicated to one procedure. At the top, there is the holon responsible of the procedure, which contains the first level rules and the whole procedure model. Then there are middle holons, each responsible of a workspace, which contain local and personal rules, and also a local representation of the procedure. At the lower level, executing holons are responsible of basic tasks on data. Figure 2 (a) illustrates the notion of multi-holon society. The holons can communicate between them by messages.

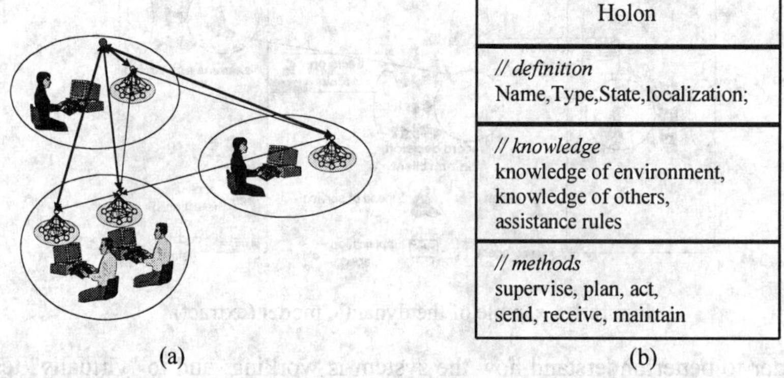

Figure 2. Holonic concepts: (a) multi-holon society, with holonic systems helping data management, (b) global architecture of a holonic agent

The multi-holon society can been designed in two steps. First, general characteristics are described with a functional analysis grid adapted from (Ferber 1995). This grid allows, for each type of holon, to define the working of the major functions (relative to the representation, to the planning, to the interactions, to the production and to the conservation) with regard to the society, the environment and itself. For the second modeling step, we have used the complex administrative system modeling method. In fact, the multi-holon society can be considered as a system composed of actors which communicate data. All the holons have the same description, cf. figure 2 (b); only implementation of their function and their knowledge change.

5 Conclusion

The proposed method is supported by the holonic concept and includes existing (adapted or not) models. This method has been defined according to the system to study. For another type of system, the method should be different. So, the method has been successfully applied in a department with procedural working mechanisms, in which cooperation plays an important role. In fact, some consequent results have appeared during the reorganization step, "coming down" from a hierarchical structure to a holonic one. From this modeling and simulating exercise, a multi-holon society has been designed to help cooperation, coordination and communication in the system. A multi-holon society is currently under test, in a procedure that has been reorganized.

6 References

Adam, E., Mandiau, R. & Vergison, E. (1998). Parameterized Petri nets for modelling and simulating human organisations in a workflow context. *Proceedings of the "Workflow Management" workshop*, within the 19th Int. Conference on Applications and Theory of Petri Nets, Lisbon, June.

Adam, E., Kolski, C., Vergison, E. (1998). Méthode adaptable basée sur la modélisation de processus pour l'analyse et l'optimisation de systèmes coopératifs dans l'entreprise. In Barthet, M.-F., (Ed.*), Actes du sixième colloque ERGO'IA 98* (pp. 270-279), Biarritz, France.

Adam, E. & Kolski, C. (1999). Etude comparative de méthodes du génie logiciel en vue du développement de systèmes interactifs dans les processus administratifs complexes. *Génie Logiciel*, (to appear).

Ferber, J. (1995). *Les systèmes multi-agents, vers une intelligence collective.* Paris: InterEditions.

Koestler, A. (1969). *The Ghost in the Machine*. London: Arkana Books.

Mandiau, R., Le Strugeon, E. & Agimont, G. (1999). Etude de l'influence de la structure organisationnelle sur l'efficacité d'un système multi-agents. *Networking and Information Systems Journal* (to appear).

Pascot, D. & Bernardas, C. (1993). L'essence des méthodes : étude comparative de six méthodes de conception de systèmes d'information informatisés. In Proceedings *INFORSID'93*, Lille, 11-14 Mai.

Work-Oriented IT Tool Design for Dynamic Business Processes and Organization Structures

Mechthild Wolber
Institute for Human Factors and Technology Management (IAT),
University of Stuttgart, Germany
E-mail: Mechthild.Wolber@iao.fhg.de

1 Concepts of Adaptation

Continuously changing business processes require adaptive and adaptable IT (information technological) tools which offer information to support the respective actions. The concepts of adaptation presented in this article are deduced from organizational as well as from human centered demands, e.g. creativity.

1.1 Information Supply to Support the Various Tasks

From the human factors' point of view, precise actions should be supported by an information supply supporting the various tasks (Hacker 1994). This kind of information supply has to support creative as well as non-creative tasks of the users. But how have information technological methods to be designed for the advanced use of an interdisciplinary information space, promising a shared mental model which is needed for successful teamwork? How can we reach a task-oriented instead of technology-centered information presentation and ordinance? How can we establish an environment which is easy to comprehend and integrates all needed or available information, especially if information is structured heterogeneously, like in product development? How can an information supply achieve the support of creative as well as non-creative tasks?

1.2 Creative Thinking and Mental Discharge

The exclusive use of technology does not suffice for the better support of business processes. In the product development process, e.g. , creative thinking includes the identification of problems and the specification of targets in open, complex, and dynamic contexts (Hacker, Sachse and Schroda 1998). Creative tasks demand for the gleaning, not always precise working knowledge acquisition. Because of the limited capacity of consciousness of every human being, mental discharge has been discovered as an extremely important factor for the ability to creative thinking. The users should be enabled to gather information during information handling. One goal therefore is to combine explorative and precise elements in a common IT system in order to reach creativity. The following questions have been considered in this context: How can explorative and precise elements working together in an IT tool? How can the gathering of interdisciplinary information be supported?

1.3 Continuous Learning and Adaptation of Processes

The demands of the various tasks can be prognosed in the context of dynamic processes only in a fuzzy way. But it is sure that the requirements concerning the co-workers respective users invoked by the dynamics itself get more and more important. Continuous changing demand for adaptive or adaptable IT tools. Hereby, the adaptation is characterized by an organizational and a human centered dimension. How can be guaranteed that the utilization of information technology is adjusted according to the requirements of the work or task design? How can we design the adaptation process to suit cooperation requirements? How is the structure and the course of the dialogue between human being and IT tool adapted to the changed abilities of the users?

2 Requirements on IT Tools

In this section, we will examine which additional requirements on information technological aids are caused by the demand for an information space supporting the respective actions, and how these requirements can be met by the use of suitable information technology.

2.1 Information Management for Non-Creative Tasks

In this context, tasks are parts of a complete action which has the target to reach a complex goal in a precise way. Therefore, team-related information are processed. Tasks belonging to non-creative actions are characterized by the need for processing and transforming information. In contrast to that, parts of creative tasks distinguish by the gathering of information. Additional demands are caused in the team-oriented processing of interdisciplinary information. Successful teamwork necessarily needs a shared mental model. This kind of interdisciplinary knowledge integration is the basis for the compensation of domain-specific barriers.

Firstly, an information supply to support the various tasks has to provide a team-related knowledge integration in an economical way. Routine work has to be supported by a precise information management. Secondly, methods for the action-related information ordinance by information technological media will be described: generic presentation media, virtual labs and spatial metaphors.

2.2 Knowledge Exploration for the Support of Creative Thinking

The intuitive knowledge exploration is closely coupled with the support of creative thinking (Lee, Ong and Quek 1995). Because of the limited capacity of consciousness of every human being, the importance of mental discharge has been acknowledged as an important factor for achieving creativity. Various concepts for mental discharge, e.g. the use of visualization as mental model, double-coding in form of images and text, as well as storage and de-storage of internal into external representations, are immediately transformable into information technological tools. For the implementation of these concepts of human factors' into IT methods, we have developed a method belonging to information science as well as to human factors: the establishment of an overlapping, jointly used information space.

2.3 Adaptation as Learning in Organizations

The continuous change of business processes requires the use of adaptive information technological aids that meet this dynamics (Wolber and Sachse 1997). So the adaptation has human oriented and furthermore organizational dimensions:

- flexible division of work in the team,
- adaptation of IT tools to the design of work,
- cooperation- and team-related adaptation.

One aim of a human centered functional division between human being and machine is an interdisciplinary qualification during work. Information technological aids have to be designed in a way that learning during work is supported. This will provide a flexible division of work in the team. Therefore, a didactic storybook for the interactive application is required. The dialogue between human being and information providing tools has to be adapted to the changing processes and tasks. Object, time and responsibility of the adaptation must be transparent to the user. The tools have to be adjusted to the changing tasks which have to be carried out by the users as well as to the changing abilities of the prospective users. Therefore, we have developed a process-oriented and team-related adaptation model.

3 Concept of Implementation and Evaluation of Results

The framework we have developed integrates human centered as well as information technological requirements. It consists of methods for the design of an information supply to support the various tasks. Furthermore, the framework includes methods for the action related knowledge integration in order to achieve an overlapping use of the information space as well as creative thinking.

3.1 Application Oriented Information Structuring

Using technology oriented information presentation methods, information is structured according to different kinds of information structures, e.g. 2D or 3D. Because of the heterogeneous nature of information, this way of structuring information has to be applied in many business processes. The disadvantage of this technology oriented information presentation is that the perception of connections between information of independent information structure is made difficult. The method of application or respective task oriented information structuring we developed in order to eliminate this problem consists of the structuring of information with spatial metaphors. In domain specific spatial units, we present semantically connecting, but structurally different information. Each of these units contains generic presentation media as structuring elements for domain specific information and tools.

3.2 Tools for the Creative Information Gathering

By the use of methods of information management and knowledge exploration, explorative as well as precise elements can be included in a common IT system. For interactive knowledge exploration, we developed methods for the gleaning knowledge acquisition. The virtual labs can be regarded as a tool for a human centered kind of knowledge discovery. According to Dutke (1994), the use of hierarchical or spatial metaphors for information structuring supports effective learning.

Mental discharge will be provided by the use of visualization as mental model and as communication aid. Furthermore, the support of storage and de-storage of mental representations into external representations will provide mental discharge. Another method for an optimal use of mental capacity is the double-coding of information with visual and notional elements.

3.3 Adaptive IT Tools for the Qualification During Work

The importance of qualification during work is indisputable in the context of modern information society. The matter of qualification is characterized by a self- or group-motivated learning without any teacher. In this context, adaptability and adaptivity of IT tools have to support following criteria:

- The methods for learning depend on a potential cooperation and on the allotment of internal or external control. Exploring kinds of learning demand for a minimum of previous knowledge. A suitable method for the learner friendly presentation makes learning from experts of other domains easier.

- A human centered functional division enables the user to choose between creative and non-creative tasks. Knowledge exploration as a creative task has to be done by the user, not by the machine.

4 Conclusion

New forms of organization, e.g. flexible division of work in a team, have to be supported by an innovative information management. The continuous change of business processes requires the use of information technological aids that meet this dynamics. However, today's IT tools support dynamic processes and organization structures only insufficiently. Being strongly related to the work design of the concerned co-workers, we have especially illuminated the human factors of system design. From that we developed the requirements on tools for the support of dynamic processes.

5 References

Dutke, S. (1994). *Mentale Modelle: Konstrukte des Wissens und Verstehens*. Verlag fuer angewandte Psychologie, Goettingen, Germany.

Hacker, W. (1994). Arbeits- und organisationspsychologische Grundlagen der Software-Ergonomie. In: E. Eberleh et al. (Hrsg.), *Einführung in die Software-Ergonomie*, 53-93, Berlin: de Gruyter, pp. 53-93.

Hacker, W., Sachse, P. & Schroda, F. (1998). Design Thinking - Possible Ways to Successful Solutions in Product Development. In: E. Frankenberger et al. (Eds.), *Designers - The Key to Successful Product Development*, Springer, pp. 205 - 216.

Lee, H.-Y., Ong, H.-L. & Quek, L.-H. (1995). Exploiting visualization in knowledge discovery. In: *Proceedings of the First International Conference on Knowledge Discovery & Data Mining*, Menlo Park - California, AAAI Press, pp. 198 - 203.

Wolber, M. & Sachse, P. (1997). Using Psychological Aspects for the Design of Adaptive IT-Tools for Cooperative Engineering. In: G. Salvendy, M.J. Smith, R.J. Koubek (Eds.), *Design of Computing Systems: Cognitive Considerations, Proceedings HCI International '97*, San Francisco, August 24-29, Elsevier, pp. 781-784.

About the Role of Intelligent Assistants in the Control of Safety-Critical Systems

Guy A. Boy
European Institute of Cognitive Sciences and Engineering (EURISCO)
4, avenue Edouard Belin, 31400 Toulouse, France
E-mail: Guy.Boy@onecert.fr

1 Introduction

In the control of complex dynamic systems, and furthermore safety-critical systems, human operators need to be aware and in control of the situation at all times. Situation awareness and control are key issues that lead to the study of human errors, vigilance, workload, performance, competence, work representation, and so on. Automation has been a crucial research topic in human factors for several decades. The nature of automation has tremendously evolved over the last few years. Software is now an integrating part of almost any complex system, to the point that it is more important both quantitatively and qualitatively than the rest of the system. Most current automated systems include software that has become more autonomous and includes more intelligent features within specific contexts. This paper tries to investigate why these intelligent assistants have become so important in industry, as well as in our everyday life. The argumentation is based on real-world experience in the field of automation and human-computer interaction applied to the aeronautics and space domains.

2 Investigating the Role of Intelligent Assistants

An intelligent assistant may play different roles according to its own level of automation. Four classes of assistance may be distinguished: a *classical technical documentation* that is either paper-based or computer-based; a *context-sensitive computer integrated documentation* that is used as an advisor where the human controls the machine directly (Boy 1991); note that in this paper, the machine is typically a complex dynamic system; an amplifier of user capacities that is located between the human and the machine, and which serves to amplify human action or perception; Hollnagel defines this role as the *embodiment* relation where germane aspects of the task are highlighted while simultaneously others are reduced or excluded (Hollnagel 1991); and an interpreter of machine capabilities that is also located in between the human and the machine, and which serves to interpret machine input or output; Hollnagel defines this role as the *hermeneutic* relation where the user has moved from an experience through the machine to an experience of the machine (Hollnagel 1991).

3 The Operational-Documentation/User-Interface Duality

Paper-based operational documentation was always an important and necessary addition to user interfaces in the control of safety-critical systems. It includes procedures, checklists and systems descriptions. It cannot qualify as an intelligent assistant since most of the intelligence is in the reader of such documentation. However, there are intrinsic factors that characterize an easy-to-use documentation, such as consistency, completeness, structure, and presentation. Technical documentation can be descriptive for training or understanding the features of a system, it can be operational for using the system efficiently. In aeronautics, operational documents such as procedures have four main goals: efficiency, safety, additional training, and legal reference.

During the mid-eighties, I faced the difficult problem of representing operational procedures in the aerospace domain. My research agenda focused on the use of *Knowledge-Based System* technology to enhance procedure following. In the aerospace domain, operational procedures are used in either normal or abnormal situations. Operational procedures are supposed to help operators during the execution of prescribed tasks. It is usually assumed that people tend to forget to do things or how to do things in many situations. Procedures are designed as memory aids. In abnormal situations, pilots need to be guided under time-pressure, high workload, and critical situations that involve safety issues. Procedures are available in the form of checklists that are intended to be used during the execution of the task (shallow knowledge that serves as a guideline to insure an acceptable performance) and operations rationale that needs to be learned off-line from the execution of the task (deep knowledge that would induce too high a workload if it was interpreted on-line.)

The main problem with this approach is that people may even forget to use procedures! Or they anticipate things before the execution of a procedure. People tend to prefer to use their minds to recognize a situation instead of immediately jumping on their checklist books as they are usually required to do in aviation for instance (Karsenty et al. 1995, De Brito et al. 1997). In other words, people are not necessarily good procedure followers. They need to be in control (Billings 1991).

Expert users build complex *situation patterns* that help them index appropriate responses to problems encountered during operations (reactive behavior). I claim that expertise is not so much a question of problem solving, but more in the way experts recognize problems. They are more efficient than others because they have compiled a large spectrum of situation patterns that they quickly pattern match with observed situations. The usability of situation patterns is restricted to the domain of expertise. If these situation patterns can be identified and made explicit, they can be used to redesign appropriately the user interface since they represent functional requirements of an operational interface. Ultimately, if the user interface includes the right situation patterns that afford the recognition of the right problems at the right time, then procedures are no longer necessary. In this case, people interact with the system in a symbiotic way. The better the interface is, the less procedures are needed. Conversely, the more obscure the interface is, the more procedures are needed to insure a reasonable level of performance. This is the *operational-documentation/user-interface duality* issue.

That is to say, representing procedures is also a matter of representing interface artifacts that afford the interaction. This confirms that the definition of interface objects needs to be concurrently associated with the elicitation of appropriate users' cognitive functions (Boy 1998).

4 Cognitive Function Allocation

A major distinction needs to be made between cognitive functions that are devoted to performing a high-level task such as flight planning for example; and cognitive functions that are mobilized to use the machine, such as the interface of the flight management system, that enables the performance of the task. The former are defined as T-CFs (task-dependent cognitive functions), and the latter as A-CF (artifact-dependent cognitive functions). From an operational documentation viewpoint, T-CFs are supported by operational procedures and checklists, whereas A-CF are supported by user guides. The main purpose of human-centered automation is to define the appropriate allocation of cognitive functions among humans and computerized machines (Figure 1).

According to this distinction, there are two types of automation: A-CF automation that facilitates the use of the machine; and T-CF that allocates part of task performance to the machine. The user controls the machine and performs the task in the former case. In the latter case, the user delegates the task (or part of it) to the machine and manages a high-level task. These two approaches to automation associate two different types of technological developments that lead to two different kinds of interaction: direct manipulation and artificial agents management. The first approach increases situation awareness to some degree of complexity of the task. The second approach tends to simplify the performance of the task, because the machine does it for the user. The user thus needs to learn a new job of managing the artificial agents that perform the task for him or her.

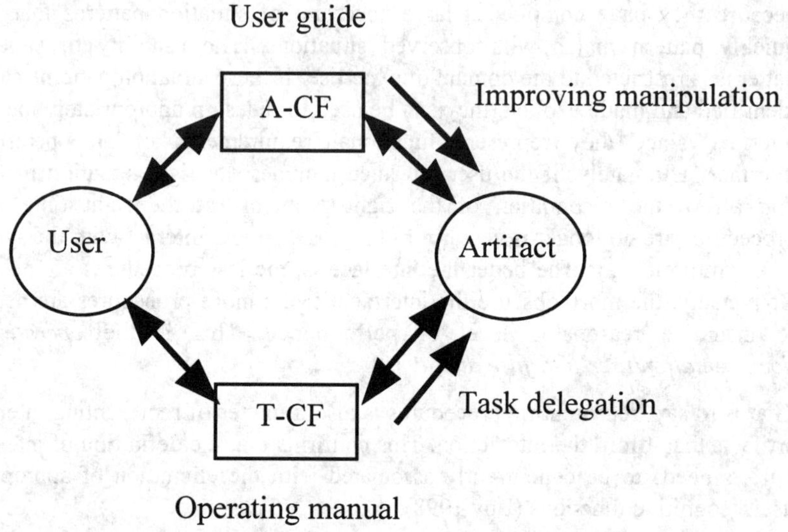

Figure 1. User-artifact interaction. A-CF are currently supported by user guides, and T-CF by operating manual. Transferring A-CF to the artifact improves manipulation (direct-manipulation -oriented automation).Transferring T-CF to the artifact increases task delegation (agent-oriented automation).

5 Training and Intelligent Assistance

High turn-over of personnel, rapid technology evolution and economical issues lead to the emergence of new practices in task performance. In particular, human operators have basic training and keep learning during task performance. They need to access the right information at the right time in the right format. Highly trained people, i.e., human operators who know how to perform the task well, need tools that they can manipulate directly, and then A-CF automation is usually the right solution. Poorly trained people need more intelligent assistance during task performance, and thus T-CF automation is an appropriate solution. However, it is not as simple as it may appear in the real world. Highly trained people often need intelligent assistance in abnormal and emergency situations. Unfortunately, it is very difficult so far to provide the right tool at the right time in such situations. I believe that such situations should be categorized using experience feedback in order to define context-sensitive intelligent assistance.

6 References

Billings, C.E. (1991). *Human-Centered Aircraft Automation: A Concept and Guidelines*. NASA Technical Memorandum 103885, August.

Boy, G.A. (1991). *Intelligent Assistant Systems*. Academic Press, London.

Boy, G.A. (1998). *Cognitive Function Analysis*. Ablex Publishing Corp., Stamford, CT.

De Brito, G., Pinet, J. & Boy, G. (1997). *L'utilisation des checklists dans un cockpit d'avion de nouvelle génération*. EURISCO Technical Report no. T-97-042.

Hollnagel, E. (1991). The phenotype of erroneous actions: Implications for HCI design. *Human Computer Interaction and Complex Systems*: ISBN 0-12-742660-4. Academic Press Limited, London.

Karsenty, L., Bigot, V. & De Brito, G. (1995). *Checklist use in new generation aircraft*. EURISCO Report no. T-95-23.

Multi-Agent Systems for Adaptive Multi-User Interactive System Design: Some Issues of Research

André Péninou, Emmanuelle Grislin-Le Strugeon, Christophe Kolski
L.A.M.I.H., UMR CNRS 8530, Le Mont Houy, BP 311,
University of Valenciennes, 59304 Valenciennes cedex, France
E-mail : {strugeon, kolski}@univ-valenciennes.fr, peninou@univ-lille3.fr

1 Introduction

It is well-known that the quality of Human-Machine interfaces (HMI) is a crucial element in order to design efficient interactive systems. It is particularly true in complex industrial systems where use constraints could be severe for safety, economic or production reasons. In this context, and according to (Browne, Totterdell, and Norman 1990), the reasons for developing Adaptive User Interface (AUI) are: "extend systems lifespan, widen system's user base, enable user goals, satisfy user wants, improve operational accuracy, increase operational speed, reduce operational learning, enhance user understanding". Within the frame of AUI in multi-user complex industrial systems, many subgoals may appear such as: be easy to learn and use by the different users, be efficient at use (according with the various tasks and objectives of the users), be tolerant to certain human errors or decrease the ratio of errors (for each type of user), enhance cooperative team work, and so on.

The paper first presents some criteria for adaptive behavior of interactive systems. Then, some issues for adaptivity from a Multi-Agent Systems (MAS) standpoint are discussed.

2 Objectives of Adaptive Interactive Systems

The state of the art in AUI does not permit, for example in the same way as in software ergonomics, to use some predefined common quantifiable criteria that enable either to decide adaptivity or to evaluate it. The criteria are always linked to potential users, to application domain, to requirements to be met by the system and to application limits. Browne, Totterdell and Norman (1990) stress on the necessity to answer four major questions during AUI development: (1) what is the objective of building adaptation into the system, (2) what are the differences or changes within the user population that justify adaptivity, (3) how can these different users categories be identified, and (4) what is the effective relationship between users variations and adaptive means. The analysis of literature about AUI permits to point out factors or criteria which are common to several AUI. Specifically, we can find general and recurrent criteria which characterise some dimensions of adaptivity (Schneider-Hufschmidt, Kühme and Malinowski 1993): the goals, the different classes of adaptivity, the stages of adaptivity, the strategies of adaptivity, that is the time where adaptivity occurs, the parts of the system which are adapted, the methods of adaptivity, the levels of adaptivity.

3 Possible Contribution of Multi-Agent Systems

In the multi-agent systems research field, studies can be found about the architecture of autonomous entities and the distributed mechanisms that can be implemented in the aim to enable a cooperation among them (see for example, (Demazeau and Müller 1991)). The autonomous entities are called agents. An intelligent cognitive agent has knowledge about itself and about its environment made of the software system and other entities that act in this environment. These knowledge enable it to act not only according to its own goals, but also in the aim to cooperate with other actors. This lead to distributed planning, to coordination mechanisms, and so on. We think that such points could contribute to solve some issues of research in multi-user interactive systems field.

Concepts of agents have yet been used in the domain of HMI. The well-known PAC model, proposed by J. Coutaz, decomposes the application and its HMI into a hierarchy of interactive agents, thus offering a recursive description of the interactive system. The benefits from the properties of agents are: modularity, levels of abstractions, parallelism in interaction handling. This distributed architecture has been extended by combining it with the components of the ARCH model into a model called PAC-AMODEUS (Nigay and Coutaz 1991). In another domain of research, work is done on interface agents and autonomous agents. These works do not directly relate to MAS concepts but make explicit interactions between humans and some autonomous entities (Lieberman 1997).

Finally, interactions between users and MAS have been studied carefully from the multi-agent systems point of view, that is, making a multi-agent system, built for application-specific reasons, interacting with end-users (Avouris and Van Liedekerke 1993) (Grislin-Le Strugeon and Péninou 1998).

In the aim to obtain a high level of adaptivity, it could be interesting to include different models in the adaptive multi-user interactive system: models of the other artificial agents (included in other adaptive or not systems, applicative agents or modules, assistance modules), users models and a model of the interface itself. Such an adaptive system, also called an intelligent agent, can reason to plan its actions in relation to goals, desires and promises of its own or/and of the user. The intelligent agent communicates with its human or artificial collaborators (users or other intelligent interfaces) to perform its tasks, requiring some knowledge or specific abilities it does not have. It must be able to decide the appropriate agents to invoke to interact with them and that leads to the need (1) for interaction protocols between the agents and (2) for organisational schemata.

One of the major purposes of intelligent agents is to show some adaptive features, in an automatic or semi-automatic fashion. For this, criteria (such as those listed above) have to be integrated within intelligent agents. From a MAS point of view, the foundation of the architecture of cognitive agents is knowledge bases use and planning of actions. For adaptive behavior, such plans should deal at least with goals, stages, strategies and methods of adaptation. Moreover, particular criteria should be used during the evaluation of the execution plans (work load, safety of the system, and so on). The multi-agent paradigm permits to use a broad range of planning techniques (hierarchical planning, planning revision, ...) which can be expected to integrate such criteria.

The use of agents concepts for both the interface and the system seems to be a promising framework. For example, in the well-known ARCH model, the agents can appear both in the interface and in the global model of the system as in (Kolski and Le Strugeon 1998) (see fig. 1 below) : (a) an intelligent agent could made up a complex dialogue component, including different models of its own and the others' activities and motivations, and having perception/action exchanges with the adjacent components (the domain-adaptator and the presentation components) ; (b) the interactions of the interface with other acting entities (assistance modules, other HMIs or human agents) can be modelled as multi-agent interactions. Indeed, the link between the interface and the application can be supported by cooperation and communication models and/or protocols which are being studied and validated in the MAS field. In addition to the architecture of an intelligent agent, these concepts and techniques could be used to give to the interfaces the means to cooperate. The results of their activities must be transmitted to the users, but at the same time, they would perform a large number of other operations, for example, the control of the interface itself.

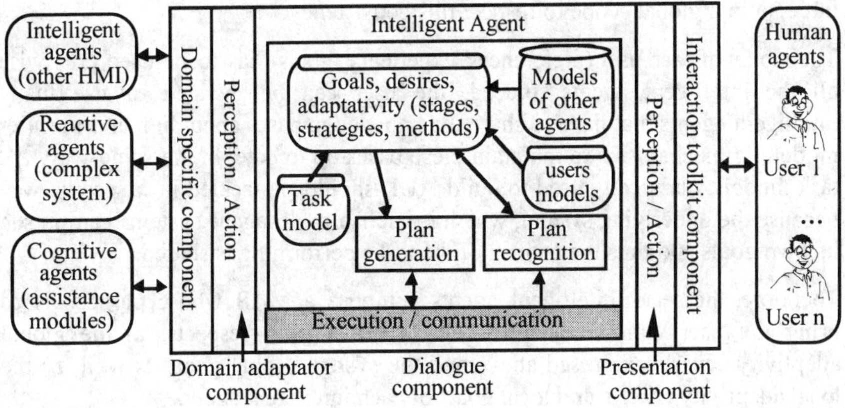

Fig.1: Intelligent agents in relation to the ARCH model, *adapted from (Kolski and Le Strugeon 1998)*

4 Research Ways

From a multi-agent standpoint, interacting with the user may bring to the fore specific problems to deal with (Grislin-Le Strugeon and Péninou 1998). Firstly, two levels of communication have to be distinguished : interacting about problem solving (task sharing, knowledge sharing), interacting about running tasks (results, explanations). These two levels, which are common to interaction analysis field, have to be studied carefully because they impact greatly the development and architecture of the MAS itself. Secondly, due to the distributed nature of multi-agent systems, does the user may see and interact with the system as a whole, or directly with each agent ? Since for complex and distributed multi-users systems, the second solution seems more promising, this aspect has a great influence on adaptive behavior of the system, and thus, on the system itself (knowledge sharing about the user, sharing of adaptive capacity of all the agents, and so on).

In a multi-users context, additional elements should be used to decide or to qualify the adaptivity. For example, the application would limit itself the extend of the adaptivity because the users work together and thus must have coherent common elements to share about the application (knowledge, exchanges about the application, and so on).

The users should have at their disposal the same adaptivity level whatever the tool they use, i.e. whatever the intelligent agent they interact with. Therefore, it seems important that the adaptivity should be not only studied in a local scope but also in a global scope to respect this consistency.

The model of each user (preferences, experience, goals) has to be used crosswise all the intelligent agents. Indeed, the user can be induced to use many intelligent agents that have to share the knowledge bases about him and the user modelling mechanisms to maintain these bases up to date. In the same way, the task models, classically used to build AUI, should be tackled in a specific way because the underlying system, which is itself a multi-agent system, can pursue its own goals and thus the control of the tasks performing is shared.

Therefore, the set of intelligent agents compose a MAS whose purpose is to bring the user with an adaptive interface that both respects: a) the global adaptivity criteria expressed above and the system global goals as well, b) the local adaptivity criteria and local goals of each intelligent agent.

5 References

Avouris, N.M. & Van Liedekerke, M.H. (1993). User interface design for cooperating agents in industrial process supervision and control application. *Int. J. of Man-Machine Studies, vol. 38*, pp. 873-890.

Browne, D., Totterdell, P. & Norman, M. (1990). Adaptative User Interface. Computers and People Series - Academic Press.

Demazeau, Y. & Müller, J.P. (eds.), (1991). *Decentralized Artificial Intelligence*. Vol. 2, Elsevier Science Publishers, North-Holland.

Grislin-Le Strugeon, E. & Péninou, A. (1998). Interaction Homme-SMA : réflexions et problématiques de conception. In *Systèmes Multi-Agents, de l'interaction à la socialité, JFIADSMA'98*, Paris: Hermès, pp. 133-146.

Kolski, C. & Le Strugeon, E. (1998). A review of intelligent human-machine interfaces in the light of the ARCH model. *International Journal of Human-Computer Interaction*, 10 (3), pp. 193-232.

Lieberman, H. (1997). Autonomous Interface Agents. In *Proceedings CHI'97, Human Factors in Computing Systems*, New-York: ACM Press, pp. 67-74.

Nigay, L. & Coutaz, J. (1991). Building user interfaces : organizing software agents. In *Commission of the European Communities, Directorate-General, Telecommunications, Information Industries and Innovation (Ed.), Proceedings of the Annual ESPRIT Conference, ESPRIT'91* (pp. 707-719). Luxembourg : Commission of the European Communities.

Schneider-Hufschmidt, M., Kühme & T., Malinowski, U. (1993). *Adaptive User Interfaces - Principles and Practice*. Amsterdam: North-Holland.

Adaptive Generation of Graphical Information Presentations in a Heterogeneous Telecooperation Environment

Thomas Rist, Matthias Hüther
DFKI Saarbrücken, Germany. Email: {rist, huether}@dfki.de

1 Motivation and Problem Description

Networked computers, multimedia, and mobile communication and computing have already begun to fundamentally change working places and private telecommunication. This progress together with powerful but nevertheless affordable computing devices help to overcome limitations such as uni-modal communication channels, connections only by pairs, and immobile communication partners. Desktop conferencing systems, tools for computer supported collaborative work (CSCW), chat corners and virtual meeting places on the world-wide-web (WWW) are the apparent messengers of this new era. However, the increasing quest for mobility together with a large variety of new portable computing and communication devices including PDA's, Palm computers, mobile phones with build-in micro computers add another level of complexity on communication and collaborative systems, since one has to take into account that the different users may not be equipped equally footed in terms of output and input capabilities. In the context of the project Magic Lounge* we are developing a number of new communication services which are to overcome such communication barriers and which aim at hiding complexity as much as possible from the communication participants. The project name "Magic Lounge" also stands for a virtual meeting place in which the members of a geographically dispersed community can come together, chat with each other and carry out joint, goal-directed activities. For example, to plan a travel they may jointly explore information sources on the WWW, such as time tables of flights and public transportation services. Unlike many other approaches (e.g., chat corners on the web (Dourish 98) or CSCW platforms (Yong & Buchanan 1996)) we do not assume that the users enter the virtual place via exactly the

* Magic Lounge is funded under the Esprit Long-Term Research pro-active initiative i3 and involves as partners: DFKI, Saarbrücken Germany, Odense University, Denmark; LIMSI-CNRS Paris France, Siemens AG, München Germany and The Danish Isles - User Community, Denmark.

same communication channels. Rather, we imagine a scenario as illustrated in Fig. 1. where two users are equipped with fully-fledged standard multimedia PCs, another with a handheld PDA that allows the display of text and graphics on a small medium resolution screen, while the fourth user is connected through a mobile phone that - apart from the audio channel - has a tiny LCD display which can be deployed for displaying minimalistic graphics.

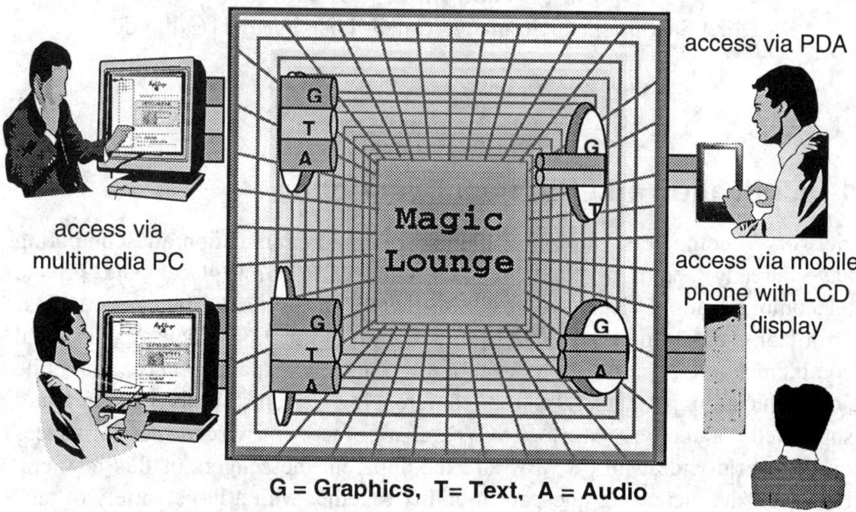

Fig. 1: Accessing the Magic Lounge virtual meeting place via heterogeneous communication devices

2 Application Scenario: Joint Solving of Localization and Route Planning Tasks

Consider the situation in which a group of three users U1,U2, and U3 meet in the Magic Lounge to discuss the details of an impending trip to downtown Saarbrücken. U1 accesses the virtual meeting place via his PC, U2 via a PDA, and U3 via a phone with a tiny LCD display. An example, how the Magic Lounge system supports the users in localization and route planning tasks is illustrated in Fig. 2. In order to clarify how to get from a certain location to another the participants want to consult a map which may be retrieved from the web. U1 is now in an advantageous position as his PC can easily display even highly colored and detailed maps. But what about the two other communication partners? As far as usability is concerned, it doesn't make much sense to output a complex graphics on a small PDA screen. Therefore, U2 receives a more abstract graphical presentation, in this case a network-style map that essentially encodes topological information about the objects in focus. The phone user U3 is certainly in the weakest position as there is only a 100x60 pixels display available on his phone to output graphics. In order to keep all communication partners in the loop, we need a service (named ML Server in Fig. 2) which provides each partner with an individual view on the underlying information. Taking into account the available communication capabilities of the communication devices the phone user U3 receives on his display a simplified graphical representation of the street which is in the current focus of interest. Other details, such as the street name or attributes which can not be shown are provided verbally using a text-to-speech synthesizer on the server side.

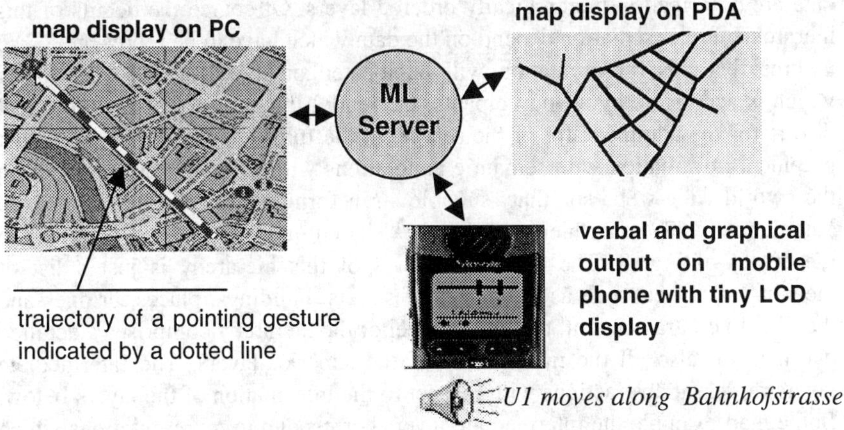

Fig. 2: The Magic Lounge Server provides different views on a shared shared source of geographical data.

Assuming that all three users have also available an audio channel for verbal conversations, the map presentations – though each being different from the others – provide some additional support for solving localization or route planning tasks. For example, verbal references to domain objects (e.g. names of streets, places, buildings) become usually easier to resolve in the presence of a graphical object representation. However, to obtain a more substantial benefit the user's should be allowed to interact with the graphical representations and to share these interactions among them. For example, U1 may perform a pointing gesture on his map in order to show the others how they could go from one location to another. To illustrate this use case, a certain street has been overlaid with a dotted line in the PC map display of Fig. 2. To a certain extend, such a pointing gesture can be transferred from the detailed map display on the PC to the more abstract map display on the PDA, or the other way round, from the PDA to the PC. However, the system has to take into account that the exact course of the pointing gesture can not be transferred using a 1:1 mapping of locations in terms of screen coordinates. Rather, the system has to translate the recorded trajectory of the gesture into a more abstract gesture which can be shown in the graphics that is being displayed on the PDA. For the phone user

U3 the system will provide a short verbal description of the gesture, such as "U1 moves along Bahnhofstrasse".

3 Approach and Technical Realization

Our approach to provide the users with appropriate information displays is to generate presentation variants (i.e. different views) from a common formalized representation of the information to be presented. We use a data model in which data are described in hierarchically ordered layers. Of course the details of this hierarchical representation depend on the data which have to be represented. We assume that this representation will be located on a centralized ML Server which is accessible by all participants. Currently, this requires some additional efforts for the administrator of the server. In the future, however, we hope that graphical information with semantic annotations will become more common in the world-wide-web so that suitable transformations can be performed automatically. The data hierarchy for the "shared map" application is shown in the center of Fig. 3. The uppermost level of this hierarchy is just a list of meaningful objects such as the names of streets, buildings, places, bridges and the like. In contrast to that, the lowest level of the hierarchy comprises a detailed pixmap, but also all the information of the superior layers. The intermediate layers represent abstractions with respect to the information of the layers below. In the map example the intermediate layers correspond to representations which subsequently abstract from geometric aspects such as shape, orientation, and distance between objects. In addition to the generation of different views on geographical data, the data hierarchy plays also a major role for mapping or mirroring of markings and pointing gestures from a source view (on a PDA) to another target view (on a PC) is illustrated in Fig. 3.

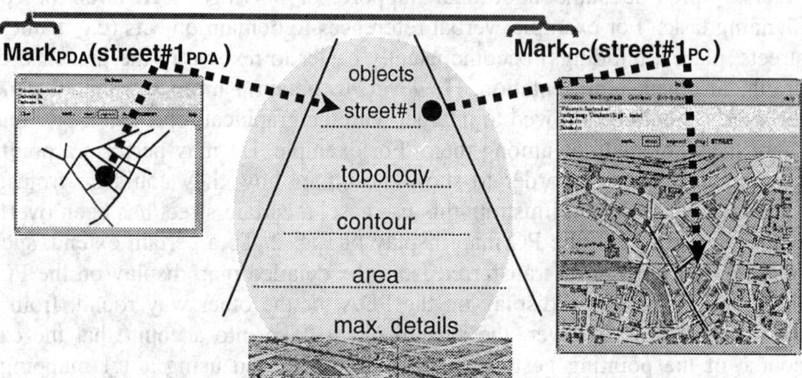

Fig. 3: An object marking on the PDA display is mirrored on the PC display. References to objects are resolved over the hierarchical data representation at the ML Server.

The mapping involves the following steps: First, determine the interaction type that is performed on the source view. In the example of Fig. 3, the user has performed a *Mark* action on his PDA. Next, the system determines the object(s) which are affected by the interaction (in the example, the graphical representation of street#1 has been marked). The system also determines which action on the PC side corresponds best to the action that has been executed on the PDA display (in the example, the appropriate action is named *Mark_PC*). Finally, the system determines the graphical representation of street#1 on the PC display and applies the action *Mark_PC* to it. In order to increase and improve the functionality of our current demonstrator we intend to equip the different viewer components with additional interaction facilities, for example, in order to enable zooming and scrolling of displays.

4 Evaluation of Usability and Acceptance

So far, we only conducted some informal usability tests with our demonstrator. Based on these observations, we got a clearer idea on the sort of appropriate information displays and the type of collaboration modes which are likely to support the users in solving their tasks. For example, it tuned out that collaborations without any verbal (or textual) communication channel does not work. On the other hand, the availability of graphical representations – though being different from each other – help to facilitate collaborations on localization and route planning tasks. Another observation concerns the way how the users exchange markings among each others. It became apparent that one should prefer a collaboration principle which leaves the decisions to the individual users when and from where they want to "import" other views. Also, it seems not advisable to import markings from several other users at the same time since this can result in confusing displays. Our ongoing evaluation work includes a comparison of the kind of interactions and conversational exchanges that occur when a group of users have to solve a route planning task together in two different settings. In the first setting, the group members will meet physically in a room, while in the second setting the group members are at different locations but will use the system to jointly solve the task.

5 References

Dourish P. (1998). *Introduction: The state of play. Special Issue on Interaction and Collaboration in MUDs.* Computer Supported Cooperative Work: The Journal of Collaborative Computing 7(1-2), 1-7.

Yong C., Buchanan C. (1996). *Evaluation of Collaborative Classroom Tools: Groupware.* CSIS Department, Kennesaw State College. Report available under: http://csis3.kennesaw.edu/~groupwre/index.html

Combining User and Usage Modelling for User-Adaptivity Systems

Alexander Nikov*, Wolfgang Pohl**
*Technical University of Sofia, PO Box 41, BG-1612 Sofia, Bulgaria, email: nikov@vmei.acad.bg. **GMD-FIT, HCI Research Group, Schloss Birlinghoven, D-53754 St. Augustin, Germany, email: wolfgang.pohl@gmd.de

1 Introduction

User-adapted interaction has been a research goal for many years now. In research on natural-language dialog systems, user models were introduced as explicit representation of assumptions about the individual characteristics of users. Systems would consult these models to decide how to adapt their behaviour to each user. Traditional user models contain *mentalistic assumptions* (MA), i.e., assumptions about knowledge, goals, interests and other mental attitudes. They are typically represented explicitly in some symbolic or numeric format (Pohl 1998). The main problem in this traditional approach is how to acquire relevant assumptions about the user. In most systems, heuristics were used to control user model acquisition, which often led to unreliable system adaptivity.

In recent years user-adaptive systems like interface agents and personal assistants (Maes 1994) have been developed that use a different approach: They analyse observations of user behaviour applying machine learning techniques, and adapt to the user based on usage patterns detected. Thus, these systems form behaviour-oriented assumptions (BOA) about the user in a systematical and reliable way, but without representing them explicitly (see Davidson and Hirsh 1998 for a recent example). Hence, the user cannot inspect and control the assumptions the system holds. Moreover, mentalistic assumptions needed by many adaptive applications are not constructed.

Our idea is to show that both mentalistic user modelling and behaviour-based usage modelling can be beneficially used to develop user-adaptive systems. We propose the MBAUM (mentalistic and behaviour-oriented user modelling) framework that combines both approaches and is based on ideas described in

(Pohl 1997) and (Nikov et. al 1997). Behaviour analysis is employed for systematic acquisition of BOA. MA are acquired from these BOA, hence also being based on reliable behaviour analysis. The framework has been applied to a server for information on research grants with the goals of both optimising the user interface in an individual way and implementing personalised suggestions of new research grants to users.

2 A Framework for Usage and User Modelling

The aim of MBAUM (cf. Fig. 1) is to adapt interactive systems to the user based on a user model containing both BOA and MA. For this purpose, logfile data of user-system interaction, interviews with users and domain knowledge (which is user-independent) are used. An adaptation of interaction structures of the system based on a neuro-fuzzy algorithm – Fuzzy backpropagation (FBP) algorithm (Stoeva and Nikov 1999) is carried out. FBP algorithm combines neural networks and fuzzy logic. The results of FBP algorithm are also used to construct an explicit user model with BOA and MA that supports further adaptation.

In Fig. 1, the *data* collected *from user and interactive system* consist of logfile, user interviews, and domain knowledge. *Logfile* data provides records of observed user (inter-) actions. *Domain knowledge* mainly contains rules that describe conditions for user satisfaction of work with the system. The initial interaction structures and their application functional interaction points (*AFIP*) are also represented within domain knowledge. For carrying out the *neuro-fuzzy adaptation*, the structure of training patterns and neural networks for processing these patterns are defined. The actual patterns are taken from logfile data using the transitions between each AFIP pair. From these patterns, the Fuzzy backpropagation algorithm (Stoeva and Nikov 1999) learns AFIP weights, developed by one of the authors, which are used to create optimal hierarchical interaction structures. These structures have minimal sum of weighted hierarchy path lengths.

Learning results (AFIP weights and interaction structures) implicitly contain information about user behaviour, used to form the explicit BOA (interaction preferences and generic action sequences) within the *user model*. MA will be mainly constructed from BOA, but can also be derived from observations, both based on heuristic rules.

The following adaptations are possible: Using domain knowledge, hierarchical interaction structures are enhanced to networked interaction structures; BOA are used for action prediction; and, based on MA, help and information are provided individually.

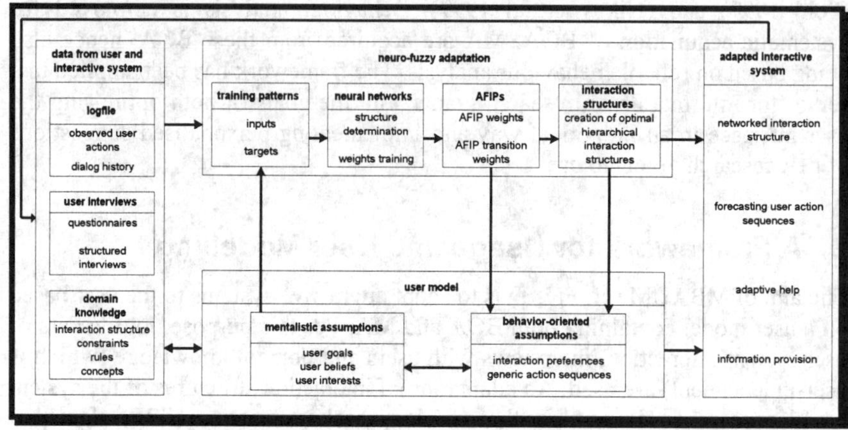

Figure 1. MBAUM framework.

3 Application

The Electronic Funding Information server ELFI [http://www.elfi.ruhr-uni-bochum.de/elfi/] provides web-based access to information on research funding (cf. Fig. 2). Detailed descriptions of funding programs are maintained in a central database. The user retrieves needed information from this database using selection trees. When the user selects a tree item, appropriate funding information is listed, from which the user can choose. All user interactions with ELFI are recorded into a logfile, which provides the basic information for adaptivity.

The selection trees represent interaction structures that can be adapted to the user. Fig. 3 shows a subtree of the selection tree 'Funding organisation' for selections based on the regional scope of funding, before and after adaptation. The number of selectable items (15) in the menu European countries was reduced to 3 menus with maximum 7 items each. The number of items (7) was chosen based on Miller's number 7±2 (Miller 1956). Stating preferred country selection as explicit BOA further leads to the MA "interest in funding from country X". Similar assumptions can be formed from the usage of selection trees concerning research topics and kind of grant (project, fellowship, travel grant, etc.). Thus, an explicit model of user information interest is formed. ELFI makes use of this model for information recommendation: Whenever a new research grant description is entered into the database, the model is consulted to determine whether the grant is of particular relevance to the user. If so, the user is informed about the new grant via email.

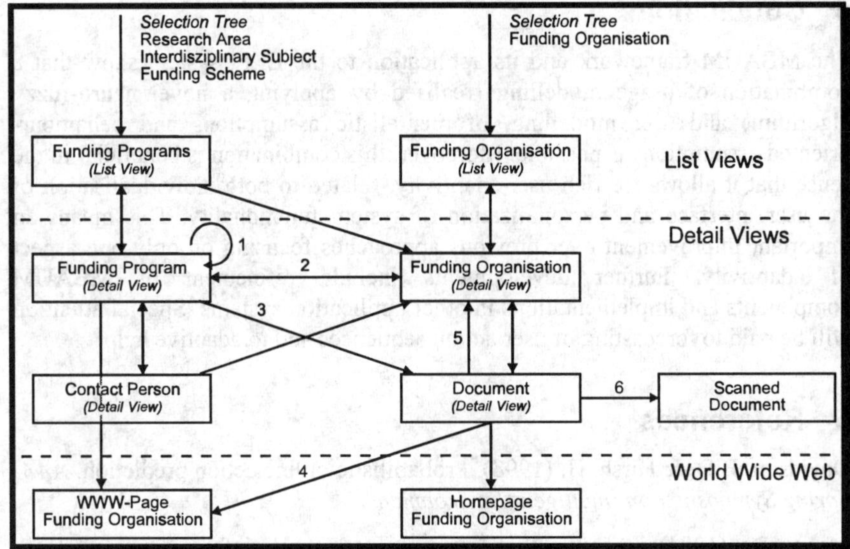

Figure 2. ELFI interaction flow.

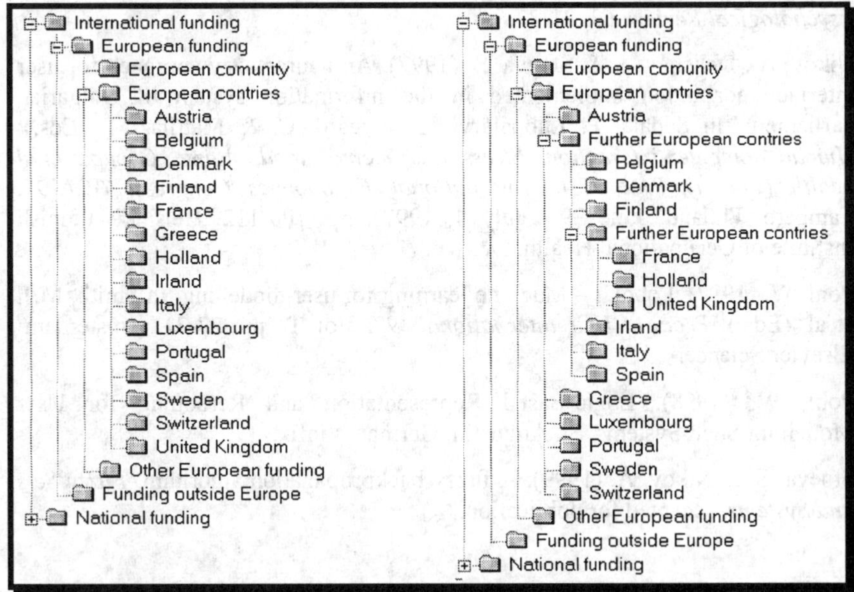

Figure 3. Initial (left) and adapted (right) ELFI selection subtrees of the selection tree 'Funding organisation'.

4 Conclusions

The MBAUM framework and its application to the ELFI system show that a combination of usage modelling (realised by applying a novel neuro-fuzzy algorithm) and user modelling, of mentalistic assumptions and behaviour-oriented assumptions is possible. Moreover, this combination is beneficial in the sense that it allows for rich user-adaptivity, related to both individualisation of the user interface and personalization of system functionality. This means an important improvement over previous approaches focusing on only one aspect of adaptivity. Further developments include refinement of MBAUM components and implementations in other application systems. Special attention will be paid to forecasting of user action sequences and to adaptive help.

5 References

Davidson, B.D. & Hirsh, H. (1998). Probabilistic online action prediction. *AAAI Spring Symposium on Intelligent Environments*.

Maes, P. (1994). Agents that reduce work and information overload. *Communications of the ACM*, 37(7), 31-40.

Miller, G. A. (1956), The magical number seven plus or minus two, *Psychological Review*, 63, 81-87.

Nikov, A., Delichev, S. & Stoeva, S. (1997). An neuro-fuzzy approach for user interface adaptation implemented in the information system of Bulgarian parliament, In Sepälä, P. Luopajärvi, T. Nygård, C. & Mattila, M. (Eds.): *Human computer-interaction, Stress and mental load, Aging, Occupational health (Vol. 1) Proc. 13th International Ergonomics Congress, (*IEA'97, Tampere, Finland, June 29 - July 4, 1997), pp. 100-112. Helsinki: Finnish Institute of Occupational Health.

Pohl, W. (1997). LaboUr – Machine learning for user modelling. In Smith, M.J. et al. (Eds.): *Proc. of HCI International '97*, Vol. B, pp. 27-31. Amsterdam: Elsevier Science.

Pohl, W. (1998). Logic-Based Representation and Reasoning for User Modelling Shell Systems. St. Augustin, Germany: infix.

Stoeva, S. & Nikov, A. (1999). A fuzzy backpropagation algorithm. *Fuzzy Sets and Systems*, accepted for publication.

A graphical user interface shifting from novice to expert

Makoto Oka and Morio Nagata
Dept. of Administration Engineering,
Faculty of Science and Technology, Keio University
3-14-1 Hiyoshi, Yokohama, 223-8522 Japan

1 Introduction

Nowadays ordinary people use computers every day as his/her personal tool. Though the beginners are confused to use computers, experts are using them conveniently. There may be many differences of using computers between the novice users and the experts.

Firstly, this study inspects the differences of the ways to use computers between the novice users and experts. Then, we propose a graphical user interface shifting from a novice user to the expert. Novice users and experts on the topics of programming and character based user interfaces have been already studied [Mayer (1997), Nagata and Nakashima (1988)]. However, there are few proposals of the graphical user interfaces from this point of view.

In this study, we assume a novice user and an expert as follows: A novice user is a user who uses computers less than 6 months. Moreover, the novice user is learning knowledge and skill in using a computer. An expert is a user who has a literacy of the computer and uses it more than one year. Moreover, the expert uses the computer for his/her everyday life.

2 Experiments

In order to find out the characteristics of novice users and experts, we have distributed questionnaires and conducted some experiments.

2.1 Questionnaires

The same questionnaires have been distributed to 147 novice users and 21 expert users. The questionnaires have asked them about icons, menus, the shortcut operations and on-line helps of Microsoft Word. The different answers between the novice users and experts are the following; the novice users do not understand the deep meanings of functions of icons, menus and on-line helps. Experts use the shortcut operations and the on-line helps frequently.

This result shows that we should provide icons and menus which are different from existing ones to the novice users. Notice that this does not mean for the novice users to use these interfaces forever. This paper will show an interface shifting a novice user to the expert.

2.2 History of keyboard and mouse operations

We have written a program to record the history of the keyboard and mouse operations of each user. This program has recorded the processes using Microsoft Word by novice users and experts.

We have obtained the following result from the records. Novice users have many mouse operations. Moreover, each mileage of the operation is long (Figure 1. Subject B). Experts have a few mouse operations. Each mileage of the operation is shorter than the novice's one (Figure 1. Subject A).

These results show that a software system can estimate the level of skill of a particular user by watching the mouse operations.

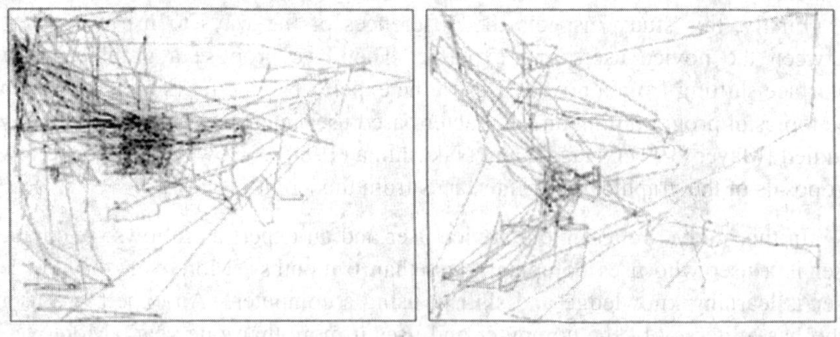

Test Subject A　　　　Test Subject B

Figure 1. Records of the mouse operations

2.3 Icons

We have recorded the users' transitions of clicking the icons of the desktop screen of Microsoft Windows by using the program recording history.

The results show that novice users tend to recognize the icons with labels. On the other hand, the experts use own experiences for selecting icons [Baecker, R. (1991)].

2.4 Menus

25 novice users and 22 experts were asked to classify 40 pieces of cards of menu items (Figure 2) with their own ways. The result shows that the classifications of the experts present the current menu structure. On the other hand, the novice users classify the menu items by the meanings of words.

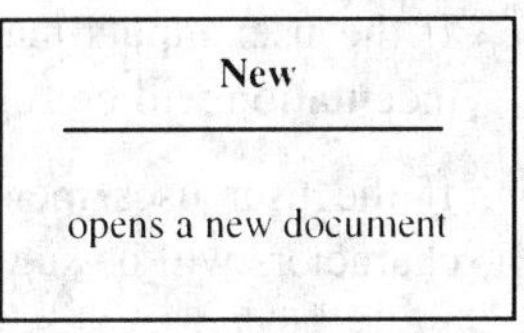

Figure 2. The card of menu items

3 Proposals

From above results, we propose a graphical user interface shifting a novice user to the expert. We have implemented a prototype system of the interface.

3.1 From redundant menus to current menus

From the result of the experiment described in section 2.1 and 2.4, we propose redundant menus that are easy to be used by the novice users. As the novice user uses the redundant menus, our system reduces the redundancy gradually. Finally, the redundant menus will become the current menus of Microsoft Windows (Figure 3).

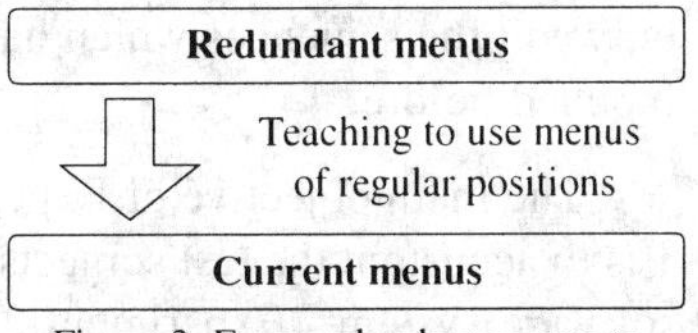

Figure 3. From redundant menus to current menus

3.2 Showing the shortcut operations to users

From the result of the experiment described in section 2.2, we propose a system to show the shortcut operations for novice users (Figure 4).

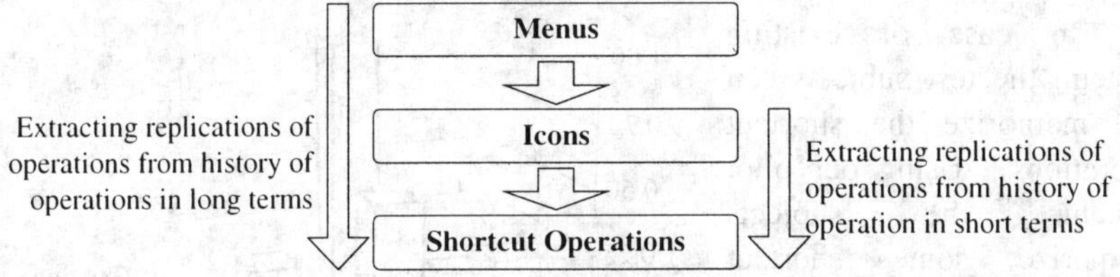

Figure 4. Showing the shortcut operations to users

3.3 Presenting better operations

From the result of the experiment in section 2.2, we propose a system to present better operations by watching the operations of a particular user.

Some examples are:

◊ If the user inputs many spaces in the head of a line, the command of the indentation will be suggested.

◊ If the user uses menus and icons for decorating characters, a dialog for characters will be suggested.

4 Evaluation

We conducted some experiments to compare our prototype editing system with the existing one. The experiments were performed in the order shown by figure 5.

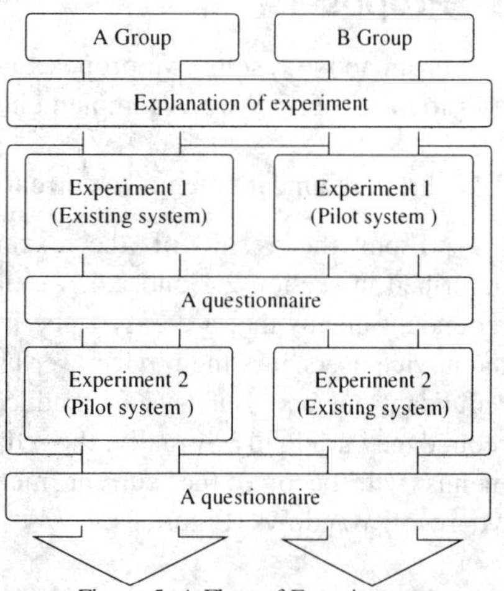

Figure 5. A Flow of Experiments

In Experiment 1 some sentences are inputted and edited. On the other hand, in experiment 2, the test subjects only edit the sentences which has been inputted before.

The main objective of Experiment 1 is to accustom the test subjects to the software system. Experiment 2 is for analyzing the effectiveness of our system. All processes of the test subjects were recorded by using the program recording the history and ScreenCam (made by Lotus).

4.1 Using rates of the shortcut operations

In case of existing system, the test subjects can not memorize the shortcut operations. Using our pilot system, the subjects memorize some shortcut operations. About 30% of their operations were shortcut ones (Figure 6).

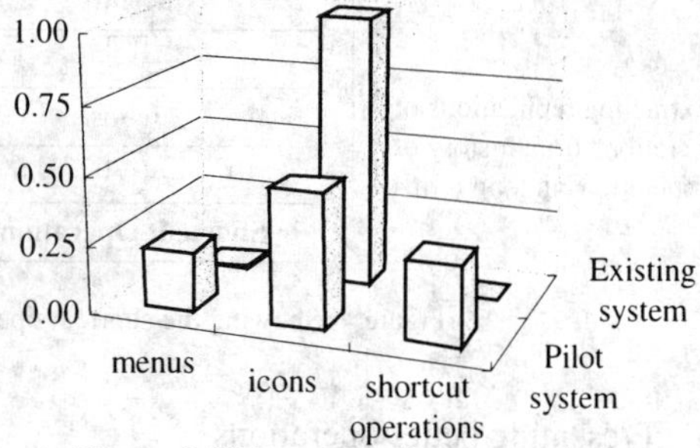

Figure 6. Using rates of the operations

4.2 Effect of presenting better operations

A test subject told us following comment

◊ When I do not hurry and I want to learn operations, I welcome the message.

◊ I am glad that the software system shows me the shortcut operations.

4.3 From redundant menus to current menus

Our system provides redundant menus for the novice users. For example, the file insertion command exists in both [file] and [insert] of the route menu. Substance is [insert], and [file] is a link. The result of two types of subjects is shown in Table 1.

Table 1. Using menu

Before experiment	Using commands	After experiment
File	Insert	Insert
File	File	Insert

5 Conclusions

In this Study, we find the main differences of using GUI between the novice users and experts by performing some experiments. Using the results of those experiments, we have implemented a graphical user interface shifting from the novice users to experts. Our prototype editing system shows that the interface near to novice's image is used easily by the novice. The pilot system shows the advanced methods appropriately. Thus it promotes skill of the user.

We evaluated the effectiveness of our proposal by conducting some experiments using our prototype editing system.

This study assumes that the current GUI is suitable for the user. In the future, we will investigate what is a good GUI.

6 Reference

Baecker, R. et al. (1991), Bringing Icons To Life, Proceeding of CHI '97, ACM, pp. 1-6

Mayer, R. E. (1997), From Novice to Expert, Helander, M., Landauer, T. K. and Prabhu, P. (Eds), Handbook of Human-Computer Interaction, Second. Completely Revised Edition, Elsevier Scince B. V., pp. 781-795

Nagata, M. and Nakashima, S. (1988). Amethod for Generating Messages of the Interactive Software Based on the Individual User Model, System and Computers in Japan, Vol. 19, No. 10, pp. 80-87

ATHENA: AN USER-CENTERED ADAPTIVE INTERFACE

Sandra de A. Siebra and Geber L. Ramalho
Universidade Federal de Pernambuco
Departamento de Informática
Caixa Postal 7851 - CEP:50732-970 - Recife - PE – Brasil
E-mail: (sas, glr)@di.ufpe.br

1 Motivation and Goals

Computers are increasingly more pervasive in all sectors of daily life and, consequently, they are being used by people with a wide variety of backgrounds, preferences, abilities and goals. This state of affairs demands a significant research effort in interface design [Preece 1994, Shneiderman 1992].

The emergence of *adaptive interfaces* [Benyon 1993, Benyon & Murray 1988] provides a satisfactory framework for taking into account users heterogeneity. In adapting to the user's characteristics, these interfaces could, in principle, minimize training and support costs, as well as improve user's satisfaction and productiveness. However, so far, there is no general agreement on adaptive interface models. There is neither a widely accepted enumeration of relevant types of adaptations the interface should be able to undertake (not only cosmetic ones), nor a definite study on the impact of these adaptations in users' learning and performance.

Motivated on a five-year experience in dealing with users having different backgrounds, we have undergone a systematic study of the main human-computer interaction problems. Once identified the potential helpful changes an adaptive interface should perform, we defined the guidelines for its development. Following, we proposed an adaptive interface model, called ***Athena***, based on a distributed problem solving architecture. A prototype of Athena has been implemented in order to evaluate properly our initial working hypothesis.

2 Identified Difficulties and Adaptations

We observed users working and also asked them to fill in forms reporting their difficulties in interacting with computers. This study revealed 15 classes of computer interaction difficulties, most of them concerning novice users. We classified the difficulties into two groups: *motion ability difficulties* (for instance, clicking in objects like icons and buttons, drag and dropping, changing the dimensions of objects and selecting them) and *conceptual difficulties* (for instance, understanding special keys, computer terminology and menu operations).

In order to minimize these difficulties, depending on the user's level, we proposed different kinds of adaptation strategies. For example, when the user is not able to click in a valid area with a mouse, the possible solutions are: (a) to increase the icon or button size; (b) to consider a valid area around the button or icon, or (c) to propose training session for the user, in the form of a "shot the target" game.

3 Athena principles

Athena was built according to four main design principles. The *Regressive Adaptation* principle states that the more expert the user, the less suggestions for automatic adaptation the interface will make, to avoid disturbing him/her. According to the *Negotiation* principle, significant modifications in the user model must be negotiated with the user before being actually registered in the system. This principle is based on recent work on Intelligent Tutoring Systems (ITS) [Paiva et al. 1995, Kay 1997, Bull & Pain 1995]. The *Control* principle assures that the user is responsible for the actual control of the interface and its adaptation. Finally, the *Training* principle declares that, in some cases (for instance, mouse handling problems), if the adaptation has no success, some specific training can be proposed to the user.

4 Architecture

In order to reduce the complexity of the automatic adaptation and user modeling processes, we adopted a Distributed Artificial Intelligence architecture consisting of four agents, as follows (Cf. Figure 1).

The *Perception Agent* receives and processes inputs from the user and the main system to which the interface is attached.

The *Modeling Agent* is responsible for the initialization and updating of the *user model*, which contains information about three generic stereotypes (beginner, intermediate and expert users) plus an individual model for each user. This information is represented by a hybrid formalism combining production rules and objects [Pachet 1995]. The user is characterized by static (e.g., user login) and dynamic (e.g., user abilities) features and his/her stereotype is dynamically updated by means of production rules.

Figure 1 – Athena's Architecture

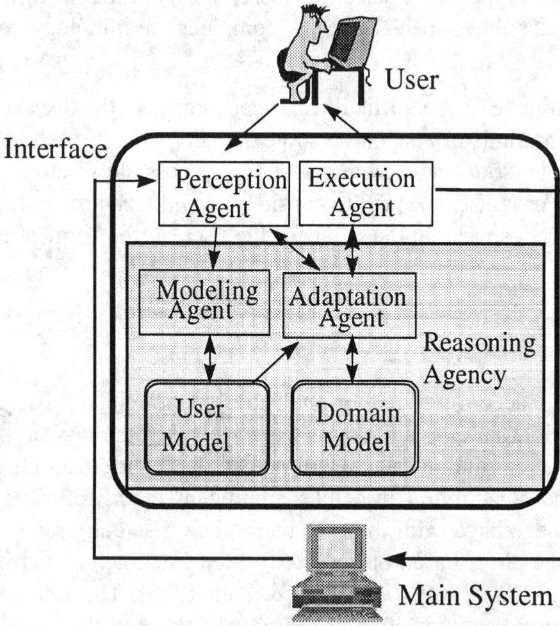

The *Adaptation Agent* has three basic functions: adapt the interface, fix anomalous actions and set training sessions to the user. The knowledge necessary to accomplish these tasks is represented in the *domain model*. It contains the interface description (the interface objects, such as windows, icons, buttons, and menus), as well as generic adaptation strategies, including bug library, advising messages, etc. The adaptations are implemented as production rules of the type *IF an error F occurs AND the user level is N THEN execute adaptation A* (c.f. Figure 2).

The *Execution Agent* implements the execution of actions and presentation of help, advising, error messages and information to the user.

Figure 2 – Athena's Java-Based Production Rules Example

```
R1: If  Perceptor.mouse_click_in_invalid_area( )
        Adaptor.user_level = = 1
        ! Perceptor.big_button(button_code)
    Then Executor.inform_user_about_event(level,event,user);
        reply_user = Executor.ask_user_decision(options, user);
        Executor.expand_button_size(button_code,reply_user);
R2: If  Perceptor.mouse_click_in_invalid_area( )
        Adaptor.(user_level = = 1)
        Perceptor.(big_button(button_code)
    then Executor.consider_valid_region();
R3: If  Perceptor.mouse_click_in_invalid_area( )
        Adaptor.(user_level = = 1)
        Perceptor.big_button(button_code)
        Perceptor.valid_region_on(scenario_code)
    then reply_user = Executor.offer_training(number_training,
                          user);
        Executor.active_training(number_training, reply_user);
```

5 Implementation and Results

We have implemented a prototype of Athena in Java to assess the impact of some of the adaptations we have identified, as well as the adequacy of the interface design principles. We performed tests with two user classes (beginners and experts) and used an "interface descriptor browser" as the main system. The first results of qualitative evaluations were quite encouraging.

We are now extending the implementation to include all identified adaptations, and setting up a more precise assessment methodology with the aid of cognitive psychologists.

By using Java and Distributed Artificial Intelligence techniques, we built Athena as a modular, reusable, extensible and portable interface. Due to this, Athena can be easily extend and attached (plugged in) to different systems.

6 Conclusions

We have presented Athena, a model of an adaptive interface designed to assess certain adaptation principles. Since interface changes can either improve or disturb the user performance, it actually deserves a systematic study of what are the potentially helpful adaptations. This is an important issue for future research in human-computer interaction.

7 References

Alma, J., (1996). *Providing an Interface for a Diverse Range of Users.* Technical Report, School of Computer Science, University of Birmingham, January.

Benyon, D. & Murray, D., (1988). Experiences with adaptive interfaces, *The Computer Journal*, 33, No. 4, pp. 453-461.

Benyon, D., (1993). Accommodating Individual Differences through an Adaptive User Interface. In *Adaptive User Interfaces.* M.Schneider-Hufschimidt, T. Kühme & U. Malinowski (Editors), Elsevier Science Publisher B.V.

Bull, S. & Pain, H., (1995) "Did I Say What I Think I said, and Do You Agree With Me?": Inspecting and Questioning the Student Model. In Proceedings of *World Conference on Artificial Intelligence and Education.* Washington DC, August.

Kay, J., (1997). Learner know thyself: a student models to give learner control and responsability. In Proceedings of *ICCE'97*, Kuching, Australia.

Kühme, T., (1993. User-Centered Approach to Adaptive Interfaces. *Knowledge-Based Systems*, 6(4):239-248.

Preece, J., (1994). *Human-Computer Interaction.* London. Addison-Wesley Pub.

Norman, D., (1988). *The Psychology of Everyday Things.* New York: Basic Books,Inc. Pub.

Paiva, A. & Self J., Hartley R. (1995). Externalising Learner Models. In J. Greer (ed.), Proceedings of *International Conference on Artificial Intelligence in Education*, Washington, DC., AACE.

Pachet, F., (1995). On the Embeddability of Production Rules in Object-Oriented Languages. *Jornal of Object-Oriented Programming*, Vol. 8, No. 4.

Shneiderman, B., (1992). *Designing the User Interface - Strategies for Effective Human-Computer Interaction.* 2nd Edition. London. Addison-Wesley Pub..

A Control Structure for Adaptive Interactive Systems

Alexander Nikov, Tzanko Georgiev
Technical University of Sofia, PO Box 41, BG-1612 Sofia, Bulgaria
email: nikov@vmei.acad.bg
Hans-Günter Lindner
humanIT Human Information Technologies GmbH,
GMD, TechnoPark, Rathausalle 10, 53754 Sankt Augustin, Germany
email: lindner@humanit.de

1 Introduction

The adaptive interactive systems are still in the beginning of their practical use. There are many scientific prototypes but only few real-life applications using adaptive technology. A lot of Internet applications use profiling techniques but they are still parameter scheduling systems, which are not really adaptive. The adaptive interactive systems have large-scaled very complex structures with high number of parameters. Great number of these parameters can be biased and vague.

There is a lack of a framework that supports the structured development of adaptive systems. The control theory is appropriate to model the structure and parameters of such systems. It allows the study of the system stability and the determination of parameter values for which the interactive systems are stable and optimal. The objective of this paper is to present an adaptive control structure that is implemented in a MATLAB environment (Pärt-Enander and Sjöberg 1999). Its neuro-fuzzy adaptation feedback is very appropriate for modelling the uncertain system parameters.

2 Description of Self-tuning Control System of Interactive Systems

We propose the presentation of adaptive interactive systems as adaptive control systems. There are four classical adaptive control system structures (see Aström and Wittenmark 1989, Unbehauen 1988). Most appropriate for structured modelling of adaptive interactive systems is the self-tuning control system shown in Fig. 1.

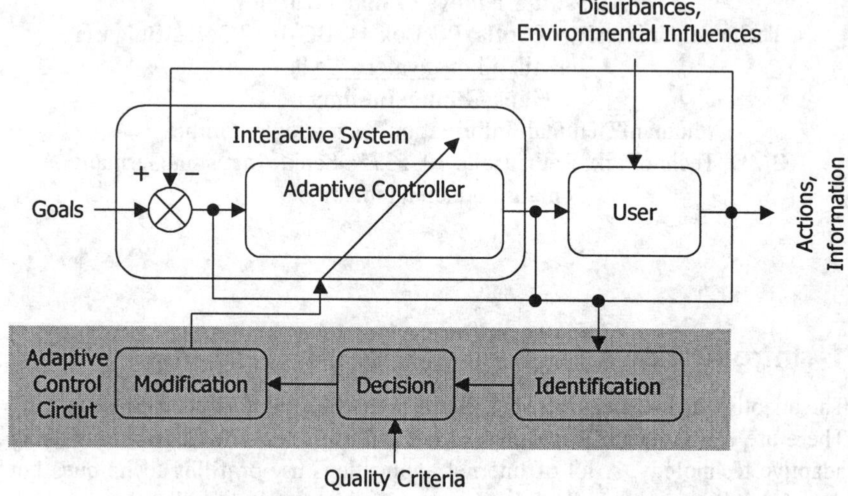

Figure 1. A self-tuning control system (cf. Unbehauen 1988)

It contains an inner and an outer control loop for adaptation. The *inner loop* models the interactive system, the user and the signal flow between them. The interactive system is presented as a regulator. It contains an adaptive controller that is responsible for the adaptation of the interactive systems and its interface. The user is modelled as controlled 'plant' that can be influenced by external disturbances or influences determining the changes of the user's behavior. Goals guide the system's behavior.

The difference between goals and actual output actions as well as information is a part of the interactive system and an input of the adaptive control circuit – the *outer loop*. The identification module collects data from the goals' deviation as well as user input and output. It estimates parameters and performance and models the process online. The decision and modification modules redesign or synthesise the controller. Quality criteria influence the decisions.

3 Matlab-based Control Structure of Adaptive Interactive Systems

The presented self-tuning control system is further developed by introduction of a neuro-fuzzy adaptation feedback. The MATLAB-based control structure of this system is shown in Fig. 2.

The module "Interactive System" is presented as a discrete transfer function derived by MATLAB toolbox "System Identification". It includes four ARMAX models (Pärt-Enander and Sjöberg 1999), which output contains a vector representing the system response to control action. The block "User" is simulated by random noise generator. The module "System Developer" presents

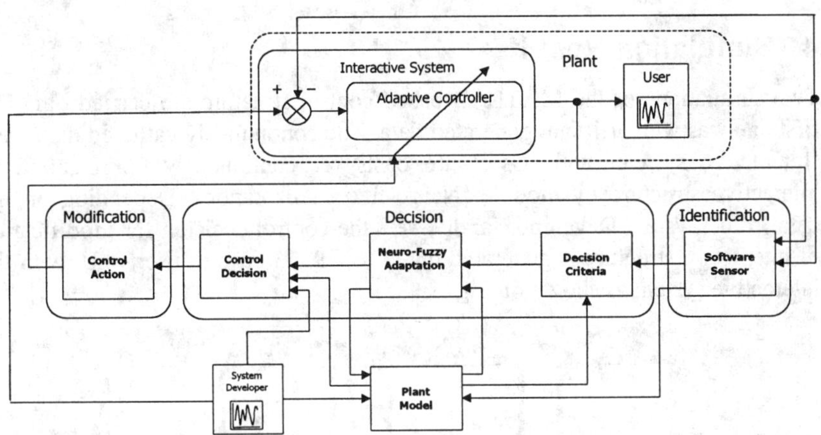

Figure 2. The MATLAB-based control structure.

the "Goal" from previous control structure (cf. Fig. 1). It concerns the supervisor functions to control decision and plant model. The role of the system developer is especially important at the development stage of an interactive system. It is simulated by random noise generator.

The block "Plant Model" presents an inner model of real-life interactive system and user. It supports the learning process of neuro-fuzzy-adaptation, determining the decision criteria and defining the control decision. This block is realised by one ARMAX model. The above-described modules: interactive system, user, system developer and plant model are the first simplified attempt to simulate their behavior.

In the following the modules of the adaptive control circuits are described (cf. Fig. 1). The module "Software Sensor" calculates the transition matrices. They contain the frequencies of transitions between the different interaction points following user work with the interactive system. The block "Decision Criteria" builds the training patterns for module "Neuro-Fuzzy-Adaptation". They include the input and target values for the learned neural network. The block "Neuro-Fuzzy-Adaptation" presents the kernel of adaptation. Here a neural network is trained by the fuzzy neural network (Nikov and Georgiev 1999). The network structure is determined based on users work with interactive system. In the module "Control Decision" by optimization a new adapted interactive structure of the interactive system is determined. For this purpose the trained weights of the neural network are used. The change of interaction structure is carried out by block "Control Action". This is done after paying attention to user expectations.

4 Simulation Results

Two simulations of the MATLAB-based control structure are carried out. The first one was with artificial generated data. The continuously gathered data from 'Interactive system' and 'User' are basis for continuously optimization of interactive structure by module 'Neuro-fuzzy Adaptation'. Depending on the data from 'System Developer' and 'User' the control decision for modification of interactive structure is generated on 7, 16, 19, 23 days of user work with the interactive system as shown on Fig. 3.

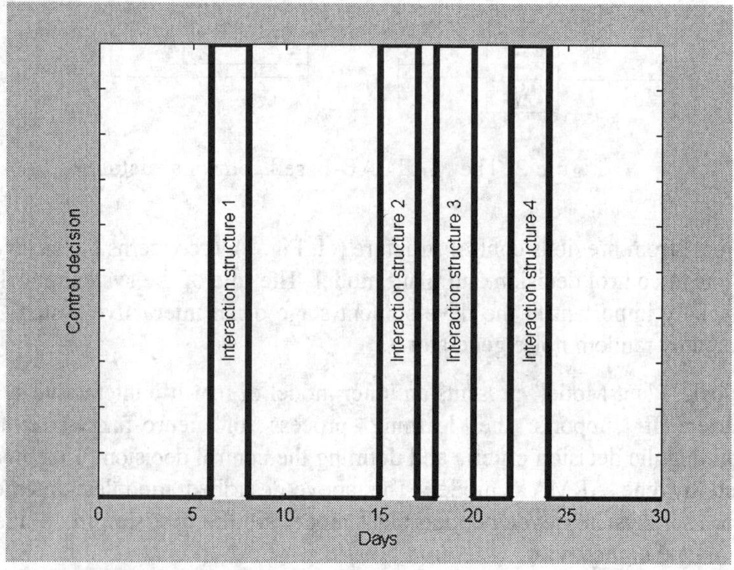

Figure 3. Control decision for modification of interaction structure

The second simulation was applied only to the adaptive control circuit. Input was logfile data from the Electronic Research Funding Information server ELFI (http://www.elfi.ruhr-uni-bochum.de/elfi/). The output was an optimal interactive structure as a selection tree. A part of it is shown in Fig. 4.

5 Conclusions

Based on a classical self-tuning control system a MATLAB-based control structure for adaptive interactive systems is proposed. By artificially generated data and real-life data the feasibility and functioning of control structure are illustrated and confirmed.

The *advantages* of such presentation are:

1) investigation and design of optimal structure of adaptive interactive systems;

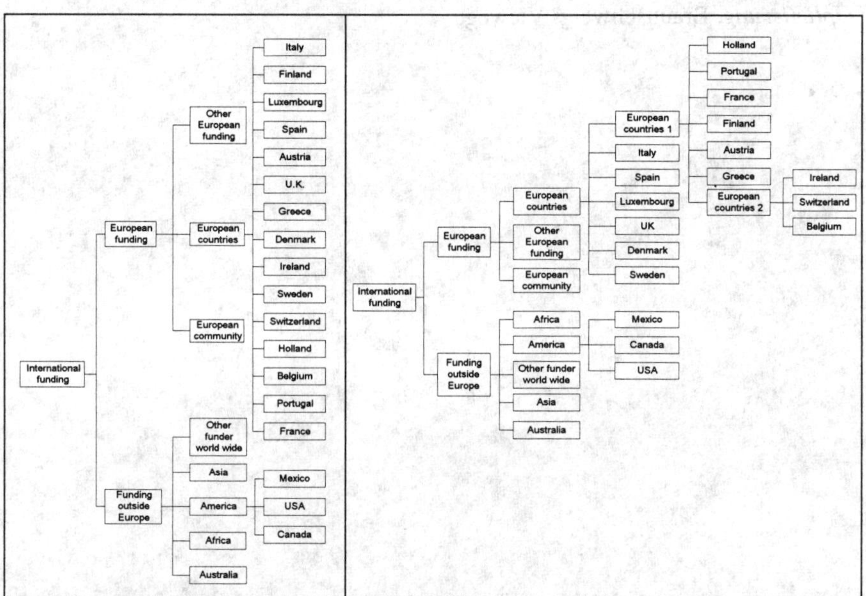

Figure 4. Input (left) and output - adapted (right) ELFI selection tree 'International research funding' with maximal 7 items in a menu

2) selection and optimization of parameters of interactive systems;
3) investigation of stability of interactive systems and design of stable interactive systems;
4) introduction of novel neuro-fuzzy adaptation approach.

The further development of the control structure proposed concerns the decomposition and refining of system modules and their application to ELFI System.

6 References

Aström, K.J. & Wittenmark, B. (1989). *Adaptive Control*. Reading: MA: Addison-Wesley.

Pärt-Enander, E. & Sjöberg, A. (1999). *The MATLAB 5 Handbook*, Addison-Wesley.

Nikov, A. & Georgiev, T. (1999). A fuzzy neural network, *Proc. 21st Int. Conference on Information Technology Interfaces* (ITI '99, Pula, Croatia, June 15-18, 1999).

Unbehauen, H. (1988). *Regelungstechnik, Bd. III - Identifikation, Adaption, Optimierung*. Braunschweig: Vieweg.

Harnessing New Dimensions for Creating Informed Interfaces

Carl Farrell , Russell Alfonso, and Frank Tillman
Hawaii Pacific University

1 Introduction

The intent of this project is to show how the quality and success of human-to-computer communication can be significantly improved by new uses of graphic dimensions. Our interest is in developing interfaces that are informative by the way they are visually structured. In the present paper, this type of visual enhancement will be applied to spreadsheet design. The idea of an "informed interface," which is referred to in this paper, was developed by Tillman and Farrell under a National Science Foundation grant entitled "The Hawaii Education and Research Network (HERN) Project."

Human interaction with a computer is, in an important respect, like human-to-human communication. The success of human communication depends on the flow of inferences that one person makes on the basis of the verbal and bodily behavior of another person. The success of human communication is measured in terms of the fluidity of interpretive inferences. The same measure of success applies to computer-to-human communication. Like human communication, the ideal in computer-to-human communication depends upon a flow of inferences that is both powerful and free of misinterpretations. But there are serious weaknesses in the flow of inferences fostered by most computer systems.

The success of interface design can be measured according to a number of scales: power, clarity, ease of learning, and ease of use. One persistent problem that runs across all of these measures is the limited scope of two-dimensional display screens.

Decisions in a dynamic interactive environment must often be made quickly. However, those decisions must also be informed. This means that what appears on the screen must also be informed. But current graphic-supported interface design appears to be locked into a tight circle of limited possibilities that depend on icons and pictures. The conventional strategy of using picture-based tools is being replicated over and over with ever increasing difficulties and diminishing advantages. In this project we focus on improving the graphic clarity of the interface, realizing that it has far-reaching implications for the power and ease of using computers. What we wish to do is to build more powerful and informative inferences into the design of the interface.

2 Our Strategy

Our strategy in addressing the issue of successful interface design is to draw on the inferential power of structural attributes within a visual interface. This is the area of interface design that is least noticed and yet powerfully informative. Informational power lies not just in what is manifest, but also in what is latent; and much of the latent power of an interface lies in the way it is structured. Structural properties and graphic dimensions are the keys to an innovative rethinking of interface design.

Interfaces have three primary functions: (1) to permit and effect actions, (2) to evoke interest or aesthetic appreciation, and (3) to create and communicate meaning. These functions are not, of course, independent, but the vast majority of previous work on interface design has concentrated primarily on the first two of these functions -- interfaces have been structured to foster action-based capabilities or to offer entertainment and artistic values. Only recently have we seen the beginnings of interface designs that attempt to tap into the latent potential of graphic structure and elaborate visual metaphors to create and communicate meaning. However, clarity is frequently impeded by problems of ambiguity, misinterpretation, and the different world-views that users bring to the computer. Our primary interest is in creating and clarifying the meaning of objects in an interface so that computer-to-human communication may occur without distraction or delay.

The success of computer-to-human communication is directly related to the success of interface design since it is the interface that frames, filters, and embodies the entire content that is available in a system. Strategic placement and combinations of graphic properties, such as color, size, etc. can substantially reduce problems of ambiguity. Within any system, *information management* is important -- within an interface, *inference management* should be paramount.

The approach we propose and demonstrate introduces the notion of an "informed" interface with self-interpreting properties and multimodal representations. Many of the difficulties of power, clarity, and ease of use arise from limited screen space. By incorporating additional dimensions of graphics, and employing the inferences supported by them, we show how interface designers can inform and empower users in new ways, thus reducing problems of ambiguity and misinterpretation.

3 New Ideas for Spreadsheet Design

A common spreadsheet interface may be used to illustrate current weaknesses and failings as well as revealing avenues leading to significant improvements. Spreadsheets are very rich in content, yet the task of finding or grasping much of the important information is usually cumbersome and time-consuming. For example, in a spreadsheet it is not concurrently possible to see calculated values and also to be able to tell, for more than one cell at a time, which of the values displayed are direct data constants and which are derived from calculations. Furthermore, when ranges are used in formulas, users currently have no easy way of seeing how these ranges may or may not be affected by subsequent insertions of rows or columns into the spreadsheet. We demonstrate how a few simple structural additions to a spreadsheet interface can greatly facilitate understanding and potentially reduce errors.

It is difficult to illustrate on paper printed in black-and-white all the techniques that could beneficially be employed to provide informed assistance in a spreadsheet, especially the use of color, depth, and interactivity. Yet, this is exactly why progress has been impeded for centuries in our use of graphic dimensions for information representation. Many graphic dimensions are available for enriching and improving the communication of information; yet because of the technologies of the time, we have generally been restricted to using only those dimensions that can easily be printed in black-and-white, and we have rarely considered employing other dimensions. We have also often ignored the fact that more information-rich interfaces could greatly enhance users' efficiency and the reliability of their work.

We can identify three basic modes of informed assistance. These are:

(1) passive information -- information that is presented without any user activity or action;

(2) active/selective information -- information that is presented when the user "selects" an item or area in some way: e.g., a mouse click, a cursor movement to the area, or an eye movement to the area;

(3) interactive information -- a two-way exchange of information; e.g., the computer provides explanations or warnings, and requests clarification or confirmation from the user before initiating some action.

Ideally, all three modes of informed assistance should be available in a system. Each type of information could be controlled through system parameters, and the normal system default should probably be to have all three modes turned on.

By drawing attention to one aspect of common spreadsheet interfaces, we can illustrate these modes of informed assistance. Figure 1 illustrates the way a spreadsheet usually looks.

Projected Revenues	January	February	March	**Total**
Receipts from sales	825	829	833	**2487**
Expenses for rebates	83	83	83	**249**
Net Revenue	**742**	**746**	**750**	**2238**

Figure 1: Example of a typical small spreadsheet.

4 The Data Type Problem

A persistent problem with spreadsheets is the determination of which cells contain raw data values and which ones contain results that are derived from formulas referring to other locations. Also, it is frequently important for a user to know whether a cell is referred to by some other cell(s) or not. We will call these attributes of a cell its Data Type. If a user mistakenly interprets the value in a cell as being raw data, and changes it to a different value, this removes any formula that was there previously and can destroy the entire integrity of the spreadsheet for all future changes. Similarly, when new rows or columns are added to a spreadsheet, it is critically important for a user to be able to tell whether the new entries are correctly included or not in the ranges referred to by the formulas in other cells.

Passive information regarding the Data Type of a cell in a spreadsheet could be presented through the use of color (e.g., derived values could be colored red or a lighter shade of the color that is used to represent fixed data values, or the background color could be made a bit darker for cells that contain derived values), underlining (e.g., derived values could have wavy underlining similar to what Microsoft Word does for words with questionable spelling or grammar), italics (e.g., values that are referenced by other cells could be shown in italics), depth (e.g., derived values could appear somewhat recessed), or captioning (e.g., a small copy of the word "DERIVED" could appear in the bottom of all cells that have derived values).

What is important is not which style is chosen for providing passive information to users, but simply that some method is employed. Microsoft is already using color to provide information about different types of objects in some of its programming software (although not in its spreadsheet applications), and wavy lines are being used in Microsoft Word systems, so we can see that the importance of this general problem of clarity is beginning to be recognized.

Figure 2 illustrates the way captioning could be used to identify derived values and italics to identify values that are referenced from other cells.

Projected Revenues	January	February	March	**Total**
Receipts from sales	*825*	*829*	*833*	**2487**
Expenses for rebates	*83*	*83*	*83*	**249**
Net Revenue	***742***	***746***	***750***	**2238**

Figure 2: Example of a spreadsheet with passive information.

Possible ways to provide active/selective information regarding the Data Type of a cell might involve any of the strategies described above for passive information, but the cell would only display the Data Type information when it was selected in some way.

Interactive information could be provided by the system through warnings and explanations which would require further input from the user before an action could be taken. This would be triggered by an event such as an attempt to enter a constant data value into a cell that previously had a derived value.

Of the three modes for informed assistance, we would like to emphasize the importance of the passive approach by pointing out the inferential potentials of graphically informed interfaces. The advantage of having passive information built into the interface of a spreadsheet is that inferences can be made quickly and easily by users, allowing them to focus primarily on entering or updating data rather than struggling to determine what is already there.

5 Summary

The spreadsheet illustration is but one exemplification of using graphic properties or structures to aid in the design of efficient interfaces.

A user's decisions in a dynamic interactive environment must be informed. Therefore, what appears on the screen also should be informed. The increasing necessity for information and meaning in the interface can only be met by inventing new ways of providing meaning. New dimensions of visual structure allow meaning to be added to an interface without increasing the density of objects on the screen.

The Effects of Representational Bias on Collaborative Inquiry

Daniel D. Suthers
Department of Information and Computer Sciences
University of Hawai'i at Manoa
1680 East West Road POST 303A
Honolulu HI 96822
suthers@hawaii.edu

1 Introduction

For a number of years, the author and his colleagues (see acknowledgments) have been building, testing, and refining a diagrammatic environment intended to support secondary school children's learning of critical inquiry skills in the context of science (Suthers *et al.* 1997). During this time, a refocus on *collaborative* learning led to a major change in how we viewed the role of the interface representations. Rather than being a medium of communication or a formal record of the argumentation process, we came to view the representations as resources for conversation (Roschelle 1994).

These observations, coupled with the fact that other projects with similar goals were using radically different representational systems, led the author to propose a more systematic study of the ways in which these different representational systems can influence collaborative learning discourse. The differences in representational notations that are provided by existing software for critical inquiry are striking. The range includes hypertext/hypermedia systems (Guzdial *et al.* 1997, O'Neill & Gomez 1994, Scardamalia *et al.* 1992), argument mapping environments (Ranney *et al.* 1995, Smolensky *et al.* 1987, Suthers *et al.* 1997), containment representations (Bell 1997), and matrices (Puntambekar *et al.* 1997). Yet there is a lack of systematic studies *comparing* the effects of external representations on collaborative learning discourse. Given that these representations define the fundamental character of software intended to guide learning, a systematic comparison is overdue.

Substantial research has been conducted concerning the role of external representations in individual problem solving, generally showing that the kind of representations used to depict a problem may influence problem solving efficiency (Kotovsky & Simon 1990, Larkin & Simon 1987, Zhang 1997). One might ask whether this research is sufficient to predict the effects of representations in collaborative learning. A line of work undertaken in collaborative learning contexts is needed because the interaction of the cognitive processes of several agents is different than the reasoning of a single agent (Perkins 1993), so may be affected by external representations in different ways. Shared external representations can be used to coordinate distributed work, and will serve this function different ways according to their representational biases. Also, the mere presence of representations in a shared context with collaborating agents may change each individual's cognitive processes. One person can ignore discrepancies between thought and external representations, but an individual working in a group must constantly refer back to the shared external representation while coordinating activities with others.

2 Hypothesized Effects of Representational Bias

Representational tools are artifacts (such as software) with which users construct, examine, and manipulate external representations of their knowledge. A representational tool is an implementation of a *representational notation* that provides a set of primitive elements out of which representations can be constructed. Developers choose a representational notation and instantiate it as a representational tool, while the user of the tool constructs particular *representational artifacts* in the tool. The present analysis focuses on interactions between learners and other learners, specifically verbal and gestural interactions termed *collaborative learning discourse.*

Each given representational notation manifests a particular *representational bias,* expressing certain aspects of one's knowledge better than others do (Utgoff 1986). Representational bias manifests in two major ways: *Constraints:* limits on logical expressiveness; and *Salience:* how the representation facilitates processing of certain knowledge units, possibly at the expense of others. Representational tools mediate collaborative learning discourse by providing learners with the means to articulate emerging knowledge in a persistent medium, inspectable by all participants, where the knowledge then becomes part of the shared context. Representational bias *constrains* the knowledge that can be expressed in the shared context, and makes some of that knowledge more *salient* and hence a likely topic of discussion.

Stenning and Oberlander (1995) distinguish constraints inherent in the logical properties of a representational notation from constraints arising from the architecture of the agent using the representational notation. This corresponds roughly to the present author's distinction between "constraints" and "salience." Constraints arise from logical limits on the information that can be expressed in the representational notation, while salience arises from how easily the agent recovers the information (via perception) from the representational artifacts.

Information that is recoverable from a representation is salient to the extent to which it is recoverable by automatic perceptual processing rather than through a controlled sequence of perceptual operators (Zhang 1997).

2.1 Notations have Ontological Bias

The first hypothesis claims that important guidance for collaborative learning discourse comes from ways in which a representational notation *limits* what can be represented (Reader unpublished, Stenning & Oberlander 1995). A representational notation provides a set of primitive elements out of which representational artifacts are constructed. These primitive elements constitute an "ontology" of categories and structures for organizing the task domain. Learners will see their task in part as one of making acceptable representational artifacts out of these primitives. Thus, they will search for possible new instances of the primitive elements, and hence (according to this hypothesis) will be biased to think about the task domain in terms of the underlying ontology.

2.2 Salient Knowledge Units are Elaborated

This hypothesis states that learners will be more likely to attend to, and hence elaborate on, the knowledge units that are perceptually salient in their shared representational workspace than those that are either not salient or for which a representational proxy has not been created. The visual presence of the knowledge unit in the shared representational context serves as a reminder of its existence and any work that may need to be done with it. Also, it is easier to refer to a knowledge unit that has a visual manifestation, so learners will find it easier to express their subsequent thoughts about this unit than about those that require complex verbal descriptions (Clark & Brennan 1991). These claims apply to any visually shared representations. However, to the extent that two representational notations differ in kinds of knowledge units they make salient, these functions of *reminding* and *ease of reference* will encourage elaboration on different kinds of knowledge units.

2.3 Salience of Missing Units Guides Search

Some representational notations provide structures for organizing knowledge units, in addition to primitives for construction of individual knowledge units. Unfilled "fields" in these organizing structures, if perceptually salient, can make missing knowledge units as salient as those that are present. If the representational notation provides structures with predetermined fields that need to be filled with knowledge units, the present hypothesis predicts that learners will be more likely to search for and discuss the corresponding information.

3 Empirical Studies

The author has begun studies that test the effects of representational notations on collaborative discourse and learning. Subjects are presented with a "science challenge problem" in a web-browser. A science challenge problem presents a phenomenon to be explained, along with indices to relevant resources. These are relatively ill-structured problems: at any given point many possible knowledge units may reasonably be considered. One side of the computer screen contains the representational tool, such as Threaded Discussion, Containment, Graph, or Matrix. The other side contains a web browser open to the entry page for the science challenge materials. Students seated in front of the monitor are asked to read the problem statement in the web browser. They are then be asked to identify hypotheses that provide candidate explanations of the phenomenon posed, and evaluate these hypotheses on the basis of laboratory studies and field reports obtained through the hypertext interface. They are asked to use the representational tool to record the information they find and how it bears on the problem. Analysis is based on transcripts of subjects' spoken discourse, gestures, and modifications to the interface; as well as measures of learning outcomes. A pilot study was conducted comparing MS Word (unstructured text), MS Excel (tables), and Belvedere (graphs), with two pairs of subjects run in each condition. The data is currently under analysis. Preliminary results are encouraging, and will be presented at the conference. Full studies have been funded and are in the planning stage.

4 Acknowledgments

The author is grateful to Micki Chi, Martha Crosby, and John Levine for discussions concerning the role of representations in learning, visual search, and social aspects of learning, respectively; to many LRDC colleagues, including Alan Lesgold, Eva Toth, Arlene Weiner, for collaborations on the design of Belvedere; and to Cynthia Liefeld for assistance with the pilot studies. This work was funded by DoDEA's Presidential Technology Initiative and by NSF's Learning and Intelligent Systems program.

5 References Cited

Bell, P. (1997). Using argument representations to make thinking visible for individuals and groups. In Proc. Computer Supported Collaborative Learning '97, pp. 10-19. University of Toronto, December 10-14, 1997.

Clark, H.H. & Brennan, S.E. (1991). Grounding in Communication. In L.B. Resnick, J.M. Levine and S.D. Teasley (eds.), *Perspectives on Socially Shared Cognition,* American Psychological Association, 1991, pp. 127-149.

Guzdial, M., Hmelo, C., Hubscher, R., Nagel, K., Newstetter, W., Puntambekar, S., Shabo, A., Turns, J., & Kolodner, J. L. (1997). Integrating and guiding collaboration: Lessons learned in Computer-Supported Collaborative Learning research at Georgia Tech. *Proc. 2nd Int. Conf. on Computer Supported Collaborative Learning (CSCL'97),* Toronto, December 10-14, 1997. pp. 91-100.

Kotovsky, K. and H. A. Simon (1990). "What makes some problems really hard: Explorations in the problem space of difficulty." *Cognitive Psychology* 22: 143-183.

Larkin, J. H. & Simon, H. A. (1987). Why a diagram is (sometimes) worth ten thousand words. *Cognitive Science 11(1):* 65-99. 1987.

O'Neill, D. K., & Gomez, L. M. (1994). The collaboratory notebook: A distributed knowledge-building environment for project-enhanced learning. In *Proc. Ed-Media '94,* Vancouver, BC.

Perkins, D.N. (1993). Person-plus: A distributed view of thinking and learning. In *G. Salomon (Ed). Distributed cognitions: Psychological and Educational Considerations* pp. 88-111.Cambridge: Cambridge University Press.

Puntambekar, S., Nagel, K., Hübscher, R., Guzdial, M., & Kolodner, J. (1997). Intra-group and Intergroup: An exploration of Learning with Complementary Collaboration Tools. In *Proc Computer Supported Collaborative Learning Conference '97*, pp. 207-214. University of Toronto, December 10-14, 1997.

Ranney, M., Schank, P., & Diehl, C. (1995). Competence versus performance in critical reasoning: Reducing the gap by using Convince Me. *Psychology Teaching Review, 1995, 4(2).*

Reader, W. (Unpublished). Structuring Argument: The Role of Constraint in the Explication of Scientific Argument, manuscript dated November 1997.

Roschelle, J. (1994). *Designing for Cognitive Communication: Epistemic Fidelity or Mediating Collaborative Inquiry?* The Arachnet Electronic Journal on Virtual Culture, May 16, 1994. Available: ftp://ftp.lib.ncsu.edu/pub/stacks/aejvc/aejvc-v2n02-roschelle-designing

Scardamalia, M., Bereiter, C., Brett, C., Burtis, P.J., Calhoun, C., & Smith Lea, N. (1992). Educational applications of a networked communal database. *Interactive Learning Environments*, 2(1), 45-71.

Smolensky, P., Fox, B., King, R., & Lewis, C. (1987). Computer-aided reasoned discourse, or, how to argue with a computer. In R. Guindon (Ed.), Cognitive science and its applications for human-computer interaction (pp. 109-162). Hillsdale, NJ: Erlbaum.

Stenning, K. & Oberlander, J. (1995). A cognitive theory of graphical and linguistic reasoning: Logic and implementation. *Cognitive Science 19(1):* 97-140. 1995.

Suthers, D., Toth, E., and Weiner, A. (1997). An Integrated Approach to Implementing Collaborative Inquiry in the Classroom. *Proc. 2nd Int. Conf. on Computer Supported Collaborative Learning (CSCL'97),* Toronto, December 10-14, 1997. pp. 272-279.

Utgoff, P. (1986). Shift of bias for inductive concept learning. In *R. Michalski, J. Carbonell, T. Mitchell (Eds.) Machine Learning: An Artitificial Intelligence Approach, Volume II,* Los Altos: Morgan Kaufmann 1986, pp. 107-148.

Zhang, J. (1997). The nature of external representations in problem solving. *Cognitive Science*, 21(2): 179-217, 1997.

Redesign of Distributed Computer-Mediated Collaboration Systems

Rita M. Vick
University of Hawaii, USA

Uncertainty and change have always been part of everyday problem solving. However, the steadily increasing pace of implementation of new technologies demands that workers engage in continuous learning cycles. The globalization of economies has fragmented the workplace into a series of mini-worlds where workers take on new roles as agents for business, education, or service organizations that operate in locations distributed around the world. This kind of operating environment requires use of various kinds of teams of individuals who work in different times and places linked by electronic media. The author has created a taxonomy of kinds of teams, a general model of the team work process, an analysis of available cultural models and a synthesis of factors representing contextual, composition, process, performance, task and team member variables (Vick 1998). The objective now is to extend this prior work to: (1) analysis of specific cognitive aspects of team decision making judgment biases and heuristics (Kahneman, Slovic, and Tversky 1982), (2) evaluation of how electronically-mediated decision-making and problem-solving processes can be enhanced through synchronous determination of user preferences (Mittal 1999; Kobsa and Wahlster 1989), and (3) development of a diagnostic decision analysis model that combines decision theory and probability theory in the form of a causal belief network (Horvitz, Breese, Heckerman, Hovel, and Rommelse 1998; Druzdzel 1996). The influence diagram shown in Figure 1 illustrates a possible configuration for such a network.

Although electronic media enable collaborative work for distributed teams, problems are encountered in a number of areas (Grudin 1994, 1988). There are three general areas of concern. First, the team, itself, may suffer from problems due to the highly focused nature of teamwork as well as problems related to organizational environment, personality conflict, lack of team-organization goal alignment, and friction due to the presence of differing professional perspectives. Second, the design of the collaborative technology must provide an environment that sustains team productivity with communication, feedback, and collaboration tools that fully support applicable problem-solving and decision-making processes. While e-mail, audio/video/text conferencing, and various brainstorming, list building, ranking, outlining, writing, survey, and analysis tools are made available by many meeting support systems, something more is needed to make computer-mediated collaborative work systems usable.

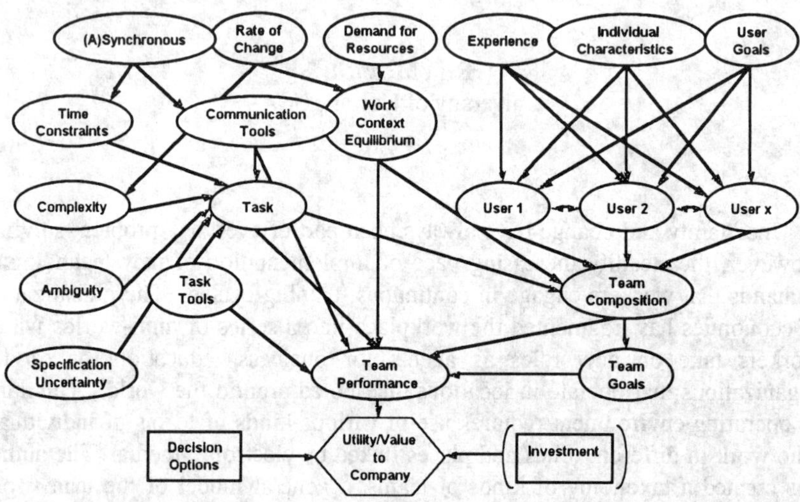

Figure 1. Influence diagram showing a belief network that might support decision options based on the utility derived from a company's hypothetical investment in a CSCW implementation. States of events or random variables can be inferred based on knowledge or on belief regarding their likelihood of occurrence. Lines indicate possible dependencies among events or variables.

Finally, the concurrently-used points of interface involved in collaborative support systems comprise a matrix of interactive nodes that must be in harmony with the task and with the work style needs of team members and the functions they serve within the organization. Poor or marginal usability of these interfaces can, in itself, lead to an unsuccessful team effort. The interface must simultaneously satisfy the interaction needs of a variety of users performing a variety of tasks. At the same time it must be "intra-active," interactive within itself, through maintenance of group memory and integration of information. Insight into the problem at hand is often the main product of parts of the complex computer-mediated collaboration (CMC) and interaction matrix. If the collective mental model that arises from use of the system is not cohesive and functional, work process and outcome will suffer. Given the emergence of a collective mental model that changes as required once trust becomes reciprocal among team members, the cyclical and open nature of a functional team process will become optimally effective despite different time, place, organizational, behavioral, and task constraints.

Having spent more than two years coordinating and facilitating meetings for business, community service, and educational groups in a face-to-face electronic meeting facility, it has been the author's experience that such facilities produce uneven results. Meeting participants are often uncomfortable with the degree of change in meeting process and outcome provided by currently available systems. Remote synchronous discussions tend to take on either too formal or too informal a character so that little of real substance can be accomplished. Asynchronous communication systems, too, are often burdened by communication problems related to resistance to change, information overload due to lack of system-provided structure for information management, storage, and retrieval, and the ubiquitous and unceasing requirement for learning new knowledge. The many advantages of computer-mediated collaborative systems include structuring of the team work process, systematic outcome evaluation, team learning, enhanced group memory, and the accumulation of a knowledge base that can be leveraged to enhance the work process. However, implementation of the majority of such systems still has not resulted in institutionalization of computer-mediated ways of working and learning (Orlikowski 1992).

How can this kind of work environment be taken apart and analyzed to see where changes need to be made to adapt electronic mediation to the needs of groups of users? Further, how can the dynamic responsiveness of the system be made to accommodate and anticipate the needs of users? Attempting to create system responses that mimic human thought patterns has been only minimally successful. If, instead, the system supports human reasoning processes by supplementing human efforts at information processing, increased productivity is more likely (Moore 1998).

Because it encompasses so many aspects of human-computer interaction, CMC-enhanced team decision making for response within complex systems (e.g., global corporate operations, military command and control, medical diagnostic and administrative functions, global/local emergency response operations) that operate under conditions of uncertainty presents an ideal environment for analysis. Creation of a model of group decision making that accounts for the cognitive diversity inherent in group reasoning processes (individual judgmental bias, heuristics, intuitive processes, subjective probabilistic knowledge) as well as behavioral and interaction diversity (attitudes, personalities, task skills, professional differences) will indicate ways in which group support system interfaces can be intelligently designed to accommodate creation of a collective mental model without destroying individual creativity. Imbedding flexible user-support intelligence into the CMC system interface will mitigate many of the problems that currently cause ineffective/non-use of such systems (e.g., awkward interfaces that require users to learn arcane methods to make the system operant for the problem at hand; or, request formats that have too many parameters that require active user intervention to tailor necessary features before productive work can begin).

A significant process-related problem lies in the area of the kind of group memory made available by the CMC system. Many systems capture process, procedural, and subject content in categorical form and spew this "history" back to users in linear, non-synthesized form at the end of each work session. What is needed, instead, is a memory system that can be referenced during a work session as well as across work sessions (Chen, Hsu, Orwig, Hoopes and Nunamaker 1994). The effectiveness, flexibility, reliability, and cognitive simplicity of each concurrently-used point of interface in the matrix of interactions making up a given collaborative support system can make or break effective use of the system.

Evidence from research on learning and problem solving in a number of disciplines (e.g., organizational behavior, economics, psychology) indicates that use of intuitive judgment in the case of experts (as well as novices) in any given field is sometimes less than optimal (de Jong and Ferguson-Hessler 1986; Chi, Feltovich, and Glaser 1981). Modeling decision making to arrive at optimal solutions lacks, of course, the degree of uncertainty present for solution of everyday problems and implementation in natural settings. However, it is evident that structuring decision problems and presenting alternative responses in critical complex domains where the amount of available information is too voluminous for unassisted human processing and the cost of decision errors is high will enhance chances for optimal outcomes. Use of imbedded models of user preferences can enhance decision making through use of statistics, reliability analysis, measuring goals and preferences, presentation of options, and maximization of utility.

A decision analysis model that combines decision theory and probability theory with qualitative and quantitative information about specific computer supported collaborative work (CSCW) processes can provide a useful system design tool for work process problem simulation. Productivity information and observation of everyday use can serve to identify unique intersections of user need and system capability. The ability to quantify subjective user opinions (beliefs) and determine the impact of mediating and conditioning variables will help make meaningful system evaluation possible.

References

Chi, M. T. H., Feltovich, P. J., and Glaser, R. (1981). Categorization and representation of physics problems by experts and novices. *Cognitive Science*, 5, 121-152.

de Jong, T. and Ferguson-Hessler, M. G. M. (1986). Cognitive structures of good and poor novice problem solvers in physics. *Journal of Educational Psychology*, 78 (4), 279-288.

Druzdzel, M. J. (1996). Qualitative verbal explanations in Bayesian belief networks. *Artificial Intelligence and Simulation of Behaviour Quarterly*, special issue on Bayesian belief networks, 94, 43-54.

Grudin, J. (1994). Groupware and social dynamics: Eight challenges for developers. *Communications of the ACM* 37, 1, 93-105.

Grudin, J. (1988). Why CSCW applications fail: Problems in the design and evaluation of organizational interfaces. In *Proceedings of the Conference on Computer Supported Cooperative Work* (Portland, OR, Sept. 26-28), pp. 85-93. New York: ACM, Inc.

Horvitz, E., Breese, J., Heckerman, D., Hovel, D., & Rommelse, K. (1998). The Lumière Project: Bayesian User Modeling for Inferring the Goals and Needs of Software Users. In *Proceedings of the Fourteenth Conference on Uncertainty in Artificial Intelligence* (Madison, WI, July 24-26), pp. 256-265. San Francisco: Morgan Kaufmann Publishers.

Kahneman, D., P. Slovic, and A. Tversky (Eds.) (1982). *Judgment under uncertainty: Heuristics and biases*. New York: Cambridge University Press.

Kobsa, A. & Wahlster, W. (Eds.) (1989). *User models in dialog systems*. Berlin: Springer-Verlag.

Moor, J. H. (1998). Assessing artificial intelligence and its critics. In Bynum, T. H. & Moor, J. H. (Eds.): *The digital phoenix: How computers are changing philosophy*, 213-230. Oxford: Blackwell Publishers Ltd.

Orlikowski, W. J. (1992). Learning from Notes: Organizational issues in groupware implementation. In *Proceedings of the ACM 1992 Conference on Computer-Supported Cooperative Work* (Toronto, Canada, Oct. 31-Nov. 4), pp. 362-369. New York: ACM, Inc.

Vick, R. M. (1998). Perspectives on and problems with computer-mediated teamwork: Current groupware issues and assumptions. *The Journal of Computer Documentation*, 22 (2), 3-22.

Hypermedia Enhancement of a Print-Based Constructivist Science Textbook

Thomas W. Speitel, Marie Iding & E. Barbara Klemm
University of Hawaii at Manoa

Introduction

An instructional computer system to enhance a high school textbook series for physical and biological oceanography is being researched, designed and developed to meet the needs of students, instructional designers, and content experts (Speitel, Reed, Shea and Inouye 1999). Often in traditional textbook and laboratory manuals, subject experts and teachers control both what is learned and how it is to be learned. More recently, instructional programs like the Hawaii Marine Science Studies (HMSS) program (Klemm, Pottenger III, Speitel, Reed and Coopersmith 1991; Klemm, Reed, Pottenger III, Porter and Speitel 1995) have shown how print-based instructional systems can be designed and taught based on principles of constructivism. Here we briefly look at the constructivist design of HMSS and at its limitations in its printed text-based format, which gave rise to the present study that we report.

In its original print-based form, HMSS is designed to provide a structure for active student learning in the constructivist sense. To learn content, students engage in active hands-on investigations of phenomena, then think about what they are doing to construct their own understandings of events. Unlike traditional texts that tell students what they are learning, HMSS initiates learning of new topics with short, focused introductions that establish a need for learning more about a problem or situation. To learn more, HMSS students engage with hands-on activities, guided to think about what they are doing and to generate their own understandings based on what they have observed. They formulate and think about activities in which they encounter concrete, real-world situations.

Key to this is that in the HMSS, students are not passively reading text to obtain information, but instead are actively testing what they already know and extracting, interpreting and applying new information to concrete hands-on problems. They work in groups to record and discuss observations.

Design Considerations

Design considerations for the computer instructional supplement for HMSS revolved around two audiences: 1) students and 2) content and pedagogical experts.

Several heuristics guide development of the computer supplement. The student is not to be withdrawn from the social constructivist context of the classroom investigation for more than a few minutes at a time. The assistance and instruction provided by the computer system therefore needs to be closely tied to the investigation and the textbook. Modes of assistance for the student are to include video, pictures, sound, and text. Types of assistance include examples, definitions, hints, pronunciations, rewording of text, cross-referencing of textbook figures and textbook text, and simple pop-quizzes.

Content and pedagogical experts were observed to not necessarily have the computer skills needed to hard-code instruction into a computer program. This was also not a good use of their time. A human computer interface was needed to facilitate rapid linking of multimedia content and instruction to extant textbook text and figures by the human expert. A "shell" was created by the programmer to easily incorporate "rich text format" text and figures from the textbook and lay them out in a manner identical to the textbook. A control panel for the expert was created to free them from needing a programmer. The expert selects a textbook figure or text segment shown on the computer screen, and then clicks on a few buttons to specify the media mode and type of assistance to be given. Audio and video digitizers enable immediate speech and sound addition, as well as digital and still video. All filing by the system and linking to extant selections is automatic. External files can also be linked. This capability was also given to teachers and students.

How does research on reading affect the computer delivery system? Some of the same basic textual adjuncts, for example, illustrations, examples, questions, appear in both traditional and computer texts as ways to facilitate understanding, especially of processes that can be easily visualized, or expressed more clearly diagrammatically than verbally. Computer-based texts have additional opportunities for "remediations" or adjuncts that students can select when their comprehension is weak, when they are particularly interested in a topic and wish to seek further information, or when they prefer to consult another form of representation (e.g., pictorial rather than verbal) in constructing their own mental models of scientific entities or processes.

In setting the foundation for describing characteristics of effective textual adjuncts for learning science, two compatible theoretical frameworks related. First, Mayer, Steinhoff, Bower, and Mars' (1995) generative theory of multimedia learning is particularly applicable for considering how learners coordinate visual and verbal information while creating a mental model of a phenomenon based upon learning from text. This model has been developed further in studies of short-term memory that provide support for separate visual and auditory stores (Mayer and Moreno 1998). These dual coding or dual processing theories are compatible with notions of situation models, as part of Kintsch's (1998) broader theory of text comprehension.

In addition to fostering conceptual change, other aspects of the HMSS text design are supported by reading research. For example, illustrations are central to the text and designed to be interacted with, not merely to decorate. It is well known that illustrated text is more memorable than unillustrated (Peeck 1989; Levie and Lentz 1982). The interactivity aspect between text and illustration should support dual coding of material into verbal and imagery stores, thereby fostering more accurate and memorable mental models of material.

Finally, hints are embedded into the computer text, often taking the form of questions, which is supported by reading research. For example, higher level questions can facilitate higher level and factual learning (Andre 1979; Hamaker 1986). Unfortunately most questions that appear in science texts are rote level (e.g., Pizzini, Shepardson and Abell 1992) and the insertion of questions about illustrations in science texts can increase cognitive load in ways that appear to be detrimental (Iding 1997).

In addition to addressing these design considerations, preliminary data was collected. There is a recorder built into the program that keeps a log of the time and type of user interactions. Data collection and review indicates that correlation with actual classroom topics being covered was high for the two high school classes observed. The instructor did not manage the student use of the computer. Most of the enrichment computer use was by boys in the coed class. Some of the interactions were rapid—up to 15 per minute. The instructor commented that the program appeared easy for the students to use. Often four or five students would be gathered around the one computer in the classroom. Data is currently being collected from seven schools. We do not have reports yet from teachers who entered hypermedia content themselves. A content expert commented that it was a simple process to link hypermedia text and graphic examples and hints, as well as quizzes.

Considering the most effective format for constructivist computer multimedia science texts and addressing questions for effective incorporation of expert information are important components in the ongoing formative evaluation of the HMSS multimedia format. Questions related to authorship recognition and validity of information need to be examined. Truly, the ability of the student and teacher to add content to the multimedia text makes this a constructivist adventure.

References

Hamaker, C. (1986). The effects of adjunct questions on prose learning. *Review of Educational Research*, 56, 212-242.

Iding, M. (1997). Can questions facilitate learning from illustrated science texts? *Reading Psychology,* 18(1), 1-29.

Kintsch, W. (1998). *Comprehension: A paradigm for cognition..* New York: Cambridge University Press.

Klemm, E. B., Pottenger III, F. M., Speitel, T. W., Reed, S. A. & Coopersmith, A. E. (1991). *The Fluid Earth: Physical Science and Technology of the Marine Environment.* Honolulu: Curriculum Research & Development Group.

Klemm, E. B., Reed, S. A., Pottenger III, F. M., Porter, C. & Speitel, T. W. (1995). *The Living Ocean: Biology and Technology of the Marine Environment.* Honolulu: Curriculum Research & Development Group.

Levie, W. H. & Lentz, R. (1982). Effects of text illustrations: A review of research. *Educational Communication and Technology Journal,* 30, 195-232.

Mayer, R. E., Steinhoff, K., Bower, G. & Mars, R. (1995). A generative theory of textbook design: Using annotated illustrations to foster meaningful learning of science text. *Educational Technology Research and Development,* 43(1), 31-43.

Mayer, R. E. & Mareno, R. (1998). A split-attention effect in multimedia learning: evidence for dual processing systems in working memory. *Journal of Educational Psychology,* 90 (2), 312-320.

Peeck, J. (1989). Trends in the delayed use of information from illustrated text. In H. Mandl and J. R. Stevens (Eds.), *Knowledge acquisition from text and pictures*, (pp. 263-267). North-Holland: Elsevier.

Pizzini, E. L., Shepardson, D. P., & Abell, S. K. (1992). The questioning level of select middle school science textbooks. *School Science and Mathematics*, 92, 74-79.

Speitel, T. W., Reed, S. A., Shea, A. M. and Inouye, B. K. (1999). *Hawaii Marine Science Studies Multimedia Draft.* Honolulu: Curriculum Research & Development Group.

Acknowledgment

This research is sponsored in part by the Space and Naval Warfare Center in San Diego. The content of the information does not necessarily reflect the position or the policy of the United States government, and no official endorsement should be inferred.

A Picture is Worth More than Two Lines

Catherine Sophian & Martha E. Crosby
University of Hawaii

User interfaces often present quantitative information visually. We focus here on the visual representation of ratios, an aspect of interfaces that may be of particular importance since people often find ratios confusing. Most often, interfaces present quantitative information in linear form; for example, the scroll bar in computer windows uses the ratio between the portion of the scroll bar above versus below the marker to represent the ratio between the portion of the document that is above vs. below the cursor's current position. However, because ratios are fundamentally two dimensional--they represent a relation between two quantities and their values are affected by alterations in either of those quantities--we hypothesize that two-dimensional representations of them, in which the two components of a ratio are mapped onto the height vs. width of a spatial figure--will be easier to process than linear ones. Experiment 1 tests this hypothesis by comparing judgments about the relation between pairs of ratios presented either linearly or two-dimensionally. Experiment 2 examines patterns of eye fixations in performing a ratio-based shape comparison task in order to learn more about the perceptual and cognitive processes underlying the interpretation of spatial representations of ratio information.

Experiment 1

In Experiment 1, adults were asked to judge whether two visually presented ratios were the same. Linear problems presented ratios in the form of pairs of vertical line segments, positioned one directly above the other. Both of the line segments comprising one pair were longer than those comprising the other, but the ratio between the upper and lower line segments might or might not be the same for both pairs. For the two-dimensional problems, the stimuli were rectangles whose heights and widths matched the lengths of the two line segments in corresponding linear problems. It was predicted that, even though exactly the same ratios were embodied in both kinds of stimuli, the two-dimensional stimuli would elicit more accurate ratio comparison judgments.

Method

Seventy-three students responded to 16 line and 16 rectangle problems. For same-ratio problems, the ratios embodied in the stimuli (height: width for rectangles, shorter: longer line for line pairs) were either small (.5:1), medium (.61:1), or large (.74:1). The two stimuli to be compared differed on both dimensions by a factor of 1.5. Different-ratio problems contrasted either small vs. medium ratios or medium vs. large ones. The two stimuli again differed in their longer dimension by a factor of 1.5 but in their shorter dimensions they differed by a factor of either 1.23 or 1.83 Line problems and rectangle problems were grouped into separate blocks that were presented in counterbalanced order .

Results

Figure 1 summarizes performance. Judgments were more accurate with rectangles than line pairs when the two ratios were different, $F(1,71) = 6.27$, $MSe = .031$, $p < .05$, but not when they were the same ($F < 1$).

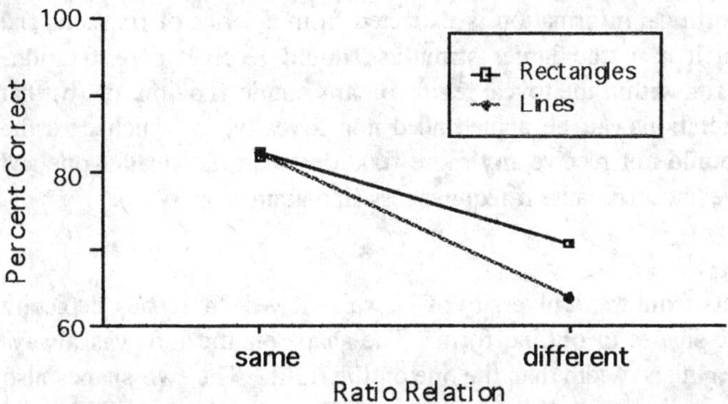

Figure 1. Percentage correct same/different judgments (Experiment 1).

Discussion

These results provide clear evidence for the expected facilitation of proportionality judgments when the ratios were represented two-dimensionally rather than linearly. The interaction between this effect and ratio relation (same/different) appears to reflect a response bias toward "same", which was especially pronounced on the line problems. Across both equal and unequal problems, adults responded"same" fully 60% of the time when the stimuli were line pairs, versus 56% of the time when they were rectangles. Because participants were more likely to judge pairs of line stimuli to be proportionally the same than to judge rectangles that way, their performance was worse on line problems, relative to rectangle problems, when the ratios in those problems were unequal (and so a "same" response was incorrect) than when they were equal (and so a response of "same" was correct).

Experiment 2

In Experiment 2, we monitored the eye movements of a sample of adults as they performed a shape comparison task which entailed deciding which of two shapes that differed in overall size was "fatter" in the ratio of its width to its height. Two questions were of particular interest. First, would the fineness of the discrimination a problem required affect viewers' patterns of fixation? In visual search tasks the discriminability of targets from distractors affects the extent to which visual attention is demanded. Correspondingly, we hypothesized that the more similar the two ratios were that viewers had to compare, the more looking they would need to do to choose between them. Second, we were interested in how fixations would be distributed across the larger vs. the smaller of the two stimuli. This again was expected to shed light on the role of attentional processing in task performance. The stimuli were clearly too large to be encompassed in a single fixation. Two alternative possibilities then, are, (a) that the requisite proportional information is extracted from a series of fixations and integrated, in which case the larger stimulus should receive more fixations because less of it fits within the foveal region on any single fixation, or (b) that the proportional relations can be apprehended non-foveally, in which case the larger stimulus should not receive any more fixations than the smaller one and might even receive fewer (because it requires less fine-grained analysis).

Method

Thirty-one students from the University of Hawaii viewed 24 scenes depicting pairs of geometric shapes in outline form. The shape on the left was always greater in both height and width than the one on the right. The two shapes also differed in that one was fatter, that is, greater in the ratio of its width to its height, than the other. The scenes varied in whether the fatter shape was the larger one on the left or the smaller one on the right, in the general shape of both figures (hexagonal, oval, or rectangular), in the relative elongation of those shapes (narrow vs. medium or medium vs. squat), and in the degree of contrast between the ratios embodied in them (close vs. far).

Eye movement data was collected using an ASL eye movement monitor. A reflection from an infrared beam was used to automatically compute the location coordinates on the array where the eye fixated. Fixations were defined by the occurrence of at least 3 consecutive point locations within a 10 by 18 pixel area. Participants sat 5 feet from the 20-inch monitor on which the scenes were displayed. They responded verbally, "top" or "bottom."

Results

Accuracy was greater for widely separated ratios, M = 94%, than for closer ones, M = 80%. It was also greater when the correct choice was the alternative on the left (M = 95%) rather than the one on the right (M = 78%).

Fixations were tallied for all problems to which viewers responded correctly. The screen was partitioned into left and right regions (a 2/3 vs. 1/3 split, as the figure on the left was substantially larger than the one on the right). Numbers of fixations to each region were then compared across problems varying in the degree of contrast between the two ratios and in whether the correct stimulus was on the left or the right. Means are presented in Figure 2.

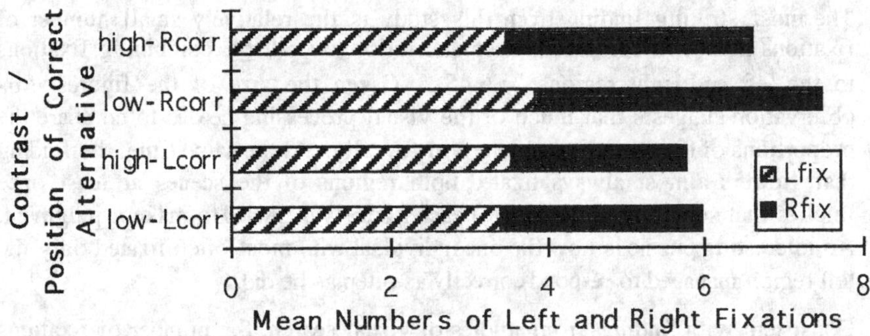

Figure 2. Patterns of eye fixations in ratio comparison task (Experiment 2).

Effects of contrast were consistent with the discriminability hypothesis, although differences between low- and high-contrast problems varied for left vs. right fixations (significantly more right fixations on low- than on high-contrast problems, Ms = 3.10 vs. 2.69, $F(1,30) = 18.87$, $MSe = .617$, $p < .001$; left fixation Ms = 3.60 vs. 3.53, n.s.) and across problems involving more vs. less elongated shapes (more fixations on low- than on high-contrast problems only when the shapes were relatively elongated; Ms = 3.72 vs. 3.24, $F(1,30)= 8.43$, $MSe = 1.91$, $p < .01$; for less elongated shapes Ms = 2.99 vs. 2.95).

Not surprisingly, viewers fixated the larger left region substantially more than the smaller right region overall, Ms = 3.56 vs. 2.86, $F(1,30) = 14.19$, $MSe = 4.27$, $p < .001$. However, almost all viewers fixated both the left and right regions prior to making correct responses. One individual responded correctly 75% of the time yet fixated only one region (always the left) on a majority of those trials (55%). All of the other participants, however, fixated both regions on virtually every trial; across these 31 individuals there were only 15 instances in which no fixations were recorded for one of the two regions of the scenes.

The distribution of fixations across the left vs. right regions varied with contrast and with the position of the correct alternative. There was a greater difference between the two regions on high- than on low-contrast problems, though left fixations predominated in both cases; for high-contrast problems Ms = 3.50 vs. 2.69, $F(1,30) = 15.33$, $MSe = 3.14$, $p < .001$; for low-contrast problems Ms = 3.61 vs. 3.10, $F(1,30) = 9.80$, $MSe = 1.69$, $p < .01$. Significantly more fixations fell on the left vs. right regions of the scenes when the correct alternative was on the left, Ms = 3.47 vs. 2.40, simple effect $F(1,30) = 25.68$, $MSe = 3.003$, $p < .001$; but fixations to the two regions did not differ significantly when the correct alternative was on the right, Ms = 3.64 vs. 3.40.

Discussion

The most striking finding from this study is the relatively small number of fixations needed to compare two spatial ratios (overall M, combining fixations to the left and right regions, = 6.45). Given the size of the figures, this observation suggests that much of the visual processing needed to compare the proportions of the stimuli was done non-foveally. At the same time, the finding that viewers almost always fixated both regions of the scenes at least once implies that some foveal viewing of each figure was needed to make a judgment. An interesting puzzle is how the one individual who most often fixated only the left region managed to respond correctly as often as he did.

Consistent with findings from studies of visual search, the number of fixations viewers made increased with the fineness of the discrimination a problem required. The finding that this pattern held only for right fixations suggests that viewers tended to focus on the left stimulus (perhaps because of its size, and/or because of a left-right scanning pattern) and only look at the right as much as they felt necessary to make their judgment. The attenuation of the left-right difference when the correct alternative was on the right may reflect a tendency to fixate the selected alternative as a choice was made. When this alternative was the one on the right, the resulting increase in fixations to that region would offset the usual predominance of left fixations.

Conclusions And Practical Implications

The comparative findings of Experiment 1 along with the fixation data of Experiment 2 support the conjecture with which we began, that two dimensional representations may facilitate the extraction of ratio information from computer displays. Effects of contrast, however, suggest that there may be limits on the precision of judgments based on these representations.

This research was supported by ONR grant N00014-97-1-0578 to the second author.

Information Gathering from Hypermedia Databases

Joan C. Nordbotten
Dept of Information Science, University of Bergen,
N-5020 Bergen, Norway
hhtp://www.ifi.uib.no/staff/joan

1 Hypermedia Databases

The Internet provides an opportunity to reach the general public with information traditionally disseminated through libraries, and museums. One consequence is that these institutions have begun to make large quantities of information available as hypertext/hypermedia databases of inter-linked images and/or documents.

The hypermedia format is particularly useful for presentation of museum data where information about a topic or artifact is normally presented using multiple media, such as images, text, tables, charts, film, sound tracks, and/or video. The French Ministry of Culture has perhaps the most ambitious museum project. It will contain more than 22.5 million documents (Mannoni 1996, 1997). J. Bowen maintains an updated index to museums with web pages (Bowen 1997). Off-line exhibits can be found in museums and other public buildings.

Hypermedia technology supports associative information retrieval and can facilitate information gathering (Bush 1945, Nelson 1967, Shneiderman 1992). However, researchers anticipate a number of problems with hypermedia system usage. As the number of inter-linked documents and path selections increases, *user disorientation* and *cognitive overload* may hinder users in gathering information (Conklin 1987, Preece 1994). Link structures may actually hinder location of specific information (MacKenzie 1996). And, it is uncertain if museum databases reach their intended public or whether their user's information requirements are satisfied (Day 1995, Futers 1997).

For the general public, gathering information entails location and retrieval of *interesting* document sets (Futers 1997). One measure of interest in a document set can be the time spent viewing/reading it. Providing effective support for information gathering from hypermedia databases requires an understanding of how the intended public finds and accesses information, including:

- How much time the information gatherer is willing to invest.
- How many page selections the information gatherer is willing to use.
- User preference for serial or associative search strategies.
- How page layout affects the user's search for information.

As can be quickly observed, there is little standardization in the structure and presentation formats for hypermedia databases, indicating a lack of consensus as to what constitutes good information presentation.

2 Information Gathering from a Museum Exhibit

Museum visitors come from the general public. They can be considered similar to web users in the sense that both groups contain casual browsers, looking for something interesting, as well as goal oriented seekers of specific information. Thus studying usage patterns for off-line hypermedia exhibits, should provide information that can also be used to design effective web databases.

We have studied usage patterns for a small hypermedia exhibit presenting social sciences projects, developed as part of the 50th anniversary celebration for the University of Bergen (Nordbotten & Nordbotten 1999). The exhibit was implemented on an off-line PC with touch screen input using a WebSite™ server with a Netscape™ browser. The exhibit content, shown in Figure 1, was given in an overview page accessible from each exhibit page.

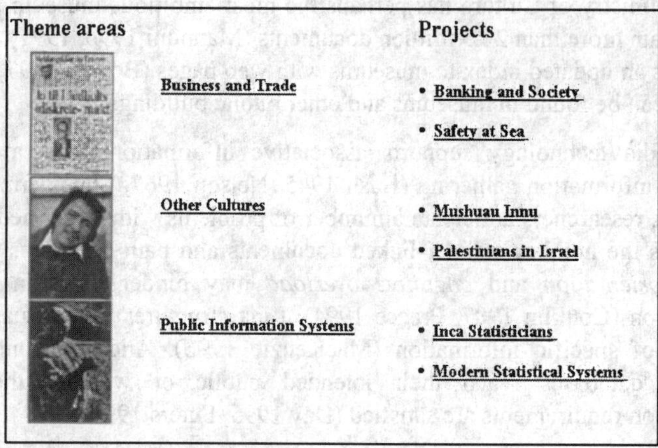

Figure 1. Exhibit content

The exhibit was first located as part of a traditional museum presentation of university research projects, made available to the general public by the Bergen Museum, in Bergen, Norway. It was later moved to the information area of the School of Social Science, at the University of Bergen. An English version is currently available at http://129.177.34.238/eng_museum/00sv-ut.htm.

Usage data has been collected using the server log. A summary, for 4 week periods for each of 3 populations, is given in Table 1. In these samples, 22-50% of the sessions contained only index pages, indicating that the visitor found nothing of interest in the exhibit. The last 3 columns of Table 1 give usage characteristics for those sessions containing at least 1 project selection.

Table 1: Session profile

Data Period	**Dominant visitor group in sample**	**Exhibit starts**	**project sessions #**	**projects/ session Avg.**	**Session length. Page #**	**session time Sec.**
Fall '96	School classes, University students	331	225	1.1	5.75	44.9
Summer '97	Adult visitor/tourist	374	187	1.6	7.0	54.6
Fall '98	Social science Student	322	250	1.4	8.2	78.8

As expected, the exhibit appears to be most interesting for the social science students, when interest is measured by length of time spent and number of pages viewed. The slight drop in projects selected can indicate that these visitors are more representative of goal searchers than general information gatherers.

We had assumed that the exhibit projects would be equally interesting to a varied, general public. As shown in Table 2, this was not the case. For each of the sample populations, project choice is highly correlated, >0.8, to its place in the index set. Note that, after the analysis for the 1st population, the theme index order shown in Table 2, was changed to the order shown in Fig.1, to test this observation.

Table 2: Project Selection

Index level 1 *Theme area*	*Index level 2* *Projects within theme*	*Pop.'96* *%*	*Pop.'97* *%*	*Pop.'98* *%*
Other cultures	Mushuau Innu	35	28	22
	Palestinians	14	12	14
Business and Trade	Banking & Society	22	23	24
	Maritime safety	8	8	8
Public information systems	Inca statisticians	18	20	18
	Modern IT systems	4	10	13

Project presentation length varies from 2 to 8 pages. Each initial page contains text links to its detail pages, which give different information about the project. Fig.2a shows that detail page selection falls from nearly 80% to 30% from selection of the 1st to 4 detail pages. Figure 2b shows a marked difference in the use of the embedded links for page selection. Here, the social science students were nearly twice as likely to choose an associative link, than the other museum visitors who selected exhibit pages using the directional buttons on each page. Possible explanations include:

- familiarity with hypermedia presentations.
- the longer time spent made it more likely that the links were seen.
- familiarity with the subject matter allowed preference selection.

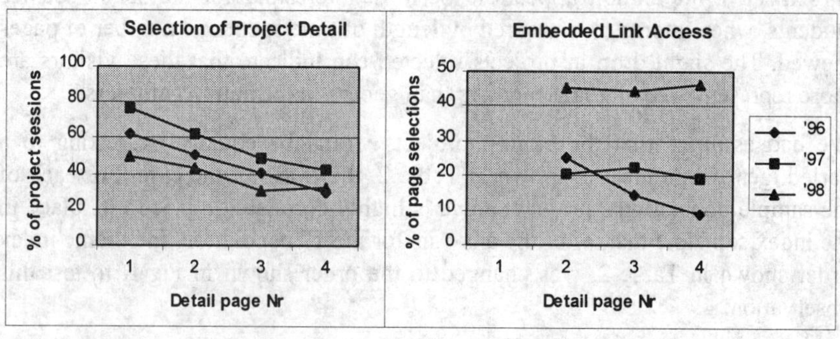

Figure 2. Detail page selection: a) by number b) use of embedded link

3 Summary and Conclusions

We have studied viewing sessions for an exhibit of social science projects, for 3 populations, dominated by school and university students, adults, and social science students, respectively. Though social science students spent the most time and viewed most pages, they selected both fewer projects and detail pages than adult museum visitors, indicating index set review and interest selection.

Our visitors spent, on average, about 1 minute viewing 7 pages on less than 2 topics. Session length in time and page number varied from 4 to 103 seconds and 3 to 36 pages, increasing with the age and topic 'nearness' of the viewer. Index placement dominated topic selection for all groups, while use of embedded links appears related to familiarity with hypermedia presentations.

In conclusion, we would advise that hypermedia exhibits be organized as short theme presentations and that indexes be ordered by theme importance.

4 Acknowledgements

This project has been part of the celebration activities for the 50th anniversary of the University of Bergen and the 25th anniversary of the School of Social Sciences. Thanks are extended to staff, faculty, and students of the Bergen Museum, School of Social Science, and the Department of Information Science for their help in the construction and test of the electronic exhibits. Special thanks are extended to Professor Svein Nordbotten for all project support.

5 References

Bowen,J. (1997). Virtual Library museums pages (VLmp). http://www.icom.org/vlmp/.

Bush, V. (1945). As we may think. *Atlantic Monthly* July 176:1, 101-108.

Conklin,J. (1987). Hypertext: An introduction and survey. *IEEE Computer,* 20:1,17-41.

Day,G., (ed.) (1995). Discussion. *Proceedings. Museum Collections and the Information Superhighway.* Science Museum, London. http://www.nmsi.ac.uk/infosh/discuss.htm.

Futers, K. (1997). Tell Me What You Want, What You Really, Really Want: a look at Internet user needs. *Mda.* http://www.open.gov.uk/mdocassn/eva_kf.htm.

MacKenzie, D. (1996). Beyond Hypertext: Adaptive Interfaces for Virtual Museums. http://www.dmcsoft.com/tamhpapers//evaf.htm.

MacKenzie, D. (1996). Beyond Hypertext: Adaptive Interfaces for Virtual Museums. http://www.dmcsoft.com/tamhpapers//evaf.htm.

Mannoni,B. (1996). Bringing Museums Online. *Communications of the ACM,* 39(6), 100-105.

Mannoni,B. (1997). A Virtual Museum. *Communications of the ACM,* 40(9), 61-62. See also http://www.culture.fr/lumiere/documents/files/imaginary_exhibition.html

Nelson, T.H. (1967). Getting it out of our system. *Information Retrieval: A Critical Review.* G.Schechter, ed. Thompson Books, 191-210.

Nordbotten,J. & Nordbotten,S. (1999) Search Patterns in Hypertext Exhibits. *Proc. 32nd Hawaii Int'l Conference on System Sciences - HICSS'32.* Maui, HI. Jan.4-9, 1999. IEEE. 1999. ISBN 0-7695-0001-3. 7p.

Preece, J. et.al. (1994). *Human-Computer Interaction.* Addison Wesley.

Shneiderman, B. (1992). *Designing the User Interface - Strategies for effective Human-Computer Interaction,* 2nd ed. Addison-Wesley.

Gender and Computer Expertise

Laurel King
University of Hawai'i

1 Introduction

There is an underlying social construction that defines technology as „masculine" and continues to act as a psychological and social barrier to females being interested in computers and other machines. This difference in expectations, in turn, encourages fewer women than men to enter into computer science and technology fields, despite the absence of any gender-based differences in intelligence. Various studies have shown that males continue to show higher computer proficiency levels and greater representation in computer-related fields than females. This tendency to exclude women from physical science and technology fields not only affects the women who are denied the higher pay and prestige associated with these areas, but also adversely affects the fields themselves (Turkle and Papert, 1990).

Studies have found that girls learn better through cooperative rather than competitive activities (e.g., Scott et al., 1991). Software company *Purple Moon*'s marketing research shows that girls favor „role-playing games with real-life characters, adventure games in which having new experiences—not winning—is the goal, and drawing, creative writing, and clue-based games." (Yovovich, 1997). Joiner et al., 1996, found that software theme affected girls' computer performance but found no relationship between theme preference and performance for boys. Elkjaer, 1992, showed that despite any innate differences in ability, girls are often treated and act as „guests" and boys as „hosts" in computer related activities. Female teachers are often untrained in computers and reinforce gender expectations because they are usually dependent on male teachers that do have computer skills and male principals to help introduce computer skills in their classrooms (Apple et al., 1990, Lee, 1997, or Reinen et al., 1997). There is also much evidence to support the claim that boys are getting preferential treatment in the classroom, particularly in subjects considered to be „masculine", such as math and physical science. Corston et al., 1996, found a social interaction effect in that females performed much better on a simple computer test in the presence of another female than they did alone or with a male present. This is significant given the high proportion of males in science and technology fields. Females have been found to be less confident and have more anxiety than males regarding computers. Also, studies have shown that males are more likely to own a computer and that owning a computer increases positive attitudes toward computers. Women often report having less computer experience, liking computers less, and seeing less value in them than men. There are indications that females are often more attracted to social and interpersonal interactions than to human-machine interactions. All of these factors have an effect on one's motivation and interest in excelling at computer technology. This effect is apparent in the United States, because in 1995 women earned only 36.3% of the doctorates in science and engineering and 30% in computer science, according to the National Science Foundation.

2 Research Questions and Methodology

Given the common perception of technology as a male domain, the following questions were studied to find potential reasons for fewer women finding computer technology interesting and pursuing technology-related fields. Are females still less likely to have a computer in their home than males? Do females have less confidence in their computer skills? Do males like fast-paced computer games more than females? Are females less likely to pursue a technology-related major than males? To investigate these questions, a one-page questionnaire was given to 112 undergraduate students. They varied in academic major and were distributed equally by gender. The largest breakdowns of majors were 20% undecided, 19% business (compared to 21% actual in 1996), and 9%-engineering majors (compared to 5% actual in 1996). The average self-reported GPA for those completing this question was 3.27 for females and 3.00 for males (P= .09). The survey was designed to look at the students' attitudes, usage, and background experience with computers.

3 Results

The data in tables 1 and 2 show the main results by gender. The students did not differ significantly in their use of e-mail, having computers at home, being the primary user or wanting to be the primary user of a computer, or on the main factors influencing their choice of major. Factor analysis of 12 questions ranking their computer skill levels yielded two clusters, one representing „computer competence" (rankings of 6 questions on attitudes about computers, technology careers, and word processing skills) and the other „computer expertise" (rankings of 5 questions on confidence in computer skills, abilities in games, spreadsheets, web page design, and programming). The mean scores for the computer competence factor did not differ significantly by gender. However, gender differences were significant for the computer expert factor with females averaging 2.5 and males 3.0 for this factor. Other significant differences showed males to be much more likely to choose technology-related majors, have had computers in their homes for more years, spend more time surfing the Web, and surf the Web for more reasons. Males liked fast games more (p = .053).

Table 1 Cross-tab Results for Selected Variables

Question	overall %	% of female	% of male	Likelihood Ratio	DF	Significance
Like fast games	61.8%	52.8%	72.9%	5.86	2	**.053**
Use e-mail regularly	83.6%	81.5%	84.6%	.184	1	.667
Computer at home	88.4%	85.2%	94.4%	2.61	1	.106
Have modem at home	76.8%	67.4%	86%	7.85	3	**.049**
Computer primary user	61.9%	58.7%	63.3%	.208	1	.648
Wants own computer	90.4%	88.6%	91.4%	.159	1	.690
Tech-related major	30.6%	18.5%	42.6%	7.52	1	**.006**
Factor in major choice:						
Family or teachers	13.1%	17%	9.3%	6.28 (Mantel-Haenszel 5.45)	3 (1)	.099 (.02)
Interest	64.5%	68%	61.1%	"	"	"
Salary	21.5%	13.2%	29.6%	"	"	"

Table 2 Independent t-Tests for Selected Variables

Question	overall mean	mean for females	mean for males	t-test Value	DF	Significance
Years computer at home	5.823	5.011	6.67	2.07	92	**.042**
GPA (self-reported)	3.121	3.273	2.996	-1.71	55	.093
Surf hours per week	3.708	1.927	5.335	2.64	74	**.010**
Total reasons for surfing	2.36	1.870	2.868	3.14	105	**.002**
Computer Competence (Factor 1) α=.82	4.1967	3.917	3.915	-.02	105	.986
Computer Expertise (Factor 2) α=.83	2.777	2.467	3.055	3.29	106	**.001**

Correlation between expert level and years computer owned (.29, p=.004) and expert level and surf hours (.51, p=.000) was found, but there was no interaction with gender. Factorial ANOVA showed interaction effects between expert level by gender, liking fast games, and whether or not one was a technology-related major. The Scheffe's test showed that male tech-majors that do not like fast games had the highest expertise mean (4.53) and differed significantly from non-technology major males (2.10) and females (2.11) that did not like fast games. The technology major males that liked playing fast games (3.31) significantly differed only from non-technology major females that did not like fast games. Table 3 shows the computer competence and computer expertise scores for the eight groups of the 2x2x2 ANOVA. It is clear that computer competence is well distributed while the technology majors and those that like fast games have higher expert scores. Oddly enough, this is not true for the male tech-majors as the expert and competence scores were higher for those that did not like fast games. Furthermore, competence scores for female tech-majors that liked fast games were lower than for those that did not.

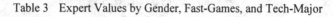

Table 3 Expert Values by Gender, Fast-Games, and Tech-Major

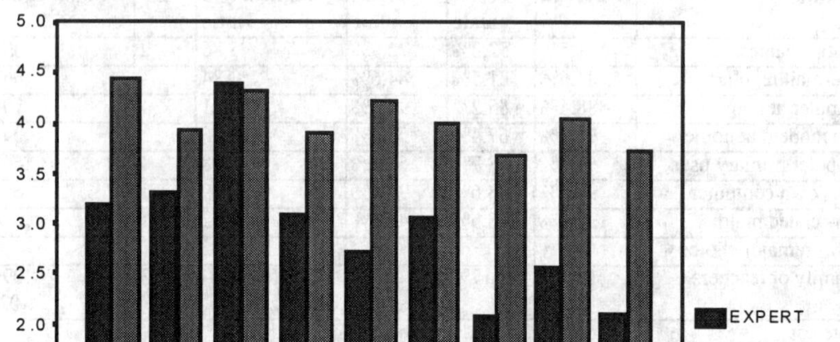

fg- likes fast games	**m-** males	**t-** technology major
nfg- no fast games	**f-** females	**nt-** not technology major

4 Discussion

This study shows that female students are as likely to have a computer in their homes as males, although males reported having had computers in their homes longer. No difference in computer competence was found, but female students reported less expertise. They are less confident about their computer skills and rate their abilities in spreadsheets, web page creation, computer games, and programming lower. Therefore, females may be losing their anxiety toward using computers, in particular word processing and e-mail, but they still do not appear on average to be as confident or skillful as their male counterparts. Male students liked fast-paced computer games much more and this was significant to the .053 level. As expected, female students were much less likely to pursue a technology-related major than male students. However, the relationship between liking fast games and having computer expertise is weak unless it is a positive factor for females and non tech-major males but a negative factor for tech-major males. It is possible that they liked games in the past and simply grew tired of them. The relationship between Web-surfing hours and expertise raises the question whether the effect is due to the added time spent on the computer or from the activity itself. The higher mean for male surfing-hours may also be related to the preponderance of „male" domain topics available on the Web.

Although the growing parity in computer competence is promising, more effort is necessary to ensure that young women have equal access to technology-related training. Efforts must be made to give girls more positive role models and encouragement to learn about computers and eventually pursue technology-related careers. In particular, computer activities that are not motivated by competition and fast-pace should also be offered in classes.

5 References

Apple, M. & Jungk, S. (1990). You don't have to be a teacher to teach this unit–Teaching, technology and gender in the classroom. *American Educational Research Journal,* 27, 227-251.

Corston, R. & Colman, A.M. (1996) Gender and social facilitation effects on computer competence and attitudes toward computers. *Journal of Educational Computing Research,* 14(2), 171-183.

Elkjaer, B. (1992) Girls and information technology in Denmark: An account of a socially constructed problem, *Gender and Education,* 4, 25-40.

Joiner, R., Messer, D., Littleton, K. & Light, P. (1996) Gender, computer education and computer-based problem solving. *Computers and Education,* 26(1-3), 179-187.

Lee, K. (1997). Impediments to good computing practice: some gender issues. *Computers and Education,* 28(4), 25-259.

Reinen, I.J. & Plomp, T. (1997). Information technology and gender equality: a contradiction in terminis? *Computers and Education,* 28(2), 65-78.

Scott, T., Cole, M. and Engel, M. (1991). Computers and education: a cultural constructivist perspective. *Review of Research in Education,* 18, 191-251.

Turkle, S. & Papert, S. (1990). Epistemological pluralism: styles and voices within the computer culture. *Signs: Journal of Women in Culture and Society,* 16(1) 128-157.

Yovovich, B.G. (1997). Girls in cyber space. *Marketing News,* 31(25), 1, 12.

Calculation of Totally Optimized Button Configurations Using Fitts' Law

Alfred Schmitt and Peter Oel
University of Karlsruhe
Institute for Operating and Dialogue Systems
Kaiserstr. 12, 76128 Karlsruhe, Germany
Phone: +49 (0)721 608-3965, Fax: +49 (0)721 696989
Email: {aschmitt,oel}@ira.uka.de

1 Introduction

This paper addresses the problem of optimal placement and configuration for a set of buttons on a graphical user interface (GUI). We present an optimization algorithm that reduces the average time necessary for pointing device motions. Fitts' law (Fitts 1954) is used to estimate the transfer time for cursor movements between buttons. Numerous studies in the literature focus on the verification and application of Fitts' law in a rather direct way (MacKenzie 1992). Our work differs in its goal: The problem of finding totally optimized button configurations is a non-trivial optimization task and is still an open issue.

Our problem can be stated as follows: Given is the number n of buttons and corresponding absolute probabilities $w(i,j)$ for user induced cursor movements from button i to button j. We are looking for a button configuration under the following conditions and constraints:

- the width of each square button may vary,
- each button may be positioned freely within a rectangular area,
- no two buttons may overlap each other,
- the average time calculated by Fitts' law for lengthy button clicking sequences is minimal.

The suggested approach is rather theoretical but its results will provide some practical insight into the placement and relative size of buttons if a total optimization based exclusively on Fitts' law is aspired.

The rest of this paper will shortly describe our solution to this optimization problem, discuss the results, and draw some conclusions.

2 Button configurations and Fitts' Law

Each button B_i is described by the coordinates of its central point (x_i,y_i) and its width $2r_i$ (see figure 1). For a concise approach to the problem we only consider square buttons. A tuple of buttons is called a button configuration. The desired configuration $\boldsymbol{C}$ consists of n buttons whose x_i, y_i, r_i have to be determined. According to Fitts' law the time to move to and to select a target of width W at a distance A is $TM = a + b \cdot \log(2A/W)$ where a and b are empirically determined well-known constants. For two buttons B_i and B_j the time $T_{pos}(i,j) = \log(d_{ij}/r_j)$ is calculated using Fitts' Law with target width $W = 2r_j$ and distance $A = d_{ij}$ (figure 1). We only consider the term $\log(d_{ij}/r_j)$ in Fitts' law, since constant factors have no influence on the results of our optimization process. Variations of Fitts' law (Card et al. 1983) like the ones proposed by Welford (Welford 1968) and MacKenzie (MacKenzie 1989) produce only very small divergences.

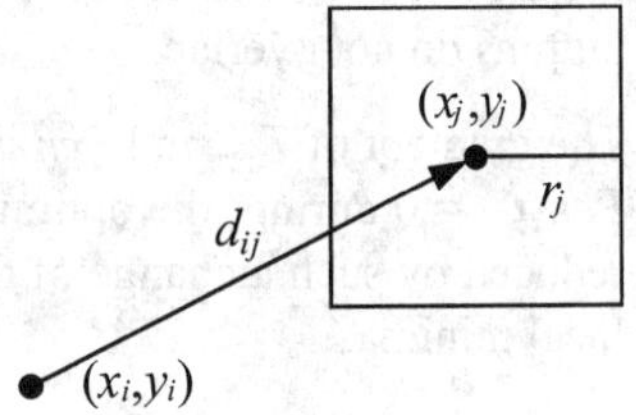

Figure 1: Movement of a user's hand to a target.

The user's behavior and his preferred traversal between the buttons induced by the graphical user interface depends on the application specific tasks to be solved. The absolute probabilities $w(i,j)$ for a movement from button B_i to Button B_j define the specific input data necessary to determine the average time $T_{avg}(\boldsymbol{C})$ for a given button configuration $\boldsymbol{C}$:

$$T_{avg}(\boldsymbol{C}) := \sum_{i=1}^{n} \sum_{j=1}^{n} w(i,j) \cdot T_{pos}(i,j)$$

3 Optimization

Finding a minimum of $T_{avg}(\boldsymbol{C})$ is a classical optimization problem. The search for optimal configurations is accomplished by simulated annealing (Kirkpatrick et al. 1983), a probabilistic approach, combined with a gradient method and a specific penalty function. The penalty function ensures that only non overlapping configurations are considered. The penalty for a configuration $\boldsymbol{C}$ is defined as

$$Pen(\boldsymbol{C}) := \sum_{i=1}^{n-1} \sum_{j=i+1}^{n} ov(i,j)$$

where $ov(i,j) := max\{0, r_i + r_j - |x_i - x_j|\} \cdot max\{0, r_i + r_j - |y_i - y_j|\}$ is an approximation of the area of intersection of the buttons B_i and B_j. The term $ov(i,j) = 0$ holds, if the buttons do not overlap.

The gradient of T_{avg} and *Pen* is used to get better values for $T_{avg}(\boldsymbol{C})$ and to obtain $Pen(\boldsymbol{C}) = 0$ during the optimization. Furthermore buttons are swapped if T_{avg} is reduced by such a change of the configuration. This ensures not getting stuck in local minima.

Another condition is of a more practical nature: The total area of all *n* buttons is limited to a predefined value. If one button grows, others must shrink. Fixing the total area of all buttons does not influence the result of the optimization. $T_{avg}(\boldsymbol{C})$ is invariant to scaling and moving in 2D since scaling factors are lost in the term d_{ij}/r_j and translation offsets sum up to 0 in d_{ij}.

4 Optimization Results

Figure 2 shows two optimized configurations of a set of 30 buttons. The probabilities w(*i,j*) used in this scenario reflect the user's behavior during a typical dialogue. The buttons numbered by 2, 3, and 5 have the highest probabilities of being visited by the cursor, whereas button 1, 16, and 17 have the lowest. Optimization for this scenario leads to the following numerical results (note that values for T_{avg} do not represent time in the sense of Fitts' *TM*):

a) $T_{avg}(\boldsymbol{C}_a) = 1.6566$ (equals 100%). No optimization: All buttons have the same shape and are arranged sequentially in rows of six buttons.

b) $T_{avg}(\boldsymbol{C}_b) = 1.2403$ (equals 74.9%). Buttons are exchanged pairwise to reduce T_{avg}. The size and row-column arrangement of buttons remains fixed. The result of this standard optimization using Fitts' law is shown in figure 2 on the left side.

c) $T_{avg}(\boldsymbol{C}_c) = 1.1023$ (equals 66.5%). Total optimization with the final configuration shown in figure 2 on the right side.

The relative improvement from b) to c) is 11.35%. These results show the superiority of totally optimized button configurations to standard optimizations with fixed button size.

Why do buttons at peripheral positions grow so significantly?

The totally optimized configuration looks quite confusing. We first thought of an error in the optimization program, but looking closer at the calculation of T_{pos}

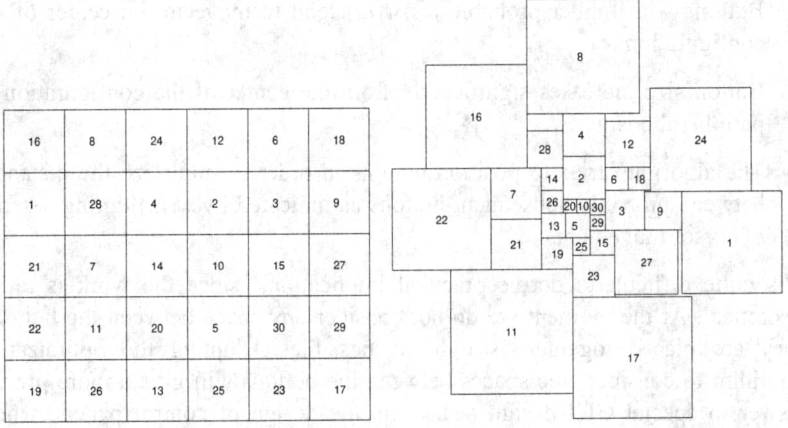

Figure 2: Two optimized button configurations, both consisting of 30 buttons. Each button is shown as a square containing the button number. In configuration C_b, on the left side, all buttons remain with same width. The optimization is done only by changing the position of the buttons. The configuration C_c, on the right, is totally optimized.

for buttons in the outer region will give the answer to the question. Let B_j be such a button and B_i a button where cursor movements to or from B_j will start or end. Terms of interest are $T_{pos}(i,j) = \log(d_{ij}/r_j)$ and $T_{pos}(j,i) = \log(d_{ij}/r_i)$, with distance $d_{ij} = r_j + d$. Finding a minimum for $T_{pos}(i,j) + T_{pos}(j,i)$ needs solving

$$\frac{d}{dr_j}\left(\log\left(\frac{d+r_j}{r_j}\right)+\log\left(\frac{d+r_j}{r_i}\right)\right)=0$$

which leads to $r_j = d$. The greater the distance between B_j and B_i, the larger the width of button B_j will be. Note that this also stops buttons in the interior region from getting bigger. This can be seen in figure 2 on the right side. Button 27 does not use the space on its right or beneath itself, because there are no possible traversals from or to the buttons 1, 11, 17 and 23.

5 Conclusions

In this paper we showed that non grid oriented placement of buttons and the variation of the button width depending on the probabilities for button traversals results in a better average time calculated using Fitts' law. The following list summarizes the main results derived from the structure of the totally optimized configuration:

- Buttons with higher probabilities $w(i,j)$ tend to move to the center of the configuration.
- Button size increases significantly from the center of the configuration to peripheral positions.
- The algorithm tries to position buttons in order to minimize the distances between probably subsequent buttons as indicated by $w(i,j)$ during a series of cursor movements.

It is quite difficult to derive practical implications, since the work is rather theoretical. At the moment we do not consider any space between the buttons. They are placed together as tight as possible. Adopting the optimization algorithm to consider free spaces between the buttons will give a more precise answer to special GUI design tasks, e.g. the design of control panels, where detailed characteristics for usage are known.

6 Acknowledgements

Finally we want to thank our students Jutta Klein, Robert Henßen, and Roland Heinemann for their implementation of the optimization algorithm.

7 References

Card, S., Moran, T. & Newell, A. (1983). *The Psychology of Human-Computer Interaction*. Hillsdale (NJ): Erlbaum.

Fitts, P. M. (1954). The information capacity of the human motor system in controlling the amplitude of movement. *Journal of Experimental Psychology*, 47, 381-391.

Kirkpatrick, S., Gelatt, C. D. & Vecchi, M. P. (1983). Optimization by simulated annealing. *Science*, 220(4598), 671-680.

MacKenzie, I. S. (1989). A note on the information-theoretic basis for Fitts' law. *Journal of Motor Behavior*, 21, 323-330.

MacKenzie, I. S. (1992). Fitts' law as a research and design tool in human-computer interaction. *Human-Computer Interaction*, 7, 91–139.

Welford, A. T. (1968). *Fundamentals of skill*. London: Methuen.

3-D Input Device using a Ball Rotation Interface

Tomoichi Takahashi **Mikio Kuzuya**
Chubu University, 1200 Matsumoto Kasugai, Aichi 487-8501 JAPAN

1 INTRODUCTION

This paper presents a new compact three-dimensional (3-D) input device. Typical 2-D input devices, such as mouses or trackballs, can input movements of two degrees of freedom (DoF) and cannot handle 3-D with six DoF at the same time. The more programs, such as computer-aided design systems and virtual reality systems, support 3-D presentation, the more a handy 3-D input device has been required (Balaguer and Gobbetti, 1996) (Vince 1995).

The Polhemus™ magnetic position and orientation sensor, the Logitec™ sonic sensor, and the SPACEBALL™ are well known devices. These 3-D input devices can directly input 3-D movement and have a more natural interface than 2-D input devices. However, they have some problems. Magnetic sensors or sonic sensors are influenced by environmental conditions and tire the operators who hold them in the air. SPACEBALL™ solves this holding problem. But, it is difficult to input spiral movement, which is composed of rotational and translation movements. Mechanical link type devices have singular postures where the number of device's DoF is reduced.

Our aim is to produce a 3-D input device which can be used beside a computer in daily life. A good 3-D manipulation interface must provide a natural way to input 3-D motion as well as support an appropriate response. A good 3-D pointing device should

(1) input 3-D motion in natural ways analogous to real life,

(2) respond effectively in noisy environments,

(3) change operation modes smoothly, and

(4) provide feedback to the operator from the computer.

We propose a 3-D rotation input interface using a ball rotation. Using a prototype device, the input interface was evaluated with several tasks. The experiments show that our 3-D device provides good manipulation of an object in 3-D space.

2 3-D INPUT METHODS and PROTOTYPE DEVICE

Fig. 1 shows the input interface and a prototype device (which we shall henceforth call "3-D device"). The device consists of a rotating ball and a joystick. It can handle six DoF movement through one-hand operation and satisfies the first three of the above good-device-conditions.

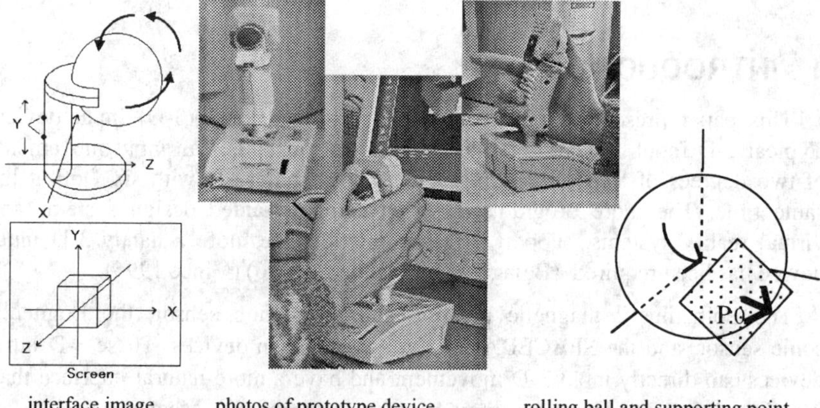

Figure 1: 3D input interface and prototype device.

The user grabs the device and rotates the ball with his/her thumb. Rotating the ball inputs 3-D rotation movement around the *x*, *y* and *z* axis. Inclining the joystick inputs traditional 2-D translation movement along the *x* and *z* axis. Pushing the buttons on the back of the joystick input positive/negative translation movement along the *y* axis. The device is position-controlled and an isotonic device (Lipscomb and Pique 1993).

A ball supported with more than four points rotates in a stable fashion to any direction. The prototype device supports the ball with two trackballs and plastics plates. Fig. 1 shows the contact plane between the ball rotating ball and a supporting point P_0. The rotation in 3-D space is observed as 2-D movement on the contact point (Goldstein 1980). The movement is sensed by the track ball which is used to support the ball. The data from the two trackballs are sufficient to calculate 3-D rotation vector $(\omega_x, \omega_y, \omega_z)$ corresponding to the ball rotation.

3 PERFORMANCE EVALUATION

Performance of the prototype is measured by manipulating objects in the computer simulation.

3.1 Experiment for 3-D basic rotational movement

The first task was to rotate a cube until the cube coincided with the displayed goal posture. All faces of the cube were colored differently. The goal direction was randomly displayed each time (Fig. 2). Seven subjects were asked to perform the task five times.

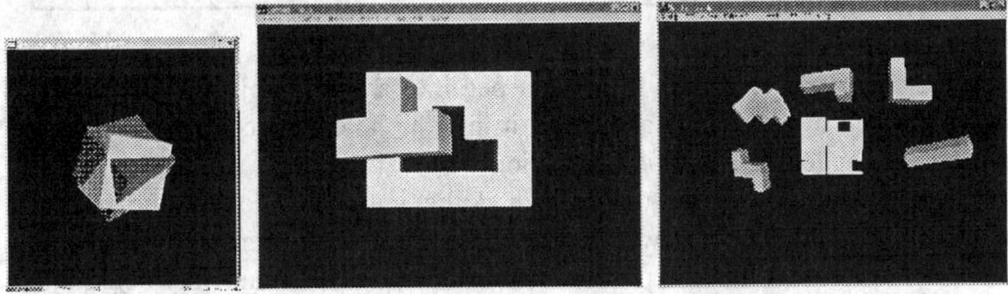

Task 1 (Cube Rotation) Task 2 (Peg in Hole) Task 3(Pentomino)

Fig 2. Tasks used for performance evaluation

Table 1 shows the result of the experiments in which the subjects completed the task with a 3-D device and a 2-D mouse. The subjects were students who use a 2-D mouse every day, but used the 3-D device for the first time. 3-D rotational input by the 2-D mouse is completed by dragging the mouse by clicking the mouse buttons. Clicking the left/right button is interpreted as input rotation about the *x* and *y*/z axes.

The table shows that average time and the standard deviation for 3-D device are smaller than those for the 2-D mouse. It indicates that the 3-D device rotates about an arbitrary 3-D axis more speedily and stable than the 2-D mouse. The superiority of the 3-D device over the 2-D mouse was tested by *t* Test with a level of significance of 0.05. The null hypothesis, H_0: $\mu_{2D} - \mu_{3D} = 0$, was rejected with T=16.94 as acceptable values are $|T| > 1.71$. The 3-D device is superior to a 2-D mouse in 3-D rotation about an arbitrary rotational axis.

3.2 Experiments for comparison with other 3-D devices.

Two kinds of task, *Peg in hole task* and *Pentomino*,were used to evaluate 3-D input interface. Four subject were asked to perform the tasks with three devices, our 3-D device, SPACEBALL^{TM} and Magellan^{TM}. The two other devices are commercial products and isometric.

Table 1. Operation time of Task 1

	average(m)	deviation
3-D device	17.3	7.7
2D mouse	33.9	25.9

Table 2. Number of Pegs insertion

	average	deviation
condition 1 (only translation)		
3-D device	3.1	5.9
SpaceBall	2.8	4.0
Magellan	38.3	12.9
condition 2 (translation & rotation)		
3-D device	7.3	2.5
SpaceBall	4.0	2.1
Magellan	6.8	2.9

Table 3. Number of pentominos

	average	deviation
3-D device	3.8	1.1
SpaceBall	2.0	1.2
Magellan	4.5	0.9

Peg in hole task: The task is to insert a peg into a hole. Subjects were asked to move the peg over the hole and insert it into the hole. The tasks were executed under two condotions. The first condition was that the peg's orientation was set as the same orientation as the hole, and the initial position of the peg was randomly located. The peg was inserted into the hole by translations only. The other condition was that the position and orientation were randomly determined. The peg must be manipulated by both translations and rotations. Performance was measured by how many times the peg was inserted within five minutes. Table 2 shows that Magellan™ was the best at condition 1, though its deviation was the worst. The 3-D device showed the best performace at condition 2, so the rotation interface might be best considering condition 1's result.

Pentominos: Pentominos is originally a puzzle game of thinking how to build a large cube by pieces of pentomino. Subjects were asked to manipulate pentominos in the same way as demonstrated on the screen. Performance is measured by how many pieces of pentomino are placed to the demonstrated places within five minutes.Table 3 shows that Magellan™ deals with the most pieces and 3-D is next.

Despite the prototype device made by our students, the experiments showed that our 3-D input device is comparable to the commercial devices. The ball rotation interface is promising for 3-D rotation input for desktop use. We think 3-D manipulation is dependent on operators' 3-D world sensing and applications. We are planning further field tests from ergonmics.

4 References

Balaguer, J.F. and Gobbetti, E (1996) 3-D User Interfaces of General-Purpose 3D Animation, *COMPUTER*, Vol. 29, No. 8, 71-78.

Vince, J(1995): *VIRTUAL REALITY*, Addison-Wesley.

Goldstein, H.(1980) *Classical Mechanics*, Addison-Wesley

Lipscomb, J.S., and Pique (1993), M.E. Analog Input Device Physical Characteristics, *SIGCHI Bulletin*, Vol.25, No.3, , 40-45.

Pointing Devices for Industrial Use

Prof. Dr.-Ing. D. Zühlke, Dipl.-Ing. Lutz Krauß
Institute for Production Automation, Kaiserslautern, Germany

1 Objective of the work

Windows-oriented software systems, such as WINDOWS and MAC OS have meanwhile become the state-of-the-art in the field of home and office computers. These systems are also used more and more in the field of industrial applications. Despite the economic advantages of this development existing disadvantages are still not taken into account.

A characteristic feature of windows-systems is the direct manipulation of visual objects. It enables users to operate almost entirely by means of pointing actions. Thus users get the feeling to work with real objects, which can be moved (Drag-and-Drop), reduced or increased in size. The precondition is the availability of a suitable pointing device, e.g. a mouse. The user clicks on the desired object with the mouse key (selection) and certain commands are executed by certain mouse actions (function activation).

However, the mouse is not always the most appropriate device to interact with computers. In industrial applications the mouse can not very often be utilised due to dirt and the lack of a horizontal surface for moving it. An effective utilisation of windows-systems in this field requires the development of a suitable alternative.

Therefore, tests were conducted with machine operators under industrial environments to determine the suitability of different pointing devices like mouse, mousebutton, joystick, trackball, touchscreen, touchpad, etc. The paper will present the used methods and the research results indicating the suitability of the tested pointing devices.

2 Conducted Tests

A comparative investigation of pointing devices with 32 persons has been carried out at the Institute of Production Automation and has provided some interesting results. The group of test persons is divided into experienced users and beginners. Users were classified as experienced if they work repeatedly with windows-oriented systems (mostly WINDOWS with a standard-mouse as pointing device). Users are classified as beginners, if they have no or little experience with windows-oriented systems and corresponding pointing devices.

The tasks which had to be performed by the test persons are standard WINDOWS-tasks. The tasks were carried out by means of the software *DEVICE* (the software *DEVICE* is available for free under http://pak.mv.uni-kl.de or by order with the Institute of Production Automation). They can be distinguished in simple tasks like clicking (CLICKING-tasks) on an object, in complex tasks like drag & drop (DRAG & DROP-tasks) an object and in sensitivity tasks like positioning the cursor very precisely on an object (POSITIONING-tasks). The operating tasks are visualized in figure 1 and described in table 1.

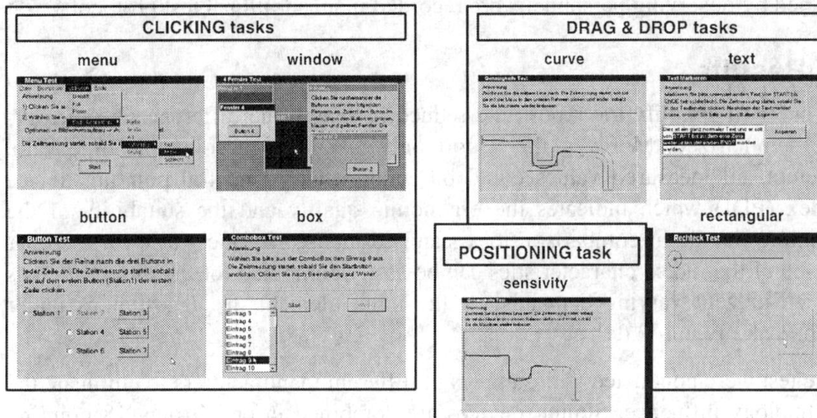

***Figure 1** Operating tasks*

task group	operating task	description
CLICK	menu	Select a menu and pull-down to defined item
	button	Click on several buttons in defined order
	window	Select large-surface windows in defined order
	box	Select a defined entry from a list box
DRAG & DROP	rectangular	Drag a rectangular to a given size and position
	text	Select a text-string within a text-page
	curve	Track a given curve
POSITION.	sensitivity	Track a given curve very precisely (different areas of precision)

***Table 1** Description of operating tasks*

A computer automatically recorded in a logfile the times needed by the test person to perform the tasks as well as the error rate. The measured values are in detail: time valid attempt, total time, numbers of attempt, points valid, points invalid and attempt canceled. Time valid attempt means the time, a test person needs to complete the task correctly in one attempt, mostly the last attempt. Total time means the time, a test person needs to fulfill the task completely inclusive incorrect attempts. If a test person is unable to solve the task, a test stop is possible and a check mark in attempt canceled is set.

It can be assumed that the data is complete and not influenced by any disturbing factors. Since the logfile only shows times and does not give any information about the users intentions an additional questionnaire was handed to the users. It provides a subjective evaluation of each pointing device. Questions like „Do you have any difficulty in learning to operate the pointing device“ have been asked among others. The suitability of the pointing devices for each test task was determined by a five-step ranking scale (Bortz and Döring, 1995). Impressions gained by observing personnel were recorded in an additional test protocol.

3 Results

In order to simplify the report and reduce its length the different kind of tasks are summarized. Morever, this report only takes into account the time valid attempt, all measured values could be combined to a neutral pointing device index (PDI) which indicates the ergonomic quality and the suitability of the pointing device in comparison to a standard-mouse. By the way of a modular set-up of the index, characteristics can be easily combined for the different goals which present varying demands for the input element (a suggestion is under http://pak.mv.uni-kl.de).

The test were conducted with samples of different manufacturers to eliminate the technology influence. Similar devices are combined in one Group. 18 pointing devices were tested and summarized in 9 groups: Mouse, mousebutton, mousestick, trackball, joystick, mousepad, keyboard, touchscreen and digitizing tablet. Figure 2 shows the comparison of measured times between the different pointing devices for each user group.

It stands out that in all tasks the experienced users on average achived better results than the beginners. Both lines have nearly the same course, on average the beginners need 10 seconds more to solve the task. But there are only minor differences between beginners and experienced users when tasks are performd with the same pointing device - for example the touchscreen.

In case of an appropriate object display (object size adjusted to finger size) the **touchscreen** proves to be very suitable for quick pointing, due to an optimal hand-eye coordination (selection tasks). Its operation can be learned within a relatively short time, thus it is an optimal input device for beginners (better then mouse). Major disadvantages become obvious if the touchscreen is utilized for movement tasks, i.e. if an object must be moved on the screen surface, a text must be highlighted or a line must be tracked. It is also unsuitable for exact positioning like on small standard Windows-elements (radio-buttons, check-boxes).

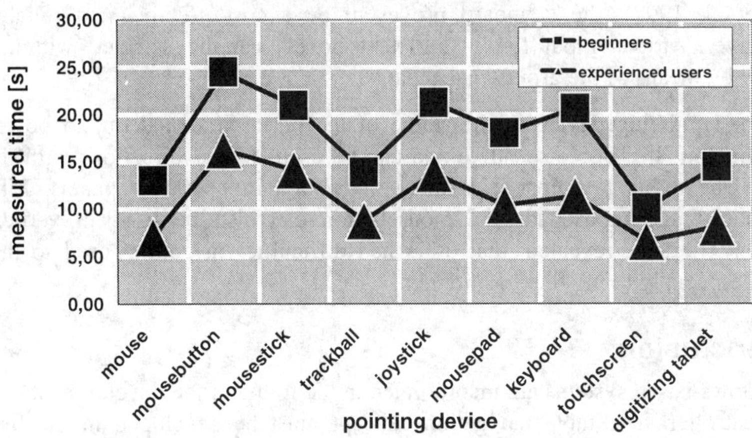

Figure 2 *Measured average time valid attempt*

As a pointing device which is very similar to the mouse the **trackball** achieved good results. It only shows disadvantages when it comes to exact tracking, which is reasoned by the finger position. A key must be pressed with the thumb or the index finger while the remaining fingers must move the ball at the same time. Most people have difficulties in performing the exact finger coordination required for this task. It proves to be positive for beginners if the ball of the trackball is heavy and rolls sluggishly. If their hand releases the ball, the cursor stops immediately (on the other hand some beginners had problems with the mouse because the tested sample moved very easily).

The **touchpad** also achieved good results in simple pointing tasks. However, it shows a drawback at movement tasks, when cursors must be moved while pressing the ENTER-key. The human finger coordination is unsuitable for the unfavorable construction. The index finger is utilized to select the desired object (e.g. painter or text marker), but the object can only be activated by pressing the ENTER-key with another finger. In most cases the thumb is used for this purpose. An optimized construction in terms of ergonomics (touch sensation of the fingers, etc.) offers new solutions to this problem.

Joysticks are a useful tool when e.g. a moving object must be tracked on the screen, but exact positioning and selection tasks can not be performed. Due to the usual way of operating a joystick (upright with loose fist) test persons considered it as very difficult to position the cursor in a reasonable time on a desired object.

The **keyboard** enables to select and activate single objects in a Window-surface but impedes spatial manipulation, i.e. movement of objects on the surface. However, if the system design has been adjusted especially to keyboard interaction as in the German Open CNC Controller Project OSACA/HÜMNOS (Boll et al. 1997), the keyboard proves to be a cost-effective and reliable alternative. Various input fields (e.g. text boxes) can be selected within a window by means of the group-change keys.

The worst pointing device in the frame of the research turned out to be the **mousebutton**. It is force-sensitive and can be used in four directions with the finger. Due to its low cost it is often embedded in operating panels. Like mousesticks and mousepads the mousebutton can also be used in vertical position. The tests revealed, that users had difficulties in getting used to this device, especially to the fine positioning.

4 Conclusion

If Windows-based systems are incorporated in the frame of the development of a new controller, a suitable navigation concept must be developed in the first place. The well-known techniques from office applications, such as mouse operation or drag-and-drop can not simply be adopted. New developments in the field of control technique will demand new solutions in the interaction with users. This study may facilitate to select an appropriate device. Further investigations need to follow resulting in concrete design guidelines for developers of control techniques. An individual final decision for an industrial pointing device should only be taken after comparative tests with several test persons have been carried out under realistic process conditions.

5 References

Boll, J. et al. (1997). OSACA / HÜMNOS Style Guide for Machine-Tool-GUI´s : Ein Handbuch zur Gestaltung von Benutzungsoberflächen für Werkzeugmaschinen. Fraunhofer IRB, Stuttgart.

Bortz, J., Döring, N. (1995). Forschungsmethoden und Evaluation. Springer, Berlin.

6 Contact

Prof. Dr.-Ing. Detlef Zühlke, Dipl.-Ing. Lutz Krauss
University Kaiserslautern, Institute for Production Automation
67653 Kaiserslautern, Germany
email: pak@mv.uni-kl.de, krauss@mv.uni-kl.de
internet: http://pak.mv.uni-kl.de

Man-Machine Interaction Using Eye Movement

K.Kohzuki,T.Nishiki,A.Tsubokura,M.Ueno,S.Harima and K.Tsushima

(Konami Co.,Osaka Electro-Communication Univ.)

1.Introduction

There are many idea processors which assist a user in his thinking process. Many of them help a user to obtain document. Some of them try to stimulate a user interacting with system in his associative thinking about given concept . We are interested in a idea processor called Mandal-Art which strongly assists a user in his associative thinking about given theme.

2.Mandal-Art

There is an idea processor called Mandal-Art which is widely used in Japan.(Mandal-Art 1984)

There are a matrix called Mandala which has nine frames in which items are typed and a handle matrix which is used to select a frame from a mandala on the screen of Mandal-Art by pointing the handle corresponding with a frame using mouse(Fig.1).

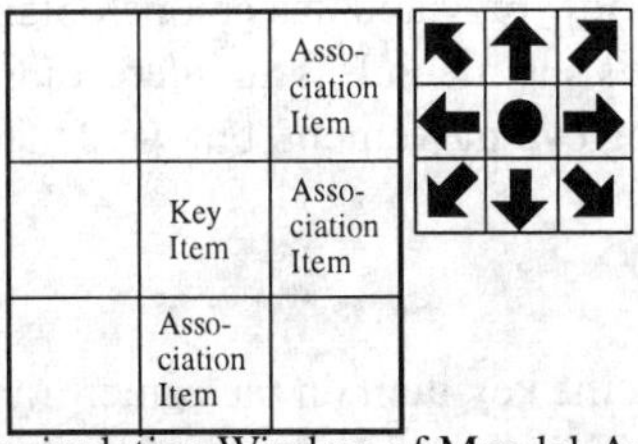

		Asso-ciation Item
	Key Item	Asso-ciation Item
	Asso-ciation Item	

Fig.1 Manipulation Window of Mandal-Art

The interaction between a user and Mandal-Art makes a progress in the following three stages,Input Stage,Edit Stage and Retrieve Stage.

A.Input stage

The user can select a subdivided square called frame by using mouse on a mandala directly in input stage , and then he can type a word or sentence as item in a frame. At first he types key-item into central frame , then he may type associated item into any surrounding eight frames.By selecting and clicking the corresponding handle which specifies an item , the item becomes the key item in the new

mandala.He can make large-scaled linking of items which is thought by himself on a series of mandala by generating mandalas repeatedly.
As a result , this linking of items which express the present subject as a whole is divided into several mandalas in which only eight links can be seen on the mandala at a time.(Fig.2)

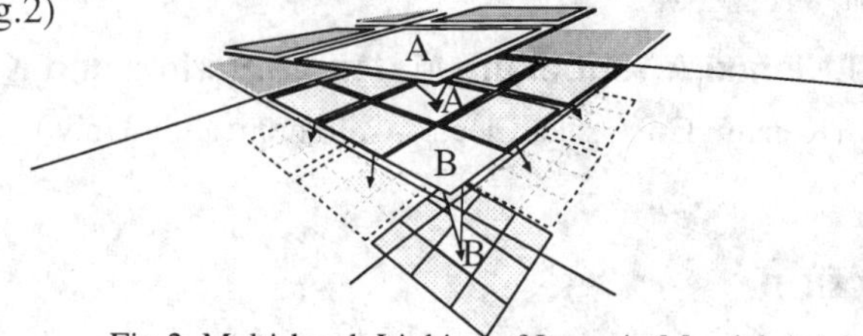

Fig.2 Multi-level Linking of Items in Mandal-Art

Mandal-Art is effective to express subject which contains many items which does not have clear linkage with each other at first. At first , the user are not aware the whole structure of subject. But he can be aware of the linkage of items gradually by interacting with Mandal-Art , and he adds a new item ,and then the new linkage is derived.

Thus a new linkage between old item on the central frame and new typed item on the surrounding frame is introduced.

Thus ,he can discover the whole structure of the given subject by communicating with Mandal-Art.

B.Edit stage

A user of Mandal-Art can edit the contents of Mandala by pointing sub menu using mouse. All over view of Mandalas can be obtained and exchange of items can be possible using this sub menu.It is an interesting theme to modify edit function of Mandal-Art using eye movement, but we don't report it in this paper.

C.Retrieve stage

At first , a user may see the key item on the center frame of mandala ,and then on the surrounding eight frames he is concentrated in a special item on the surrounding eight frame which is seemed to associate with key item. And he selects new mandala in which special item becomes the new key item by pointing the corresponding handle on the screen with it. A user can navigate in the world of items in his mind using pointing at items successively. But, pointing item in this way is not direct manipulation but indirect manipulation because of using handles. So, the association process between items in his mind is interrupted in manipulating handle on the screen to obtain another mandala, so a user feels this indirect pointing process is not comfortable in his navigation in the item world.

So,we intend to develop several modified Mandal-Arts in which the user can select frame without using handle. The user can select an item directly on a mandala by pointing it using mouse.

3.Modified Mandal-Art

As is mentioned in chap.2 , Mandal-Art has a few defects about manipulation to select frame from mandala on the CRT . So, We have developed the following three modified Mandal-Art which does not have above defects.

A) Mouse Mandal-Art
B) Eye-Mandal-Art
C) Voice-Mandal-Art

A. Mouse Mandal-Art

We have developed Mouse Mandal-Art in which a user can select an item on a mandala directly by pointing it using mouse in retrieve stage.The user does not need to be conscious of handle but only items in itself, so he can concentrate into items.

B.Eye Mandal-Art

We have already developed a computer game called modified Flipper Game in which a player can play game using not mouse and keyboard but eye movement. The details of this game is shown in appendix of this paper. We have developed Eye Mandal-Art using calibration system which had been developed for modified Flipper Game.

The user can change present mandala into the new mandala in which present associate item becomes key item by gazing at it. Conversely, if the user gazes the center frame , then old mandala in which the present key item become associated item emerges.

C. Voice Mandala

In the above improvement of modified Mandala , nothing is improved for input stage.We have developed new Mandal-Art called Voice-Mandal-Art in which a user can input item using his voice in a frame gazed by himself. So, a user can make mandalas without keyboard but with gaging and voice.

In the present , cognition capability of voice recognition is relatively poor , so completely free interaction between user and system is not allowed.

Our users report that they can more concentrate into the ideas using Mouse-Mandal-Art and Eye-Mandal-Art than using original Mandal-Art. As they don't have to use mouse and handle , they are not disturbed. So , they feel as if they were treating items directly.

4.Conclusion

We have succeeded in developing idea processor called Modified Mandal-Art in which a user can input item using voice and retrieve items using eye movement. A user does not need to use key-board and mouse in interacting with system. High speed communication with system is achieved and direct manipulation environment for associative thinking is obtained.

Appendix. Modified Flipper Game

We intend to add the capability of detecting cognitive information from learner to educational system on the computer such as CAI,ITS and ILE.This is also the trial to develop new man-machine interface in which a user can communicate with system using eye movement.

We used two Eye Mark Recorders (hereafter we call it EMR.) ,NAC EMR-600 and TAKEI T.K.K.2901 in this research. These introduced EMRs can detect eye mark of a user 600 times to 1000 times per second. But , these data can be used only after the experiment is finished.

To allow a learner communicate with system in real time, it is necessary to develop a interface to detect eye mark of a learner in short time range from 10mSec to 100mSec and to respond the learner in real time.(Mandal-Art 1984)

We developed a new hardware memory board called CINDY and a software calibration system for EMR which can detect 600 eye mark data per sec to analyze the movement of a human eye and can use them in real time(Fig.3,Fig.4).

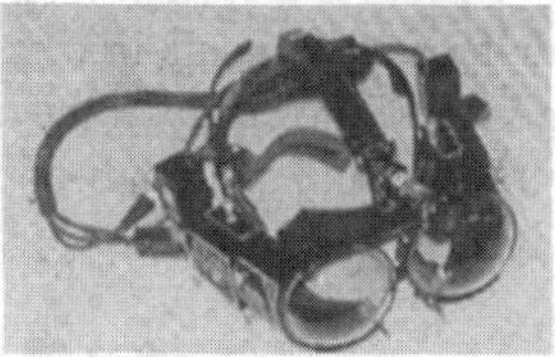

Fig.3:The used Eyemark Recorder: NAC EMR-600

Fig.4: The used Eyemark Recorder: TAKEI T.K.K.2901

Using the real time data obtained from the calibration system , the computer can quickly vary the objects which are seen by the testee on the CRT . As a result , a new man-machine interface which can respond quickly to eye movement of testee was developed in our research.

To obtain the basic feature of this communication process of testee with system using eye movement , we developed computer game called modified flipper game

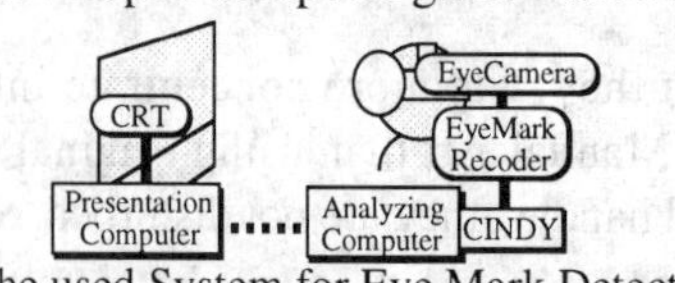

Fig.5 The used System for Eye Mark Detection

in which a learner can play game using eye movement(Fig.5). (K.Tsushima et al 1996)

There is a computer game called Flipper.A player of this game moves mouse to select the squares divided on the screen, and then he can vary the color of a selected square by clicking mouse. Several colors are used in this game. The color of selected square changes in the cyclic sequence by player's mouse clicking.

On the other hand , the color of each square varies randomly in time on the CRT .When the colors of all the squares coincident , then the game is cleared. So , a player of this game must vary the color of squares as fast as possible.

The player can confirm which square is selected by seeing a mouse pointer controlled by himself in this original Flipper game.

We modified the original Flipper game into one which the player can selects square and changes the color of selected square using eye movement.

We analyzed in detail the behavior of player playing modified Flipper Game from 2~2 division to 6~6division.(Fig.6)

The detail of this analysis can be shown in our preceding paper.(T.Nisiki et al 1997)

We obtained knowledge about player's cognition-motion process in this interface using eye movement.

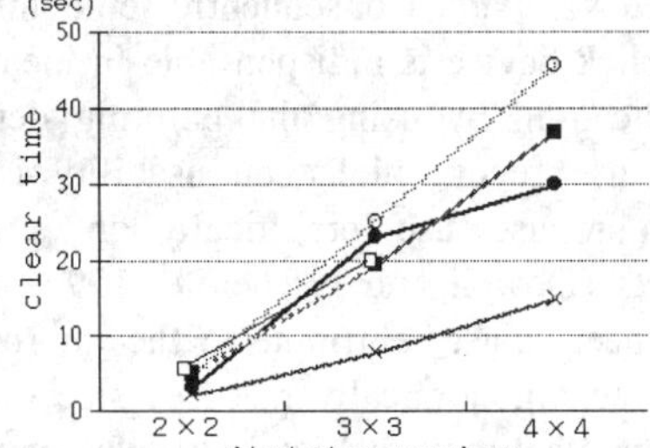

Fig.6.Used time to clear Flipper games

References

Mandal-Art (1984) Mandal-Art is the product of Hiro Art Directions.

K.Tsushima et al(1996), K.Tsushima,M.Ueno,N.Nomura,Imai,"Intelligent Learning Environment using Eye Mark",IEEE Second International Conference ---A Conference on Multi Media in Education ,534-540

T.Nisiki et al(1997),T.Nisiki,A.Tsubokura,M,Ueno and K.Tsushima:"Man-Machine Communication using Eye Mark",Proc. of International Conference ED-TELECOM,ED-MEDIA97

T.Nishiki et al(1998),T.Nisiki,A.Tsubokura,M,Ueno,K.Higasino,N.Nomura and K.Tsushima:"Human Communication using Eye Movement",Proc.of International Conference ED-MEDIA98

A.Tsubokura et al(1998),A.Tsubokura,H.Imai,T.Nishiki,T.Oshima,M.Ueno and K.Tsushima,"The Modeling of Cognition-Motion Process Using Eye Mark Tracing Data",Trans.of JSICE,14,5,191-200

Relationship between Size of Icons and Mouse Operating Force with Complex Actions during Pointing Tasks

Kentaro KOTANI and Ken HORII
Department of Industrial Engineering, Kansai University, Osaka, JAPAN

1 Introduction

A mouse is now an integral part in almost all computer tasks due to the great popularity of PC-based software. Consequently, epidemiological study on the use of this point-and-click device is indispensable in identifying the risk factors of physical disorders caused by using the pointing device. However, recent researches on mouse use focus chiefly on usability issue, and only a few researches deal with mouse use empirically in terms of physical and epidemiological aspects (Dowell and Gscheidle, 1997). Apparently, obtaining knowledge of such aspects must contribute to the improvement of the display design in determining an ergonomically risk-free size of buttons to click with and in arranging icons that ensures less stressful mouse movements. The objective of the research is to obtain empirical data on the operating force applied to the mouse in relation to the characteristics of such display layout as target areas and approaching angles to the target. In particular, the present report focuses on the results and their inferences of the relationship between the size of icons and mouse operating force.

2 Methods

A 5 mm-diameter force-sensitive resistor (FSR) was housed on the top of the microswitch inside the mouse for the measurement of operating force. The voltage signals were transmitted to a programmable A/D converter, controlled by a PC. Prior to the experiment, calibration testing was conducted and the regression equation with R^2 of 0.999 was obtained for the relationship between FSR voltage and the force applied to the mouse.

Independent variables used in the study were three types of pointing actions (click, double click, drag), three sizes of target icon (10×10mm, 5×5mm, 2×2mm) and two testing sessions. Following assumptions were made to determine the size of icons: The 10×10mm icon represents the size of desktop icons used for Windows. The 5×5mm icon represents the height of default cell in spread-sheet applications, or the size of regular 12 point font, used for word-processing software. Pointing the 2×2mm icon is a typical initial process of "rubber-banding", where the size of the object is modified by mouse pointing operations in graphic packages. A total of five subjects were chosen from the population of the engineering majors, who have used a mouse more than one year on a daily basis.

The experimental paradigm consisted of a practice session (approximately 15 minutes) and two testing sessions. Each testing session consisted of 12 trials, which were fully-crossed combined with independent variables of the pointing action and the target size. The order of trials was randomized in the session.

3 Results

Figure 1 shows typical pointing force data in time series. The grand average of mouse operating forces for all pointing actions was 142gf, which was twice that of the minimum mouse operating force (75grams, catalog data). The ANOVA showed significant main effects on the mouse operating force for the types of pointing action ($F(2,810)=138.68$, $p < .01$), subject ($F(4,810)=201.32$, $p < .01$) and session ($F(1,810)=52.61$, $p < .01$), whereas the effect of the target icon size was not significant. Interestingly, the interaction of target icon size and the type of action was significant ($F(4,810)=21.93$, $p < .01$). Figure 2 depicts the change in pointing force by the interaction of target icon size and the type of action. As the target icon size was increased, clicking and double clicking forces were increased, whereas the dragging force was decreased.

4 Discussion

Our results for the relationship between mouse operating force and the task requirement on the basis of target icon size supported our former study (Kotani and Horii, 1998), where we hypothesized that the total amount of muscle contraction employed in pointing mouse may be related to the task requirement. In other words, some amount of muscle contraction was used for pointing the mouse by flexing the index finger, and the rest of the force was merely used for developing the muscle tension prompted by the task requirement. On the other hand, the results showed that dragging force was not in the case, that is, as the size of the target icon increased, the dragging force was significantly decreased.

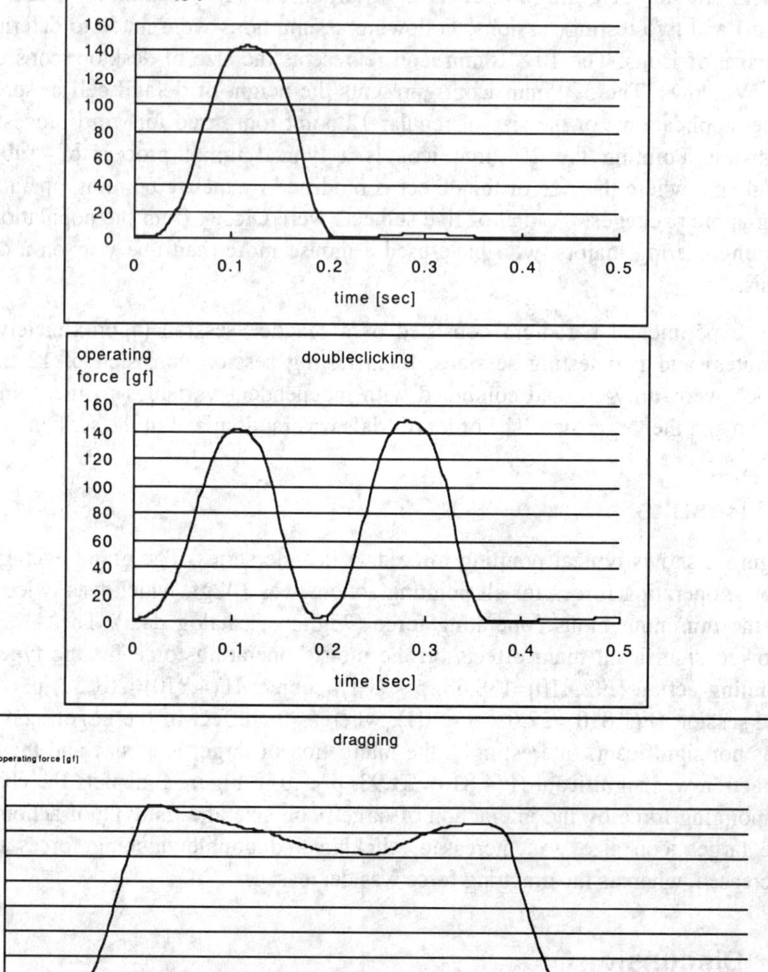

Figure 1 Mouse operating force for different type of pointing action

Currently, we hypothesized that the subjects tend to maintain the force to complete drag motions by pressing extra force, which was built up by the finger flexing muscles. The hypothesis will be enhanced by the further study including the measurements of surface EMGs during the task.

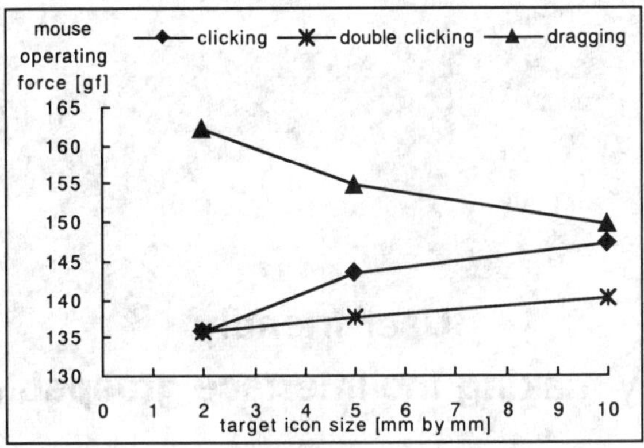

Figure 2 Peak force changes in size of target icons

In terms of the risk of musculoskeletal disorders, the most forceful motion was observed when subjects used drag motions to point the smallest target. Apparently, this type of task is frequently experienced in operating graphical packages. Some GUI design consideration may be applied for reducing such risks effectively. For example, the effective pointing area should be enlarged for not only reducing the risk of pointing errors but also reducing such forceful motions. Also, providing a priori-visual feedback to show the effective area for dragging before pressing the mouse button may be useful to realize whether or not the users can proceed drag motions before they start the rubber-banding task.

In summary, mouse operating force during various mouse actions and target icon size differed substantially. Combined with operating posture and repetitiveness of the task, the quantitative and long-term study of mouse operations should be concerned to fully analyze the epidemiological risk of musculoskeletal disorders during mouse operations.

5 References

Dowell, W.R. and Gscheidle, G.M. (1997). The effect of mouse location on seated posture, *Advances in Human Factors/Ergonomics, 21A Design of Computing Systems: Cognitive Considerations,* (eds. by Salvendy, G., Smith, M.J., Koubek, R.J.), pp.607-610, Elsevier.

Kotani, K. and Horii, K. (1998). A fundamental study on pointing force applied to the mouse in relation to approaching angles and the Index of Difficulty, *Proceedings of the 5th Pan-Pacific Conference on Occupational Ergonomics,* pp. 104-107.

User-friendly by making the interface graspable

F. A. Voorhorst & H. Krueger

Institute for Hygiene and Applied Physiology, ETH-Zürich

Clausiusstrasse 25

CH-8092 Zürich

Introduction

Human Computer Interaction (HCI) remains ignorant to the perceptual-motor skills of humans when it comes to most hardware and software interface design. Adapting to these skills may improve user-friendliness and will make computers adapt to natural ways of interaction. Compared to conventional desktop computers and Virtual Reality (VR) systems, newly emerging Augmented Reality (AR) systems show a promising medium for exploring new interaction techniques. This research investigates the posibility to design user-friendly interfaces by making them graspable. This paper describes a preliminary design of a user-friendly interface for an AR system based on the concept of a graspable user-interface.

Defining user-friendliness

User-friendliness may improve by adapting user-interfaces to the perceptual-motor skills of human. One of the major differences between interaction with the physical world and interaction with a computer is that within the physical world the difference between functions is reflected by the actions with which to perform them. When interacting with the physical world there are numerous actions, each of which unique in the function for which they are used. For example, placing a document on a table requires actions completely different from trashing a document. In interaction with a computer the richness in operation, such as e.g. the ability to hold grasp push pull, is reduced to the click of a mouse button and the selection of the mouse position. Of these, only the mouse click might be informative. The selection of the mouse position is not informative. Because there is no rigid connection between mouse and the cursor the position may be different each time, even when it concerns the same function. As a result, users remember they had to click somewhere, but may always wonder exactly where.

AR is a new type of VR systems emerging on the reality-virtuality continuum (Milgram, 1994). Until recently research in HCI was constraint to the software part, mostly because hardware is more difficult to adjust and there was not much hardware to choose from. With the coming of VR techniques this situation slightly improved as it became possible to operate within virtual environments without most limitations of desktop computer hardware. VR systems commonly consist of a head-mounted display and special input device, e.g. data-glove, which allow users not only to observe but also act in a virtual environment. VR opens new possibilities such as e.g. HCI based on gestures (Hummels *et al.* 1998). However, this new technology introduced new limitations, such as lacking tactile information and lacking force feedback from (virtual) objects. Virtual objects appear, with respect to their handling, indistinguishable from real objects and invite to be handled in a similar manner. But since they are not the same this will lead to unnatural interaction. Therefore, virtual reality may suggest natural interaction, but without tactile information and force feedback it stil is artificial. AR offers solutions that do not have this limitation as they combine physical objects with computer generated objects. For example, a physical room can be augmented with computer images of newly designed furniture with the purpose of evaluation before the actual purchase.

Build-It (Rauterberg *et al.* 1997, Fjeld *et al.* 1998), the system for which the interface described in this paper was designed, is an AR based planning tool based on intuitive manipulation for the support of complex planning and configuration tasks. Build-It presents the user with two views: a table view showing the virtual environment in plan, and a wall projection, showing the virtual environment in perspective. Users operate on the table, using real rectangular bricks with which, e.g., the configuration of a factory floor can be planned. The plan view also has a (virtual) camera. With this camera the perspective wall view can be altered. The Build-It system is set-up to allow for fast and easy interaction with objects within a virtual world. Also, the set-up is such that it supports group interaction which, combined with the ease of operation, makes it a usable tool for communication between group members and for development of ideas.

With respect to interface design AR has the advantage over desktop computers and VR that it allows for a graspable interface. An interface is called graspable when each functions of the interface is operated by means of a dedicated physical device. The device the function is operated with acts as a graspable function (Fitzmaurice *et al.* 1997). Two reasons suggest the advantage of a graspable user-interface for the development of user-friendly HCI compared to a normal graphical user-interface approach. First, a physical object naturally invites users to act. For example, a small ball invites to be squeezed or a large ball invites to be kicked. Such natural invitations to actions can, when applied properly, be used to the advantage of a user-friendly interface. Second, the use of physical objects make it possible to make explicit distinctions between (or similarities of) functions by linking them to different (or similar) actions.

A graspable user-interface for build-It

Based on these considerations, a graspable user interface was designed for the AR-system called Build-It. This interface unifies physical and computer generated objects, with an emphasis on direct manipulation and user-friendliness. The main purpose of the interface at this stage is to set a general framework for developing applications for and working with Build-It-like systems. Therefore, the functions included at this point are limited to those needed to perform conventional desktop manipulations like e.g. opening and closing a file, and starting and ending a program. The next part gives a brief overview of the interface.

The software part of the interface is organized in layers. This is done in a recursive way. This means that each element on a layer potentially can be a layer itself, each of which can consist of layers and so on. This makes all objects equal in way the user they are manipulated on. The hardware part of the interface includes a toolbox in which the tools can be placed. For now, these tools are a pen-like element, for precision manipulation tasks, a rectangular brick, for global manipulation tasks, a clipboard-like selection tool and for 3D visualization a ruler-like tool.

Given the length of this paper only a small part of the interface is described. Nevertheless, it gives an idea about which direction the project is evolving.

Figure 1: Opening the toolbox starts the application, and makes the working area appear.

Opening the toolbox starts the application (Figure 1). This will make visible the files and layers with objects belonging to the application. The position of an object can be changed by simply selecting it with either the pen or with the brick tool. To select several objects a special tool is needed: the clipboard-tool. When this clipboard tool is placed onto the table a semi-transparent layer will appear. Objects that are selected through this will be locked to it and thus selected (Figure 2).

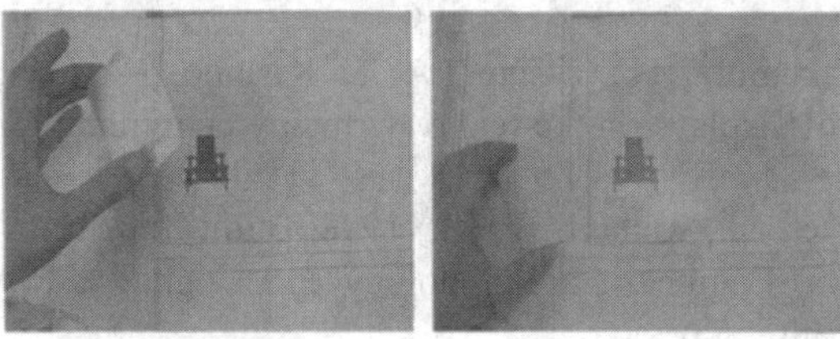

Figure 2: Selecting multiple objects: attach the objects to a virtual layer.

To copy an object it is selected with the brick, and then dragged from under the brick with the pen (Figure 3). To scale an object it is selected it with the brick, after which the object can be scaled by dragging one of its corners with the pen.

Figure 3: Copying: drag the object from underneath the tool with which it is selected.

Discussion

User-friendliness may improve by adapting user-interfaces to the perceptual-motor skills of human. For this, AR has the advantage over desktop computers and VR that it allows for an interface that is graspable. The advantage of having separate input devices for different functions is that these input devices become representatives for these functions. This will help display the structure of the interface's functionality and to emphasize the difference between functions.

In this paper the ideas for a design of such a graspable user-interface were described. Currently these ideas are being implemented in a working prototype. Future research will focus on comparing non-graspable interfaces e.g. menu based on structures, where functions are differentiated only through the selection of a menu item, with graspable user interfaces, where functions are differentiated through actions and tool selection.

References

Milgram, P., Takemura, H., Utsumi, A. & Kishino, K., (1994) Augmented Reality, a class of displays on the reality-virtuality continuum: *SPIE*, v. 2351.

Fjeld, M., Bichsel, M., Rauterberg, M., 1998, Build-It: A brick based tool for direct interaction: Engineering, Psychology and Cognitive Ergonomics, v. 3: Ashgate, Hampshire, 5 p.

Fitzmaurice, G. W., Buxton, W., 1997, An empirical evaluation of graspable user-interfaces: towards specialized space-multiplexed input: CHI 97: Looking to the future.

Hummels, C. J., Stappers, P.J., 1998, Meaningful gestures for human computer interaction: beyond hand postures: Proceedings of the 3rd International Conference on Automatic Face & Hand recognition (FG'98), p. 591 - 596.

Rauterberg, M., Fjeld, M., Krueger, H., Bichsel, M., Leonhardt, U. & Meier, M., 1997, BUILD-IT: a video-based interaction technique of a planning tool for construction and design.: Proceedings of Work With Display Units -- WWDU'97, p. 175-176.

Influence of Delay Time in Remote Camera Control

Katsumi TAKADA, Hiroshi TAMURA, Yu SHIBUYA
KYOTO INSTITUTE of TECHNOLOGY

1 Introduction

Although recent technological advances promise new possibilities for image communication, its full potential has not yet been adequately measured. It is vital to determine how users actually respond to interactive systems by studying the operating procedures of image communication.

In studying the image communication process, it is essential to examine camera control methods. Camera control in TV conferencing will allow participants to make variety of interaction among them [H. Tamura, et al. 1997]. In remote camera control, however, delay time makes visual feedback operation difficult [Hoffman, E. R. 1992]. In order to overcome this problem, camera control is made constant, and time duration of motion is quantumized (pulse operation).

Primarily, the combination of the delay time and camera movement speed determines an operator's effectiveness. Reducing superficial speed of camera view and using global view are effective way of enhancing remote camera operation were mentioned [M. Murata et al. 1997].

In this study, the remote camera control system is built on the Ethernet. Then the influence of delay time on remote camera is examined.

2 Experiment

In order to clarify the influence of delay time in remote camera control, three kinds of experiments were done. In each experiment, 13 examinees operate a remote camera with clicking a button on CRT.

2.1 Experiment 1 : Camera adjustment

Examinees were directed to adjust a camera direction to fix an object on camera view. (Figure 1)

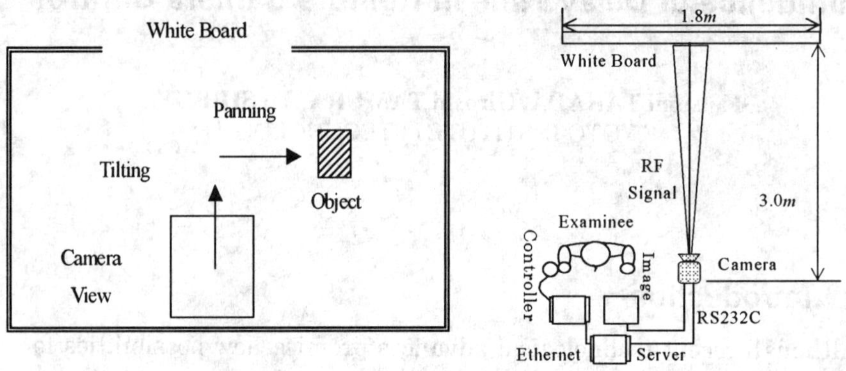

Figure 1. System Configuration of Experiment 1

2.2 Experiment 2 : Recognition objects

Examinee's task was to recognize five objects in the room by adjusting direction of a camera. (Figure 2)

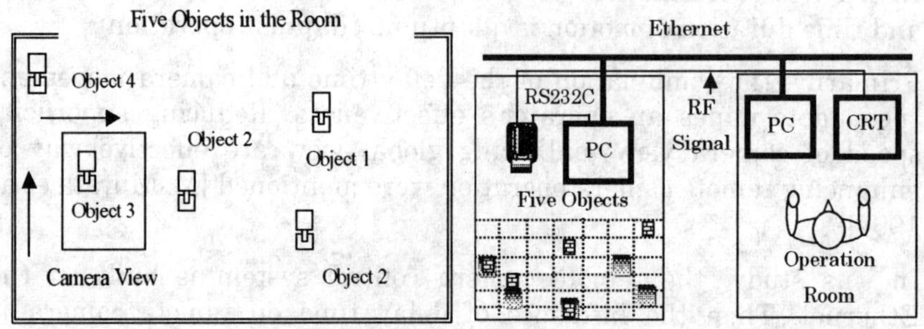

Figure 2. System Configuration of Experiment 2

2.3 Experiment 3 : Following up an object

Examinee tried to follow the moving object on board by adjusting direction of a camera. A video projector projected the moving object. (Figure 3)

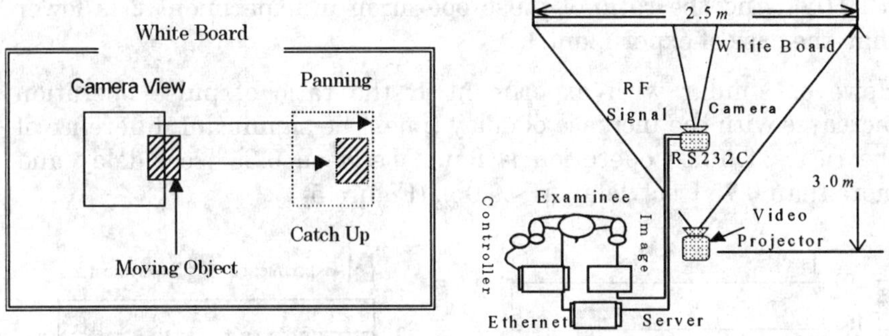

Figure 3. System Configuration of Experiment 3

3 Result

3.1 Results of Experiment 1

There was not special relation between the delay time and the task achievement time. The ratio of pulse operation increases with increase of the delay time. The significant difference of the ratio of pulse operation is shown between 0.35 [sec] delay and more than 0.75 [sec] delay ($p<0.05$). (Figure 4)

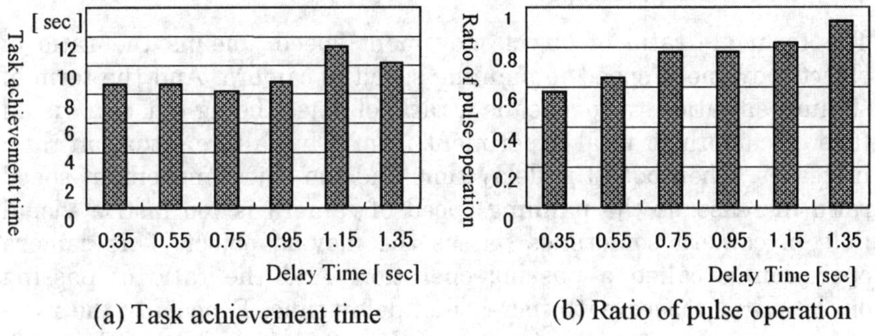

(a) Task achievement time

(b) Ratio of pulse operation

Figure 4. Result of Experiment 1

3.2 Result of Experiment 2

The task achievement time increases with increase of the delay time. The significant difference of the ratio of pulse operation is found between under 0.35 [sec] delay and more than 0.95 [sec] delay. (P<0.05). And the ratio of pulse operation in experiment 2 is lower than the case of experiment 1.

However similar with experiment 1, the ratio of pulse operation increases with the increase of delay time. The significant difference of the ratio of pulse operation is found between 0.35 [sec] delay and more than 0.75 [sec] delay ($p < 0.05$). (Figure 5)

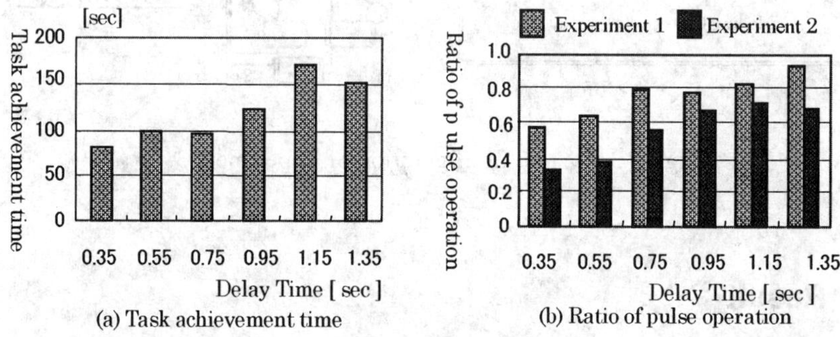

Figure 5 Results of Experiment 2

3.3 Results of Experiment 3

The term of "ratio of object movement speed" means the ratio of object movement speed to panning speed of camera. And the term of "failure situation ratio" means a ratio of object being out of a visual field of camera to total experiment term. The failure situation ratio increases, when both the delay time and the object movement speed ratio increase. If the panning speed of camera is too fast, a visual field of camera sometimes passes the moving object. This camera operation is called a passing operation. And the ratio of passing operation increases with increase of delay time. Especially the ratio of passing operation increases more than 0.95 [sec] delay. (Figure 6) As for the pulses operation, a lot of it appeared much at the low ratio of object movement speed (33% > 50% > 67%).

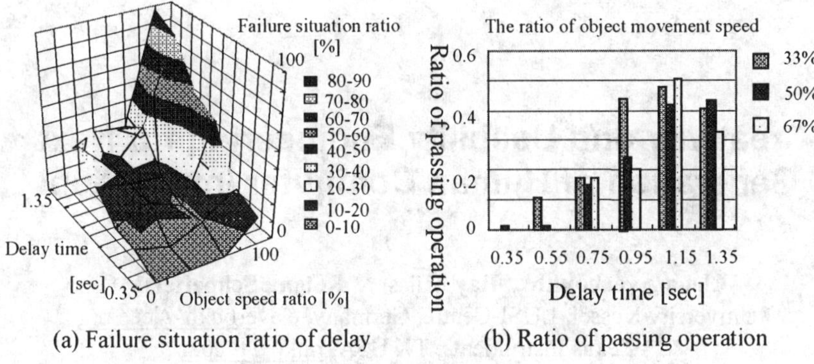

(a) Failure situation ratio of delay (b) Ratio of passing operation

Figure 6 Result of Experiment 3

4 CONCLUSION

In this study, the remote camera control system is built on the Ethernet, and the ratio of pulse operation and delay time of operation is compared. In the remote camera control, an examinee was doing the operation to have combined with the pulse operation and the visual feedback operation. And the ratio of pulse operation increases with the increase of delay time. On more complex operation, such as camera adjustment operation, increase of delay time makes camera control difficult, and operators have to adopt the pulse operation of a remote camera. In this experiment, it is mention that the use limit of the visual feedback about delay time is among from 0.75 [sec] delay to 0.95 [sec] delay.

5 REFERENCE

Hoffman, E. R (1992). Fitts' law with transmission delay, Ergonomics, 35, pp.37-48

H. Tamura, R. Zhang (1997). Remote and Local Camera Works in Media Conference, advance in human factors/ ergonomics, vol. 21B, 1997, pp.419-422

K. Murata, K. Horiuchi, K TAKADA, H TAMURA, Y SHIBUYA. (1998) .The limits of visual feedback operations on the network based remote-camera control, 14th Proceeding of 14th Symposium on Human Interface, pp.737-742

Creativity and Usability Engineering for Next Generation of Human Computer Interaction

Elena Averboukh*, Ray Miles, Roland Schoeffel*****
*University Kassel, LUSI-Centre, Germany, e.averbukh@ieee.org
** Texas Instruments, TX USA, rajko@ti.com
*** Siemens Design & Messe GmbH, Germany,
roland.schoeffel@siemens-d-m.de

1 Introduction

Increasing competition and demands to the flexibility and speed in designing interactive products and systems and to their innovation and usability on the one side, as well as rapid technological advances and new interaction media on the other side challenges

- opening new market sectors, businesses and industries,
- developing and systematically using within the business-chains innovative and user-centered design and evaluation technologies, and
- providing relevant competence development of the design and development teams.

This paper summarizes different views on these problems and advanced engineering solutions regarding different applications which are presented by all three authors in the form of penal discussion (see also Kitamura 1999)

2 Different Views on Innovation and Usability as Quality Criteria

As far as ergonomics has become a certified and licensed profession in many countries, as many relevant professional associations exist, and as there are many standards on ergonomics, one can conclude that design of usable interfaces is a trivial task. Unfortunately it is not a case, neither for complex and risky human-machine systems where preliminary training of users is considered but occurs to be nearly never sufficient, nor for every-day products like washing machines, videorecorders etc. which users are seeking to use without reading any written instructions independent of their age, professional skills and culture.

These problems become even more „severe“, e.g., in advanced internet applications for everyday family-use where especially elderly people have not only limited or no skills in using computers, but are also fearing and suspicious about them (Averboukh 1999a). Hence, rather significant part of potential users especially along with aging of our societies are not impressed with innovation which they do not understand or when they do not have yet a vision or *image* and *habit of it's everyday use.*

Users are interested in and „buy“ *functionalities* which they can easily and efficiently use for their typical everyday life-needs, for work or for fun. Product or system per se do not really matter. Availability of function and/or service which suites to the user's tasks and wishes, as well as to their perception of comfort and of „functionality-image“ (i.e. Kansei in Japanese, see Nagamachi 1995) in specific use- or task-situations influence their decision to buy and to use one or another product or system. In risky human-machine-systems safety and usability criteria are melt together, as safety of the whole system is to the most extent determined by its usability (Kitamura, 1999). Nevertherless innovation and/or so called „cool“ features can attract users and significantly influence decision to buy a new product.

To overcome especially psychological as well as user-competence barriers, any technological innovation, particularly in the field of advanced interactive products and systems have to be

- first ! *easy-to-recognize* and *easy-to-identify* in well-known to the particular user-population *terms, roles and relations,* and
- second, *easy-to-use.*

Current studies on barrier-free design of every-day products are reported and discussed.

3 Creativity- and Usability Engineering

In spite of the evidence of the above problem, a lot of difficult- or even impossible-to-use products and systems are produced worldwide. Usability tests of vending machines in the subways, of home appliances etc. show that *more than half of these products* have functionalities which users have difficulties to operate without help (Schoeffel 1999). With some apparels users have difficulties even to find the on/off button and to switch the appliance on, with others they can not select desired program etc.

The problem of producing innovative easy-to-recognize and -to-use products becomes rather sophisticated system-engineering problem. It is especially difficult because usually interdisciplinary teams of professionals at different locations within business chains have to work cooperatively under strict time and budget limits. Moreover

- design and development teams have to react very quickly and creatively to the rapidly changing market demands and end-users needs and preferences and to the technological engineering- and development platforms available,
- end-users although involved in relevant design activities and usability studies usually know what they want or need only when they try already available solutions, i.e. traditional usability engineering costs are rather high,
- large-scale technical systems are usually designed for thirty-forty years of operation life-cycle, i.e. usability has to be *modeled and predicted*!
- enormous and steadily growing amount of user-data and -knowledge has to be properly gathered, processed, interpreted and maintained across development teams and company members and/or subsidiaries,
- Innovation- and Usability-requirements to the end-products have to be interpreted (and understood!) into the relevant quality- and technical requirements for the product- or system-components etc.

Existing usability engineering methods (Nielsen 1993) as well as innovation management techniques are far not enough formalised and computer-aided to support efficiently

a) flexible and chain organizational product-development structures,

b) relevant (re-usable!) innovation- and usability-knowledge management, and

c) quick and reliable interpretation of usability requirements along with business- or technological chains.

Revisiting of software-, control-, usability- etc.-engineering and their systematic integration is discussed. Such systematic integration of these engineering methods and relevant engineering processes in design and development life-cycle is regarded as *Creativity and Usability Engineering.* Alternative system's life-cycle paradigms (see Fig.1) as well as practical experiences in the field of industrial product design and high-edge internet applications are presented.

4 Computer-Aided Platforms and Applications

Design teams permanently need advanced and up-to-date *methods and tools to analyse, forecast, evaluate and maintain* the innovation- and usability-values of new design solutions. Critical for the success and increasing productivity is *merging* mentioned above diverse engineering processes and their efficient co-management. Relevant innovative and practical competence development of interdisciplinary teams grows to a challenge and apparently to a new industry.

Recently launched in Germany, Hessen industrial R&D Programme DAKUMI „Development of Data-Banks for User-Analysis, Modelling and Interpretation" and first results are discussed (Averboukh 1999b). This project is focused particularly on the development of the internet-based engineering and evaluation platforms for efficient gathering, processing, modeling, evaluation, interpretation for different product-development-decision-makers and management of the knowledge about end-users, their goals, wishes, tasks, functions etc. and easy-to-use interaction schemes for different application areas.

Advanced approaches and development platforms of advanced Home-TV-Internet applications „for everybody" are presented. Following interpretation schemes of usability *wishes/images (Kansei)* into system functionalities and further into „technical solutions" are presented and discussed:

** *adrenalin*

entertainment

** *feel-good*

communication -> including video telephony with microphone arrays; media content can be created by content makers like movie studios, but for most humans equally relevant content are fellow humans (e.g., kids) talking live off the TV screen,

information -> including software agent services,

** *gray brain matter*

education -> interactive, web-enhanced, viewer-participatory TV engages the learner.

Easy-to-use home appliances and relevant industrial usability engineering practices are presented.

5 Standardisation and Competence Development

International activities and particularly initiative of a new German standardization group on reaching consensus what easy-to-use products mean and on issuing relevant international standard are discussed.

Advanced curriculums and practical experiences of innovative team training in cross-disciplinary areas of Creativity and Usability Engineering are presented.

6 References

Averboukh, E.A. (1999a) *Usability-Market-Study „Barrier-Free Car Design"* Kassel: LUSI-Centre.

Averboukh, E.A. (1999b). DAKUMI-Technologies: Bridging Information, Technology and Industry for Human Wellness: In: *Proceedings of International Workshop on Harmonised Technology with Human Life*, (HarTHuL, Takarazuka, Hyogo, Japan, March 3-5, 1999), pp.101-106.

Kitamura, M. (1999). Creativity and Usability Concepts in Designing Human Computer Interaction for Management of Complex Dynamic Systems, *Proc. 8th International Conference on Human-Computer Interaction* (HCI'99, Munich, Germany, August 22-27, 1999).

Nagamachi, M.(1995) Kansei Engineering: A New Ergonomic Consumer-Oriented Technology for Product Development, *Industrial Ergonomics,* 15, 1995.

Nielsen, J. (1993). *Usability Engineering,* Academic Press, Inc.

Schoeffel, R. (1999). Usability Design for Washing Machines. In*: Proceedings of International Symposium on Laundry/Washing'99* (Kobe, Japan, March 16-18, 1999).

Creativity and Usability Concepts in Designing Human Computer Interaction for Management of Complex Dynamic Systems

Masaharu Kitamura
Department of Quantum Science and Energy Engineering,
Tohoku University, Aoba-ku, Sendai, 980-8579 JAPAN

1 Introduction

The present paper discusses the issue of creativity and usability in conjunction with human-machine interactions in potentially hazardous systems such as nuclear power plants, chemical plants containing explosive and/or poisonous materials, high speed transportation vehicles, etc. The concept of creativity might be considered somewhat irrelevant to such industrial installations, but the concept, to be materialized with the concept of usability in practical situations, is closely related to functional requirements imposed on human-machine interface designed to support operators under certain operational conditions. The functional requirements and the currently available techniques to meet the requirements are briefly described in this paper.

The most important and typical situation where the role of creativity and usability becomes evident is when the dynamic system itself and its operators as well are facing to a safety-critical emergency situation. In the emergency operation, the operators are required to manage the situation by properly utilizing available resources. Since the operators are intensively trained and highly motivated, they can usually manage the situation without serious difficulty. However, the situation becomes extremely difficult to handle when the trigger event is rare and unforeseen, magnitude of the resultant transient is large, influence range of the transient is wide, multiple faults are ongoing, etc. The probability of occurrence of such a situation might be quite small, but not negligible. In fact, this is the situation when the operators are really in need of external support. It is not surprising that the once-popular expert system technology failed to support operators of these complex systems since the essence of the technology is to utilize expert knowledge established through ample experiences of problem-solving. The technology is usable to solve problems frequently experienced, but not applicable to support operators facing to a rare and unforeseen event. Therefore, even for a single subtask included in the emergency situation management, e.g. fault diagnosis, the expert-system-based problem-solving approach was ended up at the phase of far from practically dependable. An alternate option, i.e. human-machine cooperative problem-solving, seems to be an only realistic approach to meet the real need. The function to stimulate the creativity of operators is of crucial importance in this regard. Furthermore, the support of creative thinking is effective if and only if the human-machine interface can offer necessary functionality with high level of usability. Heavy operational workload to retrieve necessary information will strongly restrict the mental capability of the user to figure out the true cause of the troublesome situation.

2 Descriptions of the Problem

The feature of actual working situations we are discussing is described more specifically. As mentioned earlier, even a single task such as fault diagnosis is often complex enough to nullify the use of conventional support technology. As clearly pointed out in preceding studies, the problem solving is considered complex when the objective world is "high" in the four dimensions characterizing the cognitive demands imposed to the operators (Woods, 1988).

(1) Dynamism: When the world is dynamic, problem-solving incidents unfold in time and are event-driven, i.e. events can happen at indeterminate times.

(2) Number of parts and their interconnections: When a world is made up of a large number of highly interconnected parts, one failure can cause multiple consequences. Also, one disturbance could be due to multiple causes and thus can have multiple potential remedies.

(3) Uncertainty: When available data from the objective system are ambiguous, incomplete, erroneous, or imprecise with respect to the state of the world, problem-solving about the system can be heavily influenced by these uncertain factors.

(4) Risk: When there is risk, possible outcome of decision-making can cause huge economical losses. The presence of risk means that one must be concerned with rare but catastrophic situations as well as with more frequent but less costly situations.

The large-scale complex systems discussed in this paper are known to be high in most of these dimensions. Furthermore, it is often the case that the situation is much more complex in the sense that the operators must solve multiple and mutually related tasks such as fault detection, diagnosis, situation comprehension and assessment, selection and implementation of remedial actions, managerial and political message transfer, etc. (J-M Hoc et al., 1995). Since the most important mission of the operators is to solve such complicated problems, more attention should be paid on capability of interfaces in some ways to reduce the cognitive complexity of the world in these dimensions. The issues of creativity and usability are tightly related to this requirement.

3 Principles and Implementations

The author proposes that the complex problem of emergency management mentioned above can be decomposed to two sub-problems; (1) situation understanding and assessment, and (2) resource selection and implementation. Each of the sub-problems can be further decomposed if the problem world is organized to have certain hierarchy. The decomposition is generally an efficient strategy to reduce the complexity of the problem. The dynamism as well as the number of interconnected part, for instance, can be reduced drastically by proper decomposition of the problem world. However, the decomposition is not always the dependable strategy. The original problem can be hard to decompose, or the resultant problem can be still complex and thus direct solving of the problem by some means of computational intelligence such as mathematical programming, artificial intelligence, model-based computation, etc. is not feasible. Support of creative thinking with different perspective is still necessary.

It should be noted here that support of problem-solving does not always mean provision of candidates of solution or such; the creativity of human operators can be inspired or stimulated by providing an appropriate sign or cue which has never be highlighted in the thinking disturbed by seemingly chaotic observations. The provision of a key could be informative and valuable enough for the puzzled operator to be released from a cognitive trap or confusion and to be guided to a valid solution.

Based on these considerations, the author would like to stress that creativity and usability requirements are certainly central to the design of human-machine interface of complex dynamic systems with potential hazards. Some of the design features proposed in these industrial sectors are; adaptation to task execution mode, functional decomposition and visualization of the dynamic system, provision of multiple cognitive viewpoints, and plant history database with user-customized search scheme (M.Takahashi et al., 1997, 1998).

(a) Adaptation to task execution mode: The operator facing to a complex problem is thought to be trying to properly manage the situation represented by the multiple symptoms/observation. The task execution mode in this situation is categorized to be a "reactive" mode, since the main cognitive activity of the operator is to respond to the external signs. This has been a basic perspective in designing and evaluating interface. Actually, however, considerable portion of the activity could be "proactive" in the sense that the major tasks are to conduct active thought experiments to generate a hypothesis to make the seemingly contradictory symptoms coherent, recollect previous events with similar set of symptoms, predict possible consequences of a remedial action, etc. The design of the interface could be adaptively modified depending on the mode of task execution.

(b) Functional decomposition: Decomposition and visualization of the system in terms of its configurational structure has been a standard practice of interface design. However, the decomposition and visualization of functionality of the system is less popular since understanding and consensus about the way to fulfill the requirements are not reached yet. As one of the goals of the operator is to detect and restore the degraded function of a component or a subsystem, provision of information directly related to the function is considered to be definitely useful.

(c) Provision of multiple cognitive perspective: This function is expected to be highly helpful to support the operator in cognitive trap and/or confusion. Even if the contents are almost the same, different forms of information provision can be effective to stimulate the operator to reach the valid situation comprehension.

(d) Plant history database: Excellence of human wisdom in complex problem-solving is often attributable to memory recollection and utilization. In the past, no efficient support has been given to the operator to conduct the task of memory recollection because computer power was far less than adequate to make such support feasible. Development of a plant database which can store relevant operational records of various modality is now becoming feasible and therefore highly promising as an essential constituent of interface in near future.

It is clear that more efforts must be focused on these important aspects in design of human-computer interactions for large-scale, complex artifacts. Some of the prototypical realizations of these functions together with demonstrative results can be found in our previous reports (M.Takahashi et al., 1997, 1998).

4 Concluding Remarks

The necessity of considering the creativity and usability aspects in designing and evaluating human-machine interface of potentially hazardous artifacts where safety requirement is more rigorous than everyday products is stressed in this paper. In near future, more emphasis will be placed on even higher safety. To meet the need, efforts are expected to be required to facilitate extensive surveillance and efficient detection of mechanical failures in their very incipient phase. This task requires another type of creativity; envisioning of unforeseen events via integration of subtle symptoms.

The interdisciplinary discussions about these aspects will be stimulating and fruitful for reaching general consensus and also for further evolution of domain-specific human-machine interaction techniques.

5 Acknowledgment

The author would like to express his sincere gratitude to Y.Niwa of INSS for stimulating discussions about user requirements in dealing with complex problem-solving in nuclear power plants. He also wishes to extend his gratitude to E.Averboukh of University of Kassel for her initiative and efforts to make this valuable information-exchange possible.

6 References

Hoc, J-M., Amalberti, R., Boreham N.(1995). Human Operator Expertise in Diagnosis, Decision-Making and Time Management. In Hoc, J-M., Cacciabue, P.C.., Hollnagel, E. (Eds.); Expertise and Technology, Lawrence Erlbaum Assoc. 19-42.

Takahashi, M., Takei, S. and Kitamura, M. (1997). Multimodal Display for Enhanced Situation Awareness Based on Cognitive Diversity, Advances in Human Factors/Ergonomics 21B, Elsevier, 707-710.

Takahashi, M., Catur Diantono and Kitamura; M.(1998). Life Cycle Integrity Monitoring of Nuclear Plant with Human Machine Cooperation, Proc. 7th IFAC/IFIPIFORS/IEA Symposium on Analysis, Design and Evaluation of Man-Machine Systems, Kyoto, Japan, 431-436.

Woods, D.D.(1988). Coping with complexity: The psychology of human behavior in complex systems. In.Goodstein, L.P. Andersin, H.B. &Olsen, S.E.(Eds.), Mental models, tasks and errors, London: Taylor & Francis, 128-148.

The effect of auditory accessory stimuli on picture categorisation; implications for interface design

Myra P. Bussemakers, Abraham de Haan and Paul M.C. Lemmens
Nijmegen Institute for Cognition and Information, P.O. Box 9104 6500 HE Nijmegen, The Netherlands, bussemakers@nici.kun.nl

1 Introduction

In human-computer interfaces the analogy of human-human communication seems to be a good starting point to optimise interaction (Brennan, 1990). When two persons are communicating in a face to face situation, a whole range of modalities is used to get the message across. Besides speech and non-verbal auditory cues ('uhu'), people utilise facial expressions and gestures in their attempt to communicate all aspects of an utterance. This can result in a more effective interaction, as studies for instance have shown that carefully manufactured auditory feedback elements seem to improve the effectiveness of computer usage (Brewster, 1994).

When information is presented in multiple streams, users need to integrate these informational elements into a single experienced unity (Bussemakers and de Haan, 1998). Every occurrence in an information stream consists of a number of contingencies like colour, location, loudness and mood. In the process of integration, we suppose that some relevant aspects of the contingent information in the modalities seem to be combined. This because the mind expects these modalities to convey related information. The integration occurs for instance when you are listening to a lecture in a big hall. The lecturer is speaking into a microphone and the sound is transmitted through loudspeakers that are located on either side of the hall. Although the sound actually is coming from your left and right, you integrate the movement of the lecturer's lips with the sound in such a way that it seems as if the sound is actually coming from the lecturer (i.e. the ventriloquism effect (Howard and Templeton, 1966)).

In some situations this integration will take more time than in other situations and it seems relevant to find out when having information in more than one modality can facilitate the interaction and/or when it might even hinder the interaction.

2 Picture categorisation with accessory auditory stimuli

The study presented here was an extension of earlier experiments, where a picture categorisation paradigm with auditory presented distractors was used (Bussemakers and de Haan, 1998; Bussemakers and de Haan, submitted). It is expected that users, when working with an interface, have to do a similar task, when they work with icons representing different categories like a file, program or folder. Often these categories in a multimedia setting are accompanied by sounds. The combination of the type of sound, integrated with the visual icon will attempt to provide the users with all relevant information so that (s)he knows what the next action should be.

In these earlier experiments, subjects were instructed to categorise line-drawings of animals (for instance a cat) and non-animals (for instance a boat). The question the subjects were presented with, was: “Is this a picture of an animal?”. Subjects were asked to press a button labelled ‘yes’, or a button labelled ‘no’ to indicate their response. In some trials, the line-drawings were accompanied by one of four chords: a high or low minor chord and a high or low major chord. In other cases the drawing was presented without an additional auditory stimulus.

All pictures were shown with all chords in blocks. For instance in the first block, all drawings of animals, when accompanied by a sound, were displayed together with a high major sound and the drawings of non-animals with a low, minor sound. These contingencies in the figures were labeled ‘congruent’, because when the mood of the sound is integrated with the category of the drawing, both suggest a positive answer. A block with the drawings of animals with a minor sound and the non-animals with a major sound was similarly labeled ‘incongruent’. Blocks with both animals and non-animals presented with one sound is referred to as neutral. This presentation-strategy was adopted to enable subjects to relate the type of sound to the type of drawing more easily. Both the response times (in msec.) and the type of response, i.e. ‘yes’ or ‘no’, were registered.

Results of these experiments indicated a significant delay in responses in the conditions with sound compared to the condition without sound (see figure 1). Furthermore, between the conditions with sound, a significant effect of incongruency of sound and picture (e.g., an animal with a minor chord) was found.

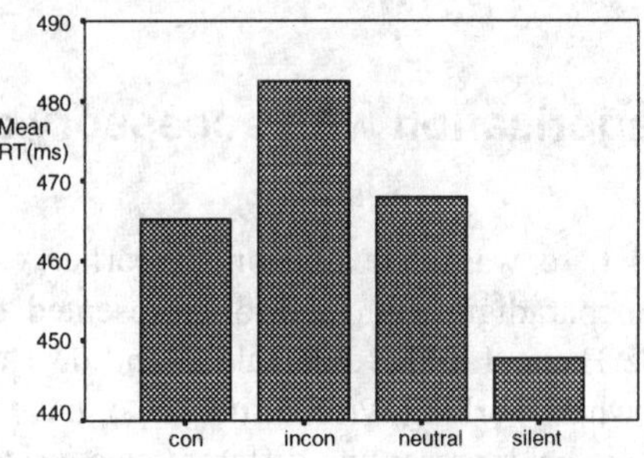

Figuur 1 Results categorisation first series of experiments.

However, no significant differences in reaction times between the congruent and the neutral condition were found (Bussemakers and de Haan, in press). These results indicate that for the process of integration it does not matter whether the information in the other modality is unrelated and therefore has no contingencies with the pictures, or the information in the other modality is the same as the pictures. Only when the auditory information is different from the visual information, when there are conflicting contingencies, so to speak, there is a significant effect on the reaction times.

In the current study these experiments were extended by adding a Stimulus Onset Asynchrony (SOA) of –500 ms., so that the sound was presented earlier than the picture, to provide an extra timeframe for the subjects to process the auditory stimulus and to relate the sound to the pictorial stimulus. It was expected that the SOA would lead to a greater difference in reaction times especially between the congruent condition and the neutral condition. Another extension to the earlier experiments was the differentiation between musically skilled and musically unskilled subjects, since it is possible that they respond to the major and minor chords differently.

The participants in the study presented here were students in psychology or cognitive science at the University of Nijmegen. Of the 14 participants, 7 participants were musically skilled (6 years or more experience in playing a musical instrument) and 7 participants were musically unskilled (no experience in playing a musical instrument). Participants were presented with all conditions.

The results again show a significant delay in reaction times for the conditions with sound, compared to the silent condition (see figure 2). In the conditions with sound, again a significant effect of incongruency was found.

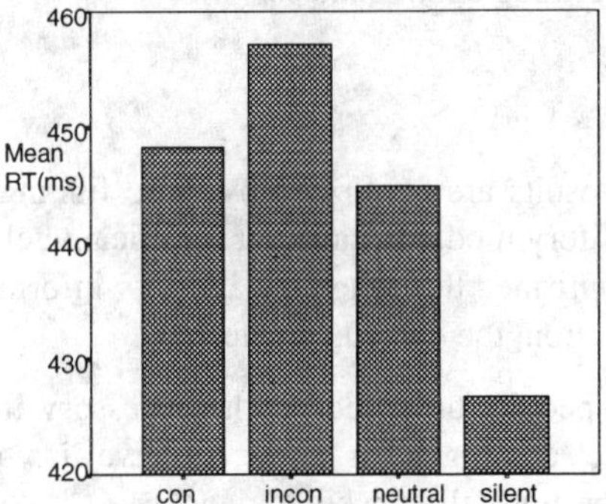

Figure 2. Overall results categorisation with SOA

However, the addition of the SOA did not result in a clearer difference between the congruent condition and the neutral condition. A reason for this may be that by adding a longer time between the stimuli, there is no change in the contingencies between both signals. Sound can be processed more quickly than visual information and having extra time might not result in the processing of more contingencies.

Furthermore, the differentiation on musical skill showed a trend, though not significant, in differences in reaction times between musically skilled and musically unskilled participants (see figure 3).

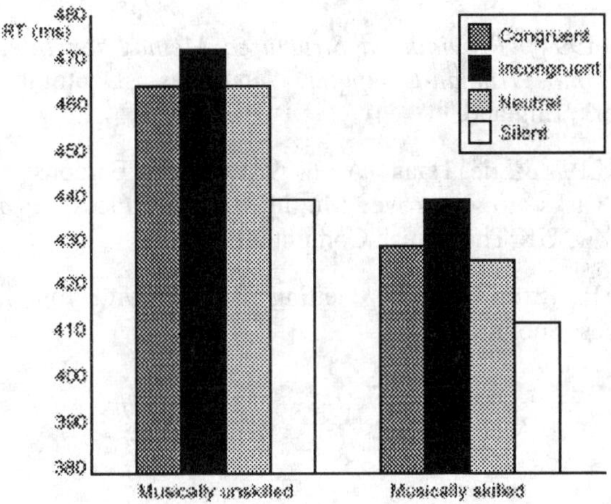

Figure 3 Results separated for musically skilled and unskilled people.

These results indicate that musically skilled participants seem to be able to process the musically originated sounds more quickly. A possible explanation is that for them there is less interference, because they are able to use the musical aspects instead of integrating contingencies.

3 Conclusions

Earlier experimental results are confirmed by these findings: having accessory information in the auditory modality causes a significant delay in reaction times. Also the results are confirmed that when the auditory information is incongruent with the visual information, the delay is greatest.

For musically skilled people, the effect of the accessory information could be different, because the contingencies they perceive in both modalities are different from those for musically unskilled people.

In designing interfaces having additional sound may not always facilitate the interaction and the effect may not be the same for everyone. It is our goal to provide a theoretical framework for these findings.

4 References

Howard, I.O., & Templeton, W.B. (1966). *Human Spatial Orientation*. London: Wiley.

Brennan, S.E. (1990). Conversation as direct manipulation: An iconoclastic view. In B.K. Laurel (Ed.), *The Art of Human-Computer Interface Design. Reading*, MA:Addison-Wesley.

Brewster, S.A. (1994). *Providing a Structured Method for Integrating Non-Speech Audio into Human-Computer Interfaces*. Doctoral dissertation, University of York, England.

Bussemakers, M.P., & de Haan, A. (1998). Using Earcons and Icons in Categorisation Tasks to Improve Multimedia Interfaces. *Proceedings of ICAD'98, Glasgow,* UK:The British Computer Society.

Bussemakers M.P., & de Haan, A. Getting in touch with your moods: using sound in interfaces. Submitted.

A New Method to Synthesize Japanese Sign Language Based on Intuitive Motion Primitives

Shan Lu[a], Hiroyuki Sakato[b], Tsuyoshi Uezono[c], and Seiji Igi[a]
[a]Communications Research Laboratory
4-2-1, Nukui-Kitamachi, Koganei, Tokyo 184-8795, Japan
[b]The University of Electro-Communications
[c]Kagoshima Prefectural Institute of Industrial Technology
Email: lu@crl.go.jp

1 Introduction

Sign language is the most important communicational tool for the hearing-impaired. Currently there are some services to assist the hearing impaired such as TV news programs that use sign language, and translation services (via phone companies) that translate spoken language into sign language. With the dramatic growth of the Internet, even more sign-language services are expected. However, these services can be difficult to develop since sign language contains many visual and dynamic factors such as hand motion, facial expression, and gesture.

Since sign language grammar is different from spoken language grammar, the synthesis of sign language is a two-step process. The first step is the linguistic translation from text to signs, and the second step is the visual synthesis of the translated signs, which is the focus of this paper.

There have been some methods that synthesize sign language by computer. There are basically methods: the first method uses sign-language motion data acquired from the motion capture system (Lu 1997b, Ohki 1994), and the second method divides a sign into more basic elements such as the phoneme in spoken language (Nagashima 1996, Kurokawa 1992). Since the first method generally uses the actual movement of actors speaking a sign word as a unit, it is not easy for an user to edit and modify the motion. Acquiring the motion data by using the motion capture system is also difficult. The second method adopts the similar phoneme to spoken language, but it is not intuitive to represent the visual factors of sign language.

2 Description of Sign Language

This paper proposes a new method to encode and synthesize the words of Japanese Sign Language (JSL). The method uses some basic motion primitives (simple movements) to encode the movements of signs. It is not only intuitive and easy-to-understand, but can accurately describe the hand movements. Since our aim is to develop a method and tool to provide more sign-language information to the hearing impaired, an easy-to-use method is very important. Although the method described here focuses on sign animation synthesis, motion primitives also reflect the general features of the sign and can also be used in sign-language machine recognition.

Sign movements are mainly represented by three factors for each hand: hand position, motion trajectory, and hand shape. Our method uses four basic parameters to describe these three factors: The start and end positions, motion primitives, hand shape, and the relationship between the two hands. The main point of this method is that it identifies the motion trajectory as the wrist movements, and classifies those movements into a dozen motion primitives.

Start and end positions. A relative coordinate system corresponding with the human model is used to define the position as shown on Figure 1. The X-Y plane is divided into 35 small areas, and the distance (Z-axis) from the body is represented by three levels. However, there are many signs which touch a specific part of the body, the head, the mouth, the nose, and the right ear, etc. So the specific part name is used to identify the position.

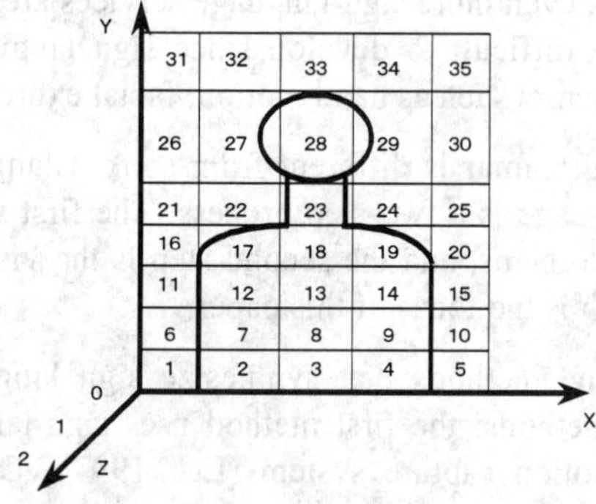

Figure 1. Coordinate system for determining the start and end positions in a sign.

Motion primitives. The motion primitives include the line, the circle, the half-circle, the arc, and the wave, etc. The motion trajectory of a sign word consists of one motion primitive or a combination. These motion primitives present the abstract movements and are independent from the other three parameters.

Hand shape. The number of basic hand shapes in any sign language has limitations. In our system, 58 kinds of hand shapes are used to encode the JSL words. Only the indication of the start and end hand shapes are needed. The intermediate hand shape is interpolated by the key-frame method during the movement of a sign.

Relationship between the two hands. If the motion primitives and the hand shapes of the left hand and right hand are symmetrical, we define the relationship as symmetry. If the two hands make the same motion trajectory together, their relationship is defined as a link. A link means that the left and right hand move together in the same direction and speed at the same time.

The authors analyzed videos of Japanese sign language and obtained twenty-one kinds motion primitives. By using these twenty-one motion primitives as an encoding method, 180 JSL words were successfully encoded.

3 Synthesis of Sign Language

We developed three-dimensional animation synthesis system to implement and evaluate the method. The system applies the four parameters described previously onto a 3-D articulated model in order to generate real-time sign-language animation.

In sign language, there are a lot of words that point to or touch a specific position on the face or the body. It is difficult to avoid the problem which the fingers are inserted into the surface of the face or the body in 3-D computer animation. In this system, the authors developed a new approach to help solve this problem. First, the names of the facial features and the body region are used to identity the start and end positions of a sign so that the parameter is independent from the 3-D articulated model. Then, animation that shows the touching of the mouth and the pointing to the eye is archived by calculating the adaptive movement direction and range of the wrist recursively.

In addition to hand motion, the facial expression and the mouth shape play the most important roles in sign language. Our other research has shown that the hearing-impaired person gazes at mouth shapes while they read sign language (Lu 1997a). The mouth shape is able to clarify some signs. Our system synthesizes the mouth shapes according to the Japanese alphabet rules. The Japanese alphabet has five vowels, and the basic mouth shapes are determined by these five vowels. A Japanese alphabet label is added to the definition of each sign word, then the vowels included in this Japanese label are used to automatically determine the mouth shape. Figure 2 shows the normal mouth shape and the shape of five vowels.

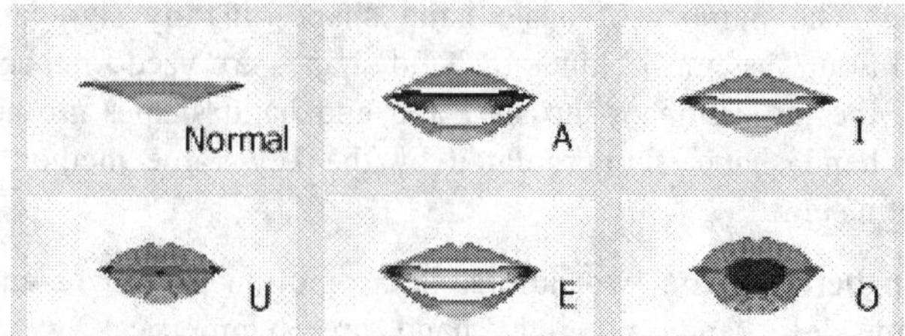

Figure 2. Images for animating the mouth shape of five Japanese vowels.

4 Evaluation

The sign-language animation created by this system was evaluated by five hearing-impaired persons and an adept JSL translator. In Table 1, Subjects A and B are female, and Subjects C, D, E are male, and Subject F is a female translator. This evaluation experiment used 118 sign words, and it did not consider the context of each sign. There were two types of evaluation. One evaluated whether the subjects could correctly understand the sign words (the recognition rate = number of correct words / number of all words x 100), and the other one evaluated how natural (the natural degree, ND) the arm-motion and hand-shape animation seemed to the subjects.

The experiment required that the subjects give one single Japanese word to represent the sign that they watched. Meanwhile, the subjects were required to evaluate how easy it was to understand. There are three levels: 1) the sign is very easy to understand, 2) there is some difficulty to understand the sign, 3) they can infer the meaning of the sign. The first four columns in Table 1 show the recognition rates in the different levels and the total rate.

Table 1. The evaluation results of JSL animation by hearing-impaired persons.

Subs	1st Lvl.	2nd Lvl.	3rd Lvl.	Total	ND of motion	ND of shape
A	66.1	22.9	3.4	92.4	4.00	4.25
B	72.0	6.0	14.4	92.4	3.36	3.07
C	79.0	6.7	5.0	90.7	4.33	4.32
D	67.2	17.6	0.8	85.6	4.27	4.32
E	58.9	27.7	3.3	89.9	4.30	4.19
F	70.3	15.6	4.2	90.1	4.25	4.32
Avg.	68.9	16.1	5.2	90.2	4.08	4.07

In order to evaluate the natural degree of the hand motion and hand shape, we define five categories to describe the different natural degrees. Each of category is assigned a score from 5 to 1. The five categories are: 1) Very natural and very easy to understand, 2) Natural and easy to understand, 3) Unnatural, but can understand, 4) Very unnatural and difficult to understand, 5) Very unnatural, and unlike human motion. The last two columns in Table 1 show the average natural degrees of motion and hand shape of all the signs.

5 Conclusions

We described a new method to encode and synthesize JSL. We also performed an experiment to evaluate how correctly a hearing-impaired person is able to understand the sign-language animation. The obtained results from five hearing-impaired subjects and a translator show that the synthesized animation has a 90.2% average recognition rate. In addition to the recognition rate, the quality of animation is also an important factor for transmitting the information to the signers. Although the natural degree depends on the personal feeling of subject, the experimental results give a 4.07 and 4.08 score (maximum is 5.0) on average. These results demonstrate that this is an effective method that can be used to accurately synthesize JSL.

Our future work will focus on the synthesis of continuous sign sentences based on this method. Furthermore, we hope to make the sign-language animation more natural by integrating the facial expressions and the body gestures.

References

Kurokawa, T. (1992). Gesture dictionary and its application to Japanese-to-Sign language translation. *Proc. the 3rd Advanced Symposium on Human Interface*, 51-58.

Lu, S., Imagawa, K., & Igi, S. (1997a). An Active Gazing-Line Generation System for Improving Sign-Language Conversation. *Proc. 7th Int. Conference on Human-Computer Interaction,* 283-286.

Lu, S., Igi, S., Matsuo, H., & Nagashima, Y. (1997b). Towards a Dialogue System Based on Recognition and Synthesis of Japanese Sign Language. *Proc. International Gesture Workshop*, Bielefeld, Germany, 259-271.

Nagashima, Y., Kamei, S., & Sugiyama, Y. (1996). Sign Language Animation System of Morpheme Information Driven Type. *Technical Report of IEICE*, ET96-86, 73-78.

Ohki, M., Sagawa, H., et al. (1994). Pattern recognition and synthesis for sign language translation system. *Proc. ASSETS '94*.

Empirical data on the use of speech and gestures in a multimodal human-computer environment

Noëlle Carbonell†, Pierre Dauchy‡

† LORIA, Campus Scientifique, BP 239
F54506 Vandœuvre-lès-Nancy Cedex, France
tel.: +33 (0)3 83 59 20 32, fax: +33 (0)3 83 41 30 79
e-mail: Noelle.Carbonell@loria.fr

‡ IMASSA-CERMA, BP 73, F91223 Brétigny sur Orge Cedex, France
tel.: +33 (0)1 69 88 33 69, fax: +33 (0)1 69 88 33 02, e-mail: pdauchy@imassa.fr

1 Context, motivations and objectives

Thanks to recent advances in automatic speech recognition and natural language processing, it is now possible to consider the integration of speech into user interfaces for the general public. Speech controlled human-computer interaction (HCI) will prove useful in contexts where the use of keyboard and mouse is awkward or impossible, such as home automation, online services, pocket devices, road navigation, etc.

In addition, speech associated with 2D gestures can provide an attractive alternative to direct manipulation in these and many other contexts of use, since:

- robust multimodal [1] HCI can now be achieved; present continuous speech decoders are capable of processing accurately large subsets of natural language; 2D designation gestures, which can also be interpreted reliably, may be used for expressing references to graphical objects or locations on the screen, thus making up for the present limitations of natural language interpreters as regards spatial linguistic references;
- spoken natural language associated with gestures on a touch screen seems more straightforward, hence usable, than direct manipulation (*cf.* menu selection through the mouse); in addition, designation gestures can be readily linked to utterances thanks to deictics (*cf.* "here/there", "this", etc.).

---

[1] Throughout the paper "multimodal" and "multimodality" refer to the combined use of speech and gestures for interacting with software.

Software issues relating to the design and implementation of robust multimodal user interfaces have been investigated extensively, *cf.* (Neal and Shapiro 91), (Binot et al. 92), (Koons et al. 93), (Nigay and Coutaz 93), among others. On the other hand, much less attention has been paid to the usability of multimodal HCI, *cf.* (Hauptmann and McAvinney 93), (Guyomard et al. 95), (Oviatt et al. 97). This is not surprising as many researchers from both the human-factors and the software-engineering communities take it for granted that HCI should emulate human communication, in order to achieve full transparency. However, various empirical and experimental findings are at variance with this judgement. It has been observed (Harris et al. 80), (Amalberti et al. 93) [2] that participants in human or human-computer oral dialogues spontaneously adapt the contents and forms of their utterances to the capabilities and to the knowledge of their interlocutors as they are perceived or inferred. The *a priori* assumption that speech and gesture HCI should emulate multimodal human communication seems then questionable or, at least, in need of empirical validation.

These observations motivate the empirical study of multimodal HCI reported here, which aims at: (i) investigating the actual use of unconstrained speech and gestures for interacting with computer applications for the general public; (ii) providing designers with an accurate characterization of the likely behaviors, needs, preferences and attitudes of future users of multimodal input facilities.

2 Experimental setup and protocol

Eight subjects of different professional profiles interacted with a simulated [3] multimodal user interface during three weekly sessions [4], using freely any of the following input modes: spontaneous speech, unconstrained 2D gestures (on a touch screen), multimodal commands. The task domain, furniture arrangement, was chosen so that all useful commands could be expressed easily and accurately using speech or gestures; see (Robbe et al. 97) for a detailed description of the setup.

The subjects performed simple design tasks. They had to modify the layout of various furnished rooms according to instructions specified in eight scenarios of increasing complexity. Room layouts (and additional pieces of furniture) were displayed on the screen in the form of 2D plans. Written instructions also specified available interaction modes. After a short oral presentation of the setup by the experimenter, subjects were left alone to process the scenarios; however, the experimenter's help was available on request throughout the experiment.

---

[2] (Harris et al. 80) concerns human communication, (Amalberti et al. 93) HCI.

[3] We used the Wizard of Oz experimental paradigm (2 wizards) for simulating some functions of the interface, namely the interpretation of the subjects' inputs and the system multimodal feedback.

[4] Each session lasted about half an hour per subject.

All subjects had previously used software intended for the general public; most of them were familiar with direct manipulation, but none had extensive computer knowledge or programming experience.
Subjects and screens were videotaped. Written descriptions of the recordings include: orthographic transcripts of subjects' utterances and system messages, together with coded descriptions of subjects' gestures and system actions. In addition, subjects' commands were labeled manually using existing taxonomies of multimodality (Coutaz and Caelen 91), (Coutaz et al. 95); they were also characterized according to their information contents. Our analyses of the subjects' commands are based on these categorizations.

The originality of this experimental setup is twofold, at least to our knowledge:
- first, most empirical studies of speech and multimodal HCI focus on situations where subjects have to comply with expression constraints [5]; such situations are inadequate for gaining an insight into subjects' preferences and attitudes regarding multimodality;
- secondly, all such studies confine the observation of subjects to one session only, thus making it impossible to study the evolution of the use of multimodality under the influence of practice.

3 Results and interpretations

Subjects in our experimental environment resorted to multimodal commands [6], *i.e.* synergic multimodality, less frequently than participants in empirical studies of multimodal HCI reported in the literature. Two main reasons may explain this difference. The task domain chosen by S. Oviatt (interaction with a map system) involved spatial location commands, the multimodal formulations of which are simpler and/or more precise than their monomodal counterparts. In the context of the other studies, the multimodal formulation of commands was easier than their monomodal expression, due to the drastic constraints which were imposed on the subjects' oral expression. Unlike these experimental setups, our setup required that all commands (spatial location designations included) be expressed as easily and as precisely through spontaneous speech or unconstrained gestures, or using free combinations of both modalities.
These comparisons suggest that monomodality is "preferred" to multimodality whenever the monomodal expression of a command is as precise and as simple as its multimodal formulation. In other words, it seems that in the context of HCI, the use of synergic multimodality is less spontaneous and/or more costly (in terms of cognitive work load) than monomodal expression.

---

[5] (Oviatt et al. 97) reports one of the few meaningful empirical studies of spontaneous multimodal HCI.

[6] *e.g.* "Put the piano here." together with a pointing gesture (indicating a location on the screen).

The evolution of the subjects' use of synergic multimodality over the experiment supports this interpretation, inasmuch as five subjects out of eight resorted increasingly to multimodal commands in the course of the experiment. The percentage of multimodal commands over the total number of commands per session increased from 37% (first session) to 47% (last session) despite marked inter-individual differences. Two subjects used oral commands extensively during the three sessions, although one of them resorted to multimodality increasingly; two others used speech and gestures alternately (according to the type of action they wanted to perform) throughout the experiment; as to the remaining four subjects, their use of multimodality increased steadily from the first session to the last. A close analysis of recorded gestures and utterances suggests that inter-individual differences in the use of modalities result mainly from differences in the subjects' mental representations of their interaction with the system, namely: manipulation of graphical representations of the application (as in direct manipulation), or cooperative manipulation of these representations (*i.e.* cooperation with the system), or communication with the system (as with a human interlocutor).

These empirical findings suggest that, contrary to current assumptions, synergic multimodality will not be used spontaneously by all users in all HCI contexts. In the first stages of interaction, multimodality will be used mostly for expressing information which cannot be formulated adequately using exclusively either speech or gestures; experienced users will resort to multimodality more often, whenever its use contributes to optimize the interaction (*i.e.* results in shorter and more precise formulations).

4 Conclusion

We have presented an empirical study of multimodal HCI which gives an insight into user preferences and behaviors regarding the use of speech and gestures in HCI environments. Interface designers should be aware that the synergic use of spontaneous speech and gestures in such contexts seems to increase the user's cognitive workload. This form of synergic multimodality may then be considered as an efficient and usable substitute for direct manipulation, thanks respectively to the simultaneity between speech and gestures and the absence of expression constraints; however, one may question whether speech associated with gestures is a more natural/transparent interaction mode than direct manipulation.

5 References

Amalberti, R., Carbonell, N., & Falzon, P. (1993). User representations of computer systems in human-computer speech interaction. *International Journal of Man-Machine Studies*, *38*, 547-566.

Binot, J.-L., Debille, L., Sedlock, D., Vandecapelle, B., Chappel, H., & Wilson, M. (1992). Multimodal integration in MMI2: anaphora resolution and mode selection. In H. Luczal, A. Çakir, & G. Çakir, (Eds.): *Proc. Work with Display Units (WWDU'92)*, pp. 335-339. Amsterdam: North Holland.

Coutaz, J., & Caelen, J. (1991). *A Taxonomy for Multimedia and Multimodal User Interfaces. Proc. First ERCIM Workshop on Multimodal Human Computer Interaction.* (Lisbon, November), Lisbon: INESC.

Coutaz, J., Nigay, L., Salber, D., Blanford, A., May, J., & Young, R. (1995). Four Easy Pieces for Assessing the Usability of Multimodal Interaction. In I. Katz, R. Mack, & L. Marks (Eds.): *Proc. CHI'95,* pp. 115-120. New York: Association for Computing Machinery Press & Addison Wesley Publishing Company.

Guyomard, M., Le Meur, D., Poignonnec, S., & Siroux, J. (1995). *Experimental Work for the Dual Usage of Voice and Touch Screen for a Cartographic Application. Proc. ESCA Tutorial and Research Workshop on Spoken Dialogue Systems* (VigsØ, Danemark, June). Grenoble, France: ESCA.

Harris, G., Begg, I., & Upfold, D. (1980). On the role of the speaker's expectations in interpersonal communication. *Journal of Verbal Learning and Verbal Behavior, 19*, 597-607.

Hauptmann, H.G., & McAvinney, P. (1993). Gestures with speech for graphic manipulation. *International Journal of Man-Machine Studies, 38*, 231-249.

Koons, D.B., Sparrell, C.J., & Thorisson, K.R. (1993). Integrating simultaneous input from speech, gaze, and hand gestures. In M. Maybury (Ed.): *Intelligent Multimedia Interfaces*, pp. 257-276. London: MIT Press.

Neal, J.G., & Shapiro, S.C. (1991). Intelligent Multi-Media Interface Technology. In J.W. Sullivan, & S. Tyler (Eds.): *Intelligent User Interfaces*, pp. 11-43. New York: Association for Computing Machinery Press.

Nigay, L., & Coutaz, J. (1993). A Design Space for Multimodal Systems: Concurrent Processing and Data Fusion. In S. Ashlund, K. Mullet, A. Henderson, E. Hollnagel, & T. White (Eds.): *Proc. INTERCHI'93*, pp. 172-178. New York: Association for Computing Machinery Press, Addison Wesley Publishing Company.

Oviatt, S., DeAngeli, A., & Kuhn, K. (1997). Integration and Synchronisation of Input Modes during Multimodal Human-Computer Interaction. In S. Pemberton (Ed.), *CHI'97 Conference Proceedings*, pp. 415-422. New York: Association for Computing Machinery Press & Addison Wesley Publishing Company.

Robbe, S., Carbonell, N., & Dauchy, P. (1997). Constrained vs spontaneous speech and gestures for interacting with computers: a comparative empirical study. *Proc. INTERACT'97*, pp. 445-452. London: Chapman and Hall.

A Multimedia System to collect, manage and search in narrative productions

Gabriella Spinelli
School of Cognitive and Computing Sciences,
University of Sussex, Falmer, Brighton U.K.

This research shows how new information and communication technology can support the community's knowledge base, providing the community members with non-intrusive tools and suitable procedures for saving, improving and transmitting cultural identity. The prototype of the system I'm going to discuss is a multimedia archive where oral memory plays the main role. The term oral memory stands for a set of knowledge which belongs to the community, has been augmented by time, and is essentially the users' history.
The oral memory has been elicited and transmitted through face-to-face interactions for three centuries. The only external signs of this implicit knowledge are the physical objects collected by the community. They don't have any meaning if not linked to that knowledge even if they are the unique guides through which people out of the community or new members of it can be connected to the socio-cultural heritage.

1 The Context

Sienna is a town in Tuscany (Italy) divided into seventeen neighborhoods (contrade7. The Nicchio, one of them, consists of 3000 people and it was the context of this research. As a matter of fact social events are the main communication channel the contrada's members use, chiefly because the direct interaction among the people o the contrada allows an effective forum for the exchange of the information and preserves and improves the community's ethnic identity. The majority of the cultural heritage is stored in the form of oral tales, and mainly by the oldest Nicchio's members.
The potential value of the system prototype consists in two functionalities it is able to support within this environment:
- information management, allowing collection, storage and search activities through the cultural and historical heritage the community members have amassed and wish to honour over time;

- usage of external representations, objects collected in the museum, and archive in order to elicit the implicit knowledge.

2 The theoretical framework

The need to study the role of external representations and cognitive artefacts from both social and cognitive perspectives has driven the choose of theoretical approach supporting the research. The adopted framework is compounded of Activity Theory, (AT) and Distributed Cognition (Dcog) [4]. According to the socio-cultural perspective of AT we should pay more attention to entities such as the community to which the acting person belongs, as well as the roles and the job distribution of the community [3].

Taking into account a larger system even compounded of the entities above, we might have a better comprehension of the interactions between the community members and be able to support them more efficiently. The relevance of this perspective to this research emerges if we analyse the problem the community seems to have. The knowledge transmission faces an obstacle in the adopted communication channel and the different roles members play within the community. The creation of an external archive which allows the elicitation of people's memory aids the dynamic of the interactions between the community members: they are supported by an alternative communication channel which improves the flexibility of the interactions and the knowledge learning for those members are first entering the community.

An archive so designed is, further, a support for the increase of the collected knowledge: from a social perspective infact the archive is not just the channel through which people can receive this knowledge, but also the place where new knowledge can be created through the interactions of the members. As far as it is possible to predict the archive might store the actual knowledge and support the creation of additional knowledge due to the interactions' traces left inside the system and accessible to the whole community.

From the computational point of view of Dcog, the unity to be analysed is the socio-technical system compounded of people performing internal and external cognitive processes, and artefacts that have the feature of representing information and treating data [1,2].

Concerning the cognitive artefacts [5] they are able to support human activities in different ways:

-changing the nature of the task that has to be performed;

-distributing the activity between time;

-distributing the activity between space;

-distributing the activity between people.

The multimedia archive's prototype can fit the description of a cognitive artefact above.

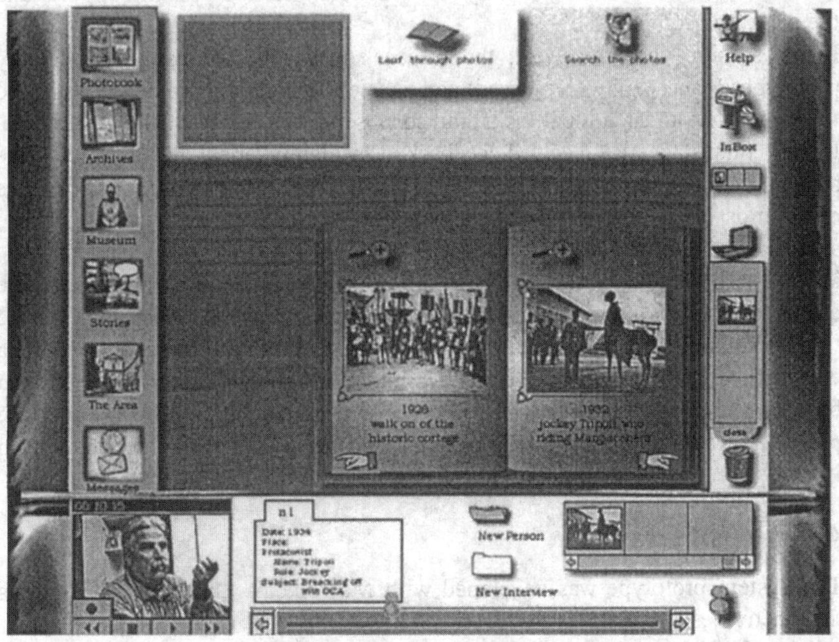

Fig 1: The Interface of the Archivist

The contrada's memory is deeply linked to the documents, objects and places belonging to and stored by the contrada in the museum, library and in the social sites. The point is that none of the collected material carries any meaning without the information handed down in oral tradition which exists alongside it. The links between oral tales and objects are a net where the knowledge is externalised and the access to it is available.

3 The design methodology

The applied methodology was User-Centred Design, (UCD), a process designed to develop artefacts that achieve usability, effectiveness, efficiency, learnability and satisfaction during user interaction [6]. The process consists of the repetition of many design cycles: prototyping-evaluation-implementation, until a suitable artefact is in place. The UCD stresses the importance of usability evaluation even if the system is under construction and the test results would be only informal evaluations.

During this research the prototype was tested with the three users groups involved (the archivists, the people who tells the stories and the people interested in browsing the materials). Observations and contextual analysis have

been carried out in order to identify the functionalities the system supports and the parts it is composed of.
In the session with the users, the feedback was collected using low-fidelity prototypes (paper- and video-based) and later on, a software prototype was used to support the most critical tasks and to observe the nature of the interaction. During this stage the received feedback was mainly about the design of the system architecture, the usage and the understanding of the icons, and the choice of the most suitable input devices concerning the different environments of the system.
When the high fidelity software-based prototype was designed the users were called to test it. Quantitative usability evaluation was not feasible due to the limited amount of data entered into the database. However, the amount of data was sufficient to support a realistic simulation of the task performed and browsing in-depth within it. The focus of the user evaluation with the software prototype was mainly the interface design and the potential interactions within it.

4 The Interface

The system prototype was developed with Macromedia Director and supports three activities:
-elicitation of the oral knowledge from the contrada's oldest members;
-knowledge recording and the management of the links between it and the historical objects belonging to the contrada;
-access to the stored information and the search in the improved relations' net.
The system prototype was developed respecting the features of cognitive artefacts Dcog has identified.
Within the archivists' interface there are tools able to decrease the user's cognitive overload anticipating and postponing some tasks of the activity. Doing this they could be performed earlier and later or even by different users; for example, the preparation of the relevant material for an interview, the filling of the references for each bookmark belonging to the interview.
Concerning the cognitive artefact's feature of distributing the activity in space, the users are supported in the virtual navigation of the real environments by three different modalities: navigation by topic, direct navigation, and exploration.
In order to change the nature of those tasks that seem to be difficult to be performed currently, for example searching within oral records without indications of the content of collected interviews, the system offers a different representation of them. The whole interview is displayed as a strip where the bookmarks stand for the beginning of the meaningful units the users might use to start their search from.

Moreover, an oral archive where such innovative tasks like oral elicitation and searching in the narratives are performed should allow the exchange and increment of community knowledge. People who search within the archives become, from the point of view of the Dcog approach, an additional representation of the collected information and, at the same time, a further and enriched knowledge representation due to the additional value individual experience might give.

In the prototype there are six different environments reciprocally interwoven: the photobook; the archives; the museum; the stories; the area; the messages.

5 References

[1] Cole M., Engeström Y. (1993) A cultural-historical approach to distributed cognition. In Salomon G. (Ed) *Distributed Cognitions: psychological and educational considerations*, Cambridge University Press.

[2] Hutchins E. (1994) *Cognition in the Wild*, Cambridge, MA: MIT press.

[3] Kuutti K. (1996) Activity Theory as Potential Framework for Human-Computer Interaction Research. In Nardi B.A. (Ed), *Context and Consciousness, Activity theory and Human-Computer Interaction*, Cambridge and London, MIT press.

[4] Nardi B.A. (1996) Studying context: a comparison of Activity Theory, Situated Action Models, and Distributed Cognition. In Nardi B.A. (Ed), *Context and Consciousness, Activity theory and Human-Computer Interaction*, Cambridge and London, MIT press.

[5] Norman D.A. (1991) Cognitive Artifacts. In Carrol J. (Ed), *Designing Interaction: Psychology at the human-computer interface*, New York: Addison Wesley, Publishing Company, Inc.

[6] Preece J. (1993) *A Guide to Usability*, New York: Addison Wesley, Publishing Company, Inc.

Multimedia: The Effect of Picture, Voice & Text for the Learning of Concepts and Principles

S. Guttormsen Schär, J. Kaiser, H. Krueger
Institute of Hygiene and Applied Physiology
ETH Zentrum 8092 Zürich
guttormsen@iha.bepr.ethz.ch

Introduction

Multimedia is a powerful information tool, particularly for complex information. It has become a key concept in contexts such as Web design and computer-aided learning (CAL). The potentials coupled with *multimedia* is however no guaranty of quality whenever the concept is used for marketing. The need for guidelines for multimedia design is obvious in many CAL products, also Internet is a live example of the experimentation going on in this field. This paper addresses the employment of multimedia in these contexts, i.e. how to represent and present complex information for the purpose of learning. There is no doubt about the great potential of multimedia in this field. It relates to the broad variety of possibilities in expressing different learning contents by employing new information technology. The big threat to this potential is the amount of knowledge in different fields that is necessary in order to achieve a good product. Beside expert knowledge in the topic to be presented, also knowledge in computer science, psychology, pedagogic, didactic, instructional design, etc. is necessary.

Background

Traditionally *multimedia* has referred to the employment of various technical tools and equipment for the storage of information, which should be presented in combination. In this paper multimedia refers to the employment of the concept which has been common with the possibility of producing and presenting information in various ways with a normal computer. According to Kerres (1993) there are four categories of multimedia features: text, voice, picture and moving pictures (e.g. video, animation, simulation). In a scientific context this is a rather elementary definition because the features are redundant as a source of cognitive information. In order to get a better understanding of their impact on learning the cognitive aspects of the multimedia features must be addressed in more detail. A cognitive framework for the investigation of multimedia is given by classifying the features as verbal vs. non-verbal, visual vs. auditory, or stable vs. variable (Paechter, 1996).

The verbal vs. non-verbal criterion represents a differentiation between language based (voice and text) and visual (picture, video) based representations. Verbal and non-verbal representation forms are different in many ways (Paechter, 1996): Verbal representations can impart basic, abstract meaning about logical relations. Non-verbal representations can impart specific information about an object, and the relation of this to other objects. Verbal representation can impart information about special attributes, and can have a referring function. On the other hand non-verbal representations can present on one glance information which only with much effort could be explained by words. It is expected that a combination of verbal and non-verbal representations can increase the learning effect (Anderson, 1996).

The visual vs. auditory criterion applies to the sense modalities involved in the learning process. Text, picture and video activate the visual system, and voice and sound the auditory system. This classification enables the design of CAL systems with a balanced load on the sense modalities, i.e. that both the visual and the auditory system are stimulated simultaneously or that both systems are used sequentially. A positive effect of this is reduced load on the senses (Paechter, 1996; Engelkamp & Zimmer, 1990). The combination of visual and auditory information can have a structuring function. The combination of picture and voice enables the direction of the attention to the essential information in the picture by the voice.

The stable vs. variable criterion refers to whether or not a student can perceive the information with a self-chosen speed. A representation is stable when it enables a student to decide how long the information shall be presented (text and picture). A representation is variable when this is not possible (voice, video). A stable representation is important for students with little previous knowledge in a subject, i.e. they can adapt the perceptual speed to the level of difficulty (Weidemann, 1993). Experts in a field can chose to read certain passages in a text only, and to skip those, which are trivial. In this way they profit from the stability of the material by increasing the tempo of perception (Kozma, 1991). Unstable representation forms can be a benefit when the presentation time has a structuring effect (Paechter, 1996). Variable representations can causes cognitive overload for students with little experience in a field. The negative effects of this can be reduced by letting the students control when and how often the unstable information units can be replayed.

Effective learning depends on the optimal combination of learning content and multimedia feature. Hence, good classification criteria of learning contents are needed. The Component Display Theory (CDT) suggests four different categories of learning contents: concepts, principles, processes and facts (Merril, 1983). Even more complex teaching sections can be sub-divided into those four categories. The benefit of the CDT is the coherence and thoroughness which is not present by other similar theories e.g. the elaboration theory (Reigeluth & Stein, 1983). According to the CDT, *concepts* are groups of objects, events or symbols, which within the group have the equal function or belong to the same category. E.g. Telephone (handy, telephone-automate, table telephone etc.). *Principles* refer to cause / effect relationships or correlation between two or more concepts, synonymous to rule. A principle explains the co-function of several concepts. One example is the principle of the population growth of the earth: i.e. the relationship between mortality and birth rate causes rapid population growth, which results in starvation and ecological problems. *Processes* are sequential actions by which a goal can be reached. A sequence of actions can result in branching of new possible actions. The number of possible actions and decision alternatives defines the complexity of a process. *Facts* are the definition of a special object, an event or a symbol. Facts do not define general aspects, but describe an example from a concept, a procedure or a principle.

On a cognitive level a coupling between multimedia and learning content is feasible. Hence, a connection between stable multimedia features (text, picture) and static learning contents (concepts and principles) seems natural, as between variable multimedia features (video, sound) with process oriented learning contents. A coupling between learning content and multimedia features classified as visual/auditory or verbal/non-verbal is not immediate obvious, but these classifications provide a useful foundation for the interpretation of the results.

An experiment was designed to investigate how to support the learning of principles by presenting combinations of text, voice and picture. Text and picture support the coupling of a static learning content with a stable presentation. Voice, which is a variable feature, is of particular interest in this context. The combination of voice and picture represents the visual auditory distinction, and picture vs. voice and text represents the verbal / non-verbal distinction. Both the verbal/non-verbal and the visual/auditory criteria advocate a combination of the modalities. The experiment can therefore illuminate the relative importance of these classifications for the learning of static information (principles). The following hypotheses were tested:

1. The combination of text and voice has a negative learning effect.
2. The combination of picture and voice has a positive learning effect.
3. The combination of text, voice and picture has a negative learning effect.

Experiment

42 fourth grade primary school children (mean age 10 years) participated in the experiment. A 2 x 7 mixed design was applied. The between group factor represented the presentation of a different set of tasks between two randomly selected groups. This was applied as a control for the effect of the tasks, and to control for task knowledge spreading between the subjects. The subjects were tested sequentially, randomly selecting subjects from either group. The within group factor represented a combination of voice, text and picture including all singular, all double and one triple combinations. Both task order and presentation categories were randomised. Hence, all subjects were given 7 tasks in a randomised order, and also the coupling between task and presentation category was randomised between the subjects. The presentation time of each task was task specific and computer controlled. After the presentation of each task, all subjects were given three questions. The questions reflected visual, verbal and inference knowledge about the principles. The visual question was only asked when the presentation category included a picture. After solving all the 7 tasks, the subjects were asked to indicate their preference for the presentation categories. A log-file registered total time, test time per task and resting time between the tasks.

Results and Discussion

There were no effects of group, the effects of presentation categories were therefore not influenced by the fact that the groups learned different principles. Consequently, the effects of the presentation categories were analysed with the complete sample. A brief overview over the result is given below (all below .05 level). Hypothesis 1 was clearly supported by the subjective preference measure. The stable/variable classification predicts that voice can disturb individual perception speed together with text. On a subjective level the subjects also perceived this. The objective measures were not clear in showing a disadvantage of the text & voice combinations. Hypothesis 2 was supported for visual and inference knowledge. Both question categories resulted in an overall preference for picture and voice (verbal / non-verbal combinations) over other combinations. This supports that voice can have a structuring effect in combination with picture. On the other hand, this shows an inconsistency in the effect of features classified as stable / variable, as the effect vas opposite for the text and voice combination. The verbal knowledge question did not produce coherent effects on this relation. Hypothesis 3 reflects an expected negative effect of information overload. This was supported on a subjective level and for visual knowledge. The subjective preferences showed that mono presentations were preferred. This shows that information overflow is to be taken seriously. Internet demonstrates the opposite trend, the production algorithm there seems to be that "more is better".

The results demonstrate that the application of cognitive descriptions (stable vs. variable and visual vs. auditory) of multimedia effects do not produce coherent results. These cognitive descriptions may be too general, and the effects of each multimedia feature must be evaluated according to the context in which it is presented. The inconsistency between visual and auditory effects may be caused by that the sense modalities interact with the structure of the representation forms. The auditory representation and reception of voice is basically different from visual representations.

This study shows that the combination of multimedia features is relevant. Further studies should continue to investigate the cognitive impact of multimedia with the aim of practicable guidelines for optimal combinations of learning content and multimedia features. This experiment should also be repeated with adult subjects, as their perception of text may be different from that of children.

References

Anderson, J. R., (1996) *Kognitive Psychologie: eine Einführung. Heidelberg*: Spektrum Akademischer Verlag GmbH.

Engelkamp, J. & Zimmer, H.D. (1990). Unterschiede in der Representation und Verarbeitung von Wissen in Abhängigkeit von Kanal, Reizmodalität und Aufgabenstellung, in: Böhmer-Dürr, K., Emig, j., & Seel, N. (Hrsg.). *Wissensveränderung durch Medien.* München: Saur.

Kerres, M. (1993). Software-Engineering für multimediale Teachware, in: Seidel, C. (Hrsg.), *Computer based training*, Göttingen: Hogrefe, p87.102.

Kozma, R. G. (1991). Learning with Media. *Review of Educational Research*, 61, pp 179-211.

Paecter, M. (1996). *Auditive und visuelle Texte in Lernsoftware* Münster, u.a.: Waxmann Verlag

Merrill, M. D. (1983). Component Display Theory, in: Reigeluth, C.M. (Ed), *Instructiona-design Theories and Models*. Hillsdale: Lawrence-Erlbaum.

Reigeluth, C.M. & Stein, F.S. (1983). The Elaboration Theory of Instruction, in: Reigeluth, C.M. (Ed.), *Instructional theories and models. An overview of their current status.* Hillsdale: Lawrence Erlbaum, p. 335-381.

Weidemann, B. (1993), Instruktionsmedien, in *Arbeiten zur Empirischen Pädagogik und Pädagogischen Psychologie*, Gelbe Reihe, Nr 27, München: Universität der Bunderswehr.

Design of Multimodal Feedback Mechanisms for Interactive 3D Object Manipulation

Roland Steffan, Torsten Kuhlen*, and Frank Broicher

Department of Technical Computer Science
*Computing Center
Aachen University of Technology
Ahornstr. 55, D-52074 Aachen, Germany
Phone: +49-241-8026105 Fax: +49-241-8888308
[steffan, broicher]@techinfo.rwth-aachen.de
*kuhlen@rz.rwth-aachen.de

1 Introduction

During the last years, Virtual Reality (VR) has proven its potential for the visualisation and manipulation of complex data like 3D geometries generated by means of CAD applications. Though virtual environments can already profitably be integrated into engineers' daily work (Gomes de Sa and Baacke 1998), there are still a lot of disadvantages in existing simulation systems, requiring trained engineers for performing interactive simulations. Especially, the lack of satisfactory feedback mechanisms in existing VR-based simulation environments complicates execution of interactive manipulations of virtual objects, as, e.g., necessary within VR-based assembly simulations. In detail, suitable modelling of user-object interactions, realistic simulation of object behaviour and intuitive presentation of information are missing.

Hence our research focuses on "physically-based modelling", "artificial support mechanisms" and "artificial sensory feedback" for interaction with virtual objects, which solve the described problems and increase user acceptance of and performance in virtual environments. A system called Virtual Workplace is used as evaluation environment, which integrates multi-plane non-immersive user-centred 3D visualisation, force-feedback technology and 3D sonification equipment (Steffan, Kuhlen, and Loock 1998).

To perform a realistic and efficient simulation, object behaviour and information presentation in virtual environments need not necessarily be an exact imitation of real environments. Instead, the simulation software of the Virtual Workplace employs mechanisms, which are abstractions of events and actions occurring in real environments and support users while performing special tasks, e.g., by using a rule-based description of interaction processes. This is indispensable because exact simulation would require a tremendous computational effort and could not be executed in real-time.

2 Physically-based modelling

The behaviour of objects within virtual environments in accordance to physical properties and laws is called physically-based modelling (Buck 1998). The Virtual Workplace software provides elementary properties of real environments, especially gravity, elasticity and friction. Modelling of gravity allows to feel the weight of virtual objects and causes automatic object movements, e.g., falling. Additionally simulating elasticity of objects and friction during object-object interactions, users get a more realistic impression when interacting with the virtual environment.

3 Artificial support mechanisms

Object handling can be difficult in goal-directed movements (Schull, Kortwinkel, Waffenschmidt, and Steffan 1998), especially when collisions occur due to confined spatial constellations, like in peg-in-a-hole situations. Although the user is supported during these movements by collision-based force-feedback, a more sophisticated rule-based mechanism called "sensitive polygons" has been invented. During the import of 3D graphical models into the environment, the user can add single or groups of sensitive polygons to special parts of objects, e.g., holes or slots. These polygons perform different object manipulation tasks if being activated by collisions with or approaches of other, manually moved objects. Sensitive polygons can adjust or position objects automatically, e.g., to simplify the screwing of a nut onto a bolt.

Besides, goal-directed movements are supported by so called "virtual guide sleeves", which for example can be positioned around drillings. If a user tries to insert a bolt into the drilling, she is guided towards the base of the virtual guide sleeve by means of multimodal feedback mechanisms. Non-interactive, artificial object movements are caused by simulation of virtual fields of force. If the virtual force fields are activated, an algorithm determines which objects are in a certain, adjustable distance to each other, moves the smaller object towards and attaches it to the larger object, so that object's surfaces are adjacent.

4 Multimodal Information Presentation

The simulation software processes and encodes data generated within the virtual environment in a way that they are intuitively comprehensible for users. For this reason, visual, acoustic, tactile and proprioceptive senses are addressed. The procedure in principle is depicted in fig. 1.

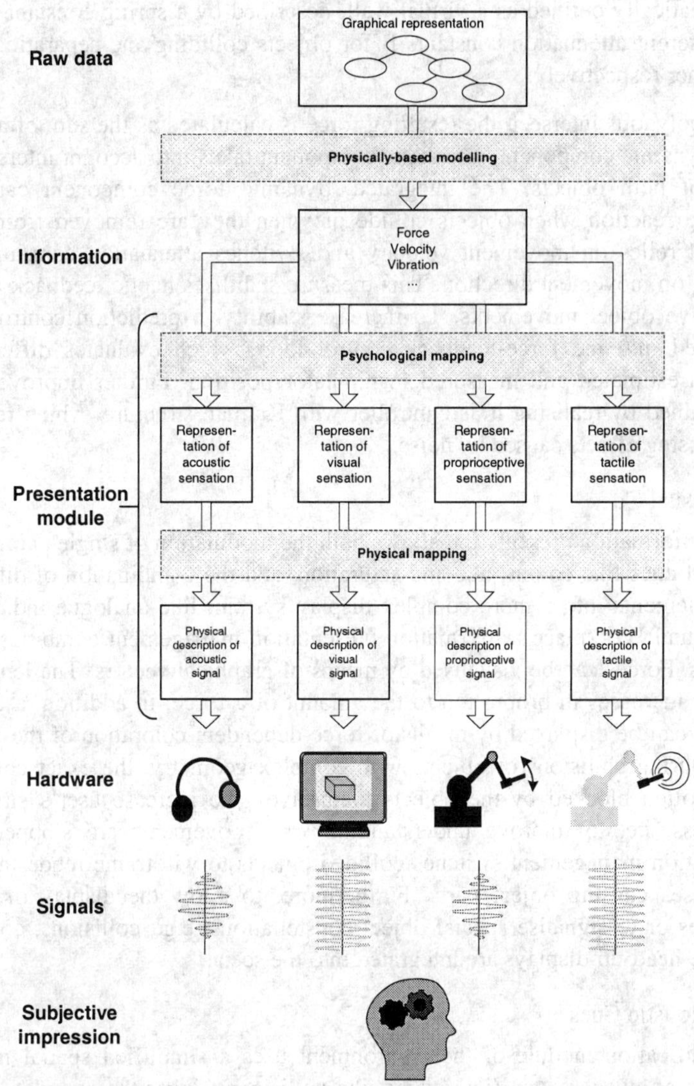

Figure 1: Procedure for multimodal information presentation.

4.1 Haptic Cues

The calculation of forces and tactile sensations and the presentation by means of adequate output devices is known as "haptic rendering" (Gillespie and Cutkosky 1996). The haptic rendering integrated into the Virtual Workplace is based on the detection of intersections of graphical objects' polygons. Every polygon is mathematically defined as a virtual wall, described by a spring constant K and two different attenuation constants B for objects colliding and separating from each other respectively.

If two polygons intersect, the resulting force is calculated as the sum of a static and a dynamic component. The static component takes into account intersection depth of both objects. The integrated, dynamic force component causes a damping reaction when objects collide or when they are removed from each other. It relies on movement velocity and switches attenuation constants depending on movement direction. This measure stabilises haptic feedback during interactive object movements. To increase stability, a prediction controller is integrated into the force-feedback control loop, which evaluates differences between estimated and measured manipulator position. Further improvements are obtained by realising it as controller with Kalman structure, which reduces destabilising effects caused by noise.

4.2 Visual Cues

Visual information presentation allows both the modulation of single parameters of visual cues like colour, hue and saturation, and the combination of different visual elements into a more complex display system, like analogue indicators. The Virtual Workplace's information presentation management combines these methods. Forces can be visualised by means of graphical vectors. The length of a vector increases in proportion to the amount of a force; in addition, a force's strength can be displayed by means of force-dependent coloration of the vector itself. During collisions of objects with complex geometry, the exact collision area is often blocked by the objects themselves. To increase user's situation awareness, i.e. to improve understanding why movements are stopped, the presentation management switches collided objects to wireframe rendering and emphasises causing object parts. Furthermore, to allow the display of exact quantities or to signalise special object constellations, e.g., collisions, context-sensitive head-up-displays are integrated into the scene.

4.3 Acoustic Cues

The sonification module of the environment uses a simplified spatial model, based on an event-driven (i.e. events like collisions trigger the sounds), direct sound approach and neglecting phase and runtime relationships. Positions and orientations of user and objects can be determined using measurements of the tracking system and data of the graphical environment, respectively. These information are used as parameters for the sonification process. As a result, lateral locating of sound sources in combination with forward-backward and near-far distinction is possible.

Object interactions like collisions and friction are associated with predefined sounds, which are modulated depending on interaction parameters, e.g., strength of a collision.

For redundant coding of force information, but also for force-feedback substitution, a force amount can be displayed by modulating the pitch of a synthetic sound signal. Besides, strength of a collision is encoded within the sharpness, i.e. a special kind of a listener's subjective impression of a sound. For this reason, a sinusoidal audio signal is mixed with a rectangular signal in the time domain. The application of force-dependent scaling functions avoids a rise of the sound's volume when raising the sharpness.

5 Future Work

At the moment, evaluation of the system is carried out using a testbed consisting of manipulation tasks with varying complexity. As a result, a judging of the implemented methods and a guideline for designing information feedback mechanisms for interactive object manipulations will be achieved.

6 References

Buck, M. (1998). Immersive User Interaction Within Industrial Virtual Environments. In Dai, F. (Ed.): *Virtual Reality for Industrial Applications*, Springer Verlag Berlin (Germany).

Gillespie, R. B. & M. R. Cutkosky (1996). Stable User-Specific Haptic Rendering of the Virtual Wall. In: *Proc. of the 1996 International Mechanical Engineering Congress and Exhibition*, pp. 397-406, Atlanta (USA).

Gomes de Sa, A. & P. Baacke (1998). Experiences with Virtual Reality Techniques in the Prototyping Process at BMW. In Dai, F. (Ed.): *Virtual Reality for Industrial Applications*, Springer Verlag Berlin (Germany).

Schull, U., W. Kortwinkel, E. Waffenschmidt, & R. Steffan (1998). Development of a Process for the Simulation of the Assembly of a Digital Mockup in a Virtual Environment. In Baake, U. (Eds.): *Proc. of the 5th European Concurrent Engineering Conference*, April 26 - 29, Erlangen-Nuremberg, pp. 70-74, SCS Publication, Ghent (Belgium).

Steffan, R., T. Kuhlen & A. Loock (1998). A Virtual Workplace including a Multimodal Human Computer Interface for Interactive Assembly Planning. In Kopacek, P. (Eds.): *Proc. of the IEEE Int. Conference on Intelligent Engineering Systems INES '98*, September 17-19, Vienna (Austria), pp. 145-150, IEEE Press.

A Distributed System for Device Diagnostics Utilizing Augmented Reality, 3D Audio, and Speech Recognition

Venkataraman Sundareswaran, Reinhold Behringer, Steven Chen, and Kenneth Wang
Rockwell Science Center, Thousand Oaks, CA 91360, USA

1 Motivation

Human Computer Interaction can be greatly enhanced if information from the computer can be projected directly in the field of view of a user as he goes about performing his tasks in the real world. This is the goal of Augmented Reality (AR): to display computer-generated information to appear embedded within the view of the real world.

Consider this future application scenario: a maintenance worker, wearing a light-weight see-through head-mounted display (HMD) approaches a complex machine; a CAD model of the machine appears in the HMD visually superimposed on the machine. As the maintenance worker moves around the machine, the pose and orientation of the display is dynamically modified so that the model appears to cling to the real machine. Using speech commands the maintenance worker asks for the machine error and diagnostics to be displayed. The error and diagnostics are displayed at 3D locations seemingly tethered to the problem spot, and 3D audio played over the headphones directs the user's attention to an adjacent machine that is part of the problem. At Rockwell Science Center (RSC), we have developed a system which is a conceptual prototype for HCI-rich AR applications. The novelty of our conceptual prototype is the integration of a new video-based AR technique, speech recognition, and 3D audio in a networked PC environment. The system demonstration is geared to address the needs in maintenance and training, to provide information not only *just in time*, but also *just in place*.

2 System Components

2.1 Head Tracking: Visual Servoing with Fiducial Makers

Head tracking using video-based methods is passive, and can achieve high accuracy in the alignment of graphical information display with the view of the real world. Head tracking in our system is based on *Visual Servoing* (Espiau et al. 1992). Visual servoing is to control a system -typically, a robot end-effector with a camera – based on processing visual information. In this approach, the error between measured image positions of visual features and their predicted image positions is minimized with respect to the camera motion parameters. The camera is moved in closed loop using the calculated motion parameters to reduce the error measure. For the head tracking application, the images from a video camera (mounted on the head) are processed to measure feature positions, and the "virtual" camera (which renders the graphics) is controlled. We use concentric ring fiducial markers as features. These markers were chosen for easy detection, in clutter and under a wide range of viewing angles. Each marker has a unique ring structure for identification.

We believe that our closed-loop, vision-based head tracking method (Sundareswaran and Behringer, 1998) solves the visual alignment problem directly, and is robust due to the use of specially constructed markers.

2.2 Visualization

The system speed and capabilities of the rendering engine set the limits for the complexity of information that can be displayed. In our prototype, we display wireframe and Gouraud-shaded 3D objects and colored text. The graphical display is overlaid on to the live video stream, using the head tracking technique described in Section 2.1. The result is a form of "X-ray" vision that enables the user to visualize internal components of the viewed object. We use simple animations (flashing) to indicate selection when the user queries about a specific component. Textual annotations (see Section 2.3) can be displayed for the selected component.

2.3 Speech Recognition Server

Rockwell Science Center's Automatic Speech Recognition (ASR) Server software provides an easy way to rapidly prototype speech-enabled applications regardless of the computing platform(s) on which they execute. The ASR server provides both automatic speech recognition and text-to-speech synthesis (TTS) capabilities. The ASR capability is obtained through abstraction of a commercially available off-the-shelf speech recognition technology, IBM ViaVoice. The TTS functionality provided with the ViaVoice engine is likewise abstracted and exposed to client applications. The ASR server's architecture provides for the future addition of other vendors' speech recognition technologies as needed. A client application connects to the ASR server over an IP network using TCP sockets. Although the ASR server runs on a Windows 95/Intel Architecture PC, the client applications may run on any operating system that supports TCP/IP networking.

In the AR system, we use speech to control the system (e.g., "Start the demo"), to query the state of the system (e.g., "What is the frame rate?"), to query the location of components (e.g., "Where is the power supply?"), and to dictate "virtual notes" that can be attached to the components and retrieved at a later time. The TTS is used to provide informative audio responses and alerts to the user (e.g., "Please repeat command").

2.4 3D Audio Auralization

A three-dimensional (3D) audio system provides an auditory experience in which sounds appear to emanate from locations in 3D space. Our particular interest is in the application of HRTF (Head-Related Transfer Function)-based 3D audio to provide cues about objects outside the field of view. Our 3D audio server is built upon an API for a chip from Aureal that implements Microsoft DirectSound3D functionality and hardware HRTF processing. The server is based on TCP/IP sockets. The sound source signals are stored as wave files (.wav) on the server. A client application can connect to the server, designate the sound source signals to be played, and stream position/orientation information for the listener and objects to which the sounds are attached in 3D. The server operates at 30 frames per second (for position and orientation updates) and can simultaneously play up to three 44.1 kHz spatialized sound sources.

3 System Integration

All the system components of the previous section have been implemented on PCs running Windows 95 or NT. A schematic of the system is shown in Figure 1 (also see http://hci.rsc.rockwell.com). The servers (ASR, 3D audio) are hosted on a Windows 95 platform with a 300 MHz Intel Pentium II processor, and the AR application (visualization and image processing) is implemented on a Windows NT machine with a 200 MHz Intel Pentium processor. The PC is equipped with an Imaging Technology color frame-grabber that digitizes video signal from a Cohu 2200 CCD camera. The image processing is carried out on the CPU, to detect and identify fiducial markers (Section 2.1). The AR visualization is implemented as a World Tool Kit (WTK) application. All user interactions are carried out in the main action loop of the WTK application.

The AR technology is demonstrated by superimposing the WTK graphics on top of incoming video signal (see Fig. 2). The display can be easily modified for a see-through system. Using a speech command the user can ask the system to highlight a specific component. If the component is within view, the selected component is flashed on the display. If the component is outside the field of view, a 3D audio icon is played and moved towards the location of the component (from a canonical location in front of the user) to direct the user's attention. Using dictation recognition, the user can compose a virtual note and attach it to a component. Previously attached virtual notes may be viewed or deleted using speech commands.

4 Discussion

We have demonstrated a distributed Augmented Reality (AR) system whose display is based on video processing, and with speech and 3D audio interaction. The powerful AR technology allows information to be displayed where it is most relevant: visually superimposed on the real world. By displaying information in 3D in the user's natural environment, HCI is enhanced to a new level in which the user's interaction with the computer is mostly transparent. The distributed architecture of the system ultimately allows for scalability and mobility (user equipped with only a light wearable computer and display). As an example, a lightweight system with AR technology can replace service manuals for maintenance applications by providing on-line information just in time and just in place.

There are several challenges to make this technology serve the vision described above and in the Introduction. First, video processing is computation-intensive, and requires fast, dedicated processors. However, with increasing demand for multimedia information, we believe that this need will be heavily addressed in the near future. Second, reliable tracking is crucial to this technology. To increase the reliability of tracking, we are currently engaged in research to supplement the video tracking with other tracking technologies (inertial and magnetic). Finally, high bandwidth wireless communication will hasten the adoption of this technology.

5 References

Espiau, B., Chaumette, F., and Rives, P. (1992). A new approach to visual servoing in robotics. *IEEE Trans. On Robotics and Automation*, 8 (3), 313-326.

Sundareswaran, V. and Behringer, R. (1998). Visual Servoing-based Augmented Reality. In *Procs. Intl. Workshop on Augmented Reality*, San Francisco, USA, Nov 1, 1998.

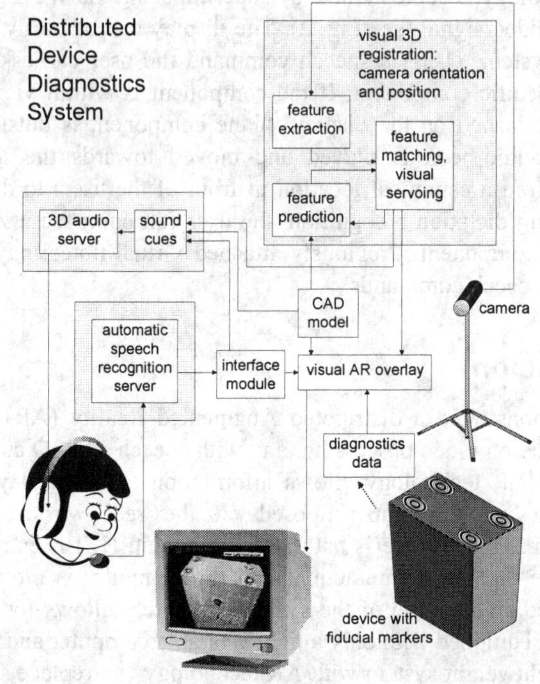

Figure 1. Schematic of the Distributed AR system

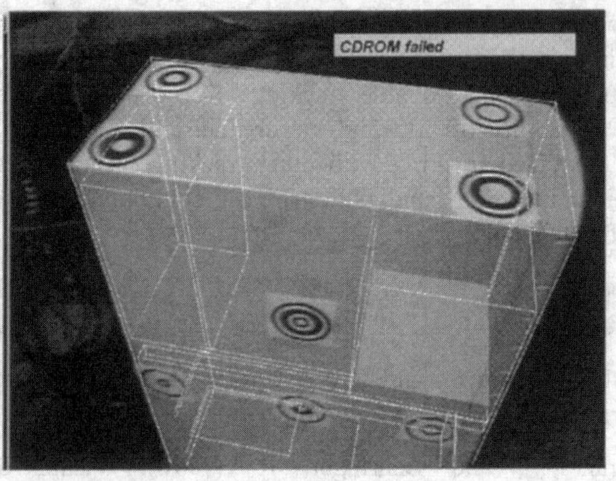

Figure 2. Sample frame with overlaid graphics

On the Visual Quality of Still Images and of Low-motion Talking Head Digital Videos

Peter J. Haubner
Institute of Applied Informatics and Formal Methods (AIFB)
University of Karlsruhe, Germany

1 Introduction

In order to achieve high fidelity performance when processing and displaying multimedia information consisting of text, still images, animations, video and audio, high data rates are needed in the order of some Mega-Bytes per second. This causes considerable transmission problems for the fastest computer devices and for today`s networks. Therefore, the data rates have to be reduced, which, however, also reduces the perceived quality of the displayed information and leads to a loss in user acceptance.

Reduction of data rates can be achieved by compression techniques or by scaling the multimedia file to a smaller window size, lower frame rate, lower resolution, lower colour depth or even by combinations of all of these measures. In an empirical study on tele-conferencing systems these possibilities of reducing data rates were evaluated in terms of perceived Visual Quality (VQ) and user acceptance. The goal of our study was to systematically investigate the influence of data rate reduction on perceived visual quality VQ and to gain statistically significant data and practical experiences from which guidelines for the ergonomic layout of „Talking Heads" and for other equivalent situations could be derived.

2 Empirical Study

2.1 Experimental Setup and Design

The experiments were carried out at a multimedia workplace in our Human-Computer Interaction Laboratory (HCI-Lab), which enables to generate, to process, to measure, to present and to transmit multimedia stimuli as applied in our investigation.

Workplace and environment had been designed to comply with relevant issues of parts 3, 6, 7 and 8 of ISO 9241 „Ergonomic requirements for office work with display terminals (VDTs)“ (International Standards Organisation 1998).

Strategies reducing data rates and hence also reducing memory requirements as well as enhancing transmission flow through networks are the following:

- compressing the video / audio files
- reducing image dimensions (window size)
- reducing screen resolution
- reducing colour depth
- reducing frame rate

For compression MJPEG was used; appropriate adjustments were found out in pretests, compression quality was set 75 % and kept constant all over the study. Still images were captured as single frames. Audio consisted of text spoken into a microphone and was captured mono with 11 kHz, 16 bit.

Three presentation modes were applied in the course of the experiments (still image, video, video & speech). Presentation mode, image dimensions, resolution, color depth and frame rate were treated as the independent variables of the experimental design.

As pointed out in the literature, visual quality is indirectly influenced also by the degree of desynchronization (asynchronity) of the video and audio tracks (Ozer 1995). It can be supposed that the acceptable degree of asynchronity depends on the content of the video and on the type of sound, too. The most critical case seems to be that of talking heads in tele-conferences for tele-teaching, tele-learning or for tele-services (e.g. tele-consulting).

Therefore, desynchronization was taken into account as a further independent variable. The only dependent variable was perceived „Visual Quality VQ“. VQ was operationalized by means of some threshold of perception, the „Borderline between Comfort and Discomfort“, a VQ-criterion well established in illuminating engineering research and practice in the context of discomfort glare assessment.

Furtheron, VQ was jugded on a 7-point rating scale reaching from very low satisfaction of the users to very high satisfaction. Data were treated according to common psychometric scaling procedures including tests of the internal consistency of data with regard to the scaling model. (see e.g. Torgersen 1958).

In principal, the layout of the experimental design was such in each experimental series as to enable a non-parametric two-way analysis of variance for related samples (a Fiedman-test in the k-sample case and a Wilcoxon-test in the two-sample case) (Siegel 1956).

The experimental design is contained in table 1:

Table1: Experimental design - treatment and operationalization

Stimulus: „Talking Head", viewed from a distance of 500 mm

Experimental design / fixed factors (independent variables):

Presentation mode:	Still image, video, video+speech (spoken text)
Window size:	38 x 50; 64 x 48; 75 x 100; 96 x 72; 160 x 120, 150 x 200 mm^2
Resolution:	4 pixels / mm^2 (640 x 480); 10 pixels / mm^2 (1024 x 768)
Color depth:	8 bit; 8 bit (image palette);16 bit; 24 bit colour; 8 bit greyscale
Frame rate:	5; 15; 30 frames/s
Asynchronity (AS):	20; 40; 80; 120; 240 ms (audio ahead; video ahead, respectively)

Operationalization of perceived Visual Quality VQ

- **BCD-threshold AS:** **B**orderline between **C**omfort and **D**iscomfort
- **Acceptance A:** Degree of satisfaction of the users (7-point rating scale)

By statistical inference, the results described in the next chapter as well as some derived layout guidelines were obtained.

2.2 Experimental Results and Design Guidelines

The „Null-Hypothesis" of every test was rejected at a level of significance ε equal to or smaller than 0.05. All fixed factors, i.e. the stimulus variables, except color against grey-scale, were of significant influence on the users` subjective judgements. This is true for the acceptance rating (user satisfaction) as well as in the case of BCD-assessment (borderline between comfort and discomfort) (see Table 2). Using the same number of steps in the colour scale and in the grey scale, visual quality VQ mainly depends on the number of pixels per area unit and on the luminance contrast for a given viewing distance, a given window size and a given frame rate, colour turns out to be of less importance for visual performance. Nevertheless, it is well known, however, that colour is preferred emotionally. Derived guidelines are shown in table 3.

Table 2: Summary of the main statistical results (testing of significance)

Variable	Statistical test	Level of significance ε / result	
Window size	Friedman	< 0.05	significant
Resolution	Wilcoxon	< 0.05	significant
Colour / Grey-scale	Wilcoxon	> 0.05	non-significant
Colour depth	Friedman	< 0.05	significant
Frame rate	Friedman	< 0.05	significant
Asynchronity	Friedman	< 0.05	significant

Table 3: Ergonomic design recommendations derived from the experimental results

Variable	Guidelines / Recommendations
Window size	minimum 50 mm x 50 mm; 60 mm x 40 mm or 40mm x 60 mm with a preferred aspect ratio (width to height) of 3:4 or 4:3 (upright or oblong format, respectively; optimum / maximum depend on sreen size and tasks
Resolution	minimum 4 pixels / mm^2 ; recommended > 10 pixels / mm^2
Colour / Grey-scale	minimum 8-bit image palette; minimum 10 bit system (windows palette), otherwise better use greyscale; recommended colour depth for still images and videos 16 bit or higher
Frame rate	Number of frames per time in the video stream: minimum 10 frames / s for minimum resolution, minimum colour depth and minimum image size, otherwise better freeze the video; better capture and replay at 15 frames / s for image sizes up to 320 x 240 pixels, recommended are 30 frames / s for a capture size of 320 x 240 pixels and more
Asynchronity	Time difference between video and audio: AS** < 100ms (90 percent of the users satisfied)

3 Concluding Remarks and Forecast

The present study aimed at low motion talking head videos, the effect of dynamics in moderate motion and in high motion scenes remains open; this, of course, could also be the subject of further research.

Some of the guidelines, elaborated in this experimental study, are of absolute nature, inasmuch as they relate to characteristics of the sensory system and do not depend on features of the test equipment given by current technology; others, however, are relative inasmuch as they are constrained to the specific experimental conditions of our study; e.g., optimum size depends on the screen size. Thus, the influence of smaller and larger screens than that used in our study should be investigated, too. Most of the minimum requirements are absolute in the above mentioned sense. Compared to it, all optimum and maximum requirements are relative recommendations.

Technology develops in the field of multimedia rapidly; therefore, some of the guide-lines have to be matched with these changing conditions, too.

The focus in our study was on perceptibility issues, mainly. If ergonomically well designed human-computer interfaces are aspired , the whole information process from strimulus to response should be considered including perception, cognition, action planning and activity regulation as well as sensu-motoric control.

That means, guidelines are needed for information content design, for task-specific multimedia information coding, for organization of information on the screen (formatting) and for navigation in the multimedia hyperspace, i.e., lexical, syntactic, semantic as well as pragmatic aspects of information exchange have to be taken into consideration.

4 References

International Standards Organisation ISO (1998). *Ergonomic requirements for office work with visual display terminals (VDTs), parts 1-17.* International standards, European standards, National standards.

Ozer, J. (1995). *Video Compression for Multimedia.* London: Accademic Press.

Siegel, S. (1956). *Nonparametric Statistics for the Behavioral Sciences.* Tokyo: McGraw Hill Book Company.

Torgerson, W. S. (1958). *Theory and Methods of Scaling.* New York: John Wiley & Sons.

Spatio-Temporal Perspectives: A new way for cognitive enhancement

Andreas Goppold
c/o FAW: goppold@faw.uni-ulm.de

Spatio-Temporal Perspective is about gaining vantage points over time and space. We are all aware of the Y2K problem, caused by the impending turn of the millennia according to the count of the Christian calendar. Due to shortsighted programming, computer systems world-wide are threatened with malfunction in the year 2000. Programmers world wide scramble to patch the sometimes 30-odd year old programs. This may not help as expected, as one forgets an old wisdom from the tailor trade: If you mend a garment too much, it will come apart in exactly those places that didn't need mending. The problem prompts us for a more profound consideration: That it may be time to overhaul our calendar system altogether. As the world is drawing together into a global civilization, the standard calendar system of Christian reckoning smacks of the parochial euro-colonialism of a by-gone era.

From a global, culture-neutral spatio-temporal perspective, the Jewish, Hindu, and Buddhist calendars offer some striking advantages over the Christian one, as they allow us to relate the present media-technological revolution to the evolution of human symbolic intelligence in a novel way. The Jewish and Hindu, as well as the Maya calendars position the zero-count at some time before 3000 BCE. By the Hindu reckoning, the beginning of the present era, the Kali Yuga (KY), was at Feb. 18, 3102 BCE, such that we are now in the year 5101 KY. The Maya calendar starts at almost exactly the same time (3114 BCE), and the Jewish calendar some time earlier (3761 BCE). The Buddhist calendar is around the 2500 mark now, which is about half the KY reckoning. A profound aspect of this numbers game turns up when compared to media technology: Our present computer technology is about 50 years old now, printing technology about 500 years, alphabetic writing has 2500 years, and writing, as it was developed in Sumer, arose quite exactly around 3100 BCE (Amiet 1966, Cohen 1958). Moreover, we can date the era of graphical symbolization around 50,000 years (Semiotica 1994, Anthro). There is nothing magic or otherwise special about the number five, except perhaps, that we have five fingers on one hand to start counting with, and that the Jewish jubilee is 50 years, which is about two human generations. This was an exercise in gaining perspective. A perspective is, as everyone will agree, entirely observer-dependent. It is a vantage point that lets the observer arrange things so that s/he can overlook them most conveniently, nothing more, nothing less.

The above projection roughly demarcates cultural eras and their dominant media (and) technology, a way of looking at history pioneered by Innis and McLuhan (Chandler). Human socio-symbolic-technological development had seen an innovation step-up in synchronicity with these eras. As to the innovation speed of computer technology, we are now in the third generation, micros, after mainframes, and minis, with the fourth already in the waiting (Landauer 1995, Norman 1993, 1998, Memex). Development of computer systems has so far been driven by technology- and marketing concerns, and not by human-potential issues, as critics point out (Landauer, Norman, Businessweek, Common, Engelbart, Karn). We can even go one step further and postulate that computer technology is constrained by an unexamined, outdated Kuhnian paradigm (Kuhn 1962): So far, it has mostly been used to mechanize the symbolics that humans had been using in the last 5000 years (Bolter 1991, Krämer 1988, Landow 1992, 1994).

Spatial perspective arose from the survival related developments of human evolution, our origin in the biosphere (Anthro, Calvin, Skoyles). As Calvin points out, the superior human facilities of spatial orientation and action were essential for the evolution of intelligence. The ability to throw objects with precision at moving targets (meaning the immense amount of neuronal computation necessary for orientation, self-stabilization, target-tracking, and trajectory projection), has been decisive in shaping the neuronal infrastructure that made us human. Using the human body as a ballistic propulsion subsystem is a neuronal computation achievement of much higher requirement than the linear force-translation of, say, a bow-and-arrow system, or simpler still, a gun. It means force-coupling a ballistical mechanical device to a human body - the arsenal of paleolithic weapons: boomerangs, propulsors, atl-atls, bolas, and slingshots (Bellier 1990). These were the greatest feats of neuronal interface technology of the last one million years. In present-day applications, we are not surprised to find the most advanced neuronal interface cybernetics in high-grade weapons systems: aimed to perform essentially the same purposes as a million years ago, but with "a bigger bang for the buck". Still time is essential: he who shoots first, and most accurately, is going to win. In military parlance, this is called the OODA-Loop: Observation, Orientation, Decision, Action (Stein 1998).

In the more refined *higher-order human symbolic activities*, this basic neuronal computation infrastructure found its appropriate re-use. The Renaissance usage of perspective in art has re-introduced and re-formulated these neuronal cybernetics as a general symbolic ordering principle. Kim Veltman has coined the metaphor of *Conceptual Navigation* for overlooking and zooming into the vast knowledge spaces of our cultural heritage with multimedia systems (Veltman). A wider meaning of *perspective* involves the utilization of our rich neuronal potential of spatial metaphors (Benking) embedded in our bodies for traversing "The Global *Semiosphere*" (Hoffmeyer 1997). While phylogenetic evolution has fitted us with an optimal equipment for dealing with spatiality, the issue of temporality is dominated by cultural evolution. The human temporal horizon of personal memory and experience is limited by our lifetime: about 75 years. All cultural evolution of the last 500,000 or so years was a product of the accumulation and re-organization of the cultural materials of the *Semiosphere*. Thus, the ability to overlook and traverse the depths of the cultural memory of humanity constitutes the *Spatio-Temporal Perspective*.

The crucial factor of *Spatio-Temporal Perspective* will be called *Neuronal Resonance*. The depth-time structure of culture, our symbolic "deep space", was treated for a long time with a purely anthropocentric approach, and only now are we coming to appreciate its neuronal infrastructure details (Brock, Skoyles). Symbols are the most recent, but certainly not the last evolutionary step in the long transmission of behavioral complexes among organisms, which started right with the first bacteria 4-5 billion years ago, and evolved in parallel with genetic evolution (Biosem, Bloom). In higher animals, all perception and behavior is mapped onto neuronal excitation fields in their brains, and all communicative and manipulative acts lead therefore to neuronal resonance fields. An ergonomically optimized tool or instrument will translate into an optimized neuronal resonance for its user. Music is the art system for producing and appreciating *neuronal resonances*. Invisibly hidden beneath the familiar complaints about computers (Landauer), and generally motor-driven machinery is their basic incompatibility with human neuronal resonance rhythms. Tognazzini (1993) shows us examples in HCI, where he details the working methods of stage magicians as "manipulations of time" (p. 359). Of course, it is not the flow of newtonian time that is manipulated, but the working of the human nervous system, whose rhythmics generates our perception of (subjective) time. For "new frontiers of cognitive enhancement", the factor of *neuronal resonance* will be essential, but this area has so far seen very little research (Halang 1992, Innis, McLuhan). There is some work around "Flow" (Csikszentmihalyi 1990, Karn 1997: 64). This denotes hard-to-define intellect-augmentation effects (Engelbart) that can occur, when expert work is able to proceed in uninterrupted sequences of cumulative efficiency. Time factor is critical, since it is interrelated with the human attention span and capacity of the short term memory. *Neuronal resonance* effects are enhanced when (parts of) the human body enter a dance-like rhythmics. Therefore, a short time lag of about 1/10 sec seems essential. Noticeable augmentation effects are attained mainly when a high level of user training and expertise is started with at the beginning, which severly limits the systematic application and testing (Karn 1997).

For *practical* HCI *applications of neuronal resonance*, we can cite systems designed for the former generation of mini computers: APL and MUMPS. These were renowned as the most powerful programming systems ever created by man. In terms of the OODA loop metaphor, they yielded maximum power for observation, orientation, decision, and action on the base of careful fine-tuning of the software to the rather minimal technology that was available then: Winchester hard disk, 80*25 CRT alphanumeric display, and 32-64 K RAM Processor. Since APL and MUMPS were virtual machine codes disguised as programming languages (and hand-crafted in native assembler), they offered extremely powerful command facilities with a few quick key-strokes (which the programmer, of course, had to memorize). The HCI "secret" of these systems was the *neuronal resonance circuit* thus created, of tight-fitting incremental loops of *code-viewing*, *understanding*, *modifying*, *testing*, and *evaluating*, the DP equivalence of the OODA loop. The later generation of mouse-driven HCI (WIMP) has sacrificed speed of interaction in favor of mass market access, leaving the power users out in the cold. This is financially understandeable, but it poses an insidious cul-de-sac for human symbolic evolution that could be possible with multi media symbol systems. Some avenues for further development are probed in (Goppold).

References

Due to space constraints, this is an abbreviated list. The full version is to be found under: http://www.uni-ulm.de/uni/intgruppen/memosys/hci.htm

Amiet, P. (1966). Les elamites inventaient l'ecriture. *Archeologia* 12, 16-23.

Anthro. http://history.evansville.net/prehist.html
http://www.massey.ac.nz/~ALock/hbook/frontis.htm
http://www.geocities.com/Athens/Acropolis/5579/TA2.html

Bellier, C., Chattelain. (1990). *La chasse dans la prehistoire*. Treignes: Ed. du Cedarc.

Benking. http://www.ceptualinstitute.com/genre/benking/homepageHB1.htm
http://newciv.org/cob/members/benking/

Biosem. http://www.gypsymoth.ento.vt.edu/~sharov/biosem/welcome.html

Bloom, H.: *History of the Global Brain*
http://www.heise.de/tp/deutsch/special/glob/default.html

Bolter, J. (1991). *Writing Space*. Hillsdale: Erlbaum.

Brock, B. *Neuronale Ästhetik*
http://www.uni-wuppertal.de/FB5-Hofaue/Brock/Projekte/NeuroAe1.html

Businessweek. http://www.businessweek.com/1998/42/b3600052.htm

Calvin, W. H. *The throwing madonna, The cerebral code.* http://www.WilliamCalvin.com/

Cohen, M.(1958). *La grande invention de l'écriture et son évolution*. Paris: Imprimerie nationale.

Chandler, D. Media Theory Web Site.
http://www.aber.ac.uk/~dgc/influ05.html
http://www.aber.ac.uk/~dgc/about.html

Common (1993). Common Elements in Today's Graphical User Interfaces: *INTERCHI '93, ACM*, p. 470-473.

Csikszentmihalyi, M. (1990). *Flow*. New York: Harper Perennial.

Engelbart, D. http://www.bootstrap.org/biblio.htm

Goppold, A. http://www.uni-ulm.de/uni/intgruppen/memosys/symbol.htm
HCI99: http://www.uni-ulm.de/uni/intgruppen/memosys/hci.htm

Halang, W. (1992). Zum unterentwickelten Zeitbegriff der Informatik. *Physik und Informatik*. Berlin: Springer, 30-40.

Hoffmeyer, J. (1997). The Global Semiosphere. In: Rauch, I., Carr (eds.): *Semiotics Around the World*. Berlin: Mouton, pp. 933-936.

Karn, K. S.; Perry, T. J.; Krolczyk, M. J. (1997). Testing for Power Usability. *SIGCHI Bulletin*, 29 (4), Oct , p. 63-67.

Krämer, S. (1988). *Symbolische Maschinen.* Wiss. Buchges. Darmstadt.

Kuhn, T. (1962). *The Structure of Scientific Revolutions.* Chicago: U of Chicago Pr.

Landauer, T. (1995). *The trouble with computers*. Cambridge: MIT Press.

Landow, G. (1992). *Hypertext.* Baltimore: Johns Hopkins.

Landow, G. (ed) (1994). *Hyper / Text / Theory*. Baltimore: Johns Hopkins.

Memex. http://www.cs.brown.edu/memex/bibliography.html

Norman, D. A. (1993). *Things that make us smart*. Reading: Addison-Wesley.

Norman, D. A. (1998). *The invisible computer*. Cambridge: MIT Press.

Semiotica (1994). Special issue on paleosemiotics. Vol. 100-2/4

Skoyles. http://www.users.globalnet.co.uk/~skoyles/index.htm
http://www.users.globalnet.co.uk/~skoyles/od0.htm

Stein, G. (1998). Talk at Ars Electronica Infowar Symposion. http://www.aec.at/infowar/

Tognazzini, B. (1993). Principles, Techniques, and Ethics of Stage Magic. *INTERCHI '93*, pp. 355-362. New York: ACM

Veltman, Kim. http://www.sumscorp.com/articles/

A Study on How Depth Perception Is Affected by Different Presentations of 3D Objects

Hanqiu Sun, Kwong Wai Chen, Pheng Ann Heng
Dept. of Computer Science & Engineering,
The Chinese University of Hong Kong

1 Introduction

To achieve high-level realism of virtual reality (VR) system, the virtual environment should reproduce sufficient depth cues the same as human's perception in the real world. These include stereopsis, perspective, occlusion, shadows, texture and the like. Conventional graphics workstations can realize some of the depth cues such as occlusion, perspective, lighting and shade, atmospheric effect (e.g. fog). More sophisticated VR system can also render binocular disparity, motion parallax, interposition, and convergence. A virtual reality application, however, should provide high refresh rate so the user can interact with the application in real time. To achieve high frame rate, the application should be tuned and optimized for real-time performance. It is usually better to reproduce more depth cues, but it can take the computer with limited resource more time to render and thus degrade the performance. For instance, stereoscopic vision requires two display channels to be rendered in a frame interval. Since humans have only limited vision, more details may be at times unnecessary. It is important to found out what depth cues are really required and what can be ignored if a high frame rate is to be achieved.

In this project, we investigated how the depth perception is affected by different representations of 3D objects in static virtual scenes. We studied depth cues in human's perception, human stereo viewing, and stereo-graphics VR system (hardware & software components). Based on the study, a set of experimental trials was designed and conducted to evaluate the factors that may influence a subject's ability to estimate the depth of objects in a static scene. The experiments also measured how advanced rendering is really needed when presenting 3D objects in static virtual scenes.

2 Related Work

Many authors in the literature have discussed general issues on depth perception. There are, however, few specific studies in the computer graphics community. These include the study for the need of rendering quality and the use of shadows. The study (Lars and Martin 1999) shows that the placement and distance of objects are much more important factors on depth estimates than the type of objects or the method of rendering. An experiment in which subjects were asked to rotate a pipe presented with different rendering qualities is described in (Barfield, Sandford and Foley 1998) and (Foley 1987). The results show that colored and highlighted presentations give a 20% better performance than simple wireframe presentations. But, a more sophisticated shaded rendering did not give any further improvement. Another study (Bulthoff 1991) performed a set of experiments regarding shape and depth perception, which found that Phong shading instead of Lambertian shading did not change perceived depth significantly.

3 Depth-Cues Perception

The primary goal of this project is to measure and evaluate the correctness of depth perception affected by different presentations of 3D objects in static VR scenes. 20 Computer Science students (most aged 18 to 22) were involved in the experimental trails. The experiments were carried out on a Silicon Graphics® Onyx™ workstation with a color display. The rendered, wireframe and textured images were created using OpenGL™. The stereopsis was implemented using Crystal Eyes® system provided by Stereographics® on stereo-supported SGI workstations.

Two types of objects, a cube and a teapot, are presented, and each of them is represented in five different configurations. The subjects were asked to estimate depth in the resultant 20 images (10 monoscopic and 10 stereoscopic) presented in a particular order during 30 minutes testing. Subjects were given an instruction sheet on how to carry out the experiments.

The scene were kept as similar as possible for all the images, with the identical rendering method, lighting model, and textures used. The illumination was chosen as a combination of ambient and two directional lights to make the scene appear pleasant. Perspective projection was used, and a floor was added to make the scene slightly complex.

There were three identical objects presented at different depths in each image. Counting the depth between the centers of the two closest objects as one unit, the subjects were asked to estimate the depth ratio (2,3,4,5, or 6) between the two farthest objects. Only the depth value in the Z dimension was evaluated. After the ten monoscopic images were first presented, the environment was changed to the stereoscopic ones in turn.

In addition to the depth ratios, the subjects were asked about their views of which image they felt being the easiest presentation to make depth estimation and whether stereoscopic image helped in judging depths.

4 Experimental Analysis

The test data were measured and analyzed based on the mean percentage errors. We have analyzed the estimating errors related to the factors of shape (teapot and cube) and placement (two horizontal and two vertical sets), shadowing, rendering methods (fully shaded, wire-frame, or texture mapped), and stereoscopic viewing.

4.1 Shape and Placement

The significant shape of the object depends on the placement of the object. The mean estimating errors of horizontally aligned sets are 7 and 8 percents while the vertically aligned sets are 19 and 31 percents. This would be due to the fact that the vertical configuration of objects appeared to be floating in the air and it was very difficult to relate them with a floor.

4.2 Shadowing

The experiments showed that shadowing was significant in judging the depth. The mean estimating error of objects with shadowing is only 3 and 4 percents. This outcome indicates that subjects have the difficulty measuring the objects freely suspended in space, rather than in reference to a familiar object. Shadowing is a useful information in judging depth because it can link the freely suspended objects to a ceiling or a floor.

4.3 Rendering

Rendering of the objects was not significant in depth estimation. The solid, wire-frame, and texture mapped scenes carry error means from 7 to 10 percents. These results showed that the rendering methods did not give the subjects further information to judge the depth more effectively.

4.4 Stereoscopic

The experiments showed no significant improvement in stereoscopic environment. This can be due to two reasons: first, subjects focused on a wrong position and the lack of blurring effect. The focus point of all the images presented in the experiments are at the center of the screen. However, during the trials subjects may focus on a wrong position, which may weaken the depth effect produced by the stereoscopic environment. Second, the depth-of-field effects that caused the blurring of the objects outside the focal field are not simulated in this experiment.

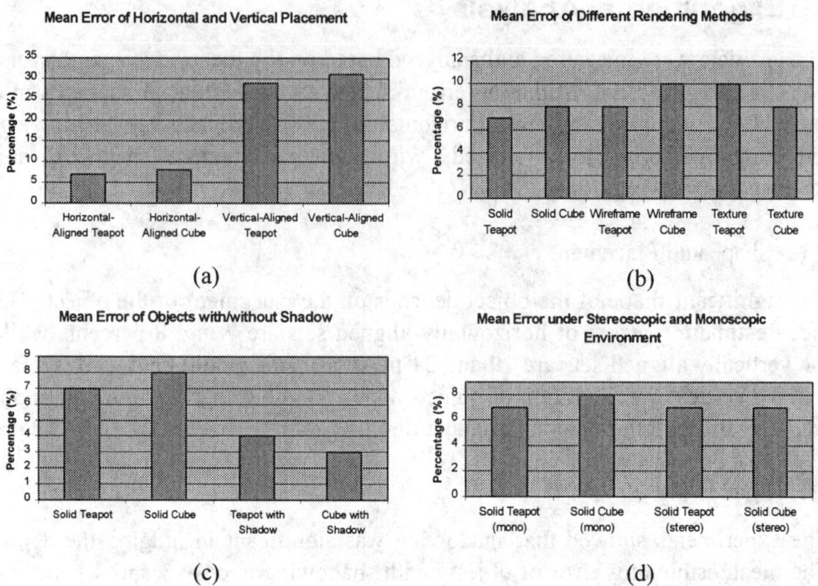

Figure 1. The Mean Errors of Depth Estimation under Different Configurations

In responding to the questions "which are the easiest and the most difficult configurations in estimating depth", the most subjects rated the shadowed objects as the easiest presentation (12 out of 20) and vertical aligned objects as the hardest one (13 out of 20).

5 Conclusion

In this paper, we investigated how the depth perception is affected by different representations of 3D objects in static virtual scenes. The experimental data are analyzed and evaluated based on the mean errors in depth estimation of each configuration, whose statistics results are shown in Figure 1. The presentation cases in the monoscopic display are shown in Figure 2. Our experiments showed that the placement and shadowing are much more important factors on depth estimation than the type of objects or the method of rendering. The stereoscopic display only assists in estimating the depth a small scale of significance. From the survey of subjects participated in the experiments, shadowed presentation is rated the easiest configuration in depth perception, and vertical-alignment presentation is the hardest one among other alternatives.

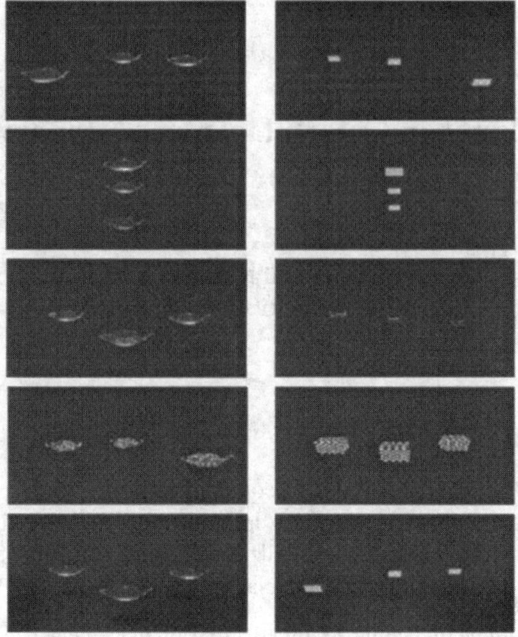

Figure 2. The screen layout of the experiments

6 References

W. Barfield, J. Sandford, and J. Foley, The mental rotation and perceived realism of computer-generated three dimensional images. *Int. J. on Man-Machine Studies*, 29, p. 669-684 (1988).

H.H. Bulthoff, Shape from X: Psychophysics and Computation. In *Computational Models of Visual Processing*, M. S. Landy and J. A. Moushon (Eds.), MIT Press, Cambridge, MA (1991).

J. Foley, Interfaces for advanced computing. *Scientific American*, 4, p. 82-90 (1987).

Lars Kjelldahl and Martin Prime, A study on how depth perception is affected by different presentation methods of 3D objects on a 2D display, *Computers and Graphics*, 19(2), p. 199-, March - Apr 1995.

R. Parslow, Spatial concepts in 3D. *In Fundamental Algorithms for Computer Graphics*, R. A. Earnshaw (Ed.), Springer. Berlin, p. 883-893 (1985).

Grigore Burdea, Philippe Coiffet, *Virtual Reality Technology*. Wiley, Interscience publication, 1994.

John lovine, *Step Into Virtual Reality*, Windcrest@ / McGraw-Hill, 1995.

Color Balance in Color Placement Support Systems for Visualization

Eric Cooper and Katsuari Kamei
Computer Science, Ritsumeikan University

Introduction

Nearly all machine-to-human communication is visual. The ability of the human designer has the greatest influence on the effectiveness of this communication. In scientific visualization, color placement performs an essential role in differentiating and associating visual data (Tufte 1990).

The number of factors involved in color placement can cause the task to be overwhelming (Silverstein 1987). Many programmers have attempted to dispose of the problem by forcing users to use the same color scheme in every situation, by limiting the user's palette, or by providing default colors.

In visualization, these solutions have had very limited success. Color schemes that work well for one particular visualization may be completely unsuitable for another. Users who encounter problems with color placement find that they either have to accept the colors provided or quickly become experts in color placement. Applications desperately need color placement support for visualizations.

The Color Balance Problem

Color balance means the relative conspicuity of each part of a design. Although color balance is clearly influenced by other design factors, such as shape, form, etc., we have chosen to ignore these for the time being and attempt to work with only the color placement aspects of design.

We propose the following problem for color balance support. On an interactive color placement system, allow the user to place colors on flatly colored, two-dimensional visualizations. Allow the user access to real-time advice on color balance. Finally, evaluate the system to prove that the system is able to improve the usefulness of the designs in this manner.

In order to explore some of the issues in constructing a color support system, we constructed a prototype using experimental data. The study gives a direct overview of the problems color support systems will encounter at all stages.

Color Balance Inference

Color support must be based on quantification of experimental data, not on any one designer's preferences. In order to quantify the sensation of color balance (relative conspicuity), we conducted color balance surveys. The central principle of color balance is that color conspicuity is related to color area (Ashizawa and Ikeda 1981). Figures with large areas are said to be "heavier," more conspicuous, than smaller figures of the same color.

In our experiment, the screen of a personal computer showed two rectangular figures on one ground as shown in Fig. 1.

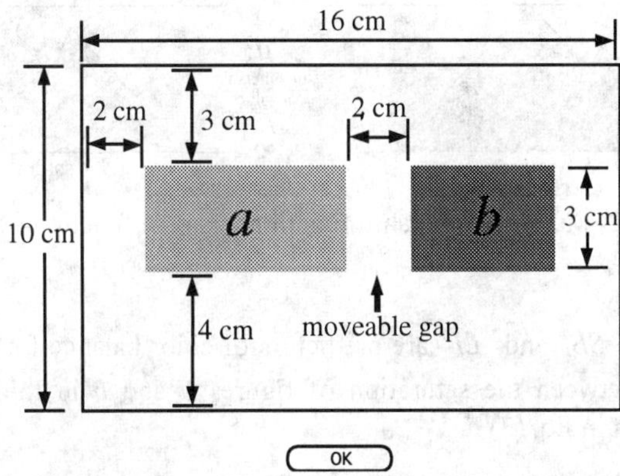

Fig. 1. Arrangement of the color balance experiments.

The subject of the experiment used the mouse to choose the figure that “stood out” more. When clicked, that figure became slightly smaller and the other became slightly larger. Repeating this, the subject could find the ratio of nine possible ratios that he/she found most balanced.

In the inference system we used for quantification, functions are used to infer color balance responses. These functions are drawn from color theory and are composed of elements of design and fuzzy subset memberships. The functions we used for implementation are of the general form:

$$f_i^j = \left[\left(\mu_i^1 + \mu_i^2 ... \mu_i^m\right) \cdot B_i^1\right] + \left[\left(\mu_i^{m+1} + \mu_i^{m+2} ... \mu_i^n\right) \cdot B_i^2\right] \tag{1}$$

Here, each function f_i^j, where j is the function number, is used to predict response of the color balance experiment set i. Each B represents an element of color design, such as the difference in lightness or saturation, etc. μ_i^x is the membership value of set i in fuzzy subset x.

As an example, we give f_i^2, function number 2 in the implemented set of rules:

$$f_i^2 = \left[\left(\mu_i^{S\_high} + \mu_i^{S\_medium}\right) \cdot SB_i^1\right] + \left[\mu_i^{S\_low} \cdot LB_i\right] \tag{2}$$

Since *S_medium* and *S_high* are consecutive and converse fuzzy subsets, we use the addition operation to form a union of these sets, as shown in Fig. 2.

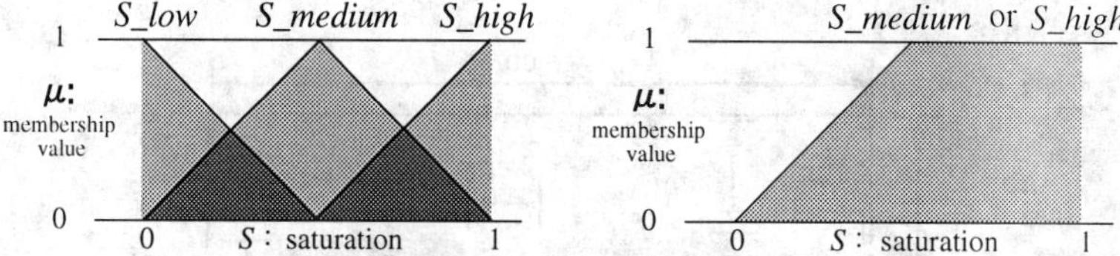

Fig. 2. Fuzzy membership functions used in function f_i^2.

In f_i^2, SB_i^2 and LB_i are predictors of color balance for set i. SB_i^2 is the difference between the saturation of figures a and b in the color balance experiment given in Eq. (3).

$$SB_i^1 = S_i^b - S_i^a \tag{3}$$

All functions use HLS color system values. LB_i is the difference in the lightness of figures a and b as compared to the ground g, as shown in Eq. (4).

$$LB_i = \left|L_i^b - L_i^g\right| - \left|L_i^a - L_i^g\right| \tag{4}$$

This method of constructing rules allows us to construct new functions in a linguistic manner. f_i^2 could be stated, "To the extent that the overall saturation is high or medium, use the saturation balance. To the extent that the saturation is low, use lightness balance."

Each function is used to estimate the balance of a set of figures using its least-squares regression line. The prediction for the set of rules is calculated from the weighted average, with the weight of each function determined by its correlation coefficient with the experimental data.

By splitting the 7993 responses in half, we were able to evaluate the validity of this color balance inference. For any given color set, the system is able to predict the mean response to a correlation coefficient of 0.97 with just 4 rules. For any one individual's response to the color balance experiment, this system can achieve a correlation coefficient of 0.67. This shows us that not only is color balance a commonly perceived notion but that it can be predicted reasonably well. Applying this knowledge to visualizations is the next and most difficult challenge.

Color Balance Support System

Using the inference described, we constructed and tested a prototype color balance support system that works in real-time. A color support system was implemented to generate color placement warnings for the user. The system calculates the difference between the predicted area ratio and the actual ratio of each color plane. If one color plane's average predicted area is much smaller than it's actual area in pixels, then the system generates a warning that the color plane in question is too conspicuous. If the predicted area is much larger than expected, a warning that the color is too inconspicuous is generated.

As a foil for balance, the prototype contains a visibility algorithm, which basically calculates the distance between each color. The system generates warnings for when two colors are indistinguishable and for when any color plane has a low visibility.

A palette interface was constructed to give the user freedom to set the colors of flatly colored visualizations. The color of a button icon showed the color support warnings by red-yellow-green signal. By pressing the button, the user could read the warnings in the form of canned sentences.

We chose a variety of graphs, charts, and maps that were about the same size to test the prototype. Ten subjects used our color placement system to design the ten visualizations, giving us 100 designs in all. Each user (and each visualization) was allowed access to the support system five times.

Ten judges rated the ten sets of ten visuals each, making 100 answers for each visualization. Each visualization was scored from 0 to 9 in order, as compared with other color schemes for the same design. Subjects were instructed to rate visualizations by their ease of use.

The prototype's warnings were not shown to have any meaningful effect on the rating for visualizations. However, the balance and visibility values generated for each visualization were shown to be good predictors of ease-of-use rating. This study also confirms that visibility is the most important factor in ease-of-use and that the overwhelming factor in visibility and balance is lightness contrast. (Cooper and Kamei 1998)

Conclusions

The goal of this study was to examine some problems and issues of color balance support system development. We found that color balance is quantifiable and plays a role in ease-of-use but that canned sentences are too simple and vague for meaningful color support.

We also found that "ease-of-use" is a commonly perceived notion for visualizations and a legitimate direct evaluation for color placement surveys. The importance of this point is emphasized by the findings of previous researchers that color placement without regard for ease-of-use has been shown to be a major detriment to visualization performance.

Automation of color placement will probably be preceded by color support systems of all kinds. These systems will have to deal with a more difficult problem than deciding how to improve the design. They will have to communicate this to the user. This most difficult challenge is our next step in the development of color support systems.

References

Cooper, E. and Kamei, K. (1998). "Development of a Color Balance Support System", *Proceedings of the 14th Symposium on Human Interface*, pp. 705 – 710.

Ashizawa and Ikeda (1981). “Size Effect in Color Conspicuity”, *Journal of the Color Science Association of Japan*, Vol. 18 No. 3, pp. 200 – 204.

L. Silverstein (1987). "Human Factors for Color Display Systems: Concepts, Methods, and Research", *Color and the Computer*, Academic Press, Chapter 2, pp. 27-61.

E. R. Tufte (1990). *Envisioning Information*. Chap. 3, 5. Graphics Press, Connecticut..

Cognitive architectures — A theoretical foundation for HCI

Dieter Wallach[1] and Marcus Plach[2]
[1]University of Basel, [2]Saarland University

1 Introduction

Recent years saw the development of a number of comprehensive theories of human cognition termed *cognitive architectures* that start to have an increasing impact on cognitive modeling applied to research questions in HCI. Gray, Young and Kirschenbaum (1997) regard cognitive architectures as the most important contribution to a theoretical foundation for HCI "since the publication of *The Psychology of Human-Computer-Interaction* (Card, Moran, and Newell, 1983)". The authors speculate that architectures are in the process of becoming the preferred route for bringing cognitive theory to bear on HCI. While applications of architectures to HCI issues are illustrated in only 11 articles published in the period from 1983-1993, more than twice as many have been published in the last five years, indicating an ever increasing interest in practical and theoretical implications of cognitive architectures

2 What is a Cognitive Architecture?

A cognitive architecture embodies a comprehensive scientific hypothesis about the structures and mechanisms of the cognitive and perceptual/motor system that can be regarded as (relatively) constant over time and independent of a task (Howes and Young 1997). On the one hand, these approaches provide an integrative theoretical framework for explaining and predicting human behavior on a wide range of tasks. On the other hand, architectures are theoretically justified, implemented software systems that allow for the computational modeling of different phenomena. By postulating a core system of theoretically motivated constructs, architectures ground HCI models in cognitive theory and provide a vehicle for our understanding of human behavior in operating interactive systems.

While early approaches in specifying an architecture concentrated on higher-level cognition, most of such theories turned out to be "brains in a box" that were typically ignorant with regard to the modeling of perception and motor behavior. Consequently, these approaches were of only limited interest for the modeling of interactive tasks in HCI. With the advent of cognitive architectures like EPIC (Kieras and Meyer 1997) and ACT-R/PM (Anderson and Lebiere 1998) theories of human perceptual-motor capabilities were incorporated to provide a more complete picture of human cognition and interaction with the environment. These architectures are concerned with the orchestration of different aspects of cognition with perception and action and allow for the modeling of event-based interactive tasks.

ACT-R/PM can be regarded as the most comprehensive approach to date which grew out of detailed experimental data on human problem solving, memory and learning (Anderson and Lebiere 1998). The ACT-R/PM architecture is divided into two layers: a *cognition layer* that handles all aspects of higher level cognition and a *perceptual/motor layer* that mediates the interaction between cognition and the environment. The cognition layer is based on a production system that distinguishes between a permanent procedural and a permanent declarative memory. Procedural knowledge is encoded in procedural memory using modular *condition-action-rules* (productions) to represent actions to be taken when certain conditions are met. So-called *chunks* are used to store knowledge about facts in declarative memory. Sub-symbolic activation processes control which productions are used and how they apply to chunks. A hierarchical representation of goals is utilized to control information processing. ACT-R/PM comprises mechanisms for the acquisition of new productions and chunks that allow for the modeling of learning processes.

ACT-R/PM's perceptual/motor layer is operating in parallel to the cognitive layer and comprises a detailed theory of human vision and motor behavior, as well as processors for modeling human hearing and speaking. Models created within this framework are *generative* and *reactive:* the model *generates* actions using the knowledge encoded in the architecture and *reacts* to events initiated by the task requirements (Gray et al. 1997). The constraint to formulate a truly *complete* model using structures and mechanisms of the architecture guides theorizing in HCI and closes the gap between cognition and overt behavior.

3 Cognitive architectures and GOMS

Before the advent of cognitive architectures, GOMS (Card, Moran and Newell 1983) was regarded a prime candidate to provide a theoretical foundation for the field of human-computer interaction. GOMS served as a framework for the systematic analysis of the goals, operators, methods and selection rules, the "how-to-do-it" knowledge comprising routine human-computer-interaction. The GOMS framework is deservedly credited to be the first approach in HCI that seriously addressed the cognitive structures underlying manifest behavior and which integrated many components of skilled performance to produce predictions about real-world tasks.

There are striking similarities between structures in GOMS and architectures like ACT-R/PM, starting from the idea of goal-oriented symbolic decomposition of human skill and ranging to central timing assumptions for perceptual/motor processes. Contrasting GOMS with cognitive architectures, however, reveals three major advantages of the latter approach: (1) GOMS models do not adress learning processes in dealing with interactive systems — although learning to operate an interface is one of the central issues in HCI; (2) GOMS models are concerned with the description and explanation of routine behaviour in system interaction, abstracting from slips or lapses of attention — modeling non-routine problem solving or error-prone behavior seems to be beyond the scope of GOMS, but even early studies showed that new users spent up to a third of their time in error recovery; (3) GOMS models are labour-intensive to construct. Cognitive architectures like ACT-R/PM, on the other hand, provide learning mechanisms to model knowledge acquisition processes, can model problem solving using empirically well supported mechanisms and are comparatively easy to construct because constraints of the architecture aid in identifying task strategies. Moreover, exploiting the generative nature of a model implemented in a cognitive architecture allows to determine its predictions by running it on the respective task — an advantage that promises to scale up to tasks where performing a laborious GOMS analysis on a larger number of benchmarks is not practical.

Cognitive architectures have served as analytical tools in the evaluation of interfaces (Anderson, Matessa & Lebiere, 1997), supported the design of manuals (Mertz, 1995), and have been used to predict the performance of users on new systems (Kieras and Meyer 1997). In contrast to empirical user studies, models developed in a cognitive architecture do not only predict *that* one system design is better than another, but also *why* this is the case, opening up directions for redesign. Alternative designs can be compared cost-effectively by using architectures on prototypes and reducing the necessity of empirical studies, especially in early phases of the design process. Applying cognitive models as "artificial users" thus allows to ask *what-if questions* (Howes and Young, 1997) about user interfaces by investigating design variants and analyzing the resulting effects on task performance.

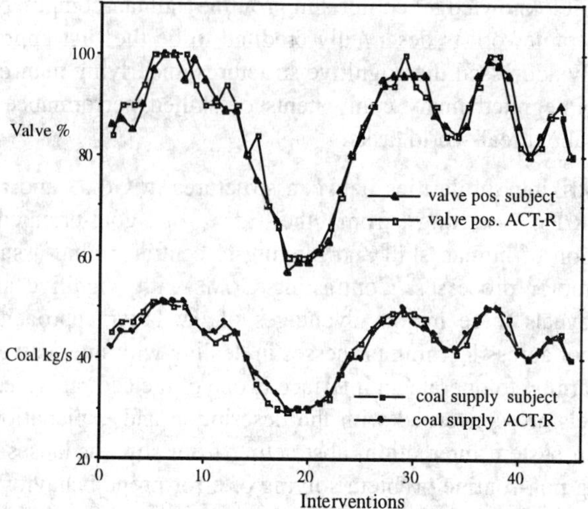

Figure 1 Control behavior subject vs. ACT-R/PM

As an example, Figure 1 illustrates the prediction of an ACT-R/PM model in comparison to the actual control behavior of a user when operating a simulation of an existing coal-fired power plant. To control the power plant, users had to fix critical variables (supply of coal in kg/s, opening of the turbine valve in %) on a graphical user interface to follow given time-dependent power curves while maintaining a constant pressure in the steam generator. Close analysis of the knowledge the model acquired strongly suggests to maintain a numerical representation of system variables in addition to a graphical display of the system trajectory on the user interface since successful performance of the model presupposes *both* information sources. For details of the model see Wallach (1998).

To be generally useful for practical HCI purposes a cognitive architecture must provide accurate qualitative and quantitative a priori predictions, must be cost-effective to use and — most important — models developed in a cognitive architecture must be understandable to system designers (John, 1996). From the viewpoint of cognitive architectures, applications in HCI provide a strict and demanding test bed for the theory underlying an architecture — HCI and cognitive architectures thus live in a mutual beneficial relationship.

4 References

Anderson, J.R. & Lebiere, C. (1998). *The atomic components of thought.* Mahwah, NJ: Erlbaum.

Anderson, J.R., Matessa, M. & Lebiere, C. (1997). ACT-R: A theory of higher level cognition and its relation to visual attention. *Human-computer interaction,* 12, 311-343.

Card, S.K., Moran, T.P., & Newell, A. (1983). *The psychology of human-computer interaction.* Hillsdale: LEA.

Gray, W.D., Young, R.M. & Kirschenbaum, S.S. (1997). Introduction to architectures and human-computer interaction. *Human-computer interaction,* 12, 301-309.

Howes, A. & Young, R.M. (1997). The role of cognitive architectures in modeling the user: SOAR's learning mechanism. *Human-computer interaction,* 12, 345-389.

John, B.E. (1996). *Cognitive Modeling for Human Computer Interaction.* Paper presented at the 1st European Conference on Cognitive Modeling, Berlin.

Mertz, J.S. (1995). *Using a cognitive architecture to design instructions.* Unpublished doctoral dissertation, Carnegie Mellon University: Pittsburgh, PA.

Kieras, D.E. & Meyer, D.E. (1997). An overview of the EPIC architecture for cognition and performance with application to Human-Computer Interaction. *Human-computer interaction,* 12, 391-438.

Wallach, D. (1998). *Komplexe Regelungsprozesse.* Wiesbaden: DUV.

Computer-based Screening of the Visual System

Andreas Hoffmann
UBS - Union Bank of Switzerland
Web Application Development Support
P.O.Box
CH-8098 Zurich
e-mail: andreas-za.hoffmann@ubs.com

Marino Menozzi
ETH Zurich
Institute of Hygiene & Applied Physiology
CH-8092 Zurich
menozzi@iha.bepr.ethz.ch

1 Introduction

Nowadays, medical screening is used in various fields, such as general medicine or occupational health screening. The objective of most applied screening examination is to discriminate a small number of conspicuous subjects from usually asymptotic individuals. Such conspicuous subjects might show high risk factors for having a disease or health problems.

The idea of the presented screening approach is to reduce the resources involved. That means, that the tests should be performed without any administrator involved during the tests.

In Germany a well known screening approach G37 is established to detect subjects showing visual symptoms for people working at VDU workplaces. For examinations as required in G37 examination a Computer-Based Screening System (CBSS) is presented in this paper covering visual screening examinations such as visual acuity for far and near viewing distance, color vision, stereo vision and heterophoria. To retrieve information on how G37 are carried out in the field, a questionnaire was sent to occupational health departments in Germany to analyze whether the G37 procedures is nowadays used as defined.

Applying visual tests to visual display units (VDU) arises technical restrictions. The size and the resolutions of the VDU used leads to discrete sizes of visual stimuli and the presentation of colors using VDUs shows that not all perceivable colors can be adequately represented. Those restrictions are to overcome to assure that discrimination of the individuals is reliable using the CBSS for visual screening tests.

2 Self-Screening Realisation

The idea of the self-screening is the reduction of resources involved in the screening test using the CBSS. In such a self-screening scenario, a subject approaches the CBSS, follows its instructions and performs the examinations by interacting with the CBSS. After finishing the test, the CBSS can show quantitative results with explanations to the subjects or a boolean qualitative result can be presented to the subject, such as *no deficiencies detected* or *please contact your physician for further examinations.* In such a scenario the medical staff can be relieved from the discrimination task of a screening process to a wide extent and the gained time can be applied to discriminated conspicuous individuals or on more efficient primary preventive tasks.

The objective to achieve for the CBSS was that computer-skilled subjects can self-examine their visual system without further instructions by an administrator. Three options for interacting with the CBSS are available: (1) Pressing an OK-Button to proceed in the dialog, (2) Pressing a BACK-Button to proceed backward in the dialog, (3) Pressing a HELP-Button to receive support.

The interaction with the CBSS during the vision test were carried out using the keyboard or a remote control.

3 Method

3.1 Questionnaire

A questionnaire was developed and sent to occupational health services in Germany. The questionnaire covered questions to evaluate which examinations, resources and devices are currently used in occupational health services to assess visual performance.

3.2 Visual acuity

The visual acuity represents a threshold that reflects the smallest angle in which a visual stimulus can be detected. One accepted approach to assess visual acuity for far and near viewing distance is the assessment using the German DIN standard DIN 58220. Technical restrictions of currently used VDUs does allow to present stimuli to assess visual acuity for far viewing distance (5m). An assessment of visual acuity for near viewing distance (0.4m) is not possible due to the resolution of current VDU (dot pitch 0.28mm). To overcome this restriction an 1'' LCD (dot pitch 0.03mm) was applied to the CBSS. Using the CBSS required the presentation of the Snellen E-hook to assess visual acuity.

A pool of 25 subjects was tested using the CBSS and a standard procedure used to analyze the reliability of the CBSS when assessing visual acuity. Further details can be taken from Hoffmann (Hoffmann 1997).

3.3 Color Vision

A frequently performed test to evaluate red/green color deficiency is the Ishihara test. Color normal and color abnormal subjects detect the presented stimuli on Ishihara plates differently. Despite the known difficulties when presenting colored stimuli using a VDU, digitized Ishihara plates were used in the CBSS to assess color deficiency. The technical restrictions when presenting the plates using a VDU and daylight were evaluated by determining the colors coordinates of different color of the plates using CIELUV. A pool of 20 subjects was tested concerning color deficiencies using the CBSS and the Ishihara test using 14 plates representing numbers. Further details can be taken from Hoffmann (Hoffmann and Menozzi 1999b).

3.4 Stereo Vision

When looking at objects, differences in distances between the objects can be distinguished by different crosswise disparities among the eyes. This crosswise disparate depth perception is called *stereo vision*. A device often used to assess stereo vision is the TNO test. The TNO test applies random dot stereograms (RDS) and the TNO test is based on retinal discrimination of stereo images using anaglyphs. Anaglyhs use colored filters to allow presentation of different stimuli to each eye. This approach was applied the CBSS. RDS were generated and anaglyphs were applied to assess stereo vision. The colors of the RDS were adjusted to the anaglyphs. A pool of 25 subjects was tested using the CBSS and the TNO test to evaluate the reliability of the CBSS to assess stereo vision. Further details can be taken from Hoffmann (Hoffmann and Menozzi 1999a).

3.5 Heterophoria

Heterophoria is a latent deviation of the visual axis of the eyes, manifested by the absence of all stimuli to fusion. Heterophoria can be an indicator that patients have asthenopic complaints. The approach applied to the CBSS to assess phoria uses two stimuli, one of which was a vertical line shortly presented to the subject. Using a psychophysical threshold procedure the subjects adjusted the vertical line to the center of a horizontal line. The colored stimuli were perceived through anaglyphs in a dark experimental environment in the absence of any fusion stimulus. The viewing distance was 0.5m which is approximately the distance when working with VDUs. 20 subject were tested for phoria using the CBSS and those results were compared to a standard assessment of phoria using the Titmus Vision Screener. Further details can be taken from Hoffmann (Hoffmann 1998).

4 Validation

4.1 Questionnaire

After sending out 40 questionnaires, eight replies were received that show similar answers indicating that current visual examinations are carried out in Germany according to G 37. The respondents claim between 6% and 33% of all employees of the companies were examined during one year which was judged to be resource intense. The mean time to perform the visual examination was estimated at approximately 20 minutes. Further details can be taken from Hoffmann (Hoffmann 1998).

4.2 Visual acuity

The technical restrictions of the standard VDU used allows the assessment of visual for far viewing distance between 0.032 and 1.6. The applied small LCD allows an assessment of visual acuity at near viewing distance between 0.063 and 1.25. The variations between the required size according to the DIN standard and the achieved sizes of each stimulus when presented using the VDU can be neglected for screening tests. Comparing the experimental results for the pool of subjects using the CBSS and the standard procedure showed not significant difference after applying the student t-test (p=0.24, near viewing distance; p=0.41, far viewing distance). The results showed that despite the technical restrictions of the CBSS the assessed results are reliable. Further details can be taken from Hoffmann (Hoffmann 1997).

4.3 Color Vision

Despite the differences of the representation of colors of the Ishihara plates when presenting the plates on the VDU to representation of the colors using daylight, the CBSS show valid results in the experiment. The pool of subjects showed 10 color normal and 10 abnormal subjects when using the Ishihara test. Using the CBSS all color normal subjects were correctly detected and 9 out of 10 color deficient subjects were discriminated. That showed a non-significant difference after applying the χ^2-text (p=0.65) between the assessment using the Ishihara test and the CBSS. The evaluation showed eight plates using the CBSS to be well suited whether six are not. Further details can be taken from Hoffmann (Hoffmann and Menozzi 1999b).

4.4 Stereo Vision

Technical restrictions when presenting colors and the resolution of the standard VDU used required viewing distance of 2m to assess stereo vision whether the assessment of stereo vision using the TNO-test applies a viewing distance of 0.4m. Comparing the experimental results of the pool of subject after assessing stereo vision using the TNO-test and the CBSS showed no significant difference after applying the Wilcoxon Ranking test (p=0.72). Further details can be taken from Hoffmann (Hoffmann and Menozzi 1999b).

4.5 Phoria

Applying anaglyphs to the assessment of phoria was possible after adjusting the colors of each stimulus. Comparing the results of the pool of subject after assessing phoria using the Titmus Vision Screener and the CBSS showed no significant difference after applying the Wilcoxon Ranking test (p=0.6). Further details can be taken from Hoffmann (1998).

5 Discussion and Conclusions

The respondents to the questionnaire established visual examinations due to G37 in Germany as the current used procedure to test the human visual system. The respondents established as well, that the current procedure is resource intense. The visual tests defined in G 37 were applied to a CBSS and each test was examined concerning technical restrictions arose by using VDUs to present visual stimuli. The evaluated technical restrictions were overcome. Experimental validation showed that using the CBSS to examine the human visual system due to G 37 leads to reliable results despite technical restrictions. That showed the self-screening approach to assess visual performance to be valid.

The approach of applying human visual test to a CBSS reduces the resources involved in the screening process. The number of tests can be extended not only covering the tests defined in G37. Work within the CBSS project started to apply test for visual search tasks, constrast sensitivity, tracking and human audio performance.

6 References

Hoffmann, A. & Menozzi, M. (1997), Using a high resolution display for screening visual acuity for near vision, *Displays*, 18(1), 11-20.

Hoffmann, A. (1998), Computer-based screening of the visual system, ETH Zurich, PhD. Thesis, Diss Nr. 12781.

Hoffmann, A. & Menozzi M. (1999a), Applying anaglyphs for the assessment of stereopsis to a PC-based screening system, *Displays*, 20(1), 31-38.

Hoffmann, A. & Menozzi M. (1999b), Applying the Ishihara test to a PC-based screening system, *Displays*, 20(1), 39-47.

Studies on Classification of Similar Trademarks Corresponding to Human Visual Cognitive Sense

Taki Kanda
Bunri University of Hospitality
Hideyo Nagashima
Kogakuin Umiversity

1. Introduction

In this paper we are concerned with classification of similar trademarks corresponding to human visual cognitive sense. When similar trademarks are classified or retrieved using computer, physical features are extracted from images of trademarks, physical feature space is constructed based upon the extracted physical features and similarity between trademarks is evaluated by the distance in the physical feature space. The results of retrieval or classification however do not always coincide with human visual cognitive sense of similarity. We have been therefore studying methods to reflect human visual mentality in retrieval of similar trademarks. In these methods a questionnaire is set out to obtain the data on human evaluation of similarity between trademarks, the physical feature space is transformed to the subjective feature space which reflects human visual mentality using the data on human evaluation obtained by questionnaire and similar trademarks are retrieved in the subjective feature space. The efficiency of this transformation has been verified but here in order to consider a method to reflect human visual mentality in classification of similar trademarks without questionnaires a function is used to express human visual cognitive sense, which is called "autocorrelation function". Since it is considered that human judgement on similarity is influenced by the pattern of trademarks and the autocorrelation function is able to express the pattern of trademarks, it is studied how to classify trademarks according to human visual cognitive sense of similarity by using the autocorrelation function.

2. Aotocorrelation Function

Autocorrelation function expresses the degree of correlation when an image is moved by a certain length of correlation to the horizontal and vertical direction. Let x, y be abscissa and ordinate and l_x, l_y be the width and length of circumscribed rectangular of trademarks respectively. Then the autocorrelation function is given by

$$Rh(u) = \frac{1}{l_y} \sum_{y=1}^{l_y} \frac{1}{l_x - u} \sum_{x=1}^{l_x} \{g(x,y)g(x+u,y)\} \tag{1}$$

[horizontal direction]

$$Rh(v) = \frac{1}{l_x} \sum_{x=1}^{l_x} \frac{1}{l_y - v} \sum_{y=1}^{l_y} \{g(x,y)g(x,y+v)\} \tag{2}$$

[vertical direction]

where u,v are the length of correlation and $Rh(u), Rh(v)$ are the values of correlation in the horizontal and vertical direction respectively, and $g(x,y)$ is density at a point (x,y) and given by the value of 1 or 0. In order to find out whether the autocorrelation function expresses human visual mentality, we calculated the values of the autocorrelation functions of 62 sample trademarks for each pixel, graphed the autocorrelation functions, classified these sample trademarks with respect to similarity of trademarks judged by persons' visual cognitive sense and examined the relation between the shape of the graphs of the autocorrelation functions and similarity of trademarks. It has been found that the trademarks the graphs of the autocorrelation functions of which resemble each other in shape fall into a same group. It is therefore expected that the autocorrelation function can be used as a function to express human visual mentality.

3. Function to Classify Trademarks

Since the autocorrelation function given by Eqs.(1) and (2) is too complicated, a method is needed to treat the function easily. As the methods to replace a function by another function so as to be able to easily treat functions, least squares approximation, interpolation or minimax approximation etc. can be considered. Here a function is concerned, the graph of which is similar in shape to the graph of the autocorrelation function. Now letting t and $f(t)$ be the length and values of correlation respectively, the function is defined as

$$f(t) = (1 - H)e^{-\alpha t} + H - A\sin\frac{2\pi}{T}t \tag{3}$$

(horizontal direction: $0 \le t \le l_x$)

(vertical direction: $0 \le t \le l_y$)

where $\alpha, A,$ T and H are parameters : α is attenuation function, A and T are the amplitude and period of oscillation respectively and H is the value of convergence. These parameters are extracted from the graphs of the autocorrelation functions as follows:

- α is obtained by the method of least squares for exponential functions.
- A and H are obtained by finding the extrema of waves of the graphs of the autocorrelation functions: A is half of the

average of the difference of the values of autocorrelation at two adjacent extrema and T is twice of the average of the interval between extrema.

- H is obtained as the average of the values of autocorrelation at middle points between extrema.

We calculated the values of these parameters, graphed the function given by Eq.(3) and the autocorrelation function for 1843 trademarks which are using in our studies, and compared the graphs of the autocorrelation function with the graphs of the function defined in this paper. For the results it has been seen that the graphs of the autocorrelation functions and the defined functions of same trademarks resemble each other.

4. Classification by Parameters

Since it has been seen that the graphs of the autocorrelation function and the function defined by the parameters extracted from the autocorrelation function resemble each other, it is considered that trademarks can be classified corresponding to human visual cognitive sense of similarity by using the parameters of the defined function. Therefore we classified trademarks according to the values of each parameter and found that classification by the value of convergence is effective compared with the rest.

5. Final Remarks

The autocorrelation function has a characteristic in shape of the graph depending on the pattern of trademarks. We therefore considered that human visual mentality can be reflected in classification of similar trademarks by using the graph of the autocorrelation function, defined the function the graph of which is similar in shape to that of the autocorrelation function and classified trademarks based upon the parameters of the defined function. It has been found that the parameters of the defined functions extracted from the autocorrelation functions are useful to classify similar trademarks corresponding to human visual cognitive sense.

6. References

Kurita, T., Shimogaki, H. & Kato, S. (1990). A personal interface for similarity retrieval on image database system. Transactions of Information Processing Sociaty of Japan, 31(2), 227-237.

Tanabe, K., Ohya, J. & Ishii, K. (1992). Similarity retrieval method using multidimensional psychological space. The Transactions of the Institute of Electronics, Information and Communication Engineers, J-75-D-II(11), 1856-1865.

Kanda, T. & Nagashima, H. (1997). Evaluation of reflection of human subjectivity in retrieval of similar trademarks. Abridged Proceedings. 7th Int. Conference on Human-Computer Interaction (HCI'97, San Francisco, USA, August 24-29, 1997), p.39.

Defending and Offending Cultures

Nuray Aykin
AT&T Labs

1 Introduction

Two French Language associations file suit against Georgia Institute of Technology for presenting information only in English about the Georgia Tech's educational programs in France. The organizations claimed that this violated a 1994 French law which outlawed the use of "foreign words in any advertisement for goods and services if an approved French word could be used instead." The suit was dropped, but the Web site now has the information in French and English (Petrovich, 1998). — Are we still insisting on English being the only language for displaying information on the Internet?

The Office de la Langue Francaise (OLF) informed the operator in a computer store in Montreal that his Web site violates the Canadian language laws requiring the use of French. The law dictates that even the corporate logos cannot be displayed unless French translations were also displayed, with the French letters at least twice the size of the English ones (Petrovich, 1998). — Although the majority of the Web content is in English, is it necessary to publish multiple language Web sites for certain locales?

Reebok names their new sneaker line 'Incubus' which means an evil spirit that lies on persons in their sleep. — Are we naming our products properly? Are we offending some cultures?

The gesture for "O.K.!" in the United States has many meanings in different cultures. For example, in Tunisia it is a threat of murder; in Japan and Korea it can mean "money"; in Germany it is a sign for "Great!"; and in Argentina, Belgium, France, Italy, Greece it means "nothing." (Transimage, 1998)— How can we know all the cultural issues that can have a big impact on our design? How do we internationalize our work?

German police arrested the head of the German division of CompuServe on charges that CompuServe was trafficking in child pornography and neo-Nazi propaganda. In 1997, Germany adopted a new law that purports to extend German legal jurisdiction to any Web site accessible from Germany (Petrovich, 1998).

Then, how do we design for multiple cultures? How do we address the users in different parts of the world? How do we attract visitors around the globe to our Web site? How do we create services or Web sites for the global market? And, how do we manage internationalization and localization processes within an organization?

When designing a software application, or designing a Web site for global markets, we need to include three steps in the design process:

1. Eliminate cultural content from a product (internationalization): Develop a "core" product that can be easily configured for any culture. Keeping the overall international market in mind provides flexibility while customizing the product for a particular locale.

2. Enhance the product to fit the users of a particular locale (localization): Design the product to accommodate the locale requirements, such as language, country related information, writing directions, date/time/address formats, cultural preferences, etc.

3. Usability Testing: The basic purpose of usability testing is getting empirical information about users into the product design cycle. The same is true for international usability testing— methodologically they are the same. The main differences are related to logistics (such as time and date differences, holiday schedules, hardware compatibility), usability test procedures adapted for culture, language and cultural differences and communications barriers.

Many global companies develop strategies and implement procedures to ensure their global investments are well thought out and studied. However, they all face problems that range from a minor headache, to costly redesign, to losing the market. Many of these problems arise from not understanding the global market, and not managing the internationalization and localization processes within the company.

One of the design decisions related to any global design process is to determine the target market. As in the U.S. market, we need to know the needs and preferences of our customers in the markets outside the US. Customer requirements should drive user interface and software requirements. Designing products for international markets requires understanding internationalization and localization concepts. The core product should be designed with internationalization in mind, and has to be ready for localization.

The identification of the market will help determine the internationalization/localization process and help decide the extent of localization needed. The cost associated with different internationalization/localization processes and levels of localization need to be justified with market data.

2 Managing Internationalization and Localization

If you are planning an internationalization/localization process for your applications or Web site, you need to educate everyone from junior staff to developer to system engineer to CEO. Everybody on the process must think globally on a daily basis. Planning for internationalization/localization involves (Luong, et al., 1995):

- Developing a globalization plan and determine the extent of localization: These involve understanding the target markets, considering business needs, identifying legal requirements and international standards/policies. The extent of localization brings up a list of questions one needs to answer before starting the design cycle:

  – Should we localize or not? – The answer depends on the comparison between incremental revenue due to increased sales of localized products and the cost of hardware, software, translation, and management resources.

  – Should we localize everything or just some parts? – With full localization, customers will have a product that fully meets their needs, however this is more expensive than providing partial localization. If customer accepts English and the market size is small, it will be cost-effective to do partial localization.

  – Should we localize in-house, or through an agency? – This depends on the translation cost and the expected length of translation needs. Many companies form long-term relationships with a localization agency to handle their localization needs.

  – Should we localize here in the U.S. or in the target country? – This depends on cost, quality, knowledge, schedules, and support functions.

- Assembling a globalization team: You need to organize the development teams to do the internationalization and localization work. There are alternative organizational structures one can choose to organize the development teams. One possibility is to have one centralized development team that is responsible for the U.S., internationalized and localized versions of the product. Another approach can be to have a core development team that is responsible for the U.S. release and the internationalization. Then regional teams are responsible for the localization for their marketplace. A third approach is to have a core development team that is responsible for the U.S. release. And a centralized international development team is responsible for internationalization and localization.

For example, in developing a multi-lingual Web site for a corporation, the core/headquarters can create the core content and framework, glossary, style guides, and translation of core messages. The local regions can provide the region-specific Web content, regional updates, and translation of updates, review translation for headquarters, and provide customer services. The core headquarters can monitor the local Web content.

Today, many companies provide tools and platforms for localization. Microsoft, Macintosh, Hewlett Packard, Sun, Netscape, Alis Technologies, etc., offer tools and processes to internationalize and localize software products. The World Wide Web Consortium provides standards for internationalization of the Web. HTML, XML, and Java support Unicode for multilingual support, and provide code for international formats. Understanding and using these tools provides the basis for the right architecture for the software. However, they will not guarantee high-quality and compelling localized products that are easy to learn and use. They will not provide guidelines on cultural preferences. This responsibility belongs to developers, user interface designers, systems engineers, writers, translators, project managers, and locale experts.

3 Legal Issues

Although it is almost impossible to educate yourself on international and state/country laws, there are ways to prevent legal actions against your product. Understanding the legal issues could help you proactively take steps to prevent unexpected legal actions. It is important to assume "anything can happen" to protect your rights and prevent damages. For example, one of the most crucial steps when publishing a Web site is to provide a legal page that includes legal statements related to trademarks, copyrights, privacy, terms and conditions on downloading software, use of chat rooms, etc. The Web site visitors should be required to accept these terms. Providing a legal page is just a preventive mechanism, and it does not guarantee protection against all possible legal actions. There still could be legal problems related to your Web site, so you should hope that your company does not have a physical presence at the country at issue. In this case, you would be dealing with a totally unfamiliar legal system.

Similar legal problems can exist for products that are designed for markets around the world. It is very important to know the internationalization and localization issues when designing products and services for other countries. Not only may your company's sales depend on these issues, but the company's image may also be associated with the successes and failures in understanding the different cultures.

4 References

Luong, T.V, Lok, J.S.H., Taylor, D.J., Driscoll, K. (1995). Internationalization: Developing software for global markets. New York: John Wiley & Sons, Inc.

Petrovich, J. (1998). Playing by the international Internet rules. Multilingual Web Sites Conference. San Diego, CA.

Transimage Inc. (1998). Examples from Transimage Inc Web site (http://www.transimage.com)

Another Software for Another Society?

Apala Lahiri Chavan
ZindaGUI
G412 Lokdarshan, Military Road
Mumbai
India

1 Globalisation -What does it mean?

We are moving towards the twenty first century with uncertainty. Globalisation, as a kind of myth of our times, is hard to grasp. The word 'globalisation' can be interpreted in many ways. In its most uncontroversial sense, globalisation refers to the 'rapidly developing process of complex interconnections between societies, cultures, institutions and individuals worldwide. It is a process which involves compression of time and space, shrinking distances through a dramatic reduction in the time taken - either physically or representationally - to cross them, so making the world seem smaller and in a certain sense bringing human beings 'closer' to one another. But it is also a process which stretches social relations, removing the relationships which govern our everyday lives from local contexts to global ones'.(Tomlinson, 1997)

It is in fact this shift in context that becomes an important issue for Third World countries, when dealing with globalisation and its effects, especially where technology is concerned. The foremost consideration of any technology imported from foreign countries, including computer software, is how well it integrates with local conditions. Here the question of dependency cannot be ignored because one source of dependency problems is the importation of a foreign technology into a cultural environment where the values are at variance with those of the exporting country.

The bias of the technology itself is complicated by the inability of the new environment to put the technology to use with its own software and instead having to use 'one size fits all' software. Another concern with the kind of globalisation that leads to one size fits all software is the reduction of local cultural space. Whether a community is able to develop its own cultural identity largely depends upon local cultural space people can control. If people are to be 'beings for themselves' (Freire, 1972), they need sufficient cultural space to define their identity autonomously. If this space is not adequately provided or acquired, people become 'beings for others.

2 Need for Culturally Localised Software?

As a software user and a user-interface designer living and working in fin-de-siecle India, one finds that people in this country (the same is probably true of other countries in Asia, Africa, etc.) belong to a predominantly pre-industrial oral culture. They face the challenge of leaping into the post-industrial world without passing through all the stages through which the West passed in its socio-economic transformation from the pre-industrial stage. Software and user-interface designers in this country must respond to many unique challenges: situations that have no precedent in past history; the economic disparity between developed and developing countries and the resultant difference in lifestyles; and religious/cultural considerations. For example, consider the question of whether a particular Hindu's obsessive ritualism and unbuttoned polytheism impacts his/her interaction with the computer differently from that of a particular Sunni Moslem who is uncompromisingly monotheistic and austerely legalistic? And how do 32 official languages spoken in the country apart from numerous dialects, affect usage of one size fits all software?

What about the fact that the emphasis on analytic communication through words is counter to the dominant Indian idiom in which words are only a small part of a vast store of signs and semiotics. Or lets look at a specific function. How does one structure the Help function for the vast majority of hesitant new users and a small minority of power users? Is the Guru metaphor appropriate? The most respected gurus in this country have always held that the guru, too, is only an aspect of the self. The guru is the disciple, but perfected, complete. At the end of your search, burn the guru, say the tantriks; kill the Buddha if you meet him on the way, is a familiar piece of Zen Buddhist wisdom. Could an understanding of the concept of the guru help us in designing an essential function like Help in a uniquely Indian way, to suit Indian users better. Even a cursory consideration of such issues seems to indicate that we need "another software for another society."

Yes indeed there is a messy trail of interface issues to be sorted out, but the question is are we looking? In this country we need a familiar interface persona that has an existence in different orders of being, like an average Indian, as a 'body, a psyche and a social being'. As Indians, our cognitive space in all matters is incredibly cluttered. Gods and spirits, community and family, all seem to be somehow intimately involved. Successful products and services have all recognised this aspect and adapted their offering in the Indian market. Software products need to learn some lessons from these success stories.

We need to study software products as artifacts-in use, and not in isolation. And we need to study specific contexts of use, where the users have certain praxis. In the context of the user interface, we need to study the use of the interface in the hands of users from different cultures.

3 Issues for Research

And if we indeed want to sort out the messy trail then we need to examine the following user-interface development issues:

- Should the development process include inputs from users across the continent? If so, how can this be implemented in a cost-effective and practical manner?
- Should the user interface, or some parts of it, be customizable according to culture/continent/groups of countries? If so, should the customizable parts of the user interface be completely designed and implemented in the countries concerned?
- Should the local SIGCHI organizations be the resource for dealing with issues of cultural customization for the country concerned? If so, how could this responsibility be established?

User-interface research issues include the following:

- How is it possible to initiate research about "another software for another society"? Could some kind of research project be undertaken under the aegis of SIGCHI, with industry sponsorship?
- Could user-interface designers/companies across the continent be involved in research about this topic and then combine research results in the form of a collection of Websites, books, CDs, and videos produced by a sponsoring organization?

Regarding information resources produced by collective research, the Websites, books, CDs, and videos could be available to all those designing software products. The materials would contain the following:

- Detailed "thick descriptions" of users studied across the world with regard to an exhaustive list of topics.
- User-interface design guidelines keeping in mind the 'thick descriptions'
- User-interface design ideas regarding cultural customization
- Recommendations, if any, regarding changes in the current development process
- Resources (individuals, companies, academic institutions, etc.) across the world that could assist product development companies with cultural customization.

4 International Comparative Research - Wanted, a new approach

If research projects were to be initiated to explore and gather data regarding users and user interface from across the world, the research methodology would be crucial. Given the nature of the research, extensive reformulation may be required of existing methodologies.

Research is not conceived, designed, executed and interpreted in a social/political vacuum. On the contrary, It reflects the system in which it is embedded. In order to successfully implement any kind of international comparative research, existing methodologies that have been taken from the natural and physical sciences, may be inadequate to deal with social/cultural relationships and situations. Existing methodologies often emphasise validity and reliability without any apparent realisation that in this kind of research, in certain situations the two may not be compatible, Reliability (which receives much more attention) is often achieved at the cost of validity.

International comparative research projects carried out in the field of communication research reveal some new approaches. There seems to be an agreement among cultural studies researchers that there is no need for all the participating countries to rigidly and mechanically apply the same agreed instruments and that a more flexible, sociological orientation which recognises national, cultural, social and linguistic differences, and the implications of these differences for design and data collection, has its advantages. In other words, it is recognised that, except in the most simple of categories, different questions, or at least differently presented questions (reflecting differences in culture, language, etc) - not the same questions- would be necessary to evoke the same type of information in the different circumstances. Moreover, it is also accepted that, in order to do justice to the complexity of the subject matter, it would be necessary for each country to include special segments in their inquiries, which, among other things, would facilitate a deeper understanding of national differences

One final point in this connection is that the hard/soft quantitative/qualitative, hierarchical dichotomy with regard to data seems to have been rejected in the sphere of communication research. Wherever possible, the two have been blended. Thus, for any research on user interface issues, quantitative data could be enriched and refined by data gained from more ethnographic and anthropological approaches - both being equally valid and useful.

Further, for the type of research being discussed here, the carbon copy blue-print application in different societies of pre-tested questionnaire items, may not be capable of providing data suitable for genuine comparative analysis. Except where the analysis is confined to very simple categories.

As in any kind of comparative research, in the kind of research being discussed here too, it will not be very helpful to provide an item-by-item inventory of similarities and differences from the various participating countries. The studies from the various countries need to be integrated or compared in relation to some model or set of guidelines. Since several models and principles exist, appropriate choices will have to be made.

It is clear that for any any such international research project to be successful, they will have to be carried out within an eclectic framework, and assessed accordingly. This does not imply the end of methodological rigour, but an acceptance of an approach, which is capable of doing justice to the complex set of relationships, structures and processes that are involved. The so-called scientific approach is incapable of doing this on its own. The adoption of complementary perspectives is essential.

Another point, very closely related to this, centres on the question of such research in Third world countries. Perhaps one needs to look at what is being exported from the 'developed' world in research? How suitable are these exported models, theories, concepts and methods derived from Western industrial experiences, when applied to completely different conditions. 'Indigenous research methodology' may be required.

5 Conclusion

Investigating and recognising the diverse needs, capabilities and expectations of culturally distinct users may hasten the process of 'glocalisation'. This would happen when the global market adapts to the local conditions while employing them to gain competitive advantage. The top down process of globalisation would then work with the bottoms up process of localisation. Glocalisation would then certainly be the tsunami of the future. ...And then users and their interfaces lived happily ever after...

6 References

Albrow, M. and King E. (1990) Globalisation, Knowledge and Society. London:Sage.

Appadurai, A (1990) 'Disjuncture and difference in the global cultural economy', in M. Featherstone (ed.), Global Culture. London:Sage.

Bagdikian, B. (1989) 'The lords of the global village', The Nation (special issue), 12 June.

Del Galdo, E. and Nielsen, J.(ed.), (1996) International User interfaces, New York: Wiley Computer Publishing.

Freire, P. (1972) Pedagogy of the Opressed. Hammondsworth: Penguin.

Hall, S. (1991) 'The local and the global: globalisation and ethnicities', in A.D. King (ed.), Culture, Globalisation and the World System. London: Macmillan.

Jussawalla, M., et al. (1986) Information Technology and Global Interdependence. New York: Greenwood.

Tomlinson, J. (1991) Cultural Imperialism: a Critical Introduction. London: Pinter.

You Don't Have to Be Jewish to Design Bagel.com, but it Helps… and Other Matters

Aaron Marcus, John Armitage, Volker Frank, and Edward Guttman
Aaron Marcus and Associates, Inc., 1144 65th Street, Suite F, Emeryville, CA 94608-1053 USA;Tel:+1-510-601-0994x19, Fax:+1-510-547-6125
Email: Aaron@AmandA.com, Web: http://www.AMandA.com

1 Introduction

The title of this paper intends to imply that user-interface and information-visualization (UI+IV) designers need not be members of the user-group, but having subject-matter experts (SMEs) and/or user-representatives on the development team, significantly increases the chances for successful design. Several project experiences lead the authors to that conclusion. In 1978, at the East-West Center, Honolulu, HI, Marcus headed an international team of visual communicators to design a non-verbal, pictographic-ideographic narrative about global energy interdependence for a multi-cultural audience of decision-makers. The project proved a multi-cultural design team successfully could "debug" cultural biases (Marcus 1981). In 1989-92, for Motorola, Aaron Marcus and Associates, Inc. (AM+A) designed UI+IV prototypes for an intelligent vehicle-highway navigation system to account for cognitive preference-differences among users evenly divided among those who preferred directions given by maps, arrow indicators, and words. After many reviews, an optimum design emerged that resolved most of the marketing and technical requirements. In 1994-97, for SABRE, a travel-agents' online-information system and one of the world's largest extranets, AM+A designed UI icons to account for racial and gender differences among international users. The UI+IV design was achieved with the significant participation of travel agents in the development process.

High-quality UI+IVs must be useful (easy access to data and functions that are easy to comprehend, remember, and use) and appealing. Their content and form depende on optimum visible language (e.g., typography, color, symbolism, layout, etc.) and effective interaction with controls to make users more productive and satisfied with products for work, home, and travel. Based on the above project experience, AM+A concludes that professional designers of specific cultures can design successfully for other cultures given significant input about the target cultures and time to do iterative improvement based on user evaluations. To achieve success, designers must analyze UI+IVs in relation to cultural diversity. The dimensions for cultural diversity are traditional for UI+IV development: the user-centered, task-oriented UI+IV development process includes planning, research (user demographics, tasks, and context, plus marketing/engineering requirements), analysis, design, implementation, evaluation, documentation, and training. The UI+IV design components are metaphors, mental models, navigation, interaction, and appearance. This paper comments on likely changes and challenges.

2 UI+IV Development Process

Planning, Research, and Analysis: The development process for culturally diverse products may differ significantly from that for more traditional client-server productivity tools, office Web applications, and home consumer products. For example, the use contexts for desktop applications in offices of developed countries is more established and uniform. In other cultures the physical environment and the structure of groups and tasks may differ significantly. The social, motivational, educational, and personal cultural characteristics of user groups also may differ significantly. In the past, UI+IV developers made assumptions about user communities that increasingly seem worth reconsidering. For example, the development process for some Web applications consists of continuous rollouts of partially finished products. If international, partial rollouts occur, the variations required must be planned to enable successful deployment. Professional and budget "turf wars" may drain valuable time of development teams unless management can establish a rational, humane, effective development process that incorporates cultural diversity early, acquires expertise, and mandates inclusion of new ideas and techniques.

One example of steps in this direction seems to be the work at Eastman Kodak (Prabhu 1997) which investigated cultural diversity parameters for a limited product range (camera-like devices) in a limited set of cultures (India, China, and Japan). More recently, Kodak announced it was targeting a particular age and gender group (teen-age girls between the ages of 11 and 15) for certain products (*San Francisco Examiner* 1999). Microsoft has announced a Website specifically for women (*New York Times* 1999). This kind of specificity of targeted consumers implies culture-specific planning, research, and analysis activities. One aspect still to be considered is the possible role of these products in accepting or changing cultures. Note, for example, that Lego sets in Germany promote female roles that are more conservative than in the USA, and that electronic "pets" in Japan "die," while those in the USA "disappear."

Design, Implementation, and Evaluation: SMEs and user representatives must participate in design and evaluations; they can be cost-effective by avoiding more extensive redesign after later evaluations. While product design may need to account for cultural diversity, specific tools supporting development are still imperfect. Accounting for international/localized products will affect not only labels on icons, but also may impact all UI+IV components. Multiple "assets" will need to be partially designed, implemented, evaluated, documented, and stored as the product moves toward completion. These libraries may change the implementation process significantly by increasing the number of parallel implementation tracks. Because of the interdisciplinary teams, frequent user evaluations, and the rapid development schedules, multiple-culture, detailed interactive prototypes oriented to use scenarios may become more important. Developers will have to plan for the of user-testing costs on the business model.

Documentation and Training: Designers can annotate culturally diverse UI+IV prototypes and transform them efficiently into interactive guidelines and specifications documents, on which software engineers can rely for the latest definitions, appearance, interaction sequences, etc. Training in cultural diversity topics of UI+IV design specific for the development team will become important. Most major conferences do not specialize enough yet in research and applied techniques to assist the UI+IV development community. Some resources are listed in a cultural diversity sub-site originated by AM+A and other contributors following a special-interest-group meeting at CHI99 (see http://www.AmandA.com). Some texts have begun to appear (see References).

3 User-Interface Design Components

UI+IV designers may introduce novel approaches to many UI+IV components:

Metaphor: Metaphors (Marcus 1998) may need to be adjusted for particular communities. Chavan (Chavan 1995) noted that Indian computer users relate to metaphors referring to books more readily than to desktops. Fundamental controls linked to widely different usage contexts also may vary widely according to different cultures. Subject matter experts and user representatives can provide guidance early on to prevent the design team from wasting effort.

Mental model: The initial display of content may differ significantly across cultures. For example, marketing decisions may cause changes of emphasis in order to convey the right level of complexity and to focus the user's attention on primary desired features. Recent news articles note that Hispanic and African-American consumers seem to value the quality of assistance provided by vendors of telephone service, and are willing to pay more for this feature. This phenomenon, related to online help, suggests cultural differences in what is defined as user-friendly.

Navigation: Widely varying culture groups use different navigation strategies. Until products can exploit these differences by building variations that can be easily and efficiently tuned to a particular culture's needs and desires, UI+IVs will not have achieved their full potential. Some products may require "simple" dashboard controls rather than "complex" conventional computer navigation.

Interaction: PDA devices with pen or stylus and improved voice input/output seem likely to increase in use. Cultural differences in interaction styles and templates oriented to these differences remain to be built into commercial authoring or design applications.

Appearance: Early UIs+IVs seem to have assumed all users would need, desire, and enjoy the same kind of display. Today, variations on the Web are built for single user's preferences. One future challenge for cost-effective design/maintenanace will be to isolate which controls and/or aesthetic elements should be standardized for re-use across groups and applications, and which should be redesigned for specific target markets. For example, a Web-based message-manager might consistently show core functions, but also a significant portion of the screen might devoted to labels and graphics oriented to specific users, e.g., teen-aged girls in Spain vs. parents in Hong Kong. One other aspect is the cultural implications of increased use of acoustic cues, or "earcons," to denote classes of objects and processes, their attributes, and, in particular, their states. Increased use of acoustic cues frees up visual areas for other semantics but immediately impacts different cultures' attitudes toward noise, music, social interaction, etc. Other challenges include being able to maintain desirable levels of legibility and readability under extreme conditions of cultural variation, and being able to assist users by viewing their differing UI+IVs.

4 Conclusion

Variations in UI+IV design will continue to increase in number for ever smaller targeted audiences (e.g., educated French-speaking Canadian girls between 11-15), while the number of functions and amount of data continues to increase. Multi-disciplinary teams in UI+IV development and good communication among team-members from different professional disciplines, including SMEs and user representatives will be indispensable. The importance of accounting for different cultural preferences for absorbing information will become apparent. Connected to the Internet, products will challenge designers to invent or re-apply UI+IV components in novel ways that break through to new levels of simplicity, clarity, and consistency for all cultures. By considering the dimensions of change outlined above, designers can continue to accommodate marketing and engineering requirements while providing the next generation of culturally diverse consumers with products that make life more humane.

5 References

Chavan, Apala Lahiri, "A Design Solution Project on Alternative Interface for MS Windows," Masters Thesis, New Delhi, India, September 1994.

DelGaldo, Elisa, and Jakob Nielsen, ed, *International User Interfaces*, John Wiley and Sons, New York, 1996.

Fernandes, Tony, *Global Interface Design: A Guide to Designing International User Interfaces*, AP Professional, Boston, 1995, ISBN: 0-12-253790-4 (paperback), 0-12-253791-2 (CD-ROM), 191pp and CD-ROM.

Marcus, Aaron, "A Rhetorical Analysis of a Pictographic-Ideographic Narrative," *Proc.* Third Int. Conf. on Semiotics, Vienna, 1981, pp. 1035-1053.

Marcus, Aaron, "Principles of Effective Visual Communication for Graphical User Interface Design," *Readings in Human-Computer Interaction*, 2nd Ed., ed. Baecker *et al.*, Morgan Kaufman, Palo Alto, 1995, pp. 425-441.

Marcus, Aaron, "Cultural Diversity," Panel at HCII-97, San Francisco, August 1997, Proc., 1997.

Marcus, Aaron, "Metaphor Design in User Interfaces," *The Journal of Computer Documentation*, ACM/SIGDOC, Vol. 22, No. 2, May 1998, pp. 43-57.

Marcus, Aaron, "International and Intercultural User Interfaces," *in User Interfaces for All*, ed. Dr. Constantine Stephanidis, Lawrence Erlbaum Associates Publishers, New York, 1999 (in preparation).

Marcus, Aaron, "Finding the Right Way: A Case Study about Designing the User Interface a Motorola Vehicle-Navigation System," in Bergman, Eric, ed., *Beyond the Desktop,* Morgan Kaufman, Palo Alto, 1999 (in press).

New York Times, "Microsoft is Starting Web Site Aimed at Big Audience: Women," New York Times, 8 February 1999, p. C2.

Nielsen, Jakob, Editor, *Designing User Interfaces for International Use*, Volume 13 in the series Advances in Human Factors/Ergonomics, Elsevier Publishers, Amsterdam, 1990, ISBN 0-444-88429-9 (Vol. 13), 230 pp.

Prabhu, Girish, "Cultural Diversity Research at Eastman Kodak", panel presentation, HCII-97, San Francisco, CA, August 1997.

San Francisco Examiner, "Kodak Focusing on Teen Girls," 18 February 1999, p. A-3.

GUI Design Preference Validation for Japan and China - A Case for KANSEI Engineering ?

Girish Prabhu and Dan Harel
Eastman Kodak Company, Rochester, NY 14650, USA

1 Background

Previous research (Choong and Salvendy, 1998) has indicated that designing user interfaces based on cognitive and cultural differences improve the overall usability indicated through increased performance and reduced error rates. Along with the typical physical/physiological and perceptual/cognitive aspects of usability evaluation, Kurosu (1997) suggests that emotional and motivational aspects (KANSEI engineering) need to be considered in Japan. The Kodak cultural research attempted to validate the importance of KANSEI engineering in user interface design for Asia. A methodology derived from cultural anthropological principles of "etic" and "emic" perspectives was utilized to study and understand users' needs and preferences for digital imaging products in Japan and China (Prabhu and Harel, 1998). The research findings were then utilized to design graphical user interfaces that are efficiently and successfully localized for China and Japan to "speak the universal language of photography". The process is based on participatory and iterative design that had "cultural research" as a prime component and is a proposed method at Eastman Kodak Company (Prabhu, Chen, Bubie, and Koch, 1997). Initial user reactions to the GUI interfaces, gathered using a modified focus group method, indicated that covert cultural issues do play an important role in overall user preferences.

2 User Interface Design

Localized user interface screens were designed for desktop user interface, back-of-the-camera user interface, and kiosk user interface. The desktop and back-of-the-camera user interfaces along with their baseline North American designs are shown in Table 1 below. Desktop user interface designs for both Japan and China utilized the Picture Easy user interface as the North American baseline.

The following variations were incorporated in the Japanese and Chinese desktop user interfaces to evaluate the six key design elements. To evaluate the preference for **typography**, the help text messages were varied on fonts size (small 10 x 10 versus large 16 x 16), usage of different font sizes for different levels of help, color of fonts, and emphasis (shadow). Different types of **screen layout** with or without specific griding, layout with harmony, and griding based on golden means were used. Three different **color** combinations were used

Table 1: Sample User Interface Designs

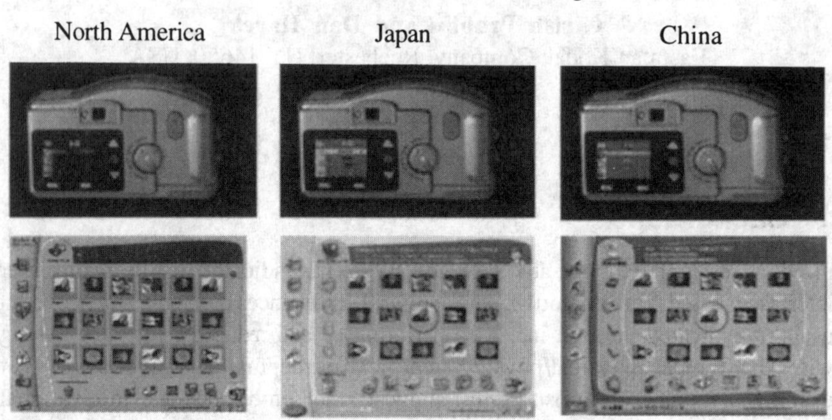

for testing: US bright colors, Chinese colors, and Japanese pastel colors. Different **icons and metaphors** (see Table 2) were tested with variations on designs to suit Chinese and Japanese users.

Microsoft Picture It!, PictureWorks PhotoEnhancer, and Picture Easy screens were used to compare different interaction styles. Picture It! interaction provided main operations as buttons and a menu-based access to other activities, and a dedicated area for image display. PhotoEnhancer interaction provides all the operations through menus and multiple windows. Picture Easy interaction was non-windows, with dedicated buttons for actions, dedicated space for image display.

3 ValidationMethodology

User interface designs developed based on the appearance characteristics generated from the anthropological research were refined through a design review session using ethnic end users. Four focus groups were conducted in China (two in Beijing and two in Shanghai), and two focus groups were conducted in Japan (both in Osaka). Though the anthropological study targeted home consumers, office consumers, professional imaging and educational institutes, the final validation research was done using home and office-based consumers. In each location, one group was composed of male respondents, and one group was composed of female respondents (in both cultures, single-sex focus groups helped to elicit fuller responses). All respondents were screened to ensure that

Table 2: Sample Icons and Metaphors

	USA	Japanese	Chinese	Standard
Trash Can				
Exit	EXIT	退出	退出	
Help				
Arrow				
Selection				
Radical				
Product Icons				

they were regular camera users, and were familiar with digital imaging equipment. These criteria were used to select individuals who would belong to the target group of potential Kodak customers for digital camera equipment. The male groups were mostly business users who used digital imaging at home, whereas participants in the female groups used digital imaging only at home or were familiar with digital imaging.

Based on the anthropological study certain hypotheses were generated regarding general preferences for Chinese and Japanese users towards different user interface elements. The user interfaces were evaluated against these hypotheses in a modified focus group setting. For each evaluation, participants were required to mark their preference on a scoring sheet first, without group discussions. The moderator would then tally the responses and lead the group into discussions. This method allowed for reduction of bias towards group conformity in Japanese and Chinese culture. However, as the sample size was very small, the validation was based on purely qualitative analysis.

4 Results

The validation research was expected to provide a clear-cut preference along the cultural line—each culture preferring their own designs. Most of the preferences for Chinese and Japanese were inline with the expectations, with some modifications.

- Japanese men preferred Microsoft Picture It! interaction style due to familiarity, whereas Japanese females and Chinese men and women preferred Picture Easy interaction style due to the intuitive design.
- Japanese preferred Picture Easy Japanese screen layout, due to harmonious, balanced, and sensuous design, whereas Chinese had mixed preferences for Chinese and Japanese screen layouts.
- Both Chinese and Japanese preferred larger size fonts. Japanese men preferred single color fonts, simple fonts without emphasis on all the three lines of help, whereas Japanese women and Chinese men/women preferred multiple colors, highlighted or emphasized fonts.
- Japanese preferred pastel colors for both the welcome screens and the interaction screens. Though Chinese men preferred Chinese colors, preference for women was mixed between Chinese and Japanese pastel colors.
- Japanese men preferred either the standard Macintosh or standard Microsoft Windows trash cans whereas women preferred the Japanese trash can. For all other icons and metaphors tested, both men and women preferred the

Japanese designs. Both Chinese men and women preferred Chinese icons for all the icons and metaphors.

5 Conclusions

The validation process concluded that there is a strong case for pan-Asian approach based on KANSEI engineering to user interface design for digital products. The process validated most of the expectations from earlier anthropological research and modified some. In doing so, significant similarities between design preferences in China and Japan were noticed as well. It should be emphasized that the results of the focus groups were qualitative, not quantitative, and therefore must be considered in the context of the entire research.

6 References

Choong Y., Salvendy, G. (1998), Designs of icons for use by Chinese in mainland China. In *Interacting with Computers: The Interdisciplinary Journal of Human-Computer Interaction, (Vol. 9 #4*), February 1998, pp. 417-430. Amsterdam: Elsevier.

Kurosu, M. (1997), Dilemma of usability engineering. In Salvendy, G., Smith, M., & Koubek, R. (Eds.): *Design of Computing Systems: Social and Ergonomics Considerations (Vol. 2), Proc. 7th Int. Conference on Human-Computer Interaction* (HCI International '97, San Francisco, USA, August 24-29, 1997), pp. 555-558. Amsterdam: Elsevier.

Prabhu, G. V., Chen, B., Bubie, W., and Koch, C. (1997), Internationalization and localization for cultural diversity. In Salvendy, G., Smith, M., & Koubek, R. (Eds.): *Design of Computing Systems: Cognitive Considerations (Vol. 1), Proc. 7th Int. Conference on Human-Computer Interaction* (HCI International '97, San Francisco, USA, August 24-29, 1997), pp. 149-152. Amsterdam: Elsevier.

Prabhu, G. V., and Harel, D. (1998), Global User Experience (GLUE): Designing for Cultural Diversity. Unpublished.

Isomorphic Sonification of Spatial Relations

Alistair D.N. Edwards
Department of Computer Science, University of York
WWW: http://www.cs.york.ac.uk/~ ...York, England

Evreinov G.E., Agranovski A.V.
Lab for Designing of Info Image Systems
SPECVUZAVTOMATIKA Design Bureau
email: asni@ns.rnd.runnet.ru Rostov-on-Don, Russia

1 Introduction

The perception of size and form of tactile objects relies on visualization, at least for people who are sighted or those who have had longer experience of seeing ('late blind', Warren 1994). Non-visual alternative representations are 3D tactile or acoustic representations. This paper describes an auditory display (Agranovski et all 1996), that provides simultaneous playback and management by eight independent virtual sound sources (m, n) or by cursors formed by four speakers located in front of the listener as shown in Figure 1.

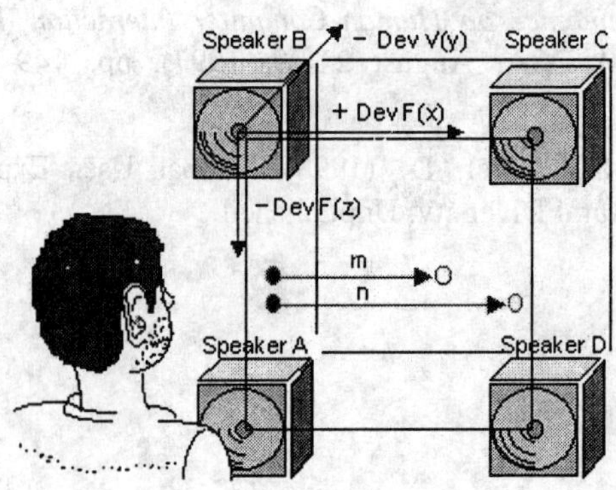

Figure 1 Formation of the auditory imaging space

2 Image Spatialization

Under certain conditions, a contour of an object designates a border of heterogeneity or interruption of continuity and could be more informative to the perceptual system with regard to other properties, allowing concentration on the geometrical relations between elements of the image.

To indicate the spatial position of specific points on the object's surface (or its edge), it is necessary to specify the additional attribute of their remoteness with respect to an observer or a frontal plane.

2.1 Linear Prospect

A resolution of sonified spatial positions within the virtual acoustic plane does not provide any cues as to the distance between the observer and the sound object. While sonification of the prospect can be produced in a number of ways.

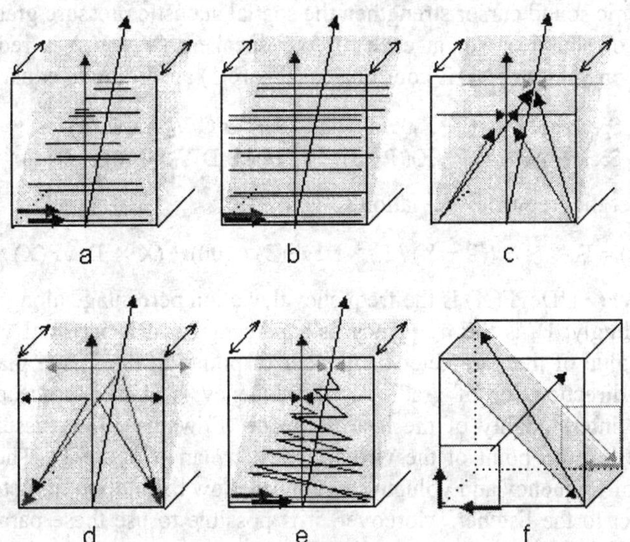

Figure 2 Some approaches toward isomorphic sonification of spatial relations: imaginary size, relative displacement and perceived distance

→ V = 40% (of Vmax), Dev V(y) = 0%; → V = 40%, Dev V(y) = -75%;
→ V = 10%, Dev V(y) = 0%; → V = 10%, Dev V(y) = +300%

Modelling by using displacement of sound tracks in virtual plane alone (similar to the visual display) is not at all interpreted by listener as a perception of 3D composition (Figure 2a, 2b). Frequency deviation of the sound cursor is more natural and effective for marking a point's remoteness from the horizon in the vertical plane – as long as other shifts of sound cue (e.g. of volume) are absent.

(See below). But increasing the gradient of this parameter causes perceptual distortion of the form of the subjective images or an infringement of sound stream continuity. At simultaneous playback of two or more sound cursors a merger of sound streams and the formation of an impression of motion of the sound front take place (Figure 2c, 2d). Since a length perception of virtual trajectory depends on the entire playback duration, linear size control can be carried out by changing the displacement of points as well as by altering the duration shift of their sonification. Volume is a natural remoteness attribute of sound sources and therefore it influences the perception of size and spatial position of frontal projections of trajectories (Figure 2).

2.2 Gradient of Surface's Texture

There is an opinion (Gibson 1950) that a textured gradient is a basic information source about the surface's slope. In particular, deviations of frequency and amplitude which we use in the creation of virtual acoustic objects based on a polyharmonic sound cursor strengthen the spatial acoustic pressure gradient. The amplitude of sound cursor at each of four speakers (V_A,..V_D) is redistributed depending on values of X, Z coordinates (Figure 1) according to equations:

$$\text{Speaker B: } V_B = (P - X)(Z / P) \qquad \text{Speaker C: } V_C = X(Z / P)$$
$$\text{Speaker A: } V_A = (P - X)((P - Z) / P) \qquad \text{Speaker D: } V_D = X((P - Z) / P)$$

while we set the frequency deviation so that:

$$F(X, Z) = F_0 \times (1 - ((P - Y) / P) \times \mathrm{DevF}(Z) / 100) + (X \times \mathrm{DevF}(X) / 100)$$

where, DevF(Z), DevF(X) is the frequency deviation percentage along Z, X axes correspondingly; F_0 is the frequency as a percentage of a maximal value; P is maximal value of the parameter or number of points in the image plane (256). Deviation direction (or of gradient) of frequency is chosen proceeding from perception intermodality of the hearing space (Edwards 1989), assuming that zero is the left top point of the virtual acoustic plane (Figure 1). The gradient dynamics of frequency and volume can indicate how the motion trajectory varies with respect to the listener. Moreover, it is possible to use these parameters to correct spatial distortions associated with acoustic systems, hardware errors or individual physiological features of the listener.

2.3 Playback of Several Sound Objects

Figure 2f represents the playback of two equal squares using different but constant volume levels, divided by an event, the perception of which can emphasize or deform the three-dimensional impression. A subjective sensation of size and spatial position of planar two-dimensional square projections depends on the sonification mode of the intermediate object, be it a line, plane or another figure.

2.4 Turn of Plane

A sequence of volume modulations along the sound track during sonification of some spatial sections of a virtual cube or appropriate projections onto its edges is shown in Figure 3a-3f. Thus, it is possible to sonify a rotation of the virtual plane image of a circle (3a), a rectangle (3b), a triangle (3c), or of any trajectory in 3D acoustic space in front of listener. However, the angle deviation of the spatial position of arches, edges or apexes with respect to the frontal plane depends also on the speed of volume shift along the trajectory and frequency deviation in a perpendicular direction with respect to the chosen axis of rotation.

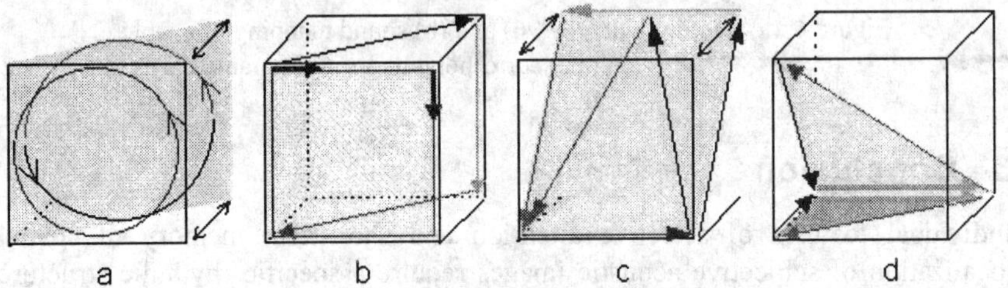

Figure 3 Sound mapping of planes rotation (a-c) and imaginary shadow (d) density of arrows (colours) correspond to parameters of Figure 2

2.5 Shadow

In an abstract sonification as a musical image, shadow can be simulated via acoustic reverberation. But the shadow of real object is a two-dimensional projection on a plane closed by this object from direct beams of the light source. The parameters governing the shadow's sonification must take into account the subordinate status of these projections (Figure 3d).

2.6 Isomorphic Sonification of Real Objects

One of problems of two-dimensional tactile graphics is that the third dimension can be displayed on a flat surface only within limits. Tactile objects should contain information concerning all elements of a spatial composition and should be easily recognizable. Therefore a two-dimensional tactile image of a table (Figure 4a) can differ significantly from the real object or its visual stereometric projection (Kurze 1996). The graphic prototype contains all the necessary information about the elements of the spatial composition and their connections. Hence, the transformation process should at least not reduce a quantity of this information about the prototype. In fact, the image is already fragmented. So the task is: How to make a sonification most naturally? How to use a suitable modulation of sound parameters for the mapping of the spatial position of these fragments? Which fragments to sonify sequentially and which ones simultaneously? How to mark attributes of object surface (colour, texture) and so on. Figure 4b shows a variant of the table sonification.

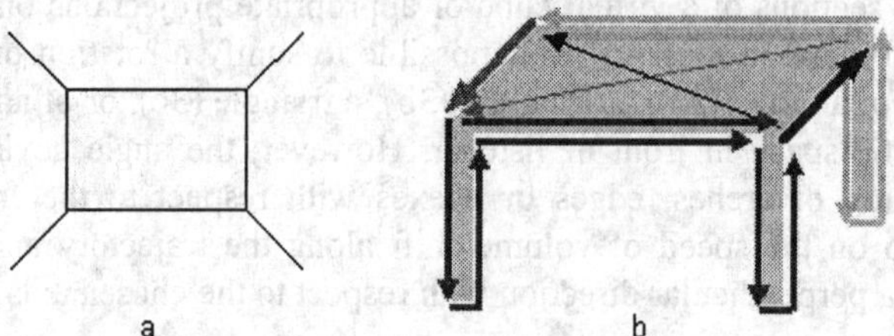

Figure 4 (a) Tactile (Kurze 1996) and (b) sound mapping of a table
, , , - colours correspond to arrows' parameters as in Figure 2

3 Conclusion

Individual features of selective attention and short-term memory of spatial localization of subjective acoustic images require a specific rhythmic structure (cadenced picture) in the creation of a sound stage of objects. Special caesurae and pitch alteration signals are necessary for the exact representation of extreme points, for switching and fixing of attention, for element separation or, conversely, for a merger of sound streams. Essentially, in these points a moving source must be rendered as a quasi-static object. In the future we wish to extend control by spatial tonal modulation and volume of sound cursors to specially analysed gradients of time intervals and playback velocities within different image fragments.

4 References

Agranovski A.V., Evreinov G.E., Yashkin A.S. (1996). Graphic audio display for the blind. In: Klaus, J., Auff, E., Kremser, W., Zagler, W.L. (Eds.): *Proc. 5th Int. Conference* ICCHP'96, (Linz, Austria, July 1996), pp. 539-542.Verlag R. Oldenburg, Wien Munchen.

Edwards A.D.N. (1989). Soundtrack: An auditory interface for blind users. *Human-Comp. Interaction* 4 (1), 45-66.

Gibson J.J. (1950). The perception of visual surfaces. *American Journal of Psychology* 63, 367-384.

Kurze M. (1996) TDraw: A computer-based tactile drawing tool for blind people. In: Proc. of ASSETS'96, (Vancouver, BC, Canada 11.-12.04 1996), pp. 131-138.

Warren, D. H. (1994). *Blindness and Children: An Individual Differences Approach.* New York: Cambridge University Press.

Virtual Tactile Maps

Jochen Schneider and Thomas Strothotte
Otto-von-Guericke University of Magdeburg

1 Introduction

Tactile maps constitute the media most commonly used by blind and partially sighted people for the exploration of spatial information. However, compared to maps for sighted people, they are inconvenient to produce and contain less information. We have developed methods and implemented them prototypically, by which visually impaired people can explore a geographical area flexibly with the help of a computer through *virtual tactile maps*. The design of a system is presented which captures hand positions through a video camera and produces acoustic output. It uses digital map data which is adapted to the requirements of the users. One of the most compelling problem is how to deal with the scale of a map, in particular the non-linearity of maps. In this paper, the development of the prototype and initial experience with it are presented.

2 A Challenging Vision

A man, recognizable by the long white cane hanging from his arm as being blind, is standing on a sidewalk. He is holding his hands in front of him and moving them with concentration. He is in the middle of planning the rest of his walk through the down town area with the help of a virtual tactile map. He had explored the area and chosen a route through it at home. Now he is checking if he still remembers the rest of the route correctly. The man is wearing a baseball hat. There is a small camera in the screen of the hat, looking downwards at his hands and picking up their positions. The camera is connected to a small computer, which emits information through small earphones.

The scenario described is a vision. We report on the development of a system for the exploration of virtual tactile maps as a wearable computer (cf. Mann 1997). In this paper, the concept of virtual tactile maps and a prototypical implementation will be presented.

3 The Virtual Tactile Map Concept

Virtual tactile map systems are digital map systems which emit speech and sound when their maps are explored by hand movements. Virtual tactile maps enable blind and partially sighted users to explore an unknown geographical space and learn routes in it, similar to tactile maps, which they are inspired by and named after. The first aspect of the user interaction with virtual tactile maps is the interpretation of hand gestures and their mapping to the geographical space; the second aspect the selection and acoustical presentation of the geographical elements on the map; the third aspect the preparation of the digital map data to a representation suitable for hand input, which is relatively coarse.

Virtual tactile maps serve to aid orientation, defined by Jansson as "perceiving/knowing the spatial relations between the traveler's current position and direction of body and significant features of space" (Jansson 99). Virtual tactile maps belong to a certain kind of electronic travel aids, namely orientation systems for larger space.

4 Related Work

The "KnowWhere" system is a hand gesture recognition system which conveys geographical information to blind people (Krueger & Gilden 1997). The system presents zoomable geographical objects (e.g., outlines of countries) by emitting a sound when the hands of a user touch their imaginary image on the desktop. The hands lie on a tactile grid. Their movements lead to speech or sound output in accordance to the kind of the geographical element which they have touched. Krueger and Gilden have conducted a test of their system with five congenitally blind subjects. The test consisted of exploring large geographic objects. The subjects were able to find easily absolute positions and recognize puzzle pieces of the shapes explored afterwards. Two were also able to produce "acceptable" drawings of the outlines of the geographical objects.

Although they both deal with presenting geographical objects to the same user group through video gesture input and acoustical output, KnowWhere and virtual tactile maps differ in some important aspects. The hardware used for KnowWhere consists of a light table and special-purpose hardware, which facilitates the gesture recognition, but also impedes portability and the widespread use of the system. Even more important, virtual tactile maps serve to convey information of an urban area, as a street map does, whereas KnowWhere conveys large-scale geographical information, as an atlas does.

Starner, Weaver and Pentland (1998) describe a wearable computer, which recognizes American Sign Language gestures through video input. The system is being developed to eventually translate the gestures into spoken language. A camera in a cap the user is wearing which looks down at his hands picks up the gestures. This approach shows the feasibility of the vision of a wearable system implementing virtual tactile maps.

5 The Design of a Virtual Tactile Map System

In this section, requirements of virtual tactile maps and the design of an actual system are presented.

5.1 Requirements

Requirements for virtual tactile maps can be elicited by asking how blind pedestrians prepare themselves for a walk in an urban area not fully known to them. Blind people who wish to navigate independently have to memorize the layout of the area given, learn segments of a path and angles between them and have to recognize them during walking (Golledge et al. 1996). A computer system can support this process by presenting the information needed. In (Strothotte et al. 1996), a study is described in which blind pedestrians and orientation and mobility trainers were asked about information an electronic travel aid should emit through speech synthesis. Information identified as essential was the name and the type of road, distances and obstacles on it; information identified as desirable was details about objects such as shops, public buildings, etc. Finally, information identified as "nice-to-have" was those on temporary obstacles such as roadworks and diversions.

5.2 System Design

Based on the requirements described above, methods and tools for the exploration of virtual tactile maps were designed. The design was implemented as a prototype, and a first test was conducted. The most important aspects of the design were the interaction and the preparation of the map data, which will be described in the following.

Interaction with cartographic objects on virtual tactile maps is done through pointing. A pointing device for a virtual tactile map system has to be large, portable, affordable, and allow absolute setting of more than one position. There are no tactile displays or touch tablets available meeting all of these requirements. Therefore, a video camera is used as an input device for the virtual tactile map approach presented here. The hand movements result in information about geographical objects under the hand and their relation to others being emitted as speech and sound.

The virtual tactile map system presented here supports both map exploration and learning of a route. Exploration is done by moving hands freely on the map, which results in the system speaking information about the objects under the hands. Route learning is done by first selecting a route. The system then guides a finger on the route through sound. Currently, pitch and balance are used to convey the distance of the finger from the route.

Commercially available map data is used in the system, enriched with positions of street car stops, etc. The system manages the map data in a small Geographical Information System (GIS), which is able to answer requests on geographical elements at a certain position or the distance between points.

6 Implementation and Test of a Prototype

The requirements above led to the design and implementation of a prototype. The prototypical setup is stationary. The camera is mounted on a tripod above a table. The stationary setup makes it possible to use a tactile grid pad on the desk the camera is aimed at, which serves to orient the hands and to facilitate image processing by providing a constant background. In addition, small markers can be employed as interaction devices, e.g. to place landmarks on the map. Finger tips and markers are recognized through color segmentation, in the case of the index fingers by finding the color of a ring each finger tip is wearing (see Fig. 1).

A prototype first prototype of a virtual tactile map system with hand gesture recognition for a single hand and without markers was tested by a blind male of about forty years, experienced in using tactile maps. Both the exploration of the whole map and the learning of a route were tested. For the former, the test subject moved his hand on the pad. When touching a street, the system emitted its name. The street exploration mode was tested as well: A long street running roughly diagonal through the map was selected. The subject then tried to follow the street with the index finger, guided by a tone changing in pitch and balance.

The map exploration feature was felt to be useful, the coding of the distance to a certain street through balance and pitch too rough, which can partly be attributed to the speakers used. Also, the gesture recognition subsystem was processing black-and-white pictures, resulting in jumps of the calculated hand position.

7 Future Work

The gesture recognition, the preparation of the geographical data and its acoustical coding still need to be refined. The extent of a map can then be chosen by the user by placing tokens of different forms on the table. It remains to be investigated how these tokens can be replaced when moving from the stationary prototype implemented thus far to a truly mobile system.

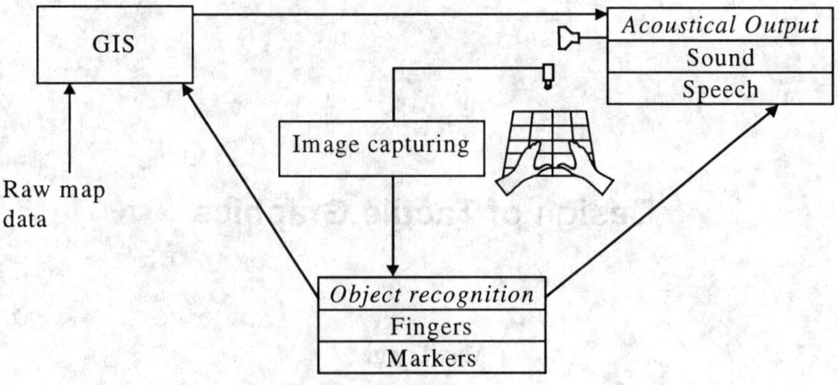

Fig. 1: Architecture of the prototype

8 References

Mann, S. (1996). Wearable computing: a first step towards personal imaging. *IEEE Computer*, 30 (2), 25–32.

Jansson, G. (1999). Spatial orientation and mobility for the visually impaired. In Silverstone, B., Lang, M. A., Rosenthal, B. & Fraye, E. E. (Eds.): *The Lighthouse Handbook on Visual Impairment and Rehabilitation.* New York: The Lighthouse and Oxford University Press. Forthcoming.

Krueger, M. W. & Gilden, D. (1997). KnowWhere™: an audio/spatial interface for blind people. *Proc. ICAD '97*. Xerox PARC: Xerox. (Available online under http://www.santafe.edu/~kramer/icad/websiteV2.0/Conferences/ICAD97/ Kruger.PDF).

Starner, Th., Weaver, J. & Pentland, A. (1998). Real-time American Sign Language recognition using desk and wearable computer based video. IEEE Transactions on Pattern Analysis and Machine Intelligence, 20 (12), 1371-1375.

Golledge, R. G., Klatzky, R. L., & Loomis, J. M. (1996). Cognitive Mapping and Wayfinding by Adults Without Vision. In Portugali, J. (Ed.): *The Construction of Cognitive Maps*. Dordrecht: Kluwer, 215–246.

Design of Tactile Graphics

Max Lange
Blista-Brailletec gGmbH

1 Introduction

Access to graphic information is becoming more and more important for the Blind. This is due partly to developments in the workplace - particularly the necessity to use graphic user interfaces to software - and partly a result of the increased freedom to integrate into everyday life, education, and work - which involves comprehending information as diverse as town maps, pictorial signs, or technical drawings. Obviously, a need therefor exists to create useful graphic output for the Blind. It is not possible, however, to simply print a graphic on a tactile medium and expect the Blind to be able to use it. For one thing, while structures as small as 0.2 mm can be perceived tactually, the resolution at which the Blind can distinguish individual dots or features is typically about 2 mm. For another, neither perspective nor color information can be depicted one-to-one in a tactile image. Finally, the way the Blind extract information from a graphic differs substantially from visual perception. This is why certain procedures and informal standards have evolved for the representation of information, based on best practice on the one hand, and on available technology on the other hand. However, recent technology development significantly extends the possibilities for printing tactile graphics, so a reassessment may be helpful.

2 Differences between tactile and visual output

In general terms, "reduce to the max" (rttm) is the dogma of tactile graphics. This is especially important when discussing computer-controlled embossing, where raster images with a resolution of typically 10 or 20 dpi are printed. The possibilities for displaying abstract graphic information such as curves, diagrams, or schematic drawings are greatly reduced as compared to visual output, and the same is even more true for artistic representations. Therefore, images need to be stripped down to their essential information before being printed. This is necessary not only because any redundant or non-essential information makes it more difficult to tactually perceive the content, but also because real estate on a sheet of paper becomes a valuable commodity at 20 dpi.

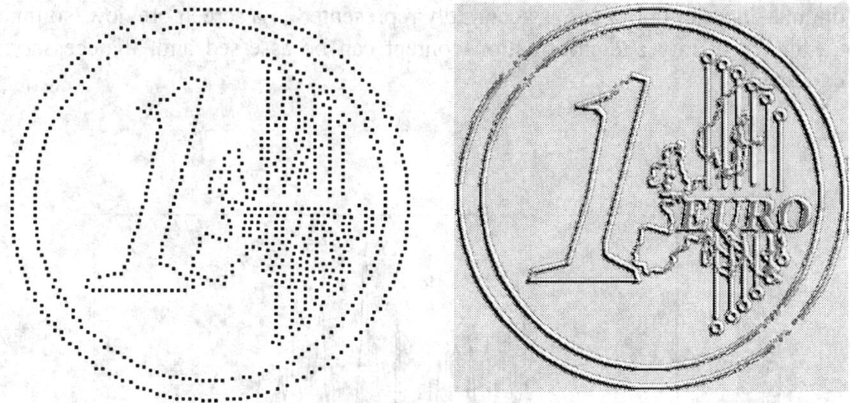

Figure 1: A Euro coin embossed with 2.5mm equidistant dots or as a relief image

In some cases, it is all but impossible to accurately translate a visual image into a tactile one. One example of this is chemical analysis spectra which tend to have a few isolated, sharp peaks. It is therefore not usually appropriate to simply take an image created for sighted users and print it on a tactile medium. Rather, tailor-made images must be created that confer the same basic information in a different way.

3 New technology

With the development of new software and hardware, tactile images can be created in a much easier, less time-consuming way (Chan 1998). A number of devices have been developed to produce tactile graphics (Vanderheiden 1994). I will mention two software products and one hardware product. On the software side, two products that help in the preparation of tactile images are TacFax

(Puck 1998) and BrailleGraf. The former, TacFax, was developed as a tactile fax engine. With its content-analysis capabilities and a possibility to detect and zoom pictures from within a text, TacFax simplifies the representation of combined text and graphics. What is even more important, TacFax can identify and mark-up text and graphic zones in a document, thus guiding the blind reader through the text, and helping to simplify the reading. BrailleGraf, the second software product, allows importing black-and-white bitmap files which are then "scanned" and translated to a tactile raster image. This procedure lets the sighted operator check and review the output on screen before printing it. The reduction of information density from a 72 dpi screen image to a 20 dpi raster image or 16 dpi out-of-raster dot image is accurately represented in a screen preview, so that the image quality and information content can be assessed and, if necessary, corrected.

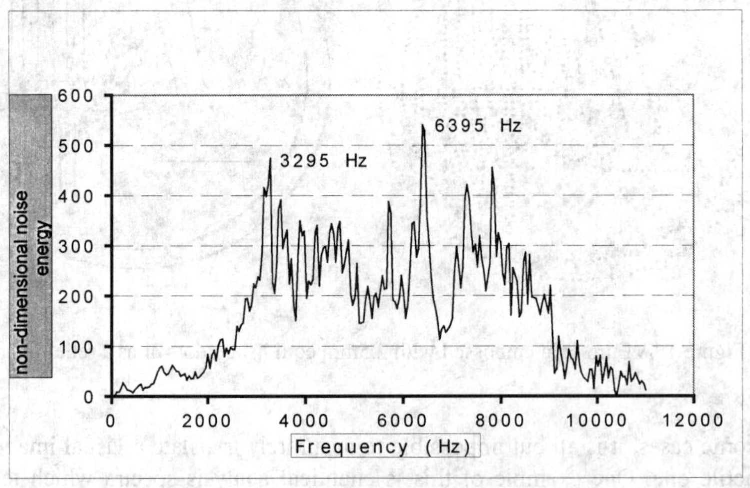

Figure 2: Noise spectrum as a relief image with PRINT

On the hardware side, an important new technology is the tactile inkjet printer developed in the PRINT project (Lange 1998). This printer, dubbed the Braillejet, prints at a 200 dpi resolution with tactile structures up to 1 mm high. thus it becomes possible to print small features such as the aforementioned sharp peaks, in a tactually accessible way. It is also possible to use different line heights and shapes, as well as hatched surfaces, so that for the first time, it becomes possible to represent an equivalent for color information. In graphs, one line style can be used for axes and another for values, for example. Simple as it sounds, this sort of image was impossible to print with a direct-output tactile printer previously. The application of this technology gives teachers, office workers, and other users of tactile graphics a powerful new tool to provide graphic information to the Blind.

As with all new technologies, though, this one, too, involves some risks. The most obvious one is that as it makes it so easy to print tactile graphics, users may just want to print any color image with it and forget about the "rttm" principle. This would be a bad service to the Blind, as tactile images aren't automatically accessible just because they can be produced. It is still necessary to strip an image of any information noise that could prevent the Blind from grasping its meaning. It is also still necessary to remember about the number of dimensions. While the world the Blind perceive is strictly three-dimensional, it is also strictly non-perspective. This limits the possibilities for imaging. Some things simply still need to be grasped in actual, rather than virtual reality. And where text and graphics are combined, it should also be remembered that braille inherently takes more space than blackprint, so the page layout may need to be different.

For all that, the new technological possibilities developed with PRINT and TacFax have created the possibility to print computer-generated images which until now could not have been printed satisfactorily. They have introduced surface texture, variable line shapes, and height graduation as possibilities in tactile printing, three tools that until now were only available in toolkits for manually crafted tactile graphics stereotypes. It is up to users and user organizations to make the most of the technology, and develop expertise in using them.

4 References

Ying H. Chan (1999). An Assessment of the Methods of Producing Raised Diagrams for the Blind. Dept. of Chemistry, University of Southampton.

M. Lange (1998). PRINT - Non-Impact Printer and Plotter for Braille/Moon and tactile Graphics. *Proceedings, CSUN Conference on Technology and Persons with Disabilities*, Los Angeles, 1998.

Greg C. Vanderheiden (1994). Dynamic and Static Strategies for Nonvisual Presentation of Graphic Information. Transcripts, Symposium on high resolution Tactile Graphics, *CSUN Conference on Technology and Persons with Disabilities*, Los Angeles, 1994.

Thorsten Puck (1998). Tactile Image Processing with Tacfax. *Proceedings, CSUN Conference on Technology and Persons with Disabilities*, Los Angeles, 1998.

Deriving Tactile Images from Paper-based Documents

Thorsten Puck, Gerhard Weber
Harz University of Applied Studies and Research, Germany

1 Introduction

Paper-based documents become accessible to blind people using automatic transcription services unless they contain graphics. But there are several classes of documents like letters, faxes or Web pages which rely on graphics to convey information and whose textual contents remain inaccessible. Like web pages may be "repaired" through alternate text descriptions, paper-based documents need to be enhanced for better automatic transcription including a review of images by the blind reader.

Through a novel process of printing high-resolution tactile images developed in project PRINT (Puck 1998) a combination of tactile notation (Braille or Moon) and high quality presentation of graphics can be achieved.

As an example application we have implemented a fax system called TACFAX. As many blind people work as telephone operators, handling of faxes could be one of their tasks if they become accessible. TACFAX can be used for several tasks related to document processing by blind people including coloring of tactile images for people who have still a small sense of vision left. As Optical Character Recognition (OCR) cannot easily distinguish text from graphics a method for characterizing documents is needed without relying on help by sighted people. Zooming and resizing of parts of the document may improve the reader's understanding of a tactile printout.

2 Non-visual User Interface of TACFAX

With TACFAX faxes can be translated into Braille or Moon and printed on paper. Most kind of graphical elements in fax documents will be recognized automatically and be printed at the end of the Braille- or Moon-printout as an tactile image. For manual interaction with TACFAX we have implemented a non-visual user interface allowing keyboard input, speech synthesis and sound feedback. Feedback through speech synthesis allows the user to use a novel menu-driven input method and to assign arbitrary categories of subject and general subject "keywords" to each fax in order to avoid lengthy title lines on the printed fax. Paper handling is thereby improved. Nonverbal sound feedback informs about the status of the document processing. For example playback of a ringing phone sound indicates a newly arrived fax even while the user is reviewing another documented.

The user interface of TACFAX is designed for three levels a user can reach: novice, advanced and expert.

The *novice user* uses the fax system as stand alone fax device. Received faxes will be printed out immediately in one of the selected notations: Braille, Moon or regular but re-sized ink print. Graphics will be printed automatically at the end of the document.

The *advanced user* considers the fax system as an electronic fax device. He has to accept the concept that a fax document is not longer only a paper-based document but an electronic one. According to his reading efforts the advanced user can make use of mark-up facilities. Therefore a tactile printout of an image of the complete fax document is printed. This tactile fax copy will be printed with zones recognised by OCR including their type (text, table or graphic) and their ordering (see next chapter). Furthermore, an orientation column is generated which allows the user to explore the document. The mark-up facilities allow the user to change OCR-settings and the layout of the printout.

Basic OCR settings are defining the page layout of a fax document which includes:

- sequence of text recognition (single columned – multi-columned),
- zone attributes: type of zone (text, graphic, table), and ordering.

Text and graphical settings are defining the newly printed document:

- text has font, size, color,
- graphics have size (zoom factor), foreground and background color, noise reduction, and title of graphic (caption).

The advanced user identifies faxes by type and assigns two keywords to each fax (a general subject and a subject). These keywords will be printed at the top of each page. With this method, faxes from the same phone number will be assigned the same keyword (general subject). In addition, TACFAX offers the possibility of forwarding received faxes. Faxes will be forwarded as text file and/or image file.

In addition to these facilities, the *expert user* makes use of the file based handling of faxes through HTML for export of electronic documents. The HTML file includes several hypertext links to a text file and to several image files of different formats all describing the original fax in different formats. File based handling of fax documents offers the possibility to retrieve and reprint older faxes by name. Furthermore HTML code includes all defined settings, whereby re-loaded faxes can be processed with the same settings again.

3 Mark-up for Document Analysis

Recognized zones will be printed within the tactile copy of fax. A rectangular shape in the context of an image of ink-print text allows the user to identify zones through their type, size and position. By touching (not reading) the tactile fax copy the user can isolate and distinguish text and graphics from its tactile perception. Figure 1 shows different kind of zones in a mark-up generated automatically.

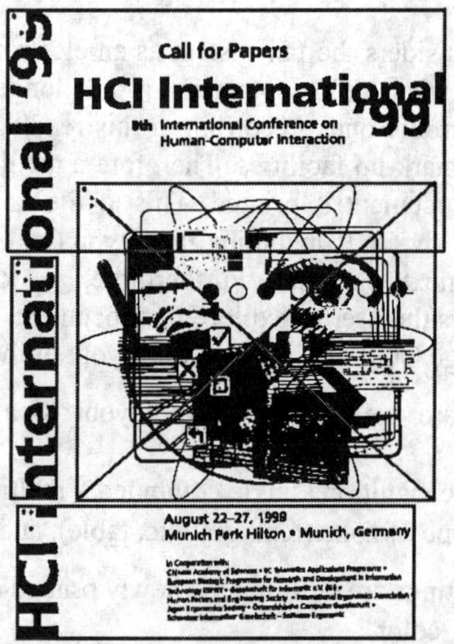

Figure 1: Graphical mark-up of a document

Text is shown with an empty box, graphical elements by a cross and tables with a grid. We assume, that the spatial properties of text or the context of text can be determined from the tactile printout, although text itself cannot be read.

Mark-up in TACFAX is an interactive process using a touch tablet as input device. The tactile document is placed on the touch tablet to identify and work on zones by spoken feedback. The continuously numbered zones in sequence of reading will help the user during mark-up. Input is formed by gestures within the selected zone. Gestures are an interaction method which require no immediate visual feedback. They are therefor suitable as an input method for blind users (Weber 1987).

We have implemented only a few gestures to avoid unnecessary training. With these gestures the user will be able to select a zone, assign the type to a zone, create a new zone or delete a zone. Mark-up of zones is a combination of a pointing task with straight linear movement and sequences of linear movement. The following table 1 shows some examples:

gestural input	**operation**
a line within a boxed zone	Change type of a zone (text – table - graphics)
a line from inside to outside a boxed zone	Change size of a zone
Four lines which are either horizontal or vertical and whose start is „close" to the end (a box)	create a zone
a diagonal line crossing the upper left corner of a boxed zone	delete a zone

Table 1: Gestural input and operations

The gestural interface includes speech feedback which identifies zones verbally by type and number as the user reaches it. Thereby the effect of „getting lost" after crossing the box boundaries (which are marked tactile) is counter-acted, although speech finishes after a while and continues only to announce leaving the zone. Graphical zones will be described verbally with either a predefined given caption or a caption assigned by the reader .

Each page of the tactile document is marked with a barcode. This offers an easy way to re-load a document's page via standard barcode scanner and synchronize the tactile paper printout with the active document used by TACFAX.

4 Further Work

Initial evaluations of sample output from TACFAX have shown the need for handling another class of documents. Diagrams which contain textual annotation are useful to blind people if the graphics is simple or schematic such as in technical diagrams and sometimes on graphs of mathematical functions.

The tactile mark-up needs to be modified such that speech feedback is generated during exploration of the tactile document conveying the recognized text included in each text zone. Furthermore TACFAX conversion of text into Braille or Moon has to ensure that the original position of the text in the document is retained. Figure 2 shows an example of Braille printing on a document retaining the text position. Thereby the user will make use of the mark-up facilities as described in the upper chapter.

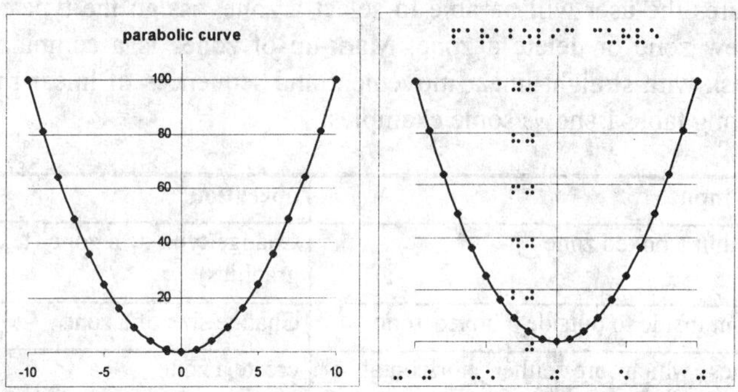

Figure 2: Braille printing on a document retaining the text position

5 Conclusions

Document processing for non-visual presentation can be based on a combination of both auditory feedback, tactile notation and high-resolution tactile presentation. Through interactive analysis of documents recognition of text and graphics can possible be improved by a blind person. Instead of an automatic method a semi-automatic method can increase accessibility to fax documents.

An ongoing field study by various user organisations throughout Europe allows now to evaluate our assumptions using a prototype setup. This will help to get a clearer understanding of the speed of learning the tactile mark-up procedure.

6 Acknowledgements

This work has been supported by the European Commission, DG XIII (TIDE), and project PRINT. We specially want to thank Prof. Giessner, FH Augsburg for preparing initial tactile printouts.

7 References

Puck, T.: *Tactile Image Processing with TACFAX*, Proceedings of CSUN 13th Annual Conference (Los Angeles, March 17-21, 1998), in print.

Weber, G. (1987). *Gestures as a means for the blind to interact with a computer*, in Bullinger, H.-J. (ed.) Human-Computer Interaction INTERACT'87, North Holland:Amsterdam, pp.593-595.

Evaluating Stress in the Development of Speech Interface Technology

Alex W. Stedmon* & Dr. Chris Baber
*Centre for Human Sciences, Defence Evaluation & Research Agency, UK.

1 Introduction

This paper offers, by way of an introduction, some of the problems in defining stress and its effects on the production of speech in the development of Automatic Speech Recognition (ASR) systems. The main focus is the findings of a recent experiment which examines the cognitive basis of speech production by implementing a paired-associates learning paradigm.

The impetus for research in this area comes from a belief that speech exists as an untapped medium for user interfaces. This, in itself, is a highly contentious issue, for as Linde and Shively argue speech is already an active or semi-active mechanism (Baber & Noyes, 1996). Taylor, on the other hand, argues for using "the potentially spare speech modality" (Baber & Noyes, 1996). Whatever the outcome of the debate, speech, just like other aspects of human performance, is prone to alter when related to other demands and stressors.

2 Stress

The concept of stress is highly problematic to define. A recent ESCA-NATO Workshop failed to provide a single standard definition, and that it arrived at six definitions illustrates, as Cox states, how "elusive ... [and] ... poorly defined" stress can be (Murray et al.,1996). One of the Workshop definitions, which shall be used for the purpose of this paper, is "an effect on the production of speech (manifested along a range of dimensions), caused by exposure to a stressor" (Murray et al., 1996). As such, stress may incorporate such diverse aspects as acceleration, vibration, temperature, noise, workload, mental load, and even emotional aspects of operator performance. Each may arise in a variety of ways, manifestations and effects and, more generally, have a negative effect on performance. From the Workshop a taxonomy of stress factors affecting speech were drawn up, with four categories: Physical, Chemical, Physiological and Psychological. Whilst these categories are distinct, they are neither exclusive nor exhaustive, and, for the sake of brevity, we shall consider only those which relate to the experimental findings described below. A more thorough appreciation of stress can be found in Baber & Noyes (1996) and Murray et al. (1996).

2.1 Noise

From a review of ambient noise levels and speech production, it is apparent that ASR performance can be enhanced if enrolment and training occur within the same environment as the operational setting (McCauley, 1984). This is due, in part, to the Lombard Effect, where increased vocal effort arises to "project through ambient noise" which acts as a stressor (Murray et al., 1996). Noise may also have the effect of depriving the speaker of feedback and self monitoring, perhaps increasing the cognitive load and affecting performance accordingly.

2.2 Memory Load

The Lombard Effect is a clear example of adaptation to stress. However, as Simpson (1986) observed, although it may be possible to mediate the effects of some, particularly physical, stressors, it may not be so simple when faced with a stressor that threatens cognition. In Klapp et al's. (1973; 1976) and Sternberg et al's. (1978) work, significant increases in response time for encoding and retrieval of stimuli and responses, illustrate fundamental effects at a cognitive level. Memory load might be a crucial component in this equation and therefore needs to be examined further.

3 Experimental Outline

Using Klapp et al's. and Sternberg et al's. findings as a basis, it was hypothesised that under increased memory load and/or ambient noise, overall reactions times and accuracy to the task would be significantly effected. By using a speaker-dependent ASR interface, idiosyncratic changes and nuances in the participants' speech patterns could be captured. The paired-associates used were the first six letters in the ICAO alphabet: alpha, bravo, charlie, delta, echo and foxtrot. A 2 x 2 (within) factorial, fully repeated and completely counterbalanced design was used. Controlling the variables in this way kept the task content homogenous, across subjects and conditions, and minimised error variance through order, practise, learning and/or fatigue effects.

The independent variables were memory load (simple or complex) corresponding, respectively, to congruent paired-associates where the stimulus word was repeated and incongruent paired-associates where the word after the stimulus word was required; and ambient noise levels (quiet or noisy) corresponding, respectively, to 50db and 85db of white noise generated through headphones. Objective measures of the dependent variables (response time, system accuracy, user accuracy and speech level) were taken and analysed.

4 Results

All data were tested and, where necessary, transformed to meet the assumptions for parametric analysis. Two-way ANOVAs were carried out on all the measures.

4.1 System Accuracy

No significant main effects were observed for either task ($F=0.11$, df.1,92; $p>0.05$), or noise ($F=0.48$, df.1,92; $p>0.05$) and no interaction was observed ($F=0.37$, df.1,92; $p>0.05$). No further analysis was carried out.

4.2 User Accuracy

A significant main effect was observed for task ($F=11.81$, df.1,92; $p<0.001$). No other significant effects were observed for noise ($F=0.24$, df.1,92; $p>0.05$) or for any interaction ($F=0.03$, df.1,92; $p>0.05$). Comparing the mean task scores illustrated that participants were more accurate in the simple task (23.96) than in the complex task (23.52).

4.3 Response Times

A significant main effect was observed for task ($F=89.62$, df.1,92; $p<0.0001$). No other significant effects were observed for noise ($F=1.15$, df.1,92; $p>0.05$) or for any interaction ($F=0.10$, df.1,92; $p>0.05$). Comparing the mean task scores illustrated that participants responded more quickly in the simple task (1345.24 m/s) than in the complex task (1811.75 m/s).

4.4 Speech Level

No significant main effects were observed for either task ($F=0.72$, df.1,92; $p>0.05$), or noise ($F=2.61$, df.1,92; $p>0.05$) and no interaction was observed ($F=1.33$, df.1,92; $p>0.05$). No further analysis was carried out.

5 Discussion

The non significant data for system accuracy was to be expected as the participants trained the interface for each noise condition. As such, and in keeping with McCauley (1984), the ASR used optimised templates in each condition and would be expected to maintain similar levels of accuracy across the conditions.

The significant main effect for task in user accuracy indicates that the complexity of the task duly affected performance. Whilst no other effects were observed for either noise or an interaction, it would appear that memory load was a primary stressor in the experiment. This is further supported by the similar findings for response times where the same main effect appeared significant whilst noise and any interaction effects did not. The participants were able to complete the task more accurately, and in a quicker time, when the task was a simple repetition task rather than anything more complex.

The lack of significant data for speech levels is remarkable, and does not support the Lombard Effect. However, as Junqua (1995) states, it is highly variable, and difficult to simulate. Whilst an increase in the speech levels was observed, the measurements were approximate and may not have been sensitive enough to illustrate the effect. Similarly, the lack of significance for ambient noise levels was also unexpected. However, this may be explained by the work of Miles et al. (1991) and their work on irrelevant speech. As such the white noise may not have been sufficiently loud enough (even at 85db) to effect speech, or similar enough to speech (in that it is a constant noise) to cause interference.

Relating these findings back to Klapp et al. and Sternberg et al. the results support a cognitive basis to speech production under stress. This will be examined further using the data, and audio recordings of the participants, and partitioning the response time into component parts of articulation onset time (AOT), and utterance duration. Already from preliminary analyses it would appear that AOT is the fundamental factor that increases under stress.

If this is validated, it will further reinforce the argument and support the earlier work. Applications of the paired-associates paradigm are already being considered in ASR form-filling tasks. A simple extension of the one word stimulus-response format to a question and a number of possible responses could be developed so that the theory builds the basis for more applied research and the intermediate research hopefully provides a synthesis of the two perspectives.

6 References

Baber, C., & Noyes, J., (1996) "Automatic Speech Recognition in Adverse Environments". In, *Human Factors*, 38(1), pp.142-155.

Junqua, J-C., (1995) "The Influence of Acoustics on Speech Production: A Noise-Induced Stress Phenomenon Known as the Lombard Effect". In, I. Trancoso & R. Moore (eds.), *Proc. of ECSA-NATO Workshop on Speech Under Stress*, Portugal 1995.

Klapp, S.T., Anderson, W.G., & Berrian, R.W. (1973) "Implicit Speech in Reading, Reconsidered". In, *Journal of Experimental Psychology,* 100, pp.368-374.

Klapp, S.T., & Wyatt, E.P., (1976) "Motor Programming Within Sequences of Responses". In, *Journal of Motor Behaviour*, 8, pp.19-26.

McCauley, M.E., (1984) "Human Factors in Voice Technology". In, F.A. Muckler (ed), *Human Factors Review 1984*. Santa Monica, CA. Human Factors Society.

Miles, C., Jones, D.M., & Madden, C.A., (1991) Locus of the Irrelevant Speech Effect in Short-Term Memory. In, *Journal of Experimental Psychology: Learning, Memory & Cognition*, 1991, Vol.17, No.3, pp.578-584.

Murray, I.R., Baber, C., South, A., (1996) "Towards a Definition and Working Model of Stress and its Effects on Speech". In, *Speech Communication*, 20, (1996), pp.3-12.

Simpson, C.A., (1986) "Speech Variability Effects on Recognition Accuracy Associated with Concurrent Task Performance by Pilots". In, *Ergonomics*, 29(11), pp.1143-1357.

Sternberg, S., Monsell, S., Knoll, R.L., & Wright, C.E., (1978) "The Latency and Duration of Rapid Movement Sequences: Comparisons of Speech and Typewriting". In, G.E. Stelmach (ed), *Information Processing in Motor Control and Learning*. New York: Springer.

Integrated Recognition and Interpretation of Speech for a Construction Task Domain

Hans Brandt-Pook, Gernot A. Fink, Sven Wachsmuth, Gerhard Sagerer*
Technische Fakultät, Angewandte Informatik Universität Bielefeld
Postfach 100131, 33501 Bielefeld, Germany
http://www.TechFak.Uni-Bielefeld.DE/~gernot/

1 Introduction

The development of speech processing front-ends for the controlling of complex systems has received more and more interest during the last years. Usually this task is divided in two subtasks. The *speech recogniser* records the utterance and puts out a corresponding text, and the *speech understanding module* tries to extract an internal representation of the meaning of the utterance. As shown in Figure 1 the recogniser uses an acoustic model as well as a statistical language model, which are both generated from a corpus. The speech understanding uses linguistic and domain specific knowledge designed by experts.

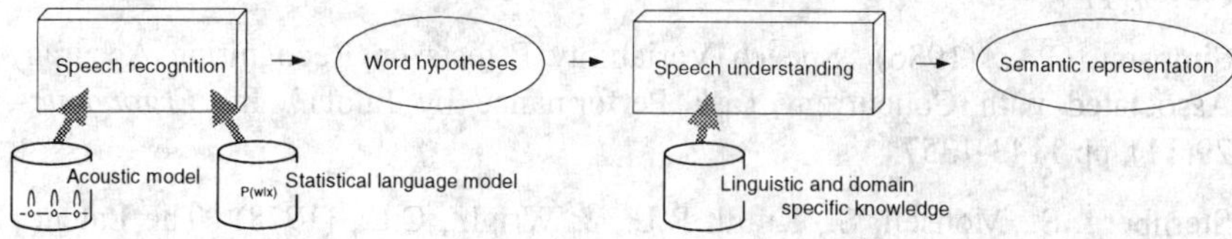

Figure 1: General architecture of speech processing systems

Traditionally speech processing systems are organised strictly horizontally under two aspects. Firstly, the knowledge represented in the speech recogniser is completely independent from the linguistic knowledge used in the understanding component. Secondly, within the linguistic analysis the utterances are processed sequentially in several layers of abstraction (syntax, semantics, pragmatics). These general principles can be found in most speech processing projects (see e.g. (Eckert et al. 93, Lavie et al. 97)). This approach seems to be motivated

* This work was partially supported by the German Research Foundation within SFB360.

by different paradigms used for the components. Whereas the speech recognition usually uses statistical methods, the models appropriated for speech understanding are often knowledge based. The advantage of this approach is that each component can be optimised for itself. On the other hand an interaction of the components increases the robustness of speech processing systems (McKevitt 94). So we propose in this paper a technique that tightly couples the speech recogniser and the speech understanding. We first give an overview of the system architecture and then we focus on the language modelling and the linguistic analysis respectively. Finally evaluation results are presented.

2 Integrated System Architecture

The general idea of the approach is to overcome the horizontal organisation of the system architecture by using a vertical approach for the knowledge representation and an integrated processing of speech input. Figure 2 shows the tightly coupled components. In contrast to traditional systems, the speech recognition process is influenced by the linguistic and domain specific knowledge, which is also used by the understanding module. Due to the use of a declarative *domain specific grammar* the performance of the speech recogniser increases and it produces structured results, that minimise the effort of the understanding component. Furthermore, our approach works incrementally, i.e. the recognition and the analysis of the human instruction deliver results while the human instructor has not yet finished the utterance.

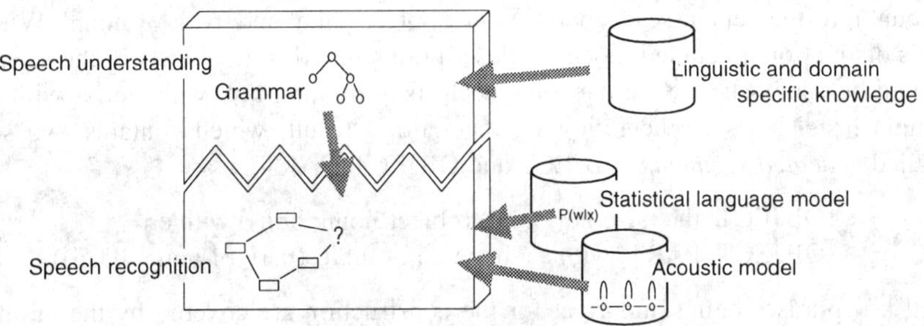

Figure 2: Overview on the proposed architecture

3 Speech Recognition and Language Modelling

In speech recognition the dominant paradigm is the statistical approach. For the acoustic description of individual speech sounds and words Hidden Markov Models are employed. In order to capture language regularities statistical models, which provide a probability for any given word sequence, are combined with the acoustic models. A joint optimisation of the probability for observing a sequence of acoustic events together with the sequence of spoken words yields the result of the recognition process.

As both types of statistical models have to be estimated on large amounts of training data that might not be readily available, we developed an approach to make use of declarative language constraints. A parser integrated with the recogniser robustly combines the restrictions of a context free grammar with the acoustical scores and the statistical language model (Wachsmuth et al. 98).

Due to the goal 'increasing robustness' the integrated grammar differs from a strict linguistic grammar definition. We do not use a sentence grammar that decides after the end of an utterance about its validity. Instead the development of the grammar for the approach proposed here was based on two aspects: It has to support the speech recognition process as explained above, and it has to define well formed parts-of-speech as basic units for the speech understanding component. These basic units must cover syntactic knowledge as well as semantic information. Due to this vertical organisation of the knowledge a major part of the complex syntactic-semantic analysis is already done by the speech recognition process.

For the modelling of the grammar we studied a corpus of simulated human-machine-interaction and a corpus of human-human-interaction. Both corpora contain instructions that guide a human constructor and a robot respectively. Thus we were able to respect the special features of instructions in this domain in the design of the grammar. For instance we divided the traditional word class 'noun' into the subclasses 'general-noun', 'object-noun' and 'part_of-noun'. With this distinction a detailed model of descriptions that denote objects in the scene could be established. So a part of speech like 'the long bolt with - err - with a round head' leads to the following recognition result, which contains two so called *accepted segments* (OBJECT and OBJECT_PART):

> (OBJECT: the (size-adj: long) (object-noun: bolt)) with err
> (OBJECT_PART: with a (shape-adj: round) (part_of-noun: head))

All key phrases being important for the construction are covered by the grammar, e.g. objects, intended actions, instruments needed for the execution of actions, ports of objects, and interventions to stop the execution immediately.

4 Linguistic Analysis

The analysis of the structured recognition results is realised within a semantic network language (Sagerer & Niemann 97). The semantic net is divided into three levels of abstraction. On the *interface level* the different segments modelled in the grammar are covered by corresponding concepts. The special contribution of the *interpretation level* to the entire processing is the combination of the segments to relevant constituents. Furthermore on this level for each of the obtained constituents the semantic representation is extracted and these are combined to a representation of the whole utterance. Figure 3 shows the incremental processing of an utterance. Finally on the *dialogue level* of the semantic

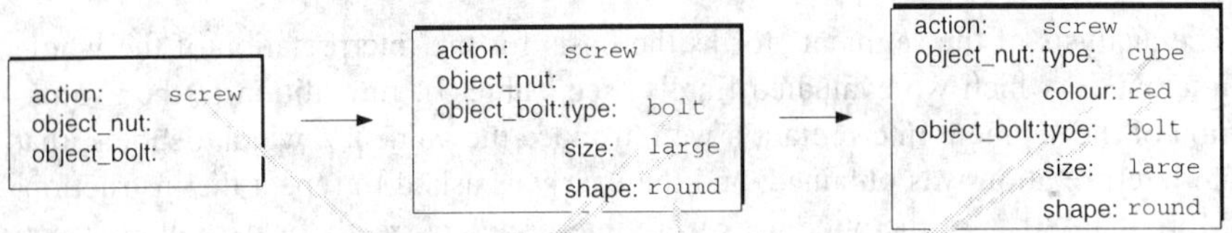

Figure 3: Incremental generation of a semantic representation

net the semantic representation of a single utterance is analysed with respect to the dialog context. Based on this, an adequate system reaction is chosen and a feedback for the users instruction is generated.

5 Results

The approach was evaluated on two sets of altogether 627 independent instructions given spontaneously by speakers who were familiar with the system. Their task was to ask the robot for taking an object presented on an image of the construction scene. The coverage of the grammar on this test set is about 74 %. The influence of the statistical language model (LM) and the domain specific grammar (DSG) on the speech recogniser is given in Table 1. The best results were achieved by combining the LM and DSG, which shows the adequacy of the approach concerning the speech recognition.

without LM and DSG	with LM	with DSG	with LM and DSG
49.1 %	31.3 %	32.7 %	28.8 %

Table 1: Recognition results (word error rate)

Secondly, we evaluated the semantic analysis of the segments. In Table 2 *correct* denotes segments, that are interpreted completely correct, whereas an interpretation was *partially correct*, if some features (e.g. the colour of the object) were not detected. The remaining segments were processed wrongly or completely ignored. We obtained the best result by using only the DSG. Regarding Table 1 and Table 2 it can be seen, that the speech processing can be optimised with respect to different goals: to achieve the lowest word error rate, the combination of the LM and DSG is the best solution, but under the perspective of the following interpretation it is better to use only the DSG, because this method focuses on the key phrases relevant for the domain.

	Correct	Partially correct.
without LM and DSG	34.1 %	4.6 %
with DSG	61.2 %	10.9 %

Table 2: Semantic interpretation of the segments

The analysis of the segments forms the basis for the interpretation of the whole utterances, which we evaluated finally (see Table 3). In addition to the evaluation of the segment interpretation we introduce the value *no*, which denotes, that no interpretation was obtained, and the user was asked to repeat the instruction. More than 80 % of the utterances were interpreted correctly or partially correct, which demonstrates the strength of the approach.

correct	partially correct	False	No
42.7 %	39.4 %	7.3 %	10.5 %

Table 3: Interpretation of the utterances

6 Conclusion

In this paper we presented an incremental speech recognition and interpretation system. The proposed architecture differs from traditional ones under two aspects: the knowledge is organised vertically and the modules work together in an integrated manner. The use of a domain specific grammar improves the recognition results as well as the interpretation performance of the system.

7 References

(Eckert et al. 93) W. Eckert, T. Kuhn, H. Niemann, S . Rieck, A. Scheuer, E. G. Schukat-Talamazzini: *A Spoken Dialogue System for German Intercity Train Timetable Inquiries.* Proc. European Conf. on Speech Communication and Technology, Berlin, 1993, pp. 1871-1874.

(Lavie et al. 97) A. Lavie, A. Waibel, L. Levin, M. Finke, D. Gates, M. Gavalda, T. Zeppenfeld, P. Zhan,: Janus III*: Speech-to-Speech Translation in Multiple Languages.* Proc. Int. Conf. on Acoustics, Speech and Signal Processing, München, 1997, pp. 99-104.

(McKevitt 94) P. McKevitt (ed.): AAAI-94 Workshop Program: Integration of Natural Language and Speech Processing, Seattle, Washington, 1994.

(Sagerer & Niemann 97) G. Sagerer, H. Niemann*: Semantic Networks for Understanding Scenes.* Plenum Press, New York, 1997.

(Wachsmuth et al. 98) S. Wachsmuth, G. A. Fink, G. Sagerer*: Integration of Parsing and Incremental Speech Recognition.* Proc. European Signal Processing Conf., Rhodes, 1998, pp. 371-374.

Exploring the Learnability of Structured Earcons

Edwin Pramana and Ying K. Leung,
School of Information Technology,
Swinburne University of Technology,
P.O. Box 218, Victoria 3122,
Australia

1 Introduction

The sound card is now becoming a standard component of the personal computer. Increasingly, software applications are exploiting the audio channel to enrich the human-computer interface, using sounds to complement and supplement the visual information presented on the screen. In recent years, two approaches in non-speech auditory interfaces have emerged – earcons and auditory icons. Earcons and auditory icons are distinct in their respective approaches to the representation problem. Earcons are totally symbolic and incorporate a musical approach while auditory icons may be symbolic, metaphorical or nomic and incorporate everyday listening.

Despite the wide range of research conducted on earcons by researchers, systematic application of earcons at the user interface is scarce. This is perhaps attributed to the fact that humans have limited capacity in learning arbitrary sounds (Patterson and Milroy, 1980). Whilst technology enables families of earcons to be easily created, their usability is primarily determined by how easily they can be learned (Leung, Smith, Parker and Martin, 1997). It is demonstrated by the fact that the research conducted on earcon's learnability generally involved a small set of earcons, typically not more than eight.

This paper describes a study to explore the learnability of structured earcons. It is hypothesised that (1) structured earcons are more easily to learn if the user knows how the timbre and pitch of the earcons are mapped to the objects or events they represent and that (2) user's prior knowledge in music affects the learnability of structured earcons. The experiment was conducted in a simulated multi-tasking environment. The results of this study will provide useful information about how structured earcons should be designed and how we should train users to learn a large set of structured earcons.

2 Experiment

2.1 Subjects

Forty-eight undergraduate and postgraduate students from Swinburne University, males and females aged between 19-38 years, participated in this experiment. Prior to the experiment, subjects were asked about their musical knowledge. Subjects were categorised as having musical knowledge if they knew the basic music notations and/or could play an instrument.

None of the subjects had heard the earcons used prior to the experiment. All subjects had used computers before and were familiar with Microsoft Windows 3.1, which was used as the multi-tasking environment in this experiment.

2.2 Materials

Sixteen structured earcons, representing four different processes, each of which with four possible outcomes, were created. The four processes, running in background, were downloading files, compiling large programs, checking the arrival of an e-mail message and data back-up operation. Each of these four processes has four possible outcomes: (1) successful completion, (2) failed completion (disk full), (3) failed completion (disk read error) and (4) failed completion (disk write error). The earcons were carefully designed by following Brewster's Earcon Design Guidelines (Brewster, 1993) and created using the Roland MIDI keyboard. Timbre and rhythm were the two dimensions of sound used in the earcons, mapping onto the process and the specific outcome respectively (see Tables 1 and 2). Each earcon was composed of three to four pitches.

Process	**Timbre**
Download	Vibraphone
Compile	Trombone
Back-up	Hammond Organ
E-mail	Acoustic Grand Piano

Table 1. Process-Timbre Association

An application was developed using Borland Delphi (version 1.0) to run under Microsoft Windows 3.1 for the participating subjects to learn the earcons, and a multi-tasking environment was simulated for testing each of the subjects on the earcons learned. Two versions of the application for the subject to learn the earcons were made, one included an explanation about the structure of the sixteen earcons and how their timbre and rhythm map to the processes and outcomes, and the other without.

Event	Rhythm
Successful	
Disk Full	
Disk Read Error	
Disk Write Error	

Table 2. Outcome-Rhythm Association

2.2 Procedure

Subjects were divided into two main groups. One group of subjects (20) was given an explanation about the earcons at the start of the experiment and the other group (28) was not. Of these two groups, 7 and 16 subjects respectively were classified to have musical knowledge. The experiment consisted of four stages and each stage consisted of two phases. During each stage, the subject was required to learn four additional new earcons, representing a new process and its four outcomes, while the testing phase would involve these earcons as well as all previous ones. This incremental approach was used to simulate a more realistic way of learning.

During the learning phase, each of the earcons to be learned was played to the subject once and the name of the event that was to be associated with it displayed on the screen. In order to minimise the order effect during the learning phases, the order of earcons to be learned was rotated using a Latin Square arrangement.

On completion of the learning phase, the subject proceeded to the test phase. During the test phase, the subject was asked to perform a simple typing task. The purpose of this was to simulate a typical work environment where the earcons were played to indicate the status of a process. Each of the earcons learned was played randomly while the typing task was in progress. The subject was asked to identify the earcons as quickly as possible via a pop up window.

During each stage of the experiment, the subject was given a maximum of three trials to achieve the specified criterion level before progressing to the next stage. The number of correct identifications required for the four stages with 4, 8, 12 and 16 earcons respectively were 4, 7, 10 and 13. If the subject failed to achieve the specified criterion after three trials in a particular stage, the experiment would be terminated.

	Without music knowledge	With music knowledge
Without description	41.67%	50.00%
With description	53.85%	71.43%

Table 3. Percentage of subjects successful in learning the earcons

3 Results and Discussion

The percentage of subjects successful in learning the earcons is presented in Table 3. Successful completion of the test means that the subjects have learned the whole set of earcons (16 earcons) and have been successfully tested. There are two independent variables used; explanation of the mapping of earcons (Factor A) and subject's prior experiences in music (Factor B). The results of the experiment are elaborated in Table 4 and Table 5.

Using one-way analysis of variance (ANOVA) procedure, both Factor A and Factor B were found to be significant in user performance of learning the earcons; for factor A, $F(1,1)=13.2$, $p<0.05$, and Factor B, $F(1,1)=7.85$, $p<0.05$.

Further analysis of the results showed that subjects are better in identifying the process (represented by the timbre), compared to the outcome (represented by the rhythm); the difference in user performance is statistically significant (t=18.2, df=3, $p<0.05$). An analysis of user reaction time also showed that both Factor A and Factor B are also significant; Factor A, $F(1,1)=10.18$, $p<0.05$ and Factor B, $F(1,1)=0.65$, $p<0.05$.

	Process	Outcome
Without description No music knowledge	97.22%	79.55%
Without description With music knowledge	94.78%	80.54%
With description No music knowledge	95.45%	80.99%
With description With music knowledge	95.43%	81.10%

Table 4. % of subjects successfully identified the processes and outcomes

	Without music knowledge	With music knowledge
Without description	5.57 seconds	5.61 seconds
With description	5.11 seconds	4.73 seconds

Table 5. Average time to recognise the meaning of the earcons

Learnability is the major limiting factor constraining the usability of earcons. In many applications, earcons can be designed in a structured way, thus allowing large number of earcons to be used. The results of this study showed that subjects learn a large number of structured earcons significantly faster and with less errors if they know the association between the abstract sounds and the objects/events they represent. An implication of this finding is that earcon designers need to give due consideration in terms of mapping between the sound dimensions and the events they represent. This consideration is particularly important when more dimensions of sounds are used to construct a large family of structured earcons.

4 References

Brewster, S.A., Wright, P.C. & Edwards, A.D.N. 1993, An Evaluation of Earcons for Use in Auditory Human-Computer Interfaces, *Proc of ACM INTERCHI '93*, Amsterdam, pp. 222-227.

Leung, Y.K., Smith, S., Parker, S. and Martin, R. (1997), Learning and retention of auditory warnings, *Proc of the 4th ICAD*, Palo Alto, pp.129-133.

Repeats, Reformulations, and Emotional Speech: Evidence for the Design of Human-Computer Speech Interfaces

Kerstin Fischer
University of Hamburg

1 Introduction: The Problem

At human-computer speech interfaces, irritations caused by system malfunctions cannot be completely avoided. These irritations are not just local problems which can be easily overcome; they constitute severe problems for human-to-computer communication in at least two ways:

Firstly, the acoustic characteristics of the users' utterances have been found to be very different if they constitute repetitions or reformulations of previous utterances. That is, if the system claims not to have understood a contribution by the speaker, the speaker will repeat his utterance, however, possibly with a different stress pattern, different phrasal intonation, with a strong emphasis on exact pronunciation or even hyper-articulation, and short pauses between the words. Some of these properties may cause severe problems for current automatic speech processing systems; for instance, Levow (1998) finds that the error rate in speech recognition rises from 16% to 44% for repetitions. That means that the characteristics of an utterance are very different if it constitutes a repetition of a previous contribution, and that these differences cannot be neglected in human-computer interaction (HCI).

The second problem concerns the fact that speakers may become emotionally involved when they are repeatedly confronted with errors by the automatic speech processing system such that the system's malfunctions may provoke emotional responses in the user; thus, the speakers' attitude towards the system may change over time. This change in attitude may have global consequences for the prosodic, lexical, and conversational properties of the speakers' utterances. For instance, the average pitch may rise, the local properties as the above may occur also when no irritation directly precedes the current utterance, people may start talking to themselves, and words (e.g. four-letter words) may be used the system has not been trained for. Huber et al. (1998) show that if a speech recognizer was trained on normal speech and tested on emotional speech or vice versa, the speech recognition rate decreases significantly. Like the local changes observed in direct reaction to system malfunctions, these linguistic properties (prosodic, lexical, and conversational) thus constitute great problems for current automatic speech processing systems which need to be addressed if HCI speech interfaces are to be successful. This paper will show which irritations can be found in reaction to system malfunction and how these can be addressed.

2 Method

In order to get data for the analysis of these speaker reactions, a corpus has been designed especially to provoke reactions to probable system malfunctions. In that scenario, the speakers are confronted with a fixed pattern of (simulated) system output which consists of sequences of acts, such as messages of non-understanding or insufficient perception, and rejections of proposals, which are repeated in a fixed order. This allows to compare the speakers' behaviour through time and therefore to analyse their strategies in repetitions, reformulations, and in situations of emotional involvement. For instance, in the dialogues a sequence of a rejection of a date, a misunderstanding and a request to propose a date occurs three times in each dialogue and allows to compare how each speaker's reactions to the system's utterances change over time. After it is clear how speakers, whose perspective is a system which is not functioning properly, react in general, it can be experimented with certain de-escalation strategies. Thus it is possible to initiate clarification dialogues if the system encounters problems with the user's speech, or to generate utterances which may possibly calm down an angry user. Therefore, the fixed dialogue structure not only allows to control for local and global changes in speaker behaviour, but also to experiment regarding speaker behavior by varying the system output systematically. The dialogues are finally analysed regarding their lexical, conversational, and prosodic properties.

3 Local and Global Reactions to System Malfunction

One obvious effect of speakers' getting angry is their use of vocabulary which expresses a negative evaluation of the system. In the corpus described above, this concerns four-letter-words and, for instance, the interjection *hm* which indicates dissatisfaction and divergence. However, more often the speakers become ironical, using vocabulary like *brilliant* or *very interesting*, which do not belong to standard word lists, either. Furthermore, metalinguistic vocabulary is frequently used, such as *this is not a proposal*, or *what I mean is....* Designing a system on the basis of human-to-human communication (HHC) causes wrong predictions on the probability of these lexical items if they are included in the system's lexicon at all.

Regarding discourse strategies, the dialogues recorded diverge from natural conversation in a number of ways; a property which may cause problems for current speech processing systems is the use of greeting acts in the middle of the dialogue, possibly to attempt a restart of the system, which is in contradiction with any dialogue model constituted on HHC data (Schmitz and Quantz 1995). Furthermore, speakers' metalinguistic statements of what they said and not said may include aspects which in natural dialogues do not occur and which may not have been accounted for in the linguistic models of the domain.

Finally, the acoustic properties of utterances have been found to change considerably if speakers have to repeat their utterances; if they are getting angry, even more changes occur. These changes can be attributed to attempts to make understanding easier for the automatic speech processing system, for example, by hyper-articulation, or to the changing attitude towards the system when the system is unexpectedly unsuccessful. The acoustic properties of repeated, reformulated and emotional speech include the lengthening of syllables, e.g. *mo:::nday*, increasing loudness, hyper-articulation, pausing between syllables, systematic variation of stress, and the occurrence of laughter, sighing, and audible breathing.

4 Addressing the Problem

Designing HCI systems always involves finding a way between adapting the system to the users' habitual verbal behaviour and imposing specific requirements on the user (Ogden and Bernick 1997); it is constrained by current technological possibilities on the one hand and by the adaptability (and willingness to adapt) of the speaker on the other. Likewise there are two ways to approach the problems described above: On the one hand, the system can be adapted to understand in spite of the properties characteristic of repetitions, corrections and emotional speech. Thus it need to be investigated what exactely the linguistic features are which differ from the normal training data and how recognition and processing can be adapted to be capable of dealing with these changes. Preparing the system for these peculiarities may mean to train a speech recognizer on data elicited under conditions similar to those described above; likewise, lexicon and other linguistic knowledge resources like a dialogue model need to be constituted on the basis of such data.

On the other hand, methods may be developed to prevent these irregularities. This may be done in two ways: firstly speakers could be taught to behave in a particular way; this may include instructions regarding syntax, vocabulary and certain conversational strategies (Ogden and Bernick 1997). However, there may still be aspects which are hardly controllable, and the usability of such a system may be affected since speakers need to be trained before they can use the system, which is not suitable, for instance, for telephone use. Furthermore, it is not unlikely that instructing the user about the system's restrictions may even trigger behaviour such as hyper-articulation and syllable-lengthening.

Besides explicitely instructing the speakers, preventing the occurrence of the above peculiarities can also be attempted by subtly guiding the speakers' behaviour and influencing their attitude towards the system, preventing them from getting angry. Here it may be useful to see which strategies speakers employ in conversation to ensure a harmonious flow of information and how these can be applied to HCI design.

5 Using De-Escalation Strategies from Human-to-Human Communication

Speakers devote almost every tenth word in natural conversation to anchor their utterances in the communicative situation (Fischer 1998). Thus speakers constantly provide feedback for their partners on the one side and make sure that their partners have understood on the other, by means of discourse particles, tag questions, and speech routines. Furthermore, with every turn, speakers display their understanding of the previous turn to each other (Sacks et al. 1974). Normally, they also make sure themselves that their utterances are understandable, for instance by self-initiating repair (Schegloff et al. 1977). If problems occur, such that the speaker has to reject a proposal, these possibly face-threatening acts are presented very carefully and are usually accompanied by accounts of this behaviour (Brown and Levinson 1987). Consequently, misunderstandings, caused by, for instance, recognition errors, which are very frequent in HCI, are very rare in HHC because speakers devote much attention to their prevention.

The transfer of some of these practices to HCI design is not trivial; for instance, the employment of discourse particles in system output demands not only an explicit description of their use in spontaneous spoken language dialogues and a system which can process the relevant higher level dialogue information, but also a speech synthesizer which is capable of generating the appropriate intonation contours for these lexical items. Furthermore, it is questionable whether signalling perception and understanding by the system is useful if it actually has not understood; thus those strategies meant to avoid (rather than repair) misunderstandings may be useless in HCI. In contrast, direct explicit accounts, such as explanations of the system's malfunctions, are more straightforwardly employed and only presuppose the recognition of critical situations. For instance, comparable to speakers' accounts of face-threatening acts in conversation, if the system detects changes in speaker behaviour which may be caused by a changing attitude towards the system, it may calm down the user by explaining its shortcomings or by apologizing for them. It may also be very effective to sum up the current state of the discussion, but this again requires elaborate capabilities including dialogue memory and the ability to compare the current state to a projected goal. Consequently, considering the difference between HHC and HCI, the least costly and most efficient way to influence the speakers's attitude towards the system, which also has been proven to be effective in the Wizard-of-Oz experiments, may be to use explicit accounts which can be previously generated and which only require that the system recognizes situations which make such accounts necessary.

6 Conclusion

Three ways of addressing the problems caused by speakers' reactions to system malfunction which are manifest in repetitions, reformulations, and emotional reactions have been presented: Automatic speech processing systems can be adapted to the peculiarities of utterances of this type, speakers can be explicitely instructed, and they can be subtly guided by means of features of natural conversation. For the latter alternative it was examined in how far strategies from HHC can be useful for HCI design, how costly an implementation would be and what could be gained. It can be concluded that while it is often impractical to instruct speakers before they begin their interaction with an automatic speech processing system, both alternatives, the adaptation of the system to real conversational conditions such as reformulations and emotional involvement and the employment of some of the less costly strategies speakers use in HHC, should both be followed.

7 References

Brown, P. & Levinson, S. (1987). Politeness. Some Universals in Language Usage. 2nd edition, Cambridge: Cambridge University Press.

Fischer, K. (1998). A Cognitive Lexical Pragmatic Approach to the Functional Polysemy of Discourse Particles. PhD Thesis, University of Bielefeld.

Levow, G.-A. (1998). Characterizing and Recognizing Spoken Corrections. *Proceedings of Coling/ACL '98*, Montreal, Canada.

Huber, R., Nöth, E., Batliner, A., Buckow, J., Warncke, V., & Niemann, H. (1998). You BEEP Machine - Emotion in Automatic Speech Understanding Systems. Proceedings of TDS '99, Brno, Czech Republik, pp. 223-228. Masaryk University Press.

Ogden, W.C. & Bernick, P. (1997). Using Natural Language Interfaces. In: Helander, M., Landauer, T. & Prabhu, P. (eds.). *Handbook of Human-Computer Interaction*. 2nd edition, pp. 137-161. Amsterdam et al.: Elsevier.

Sacks, H., Schegloff, E. & Jefferson, G. (1974). A Simplest Systematics for the Organization of Turn-Taking for Conversation. Language, 50 (4), 696-735.

Schegloff, E., Jefferson, G. & Sacks, H. (1977). The Preference for Self-Correction in the Organization of Repair in Conversation. Language, 53, 361-382.

Schmitz, B. & Quantz, J (1995). Dialogue Acts in Automatic Dialogue Processing. Proceedings of TMI '95, pp. 33-47. Leuven.

Task Oriented and Speech Supported User Interfaces in Production Technology

Rainer Daude, Hendrik Hoymann, Manfred Weck
Laboratory for Machine Tools and Production Engineering (WZL),
RWTH Aachen University of Technology

1 Difficulties in Machine Operation

Machine tools play a primary role in production technology, paying a significant contribution to industrial economics. Due to the aggravated market conditions and customer needs, manufacturing companies as well as their machines are required to be highly flexible. Unfortunately, more flexible machines are often more complex and thus complicated to use.

Nowadays, classical input/output (I/O) elements such as monitors, single buttons or rotary switches are fixed to the machine so that generally a simultaneous observation of the manufacturing process and the input unit is not possible. Additionally, operating procedures are often characterised by rigid, hierarchically structured dialogues that diminish the worker's productivity.

New HMI developments in manufacturing therefore have to solve two problems: mobility – supported by suitable mobile I/O media – and task orientation. Mobile operator units offer close and simultaneous observation of the manufacturing process and the control pendant. Task oriented HMI's will result in a clear, flat menu structure with flexible dialogues allowing direct access to functions that have to be released in the actual working situation.

2 Speech Control in Manufacturing

Speech based operating elements are a good solution for improved I/O media. Due to size and weight limitations, mobile systems must not be equipped with many buttons. Speech, however, offers high functionality with only few I/O elements as well as a very intuitive way of communication. Thus, mobility can be easily supported by speech, using wireless head-sets or microphones.

At WZL, speech recognition is used for activating functions of the control which are usually released through buttons at the machine operating desk. The user is enabled to activate commands like “coolant on” or “spindle start” from his current working position with his hands and eyes free for other tasks. The speech system is also employed to control the user interface of a head-mounted display (HMD) which provides a visual display of selected control, machine, and help information. Commands like “feedrate” make the corresponding data appear on the display. The ‘see-through’ capability of the HMD enables the user to pay visual attention to control information as well as to the machining process (Daude and Weck 1997). In addition to the speech recogniser it is possible to emit acoustic information, e.g. hints for error detection or troubleshooting.

Experiences and user evaluations show good results, but there is an even bigger potential in speech recognition than just „replacing buttons“, that should be exploited. Speech can serve as a vital means to overcome the problems of function oriented user interfaces as described above and boost the development of task oriented, clear, flat, and flexible dialogue structures. Fig. 1 shows the multiplying advantages of combining speech and task orientation.

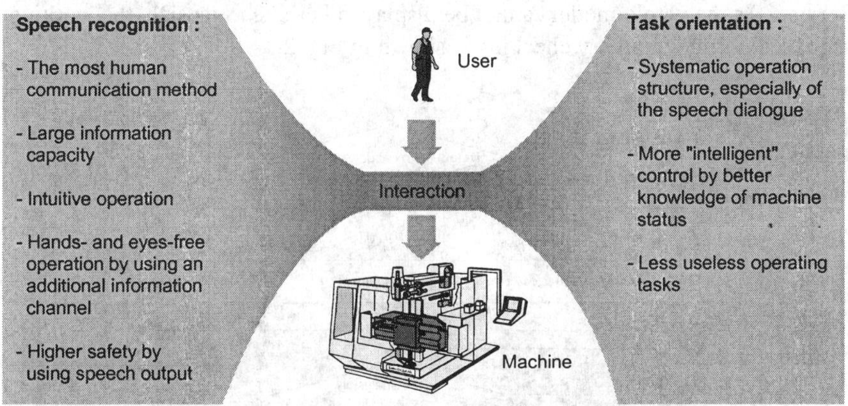

Fig. 1: Improving HMI’s by combining speech processing and task orientation

3 Task Oriented Operating Structures

Challenges in the area of task orientation have been found in the examination of workers’ tasks in production technology and the determination of an equivalent structure of the user interface. According to related literature e.g. in robotics (Dubey and Nnaji 1992), task orientation is defined as the structurisation of the user interface according to the users’ tasks, classifying activities that belong together either by theme or by time, and not referring to any functionality. *Tasks* are understood as those activities that help contribute directly to reach the given *goal*. For example, the superior goal may be the production of 100 work pieces of a certain type. According to this goal a task could be the programming of the workpiece geometry. Tasks are divided into *sub-tasks* that are characterised by some computational support; they may include other sub-tasks or mere *functions* which are defined as low-level activities with neither significant intellectual nor computational effort. E.g., this computer support could be a data base containing all possible tools, which helps the worker to select the right tool.

In order to transfer this definition to the practical application of machine tool operation and to adapt the dialogue structures to the workers activities, operators' tasks in manufacturing have been modelled. Basing on previous works by (Schlick and Luczak 1995) for milling machines, the task model set up here comprises all activities at a lathe and a handling system including NC programming as well as necessary incoming and outgoing information. The model is depicted as a flowchart according to (Boillot 1995). A typical working sequence is as follows: (1) Programming; (2) Chuck assembly; (3) Tooling; (4) Insert workpiece; (5) Determine raw part position; (6) Operation; (7) Check geometry; (8) Disassemble workpiece; (9) Disassemble chuck; (10) Diagnosis & Service. As the whole model cannot be displayed her due to its size, the working sequence during geometry checking is shown in Fig. 2.

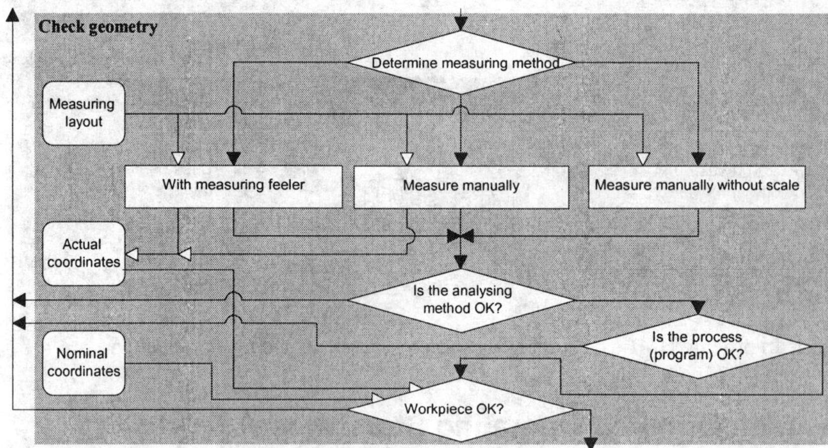

Fig. 2: Exemplary working sequence "Check geometry"

4 User Investigation

For the creation of the task oriented HMI, the work described here relies on a participative design approach. Therefore, a detailed user investigation with 29 workers (including operators, foremen, service technicians, programmers and product engineers) from six companies has been carried out[1]. Their age varied from 19 to 55 years, their work experience from one to 32 years.

As investigation method, guided interviews had been chosen as these are more suitable for asking background information (Nielsen 1997). The interviews had been divided into three parts. The first one referred to production surroundings including organisational aspects; these questions were asked to engineers or foremen and make the results of different companies comparable. The second part focused on the machining environment: The potential of speech processing in task oriented machine operation should be discovered, describing the areas with the biggest decrease of operation time and mental load. The third part should verify and refine the task model described in chapter 3.

According to the interviewed machining experts, a speech driven, task oriented HMI would improve productivity best in the areas of programming, set-up (e.g. tooling), approach, diagnosis and service. The investigation confirms that different production environments, i.e. single (or small batch) and mass production, focus on different areas with high potential to HMI improvement, Fig. 3.

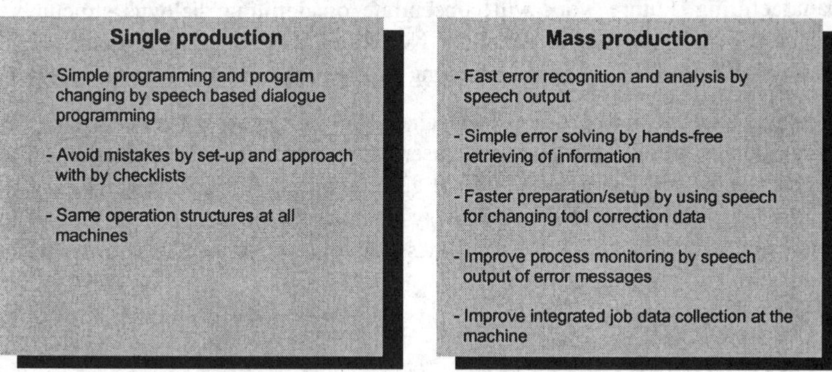

Fig. 3: Main areas of operator support in single and mass production

According to the operators, a real speech based dialogue with the machine would help in most of these areas. E.g. the machine control comes up with a (spoken) error message. Nowadays, the service technician often has to move between the control unit and the switch cabinet that is attached to the back of the machine. With speech, he could just ask the machine for error codes, an I/O setting, additional information or for some repair instruction from everywhere around the machine, and the machine answers acoustically.

---

[1] The work described here is funded in part by the European Commission through the TOSCA project (ESPRIT 29165). This contribution is gratefully acknowledged.

Very often, the experts mentioned the input of measured tool data as an excellent example of a speech driven, task oriented human-machine communication. In addition to measuring some characteristical points of the finished workpiece, the operator measures in between operating steps to protect further processing. From measuring the specified positions and calculating the difference to the nominal size, the tool data is derived. This working sequence takes several steps and navigation activities in function oriented HMI's. With the HMI targeted here, the user would just say e.g. "Change length correction of actual tool to 0.04", while measuring the next point. The speaking of this one sentence is the perfect description of the task to do: the machine is intelligent enough to know which is the "actual tool" and processes all information provided without the necessity of navigation within the menu tree.

5 Conclusion

The user investigation described here proves that the combination of task orientation and speech processing is a promising way for improving HMI's in manufacturing. Future work will concentrate on defining dialogue structures, vocabulary and grammar in detail for NC lathes with handling systems. This work will be carried out with a continuing strong participation of the end users.

A big challenge with the targeted speech controlled system is the occurrence of recognition errors. For this reason the user has to understand the limitations of a speech recognizer. The user should not get the impression that he is the reason for troubles. Therefore the technical system has to be adapted to the work station and the user itself. Noise robustness, dialogue design and task orientation could help a lot to reach that goal.

6 References

Boillot, M.H. *et al.* (1995). *Essentials of Flowcharting*. New York: McGraw-Hill.

Dubey, A. & Nnaji, B. (1992). Task-Oriented and Feature-Based Grasp Planning. *Robotics & Computer-Integrated Manufacturing*, Vol. 9, No. 6, 471-484.

Daude, R. & Weck, M. (1997). Mobile Approach Support System for Future Machine Tools. *Proc. Int. Symposium on Wearable Computers* (Cambridge, USA, October 13-14, 1997), pp. 24-30. Los Alamitos: IEEE Computer Society.

Nielsen, J. (1997). Let's Ask the Users. *IEEE Software*, 14, 3 (May), 110-111.

Schlick, C. & Luczak, H. (1995). Facharbeiteranforderungen an Mensch-Maschine Interaktionsflächen von CNC-Werkzeugmaschinen. *Jahresdokumentation 1995 der Gesellschaft für Arbeitswissenschaft*. Köln: Verlag Dr. Otto Schmidt KG.

Compatibility Problems in e-Commerce Interfaces

Horst Henn, Thomas Stober
IBM Deutschland Entwicklung GmbH

1 The Changing Environment

Classical closed shop information technology was concentrating on creation of a stable environment for employees. Central management of equipment, applications and user interfaces was key to overall productivity. Customers or business partners were hardly involved in the daily information technology operations. Tightly controlled data interfaces or satellite systems were the only interfaces to the outside world. Users would be faced with a limited set of functions and user interfaces, which typically have been customized by every application provider. Proper selection and education of user groups was required to manage the system complexity. User education was a prime investment for every system being introduced.

Today's organizations are faced with the challenge of interfacing directly with business partners and customers. Competitive services must be accessible directly by the customer not only via employees in the own organization. Typical examples are e.g. Internet based information services, online shopping or web based parcel tracking systems. Services must be applicable without any training, since the users expect them to come with "standardized user interfaces". It is not acceptable any more that users spend significant portions of their time learning to work with user interfaces needed for their daily work. The acceptance of user interfaces will have a major impact on the acceptance of services and products offered in the Internet environment.

In addition to the common personal computer an increasing number of new devices like mobile phone, hand held organizer, web TV or screen phone must be integrated into the IT organization. These devices have limited bandwidth, display and key entry capabilities which require tailored user interfaces. Users will only accept easy to use, familiar interfaces.

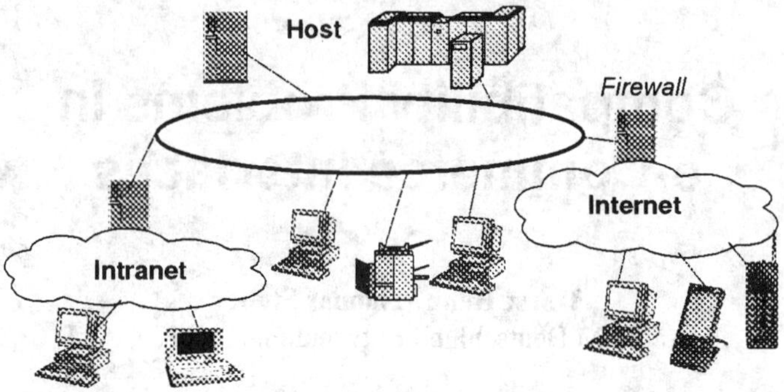

Figure 1: An e-Business system environment

Additional security functions must be provided in order to guarantee integrity as well as privacy of communication and to identify communication partners unambiguously. The following requirements are inevitable for interfaces which are applicable for e-Business:

- independent from a specific application context
- simple and intuitive to integrate and use
- connectable and able to communicate with network services

2 Interface Compatibility

Operation standards for hardware, the basic communication layers and the presentation layers are defined reasonably well in the Internet environment. A deficit in standardization can be observed when regarding the application levels. To close this gap common standards for seamless integration of hardware and software components must be provided. There are different major views to be considered when talking of interfaces:

- User interfaces
- Security interfaces
- Developer interfaces
- Services and contracts

2.1 User interfaces

There is a number of well-defined standards defining basically how user interfaces should be arranged and designed. These standards are key to the widespread use and acceptance by the user community.

Browsers provide a common base for platform independent interfaces. They can be used with minimum hardware and software prerequisites. Browsers are simple and common available tools to display and manipulate multimedia data from almost all kinds of information sources. They provide the basic navigation, presentation administration and security services.

Especially in e-Business most applications are based on Internet browsers, since the Internet service World Wide Web and its protocol HTTP offer a wide range of applications and are open for a large number of potential users. However, there are hardly any application specific standards established. Even simple tasks like searching or navigation within a HTML document are handled differently from website to website.

2.2 Security interfaces

Traditional personal identification and authentication by user id and password are replaced by certificates and digital signatures. Smart cards will play an important role in these security issues as well as the emerging trust centres which provide a public/private key infrastructure.

Today's smart cards offer access to a variety of different applications. Very reliable authentication, electronic signature and cryptography are just a few examples of typical smart card applications. Smart cards play an important role especially in e-business and banking. Other targets will be a new generation of electronic devices: mass produced low cost computers like screen phones which have been designed for a variety of all-day purposes (figure 2).

Modern smart card applications involve far more than just several card accesses. The Smart Card Solution business comprises the integration of multiple applications (e.g. adding a loyalty application to a credit card application on the same card), the connection to network and back-end systems (e.g. to the reservation system of an airline) as well as to complex card management systems of the card issuer. Accessing these systems is a very important issue in smart card projects.

2.3 Developer Interfaces

Heterogeneous platforms hinder information flow. It is a major task for today's IT to write, provide and use software for all operating systems. Java is one approach to implement platform independent software. Java developers can take advantage of a variety of class libraries providing objects out of which applications can be assembled (e.g. using the JavaBeans technology). Due to the inter operability of Java components, a large number of devices and systems can be integrated into an application (figure 3).

Regarding the increasing complexity of e-Business solutions it is an important issue to make application development as easy as possible. One approach to accelerate programming is the visual programming concept. The application logic can be implemented almost intuitively by connecting components with each other in a graphical editor.

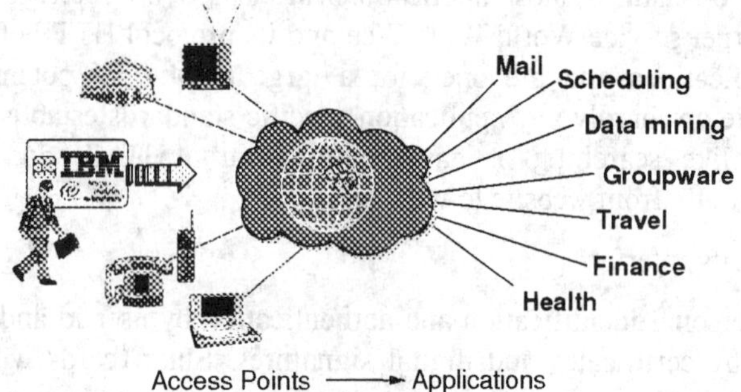

Figure 2: Smart cards are a key to access and exploit network based services

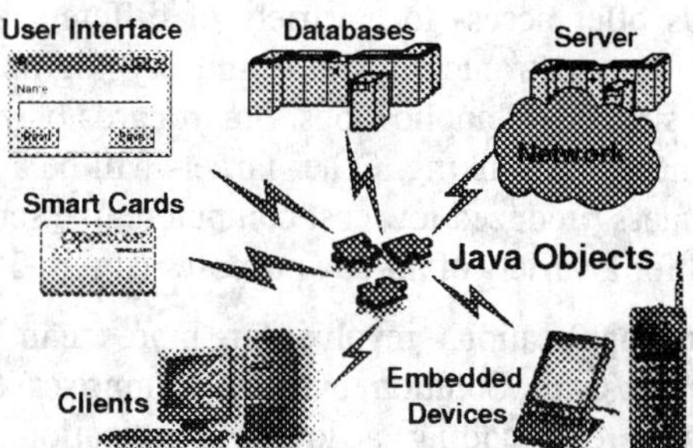

Figure 3: Using Java objects to access heterogeneous platforms and Systems

The following example describes OCEAN, an IBM development environment for smart card applications. This sample illustrates how modern development interfaces can look like (figure 4):

By using Java and the JavaBeans technology OCEAN allows to assemble even complex smart card solutions out of flexible components without having detailed smart card knowledge. With the help of a visual programming environment like VisualAge for Java, the application logic is easily implemented by connecting the OCEAN components with user interface components, databases and servers. The resulting applications are generated by VisualAge. They are independent from card manufacturer specific issues. Solutions are able to work with a variety of different cards and card readers - from today's file system oriented card to the latest generation of smart cards like JavaCard, Multos or Windows for Smart Card. Making smart card application development easier and faster, OCEAN has a significant impact on the development process of smart card based secure e-Business solutions.

2.4 Services and contracts

In today's information technology there is a significant shift from data to application exchange. There is an urgent need for common services and contracts. Standardized models for information gathering, contracting, payment and delivery are required by end-users who want to have the same look-and-feel as well as by developers who want to improve the reusability of their software modules.

One approach to achieve a "Convergence of Computers, Communication, Consumer Electronics, Content and Services" (Sun 1999) is Jini. Jini enables components of network systems to communicate with each other and share their resources (Sun 1999). The goal is to provide flexible components which can be used easily by other components, users, software or hardware. Devices make themselves available, they remove themselves and make an explicit device administration and configuration obsolete (figure 4). Easy access is provided by a common infrastructure. With help of Jini, the network turns into a dynamic and distributed system. Creating, using and maintaining of devices is simplified significantly.

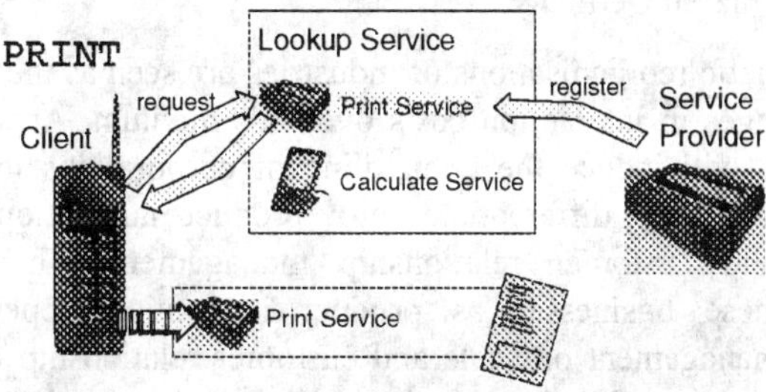

Figure 4: A printer device registers itself to a Jini lookup service. This print service can be used now by another device.

3 References

Sun (1999): *Jini Architectural Overview*. http://www.sun.com/jini

Stober, Thomas (1998): OpenCard Application Framework (OCEAN). In: Cap, C.H.: *JIT'98, Proceedings, Java Information Days*, Frankfurt, December, 4.-5., 1998, Berlin: Springer.

Customer Segmentation and Interface Design

Hans Schedl, Horst Penzkofer, Heinz Schmalholz
Ifo Institute for Economic Research, Munich (Germany)

1 Introductory Notes and Definitions

More and more business interactions are moving onto electronic networks like the internet. Forrester Research sees the increase in online retail sales in the US expanding from $2.4bn in 1997 (1998 estimates $7.8bn) up to $108bn by 2003. Similar growth rates have been estimated for Europe.

In comparing the number of internet users, the US is in the lead with: roughly 50m users in 1997 compared to 25m in Western Europe or 6m in Germany – according to EITO figures. Even corrected for population size the use rate in the US was double that in Germany.

Rapid and dramatic reorganisations of industries are seen as the consequence of significant changes in interaction costs over this medium. According to Hagel and Singer this will induce the unbundling of corporations into one of their businesses which they differentiate into: product innovation, infrastructure management and customer relationship management. They see inherent conflicts in these businesses as product innovation depends on speed, infrastructure management on scale and customer relationship management on scope. Specialised networked competitors are thought to outperform "bundled" corporations. They cite the computer industry, regional Bell operating companies, and newspapers as examples where huge vertically integrated companies specialise or will have to specialise on one business.

Contacts between the enterprise, its business partners and customers should – as a consequence – change significantly. In order to classify this change we differentiated between three simplified communication types:

- personalised contacts (visits, conferences, calls, faxes, letters or e-mails) with known persons,

- depersonalised contacts (e.g. via call centres, newsletters, mailings [physical or electronic]) and
- depersonalised human-to-computer contacts (via internet sites or proprietary networks to a firm server).

We wanted to ascertain induced changes in customer relations management: if personal contacts were still the standard in business-to-business customer relations and how the development towards the second or third contact option might develop in the future.

For this purpose we conducted 40 in-depth interviews with German firms. The limited number of interviews led us to a selection from three typical sectors of German industry: machinery, electrical machinery and electronics, and chemicals. We have classified these firms into four types: *Type one* includes small-to-medium-sized firms without own customer relations management with end users; these function and distribution logistics are transferred to trade organisations.

Type two comprises small-to-medium-sized firms with a small scope of products or small scale of production, strong global orientation (export shares above 50%), own customer relations management and own distribution. Markets and competitors are usually well-known by these firms. Usually long standing relations exist with (industrial) end-users. Products tend to be complex and necessitate extensive explanations. We might call this group small-to-medium-sized specialists.

Type three contains firms with more than 500 employees and a medium-to-large scale of production. These firms tend to profit from automated communication processes. Six of the nine firms in this group do not reach the end user of their product in pre-sales transactions. Call-centres or the internet would be solutions for them to gain proximity to the end-user.

Type four consists of firms of more than 1000 employees with at least an important share of complex and/or expensive products that are customised. These firms, though big in size, are specialists in transparent, competitive and global markets.

For a more detailed analysis of communications-media use in customer contacts, we defined six different transaction types in order to characterise the typical customer relation transactions. The first three cover standard pre-sales activities, the second standard after-sales transactions:

- information on the firm and its products,
- the handling of requests and offers,

- ordering,
- the handling of complaints,
- classical service (e.g. installation, repair), and
- professional services (e.g. consultancy, process optimisation, maintenance).

In a next step we tried to identify factors which may have led to the observed developments. We assumed three factor-groups: the first were phases of use, planning and experience, the second size-related factors and the third product-related factors. The outlook evaluates further developments.

2 The Use of Communication Types

The use of depersonalised media in customer contacts was the exception in our sample: very few firms actually used call centres or computer interfaces in one of the typical customer relations transactions. If these technologies were used, their installation was mostly recent. Similar statements could be made for the use of modern messaging technologies with the exception of e-mail, where use is relatively stronger compared to voice-mail or unified messaging.

3 Factors Shaping the Development

One important factor for this development has been the orientation of IC investment on intranets aimed more at productivity increases than communication purposes. In many of the firms the convergence between telecoms and computers is just starting.

Another important factor was firm size: for the firms of type one and two the use of depersonalised communication options was ruled out. In the case of type two firms this was combined with the third factor: specialisation on complex products. In these cases call centres were judged as not being appropriate. This opinion was shared by most of type four firms.

4 Outlook

Will traditional personalised contacts prevail in customer relations in the near future? The answer is 'yes' with exceptions. These exceptions depend on customer segmentation, product type, distribution system and transaction type. In general personalised contacts will prevail with

- important customers (e.g. A and B groups regarding an ABC classification) and, to a higher degree, with European customers (US customers, as far as cited in our sample, were more open to depersonalised or human-to-computer contacts),

- specialised products demanding a high degree of explanation and/or adaptation; this is quite often coupled with a distribution system that reaches a business end-user of the product or service,
- the handling of requests and offers, classical service and professional services.

The use of personalised communication media will change. E-mail will increasingly substitute traditional mail and fax. Video conferencing will reduce (though, at a significantly lower speed) face-to-face communication with big customers in distant locations.

Depersonalised human-to-human communications will, given the restrictions above on customer groups and specialisation, increasingly gain importance in information on the firm and its products, ordering, and in the handling of complaints for commodity producing enterprises. The technology of choice will be call centres. Due to their cost, the diffusion in SMEs may depend on the success stories of other SMEs. Especially if the firm has no direct contact to the end user (product needs installation or the scope of production is limited), call centres offer the possibility of improving the contact to end-users and a better perception of the market.

Human-to-computer contacts with the customer were seen as limited to information and ordering. An increase in this domain is seen to be correlated with generational change.

User-Side Web Page Customization

Yoshinori Aoki and **Amane Nakajima**
IBM Research, Tokyo Research Laboratory
1623-14 Shimotsuruma, Yamato, Kanagawa 242-8502, Japan
{yoshia, amane}@jp.ibm.com

1 Introduction

The World Wide Web (WWW), while undeniably a wonderful mass medium, does not support individual adaptation very well. Some Web sites have therefore started to provide a personalization service, which customizes Web pages for individual users [1]. At the Web site, a server-side program generates a customized Web page according to its policy, strategy, and the user's preference. In other words, a personalization service is a provider-side Web page customization service. On the other hand, users still have no mechanism for customizing Web pages by themselves without editing the original HTML files. It would be very useful if one could customize Web pages and share them with a group, since that would make it possible not only to add personal memos or hyperlinks to a Web page, but also to exchange opinions about the page. The ability to form communities around Web pages would give new value to the WWW.

This paper describes a mechanism for user-side Web page customization. The mechanism provides three functions, allowing users to (1) add their own objects such as text, images, hyperlinks, and Java applets to a Web page, (2) change the destination URLs of hyperlinks, and (3) share their customized pages with other users. Users can add text and hyperlinks to Web pages as personal memos. Next time they visit the same Web pages, they will see only their own memos, not the other people's memos, on those Web pages. Thus, they can use a Web page like a notebook to make annotations. If a user makes a memo public, other users can see the memo on the Web page. An administrator can hide unsuitable hyperlinks, and can also change their destination URLs. Figure 1 shows an example of a customized Web page.

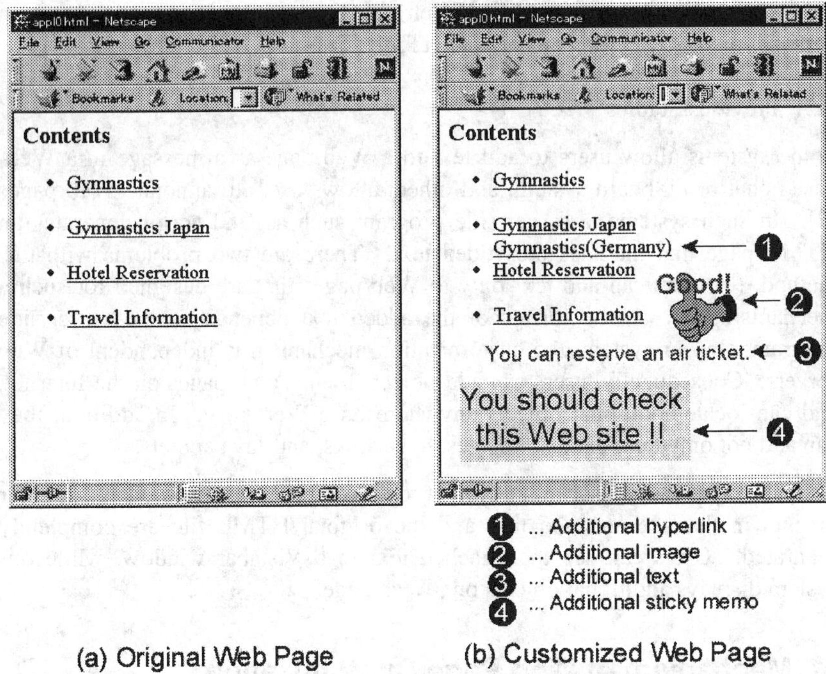

Figure 1 Example of a Customized Web Page

2 Approaches to Customizing Web Pages

2.1 Our approach

One of the major advantages of our mechanism is that it does not modify original HTML files. By separating original HTML files and customization data, it can handle existing HTML contents on the Internet. Figure 2 shows an overview of the mechanism. Customization data are stored in a *Customization Database* (CDB). The CDB is independent of the Web servers that hold the original HTML files. *Customizer* is a program written in Java and JavaScript that is embedded into an original HTML file at a proxy server. It interacts with a user and CDB, and then customizes a Web page on a Web browser. We separate original HTML files from customization data for the following reasons. First, a user usually has to edit an HTML file to add text or images to a Web page for customization, but it is impossible to edit another person's HTML file. Second, modification of an original HTML file affects other users' browsing, since they will all see a modified Web page. Hence a user cannot create a personal memo by editing an original HTML file.

2.2 Previous Studies

Some systems allow users to add text to a Web page as a message in a Web-based chat or billboard system, and others allows users to annotate Web pages [2]. In such systems, a server-side program such as CGI script generates an HTML page that includes the added text. There are two problems with this method: (1) users can add text only to Web pages that are designed for such a mechanism, and (2) the format of the added text depends on the server-side program. In our system, the customization mechanism is independent of Web servers. Consequently, users can add objects to any Web pages on the Internet, and can locate additional objects anywhere on a Web page. In addition, they can add not only text, but also images, hyperlinks, and Java applets.

Vistabar [3] allows a user to attach text to a Web page. The approach is similar to ours in that the attached text and the original HTML file are completely separated. Users can see the attached text in a Vistabar window, while our system displays additional objects on a Web page.

3 Mechanism of Web Page Customization

In Figure 2, when a user requests a Web page, a program called Customizer is embedded into an HTML file on the fly. Customizer is written in Java and JavaScript, and sends the URL of the Web page and the user ID to CDB. Customizer can get the URL by checking the href property [4], and the user ID by using a cookie or through interaction with a user. It then receives the customization data prepared for the Web page, and immediately reflects the customization in the user's Web browser. Customizer uses functions of Dynamic HTML to add objects to a Web page or change the destination URL of a hyperlink. For example, to add a hyperlink to a Web page, it creates a layer and puts a hyperlink on it. Users can add text, images, hyperlinks, and Java applets, because any HTML statements can be put on a layer. The destination URL of a hyperlink can be changed by overwriting the href property of a link object with a URL specified by a user [4].

Customizer opens a control panel in which a user can give instructions on how he or she wants to customize the Web page. After getting a user's instruction, Customizer immediately reflects it in the Web browser window by using Dynamic HTML functions. Customizer stores each operation in CDB to share with other users. It also allows the user to specify access controls determining which other users can share his or her customization.

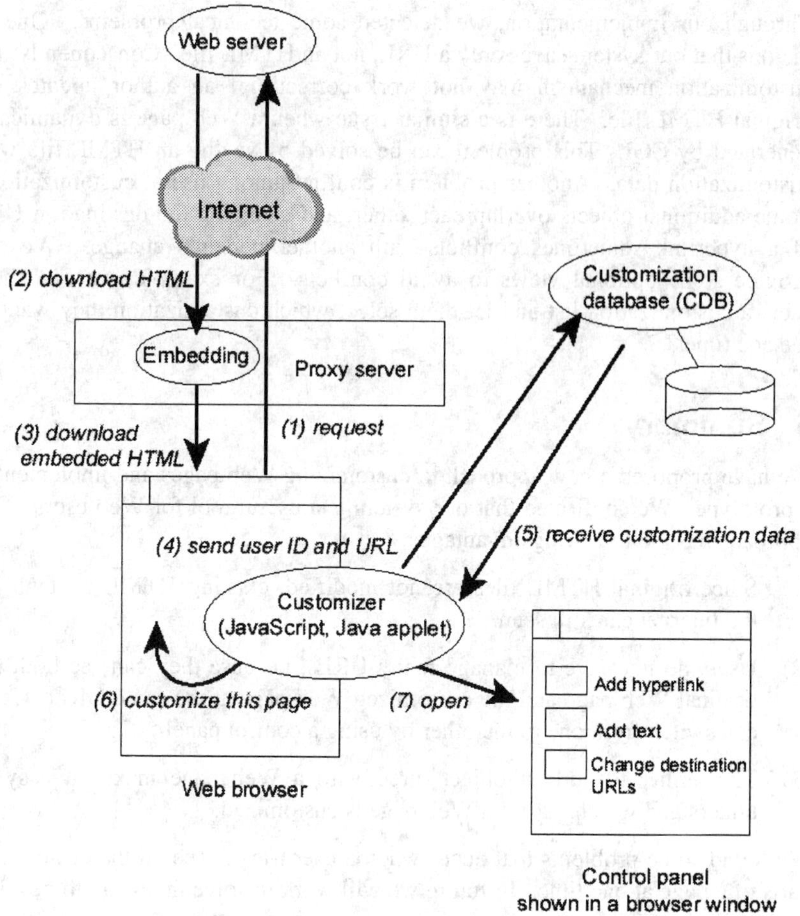

Figure 2 Mechanism of Web Page Customization

4 Discussion

We have implemented a prototype of our Web page customization system that does not include CDB. After adding an object to a Web page on a Web browser, a user can immediately see how it looks after customization. To reduce the time needed to customize a Web page, our system provides users with the WYSIWYG (What You See Is What You Get) environment for customization on a Web browser.

Through our implementation, we detected some technical problems. One of them is that our system saves only a URL, not an HTML file. Consequently, the customization mechanism may not work correctly if an author updates the original HTML file. There is a similar issue when a Web page is dynamically generated by CGI. This problem can be solved by saving an HTML file with customization data. Another problem is conflict among users' customizations. Some additional objects overlap each other, and changing the destination URL of a hyperlink sometimes conflicts with another person's change. We can provide some optional views to avoid conflicts. For example, we can show users a customization list and let them select which customization they want to see at a time.

5 Summary

We have proposed a new approach to customizing Web pages and implemented a prototype. We confirmed that our system is a useful tool for Web users. Our approach has the following advantages:

(1) Since original HTML files are not modified, existing HTML contents on the Internet can still seen.

(2) Users do not have to manage many URLs, because they can see both the original Web page and the customized Web page at the same URL. Users can switch from one to the other by using a control panel.

(3) The ability to add an object directly to a Web page makes it easy to understand which part of a Web page is customized.

We found some problems that occur when a user tries to see all the customized parts of a page at one time. In future we will work to solve them and to develop applications on the prototype.

6 References

[1] Yahoo!, http://www.yahoo.com.

[2] WebCT Educational Technologies Corporation, WebCT, http://www.webct.com.

[3] Marais, H. and Bharat, K., Supporting cooperative and personal surfing with a desktop assistant. In *Proceedings of UIST '97* (Banff Alberta, Canada, Oct. 1997), ACM Press, NY, pp. 129-138.

[4] World Wide Web Consortium, Document Object Model, http://www.w3.org/DOM.

A Direct Manipulation User Interface for a Telerobot on the Internet

Peter F. Elzer, Harald Friz, Ralf Behnke
Institute of Process- and Production Control Technology (IPP)
Technical University of Clausthal (TUC)
Julius-Albert-Str. 6
D-38678 Clausthal-Zellerfeld, Germany

1 Introduction

In the framework of a master«s thesis a user interface for an Internet telerobot has been developed. The work was a joint effort of the UWA and the IPP and mainly done in Australia. The authors are therefore indebted to Barney Dalton and Ken Taylor at the Department of Mechanical Engineering at The University of Western Australia, for providing the robot, lab room and advice. The interface overlays multiple monoscopic images of the robot's environment with a three-dimensional stick figure visualization of the robot's 5 degrees of freedom (DOF). The operator drags the elements of this visualization with a pointing device to obtain additional depth cues, measure size, location and orientation of objects, and to control the manipulator. This paper describes the interface design from a human-computer interaction perspective. The user interface was designed by applying the ecological approach of visual perception to visualize 5 degrees of freedom. This leads to a simple stick-figure-like structure that can be used to specify a pose in 3D space using only two-dimensional input devices. The user interface is implemented as a Java applet and can be used to control Australia's Telerobot on the Web at http://telerobot.mech.uwa.edu.au/.

2 Motivation

The Internet is becoming the standard medium for digital communication and is also increasingly used to control devices at remote locations. Using standard Internet technology for telerobotic applications offers the advantage of low-cost deployment. Disadvantages are limitation of varying bandwidth and time delays.

Human-computer interaction in Internet telerobotics is determined by the configuration of a typical personal computer, i.e. keyboard, mouse (or some other pointing device), video screen and audio speakers. The design of user interfaces is limited by the capabilities of web browsers and the programing language Java.

3 Australia's Telerobot on the Web

Australia's Telerobot on the Web (Taylor and Trevelyan 1995) was one of the first industrial robots connected to the Internet. It is a 5 DOF industrial manipulator that is set up in a laboratory environment with a table and some wooden blocks for pick-and-place tasks.

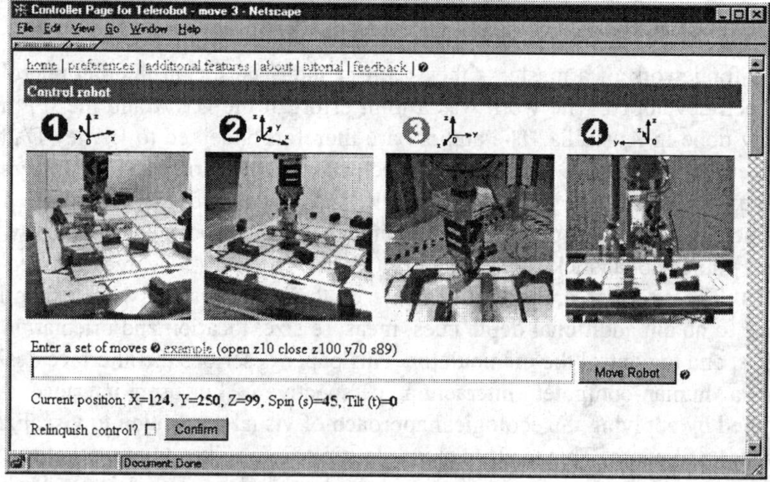

Figure 1: Standard user interface of Australia's Telerobot on the Web with multiple monoscopic views.

The original user interface (fig 1) is an HTML form that presents four images of the current workspace. Operators estimate the coordinates that the gripper is supposed to move to and enters them explicitly. Next the robot moves to the specified position. New pictures of the environment are taken and returned in an updated version of the form.

Basically, this interface supports all actions neccesary to control the robot, but it does not support direct perception and manipulation and therefore, requires much skill and training.

3.1 Augmented Reality

One way to overcome these problems is to apply Augmented Reality (AR) techniques. Computer-generated visual enhancements added to the monoscopic images would provide operators with additional visual cues and make it easier to judge both depth and pose and to visualize the gripper's next position. Controlling the robot by manipulating these enhancements with e.g. a mouse would relieve operators of the necessity to "think in numbers".

3.2 5 Degrees of Freedom

The key element of Usher is to visualize the robot's 5 DOF. Three DOF define the gripper's position in world coordinates (x, y, z), another two its orientation. Restricting the gripper's orientation to spin and tilt rather than roll, pitch and yaw is part of the robot's original design.

3.3 Data Visualization

The minimal visualization required to perceive depth and height correctly is a vertical reference line from the gripper reference point to the ground surface. This line relates the position of the gripper to the surface of the environment. Depth and height are directly perceived by relating the length of the reference line with the texture of the surface (a black grid drawn onto a white table) and the size of objects visible in the image (Friz 1998).

The design of these additional cues is based on Gibson«s ecological approach to visual perception (Gibson 1986) and results in the so-called "stick cursor" (fig 2). It is composed of 6 lines: the vertical reference line, two horizontal reference lines, two "needles" that visualize angles and the tilting axis, respectively.

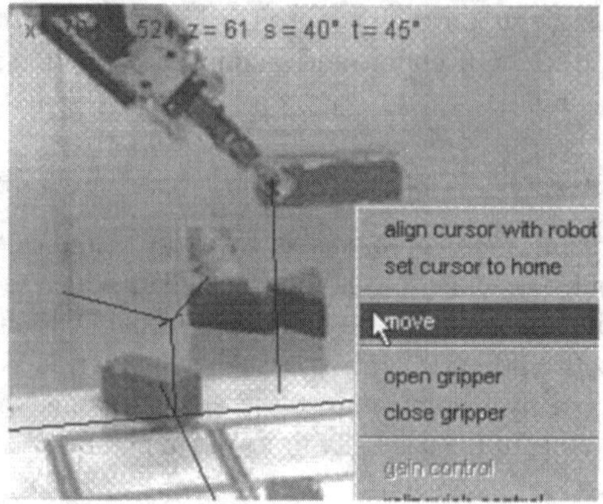

Figure 2: Usher, the augmented reality user interface for Australia's Telerobot on the Web (Friz 1998)

4 Interaction and Control

The stick cursor does not only provide static visualization but it can also be interacted with.

Each of the elements, like the reference lines or the needles, can be manipulated with a mouse. The three-dimensional model of the stick cursor is manipulated by dragging the two-dimensional projections of these lines.

Direct manipulation of the graphical elements offers several benefits:

- Additional cues for perceiving depth and solid angles.
- A tool to measure dimensions and orientations of objects.
- A user interface that allows to control the robot.

4.1 Additional Cues

Each element of the stick cursor can be manipulated directly. This increases the operator«s perception of kinematics and angles.

4.2 Measuring Objects

Usher can be used in many ways to directly measure location, height and orientation of object features.

4.3 Controlling the Robot

The stick cursor provides all features necessary to control the robot. For additional commands such as "gripper open/close" and "move to specified position" Usher uses a pop-up-menu (fig 2).

5 Conclusions

Usher, the Augmented Reality user interface (fig 2) for Australia«s Telerobot on the Web, gives an example of how visual enhancements can compensate for most of the disadvantages of multiple monoscopic views.

The stick cursor is a simple and versatile tool to visualize and specify locations in three-dimensional environments with a known surface topology. It also allows visualization and manipulation of two rotational angles. The stick cursor does not only present data to the operator but it can be interactively manipulated with a mouse. DOF are represented by one element of the stick cursor each. Hence they can be controlled independently. Consequently, users can manipulate 5 DOF with a two-dimensional mouse graphically.

Applications of the new interface are not limited to telerobotics. Almost any task that includes manipulation of elements in real or virtual three-dimensional environments could integrate the stick cursor as an additional tool, e.g. path planning for indoor cranes and blimps, manipulation of camera or light positions for theater or TV productions, telemaintenance, CAD and animation software.

6 References

Australia's Telerobot on the Web, http://telerobot.mech.uwa.edu.au/.

Friz, H. (1998). „*Design of an Augmented Reality User Interface for an Internet based Telerobot using Multiple Monoscopic Views*", Diplomarbeit, The University of Clausthal, Institute of Process- and Production Control Technology, 1998

Gibson, J. J. (1986). *"The ecological approach to visual perception",* Hillsdale, NJ: Lawrence Erlbaum, (reprint of the original edition from 1979)

Goldberg, K., Mascha, M., Genter, S., Rothenberg, N., Sutter, C., Wiegley, J. (1995). "Desktop Teleoperation via the World Wide Web", *Proc. IEEE International Conference on Robotics and Automation,* Nagoya, Japan, May 19-26

Taylor, K., Trevelyan, J. (1995). "Australia's Telerobot on the Web", *Proceedings of the 26th International Symposium of Industrial Robotics,* Singapore

Taylor, K., Dalton, B. (1997). "Issues in Internet Telerobotics", *Proceedings of the International Conference on Field and Service Robotics (FSR 97),* Canberra, 1997, pp. 151-157.

A new Concept for Designing Internet Learning Applications for Students of Electrical Engineering

Dirk Thißen, Dr. Birgit Scherff
FernUniversität Hagen

1 Introduction

At the beginning of the new millennium, the special demands of information society, resulting from rapidly growing information databases and the more and more decreasing information half period lead to what is called lifelong learning (Höfling 1996). Against this background, qualification concepts nowadays have to offer a kind of learning, that is independent of time and place and orientated towards the learner's individual needs (Zimmer 1995). Therefore, in the course of an interdisciplinary research and development project **InMedia**, which is supported by the Ministry of Economics and Research in NRW (a federal state of Germany), research has been carried out to develop a mediadidactic concept for designing computer-assisted learning for students of electrical engineering. Up to now, the concept has exemplarily been realized in form of an HTML-based internet learning application for *mechanical engineering basics in automation technology.*

2 Objectives and Strategy

The new concept has got two main objectives. First, with regard to today's high development expenses for computer based learning materials (Kerres 1995) and the desire to offer students a unique user-interface, it shall be applicable to many subjects of electrical engineering's curriculum. Second, the concept is to provide some additional value as compared with written material and especially internet learning applications, that simply put a single subject into multimedia form. To achieve these objectives a strategy for the concept's development was chosen, that is based on three main working fields.

By the analysis of curriculum's courses, structural similarities between the single subjects could be discovered. For example all investigated courses present the contents in theory, provide exercises and include examples in reality and/or professional life.

Then, the study of managers' demands with regard to collaborators' qualifications give information about professional life's special requirements for electrical engineers.

Last but not least, the questioning of students at an early point provides the opportunity to consider learners' experiences with computer assisted learning and their expectations for future developments.

3 The concept's basic approach

The result of the above listed investigations is a concept, that recommends a structure for internet learning applications for electrical engineers, that consists of eight different modules and six tools.

3.1 Modules

All of the eight modules have two things in common. Each of them is in a way related to a special subject's contents and second, they try to combine computer assisted learning with traditional learning methods, e.g. solving exercises by writing on a piece of paper. But the modules differ with regard to the learning objectives, didactic approach, and the use of media elements.

Course. Offers the contents of special subjects as it is done in specialist books, here however in multimedia form.

Exercise book. Offers a number of exercises, derived from a typology of exercises, especially elaborated for engineering subjects.

Media corner. Presents strategies, hints, and tools for autonomous information acquisition.

Base knowledge. Gives a brief summary of other subjects' contents necessary to comprehend the special subject.

Real life. Points out the actual use of a special subjects in reality and professional life.

Laboratory. Offers simulation environments for special items.

Formulary. Presents a combination of single subjects' formulas.

Cup of coffee. Offers opportunities for recovering.

For each module, the concept recommends a division into smaller information units (for example in *exercise book* a single exercise represents an information unit), which are to be connected by a linear navigational scheme. Beyond that, cross references are to connect all modules with each other. This allows guided as well as self-determined exploration of the single information units.

For each module, beside a detailed description a set of functions is specified, which serves as a guideline for the developer to facilitate an implementation of the modules in the concept's sense. Additionally, the development of internet learning applications for other single subject can partly be based on certain templates derived from the HTML-based realization of the concept for *mechanical engineering basics in automation technology*.

In dependence of the typical demands of a single subject, the developer can decide for his current internet learning application, whether to stress one or a few of the eight modules, perhaps leaving out one and so on. Besides, it is still left in the hands of the special subject's expert to determine the composition of a certain animation, figure, video, etc..

3.2 Tools

In addition to the modules, the new concept recommends six tools for coordinating and supporting the student's learning sessions to become more efficient ("to do things in the right way") and effective ("to do the right things").

Flipchart. Gives an overview of the user's navigational schemes, comments, times for learning, test ratings, etc..

Calendar. Offers to plan a single learning session and to elaborate learning schedules.

Plant. Motivates the student using the Tamagotchi-effect[1].

Search engine. Offers quick access to special information units or parts of them.

Calculator. Offers a scientific, interactive calculator.

As the tools *search engine* and *calculator* are standard in highly sophisticated computer based learning, the other four tools are new. They are all intended to take advantage of the medium computer, that is besides others the general possibility of recording learner's activities.

---

[1] In 1997 a popular and controversial discussed electronic toy for children. Children have to keep alive a figure in a mini computer by regular care.

3.3 Adaptability

Many of today's computer based training courses take into consideration the student's knowledge of a special subject. For example, based on a test in the first learning session the student is advised to leave out or to repeat certain parts of the course. The new concept follows a different approach. It recommends for internet learning applications to be adaptable to different strategies in learning as well as to a student's learning success.

Adaptability to different learning strategies. Based on empirical research two fundamental strategies in learning have been detected. These strategies have to be seen against the background of traditional university's curricula, characterized by the distinction between *lectures* and *exercises.*
In a first mode, the internet learning application presents itself primarily like a specialist book. Now the module *course* is in the focus of attention; which becomes obvious as it is the only module that is combined with a table of contents, allowing the students to directly access each information window. The use of the other modules is only possible if using the cross-references within *course.*
In a second mode the internet learning application is characterized by the *exercise book* module, that means learning primarily occurs by dealing with exercises. Now the student can easily access the single exercises using a table of contents. Analogous of theory based learning, in this mode exploring of the other modules' contents depends on the cross references within the module *exercise book.*
Additionally, the whole internet learning application can also be used in an independent mode, where all modules are treated in the same way. The choice of one of these modes can be made by the student and can be changed at any time of a learning session.

Adaptability to a student's learning success. For the modules *course*, *exercise book* and *base knowledge* the students can classify each information unit as easy, medium or difficult, which will be recorded. Depending on that classification, a special strategy for using the information unit is suggested to the student, which he can either accept or ignore. If a student didn't work on a certain information unit so far, it would be classified as difficult.

4 Conclusions

The concept's intention is to provide a didactic framework for designing internet learning applications for electrical engineers. This framework is enriched by a partly verbal and partly flowchart based description of the eight modules specific for engineering subjects and six more generally usable tools. The use of the new concept offers three great advantages:

With its modular structure the human learning process (adopting knowledge up to using it) cannot only be supported on the whole, but due to the clear separation, it is in the student's hands to choose his own aspiration level at any time of his learning session.

With the integration of traditional methods in learning and computer assisted learning in a didactic concept a pedagogical surplus value could be achieved.

Assimilating the tools *learning optimizer*, *flipchart*, *calendar*, *plant* further contributes to the student's ability to learn, which can be regarded as a presumption of a successful professional life nowadays.

5 References

Zimmer, G. (1995). Mit Multimedia vom Fernunterricht zum Offenen Fernlernen. In Issing, L.. Klimsa, P. (Eds.): *Information und Lernen mit Multimedia*, pp. 337-352. Weinheim: Pschologie-Verl.-Union.

Kerres, M. (1995). Technische Aspekte multimedialer Lehr-Lernmedien. In Issing, L.. Klimsa, P. (Eds.): *Information und Lernen mit Multimedia*, pp. 25-46. Weinheim: Pschologie-Verl.-Union.

Höfling, S. (1996). Informationszeitalter, Informationsgesellschaft, Wissensgesellschaft. Hans-Seidel-Stiftung, Akademie für Politik und Zeitgeschehen.

Design strategy for an HTML administration and management user interface

Thomas Wolff
SIEMENS ICN – Intelligent Networks
Thomas.Wolff@icn.siemens.de

1 Introduction

This is to describe design aspects that are relevant in the definition of a user interface design strategy for a specific large system.

The system is a platform to host multiple telecommunications services in the area of Intelligent Networks. Various telecommunications providers can run various call processing or Internet based services on this installation. Several kinds of user interfaces are involved, such as system administration, operation and maintenance, service administration, and service use by numerous customers of some of the installed services.

Planning the user interface design approach had to consider two aspects:

- Compared with legacy systems in this area, a substantial improvement in user interface quality should be achieved.
- It was decided to use web technology and especially pure HTML for most parts of the user interface. This imposes certain technological limitations that have to be dealt with but also opens chances to approach the design task in a fresh way, meeting a large part of the users in an environment that is familiar to many people nowadays.

The paper

- describes the selected approach to setting up a user interface design strategy in an industrial project environment,
- explains why plain HTML technology is sufficiently well suited for the project,
- illustrates how HTML can be used to achieve an optimum of interface quality.

2 Achieving quality design goals

In a development environment which has traditionally been focussed on high-tech hardware and software components, with extreme reliability ("high availability") being one of the predominant design requirements, it is not always easy to propagate the importance of high user interface quality as other design criteria are more evident within the development process. Two issues have promoted taking special and centralised care of this topic:

- Evolving expectations of customers even for system operation by dedicated personnel.
- Adding web based services to a system concept which had previously focussed on telecommunication call processing, having been more like a real-time control system.

2.1 Design goals

Design of the user interfaces for the system shall achieve a high degree of user acceptance by assuring

- ergonomic user interface components designed for optimal usability,
- uniform visual and interactive design throughout the interface components ("common look-and-feel"),
- product design to suggest product recognition and identification to users.

2.2 How to approach these goals

Many companies and institutions maintain their guidelines for GUI design (e.g. Stüper 1998) and quite a number of such guidelines are available publically. These guidelines do often not reflect the state of the art in user interface design (as covered in Shneiderman 1998 or Mandel 1997). They tend to be collections of mixed-up aspects of interface design, often focussing on layout properties which everyone is aware of as they can be seen immediately on the interface. Rules to handle these aspects are much more evident and easy to handle and there is a risk that crucial issues for the interactive behaviour are neglected – properties or structure and contents of the interface are harder to describe in rules. (Here lies some similarity with "programming rules" which are also often a mix of weakly profounded rules and arbitrary personal taste about source code layout.)

This is not saying that layout properties would be of subordinate importance. But a strict separation of concerns should be enforced in order to achieve the design goals in all relevant aspects. Modern guidelines do provide a well-sorted collection of established principles and advices (e.g. Weinschenk et al 1997).

2.3 Actual design aspects

The following structuring of design aspects was devised to prepare the design strategy established for the project.

Behaviour, structure, contents: For an interface to successfully provide good usability, the following structural aspects need to be observed:

- User-controlled and intuitive interaction – typical work-flows must be easily approachable.
- Users must be presented interface entities (pages) of comprehensive contents; functions should be attached to these pages in an object-oriented way.
- Easy orientation and navigation within the interface structure must be intuitive and comprehensible.
- Enterface components should be interconnected à la web-style, offering quick access to any related task even if it is logically located in a different system component or function group – this means a break with traditional hierarchical interface menu structures.
- The problem of keeping the orientation while browsing non-hierarchically through web links (a mild form of the "lost in cyberspace" syndrome as known from the inhomogeneous Internet world) is compensated by structuring the interface into homepage areas. They focus a collection of related interface pages on a common introductory or overview page – the home page – and share an associated common navigation framework.
- Adequate feedback and task-oriented help and assistance complement the user orientation of the interface.

Presentation, layout, style: To support usability and user acceptance, an adequate layout is equally important:

- Comprehensible and moderate layout of the interface should support usability and orientation, and not obstruct the user.
- Properties of interaction, style, language need to be consistent.
- Aesthetic layout provides a comfortable feeling while handling the interface.

2.4 Development process impacts

The user interface is the users' man-machine interface. Its purpose is not to just map a system component's functionality but rather to serve the users to perform system functions. Thus system developers must be acquainted with the "keep the user in mind" principle while constructing their part of the user interface.

In order to meet the point of view of the users and finally acquire their acceptance for an interface, it is not sufficient to identify or adopt design principles and realise design rules. Rather an evolving process is required getting feedback on early system versions and taking this for tuning or adapting the interfaces. While actual customers of a product are not yet available, system test department members and experienced support personnel acting as users may hopefully be able to direct the development into the intended direction during the initial feedback phase.

Systematic acquirement of feedback is part of the *usability engineering* process which should be included in development plans well in advance.

2.5 Customisation requirements

As an additional system requirement, interfaces must be customisable to numerous international and other customer-specific preferences. This covers all kinds of data representation, presentation language, layout, visual aspects (like, e.g., colours and button forms) as well as structural properties or functional assortment of individual pages.

3 HTML aspects of the management interfaces

3.1 HTML suitability of a system

The product being developed has a lot of data administration interfaces:

- Administration, configuration of installed service software, many tasks of daily operation are all tasks that modify a couple of data just like an elementary database access. This can easily be presented on an HTML page. Additional help information which is often very useful with more specialised tasks can optimally be placed wherever appropriate directly on an interface page or quickly available from it.
- Service use as intended for web service customers would basically also work on data. Finding and managing personal data records can easily be handled; radio buttons, check boxes, popup menus, and forms yield a sufficient collection of interaction paradigms to attach the required functions to.

For the kind of interaction just described no versatile interactive graphics is needed. Neither would the usual large number of functions sorted in various interface menus be helpful in most situations – this is also a positive effect of object-oriented interface page assortment which focusses interface pages around data rather than presenting all available functions in a global menu set.

3.2 Interface adaptability

A high level of interface adaptability is needed for two reasons:

- to implement the customisation requirements, and
- to facilitate the interface optimisation in the course of the usability process.

The software components of the system are of such a high-reliability nature that kernel software modifications can barely be afforded just for interface adaptation. This is a point where HTML comes in handy, or rather an interface generation approach which makes use of HTML. Actual HTML pages are only produced from application-provided data using HTML templates to add common elements and layout. This way, common look-and-feel throughout the system interfaces is ensured in a straight-forward, automatically handled generation technology. Also, various degrees of adaptation, customisation and even user preferences can be achieved with this approach.

3.3 Benefits of full HTML page customisation

In addition to the dynamic properties implemented by the page generation approach, also the HTML templates can be completely exchanged. This enables extensive but still easy customisation of all interface aspects; the interface structure (the hyper-linked network of interface pages) and all the kinds of presentation for certain elements (link/button/icon, graphic images, specific layout attributes).

Flexibility: HTML allows experimentation, modification, and layout tuning (within limits) with an amazing flexibility. A large number of tools is available to construct HTML pages, including working on the plain source which is still much easier than fiddling with a cumbersome GUI programming interface. All HTML elements including graphics are always available for immediate integration and test.

Acceptance: Finally, the resulting interface provides a familiar feeling to many customers, not only computer experts. Also, HTML interfacing works seamlessly without client installation or applet download delays even for home customers which is an important aspect for the intended application domain of the system.

4 References

Mandel, T. (1997). *The Elements of user interface design.* Wiley.

Shneiderman, B. (1998). *Designing the User Interface.* Addison Wesley.

Stüper, U. (1998). *Style Guide for Development of Graphical User Interfaces.* Siemens ICN Handbook.

Weinschenk, S., Jamar, P., Yeo, S. C. (1997). *GUI Design Essentials.* Wiley.

A Visual Approach to Internet Applications Development

Mauro Mosconi – Marco Porta
Dipartimento di Informatica e Sistemistica – Università di Pavia
Via Ferrata, 1 – 27100 – Pavia – Italy
mauro@vision.unipv.it – porta@vision.unipv.it

1 Do Advanced Users Wish to Build and Control Their Own Internet Applications ?

To allow Internet users to find, collect and manipulate information available on the Web, different solutions have been developed (IEEE 1997) by researchers and software companies which aim at simplifying the interface as much as possible (search engines) or even acting on the users' behalf (software agents). Nevertheless, there is a range of applications where the overload involved in training an adaptive intelligent system would be unacceptable, while a traditional browsing approach would result in tiresome, time-consuming effort.

To tackle this class of applications, we propose a data-flow visual environment in which non-naive users can accomplish their goals better and more rapidly by directly composing simple visual programs themselves, while preserving a sense of control over the system. In the following, a simple example application is discussed to highlight the potential of our approach.

2 The Development Platform

The system we use for building our visual Internet applications is VIPERS (Mosconi and Porta 1998), a general purpose, visual programming environment based on an augmented data-flow model (Hils 1992) and developed at the University of Pavia. VIPERS uses a single interpretive language (Tcl) to define the elementary functional blocks (the nodes of the data-flow graph). Each block corresponds to a Tcl command (or procedure): such a command may itself invoke the execution of other programs as subprocesses.

VIPERS elementary modules have a square shape and present connection points, or ports, on their lateral sides; programs are assembled through direct manipulation, by positioning and properly connecting the available modules: entire programs can therefore be built without typing any line of code.

3 The Visual Approach

Figure 1 shows an example of a visual program built through VIPERS. Its purpose is to explore the Web to find a set of E-mail addresses (to which a message will be sent later on), starting from one or more words and/or sentences. To avoid sending messages to persons out of target, the program also generates a form, to be filled in by the user. For each address, this form shows its context (the ten words preceding it and the ten words following it) and a check box, used to select or exclude the address. The form is subsequently processed by a CGI program. In the example, you may note both ad-hoc blocks (purposely prepared for this kind of application) and more general blocks, which can be used in other application domains as well, for building general-purpose programs.

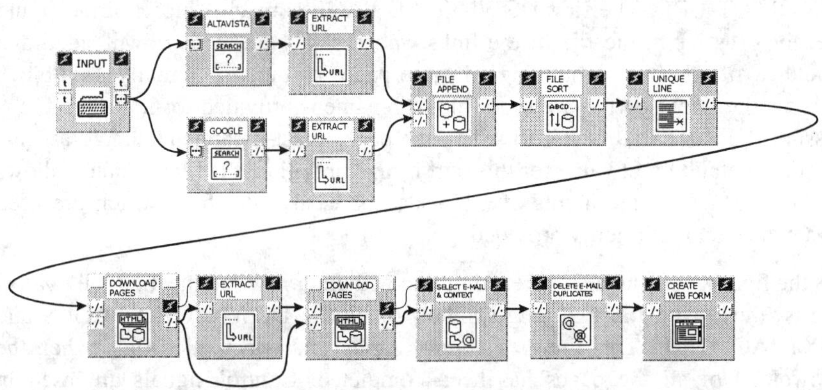

Figure 1: an example of visual program built through VIPERS

Each input/output port in VIPERS is characterized by a special icon indicating the corresponding data type. Data types used in our example are: t (text), [...] (list) and •/• (file path). It is to be noted that (at least in this example) when a module gives out a file path name, this path refers always to a temporary file.

The first module in the figure (block INPUT) allows the user to enter the words/sentences to be searched for. These words/sentences are then provided, in parallel, as inputs, to blocks ALTAVISTA and GOOGLE, which connect to their relative search engines and produce HTML pages (local temporary files) containing the search results. Blocks EXTRACT URL analyze these results, locate the links they contain and create text files with one link per line. The purpose of block FILE APPEND is to link together the files it receives as inputs and to produce a single file. In the example, therefore, it generates a file holding all the links present in both the Altavista and the Google search results. This file is then alphabetically sorted by block FILE SORT and "cleaned up" by block UNIQUE LINE, which eliminates any duplicated lines (links).

At this point, a quite meaningful set of results should have been obtained (which could be exploited to automatically download the sites' pages, through already available software tools). In our program, now the information access phase starts. For each link in the temporary file at the output of block UNIQUE LINE, the following operations are performed:

- the page corresponding to the link is accessed (downloaded);
- the page is appended to a buffer;
- the page's links are in turn extracted and the relative pages are then accessed and appended to the buffer (*two-level spidering*).

These operations are accomplished by blocks DOWNLOAD PAGES and EXTRACT URL. The first DOWNLOAD PAGES (macro) block in the figure receives, as input, the file of the links whose pages are to be downloaded and yields a file (buffer) containing all these pages as output. Actually, this block has another input port as well: if a file path name is provided for it, such a file is assumed as an initial buffer to which the pages must be appended. We assume that the default input value for this port is an empty file: VIPERS, in fact, allows default values for the inputs of a block to be easily specified, which are used when no data is explicitly provided.

In the figure, you can also note a thin line without arrows connecting the *control* ports (those with the lightning symbol) of blocks DOWNLOAD PAGES and EXTRACT URL. This line is a *control signal* and is activated only when the download of all the pages has been completed. Control signals are used in VIPERS to achieve correct synchronization: if a signal exists between an output control port (on the right) of block A and the input control port (on the left) of block B, then the execution of block B can not occur before the whole execution of block A.

Returning to the example, when all the first-level pages have been downloaded, the links they contain are extracted by block EXTRACT URL and provided for the second DOWNLOAD PAGES block. This block also receives the output of the first DOWNLOAD PAGES block as an initial buffer and produces a single file containing both the first-level and second-level downloaded pages. The specialized module SELECT E-MAIL & CONTEXT receives this buffer as input and singles out the E-mail addresses contained in it. For each address, it then yields, besides the address itself, the title of the page in which it was found and its context as output (the ten words preceding it and the ten words following it). Lastly, block DELETE DUPLICATES eliminates any duplicated E-mail addresses and the block CREATE WEB FORM generates the web form to be checked by the user.

As already stated, block DOWNLOAD PAGES is a *macro* block, that is, it is composed in turn of other blocks. Figure 2 shows its internal structure.

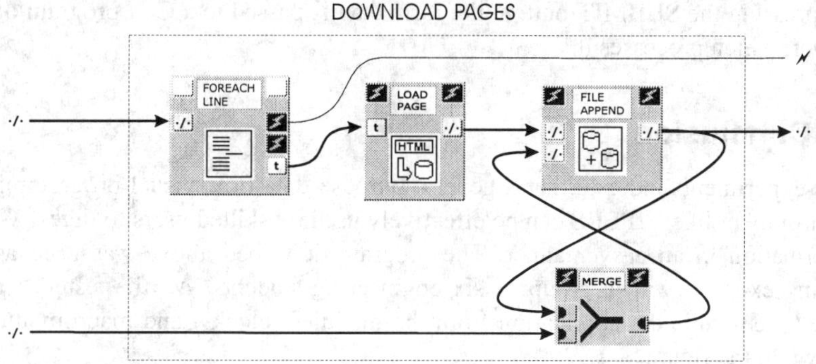

Figure 2: the internal structure of block DOWNLOAD PAGES

As one can see from the figure, a file path name is provided as input to block FOREACH LINE. This block, in turn, is a macro block implementing the *foreach* control construct, allowing elements of a sequential data structure (in this case, a file) to be analyzed and made available, one after another. Block FOREACH LINE emits the various lines (links) of its input file, as they are taken out, but only at the end of this process does it activate its output control signal. Block LOAD PAGE connects to the URL it receives as input and downloads the corresponding HTML page on a temporary file, whose path name is made available as its output. The MERGE block fires when either of its two input ports receives a new data item, which is then emitted as an output. Here, it allows an initial buffer path name to be specified, to which, during the subsequent iterations, the various HTML page contents will be added by block FILE APPEND.

The whole output of the program of figure 1 is a Web page (a form) like that shown in figure 3, which could be delivered via E-mail to the user. By displaying the form in his/her browser, the user will then be able to select the addresses, excluding, for example, those of the various webmasters, etc..

mauro@vision.unipv.it	include ☐
http://**mauro.mosconi.it**/university.html ...presso il Laboratorio di Visione Artificiale (piano D). Sono raggiungibile **via e-mail** Argomenti per Temi e Tesi di laurea e di diploma...	
porta@vision.unipv.it	include ☐

Figure 3: an example of form generated by the program of figure 1

By pressing the SUBMIT button, the form data is passed to a CGI program on a server, which processes it.

4 Conclusion

Our experiments indicate that a general purpose data-flow visual programming environment like VIPERS can be effectively used by skilled users to digest Web information in an easy manner. The program described above was used as a testing exercise with a group of six engineering students. All the testers were able to assemble the application from the modules (blocks) and macromodules shown in the figure.

It is worth noting how the visual programming approach described here could be used for rapid program prototyping as well. Indeed, once a program has been built, it could be advantageously translated into a single script (for example in Tcl), which can be used without need for the VIPERS visual interface.

We are now developing an on-line version of the VIPERS system, which will allow to build visual Internet applications by simply connecting to a proper site and assembling ready-made blocks taken from various libraries. Block execution will occur both locally or, when needed, on remote machines.

References

Hils, D. D. (1992). Visual Languages and Computing Survey: Data Flow Visual Programming Languages. *Journal of Visual Languages and Computing*, 3, 69-101.

IEEE Internet Computing, vol. 1, n. 4, Jul-Aug. 1997.

Mosconi, M., Porta, M. (1998). Designing new Programming Constructs in a Data Flow VL. *Proceedings of the 14th IEEE International Conference on Visual Languages* (VL'98, 1-4 September 1998, Nova Scotia, Canada).

PART 3

USER INTERFACE DESIGN

Systematic Design of Human-Computer-Interfaces as Relational Semiotic Systems

Rudolf Haller, M.A.
Forschungsgruppe für Semiotik und Wissenschaftstheorie, Universität Stuttgart

1 The Development of Relational Semiotics

The semiotic concepts presented here are based on the work of Ch. S. Peirce and its further development by Max Bense, Elisabeth Walther and their coworkers during the last 40 years at Stuttgart University. For details see (Bense 1983), (Walther 1974). For relational semiotics see (Haller 1999).

2 Some Basics of Semiotics

Red light at a crossing of streets means or signifies a 'Stop', not a 'Stop, if you like to'. How can the red light do so? It doesnt. We do, as we have a condition by having learned a rule giving the red light of a certain design and place the value of a light used for traffic regulations. And we have among the traffic regulations a rule giving the value when it is red. The rule is: if the traffic light is red, then stop. The rule is of well known form and obtaines a single value. 'Stop' is the answer to the question 'What is to do if the traffic light is red?'. Note, that we cannot find the rule by investigating the red light or the signified stop. The light at the crossing is a traffic light only, if we realize the governing rule for it. That is: the light at the crossing without its rule is no traffic light; nor do have we the traffic light when we have the rule alone.

Semiotics deals with the recognizable by signs. Hence any representation by signs used in the sciences may be used to show the basics of semiotics.

We may start with the denotation of a function

$$f: x \dashrightarrow f(x),$$

where x denotes a variable in domain X, f(x) a variable in range f(X), and f denotes a defining rule, also called the function choice connecting x and f(x). We remind, that X, f(X) and the defining rule can be given in several different ways. We may conclude to any of the f(x) from x under the defining rule f. A function has three components.

We generalize the concept of the function to a multi-valued relation

R : x --> y

where R now denotes the defining rule, the relation choice connecting x and y, each is variable on some repertoire. The relation sign --> may be given in some different representations. Wether an attribute, e.g. 'single-valued' or 'multi-valued', applies to the relation, depends on the defining rule, the relation choice R of the relation only.

In general: any **semiotic relation** is written in the form

interpretant : the signifier --> the signified

where interpretant is a generalizing term for the defining rule for the connection of the signifier to the signified. The signified is also called object. Every question and answer pair belongs to a semiotic relation ruled by its interpretant. From two of the components of the semiotic relation we may conclude the third of its components in semiotic terms. Increasing precision of its representation is obtained by increased recognition of the sign components.

The relation interpretant : signifier --> object is sometimes written as a **ordered triple**: (signifier, object, interpretant).

This triple is a **sign** and may be signified.

We refer to some results :

A sign is recognized only, if its components are recognized.

Every sign consists of three components in triadic relation; any combination of less components cannot be a sign.

Signs may be checked: if two components of the signs are different, then the signs are different; if one component is different, then this component is faulty. The acceptance of the denotational representation of the components of a sign is obtained by convention in discourse, that is: by continued relating the components of the sign by the participants in communication.

Any sign may be named, but it may be named wrong.

Interpretant, signifier and object may be represented in different ways without the sign being changed.

There is no language, where signs get named, without dictionary. The lexical rules of a language, as part of its syntax, signify objects not belonging to, but signified in the language. Hence lexical rules are protocol sentences.

Facts are set according to the signs possible to set.

Any sign is in present time. The repeated sign is a sign actualized again using memory. The presentation of a sign from the past shows the signifiers we give its components in the present time.

We observe the construction of **connected signs**:

(i) In any representation of a sign in its components, each component is represented by a signifier. In the represention of a sign we present signs of its components.

(ii) A signified sign represents a sign component.

According to (i) and (ii) signs may be connected stepwise. Any pair of such connected signs, actualized in the class of its representations, is called a sign **morphism**. The definition of the mathematical category applies to sign morphisms.

The components of a sign are signs having three components each. By bringing a component of the signifier, a component of the object and a component of the interpretant each in triadic relation a **characterization** of the sign is obtained. The components of a characterization are named in Peircian terminology consisting of three parts: according to the component of the characterization being a component of the signifier, the sign is characterized as qualisign, sinsign or legisign respectively; according to the component of the object the sign is characterized as icon, index, or symbol respectively; according to the component of the interpretant the sign is characterized as rheme, dicent or argument respectively. Rhematic iconic sinsign is of course the usual characterization of the icon on a computer screen as we know it.

By characterizing a sign we set the **focus** onto the specified sign components. During usage of a sign the change of its characterization is possible. In designing the representation of a sign we may take care of the recognizability of the sign with the characterization we wish it should show.

3 Application to Design and Review

Taking interfaces of humans to computers as working example, we denote:

(1) interface : human --> computer

where we want to analyse the signified sign components in its infrastructure.

We have the denoted sign relation as a starting point: humans use or signify computers according to the rule or choice of an interface relating them.

A first goal is: to conclude from the signifying human and the signified computer to the infrastructure of the interface. If we resume the usual objectives of humans, then the infrastructur of the interface has to represent the available objectives of the infrastructure of the signified computer. A question to find the signifier in the sign relation named computing infrastructure is: what media is used under the concept of computing infrastructur? To find the signified: what are the objects pointed to? To find the interpretant: what connects media and objects?

Thus a computing infrastructure may be:

(2) lookup, granted access, usage :

devices (network of components and computing devices) -->

tasks (collected services)

As usual, we understand tasks as the collected services of computation, storage, communication channel, software filter, hardware device, other services and users. Please note, that any sign and sign component is understood better, when represented or described in other terms too, but this is not to spoil a clear understanding of the concepts in discours. In our example every description belonging to tasks should however answer a question of the form: to what purpose may one of the devices looked up, accessed or used?

Any of the sign components may consist of a list. After few steps we will come close to the details of an actual project.

Taking any one of signified services as a sign and trying to find its components, we may obtain:

(3) transaction manager (enables groups of objects to participate) :

input/output of computing -->

the communication or storage of related groups of objects

For the components of lookup, access and usage we may write:

(4) performing operation :

invocation of services --> coded object (may be distributed)

where

performing operation may be

method : parameter --> result (or exeption)

invocation of services may be

interfacing : method call --> coded client

coded object may be

behavior : state --> services

For interface design according (4) we may check, if the service specification underlying their invocation signifies its implementation as coded object such, that the intended items become public and visible, this is: signified, as far and only as far as the performing operation requires.

Looking for the sign components for a coded object by applying questions accordingly, we may obtain

(5) Apply the operation needed :

set up any of its properties -->

creation of the required object

The creation of a required object should answer the question: what properties are to set up if an operation needed is applied? Any components may be checked: is their relation to other components realized clearly and distinctly when composing a triadic sign relation? Are their concepts recognized at all?

By applying the sign characterization to any of the determined components their realizability and recognizability are categorized. Sign characterization may be applied to all components including the aesthetic design of interface components. For further details of applying characteristics see (Haller, 1999).

4 References

Bense, M. (1983). *Das Universum der Zeichen*. Baden-Baden: Agis-Verlag.

Walther, E. (1979). *Allgemeine Zeichenlehre*. Stuttgart: DVA.

Haller, R. (1999). *Zeichen im Zusammenhang*. Universität Stuttgart: Diss.

A WANDERING PHOTOGRAPHER

Dipl. Des. Oliver Endemann
Kunsthochschule in der Universität Kassel
Product-Design

Context

This presentation of work is part of my National Degree with the theme:
'Retailers in Public Space - Smallest Markets'.
The presentation addresses the following concerns:

a) Social background of retailers within public space (their history, needs and compulsions)
b) The changing markets (rising unemployment, new work, job-sharing...)
c) The inner city and the possibilities for fluctuating markets.
d) The conceptual design of different possible jobs or trades (including the design of the relevant equipment)

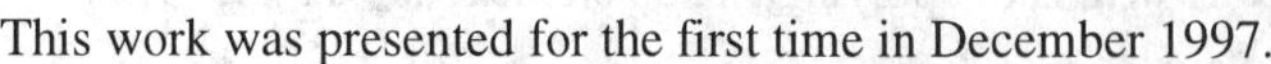

This work was presented for the first time in December 1997.

A wandering photographer

More than hundred years ago, wandering photographers started to conquer the world with their new eyes, and where ever they went, they were covered with glory. Nowadays we have a very different situation - mass production has made this technology cheap and almost everybody can afford their own camera.

But with the development of digital technology in photography- and Data-transfer technology, there is a market for photographers and their services again. Although we have had digital cameras and 'handys' for a few years, they are not easily accessible to everybody. The service for the customer could be, for example, a modern variation of the traditional postcard: A high quality image of a person or location taken somewhere in the world will be sent to an email-recipient elsewhere.

Scenario

A group of friends is in town for shopping and amusement. On their way they meet a person who makes them a nice offer. He says he is a wandering photographer and he will take some stills of their crew with his digital camera. The pictures taken can then be sent, right now, from here to any person who has an e-mail address. Our gang says: "We like this idea and we will go for that!" Now the photographer shows his ability as a host and he encourages his customers to pose in an exiting way. He takes several pictures and simultaneously he records the sound from the surrounding scene. After the session he shows the results to the gang on the little LCD-monitor on the back of the camera. The customers choose a series of pictures - four stills. Having pressed the ok-button on the camera the pictures are automatically attached to a special email-program running on the connected portable computer. With the keyboard the photographer inserts the given email address and also a little text message. The small TFT-display shows the written text and address to the customers. With the ok from the group the photographer sends the picture-mail forward to the recipient using the internal radio-sender. Finishing the business, the customers pay the photographer and he gives them a little Polaroid of the sent pictures. Where has he got this proof from? Out of the portable computer!

Comments on the Design and the handling of the Used Technologie

All the necessary functions for the above experience need to be incorporated in the design of the equipment, without losing the relation to the typical and grown aesthetics of photo technology. The main focus in the development was on the digital camera itself. She is the 'optical heart' and should be recognizable from a far distance. Maybe she looks more like an eye than other cameras do, because of the 'eye-brow', which takes over the function of a lens-hood, and an aerial for data transmission. Not difficult is the comparison of the bent 'brow' with a wave expanding concentrically. She says: 'I'm sending'.

During the daily action, digital photos are made with the help of controllable programmes. These token pictures can be viewed for assessment on the back of the camera. The selection of the image or images chosen by the customer is made with a laptop that makes up part of the equipment. This small portable computer is located on the photographer's pushcart and communicates with the camera in the same way as an external disk-drive. In this way the picture is integrated into a special e-mail programme running on the laptop. With the keyboard of the computer the service person types in the recipient's e-mail address and some optional information, say 'The weather is fine over here...'. For quicker and more comfortable handling it might be better to have some predefined options so that you can choose between several values (for example, the photographer offers his service at a music-festival, so he could generally attach a prepared small sound-file to the message. In case he is in front of a historical building, extra information to the motif could be added.) With these values the whole procedure can be performed within a minute and a few clicks. However, it always depends on how much expenditur of time and money the customer is willing to spend.

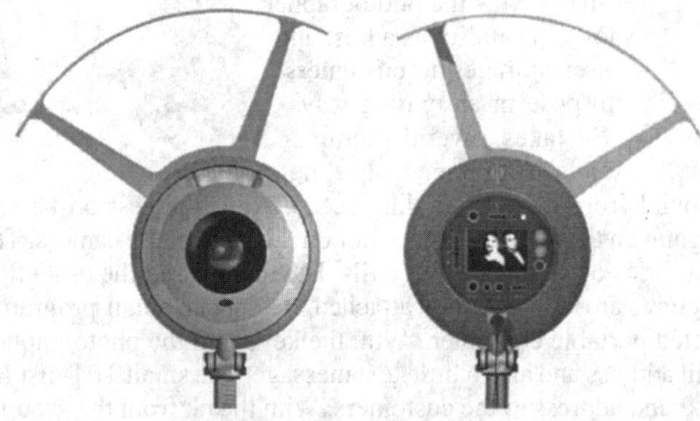

From the customers point of view...

It is really practical and useful that we don't have to buy postcards anymore. Postcards tend to show objects from past times, it's seldom that we find a brand new motif. We don't have to find a place where we can buy stamps and also we don't have to search half of the city to find a letterbox. Nowadays we go to the mobile photographer who is always there where something is going on, or you will find him at places of general interest. And he does much more than regular postcard-shops do! You can have your very unique, freshly taken picture, or you can choose from hundreds of 'postcards' saved on his disk drive and presented in a catalogue. And the best advantage is that the 'postcards' will be delivered within minutes.
Maybe the customer would like to have a more generous sized monitor to have a look at the images, but everybody will understand that this will increase the costs for both the customer and the operator. Furthermore, those monitors need a lot of power to run constantly, and this is no good when you are 'on the road'.
So the customers have to take a look at the small monitor while the photographer starts to prepare the e-mail for sending by entering the address. On the display of the laptop the customer can see the status-bar of the file-transfer. After this transfer the essential job is done and it is time to pay. But at this moment the client can choose whether he wants a little proof that the picture has been sent (at extra cost). This can be a nice remembrance of this 'immaterial' service.

Summary

So this project is a bit 'aged' now, (almost two years have passed since I started to think about the technology for the equipment) technology has been dramatically developed. Polaroid has developed a picture output for digital images, and with Noika's 'Communicator' we have a portable computer that has similar features to the pieces of equipment I described with this work.
Already we are sending movie-clips by e-mail, and everybody can imagine that we will be able to do this live, in any length and quality, from every place in the world soon. Maybe we can really see with the others eyes someday, like in the movie 'Strange days' (Directed by Cathryn Bigelow), where a little recorder records all impressions and feelings of a person and anyone can play it back.

But whatever will be possible someday, the rules of good communication won't change.

Multidimensional Orientation Systems in Virtual Space on the basis of Finder

Philip Zerweck
Universität Gesamthochschule Kassel, Fachbereich Produktdesign

The work is about orientation systems meant for data files that help users to act within the frontend of computer (MacOS: Finder). The focus is also on stereoscopic visualisation systems. It is a search for possibilities of spatial graphical user interface to furnish, establish and manage personal computer environments. In the end developements are being made like e.g. digital fixtures for deskspace or a supportive agent of character. Another developed tool is the Semantic Browser. With its help it is possible to create interactively multivariate visual sensations out of a quantity of files. It is based on principal ideas on semantic dimensions of Alan Wexelblat.

1 Deskspace

The success of WIMP–systems is not based on the superiority of one of its parts but on their coordination and consistency. These are the demands to be fullfilled to successfully create next generation interfaces for operating systems.

As said before in this work, tools were developed to show a path to a deskspace–finder. Even the building of a complete finder was forbidden because of the scope, a valid concept is the needed starting point. This is described by a few key points: Input or output does not need any put on or buckle on components. The presentation has a bottom, a back, a top, a left and a right in the way of a stage. The viewer does not change his standpoint only the content / scene of the stage gets changed. The visualisation is based on the ideas of cellular–picture–space, voxel objects and radiosity procedures.

2 Gesture Control

Gesture control has high potentials to act as an input unit for practicable daily use computersystems with spatial user interface. This is because of several reasons: Gesture control is spacial by principle. It does not need any body contact and, because of that, interferes less in user's physical constitution. It does not need any special or additional space on the real desktop. It enables concurrent input of location and standardized commands. The hand is the first tool of man. With it things that we are able to grasp or want to grasp get closer.

Whether in future the user will learn the gestures that the system understands or vice versa remains to be seen. This question does not affect the necessity and designers duty to supply a start code. To control the system I worked out a set of gestures; the zero gesture, to take hold, the pointing at / approaching, to fetch, to leaf or hop e.g. through hierarchy, setting up a sphere for the purpose of manipulate the scene, grabbing, motion to come nearer or to withdraw, accentuate, hit–and–bang as a kind of drag–and–drop, open by tickling, delete by flicking to pieces.

The user himself also needs a representation (of his hand). I suggest a specific and maybe even personal sculptural icon wich underlines users potentials instead of the common rough chopped off hands.

Fig. 1: If you point at a file, it shows its label and preview. A 2d picture has an icon but no vicon (volume–icon).

3 Structure Elements

The deskspace–finder as it is seen by the user is composed by an unclosable list of different elements. All representations / appearances of these elements can be classified in: icons, vicons (volume–icon, consists out of 32^3 voxel), label, preview and view. While the appearance of files or additional elements vary strongly and in the future even the styling of the system elements will be able to be changed by the user, it is the gestalt of those system elements that settles whether the system is suitable to the user or not.

With the designing of orientation systems it is important that order functions individually. Necessary are different ordering elements that could be used by choice. The analogy with offers for furniture is appropriate. Because I think that in future foreign providers will bring a variety of „furnitures" on the market, it was my aim to design supplies for archetypal patterns of order and safekeeping. Each tool is original, has special features and therefore it is restricting. Man needs to be able to assign specific characteristics to (even digital) things so he can remember them and therefore be able to handle them.

Besides the basic elements File and Deskspace, there are further elements, representatives of prototypical putting in order: Alias (as in principle known from MacOs), Folder (as in principle known), Galaxy (representation of other deskspace / home), World (a kind of 3dimensional folder with Semantic Browser instead of list), Container (for aim of long term storage), Cluster, Pinboard, Pile (among others a description of how unstored data gets stacked in the deskspace), Box (like a cardboard box for old bills), Bundle (like a tramp's bundle)

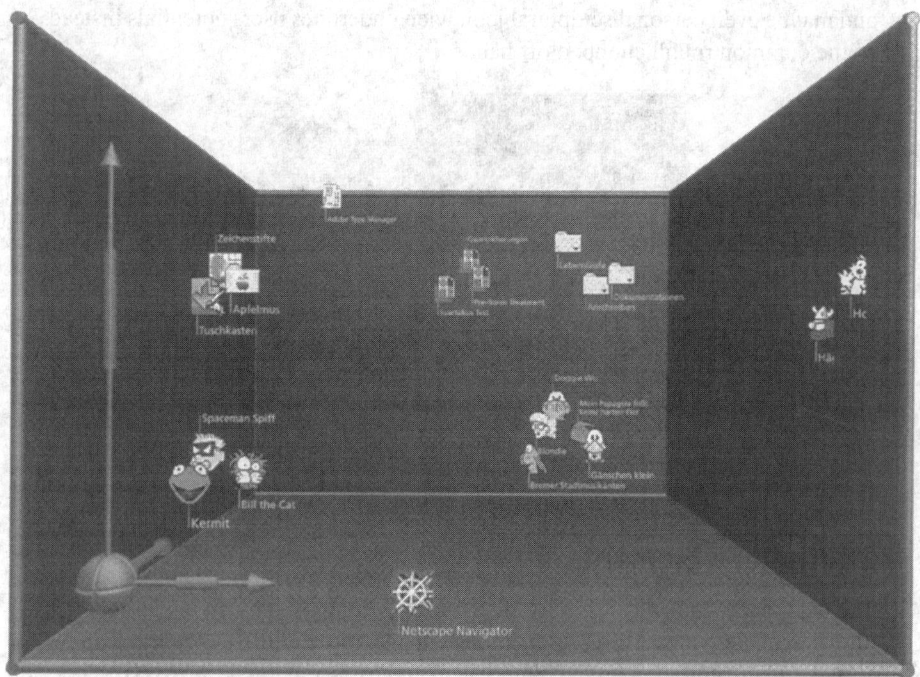

Figure 2: World with embeded Semantic Browser (origin and dimension arrows with a sleeve as choosing tool)

4 Semantic Browser

On one hand there is an absolute disintegration of locality in virtual space because there is no obvious frame of reference. On the other hand there is the anthropogenesical necessity of man to remember relations of subjects and as a result their meanings by spatial structures. Architecture and design create orientation structures, order by locating things in space. This ability which consists of defining a consistency of time and location for subjects and so memories in space comes into real conflict with the structural potentials of virtual space. There is not only a lack of any tradition and therefore skill but also for the first time the space itself included with its dimensions and frames of reference need to be arranged.

To create an artificial cartesian space out of really arbitrary dimensions and enter images in it which should transport meanings by their location, would not cope with the potentials of virtuality. Instead of that the user has to get the opportunity to manipulate not only the arrangement but also the space. For his individual needs of information and explanation he chooses semantic dimensions and form with their help that space from which he expects the most useful answers. Now the task for designers is to identify variables, transform them to sense causative dimensions and together with possibilities of manipulation condense them into a understandable intuitive operation. The Semantic Browser is such a design.

The user puts any semantic dimension of his choice at the origin in any angle he likes. The dimensions can be changed in direction and length and so control the extension of the semantic space. Each dimension has an additional tool. With its help a range out of the dimension is selected and so the appearance and spreading of the files are manipulated. The on the fly made sight shows a sworm of files. The figure is determinated by the chosen dimensions, theire directions and lengths and the scope and position of the selection tools. Another operating possibility of the Semantic Browser is to activate a dimension without putting it at the origin. So the dimension with its selection tool serves as a general filter of appearance for files without influences on theire location.

The Semantic Browser can be used like e.g. list views in different environments. According to the environment the available dimensions / criterias will change practically. The criterias can be divided by principles into three groups: criterias of the system, criterias of metainformation and criterias of files. I call those criterias of the system that are related to values / variables known by the system without any research and that are available for any file. These are for example name of file, size of file, age / date (creation, emergence, change, last use), kind / creator, tag, version. Besides that there are the three spatial dimensions X, Y and Z respectively right / left, up / down and back / forward that are needed if files should be arranged in free spatial order.

Criterias of metainformation are related to values / variables that are added on the files. Nowadays the adding is made by the creation programs like Word prompted by the user or by archive programs. That is why not every file will have values / variables for every of those dimensions. Criterias could be: title (not necessarily the same as name of file), topic, author, subject (even multiple), length of text (strokes, characters, words, pages), filling of pages (different relations), size of document (width, height, depth, plane, volume), size of picture (width of, height of, number of pixel), number and depth of colour, length by time or frames, tree of descent / geneses.

Criterias of files are understandable as such criterias that point at the content of files. With their help and their implementation into dimensions it is possible to integrate e.g. retrieval systems into the Semantic Browser. Reasonable are e.g. word or text research, thesaurus (semantic lexicography) or picture research.

The integration of criterias into dimensions is critical. The perception of the recipient has to be anticipated. With the representation of the size of file e.g. a linear interpretation has to be rejected because the perceived difference between 1 MB and 2 MB is much higher than between 100 MB and 101 MB. In addition there are much more small files than big ones. A logarithmic to the base e interpretation turned out to have advantages. Also the question of discrete or continuous appearance as with e.g. alphabetical order has to be decided or both offered. Finally there are dimensions where files or values eventually appear more than ones like e.g. author because one text could have more than one author.

5 Outlook

This has been some results of my work. It shows that spatial user interfaces are not far away, because the singular components already exist. Deskspace could be the initial point to introduce the needed systems. This is possible if the aim always is to enable any individual user his intuitive access to his own files.

Development of an Interface for Computing and Analyzing Multidimensional Perceptual Space

Lin-Lin Chen and Jeng-Weei Lin
Department of Industrial and Commercial Design
National Taiwan University of Science and Technology

1 Introduction

Product perceptual space [Green 1989; Moore 1993] is often used to visualize consumers' perception of the various products in the market, to determine consumers' satisfaction with the existing products, and to seek opportunities for new products, or directions to improve existing products. Product perceptual space can be constructed by using a number of methods, including *multidimensional scaling* (or MDS) [Kruskal 1978; Schiffman 1981].

When using MDS for similarity analysis or preference analysis, the designers often face the following difficulties [Lin 1995]:

1. Collecting similarity or preference data from surveys, and transforming the collected data into the input formats required by the various MDS programs are tedious and error-prone.

2. In the perceptual space generated by MDS programs, stimulus objects are represented symbolically, as numbers or names, rather than graphically, as icons or images. For design related applications, such a representation does not provide sufficient information for interpreting the perceptual space. To visualize the distribution of stimulus objects in the perceptual space, designers often create a graphical representation from the output of MDS programs by hand, or by using a separate drawing program.

3. To interpret the meaning of each axis in the perceptual space, designers need to visualize ideal vectors that represent subject preferences, as well as the projections of points onto the ideal vectors. Again, designers usually need to draw these vectors and to find projections of points based on the output of MDS programs.

4. If the perceptual space requires more than three dimensions, the MDS programs display the perceptual space in a series of two-dimensional projections. Such a representation can be misleading: two points that are far apart along the z dimension, for example, can appear to be quite near in the x-y plane.

WebMDS is a WWW-based system designed to address these problems. By using WWW as the common interface, WebMDS provides an integrated, intuitive, and interactive environment for performing design analysis using MDS methods.

2 Overview of WebMDS

The process of using MDS methods in design analysis consists of three stages: *data collection*, *perceptual space computation*, and *perceptual space visualization*. WebMDS contains modules for assisting the designers with each stage of the process, as illustrated in Figure 1. Details about each module and its functions are described in the following sections.

3 Data Collection

To reduce the amount of time required to collect and record proximity data, the MDS survey can be conducted online to enable automated collection of proximity data. The World-Wide Web provides an excellent environment for conducting such kind of online surveys. By basing on the WWW, the online questionnaires can be accessed through a consistent interface from any computer connected to the Internet.

WebMDS provides a questionnaire generation module, that allows the designers to interactively design a questionnaire for similarity or preference analysis. This module is implemented using Javascript and CGI programs. When the designers finish designing the questionnaire, WebMDS automatically creates the hypertext questionnaire and a corresponding program. The designers can then conduct the online survey by putting the hypertext questionnaire and the corresponding program on a WWW server. The answers to these online surveys are collected automatically by the corresponding program for the subsequent analysis.

4 Perceptual Space Computation

One of the problems associated with using MDS methods is in transforming the collected proximity data into archaic formats required by MDS programs, which are often written in FORTRAN. WebMDS hides the nuts and bolts of creating an input file for a particular MDS program by using WWW as the common interface to the various MDS programs. The designers can choose the desired type of analysis, select a particular perceptual mapping method, set the corresponding set of options, and choose survey data sets from a consistent interface. To compute perceptual space from the collected survey data, WebMDS provides a perceptual space construction module for computing the perceptual space by using the ALSCAL program. This module is implemented by using CGI programs in C and Perl.

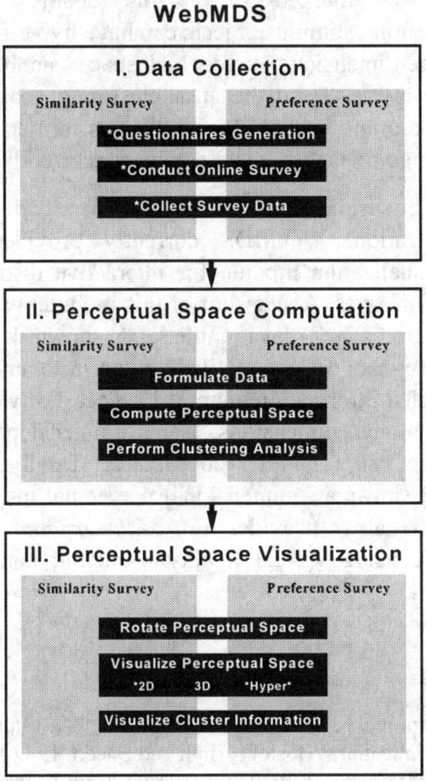

WebMDS

I. Data Collection

Similarity Survey
Preference Survey
*Questionnaires Generation
*Conduct Online Survey
*Collect Survey Data

II. Perceptual Space Computation

Similarity Survey
Preference Survey
Formulate Data
Compute Perceptual Space
Perform Clustering Analysis

III. Perceptual Space Visualization

Similarity Survey
Preference Survey
Rotate Perceptual Space
Visualize Perceptual Space
*2D 3D *Hyper*
Visualize Cluster Information

Figure 1 Structure of WebMDS

5 Perceptual Space Visualization

Interpreting the perceptual space is one of the most difficult tasks involved in using MDS programs. For design related applications, being able to visualize the clustering of stimulus objects and the bundling of ideal vectors about each axis (in the perceptual space) is essential for searching and validating an interpretation. Yet MDS programs currently provide few visualization tools for such purposes. WebMDS makes use of the multimedia nature of WWW to graphically display the perceptual space. Such a visualization as several advantages over a paper-based visualization. First, as is the case for online surveys, the stimulus objects are no longer limited to be images. They can be music or video segments, as well as three-dimensional CAD models. Second, the designers can interact with the perceptual space. For instance, a region of the perceptual space can be enlarged to reveal finer details of the clusters in that region. Icons representing stimulus objects can have hyperlinks to more detailed information. A three-dimensional perceptual space can be rotated to expose occluded stimulus objects. Third, because clustering analysis and preference mapping are integrated into WebMDS, the clusters formed by stimulus objects and the bundles formed by ideal vectors can be concurrently visualized.

Depending on the dimensions of the perceptual space, different techniques are needed for visualization. WebMDS currently provides a 2-dimensional perceptual space visualization module for interactive display of a perceptual space using orthogonal axes. A snapshot of this module, which is implemented by using CGI programs and Java applets, is shown in Figure 2. In this visualization, the designers can interactively zoom in to any region, or click on any icon to see detailed information about the object. For visualizing perceptual spaces of four or higher dimensions, a multidimensional perceptual space visualization module using parallel coordinates [Inselberg 1995] has been developed. Figure 3 shows a 2-dimensional perceptual map for sporting event mascots. WebMDS also provides a *3-dimensional perceptual space visualization module* using Virtual Reality Modeling Language (VRML), as illustrated in Figure 4.

6 References

1. Green, P. E., Carmone, F. J. Jr., and Smith, S. M: 1989, Multidimensional Scaling: Concepts and Applications, Boston: Allyn and Bacon, 1989.
2. Inselberg, A.:1985 "The Plane with Parallel Coordinates", Special Issues on Computational Geometry of The Visual Computer 1, 69-97.
3. Kruskal, J. B., and Wish, M.: 1978, Multidimensional Scaling, Beverly Hills, CA: Sage University Series.
4. Lin, Rung-Tai., private communications, 1995.
5. Moore, W. L., and Pessemier, E. A.: 1993, Product Planning and Management: Designing and Delivering Value, McGraw-Hill.
6. Schiffman, S. M., L. Reynolds, and F. W. Young, Introduction to Multidimensional Scaling, Academic Press, 1981.

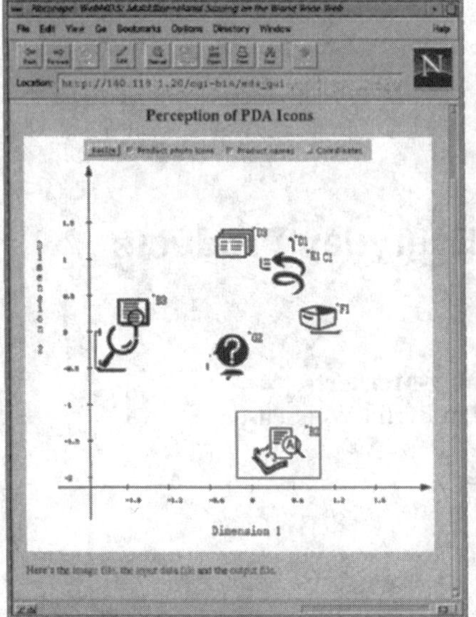

Figure 2. Two-dimensional Perceptual Space Visualization Module Based on Java

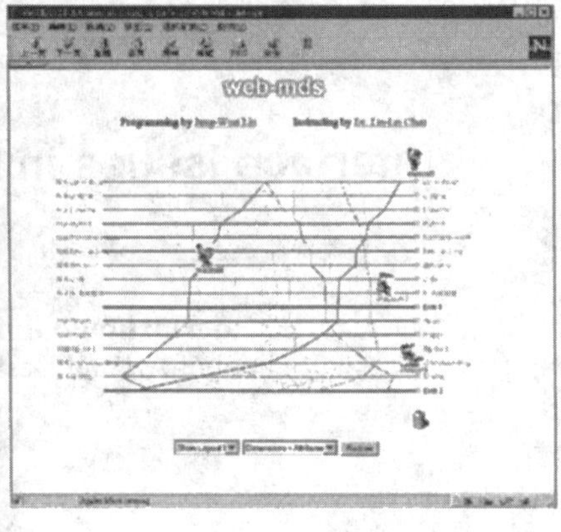

Figure 3. Higher-dimensional Perceptual Space Visualization Module Using Parallel Coordinates.

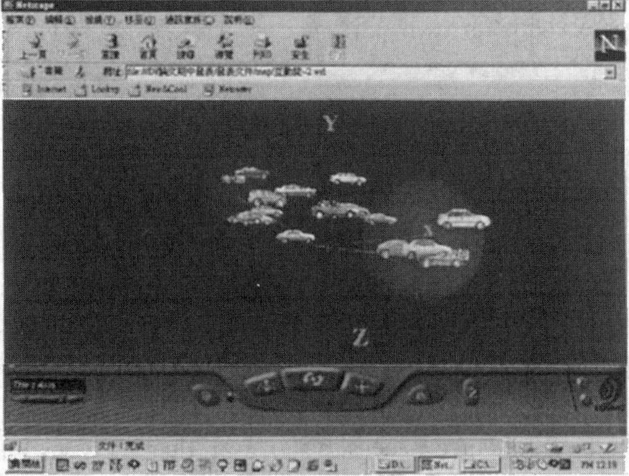

Figure 4. Three-dimensional Perceptual Space Visualization Module Based on VRML

Interface Issues in Everyday Products

Hartmut Ginnow-Merkert
School of Art and Design Berlin-Weissensee

1 Introduction

Contemporary household and consumer appliances, medical equipment or transportation products all have one feature in common: Their behavior is no longer cast in hardware but programmed into software. In this respect, nearly every electrically powered product may be considered an autonomous special-purpose computer with its own particular integrated cast of input and output devices.

With today's technological potential to program nearly any behavior into a product's circuitry, design has been liberated from the mechanical and material demands by which it had been ruled since the industrial revolution.

Along with this new freedom emerged a need for designers to equip their products with meaningful behaviors comprehensible to their human masters who find themselves still firmly held in their own sets of evolutionary conduct.

In today's complex universe of products, conventional industrial design – as it focuses merely on a product's *visual* appearance – no longer suffices to provide the information and guidance necessary for a product's joyous and harmonious interaction with its human counterpart. Anger, frustration and consumer abstinence are among the undesirable consequences of the designer's negligence.

Industrial design will need to operate from an interface design perspective, addressing a wealth of previously unexplored issues related to a product's interaction with humans.

2. Multisensory Interface

A product's human user will learn about the product's purpose, its functions, its operation, and its current status via multisensory interaction. The limited capabilities of the human being's five senses provide all that he or she will ever know about the product.

Far too often designers pay meticulous attention to a product's visual appearance while leaving its acoustic, tactile, gustatory and olfactory utterances to chance. Any mechanical engineer selecting materials and production processes based on an engineering set of criteria will potentially jeopardize the product's integrity as perceived by its human user. An expensive-looking toaster – to name a profane example – will suffer terminally from a cheap-sounding metallic clatter.

Therefore careful orchestration of a product's multisensory manifestations for optimum harmonious effect will become a mandatory part of the designer's business.

3. Mental Model

Virtually every product emits output available to the human being via the visual, acoustic and tactile senses. Often enough, the olfactory and gustatory senses are involved as well.

All the product's multisensory output will combine into a mental model in the user's mind. With the human mind's unrelenting effort to fabricate mental models from the observed input, designers will need to create meaningful mental models well before they begin to work on the product's visual appearance.

4. Learning Style Adaptive Interface

Based on our current knowledge about learning styles, the 'one-size-fits-all' approach to interface design is no longer justifiable. While any designer readily indulges the customer's yearning for a diversified scope of colors, sizes and shapes, current display and interface designers limit the user's options to a singular choice of unimaginative buttons and displays. With interaction via buttons and displays playing the crucial role in customer satisfaction, such neglect results in a market mismatch with the majority of the potential clientele. A product's user interface will need to come in many colors, sizes and shapes, just like many a product's exterior. Diversified mental models will need to cater to different learning styles.

5. Dynamic Interface

Designers will soon need to develop answers to a current dilemma: While current technology is able to provide an increasing wealth of product functions at a decreasing cost, consumers are overwhelmed and frustrated with functions they can't or don't care to learn.

Yet, when given a choice between two comparable products, consumers will choose the product equipped with the greater number of functions!

Reducing the number of functions will undoubtedly speed up the learning process. But in the face of the dilemma described above, a reduced set of functions will put the manufacturer of this product at a marketing disadvantage. Furthermore, a simple interface that is quickly mastered will as quickly lead to boredom. A complex interface will overwhelm the user in the early stage of his or her familiarization with the product.

The *dynamic* interface – an interface that evolves along with and just a bit ahead of the user's knowledge of the product – will satisfy the user's needs and provide just the right challenge in every stage of the product's life cycle.

Considering recent advances in sensor design technology, designers will need to program an evolutionary quality into the product's interface.

6. Human Evolution

Humans owe most of their behavior to their four-million-year history as participants in Nature's evolutionary process. In spite of all the technology and whatever we call civilization, modern man behaves not far from his club-wielding hunter and gatherer ancestors.

Human cognitive capabilities and intelligence all were optimized for a hunter and gatherer's environment. Whether we drive a car on the freeway, turn our faces towards the elevator door or perceive visual, acoustic or tactile information about a product's value and quality all relate to mankind's past experience with the natural environment.

Human perception of value and quality follows directly the laws of Nature, many of which are yet to be discovered. A thorough study of this field will yield additional criteria to be used in product and interface design.

7. Cross-Cultural Design Issues

When using metaphors to transport meaning, metaphorical imagery typically is taken from the individual's own cultural background. Designing for foreign cultures or with metaphors taken from a foreign culture requires sensitivity and understanding as well as new methods of analysis and user testing.

Cross-cultural design bears new opportunities. Ancient traditions, wisdom passed orally from generation to generation, teachings about aesthetics based on different views of the universe (e.g. Asian monistic philosophies), and different perspectives on design derived from our studies of foreign (e.g. Chinese) language and script all will enrich the designers' spectrum of possibilities. Cross-cultural design will allow designers to break common barriers of routine thinking.

8. Closing Statement

Industrial Design and interface design will need to merge, or at least to cooperate more closely. They also need to enter into close cooperation with scientific disciplines such as Evolutionary Psychology, Psychoacoustics and Cognitive Science – to name a few.

Only by means of a combined, integrative and multidisciplinary effort will Designers continue to provide a valuable service in the development of contemporary and future products.

9 Further Reading

Ackerman, Diane. *A Natural History of the Senses.* Random House, 1990.

Degen, Helmut. *Multimediale Gestaltbereiche als Grundlage für Entwurfswerkzeuge in multimedialen Entwicklungsprozessen.* Freie Universitaet Berlin, Institut fuer Publizistik- und Kommunikationswissenschaft,

Arbeitsbereich Informationswissenschaft, 1996. http://www.kommwiss.fu-berlin.de/~degen/publikationen/isi96/isi96.htm.

Ginnow-Merkert, Hartmut. *Beyond the Visual.* form diskurs #1. Form-Verlag, Frankfurt, 1996.

Ginnow-Merkert, Hartmut. *Metaphors in Design: A Korean Experience.* http://www.kh-berlin.de/korea, 1998.

Krippendorff, Klaus. *On the Essential Context of Artifacts or on the Proposition that "Design Is Making Sense (of Things)".* Design Issues: Vol. V, Number 2, Spring 1989.

Krippendorff, Klaus. *Philosophy of Semantics in Design.* University of Pennsylvania, Philadelphia, 1994.

Krippendorff, Klaus. *Redesigning Design. An Invitation to a Responsible Future.* Conference Proceedings "Design–Pleasure or Responsibility". Helsinki, Juni 1994.

Tversky, Barbara, et al. Spatial Thinking and Language. *How Do Words Create Pictures, and Pictures Words?* Interface Laboratory, Stanford University, 1995.

Wulf, Gabriele. *The Learning of Generalized Motor Programs and Motor Schemata: Effects of KR Relative Frequency and Contextual Interference.* Teviot Scientific Publications, 1992.

Cognition and Learning in Distributed Design Environments: Experimental Studies and Human-Computer Interfaces

Frue Cheng, Ying-Hung Yen and Holger Hoehn
Department of Industrial Design
National Taipei University of Technology

1 Introduction

The Report of this Joint Project between the Department of Industrial Design, National Taipei University of Technology, Taiwan and the Department of Product Design, University of Kassel, Germany that was to explore the possibilities and the limitations of Distributed Design. Three groups of design students from two different countries with different cultural background and different native language were to work on a specific design task and to communicate and exchange ideas via Internet.

2 Organization

From the Taiwanese side Prof. Dr. Frue Cheng directed eight students. Two Assistants, Ying-Hung Yen (National Taipei University of Technology), responsible for the technical aspects and Holger Hoehn (Kusthochschule Berlin-Weissensee, Hochschule fuer Gestaltung), who was in charge of the design issues also participated in this project.

On the German side three students participated in the joint project, directed by Prof. Hans Dehlinger, who also was the initiator of this project. He was being supported by his assistant Oliver D. Endemann.

The idea was to cooperate on a classical design task and to develop a final product by communicating over long distance via Internet. As a theme the product category of a flash light was chosen.

The first step of the project work was to form three groups consisting of German and Taiwanese students. They were to cooperate and develop ideas and concepts together by using the internet.

3 Design Process

During the whole period of the project the usual traditional means of developing concepts were being applied. Loose sketches and storyboards were being created by the students as well as full-scale orthographic and perspective line drawings and renderings in the later phase of the project. Then the two-dimensional information also needed to be translated into three-dimensional physical models.

Being supported and directed by a non-Chinese speaking assistant, the Taiwanese students were confronted with a usual situation. They had to present their ideas verbally in English or with most students having difficulties in speaking English, by drawing. - But then also those circumstances provided a good basis for presenting concepts to the German team members in Kassel.

4 Communication

In order to exchange ideas by using the internet the three teams had to transform their ideas in form of text and sketches into digital information. Text had to be typed in a word processing format and sketches had to be scanned. Also pictures of models had to be taken. Crucial issues of digitizing the information were standards for text and picture data, so that both sides with different hard- and software standards had access.

A homepage served as a forum for the cooperating teams where the project was being introduced and new ideas and concepts could be posted by the students by e-mail.

With the project proceeding also individual students were using their own homepages to present their ideas and concepts. By doing so all team members were giving the chance to present their work and update their Product ideas to everybody involved, and they could always receive feedback from other team members.

Finally also desktop videoconferencing was applied, so that the students could present their concepts and the product models and get a direct and immediate feedback.

4.1 Hardware platforms and available peripheral equipment

In this project, both Taiwanese and German side integrated the several kinds of hardware software and peripheral equipment to upload and download data efficiently. The equipment list is described in the following:

Table1 The equipment of the project

Name	Hardware	Software	peripheral equipment
Taiwan	PentiumPC200Mhz 128M RAM 17" Monitor	I-DEAS IE4.0 Frontpage98 Adobe PhotoShop Microsoft Word	CCD Digital camera A4 laser printer A4 colour scanner ISDN network.
German	PowerMac 8500 21" Monitor	I-DEAS StrataStudioPro Adobe PhotoShop Adobe Acrobat Quark Express Microsoft Word	CCD digital camera A4 laser printer A4 colour scanner ISDN network.

4.2 Formatting data for exchange

Each group is responsible for proper formatting communication material into the following formats:

Table2 Formatting data for exchange

Format		Description
Text	RTF(Rich Text File)	Use this format for exchange text-documents.(text only)
	HTML(HyperText Markup Language)	Use this format for exchange text-documents (with embedded pictures) and view them in a Web-Browser.
Picture	GIF(Graphics Interchange Format)	Use this format for Line-Art or 8-bits color (static or dynamic) pictures.
	JPEG(Joint Photographic Expert Group)	Use this format for color or grayscale photographic based pictures.

4.3 Methods of communication

There are two methods of the communication The first one is informal communication. Both sides of the students can send E-mail or post suggestion to show the idea of design and discuss the viewpoint of each other. The second one is formal communication. Each group can show the drawing on homepage and present their sketch model to the participators of the other side through desktop videoconference.

- E-Mail to your group for standard communication: If you include attachment (like a digital photo or a scanned hand drawn sketch) observe formatting.
- Forum: Read the forum to inform yourself about issues of interest available to all participants.
- Homepage: Use homepage for presenting the sketches or downloading material that is made available to all participants formally. (See figure1, http://www.uni-kassel.de/~dehl-www/DISDES)
- Videoconference: We have to test this first and will use this later. Assistant will perform the test.

5 Evaluation

During the project work both groups were facing a lot of barriers that were disturbing the workflow. As mentioned before especially the Taiwanese students had strong difficulties in translating their ideas into English, especially during the video-conference, because most of them still lacked English conversational skills, while their English reading skills were mostly good.

The most important problems though were on the technical side.

Bandwidth is unaffordable: The effect of the transmission is not well on desktop videoconference, although the data of the project is transmitted by ISDN. Especially, we can't receive the voice message in the real-time, because of the bandwidth.

Lost of the data package: When we are presenting the sketches of design by shared whiteboard in Internet. There is no reason that some sketches are disappearing, although the situation is the weak of the initial structure of Internet, the result of the project is disturbed in videoconference.

Altogether though the Distributed Design project can be considered as a success. The students earned a lot of knowledge about organizing their work to be presented on the internet by digitizing analog information and gained experience about exchanging ideas and concepts and cooperating with a remote team over long distance.

Besides improving their technical skills in the realm of traditional means of design as well as in the field of modern computer tools all team members, including German and Taiwanese students probably also started to develop a sense for developing design for a world with merging cultural barriers.

6 References

Liang Ting-Peng (1997), Decision Support System. Taipei:Unalis Corporation.

Yang Cheng-Fu (1997), Management Information System. Taipei: Chuan Hua Corporation.

Bill Gates (1999), Business @ the speed of thought using a digital nervous system. New York: Warner Books,.Inc.

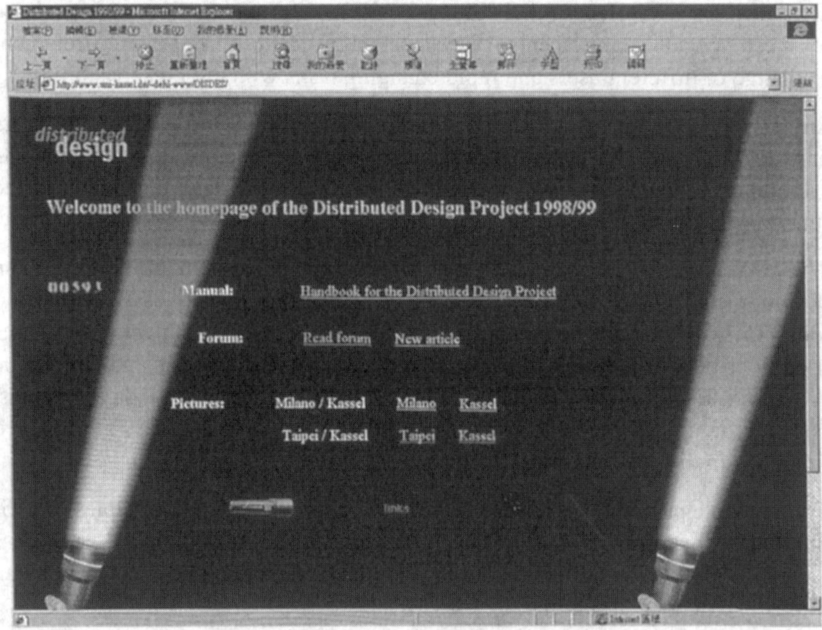

Figure1 Homepage of Distributed Design

Adaptable GUIs Based on the Componentware Technique

Dipl. Inform. Sabine Schreiber-Ehle
Research Institute for Electronics and Mathematics, Research Establishment of Applied Sciences, Wachtberg, Germany

1 Introduction

Graphical User Interfaces (GUI) of e.g. business information systems often have to meet special requirements. The most significant of these are frenquent exchange of different users and rapid changing tasks with high responsibility.

In daily system operation the users often change especially when casual users have to work with the system. They have to learn the use of the system and its GUI quickly. The users want to complete their duties, instead of exploring the functions of the GUI in order to adapt the task to the GUIs functions.

The evolution of tasks implies extensibility of the system at runtime. The development cycles of classical software engineering techniques are much too long to match todays rapid changes of the working environment. In systems used in critical processes mistakes may have fatal consequences and have to be prevented. Especially the last requirement surely is not met by todays systems, because the user is unable to locate the needed function in the mass of diplayed functions in the user interface.

These problems make it extremly urgend to apply ergonomic principles to GUI building. A modern GUI system design has to be task oriented and the interfaces have to be lean. The system has to provide task adaptable interfaces which can be easily extended at runtime.

2 Available techniques

A rather classic approach for GUI-design is the monolithic software achitecture that would be resticted to the tasks it was designed for. This concept requires the complete knowlege of the needed functionality during the design phase, but that is impossible in most cases.

The library architecture is another approach. It is far better adaptable to a changing working environment, but still needs qualified programmers to change the system functionality and still thc requirements have to be known during design.

An extension to the single application model discussed so far is the compound document concept. Here the user is automatically supplied with the appropriate application for the data he works with. He has to deal additionally with the complete functionality of a second application embedded in the controls of the original one (Balzert 1996).

Recently the componentware principle offers applicable solutions to evolve applications at runtime. This technique defines standardized protocols which independently developed programs can use to communicate (Microsoft 1998, OMG 1998).

Thus program systems can be put together by using products of different manufacturers. If this technique is used to develop complex program systems, modules can be exchanged at runtime depending on the underlying system architecture. Therefore, changes of the system´s performance and its GUI can be achieved even after the software installation process has been completed (Plasil 1998).

To create a suitable software architecture consisting of the required independent exchangeable parts the use of Design Patterns is helpful. Design Patterns structurize logigal dependencies on a high abstraction level. So they are applicable to a software design in general and they supply designers with premanufactured design solutions (Gamma 1995).

From the different applicable patterns only architectural and structural patterns are considered. Useful architectural patterns to create componentware systems are the Model-View-Controller (MVC) paradigm and the Presentation-Abstraction-Controller (PAC) pattern (Buschmann 1996). They provide principles for functional separation according to the goals of elements and for organization of independant parts working together. Structural patterns like the factory pattern provide blueprints for mechanisms to abstract from application specific details.

The advantage of components is their universality. At the same time this universality is their biggest disadvantage, the independant components do know nothing of each other. So they will not cooperate, but work separately. But the relations between the systems parts represent the essence of the users task.

In order perform tasks the user has to combine the parts acoordingly. This is even more strenious than looking for functions on a single crowded interface. To complete the users trouble he additionally has the choice between various coupling techniques for the components.

3 An improved approach for HCI design

To combine the advantages of componentware and abstraction with the optimal adapted user interface of the monolithic system one has to create a framework for the components. This framework has to have specific qualities, or it will be no better than a classic monolith.
First of all, the framework has to be generic according to the used components. The possibility to add components to the system has to be dynamic and configurable in order to have an adaptable system.
To fulfill this requirement, a control system has been develped, that configures the components, their communication paths and their relations is controlled by external descriptions. The task oriented interfaces can be formed by these configurations. New components can be added at runtime using default integration mechanisms by the user. The integration of new components can be achieved later on by the system designer who adapts the new component to the needed interface (Sane 1995).

This strategy is used to create a task oriented user interface that fulfils the requirements specified in ISO 9241/part10 (European Standard 1995, VDI 5005). The approach is focussed on suitability for the task, controllability, and suitability for individualization. The user (her)himself or some administrator prepares the users workspace by combining the necessary software tools as components of an aggregated program system. So the users task shows up in the structure of the interface. The use of the resulting lean interface is easier to learn and the user can not accidently use superfluous functions, because they are not displayed.

While working with the system the user additionally can extend the interface responding to incoming new task requirements or as an adaptation to individual needs.
To configure a system, software components can be pre-selected and applied immediately or subsequently. Each component has its own menus and controls which are normally supplied.
The designer can create dedicated GUIs to fulfil special task requirements using components that have special protocols (e.g.,ActiveX) controls. A chance for designing well structured GUIs is to use only parts of components. For that, the designer can select only suitable functions and leave out unnecessary ones.

4 Application example

For developing a Human Computer Interface (HCI) of a complex information system a situation editor with geographic reference has been designed and implemented as a component. The architecture of that HCI hosting this component is designed to enablc communication on different levels of abstraction such as DLL, ActiveX, OLE, and CORBA (Gruntz 1998).
The complete HCI is made up of commercial-off-the-shelf (COTS) components, like word processor, presentation tool, image editor, and database application, and specialized components, like situation editor (Kaster 1998) as well as geographic vector and raster map display and creation programs.
All these components can be put together at runtime according to the user‘s tasks requirements. To achieve a smooth exchangeability of the specialized component parts the underlying software design is structured by strict separation of the user interface and the remaining system functionality.
This separation especially means, that dialogs are implemented as single components as well as menu bars. They do not belong to the kernel of the system. Therefore they can easily be replaced, even by third party products if those implement the required protocols.
Furthermore well known design patterns such as the Model-View-Controller (MVC) paradigm, Presentation-Abstraction-Controller (PAC) and the factory pattern have been used to group the parts logically. This facilitates the easy adaptation of the GUI to the users task requirements by exchanging whole GUI element sets.
Because of the underlying structure of the MVC paradigm the user can choose different characteristics for presenting output data, like graphics or ASCII, and for manipulating input data, like different mouse controllers. The factory patterns, for instance, facilitate the user to choose different display elements for presenting the same data with different task context, e.g. a ship´s position by different icons.

5 Further extensions

The mentioned configuration descriptions that parameterize the interface could be generated by a intelligent GUI agent. This agent could observe environmental states or user actions to configure the interface according to the situation he analyzed. Tools for integration of components could be desinged to assist a system advisor. Such a specially trained person transforms task requirements into dependencies between components. The result is a special interface for each task.

6 Acknowledgments

I would like to thank Prof. Dr. B. Döring of the FGAN instute for his friendly advice and helpful comments, and Mary Rotman of a well known newsgroup for speeding up my english phrases.

7 References

Balzert, H. (1996). *Lehrbuch der Softwaretechnik.* Heidelberg , Spektrum.

Buschmann, F., Meurier, R., Rohnert, H., Sommerland, P., Stal,M., *Pattern orientierte Softwarearchitektur*, Addison-Wesley, 1996

European Standard, *ISO9241 Part 10 Dialogue Priciples,* 1995.

Gamma, E., Helm, R., Vlissides, J.: *Design Patterns* . Addison-Wesley, 1995.

Gruntz, D., Pfister, C., (1998). *Komponentensoftware und ihre speziellen Anforderungen an Standards.* Objekt Spektrum, 4/89 , 38–42.

Kaster, J., *Handbuch der xiris 2.1 Arbeitsstation,* FGAN,1998.

Microsoft : *Microsoft White Paper: DCOM Architecture,* 1998.

Object management group (OMG), *CORBA2.2 Specification :* OMG98-2-33, 1998.

Plasil, F., Stal, M. (1998). *An Architectural View of Distrubuted Objects and Components in CORBA, Java RMI, and COM/DCOM.* Software Concepts & Tools 1998(66), Springer.

Sane, A., Campbell, R., Composite Messages : *A structural pattern for communication between components,* 1995.

Verein deutscher Ingenieure (VDI), *VDI5005 Softwareergonomie in der Bürokommunikation,* 1990

An Approach for the Complementary Model to Improve Supervisory System

Wang, Chau-Hung* **Lin, Tsai-Duan****
* Department of Business Administration, Soochow University
** Hwa Hsia College of Technology and Commerce,Taipei, Taiwan

1. Introduction

Generally, the output information from monitors and the operating processes are two important factors for supervisors to well do their works in an automatic system. However, even through supervisors really know the operating processes, there are a lot of uncertain things, such as nervous anxiety or machine breakdown, to interrupt supervisors to do the right works. Thus, supervisors must carefully pay attention to watch monitors, and then take some suitable operations to control supervisory processes to ensure system can run well.

We can find from many studies that cooperation of human and computer is a suitable way for supervisory tasks. For example, using interactive scheduling methodology is better in the performance than using analytic approaches or using priority rules for dispatching jobs in real-time schedule works. [5,6] However, human and computer have their individual characteristics. The problem is how to allocate tasks to human and computer to produce more interactive effectiveness.

In this paper, we will initially discuss the matters from human cognitive limits, the concepts of task allocation between human and computer, and the human-computer complementary concept. After this, a human-computer complementary model will be developed to decrease supervisor's cognitive limits and to improve interactive effectiveness. Then, the complementary model was used to improve graphic interface in Environment Control System (ECS) of Taipei Rapid Transit System. Finally, an experiment was taken to evaluate the effectiveness of the new interface.

2. Task Allocation and Cognitive Limits

Man-machine system is an operational system in which the roles of human and machine are assigned to achieve a particular goal. The assignment job, also called task allocation, means that allocate tasks to human or computer according to some rules. The objective of allocation is to decide which jobs would be done by human, which done by computer, and which jobs done by human cooperating with computer. [2,3] Sanders listed a lot of elegant characteristics of human and machine. [4] However, it does not mean that which kinds of jobs must be performed by human and which done by machine. Jordan said that human and machine could not compare directly and allocate tasks according to their relative advantage, because the essences of human and machine are different. [1] The essences of human are flexible and fewer stables, but the essences of computer are nearly opposite to human. It will result failure to allocate tasks only refer to the relative advantage between human and machine. However, the concept of cooperation may be a better way allocate tasks between human and computer. The complementary model is based on the concept of cooperation. That is supervisor and computer to control the process at the same time. The tasks of computer are to display situation information on screen and collect processing data, and then compare it with prescribed data in database to find out if something out of control exists in system. If computer finds a problem and knows its solving method, then the computer could solve it by itself. Otherwise, the computer will display some information on screen to tell supervisor that it has happen some trouble things needed supervisor to help solve it. In this time, the supervisor will interrupt computer and take over authority to control the process. Meanwhile, the computer provides supervisor information from on-line help system to help supervisor solve problem. In other words, the role of computer changes from monitor to supplier of help information.

In addition complementary concept, the supervisor's cognitive limits are another important factors to affect supervisory task to be done effectively and efficiently. From the cognitive limits, Zachary listed six reasons that supervisor cannot make decision effectively: [7]

(1) inability to predict processes
(2) difficulty in combining competing attributes or objectives
(3) inability to manage information needed in the decision process
(4) problems in analyzing or reasoning about the situation
(5) difficulties in visualizing
(6) quantitative inaccuracies in heuristic judgments

3. A Model of On-line Help System for Supervisory Task

Figure 1 illustrates a model of on-line help system what is established according to the human-computer complementary model for supervisory task. The purpose of the on-line help system is to overcome the supervisor's cognitive limits and to help supervisors to improve their decision-making quality from the viewpoints of complementary concept in task allocation.

In this study, we provide three types of help information for the supervisory task:

information for system condition, analysis of system information, and methods of control. In each type of help information, in order to find as much as possible situations that supervisor will meet, it is necessary to consider content of task, supervisor's jobs, and supervisor's cognitive limits. These possible situations will be used to search methods of solving problem. After finding the possible solving method, help information will display on a screen to help solve problem. The quality of supervisor's decision making will be improved by using the help information.

A. Information for system condition

In this supervisory step, supervisor always cannot do their jobs efficiently and effectively, because many supervisor's cognitive limits exist, such as:

(1) It is difficulty for supervisor to effectively manage information needed in the decision process.
(2) It is difficulty for supervisor to pay attention all the time.
(3) It is difficulty for supervisor to react in time when meets an emergency.

In order to overcome the cognitive limits and improve the efficiency of operations, computer should provide some suitable on-line help information to help supervisor. It means that the on-line help system's main tasks are to display some useful information about system condition and try to find if it has existence of some potential conditions that will cause problems. At the same time, the supervisor's jobs are try to understand meanings of the displayed information, relationship among all displayed information, and relationship between the system's objective and displayed information. They then try to use help information to find if it has existence of some potential conditions that will cause problems. However, If it really has existence of an abnormal situation, then computer could use light and sound to call supervisor's attention, and try to take some actions to reduce this abnormal situation in a short time or display recommendation to help supervisor to solve it.

B. Analysis of system information

In this supervisory step, the jobs of supervisor are to use help information to analyze the present operational situation, to ensure the system is in control. They then efficiently use analytical method to analyze displayed information, predict the future operational situation, and detect if the system will be broken down in the future. However, it is not easy to predict the future events and to get quantitative inaccuracies in heuristic judgments, because supervisor's mental load may be overload. Thus, the main tasks of help system are to reduce the mental load of supervisor. It will be useful to provide supervisor on-line help information for the means of analysis and the meanings of analyzed results, the means of prediction and the meanings of predicted result, and the meaning of level for each measurement index at. Additionally, help system could analyze system condition and judge the system situations, then to suggest a suitable way to control system. There are many types of present methods to show the system's condition, such as light, sound, graph, figures, and characters.

C. Methods of control

In this supervisory step, the main task of supervisor is to select the most suitable control method according to the analyzed result of situation information. If the result is system in under control, then system will operate the system by computer automatically. However, if the result is system out of control, then the supervisor will interrupt computer and take over authority to control the process. In this situation, supervisor must know how to input commands to control system and what kind of result will be happen after entering commands. Beside these, to understand the sequential relationship among commands, and clearly know the control processes to stop system running are another important things for supervisor. However, it will need too much information for supervisor to efficiently use. In other words, it is not easy to effectively manage all information at a time. It is necessary to provide supervisor all information what is needed. These help information include telling supervisor how to input commands to control system and what kind of result will be happen after entering commands, the sequential relationship among commands, and the way to stop system running.

4. Experiment and Results

Environment Control System (ECS) of Taipei Rapid Transit System (TRTS) is a centralized supervisory system. The ECS have located in central control room. In the ECS, a monitor system what is a highly complicated human-machine system has been used to supervise and control the TRTS by supervisors. However, supervisors always complain that the interface on the monitor system is not clear enough. Many errors were occurred because the information on monitor screen was too complex to read or to use. Additionally, the design of graphic interface was also not friendly enough.

In this study, two major changes on the graphic interface were done in order to improve the quality of supervisory. We changed the design of graphic interface and constructed a new on-line help system into the ECS. These jobs were done according to the concept of the human-computer complementary model. These interface and help system could use different colors, sounds, lights, and alphanumerically word to provide supervisors with a lot of real-time and important information for detecting situation, comparing messages, and giving warning.

Finally, an experiment was used to test whether the new graphic interface and on-line help system could decrease supervisor's cognitive limits and improve the quality of supervisory work. The results of experiment indicated that new interface and help system could reduce supervisor's workload and enhance the performance of supervisory work. Additionally, subjects who took part in the experiment said that all components in this new interface are useful to improve the efficiency and effectiveness on supervisory work.

5. References

1. Jordan, N. (1963). Allocation of functions between human and machines in automated systems. *Journal of Applied Psychology*, 47, 161-165.
2. Kantowitz, B. H. and Sorkin, R. D. (1987). Allocation of functions. In Solvency, G. (Ed.): *Handbook of Human Factors*, New York: John Wiley.
3. Price, H. (1985). The allocation of functions in systems. *Human Factors*, 27(1), 33-45.
4. Sanders, M. S. and McCormick, E. J. (1987). *Human Factor in Engineering and Design*, New York: McGraw-Hill.
5. Sanderson, P. M. (1989). The human planning and scheduling role in advanced manufacturing system: an emerging human factors domain. *Human Factors*, 31(6), 635-666.
6. Sheridan, T. B. (1988). Task allocation and supervisory control. In Helander, M. (ed.): *Handbook of Human-Computer Interaction*, North-Holland.
7. Zachary, W. W. (1988). decision support systems: designing to extend the cognitive limits. In Helander, M. (ed.): *Handbook of Human-Computer Interaction*, North-Holland.

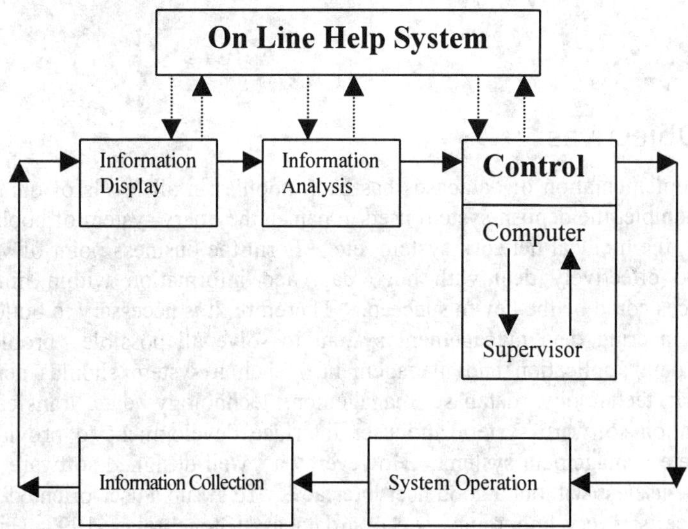

Figure 1: the concept of on-line help system in a supervisory task

An Experimental Study of User Interface in a Data Management System

Jessica Chen[1] and Sheue-Ling Hwang[2]

1. Project Engineer
   Faltien Trading Co., Ltd.
   Chungho, Taipei, Taiwan
   Fax: 886-2-3234-0389
   Email: Jchen@mail.faltien.com.tw
2. Department of Industrial Engineering
   National Tsing-Hua University
   Hsin-Chu, Taiwan
   Fax: 886-35-722685
   Email: Slhwang@ie.nthu.edu.tw

1 Objectives

The implementation of databases has been popular in all levels of our life. For example, the deposit system used in banks, the query system of books in library, the medical network system, etc. Form the business point of view, how to effectively deal with huge data and information within limited resources could be the key to succeed. Therefore, it is necessary to build up an engineering data management system to solve all possible problems about data application and management. Such a system should include network technology, database management technology, data transfer of application software system and user interface development to provide a complete management system. However, a well designed software will become useless without a good user interface. To set up a user-oriented interface is a very important issue in a database (Stonebraker, 1993; Haber, 1994). This research is to develop an interface of an engineering data according to the principles and standards of interface design. We expect to find out some strategies to prevent the operating errors of an engineering data management system, and give suggestions for the principles of user interface design.

2 Methodology

The literature in database, especially the users' requirements and interface needs of an engineering data management system, are reviewed (Levine, 1994; Chang, 1996;Ebert, 1994; IS09241, 1995). By interviewing the users and designers of engineering data management systems, and evaluating existent engineering data management systems, one may figure out the characteristics of the user interface design. Then, applying the principles of interface design, (ISO9241, 1995) the principles of dialogue control of an engineering data management system are set up. Afterwards, an experiment is conducted to verify the effects of principles.

2.1 Questionnaire Survey

A questionnaire is designed to survey the problems of using an engineering database. Nearly 100 questionnaires were sent out and the response rate was 90/100, i.e. 90% return rate. There were 79 valid questionnaires. The subjects were all users of engineering data management systems.
The average period of using database was 36.4 months which was about 3 years.

Form the results of the questionnaire, one could see the common errors that users always made were: (1) the errors of entering a system, and (2) the errors of operating tasks.

2.2 A Commercial System Evaluation

By evaluating a commercial software, one may find out the main causes of making human errors, especially the errors from questionnaire survey. In addition, the results of evaluation may give some key points to draw up the principles of dialogue control in an engineering data management system.

The engineering data management system to be evaluated is an integrated system. It is mainly divided into plan management module, material requirement management module, engineering drawing management module, documentation management, system maintenance module, etc. The user interface of the system is by direct manipulation, and the information is presented in windows, using list menu. The interface of the system is evaluated by usability evaluating principles.

From the results of the evaluation, we can find out that the display representation, such as font, button and window style, satisfy the principles of ISO9241. However, the design of the menu grouping levels of tasks, and help message should be improved according to the characteristics of an engineering data management system, users' requirements and the principles of interface design.

2.3 Experiment

An experiment was designed to verify applications of the design principles of interface in an engineering data management system. We compared the original and the improved interface designs in operating time, error rate, and the users' subjective feeling.

The interface design in this experiment mainly referred to the software architecture of Design Manager. (PDM tech,1996) It included two experimental parts, A and B.

(1) Experiment A

The hypothesis of this experiment was that menu map could help the users to find the task option when the users were not familiar with operating the system or get lost in the system. The original interface was the list menu . The improved interface presented all options in one list, and distinguished tasks by different colors

(2) Experiment B

The hypothesis of this experiment was that if the users hadn't finished the prior task, it was better that the system provided the messages as well as shortcuts between the related tasks. The original interface only presented the message to tell the users "What's the task you should finish first!" The improved interface presented the message about the task the user should finish first and the users could get into the prior task and returned by shortcuts. Each experimental part had two different designs, original and improved ones. The original interface was simulated the interface of Design Manager and the improved interface was designed by the previous survey and evaluation feedback and the principles of ISO9241.

Twenty-four students of Industrial Engineering at National Tsing Hua University participated the experiment with payment. Their ages ranged from 23 to 30 with the average age 24.6 years old. All subjects were familiar with operating in windows environment and were experienced in the database system. Subjects sat in front of a computer and were given instructions about the functions and description of the engineering data management system. After practiced for some tasks, the subject started a formal experiment. We controlled the experimental environment and factors to be the same for each subject. When the subject finished each task, the operating time, total counts of clicking buttons, and error counts would be recorded. After finishing all the tasks, the subject was told to compare the two interfaces and give the grade (1-7 points) to the interfaces subjectively.

3 Results and Discussion

The results of objective performance measures (operating time, numbers of clicked buttons, numbers of error) shows the effect of improved design are significant on all the performance measures ($P<.01$).

The mean grade of each interface indicated that the subjects preferred the improved interface in the experiment ($P<.001$).
In the experiment A, it assumes that the performance of using the improved interface is better than the original interface (i.e. operating time is shorter and error rate is lower). The results revealed that there were significant differences on performance between two interface designs. According to the mean of the operating time, it is shown that providing menu map can help the users to find out the location when they get lost in the system. Moreover, according to the results of numbers of clicked buttons and errors, we find out the users' workload is lighter in using the menu map to browse option, make decision and trial and error. It is also verified from the users' subjective feeling.

In the experiment B, the improved interface gives the message about what the unfinished task is and provides the shortcuts between the related tasks. The subjects' subjective response also indicated that they preferred the improved interface obviously. Usually, the users expect an interface that can judge if the task has been finished automatically and is simple to be operated. Providing the shortcuts can ensure the accuracy in data processing, and significantly improved the performance of the related tasks.

The results of the research revealed that exploring the users' operating problems and quoting the principles of interface design in designing the interface of an engineering data management system could increase the users' performance and prevent the users' errors.

Acknowledgments

This study was supported by the National Science Council of the Republic of China (Contract NSC 86-2212-E007-012)

References

Chang, C. M. (1996). "The development of the human-computer interface design tools for an engineering data management system", M. S. Thesis Department of Industrial Engineering, National Tsing Hua University, Taiwan.

Eberts, R. E. (1994). "User Interface Design", Prentice Hall.

Haber, E. M. (1994). "The ambleside survey: important topics in DB/HCI research", Interfaces to Database Systems, Lancaster, pp.361-361.

ISO 9241 (1995). Ergonomic Requirements for Office Work with Visual Display. Terminals, Part 10, Dialogue principles, International standard.

Levine, J. R., (1994), "Designing GUI Applications for Windows", M&T Books.

PDMtech Inc, (1996), "Design Manager Manual", Version 2.2.

Stonebraker, M, et. al., (1993), "DBMS research at a crossroads: the vienna Update", Proceeding of the 19th VLDB Conference, Dublin, Ireland, pp.688-692.

Combined Metric for User Interface Design and Evaluation

Andy Smith [1] and Lynne Dunckley [2]
(1) Department of Computing and Information Systems, University of Luton, Park Square, Luton, LU1 3JU, UK. Email: andy.smith@luton.ac.uk
(2) Faculty of Maths & Computing, The Open University, Walton Hall, Milton Keynes, MAA 3JU. Email: L.Dunckley@open.ac.uk

1 Introduction

This paper analyses methods for obtaining a combined usability metric for applications. The rationale for the development of this research emanates from the specification of LUCID, a novel interface design method which has been proposed (Smith and Dunckley, 1998). It is however suggested that the techniques proposed have generic applicability.

2 Candidate techniques

In developing tools and techniques for the determination of a combined metric there are two main issues. Firstly a range of potential metrics need to be ascertained. Secondly the major potential metrics need to be prioritised in some manner, so that by allocating weightings to each an overarching value can be ascertained. These issues will each now be further explored.

There are two options for the process of proposing a range of metrics for possible inclusion within a combined metric. Firstly it may be sufficient to rely on *developer introspection*. The evidence gained from this work and other related studies (Smith and Dunckley, 1999) would suggest however that collaboration between a number of developers is to be recommended above the introspection of a single expert alone. The alternative approach is to directly involve users and developers in a participative environment.

There are also two options for the process of prioritising the metrics that are proposed. Firstly Lin, Choong, and Salvendy's (1997) single usability index could be adapted to metric prioritisation. Essentially this would require the interface developer / evaluator to simply rate each metric which has been proposed on a Likert scale. The percentage which each metric might make to a combined metric would be based upon its score divided by the total of all scores allocated. This approach is henceforth referred to as the *Numerical Assignment* (NA) technique. Karlsson (1996) has investigated the issue of generic software requirements prioritisation and analysed the *pairwise comparison technique* (PW) which is based upon the Analytic Hierarchy Process (Saaty, 1980). It would require the developer to decide, for a number of pairs of metrics, the extent to which one of the two metrics is more important (Table1).

Table 1 Pairwise comparisons - intensity of importance

Importance	Definition
1	equal importance
3	moderate importance of one over other
5	essential or strong importance
7	very strong importance
9	extreme importance

3 Acadmin practical studies

A study based upon *Acadmin,* a commercially available student administration system, required four HCI experts to prioritise each of four candidate metrics using both the NA and PW techniques.

Table 2 Numerical assignment data

Developer Metric	D1	D2	D3	D4	Av. %	St. dev.	Av. Pr'ty
A Task Time	21%	24%	37%	31%	28%	6.2%	1
B PUT	26%	29%	11%	31%	24%	8.1%	3
C Errors	21%	29%	21%	21%	21%	5.7%	4
D Satisfaction	32%	19%	32%	27%	27%	5.2%	2
Total					Av.	6.3%	

Numerical assignment. The individual and average percentage contributions to a combined metric that were derived are shown in Table 2. Two issues appear to be significant. Firstly although the four developers differed considerably in their interpretation of the importance of each metric, there was little difference in the individual percentage contributions to the combined metric.

Pairwise comparison. In order to analyse the data provided by the pairwise comparisons it should be inserted into a square matrix of order n, where n is the number of contributing metrics. Developer 2, for example, decided that Metric B was 'moderately more important' than Metric A and therefore a '3' is inserted in row 2 column 1, and 0.333 is inserted in row 1 column 2. These and the other values for Developer 2 are shown below:

	A	*B*	*C*	*D*
A	1	0.33	0.2	5
B	3	1	0.33	5
C	5	3	1	7
D	0.2	0.2	0.143	1

To calculate the percentage contributions it is necessary to calculate the eigenvalues of the matrix. Metric A was found to have a percentage contribution of 13.62%. (Table 3).

Table 3 Pairwise comparison data

Developer **Metric**	**D1**	**D2**	**D3**	**D4**	**Av. %**	**St. dev.**	**Av. Pr'ty**
A Task Time	15%	14%	64%	51%	34%	21.9%	1
B PUT	20%	26%	4%	26%	19%	8.8%	4
C Errors	11%	56%	12%	7%	21%	19.9%	3
D Satisfaction	54%	5%	20%	16%	24%	18.2%	2
CI	0.055	0.080	0.132	0.044	**Av.**	17.2%	
CR	0.062	0.089	0.146	0.048	0.086		

As with the numerical assignment technique there was considerable variability between the developers in terms of the priority given to each metric. Indeed the variability is greater with the pairwise comparison technique (Figure 1).

Figure 1. Comparison of numerical assignment (N) with pairwise comparison (P)

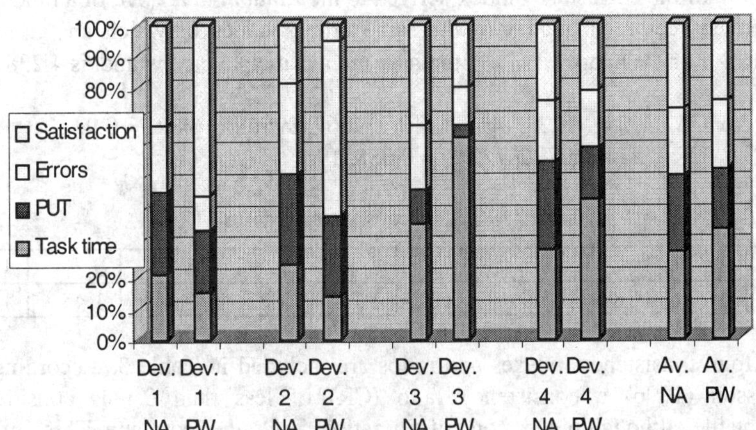

Such large differences between HCI experts may appear somewhat surprising. One of three conclusions can be drawn. One possibility is that the developers views were spurious. In fact, as outlined below when calculating consistency ratios, this is likely not to be the case. A second possibility is that the developers make different assumptions about the context of use and bring differing experiences to the evaluation process. A third is that the pairwise comparison technique is not a valid approach as it does not accurately reflect actual developer perceptions.

By comparing the percentage contributions two main findings can be noted. Firstly three of the four developers generated the same ranking of metrics using both techniques (NA and PW). The average percentage contribution of each metric over the four developers were in a very similar ranked order compared to the numerical assignment technique with a Spearman's rank correlation coefficient of 0.625. These results tend to support the proposition that the pairwise comparison technique is indeed valid. The second main finding was that the pairwise comparison technique greatly magnifies the difference between metrics.

An additional feature of the pairwise comparison technique is a method by which the significance of any judgement errors can be determined. The analytical hierarchy process (Saaty, 1980, Karlsson, 1996) allows us to calculate a *consistency ratio* to measure the acceptability of the comparisons. Firstly a *consistency index*, CI, is calculated where:

$$CI = (\lambda_{max} - n) / (n - 1)$$

where λ_{max} denotes the maximum principal eigenvalue, and n denotes the number of candidate metrics. The closer λ_{max} is to n, the more accurate the resulting percentage contributions. The consistency ratio, CR, is calculated as the ratio of the consistency index derived to the *random index*, RI, of a randomly generated reciprocal matrix of the same order, values of which are given in Table 4. For Developer 2, λ_{max} , the maximum principal eigenvalue, is 4.238 and so:

$$CI = (4.239 - 4)/3 = 0.079667, \text{ therefore as RI} = 0.9,$$
$$CR = 0.079667 / 0.9 = 0.088519.$$

Table 4. Random consistency indexes - RI

n	2	3	4	5	6	7	8	9	10
RI	0	0.58	0.90	1.12	1.24	1.32	1.41	1.45	1.49

The four consistency indexes and ratios are included in Table 3. According to Karlsson (1996) a consistency ratio (CR) of less than 0.1 is considered acceptable, although higher consistency rations are rather common. Three of the four developers generated a CR of less than 0.1 and the fourth was not greatly outside the range. These results clearly indicate that the technique provides a suitable mechanism on which to determine a combined metric.

4 Conclusions

Though this reported case study and other studies results have illustrated the strengths of the pairwise comparison technique for determining a combined metric. Interface developers on their own can generate quite widely differing percentage contribution by either method, and some mechanism for averaging the values of a group of developers is therefore necessary. Averaging differing percentage contributions generated by the numerical assignment tends to lead to little discrimination between metrics whereas doing this for pairwise compared data tends to leads to greater difference between metrics. Achieving a greater discrimination is only a good thing if it correctly represents the real views of the developers. In this respect the ability to determine consistency ratios leads to a quantitative way in which developer assessments can be analysed, individual views eliminated if necessary, and the overall metric substantiated. A key feature of the pairwise comparison technique is that developers / evaluators are forced to make trade offs between desired quality characteristics and a higher consensus about their relative importance is thereby generated.

5 References

H.X. Lee, Y. Choong, and G. Salvendy, (1997), A proposed index of usability: a method for comparing the relative usability of different software systems, *Behaviour & Information Technology*, 16 (4/5), pp. 267-278.

Karlsson, J. (1996), Software requirements prioritizing, *Proceedings of the second international conference on requirements engineering*, (ICRE'96), IEEE Computer Society Press, pp. 110-116.

Saaty, T. L. (1980), The analytic hierarchy process, Mc-Graw Hill.

Smith, A. and Dunckley, L. (1998), Using the LUCID method to optimise the acceptability of shared interfaces, *Interacting with Computers*, 9, pp. 335-345.

Smith, A., and Dunckley, L. (1999), Importance of collaborative design in computer interface design, *Contemporary Ergonomics, 99*, Taylor and Francis.

User-Interface Design for Online Transactions: Planet SABRE Air-Travel Booking

Aaron Marcus, John Armitage, Volker Frank, and Edward Guttman
Aaron Marcus and Associates, Inc., 1144 65th Street, Suite F, Emeryville, CA 94608-1053 USA;Tel:+1-510-601-0994x19, Fax:+1-510-547-6125
Email: Aaron@AmandA.com, Web: http://www.AMandA.com

1 Introduction

During 1994-99, Aaron Marcus and Associates, Inc., (AM+A) worked closely with SABRE Travel Information Network (SABRE), providers of one of the world's largest private extranets, to develop an innovative user interface and information visualization (UI+IV) for Planet SABRE ™, which is used by travel agents world-wide (in six languages: English, Spanish, French, German, Portuguese, and Italian). Approximately one third of the world's travel agents access the 42 terabytes in SABRE's databases of information about air travel, hotels, and auto rentals on 125,000 terminals. The system receives up to one billion "hits" per day, and generates approximately US$1 billion dollars per year in sales. SABRE is a twenty-year-old example of a very successful transaction system within e-commerce. Challenges and accomplishments in this long-duration, large-scale development effort of UI+IV design provide important lessons for development of applications for Web-based e-commerce. This paper reviews designing the UI+IV specifically for the primary module or application of the system, air travel booking, in 1998-99, for a user community whose characteristics and tasks were generally well-understood based on previous years of project experience and user evaluation..

Developing a UI for a system enabling access, search, retrieval, and rapid decision making about competing elements of knowledge typically has these steps: planning, research, analysis, design, implementation, evaluation, documentation, and training. Development is cyclical and may be partially repetitive. For example, evaluation may be carried out prior to, during, or after the design step. For specific users (defined by their demographics, experience, education, and roles in organizations of work or play) and their tasks, user interfaces must provide these components: *Metaphors* are essential concepts conveyed visually through words, images, and acoustic or tactile cues. *Mental Models* are organizations of data, functions, tasks, roles, and people in groups at work or play. *Navigation* is the movement through mental models afforded by windows, menus, dialogue areas, control panels, etc. *Interaction* is the means by which users input changes, and the system supplies feedback. *Appearance* consists of verbal, visual, acoustic, and tactile perceptual characteristics.

2 Developing the UI for SABRE Graphical Air

In developing a user interface for a system enabling access, search, retrieval, and rapid decision making about competing "products" typically, SABRE and AM+A followed the typical steps identified above

Plan: Defined the problems or opportunities; established objectives and tactics; determined budget, schedule, tasks, and development-team and other resources. *Research:* Investigated dimensions and techniques for all subsequent steps, e.g., techniques for analysis, criteria for evaluation, media for documentation, etc. *Analyze:* Examined results of research, e.g., problem or opportunity (SABRE conduct market research), refined criteria for success in solving problem or exploiting opportunity (write marketing or technical requirements), determine key usability criteria; and define the design brief, or primary statement of the design's goals. *Design:* Visualized alternative ways to satisfy criteria using alternative prototypes; based on prior or current evaluations, selected the design that best satisfied criteria; prepared documents that enabled consistent, efficient, precise, accurate implementation, in particular, interactive guidelines documents. *Implement:* Built or carry out the design to complete the final product, e.g., SABRE wrote code using appropriate tools. *Evaluate:* Tested results at any stage against defined criteria for success, e.g., conducted focus groups, tested usability on specific functions, gathered customer and user feedback. SABRE continued until tests showed satisfactory low levels or users' errors or confusion. *Document:* Recorded development history, issues, and decisions in interactive guidelines (Fig.3), and recommendation documents, including scenario documents and illustrated "dictionaries" of controls or widgets.

The designers considered these project attributes: The application contained significant complexity. The project spanned the entire software development cycle. SABRE wished to change the industry fundamentally to gain a competitive advantage. SABRE sought to increase the competitive expertise of travel agents who were potentially squeezed by Internet-based businesses. Travel agents felt pressure to be more productive because of new fee caps on transactions. The design needed to change a Legacy system command-line user interface, involving arcane codes in place for 20 years, into a graphical one that novice as well as power users could find attractive (Fig. 1). Current functionality was driven by data flow, not task flow. The system needed to become Internet-enabled for future evolution without overwhelming the host.

Among marketing and engineering objectives, the new design needed to accomplish the following: Accommodate beginners and power users, yet convert Legacy users painlessly. Help agents find the lowest fare quickly, yet not be biased in favor of any particular airline. Be easy to use and reduce help-desk calls to SABRE, but also had to be *perceived* immediately as usable. Result in cost savings for travel agencies, while increasing revenue from airlines. Satisfy both business (time-sensitive) travelers and leisure (cost-sensitive) travelers, the two basic market niches. Comply with international government regulations and support localization.

Because initially relatively little useful documentation was available. AM+A relied heavily upon subject matter experts who evaluated partial prototypes. The prototypes explored the three basic categories of critical information necessary for specifying a flight: fares, schedules, and rules, where rules refer to constraints airlines place on tickets. In the Legacy system, these three kinds of information were not displayed together, somewhat analogous to a grocery store that displays vegetables on one aisle, and their prices on another aisle. A major IV challenge was to combine interdependent facts into a single display while simultaneously limiting the depth of dialogue boxes to only three levels.

Another significant task was designing a seatmap module (Fig. 2), because each airplane has different seat configurations. AM+A's solution was to create a way of dynamically drawing the seatmap for any airplane, using a grid and a set of bitmaps that could be sent efficiently to the display terminals.

SABRE wanted the new system to be intuitive immediately to new users. This was not feasible, because of the extreme complexity of the system, nor appropriate, because most people would be using it full-time. AM+A believed it made sense to design for the intermediate user, with the expectation that users would receive significant training. An independent researcher did a formal task analysis of fifteen different travel agencies, and this information was used to develop the mental model.

AM+A made static and then interactive prototypes using Macromedia Director, which was useful for displaying use scenarios. The initial prototypes assumed a breadth-first interaction style of searching, going from general to specific. However, travel agents prefer to work in the opposite direction, going directly to what is likely to be the best solution. Then, if customers are not satisfied, they widen the search in close proximity to initial constraints.

One design challenge was to communicate the design to SABRE users, managers, and engineers. AM+A showed prototypes to focus groups of users to evaluate tasks, and also to managers to secure funding. Engineers objected to technical aspects of some of the designs developed via marketing; consequently, AM+A revised them. Development cycles exceeded initial budget estimates. SABRE solved this challenge by increasing airlines' fees for participation in the system, but enabling them to advertise in the displays in order to compensate for the higher fees.

3 Conclusion

The project was goal-driven, complex, and daunting, relying heavily on new metaphors and IV. The UI process was iterative, user-centered, and task-oriented. This UI+IV development process is fundamental for successful UIs of Web-based transactions and e-commerce: Metaphors, mental models, and navigation must avoid overwhelming users with content and tools and must enable them to navigate simply and quickly. Primary information-visualization displays must display effectively content tht is user and task-driven, not database driven. The application is scheduled for release in late 1999.

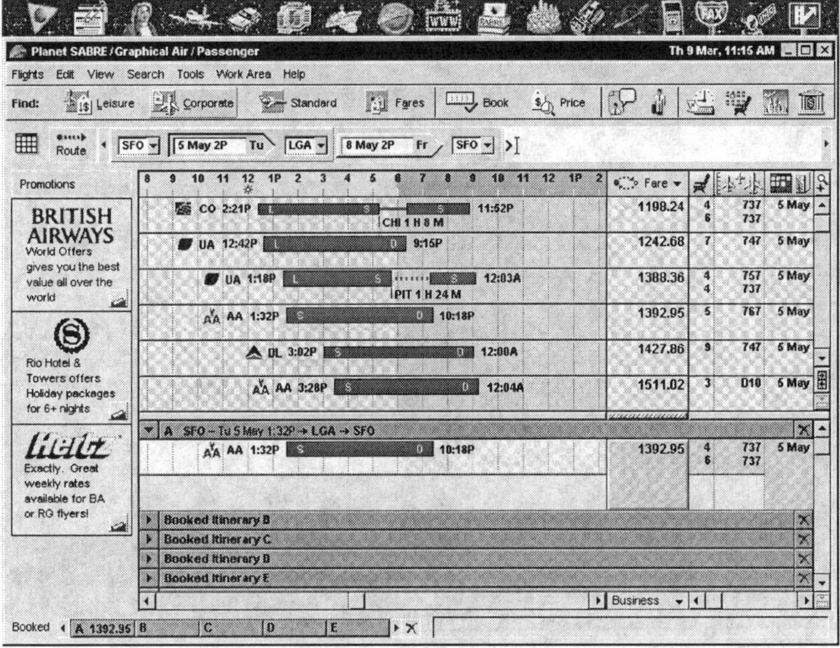

Figure 1

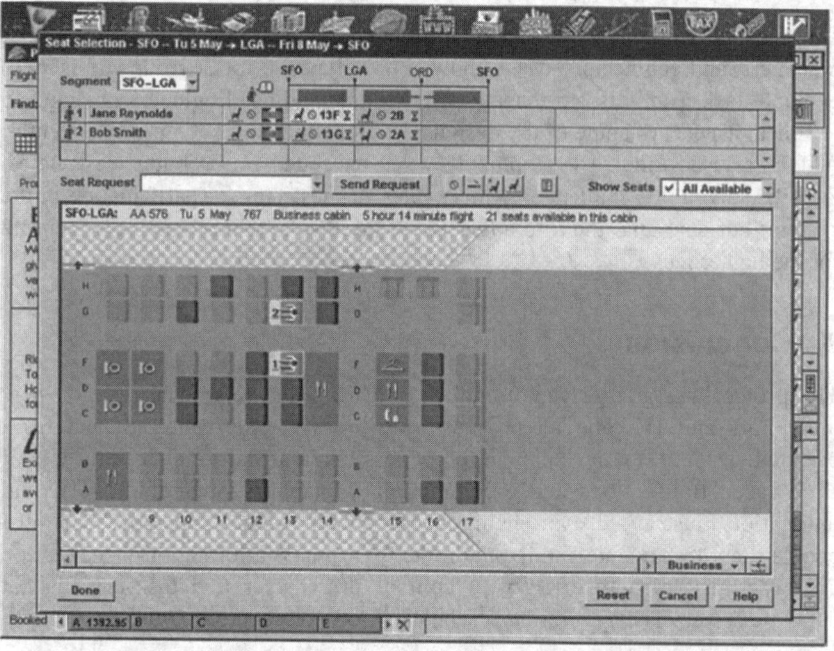

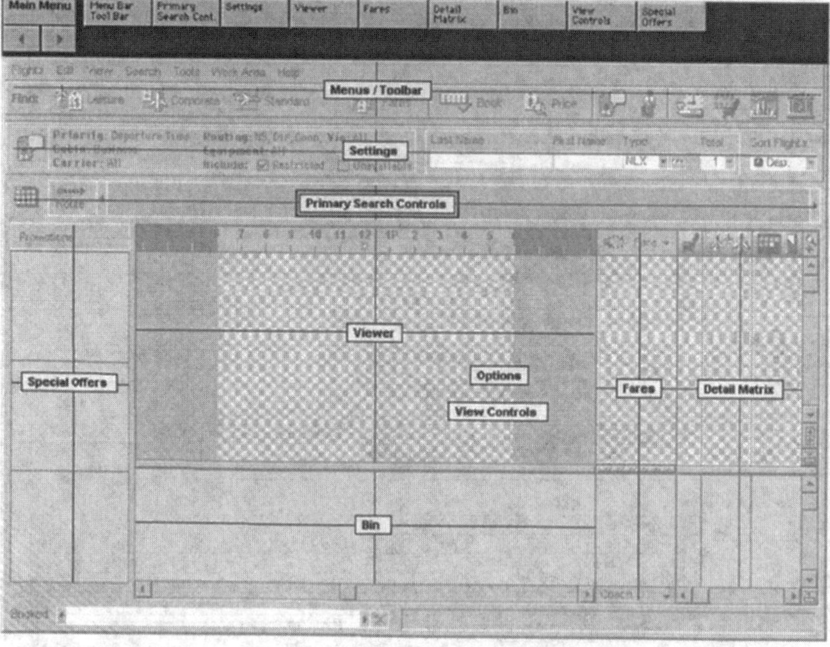

Figures 2 and 3

Interface Design Factors for Shared Multi-Cultural Interfaces

Andy Smith [1] and Lynne Dunckley [2]

(1) Department of Computing and Information Systems, University of Luton, Park Square, Luton, LU1 3JU, UK. Email: andy.smith@luton.ac.uk

(2) Faculty of Maths & Computing, The Open University, Walton Hall, Milton Keynes, MAA 3JU. Email: L.Dunckley@open.ac.uk

1 Introduction

Shared interfaces, with users from different cultural and geographical domains present designers with new dilemmas. There is the choice of overall design strategy; whether to develop an international culturally free interface or to provide a number of localised versions. Whatever the approach, effective design involves recognising the cultural elements within a given design proposal.

Designing for diverse user groups involves consideration of culture, gender, age, and physical disability. Such diversity makes it even more unrealistic for designers to rely on intuition or personal experience for interface design. However designing multiple interfaces for different user groups adds significantly to the cost of development. It is important therefore to focus on characteristics of interfaces which are sensitive to user factors in order to produce a cost effective solution. This paper describes the results of some experimental developments for shared multi-cultural interface design.

2 LUCID method for Shared Interfaces

The authors have developed the Logical User Centred Interface Design (LUCID) method for interface design that has been validated for the development of mono-cultural systems. The method (Smith & Dunckley, 1996) aims to provide a user-centred environment through which usability is enhanced with respect to identified criteria. The identification of key interface design factors is crucial to the LUCID method and occurs at an early stage in the design process. Through an experimental prototyping design strategy based upon Taguchi Methods (Taguchi 1986) the objective is to set the specifications of the contributing design factors to improve quality and reduce costs. The adaptation of the LUCID method for shared interfaces, requires the identification of user factors which will influence the usability of different cultural groups (Smith & Dunckley, 1998). In preliminary research user factors have been classified into the following types:

- *objective factors*, such as gender, age, ethnic background, mother tongue which can be objectively identified for each user subject and used to categorise different user groups,
- *subjective factors*, which cannot be directly measured or identified (e.g. cognitive style, locus of control and user attitudes) and need to be assessed by other methods for example from the user responses to questions or cognitive tests (see Hofstede, 1991).

The method aims to measure which design factors affect the user's performance and the variability of the user's response. The objectives of the LUCID shared interface design method are to identify, for a given application, the optimum combination of design factors to give:

(a) an internationally culturally free interface

(b) the localised interfaces for particular cultural groups.

The method uses a combination of inner and outer Taguchi orthogonal arrays to study the interaction between design factors and user factors, then to analyse the results by ANOVA to predict the optimum localised and globalised interfaces.

3 Experimental Studies

One of the issues we were concerned about was whether the subjective cultural factors would be significant in terms of usability, since they would be more difficult to assess. Two experimental studies of the proposed method have been carried out, one used only objective user factors while the other used both subjective and objective user factors. The lack of an established instrument for assessing subjective cultural factors in an IT context was a problem which led to development an questionnaire instrument for the subjective factor study.

3.1 Multi-media Application with Objective User Factors

In the first study, a multi-media CAL application, preliminary analysis suggested three objective user factors were significant, i.e. ethnicity, learning difficulty and gender. These three factors can be combined in eight different ways but by using an L_4 orthogonal array the number of user groups who need to be studied is reduced to four, as shown in Table 1.

Table 1 Outer User Array for Multi-media Interface

User Factor	Group 1	Group 2	Group 3	Group 4
Learning Difficulty	*Dyslexic*	*Non-dyslexic*	*Non-dyslexic*	*Dyslexic*
Ethnicity	*Afro-Caribbean /African*	*English*	*Afro-Caribbean /African*	*English*
Gender	*Female*	*Female*	*Male*	*Male*

Other user factors such as age, IT literacy and educational level were held constant across the groups. In this application the design factors studied were: navigational style, vocal hotwords, video screen style, use of status details, warning prompts, check boxes, combo boxes and sliders.

The results showed that it was possible to classify these design factors into :

(a) factors which affect the average performance of the interface – in this case - navigational style, warning prompts, the use of sliders

(b) factors which affected the variability of the performance i.e. were sensitive to the different user groups – in this case – video style and combo boxes.

Cultural effects were significant and it was possible to identify the optimum settings for the design factors across all user groups to give the specification of the best global interface. However the user group *dyslexic*, *Afro-Caribbean/African, male* was still disadvantaged compared with other user groups and a different specification was identified, the localised interface.

3.2 Large Scale Database using Objective/subjective Factors

This application was a database of transportation employment patterns for long term planning purposes. Real data was used but the users were selected from multi-cultural undergraduate student groups, who were realistic models of the target users. The system provided data entry screens, summary reports and help screens. Three design factors given in Table 3 were studied . A large group of 95 multi-cultural users completed an electronic questionnaire, the User Perception Questionnaire consisting of 26 semantic differential questions based on a 7 point Likkert scale. The cultures represented were British, USA, African, Afro-Caribbean, Greek, Chinese and Indian Sub-continent. Both objective data (gender, age, language and ethnic background) and subjective data (individualist/collectivist, power distance, femininity/ masculinity and locus of control) were collected. From the original group 32 users were selected for the usability tests.

Table 3 Design Factors with Objective User Factors

Design Factors			Objective User Factors		
F1Text Display Style	F2 Screen Design	F3 Help Provision Style	F1 Age	F2 Mother Tongue	F3 Orientation
Upper Case	Multi	Pop-up Help	Mature	Non E	Business
Upper Case	Single	Menu System	Mature	English	Computing
Lower Case	Multi	Menu System	Young	Non E	Computing
Lower Case	Single	Pop-up Help	Young	English	Business

The same optimum interface and design factor settings were identified for all user groups. However the contribution of the design factors was different across the user groups. For the *Non-English* user groups, text display style was much more important for performance metrics than help provision style, although this was a key variability factor for the different groups.

Table 4 Cultural Dimensions User Array

Optimum Level	Contribution	Favoured User	User Contribution
Lower Case	69.5	High Power Distance	69.1
Multi screen	3.25	Individualism	30.8
Help menu	27.25	Femininity	0.1

Alternative users groups based on the subjective factors Power/Distance, Individualism/Collectivism and differentiation of Femininity/Masculinity were tested. Table 4 gives the results from the subjective factors user groups, showing there was no difference in the predicted optimum interface but the contribution of the text style design factor was greater.

4 Results and Conclusions

The results of both practical applications show that it is possible to measure the variability of the user's response to the interfaces and to identify which of the design factors affect this variability. In this way it is possible to identify the best 'global' interface and those design factors, which contribute the most to optimising the 'global' interface in contrast to those design factors which have maximum affect on the user differences and should be considered when localising the interfaces.

User groups based on objective cultural factors gave clear and understandable results. Identifying and classifying users using subjective cultural factors was both more difficult and more resource intensive. The insights for interface design obtained from the results were also less clear. For example, the subjective factors indicated the significance of power distance and individualism/collectivism for the database application. This result was less obvious but could be an important insight into the way that users perceived their tasks in relation to the computer system and planning decisions. The other preliminary finding (which will need further verification) was that the subjective factors seemed much better at clustering users than the objective factors. It is common knowledge that individuals of the same age and ethnic background vary a great deal in other ways. However the results of the users in the same subjective categories showed a surprisingly small variance. This was an unexpected and intriguing result.

The usability of the multi-media application was more sensitive to cultural factors than the database application. Western culture tends to dominate computer systems at the moment and it was significant that in both applications users from non- Western cultural groups were at a disadvantage.

5 References

Hofstede,G. (1991) '*Cultures and organizations - software of the mind*' McGraw-Hill

Smith, A. and Dunckley, L. (1996), Towards the Total Quality Interface - applying Taguchi TQM Techniques within the LUCID method, in Sasse, M. A., Cunningham, R. J. and Winder, R. L. (eds.), *People and Computers XI*, Proceedings of HCI-96, Springer. ISBN 3-540-76069-5, pp 3-17.

Smith, A. and Dunckley, L. (1998), Using the LUCID method to optimise the acceptability of shared interfaces, *Interacting with Computers*, 9, pp. 335-345.

Taguchi, G. (1986), 'Introduction to Quality Engineering'. Asian Productivity Organisation.

Combining Knowledge Acquisition with User Centred Design

Hans I. Myrhaug, Nils B. Moe, Ole M. Winnem
SINTEF Telecom and Informatics, N-7035, Trondheim, Norway

1 Introduction

A Norwegian artillery school was recently relocated from Gardermoen to Rena. Rena is an agricultural area with many small villages surrounded by hills and forests. Because of this fact, the Norwegian Pollution Control Authority gave the artillery school new environmental directives for both noise propagation and chemical/physical pollution.

The new directives created a need for a noise-planning tool. The system and method requirements showed that techniques from knowledge acquisition (KA), user centred design (UCD), and software engineering (SE) had to be combined in order to create this new planning tool.

System requirements. The directives ordered the army to *monitor*, *control* and *report* their noisy activities, leading to the design of a new system called STOY - An Information System for Environmental Noise Planning. The requirements are among others: 1) monitor, control and report noisy activities, 2) recognise and separate own noise from background noise, 3) (re)plan future activities and simultaneously evaluate them against directives and past experiences and 4) review past activities regarding complaints, directives and past experiences.

Method requirements. First, the system requirements in fact ordered the use of several techniques from artificial intelligence, one of which requires the use of extensive domain modelling. Second, since no similar system exists, the development process had to rely upon user centred design. Third, since the system requirements ordered the use of complex technology, the method had to be scalable. Fourth, the method had to help designers, programmers, and users to communicate the designs. Therefore, in an attempt to find the right development method, elements from knowledge acquisition and user centred design was combined.

1.1 Knowledge acquisition

Knowledge acquisition (KA) can be defined as the act of eliciting, analysing, and modelling knowledge of human experts. The goal of the KA process is to transfer human expertise to other humans or computer systems. The KA process is evolutionary, where the final or intermediate results of the various activities are explicit models. In (Aamodt, 1991) the KA models and processes are:

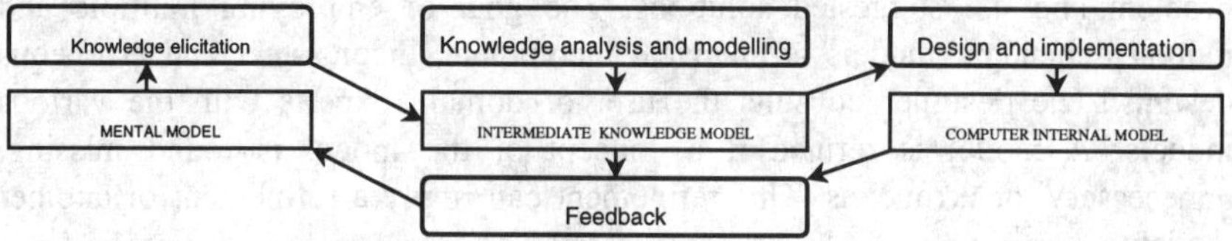

CommonKADS (Wielinga et al, 1994) is one methodology of KA. It produces an expertise model containing, among others, a task model, a domain model and an agent model. The focus is on how organizations develop, distribute and apply their knowledge resources.

1.2 User Centred Design

User centred design (UCD) can be said to consist of the *evolutionary* approach and the *model-based* approach. Both approaches emphasise the user dimension, and can be used during the complete development cycle. The difference is in how the interactive system is shaped. UCD is a task-oriented approach to system development that aims at making interactive systems more usable.

Evolutionary design is a collection of activities that meet the user needs through an iterative process of analysis, (re)-design, and evaluation phases. It is attractive because many prototypes are suggested, tested, modified, or discarded. The prototypes are close to what the user gets in the final interactive system, and it is easy to discuss the solutions with the user. Indeed, this approach is so appealing that some designers believe it is a complete solution to the system development problem, and that there is no need for specifications, since the prototypes specify parts of the final system. Consequently, the analysis phase is given less attention compared to the (re)-design and evaluation phases. Thus, it is strong at suggesting alternative solutions and modifying them based upon the evaluations. But, the approach does not focus on understanding the problem.

Model-based approach attempts to design interactive systems from explicit models. The designers use the models in order to communicate aspects of the interactive system design to users and programmers. The various model categories are user models, task models, domain models, and user interface models.

User models describe aspects of the user, task models specify the user's tasks, domain models describe the underlying concepts of the system and the environment, and finally, user interface models describe aspects of the user interface. Mock-ups, navigational maps, horizontal designs, and state diagrams are some examples of user interface models.

The models are concise and structured. Each model involves aspects of the other models. The use of several types of models provides several views to the design problem and the suggested solutions. The goal of employing multiple and mutual dependant models is to reveal incomplete, inconsistent and erroneous designs. The designer consults the user or domain experts with the various models. A model is refined if a concept of the model is found missing, unnecessary, or erroneous. This refinement can imply a refinement of another model.

Evolutionary design is strong at providing a formal strategy to assess usability. However, we argue that it is more difficult to follow when little is known about the users and the domain is complex. The model-based approach complements it because it focuses on formal specifications. Consequently, the analysis phase is given more attention in the model-based approach. It can also be said to extend the evolutionary approach because the focus is mainly on user interface models (prototypes) within the evolutionary approach.

However, we do not suggest that one approach is better than the other. Rather, we believe that a balanced synthesis of the two approaches enhances both the development process and the usability of interactive systems. A more complete discussion on this topic is found in (Hall, 97).

2 The Model Based Method

To meet the method requirements the following method was used:

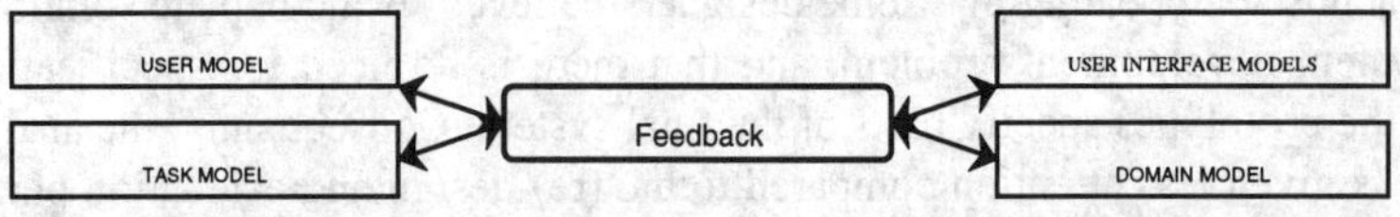

All models of the method are subject to the iterative and generic steps of analysis, design, and evaluation. A change of one model gives feedback and may cause changes of other models.

1. Brief design specification - a brief brain storming activity ended up in a textual user interface design specification.

2. Domain modelling for experience-based technology - an iterative knowledge acquisition interview technique was used to produce a domain model for outdoor noise emission, so that the experience-based system can explain why two activities are more or less equal. There were 11 iterations, where 4 acoustic researchers were interviewed. This phase gave much feedback to phase 4.

3. Domain modelling for noise directives – the rules from the textual directive was modelled in first order predicate logic (Prolog). The result was a subsystem that criticises the user's plan. This gave feedback to phase 2.

4. Task modelling - this phase resulted in a complete and hierarchic task model. The brief design specification and the domain model for outdoor noise emission served as invaluable inputs to this phase. The model was refined several times. This gave much feedback to phase 2 and phase 5.

5. Navigation design – the task model was mapped down to a navigation model. The model was represented as a graphical state diagram where each state represented a dialog box or window. There was roughly a one-to-one mapping between each subtask of the task model and each dialog box. This gave much feedback to phase 4 and phase 7.

6. Functional core design - an iterative design phase of the functional core was completed. A separation strategy for the user interface objects and the functional core objects (application core) was needed because of the complexity. Both phase 2 and phase 4 provided input to this phase. The result of this phase was a complete UML class model of the functional core.

7. User Interface design - the system's usability was defined as time effectiveness for each subtask. The task model and the navigation model served as input to the horizontal design of windows, dialog boxes, and the menu system. Several dialog boxes have been tested with users (evolutionary design). The users were given particular subtasks, and the completion time of the task was recorded. The philosophy behind this technique is that one can speed up the user interface by speeding up each subtask.

8. User interface and functional core implementation - the UML class model and the horizontal design prototypes served as input to the complete implementation.

9. Acceptance test, emission to market, and maintenance.

3 Discussion and results

The challenge of STOY was to combine techniques from KA, UCD and SE. Several of the method activities were carried out in parallel. They were also iterative.

The Model Based Method relies upon the development of a set of mutual dependant models. All models are subject to the generic and iterative modelling phase of analysis, design, and evaluation. The consequence is that the design team can leave one perspective and switch to another at any time in the development cycle, since the modification of one model implicitly affects the other models. Thus, problems abandoned in one perspective can be solved implicitly from another perspective.

The method can be conceptually heavy for the developers. The main reason for this is the employment of multiple perspectives to the interactive system to be developed. Despite this, the development team appreciated the method because the models served as a concise communication medium. Both users, other experts, and members of the development team can discuss the various models and discover unnecessary, erroneous or lacking concepts at an early stage in the development cycle - even before prototypes are developed.

The budget of STOY is 500.000 ECU. Four people have been involved in the development team. The project is on schedule regarding both time and budget constraints. The system is when writing in phase 8.

The method is now being used by two other projects. The first project aims at making a navigational aid system for blind people. The second project designs and tests prototypes and existing information- and communication systems for truck drivers. Both projects share the navigation task, but the user and the domain are slightly different. Thus, the task and domain models in the former project were reused and slightly modified in the latter project because of a different user profile.

In a third project we are searching for an effective development method for multi-modal user interface systems. The method will be reviewed in this project.

4 References

Aamodt, A. (1991), A knowledge-intensive, integrated approach to problem solving and sustained learning, Ph.D. thesis, Norwegian University of Science and Technology, May 1991.

Hall, A. (1997), Do interactive systems need specifications?, In: *Design Specification and Verification of Interactive Systems '97*, pp. 1-12, Springer-Verlag, June 1997.

Wielinga et al (1994), Expertise model definition document. ESPRIT Project P-5248 /KADS-II/M2/UvA/026/5.0, University of Amsterdam, Free University of Brussels and Netherlands Energy Research Center ECN, June 1994.

Aesthetic design – just an add on?

Michael Burmester, Axel Platz, Udo Rudolph & Brigitte Wild
Siemens AG Corporate Technology
User Interface Design,
D-81730 Munich
michael.burmester@mchp.siemens.de,
axel.platz@mchp.siemens.de
LMU Munich
Institute for Psychology
D-80802 Munich
rudolph@mip.paed.uni-muenchen.de
brigitte.wild@stud.uni-muenchen.de

1 Need for appealing user interfaces

Recently, the main focus of industrial HCI research was on improving user interfaces so that users can accomplish their tasks in an effective and efficient way in order to reduce their mental workload and to make them feel satisfied by improving product usability (ISO 9241-11, 1997). Logan (1994) divided usability into two components – behavioral and emotional usability. Emotional usability "refers to the degree to which a product is desirable or serves a need(s) beyond the traditional functional objective" (p. 61). User interfaces should be engaging, foster a sense of discovery and eliminating fear.

These demanded user interface qualities are also related to the concept of an 'event-society' (e.g. Weinberg, 1992). Human decisions and preferences are not purely rational, but rather heavily influenced by emotional contents (Koppelmann, 1997). Although today mostly consumer products are connected with such an emotional component this is getting more and more important for capital goods.

There has been little research concerning aesthetic design and emotionally appealing user interface design. Leventhal et al. (1996) compared different dialogue boxes and there preferences among different user groups. Kurosu & Kashimura (1995) and Tractinsky (1997) investigated the relationship between users' perceptions of interface aesthetics and usability. Others like Reeves & Nass (1996) looked for anthropomorphism aspects of user interfaces and their influence on the behavior of persons.

In order to increase knowledge about emotionally appealing user interfaces, Siemens has called into being the project 'ATTRACTIVE'. Questions of the project are: How people perceive user interfaces? Which features, visual design and elements of user interfaces lead to positive emotions and impression? How can this be measured?

2 Design theory

The traditional perspective in industrial design is 'form follows function' (Bürdek, 1991), which means that perfect aesthetic form is reached if the form corresponds exactly to the function of a product. This design leads to the style of aesthetic functionalism (Koppelmann, 1997). Aesthetic form increases order and facilitates communication without confusion (Arnheim, 1966).

Due to the micro controller technology the inherent functions of modern products are not visible (Bullinger et al., 1995). This generates a special challenge to design user interfaces adequately.

3 Research questions

In most cases user interfaces are designed by software developers using tools for design of graphical user interfaces (GUI tools) and style guides focusing on usability requirements. These tools provide basic support for aesthetic design such as pre-designed standard user interface building blocks and design grids for placing elements on the screen. Taking this into account, the question is whether users react differently to standard GUIs and user interfaces designed according to the principles of aesthetic functionalism by a professional communication designer.

Furthermore, different types of users are in contact with product, in this case medical equipment. The importance of a product is different for professionals and patients. On the one hand professionals have job and task know-how and doing their work with the computer tomograph (CT). On the other hand patients do not know much about the CT and are in an insecure situation. They have to trust the professionals and the device.

The research questions are:

- Which dimensions are valid for the judgement of user interfaces?
- Is there a difference in the perception of standard GUIs and designed user interfaces?
- Do non professionals (e.g. patients) react in a different way to a professional product (a CT) than professional users?

4 Methods

To answer the questions, two user interface versions for the same CT were compared. One version was developed by the traditional way (GUI Version, Figure 1) and the second version was worked over completely by a communication designer (Design Version, Figure 2).

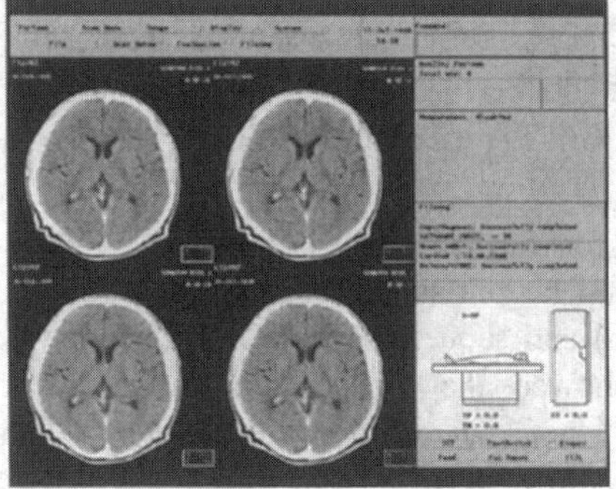

Figure 1: GUI Version

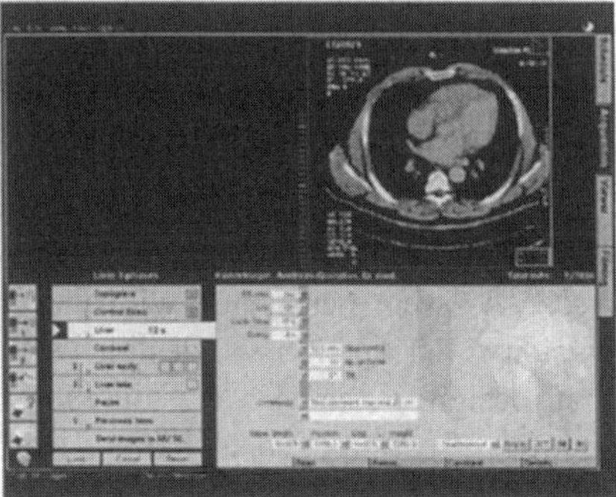

Figure 2: Design Version

To measure the impression upon the subjects a questionnaire was developed. The 23 items were constructed according to *general evaluation* of the CT as a whole, *aesthetic form* of the user interface, impression of *usability*, *professionality* and *safety*.

15 experts and 41 laymen were invited to the usability lab of the 'User Interface Design' Department of Siemens. Most of the experts were MTAs[1].

Each of the two user interface versions used as stimuli were represented by a sequence of 5 resp. 6 screen shots. Each screen shot was presented for 15 sec. Within a sequence of one version the screen shots were presented in the order of the working process. After the presentation of one version, the subjects filled out the questionnaire. The two versions were presented in random order.

5 Results

A factor analysis (principal component extraction followed by a varimax rotation) was applied in order to show the hidden concepts behind the items

[1] MTA = Medizinisch-Technische-Assistenten. MTAs assist the physicians, in this case operate the computer tomograph.

(Brosius & Brosius, 1995). The analysis indicated that the five item sets *general evaluation*, a*esthetic form*, *usability*, *professionality* and *safety* were reduced to four dimensions *quality impression, apparent usability*, *safety* and *superiority*. The most striking result was, that the aesthetic dimension vanished. What happened to it?

The first factor extracted was *quality impression*. Here about half of the original aesthetic items can be found. The second factor was *apparent usability* (see also Tractinsky, 1997). This factor was composed of all the usability items and almost half of the aesthetic items. Additionally, there was a *safety* factor, how much one trust the CT, and a smaller *superiority* factor, which compares the CT with others. 69% of the entire variance was explained by these four factors.

Out of these four dimensions, scales for further analysis were calculated. ANOVAs were performed with the group factors status (experts vs. laymen), version (GUI version vs. design version) and sequence (first GUI version vs. first design version). The outcome showed that the design version was significantly judged better at the *quality impression* scale ($F(1,52)=24.61, p<.001$), the *apparent usability* scale ($F(1,52)=7.37, p<.009$) and the *superiority* scale ($F(1,52)=4.35, p<.042$) but not at the *safety* scale. Additionally, laymen rated the GUI version more negative than the experts.

6 Discussion

The factor analysis showed that the most important dimension for judgement of the user interface screen shots were *quality impression, apparent usability, safety* and *superiority*. Aesthetic form belongs to the apparent usability AND the quality impression factor. Aesthetic design is not a dimension of its own. The close relationship of quality impression and apparent usability of a product shows that design is not an add on to make user interfaces more beautiful!

Furthermore, the analysis showed that the design version was rated higher than the GUI version with respect to positive *impression* of product quality, *apparent usability* and *superiority* of the product. This effect is stronger for the laymen.

Due to the professional background, experts have an analytical view and a good understanding of the user interface screen shots shown. The layman have had no task knowledge and therefore they did not understand the controls and information presented. They just referred to the graphical impression.

In this study, the subjects could not interact with the user interface, but this situation can be compared to showroom or fair situations. In these situations it is important to generate a good quality impression for a product and the impression that the product is easy to use.

Following studies should investigate the relationship between design styles and their influence on the user interface impression judgement and they should include the interactivity dimension of user interfaces.

7 References

Arnheim, R. (1977). *Zur Psychologie der Kunst*. Köln: Kiepenheuer & Witsch.

Brosius, G. & Brosius, F. (1995). *SPSS, Base System und Professional Statistics.* Bonn: International Thomson Publishing.

Bullinger, H.-J., Burmester, M., Mangol, P. & Vossen, P.H. (1995). *Design interaktiver Produkte - Dialog zwischen Mensch und Produkt.* In H.-J. Bullinger (Hrsg.), Design interaktiver Produkte - Dialog zwischen Mensch und Produkt (S. 13-28). Stuttgart: Fraunhofer IRB.

Bürdek, B. (1991). *Design: Geschichte, Theorie und Praxis der Produktgestaltung.* Köln: Du Mont.

ISO 9241-11 (1997). *Ergonomische Anforderungen für Bürotätigkeiten mit Bildschirmgeräten - Teil 11: Anforderungen an die Gebrauchstauglichkeit.*

Koppelmann, U. (1997). *Produktmarketing, Entscheidungsgrundlagen für Produktmanager.* Berlin: Springer-Verlag.

Kurosu, M. & Kashimura, K. (1995). *Apparent usability vs. inherent usability.* CHI'95 Conference Companion. 292-293.

Leventhal, L.; Teasley, B.; Blumenthal, B.; Instone, K.; Stone, D. & Donskoy, M. V. (1996). *Assessing user interfaces for diverse user groups: evalutation strategies and defining characteristics.* Behaviour & Information Technology, vol. 15, no. 3, 127-137.

Logan, R.J. (1994). *Behavioral and Emotional Usability: Thomson Consumer Electronics* (pp 59-82). In M. E. Wiklund (ed.), Usability in Practice. Academic.

Reeves, B. & Nass, C. (1996). *The media equation: how people treat computers, television, and new media like real people and places*. Cambridge Univer. Press.

Tractinsky, N. (1997). *Aesthetics and Apparent Usability: Empirically Assessing Cultural and Methodological Issues.* CHI'97 Conference Proceedings . 115-122.

Weinberg, P. (1992). *Erlebnismarketing.* München: Vahlen.

Keep Cool: The Value Of Affective Computer Interfaces In A Rational World

Erik Hollnagel
Graduate School for Human-Machine Interaction
University of Linköping, Sweden

1 Introduction

The standard model for human-machine interaction is based on the communication paradigm formulated by Claude Shannon in the early 1940s (Shannon & Weaver, 1969). This paradigm describes the exchange of messages between a sender and a receiver, with emphasis on the capacity of the information channel and how messages can be distorted by noise. The paradigm was eagerly adapted by a science of psychology trying to break lose from the confines of behaviourism (Attneave, 1959; Miller, 1967), and provided the foundation for the psychological models that became part and parcel of information processing psychology, cognitive psychology and cognitive engineering (Lindsay & Norman, 1977; Newell & Simon, 1972; Wickens, 1984). It also corresponded well to the endeavour to describe human actions as the result of a rational choice, exemplified by the models for human decision making (Lee, 1971; Edwards, 1954).

When the study of human-machine interaction was devoured by the swelling interest for human-computer interaction in the 1980s, the basic paradigm remained. Human-machine interaction was seen as the exchange of messages (signals and control actions) across an interface, and the information processing view reigned supreme. The processes, cognitive or otherwise, that provided the basis for the communication and interaction were all "cold" rather than "hot", i.e., excluding emotions and affect (Abelson, 1963). Information processing psychology several times tried to develop theories that included emotions and affect, but with very limited success (Mandler, 1975; Simon, 1967).

2 The Efficiency Of Human-Human Communication

As the use of computers has spread in an almost epidemic fashion to all areas of life and work, enormous efforts have been was put into making human-machine interaction as easy and efficient as possible. One example of that is the campaign to develop systems that are user-friendly and require little or no learning. Another is the desire to use multi-media to enhance the communication and comprehension. A third is the effort to build adaptive systems and interfaces that "automatically" provide the user with the right information, in the right form, at the right time.

In all of this human-human communication has often been used as a paragon. The smoothness, versatility, and efficiency of communication between people is a distant goal for human-machine interaction, and it is therefore tempting to look for features of natural communication that are missing in artificial communication. One obvious candidate is emotion or affect, i.e., the fact that people have affects and also are uncannily effective in recognising the affects and emotions of others. Since affects clearly are missing from the information processing paradigm, it is tempting to consider whether affective interfaces would provide the coveted quantum leap in human-machine interaction.

2.1 Emotions And Control

It is not impossible that affective interfaces may serve a therapeutic purpose, in the sense that they make the user feel better. The main reason for using an affective interface should nevertheless be that it improves the efficiency of the communication, hence the ability to retain control. Generally speaking, the purpose of communication is to control the behaviour of the receiver, whether it is a human or a machine, and anything that enhances the ability to control is therefore worth considering. One effective way of describing human (and machine) performance is by means of a cyclical, rather than a linear, model (Hollnagel, 1998; Neisser, 1976). In Figure 1 this principle is used to show the communication between two agents, in this case two humans.

The model describes how actions are determined by the current understanding, which in turn depends on the evaluation (or interpretation) of the feedback. In communication, the message of one agent becomes the response of another, and *vice versa*. The model emphasises that performance is both feedback and feedforward controlled. As pointed out already by Conant & Ashby (1970, p. 92), feedback (error-controlled) regulation is inferior to feedforward (cause-controlled) regulation. Despite that observation, practically all information processing models of operators exemplify error-controlled regulation.

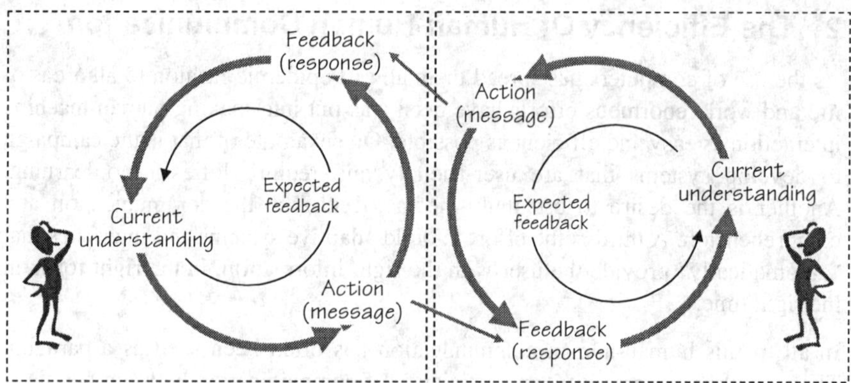

Figure 1: The cyclical communication paradigm.

2.2 Understanding The Mood Of The Other

The model in Figure 1 can easily be extended to address the role of affects by noting that one aspect of the current understanding is the assessment of the mood or emotional state of the communication partner. The emotional state of a person plays a major role in determining the reaction. "Hot" cognition may narrow the scope of interpretations and conclusions may be based on what was expected rather than what actually happened. Decisions may become coloured by highly subjective and temporary criteria, and otherwise reasonable alternatives may be disregarded. (This happens not just during affective states, but also during high stress and discomfort.) It is therefore important to take this into consideration when formulating and expressing the message. In normal life, and specifically during work, affective-emotional human-human interaction is usually not encouraged. On the contrary, we often try to counter the affective state of the other person to avoid the situation from boiling over or to prevent getting into a heated argument. The reason is simply that reciprocal affects may bring the system out of control. On the other hand, if the affective state of the other is correctly recognised, control can be regained and stability ensured.

3 Affective HMI – Dream Or Nightmare?

The model and the principles can be applied to human-machine interaction (HMI) as well as to human-human interaction. The purpose of HMI is not primarily to make the user feel good, but to enable the joint system to maintain control of a situation. The value of affective interfaces must therefor be considered in relation to this overall goal. One possibility is that the interface (or rather, the machine) is able to recognise the mood or affective state of the user. If this can be done in a reliable fashion, and if it further is possible to adjust the functioning of the HMI accordingly, mood recognition may conceivably have a beneficial effect. It is in a way a logical extension of the principles to design interfaces for everyone (Newell, 1993), although extending the scope of normal and disabled users to include calm and agitated users as well. Another possibility is that the interface also can show or represent emotions, i.e., be an affective interface. This could be seen as an extension, although a significant one, of adaptive interfaces. It would therefore also comprise the risks of adaptive interfaces, and possibly amplify them. The main risk is that two systems that reciprocally adapt to each other very easily run the risk of going into an unstable region. If the response to a recognised affect is inappropriate, which in control terms means that the damping of the system is inadequate, then this may all too easily lead to an amplification of deviations, rather than the neutralisation and homeostasis that is really wanted (Maruyama, 1963).

For this reason alone, it is highly uncertain whether there is anything gained by having an affective interface, although there may be a potential advantage in having one that can recognise affects. The question to be pondered by researchers in HMI (and HCI) is therefore not whether an affective interface, in either sense, is possible or technologically feasible, but whether it is required. Once affective interfaces are available, and this will undoubtedly happen at some time, they may join the many other technological solutions that are in search of a problem. It would be nice if, for once, a genuine need was established first. At the moment, it is my belief that the virtue of human-human communication are exaggerated, and that the risks of affective interfaces need to be understood better. This will require a sorely needed revision of the dominating paradigms for HMI.

4 References

Abelson, R. P. (1963). Computer simulation of "hot" cognition. In S. S. Tomkins & S Messick (Eds.), *Computer simulation of personality*. New York: Wiley.

Attneave, F. (1959). *Applications of information theory to psychology: A summary of basic concepts, methods, and results*. New York: Holt, Rinehart & Winston.

Conant, R. C. & Ashby, W. R. (1970). Every good regulator of a system must be a model of that system. *International Journal of Systems Science*, *1*(2), 89-97.

Edwards, W. (1954). The theory of decision making. *Psychological Bulletin, 51*, 380-417.

Hollnagel, E. (1998). Context, cognition, and control. In Y. Waern, (Ed.). *Co-operation in process management - Cognition and information technology*. London: Taylor & Francis

Lee, W. (1971). *Decision theory and human behavior*. New York: John Wiley & Sons.

Lindsay, P. H. & Norman, D. A. (1977). *Human information processing: An introduction to psychology* (2nd Ed.). New York: Academic Press.

Mandler, G. (1975). *Mind and emotion*. New York: Wiley.

Maruyama, M. (1963). The second cybernetics: Deviation-amplifying mutual processes. *American Scientist*, *55*, 164-179.

Miller, G. A. (1967). *The psychology of communication*. Harmondsworth, UK: Pelican Books.

Neisser, U. (1976). *Cognition and reality*. San Francisco: W. H. Freeman.

Newell, A. (1993). *HCI for everyone*. Invited keynote lecture at INTERCHI ′93, Amsterdam, April 27-29, 1993.

Newell, A. & Simon, H. A. (1972). *Human problem solving*. Englewood Cliffs, NJ.: Prentice-Hall.

Shannon, C. E. & Weaver, W. (1969). *The mathematical theory of communication*. Chicago: University of Illinois Press.

Simon, H. A. (1967). Motivational and emotional controls of cognition. *Psychological Review*, *74*(1), 29-39.

Wickens, C. D. (1984). *Engineering psychology and human performance*. Columbus, OH: Merrill.

Affect-Adaptive User Interface

Eva Hudlicka & John Billingsley

Psychometrix Associates, Inc.
Lincoln, MA, USA
hudlicka@acm.org

1 Introduction

As decision-support systems (DSS) and user interface (UI) technologies mature and proliferate into critical applications, and increasingly heterogeneous user populations, it becomes particularly important that they accommodate individual user characteristics. Recent research provides increasing evidence that individual differences in general, and affective states in particular, have a major impact on performance (Williams et al. 1997; Eysenck 1997; Mineka & Sutton 1992; Isen 1993; LeDoux 1992; Fallesen 1993). Specific influences include the following: 1) anxiety influences attention, by narrowing of attentional focus and predisposing towards detection of threatening stimuli (Mineka and Sutton 1992; Eysenck 1997); 2) mood influences memory, by biasing recall of information that matches the current affective state (mood congruent recall) (Bower 1981); and 3) obsessiveness influences decision-making by increasing "checking behaviors" and delaying decision making (Williams et al 1997). Current lack of accommodation of individual variations in performance in most human-machine systems can lead to non-optimal behavior at best, and critical errors with disastrous consequences at worst (e.g., USS Vincennes).

To address this problem, we developed an affective, adaptive methodology and implemented it within a software prototype: the Affect and Belief Adaptive Interface System (ABAIS). The methodology consists of: 1) sensing/inferring the individual's affective state and performance-relevant beliefs (e.g., high level of anxiety; aircraft under attack); 2) identifying their potential impact on performance (e.g., focus on threatening stimuli, biasing perception towards identification of ambiguous stimuli as threats); 3) selecting a compensatory strategy (e.g., redirecting focus to other salient cues, presentation of additional information to reduce ambiguity); and 4) implementing this strategy in terms of specific UI adaptations (e.g., highlighting relevant cues or displays); that is, presenting additional information, or presenting existing information in a format that facilitates recognition and assimilation, thereby enhancing situation awareness.

The paper is organized as follows. Section 2 describes the ABAIS demonstration prototype. Section 3 illustrates ABAIS functionality in the context of an Air Force sweep mission. Section 4 provides a summary and conclusions.

2 ABAIS Prototype

The ABAIS architecture consists of four modules, each implementing the corresponding step of the adaptive methodology (see figure 1). The *User State Assessment Module* identifies the user's predominant affective state (e.g., high level of anxiety) and likely situation-relevant beliefs (e.g., interpretation of ambiguous radar return as threat), from a variety of user-specific and task-specific data. Since no single reliable method currently exists for affective assessment, the User Assessment module combines multiple, complementary assessment methods, including: physiological assessment (e.g., heart rate); diagnostic tasks; self-reports; and use of knowledge-based methods to derive likely affective state based on factors from current task context, personality, and individual history.

The *Impact Prediction Module* predicts the influence of a particular affective state (e.g., high anxiety) or belief (e.g., "aircraft under attack", "hostile aircraft approaching", etc.) on task performance. Impact prediction process uses rule-based reasoning (RBR), and instantiates the known generic effects of the identified affect / belief within the current task context (e.g., increased focus on threatening stimuli may mean neglect of other tasks).

The *Strategy Selection Module* selects a compensatory strategy to counteract the affect- and belief-induced performance biases. Strategy selection uses rule-based reasoning, with rules mapping specific performance biases (e.g., task neglect, threat-estimation bias, failure-estimation bias, etc.) onto the associated compensatory strategies (e.g., present reminders of neglected tasks, present broader evidence to counteract threat-estimation bias, present contrary evidence to counteract failure-driven confirmation bias, etc.).

The *GUI Adaptation Module* performs the final step of the adaptive methodology, by implementing the selected compensatory strategy in terms of specific GUI modifications. Rules map the specific compensatory strategies onto the necessary GUI / DSS adaptations. The specific GUI modifications take into consideration information about the individual pilot preferences for information presentation, encoded in customized user preference profiles; for example, highlighting preferences might include blinking vs. color change vs. size change of the relevant display or icon.

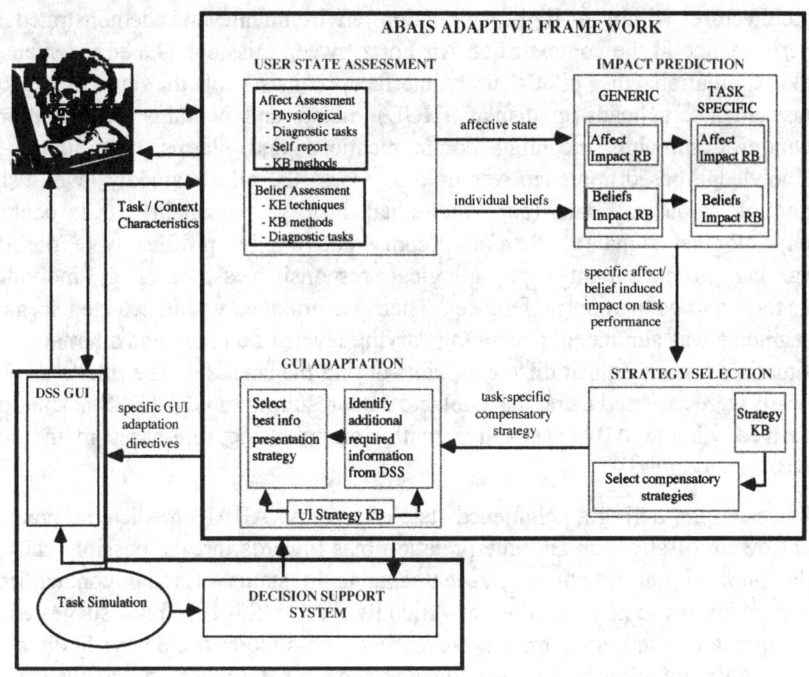

Figure 1: ABAIS Architecture Implementing the Adaptive Methodology

In general, two broad categories of adaptation are possible: *content-based*, which *provide additional information* (e.g., providing additional data about an ambiguous radar signal to prevent an anxiety-induced interpretation bias), and *format-based*, which *modify the format of existing information* (e.g., presentation of existing data in an alternative format, to enhance visibility and / or to draw attention to neglected displays). Both methods result in the modification of specific user interface (UI) attributes. Adaptation can take place at any of these four levels: *icon* (e.g., changing size or appearance to enhance visibility), *individual display* (e.g., decluttering a display to enhance assimilation of relevant data), *notification* (e.g., adding / modifying an alarm string), or entire *user interface configuration* (e.g., global changes in the overall UI such as reconfiguring existing displays). Each level affords different alternatives, and is more or less suitable for a given situation, depending on the task, task context, and the individual.

3 ABAIS Demonstration Example

To demonstrate the adaptive methodology, we embedded the ABAIS adaptive architecture within a flight simulation environment, and demonstrated its performance in the context of an Air Force sweep mission. The key element of the simulation, the pilot's user interface, consists of the critical cockpit instruments, a heads-up display (HUD), radar, and contains two additional windows showing incoming communications and alarms (see figure 2). Knowledge-based approach was used to assess the pilot's anxiety level, using multiple types of data (e.g., individual history, personality, task context, physiological signals). Several representative pilot profiles were defined, varying in personality, physiological responsiveness, training, individual history, and adaptation preferences. Their performance within selected scenario segments was simulated, generating varying levels of anxiety and alternatives in situation assessments at different points during the scenario. The pilot's anxiety levels were assessed using the available data, resulting in GUI / DSS adaptations derived via the ABAIS adaptive methodology, using rule-bases in the four ABAIS modules.

For example, during a heightened state of anxiety, ABAIS predicted a possible narrowing of attention and interpretation bias towards threats, possibly causing the pilot to fail to notice a recent change in status of radar contact from unknown presumed hostile to friendly. ABAIS therefore suggested a compensatory strategy aimed at preventing a possible fratricide by: 1) directing the pilot's attention to the radar display showing the recent status change, by displaying a notification string across the HUD display, presumed to be the current focus; and 2) enhancing the visibility of the relevant signals on the radar display. Specifically, the standard contact icon on the HUD display was modified by enlarging its size, changing its color, and blinking, to draw the pilot's attention. In addition, a blinking "RADAR" string was displayed on the HUD, the pilot's presumed current focus, directing him to look at the radar display below, which shows an enhanced contact icon indicating a change in status, with additional details provided in a text box in lower left corner of the display.

4 Summary and Conclusions

Development of the ABAIS proof-of-concept prototype demonstrated feasibility of the overall adaptive methodology and its implementation. Specifically, ABAIS successfully *assessed the user state* using a knowledge-based approach and information from a variety of sources (e.g., static task context, dynamic external events occurring during the scenario, pilot's individual history, personality, training and proficiency, and simulated physiological data), *predicted the effects* of this state within the constrained context of the demonstration task, and *suggested and implemented specific GUI adaptation* strategies based on the pilot's individual information presentation preferences (e.g., modified an icon or display to capture attention and enhance visibility). An empirical study would be required to assess the effectiveness of the adaptations in an operational context.

As a result of the ABAIS prototype demonstration, a number of specific requirements were identified for developing an operational affective adaptive interface. These include: 1) limiting the number, type, and resolution of affective states; 2) using multiple, complementary affect assessment methods; 3) providing highly individualized user data (e.g., past performance history and physiological data; 4) constraining the context in terms of situation assessment and behavioral possibilities; 5) fine-tuning the rule-bases and inferencing to "personalize" the system to the individual user-task context; and 6) implementing only 'benign' adaptations, that is, GUI / DSS modifications that at best enhance and at worst maintain current level of performance (e.g., adaptations should never limit access to existing information).

References

Bower, G.H. (1981). Mood and Memory. *American Psychologist,* 36, 129-148.

Eysenck, M.W. (1997). *Anxiety and Cognition: A Unified Theory.* Hove, East Sussex: Psychology Press.

Fallesen, J. J. (1993). *Overview of Army Tactical Planning Performance Research* (Technical Report 984): Washington, DC: US ARI.

Hudlicka, E. (1998*). ABAIS: Affect and Belief Adaptive Interface System. Report No. 9814.* Lincoln, MA: Psychometrix Associates, Inc.

Isen, A.M. (1993). Positive Affect and Decision Making. In *Handbook of Emotions*, J.M. Haviland and M. Lewis, eds. New York, NY: The Guilford Press.

LeDoux, J.E. (1992). Brain Mechanisms of Emotion and Emotional Learning. *Current Opinion in Neurobiology*., 2(2), 191-197.

Mineka, S. and Sutton, S.K. (1992). Cognitive Biases and the Emotional Disorders. *Psychological Science*, 3(1), 65-69.

Williams, J.M.G., Watts, F.N., MacLeod, C., and Mathews, A. (1997). *Cognitive Psychology and Emotional Disorders.* NY: John Wiley.

Measurement in Synthetic Task Environments for Teams: A Methodological Typology

Michael D. Coovert
University of South Florida

Linda Elliott
Brooks Air Force Base

Lori L. Foster
University of South Florida

J. Philip Craiger
University of Nebraska at Omaha

Dawn Riddle
University of South Florida

Work in organizations today has been moving from the autonomous worker to various team based structures. Given that the product of work is no longer dependent upon just one individual, methods have begun to emerge regarding the role of the performance unit (e.g., individual, dyad [sub-team], whole team) and the type of interactions required (process vs. outcome). Depending on the purpose of measurement (describe, evaluate, or diagnose behavior), tradeoffs can be made relative to the importance of these dimensions. Figure 1 summarizes this perspective,

	process measures			outcome measures		
	individual	dyad	team	individual	dyad	team
description						
evaluation						
diagnosis						

Figure 1. Global issues in measurement.

Of course, these global issues must not be considered in a vacuum as different criterion issues come into play, depending on the level of the measurement. As Elliott, Schiflett, and Dalrymple (1998) have pointed out, this level can be very low -- such as individual task proficiency; moderate -- such as the individual or team; versus very high -- such as at the mission/organizational level.

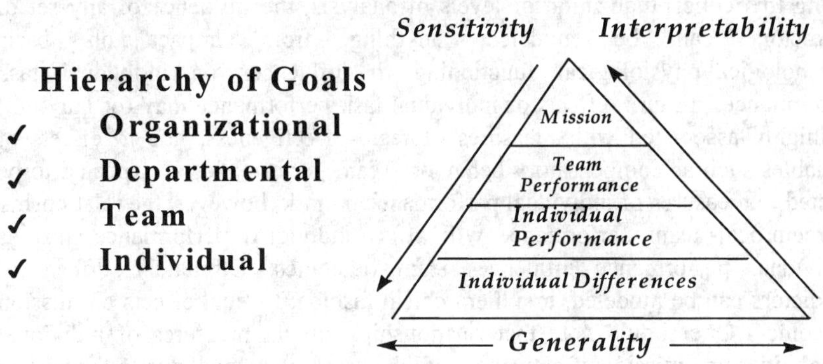

Figure 2. Criterion issues related to level of analysis.

Figure 2 illustrates four broad categories of performance measurement and tradeoffs inherent in their measurement (Elliott, Schiflett, and Dalrymple, 1988). Tradeoffs exist relative to interpretability (performance as it relates to the mission), sensitivity (performance measures as functions of the performer as opposed to context), and generality (relationships to other teams and/or situations). At the tip of the triangle are measures related to overall mission effectiveness. These measures are interpreted in light of the context and goals of the team. However, they are less straightforward with regard to assessing individual or teamwork effectiveness, due to the increased sources of error due to external variables (e.g., overwhelming hostile forces, unfavorable timing of events, etc). For example, the team may have coordinated information as efficiently as possible, but still fail in overall outcomes.

While mission outcomes reflect a bottom line criterion measure of effectiveness, other measures of performance are required, which are more fine-grained and interpretable from the perspective of construct validity. At the bottom of the triangle are the measures of individual task performance, and of individual differences in capabilities. As one moves down the triangle, measures become more defined as constructs and thus more generalizable to other subjects or situations. In contrast moving up the triangle, to measures of mission effectiveness, measures increase in interpretation of effectiveness within that specific context—measures are interpreted in light of situational and scenario-based factors. In moving down the triangle, measures are more sensitive to experimentation—for example, more fine-tuned insight with regard to effects of fatigue on cognition is gained when criterion measures are those of a cognitive test battery than evaluations of overall outcomes.

Using this conceptualization of levels of analysis, the influence of any set of predictors can be modeled, anything from impact on basic psychological/physiological functioning to influences on individual task performance. In turn, effects of individual task performance may (or may not) be highly associated with measures of team effectiveness, due to effects of variables such as compensatory behavior. Team performance is expected to be related to measures of individual performance on task; however, the relationship is reciprocal--team performance will affect individual performance, just as individual performance influences team outcomes. Ultimately, effects of predictors can be modeled, in a hierarchical fashion, to final effects on mission outcome. One should note that relationships among measures of individual capabilities and mission effectiveness will be most attenuated, due to the variety of contextual variables which can affect mission effectiveness.

These meta-issues are relevant when we consider performance in a synthetic task environment. Cannon-Bowers and Salas (1997) took a first step in considering the issues by generating a typology similar to that presented in Figure 3. Focused on the concepts depicted in Figure 1, the typology is useful to guide issues from a particular theoretical perspective but unfortunately, does not provide guidance regarding human –computer measurement issues that are central to an individual or team-oriented synthetic task irrespective of theoretical orientation.

Our work our work builds upon that framework by generating a strategy that is useful to guide human-computer interaction measurement for individual and team oriented synthetic tasks, even when those tasks are used for a variety of purposes – such as training versus research.

The measurement typology considers five central issues that become crossed with the level issue presented in Figure 2. These are:

1. **<u>Variable/construct</u>** - what it is we are trying to measure (e.g., stream of a Weapon's Director mouse clicks).
2. **<u>Source</u>** - how the information is to be obtained or where the captured information should reside.

3. **Purpose** - why it is important to measure the variable or construct (e.g., theoretical or practical considerations).
4. **Time of measurement** - when the measure should be gathered (e.g., continuously vs. after a critical event).
5. **Design implications for the synthetic task** - implications relative to the capability of the task in order to gather the measure.

	Team	Individual
Process	Shared Task	Assertiveness; Flexibility
	Compensatory	Information processing
	Collective	Task completion procedures
	Dynamic	Mutual performance monitoring
Outcome	Mission/Goal accomplishment	Accuracy
	Aggregate latency	Latency
	Error accuracy	Error
	Error propagation	Safety

Figure 3. Typology: Level by type. (Adapted from Cannon-Bowers and Salas 1997)

Our typology provides guidance to those interested in using a synthetic task environment for a variety of purposes.

For example, the following TEAM variables might be investigated.

- Overt team interactions
- Judgment accuracy
- Performance scores of various types
- Team preplan (e.g., mission planning) behavior
- Computer mediated team communications
- Shared mental models
- "Teamness" measures (cohesion, morale, resource allocation)
- Team situational awareness

Individual difference variables might also be investigated so their relationship to team performance can be explored, some examples:

- Intelligence
- Experience

- Aptitude
- Personality

System parameters of the synthetic task that should be modifiable, examples include:

- Specification of scenario pauses
- Recording of operator actions
- Scoring
- Software agent editor
- Cognitive requirements
- Pop-up windows
- Recording of operator input
- Scenario editor
- On-the-fly manager input
- Computer-user task allocation

While building on our experience in a variety of synthetic task environments (e.g., individual human-computer task performance to full team computer-cooperative work environments), our typology provides conceptual model for identifying those constructs that need be considered in a variety of human-computer contexts. These range from the individual level, through dyadic groups, to full team computer-supported cooperative work environments. By filling in the cells of Figure 1, the typology provides guidance on obtaining process and outcome measures given the purpose of the measurement (describe, evaluate, vs. diagnose) behavior. Tradeoffs in measurement should be assessed via the hierarchy of goals as depicted in Figure 2, and an assessment of the measurement process can be determined relative to sensitivity, interpretability, and generalizability of conclusions while considering issues from level of analysis and process versus outcome assessment (Figure 3). Taken as a measurement typology, our framework provides both researchers and practitioners specific guidance on human-computer interaction measurement issues, regardless of theoretical orientation.

1 References

Cannon-Bowers, J. A., & Salas, E. (1997). A framework for developing team performance measures in training. In M. T. Brannick, E. Salas, and C. Prince (Eds.), *Team performance assessment and measurement* (pp. 45 - 62) Mahwah, NJ: LEA..

Elliott, L., Schiflett, S. G., & Dalrymple, M. A. (1998). Modeling the decisionmaking process and performance of Airborne Warning and Control System (AWACS) weapons directors. *NATO Symposium on Collaborative Crew Performance in Complex Operational Systems*, Edinburgh, UK.

Countermeasures Against Stress: Dynamic Cognitive Induction

Robert P. Mahan
University of Georgia, U.S.A.

Christopher J. Marino
University of Georgia, U.S.A.

Linda Elliott
Brooks Air Force Base

Eric Haarbauer
University of Georgia, U.S.A.

Philip T. Dunwoody
University of Georgia, U.S.A.

Command-level decision making is not deterred by the uncertainty associated with the problem of forecasting future events. Faced with irreducible levels of uncertainty, commanders often use expert intuition as a means to assess and implement particular military options. Expert intuition in military, command-level decision making is configured around a number of knowledge domains including the behavior of military systems, the behavior of personnel in those systems, tactical doctrine, enemy doctrine, resources, orders, and accepted military practices. Further, the application of command-level expertise often occurs under tactical conditions that induce significant levels of stress in the decision makers. The most common of these occupational stressors are associated with continuous and sustained operations.

This paper represents a heuristic model and description of research efforts designed to develop dynamic methods, implemented in an interface, to help counteract the effects of environmental stressors on decision performance. The decision support theory expressed in this paper comes from the convergence of three areas of research: (1) judgment and decision making, (2) display engineering, and (3) occupational stress. The thematic features that unite these unusually disparate literatures (see Hammond and Doyle, 1991 for a review) can be summarized in the three global, scientifically validated premises below.

(1) Tasks contain structural properties that induce a particular processing response and these properties directly constrain expert decision making. Before one can ask any questions about the nature of information processing (i.e., what is going on inside the head of the decision maker), there must be a degree of clarity concerning the characteristics of the problem and the demands that are being placed upon the expert. This calls for a model of the decision task, or more generally, a methodology directed at modeling the ecology in which the decision maker operates.

(2) The manner in which information from a decision environment is presented will induce a particular processing response in the decision maker. This idea is analogous to the task property premise (#1) above. The system interface is a window to the task world. It can be viewed as representing the surface features of a system that connects (in various ways) to depth or hidden properties of the complex task environment. That is, our knowledge of the covert properties of the task world is provided through surface information. If the window distorts the underlying task structure, then the responses to task demands will also be distorted.

(3) Decision makers exhibit individual differences in information processing characteristics. The variability is clearly associated with task demands (premise #1), mediation characteristics (premise #2), as well as many individual difference variables. However, this variability is also caused by a systematic response to particular operational factors, such as stress. The premise concerns the nature of this variation, and the identification of selected countermeasures to compensate for its effects on decision performance. Stress causes drift in complex judgment behavior. The drift in cognitive functioning is viewed as fatigue-induced changes in decision making strategies for operating on surface information. Fatigue-induced strategy changes are often deleterious to the courses of action implemented by decision makers.

Using the three premises outlined above, we suggest a means to create an intelligent agent mechanism that maintains real-time congruence between the Task, Mediation and Cognitive systems. Cognitive responses to a task depend on task properties. Further, these properties must not be distorted by the mediation technique. Thus, cognitive responses are derivative of mediation techniques mapping the task world. A display, for the most part, must be an accurate representation of the world to be judged.

As an example of task properties inducing cognition, consider an expert economist using a mathematical model to predict future states of the economy. The economy is represented by a series of equations that provide the user with a prediction model that is retraceable and coherent. The prediction will be precise and because of these properties will induce the user to become analytical (Hammond, pp. 181; 1996). However, as the history of economic theory demonstrates, if a task contains irreducible uncertainty, then the application of precision-geared analytical cognition may lead to predictions that deviate widely from the true state of affairs. In this instance, the use of mathematical models induces analysis, yet expert intuition may be more compatibly (or congruent) with the task (Dunwoody et. al., in press).

In order to assess congruence, one must first develop a methodology that allows quantification of states of (1) task demands, (2) display (mediation) properties, and (3) cognitive functioning of the user. This requires a set of continua (one for each system) to identify processing characteristics that demarcate the range in states that can exist in a given context. Analysis and intuition have traditionally bounded the decision sciences information processing domain (albeit as a strict dichotomy) (cf., Kahneman, Slovic and Tversky, 1982; Hammond, et al., 1987, 1996; Anderson, 1991). Thus, each continuum is anchored by Analysis at one end and Intuition at the other. For example, the task continuum will define an ordered set of task properties that induce a particular cognitive response in the operator, ranging from analysis to intuition. Similarly, the display continuum will define an ordered set of representational properties that map the task continuum and also induce particular cognitive properties in the operator, ranging from analysis to intuition. Finally, the cognitive continuum will define an ordered set of cognitive properties, ranging from analysis to intuition that reflect the mode of cognitive processing that an operator is using at any given moment in time.

The "band of congruence" quantifies the degree to which the separate systems are aligned. Deviations from this alignment state will represent potential shortfalls in performance. Through implementation of a real-time monitoring process for deviations in congruence discussed below, certain adaptable countermeasures can be automatically invoked. For example, drifting operators (operators not in congruence with task demands due to the effects of stress) can be brought closer to the region of optimal cognitive performance by the representational structure of the interface (e.g. shifts from tabular information presentation to graphically displayed information). For example, assume a decision maker is using a mode of cognition that is optimal for conducting an off-line (no time pressure) assessment of a single information source. However, the task requires a mode that supports combining multiple uncertain indicators in real-time. Based on dynamic cognitive induction, representational variables can be implemented to achieve this modification in processing mode (discussed below).

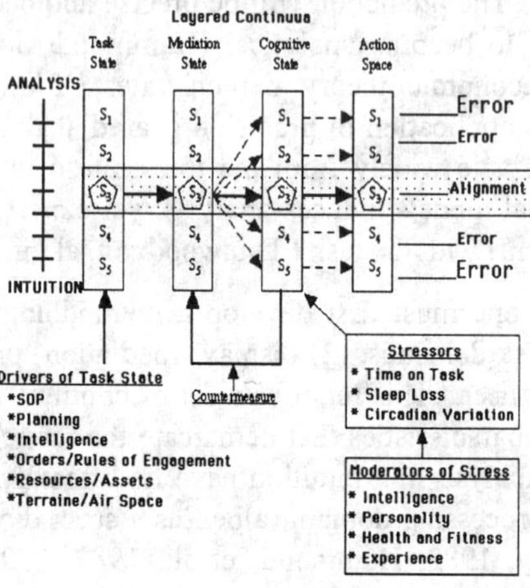

Figure 1. The DCI framework model conceptually illustrates the "band of congruence" principle.

In Figure 1, a particular task configuration (S3 in Task State vector) will require a unique representation (S3 in Mediation State vector) that in turn activates or causes a given mode of cognition in the decision maker (S3 Cognitive State vector). This mode generates a particular course of action or decision (S3 Action Space vector). All systems must be in congruence to generate optimal performance. Stressors (fatigue and sleeploss) and individual differences (training, intelligence, health and fitness) introduce fluctuations in alignment that alter cognitive functioning (the dashed arrows emanating from the mediation vector) and cause poor performance (Error). The Cognitive state shares a feedback loop with the Mediation vector which allows the Mediation vector (interface) to guide cognitive behavior to higher levels of alignment via changes in the Mediation state of the system (countermeasures applied to the representation of task). As an example, if S1 in Cognitive State is activated from an S3 Mediation value due to external factors (e.g., fatigue), this will produce Error. A countermeasure might be to implement an S5 Mediation value in an effort to raise the cognitive state to a value more homogenous with the S3 Task State being assessed by the decision maker.

The above discussion argues that displays can be created that drive an individual decision maker toward a particular kind of cognition. Anderson (1991) has written extensively on the adaptive qualities of cognitive functioning. Further, he notes that cognition is profoundly influenced and constrained by the properties of the environment in which the decision maker is embedded. This position is echoed throughout the above discourse in views that provide a formal approach to support individuals in specific operational contexts.

1 References

Anderson, J. R. (1991). Is human cognition adaptive? *Behavioral and Brain Sciences, 14*(3), 471-517.

Dunwoody, P. T., Haarbauer, E., Mahan, R. P., Marino, C. J., & Tang, C. C. (In Press). Cognitive adaptation and its consequences: A test of cognitive continuum theory. *Journal of Behavioral Decision Making.*

Hammond, K. R., Hamm, R. M., Grassia, J., & Pearson, T. (1987). Direct comparison of the efficacy of of intuitive and analytical cognition in expert judgments. *IEEE Transactions on Systems, Man and Cybernetics, SMC-17*(5), 753-770

Hammond, K. R. (1996). *Human judgment and social policy: Irreducible uncertainty, inevitable error, unavoidable injustice*. New York: Oxford University Press.

Hammond, K. R., & Doyle, J. K. (1991). *Effects of stress on judgment and decision making* : Boulder: University of Colorado, Center for Research on Judgment and Policy.

Kahneman, D., Slovic, P., & Tversky, A. (1982). *Judgment Under Uncertainty: Heuristics and Biases*. Cambridge: Cambridge University Press.

The Social Psychology of Intelligent Machines: A Research Agenda Revisited

A. Rodney Wellens and Michael D. McNeese
Department of Psychology, University of Miami, Coral Gables, Florida U.S.A.
and
United States Air Force Research Laboratory
Wright-Patterson Air Force Base, Ohio U.S.A.

1 Introduction

Over a decade ago Wellens and McNeese (1987) proposed a research agenda for the social psychology of intelligent machines. The term "intelligent machine" was used to describe any information processing system designed as a social surrogate that could respond meaningfully to human action along communication channels previously reserved for human contact. Of special interest was a class of "socially skilled" machines that used human-like interaction styles (e.g., employing natural language, voice synthesis, animated facial expressions) to interface with humans. The interaction of humans with intelligent machines was seen as a new form of dynamic social interaction. The "social psychology of intelligent machines" represented a proposed interdisciplinary approach to understanding the impact of artificially intelligent systems upon human cognitive, affective and behavioral functioning. Observations from the disciplines of social, cognitive and clinical psychology, sociology, communications and human factors engineering were brought to bear upon the problem of assessing the influence of intelligent machine systems on human functioning. Within this paper we will draw upon those early reports as well as more recent observations to re-explore how social influence can occur via exposure to intelligent machines and what consequences this might have for system users and future system design.

2 Background

Wellens (1990) examined the effects of communication media (email, voice intercom, two-way television) upon collaborative problem solving within hybrid teams composed of human and electronic teammates as well as all-human teams. The electronic teammates used were actually rule-based expert systems that had been augmented with digitized faces, voice synthesizers and text generating software so that they could transmit information using varying degrees of "social presence" associated with the communication media employed (a rudimentary form of avatar). While results showed that humans could readily accept machines as teammates, it was also discovered that task success did not necessarily increase with increased communication bandwidth. While reported "team feelings" increased with communication, objective measures of performance did not increase proportionately. Having to manage communications with a remote teammate (human or machine) seemed to reduce attention to the fairly complex primary task at hand.

This initial venture into creating human-computer teamwork coincided with developmental efforts on the part of others to create computer "associates" to help pilots manage the increasingly complex array of information systems packed into modern cockpits. Already there was some evidence (Wiener & Cury, 1980) that automated systems could have some unexpected side effects (e.g., skills loss, lapses in attention, devalued sense of self) on system operators. The time seemed ripe to assess additional psycho-social effects expected with further exposure to advanced information systems. Since these systems were being touted as substitutes for human partners, the social psychology literature seemed like a good place to look for ideas. Traditional areas of social psychology, such as self perception, affiliation, attraction, conformity, social facilitation, group structure, leadership and communication were all selected as potential candidates. The relevant theoretical and empirical literature was reviewed and speculations were made regarding potential organizing principles (Wellens & McNeese, 1987).

The world of humans and machines has changed dramatically since our original proposal. The off-the-shelf systems that were used to create simulated "smart socially skilled" machines in the 1980's are primitive by today's standards. What was once relegated to the realm of science fiction has now become technological reality (Stork, 1997). The relative sophistication of human users has also increased dramatically. Individuals born in the 1980's have grown up with computers and depend upon them to navigate information highways that didn't exist for prior generations. The entertainment industry has exploited the power of interactive devices and used the newest technologies to capture the leisure time of many individuals who have been all too willing to enter the world of cyberspace and virtual environments.

Along with these rapid changes in technology and culture has come a recent call to take the next step in techno-evolution and imbue computers with the ability of sharing human emotions. Picard (1997) has detailed the technologies needed to sense and express emotional signals between humans and machines. "Wearable" computers have been created that maintain close contact with their human hosts. Picard is in the vanguard of a group of computer scientists who are promoting the new field of "affective computing." The idea comes in part from the concept of "emotional intelligence" popularized by Goleman (1995). Many of the ideas that lead up to the notion of affective computing had their scientific roots in the still emerging field of "affective neuroscience" (Panksepp, 1998) that points to emotions as being part of, rather than separate from, cognitive functioning.

Now more than ever there is a need to step back and assess the costs and benefits of widening even further the bandwidth connecting humans and intelligent machines. What has been learned thus far and what does the future hold?

3 Lessons from the Social Sciences

It has been suggested that ideas about how people will respond to artificial social agents and other new forms of electronic media can be found in the social science section of the library. Reeves & Nass (1996) point out that there is over a century of research about how people respond to each other, as well as a rich legacy of methods used to study social phenomena, that can be consulted when designing experiments assessing human responses to artifacts that display human-like qualities. This insight is similar to the one espoused by Wellens and McNeese (1987) when they introduced the phrase "the social psychology of intelligent machines." Reeves & Nass go on to point out that human brains evolved in a world where only humans exhibited rich social behavior; what appeared to be real *was* real. Machines that now generate and display social cues engage the old brains of their human interlocutors. Social schemas derived from human-to-human contact are readily transferred to human-machine interactions.

Reeves & Nass (1996) describe a series of experiments that demonstrate how people will use social schemas in interacting with computers that display only minimal social cues. For example, gender stereotypes have been elicited by manipulating synthesized speech used by a computer to appear either "male" or "female". Perceptions of computer "personality" (e.g., "dominance" vs. "submission") have been successfully manipulated by changing the number of assertions and commands used by a computer instead of using a more tentative, suggestive verbal style. When the computer's personality style was matched to that of its user, the computer was judged by the user as more intelligent, knowledgeable, insightful and helpful. Furthermore, humans readily formed team relationships with computers and subsequently accepted more influence from them. Collins (1997) points out that humans are especially forgiving of imperfect machine socialization when interacting with artificial intelligence and engage in considerable repair and attribution work in order to make sense out of the interaction. The tendency for humans to use social schemas to interact with computers can be (and has been) exploited to increase user acceptance. Computers don't have to be especially "smart" in order to engage their human users, just "clever" in the way they (or their programmers) present themselves.

It has been assumed that developing “friendlier” interfaces for computers would make them easier to use and this would be a “good” thing. Part of the motivation for bringing humans and machines even closer together via affective links is to make machines even better at what they do. However, there can be some potentially significant psychological costs to the human user over time. Early reports of "computer addiction" (Ingber, 1981; MacHovec, 1984; Turkle, 1984) and the preference for machines over people in adults (Simons, 1985) and children (Selnow, 1984; Brod, 1984) have been followed up with surveys and in-depth interviews with "computer-dependent" individuals (Shotton, 1991) as well as novice internet users (Kraut, et al, 1998). It appears that for a growing number of people, increased computer use is replacing direct human social contact. Since humans depend upon social feedback for the development and maintenance of the self concept, they are especially vulnerable when isolated from other people. Affiliation and the self-concept are at the root of many social behaviors (e.g., attraction, group membership, conformity) so changes in these key processes may be expected to have ripple effects throughout the social fabric. As we expend effort and resources to make machines smarter, we need to focus at least as much attention on understanding the impact they may be having on their creators.

4 References

Brod, C. (1984). *Technostress: The human cost of the computer revolution.* Reading, Massachusetts: Addison-Wesley.

Collins, H.H. (1997). Rat-tale: Sociology’s contribution to understanding human and machine cognition. In P. Feltovich, K. Ford & R. Hoffman (Eds.): *Expertise in context: Human and machine*. Cambridge, Mass: MIT Press.

Goleman, D. (1995). *Emotional intelligence*. New York: Bantom Books.

Ingber, D. (1981). Computer addicts. *Science Digest*, 89, 74-114.

Kraut, R., Patterson, M., Lundmark, V., Kiesler, S., Mukopadhyday, T. & Scherlis, W. (1998). Internet paradox: A social technology that reduces social involvement and psychological well being? *American Psychologist*, 53, 1017-1031.

MacHovec, F. (1984). Love at first byte: a terminal affair. *APA Monitor*, 15, 5.

Panksepp, J. (1998). *Affective neuroscience: The foundations of human and animal emotions*. New York: Oxford University Press.

Picard, R. (1997). *Affective computing*. Cambridge, Mass: MIT Press.

Reeves, B. & Nass, C. (1996). *The media equation: How people treat computers, television, and new media like real people and places*. New York: Cambridge University Press.

Selnow, G.W. (1984). Playing videogames: The electronic friend. *Journal of Communication,* 34,148-156.

Shotton, M.A. (1991). The costs and benefits of "computer addiction." *Behavior and Information Technology*, 10, 219-230.

Simons, G. (1985). *The biology of computer life: Survival, emotion and free will.* Boston: Birkhauser.

Stork, D. (1997). Hal's legacy: *2001's computer as dream and reality*. Cambridge, Mass: MIT Press.

Turkle,S. (1984). *The second self: Computers and the human spirit*. New York: Simon and Schuster.

Wellens, A. R. (1990). *Assessing multi-person and person-machine distributed decision making using an extended psychological distancing model*. AAMRL-TR-90-006. Wright-Patterson Air Force Base, Ohio.

Wellens, A. R. & McNeese, M.D. (1987). A research agenda for the social psychology of intelligent machines. *Proceedings of the IEEE National Aerospace and Electronics Conference*, 4, 944-950.

Wiener, E. & Curry, R. (1980). Flight-deck automation: Promises and problems. *Ergonomics*, 23, 995-1011.

Matrix Evaluation Method for Planned Usability Improvements Based on Customer Feedback

Konrad Baumann
Philips Business Electronics

"The last decade has witnessed significant world-wide growth of products with small screens and other limited characteristics of appearance and interaction." (Marcus 1999)

"Are your help lines too expensive? The fact that customers need so much help is a sign of poor products. Want a better product? You probably have to reorganize your company and change the product process." (Nielsen, Norman 1999)

1 Usability in Telecommunication

The matrix evaluation method described in the following is based on information about customers' usability problems gathered in a call center.

Call center data can provide information on "usual" usability problems having as root cause the insufficient consideration of usability principles during development. Usability principles include self-descriptiveness, consistency, simplicity, compatibility, error tolerance, and feedback. (Baumann 1998)

More than this, call centers for telecommunication products can help detecting usability problems that arise during installation and setup of the product, connecting it properly to other devices, and registering for services and accounts.

These problems do not necessarily stem from a poor user interface or user manual of a single appliance but from the necessity to setup and use two or more appliances or services together in order to make them fulfill a specific task.

The circumstances leading to a usability problem may for example include a telephone or internet appliance owned by a specific customer, other devices the customer has already installed, the services offered by his telephone and internet service providers, as well as the appliances and service providers of the sending or receiving counterparts.

Thus it may be difficult to eliminate all possible pitfalls in the test phase because a product can hardly be tested in all possible technical environments, or simply because some of the involved appliances or services are outside of the own company's responsibility.

So in the context of telecommunication we consider it a usability problem when a user is not able to have a specific task fulfilled by his appliances and services while all of them, however, perform according to specification.

2 Usability Problem Clusters and Improvement Areas

The variety of possible customers' usability problems may be clustered in a set of problem areas depending on the type of appliance. A fictional example for a set of possible problem areas is:

1) Connection of device to power and communication lines
2) Setup of device to match communication environment
3) Interaction with the user interface (e.g. data input)
4) Transmission and reception of data
5) Use of special features and other problems

Usability improvement areas for telecommunication appliances or services may be defined as shown by the following fictional example:

1) operation panel elements (key grouping and naming, display size, availability of softkeys or a touchscreen),
2) user interface software design (interaction design, definition of menu size and menu items, screen graphics),
3) user-relevant technical characteristics (selection and presentation of features and functionality according to the users' needs),
4) acoustical interaction elements (audible feedback, speech input or output),
5) built-in help functions, and
6) directions for use.

The applicable usability improvement areas for telecommunication services form a subset of the described areas.

3 Field of Application and Necessary Tools

Field of application. The evaluation method described in the following has been developed and is used for telecommunication products and services. Products include telephones, modems, fax units, answering machines, cellular phones, internet appliances, and video conferencing systems. Services include mailbox and other services offered by telephone providers, internet access services, facsimile forwarding, telephone and video conferencing.

The method may also be used for other appliances "with small screens (baby faces)" or "other limited characteristics of appearance and interaction." (Marcus 1999)

Necessary tools. Use of the evaluation method is possible if a certain infrastructure is available that includes a call center with well-trained employees, a database for storing customer and problem data, a dedicated problem classification system, and a multidisciplinary team that performs the evaluation and closes the feedback loop back to the rest of the company.

The problem classification system assigns a name and code to each problem. It is defined like a tree structure starting at the top level with "customer problem call" as opposed to "marketing call" or other friendly call reasons.

At the next level the branch "customer problem call" has to be split into "usability problem call" and "technical problem call". The definition of a usability problem is that it can be solved by giving advice, while technical problems can only be solved by hardware exchange, repair, software reset, or software update.

At the next tree level the problem areas are defined as shown earlier in the example. These problem areas and all lower tree levels specify the problems in more and more detail.

The output of the problem database is a report indicating the rate of usability problem calls for each detailed problem or for each problem area. The usability problem call rate is a figure indicating the number of related customer calls divided by the average number of sold products (appliances or service packages) in the same time frame.

Detailed improvement action proposals can be collected by means of heuristic evaluation, benchmarking, brainstorming, or application of known usability principles to product aspects.

4 The Matrix

"Consumers can't tell you what they will want tomorrow. They can tell you what makes them unhappy today." (Philips 1999). This statement can be considered as the essential idea that led to the matrix evaluation method.

It is evident that customers' problems and possible improvement actions do not correspond to each other via a one-to-one relationship (1 problem - 1 root cause - 1 solution).

The relationship of improvements and problems can be represented in the form of the following matrix. A dot stands for a positive influence of an improvement on a problem.

Usability problems: / Improvement areas:	Connection	Setup	Interaction	Transmission, reception	Special features, others
Operation panel		•	•	•	•
User interface software		•	•	•	•
Technical characteristics	•	•	•	•	•
Acoustical interaction		•	•	•	•
Help functions	•	•	•	•	•
Directions for use	•	•	•	•	•

This matrix shows the basic relationship of improvement and problem areas. However, as most of the matrix elements are marked with a dot, we cannot use it for gathering more detailed information.

For practical use the matrix has to be more detailed in both dimensions. A matrix with 20 problem clusters and up to 100 improvement proposals can still be handled. Of course a single-line matrix can be used to evaluate a single improvement proposal.

	Usability problems (detailed):	Text input	Status setting	Memory	Text legibility	...	Others
	Problem call rate (%):	0,8	1,1	0,4	0,6	...	0,5
No.	**Improvement actions:**						
1	Redesign keys	0,7	1	1	1	...	1
2	Rename keys	1	0,9	1	1	...	1
3	Bigger display	0,8	0,8	1	0,2	...	1
4	Redesign feedback tones	0,9	1	1	1	...	1
...	etc.	...	...	...	...	...	...

The matrix elements are figures indicating the expected influence of an action on a problem. If the element is equal to 1 the problem is expected to remain unchanged by the action. If the element is smaller than 1 a problem call rate reduction is expected. For example the matrix element "0,7" indicates that the action "redesign keys" is expected to reduce the problem "text input" by 30%. (Problems, actions, and figures of these examples are fictional.)

The matrix elements are defined by several persons (ideally more than 10 persons). These persons heuristically evaluate every proposed action and fill in the complete matrix according their personal opinion. In order to reduce guessing error the standard deviation of all matrix elements is calculated. The elements having the highest standard deviation are then discussed in an evaluators' meeting. In this meeting the evaluators having done the most untypical estimations can still change them.

Then the average values of the estimated matrix elements by all evaluators are calculated. This average values matrix is used for further processing.

5 Calculation of the Improvement Potential

Every problem has an associated problem call rate as defined earlier. This figure is indicated in the matrix representation below the name of each problem.

The improvement potential of an action is calculated as follows: Multiply all matrix elements in the matrix line associated to the action by the problem call rate of the corresponding matrix columns. Add up the results.

The improvement potential of all actions for a specific problem is calculated as follows: Multiply all matrix elements in the matrix column associated to the problem. Multiply the result by the problem's call rate. (This calculation is valid under the assumptions of independent or non-overlapping actions and equal distribution of problems within the pool of calling customers. The latter assumption simplifies reality and hence limits accuracy of the calculation.)

The total improvement potential of all actions and problems is calculated by adding up the single improvement potential figures for all problems.

If replacing the percent call rate figures by call center costs the improvement potential becomes a realistic figure that can be compared to the change costs.

Accuracy of this figure is best if the problem clusters and the actions have been defined in the most detailed way that still allows easy handling of the matrix.

Ranking. The single action proposals can now be ranked according to their improvement potential. This ranking can be used as an action priority list for running changes or development of new user interfaces.

It closes the overall feedback loop about customer satisfaction and product usability and can be used as a valuable quality process tool and decision aid.

6 References

Baumann, K., Lanz, H. (1998). Mensch-Maschine-Schnittstellen elektronischer Geraete. Berlin, Heidelberg: Springer.

Marcus, A. (1999). Finding the Right Way: A Case Study about Designing the User Interface for Motorola's ADVANCE Vehicle-Navigation System. Unpublished case study.

Nielsen, J., Norman, D. (1999). Nielsen Norman Group: Company Backgrounder.

Philips Consumer Communications (1999). FWD>> Calendar 1999.

Will Baby Faces Ever Grow up?

Lars Erik Holmquist
PLAY: Applied research on art and technology
The Viktoria Institute, Box 620, S-405 30 Gothenburg, SWEDEN
http://www.viktoria.informatics.gu.se/play/
leh@viktoria.informatics.gu.se

1 Introduction

For a discussion about user interfaces for small screens, so-called *baby faces*, it might be interesting to start with considering another type of "baby faces" – namely, real ones. Until quite recently, human children were perceived as being basically smaller versions of adults, who were just waiting to become strong and educated enough to take their productive place in the grown-ups' society. But with the increasingly high living standard of the industrial age, coupled with a more developed education system, it became apparent that children are in fact unique individuals and quite different from adults in their needs and capabilities. Nowadays, there is an understanding that children are in fact inherently different from adults in many respects.

At the moment, baby-faced computers are perceived perceived as smaller versions of their desktop "parents". But there are many factors that make baby-faced computers inherently different from desktop computers, just like children are different from grown-ups. It is quite possible that with baby faces, many new uses for computers will be found, and completely new markets will open up, with baby-faced computers reaching user groups that are different from anything that has come before. These baby faces may be used for very different tasks and in very different contexts than that of the comparatively well understood desktop setting. Users then will want to bring their baby faces on buses or trains, use them in cars or while taking a walk, completely changing many of the parameters that have previously dictated human-computer interface development. Therefore, can not go on thinking of baby faces as if they are basically desktop computers that just happen to be very small!

2 Why Baby Faces Are Different

Whereas a desktop computer is very much identifiable as a "computer", and therefore approached by users with an appropriate mindset and often a great deal of respect, tomorrow's baby-faced computational devices may not be perceived as computers at all. These will be the future versions of mobile phones, watches and notebooks, aimed at everyone from schoolchildren to grandmothers. Therefore they may require quite a different approach to usability compared to what currently dominates the field. We may very well find that the conventions of user-interface design that have been defined and refined during the last twenty-odd years of desktop computing do not apply to these new computing devices.

The shift from mainframe computers to desktop computers gave rise to a radical change in how computers were perceived and used. If there is a similar shift to even smaller computers, it should be safe to assume that this will make computing even more accessible to the general public. If this will be the case, the challenge of creating user interfaces for these ubiquitous "baby faces" will be much greater than the one that faced designers when the first so-called user-friendly computers were constructed. While the first machines based on graphical user interfaces were a revolution in their day, tomorrow's mass-market screen-based devices may reach people who would never think of using a "real" computer. A majority of baby-face users might well be completely baffled by traditional user-interface conventions like folders, scrollbars and menus. Should we then force them to accept these conventions, or should we see this as a chance to come up with something better?

Even current baby faces have many properties that can pose unexpected problems. When exploring novel user-interface designs on small screens, we have found that it is far from straightforward to adapt techniques developed on traditional screens to their smaller counterparts. For instance, the shape of the screen is often different. Whereas almost all PC screens have the proportions 4:3, a typical keyboard-based PDA has "wide-screen" proportions, approximately 5:2. Pen-based PDAs, on the other hand, have proportions more closely resembling that of a traditional monitor turned on its side. This change in proportions can make some pixels more precious than others; for instance, on the wide screen, vertical space is much more scarce than horizontal.

Anyone who finds this to be a trivial observation should try the following experiment: take a piece of cardboard and cover the lower half of your computer screen. Then try working as normal. You will probably soon find the generous horizontal icon bars on your web-browser and word processor to be a major distraction, since they take up a lot of valuable (vertical) screen estate. The thick borders and title bars on application windows will also soon seem like an intrusion. Indeed, it seems that window-based user interfaces are not very suitable to small screens, and on many current PDA models, software applications are usually allowed to occupy the entire screen with no window borders. Interestingly, the makers of a major operating system for small devices do not necessarily seem to agree with this approach, since they are even using the term "Windows" as a selling point! Time will tell if the flexibility of having a true windowing system will be enough to offset the loss of valuable screen space to borders and title bars.

3 Size Is Not All That Matters

Besides the size and shape of the screen, there are many other factors that should be taken into account when designing interfaces for baby faces. For instance, small devices are not on the same curve of increasing processing power from which stationary and laptop PCs have long benefited. For pocket-sized devices, battery-life is often considered more important than speed, which means that it is rarely possible to perform advanced graphical transformations or content filtering. Researchers working with solutions for traditional screens can often safely assume that even if requiring an expensive super-computer today, their techniques will run effortlessly on consumer PCs a few generations down the line. Designers for baby faces can not always afford this luxury.

Another aspect worth noting is that the user-interface components that we take for granted when designing for desktop computers may behave completely differently on small devices. For instance, the author once used a pen-based hand-held computer with a touch-screen to look at a page of text. Only a few lines were visible on the display and he naturally pointed the pen to the scrollbar and started to drag it to reveal more text. But since the author is left-handed, and the scrollbar was on the right-hand side of the screen, he covered the whole display with his hand in the process! As computers move off the desktop and into the hands of users, designers will have to take the whole interaction model into account, not just the size of the screen.

4 A Challenge for User-Interface Designers

There is no question that from a purely technical viewpoint, most computing functions can be miniaturized enough to allow them to be crammed into a pocket-sized device. But the question remains whether these functions can be accessed in a meaningful way. If the only way to make all this functionality accessible is through an extremely small screen, it will put extremely high demands on the design of the graphical user interface, demands some designers currently seem unable and even uninterested to meet.

As a case in point, mobile phones are perhaps the most ubiquitous "baby faced" devices currently in use. The penetration has already passed 50% in several Scandinavian countries, and soon mobile phones may be as common as stationary phones. Modern models give the user access to a multitude of functions, most of which are hardly ever used. It seems that the design of these phones give rise to a form of "functionality hiding", in that the limited size of the display implies that all functions will have to be accessed by wading through endless menu options. Only the most persistent users will ever have the patience to explore these options, leaving most of the potential functionality untapped.

This author has personally spent many frustrating hours navigating the menus of his new, top-of-the-line phone. The device will supposedly handle everything from fax messages to databases and shopping list, throwing in the occasional action game in the process. Despite this, the author has barely been able to use it even for basic tasks such as entering often-used phone numbers. We can only imagine what a usability nightmare it will be when all phones come with built-in e-mail clients, web browsers, and word processors. Will the designers and marketers of these devices expect all new customers to have a degree in computer science before making their first call?

5 Conclusion

Designing user interfaces for small screens is a difficult problem, much more difficult than it may seem at first glance. We can not simply take established interface conventions and "shrink" them to baby face size, because just like children have a unique way of life, baby faces are different to desktop computers in ways that we are only beginning to comprehend. But this difference also presents great opportunities for interface designers to find new and valid interface paradigms, paradigms that will be relevant not just for baby faces, but for mass-market computing devices in general.

6 Acknowledgements

The basis for the discussion in this paper comes from experiences gathered in the project *Effective Display Strategies for Small Screens,* which was conducted at the Viktoria Institute as part of the Mobile Informatics research program sponsored by SITI, the Swedish Institute for Information Technology. For this I am grateful to have worked with colleagues at the Viktoria Institute, including Staffan Björk, Johan Redström and Peter Ljungstrand, and our project partners at Telia Research, Ericsson Radio Systems, and the Swedish Institute of Computer Science.

Small interfaces -- a blind spot of the academical HCI community?

Kari Kuutti
University of Oulu, Finland

1 Small interfaces everywhere

Small interfaces are ubiquituous: for example, there are 11 in my own household: a PDA, 2 CD-players, a videorecorder, a television set, a calculator, a microwawe oven, 2 mobile phones, a clock and a radio in the car. In an informal poll among my university colleagues there was a range from 6 to 20, and my count was clearly lower than the median... We are practically immersed in small interfaces, and if we take the whole population it is not a bold guess to say that they are used more than those in "real" computers.

2 But where is the academical research?

Surprisingly enough, for the academical HCI-community these small interfaces do not exist, or at least the community has not published practically anything about them in CHI-conferences or in major HCI journals. This is surprising, to say the least, when compared against their penetration in our lives, and the non-standard nature of them -- each of them different. One can imagine several potential reasons for this glaring omission:

a) the problems related to small interfaces are so trivial that they are not worth of mentioning,

b) there are no reasons to talk separately about small interfaces because the problems are so similar with those in the PC realm,

c) the problems related to small interfaces are transitory and will rapidly vanish when technology developes,

d) in general, there is no interest and pressure to develope small interfaces further, and thus there is no market for research either,

e) for some other reason researchers simply do not think small interfaces worth of any effort.

The option a) does not seem reasonable. On the contrary, it can be claimed that the limitations set by small dimensions and limited ways of interacting make small interfaces more problematic to design than their large counterparts. This is exaggravated by the continuing trend and possibilities to squeeze more and more functionality into small devices.

The option b) does not sound plausible, either. It is true, of course that many of the "golden rules" of interaction design apply to small interfaces as well as larger ones. Because of the general nature of rules their utility may be even less with the small interfaces, however. There are also apparent differences. For the first, the traditional ergonomics has practically totally disappeared from HCI discussions - who remembers anymore that The International Journal of Human-Computer Interaction, fromerly the International Journal of Man-Machine Studies was founded in 1940s as a purely ergonomical forum? With respect to small interfaces this has to change and ergonomical considerations must again become an essential integral part of interface design when properly done.

For the second, the major share of the HCI research has been devoted to the work at the conceptual-cognitive level, understanding and improving the development and use of mental models about the computer system and objects to be manipulated through it. Although the interest in contexts where use acts and thus also the mental models are always embedded has been increasing during the 1990s, this far it is still only a minority of researchers that try to take them seriously into account in practical research. With small interfaces it is impossible to forget the context. Electronic games notwithstanding, nobody will sit down and "use" a small interface device in the way we use computer applications. Small interfaces are occasionally visited in a chain of actions and something is enacted during the visit, something that will happen or whose purpose and meaning lies outside the interface itself, in the chain of actions. It is not possible to even discuss about mental models with respect to small interfaces without the context where they are embedded. Thus design for small interfaces will force us towards more integrated approaches -- where ergonomic, cognitice and contextual issues are designed together -- than those currently mostly used with WIMP interaction.

With respect to option c) one reviewer of a proposal on the "Baby faces" panel on small interfaces for the CHI'98 conference suggested indeed that it is not worthwhile to discuss at all about them, because the technological development will soon make the problems disappear... While it is true that the screens of small interfaces will probably grow larger and it will be possible to put more elements, even in color, on them, there are limits set by human perception and manipulation abilities, and the practical size of hand-held devices, for example, that technological development cannot easily override. Furthermore, as noted in previous paragraph, the nature of using small interfaces is characteristically different than that within the PC realm. It cannot be assumed that any technological advances that will facilitate large interfaces, will automatically offer some benefit for small ones.

There may be some truth in possibility d), at least what comes to the contacts between industry and the academical HCI community: pessures cannot develope where mutual ignorange reigns. In a lucky case small interfaces are designed by industrial desigers - a community of our colleagues, with whom we in academical HCI have not been able to establish much contacts. In a less lucky and perhaps more typical case small interfaces are designed by the electronics engineers building and coding the logics of the device. This group is even further apart and usually does not know or care much about HCI research. And, if we think what HCI can currently offer, they have a good reason in doing so. My own experiences hint, however, that the lack of pressure from the industry side is deceiving and more a problem of communication than lack of respect: people involved in small interfaces design in industry have usually been delighted to find out that someone in the university is interested in their problems.

Finally, there is the possibility e), that for reason or another the HCI community has developed a blind spot towards small interfaces. This opens up a range of potential interpretations, and I'll follow here only one of them: one of the most central characteristics of small interfaces is that they belong to the realm of mundane and everyday. Perhaps it is mundane nature of small interfaces that keeps academical HCI researchers off?

Instead of working at leading edge of the high technology one is forced to struggle with all kinds of size, economical etc. limitations -- that does not make easy to look glorious. It is even difficult to make a spectacular demo about the charcteristics of a small interface, for example when compared to a VR demo. With respect to funding, it might be more complicated to convince a funding body that a study where the technological edge is not so much advanced but a reasonable compromise is sought instead, is really worth of funding. There is at least a grain of truth in this speculation: the majority of systems used as examples in CHI conference papers and demonsrations, for exaple, are either very complex or technologically very advanced: they shy away from everyday. Definitely the CHI panel reviewer mentioned above would do so... To be honest, also the most profitable research contract in my own laboratory is about "virtual prototyping" -- trying to push the leading technological edge in a couple of respects, and definitely distanced from everyday practices. One must indeed ask, if we as a community have developed a blind spot for everyday issues?

Curiously enough, that is one of the accusations feminists have thrown towards the academical research in general. They may have a point here....

3 What is going to happen, and what should we do?

As mentioned above, when studying small interfaces seriously it is necessary to integrate ergonomical, cognitive and contextual perspectives in the design process. This is something new for HCI research. It is true that during the last years there has been growing CSCW-inspired interest in contextuality of use acts, and correspondingly new methods to study and understand it have been developed, but since the GUI paradigm has won and more or less stabilized for mass markets, the experience and understanding of ergonomical perspective has grown thin within the HCI research. We are no more familiar with any degrees of freedom at the hardware level: how to do interface design, when you must shape not only the program inputs and outputs, but hardware inputs and outpus as well?

Luckily enough, there is a community that is well accustomed to that: industrial designers mentioned above, and HCI has a lot to learn from them in this respect. Thus it seems obvious, that to cope with small interfaces, we as HCI researchers must start to open connections to two directions: towards CSCW rersearch for better understanding of contexts and towards industrial design for better understanding of ergonomical and aestethical product design. But interdisciplinary research is demanding and expensive – will there ever be funding for that?

Signs of a change are already visible: the area where the small interface research has great possibilities is wireless mobile services. The PalmPilot has finally made PDAs a viable alternative, wireless phones have already become a commodity, and the two are just about merging, Nokia Communicator as the first example. But a phone + computer is not the only alternative: with the possibilities of wireless communication channels expanding with Bluetooth in the low end and UMTS in the high end, a wide range of new kind of mobile services becomes possible. These services are typically place-dependent and may use interaction between devices in the environment and the hand-held device. Services must be activated, navigated and so on, but for that the complexity of current computers is an overkill: something closer to a wireless phone may be more appropiate. Moreover, many of the services are useful only if there is a critical mass of other users using them. Thus the ease of use becomes a vital factor: the complexity of activation and navigation actions must be somewhere at the level of making a wireless call or sending a text message. For this, the whole chain from the small interface to use situations to services themselves must be analysed and designed. The demand for that will be huge, and it will take the design of small interfaces into the new millenium.

Good Things in Small Packages: User-Interface Design for Baby Faces in Devices and Appliances for Y2K and Beyond

Aaron Marcus, John Armitage, Volker Frank, and Edward Guttman
Aaron Marcus and Associates, Inc., 1144 65th Street, Suite F, Emeryville, CA 94608-1053 USA;Tel:+1-510-601-0994x19, Fax:+1-510-547-6125
Email: Aaron@AmandA.com, Web: http://www.AMandA.com

1 Introduction

User-interface and information-visualization (UI+IV) development for small displays ("baby faces", Marcus 1998) attempts to solve constraints that are relevant for many consumer devices and appliances. This equipment must be useful (easy access to data and functions that are easy to comprehend, remember, and use) and appealing. High-quality UI+IVs, whose content and form are so much dependent upon optimum visible language (e.g., typography, color, symbolism, layout, etc.) and effective interaction with controls, improves the likelihood that users will be more productive and satisfied with computer-based products in many situations beyond the traditional office desktop. The success of products with "baby faces" for home, work, and travel seem to make it imperative that UI+IV designers analyze the subject of baby-face design carefully. The dimensions of this analysis are the traditional ones of UI+IV development: The user-centered, task-oriented development process for UI+IV design includes planning, research (user demographics, user tasks, user context, and marketing/engineering requirements), analysis, design, implementation, evaluation, documentation, and training. The UI+IV design components are metaphors, mental models, navigation, interaction, and appearance. This paper comments on some of likely changes and challenges in these topics.

2 Development Process

Planning, Research, and Analysis: The baby-face development process routinely may differ significantly from the development process for more traditional client-server productivity tools and office desktop Web applications for several reasons. The use contexts of desktop equipment is more established, and the information processing equipment, i.e., desktop or portable PCs, is in place and more stable. In the world of devices and appliances, the cost often may be lower, implying a more casual approach to users' original purchases, their turning over their equipment for a newer models, and thus potentially greater competition among equipment providers. In the past, industrial designers, or product designers, and engineers, including human-factors specialists, have routinely directed the design of devices and appliances. Each of these two groups sometimes has different philosophies of function and form that compete in the minds/eyes of both management and users. The development process for devices and appliances currently differs significantly from the emerging Web paradigm of continuous rollouts of partially finished products. For example, equipment is physical hardware that must be physically manufactured, warehoused, and distributed. Now, the philosophies, principles, techniques, experience, and team members of these two groups must be integrated with those coming from the UI+IV design communities, who have more than two decades of relevant, crucial, but *different* expertise. It seems likely that intellectual and budget "turf wars" will pre-occupy the time of the development team unless upper management can establish a rational, humane, effective development process that makes use of existing expertise, but mandates clearly the incorporation of new ideas and new approaches.

Design, Implementation, and Evaluation: To the extent that devices and appliances with significant UI+IV components are distributed to worldwide markets, the equipment must be designed with globalization in mind. Accounting for international/localized products may affect not only labels on hard buttons, but also may impact other UI+IV components, e.g., metaphors (Chavan 1995, Marcus 1998). Devices and appliances are imbedded in more complex social, cultural, and physical contexts than the traditional office product, which tended to have more universal, limited contexts of use. This new equipment may change social and leisure-time experiences significantly. Consider the Sony Walkman in the 1980s, a radio without a speaker like the first crystal-radio sets in the early twentieth century, which reintroduced the solitary-listener ritual into the mass-market mainstream of music appreciation and social interaction. Because of the interdisciplinary team, the small scale, the intimate relation to hardware components, the importance of frequent user evaluations, and the rapid development schedules, detailed interactive prototypes oriented to use scenarios become especially important.

Documentation and Training: UI+IV design for baby faces requires innovative solutions to all components. Designers can annotate interactive prototypes and transform them into interactive guidelines and specifications documents, upon which software engineers can rely for the latest definitions, labels, interaction sequences, etc. Training in topics of UI+IV specific for the baby-face development team will become important. Most major UI+IV conferences do not specialize enough yet in research and applied techniques to assist the development community, giving rise to special-focus conferences and workshops. Some of these are listed at the Baby Face Website originated by Kuutti, Marcus, and other contributors to an ACM/SIGCHI-99 panel (http://www.tol.oulu.fi/projects/hci/babyfaces/, Marcus 1998). Case studies and texts also have begun to appear (Marcus 99).

3 User-Interface Design Components

Because equipment and use context are different from desktop paradigms, baby faces are introducing novel approaches to many UI+IV components.

Metaphors: After the Apple Newton debuted, personal digital assistants (PDAs) promoted a different set of fundamental nouns and verbs than previous desktop productivity tools (files, folders, applications, etc.) or the emerging Web browsers (documents, pages, etc.). Concepts like to-do lists, contact lists, calendars, planners, schedulers, and their associated sub-functions, became top-level items in menus. As another example, essential metaphorical references in navigation devices typically include maps, with roads, scales, north arrows; trips, with planners, destinations, route preferences; and keyboard and telephone keypads, with buttons and labels. Home control devices emphasize scheduling calendars, process controllers, and monitoring gauges. Device and appliance controls are linked to widely different usage contexts that also may vary widely according to individual cognitive preferences and cultures. For example, in trip navigation, research has shown that typical users are evenly divided among those who prefer signs, maps, and words for guidance (Marcus 1999).

Mental model: Many baby faces for devices and appliances vary in their mental models. One significant challenge is the initial display contents upon turning on the device. In general, to convey a level of complexity that is not daunting to the novice or occasional user, the initial screen should have 7±2 items to denote and connote a simple product. This guideline contrasts with the established remote control conventions for televisions and videocassette recorders, which have a reputation for being user-unfriendly. With almost all devices and appliances likely to be connected to communication networks, retaining simplicity, clarity, and consistency within and across equipment will become a primary challenge for baby-face UI+IV design.

Navigation: Physical devices and appliances use “simple” dashboard controls rather than “complex” conventional computer navigation. Baby faces seem likely to change to complete windows or panes switching. In AM+A’s experience for desktop applications, navigation with multiple windows, such as Microsoft’s multiple document interface (MDI) paradigm, almost always seems an undesirable approach. Baby faces seem likely to need very rapid navigation with a minimum amount of scrolling and maximum stability to locations. This navigation paradigm differs from typical desktop personal computer interaction with detailed internal hypertext-like navigation in which any part of the display might be interactive and show small-scale pop-up controls, or widgets.

Interaction: PDA devices have favored a pen or stylus for interaction, including picking and character entry. Improved voice input and output seem likely to increase in use, especially for limited display real estate. Research (Pooon, 1999) has shown the value of stable locations for pick-lists over scrolling for medical-symptom selection; this paradigm and its benefits seem generalizable.

Appearance: The designers of the earliest graphical UIs designed visual components with the assumption that screen-space was not severely limited. Even the earliest Macintosh computers with 72 pixels per inch could appear on screens with approximately 20-cm diagonals. Menu bars, scroll bars, title bars, and icons, among other items, all consume considerable space, often 10 to 25 per cent of available area for screens with approximately 30 cm diagonals. This level of inefficiency seems intolerable in baby faces that might measure 3 to 12 cm on the diagonal, even with higher pixel density. In addition, the demands of color, varieties of ambient light characteristics, selection target-size, and requirements for product and corporate branding, are likely to make achieving success an extremely demanding challenge for designers. One other aspect is the possibility of increased use of acoustic cues or "earcons" to denote and connote classes of objects and processes, their attributes, and, in particular, their states. The increased use of these acoustic cues frees up visual real estate for other semantics. Finally, one special challenge will be to maintain desirable levels of legibility and readability for miniature displays of people's faces, charts, maps, and diagrams under extreme conditions of reduced numbers of pixels.

4 Conclusion

Baby faces for devices or appliances continue to decrease in size, while their functions and data-access continue to increase. For example, Samsung recently announced a wristwatch that also is as a telephone and pager. Baby-face UI+IV design is likely to bring changes. Multi-disciplinary teams in UI+IV development and good communication among team-members from different professional disciplines will be indispensable. Interactive design-space exploration tools that enable clients to understand what variations are possible and to appreciate the professional skill of designers in synthesizing solution will also be crucial. The importance of accounting for different cognitive preferences for absorbing information will become apparent. When designers are targeting "wide-band audiences" with multiple demographic characteristics of age, gender, education-level, cultural background, etc., even if the users are united in their immediate tasks, it seems important to account for diversity in the components of the UI. Developers will have to plan for the impact of user testing on the development process, especially on the cost of user-testing's impact on the business model. Without planning, the cost of these efforts may defeat the business case for product development. Over the last twenty years, the author has noted hesitation on the part of some clients to undertake appropriate focus group investigation and usability testing during development. The challenge in the design community is to gather valid, precise, accurate, clear statistics of success cases and to make these available to the industry.

Connected to the Internet, devices and appliances with baby faces challenge designers to invent or re-apply metaphors, mental models, navigation, interaction, and appearance in novel ways that break through to new levels of simplicity, clarity, and consistency. By considering the dimensions of research, analysis, and design outlined above, designers can continue to accommodate marketing and engineering requirements while providing the next generation of consumers with devices and appliances that make life more humane.

5 References

Chavan, Apala Lahiri, "A Design Solution Project on Alternative Interface for MS Windows," Masters Thesis, New Delhi, India, September 1994.

Clarke, Cathy, and Lee Swearingen. "Motorola Smart Car," Chapter 2.3, in *Macromedia Director Design Guide*. Hayden Books, Indianapolis, IN, 1994.

Marcus, Aaron, "Principles of Effective Visual Communication for Graphical User Interface Design," *Readings in Human-Computer Interaction*, 2nd Ed., ed. Baecker *et al.*, Morgan Kaufman, Palo Alto, 1995, pp. 425-441.

Marcus, Aaron. "Baby Faces: User-Interface Design for Small Displays." Panel Description, *Conference Summary*, CHI-98, National Conference of ACM/SIGCHI, 18-23 April 1998, Los Angeles, CA, pp. 96-97.

Marcus, Aaron, "Metaphor Design in User Interfaces," *The Journal of Computer Documentation*, ACM/SIGDOC, Vol. 22, No. 2, May 1998, pp. 43-57.

Marcus, Aaron, "Finding the Right Way: A Case Study about Designing the User Interface a Motorola Vehicle-Navigation System," in Bergman, Eric, ed., *Beyond the Desktop,* Morgan Kaufman, Palo Alto, 1999 (in press).

Poon, Alex, "Pen-based System for Structured Data Entry", in plenary lecture by Dr. Edward H. Shortliffe, Ted.Shortliffe@Stanford.edu, Medical Informatics, Stanford Univ., 33rd Hawaii Internat. Conf. on Systems Science, January 1999.

An Integrated CAI/CAD System for Graphic Design

Rungtai Lin, Yen-Yu Kang

Department of Industrial Design
Mingchi Institute of Technology, Taipei, Taiwan
Email: rtlin@ccsun.mit.edu.tw, yen@ccsun.mit.edu.tw

1 Introduction

Recently, The graphics design have been enhanced and expanded with the integration of computer graphics and CAD technology. Many research involved in using mathematics and computers for artistic purposes, however, the wide range of current research do not seem to include any methods which makes the design process of early externalized easier. In addition, present CAD systems in graphics design only provide graphic representation of idea recording and final solution in design process. It is very desirable to extend current CAD systems to support the activities in the entire graphic design process by integrating the concept of symmetry groups and CAD technology.

2 A new approach in graphic design

In graphic design, like other design fields, there is a systematic approach toward the solution of a graphic problem. Many combinations of steps have been suggested to enable graphic designer to achieve his or her design. one possible problem-solving model is the following five-step procedure (figure 1):

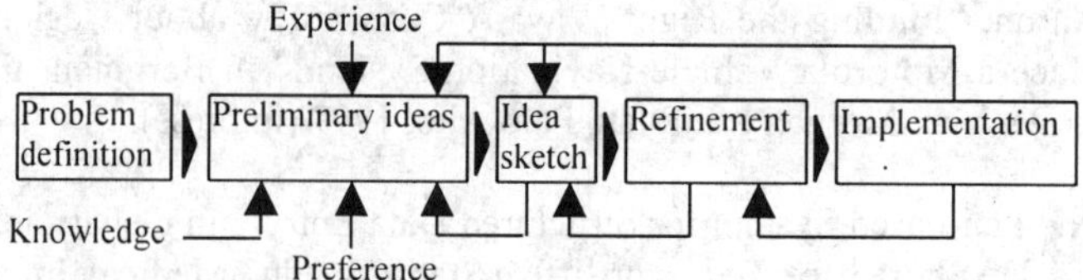

Figure1 A traditional procedure of graphic design

The design procedure is a rational decision-making and problem-solving process in which graphical representation is a necessary vehicle for recording ideas and different design stages. Graphical representation in the tradition design process falls into two main categories: the representation and recording of design ideas during the design process and representation of a finalized design solution. If creativity can be defined and explained as an activity based on rationality, it is possible to develop a tool by using computer graphic and CAD technology to support creativity.

Figure 2 shows the new approach to graphic design. Since conceptual frameworks for the use of graphic representation can be served of major benefit in the graphic design process, a new tool is developed for designer in order to generate and explore ideas easier in the initial design stage. In the latter stages of design process, the current CAD system can serve as a powerful tool for designing the repeating patterns through design process. The greatest benefit of the new approach is not only the approach provides a dynamic method for presenting visual and graphic information in the entire design process, nut also the new approach provides a real-time interactive for generating and exploring new ideas in the earlier design stages.

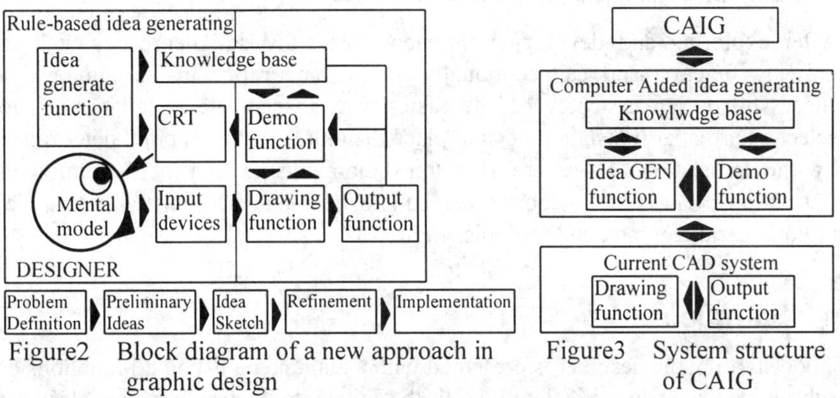

Figure2 Block diagram of a new approach in graphic design

Figure3 System structure of CAIG

3 System structure and implementation

The system consists of two major parts: CAIG subsystem and current CAD system. CAIG subsystem is composed of a knowledge base for storing the rules of symmetry groups. CAD system contains drawing function and output function for refining and graphical representation. These two major parts including five functions are logically connected to each other and with Autocad as shown in (figure3).

At the very beginning, the designer attempts to develop new concepts through searching his store of knowledge, experience, past work and researching from sources of references. Demonstration function is to show good examples of tessellation for designers as a source of reference. The idea generate function can produce all the types of tessellations by using a given unit cell.

Upon the ideas developed and a desired unit cell selected, the drawing function connected with Autocad serve as a tool for modifying and refining the designs until a successful design is achieved. In the final stage, output function based on Autocad's original utility functions can serve as a tool for graphical representation.

The idea generate function contains three subroutines: design subroutines for giving a unit cell, display subroutines for displaying the procedures and results, generate subroutines for generating all the repeating pattern. First, the design subroutines will ask the designer input a unit pattern into a unit cell frame. Then, the generate subroutines will automatically generate all the possible repeating patterns by using the given unit cell. Every repeating pattern will be generated step by step and the procedures will be displayed on the screen by display subroutines. After every individual repeating pattern is produced, all the repeating patterns will be displayed on the screen at the same time for designer's checking. One example is shown in (figure 4).

After exploring and developing the new ideas, the designer can switch to drawing function and can continually modify and refine his design until a successful design is achieved. The easiest way to generate tessellations is to select a particular translation net and place the unit cell at each point determined by the translation. There are five translation nets to control the unit cell translations. The five categories are square, rectangular, equilateral triangle, oblique-parallelogram, and rhombic (figure 5).

4 Sample run

Upon start up, the designer is presented with a main menu listing all functions or subroutines as shown in (figure 6). In the initial stage, designer can select the demonstration function for searching good ideas from good design examples, or past works. In the stage of preliminary idea, the designer can switch to idea generate function for exploring the new ideas. After switching to idea generate function, the function will ask designer input a unit cell. Then, the function will automatically generate all the possible repeating patterns by using the given unit cell. Every repeating pattern will be generated step by step and the procedures will be displayed on the screen. After every individual repeating pattern is produced, all the repeating patterns will be displayed on the screen at the same time for designer's checking. Two examples are shown in (figure 7 and 8). For practicing and forming designer's own works, he can switch to drawing function for designing, modifying, and refining his own designs. After the function is running, the designer will be asked to input three parameters, a and b are the distances in the two axes, the angle between these axes. Then, a frame of unit cell will be displayed on the screen. At this moment, designer will give the unit pattern in the frame and select the different generators the desired tessellation will generate for designer. An example is shown in (figure 9).

5 Conclusion

In this study, the development of CAIG system is to provide an efficient way for graphic designers to explore and develop ideas in the early stages of design process, and to modify and refine their works in the final stages. The CAIG system, which integrates knowledge base and CAD into graphic design, offer several major advantages over the conventional approach. The computer is designers' instrument producing the final results, and the mathematics are the rules of tessellation groups that control the type or structure of the tessellation produced. By integrating AI and CAD technology, the designer can produce anything that he used to design in the traditional design process, and like conventional the designer is limited only by his imagination. With this in mind, this study is to develop an efficient way for designers to inspire his imagination. Working with CAIG system, it is worth emphasizing that the, designer is only limited by his imagination and by the initial unit cell that he choose.

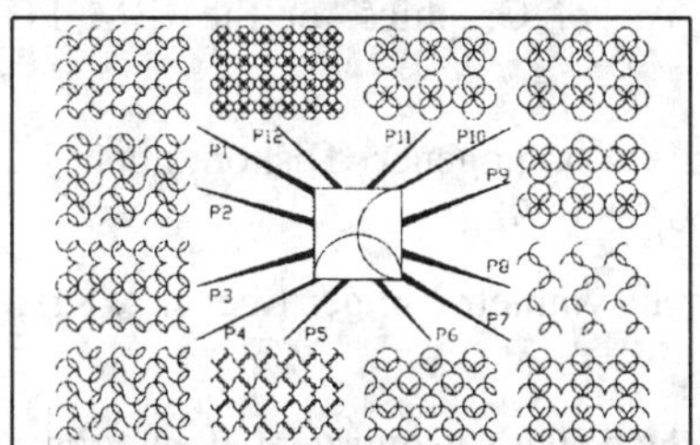

Figure 4 One of idea generating using a given unit

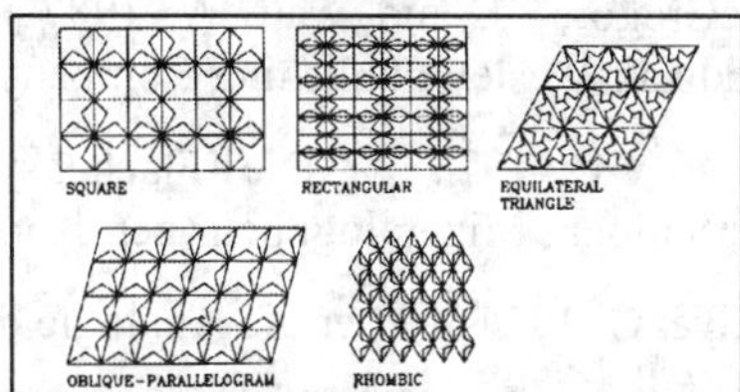

Figure 5 The five translation nets for unit cell placement

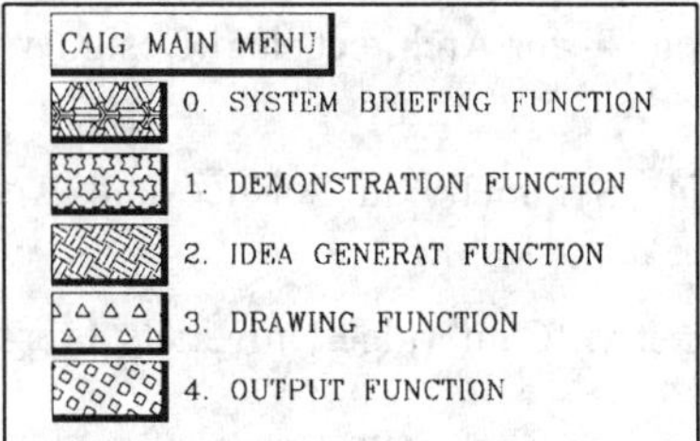

Figure 6 Main menu of CAIG system

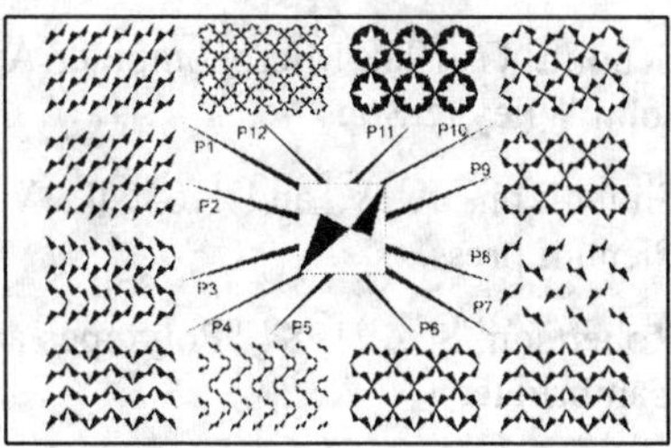

Figure 7 One example of idea generating using a given unit

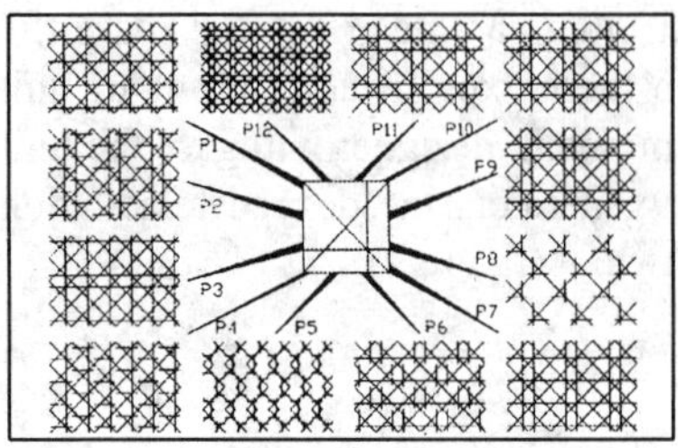

Figure 8 A example of idea generating using a given unit

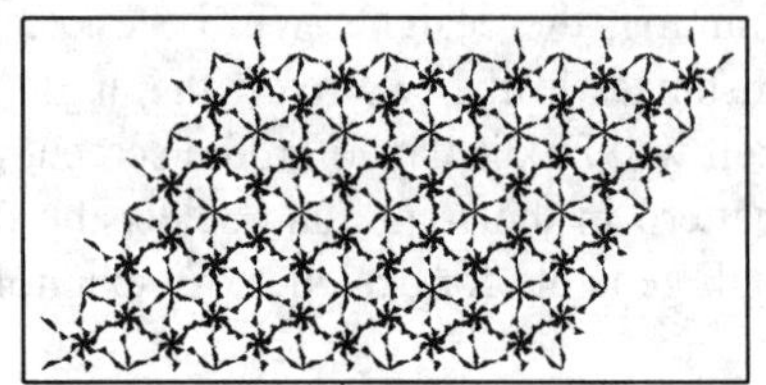
Figure 9 One of tessellations generated by drawing function

Reference

Chern, S. S. 1987 "CAD/CAM, What is Next", 1987 ASME International Computers in Engineering Conference, New York.

Coxeter, H. S. M. and Moser, W. 0. J.1957 Generators and Relations for Discrete Groups, Springer-Verlag.

Fricke, R. and Klien, F. 1897 Vorlesungen Uber die Theorie der automorphen Funktionen, Leipzig.

Locher, J.L. 1971 The World of M.C..Escher, Harry N. Abrams, Inc., New York.

McGregor, J. and Watt, A. 1987 The Art of Graphics for the IBM PC, Addison-Wesley Publishing Company.

Niggli, P. 1924 Die Flachensymmetrien homogener Diskontinuen, Z. Kristallogr., Mineralog. petroger, Abt. A 60, 283-298.

Polya, G. 1924 "Ueber die Analogie der kristallsymmetrie in der Ebene, 2. Krist. 60, 278.

Schattschneider, D. 1978a "Tiling the Plane with Congruent Pentagons," Mathematics Magazine, vol. 51, no. 1, January, pp. 29-44. 1978b "The Plane Symmetry Groups: Their Recognition and Notation," The American Mathematical Monthly, vol. 85, no. 6, June-July, pp. 439-450.

Schmitt, G. 1988 Microcomputer Aided Design -- for Architects and Designers, John Wiley & Sons.

Shubnikon, A. V. and Koptsik, V. A. 1974 "Symmetry in Science and Art, Plenum press.

Robertson, S. A. 1984 "Polytopes and Symmetry," Cambridge University Press, Cambridge.

Wieting, Thomas W. 1982 "The Mathematical Theory Of Chromatic Plane.

Cultural Differences in Icon Recognition

Rungtai Lin
Department of Industrial Design
Mingchi Institute of Technology, Taipei, Taiwan
Email: rtlin@ccsun.mit.edu.tw

1 Introduction

Along with the increasing availability of high quality graphic displays, the use of iconic interfaces in computer system is becoming common and accessible to users. Many users have come to view an iconic interface as a requirement for their systems and software. While icons are playing an increasingly important role in iconic interface, many cognitive human factors of icon design are not well understood (Gittins 1986, Lodding 1982). There are no specific set of rules or criteria that can be followed by the designers during the design stage. Although the International Organization for Standardization (Easterby and Zwaga 1976) provides the selection criteria and original reference principles, whether an icon has met the criteria can be known only after the test is completed. For icons are effective in an iconic interface, they need to be properly designed to be meaningful, associated, learnable, memorable and consistent. Collins (1982) listed six questions to evaluate an icon, including 1)Can the symbol be detected? 2) Can it be discriminated from all others? 3) Can it be recognized when seen in a different context? 4) Does it communicate the intended meaning? 5) Does it gain attention? 6) Does it alter behavior appropriately? This six questions can be grouped into the four categories of human factors which are physical, cognitive, social, and cultural. Cultural factors are the vaguest and the most unspecific, the least known, the most difficult to codify, but perhaps, precisely because of these reasons are the most important. Recently, the research in this field has been focusing on "cultural differences in the user perception" (Resnick et al. 1997) with an attempt to study the influence of culture in icon recognition. Therefore, the purpose of this paper intended to study the cultural factors in the preferences of pictorial symbols.

2 Method

A semantic differential rating was used to evaluate 35 icons (Figure 1) with three cognitive factors by three groups of subjects with different cultural backgrounds. The stimuli consisting of thirty-five public information symbols were adopted from the results of the ISO test series 1985/6 and 1979/80 (Brugger 1990). The three adjective pairs and their corresponding cognitive factors were meaningless / meaningful for "communicativeness," complicated / concise for "design quality," and abstract / pictorial for the "image function" (Lin, 1992). 104 students with different cultural backgrounds participated in this study. The subjects were grouped into three groups: 1) Group I were Forty-one Taiwanese students with design background from Mingchi Institute of Technology, Taiwan. They were 20 males and 21 females and aged between 16 and 22. 2) Group II were thirty-two Taiwanese oversea students in the US. They were 19 males and 13 females and aged between 26 and 42. 3) Group III were thirty-one American students with engineering background from Tufts University, MA. They were 20 males and 11 females and aged between 18 and 24. All participants were told the purpose of the study, then were asked to rate each icon with its referent according to the degree of association with three cognitive factors on a 0 to 6 rating scale (Figure 2).

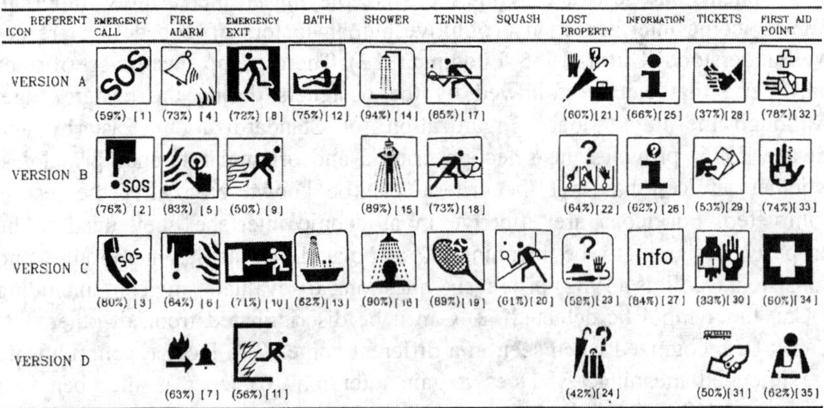

Figure 1 Referents and variants of thirty-five public information symbols

3 Results

Table 1 lists means of the ratings on three cognitive factors of 35 icons. The first two columns show the comprehension rates of the variants for each referent (icon). Variants A, B and C are the ISO test series 1985/6, while variant D is the best alternative variant from the ISO test series 1979/80. The last column gives the means of overall ratings which are weighted based on the percentage of variance each factor can explain (Lin 1992). The results show that an icon with high comprehension rate got a high rating scores on the three cognitive factors. Figure 2 shows the comparison graph of overall rating scores of 35 icons and the comprehension rates of the variants tested for each referent. The straight line indicates the ISO recommended 67% of comprehension rates as a public information symbol for international use. In figure 2, there were differences in referents and subjects, for examples, variant C of "Tennis" shows the consistence of the subjects, while variants A and B are varied among the subjects. Through Duncan's multiple range test, significant differences (p <.05) of six referents, both in subjects and variants were found. Table 2 lists the comparison of six referents with their variants and the differences among three groups of subjects.

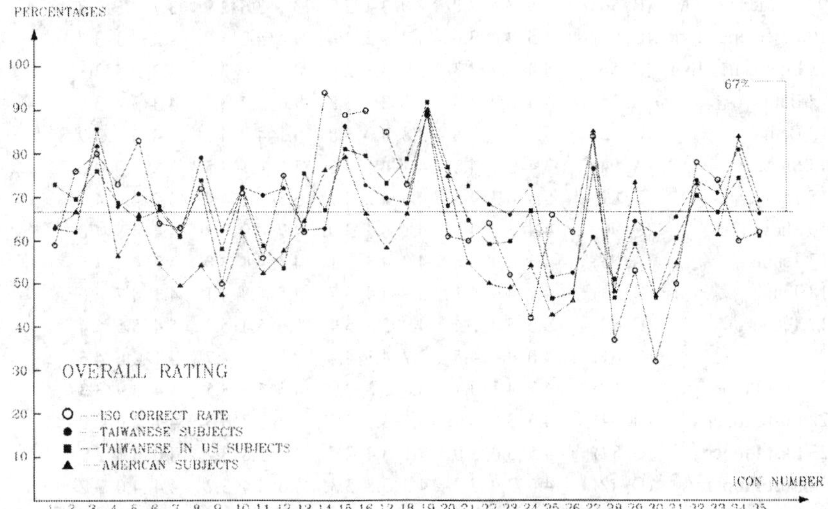

Figure 2 Comparison graph of thirty-five public information symbols

Table 2 Differences of six referents in subjects and their variants

Referent for symbols	Subject difference *	Variant Difference **
Fire Alarm	S(T) = S(A) >S(TA)	V(B) > V(A) = V(C) > V(D)
Emergency Exit	S(T) > S(TA) > S(A)	V(A) = V(C) > V(D) > V(B)
Bath	S(T) > S(TA) > S(A)	V(C) > V(A)
Tennis	S(TA) > S(T) > S(A)	V(C) > V(B) > V(A)
Lost Property	S(T) > S(TA) > S(A)	V(D) = V(A) > V(B) = V(C)
Tickets	S(T) > S(TA) = S(A)	V(B) > V(D) > V(C) = V(A)

* S(T): Subject group I; S(TA): Subject group II; S(A): Subject group III.

** V(A,B,C,D): Variants for different referents.

Table 1 Comprehension rates and mean values of 35 icons

Public Information Symbols	(Version , %)	Factor I			Factor II			Factor III			Overall		
		T	TA	A	T	TA	A	T	TA	A	T	TA	A
1 Emergency Call	(A, 59)	3.6	4.6	3.9	4.9	4.6	4.0	2.7	3.8	3.3	3.8	4.4	3.8
2 Emergency Call	(B, 76)	3.7	4.3	4.2	4.0	3.9	4.0	3.4	4.3	3.6	3.7	4.2	4.0
3 Emergency Call	(C, 80)	5.4	4.8	5.0	4.8	4.3	4.8	5.0	4.5	4.8	5.1	4.6	4.9
4 Fire Alarm	(A, 73)	4.1	4.1	3.3	4.2	4.0	3.6	3.9	4.3	3.2	4.1	4.1	3.4
5 Fire Alarm	(B, 83)	4.5	4.1	4.0	3.8	3.6	3.9	4.4	4.1	3.8	4.3	3.9	3.9
6 Fire Alarm	(C, 64)	4.3	4.0	3.2	3.9	3.9	3.4	4.0	4.2	3.2	4.1	4.0	3.3
7 Fire Alarm	(D, 63)	3.9	3.8	3.0	3.5	3.7	2.8	3.6	3.3	3.0	3.7	3.7	3.0
8 Emergency Exit	(A, 72)	4.7	4.6	3.0	4.7	4.2	3.4	4.9	4.6	3.5	4.8	4.4	3.2
9 Emergency Exit	(B, 50)	3.6	3.5	2.7	4.0	3.4	3.0	3.7	3.6	2.9	3.7	3.5	2.8
10 Emergency Exit	(C, 71)	4.5	4.4	3.6	4.2	4.3	3.6	4.4	4.2	4.0	4.3	4.3	3.7
11 Emergency Exit	(D, 56)	4.4	3.7	3.3	3.9	3.3	2.9	4.4	3.6	3.2	4.2	3.5	3.1
12 Bath	(A, 75)	4.4	3.2	3.4	4.1	3.2	3.5	4.6	3.2	3.5	4.3	3.2	3.5
13 Bath	(C, 62)	3.5	4.7	3.8	3.8	4.0	3.8	4.1	4.9	4.1	3.8	4.5	3.9
14 Shower	(A, 94)	3.4	3.9	4.6	4.4	4.1	4.6	3.5	4.1	4.5	3.8	4.0	4.6
15 Shower	(B, 89)	5.5	5.3	4.8	4.6	4.2	4.7	5.5	5.1	4.7	5.2	4.9	4.7
16 Shower	(C, 90)	4.4	5.0	4.1	4.1	4.4	3.9	4.7	4.9	3.8	4.4	4.8	4.0
17 Tennis	(A, 85)	4.3	4.4	3.6	4.0	4.3	3.1	4.3	4.4	3.8	4.2	4.4	3.5
18 Tennis	(B, 73)	4.1	4.9	4.1	3.9	4.4	3.7	4.5	4.8	4.1	4.1	4.7	4.0
19 Tennis	(C, 89)	5.6	5.8	5.4	4.9	5.2	5.4	5.6	5.4	5.3	5.4	5.5	5.4
20 Squash	(C, 61)	4.0	4.8	4.5	3.7	4.1	4.4	4.7	4.9	4.7	4.1	4.6	4.5
21 Lost Property	(A, 60)	4.7	4.1	3.4	3.8	3.5	2.9	4.6	4.1	3.5	4.4	3.9	3.3
22 Lost Property	(B, 64)	4.5	3.8	3.0	3.4	3.1	2.7	4.4	3.7	3.3	4.1	3.6	3.0
23 Lost Property	(C, 52)	4.5	3.6	3.0	3.1	3.4	2.7	4.2	3.8	3.2	4.0	3.6	2.9
24 Lost Property	(D, 42)	4.6	3.9	3.2	4.0	3.8	3.3	4.6	4.4	3.2	4.4	4.0	3.2
25 Information	(A,66)	2.5	2.2	2.2	4.7	3.8	3.4	2.0	2.4	2.0	3.1	2.8	2.6
26 Information	(B,62)	2.9	2.6	2.8	4.1	3.6	3.0	2.4	2.4	2.3	3.2	2.9	2.8
27 Information	(C,84)	2.9	4.7	5.5	4.9	4.8	5.2	3.2	4.2	4.4	3.7	4.6	5.1
28 Tickets	(A,37)	2.9	2.6	2.7	3.4	2.9	2.9	2.9	3.0	3.0	3.1	2.8	2.9
29 Tickets	(B,53)	3.6	3.4	4.3	4.3	3.8	4.4	3.9	3.4	4.6	3.9	3.6	4.4
30 Tickets	(C,32)	3.9	2.7	2.8	2.9	2.6	2.6	4.4	3.2	3.2	3.7	2.8	2.8
31 Tickets	(D,50)	4.1	3.6	3.0	3.4	3.5	3.6	4.4	3.8	3.3	3.9	3.6	3.3
32 First Aid Point	(A,78)	4.9	4.5	4.7	3.6	3.8	4.0	4.6	4.3	4.4	4.4	4.2	4.4
33 First Aid Point	(B,74)	4.4	4.3	3.9	4.0	3.4	3.5	4.3	4.2	3.6	4.3	4.0	3.7
34 First Aid Point	(C,60)	5.4	4.4	5.2	5.6	5.4	5.3	3.1	3.4	4.4	4.9	4.5	5.0
35 First Aid Point	(D,62)	3.8	3.7	4.2	4.1	3.6	3.9	4.2	3.8	4.4	4.0	3.7	4.1

4 Conclusion

Understanding how users recognize an icon is as complex as understanding visual perception itself. The factors that affect the comprehension of pictorial symbols can be grouped into four categories of human factors which are physical, cognitive, social, and cultural. In this study, the results indicate that culture differences affect user preference of icons. The cultural factor plays an important role of the comprehension of icons, and is worthy of more in-depth study.

5 References

Brugger, C. (1990). Advances in the International Standardization of Public Information Symbols. Information Design Journal, Vol. 6/1, 79-88.

Collins, B. L. (1982). The Development and Evaluation of Effective Symbol Signs. Washington DC: US Department of Commerce, National Bureau of Standards NRS Building, Science series 141.

Easterby, R. S., and Zwaga, H.J.G. (1976). Evaluation of Public Information Symbols, ISO test: 1975 series Report 60.

Gittins, D. (1986). Icon-based human-computer interaction. International Journal of Man-Machine Studies, 24, 519-543.

Lin, R. (1992). An Application of the Semantic Differential to Icon Design. Proceedings of the Human Factors Society 36th Annual Meeting, 336-340.

Lodding, K.N. (1982). Iconic interfacing. IEEE Computer Graphics and its Applications, 3, 11-20.

Resnick, M.L., Zanotti, A. and Jacko, A. (1997). Cultural Differences in the Perception of Responsibility for Child Safety. Proceedings of the Human Factors Society 39th Annual Meeting, 1015-1019.

Synthesizing Ideal Product Shapes by Image Morphing

L. L. Chen, S. H. Yang, and J. W. Lin
National Taiwan University of Science and Technology

1 Introduction

Under the current trend towards *mass customization* and *agile manufacturing* (Anderson and Pine 1997; Pine 1992), it is becoming ever more important to understand what each consumer wants, and to assist consumers express their likes and dislikes. For studying the perceptions of products by target consumers, market researchers and designers often use perceptual maps for visualizing consumers' perception of different products and the critical factors affecting their decisions.

Product perceptual map can be constructed by using a number of methods, including *multidimensional scaling* (or MDS) [Kruskal 1978; Schiffman 1981]. Based on proximities (or, conversely, distances) among the stimulus objects, MDS methods computes a (usually low dimensional) perceptual map such that each point in the map corresponds to a stimulus object, and distance between any two points match the proximity between the corresponding stimulus objects as much as possible.

By incorporating additional information about the rankings of stimulus objects according to a number of attributes, ideal vectors or ideal points, that represent the preferences of target consumer groups, can be located in the perceptual map. From proximities among stimulus objects, it is also possible to conduct clustering analysis to partition the stimulus objects into clusters, each consisting of stimulus objects that are considered to be more similar than objects in other clusters.

Visualization of a perceptual map, where the image of a stimulus object is shown at the location of its corresponding point, is very useful for interpreting the perceptual space. Because the proximities among the stimulus objects are estimated from survey data, the number of stimulus objects is usually limited. Since the perceptual map is not dense enough, there is usually no stimulus object that coincides with the ideal points or cluster center points. It is neither possible to pick an arbitrary point in the perceptual map and find out the corresponding shapes or characteristics.

2 Automobile Perceptual Map

A perceptual map of automobiles summarizes how consumers perceive the various automobiles currently on the market. In this research, such a perceptual map is the basis for determining the perceptual qualities of the automobiles so that automobiles with specific perceptual qualities can be selected for image morphing.

The side-view images of automobiles are collected in magazines from 1991 to 1997 or from photographs of actual cars. We collected a total number of 104 automobiles including some concept cars. We also collected 50 adjectives that can be used to describe automobiles.

A first survey is conducted with thirty subjects who sorted the automobiles into clusters and picked an automobile best representing each cluster. In the same survey, the subjects were also asked to choose the most often used adjectives and to cluster the adjectives according to their perceived similarity. As a result of this survey fourteen representative automobiles and ten adjective pairs were selected.A second survey is next conducted with forty five subjects in which the subjects are asked to rank each of the fourteen automobiles according to the degree each of the ten adjectives describes the automobile. We then used MDPREF program to construct the perceptual map using the preference data resulted from the survey. We also used PREFMAP program to determine the location of the ideal point corresponding to each adjective in the perceptual map. The result is an automobile perceptual map where the two major axes are interpreted as “streamline” and “width/height ratio”, respectively.

3 Image Morphing

The basic idea of *image morphing* [Gomes et. al 1999; Wolberg 1990] is to obtain a series of images that interpolate between two given images. In computer graphics, such a process is usually called *morphing* or *metamorphosis*. In practice, morphing refers to the process of an image gradually changes from a starting image to an ending image. To achieve the effects of morphing, two techniques, *cross-dissolve* and *warping*, are often used.

In this research, we use a program that is based on the mesh method for specifying the warping functions and pixel cross-dissolve to create image morphing. Our objective is to provide an interface to allow a user to interactively pick one point in the perceptual map and be able to see a visualization of the corresponding product by morphing some of the given product images. Two of the problems we encountered are determining *which* product images to morph, and, since we could only morph two images at a time, the sequence and the weights of each of the product images. These two problems are solved by using *Delaunay triangualtion* and *barycentric coordinates,* respectively.

4 Delaunay Triangulation

For given set of points in the plane, it is possible to obtain a Delaunay triangulation [Goodman and O'Rourke 1997] of the points that offers many desirable properties. In particular, it is possible to use the triangles in a Delaunay triangulation to define the "closest" three points to any given point in the plane.

The Delaunay diagram for points in the plane can be computed very efficiently, in *O(n log n)* time where *n* is the number of given points. In our example of automobile perceptual map, this number *n* is 14. The Delaunay triangulation of the automobile perceptual map is shown in Figure 1.

The circumscribing circle of any triangle in the Delaunay triangulation does not contain any other points in the set. This means that, in a certain sense, we can say that these three points are "closest" to the points in the triangle. For an arbitrary point in the plane, we simply locate the point in the Delaunay triangulation to determine the three closest points (or the product images) to use for image morphing.

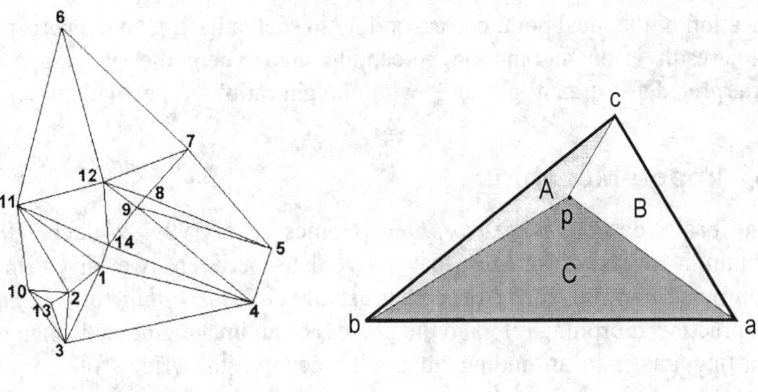

Figure 1

Figure 2

5 Barycentric Coordinates

Given a point p and a triangle Δabc, the line segments connecting the point p to the vertices of the given triangle partition the original triangle into three smaller triangles Δpbc, Δpca, and Δpab, as shown in Figure 2. The barycentric coordinate [Gomes et. al 1999; Farin 1996] of the point p within the triangle Δabc is a tri-tuple of weights (u, v, w) such that $u=\Delta pbc/\Delta abc$, $v=\Delta pca/\Delta abc$, $w=\Delta pab/\Delta abc$ are the weights for vertices a, b, c, respectively.

In this research, we will use barycentric coordinates as the weights to assign to the product images for image morphing. We observe that barycentric coordinates provide a nice generalization from the interpolation between two points to interpolation among three points. Notice that, if the point p lies on an edge of the triangle, say bc, then the area of the triangle diminishes. So, the weight u becomes 0 and p is expressed as a linear interpolation between the two vertices b and c. As the point p gradually moves away from the edge bc, the weight u for the vertex a gradually increases from 0 to 1.

We next need to determine the order in which to morph the images. Our strategy is to morph the images corresponding to the two vertices farther away from the point p first to obtain an intermediate image q, which is then morphed with the image corresponding to the vertex closest to the point p. Without loss of generality, we assume that vertex c is the farthest away from the point p, followed by vertex b and then vertex a. Then, we have:

$$\begin{aligned} p &= u\,a + v\,b + w\,c \\ &= u\,a + (v+w)\,q \\ q &= v/(v+w)\,b + w/(v+w)\,c \end{aligned}$$

6 Ideal Shapes by Image Morphing

By the image interpolation technique described above, we can compute a corresponding image for any point in the perceptual map within the convex hull [Goodman and O'Rourke 1997] of the given points.

For an ideal point that lies within the convex hull of the given points in the perceptual map, it is possible to compute a shape that represents the ideal shape for the particular adjective, as illustrated in Figure 3. It is also possible to compute a shape that corresponds to the center points of the clusters in the perceptual map, as shown in Figure 4. Here the center point is computed by finding a smallest enclosing circle of the points in a cluster.

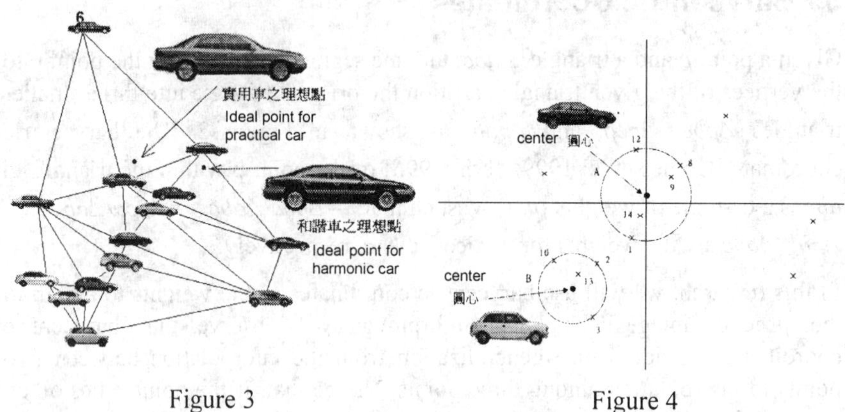

Figure 3 Figure 4

7 References

D. M. Anderson & B. Joseph Pine (1997). *Agile Product Development for Mass Customization: How to Develop and Deliver Products for Mass Customization, Niche Markets, JIT, Build-to-Order, and Flexible Manufacturing*, McGraw-Hill.

P.Gerald Farin (1996). *Curves and Surfaces for Computer Aided Geometric Design: A Practical Guide (4th Ed)*, Academic Press.

Jonas Gomes, Lucia Darsa, Bruno Costa, & Luiz Velho (1999). *Warping and Morphing of Graphical Objects*, Morgan Kaufmann.

J. E. Goodman and J. O'Rourke, editors (1997). Handbook of Discrete and Computational Geometry, CRC Press LLC, Boca Raton, FL.

J. B. Kruskal & M. Wish (1978). Multidimensional Scaling, Beverly Hills, CA: Sage University Series.

B. Joseph Pine (1992). *Mass Customization: The New Frontier in Business Competition*, Harvard Business School Press.

S. M. Schiffman, L. Reynolds, & F. W. Young (1981). Introduction to Multidimensional Scaling, Academic Press.

G. Wolberg (1990). Digital Image Warping, Los Alamitos, Calif.: IEEE CS Press.

WWW Interface Design for Computerized Service Supporting System

Thu-Hua Liu
Department of Industrial Design, College of Management,
Chang Gung University,
Tao-Yuan, Taiwan, R.O.C

Sheue-Ling Hwang and **Jia-Ching Wang**
Department of Industrial Engineering, National Tsing Hua University,
Hsinchu, Taiwan, R.O.C

1 Introduction

Recent surveys on MIS (Management Information System) issues show that effective use of the data resource is very important. For example, one survey shows that data resource is the second top critical issue [Wiederman et.al. 1991]. On the other hand, the design of effective user interfaces to the database systems has not been vigorously pursued. The major interface language is still SQL, a system that was developed almost 20 years ago [Chamberlain and Boyce 1994] and was found to be very difficult, even for trained users [Date 1990; Greenblatt and Waxman 1978; Welty 1985].

With the widespread availability of computers and data to not only MIS professionals but also increasingly to end users, data access will expectedly remain an important issue, but with some new twists. To avoid any bottle-necks caused by heavy end-user demand on MIS professionals, it has become imperative to provide general-purpose database interfaces that enable end users to perform reliable database retrieval (and occasionally design) on their own [Welty and Stemple 1981]. The user interface provides critical ability for how information in a database is absorbed, accessed, and used, because the interface would directly affect the efficiency when users use the database. If the databases are integrated with the Webs, the interface of database must be built on the Webs. But how can we utilize the characteristics, like HTML, Java and JavaScript etc., of webpages for interface design?

2 Case Analysis

A calibration service company, we call it company A, is chosen for our case system. The workflow of company A and the current interface functions of their database are analyzed. Then, the database and interface are transformed into an OODB with web based interface as shown in Figure 1. Finally, the new system was tested and evaluated. The main functions of the new system are depicted in Figure 2. Different users can access certain functions of the system through the web-based interface based on their designated types and authority.

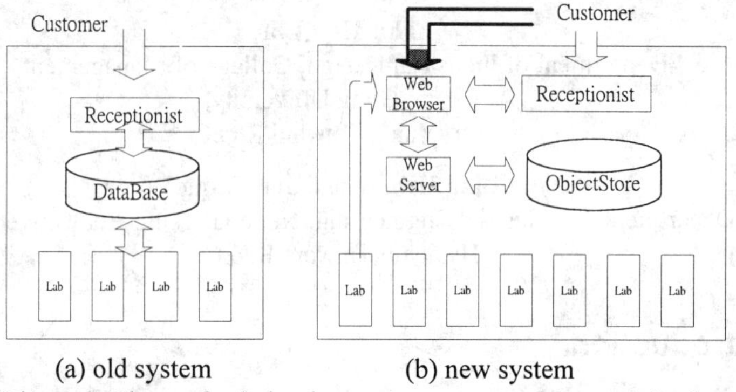

(a) old system (b) new system

Figure 1. Changed relation between customers, labs, and database

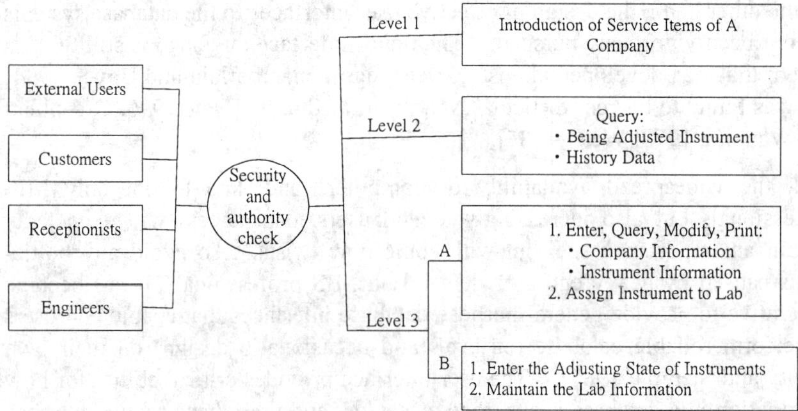

Figure 2. Main functions of the new system

3 Result and Discussion

After completing the user interface on the web for the database system of company A, the evaluation result of the interface needs to be discussed and the advantages and disadvantages about applying the web technology to the user interface are documented.

Evaluation and Result

The guidelines of designing software user interfaces as well as the evaluation result are summarized in Table 1. The outcome of evaluation shows that the interface satisfies most of the requirements. But due to the limitations and characteristics of the Webs, the interface does not satisfy some of the guidelines, such as items 3.3, 6.4, and 9.1, stated in Table 1. The column and row headings can't be maintained when scrolling are used, because the tables are fixed in the WebPages (item 3.3). The WWW browsers just provide the function to decode the documents to different languages, and there is only one type of font for each language (item 6.4). The screen must be changed from one page to another when the users select functions from the interface, hence the number of layers is possible to get over three (item 9.1).

Table 1. The Guideline and evaluation result of User Interface Design

No.	Item	Guideline	Result
1	**Areas**		
1.1	Consistent Location of Areas	Dialogue areas used in the application should be consistently located.	●
1.2	Display Density	Display density of characters in a text display should be less than 40 percent.	●
1.3	Scrolling and Paging	If the amount of information to be displayed exceeds the available area, the user should be provided an easy ways to view portion of the information not currently displayed.	●
2	**Lists**		
2.1	Alphabetic Information	Vertical lists of alphabetic information should be left-aligned.	●
2.2	Numeric Information	Numeric information representing values or quantities should be right-aligned.	●
2.3	Decimal Number	If numeric information contains decimal points, it should be aligned with respect to the decimal point.	–
3	**Tables**		
3.1	List Organization in Table	Tabular information should be arranged such that the material most relevant to the user or with the highest priority is displayed in the left-most column.	●
3.2	Accordance with Paper Forms	If paper forms are used in the task, the information display format and the paper forms should be designed to match each other.	–
3.3	Maintaining Column and Row Headings	Column and row headings should be maintained when paging and scrolling are used with tables that extend beyond the available display area.	×
4	**Labels**		
4.1	Labeling Fields, Items, Icons and Graphs	Fields, items, icons and graphs should be labeled unless their meaning is obvious and can be understood clearly.	●
4.2	Designation	Field labels should explain the content of the fields.	●
4.3	Label Position	Field labels should be consistently located adjacent to (above, or to the left) a displayed field or group.	●
4.4	Label Position for Check Box or Option Buttons	Label for check box or option buttons should be consistently located to the right.	–
4.5	Separation of Labels and Associated Information	Labels and the associated information should be distinguished clearly by e.g. space, delimiters, etc.	●
4.6	Label Format	Labels and fields should be consistently formatted and aligned. If fields are aligned vertically in columns and label lengths vary, labels should be right-justified and fields should be left-justified. If fields are aligned vertically in columns and label lengths do not differ significantly, labels and fields both should be left-justified.	●
4.7	Label for Units of Measurement	The units of measurement for displayed information should be either included in the label or as a part of each information item.	●
5	**Fields**		
5.1	Input Field Format	The format of an input field should be clearly included. This recommendation is most relevant for users not familiar with field formats.	●
5.2	Input Field Length	The length of an constant input field should be clearly indicated (except for input fields designated for security purpose e.g. password). The maximum input field length should be indicated.	●

5.3	Partitioning Long Information Items	Long information items should be partitioned into groups with a specified number of characters and consistently used for entry and display.	–
6	**Font**		
6.1	Type	The font types should be simple, familiar, and easy to read.	●
6.2	Size	There should be 3 different sizes of fonts or less.	●
6.3	Coherence	It should keep the same font type in each screen.	●
6.4	Controllability	The users can change the font type.	×
7	**Button**		
7.1	Shape	The buttons should be rectangular	●
7.2	Position	The buttons should be lined in the lower part of the screen.	●
7.3	Text on the Button	The signification should clearly show its function.	●
7.4	Layout	Buttons of the same assortment should be putted together, and different assortments should be apart.	●
8	**Identify User's ID**		
8.1	Length of Password	The length of password should not be shown out.	●
8.2	Display of Password	The inputted characters should be displayed as "*".	●
8.3	Trial and Error	There should be at least two times of trial and error.	●
9	**Layer**		
9.1	Layers	The layers of each function should not be over 3.	×
9.2	Indication	There should be some indications to show where the users are.	●
9.3	Exit	There should be an exit option to go to the forward or backward layer.	●

●: Satisfy ×: Not Satisfy –: Not Applicable

Pros and Cons of Applying the Web Technology

The advantages of applying the web technology in designing user interface of the calibration service supporting system are as follows:

(1) Quick-Design: Using the webpage editor equipped with handy features, the designers can construct webpages just like producing and formatting digital documents by a word-processing software. By just adding necessary hyperlinks to the pages, the construction of a user interface on WWW is much easier and faster then on other systems.

(2) Easy to Integrate: WebPages are written in HTML format, and some may be mixed with Java and JavaScript. Their working platforms are browsers only. All users can access the same interface under any computer if there is a web browser installed in the computer system. The connection between each department can depend on HTTP as long as the department applies the user interface on WWW.

(3) Commercial Potential: Each company will have its own homepage in the future. Besides that, the continuous improved web technologies will soon have the capability to integrate most of office works through the Webs. Hence, the business dealings between companies can be accomplished through the WWW. The ability and degree of integration of business processes on the Webs will be an important competitive force of the company.

Some drawbacks need to be considered before advanced web related technologies are matured.

(1) Security: The security problem will always be a threat for the applications through Internet. No matter how advanced the web security technology is,

the security problems occur somewhere soon or later.

(2) Transfer Speed: Internet is getting more popular in the world than before, the loads of transferring information become heavier than ever. Unless the shared band width of the internet is expended enough to overcome the slow transfer speed transferring large volume data, communication through Internet will probably still a time consuming task for a big file.

(3) Layer: Because the user interface on WWW is organized with WebPages, the screen must be changed from one to another when users select any function of it. Every time the main screen changes, users would feel that they switch to another layer. Too many layers could make the users to get lost in the system structure.

(4) Limitation: languages, such as HTML, are not enough to handle all the functions of an interface should equipped with. Hopefully, some webpage design languages will be developed with handy plug-in functions to overcome the above shortage. In the near future, a standard language, such as XML, can be fully developed to facilitate the easy webpage design.

4 Conclusion

In this research, not only the WebPages are connected with the database, but also the office work and customer service processes are integrated through the WWW by the web browsers. With the evolving of Internet, the development of web technologies will impact the interface design of software systems that enable new ways of business communications and satisfy unique user preferences.

Reference

Chamberlain, D.D. & Boyce, R.F. (1974). "SEQUEL: A Structured English Query Language," *Proceedings of the 1974 ACM SIGMOD Workshop on Data Description, Access, and Control*, Ann Arbor, Ml, May.

Date, C.J. (1990). *An Introduction to Database Systems*, Volume I, 5th edition, Addison-Wesley, Reading, MA.

Greenblatt, D. & Waxman, J. (1978). "A Study of Three Database Query Languages," in *Databases: Improving Usability and Representativeness*, B. Shneiderman (ed.), Academic Press, New York, NY.

Niederman, F., Brancheau, J.C., & Wetherbe, J.C. (1991). "Information Systems Management Issues for the 1990s," *MIS Quarterly*, 15 (4), pp. 475-500.

Welty, C. & Stemple, D.W. (1981). "Human Factors Comparison of a Procedural and a Non-procedural Query Language," *ACM Transactions on Database Systems*, 6 (4), pp. 626-649.

Welty, C. (1985). "Correcting User Errors in SQL," *International Journal of Man-Machine Studies*, 22 (4), pp. 463-477.

A study on tactile interfaces as an aid to home electronic appliance operation for the visually impaired

Wen-Zhi Chen and Wen-Ko Chiou
Department of Industrial Design
Chang Gung University, Tao-Yuan, Taiwan
E-mail: wenzhi@mail.cgu.edu.tw

1 Introduction

Much of the information human perceives is sensed through the visual organ. When the visual faculty has been impaired or the visual has been mask and no visual information is possible to be perceived, the aided perceiver of information is need. Now, when one looks at the products that are available in the market, the symbols and signs of product operation are almost is visual symbol, that is no help for visually impaired people, and sometime they may also harm themselves. People who lost their vision will have trouble in living, learning, and traveling. However, nowadays' development in auxiliary aid aims mainly at learning and traveling, particularly in information perception -- tactile display for computer user (Shinohara et al. 1998), and seldom aims at living. Therefore, in this study, we intent to help to solve the living problems face by visually impaired people, particularly in how to use home electronic appliance when one's vision is lost.

Tactile and auditory faculties are the two main channels for person has relied on in receiving outside information, since the visual has been impaired or mask. For blind people the use of auditory in products has been of benefit in several areas (Blenkhorn 1994). Making products speak is to make them more accessible to blind people. Tactile information offer some promise of being useful in certain special circumstances in which the visual and auditory senses are overburdened, and definite potential as aids to blind persons (Sanders & McCormick 1987). Chen and Yu (1997, 1998) study the tactile symbols as an aid to the visually impaired for product operation. Results showed that most widely used tactile pattern in controller of radio cassette recorders is of the solid-furrow type, with a smallest average size as compared to all other patterns of tactile symbols that were experimented. The redesigned symbols can significantly reduce the time and error in product operation.

This study will investigate the situation of product usage and use of tactile interface as an aid to help the visually impaired people operate the function of home electronic appliance. The objectives of this study are fivefold. 1) Survey the home electronic appliance that the visually impaired often used. 2) Find the insufficiency and dangerous of the home electronic appliance when used by visually impaired. 3) Investigate the auxiliary interface which original designed for visually impaired. 4) Establish a protocol in auxiliary of tactile interface. 5) Find the best type of tactile interface as an efficient aid.

2 Methods

There will be two stages in this study. In the first stage, a survey consists of 300 visually impaired people will be conducted. Subjects will be asked about the problems they encountered when using the home electronic appliances, including question such as: Is there any auxiliary interface? Ever been injured while operating the product, or Is auxiliary interface helpful? etc. These answers will be helpful to construct an overall operating model. In the second stage, the operation interface of tactility type will be used for testing. The tactile interface will be added to the home electronic appliance, which gives the most problem to the visually impaired people. Thirty people who are visually impaired will be tested against this modified appliance, and their comments and suggestions will be evaluated accordingly.

Twelve subjects were interviewed in the pilot test, which included 8 men and 4 women, with a mean age of 35 years. There were 9 total blind (4 subject born with blind) and 3 amblyopic. The questionnaire includes 4 categories: 1) about the home electronic appliance; 2) operation of home electronic appliance; 3) safety of home electronic appliance; and 4) aided of home electronic appliance.

3 Results and Discussions

3.1 About the home electronic appliance

Table 1 lists the important appliances subjects selected; there are telephone, TV, Walkman, wash machine, rice cooker and drinking fountain. The appliances of leisure are the most important, like the telephone, TV, and Walkman, for visually impaired people. Visually impaired people get information about the product is from broadcast, TV, family and friends. Function, operation, safety, texture, price and distribution channels considered when they buy it (Table 2).

Table 1 The important appliance

Appliance	Freq.	Perc. of cases (%)
Telephone	12	100.0
Television	12	100.0
Walkman	9	75.0
Wash machine	7	58.3
Rice cooker	7	58.3
Drinking fountain	6	50.0
Electronic hob	3	25.0
Audio system	1	8.3

Table 2 Requirements for buying appliance

Requirements	Freq.	Perc. of cases (%)
Operation	8	66.7
Safety	8	66.7
Speech	5	41.7
Tactual	4	33.3
Price	4	33.3
Distribution channel	2	16.7

3.2 Operation of home electronic appliance

Table 3 lists the appliances, which are easiest operation, include Walkman, rice cooker, telephone, TV, wash machine, audio system and drinking fountain. Audio system, telephone, wash machine and microwave oven are need assistance for use. About 58.3% of subjects will use tactual to guess the function and operation of product.

Table 3 Appliances that easiest use

Appliances	Freq.	Perc. of cases (%)
Walkman	7	58.3
Rice cooker	6	50.0
Telephone	5	41.7
TV	4	33.3
Wash machine	3	25.0
Audio system	2	16.7
Drinking fountain	2	16.7

3.3 Safety

Seven subjects have been harmed by use home electronic appliance. Table 4 lists the appliances make harm. The rate of harm is high by use cooking appliance. Because they produce heat. Alarm system can alter the user for safety. Speech, auto-switch, and bee-sound are the alarm method for safety subject preference (Table 5).

Table 4 Appliances that make harm

Appliances	Freq.	Perc. of cases (%)
Microwave oven	2	16.7
Oven	2	16.7
Drinking fountain	1	8.3
Blender	1	8.3
Rice cooker	1	8.3

Table 5 Alarm for safety

Appliances	Freq.	Perc. of cases (%)
Speech	3	25.0
Auto-switch	3	25.0
Bee-sound	2	17.7

Table 6 Appliances unsatisfied of operation

Appliances	Freq.	Perc. of cases (%)
Microwave oven	4	33.3
TV	3	25.0
Wash machine	2	16.7
Electronic hob	2	16.7
Oven	1	8.3
Walkman	1	8.3
Audio system	1	8.3

Table 7 Methods of operation aided

Appliances	Freq.	Perc. of cases (%)
Texture	3	25.0
Form/Shape	3	25.0
Tactile symbol	2	17.7
Braille	2	17.7

3.4 Aided

Unsatisfied on appliance operation by visually impaired people list in Table 6. Microwave oven is most unsatisfied on operation. Appliances need complexity set and process on use, unaccepted by visually impaired people. An aided can improve operation. Appreciate methods of operation aided visually impaired like are use texture, form or shape, tactile symbol, and Braille (Table 7).

Result of pilot test find the leisure appliances (telephone, TV, Walkman) are most need for visually impaired people. Wash machine, rice cooker, and drinking fountain also are importance. Jian (1997) investigated home electronic appliances use at Taiwan, find radio cassette recorder, rice cooker, telephone, TV, wash machine is the most need appliance. The results of both studies, find the need on leisure, food, clothes are height. Heating appliances, like microwave oven and oven, are easiest harm. Appliances can talk, auto-switch are acceptable by visually impaired. Most visually impaired use tactual for recognizes the function of appliance. Texture and form are the preference. About safety, most subject response safety of operation of appliance can increase by learning and practice. But need long time to learn by try and error.

Three electronic substitution techniques that have been shown to give the blind a sense of their surroundings: tactile visual substitution systems, visual cortex neuroprosthesis, and mobility/navigation devices (Shimizu 1992). These techniques have been around for at least two decades without much enthusiasm from the rehabilitation community; but are now being resurrected by a number of research laboratories that are incorporating technological advances into their designs. On the surface, all three techniques world appear to have merit and be feasible; yet, significant problems exist. The diversity of the control characteristics has made it impossible at this time to clearly analyze the shape, size, and surface finish of the controls, let alone make appropriate design recommendations. This problem will be examined further in a follow-up study.

4 Conclusions

In conclusion, many need on live of visually impaired people are unsatisfied. When design the appliance for them, must consider the learning and memories of people, not just consider operation process, function recognize. Some recommendations outlined from pilot test are restrictive from a visually impaired people point of view, rather than prescriptive. Continuous study will design best type of tactile interface as an efficient aid for visually impaired people. In any case, industrial designers are faced with a complex but challenging design problem.

5 References

Blenkhorn, P. (1994). Design products that speak: lessons from talking systems for blind people, *Computing & Control Engineering Journal*, August, 172-176.

Chen, W.Z. & Yu, W.L. (1997). The study on discriminability of tactile symbols of radio cassette recorder controllers, *Journal of Special Education (Taiwan)*, 12, 41-46.

Jian, Q.W. (1997). A study of sense for blind: analysis and application on ergonomic safety design, *Thesis of Master*, Cheng Kung University, Tainan, Taiwan.

Sanders, M.S. & McCormick, E.J. (1987). *Human factors in engineering and design*, Singapore: McGraw-Hill.

Shimizu, Y. (1992). Tactile Sense and Interface Design, *Industrial Design (Japan)*, 157, 56-57.

Shinohara, M., Shimizu, Y. & Mochizuki, A. (1998). Three-dimensional tactile display for the blind, *IEEE Transactions on Rehabilitation Engineering*, 6 (3), 249-256.

Yu, W.L. & Chen W.Z. (1998). A study on tactile symbols as an aid to the visually impaired for product operation, *Journal of the Chinese Institute of Industrial Engineers*, 15 (1), 9-18.

An Ergonomic Design of EMG Signal Controlled HCI for Stroke Patients

Wen-Ko Chiou, Ph.D.[1] and May-Kuen Wong, M.D.[2]
Department of Industrial Design[1] and Department of Physical Therapy[2]
Chang Gung University, Kwei-Shan, Tao-Yuan, Taiwan
Email: wkchiu@mail.cgu.edu.tw

1 Introduction

Stroke is an important cause of death and one of the main causes of morbidity. Many of them will live with significant sensory-motor deficits that will considerably impede their level of functional independence (Capaday and Stein 1986). In fact, between 30% and 60% of people who survive a stroke will be dependent in certain aspects of their daily activities (Olney et al. 1998). The goal for the final rehabilitation is independent walking for stroke patient. Although there are various types of preliminary training before ambulating rehabilitation, upright stepping and the exercise model of walking are probably the most similar (Werning and Muller 1992). Upright stepping devices have become commonly used pieces of exercise equipment in health clubs, rehabilitation clinics, hospitals, and physical therapy practices (Holland et al. 1990, Olney et al. 1998). Spasticity is a complex phenomenon that interferes with motor control. The electromyography (EMG) activity obtained during upright stepping always used to quantify disordered muscle activation patterns in spastic paretic stepping (Fung and Barbeau 1989).

During the active rehabilitation period after a stroke, rehabilitation interventions emphasize stimulation of the recovery of the sensory-motor function of the paretic side (Brask et al. 1992). There had poor feedback to know real-time function for stroke patients on upright stepping devices until now (Obradovich and Woods 1996). The objectives of this study were twofold. (1) Design an EMG signal controlled human-computer interface (HCI) built in upright stepping device. (2) Explore the clinical effect of EMG recognition system on the balance, functional independence, psychological responses and progress of activities of the stroke patients.

2 Methods

EMG activity (using MA-100 System, USA) from the hip adductors (HA) and hamstrings (HS) were detected with disposable bipolar surface electrodes after proper skin preparation, such as shaving, abrading, and rubbing with alcohol to reduce skin impedance. Footswitches placed bilaterally beneath the heel, the head of the fifth metatarsal and the big toe registered the temporal relations of the stepping cycle. The EMG data were normalized to the stepping cycle from one foot-floor contact to the next (0-100%) and averaged for each 0.4% interval of the cycle. A stepping index, *I*, derived from the EMG activity obtained during upright stepping in normal subjects, is therefore proposed as a functionally relevant measurement of spasticity in locomotion. The predetermined windows gave rise to 3 bins (b_1, b_2, b_3) of on-off activity for each muscle. Integrated area under the profile (a_1, a_2, a_3) could be calculated for each corresponding bin. The proposed *Spastic Stepping Disorder Index*, *I*, defined as the ratio of the integrated amplitude area in the 'off' bin(s) to that in the 'on' bin(s), i.e., $I=a_2/(a_1+a_3)$ for HS (Figure 1), and $I=(a_1+a_3)/a_2$ for HA (Figure 2). *I* represent the degree of abnormal activation of locomotor muscles from their normally relaxed state in a defined phase of the stepping cycle, as compared to the total recruitment in the active phase. The index, *I*, was calculated for each muscle in each of the 30 normal subjects and 25 stroke patients.

A new HCI based on a real-time EMG recognition system is developed. A personal computer with a plug-in data acquisition and processing board (dSPACE DS1102) containing a TMS320 C31 floating-point digital signal processor are used to attain real-time EMG classification. The recognition results are used as control commands of the HCI. This system compatible with Microsoft mouse can move the cursor in four directions and click the icon in GUI operating systems. Two channel EMG signals are collected by two pairs of surface electrodes located on the HA and HS. The integrated EMG is employed to detect the onset of muscle contraction. The cepstral coefficients, which are derived from auto-regressive coefficients and estimated by a recursive least square algorithm, are used as the recognition feature. These features are then identified using a modified maximum likelihood distance classifier with a rejecting rule.

In clinical validation section, fifty stroke patients who were referred to the rehabilitation center of Chang Gung Memorial Hospital. The EMG recognition system built in upright stepping device was implemented on experimental group (n=25) whereas control group (n=25) received conventional upright stepping treatment over a 1-year period. The dependent variables measured were of 3 types: (1) balance (14 items, score range from 0 to 56), (2) functional independence (13 items, score range from 13 to 91), and (3) psychological negative responses (22 items, score range from 22 to 110).

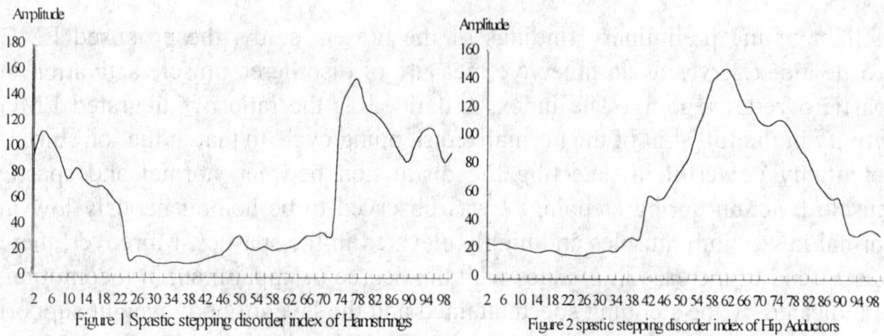

Figure 1 Spastic stepping disorder index of Hamstrings

Figure 2 spastic stepping disorder index of Hip Adductors

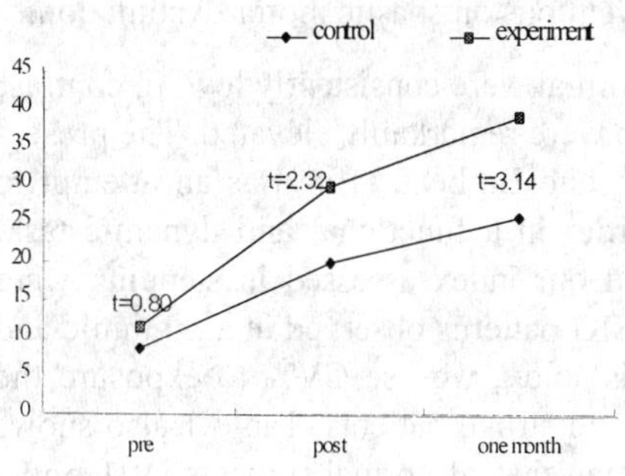

Figure 3 Clinical effect of EMG recognition system on balance

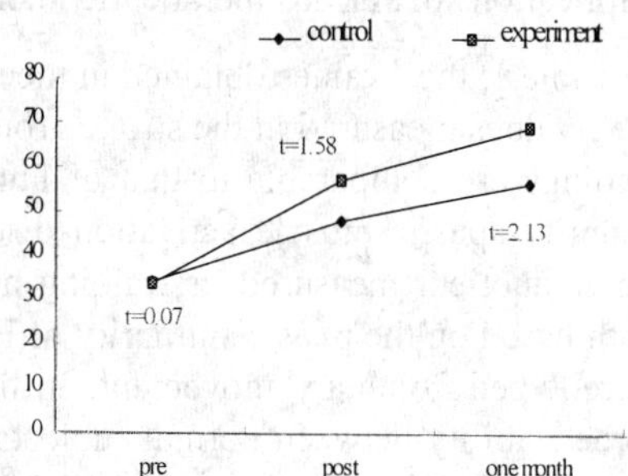

Figure 4 Clinical effect of EMG recognition system on independence

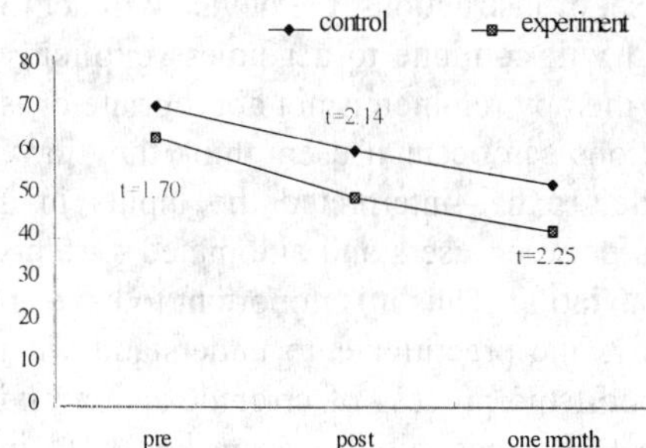

Figure 5 Clinical effect of EMG recognition system on psychology

Table 1 The means of stepping index (I) and leg force (CV%)

Lower Extremities	Normal	Patient
Hamstring muscles*	0.24	0.77
Hip Adductor muscles*	0.23	0.71
Left leg stepping*	22.4	82.6
Right leg stepping*	18.6	79.5

*: t-test of two groups, P<0.001

3 Results and Discussions

In light of the preliminary findings of the present study, the proposed EMG profile index serves as an objective measure of disordered muscle activation in spastic paretic stepping. The index, *I*, defined as the ratio of integrated EMG activity in the 'off' bin of the normalized stepping cycle to that in the 'on' bin, is potentially powerful in detecting the distinction between normal and spastic muscle function during stepping. *I* was observed to be homogeneously low in normal lower limb muscles and highly elevated in the spastics. Moreover, there was a trend of increase in *I* values with the degree of impairment in locomotion, as indicated by the stepping speed attained and the level of body weight support required. This preliminary study paves the way for a future full-scale study to validate the use of the index in quantifying spastic motor disorder in stepping. The potential application is broad, in that the index can become a great adjunct for purposes of classification, evaluation, and assessment of functional implications of various therapeutic interventions on spastic motor dysfunction.

In Table 1, the *I* values obtained in the normal were consistently low. In contrast, the *I* values measured in the stroke group were remarkably elevated. The present findings are comparable to that of Fung and Barbeau (1989) as an attempt to quantify spastic muscle activation disorder in a functional and dynamic task. Their quotient measured in walking and our index assessed in stepping were both based on the phasic similarity of EMG patterns observed in a rhythmic and stereotyped voluntary movement. In this study, we use CV% to exposure the force stability between normal subjects and stroke patients. Table 1 also shows that stroke patients were greater CV% than that of normal subjects in stepping evaluation.

Characterizing the problems and deficiencies in device use in context points to new design concepts, such as ways to make the current displays and control sequences more usable. However, understanding HCI in context can point to deeper implications for redesign. One of these is the question of how to provide more effective feedback about device status and activities. The upright stepping device is an automated system. Given a set of instructions, the device will act to implement the programmed therapy and will continue to act unless explicitly instructed to stop or change, even if that therapy regimen is not appropriate or is not what was intended. Miscommunications can occur if users think they have communicated one intention but the device has interpreted the inputs in a different way. Such miscommunications between users and automated systems have been on contributor to accidents in aviation. Thus it is important to have an effective feedback mechanism that allows the practitioner to understand what the automation is actually doing to support the process of error detection and recovery. After redesign procedure, the EMG recognition system had built in upright stepping device to validate clinical effects.

The final rehabilitation evaluation program consisted of 25 stroke subjects and 25 comparison patients. There were no significant differences in demographic data and three types of dependent variables in pretest period between two groups. In balance evaluation (Figure 3), experimental group presented better improvement of balance than control group in posttest and one-month after outpatient ($t=2.32$, $P<0.05$; $t=3.14$, $P<0.01$). In functional independence evaluation (Figure 4), there were no significant differences in posttest period between two groups ($t=1.58$, $P>0.05$), experimental group presented better improvement than control group in one-month after outpatient ($t=2.13$, $P<0.05$). In psychological responses evaluation (Figure 5), experimental group presented better improvement of psychological negative responses than control group in posttest and one-month after outpatient ($t=2.14$, $P<0.05$; $t=2.25$, $P<0.05$).

4 Conclusions

This study suggested that EMG recognition system built in upright stepping device could enhance the improvement of balance, functional independence, and psychological negative responses for stroke patients. Further research can be verified that of a useful medical instrument in a long-term rehabilitation period (three-month, six-month, and one-year) and other kinds of patients.

5 References

Brask, B., Lueke, R.H. & Soderberg, G.L. (1992). Electromyographic analysis of selected muscles during the lateral step-up exercise. *Physical Therapy*, 64, 324-329.

Capaday, C. & Stein, R.B. (1986). Amplitude modulation of the soleus H-reflex in the human during walking and standing. *Journal of Neurosciences*, 6, 1308-1313.

Fung, J. & Barbeau, H. (1989). A dynamic EMG profile index to quantify muscle activation disorder in spastic paretic gait. *Electroencephalography and Clinical Neurophysiology*, 73, 233-244.

Holland, G.I., Hoffman, W.V., Mayer, M. & Caston, A. (1990). Treadmill vs. steptreadmill ergometry. *Physical Sportmedicine*, 18, 79-85.

Obradovich, J.H. & Woods, D.D. (1996). Users as designers: how people cope with poor HCI design in computer-based medical devices. *Human Factors*, 38, 574-592.

Olney, S.J., Griffin, M.P. & McBride, I.D. (1998). Multivariate examination of data from gait analysis of persons with stroke. *Physical Therapy*, 78, 814-828.

Visintin, M. & Barbeau, H. (1994). The effects of body weight support on the locomotor pattern of spastic paretic patient. *Paraplegia*, 32, 540-553.

Werning, A. & Muller, S. (1992). Laufband locomotor with body weight support improved walking in persons with server spinal cord injuries. *Paraplegia*, 30, 229-238.

A Comparison of Multi-modal Combination Modes for The Map System*

Gao Zhang, Xiangshi Ren, Guozhong Dai
Software Institute of Chinese Academy of Sciences, Beijing, China
Department of Information and Communication Engineering, Tokyo Denki University

1 Introduction

The map system is usually used in public places, it needs more convenient user interface. Some multi-modal interactive methods, such as spoken natural language, gesture, handwriting, etc. have been introduced into the map systems to produce more natural and intuitive user interfaces(Oviatt 1996; Adam Cheyer and Luc Julia 1996).

There is still no persuasive proof, especially the quantitative comparison, to certify that the multi-modal user interface is more natural and efficient than single-modal user interface in the domain of map applications. Although research of multi-modal interface has been around for a long time, the choice of included modalities and their combinations is usually made without any quantitative consideration.

Many studies (Bekker, M.M., Nes, F.L.van, and Juola, J.F 1995; Ren, X., and Moriya, S 1997) have showed that evaluation experiment could help to solve this problem. The evaluation experiment makes it is possible to do quantitative analysis based on the statistical data of the result and get convictive conclusion about the optimal modality combination mode.

We set up a prototype multimodal map system, where users can use pen, spoken natural language, keyboard and mouse to accomplish some trip plan tasks. For

* This paper is supported by the key project of National Natural Science Foundation of China (No.69433020) and key project of China 863 Advanced Technology Plan (No. 863-306-ZD-11-5)

this environment, we designed two experiments to investigate the differences of different combination modes

2 Multi-modal Map System in Experiment Environment

The experiment environment is based on a prototype trip plan system. Users can use voice (spoken natural language, Chinese), handwriting (handwriting of Chinese character), pen-based gesture (Drawing Graphics by pen), in addition to traditional mouse and keyboard modes (typing, picking and dragging), to get the necessary information in planing their travel routes.

There are 4 classes of tasks in the map system: Distance calculation; Object location; Filtering; Information Retrieval. All these tasks can be accomplished by multiple modality combination modes.

3 Method

3.1 ANOVA (ANalysis Of VAriance between groups)

ANOVA tells us if the variation of multiple groups is significant by putting all the data into one number (F) and providing one P for the null hypothesis.

F=(found variation of the group averages)/(expected variation of the group averages). The number of degrees of freedom for the numerator is (k-1), and the number of degrees of freedom for the denominator (k(n-1)).

How big should F be before we reject the null hypothesis? P reports the significance level. Generally, if $P<0.05$, we can reject the null hypothesis.

3.2 Subjects

Twenty-four subjects (12 male, 12 female, all right handed; 12 students, 12 workers) were tested for the experiment. Their ages ranged from 20 to 35. All of them have no experience in using this kind of trip plan system.

3.3 Procedure

In order to pay more attention on the often-used combination modes, we excluded those seldom-used combinations modes by an experiment at first. We called this experiment as Selection Experiment.

In the Selection Experiment, each of the subjects had 30 minutes to learn how to use the input equipment (mouse, keyboard, pen and voice) to accomplish trip plan tasks. After they were familiar with the experiment environment, the Selection Experiment began. The system provided 10 minutes for each class of task. The 15 possible combination modes were given to subject to perform circularly. The subjects were asked to perform the task as soon as possible by use of the appointed combination mode. If the performance is successful, the combination mode is recorded by a hidden Task Agent. If the performance take a too long time (about 20seconds), system will overleap this combination as a failed combination. The rate of successful times of each combination method from the total successful times was calculated and recorded.

Based on the result of the Selection Experiment, we performed another experiment to compare the difference of different combination modes. We called this experiment as Comparison Experiment.

The subjects were asked to accomplish 24 sample tasks for each combination modes selected, i.e., keyboard, pen, voice, mouse + voice, keyboard + voice and pen + voice. And in the test of each mode, the sample tasks were given out randomly.

Data for each combination mode was recorded automatically as follows. Manipulation Time: the time taken to finish all the samples. This is the time lapsed from the beginning of the first task to the end of the last one. Subject Preference: The subjects were questioned about their preferences after they finished testing each mode of combination. They were asked to rank (on a scale of 1-10) the mode just tested according to their satisfaction and desires to use.

4. Results and Discussion

4.1 Useful combination modes for map-based system

From the Selection Experiment, we can analyze useful combination modes for accomplishing the trip plan tasks. The mean successful rate is 1/14= 7%, so we treated those combination with successful rate above 7% as useful combination modes. The statistic result of showed that: In single-modality mode, the successful rate of using mouse only is 1%, keyboard rate is 10%, and the rates of voice and pen are 15% and 12% respectively. We can find that the accurate interactive tool, such as mouse, is not suitable for the trip plan tasks.

In bi-modality mode, the successful rates of mouse + voice, keyboard + voice and voice + pen are 13%, 11% and 18%. And the successful rates of mouse + keyboard, mouse + pen, keyboard + pen are 5%, 3% and 5%. We can find voice play an important assistant role in trip plan tasks. The tri-modality modes are seldom be used with the successful rates less than 3% (see figure 1).

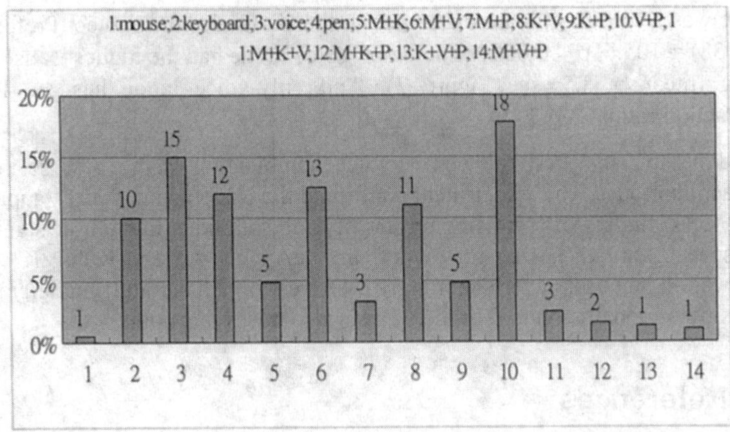

Figure 1 Successful Rates of Different Combination Modes in Selection

An ANOVA (analysis of variance) was conducted on the result of Comparison Experiment. The manipulation efficiency and subjective preference differences of these multi-modal interactive modes were compared.

4.2 Pen + Voice mode interface

Significant differences in the six multi-modal combination modes were found, in mean manipulation time, $F(5,138) = 105.6$, $p < 0.0001$. The result revealed the pen + voice combination was faster than the other five modes in total time (mean manipulation time of pen + voice is 6.8 minutes, see Figure 2). On the other hand, the only-pen interface was the slowest among the six combination modes with mean manipulation time of 9.1 minutes.

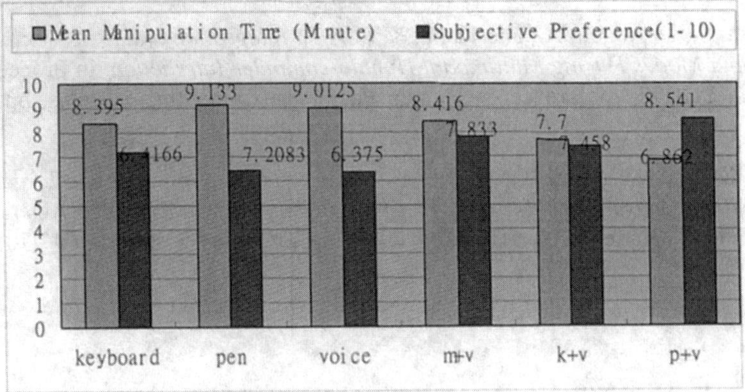

Figure 2 Mean Manipulation Time and Subjective Preference Comparison of Six Modality Combinatio

There was also a significant difference in the six modes for Subject Preference, $F(5,138) = 105.6$, $p < 0.0001$. The pen + voice mode had the highest satisfaction rating (mean = 8.5, see Figure 2). And only-voice mode has the lowest satisfaction rating of 6.3.

Based on the analyses, the pen + voice combination is the best of the six interaction modes. This experiment confirmed the Hauptmann *et al.* (Hauptmann A.G. 1989) finding of surprising uniformity and simplicity in the user's gestures and speech, and Oviatt *et al.* (Oviatt, S., DeAngeli, A., and Kuhn, K. 1997) observation that users overwhelmingly preferred to interact multi-modally rather than single-modally. Nitta (Nitta, T. 1995) also raises this point.

5 References

Adam Cheyer and Luc Julia (1996), *Multimodal Maps: An Agent-based Approach; SRI International,* 1996, http:// www.ai.sri.com/ ~cheyer/papers /mmap/ mmap.html.

Bekker, M.M., Nes, F.L.van, and Juola, J.F(1995), *A Comparison of Mouse and Speech Input Control of A Text-annotation System*, Behaviour & Information Technology, Vol.14, No.1, 14-22,

Hauptmann A.G. (1989), *Speech and Gestures for Graphic Image Manipulation,* In Proceeding of the CHI'89 Conference on Human Factors in Computing Systems, 241-245.

Nitta, T. (1995), *From GUI towards Multimodal UI (MUI) (in Japanese)*, Information Processing Society of Japan, Vol. 36, No.11, 1039-1046.

Oviatt (1996), *Multimodal Interfaces for Dynamic Interactive Maps.* In Proc. CHI '96, (Vancouver), 95-102.

Oviatt, S., DeAngeli, A., and Kuhn, K. (1997), *Integration and Synchronization of Input Modes During Multimodal Human-computer Interaction,* In Proceeding of the CHI'97 Conference on Human Factors in Computing Systems, pp.415-422, 1997.

Ren, X., and Moriya, S. (1997), *The Strategy for Selecting a Minute Target and the Minute Maximum Value on a Pen-based Computer*, In Extended Abstract of the CHI'97 Conference on Human Factors in Computing Systems, 369-370,

Multimodal Human-Computer Communication in Technical Applications

Hermann Hienz, Jörg Marrenbach, Roland Steffan, and Suat Akyol

Department of Technical Computer Science
Aachen University of Technology
Ahornstr. 55, D-52074 Aachen, Germany
Phone: +49-241-8026105 Fax: +49-241-8888308
Email: [hienz, marrenbach, steffan, akyol]@techinfo.rwth-aachen.de

1 Introduction

Human-computer interfaces represent the borderline between the computer and the user. People naturally communicate with other people and their environment by means of different modalities, e.g., speech, gestures, touch, and mimic. Therefore, the development of advanced human-computer interfaces, which are more natural, efficient and intuitive is worthwhile (Bolt 1984). In our department a research group is concerned with the development and evaluation of multimodal human-computer interfaces in technical applications. Where nowadays the most popular input modalities relies on keyboard and mouse, future interfaces will benefit from natural communication modes.

According to develop advanced multimodal human-computer interfaces, research is focussed on the following topics:

- analysis of natural communication procedure
- selection of communication modality and its context-dependent use
- suitable combination of input as well as output modalities

Multimodal human-computer communication might be appropriate for operating a wide range of applications, e.g., multimodal information browsing. Regarding technical applications, such as machine tools, it will be advantageous if the user can operate the machine in a hands-free and eyes-free manner instead of using a keyboard- and monitor-based input system. A multimodal input device can be a useful add-on or even a substitute for common input devices. It creates new functionality and comfort for controlling technical systems.

2 System Architecture

The system architecture for multimodal human-machine communication is depicted in figure 1.

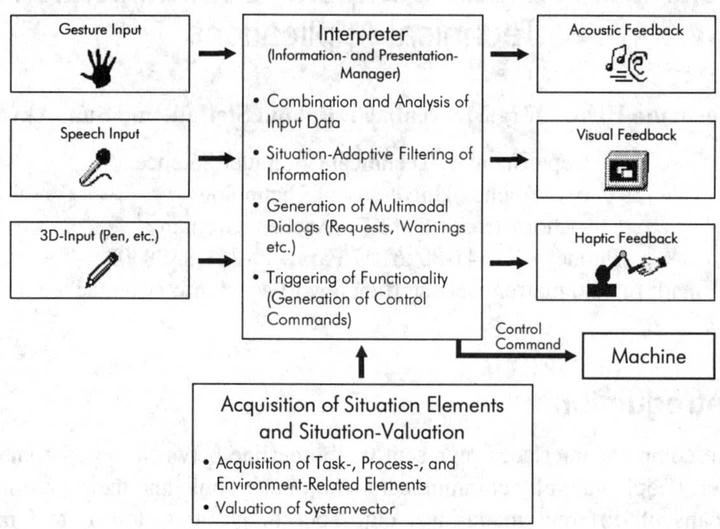

Figure 1: System architecture of multimodal human-machine communication

The system is composed of four major components: the multimodal input unit, the interpreter, the acquisition of situation elements and its valuation, and finally the output unit. On the input side data from different modes, such as, speech, gesture, and 3D-input (e.g. pen-based magnetic or optic sensor) are collected. Automatic acquisition of situation elements plays a major role within the system. This component is able to record task-, process- and environment-related elements by using different sensors. The situation-valuation module aims to classify the situation-dependent system-vector to a certain class of situation. For this reason a comparison with reference classes, which are stored in a database, is carried out. The main task of the interpretation module represents situation-adaptive filtering of information. This is carried out in a way that the system presents only relevant information to the user. Furthermore, situation-adaptive dialogs (e.g. system warnings and requests) are generated, which improves the usability compared to simple command & control systems. Finally, the interpreter is responsible for triggering the functionality of the system by generating commands for controlling the machine. The task of the output unit is to provide multimodal feedback to the user as well as sending control commands to the machine. User's feedback is based on acoustic, visual, and haptic components.

The described architecture offers the opportunity to use a different modality if a single input or output channel can not adequately be recognised and interpreted by the system and the user, respectively. As a consequence operating errors can either be avoided or corrected in an easier way.

3 Input Modalities

In addition to conventional input modalities, the use of gestures and speech is encouraged. This section describes implementation details regarding these new input channels.

3.1 Gesture-Based Input

Gestures as part of a multimodal input system can be classified as body-related and non-body-related (Akyol and Hienz 1999). Body-related gestures are performed at predefined positions, while non-body-related are performed at positions that depend on the actual context. The purpose of a gesture can be selection or control, while non-body-related gestures can be additionally divided into qualitative and quantitative control. The taxonomy of gestures described above is shown in table 1.

Gesture Type	Purpose
Non-body-related	Selection Qualitive control Quantitive control
Body-related	Selection Control

Table 1: Gesture taxonomy

According to this taxonomy examples for gestures in multimodal task-processing can be given. The non-body-related case, for example, is a pointing-gesture to select an object followed by a speech command to determine what to do with it. The body-related case is covered by examples like the "telephone-gesture" with the phone-number input by speech or buttons.

Some aspects of human gestures, that have to be regarded for the design of the gesture-recogniser, are the following: the meaning of gestures is context dependent; cultural and personal aspects belong to the context, gestures are often used in combination with other modalities; gestures are not always done intentionally and can be performed with different bodyparts. In the present approach, some restrictions have to be stated in order to consider those aspects. First, the context must be defined with respect to the technical system to be controlled. Second, to detect the users intention, a set of gestures and a gesturing area must be defined.

3.2 Speech-Based Input

The speech recognition system used in this research project is a speaker-independent recogniser for continuous german speech. Keyword spotting is used to pick out the words which are part of the vocabulary. Importance is attached to offer users the opportunity to employ a wide variety of keywords to trigger functions and to enter data in their own personal, flexible way rather than by following a rigid, prescribed procedure. It should be stated that a push-to-talk button is implemented to activate speech recognition. This approach is required in order to avoid misclassification when speech input is not used. Moreover, the influence of noisy environments is manageable.

3.3 Requirements of Gesture- and Speech-Based Input Modalities

With regards to the purpose and the restrictions specified above, the requirements of the input unit, which make use of speech and gestures, can be stated as follows:

- user independent recognition
- no calibration requirements
- environment independent recognition
- realtime recognition
- gesture-vocabulary of 10 to 15 hand-gestures that are common knowledge or can easily be learned

4 Output Modalities

Due to the increased complexity of technical systems it is necessary to provide task-dependent information to the user. Furthermore, the use of gesture and speech requires new concepts of feedback.

Utilisation of different feedback systems permits the stimulation of several human senses, especially visual, acoustic and haptic sensation. The idea of multisensory data presentation is to provide information and situation adaptive stimuli in a way that data can intuitively be accessed and understood by users (Steffan, Kuhlen, and Loock 1998). Redundant coding of information, e.g., visual and acoustic display of machinery status, can increase presentation efficiency, especially when user's communication attention varies between different senses. Moreover, substitution of senses stimuli, e.g., expressing acoustic information by means of visual signals, is indispensable if single senses are blocked, for example acoustic channel in loud environments.

5 System Evaluation

For the evaluation of usability of the multimodal human-computer interface, three criteria are considered as essential: effectiveness, efficiency, and satisfaction. These criteria determine the general evaluation framework for the investigation of man-machine systems.

For system evaluation normative user models are utilised. They describe a normalised user with usual knowledge about the devices and its purpose. These models are built up using the keystroke modelling technique and the GOMS analysis (Marrenbach and Leuker 1998). The evaluation of the device can be done by a simulated use via the user model and by tests with prospective users. Action logs are generated automatically and conclusions concerning the operability of the prototype can be drawn, e.g., errors and operating time. Furthermore, the comparison of the normative user models and the action logs of the test user supply predicates about the effectiveness and the efficiency of the device with respect to the different modalities. Additionally, a comparison between the multimodal and conventional system must be carried out in order to evaluate the advantages of the new interface.

6 Conclusion and Future Work

In this paper we suggested a multimodal human-computer interface as an alternative to existent systems. The interface is composed of four components: the multimodal input unit, the interpreter, the acquisition of situation elements and its valuation, and finally the output unit. The system is based on natural communication modes, e.g., gestures and speech, and offers different ways of user feedback, especially acoustic, visual, and haptic.

Future development will concentrate on the realization of a technical application which will be evaluated analytically as well as empirically.

7 References

Akyol, S. & H. Hienz (1999). Controlling Technical Systems with Gestures. In Kraiss, K.-F. (Eds.): *Bi-Annual Report – Department of Technical Computer Science*, Shaker-Verlag Aachen.

Bolt, R.A. (1984). The Human Interface – Where People and Computers Meet. Belmont (USA). Lifetime Learning Publications.

Marrenbach, J. & S. Leuker (1998). Konzept zur Evaluierung technischer Systeme in der Entwicklungsphase. In: *ITG-Fachbericht: Technik für den Menschen. Gestaltung und Einsatz benutzerfreundlicher Produkte*, VDE-Verlag, Berlin, Bd. 154, S. 103-112.

Steffan, R., T. Kuhlen & A. Loock (1998). A Virtual Workplace including a Multimodal Human Computer Interface for Interactive Assembly Planning. In Kopacek, P. (Eds.): *Proc. of the IEEE Int. Conference on Intelligent Engineering Systems INES '98*, September 17-19, Vienna (Austria), pp. 145-150, IEEE Press.

Extracting Baby-Boomers' Future Expectations by the Evaluation Grid Method through E-Mail

Junichiro Sanui* and Masato Ujigawa**

* Product & Technology Planning Department, Nissan Motor Co., Ltd.
6-17-1, Ginza, Chuo-ku, Tokyo, 104-8023, Japan.
E-mail: j-sanui@mail.nissan.co.jp

** Research & Development Institute, Takenaka Corporation.
1-5-1, Otsuka, Inzei-shi, Chiba, 270-1395, Japan.
E-mail: masato ujigawa@takenaka.co.jp

1 INTRODUCTION

At HCI-97, we introduced an e-mail version of our research method called Evaluation Grid Method (EGM) for eliciting users' preference structure and its first empirical work on the gas station preferences among drivers (Sanui and Maruyama 1997), which produced insightful results of satisfactory quality in light of our past experiences with the conventional EGM by face-to-face interviews. There were, however, a couple of shortcomings attributable to the difference in the communication mode. For instance, typing all the statements was troublesome to some participants, and subtle feelings were hard not only to express but also to grasp by e-mail which limits communication to literally verbal.

This paper describes the second pilot work by the EGM through e-mail with the following purposes:

1. To test whether the method is useful for future expectations beyond existing products services that have been studied so far;
2. To develop a research procedure to minimize disadvantages of the media, i.e. e-mail; and,
3. To quantify the efficiencies of the present method.

The topic chosen for the first purpose was the future life expectations of the baby-boomers'. There have been plenty of publications concerning the aging society of Japan, their predictions and proposals are mostly based on the stereotyped conceptions of the aged, i.e., becoming less able, less mobile, less out-oriented and so forth. This seems, however, hardly applicable to the baby-boomers who will enter the aged population in ten years time, since they are the generation who have actively shaped a new way of life for decades. Therefore, we believe it quite meaningful to put into light what they themselves expect about their own future.

2 Research Methods

In carrying out the research, a two-step e-mail procedure was employed to help clarify the vague present issue and to attenuate the disadvantages in e-mail, which is our second purpose. The first step started with a priming question that would help the participants to envision their future. They named two acquaintances of an older generation with approximately 10 years difference in age, and evaluated their way of life in terms of desirability. Then, the participants were asked to enlist the conditions and factors for foreseeable happy life of their own ten years later. The returned responses were edited for the questionnaire in the second step to inquire why the participant had mentioned such conditions and how he/she thinks the conditions could be realized. In addition, the questionnaire requested to estimate how likely each condition be realized and to list what the participant has done for its realization. The number of questions was kept to minimum in order to lighten participants' mental load and to better communicate.

As total, there were 19 participants (male:13, female:6, average ages:48.7). The data collection was carried out in October-December, 1998.

3 Results

Before reporting the results, we should mention the generality of our findings. There were only 19 people selected for the survey through our social-network. Although arbitrariness in sampling is undeniable, they were very cooperative to produce valuable outcome which was organized into the hierarchical links of goals (i.e., desires) and means.

Set aside the sample limitation, we gained confidence in the present approach on the topic hard to crystallize. The hierarchical structure of baby-boomers' future expectations obtained is shown in Fig.1. Of interest were some common desires shared by the majority of the participants such as "keeping social networks either by means of job or by associating with many friends", "Keeping youthfulness in

Fig.1 Obtained structure of baby-boomers' future expectations.

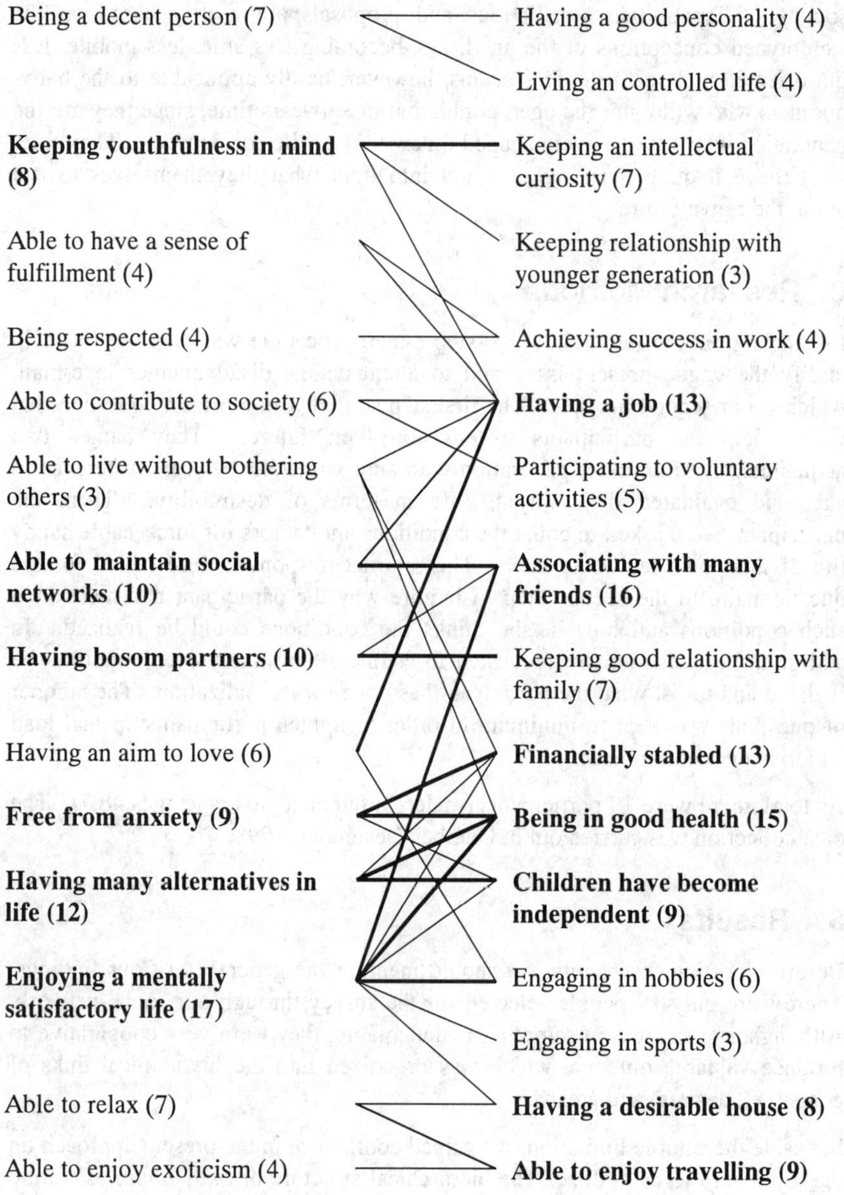

Note-1 : Figures indicate the number of participants who hold the expectation.
Note-2 : Major expectations and relations are indicated by thicker letters/lines.

mind", "having a desirable house", "able to enjoy travelling", "engaging in hobbies" and so on, besides the fundamental desires concerning health, financial stability and independence of children. We are planning to conduct large-scale survey on the generality and importance of these found desires in a fixed format questionnaire.

This outcome containing rich information could be of an support for the validity of the simplified question format that were developed to meet the second purpose of this research.

Concerning the third purpose, we measured the average time spent for composing e-mails for the first and second step. It took on the average 32 minutes as total per participant. The other measure was the number of days used to complete the two-step survey. It took on the average 7.5 days per participant. Efficiency in this regard would increase with the improvement of reminder mails and other means.

4 Conclusions

Our two pilot studies seem to indicate the practical feasibility of the EGM by e-mail in the computer-network era in which the global community is being restructured at an unprecedented rate. People have started using computers as a vital means of wide communication. E-mail or more advanced technique will soon become an everyday tool. The shortcoming in sampling should not cloud the value of our effort when there is no established theory or procedure for sampling network users.

5 References

Sanui J. and Maruyama G. (1997) Revealing of preference structure by the Evaluation Grid Method. In Salvendy, G. Smith, M. & Koubek, R. (Eds.): *Design of Computing Systems: Cognitive Considerations (Vol. 1), Proc. 7th Int. Conference on Human-Computer Interaction* (HCI International '97, San Francisco, USA, August 24-29, 1997), Vol.2, pp.471-474. Amsterdam: Elsevier.

Genetic Algorithm (GA) as a Means of Ambiguous Decision Support by AHP

MATSUDA, Noriyuki
IPPS, University of Tsukuba, Japan

1 Introduction

AHP (Analytic Hierarchy Process), proposed by Saaty (1980), is one of the most popular decision support tool in many areas. It derives relative priorities for a given alternative set from the weights accorded to the hierarchically arranged decision items. Its user-friendliness stems from the localized evaluations within a hierarchy, since our natural judgments are mostly *kanseic* rather than rational. The contrast does not mean that *kansei* is irrational. Instead, they both constitute human intelligence, but differ in orientation as follows (Matsuda, 1997, 1999):

Rationality: Intelligent capacity oriented toward unambiguity, precision, rigor, consistency,...

Kansei: Intelligent capacity that allows partial deviations from such standards.

In short, *kansei* enables us to live with practical efficiency when relevant information is partially available and/or optimal procedures are not completely known in the ever changing environments. Within reasonable limits, however, *kansei* leads us to attain, to a satisfactory extent, what rationality strives for.

Taking advantage of the hierarchical structure of AHP, its user can concentrate on localized or partitioned judgement, leaving systematic information integration to the tool. Nevertheless, we believe that the standard procedure of pairwise comparisons are too fragmented and repetitive for the user to maintain consistency even in the partitioned classes. As a solution, we suggest simultaneous multiple comparisons, making use of interactiveness of graphical computer interface (Matsuda, 1999).

To further enhance the practicality of AHP, we propose here an application of GA (Genetic Algorithm) in expectation of reaching a set of priorities among awfully many possibilities arising from ambiguities inherent in ordinary judgment. A procedural tool that bridges between GA and AHP is called a wave model that generates upper- and lower-boundaries for each judged value of the user's. GA is expected to lead us to find a prominent alternative out of numerous possibilities.

2 Method

The AHP structure in Figure 1 will be used in the following explanations for illustrative purposes.

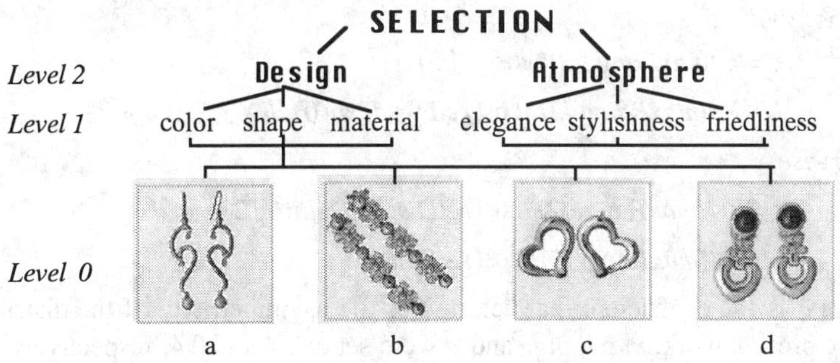

Figure 1. Evaluation structure by AHP

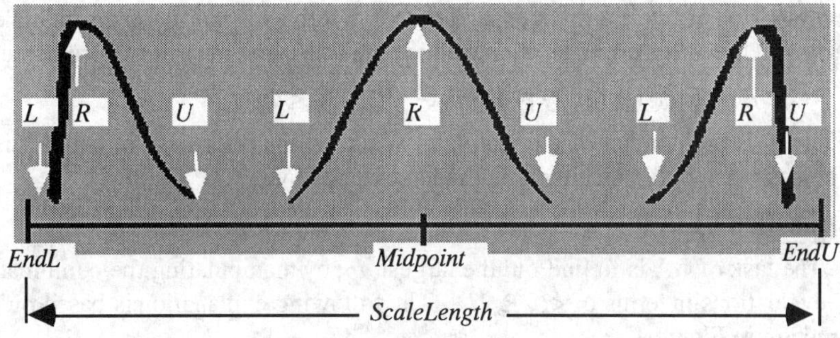

Figure 2. Sketches of the wave model

Wave model

The model (see Gärdenfors & Sahlin ,1983) is likened to waves hitting wharfs on both sides of a scale (Figure 2). The height of the distribution may reperesent the probability, the degree of confidence, or a fuzzy membership associated with a value on the scale.

For the ease of presentation, we let *R*, *L*, and *U* denote the response value, lower and upper ends of the associated interval, respectively. The distribution under the model is symmetrical when the response value (R) falls on the middle of a scale, but becomes positively (or negatively) skewed toward the lower (or upper) end of the scale. Also, its range (U-L) reaches maximum when R is on the middle of the scale. The range decreases monotonously, as R nears the scale ends. Cos() seems to be a natural choice for such function, being simple and mildly nonlinear. The argument of cos() is the distance (*relD*) of R from the midpoint of a scale, *Midpoint*, relative to the half of the scale length, *ScaleLength*.

The following are the formulae for computing U and L from R:

if $R \leq$ *Midpoint*

$$L = R - (maxRange/2) * \cos(relD)$$

$$U = R + (maxRange/2) * \cos(relD) + a * sin(relD)$$

otherwise

$$L = R - \{(maxRange/2) * \cos(relDist) + a * sin(relD)\}$$

$$U = R + (maxRange/2) * \cos(relDist)$$

where a is the coefficient that determines the asymmetricity of the distribution. In the present work, *maxRange* and a were set at 1.4 and 0.4, respectively.

Degree of prudence (dp)

AHP priorities with small differences leave the user indecisive. If, in contrast, one or two alternatives are prudent in terms of priority, decision making is easy. Under this premise, we introduce the following measure of prudence:

$$dp = (\max(\{\text{priority}\}) - 1/N)/(1 - 1/N), \quad 0 \leq dp \leq 1$$

where N is the number of alternatives. When all the priorities are equal it is 0, whereas when only one alternative receives a positive.

Estimation by GA

The task of GA is to find out the largest *dp*, by manipulating the combination of evaluations in terms of <*L*, *R*, *U*>. The following explanation is based on the structure in Figure 1.

A chromosome was created at each level to confine crossover within a level:

Level 2: * *

Level 1: *** ***

Level 0: ****,****,**** ****,****,****

where * marks a locus. Items at Level 2 are separated by a space, which is used as a delimiter among the corresponding items or alternatives at lower levels. At level 0, alternatives are to be arranged horizontally. Total number of assignment of <*L*, *R*, *U*> to each locus amounts to 3^{32} possibilities.

The first generation consists of individuals whose genes are created by biased- and unbiased-coding. The former is favors particular alternatives, while the latter has no such favoritism.

1) Biased coding: The coding is unaffected by the judgment data at Level 0, but dependendent on them at higher levels.

At Level 0, every alternative has its own gene in which *U* is assigned to the corresponding loci , while *L* is assigned to the remaining loci (see Table 1).

Table 1 Biased coding at Level 0

favored alternative	coding
a	ULLL,ULLL,ULLL ULLL,ULLL,ULLL
b	LULL,LULL,LULL LULL,LULL,LULL
c	LLUL,LLUL,LLUL LLUL,LLUL,LLUL
d	LLLU,LLLU,LLLU LLLU,LLLU,LLLU

2) Unbiased coding: This is divided into random coding inwhich an element of <*L*, *R*, *U*> is randomly chosen for every locus, and uniform codingin which all the loci at a given level share a single element . To exhaust the possibilities in the latter, three genes are created at each level (see Table 2).

Table 2 Uniform coding

Level	Coding	
2	L	L
	R	R
	U	U
1	LLL	LLL
	UUU	UUU
	UUU	UUU
0	LLLL,LLLL,LLLL	LLLL,LLLL,LLLL
	UUUU,UUUU,UUUU	UUUU,UUUU,UUUU
	RRRR,RRRR,RRRR	RRRR,RRRR,RRRR

The chromosomes of an individual in the intial generation is created from union of the direct products of the above coding schemes:

$$\{\text{random,uniform}\}_{\text{Level 2}} \otimes \{\text{random,uniform}\}_{\text{Level 1}} \otimes \{\text{random,uniform}\}_{\text{Level 0}}$$
$$\cup \{\text{biased}\}_{\text{Level 2}} \otimes \{\text{biased}\}_{\text{Level 1}} \otimes \{\text{random,uniform,biased}\}_{\text{Level 0}}$$

The 35 fittest individuals in terms of *dp* are chosen as elites to remain the the next generation. Additional individuals are created by applying crossover and mutation on the elites. Moreover, inorder to avoid stagnation by incest, 20 individuals coded totally randomly were added. The same operation is repeated for the successive generations until no forther improvement in *dp* is expected.

3 Results and concluding remarks

To obtain baseline information, we first performed blind search using 1,000 individuals created by total ly random coding. The best *dp* in 10 independent trials was only 0.157. In contrast, the aforementioed GA started with only 400 individuals produced more than one elites with *dp* above 0.200 only in three generations. Although the 20% improvement is not highly striking, efficiencies in time and in the size of individuals involed are ceratainly appreciatialbe.

References

Gärdenfors, P., & Sahlin, N-E. (1983). Decison making with unreliable probabilites. *British Journal of Mathematical and Statistical Psychology*, **36**, 240-251.

Matsuda, N. (1997). *Kansei Information Design*. Tokyo: Ohm.

Matsuda, N. (1999). Iterative use of flexible AHP as an aid to impression crystalization and decision making. In Osawa, M. (Ed.) *Toward Iimpression Engineering*. Tokyo: Maruzen Planets.

Saaty, T.L. (1980). *The Analytic Hierarchical Process*. New York: McGraw-Hill.

Corrspond to: Prof. MATSUDA, Noriyuki
Inst. of Policy & Planning Sciences, Univ. of Tsukuba
1-1-1 Tennou-dai, Tsukuba, Ibaraki 305-8573, Japan
email: mazda@shako.sk.tsukuba.ac.jp

Cyber-Messe part 2: An Operational Prototype Model

Tomoyuki Tsunoda
The Institute of Social Environment Systems, Inc.
2-5-22,Suido, Bunkyo, Tokyo, 112-0005, Japan

1 Introduction

"Cyber Messe" is a project to research the feasibility of holding industrial trade fairs on the Internet, and to explore an effective managing way of them. This project is being undertaken by the ENAA; Engineering Advancement Association of Japan. The research is based on the evolutionary prototyping (Budde 1992). In 1997 a Cyber Messe feasibility study was undertaken, and Paper Prototype Models were drawn up (Funane 1999). In 1998 a Cyber Messe Operational Prototype Model was developed, and a test-run was made on the Internet. In 1999 the model will be improved based on the results of the test-run in order to complete a practically useable system.

In 1998, "Toyama Craft Design Messe (hereinafter called TCDM)", was performed for the industries located in Toyama prefecture. The TCDM succeeded the Cyber Design Messe concept which was proposed in the previous study, and set up a handicrafts (woodwork, ceramics, glassware, etc.) exhibition on the Internet. The author participated in this project as a webmaster.

This report describes the development process and the results of the test-run of the TCDM.

2 System Design

2.1 Grand Design of the System

Considering of the outline of the TCDM, the following items were defined.

Strategy design. To succeed with the objectives and strategy of the system considered in the previous study. However, the functions offered were chosen from those suggested in the previous study. For example, the Cyber Tour and the Comment Board were done away with at an early stage.

Team design. After consideration of the team make-up proposed by Holzschlag (1996) and Sasaki (1997), a project team of eight was organized (Figure 1).

Facilities design. The computer equipment to be used was decided upon after consideration of the necessary capabilities and process capacity.

User Design. A detailed list of potential exhibitors and visitors to the TCDM was put together.

Promotion Design. Effective methods of attracting new visitors and repeat visitors were considered.

- Director
- Web Master
- Content Master
- Content Editor
- Html Coder
- Programmer
- Graphic Designer
- System Administrator

Figure 1:The project team

2.2 Design Guidelines

Policies and rules that should be followed while developing the system were specified.

Content and Site Structure. Content and site structure were defined upon while considering the ease of visitors to search, retrieve and memorize information.

Page Design. As Nielsen (1989) pointed out, one of the most important aspect of usability must be consistency in user interface. Some rules and instructions which would give users consistency were specified.

Exhibition Hall System. In order to maintain uniformity and quality control, each page of exhibitor's booth was created by the project staff in a unified page format. Also the navigation system was considered. These were discussed using the storyboard method (Holzschlag 1996) as shown in Figure 2.

Figure 2: The storyboarding

Content Size Estimate. Estimated content size was at 15 Mbytes; 200 pages.

Content Update. The results of the previous research showed that something needed to be done for attracting repeaters. To cope with this, three pages which would be renewed monthly were included.

Membership System. As suggested in the previous research, a membership system was employed. It was studied a must to place importance on the member-generated content (Hagel 1997), so some benefits for posting news and articles on the web site were set up.

User Communication. "intimate exchange" was pointed out as an important aspect of the system at the prior research. Not only Discussion Forum system on the Web, but also the news distribution services by e-mail and the off-line meetings were included in the model.

HTML Coding Standards. The TCDM HTML 4.0 coding guide was created.

3 Web page Production

The web pages were produced in the following order.

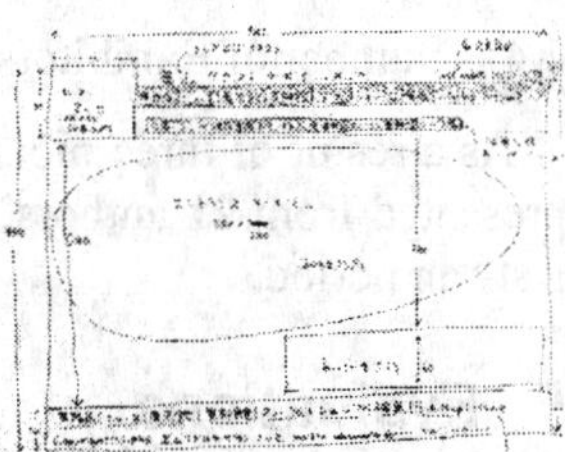

Figure 3: The design sheet

- Design sheets production (Figure 3)
- Graphic parts production and evaluation
- CGI program production and execution test
- Content production
- HTML authoring
- HTLM inspection
- Web site inhouse review (α version test)
- Web site users review (β version test)

Errors of HTML codes were examined by a check program. To prevent dead links, a link-check program was also used. With the inhouse review, many corrections were made to how the exhibit information was set-out, and to how it was navigated around. With the users review, Exhibitors strongly concerned with quality of graphic images of their products.

4 Test-Run

The test-run was made from 15 January, 1999 to the end of March, 1999. Content was finally with the volume of 10 Mbytes; 120 pages.

The URL was : http://messe.toyama-smenet.or.jp/craft/

4.1 Invitation to Exhibitors

Two meetings were held with potential exhibitors (Figure 4).

Figure 4: The invitation meeting

4.2 Communication with the Members

Two opportunities for exchange of opinions with the members were held at invitation meetings. Further opinions were also heard at receptions held by the exhibitors and the staff.

4.3 Promotion Activities

The following promotion activities were undertaken.

Newspapers. Three newspapers carried articles.

Search Engines. The TCDM was registered in the top 20 search engines .

Bulletin Boards. Advertisements were posted to four major Bulletin boards.

Email News. Articles were sent to some major Email news distributors.

Promotion Postcards. Postcards with the URL printed on them were produced for publicity purposes as shown in Figure 5.

Figure 5: The promotion postcard

4.4 Number of Exhibitors and Access Statistics

As a result of three month activities of attracting exhibitors, 53 exhibits were presented from 21 authors. There were 2,155 page views during the two-month test-run period.

5 Evaluations

5.1 Evaluations by the exhibitors

Interview surveys were carried out with the exhibitors mainly during the receptions. Some of these evaluation are presented below.
Number of Exhibitors. For a web site to be successful, the number of visitors must exceed the critical mass (Hagel 1997), but at this stage the number of exhibitors is too low. The opinion is that there should be about 100 exhibitors.
Exhibitor's Needs. Some exhibitors said the reason for exhibiting in the TCDM was not to sell the exhibits, but rather to increase the connection between people in the handicraft-related areas throughout Japan, so they proposed more should be done to extend this exchange between people.
Information to display. Not only images to show the forms and shapes of the exhibits, but there was also a call for images (animations, etc.) showing how to use the exhibits.
Participation in the Content Production. Some exhibitors expressed their desire to participate in the planning and production of content.

5.2 Evaluations by visitors

As well as also to promote the site, email was sent to the webmasters of craft design and handicraft-related web sites announcing the opening of the TCDM, and they were asked for their impressions. Followings are some of their replies.
Navigation & Guide. It was pointed out that two different orders to navigate exhibits made some users confused. It was also pointed out that some explanation of links to other pages were not enough to help users find their way.
Where Information is Placed. The words "Organization & Activities," "Membership Rules" and "Exhibit Requirements" were placed in prominent positions. These are important content for exhibitors, and for people planning to exhibit, but as far as visitors to the site were concerned they were "boring" and it was pointed out that they need not be put in such prominent positions.
Enrichment of Information. Some people suggested us that because it was a virtual space, there needed to be more information than in reality.

5.3 Evaluations by the project staff

The impressions of the project staff were collected together and analyzed. Some of these are presented below.
Expansion of the Project Staff. There was more promotional work involved than was expected. Someone needs to be in charge of the promotional activities. On

top of that, in order to enrich the content, someone deeply involved in craft design needs to be employed as a Content Master.

Content Quality Control. Quite often some pages were published without spelling mistakes, dead links and wrapped links (Spool 1997) being picked up by the person responsible.

Editorial Power. The amount of information received from each exhibitor was different. For example, some exhibitors did not tell us the weight of their products. It is a lot of work to put all the information on each booth page equally, but the quality of information may affect how the visitors trust the web site, so there is a need to enhance ability and productivity of our editorial work.

6 Conclusion

As the second stage of the Cyber Messe project, the Paper Prototype Model was developed into the Operational Prototype Model. The structured development procedures we adopted functioned extremely well. From the two month test-run, the following significant conclusions were drawn up.

- Even those people who did not display any interest with just the concept or the paper prototype model began showing interest as the web site began to appear.
- As a result, while being small, an on-line community which we thought had the potential to grow was established. Some evangelists (Cohill 1997) came forward by themselves to cooperate with the operation.
- Many areas which could only be grasped in a very abstract way with the paper prototype model, for example, what the exhibitors and visitors would be hoping for in detail, the amount of work involved and how much it would cost, the necessary make-up of an effective management team, became clear.

References

Budde, R. Kautz, K. Kuhlenkamp, K. Zullighoven, H. (1992). *Prototyping: an approach to evolutionary system development.* Berlin: Springer-Verlag.

Cohill, A.M. (1997). *Success factors of the Blacksburg Electronic Village.* In Cohill, A.M. & kavanaugh, A.L. (eds.) *Community Networks –Lessons from Blacksburg, Virginia–.* pp.297-318. London: Artech House.

Funane, K. (1999). *Cyber-Messe Part 1: Paper Prototype Models.* Proc. 8 th International conference on Human-Computer Interaction, Amsterdam: Elsevier.

Hagel III, J. (1997). *Net Gain.* Boston: Harvard Business School Press.

Holzschlag, M.E. (1996). *Professional Web Design.* California: Prima.

Nielsen, J. (1989). *Executive summary: Coordinating User Interfaces for Consistency.* In Nielsen, J. (ed.) *Coordinating User Interfaces for Consistency,* pp.1-7. London: Academic Press.

Sasaki, H. (1997). *Web Design Handbook.* Tokyo: AI Publishing. (Japanese)

Spool, J.M. & Scanlon, T & Scroeder, W. & Snyder, C & Deangelo, T. (1997). *Web Site Usability: A Designers Guide.* Massachusetts: User interface Engineering.

The Web Crusade Part I: Motives for Creating Personal Web Pages by Individuals

Gen Maruyama
Technology Research Center, Taisei Corporation

Sanken Bldg., 25-1, Hyakunin-cho 3-chome, Shinjuku-ku, Tokyo 169-0073, JAPAN

phone: +81-3-5386-7561 fax: +81-3-5386-7577

E-mail: maruyama@mb.kiku.taisei.co.jp

1 Goal

Millions of Web pages, a galaxy of "homepages", are now found on the Internet, each of which has been created with its own purpose or motive.

Our "Web Crusade" activities aim to study and analyze those Web pages to see if they are designed appropriately to meet their motives or intentions, and help them create better-looking, easier-to-read Web pages. This report gives, as the first step, a definition of purposes or goals that creators of Web pages may have when they try to design "homepages". Particularly we focused on pages that individuals created (called personal Web pages), thus determining motives or intentions that the creators of personal Web pages have. In this study, we employed a psychological method to analyze creators' motives and tried to break down the motive types into several categories.

2 Method and Implementation

With this research we employed the Evaluation Grid Method (EGM), an approach where the motives or intentions have been obtained by direct interviews with each Web page creator.

This method originally was developed by environmental psychology, a field of architecture. We reported its effectiveness in the provisos HCI session.

EGM is an improved version of the repertory grid developed for use in interviews by the clinical psychologist G.A. Kelly (1955), with the aim of elucidating phenomenologically those elements of the human cognitive system having a bearing on assessment of the environment. The method for developing the repertory grid has the following characteristics:

1) As this was originally an interviewing technique used in clinical psychology with the aim of healing patients, it is possible to use it to determine the way in which people assess their environment.

2) As the method is based on a clear cognitive psychological theory known as the Personal Construct theory, this theory can be used to provide a lucid explanation of the results of the survey.

3) While the survey technique is based on free responses by the interviewee, the systematic format imposed by the interviewing technique itself reduces to a minimum interference from the subjective opinions of the interviewer. Reliance on the interviewing ability of the researchers is also kept to a minimum, so little variation is found in the results obtained by different interviewers.

EGM proceeds in two stages. First, given a pair of design solutions, or elements, the participants are asked to state why they consider one element to be preferable to the other. Second, these reasons (i.e., requirements) are organized into a hierarchical diagram with the aid of guided questions to elicit antecedents and consequences. The procedure can be likened to laddering. By accumulating individual diagrams, we can obtain a group diagram on which the strength of links is indicated by the frequency of cases.

Strictly speaking, this method refers to a research technique to identify the structural depths of human's consciousness, or the mechanism of determinations and evaluations, by direct interviews with people. However, we have developed an approach of virtual interviews, that is, interviews via E-mail, and used it again in this research. This time, we had detailed interviews through E-mail with eight devoted Web page creators.

3 Results and Considerations

Figure 1 shows the results of interviews in the form of a structured network, obtained using an analytical method of perceptive psychology that is unique to the Evaluation Grid Method.

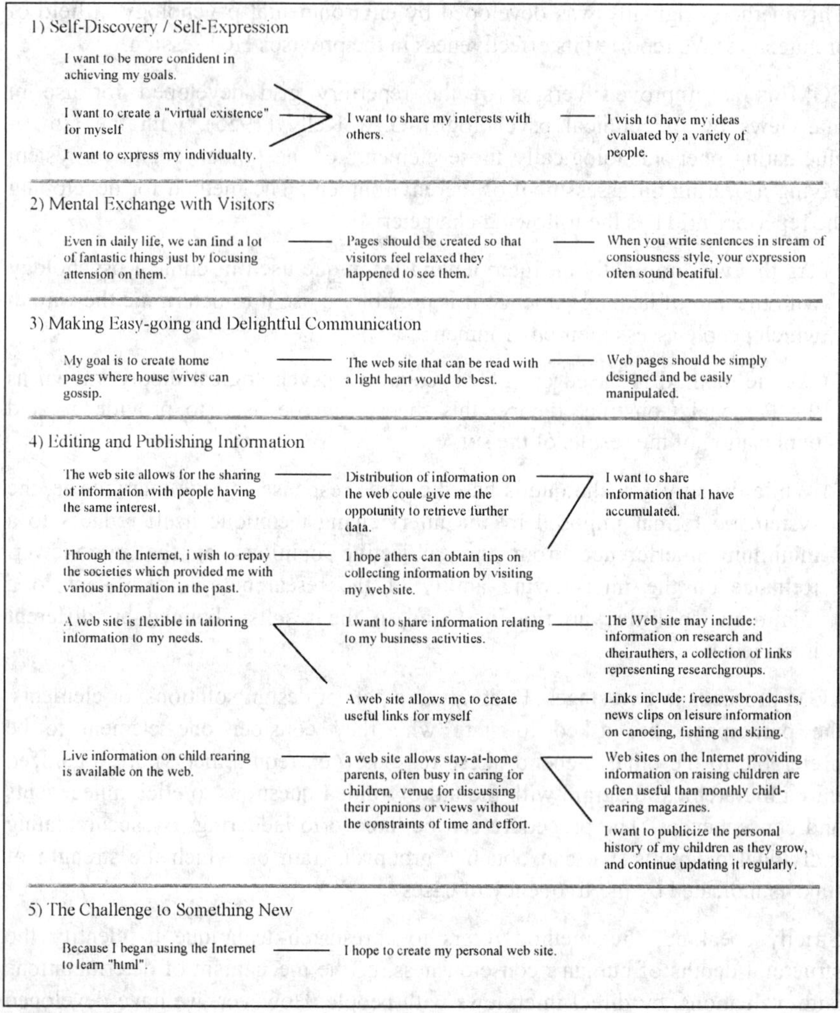

Figure 1. Psychological Flow Chart of Motives for the Creation of Private Web Pages

As is known by a motive model, shown in Figure 2, the motives of creators can be broken down into the five categories: 1) the challenge to something new (curiosity), 2) editing and publishing information (usefulness), 3) making easy-going and delightful communication (mutual understanding), 4) mental exchange with visitors (sympathy), and 5) self-discovery or self-expression (formation of ego). Of the five categories, the last motive represents the deepest consciousness level of the creators and the depth level goes up towards the first motive.

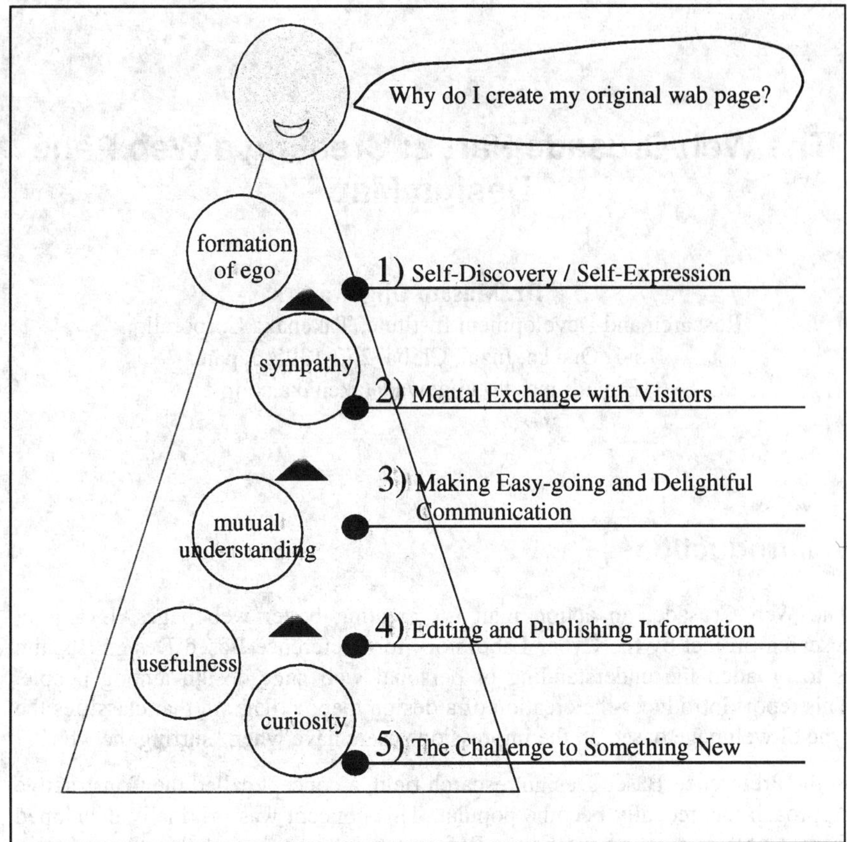

Figure 2. A Motives Model for the Creation of Private Web pages

By the psychological analysis using the structural method, we reached the conclusion that the motives of personal Web page creators varies widely, ranging from a curiosity to use the newly developed technologies (Internet), to an attempt to take advantage of the Internet as a way to foster or enlighten themselves

4 References

Maruyama, G. and Sanui, J. (1997). Revealing of Preference Structure by the Evaluation Grid Method: *Design of Computing Systems: Cognitive Considerations (Vol. 2), Proc. 7th Int. Conference on Human-Computer Interaction* (HCI International '97, San Francisco, USA, August 24-29, 1997), pp. 471-474.

The Web Crusade Part 2: Creating a Web Page Design Map

Dr. Masato Ujigawa
Research and Development Institute, Takenaka Corporation
1-5-1, Otsuka, Inzai, Chiba, 270-1395, Japan.
E-mail: masato.ujigawa@takenaka.co.jp

1. Introduction

The Web Crusade, an action plan for creating better web pages, is a joint research project by the Cyber Laboratory for Preference-Based Design. Its aim is to broaden the understanding of personal web page design among people. This report introduces the creation of a design map; a diagram that classifies the type of web page based on the impressions users have when "surfing the web".

In the Preference-Based Design research field, a concept called the Constructive Approach has recently become popular. This concept was originally developed from a golf instruction method. Before the introduction of this Constructive Approach, one standard and established style was applied to each golfer in lessons (called the Corrective Approach). However, because every golfer's goals, endurance and effort varies, it was concluded that there are limitations in this Corrective Approach where one ideal form is applied to each player. Therefore it has been recommended that professional trainers try to choose an appropriate goal for each golfer and instruct them how to achieve it. This is the Constructive Approach.

When individuals create web pages, their purposes are different, and the knowledge and time they can spend in creating them also varies. Therefore we thought it best to share techniques and information based on personal needs, that is, useful information for each individual.

Professional designers have extensive knowledge of various design types and can select an appropriate one depending on the situation at hand. General users do not usually have this knowledge, selecting only casual ideas in their design. If general users are given organized information on design types, it will be easier for them to choose the style they want and also create new designs based on a style from our selection. Therefore, we studied the impressions users got from web pages, and created a design map. This breaks down web pages by category.

2. Method

2.1 Preparation of Stimuli

In psychology, an object that is presented to the subject is called a stimulus. We created 50 web page printouts and divided them into 20 groups, which are similar in type. Then we selected one typical printout from each group. Selected printouts are all color; rectangular pictures approximately 8 x 8 centimeters in size.

2.2 Evaluation of Impressions

A semantic scale is used to evaluate impressions in psychology. An example of this scale is where an adjective is placed at one end and an adjective having the opposite meaning at the other. Across the scale, adjectives whose meaning falls within the range of the two "end" adjectives are arranged as options.

We used eight semantic scales in this survey: like - dislike, bright - dim, quiet - lively, warm - cold, strong - weak, intelligent - un-intelligent, delightful -not-delightful, and friendly - un-friendly. We picked six persons (three males and three females each), who often use home pages on the Internet, and surveyed their impressions of the 20 printouts using the eight semantic scales described above. The results were tabulated and are presented as the "Rating Value Mean", in a possible rating range from 2 to 0, e.g., like=2, moderate=1, dislike=0.

2.3 Grouping Design Types Based on Impressions

Based on the results of the survey, we correlated the information by calculating a "coefficient of correlation" between the semantic scales. Then using a statistical analysis method called Principal Component Analysis; we obtained a general trend. Finally, we placed the 20 printouts on a two-dimensional plane by using a method called factor rotation, and pasted the 20 printouts to their corresponding positions to make a design map.

3. Result

Table 1 shows the average values calculated from the results we obtained by evaluating the subjects' impressions. Table 2 illustrates the correlation matrix obtained from the responses to the eight semantic scales. We did not find a strong correlation for the semantic scale of like - dislike, however we could see a strong correlation within the other seven semantic scales.

Table 1 Evaluated Impressions (Rating Value Mean)

No.	like - dislike	bright - dim	quiet - lively	warm - cold	strong - weak	intelligent - un-intelligent	delightful - not-delightful	friendly - un-friendly
1	0.5	1.3	0.5	1.2	2.0	0.5	1.5	1.5
2	1.3	0.8	1.5	0.7	0.3	1.7	0.8	0.7
3	1.5	0.7	1.7	0.8	0.7	1.5	0.5	0.7
4	0.7	1.3	0.2	0.8	2.0	1.0	1.2	1.7
5	0.7	0.8	0.5	1.0	1.5	1.2	0.5	0.5
6	0.8	1.7	0.0	1.7	1.5	0.3	1.7	1.7
7	1.2	1.5	0.5	1.2	1.0	1.2	1.0	1.2
8	1.2	0.8	0.7	0.2	1.7	0.8	0.8	0.5
9	0.7	0.0	2.0	0.0	0.3	1.7	0.3	0.0
10	1.2	1.3	1.2	1.2	0.7	1.0	1.2	1.3
11	0.7	1.5	0.5	1.7	1.3	0.3	1.0	1.2
12	1.0	1.2	1.5	1.3	0.8	1.3	1.2	1.0
13	1.2	1.0	1.5	1.3	0.7	1.0	0.7	1.0
14	1.3	0.7	1.8	0.5	0.0	1.5	0.3	0.3
15	0.7	1.7	0.2	1.8	1.3	0.3	1.7	1.5
16	1.0	0.7	2.0	0.3	0.0	1.7	0.2	0.2
17	0.5	0.7	1.8	0.3	0.2	1.2	0.2	0.2
18	1.2	1.0	0.2	1.0	1.5	1.2	1.3	1.0
19	1.2	1.3	0.2	1.2	2.0	0.5	1.5	1.5
20	1.2	1.7	0.8	1.3	1.0	0.8	1.2	1.5

Table 2 Correlation Matrix

	bright - dim	quiet - lively	warm - cold	strong - weak	intelligent - un-intelligent	delightful - not-delightful	friendly - un-friendly
like - dislike	-0.107	0.241	-0.098	-0.296	0.389	-0.102	-0.080
bright - dim	1.0	-0.742	0.867	0.582	-0.799	0.831	0.919
quiet - lively		1.0	-0.611	-0.903	0.778	-0.815	-0.770
warm - cold			1.0	0.448	-0.710	0.744	0.811
strong - weak				1.0	-0.726	0.738	0.697
intelligent - un-intelligent					1.0	-0.753	-0.748
delightful -not-delightful						1.0	0.906
friendly - un-friendly							1.0

With Principal Component Analysis, both the first and second principal components showed a cumulative contribution rate of 84%.

Axis rotation was performed by Factor Analysis and the position of each stimulus was calculated. Table 3 indicates a factor-loading matrix of rotated components. For the first principal component, we found a high proportion of bright, lively, warm, strong, un-intelligent, delightful, and friendly factors. For the second principal component, the burden is particularly high in the like - dislike scale and relatively high in such scales as strong - weak, intelligent - un-intelligent, and quiet - lively.

Table 3 Factor Loading Matrix of Rotated Components

Principal Component	First	Second
Eigen Value	5.61	1.12
Cumulative Contribution Rate	70.1%	84.1%
like - dislike	-0.106	0.847
bright - dim	0.388	0.206
quiet - lively	-0.381	0.127
warm - cold	0.350	0.249
strong - weak	0.346	-0.258
intelligent - un-intelligent	-0.377	0.189
delightful -not-delightful	0.390	0.134
friendly - un-friendly	0.395	0.202

We created the design map by assigning the first factor to an X-axis and the second factor to a Y-axis and pasting the printouts to their corresponding stimulus positions. The map is shown in Figure 1.

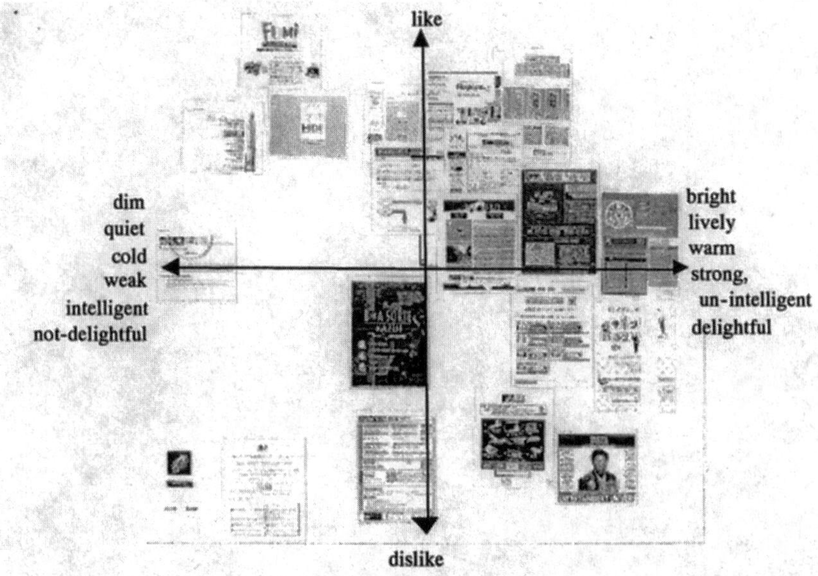

Figure 1 A Design Map of Web Pages

On the right side of the figure, impressions such as bright, lively, warm, strong, un-intelligent, and delightful are arranged. On the opposite side, impressions such as dim, quiet, cold, weak, intelligent, not-delightful are distributed. The upper area of the figure relates to impressions of like, and the lower area displays contrary perceptions.

4. Conclusion

We came to the following conclusion after having created the design map: Adjectives such as bright, lively, warm, strong, and delightful have similar meanings or concepts. We found a weak relation between the like - dislike concept and such adjectives as bright, lively, warm, strong, and delightful. Users like something intelligent and quiet, but do not like something dim, dull, or forced. However, because the number of subjects in this survey was small, one should not assume that the results are a general trend.

The Effects of Facial Expression on Understanding Japanese Sign Language Animation

Sumihiro KAWANO and Takao KUROKAWA
Graduate School of Engineering and Science, Kyoto Institute of Technology, Matsugasaki, Sakyo-ku, Kyoto 606-8585, JAPAN

1. INTRODUCTION

Japanese Sign Language (JSL) is a native language for auditory handicapped people in Japan. It is multi-modal, and it is signaled in many kinds of body actions which include head movement, eye movement, posture, facial expression as well as manual gesture. It is difficult for hearing people to acquire JSL and to communicate with the hearing impaired in it. In order to solve this problem and improve the present communication environment for the hearing impaired, the authors are developing a system which translates Japanese into JSL and vice versa [1]. In this system, the results, or JSL sentences, of translation from Japanese sentences are shown to non-hearing people by means of sign animation featuring a model human [2]. The animation is synthesized by a rule-based method. In the previous study [3][4], we analyzed movements such as nodding, blinking and gaze behavior appearing on a sign interpreter's face while talking in JSL and found their positive effects on understanding JSL sentences and their structures through an experiment using the animation. In the experiment two kinds of the JSL animation were produced; in one the model human talked in JSL with only hand gesture and in the other with facial movement as well as hand gesture. Subjects watched one of them and responded with their recognized meanings of JSL sentences. There were clear correlations between occurrence facial movement and correct recognition of the words and sentences.

But as to facial expression its effects on understanding JSL have not been known, although it is said to be an important component in JSL. The purpose of this paper is to analyze roles of facial expression in JSL and to improve the JSL animation based on the results. It will also describe the effects of the synthesized facial expression on understanding JSL sentences based on the results of the experiment for evaluating JSL animation with or without facial expression.

2. ANALYSIS OF FACIAL EXPRESSION IN JSL

In order to clarify the functions of facial expression, we analyzed a female sign interpreter's non-manual gesture on JSL TV News of NHK. The material for the analysis was videotapes which recorded 304 JSL sentences selected from JSL TV News. All the sentences were signed together with utterance in Japanese by her. We first measured movement of some parts of her face and second examined relations between facial expression and JSL sentences. They contained 1824 sign words, each of which generally corresponded to a Japanese phrase. As for facial expression we focused on opening and closing of the eyes, raising and lowering of the eyebrows, and cheek movement. They were measured for each of the sign words. Eye opening was graded on the scales of closed, half opened, normally opened and largely opened. Raising of the eyebrows was measured on a three-graded scale; raised, not moved, lowered. The cheeks were judged to be raised only when the upper cheeks markedly moved up, because they were under the influence of utterance. For the same reason, mouth movement were ignored. Although the results are not shown here, facial movements mentioned in 1. were also measured.

Table 1 Frequencies of the combined movements of eyebrows and eyelids in the words used on JSL TV News

Degree of eye opening	Movement of eyebrows: lowered	not moved	raised
closed	4.3	3.5	1.1
half opened	10.8	2.6	1.1
normally opened	2.9	53.5	9.3
largely opened	0.4	3.4	7.0

n = 1824 (%)

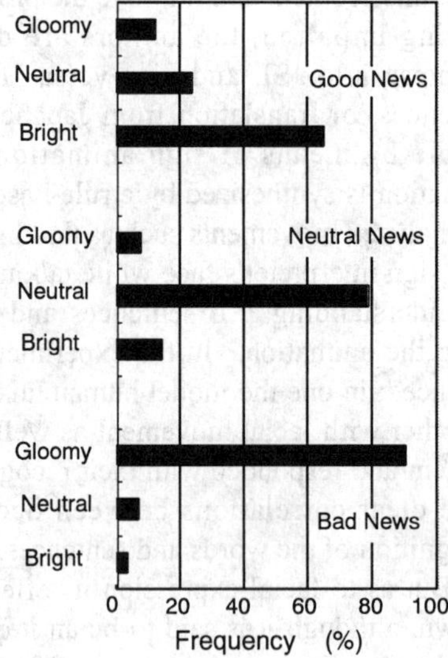

Fig. 1 Correlation between sentence image and facial expression

Table 1 shows the frequencies

in percentage of the combined movements of the eyes and the eyebrows. Although 46.5% words were accompanied by some movements of the eyes and/or the eyebrows, the facial expression was strongly correlated with the image or impression of sentences. In order to relate the facial expression to the sentence image, every sentence was classified into one of the three categories; good, neutral and bad news. Then in each sentence, we chose the most appropriate word to strongly express the image of the sentence. On the other hand, the facial expression that the chosen word was accompanied by was distinguished into the three typical expressions; bright, neutral and gloomy expressions. Bright expression was defined as one with largely or normally opened eyes and raised eyebrows, and gloomy expression as one with half opened eyes and lowered brows. Other combinations were judged as neutral. As shown in Fig. 1, opening of the eyes and raising of the brows markedly signaled the image of words and sentences.

3. THE EFFECTS OF FACIAL EXPRESSION

3.1. Methods

In order to examine effects of facial expression on understanding JSL sentences, we carried out an experiment where subjects were shown JSL animation and put the JSL sentences into Japanese. The animation consisted of three current topics: an annuity in the future, the economic growth of Japan and the next day weather, taken from JSL from TV News. They contained 101 words (76 different words). The two kinds of animation were synthesized; one had facial expression and movement as well as manual signing ("with facial expression" in the following), and the other had only hand gestures ("without facial expression"), and then they were recorded on the separate tapes A and B (A: without facial expression, B: with facial expression). The hand gesture and facial expression in the animation were designed to imitate those of the interpreter as far as possible. For comparison with the animation, tape C was

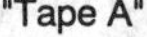

"Tape A"

"Tape B"

"Tape C"

Fig. 2 Images presented in the experiment (talking "tomorrow" in JSL)

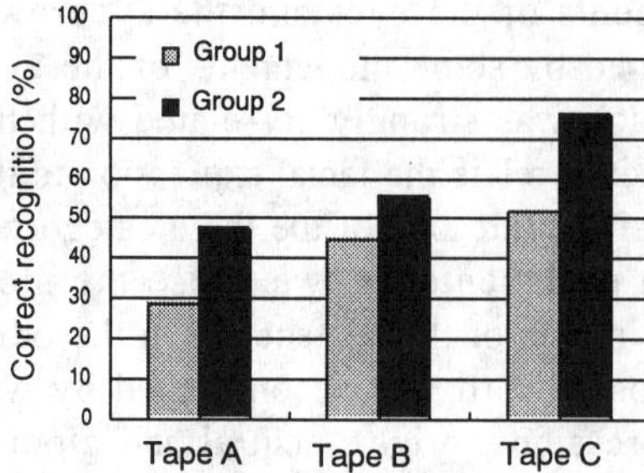

Fig. 3 Rates of correct recognition of JSL words

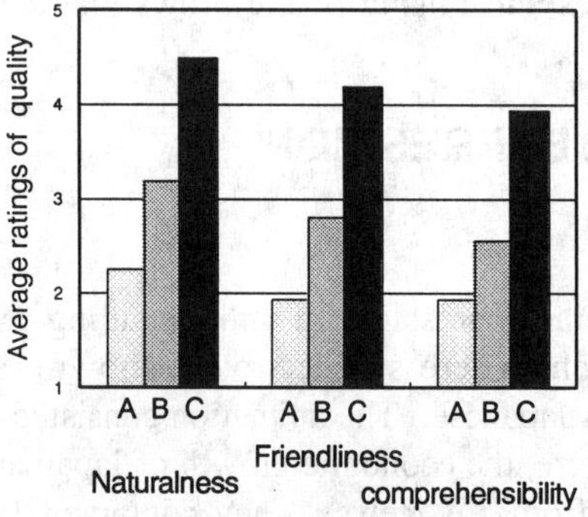

Fig. 4 Average ratings of image quality in Tapes A, B and C

made of the original TV News and contained the same sentences as the synthesized ones. 16 subjects participated in the experiment; 8 were hearing disabled (Group 1) and 8 were hearing persons (Group 2). Groups 2 subjects served as interpreters, or teachers in a school for deaf pupils. Their experience in JSL was distributed from one to ten years. After given the instructions about the methods, they were presented all JSL sentences on Tapes A, B and C. Every time one of the sentences was shown twice, they wrote down the recognized sentence in Japanese. After the experiment they answered a questionnaire for evaluating the quality in naturalness, friendliness and comprehensibility of the animation and the TV News image on a 1-to-5 scale. Finally, they were asked their opinion about JSL animation and machine translation between JSL and Japanese.

3.2. Results and Discussions

Fig. 3 shows the correct recognition rates of the JSL words in Tapes A, B and C. The rate was highest in the original news (Tape C) and lowest in the animation without facial expression (Tape A). Tape C, however, did not always get a rate high enough because they had difficult words concerning the economic growth of Japan and an annuity in the future, which do not frequently appear in everyday conversation. There was much difference in quality between the animation and the live news. The synthesized images were not natural in that actions of the model human were awkward and that movements in the depth direction were difficult to detect. We think that these were the reasons why Tapes A and B got lower rates. There was a large difference between Groups 1 and 2 in all the tapes. Maybe this fact came from the difference in experience and habituation to JSL. Fig. 3 also clarifies that there was a significant difference ($p < 0.05$) in the recognition rate between Tapes A and B. This means that facial expression and movement helped the subjects understand the JSL words and sentences.

The results of the subjective rating about image quality are drawn in Fig. 4. They are parallel to the above results. The quality of the animation synthesized by our computer was not so high as the live image because of syntactic and technical imperfections. The subjects, however, judged that the animation with facial expression was superior to that without it. They stressed the same in their opinion after the experiment. The results mean that facial expression is an important component in JSL animation. All the images got the lowest rating in comprehensibility. This is also guessed coming from that there appeared many words unfamiliar to them.

4. CONCLUSIONS

In this study we have confirmed that facial expression can convey information about the image of topics, sentences and words in JSL and that facial expression and movement are indispensable in sign language animation. We also have found some problems to solve in the future study. They include improvements in representation of the model human, synchronization among bodily actions and modulation of actions.

Acknowledgement: We thank Ms. Nakano of NHK for her permission of recording her image during talking in sign language. We also thank Mr. Itou, a graduate student of Kyoto Institute of Technology, and Mr. Sakai of Kyoto Prefectural School for the Deaf for their cooperation in the experiment. This research was partly supported by the MESCJ under Grants-in-Aid No.10480072.

References

1. Kurokawa, Morichi and Watanabe (1991) Bidirectional translation between sign language and Japanese for communication with deaf-mute people, Advances in Human Factors/Ergonomics 19B, 1109-1114.
2. Watanabe, Izuchi, Fujishige and Kurokawa (1993) Sign Language Animation Using a Synthesis-by-Rule Method, Proc. 9th Symp. on Human Interface 385-390.
3. Kawano and Kurokawa (1998) An Analysis of Characteristics of Facial Expression of a Sign Language Interpreter of TV News, Human Interface News and Report 1998 vol.13 287-292.
4. Kurokawa T, Senba K and Kawano S (1997) Synthesis of facial movement images for sign language animation and their effects, Advances in Human Factors/Ergonomics, 21, 5-8.

Small Group CSCW Using Web Resources: A Case Study

Tan Kay Chuan
Department of Industrial and Systems Engineering
National University of Singapore
Correspond to: isetankc@nus.edu.sg

1 Introduction

Computer Supported Cooperative Work (CSCW) emerged due to increasing research and development activities associated with group computer work. One area of concern for HCI specialists is in understanding how to manage both cooperation and conflict at a general and a specific level. At a general level, there needs to be awareness of principles that apply widely across different cooperative work contexts. At a specific level, there needs to be a set of concepts and procedures for analyzing particular work patterns.

Given the present state of computer and information technology, the range of CSCW systems that can be developed and deployed is limited. The development of World Wide Web (WWW) applications and internet tools provide good opportunities for CSCW design.

This paper describes a study that uses qualitative methods and the technology offered by the WWW to build an effective web-based CSCW system in a typical small-group academic environment. The Department of Industrial and Systems Engineering (ISE) at the National University of Singapore was chosen for the case study. The objective was to first analyze the current cooperation and conflict activities that occurred in the Department in the absence of computer support. Having established this baseline, various CSCW technologies widely available on the internet were adopted to develop a prototype system for use in the Department.

2 The WWW as an Enabling Technology for CSCW

The WWW offers much in terms of cooperative technology for work and information sharing. Most CSCW systems now developed are based on particular platforms and are only usable within the particular organizations for which they were developed. The diversity of machines, operating systems, and application software makes it difficulty for CSCW systems to be deployed across platforms. In contrast, the WWW offers a globally accessible and platform-independent infrastructure. It can be seen as an enabling technology for CSCW from several perspectives:

a) The WWW provides access to information in a platform-independent manner because its client programs (e.g., browsers, add-ons) are available for all popular computing platforms and operation systems;

b) Browsers offer a simple user interface and consistent information presentation across these platforms;

c) Browsers are already part of the computing environment in an increasing number of organizations, requiring no additional installation or maintenance of software for users to cooperate using the WWW; and

d) Many organizations have installed their own web severs as part of an internet presence or a corporate intranet, and have familiarity with sever maintenance and sever extension through programming the sever API.

3 Case Study in the ISE Department

3.1 An Overview of the Research Site

The research site is a typical small academic department at a university. In general, working activities in the Department of Industrial and Systems Engineering are conducted in an atmosphere of shared information. The academic staff, non-academic staff, and students communicate with one another using both formal and informal procedures. Table 1 portrays the interaction among these three groups.

Table 1 Interaction activities in the ISE Department

Groups interacting	Activities
Acad staff and students	Formal and informal reports, group discussions, occasional chatting, and lectures
Acad staff and non-acad staff Students and non-acad staff	Student and course administration, arrangement of seminars, etc

The use of formal procedures has met with limited success. For example, the Department conducts regular seminars. At these seminars, the speakers' presentations are not always fully understood. Thus, informal communication is sometimes used to clarify points brought up during the seminars. This takes place in the form of informal chats after the seminar, having a copy of the researcher's paper, or e-mailing to the speaker to discuss a specific problem or future course of research.

3.2 Questionnaire and Interview

In order to identify how cooperation and conflict occur in the Department, a questionnaire was administered to determine the kinds of cooperative work that needed to be supported and the conflicts that needed to be resolved (see Appendix A for list of questions used in questionnaire). Interviews were conducted also using a semi-structured format. All questions related to the formal and informal ways used to interact and cooperate with one another.

3.3 Results

An analysis of the questionnaire and interview results focused on the deviations and errors that had occurred and had hampered cooperation. They included:

Informal practices. Three main channels of informal communication were identified, including: e-mail, use of the telephone, and face-to-face conversation. It was interesting to note that the majority of e-mails focused on the senders' rather than the receivers' interests. There also existed much non-work related messaging. To an extent therefore, e-mail has brought about some degree of time wastage especially for the receiving end.

Absences at seminars. The ISE Department conducts an average of two seminars a month. These are compulsory for students but absences are frequent due to a host of reasons. Respondents indicated that there were instances where they wanted to attend a seminar but were unable to do so. They, furthermore, would like to have information on the seminar available (e.g., an extended synopsis). Under the current means available, only 9.5% of the respondents made any effort to follow up a seminar for more information.

Delay of post mail. 57% of the students in the Department felt that they do not receive post mail in a timely manner. This was a concern especially in light of the fact that the academic staff at times used this mode of communication with their students.

4 The CSCW Prototype System

Through the use of technology and software currently available, a prototype CSCW system was established in the Department. Two software were used, including: Microsoft NetMeeting and ICQ (Internet Pager software). Microsoft NetMeeting provides for conferencing and collaboration capabilities. It was chosen because it came with ample capabilities. It also exposes Component Object Model (COM) interfaces so that applications will be extensible with dynamic component recovery. NetMeeting can be used free of charge with no run-time royalty. The components used included internet video conferencing, internet phone/audio conferencing, and multipoint data conferencing.

The ICQ software was used in conjunction with the NetMeeting software. ICQ is a shareware software that lets users create contact lists from which messages and file can be sent real-time to many parties simultaneously. ICQ also allows one to locate those who are online and contact them for a meeting on the internet. ICQ can be configured to work with external applications. It also provides for e-mail notification.

The CGI programming technology was used to create a bulletin board in support of group discussions. This way, supervisors and students can raise topics or questions for discussion.

5 Implications for CSCW Design

Positive results. The prototype CSCW system integrated traditional data and text processing with teleconferencing technology. This turned out satisfactory results. More than 95% of the respondents thought the prototype provided support for their daily cooperative work in terms of increased flexibility in the frequency and timing of their communication. This web-based CSCW system can be built and used easily because it is based on internet technology that is widely available. It had increased that amount of formal and informal communication thereby increasing interaction among members of the ISE Department.

Unsatisfactory results. Several problems were evident also with the prototype. It should come as no surprise that these include slow transmission speeds and a lack of support for video and audio input/output. The solution to these problems lies in the maturing of these communication technologies.

6 Conclusion

In this study, an application prototype of a WWW-based CSCW system was developed. Since the internet has become increasingly commonplace, it is very useful to make good use of internet tools to build applications comparable to some commercial software. Using qualitative methods, a deeper and broader understanding of the way people interact with each other in a work setting was obtained. A CSCW prototype was developed and this turned out good results in solving conflict and in supporting cooperation. The future should see more useful internet software that support CSCW activities.

7 Appendix A. Questions used in the survey

1. Which platforms are you familiar with?
2. How knowledgeable are you on networking?
3. What formal means of communication do you use?
4. How often do you cooperate with others in the Department?
5. How do you record the work-related portions of your communication?
6. Are there alternatives if you should be unable to attend a seminar that interest you? What are these alternatives?
7. Do you use ICQ for mail notification?
8. Do you use NetMeeting for cooperation?
9. What do you think are the major issues surrounding the use of computers for both formal and informal cooperative work?

Testing Videophone Graphical Symbols in Southeast Asia

Piamonte, D. P. T.[1], Abeysekera, J. D. A.[1] and Ohlsson, K.[2]
[1]Division of Industrial Ergonomics, [2]Division of Engineering Psychology,
Department of Human Work Sciences, Luleå University of Technology, Sweden

1 Introduction

Southeast Asia is a region characterized by diversity in language and culture. Although it is slowly recovering from the recent economic crisis, it still remains as one of the most dynamic regions in the world. Economic trades and technology transfers from the west still abound in most of its countries. One palpable sign is the continued rise in the use of western products. With these also came the use of graphical symbols (icons and pictograms) as a common mode of user interface of the products. Easy learnability, less space requirements and being non-text dependent are just few of the reasons why graphical symbols have become a major means of presenting information and functions especially in travel, communication and consumer products (Zwaga, 1989 and Horton, 1994). In order to elicit the above benefits of using graphical symbols, they must be tested across their intended user groups. Symbols are often designed and tested in the west but targeted for international use. Thus, the extent of their usability and acceptability among user groups with different languages and cultures from those of the west can be questionable. Studies on how icons and pictograms in Asia are still sparse. This paper aimed to study how symbols designed and tested in Europe fared among subjects from different countries in Southeast Asia. It was part of an international study on graphic symbols involving countries from Southeast Asia (Piamonte, 1998). Specifically, videotelephone symbols designed and tested by the European Telecommunication Standards Institute (ETSI) (Böcker, 1993) were evaluated using subjects from five Southeast Asian countries.

2 Methods and Materials

Three sets of videophone symbols were derived studies by Böcker (1993) for the European Telecommunications Standards Institute (ETSI), including the set (Set 1) eventually proposed by ETSI as the standard (Figure 1). The symbols were tested in the form of questionnaires. Each symbol set contained seven symbols representing seven videophone functions.

Testing of the subjects were done around Metro-Manila, Philippines, Bangkok, Thailand, Bandung, Indonesia, Sarawak, Malaysia, and Sri Lanka, and in were and in small groups in each country, lasting for about 45-60 minutes. Subjects were randomly given one of three versions of the questionnaires. The subjects were shown an illustration of a videophone. Its general functions were then discussed. Afterwards, instructions were given on how to go about the different test parts. Questions were entertained prior to the administration of the tests. Emphasis was given on avoiding omissions in trials in order to get back to them later on.

In the spontaneous identification test (Part One), each page contained one set of the seven videophone symbols, for a total of three pages for the three symbol sets. On spaces provided for, they would write what videophone function they thought was represented by each symbol. They were also asked to rate the level of certainty for each of their answers using a seven-point rating scale (from "very certain" to "very uncertain"). Part Two was a cued response test patterned after a method developed by Böcker (1993) in testing simultaneously appearing symbols. The subjects first read a referent and its description. Then they had to choose one symbol from a set of seven symbols which they thought best represented the referent in question. Each page contained one referent-description and one set of symbols. They were also asked to rate their certainties for their answers using the 7-point rating scales. There were seven videophone referents tested on three sets of symbols rendering a total of twenty-one trials in Part Two.

3 Results and Discussion

Two hundred forty-eight subjects (127 males and 113 females) from the five Asian countries acted as respondents. The mean age was 25.12 years (SD=6.08, Range=17-52 years). There were forty-eight subjects per country with equal numbers of students and employees. In both Parts One and Two, correct responses resulted when a subject correctly identified (Parts One) or matched (Part Two) the referent in question to its right symbol. Correct responses were labeled as "1" (hits) and wrong responses as "0" (misses). Missing data or "no responses" were initially treated as also "0" during this stage of the analysis.

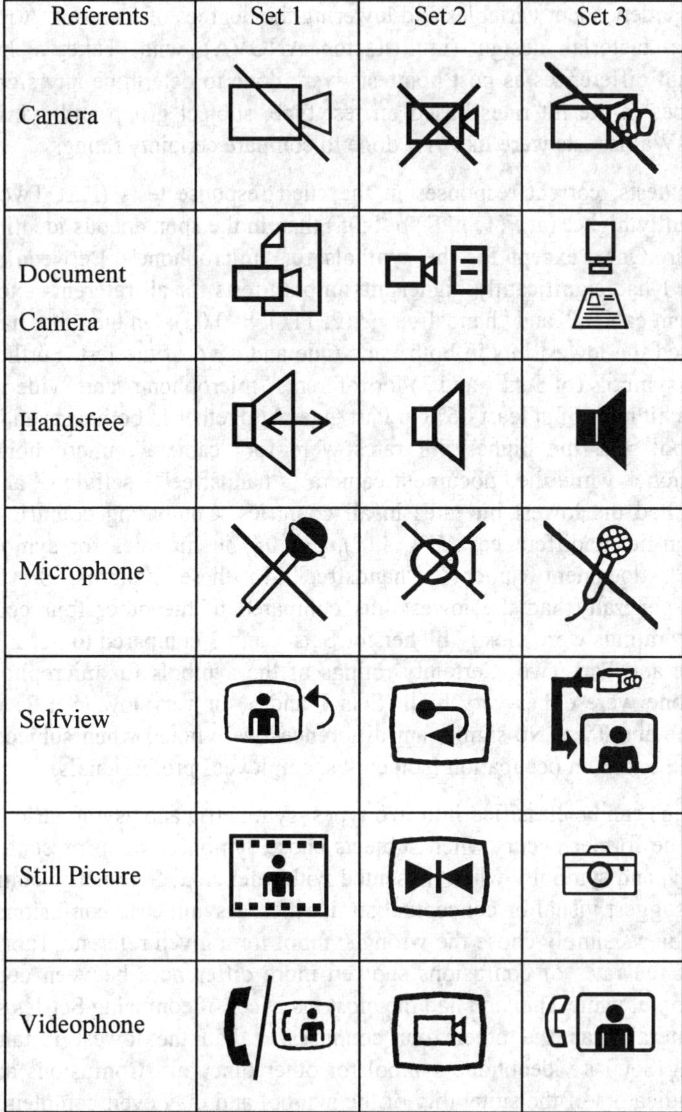

Figure 3. Graphical symbols used in the present study (Papers 2-6) as based on the studies by Böcker (1993) for the European Telecommunications Standards Institute.

The response data were thus basically binomial percentages and arcsine transformations are warranted to ensure that assumptions in normality were met. Such transformations of percentages eliminate subjects as a variable leaving only the independent variables and lowering the degrees of freedom. A repeated measures factorial design (multi-factor ANOVA) with Tukey's honestly significant differences as post hoc tests were done to determine any significant differences in the hit rates based on test type, subject group and symbol set. Kruskal-Wallis tests were likewise done to compare certainty ratings.

Between tests, correct responses in the cued response tests (Part Two) were significantly higher (at $F(1, 111)$, $p<0.05$) than in the spontaneous identification tests (Part One), except for the symbols for "microphone". Between symbol *sets*, Set 1 had significantly higher hits in both tests for all referents except for "document camera" and "handsfree" ($F(2, 111)$, $p<0.05$). On the other hand, Set 2 garnered the lowest hits in both Parts One and Two of the test. Furthermore, only the symbols (of Sets 1 and 3) for referents "microphone" and "videophone" garnered hit rates of at least 66% in Part One. Altogether in both tests and across all symbol sets, the highest hit rates were for "camera", microphone" and "videophone" while the "document camera", "handsfree", "selfview" and "still picture" had the lowest hit rates in all countries. Comparing countries, there were significant differences ($F(4, 111)$, $p<0.05$) in hit rates for symbols for "camera", document camera", "handsfree". In these symbols, Sri Lankan subjects generally had the lowest hits compared to the other four countries. Certainty ratings were mostly higher for Sets 1 and 3 compared to Set 2 in both Part One and Part Two. Certainty ratings of the symbols for microphone and videophone were either very high (Sets 1 and 3) or very low (Set 2) in both Parts One and Two. No significant differences were noted when subjects were compared based on occupation (student vs. employees/professionals).

Confusions can be classified into two types: symmetric and asymmetric (Nolan, 1989). The former occurs when subjects chose symbol x when presented with referent y, and symbol y when presented with referent x. Symmetric confusions usually suggest visual or conceptual similarities. Asymmetric confusions occur when subjects simply chose the wrong symbol for a given referent. Tabulations made on the rates of confusions showed more differences between countries. For example, while Thailand had the most instances of confusing Set 1's symbol for "camera" than the other four countries, it had the lowest instances of confusing Set 1's videophone symbol for other functions. Confusions are very useful indicators of the suitability of the symbol and may even complement the hit rates. For example, the videophone symbols of Sets 1 and 3 had very similar hit rates among the countries. Nevertheless, when symbol confusion was also considered, Set 1's version was better because of its lesser number of confusions. With regards to the microphone symbols wherein subjects from the five countries had similar hit rates also, Set 3's version performed best for Indonesia, Malaysia and Sri Lanka (lower symbol confusions). In the case of the Philippines and Thailand subjects, Set 1's microphone would be the better choice. Lastly, missing values were instances when some of the subjects gave no response or answers during the cued response tests. Between countries, Thailand had the most instances of missing values among all countries. Its missing values ranged from 2% to 23% across all 21 symbols. Sri Lanka was the opposite with no missing values in the cued response test. Missing values or subjects giving no answers as a parameter is a very good gauge of situations wherein the subjects did not know the answer because he/she couldn't understand the referent or the symbols presented. In actual user scenarios, this would be akin to the user not knowing which control would cause the target function (Böcker ,1993).

Altogether, some both similarities and differences were noted regarding symbol understanding among Southeast Asian subjects. The results in the spontaneous identification and cued response tests in testing videophone symbols showed very low rates of correct identification and even correct association of the symbols across all countries. Differences between countries lie mostly in the levels of confusing the symbols from each other and their missing values. Both results showed the potential difficulties encountered by people from this region of recognizing and using these symbols. However, it is still believed that graphical symbols undeniably are still very useful. When conditions warrant that they be used alone or as a primary interface, user testing, adequate information, training as well as other help procedures become very important.

4 References

Böcker, M. (1993). A multiple index approach for the evaluation of pictograms, In: *Proceedings of the 14th International Symposium on Human Factors in Telecommunications*, Darmstadt, Germany, May 11 - 14, Heidelberg: R. v. Decker's Verlag, G. Schenck GmbH, pp. 73-85.

Horton, W. (1994). *The Icon Book – Visual symbols for computer systems and documentation*, New York: Wiley and Sons.

Nolan, P. R. (1989). Designing screen icons: ranking and matching studies. In *Proceedings of the Human Factors Society 33rd Annual Meeting*, Santa Monica: Human Factors, pp. 380-384

Piamonte, D. P. T. (1998) *A User-based Evaluation of Candidate Telecommunication Icons*, Industrial Ergonomics Division and Engineering Psychology Division, Human Work Sciences Department, Luleå University of Technology-Sweden, ISSN: 1402-1757, ISRN:LTU-LIC, 98/12-SE, Luleå: Luleå University of Technology.

Zwaga, H. L. G. (1989), Predicting the public's comprehension of graphic symbols. In Megaw, E. D. (Ed.): *Contemporary Ergonomics*, pp. 431-436.

An Empirical Evaluation of Videophone Symbols: An International Study

Piamonte, D. P. T.[1], Ohlsson, K. [2], and Abeysekera, J. D. A. [1]
[1]Division of Industrial Ergonomics, [2]Division of Engineering Psychology, Department of Human Work Sciences, Luleå University of Technology, Sweden

1 Introduction

The widespread use of graphical symbols in the light of the current economic globalisation is undeniable. Symbols such as icons or pictograms as interfaces in modern technological devices have become very common. Although generally designed in western countries, their use can be found worldwide from computers to audio-visual appliances and other similar devices. Their use offers several potential advantages. They require minimum space, are easy and quick to recognise, and are non-text dependent. The use of graphical symbols, however, is not devoid of problems. Foremost among these is that some symbols are simply not understood well (Brelsford, Wogalter and Scoggins, 1994). The above advantages were cited as evident when the concepts being represented were well understood or concrete, and not when ideas become more abstract. Regarding older people, Davis, Haines, Norris and Wilson (1998) cited the studies of Easterby and Hakiel (1981) showing that although symbols are easily recognizable, understanding their meaning are generally poorer compared to younger people. One basic principle of effective UI design is to involve the potential users in the design and evaluation stages especially when the user groups are quite diverse. Thus, empirical tests are important to properly evaluate the symbols on their usability especially across different cultures.

This study was part of an international project on evaluating telecommunication symbols (Piamonte, Ohlsson and Abeysekera, 1999b). One of the major objectives was to test different graphical symbols (in this case, those of the videophone), designed and tested in Western Europe using subjects from different subject groups from Asia and the United States. The elderly as a special group was also included. The previous paper dealt with the study using subjects from five Southeast Asian countries (Piamonte, Abeysekera, and

Ohlsson, 1999a). This paper would discuss the results involving U.S. and Finnish elderly subjects and comparing them with the Asian results as well.

2 Methodology

The materials (symbols and questionnaires) used in the previous paper were also used in this study (Piamonte *et al*, 1999a and b). However, aside from the spontaneous identification and cued response tests, symbol and set preferences were also tested. Except for the spontaneous identification tests (Part One), the whole method was based on the multiple index approach (MIA) developed by Böcker (1993) with the European Telecommunications Standards Institute (ETSI).

In the spontaneous identification test (Part One), each page contained one set of the seven videophone symbols, for a total of three pages for the three symbol sets. On spaces provided for, they would write what videophone function they thought was represented by each symbol. They were also asked to rate the level of certainty for each of their answers using a seven-point rating scale (from "very certain" to "very uncertain"). In the cued response test (Part Two), the subjects first read a referent and its description. Then they had to choose one symbol from a set of seven symbols which they thought best represented the referent in question. Each page contained one referent-description and one set of symbols. They were also asked to rate their certainties for their answers using the 7-point rating scales. There were seven videophone referents tested on three sets of symbols rendering a total of twenty-one trials in Part Two. In the last part, the three symbols (one for each set) together with the referent they represented were shown. The subjects had to choose one symbol they thought best represent that referent. Next, with the three symbols sets presented together, they had to choose the set they prefer most. In all, the subjects would choose seven symbols to represent the seven referents and one symbol set they preferred most.

3 Results and Discussion

For purposes of extended comparisons, subjects from the United States were included, using the same method in the Southeast Asian studies (Piamonte *et al,* 1999a). The nineteen subjects were students and employees in the University of Central Florida. The age range was 19-43 years with a mean of 22.6 years (SD=5.24). All of them have used computers and computer-related products at home and at work, with 78.9% (n = 15) using both graphics-based and DOS-based programs. Similar to the previous study, the US subjects performed poorly in Part One, even failing to identify the handsfree symbols in all 3 sets.

The still picture symbols failed to score hits above 66% in both Part One and Part Two. In contrast, the symbols for camera, microphone and videophone had high recognition rates regardless of test type and set. Set 1 had the most number of symbols with recognition rates above 66% and Set 2 with the least. Certainty ratings generally were higher when the hits were higher as well.

Aside from correct identification and associations, other parameters were also measured. These were confusion, missing values, symbol and set preferences. Regarding preferences, the US subjects preferred five of seven of Set 1's symbols. These results were similar to the Southeast Asian study. However, they differ from the latter group by equally preferring Sets 1 and 3 as a group. One important factor that leads subjects or potential users to their choice of symbols is easy recognition. Set 3's hit rates were generally comparable to Set 1. Hence, it was not surprising for preference to be similar as well.

In some parts of Northern Finland, videophone-based services for the elderly are being tested. This was one of the major reasons why the current study was also tested among a group of elderly subjects. The respondents in the tests were twenty-nine (29) subjects from Finland (14 women and 15 men). The women were 61 to 81 years old (mean age: 69.6 years) and the men 60 to 88 years old (mean age: 70.9 years). The mean age for the whole group was 70.3 years. Compared to the subjects from Southeast Asia and even the USA, the Finnish elderly failed to do Part One. Majority of the subjects claimed that the tasks were either too difficult or too complicated. When compared to younger subjects (from Asia and the US), the elderly subjects had much lower hit rates in all symbols of the three sets used. Lastly, in contrast to the other groups also, the preference tests revealed the elderly favoring Set 3 to Sets 1 and 2. They commented that Set 3's symbols were more concrete and easily recognizable than the others.

The trend of demographic evolution shows that the European population, in particular, is one that is becoming old. This trend will have implications on the social structure, where respect for an independent living should be taken into consideration. It should taking into account the new services and products just available or those under development, such as telealarms and teleservices to assist people at home or elsewhere. Understanding the human factors of aging can lead to computer designs that will facilitate access by the elderly. Benefits include increased access of the society to the elderly for their experience, increased participation of the elderly in society through communication networks, and improved chances for productive employment of the elderly. (Shneiderman, 1992). In this case, the symbols were intended to improve the usability of a product (i.e. videophone). However, the results showed that the opposite might occur among the elderly. Only by a conscious and active participation of the elderly in designing and developing modern products will their needs be truly met.

As a whole, empirical tests using multiple indices are important to properly evaluate the symbols on their usability especially across different user groups. The evaluation method should also simulate actual usage scenarios to determine the suitability of each symbol alone, but together with the other symbols as well. This can be exemplified by the elderly and US subject groups who chose symbol sets quite different from the Asians. Hit rate is an important parameter, but must be tested under different user scenarios (spontaneous identification and cued responses) to determine if the symbols indeed are understood with and without cues. At the same time hits rates alone are not enough. Subjective certainties, false alarms (confusions), preferences, and even missing values are equally useful factors needed to make a deeper analysis. They enable the tester to see other often subtle but important differences (i.e. subjective biases, confusing symbols to other referents and symbols) on how users perceive and understand symbols.

4 Acknowledgements

- To Mr. Martin Böcker, of Siemen's-Nixdorf and ETSI Human Factors Group, for his generosity not only in sharing his resources on icons and pictograms, but more so for unselfishly sharing his insights and ideas about the study.
- To the National Communications Research Board of Sweden (*Kommunikations Forsknings Beredningen, KFB*) for the financial and logistical support

5 References

Brelsford, J. W., Wogalter, M. S., and Scoggins, J. A., (1994). Enhancing comprehension and retention of safety-related pictorials, In *Proceedings of the Human Factors Ergonomics Society 38th Annual Meeting,* pp. 836-840.

Böcker, M. (1993). A multiple index approach for the evaluation of pictograms, In: *Proceedings of the 14th International Symposium on Human Factors in Telecommunications*, Darmstadt, Germany, May 11-14, Heidelberg: R. v. Decker's Verlag, G. Schenck GmbH, pp. 73-85.

Davies, S., Haines, H., Norris, B., and Wilson, J., (1998). Safety pictograms: are they getting the message across? *Applied Ergonomics,* 29, 1, 15-23.

Easterby, R. S., and Hakiel, S. R., (1981). Field testing of consumer safety signs: the comprehension of pictorially presented messages. *Applied Ergonomics*, 12, 3, 143-152.

Horton, W. (1994). *The Icon Book – Visual symbols for computer systems and documentation*, New York: Wiley and Sons.

Piamonte, D. P. T., Abeysekera, J. D. A., and Ohlsson, K. (1999a). Testing videophone graphical symbols in Southeast Asia. In *Proc. 8th Int. Conference on Human-Computer Interaction* (HCI International '99, Munich, Germany, August 22-27, 1999).

Piamonte, D. P. T., Ohlsson, K., Abeysekera, J. D. A. (1999b). *Final Project Report: Across the seas-Auser - based evaluation of candidate telecommunication icons (Sponsored by the Communications Research Board of Sweden, KFB)*. Luleå: Luleå University Press (in press).

Shneiderman, B. (1992). *Designing the user interface – Strategies for effective human-computer interaction (2nd edition)*. New York: Addison-Wesley.

Connecting Culture, User Characteristics and User Interface Design

Thomas A. Plocher*, Chaya Garg*, Jacqueline Chestnut**
*Honeywell Technology Center, Minneapolis, Minnesota
**Mississippi State University, Starkville, Mississippi

At Honeywell, we have embarked on an initiative to improve our Asian market share by attending more carefully to the needs and requirements of our product users in Asia. As the user interface design experts supporting the initiative, we are frequently asked questions like, "Why should user interfaces to our products be any different for Asians than for Americans? Why would one expect to find cultural differences in user interface preferences?". The short answer is to point out the dozen or so papers in the literature that directly compare Americans and Asians (usually Chinese) on their performance with and without some user interface feature that has been culturally adapted. But what is the theoretical foundation that allows one not only to integrate these rather scattered research results, but also to predict where cultural differences will be found in other cultures and products, and how they might be addressed through design? A good theoretical model relating cultural differences and user interface design is the most compelling answer to these questions.

This paper reviews our research into the theoretical connections between culture, user characteristics, and user interface design. As a result of this work, we have developed our own model relating these variables. This model will continue to evolve, as we gain practical experience with our products and their Asian users.

1 Definition of "Culture"

Culture is not only how individuals act, behave, and respond, but also the beliefs, values, norms, and ideologies that they share and use to justify their actions and behavior to themselves and others. Geertz's (1973) definition captures this dual aspect of culture. Geertz (1973, pg. 145) defines culture as the fabric of meaning in terms of which human beings interpret their experience and guide their action.

2 Models of Culture and the Connection to User Interface Design

The meta-models of culture reviewed by del Galdo and Nielsen (1996) are of some help in organizing the important cultural variables and predicting how they might affect information processing preferences and performance. Several of these meta models are summarized below.

Stewart and Bennett (1991) make a distinction between the objective and the subjective layers of culture. The subjective layer of culture (Stewart and Bennett, 1991) consists of psychological features including assumptions, values, and patterns of thinking. Subjective culture is difficult to examine because it operates outside of conscious awareness. However, it manifests itself in the institutions and artifacts of objective culture. The objective layer consists of the institutions and artifacts of a culture, such as its economic system, social customs, political structures and processes, arts, crafts and literature. Objective culture is visible, easy to examine, and tangible.

Although subjective culture is harder to study, its importance cannot be emphasized enough. Hoft (1995) suggests in her "Iceberg" model that, just as only 10 percent of an iceberg is visible above the surface of the water, only 10 percent of the cultural characteristics of a target culture are visible, obvious, and easy to research. The unspoken and unconscious rules of a culture constitute the remaining 90 percent. Although these rules have a significantly greater impact on user attitudes, behaviors, personality, communication, and problem-solving styles, they are more obscure and are harder to research and identify. To understand these rules and their implications, one has to understand the target cultures rules regarding nonverbal communication, sense of distance and time, rate and intensity of speech, and business etiquette and protocol.

These ideas are further reinforced in Trompenaars (1993) "Onion" model. Trompenaars asserts that there are three layers of culture: Outer, middle, and core. The outer layer consists of explicit products and physical artifacts that are the symbols of a deeper level of culture. These explicit and easily observable artifacts include the language, food, buildings, markets, buildings, shrines, and fashions. The middle layer of a culture is defined by the norms and values that are shared among a group of people. The core of culture is implicit and consists of how people adapt to their environments, including how they seek and use information and make decisions.

In summary, models of culture almost universally conclude that a significant portion of what can be called "culture" lies "below the surface" in individuals. It is present in the form of the values, assumptions, and attitudes of individuals of a specific culture, as well as their problem-solving styles, preferred modes of communication, and ways of organizing the environment around them. These culturally-determined user characteristics and variables have significant implications for the design and development of user interfaces for information systems. They highlight the need to either remove the ethnocentricity from our information systems or design them in a manner that is compatible with the culture of the end user. Figure 1 summarizes some relationships that have been either hypothesized or empirically-demonstrated in the literature between fundamental cultural variables and user characteristics, and user characteristics and user interface design considerations. At Honeywell, we are using this model and the connections it depicts as the theoretical foundation for our notion of *culturally-adaptive user interfaces*. The remainder of the paper will discuss some examples from the Asia market region that illustrate this theoretical connection between fundamental cultural attributes, specific user characteristics, and user interface design features.

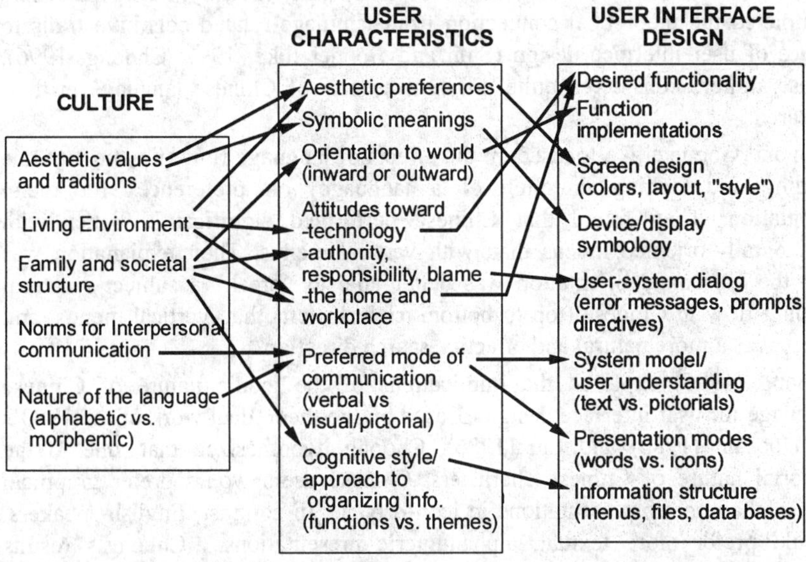

Figure 1. Connections between culture, individual characteristics, and UI design.

2.1 Aesthetic Values and Traditions

Aesthetic values and traditions are perhaps expressed most directly in cultural preferences and affective associations with colors and symbols. Fernandez (1995) provides a large amount of anecdotal evidence of these cultural variations. However, there also have been a few empirical studies, particularly on color stereotypes, among specific regional populations in China.. Courtney (1986) found that color associations or stereotypes are not absolute, but rather a matter of degree. The Mainland Chinese population he studied had weaker associations with several classic indicator and warning colors than Americans. But more recently, Luximon, Lau, and Goonetilleke (1998) failed to find significant differences in color associations with hazard conditions in a Hong Kong university student population. Hence, the issue of color preferences and associations and their translation into design guidelines may be rather more complex than we have assumed. Perhaps we can expect to find regional differences within a culturally complex country such as China, as well as interactions with education and occupation.

2.2 Nature of the Language

Numerous authors have speculated about the profound connections between Chinese language and typical Chinese cognitive traits and behaviors (Hoosain, 1986 and Liu, 1986 provide good discussions). Only recently, have researchers completed the theoretical connection from language-related cognitive traits to issues of user interface design (Shih and Goonetellike, 1997; Choong, 1996). These authors considered quite different aspects of Chinese language in their research.

Shih and Goonetellike focused on the effects of language flow (i.e. the intrinsic reading and writing direction of a language) on preferences for menu orientation. They found that Chinese performed significantly better with horizontally-oriented menus than with vertical menus. Their explanation was that the horizontal orientation was better able to "break" a subject's natural reading flow in Chinese (top to bottom, right to left) than vertical menus, and hence was a more natural and effective search direction.

Choong (1996) explored the implications of the iconic nature of Chinese language for user interface design. Based on the theoretical work of the 1980's by Hoosain (1986) and Liu (1986), Choong hypothesized that, due to the pictorial nature of Chinese characters, Chinese users would prefer graphical over alphanumeric presentations in icon-driven. In contrast, English-speakers, should prefer more textual, alphanumeric presentations. Choong's results largely supported these theoretical predictions.

2.3 Family and Societal Structure

Theory predicts that the traditional Chinese emphasis on obedience and parental authority has significant effects on a variety of variables related to cognitive style. These aspects of cognition include verbal and ideational fluency, as well as categorization and organization of information. Chui (1972) contrasted the classification tendencies of Americans with those of Chinese. Americans tended toward an inferential-categorical style, classifying objects or items of information on the basis of functions or inferences. with their relational-contextual style, Chinese tended to classify things on the basis of their thematic relationships. Choong (1996) tested these theoretical predictions by comparing

American and Chinese performance with an on-line shopping center in which the items for sale were organized either functionally or thematically. The most significant finding was that Chinese subjects consistently made fewer errors with the thematically-organized on-line shopping center. Effects on performance time were less clear.

3 References

Chiu, L. H. (1972). A Cross-cultural comparison of cognitive styles in Chinese and American children. *International Journal of Psychology*, 7(4), 235-242.

Choong, Yee-Yin. Design of computer interfaces for the Chinese population. Doctoral Dissertation. Purdue University, August, 1996.

Courtney, A.J. (1986). Chinese population stereotypes: color associations. *Human Factors*, vol. 28(1), 97-99.

del Galdo, E. M and Nielson, Jakob (1996). *International User Interfaces*.. Wiley Computer Publishing, New York.

Fernandes, Tony (1995). *Global Interface Design: A Guide to Designing International User Interfaces*. AP Professional, MA.

Geertz, C. (1973). The interpretation of cultures. New York: Basic Books.

Hoft, N. (1996) Developing a cultural model. In E. del Galdo and J. Nielsen (Eds.) International User Interfaces. 41-71. Wiley Computer Publishing: NY.

Hoosain, R. (1986) Language, orthography, and cognitive processes: Chinese perspectives for the Sapir-Whorf hypothesis. *International Journal of Behavioral Development*. 9. 507-526.

Liu, I.M. Chinese cognition. In M.H. Bonds (Ed.*), The Psychology of the Chinese People*. New York: Oxford University Press.

Luximon, A., Lau, W.C., and Goonetilleke, R.S. (1998).Safety signal words and color codes: the perception of implied hazard by Chinese people. *Proceedings of the 6th Pan-Pacific Conference on Occupational Ergonomics*.

Shih, H.M. and Goonetilleke, R.S. (1997). Effectiveness of menu orientation in Chinese. Working Paper. Hong Kong University of Science and Technology.

Stewart, E.C. and Bennett, M.J. (1991). American Cultural Patterns : A cross-cultural Perspective. Yarmouth, Maine: Intercultural Press, Inc.

Trompenaars, F. and Hampden-Turner, C. (1998). Riding the Waves of Culture: Understanding Culture Diversity in Global Business. New York, McGraw-Hall Companies, Inc.

Challenges in Designing User Interfaces for Handheld Communication Devices: A Case Study

Tham Ming Po, Ph.D.
Human Factors Engineering Laboratory
Motorola Electronics Pte Ltd
10 Ang Mo Kio Street 64
Singapore 569087
email: Mingpo.Tham@mot.com

Tan Kay Chuan, Ph.D.
Dept of Ind & Systems Engineering
National University of Singapore
10 Kent Ridge Crescent
Singapore 119260
email: isetankc@nus.edu.sg

1. Introduction

China is one of the largest markets for communication products. To stay competitive in the business, Western manufacturers are adapting the user interface of these products to meet the needs of the Chinese consumers.

Converting an English language user interface to one in Chinese is relatively straightforward for pagers displaying only numeric or simple text information. However, as the functions and features of these communication devices become more complex, for instance, a pager incorporating the capabilities of a Personal Digital Assistant (PDA) and a wireless internet browser, designing the user interface for such a device becomes a formidable task. *Formidable* because not only is there a need to customize the user interface for consumers in a foreign market but also in getting this interface to work intuitively within the constraints imposed by the hardware and software limitations of the device. This paper draws attention to some of these constraints and their impact on the design of graphical user interface (GUI) for pagers and handphones.

2. Challenges in the Design of UI for Pagers and Handphones

2.1 Lack of Guidelines and Standards

Much has been written about the internationalization of products and guidelines for designing user interface (UI) for different cultures abound in the literature (for examples, see del Galdo & Nielsen, 1996; Bonner, 1998; Shneiderman, 1992). However, the UI designer still has to adopt a trial and error approach when it comes to the actual task of designing the ideal interface for users from a different culture and language group. Furthermore, Human-computer interface (HCI) guidelines for PC and Internet applications do not address the challenges frequently encountered in the design of the UIs in pagers and handphones. Usability standards pertaining to software applications for PCs and electronic appliances are often not applicable to devices like pagers and handphones even though their UIs emulate that of PCs.

2.2 Impact of Power Supply Limitation and Display Size on Usability

Products like pagers and cellular handphones do not have the memory (RAM/ROM) size, microprocessor capabilities, power supply and the bright, big, and high resolution display screens found in laptop or notebook computers. The small liquid crystal displays (LCDs) in pagers and handphones -- typically no larger than 4 by 6 cm in size -- severely curtails the use of text labels and pop-up help-screens. This has a telling effect on usability, particularly in Chinese GUIs because, unlike English, a number of Chinese characters are often needed to represent a concept or to identify a function/feature in the product.

Conservation of power and prolonging battery life is paramount to the operation of portable communication devices. Unlike palmtop PCs and PDAs, the batteries in pagers and handphones besides providing power to drive the microprocessor and display module also has to power the transceiving and the signal monitoring/encoding/decoding processes. As a consequence, the use of colour to enhance usability (e.g., colour coding of icons), is a strategy seldom used in smaller devices like pagers because colour LCDs cost more and can consume up to 3 times more power than grey scale LCDs. Having to work around these factors makes the already challenging task of designing a GUI for users from a different culture more difficult for the UI developer.

3. The Motorola Experience

The case study in this section highlights the efforts of two design teams in Florida, USA and Singapore working independently on the GUIs of two Pager-PDA products developed by Motorola for the Chinese market. The icons created by the two teams reflect the difference in their cultural background as well as the different approaches adopted by the two teams in response to the constraining factors mentioned above. Other issues encountered in designing the user interfaces for these products are not discussed here as this is beyond the scope of this paper.

3.1 Impact of "Common look and Feel" on Usability

Team Y in Florida modified the UI of an in-production Info-Pager (a pager that allows the user to receive information on a variety of topics such as news, financial and stock markets, entertainment, weather, transportation schedules, book reviews, etc.) for the China market. Because of the Marketing Department's stipulation to maintain a common look and feel, many of the icons from the UI in the original version of the product were retained together with the dimensions of the pager physical components.

Team AB in Singapore developed the UI for a brand new product (Product AB) that combined the features of an Info-Pager with a PDA. Product AB also allowed the user to make alphanumeric keyboard and Chinese handwriting entries via the touch screen. As this was the first product of its kind for Motorola, the team had more leeway in interpreting the "'desirable size" requirement stipulated by Marketing. This allowed the team to specify the use of an LCD panel larger than any previously used in a Motorola pager product. In spite of the initial misgivings from various sources in the organization, this decision subsequently proved to be the right one. The larger display format allowed the use of text labeling for the icons, an UI feature that was found to be highly desired by users of the product.

The independent efforts of the two teams led to very different outcomes. Even though Product Y and AB have many common features/applications and functions, the icons designed by the two teams to represent these features turned out to be very different in appearance (see Figure 1).

Each icon in Product AB was labeled in Chinese. As a result of their working with a much smaller LCD and to preserve the common look of the original product, Team Y was forced to forego text labeling for their icons in order to display the same number of features/applications per screen. Because of the commonality between Products Y and AB, an evaluation of the icon designs was carried out in China and Singapore to determine which set of icons should be adopted for the Chinese UI. The participants in the surveys were asked to choose the icon that they thought best represent each feature or function in the pagers. The text labels from the AB icons were not shown during the surveys. The results from the surveys conducted in China and Singapore showed similar trends and the responses from a subset of the Chinese participants in the survey are shown in Figure 2.

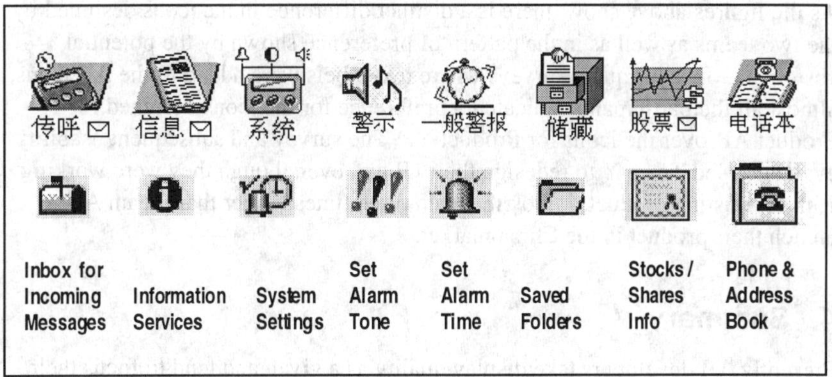

Figure 1. An example of the influence of culture on icon design: different icon designs representing the same features/functions in two similar Motorola pager products. The icons in the upper row were designed by a team in Singapore while the icons in the lower row were designed by a team in Florida, USA.

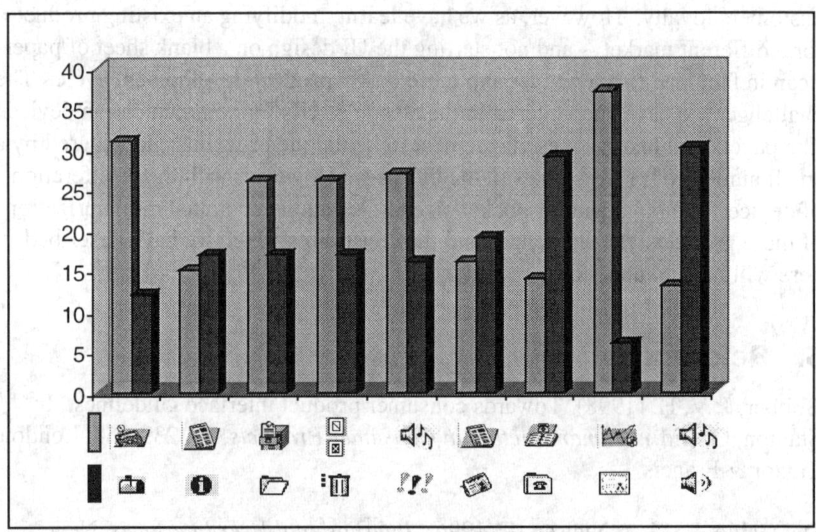

Figure 2. Frequency plot of the responses of forty-five experienced pager users in China who were asked to indicate their preference for the icon which best represent a given feature/function in the new products. The light grey columns are the responses in favour of the icons designed by the Singapore team. A similar survey carried out in Singapore displayed the same trend in users' preference.

As the figures above show, there is a distinct difference in the icons designed by the two teams as well as in the pattern of preference shown by the potential customers. In subsequent surveys where text labels were added to the AB icons, almost all the participants indicated a preference for the icons designed for Product AB over the icons for Product Y. The survey and subsequent usability evaluation led Team Y to redesign their UI and even though they were working with an existing product, it took the team three times longer than Team AB to launch their product in the China market.

4. Summary

Presently, UI developers take display quality as a given and tend to focus their effort on making the software component of their UI intuitive. However, with handheld devices, display quality becomes a primary concern because of its impact on usability. As we have highlighted above, current HCI guidelines do not address this and the other issues we have described earlier. Compounding this is the product manufacturer's desire to provide a "common look and feel" identity to its family of products. Of course, this is clearly motivated by the economy of scale as much as the need to build up brand awareness and customers loyalty. However, as we have learnt, modifying an existing product for a different market -- and not starting the UI design on a blank sheet of paper -- can in fact lead to a lengthier and more costly product development cycle. The challenges we have encountered in the design of UIs for communication devices like pagers and handphones may appear to be unique problems encountered by a small number of professionals in the business. However, with the proliferation of devices like palm and /or pocket-sized PCs and the continual miniaturization of these products, the challenges and the constraints which we have described here will be encountered more widely.

5. References

Bobber, J. V. H. (1998). Towards consumer product interface guidelines. In Stanton, G. (Ed.): *Human Factors in Consumer Products*, pp. 239-258. London: Taylor & Francis

Del Galdo, E. M. & Nielsen, J. (1996). *International User Interfaces.* New York: John Wiley & Sons, Inc.

Shneiderman, B. (1992). *Designing the User Interface: Strategies for Effective Human/Computer Interaction.* (2nd edition). Reading, Mass.: Addison-Wesley.

Design Methodology of Chinese User Interface[1] Lessons from Windows 95/98

Jian Wang, Dejiang Fu, Zhaoxia Jin and Shaosheng Li
State Lab of Human Factors, Department of Psychology
Zhejiang University, Hangzhou 310028 China

1 Introduction

With the introduction of graphical user interface and internet browser, user interface is becoming an integral part of software system and web site. User interface is facing a big challenge that never had before since software and the web site is available internationally. One thing we have to do is the localization of user interface, translating the text items (prompts, status indication and help) into local language like Chinese. But Chinese user interface is more than just the translations. A detail user interface analysis of Chinese Windows 95/98 is conducted as a case study for the purpose of exploring the design methodology of Chinese user interface. No intention is made to do a complete usability evaluation of Chinese Windows 95/98. Major user interface problems identified in Chinese Windows 95/98 are valuable for formulating design methodology of Chinese user interface. The experience from Chinese user interface design for Motorola is also briefly reviewed. The analysis and experience have general implications for the user interface design in other cultural in Asia.

3 Method

A heuristic evaluation method (Nielsen 1994) was used for the cheaper and fast analysis of Chinese Windows 95/98 user interface. Three evaluators participated to inspect the interface. No communication is allowed among evaluators before they have been completed the inspection. Their finds were aggregated in a series of discussions late.

---

[1] This work is supported by a grant from 863 Project and Science Foundation of Zhejiang Province. Author's e-mail address: jwang@mail.hz..zj.cn

To evaluate the icon and labeling design, eight subjects participate the evaluation by doing a modified version of free describing task which is very similar to the naming test of icon shape test (Bewley, Roberts, Schroit and Verplank 1983, Baeker, Small and Mander 1991). This is a little bit different from traditional heuristic evaluation. Three of them have programming experience, considered as expert users. Two of them have good knowledge of Windows 95/98 applications. The remaining three have no experience of Windows 95/98 but can use MS-DOS based applications.

For the purpose of this research, we are particularly interested in the design problems that are specific to Chinese use interface. So our evaluation focuses on Chinese character design, Chinese character input interface, icons and labeling. Total sixty icons from Windows95/98 and its corresponding labeling are used for the evaluation. The Windows® Interface Guidelines for Software Design (Microsoft 1995) was used as reference to the selection of these icons so they cover from desktop to control panel of Windows 95/98.

Due to the nature of heuristic evaluation method, the findings reported here are qualitative.

4 Evaluation Findings

4.1 Chinese characters design

Chinese character design is a very important design factor for common look and feel of Chinese user interface. True type Chinese font is used for menu and labeling screen display in Windows 95/98. Since the technical limitation of true type font, the displayed characters do not look like a regular stable and structural Chinese characters. Some local vendors provide a Chinese platform that a special hand-designed dot matrix font (Arial-like) for the menu and labeling display of Windows system to improve the visual perception of Chinese character on the screen. A better way other than dot matrix character design may needed.

It is very common to display Chinese and English in a Windows 95/98 system at same time. Current Windows 95/98 font designs are not designed for mixture display of English and Chinese. It does not provide a visually similar English and Chinese font. Chinese and English do not look like a same font. The resulting interface is inconsistent in terms of visual display. Arial-like Chinese font is suggested to reach a kind of similarity between English and Chinese font. As to other different fonts, further ergonomic research is needed to have a total solution.

4.2 Icons

It turns out in current evaluation that icon design is a big problem in Windows95/98. The key problem is that the most of icons do not provide enough meaningful information to indicate which functions they represent. It our function describing task, none of eight subject could recognize the icons of "Shut Down", "Dial-up Network", "System Information", "Run". Nobody could figure out the meaning of "Character Maker", an icon that only available in Chinese Windows95/98. For some expert user, they know they see these icons before but still have difficulty to connect these icons to the functions they represent

Some icons are cultural-dependent. Some designs used in Windows 95/98 are not very popular and well accepted in China. For example, "Recycled" is an obvious one. There is no much "Recycled" stuff yet in China. And you could see the Chinese translation is very awkward. We could also see some extra examples that is not tested in this evaluation, such as mailbox and file cabinet. Most of Chinese may never see a regular mail box that could be easily seen in US. In this case, it is inappropriate to design an American mailbox icon for Chinese user interface like most Windows 95/98 application does.

Interview with user shows that they have to use labeling of icons to understand the function even most of them have be using Windows system for a long period of time. Further exploration is needed to see if traditional icon concept is well applicable in a large system that have hundred of icons. We also notice that Windows95/98 use same icon for different application, such as sound volume control in task bar and sound recorder, while same sound volume control use a different icon when you try to access through programs menu.

4.3 Chinese character input interface

The hot issue in Chinese character input is used to be focused on the creation of input coding system and Microsoft provide its one PingYin input coding system. But now, there are may different input coding systems and user interface design of Chinese character input is becoming a big issue.

Any existing Chinese character input interface in Windows95/98 violates the principle of modeless user interface design. This result in so-called mode interface. That means users have to keep the status of keying mode (English or Chinese) in their working memory and switch between English letter and Chinese character input back and forth. Modeless interface is important for both productivity and user-friendliness. The difficulty of designing modeless interface Chinese character input should be identified before any alternative solution could be proposed. Other interaction method like handwriting and speech recognition system is a very competitive alternative to reach a modeless Chinese input interface (Wang 1996). This problem is true for most of Asia languages.

Another problem of Chinese character input interface is that user have to handle up to three different cursor position. This is a very demanding task, especially in the condition of error-correction.

4.4 Language-related design

4.4.1 Word Form

Some Chinese word could be used as both a verb and a noun. This may cause understanding of labeling sometimes. For example, the translations of "Communications" and "Communicate" are the exactly same. This will provide an incomplete information sometimes to let user know a particular operation will initiate an operation (verb) or access a property (noun) (e.g. in menu or prompt design) and reduce the usability of the user interface. Special menu and prompt design strategy, like long prompt labeling should be used in Chinese user interface design.

4.4.2 Bilingual Reading

It is typical to read text in both English and Chinese in Windows95/98. The words like "Internet", "Outlook Express", "Internet Explorer" and "Windows Update" were kept there in English. This is a very unfriendly design in some situation. For example, you can see the Chinese translation of Recycled and English word "Recycled" at same time when you use Windows Explore.

4.4.3 Prompts

Equivalent translation of Chinese text is usually short in terms of space but less informative. This makes some icon prompts difficulty to understand. Good icon design with appropriate prompt makes a better and intuitive user interface with less screen space.

These are major findings we get by using a modified heuristic evaluation method. These findings are helpful to formulate design methodology and guideline of Chinese user interface.

5 Conclusions

We have designed a Chinese user interface for Motorola's electronic device, which has been done an extensive usability test with various performance measurement. Compared with Chinese Windows 95/98 user interface, this device has a slim interface. So icon itself give enough information about the function it represent. Our experience to design small Chinese user interface for Motorola is also valuable for establishing the guidelines of Chinese user interface design. A guideline based on our work for Motorola need further comparison with evaluation findings from Windows system. Different design methodology may be adopted for small and large system.

The lessons from Chinese Windows 95/98 user interface design is an important information source for future Chinese user interface design. This also suggests that we need a Chinese user interface design guideline and methodology if we want to reach a quality Chinese user interface. Based on the experience of our user interface design and future integration with ergonomic study, we reach a better user interface than what Windows 95/98 have now.

6 Reference

Nielsen, J. (1994). Heuristic evaluation. In Nielsen, J., and Mack, R.L. (Eds.), *Usability Inspection Methods*. John Wiley & Sons, New York, NY.

Bewley, W., Roberts, T., Schroit, D., and Verplank, W. (1983). Human Factors Testing in the Design of Xerox's 8010 'Star" Office Workstation. *Proceedings of CHI '83*, 44-47.

Baeker, R., Small, I. and Mander, R. (1991) Bring icons to life. *Proceedings of CHI '91*, 444-449.

Microsoft (1997) *The windows interface guidelines for software design*. Microsoft Press Redmond, Washington.

Wang, J (1996) Multi-modal user interface based on natural human-computer interaction. *Journal of Computer (in Chinese)*, Vol. 19, 108-111.

The Emotion Mouse

Wendy Ark, D. Christopher Dryer, Davia J. Lu
IBM Research Division

1 Introduction

One goal of human computer interaction (HCI) is to make an adaptive, smart computer system. This type of project could possibly include gesture recognition, facial recognition, eye tracking, speech recognition, etc. Another non-invasive way to obtain information about a person is through touch. People use their computers to obtain, store and manipulate data using their computer. In order to start creating smart computers, the computer must start gaining information about the user.

Our proposed method for gaining user information through touch is via a computer input device, the mouse. From the physiological data obtained from the user, an emotional state may be determined which would then be related to the task the user is currently doing on the computer. Over a period of time, a user model will be built in order to gain a sense of the user's personality. The scope of the project is to have the computer adapt to the user in order to create a better working environment where the user is more productive. The first steps towards realizing this goal are described here.

2 Emotions and Computing

Rosalind Picard (1997) describes why emotions are important to the computing community. There are two aspects of affective computing: giving the computer the ability to detect emotions and giving the computer the ability to express emotions. Not only are emotions crucial for rational decision making as Picard describes, but emotion detection is an important step to an adaptive computer system. An adaptive, smart computer system has been driving our efforts to detect a person's emotional state.

An important element of incorporating emotion into computing is for productivity for a computer user. A study (Dryer & Horowitz, 1997) has shown that people with personalities that are similar or complement each other collaborate well. Dryer (1999) has also shown that people view their computer as having a personality. For these reasons, it is important to develop computers which can work well with its user. By matching a person's emotional state and the context of the expressed emotion, over a period of time the person's personality is being exhibited. Therefore, by giving the computer a longitudinal understanding of the emotional state of its user, the computer could adapt a working style which fits with its user's personality. The result of this collaboration could increase productivity for the user.

One way of gaining information from a user non-intrusively is by video. Cameras have been used to detect a person's emotional state (Johnson, 1999). We have explored gaining information through touch. One obvious place to put sensors is on the mouse. Through observing normal computer usage (creating and editing documents and surfing the web), people spend approximately 1/3 of their total computer time touching their input device. Because of the incredible amount of time spent touching an input device, we will explore the possibility of detecting emotion through touch.

3 Theory

Based on Paul Ekman's facial expression work, we see a correlation between a person's emotional state and a person's physiological measurements. Selected works from Ekman and others on measuring facial behaviors describe Ekman's Facial Action Coding System (Ekman and Rosenberg, 1997). One of his experiments involved participants attached to devices to record certain measurements including pulse, galvanic skin response (GSR), temperature, somatic movement and blood pressure. He then recorded the measurements as the participants were instructed to mimic facial expressions which corresponded to the six basic emotions. He defined the six basic emotions as anger, fear, sadness, disgust, joy and surprise.

From this work, Dryer (1993) determined how physiological measures could be used to distinguish various emotional states. Six participants were trained to exhibit the facial expressions of the six basic emotions. While each participant exhibited these expressions, the physiological changes associated with affect were assessed. The measures taken were GSR, heart rate, skin temperature and general somatic activity (GSA). These data were then subject to two analyses. For the first analysis, a multidimensional scaling (MDS) procedure was used to determine the dimensionality of the data. This analysis suggested that the physiological similarities and dissimilarities of the six emotional states fit within a four dimensional model. For the second analysis, a discriminant function analysis was used to determine the mathematic functions that would distinguish the six emotional states. This analysis suggested that all four physiological variables made significant, nonredundant contributions to the functions that distinguish the six states. Moreover, these analyses indicate that these four physiological measures are sufficient to determine reliably a person's specific emotional state.

Because of our need to incorporate these measurements into a small, non-intrusive form, we will explore taking these measurements from the hand. The amount of conductivity of the skin is best taken from the fingers. However, the other measures may not be as obvious or robust. We hypothesize that changes in the temperature of the finger are reliable for prediction of emotion. We also hypothesize the GSA can be measured by change in movement in the computer mouse. Our efforts to develop a robust pulse meter are not discussed here.

4 Experimental Design

An experiment was designed to test the above hypotheses. The four physiological readings measured were heart rate, temperature, GSR and somatic movement. The heart rate was measured through a commercially available chest strap sensor. The temperature was measured with a thermocouple attached to a digital multimeter (DMM). The GSR was also measured with a DMM. The somatic movement was measured by recording the computer mouse movements.

4.1 Method

Six people participated in this study (3 male, 3 female). The experiment was within subject design and order of presentation was counter-balanced across participants.

4.2 Procedure

Participants were asked to sit in front of the computer and hold the temperature and GSR sensors in their left hand, hold the mouse with their right hand and wore the chest sensor. The resting (baseline) measurements were recorded for five minutes and then the participant was instructed to act out one emotion for five minutes. The emotions consisted of: anger, fear, sadness, disgust, happiness and surprise. The only instruction for acting out the emotion was to show the emotion in their facial expressions.

4.3 Results

The data for each subject consisted of scores for four physiological assessments (GSA, GSR, pulse, and skin temperature), for each of the six emotions (anger, disgust, fear, happiness, sadness, and surprise) across the five minute baseline and test sessions. GSA data was sampled 80 times per second, GSR and temperature were reported approximately 3-4 times per second and pulse was recorded as a beat was detected, approximately 1 time per second. We first calculated the mean score for each of the baseline and test sessions. To account for individual variance in physiology, we calculated the difference between the baseline and test scores. Scores that differed by more than one and a half standard deviations from the mean were treated as missing. By this criterion, twelve score were removed from the analysis. The remaining data are described in Table 1.

Table 1: Difference Scores.

		Anger	Disgust	Fear	Happiness	Sadness	Surprise
GSA	Mean	-0.66	-1.15	-2.02	.22	0.14	-.1.28
	Std. Dev.	1.87	1.02	0.23	1.60	2.44	1.16
GSR	Mean	-41209	-53206	-61160	-38999	-417990	-41242
	Std. Dev.	63934	8949	47297	46650	586309	24824
Pulse	Mean	2.56	2.07	3.28	2.40	4.83	2.84
	Std. Dev.	1.41	2.73	2.10	2.33	2.91	3.18
Temp	Mean	1.36	1.79	3.76	1.79	2.89	3.26
	Std. Dev.	3.75	2.66	3.81	3.72	4.99	0.90

In order to determine whether our measures of physiology could discriminate among the six different emotions, the data were analyzed with a discriminant function analysis. The four physiological difference scores were the discriminating variables and the six emotions were the discriminated groups. The variables were entered into the equation simultaneously, and four canonical discriminat functions were calculated. A Wilks' Lambda test of these four functions was marginally statistically significant; for lambda = .192, chi-square(20) = 29.748, $p < .075$. The functions are shown in Table 2.

Table 2: Standardized Discriminant Function Coefficients.

	Function			
	1	2	3	4
GSA	0.593	-0.926	0.674	0.033
GSR	-0.664	0.957	0.350	0.583
Pulse	1.006	0.484	0.026	0.846
Temp.	1.277	0.405	0.423	-0.293

The unstandardized canonical discriminant functions evaluated at group means are shown in Table 3. As shown in this table, function 1 is defined by sadness and fear at one end and anger and surprise at the other. Function 2 has fear and disgust at one end and sadness at the other. Function 3 has happiness at one end and surprise at the other. Function 4 has disgust and anger at one end and surprise at the other.

Table 3: Functions at Group Centroids.

EMOTION	Function			
	1	2	3	4
anger	-1.166	-0.052	-0.108	0.137
fear	1.360	1.704	-0.046	-0.093
sadness	2.168	-0.546	-0.096	-0.006
disgust	-0.048	0.340	0.079	0.184
happiness	-0.428	-0.184	0.269	-0.075
surprise	-1.674	-0.111	-0.247	-0.189

To determine the effectiveness of these functions, we used them to predict the group membership for each set of physiological data. As shown in Table 4, two-thirds of the cases were successfully classified.

Table 4: Classification Results.

		Predicted Group Membership						Total
	EMOTION	Anger	Fear	sadness	disgust	happiness	surprise	
Original	anger	2	0	0	0	2	1	5
	fear	0	2	0	0	0	0	2
	sadness	0	0	4	0	1	0	5
	disgust	0	1	0	1	1	0	3
	happiness	1	0	0	0	5	0	6
	surprise	0	0	0	0	1	2	3

5 Conclusions

The results show the theory behind the Emotion mouse work is fundamentally sound. The physiological measurements were correlated to emotions using a correlational model. The model is derived from a calibration process in which a baseline attribute-to-emotion correlation is rendered based on statistical analysis of calibration signals generated by users having emotions that are measured or otherwise known at calibration time.

Now that we have proven the method, the next step is to improve the hardware. Instead of using cumbersome multimeters to gather information about the user, it will be better to use smaller and less intrusive units. We plan to improve our infrared pulse detector which can be placed inside the body of the mouse. Also, a framework for the user modeling needs to be develop in order to correctly handle all of the information after it has been gathered.

There are other possible applications for the Emotion technology other than just increased productivity for a desktop computer user. Other domains such as entertainment, health and the communications and the automobile industry could find this technology useful for other purposes.

6 Acknowledgements

Many thanks to David Koons, Florian Vogt, Frank Hoffmann, Steven Ihde and Rajat Paharia for their help on early versions of the software and hardware. Also, thanks to the BlueEyes team for their support.

7 References

Ekman, P. and Rosenberg, E. (Eds.) (1997). *What the Face Reveals: Basic and Applied Studies of Spontaneous Expression Using the Facial Action Coding System (FACS).* Oxford University Press: New York.

Dryer, D.C. (1993). Multidimensional and Discriminant Function Analyses of Affective State Data. Stanford University, unpublished manuscript.

Dryer, D.C. (1999). Getting personal with computers: How to design personalities for agents. *Applied Artificial Intelligence, 13*, 273-295.

Dryer, D.C., and Horowitz, L.M. (1997). When do opposites attract? Interpersonal complementarity versus similarity. *Journal of Personality and Social Psychology, 72*, 592-603.

Johnson, R.C. (1999). Computer Program Recognizes Facial Expressions. EE Times www.eetimes.com, April 5.

Picard, R. (1997). *Affective Computing.* MIT Press: Cambridge.

Smiling Through: Motivation At The User Interface

Nicola Millard[*], Linda Hole[#] and Simon Crowle[#]
[*]BT Laboratories and [#]Bournemouth University

1 Introduction

The concept of a 'Call Centre' is regarded as fundamental to enhancing relationships, improving the quality of contact, and maximizing the possibility of conducting business with customers. This is achieved by combining innovative technologies with skilled staff into a closely managed and controlled specialist call handling function, which supports the core area of a company's business activities. The call centre is a true socio-technical system. The agent must contend with the customer interface, maintaining a coherent conversation with the person on the other end of the phone line, whilst also interfacing with the various customer handling tools that they have to help them. The customer should be unaware of the technology whilst talking to the human agent. Enquiries can be resolved on-line, and although the volume of calls is generally very high, typically the nature of the calls is predictable and uniform. The call distribution technology automatically forces the agents to answer a call as soon as their lines are free. It tracks all agent activity, all the time, making call centres the most highly supervised and monitored environments in business today. Commonly, incentive schemes for customer service people are driven by measurements designed to assess how efficient and productive they are. However, incentives based upon quantitative measures such as call throughput, talk time and volumes reflect little on the quality of service provided to the customers.

In reality, a call centre is only as good as its people; motivating and retaining good customer service agents can be the single biggest challenge in running a call centre. This research examined how customer service agents in an inbound customer service centre could be motivated to 'smile through adversity' by incorporating motivators into their system interface, which might enhance their levels of motivation throughout the working day.

2 Studying the agents' motivation

The research team worked closely with the agents to discover what did motivate them and what could be incorporated into the interface. The study used four methods: a ranking exercise; paper-based emotion elicitation; a questionnaire; protocol analysis of the agents' conversations.

A group ranking exercise with four groups of five customer service agents investigated what the agents felt motivated them. Each team was asked to rank, in order of importance to themselves, a set of key phrases pertaining to four different views of their work. They considered what the customer wants from the company, what the company wants from its agents, the best aspects of the agents' job, and feedback they receive about the job. Once the teams had completed the ranking exercise, individuals were asked to complete a structured, open-ended questionnaire, designed to elicit individual agents' emotions. They were asked how they felt about selling, about the customers, about other team members, about themselves, and about their work sessions.

A second, structured questionnaire, using both attitude scales and bipolar scales, elicited responses from a random sample of agents. This was designed to validate the ranking exercise; there was a 54% response rate from 100 administered questionnaires. Finally, a sample of 41 agents' interactions with customers was collected, and their content examined. This was classified as either 'background activity' (handshake; general conversation; agent-system activity; customer data), or 'business activity' (service enquiry; customer problem; marketing). More than 50% of each call was found to focus on the background, 'service with a smile' activity.

3 Motivation in the Call Centre

The agents gained satisfaction from contact with both their team, and the customers, i.e. *intrinsic* motivators. However, the *extrinsic* company motivators focused on minimizing the customer contact time, so the agents did acknowledge that this was in direct opposition to the company's view. They felt that their company wants them to be efficient 'problem-solvers', providing the customers with both service and product data. They needed to use good interaction skills to maintain the customer relationship, as they felt that the customers want a pleasant and positive response to their calls, and that they prefer to obtain all the information they need within one phone-call.

They found the workload tiring, and at times, system demands caused them to break their conversation with the customer to concentrate. Although they did their best, they felt that they could improve performance. Customers' reactions affected the agents' spirits throughout their shift, but teamwork was an important factor, since fellow team members could help them in their work.

4 Having fun with technology

A stream of research has investigated whether users can enjoy their workplace interactions with computer technology. Factors for the adoption of computer technology have been identified as perceived usefulness, perceived ease of use, perceived enjoyment (Davis, Bagozzi and Warshaw, 1992), along with perceived fun (Igbaria, Schiffman and Wieckowski, 1994). Fun is an example of intrinsic motivation, where the person performs an activity for no apparent reinforcement other than the process of performing the activity *per se*. Increased ease of use of a system motivates the user to explore the system functionality, which may in turn increase intrinsic motivation, and result in greater enjoyment of the activity. Enjoyment of an activity may also be partially determined by the individual's perception of how much fun he or she is having in doing the activity (Davis et. al., 1992; Atkinson and Kydd, 1997). Kendall and Webster (1997) believe that "Fun and computers go hand in hand".

The call centre agents had no choice but to engage with the technology, so the research team attempted to improve the ease of use of the call handling interface (Millard, Hole and Crowle, 1997). Activities observed during the agents' task performance were represented in the design of performance-relevant objects featured in a Personal Office Workspace (POW). Weller and Hartson (1992) noted that, as an illusion, a user interface is itself a world in which the user can act, and by suspending a certain amount of disbelief, the user can directly engage in the world of objects. The POW user interface design offered the users such a world, and had been received enthusiastically. The rationale for the Motivational User Interface (MUI) design was to further develop the interface to continue to provide support for their task performance, but also to provide a fun, motivating environment for the agents to use (Hole, Crowle and Millard, 1998).

The MUI environment was developed to offer a more aesthetic view to the user. The agents were offered views of the *Outside world*, (a tropical beach or an Italian city) which gave them a more relaxing backdrop to focus on. The work artifacts were reproduced upon a work surface, which could be taken into any context. The POW workbooks became amalgamated into one simple *Customer book*, occupying less screen space and providing room to incorporate other items for the agents to tailor their personal workspace (e.g. clocks, pictures, etc). The peel-off notes were complemented by a *Customer capsule*, which acted as a database key, calling up relevant customer information, and maintaining the simplicity for database querying. *Clouds* in the sky represented the queued calls waiting to be taken, whilst a *window blind* provided a screen which brought the agents news of their individual or teams' successes, or company news. Such news reflected the extrinsic, company motivators, measured quantitatively.

The MUI design also sought to address the qualitative nature of the agents' work, which generated their intrinsic motivation. It provided a *Communication cube*, offering the functionality to send email or voice messages to other team members or the team leader, depending on which of the *six faces* is accessed. This would enable the agents to communicate their needs or emotions (such as customer-induced anger or frustration) to their co-workers via the interface, whilst they continue to handle customer calls. Kim and Moon (1998) observed that prior usability research has focused on the *cognitive* usability of the system. However, there is also the *emotional* usability of the system: the feelings of users as they interact with the computer system. They argue that the interface may elicit a variety of emotions ranging from the basic affective feelings such as joy or fear, to non-basic feelings such as trustworthiness or sophistication. The MUI display offered transportation objects (hot air balloon or jet) so that the agents could 'move to' a different interface environment if they became depressed or tired during their shift.

Webster and Martocchio (1992) addressed the notion of microcomputer playfulness, which relates positively to mood, involvement and satisfaction. Play has been found to be positively associated with mood satisfaction, learning, performance, and intrinsic motivation (Perry and Ballou, 1997). The monotony of the call handling work was addressed by enhancing mundane aspects such as product scripts by presenting them in the form of *script bubbles* which could be 'popped' when read. The *world face* of the *communication cube* also offered intranet access: the source of the product information.

The most frustrating aspect of the agents' work was found to be the tension that existed between limiting their call handling time and spending enough time talking to the customers to deal with their queries or concerns. The quality of their interactions with customers could not be collected via the technology, so the interface generated animated figures, or *moodies*, to provide customer images whose colours could be changed to indicate what sort of mood they had been in. The *moodies* would enable the agents to produce instant images of the types of callers they had dealt with during their shift.

5 Agent reactions

The provision of an interesting environment at the interface was very well received, as the agents liked the idea of 'looking out' at something, rather than 'looking in' on a desktop. Agents of all ages responded positively, and suggested their preferences for their own personal workspace. The agents felt that the animated figures on the interface would reduce the tension for them.

Laughter, and ideas about how much fun the agents might have during their working day, punctuated the feedback sessions. It remains to be seen whether daily use of this type of interface will truly motivate the call centre agents to enjoy achieving the company targets. The next step will involve 'proof of concept' trials in the call centre environment.

6 References

Atkinson, M., & Kydd, C. (1997) Individual Characteristics Associated with World Wide Web Use: An Empirical Study of Playfulness and Motivation. *The DATABASE for Advances in Information Systems, 28, 2, 53 -62*

Davis, F., Bagozzi, R. & Warshaw, P. (1992) Extrinsic and Intrinsic Motivation to Use Computers in the Workplace. *Journal of Applied Social Psychology, 22,14,1111-1132*

Hole, L., Crowle, S. & Millard, N. (1998) The Motivational User Interface. In J. May, J. Siddiqi & J. Wilkinson, (Eds.): *HCI'98 Conference Companion*, BCS

Igbaria, M., Schiffman, S. & Wieckowski, T (1994) The respective roles of perceived usefulness and perceived fun in the acceptance of microcomputer technology. *Behaviour & Information Technology, 13, 6, 349-361*

Kendall, J. & Webster, J. (1997) Computers and Playfulness: Humorous, Cognitive, and Social Playfulness in Real and Virtual Workplaces. *The DATABASE for Advances in Information Systems, 28, 2, 40 -42*

Kim, J. & Moon, J. (1998) Designing towards emotional usability in customer interfaces - trustworthiness of cyber-banking system interfaces. *Interacting with Computers, 10, 1-29*

Millard, N., Hole, L. & Crowle, S. (1997) From Command to Control: interface design for future customer handling systems. In S. Howard, J. Hammond & G. Lindgaard (Eds.): *Human-Computer Interaction INTERACT '97,* London: Chapman & Hall, 1997, pp. 294-300

Perry, E. & Ballou, D. (1997) The Role of Work, Play and Fun in Microcomputer Software Training. *The DATABASE for Advances in Information Systems, 28, 2, 93-112*

Webster, J. & Martocchio, J. (1992) Microcomputer Playfulness: Development of a Measure With Workplace Implications. *MIS Quarterly June, 201-226*

Weller, H. & Hartson, H. (1992) Metaphors for the Nature of Human-Computer Interaction in an Empowering Environment: Interaction Style Influences the Manner of Human Accomplishment. *Computers in Human Behaviour, 8, 313-333*

AFFECTIVE COMPUTING FOR HCI

Rosalind W. Picard
MIT Media Laboratory

1 Introduction

Not all computers need to pay attention to emotions, or to have emotional abilities. Some machines are useful as rigid tools, and it is fine to keep them that way. However, there are situations where the human-machine interaction could be improved by having machines naturally adapt to their users, and where communication about when, where, how, and how important it is to adapt involves emotional information, possibly including expressions of frustration, confusion, disliking, interest, and more. Affective computing expands human-computer interaction by including emotional communication together with appropriate means of handling affective information.

This paper highlights recent and ongoing work at the MIT Media Lab in affective computing, computing that relates to, arises from, or deliberately influences emotion. This work currently targets four broad areas related to HCI: (1) Reducing user frustration; (2) Enabling comfortable communication of user emotion; (3) Developing infrastructure and applications to handle affective information; and, (4) Building tools that help develop social-emotional skills.

2 Reducing User Frustration

Not only do many people *feel* frustration with technology, but they show it. A widely-publicized 1999 study by Concord Communications in the U.S. found that 84% of help-desk managers surveyed said that users admitted to engaging in "violent and abusive" behavior toward computers. It seems that no matter how hard we researchers work on perfecting the machine and interface design, frustration can occur in the interaction. Most HCI research has aimed to prevent frustration, which continues to be an important goal. However, there is also a need to address frustration at run-time. Affective computing can be used to address both: (1) Design-time and run-time identification of frustrating situations, and (2) Helping reduce user frustration during an interaction.

We have developed a system that gathers and analyzes two physiological signals together with mouse clicks in an effort to characterize episodes of user behavior when the user experiences problems (Fernandez and Picard 1997). Initial results were significantly better than random at detecting and recognizing such episodes in 21 out of 24 users. We are also adapting mice with pressure sensors to make it easy for people to deliberately express frustration at an application, and to have these moments of expression associated with software events. Even if the system is not smart enough to *fix* the problem that irritates you, it could (perhaps anonymously) begin to let designers know what those things are---providing a kind of *continuous* human factors analysis.

"It looks like things didn't go very well," and "We apologize to you for this inconvenience" are example statements that people use in helping one another manage negative emotions once they have occurred. Such statements are known to help alleviate strong negative emotions such as frustration or rage. But can a *computer,* which doesn't have feelings of caring, use such techniques effectively to help a user who is having a hard time? To investigate, we built an agent that practices some active listening, empathy, and sympathy, and tested it with 70 users who experienced various levels of frustration (Klein *et al.,* 1999). The agent assesses frustration and interacts with the user through a text dialogue box (with no face, voice, fancy animation or use of the pronoun "I"). Compared to two control conditions, interactions with the emotion-savvy agent led to behavior indicative of a significant decrease in frustration. These results suggest that today's machines can begin to help reduce frustration, even when they are not yet smart enough to identify or fix the cause of the frustration.

3 Enabling Communication of User Emotion

People naturally express emotion to machines, but machines do not naturally recognize it. Emotion communication requires that a message be both sent and received. In addition to the efforts above aimed at user frustration, we are building tools to facilitate deliberate emotional expression by people, and to enable machines to recognize meaningful patterns of such expression.

Emotion can be sensed in an ongoing way, or by interrupting the user for feedback. Consider a focus group where participants are asked to indicate clarity of packaging labels. If while reading line 3, a subject furrows his brow in confusion, then he has communicated in *parallel* with the task at hand, which has many advantages. Alternatively, he could stop at the end of the task and rate the label as mildly confusing on a questionnaire---*non-parallel* affective communication---occuring via interruption of or at completion of the primary task. We are working to enable both kinds of communication, e.g., via eyeglasses that sense chances in facial muscles, such as furrowing the brow in confusion or interest (Scheirer *et al.*, 1999). One advantage of these expression glasses is that they can be used in parallel with concentrating on a task or not, and can be activated either unconsciously or consciously. People are free to have a "poker face" to mask true confusion if they do not want to communicate their true feelings, and we think this is good.

We are also exploring multi-modal means of emotion communication. Current recognition rates are up to 81% in automatically detecting and recognizing which of eight emotions an actress expressed through four physiological channels (Vyzas and Picard 1999), which is at a level comparable to machine recognition of facial and vocal expressions. We are also beginning to analyze affect in speech jointly with other natural modes of expression. However, all these efforts seem to push the abilities of traditional pattern recognition and signal processing algorithms, which have difficulty handling the day-to-day and interpersonal variations of emotional expression; consequently, we are conducting basic research in machine learning theory and in pattern recognition to develop better methods.

It is important to keep in mind that some people do not feel comfortable with "parallel communication" of affect, especially with methods involving signals that people do not usually see. Users may prefer either no sensing, or non-parallel communication means such as dialogue boxes that they control, or tangible or non-tangible icons that they can "hit," "kick" or otherwise interact with to directly communicate affective feedback. People have strong feelings about if, when, where, and how they want to communicate their emotions, and it would be absurd if affective computing technology did not respect these feelings. It is important to develop a variety of means and give users choices.

4 Developing Infrastructure and Applications

Most people think it should be easy to gather data on frustration expression: Just sit a subject down in front of a computer running a certain operating system, and "voilà!." Alternatively, hire an actor or actress to express emotions, and record them. If the actor uses method acting or another technique to try to self-induce true emotional feelings, then the results may closely approximate emotions that arise in natural situations. However, these examples are not as straightforward as they may seem at first: they are complicated by issues such as the artificiality of bringing people into laboratory settings, the mood and skill of an actor, whether or not an audience is present, the expectations of the subject who thinks you are trying to frustrate them, the unreliability of a given stimulus for inducing emotion, the fact that some emotions can be induced simply by a subject's thoughts (over which experimenters have little or no control), and the sheer difficulty of accurately sensing, synchronizing and understanding the "ground truth" of emotional data.

We have developed lab-based experimental methodologies for gathering data (Riseberg *et al.*.1998). However, the best way to get realistic data may be to catch people expressing emotions to technology in everyday situations. Wearable and ubiquitous computing both offer new possibilities toward this goal. We have built "affective wearables" that sense information from a willing wearer going about daily activities (Picard and Healey 1997). Some of these wearables have been adapted to control devices for the user, such as a camera that saves video based on your arousal-response (Healey and Picard 1998), and a wearable "DJ" that not only tries to select music you like, but music that suits a feature of your mood (Healey *et al.*. 1998). We are sensing data from drivers *in situ* to learn about natural driving behaviors under stress (Healey *et al.*. 1999). We have also designed and built a wearable system to measure features of expression from professional conductors (Marrin and Picard 1998). Marrin is now adapting this "conductor's jacket" so the wearer can control the play of MIDI music in real-time while making expressive conducting gestures.

5 Building Tools to Develop Social-Emotional Skills

Autistics, who tend to have severely impaired social–emotional skills, have sometimes expressed that they love communicating by computer: computers allow for little transmission of non-verbal affective information and help level the playing field for them to communicate with non-autistics. Current intervention techniques for autistic children suggest that many of them can make progress recognizing and understanding the emotional expressions of people if given lots of examples to learn from, and extensive training with these examples. We have developed a system that is aimed at helping young autistic children learn to associate emotions with expressions and with situations. The system plays videos of both natural and animated situations giving rise to emotions, and the child interacts with the system by picking up one or more stuffed "dwarfs" that represent the set of emotions under study, and that wirelessly communicate with the computer. This effort, led by Kathi Blocher, is being tested with autistic kids aged 3-7 this month. We are also developing a stuffed animal, "Tigger," that exhibits expressive behaviors in response to how a child plays with it, discriminating potentially abusive actions like poking of the eyes from potentially playful actions like bouncing and light pulling on the tail. This work, led by Dana Kirsch, is also undergoing trials with young children.

Over the years, scientists have aimed to make machines that are intelligent and that help people be intelligent. However, they have almost completely ignored the role of emotion in intelligence, leading to an imbalance where emotions are almost always ignored. We do not wish to see the scale tilted out of balance the other way, where machines twitch at every emotional expression or become overly emotional and utterly intolerable. However, we think research is needed to learn about how affect can be used in a balanced, respectful and intelligent way; this should be the practical aim of affective computing in HCI.

6 References

These support this brief overview of HCI-related work in affective computing at the MIT Media Lab; for our references to related research not conducted at the MIT Media Lab, please see the lists in these articles.

Fernandez, R. and Picard, R.W. (1997) Signal Processing for Recognition of Human Frustration, *Proc. IEEE ICASSP '98*, Seattle, WA.

Healey, J., Dabek, F. and Picard, R.W. (1998). A New Affect-Perceiving Interface and its Application to Personalized Music Selection, *Proc. 1998 Workshop on Perceptual User Interfaces*, San Fransisco, CA.

Healey, J. and Picard, R.W. (1998). StartleCam: A Cybernetic Wearable Camera, *Proc. Intl. Symp. on Wearable Comp*uting, Pittsburgh, PA.

Healey, J., Seger, J., and Picard, R.W. (1999) Quantifying Driver Stress: Developing a System for Collecting and Processing Bio-Metric Signals in Natural Situations, *Proc. Rocky-Mt. Bio-Eng. Symp.*. Boulder, CO.

Klein, J., Moon, Y, and Picard, R. W. (1999). This Computer Responds to User Frustration. *CHI 99*, Pittsburgh, PA.

Marrin, T. and Picard, R. W. (1998). Analysis of Affective Musical Expression with the Conductor's Jacket, Proc XII Col. Musical Informatics, Gorizia, Italy.

Picard, R. W, and Healey, J., (1997). Affective Wearables, *Personal Technologies* Vol 1, No. 4 , 231-240.

Riseberg, J., Klein, J., Fernandez, R. and Picard, R.W. (1998). Frustrating the User on Purpose: Using Biosignals in a Pilot Study to Detect the User's Emotional State, *CHI '98, Los Angeles, CA.*

Scheirer, J., Fernandez, R. and Picard, R.W. (1999). Expression Glasses: A Wearable Device for Facial Expression Recognition, *CHI '99, Pittsburgh, PA.*

Vyzas, E., and Picard, R. W. (1999).Online and Offline Recognition of Emotion Expression from Physiological Data, submitted to Workshop on Emotion-Based Agent Architectures, Int. Conf. on Autonomous Agents, Seattle, WA.

Computer Assisted Education with Cognitive and Biological Feedback Loops

Evgenij Sokolov
Department of Psychophysiology,
Moscow State University, Moscow, Russia

One of the forms of human-computer interaction (HCI) is computer-assisted education (CAE). To conduct CAE in a corridor of optimal functional states avoiding frustration and drowsiness two contours of feedback are suggested: a cognitive and biological one.

A cognitive feedback has to correct the process of concept formation. The terms to be acquired are represented in a geometrical model - a semantic space. The semantic space is constructed from a matrix of subjective differences between respective terms using multidimensional scaling (MDS). Each term in the n-dimensional semantic space is characterized by specific location being represented by an excitation vector composed from excitations of long-term memory units. The semantic subjective differences are equal to distances between the ends of such excitation vectors. Two terms are synonyms if their excitation vectors are equal. It means that their representations in the long-term memory coincide. In the process of learning the student semantic space is matched against expert semantic space. The discrepancies between the spaces are used to generate instruction to pay attention to a content of respective terms to eliminate a mismatch between student and expert semantic spaces.

The vector encoding of semantics with participation of long-term memory is based on vector code at perceptual level. The study of color vision in normal trichromatic subjects has shown that colors of different hue, lightness and saturation are located on a hypersphere in the four-dimensional Euclidean space. Each color stimulus is encoded by an excitation vector having four components-excitations of four color-coding neurons: red-green, blue-yellow, bright and dark (Izmailov and Sokolov, 1991). Color working memory and color long-term memory are replica of the perceptual color space. Color names of the native language are labels of specific long-term memory units (Sokolov, 1998).

Subjective differences between native color names are closely related to subjective differences between respective perceived colors (Sokolov and Vartanov, 1987). A matrix of subjective differences between native color names treated by MDS revealed a four-dimensional space in which color names are located on a hypersphere, so that color names are located close to color stimuli that they label. Universal vector code and spherical representation found for perception, memory and semantics in color vision is operating also in the process of learning (Sokolov, 1998). Learning in the form of associations of arbitrary color names (combinations of consonant-vowel-consonant) with color stimuli was used to test gradual modification of semantic color space for arbitrary color names. Subjective differences between arbitrary color names were used to construct learning-dependent semantic color space. In the process of association of arbitrary color names with color stimuli random guesses of semantic differences were replaced by more consistent estimations. Gradually semantic space of arbitrary color names approached perceptual color space and semantic color space of native color names. At initial stage of learning arbitrary color names concentrated near red, yellow, green and blue axes. In the process of learning color names are located close to respective color stimuli constituting a sphere in the four-dimensional semantic space (Izmailov and Sokolov, 1992).

Semantic color space was used to compare the mother language with the second and third acquired languages. It has been found that an increase of the semantic knowledge is expressed in reduction of noise - decrease of a thickness of a spherical layer where color names are located. Learning of arbitrary color names as a model of semantic learning is of particular value because semantic space instead of an expert semantic space can be matched either against perceptual color space or against native color names of the same subject.

A biological feedback is used to identify user's functional state. Skin conductance, heart rhythm, brain waves can be inputted into computer to identify arousal level and emotional state of the subject. The objective data applied for modification of educational material and learning procedure to induce positive attitude, to reduce anxiety and to optimize learning. Heart rhythm is mostly convenient measure of user's functional state (Zemaityte et al., 1984). Spectral analysis of the heart rhythm reveals three peaks related to metabolic, vascular and respiratory oscillators. The magnitudes of the peaks constitute components of output excitation vectors, so that different functional states can be represented by points in the three-dimensional autonomic space (Sokolov and Cacioppo, 1997). The modifications of the functional state in the process of learning can be given by trajectories in the space. An increase of respiratory and vascular arrhythmia indicates positive attitude related to orienting responses. A depression of these components combined with an increase of the heart rate suggest anxiety and defensive responses (Danilova, 1995).

A promising approach to evaluate emotional state of user would be a computer-based identification of facial expressions. MDS of subjective differences between outlined facial emotional expressions has shown that the angles of brouttes and mouth strongly determine emotional aspects of facial expression. Different emotional expressions are located on a hypersphere in the four dimensional emotional space. Emotional tone, emotional intensity and emotional saturation correspond to three angles of the emotional hypersphere. Subjective differences between emotions highly correlate with distances between points in the four-dimensional space (Sokolov, 1992, Izmailov et al, 1999). The geometrical model of facial emotional expression implemented as a neuron-like structure in a computer might be used in parallel with autonomic indices of emotions.

Combination of cognitive and biological criteria can contribute to individual optimization of computer-assisted education.

References

Danilova, N. N. (1995). Serdechnyi ritm i informationnaya nagruzka (Heart rhythm and informational load). *Vestn. Mosk. U-ta, Seria 14, Psikhologiya*, 4, 14-27.

Izmailov, C. A., and Sokolov, E. N. (1991). Spherical model of color and brightness discrimination. *Psychological Science*, 2, 249-259.

Izmailov, C. A., and Sokolov, E. N. (1992). A semantic space of color names. *Psychological Science*, 3, 105-110.

Izmailov, C. A., Korshunova, S. G., and Sokolov, E. N. (1999). Sfericheskaya model razlicheniya emotsionalnykh vyrazhenii skhematicheskogo litsa (Spherical model for differentiation of the outlined face). *Zhurn. vys. nerv. deyat.*, vol.49, pp. 186-199.

Sokolov, E. N., and Cacioppo, J. T. (1997). Orienting and defense reflexes: Vector coding the cardiac response. In Lang, P. J., Simins, R. F., and Balaban, M. (Eds.): *Attention and Orienting: Sensory and Motivational Processes*, pp. 1-22. Mahwah - London: Lowrence Erlbaum Associates Publishers.

Sokolov, E. N. (1998). Model of cognitive process. In Sabourin, M., Clark, D. and Robert, M. (Eds): *Advantages in Psychological Science*. Vol.2. Biological and Cognitive Aspects, pp. 355-379. East Sussex, UK, Psychology press.

Sokolov, E. N. and Vartanov, A. V. (1987). K issledovaniyu semanticheskogo tsvetovogo prostranstva (Concerning study of semantic color space). *Psikhologicheski Zhurnal*, 2, 58-65.

Sokolov, E. N. (1992). Detecter mechanisms of perception and emotions. In Forgays, D. G., Sosnowski, T., and Zesnewski, K. W. (Eds.): *Anxiety: Recent Developments in Cognitive Psychophysiology and Health Research.* pp.153-165. Washington: Hemisphere Publishing corp.

Zemaityte, D., Varoneckas, G., and Sokolov, E. N. (1984). Heart rhythm control during sleep. *Psychophysiology*, 2, 279-289.

What Emotions are Necessary for HCI?

Dolores Cañamero
Artificial Intelligence Research Institute, Spanish Scientific Research Council

1 Introduction

The Human-Computer Interaction (HCI) community is showing increasing interest in the integration of affective computing in their technology. Particular attention is being paid to research on emotion recognition, since computer systems should be able to recognize human emotions in order to interact with humans in a more adaptive and natural, human-centered way. However, other aspects of emotion might be equally important, depending on the kind of interaction we are aiming at. My research has focused on two different aspects of affective computing. On the one hand, the generation of affect-driven expressive musical performances that can elicit different emotional states in humans, and that can also contribute to the understanding of how listeners perceive the expressive aspects that the musician intends to communicate. On the other hand, my work focuses on emotion synthesis for action selection and behavior modulation in autonomous agents. Both topics raise many open questions that are equally relevant within the HCI community.

2 Expressing and Eliciting Emotion with Music

One of the main problems with the automatic generation of music is to achieve the degree of expressiveness that characterizes human performances. The main difficulty arises from the fact that performance knowledge - including not only the sensible use of musical resources, but also the ability to convey emotion - is largely tacit, difficult for musicians to generalize and verbalize. Humans acquire it through a long process of observation, imitation, and experimentation (Dowling and Hardwood 1986). It seems thus natural to adopt a similar approach for the automatic generation of expressive music. This was the main motivation to develop SaxEx, a Case-Based Reasoning system that turns inexpressive performances of jazz ballads into expressive ones by imitating human performances of other jazz ballads stored in the case base of the system. In its current version (Arcos et al. 1999) SaxEx uses, in addition to expressive musical parameters (dynamics, rubato, vibrato, articulation, and attack), information concerning the affective dimensions of music in order to generate expressive performances. The system is given as input: a score (a MIDI file), which provides the melodic and the harmonic information of the musical phrase; a sound file containing an inexpressive performance of the phrase; and qualitative labels along three affective dimensions (sad-joyful, tender-aggressive, and calm-restless), through which the user can specify the expressiveness desired in the output. Values for affective dimensions guide the search in the memory of cases so that the system gives priority to performances with similar affective qualities. The selected performances are used to modify the inexpressive input file with the help of background musical knowledge (theories of music perception and understanding). The output of SaxEx is a new expressive performance (a sound file), obtained by transformations of the original inexpressive sound through imitation of the expressive resources of the selected sample cases (Cañamero et al. 1999). Readers can visit the SaxEx web site for sound examples: http://www.iiia.csic.es/Projects/music/Saxex.html. As for the uses of SaxEx in HCI, we envisage three main applications: as a pedagogical tool, as an experimentation tool for musicians, and as a module in multimedia applications to generate expressive music coupled with video images or animated cartoons.

Although the approach adopted seems to be very successful in grasping performance knowledge - listeners agree that the music generated by SaxEx sounds very natural and human-like, and they can easily identify the affective intention of the piece - a number of issues are still to be solved. The affective space is characterized in SaxEx along three dimensions (sad-joyful, tender-aggressive, and calm-restless), instead of using isolated adjectives, for two main reasons. First, the use of affective adjectives for musical purposes presents problems related with linguistic ambiguity. Second, the selected dimensions reflect three dimensions of affective meaning which seem to be culturally universal (Osgood et al. 1975): "evaluation" (good vs. bad), "potency" (powerfulness vs. powerlessness), and "activity" (activation vs. tranquillity). It is not clear, however, that these dimensions allow to express the full range of emotions that music could in principle convey. Much experimentation using very different musical styles would be needed in order to look for possible counterexamples. A major difficulty here is the fact that not much theory is available to guide experimentation. Surprisingly, as Peretz. et al. (1998) point out, although music is often characterized as the language of emotions, very few studies in psychology have been devoted to the study of music as an emotional language. Another issue that remains open is how musical elements relate to emotions. In SaxEx, musical analysis of the score and its partition into affective regions are currently performed separately. One of the extensions envisaged is to have the system learn associations between affective labels and expressive musical parameters for situations appearing recurrently in the phrases, and to use this knowledge to help assess when performers introduce expressive variations due to the logical structure of the score vs. other motives (e.g. their own expressive intention). Again, little psychological research has been devoted to this issue. It would seem that the joyful-sad dimension is to a big extent conveyed by structural properties of the music - mode and tempo of the piece (Crowder 1984) - rather than by human interpretation, but to our knowledge, almost nothing is known about the other dimensions.

3 Does HCI Need Emotion Synthesis?

The other research line that I have undertaken in affective computing is emotion generation or synthesis—computers (or in general artifacts) that "have" emotion, to put it in Picard's words (Picard 1997). My work in emotion synthesis focuses on the use of an emotional system for action selection and behavior control in autonomous agents living in a highly dynamic world, where different elements, external and internal, can threaten their lives. Emotions are in this case adaptive mechanisms related with the agent's survival, and are elicited by the presence of events that are significant with respect to survival-related goals. In my physiological model of emotions (Cañamero 1997), a motivational system drives behavior selection and organization based on the notions of arousal and satiation. Basic emotions (anger, boredom, fear, happiness, interest, and sadness) exert further control of the agent's behavior through the action of hormones that may affect the agent's perceptive, attentional, and motivational states, also modifying the intensity and execution of the selected behavior. Emotions are therefore a key element in determining what the resulting behavior of the agents, including their interactions with other agents, whether artificial or human, is going to be.

The explicit inclusion of emotional mechanisms that have direct feedback on the behavior of the system - in this case, agents - seems natural in this kind of application, but it is less clear whether every system that has to interact with humans must "have" emotions as well. Work on believable agents shows that humans easily attribute the illusion of life, personalities, and emotions, to characters that do not have explicit mechanisms for them. For some applications, one may think that the ability to recognize the emotional state of the user and to reason about it would suffice for the system to respond in a way that is adapted to the user's state. I will argue, however, that endowing the system with some mechanism that allows it to generate and show emotional behavior - putting some sort of emotions into it - can be very beneficial for most applications. Indeed, for a natural and human-like interaction with humans, HCI systems must be adaptive themselves. Emotions are essential mechanisms for adaptation, not only as far as survival is concerned, but most importantly to be able to properly interact in the social world. In social interactions, emotions are mechanisms for communication and negotiation, and they emerge from (and have an impact on) the interactions among the different individuals involved. If we devoid HCI systems of the ability to *have* emotions, we may end up developing autistic HCI systems. My claim is thus: Emotion synthesis is necessary to close the human-computer loop in HCI. Many tough problems have to be solved first, regarding for instance the choice of the emotional repertoire needed for the different applications, as well as many design choices; the solution of these problems requires the collaboration and mutual feedback of both the affective computing and HCI communities.

4 Conclusion: What Does Anthropomorphism Buy Us?

Since HCI is essentially human-centered, it seems natural to take a human-centered approach when integrating affective components in this technology. However, it is far from clear what "human-centered" means as far as affective computing is concerned. In what could be called the weak sense of the term, human-centered affective computing is intended to be adapted to the users' emotional world in order to enhance their cognitive capabilities and interactions, hiding the computer in the human-computer loop. The strong sense of the term seems to imply, in addition, some form of anthropomorphism, i.e. the attempt to integrate elements that mimic human emotion. Is the second sense of the term necessary to achieve the first one? Does affective computing need to take into account the full complexity of the human emotional apparatus? What does it mean for a computer to have human-like emotion?

I argued somewhere else (Cañamero 1998) that, if we have an engineering purpose in mind, it is neither desirable nor possible (at least for now!) to endow artificial systems with emotions of full human-level complexity (see (Picard 1997) for a list of components of human emotional systems). I also advocated for the adoption of a functional approach (as opposed to a "componential" one) to guide the design of the system. Following Frijda (1995), from a functional point of view, we have to pay attention to the properties of the structure of humans and their environment which are relevant for the understanding of emotions, and that can be transposed to a structurally different context to give rise to the functions or roles of emotions we are interested in. Not all the functions of emotions are relevant for every application - their choice, as well as of the mechanisms used to give rise to behavior fulfilling these roles, depend on the kind of interactions needed in the application. In other words, the design of the emotional system must be guided by the requirements on the system and on the interactions with the human users and their environment, so that they can show human-adapted (and in some sense human-like) emotional behavior. The other side of the coin could be that human users might tend to attribute to a system showing human-like emotional reactions the full complexity of their own emotions. This may result in the user being sometimes worried about the system, disappointed, or frustrated when s/he realizes that the system is not a human. Or perhaps this danger will simply not exist, as humans will become used to interacting with affective systems and get to know them (including their limitations). In any case, we as designers should always make clear what is "inside the system" in order to avoid promoting false illusions in the users. This is not in contradiction with the fact of trying to make emerge not only human-centered, but also human-like emotional interactions in HCI.

5 References

Arcos, J.L., Cañamero, D., López de Mántaras, R. (1999). Affect-Driven CBR to Generate Expressive Jazz Ballads. In *Proceedings of the 3rd Intl. Conf. on Case-Based Reasoning*. Berlin-Heidelberg: Springer-Verlag, LNAI (in press).

Cañamero, D. (1997). Modeling Motivations and Emotions as a Basis for Intelligent Behavior. In W. Lewis Johnson (Ed.), *Proceedings of the 1st Intl. Conference on Autonomous Agents*, pp. 148–155. New York: The ACM Press.

Cañamero. D. (1998). Issues in the Design of Emotional Agents. In *Emotional and Intelligent: The Tangled Knot of Cognition. Papers from the 1998 AAAI Fall Symposium*, pp. 49–54. TR FS–98–03. Menlo Park, CA: AAAI Press.

Cañamero, D., Arcos, J.L., López de Mántaras, R. (1999). Imitating Human Performances to Automatically Generate Expressive Jazz Ballads. In *Proceedings of the AISB'99 Symposium on Imitation in Animals and Artifacts* (AISB'99, Edimburgh, UK, April 7–9, 1999) pp. 115–120.

Crowder, R.G. 1984. Perception of the Major/Minor Distinction: I. Historical and Theoretical Foundations. *Psychomusichology*, 4, 3–12.

Dowling, W.J., Harwood, D. 1986. *Music Cognition*. Academic Press, 1990.

Frijda, N.H. 1995. Emotions in Robots. In H.L. Roitblat, J.-A. Meyer (Eds.), *Comparative Approaches to Cognitive Science*, 501–516. The MIT Press.

Osgood, C.H., May, V.H., Miron, M.S., 1975. *Cross-Cultural Universals of Affective Meaning*. Urbana, IL: University of Illinois Press.

Peretz, I., Gagnon, L., Bouchard, B. (1998). Music and Emotion: Perceptual Determinants, Immedicay, and Isolation after Brain Damage. *Cognition*, 68, 111–141.

Picard, R. (1997). *Affective Computing*. Cambridge, MA: The MIT Press.

Affective Computing: Medical Applications

Raoul N. Smith
College of Computer Science
Northeastern University
Boston, MA 02115 USA

William J. Frawley
Department of Linguistics
University of Delaware
Newark, DE 19711 USA

Introduction

Marketers have long known that a buyer's reaction to an interface is a major factor in consumer purchase and loyalty. And this reaction can be largely affective. So the interface is not just a value-added characteristic of a piece of software; it has tended to be the critical component.

The uses or even control of emotions in the design of and subsequent reactions to interfaces, and, for that matter, other computer systems or system components, is still not completely clear[1]. What emotions are appropriate in HCI? And how do emotions fit into the actual construction of practical computing systems, in addition to human-computer interfaces?

We do not think that the field should take the path of early work on natural language interfaces that tried to account for all of English grammar in one system. In particular, in this emerging research area, we should be conducting studies of actual users and see what subset of all possible emotions actually come into play in human interactions with computers. When is the knowledge of emotions appropriate to HCI and other such questions need to be answered before embarking on actual large-scale system building.

Obvious uses of emotions *in* computer systems include the following applications:

---

[1] Rich 1996 uses anthropomorphism as a guiding principle in his development of software interface agents. Emotions, as an indication of anthropomorphism *par excellence*, play a role in such agents but what, in detail, should that role be?

- computer aids to psychiatry
- text understanding
- text and speech generation
- computer art
- computer music
- robotics
- synthetic agents
- behavior animation
- artificial life
- virtual environments

Emotions in Virtual Environments for Medicine

A large class of systems that have a place for emotions is that of virtual environments, in particular, virtual environments in medicine where emotions might play a central role in applications for treating physical and mental problems[2]. These environments can be divided into two classes depending on the perspective of the users--health care professionals and patients--and hence speak to the need for designing affectively intelligent user interfaces.

Some illustrations of the use of emotions in user interfaces according to these classes of virtual environments are below:

I. Emotions in virtual environments in physical health care:

a. Health care professional as user

- enhancing physicians' empathy via interaction with a virtual patient
- designing health care facilities through virtual buildings that embody a calming environment
- reacting emotionally to a virtual patient's response to physician-caused pain such as in touching a simulated burn victim
- navigating through corporal virtual environments that cause negative affective reactions (e.g. disgust to a puncture in the intestines or pain during simulated childbirth)
- reacting affectively to a mistake in a simulated procedure
- reacting to mass casualty triage in a simulated combat environment
- emotional reacting to simulated olfactory stimuli, important in diagnosis and surgery (hence designing multimodal interfaces)

---

[2] Not that these two should be treated independently. Nor should they be treated separately from the spiritual during this evolution of research in a post-rational, post-scientific era.

- interpreting a virtual patient's medical history affectively as well as cognitively

b. Patient as user

- reducing patient anxiety during a procedure, (e.g. headmounted displays (HMD) of virtual space on a patient during dental surgery; calming scenes while having blood pressure taken)
- experiencing a facilitative virtual environment during physical therapy (inducing competition, joy, etc.)
- educating on health issues such as AIDS
- simulating a world in a disabled person to enhance social involvement so they do not become depressed, etc.
- gauging the problems and prospects of an architectural design for a disabled individual before the structure is built

II. Emotions in virtual environments in mental health care:

a. Health care professional as user

- simulating syndromes and therapies through virtual situations, e.g. fear of heights in simulated situations
- measuring success of interventions, e.g. charting a patient's mood swings to a changing environment

b. Patient as user

- identifying and expressing affective reactions to various virtual experiences
- therapeutic games to help recognize emotions and build self-esteem
- "living" a fake-it-till-you-make-it experience in a safe, up-beat virtual environment (thus replacing the traditional therapeutic strategy of the physician asking the patient to imagine their favorite place; it could be projected virtually)

Emotions In Computer Aids To Psychiatry: Two Illustrations

Here we want to describe two medical applications we have developed that rely on affectively intelligent interfaces, focused on the patient as user in a virtual environment for mental health care. The first is Dr. Bob (Smith et al. 1993), a system that retrieves stories from the Big Book of Alcoholics Anonymous and books on childhood sexual abuse. It is a memory-based reasoning system that runs on the parallel processor Thinking Machines, Inc. CM2a. It prompts the user for words that capture how they are feeling, then uses these and other words to retrieve stories. The idea behind the application is that patients will be able to read stories that are similar to their own and from these stories glean solutions to their own problems.

The other system is Alex (Smith and Frawley, to appear), which helps people suffering from alexithymia, an affective disorder in which the patient is unable to articulate his/her feelings.

Alexithymics display a variety of behaviors marked by affective reduction, from the inability to recognize, use, and verbally describe one's own emotions, and sensations, stiff posture and a lack of affective facial expressions. Alexithymia has been described variously as emotional illiteracy and has been observed in a variety of patients including those suffering from post-traumatic stress, alcoholism, and drug addiction.

Alex uses a data structure for emotions similar to a frame as used in artificial intelligence. When the patient responds to a set of questions concerning their behavior and their beliefs, the computer tries to identify the emotion which the patient is/was experiencing. As part of the system we are in the process of building a large data/knowledge base of emotions based on a word list of emotions which now contains over 7500 items. This database, when structured formally, should be usable by any of the systems in the categories mentioned in the Introduction.

Both systems lead to the following questions: What are the parameters of emotion structure that have to be codified in order to build systems that take emotions into consideration? How do these parameters bear on the design of user interfaces for practical affective computing?

Lack of emotional uptake and failure of affective cue recognition characterize a variety of syndromes (including the ones for which we designed Dr. Bob and Alex). Insofar as the cues constitute a significant part of the system-user interaction, it is important to design the interface with usable cues that are productive, that is, cues that can fully engage the system and the user. For example, lowering the eyes and the downward motion of the head are robust cues to the emotion of shame. An intelligent emotion tutor designed with a range of variations on this eye/head lowering scenario could lead individuals through these cues. It could present both canonical and ambiguous instances of shame to help consolidate and generalize the emotion and gather data on the cue uptake to trigger appropriate speech or text from the knowledge base or from the user herself.

While the tractability of these tasks might seem problematic because of the many individual differences in affective reactions to an environment, research has shown these to be a small number of prototypical, healthy reactions. Identifying these will help us build emotionally cognizant virtual environments as well other types of systems.

Emotional aspects of computer systems have generally not been attended to because of the nature of training and the state of the field which stresses the cognitive-representational side of system design. It is now time to give emotions their proper place in the development of computer systems.

References

Frawley, W. and Smith, R. N. (1998). Alexithymia and the biocybernetics of shame. *Workshop on Emotions, Qualia, and Consciousness*, Ischia, Italy, October, 1998.

Rich, C. (1996) Window sharing with collaborative interface agents. *SIGCHI Bulletin* 28.1, 70-78.

Smith, R., Chen C. C., Fang, F. F., and Gomez-Gauchia, H. (1993). A massively parallel memory-based story system for psychotherapy. *Computers and Biomedical Research*, 26, 415-423.

Smith, R. and Frawley, W. J. (to appear). ALEX: a computer aid for treating alexithymia. *Proceedings of the Twelfth International Conference on Industrial and Engineering Applications of Artificial Intelligence and Expert Systems*, Cairo, Egypt, May 31-June 3, 1999.

PART 4

USABILITY ENGINEERING

Asynchronous and Distance Communication over Digital Networks

Edoardo S. Biagioni
University of Hawaii at Mânoa

1 Introduction

Communication is the transfer of information between a speaker (sender, writer) and a listener (receiver, reader). Communication is asynchronous when the speaker and the listener perform the communication at different times. Distance communication allows the speaker and the listener to be in different places. At present the ultimate form of asynchronous, distance communication is the book, requiring the reader to be neither in the same time or place as the writer. Media that support both Asynchronous and Distance communication will be referred to as ADC media.

Traditional forms of communication between humans include written communication, including text, images, and drawings, and voice-centered communication, spanning the range from full personal interaction through voice mail. Common modes of communication include one-to-one, one-to-many (broadcasting), and parliamentary or conference call, a sequence of one-to-many communications with the same participants but different speakers.

Traditional ADC media, including print, telephone, and broadcast, each support different combinations of these forms, modes, and degrees of cost and convenience for asynchronous or distance access.

Focus on digital networks is the consequence of widespread interest in the Internet (a medium for distance communication) and the world-wide web (WWW - an ADC medium). While text and voice have been well-supported by print, telephone, and more recently e-mail, part of the interest in the WWW is its low-cost support for non-traditional forms of communication such as hypertext (text with links), images, animation, and store-and-forward sound and video. What the WWW is less well-suited for is interaction, a mode of communication supported on the Internet by chat rooms, ``talk", the MBone (Kumar 1996), and Internet and video telephony (Wu and Irwin 1998).

In this abstract we explore more advanced applications of digital networks which are currently starting to become practical due both to improving cost and bandwidth on digital networks, and to newly developed and implemented standards. These applications include interactive video-conferencing and shared whiteboard and workspace interaction, together with store-and-replay for asynchronous access and multimedia e-mail for asynchronous interaction.

2 Feasibility

In this section we focus on the feasibility of video-conferencing because video-conferencing is the most expensive of the technologies needed to support distance communication over digital networks. While video-conferencing is synchronous, the technical requirements for asynchronous delivery of on-demand stored video are similar.

On-demand, ``anytime", ``anywhere", video-conferencing has long been considered the next step for both networking and telephony. The difference now is that both cost and technology are coming within reach of the mainstream business market, educational institutions, and users of advanced technology.

For example, David Brown (Brown 1998) states that reasonable-quality bi-directional reasonable-quality video need not take up more than 400Kb/s. At the same time that advances in networking technology are providing ever-increasing bandwidth from the backbone, with ATM (Stallings 1998) and Gigabit Ethernet (IEEE 802.3), to the desktop (100 Mb/s FastEthernet cards can be obtained for under \$100 retail), advances in encoding standards, hardware, and software for video and audio transmission are reducing the load for multimedia transmission and reception on both the networks and the processors. These advances are encouraging commercial suppliers to provide encoding and decoding equipment that can use the existing Internet for video and audio transport.

While current prices for high-bandwidth long-distance communication are still fairly high, they already compare favorably with commercial travel when savings of time and travel expenses are taken into account, especially for short or frequent interactions. Without quoting specific prices, it is sufficient to point out that 400Kb/s is only about 6 times as much bandwidth as is required to carry a regular (audio) telephone call, and in the long term prices can be expected to become correspondingly low. For higher quality video, the price would be correspondingly higher, but several trends would tend to keep the price in check:

- communications capacity overall is increasing

- the cost per unit bandwidth is generally decreasing
- new compression algorithms, new hardware, and better computers can be expected to reduce the bandwidth requirements even further

The future is ultimately unpredictable, but there is every reason to believe that these trends will continue.

3 Asynchronous and Distance Education

Interest in asynchronous and distance education is partly fueled by the widespread availability of technologies such as the Internet, and partly by the perception that the rapid change of technological innovation requires lifelong learning and hence non-traditional modes of education delivery (Sloan 1999).

Our exploration is proceeding in the context of distance and asynchronous education at the department of Information and Computer Sciences (ICS) at the University of Hawaii at Mânoa (UHM). UHM has access to the Hawaii Interactive Television System (HITS), which uses analog microwave transmission, to deliver courses to other UH campuses, some on other islands. This system is expensive and only available by prior reservation. Due to these limitations, interaction for distance courses is typically limited to interaction during the scheduled lecture period or by telephone or e-mail. Asynchronous access is achieved by e-mail and by going to the local library to review videotapes. A specific example is the Maui High Performance Computing Center (MHPCC), located on an island without a major research university: professionals who work at the center and at nearby high-technology firms currently take asynchronous (web- and e-mail based) or evening HITS courses to obtain master's degrees or simply keep up with their fields.

As is true for other forms of formal interpersonal communication, the traditional delivery of education has many components, including face-to-face interaction in lectures (one-to-many) and office hours (one-to-one or parliamentary), and asynchronous communication in the form of assignments, projects, lecture notes, and testing. Of these, testing often involves trust issues that in many cases render current forms of ADC unsuitable, but the remaining forms of communication seem well-suited to asynchronous and distance delivery. To preserve quality of interaction, informal modes of interaction including assistance in labs for projects or assignments and simple questions often asked before and after lectures must also be supported when converting a class to an ADC format.

We emphasize that we are not talking about courseware, i.e. educational material that can be used independently of teachers. Instead, we are interested in providing teachers and students with better tools for reaching each other. Open-source software development has long used ADC (mostly e-mail) without reducing or removing the need for programmers; likewise we foresee ADC being used to enhance the quality and availability of education without substantially altering the model under which education is provided.

4 Capillary High-Performance Networking

A substantial improvement to HITS can be obtained by using digital networking technology with reserved bandwidth instead of the existing analog infrastructure. Compression and encoding can provide greater capacity using existing hardware, and using digital networking allows additional capacity to be purchased on a more cost-effective ``as needed" basis.

Ultimately, the need and potential is for a network as capillary as that of the telephone system, but with higher raw capacity and the same or lower cost, as well as inexpensive additional services such as voice- and video-conferencing for one-to-one, one-to-many, and parliamentary communication. The terminal can be each individual's desktop workstation, augmented with cameras and necessary software, at a cost no more than double the cost of the workstation itself. The necessary connectivity, bandwidth, and bandwidth and latency guarantees can be provided by technologies ranging from conventional telephone trunk lines to ATM and Gigabit Ethernet. While a bandwidth of 400Kb/s is sufficient for low-quality video, bandwidth reservation is required for reasonable interaction, as is a guaranteed latency of no more than a few hundred ms. By capillary we mean that every faculty's office, every conference room, and most students' offices (especially in the case of non-traditional students working full-time in the high-technology field) can be equipped at reasonable cost for video-conferencing, video production, store-and-replay, and the required network connectivity.

With capillary on-demand video-conferencing, we can provide not only lectures (with convenient and informal before-and-after lecture interaction), but also extended remote office hours, perhaps instructor-initiated for students who exhibit deficiencies. Students who exhibit similar deficiencies can be brought together into a small ad-hoc asynchronous "study group", with and without instructor interaction, something which is considerably more cumbersome in the current environment where the normal mode of interaction requires people to be in the same place at the same time, and the only realistic asynchronous alternative is e-mail.

Capillary on-demand video-conferencing with store-and-replay also provides opportunities for student interaction with other students. Instructors could periodically review (on-line or off-line) these interactions to see whether the students have an adequate understanding of the material, in much the same way that an instructor can walk into a computer lab to help students and to see what kinds of problems the students are having.

5 Automatic Video Indexing

As part of the experience of using UHM's HITS system, the author has requested feedback and suggestions on a digital version of this system. One of the suggestions (Brem 1999) was that students need a mechanism for searching quickly through stored video, with the ability to review or find specific parts of lectures without having to scan through the entire lecture.

In the ensuing conversation, we realized that any notes, slides, or overheads the instructor is using in the class can, if displayed by a computer, automatically be used as an index into the video of the lecture. The computer simply keeps track of the time at which a slide is presented, and indexes the video stream correspondingly.

We are not aware of any existing system that does this, and we therefore suggest it as a simple, practical, low-overhead way of indexing video of formal presentations.

6 Conclusions

We live in a changing world. The technologies we have examined are to varying extents already available, and most of them are incremental improvements over existing technologies. The incremental nature of these changes makes it likely that they will happen, and the opportunities for distance education are very real. These technologies can be used for most forms of asynchronous interpersonal communication, spanning the fields of higher education, business, government, and personal activities.

7 References

1. Robert Brem. Personal communication, 1999.
2. Brown. Making H.323-to-H.320 connections with two videoconferencing solutions. *Network Computing online*, http://www.nwc.com/908/908r2.html, May 1, 1998
3. Alfred P. Sloan Foundation. Asynchronous learning networks program in learning outside the classroom, http://www.sloan.org/programs.html.
4. IEEE 802.3. Originally published as IEEE 802.3z-1998. Clauses 38-42.
5. Kumar. MBone: Interactive Multimedia on the Internet. New Riders, 1996.
6. Stallings. ISDN and Broadband ISDN with Frame Relay and ATM. Prentice Hall, 1998.
7. Wu and Irwin. Emerging Multimedia Computer Communication Technologies, chapter 9-10. Prentice-Hall, 1998.

Investigating User Comprehension of Complex Multi-user Interfaces

Martha E. Crosby and David N. Chin
University of Hawaii

1 Introduction

Computers are capable of instantly providing large amounts of information to users. Unfortunately, it takes time to sift through large amounts of data to find what is needed and critical data may escape the user's notice. In time-critical applications such as crisis management, delays from search or overlooked information can have life and death consequences. To help users, computer systems can use a user model to reduce the information load by presenting only data that is relevant to the user's current task (user-model-based information-filtering). Unfortunately, there have been few, if any, experimental studies to verify the conditions under which user-model-based information-filtering leads to improved user performance in time-critical tasks. Indeed, even the utility of information filtering is as yet an unproved speculation. It may be that in (some) situations, some amount of context (which is not strictly needed for the task) helps (some) users situate themselves, thus leading to better performance. For example, in finding routes, landmarks pictured on a map are not be absolutely necessary and some people may even find them distracting, but many others find them helpful. This paper describes two controlled experiments conducted to evaluate the effectiveness of user-model-based information-filtering techniques for reducing the cognitive load of users in time-critical crisis-management situations.

2 Experiment 1

The purpose of experiment 1 was to determine the extent that different types of complexity (map or task) influenced the accuracy of the participants ability to find an optimum route (as measured by the number of intersections) from a fire station to a fire. The map background condition was either simple or complex.

The maps with the simple background showed only the roads and intersections. The maps with the complex background, however, included details such as street names and landmarks. The problems also had a simple or complex factor. A simple problem had only one fire to put out. A complex problem had multiple fires as well as distractors such as ambulances, road blocks, and icons of different sizes to indicate the severity of the fire.

2.1 Method

Sixteen students at the University of Hawaii were given booklets consisting of twelve maps. On each page were icons and roads. For each fire icon, they were to bring one fire truck to it (simple task). If there were double fire icons, they had to bring two fire trucks and one ambulance to it (complex task). For complex tasks, counting the intersections became more difficult as fire trucks and ambulances were not allowed to pass through road blocks that existed on some of the roads.

2.2 Results and Discussion

Results were analyzed with respect to the types of errors made and the strategies the students employed to solve the problem. A multiple regression showed that 3 of the variables accounted for 86% of the variance, ($F=16.8$, $p=.0008$). Figure 1 shows results of a 2 factor ANOVA between the tasks and the maps.

The AB Incidence table on Y 1 : avg % correct

	map:	simple	complex	Totals:
task	simple	4 92.109	2 93.125	6 92.448
	complex	5 75.688	1 69.688	6 74.688
	Totals:	9 82.986	3 85.312	12 83.568

Figure 1. Percentage correct by map and task complexity (Experiment 1)

Problem complexity was by far the most salient factor in the successful solution of the simulation. Although the students performed better on maps without labeled streets, the backgrounds did not contribute much to the solution variance. Icon size, which was used to designate the magnitude of a problem, was too difficult for the students to discriminate, thus many of the simulations were only partially completed.

3 Experiment 2

A User Modeling System crisis-management system, EMI (Evaluating Multi-user Interfaces), was written to support multiple team members collaborating in a crisis management situation. EMI provides team members with customized user interfaces that are adapted to the user's current role and task. For example, a fire captain (the user's role) dispatching fire trucks (the user's task) sees a map showing the locations of fires, roads, fire stations, and fire trucks. However, if the fire captain switches to the task of ensuring adequate water supplies for fighting a particular fire, then fire stations and fire trucks are replaced with water hydrants (with pressure readings) and pump trucks (with capacities). In contrast, an ambulance dispatcher selecting an ambulance to send to an accident site sees an entirely different map, one that shows the region's roads, the accident's location, hospitals, and all ambulance locations and availability. These different maps, called thematic maps, are created automatically by EMI based on the user's current role and task. The EMI crisis management system employs user task models to filter out information that is not relevant to the user's current task. In particular, irrelevant (distractor) icons are not displayed on the user's crisis map. To determine the effectiveness of such information filtering, two controlled experiments were performed using a counterbalanced repeated-measures design.

For the purpose of Experiment 2, EMI was modified to record all user interactions including keystrokes, mouse clicks, and mouse movements. The recorded interactions can be replayed by EMI at a later time. EMI also includes a server component that receives information from clients (the map-based user interfaces) such as the addition, deletion or movement of icons and rebroadcasts this information to other clients based on the current job and task of the user at each client. For Experiment 2, the server component was modified to allow adaptable filtering. In the normal filtered mode, only those icons that are relevant to a user's current job and task are forwarded to the user's client program by the server. In the control (unfiltered) mode, all icons are forwarded and displayed on the user's client program. In addition, the EMI server has the capability to record icon interactions (additions, deletions, and motions) from clients and later replay these icon interactions. This feature was used to create scenarios for the experiment. For this experiment, all scenarios put the user in the fire captain role and the dispatching fire trucks task. The dispatching fire trucks task requires the user to select which fire station to send fire trucks to fight which fire and select the shortest route to the fire from the fire station. For simplicity, each fire station was to be assigned to fight one and only one fire.

The shortest route was defined to be the path with the least number of intersections, since fire trucks must slow down before crossing intersections to avoid causing traffic accidents. Also, the route cannot pass through road block icons, which represent reported hazards such as fallen trees, floods, rock or mud slides, disabled cars, lava flows, etc. that make the road impassable.

The initial maps contain from one to three fire stations and from one to eight road blocks. The filtered versions of the scenarios only contain fire stations and road blocks whereas the unfiltered versions contain other types of irrelevant icons (distractors). In a short time after the initial screen, one to three fire icons appear simultaneously on the user's map. At this time, the user begins to select the best route(s) by pointing and clicking on roads using the mouse. Incorrectly selected roads can be deselected by clicking a second time on the road. Selected road sections are highlighted by EMI by redrawing the road as a wider line (three pixels wide instead of one). In some of the unfiltered version of the scenarios, additional distractor icons enter, leave, and move around in the map simulating a real crisis management situation in which many things are happening at the same time.

3.1 Method

Subjects were given a computer version of six scenarios in which they had to click on the best route. The six scenarios were repeated for each user both with and without user-task-model-based information-filtering. All twelve scenarios were presented to each subject in varying order of difficulty. The effectiveness of filtering was determined by comparing multiple dependent variables: the subject's performance speed and accuracy, their subjective evaluation of the difficulty of the scenarios, and their pupil dilation (which previously has been shown to be correlated with cognitive load).

Thirty eight faculty, staff, and students from the University of Hawaii participated in Experiment 2 which used the EMI system, run on a Sun Spark Station 20 with a 20 inch color monitor. The purpose of this study was to investigate the influence of filtered versus unfiltered data in complex user interfaces. Based on the results from Experiment 1, six scenarios were devised: two scenarios containing one fire station and one fire, three scenarios containing two fire stations and two fires, and one scenario containing three fire stations and three fires. These six scenarios were duplicated to produce twelve total scenarios: six filtered scenarios (with user-model-based information-filtering) and six unfiltered scenarios (without user-model-based information-filtering). Using a counterbalanced repeated-measures design, each subject was given all twelve scenarios at one sitting in one of four orders. Half of the subjects were given all filtered scenarios first and half were given all unfiltered scenarios first. Among the six filtered or unfiltered scenarios, half of the subjects were given the six scenarios in order of difficulty (from one fire to three fires) and the other half were given the six scenarios in opposite order of difficulty (starting with three fires first). The subjects were first given some simple instructions to read. Next they listened to a previously recorded tape of instructions that were synchronized to a recorded demonstration of using the EMI system to solve a two fire station and two fire practice scenario. After watching the demo scenario (which includes descriptions of the EMI map interface, the task, and the evaluation criteria for the task), the subject was given the same scenario as the demo to practice before proceeding to the actual experimental scenarios. The subject was allowed to ask questions during the practice scenario. Between scenarios, the subject was allowed to rest for as long as the subject wished. At the end of each scenario, the subject was asked on the computer screen to rate the difficulty of the scenario on a scale of 1 to 7. The independent variables in the study are the presence or absence of user-model-based information-filtering and the task difficulty (number of fires).

3.2 Results and Discussion

The results suggested that the user-model-based information-filtering differentially influences performance. There variability in the time the participants took to complete the task accounted for the greatest variance As expected the most participants gave more accurate answers if the background was filtered as opposed to unfiltered. This was particularly true for the more complex versions of the six scenarios. The problem accuracy (participants' proportion of correct answers) was not as heavily influenced by the task complexity in experiment 2 as it was in experiment 1. The user's subjective evaluation of the difficulty of the scenario..

4 Conclusion

The use of multiple dependent variables allows analysis of the consistency and relationship between variables and serve as a check on the validity of the experimental design. The findings of Experiment 1 and Experiment 2 suggest that many variables effect the speed, accuracy, and perception of even very simple tasks. In a crises or emergency situation errors and poor problem solutions are even more likely to occur. Designers of computer interfaces for complex systems need to be aware of the effects of too much detail. User-model-based information-filtering may help alleviate the problem.

This research was supported by ONR grant N00014-97-1-0578 to both authors.

An Extendible Architecture for the Integration of Eye Tracking and User Interaction Events of Java Programs with Complex User Interfaces

Jan Stelovsky
Department of Information and Computer Sciences, University of Hawaii, USA

1 Background

With the advent of Java Foundation Classes (or Swing), Java applications and applets can use virtually all the sophisticated features of modern graphical user interfaces, such as iconic buttons, combo boxes, nested menus, toolbars, internal windows with menus, tables, hierarchical displays reminiscent of the "Windows Explorer", and World-Wide Web (WWW) browsers. As a consequence, we see the emergence of advanced applications written entirely in Java - word processors, spreadsheets, databases - whose user interfaces approach quickly the complexity of the comparable standard tools for desktop computers.

The Java Development Kit now includes a set of "reflection" classes which give a developer the unique opportunity to write a program that examines all objects within a software system. In particular, we can provide a "monitor" program that traverses the hierarchy of user interface elements and attaches itself as a "listener" of the events that constitute the user's interaction with the system. Such a generic monitor program can be attached to a wide variety of new tools written in Java – whether WWW applets or stand-alone applications - and log the interactions or replicate them remotely on another computer on the network.

Eye tracking equipment provides us with a wealth of data about the focus of the user's attention. We can analyze the eye positions to filter out the eye fixations. The integration of the stream of eye fixations with the user interaction events enhances the basis for the empiric analysis and interpretation of subjects' interactions with a variety of emerging user interfaces.

2 Analysis of user interfaces in Java programs

The user interface of all Java applets and applications consists of objects derived from one common class: *java.awt.Component*. These components are nested and form a large hierarchy whose root is an object of type *java.awt.Frame* (which is derived from *java.awt.Window*). While a stand-alone application must create the root frame, an applet is embedded in a frame which is provided by the outer environment, e.g. by a WWW browser. A Java tool can use other auxiliary *java.awt.Window* objects, typically to provide dialog windows.

A typical Java program creates the entire hierarchy of user interface components at initialization time. An outside "monitor" program can then detect the root frame and extract the entire hierarchy of components using the classes in the *java.lang.reflect* package. If all the auxiliary window objects are accessible via instance variables and their content is created and composed during the initialization, then the entire user interface of such a "static" system can be analyzed by the monitor program. The monitor program provides a method that a more "dynamic" target program - i.e. one that creates the windows in local variables and/or changes the structure of components at run time - needs to call to let the monitor program analyze the new components and their structure.

The event notification framework introduced in Java Development Kit 1.1 provides two other convenient features: 1) the possibility to attach additional listener objects to each relevant component, and 2) the direct access to the event queue. These features enable the monitor program 1) to be notified whenever the user interacts with the target application, and 2) to replay these interactions.

When the target program is "static" in the above sense, our monitor can present the experimenter with a tree-like display of the user interface components and let them specify which events should be monitored during the experiment. The resulting event filter can be prepared ahead of time and read at the beginning of an experimental session. The monitor program will then install listeners only for the relevant components thus improving the efficiency of the monitoring process and reducing the size of the resulting data.

The eye tracking equipment is a source of a steady stream of data. The driver software that supports the eye tracker delivers an eye location sample every 1/60 of a second. The stream of eye location data must be analyzed to determine the actual eye fixations. The resulting stream of eye fixations needs to be combined with the user interaction events – either in real time, or in a separate processing step, or during the post-experimental analysis phase. We plan to augment our "Experimenter's Workbench" (Stelovsky, J. and Crosby, M.E. 1997) with tools that take advantage of the interleaved user interaction and eye-fixation events.

3 Unified framework for event processing

A flexible monitoring and analysis environment should support numerous data types, data formats and processing modes. The data types include:

1) raw eye location data,
2) eye fixation events, and
3) user interaction events.

This data can be stored and retrieved in various formats, such as:

1) raw binary form (e.g. as obtained from the eye tracker hardware),
2) textual human-readable form, and
3) persistent object format as defined by the *java.io.Serializable* interface.

Moreover, there are several processing modes:

a) during experimental session:
    1) real-time events arriving at irregular time intervals,
b) during post-experimental analysis:
    2) stream of events read from a local file or a network connection,
    3) randomly accessible local data structure containing events,
    4) events arriving at time intervals corresponding to data obtained via 2) or 3).

These processing modes fall into two categories:

1) "pull" type where the consumer reads data in stream-like fashion, and
2) "push" type where the client is notified whenever data is available.

The object-oriented approach to analysis and design allowed us to develop a unified framework that satisfies the above requirements and allows the programmer to combine them within tools that convert data in between formats as well as interleave eye fixation and user interaction events. This framework lets the monitoring and analysis tools - such as the "Analyzer" tool within the "Experimenter's Workbench" - consume data in event-oriented manner even though is read from a stream and vice versa.

The data formats are represented by classes derived from a superclass *TimeEvent* such as the subclasses *InteractionEvent*, *EyeLocation* and *EyeFixation*. Each of these subclasses is responsible for the conversion to a common format. The classes *InputEventStream* and *OuputEventStream* handle the "pull" type of reading and writing of any *TimeEvent* object from and to streams - such as local files or network connections. The "push" type of processing is provided using the event-notification schema as defined in the JavaBeans specification. The interface *TimeEventListener* defines the method *eventOccured()*. The clients which implement this interface and register themselves with an object of type

EventSource will be notified whenever a *TimeEvent* occurs. This framework can be easily extended by constructing new subclasses of *TimeEvent* to accommodate new data formats.

This framework has been used to construct a World-Wide-Web based analysis tool that visualises data collected during past experimental sessions (Figure 1). This tool displays the picture of the of computer screen presented to the subject and draws on top of it the subjects' eye-tracking samples (small circles) and fixations (big circles connected with lines). The user can vary the speed of the animation and the attributes that guide the computation of fixations.

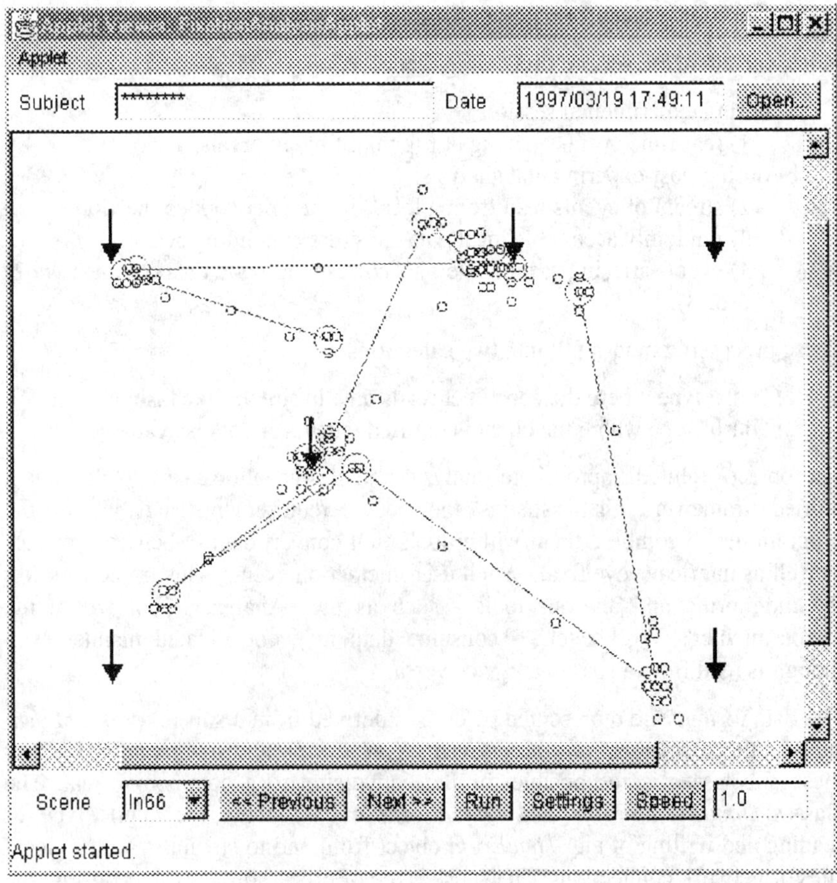

Figure 1: Visualisation of subject's eye-tracking samples and fixations.

We are currently implementing a distributed Java system that uses the proposed framework to monitor an eye-tracking experiment in real time. The server application resides on the computer that is connected to the eye-tracking equipment and runs the driver software. This application uses Java Native Interface (JNI) to convert the driver's data samples into events and Remote Method Invocation (RMI) to send these events to the client listeners. This system will enable the experimenter to monitor the subject's eye fixations with a WWW browser.

Other proposed tools will convert legacy data collected during past experiments to new formats so that they can be presented on WWW.

4 Conclusion

The proliferation of Java programs gives us new opportunities to study human-computer interaction. Our monitoring tool can analyze a wide variety of Java applets and applications, record the users' interactions with such systems and integrate them with the stream of user's eye fixations. Our framework employs object-oriented principles to ensure a flexible, easily extendible architecture that can incorporate future data formats as well as different processing schemas with minimum programming effort. This framework is used in monitoring and analysis tools whose software architecture supports WWW browsers and other computer systems distributed across Internet.

5 References

Stelovsky, J., Crosby, M.E. (1997). A WWW Environment for Visualizing User Interfaces with Java Applets. In Salvendy, G. Smith, M. & Koubek, R. (Eds.): *Design of Computing Systems: Social and Ergonomic Considerations (Vol. 2), Proc. 7th Int. Conference on Human-Computer Interaction* (HCI International '97, San Francisco, USA, August 24-29, 1997), pp. 755-758. Amsterdam: Elsevier.

Evolutionary Approach to Revolutionary Use of Software Diagrams in Client-Designer Communication

Koji Takeda
University of Hawaii

Objective

Managing communication in a software development project team is troublesome for many reasons, including the different skill levels and philosophy among designers, different backgrounds of clients and designers, the ambiguity of a client's requirements, the complexity of software due to its abstract and dynamic nature and the lack of a recording method for communication processes. This is particularly true when client involvement in the process is high. A computer support is needed.

The software specification diagrams developed in CASE (Computer Aided Software Engineering) tools are used widely by designers. The use of diagrams in software specification has proven to be helpful for designers to improve the quality of communication and elicit clearer thinking. With the aid of the interactive visual user-interface technologies available on most computers today, a diagram's expressibility and understandability can easily exceed that of natural language.

Our idea is to use those diagrams to enhance client-designer communication. However, for such an attempt, the diversity of diagram language is a burden for learners. Also, since diagrams are designed for software design activities (and most of time designed by computer experts), they lack design perspectives for a client's use. Designing a uniform diagram language system, which will overcome these difficulties, is our goal.

Significance

We have developed a diagramic language system called MERA that enhances client-designer communication and significantly reduce the stress level of project mangers. It contributes to long overdue software engineering missions of productivity and quality.

The most significant aspect of this project is that, the design of MERA and associated tools was developed incrementally and tested through actual use in seven major software development projects over a period of ten years. Each project lasted one to three years, with ten to thirty members, and with the Japanese private sector as the clients. The project members included people from computer areas such as programmers, system analysts, professors and graduate students of computer science department, and from other areas including customer service, business planning and sales, and actual users of the target system. The scale, length, number of projects and the diversity of the participant's backgrounds created an excellent environment to solicit requirements and test the designs of MERA.

Technical Approach

The diagram scheme uses a uniform syntax so that clients do not need to learn different kinds of notations for different diagrams and they can use the same tool for all the diagrams. Moreover, MERA can be used to define new diagram notations. Therefore, both end diagrams and diagram notation definitions are represented uniformly and can be edited by the same tool. This makes it easy, even for the client, to create a new diagram notation by adapting the existing definition.

Figure 1 shows an example of MERA diagram. This diagram depicts the data flow of a point-of-sale (POS) system, and is using a variation of data-flow diagram notation, a popular software specification diagram. Figure 2 is a partial definition of the diagram notation used to draw the diagram in figure 1.

Over a period of ten years, several features have been added and removed to make the language easier to use. Current version includes separation of view data, animation of diagrams, stereotype models, methodology models and common data format.

Many clients have a strong preference about the way each diagram is displayed. The separation of view data from definitional data increases the flexibility of diagram display so that the client's diagram preference can easily be accommodated.

When dynamic property of system, such as user interface, is abstracted into a static model, the client may not be able to reconstruct the original dynamism. The animation of diagrams enhances the understandability of abstract models.

The stereotype model serves as a template diagram for particular classes of software. It provides guidance for those clients who have difficulty writing a diagram from scratch. The stereotype model also implements storage and reuse of domain knowledge.

The methodology model provides additional guidance during the construction process of diagrams. By suggesting and controlling the organization of the diagrams, tools to be used and process steps to be taken by the client, completeness and consistency of the resultant diagram is assured.

The common data format allows the exchange of diagrams across heterogeneous computer environments.

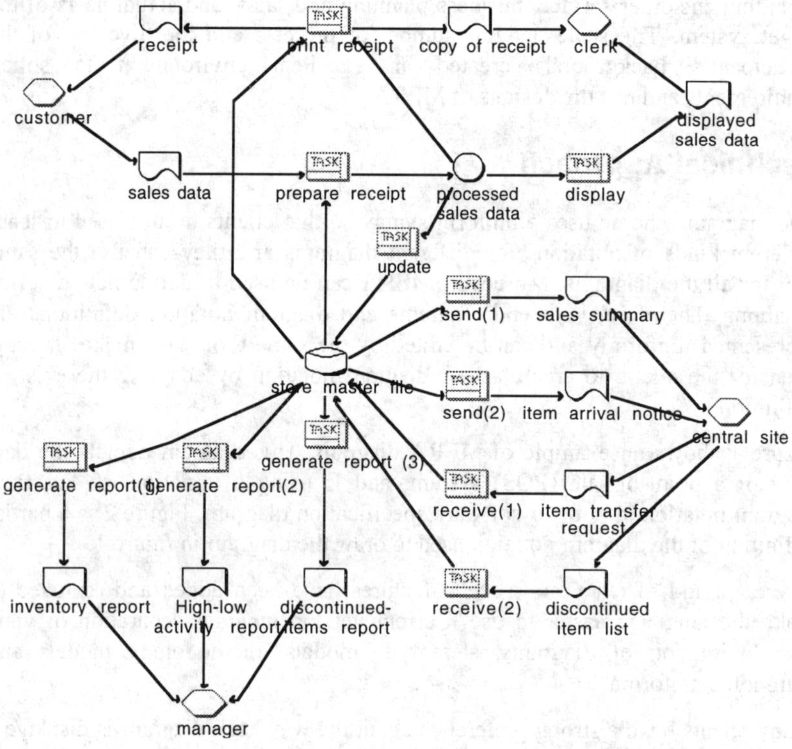

Figure 1. MERA diagram example.

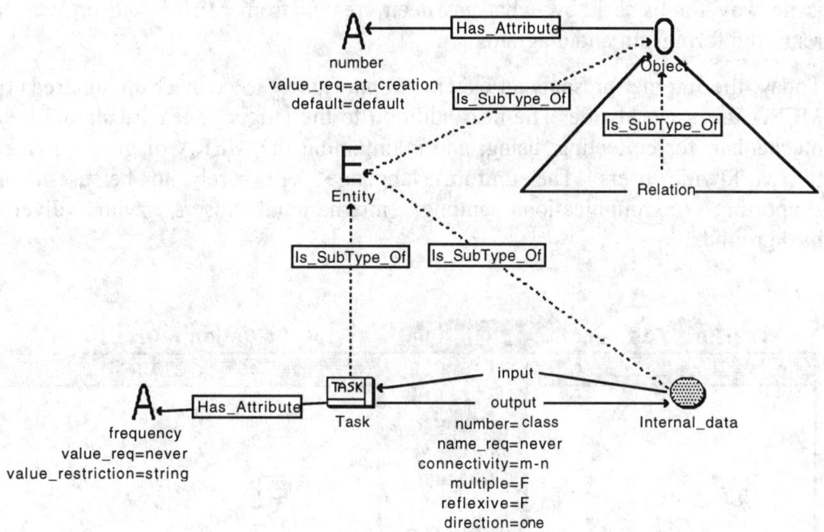

Figure 2. Diagram notation definition example.

Result

The basic editor for MERA is being implemented on Java for platform independent operation. Other versions of editors are already running on Macintosh, PC, UNIX workstations, and any Smalltalk-80 compliant platforms. The Macintosh version (figure 3) has been distributed to research institutes worldwide. We have been receiving positive responses from individual users of our editors. During the past few years, MERA was used in projects conducted by organization other than University of Hawaii. These organizations also scored high on the flexibility of MERA for representing variety of concepts, and acceptability by diverse class of users.

One way to set our future goals is to analyze the weaknesses of the current implementation. We want universal acceptance of our language. However, declaring it as a standard and providing an editor will not lead to the acceptance of the language by a large user community. We need to present the language as a way to encode knowledge by making created diagrams reusable and share-able. In this regard, a weak point of the current language design is its inadequate support for knowledge acquisition. Over 100 copies of the MERA editor have been distributed worldwide. Over 50 researchers studied at University of Hawaii and went home with lots of experience using MERA. Probably, dozens of diagram formalism have been implemented and hundreds of diagrams have been drawn using those formalisms at various sites. Yet there is no way for us to know what has been created from MERA and no way to reuse the formalism and diagrams.

Today, the Internet presents an excellent opportunity to connect up hundreds of MERA users worldwide. The next addition to the language feature should be a mechanism for collecting, using, and maintaining the MERA diagrams created by worldwide users. The uniform language will surely to be useful in supporting communication among international users with diverse backgrounds.

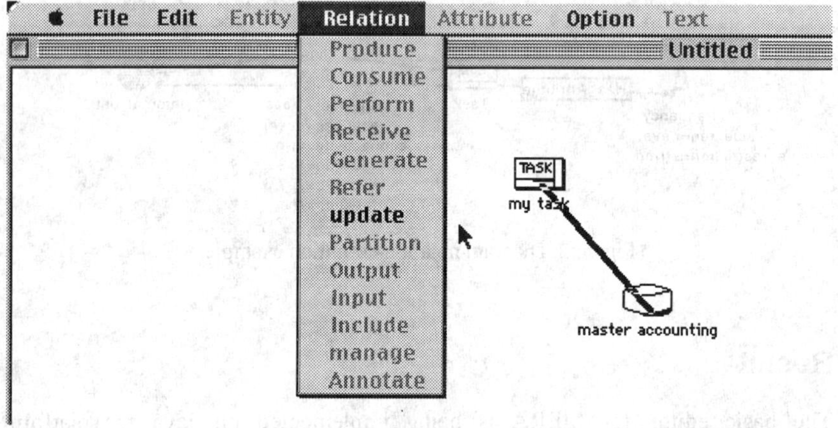

Figure 3. Syntax driven editor for MERA diagram.

References

Chin, D. (1989). Software Requirements as Variations on Stereotypical Models. *In Proceedings of the Workshop on Automating Software Design* (IJCAI-89, Boston, MA, August 1989), pp. 60-64

Takeda, K., Chin, D., Miyamoto, I. (1993). MERA: Meta Language for Software Engineering. *In Proceedings for 4th International Conference on Software Engineering and Knowledge Engineering* (SEKE-93, Capri, Italy, 1993).

Takeda, K., K, Inaba, M., Sugihara, K. (1996). User Interface and Agent Prototyping for Flexible Working. *IEEE Multimedia*. Vol. 3, No. 2, pp. 40-50.

Miyamoto, I. (1987). A Prototyping Tool for Graphical Software Engineering Tools. *ACM Software Engineering Notes*, Vol.12, No.4.

Project Leap: Addressing Measurement Dysfunction in Review

Carleton Moore
University of Hawaii, Manoa

1 A Problem with Reviews

The software industry and academia believe that software review, specifically Formal Technical Review (FTR), is a powerful method for improving the quality of software. FTR traditionally is a manual process. Recently, computer mediated support for review has had a large impact on review. Computer support for FTR reduces the overhead of conducting reviews for reviewers and managers. This reduction in overhead increases the likelihood that software development organizations will adopt FTR. Computer support of FTR also allows for the easy collection of empirical measurement of process and products of software review. These measurements allow researchers or reviewers to gain valuable insights into the review process. With these measurements reviewers can also derive a simple measure of review efficiency. A very natural process improvement goal might be to improve the numerical value of review efficiency over time.

My research group, the Collaborative Software Development Laboratory (CSDL), developed a computer supported review system called Collaborative Software Review System (CSRS) (Tjahjono 1996). CSRS is a computer supported software review system implemented in UNIX with an Emacs front-end. It is fully instrumented and automatically records the effort of the participants of the review. The moderator of a review using CSRS may define and implement their own review process. Dr. Tjahjono and Dr. Johnson used CSRS to investigate the effectiveness of group meetings in FTR (Johnson and Tjahjono 1998). Our experiences with CSRS caused us to start thinking about the issues of measurement dysfunction.

After looking closely at review metrics, we started to worry about *measurement dysfunction* (Austin 1996) and reviews. Measurement dysfunction is a situation in which the act of measurement affects the organization in a counter-productive fashion, which leads to results directly counter to those intended by the organization for the measurement. Some of the different types of measurement dysfunction that can occur in software review are: Defect severity inflation where the severity of a defect is raised because the reviewers want it fixed faster; Defect severity deflation, where the severity is lowered so the project may complete a milestone. Defect density inflation, where many minor defects are reported or the size of the document reduced to bring the document up to norms. Defect density deflation, where the number of defects are under reported or the size is over reported to make the document look better. Other types are artificial defect closure, and reviewer preparation time inflation/deflation (Johnson 1996). Reviewers or project managers may feel pressure to report good defect severity or density levels. They can miss report the severity or number of defects to affect the metric reported to management.

I have witnessed defect severity reduction in industry. An organization's policy was a project could not pass any milestone with any open defects of critical severity. The project manager talked to the developers and got them to reclassify all the critical defects as high so the project could pass the milestone on date. Since all the critical defects were now high they did not get fixed first. This reduced the quality of the overall product.

2 Project LEAP

How can we reduce the threat of measurement dysfunction in software review without losing the benefits of metrics collection? Project LEAP (Johnson and Moore 1999) is our attempt at to answer this question. It investigates the design, implementation, and evaluation of tools and methods for empirically based, individualized, software developer improvement. A major factor in developer improvement is receiving external reviews of products. Project Leap has produced a toolkit for personal process improvement that supports distributed review of documents. It also supports the generation of review and process checklists and patterns.

The Leap toolkit gives the developer or reviewer full control on the information they share with others. In small groups that they trust, developers can share all their data. In larger groups that the developer does not trust they have three options: sharing the data they are comfortable with, such as checklists and patterns, obfuscating their data, or falsifying their data. It will also provide an obfuscater. The developer runs their original data file through the obfuscater and all identifying data is changed so there is no way to associate the Leap data with the developer. Once the data is obfuscated the identity of the developer or reviewer is removed. The last option, falsifying data, is possible since Leap shows the user exactly what will be sent and allows the user to change any entry. For example, a developer can copy their accurate data and modify their productivity to match some corporate standards.

3 Evaluation

We are evaluating Project Leap in two ways. First, we are introducing the Leap Toolkit to industry and academia. The adoption of the Leap toolkit in industry will be an interesting case study. The design features that reduce measurement dysfunction allow the users to produce "good" answers to management. The case study may provide evidence that developers keep accurate personal data even while maintaining an organizational set of data containing different values.

Second, we are building the Leap data obfuscater and Web based shared data repository. The Leap data obfuscater removes all the identifying information from Leap data files, yet retains the accuracy of the data. We hope that Leap users will upload their obfuscated data files to our web data repository and share their insights into review and development. We will have a repository of checklists and patterns for software development.

Project Leap represents an explicit look at human factors issues in empirically based measurement, in the attempt to collect more accurate and useful data for personal process improvement.

4 Preliminary Results

The Leap toolkit is in use in a senior level ICS class this spring at the University of Hawaii. 19 students are using it to track their software engineering work. So far we have had good results from our students. We will track them to see if they continue to use Leap after they finish the class.

We are using Leap for document review in CSDL. We have even used the Leap toolkit to review the Leap toolkit. One of my colleagues in CSDL feels uncomfortable about sharing their time data with the rest our group, yet they are very comfortable using the Leap toolkit for reviewing since it allows them to not reveal their time data to the rest of the group. During one particular review, my colleague collected very accurate time and defect data. Leap allowed my colleague to share the defects with the author of the document while keeping their time data private.

Our use of the Leap toolkit will enable us to improve our own review practices. Another colleague made the following remark during one of our weekly meetings. "I have gathered 38 defects on the PSP journal article [using Leap]. I am interested in seeing what defects the external reviewers find, and comparing these two datasets to learn how we can improve our own review process." By using the Leap toolkit we are able to easily compare the defects we find in review against the defects the journal reviewers find. We should be able to develop checklists to help us find and remove the defects found by those external reviewers.

5 Current Status

Leap is publicly available at http://csdl.ics.hawai.edu/Tools/LEAP/LEAP.html. Since we publicly announced the release of Leap in December 1998, several individuals, not associated with CSDL, have adopted Leap.

We are looking for industrial partners who are willing to use Leap and allow us to record their experiences. We are going to use Leap to help teach a graduate software engineering class in the fall of 1999. The students will be using Leap to conduct reviews of their fellow classmate's programs.

We are developing the Leap obfuscater and designing the Leap data repository. These two tools should help with the maintenance and distribution of valuable review checklists.

6 References

Austin, R. D. (1996), *Measuring and Managing Performance in Organizations.* New York, Dorset House.

Johnson, P. M. & Moore, C. (1999). Project leap: Personal process improvement for the differently disciplined., http://csdl.ics.hawaii.edu/Research/LEAP/LEAP.html.

Johnson, P. M. (1996). Measurement Dysfunction in Formal Technical Review. Technical Report ICS-TR-96-16, Department of Information and Computer Sciences, University of Hawaii, Honolulu, Hawaii 96822.

Johnson, P. M. & Tjahjono, D. (1998). Does every inspection really need a meeting? *Journal of Empirical Software Engineering*, 4 (1), 9–35.

Tjahjono, D. (1996) *Exploring the effectiveness of formal technical review factors with CSRS, a collaborative software review system.* Ph.D. thesis, Department of Information and Computer Sciences, University of Hawaii.

Between the situation of simulation and the situation of reference : the operators' representations

A. Van Daele & D. Coffyn
University of Mons-Hainaut
Department of Work Psychology
Belgium

1. Introduction

The simulation of work situations is characterized nowadays by the diversity of the finalities that it pursues as well as by the multiplicity of the practices that it produces. It is usual to distinguish three types of finalities : research, training and design (Leplat 1992, Béguin and Weill-Fassina 1997, Rogalski 1997). In all cases, the simulation is often defined as a substitution process in comparison with reality (Weissberg 1989). A major question comes up : "to what degree is the simulated situation representative of the natural situation at work ?". This question refers to the representations of those who design the simulation. We think that even if this question is crucial, it only covers a part of the important questions to be asked. The representation of those who are confronted with it seem as important to take into consideration. That involves considering the point of view of the operators in a situation of simulation, the understanding that they elaborate from it and the meaning they give to their activity. Defining the situations of simulation from the significance that the operators develop allows to reconsider the stakes of the relation between the situation of simulation and the situation of reference (Dubey 1997, Pastré 1997, Samurçay and Rogalski 1998).

2. Context and objective of the study

In the context of the training, the didactical use of the simulators is spoken of more and more. The efficiency of a training on a simulator is not limited to the realism of the simulator. A particular attention must be granted to the management of the simulation situations by the trainers.

The study concerns a training simulation. More particularly, it is about training on full scale simulators of train driving. In this situation, every 18 months, in group of 4, it is planned that some drivers spend half a day training on a simulator. The prescribed organisation of the training foresees that the latter primarily aims at the procedures and safety. Each session must begin with a 45 minutes briefing intended to introduce the simulator, to relay information needed to perform exercises and to answer the possible questions of the drivers. Afterwards, each driver proceeds individually on the simulator and is submitted to an exercise which must last approximately 30 minutes. A debriefing follows each exercise. It analyses the incidents that have been simulated as well as the activity displayed by the driver to fare them. The debriefing must last about 10 minutes per driver.

The training situation has been analysed according to 2 axes : the didactical use of the simulator by the trainers and the drivers activity in simulation. Our main objective throughout those analyses of activities is to better point out the trainers and the drivers representations redefining the stakes of the relation between the situation of simulation and the situation of reference.

3. Methods

The didactical use of the simulator by the trainers has been analysed by an observation grid. This grid was allowed to analyse the activity of the trainers in the different phases of the sessions : briefing, exercise, debriefing. The collected data are the phases duration, the number of simulated incidents, the type of incidents, the number of drivers per session, the topics dealt with in the different phases, the actions of trainers during the exercises... This grid has been used in many sessions concerning a total of 200 drivers and 8 trainers.

The activity of the drivers in simulation has also been analysed from an observation grid. The 2 big categories of observable are the errors and the driving speed. That grid could be used for 70 drivers.

4. Main results

As for as the activity of the trainers is concerned, a certain number of deviations has been observed in comparison with what is expected.

	Prescribed	Real
Briefing	45'	21'
Exercise	30'	39'
Debriefing	10'	18'
Number of incidents	3	4

Table 1. Comparison between some prescribed variables and real variables of the training.

The number of drivers by session is sometimes higher than 4. This involves inevitably some adjustments in the management of the training : less time for each driver in the simulator, less time for debriefing...The average duration of the debriefing phase is shorter while the average duration for the exercises and debriefing are longer. As for as the briefing phase is concerned, we think the fact nearly all the drivers have already experienced the simulator a first time explain the result. As a matter of fact, those drivers who already know the simulator, ask few questions and consequently, the trainers give less information. As for as the debriefing phase is concerned, it seems that the duration scheduled for 10 minutes per driver is too short to comment the activity of each and it is, especially when the number of drivers is higher than expected. As for as the exercise are concerned, another results must be taken into consideration. That result refers to average number of simulated incidents per exercise which higher as well as the number of changes (additions or withdrawals) during the exercise. Therefore it is not surprising that the duration of the exercise is also longer than expected.

As for as the activity of the drivers is concerned, we have observed many interference from the trainers on the speed, encouraging the drivers to go faster. Therefore, we have taken an interest in this variable to find out that a majority of drivers adopt on the simulator a maximal speed inferior to the maximal prescribed speed.

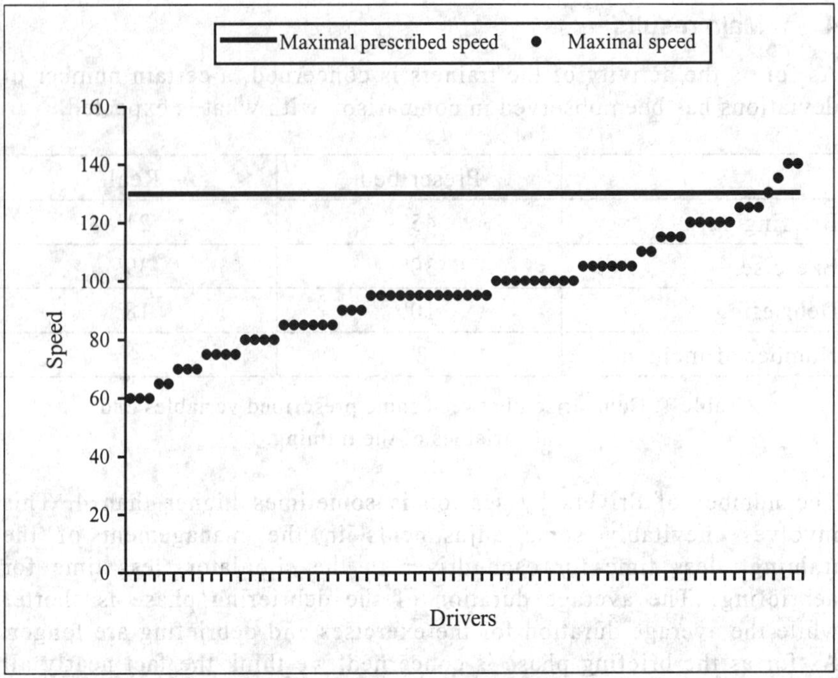

Figure 1. Deviations between the maximal prescribed speed and the maximal speed (70 drivers).

Consequently we can think that the drivers adapt their speed reducing it, which allows them to better anticipate and / or better manage the incidents introduced in the simulation. As we have already underlined in the analysis of the management of the simulation situations by the trainers, a frequent tendency is to be able to exploit the potentialities of the simulator to a maximum, multiplying the incidents within the same exercise. The driver is therefore induced to accomplish more actions in less time. Reducing the speed is a way to recuperate the time needed for action planning. This planning is all the more necessary as the driver experts to be able to manage several incidents and sometimes even several incidents at the same time.

5. Conclusions

It is know by the definition that the simulation is not the reality. Therefore it is know that any simulation introduces deviations in comparison with that reality (in terms of stakes, action time, history, complexity...). On the other hand, it is less know that the didactical use of the simulator itself can bring deviations in. Now if we want to guarantee a transfer of competence from the situation of simulation to the situation of reference, it seem very important to apprehend and to master those deviations as much as possible. In the studied situation where the interest for the technical aspects of there simulators is very much designed to the detriment of the training program, we have noticed several of those deviations. In the absence of a real training program, the management of the simulation is strongly influenced by the representation that the trainers make of a "good training on simulator". An illustrative example refers to the high number of simulated incidents. The trainers have the impression that an efficient training is one that confront the drivers with a maximum of incidents in the limited time accorded to them. Now, we have observed that this leads to an adjustment of the drivers activity consisting especially of reducing the speed of the train. This adjustment appears to be specific to the simulation. For drivers, even if no formal evaluation of their competence is planned, it is very important not to fail in the simulator, not to lose face in front of the colleagues who observe, to be stronger than the trainers who don't hesitate to make the exercises more and more difficult.

References

Béguin, P. & Weill-Fassina, A. (Eds) (1997). *La simulation en ergonomie : connaître, agir et interagir*. Toulouse : Octarès.

Dubey, G. (1997). Faire "comme si" n'est pas faire. In Béguin, P. & Weill-Fassina, A. (Eds) (1997) : *La simulation en ergonomie : connaître, agir et interagir,* pp.39-53. Toulouse : Octarès.

Leplat, J. (1992). Simulation and generalization work context : some problems and comments. In Brehmer, B. (Ed.) : *Models of human activities in work context.* Risö National Laboratory, Roskilde, Denmark.

Pastré, P. (1997). Didactique professionnelle et dévelopement. *Psychologie Française,* 42, 89-100.

Rogalski, J. (1997). Simulations : Fonctionnalités ? Validités ? Approche sur le cas de la gestion d'environnements dynamiques ouverts. In Béguin, P. & Weill-Fassina, A. (Eds) : *La simulation energonomie : connaître, agir et interagir,* pp.55-76. Toulouse : Octarès.

Samurçay, R. & Rogalski, J. (1998). Exploitation didactique des situations de simulation. *Travail Humain,* 61(4), 333-359.

Weissberg, J.L. (1985). Lecompact réel/virtuel. In Weissberg, J.L. (Ed) : *Les chemins du virtuel: simulation informatique et création industrielle, Cahiers du CCI, numéro spécial.* Paris : Edition du centre Pompidou.

From Field to Simulator and Microworld Studies: The Ecological Validity of Research

Anthony Loiselet, Jean-Michel Hoc, and Pascal Denecker
CNRS – University of Valenciennes, LAMIH, PERCOTEC, France

1 Introduction

Research in ergonomics is now using simulation because, either field study is not feasible (for reasons of cost, factor control, disturbance of operators, etc.), or it is not the actual situation which is under study but a future one to be explored. Within this context, the ecological validity of simulator studies is often questioned, particularly in regard to the confidence that can be put in the possible transfer of results from simulation to "natural situation"[1]. This paper addresses ecological validity as a possible feature of external validity, that is the transfer of knowledge from a situation under study to another situation or a class of situations where conditions are at least partly different.

Ergonomists are used to distinguishing basic research from applied research in terms of ecological validity. Basic research would aim at finding general primitives with maximal external validity, but could be irrelevant to natural situations. Applied research would be designed to intervene on specific situations with a minimal external validity but with a maximal ecological validity. This paper will show that this distinction is short-sighted. Basic research is mainly motivated by knowledge of theories and applied research by knowledge of situations, but the ecological validity problem is the same, although it could be less crucial in the former than in the latter (Matalon 1988).

---

[1] Work situations are clearly cultural. Here, the term "natural" refers to situations which are not designed for scientific purpose.

2 Theoretical Background

Two meanings of validity are usually distinguished. Logical validity refers to the consistency of a proposition with other propositions which are taken to be true. Empirical validity deals with the consistency of a theory with facts. A third and more common meaning of validity applies to an instrument that does measure what it is supposed to do. These three meanings of validity are all relevant to approaching ecological validity accurately.

Any empirical study tests (or gives some evidence in favour of) a (theoretical) proposition on the basis of some assumptions which can not be tested. It is an old philosophical aphorism that we cannot directly access "reality", despite sophisticated modern measurement techniques. Even if we observe the actual operators' behaviour in the field, we make use of "descriptors" of (overt or covert) behaviour. Such descriptors spring from no other sources than (explicit or implicit) theories. Doing observational studies in the field we take for granted macro-theories to test or elaborate micro-theories. Thus, the validity problem concerns a large range of studies, from field studies to simulator and microworld studies (Leplat 1978).

As opposed to internal validity (the property of the observational device to give evidence to the micro-theory, on that we will not argue here), external validity enables the researcher to generalize (transfer) the results of one particular study to a class of situations (Brunswick 1947). This statement refers to the "extension" of generalization, but "intension" should be also considered. If the results can be extended to a class of situations, that is because the interpretation of the results is reflecting an invariant, common to all the situations of the class. Consequently, although the study is done on a specific situation, the level of description of this situation and the level of interpretation are abstract. Such an abstraction is only possible (or understandable) if a theory is available and preferably made explicit (Rasmussen 1993).

Thus, the external validity of a particular study, performed on a necessarily specific situation, involves the three forms of validity we have just introduced. The conclusion should be logically valid, in relation to the macro-theory, that is, integrated into the current state of knowledge. It should be empirically valid, that is, consistent with the empirical facts shown by the study. However facts are not raw data, but data interpreted by means of a macro-theory. Finally, the observational device should be valid in that it gives access to the object it is supposed to do. Again, the macro-theory is used to insure this access.

Ecological validity can be seen as a possible feature of external validity. Even if the conclusions are transferred from a situation under study to a specific work situation, the assumption and the abstraction problems cannot be ignored. Besides, a work situation covers a variety of conditions of occurrence so that all of them cannot be observed during a study. The (abstract) object under study should be the same, although its instantiations are different in the different situations. But what kind of object is found in our research domain?

Training situations will not be addressed in this paper, which is devoted to the study of experienced operators (experts) in order to design efficient human-machine systems (e.g., computer assistance to blast furnace operators, human-human and human-machine cooperation in fighter aircraft cockpit, or more generally the role of process dynamic features on supervision and control strategies). At least in French-language literature in ergonomic psychology of cognition, situation refers to the result of an interaction between an operator and a task (including machines). From this point of view, the definition of the object under study involves making clear the representation of the task the operator has elaborated, and consequently the knowledge used to do that.

More specifically, we will consider that a situation under study (field, simulator, or microworld) is ecologically valid as regards a specific situation or a class at least if expert strategies can be implemented in the former and if they does not lose their heuristic properties. Based on a common general architecture, the human cognitive system shapes itself according to the requirements of the environment in which it lives. Then, the ecological validity of a situation may be understood as the degree to which it allows the operators to shape their cognitive architecture as they normally do in daily situations. Gibson (1986) introduced the notion of affordance in order to explain the way by which perception could be directly coupled to action. An observational situation then has a good ecological validity if it retains the affordances used by the cognitive system in the usual situation and the operationality (the action-oriented nature) of the coupling between the operator's cognitive system and the environment. Similarity (or fidelity) is very often stressed in this context, but it should be defined in relation to such a coupling rather than in the sole terms of surface characteristics of the task. However, some surface similarity is needed by affordances.

In any case the situation under study is simpler than any situation of the class because, either one cannot access the operators' overall activity (observational device) or some characteristics, considered as less important as regards the aim of the study and the macro-theory, are removed (simulator). Thus, any study leads to a reduction in complexity, either on the activity side (some parts of the activity are not studied) or on the task side (some subtasks are suppressed). Even if such removals can be justified by the objectives, the problem of possible interaction between the parts remains (Cook and Campbell 1979). However the primitives should be studied, even in simplified situations to enable researchers to elucidate interaction. How can we define interaction between primitives if we do not know anything of primitives?

3 Methodological Proposals

3.1 Managing a Trade-off between Research Cost and Relevance

Research cost is obviously related to observational and analysis duration (implying the operators' work duration for the study), and technical requirements (a microworld is clearly less expensive than a full-scale simulator). Relevance of research is less easy to evaluate, either in terms of long-term value conclusions or short term improvement of work conditions. In the first case, we are often interested in high level conclusions capable of guiding the design of new work situations during several years. Precise numerical values are not needed but only trends. Maintaining the work context can be sufficient to fulfil the requirements of situated cognition in a simplified situation. However, precise numerical conclusions can be necessary to satisfy designers for well-defined problems and ecological validity should be at its best.

3.2 Designing an Ecologically Valid Situation

In any case, nobody can do any research without a minimal theory to define the characteristics common to the situation under study and the target situations. The role of experts in the work domain is crucial to help the psychologist sketch an overall model of the operators' activity as a particular implementation of a psychological theory. Cooperation between the two insures that the sketchy model will not only be a technical (machine-oriented) model but a psychological one. The relation between the theory and the model will enable the psychologist to define the psychological invariants and to control the ecological validity of the situation under study.

3.3 Evaluating Ecological Validity

Experimental or observational reports sometimes make explicit the method adopted to design an ecologically valid situation, although the method is not always justified by a sufficiently detailed theory. However, accounts of ecological validity evaluation are extremely rare. The reason is that, if internal validity is easy to justify because it is a deductive inference from assumptions to results, external validity bears on these assumptions, and becomes almost unfalsifiable within the context of the study. More is needed. The situation under study should be compared to the natural situation, at least at some level. For example, performing a simulator study without any other access to the natural situation is a blind study. *A fortiori*, a microworld study without reference to a large amount of studies on similar situations (from a theoretical point of view) takes the risk of losing ecological validity. Comparisons between experimental situations (replications) and between natural situations themselves is also needed (Chapanis 1988). Certainly, the costs involved in making these kinds of comparison are high, but it is the price which must be paid to do relevant research in such domains. Researchers do not work alone, but in research communities that can provide them with such comparative data.

4 Conclusion

In this paper, we have stressed two main ideas to contribute to the discussion opened several years ago: (a) the role of theory to define ecological validity, and (b) the object in relation to which ecological validity is defined. The main difficulty is that ecological validity bears on theoretical assumptions that cannot be tested by the study in itself. In addition, there is no theoretical consensus in psychology and a particular researcher can always be criticised as regards ecological validity by another who adopts another theory. Thus, the theoretical assumptions behind the criticisms should always be made explicit. In any case, the argument that the situation under study does not reproduce the surface features of the natural situation is not allowable. In fact, the object under consideration is an abstraction that is likely to be described by inference from overt variables to covert phenomena.

5 References

Brunswick, (1947). *Systematic and representative design of psychological experiments*. Berkeley, CA: University of California Press.

Chapanis, A. (1988). Some generalizations about generalization. *Human Factors, 30*, 253-267.

Cook, D.T., & Campbell, D.T. (1979). *Quasi-experimentation.* Boston, MA: Houghton-Mifflin.

Gibson, J.J. (1986). *The ecological approach to visual perception* (orig. publ. 1979). Hillsdale, NJ: Lawrence Erlbaum Associates.

Leplat, J. (1978). L'équivalence des situations de laboratoire et de. *Le Travail Humain, 41,* 307-318.

Matalon, B. (1988). *Décrire, expliquer, prévoir.* Paris: Armand Colin.

Rasmussen, J. (1993). Analysis of tasks, activities and work in the field and in the laboratories. *Le Travail Humain, 56,* 133-155.

THE TEMPORAL STRUCTURE OF A DYNAMIC ENVIRONMENT IN A MICROWORLD EXPERIMENT

Ophélie Carreras & Hélène Eyrolle
Laboratoire Travail & Cognition, UMR 5551 CNRS-UTM, Maison de la recherche, 5, allées Antonio Machado, 31058, Toulouse Cedex 1, FRANCE.
email : carreras@univ-tlse2.fr

1 Introduction

Our researches take place in particular situations, namely dynamic environments. These situations have the characteristic of evolving partly independently of operators' actions. One prototypical situation of that kind is process control : *e.g.* nuclear or chimical plants, blast furnace… Time is a fundamental aspect of those situations in the sense that operators must anticipate the process evolution and the effects of their own actions regarding the process. Our particular question of research has to do with this management of time. We are interested in the way operators cope with the temporal properties of the environment.

Using simulation for studying such a question seems appropriate for differents reasons which we will develop further. This presentation begins with a general consideration of the different types of simulation. We will then concentrate on microworlds by presenting the one we designed for our objectives : the microworld "Epure". Finally, we will try to characterize the main interests and limits of microworlds in general, and "Epure" in particular.

2 Simulation : General Aspects

Different kinds of simulation are used for different purposes. Simulation can be used to create task environments for studying cognitive behaviour of human subjects. Another kind of simulation is the one which is used to simulate cognitive processes, it is a modelling of subjects' cognitive behaviour. We will focus here on the first kind of simulation.

We can distinguish simulators and microworlds. Simulators are constructed by reproducing the main characteristics of real work situations. They can be used for training or for studying human behaviour in some kind of "realistic situation" with a minimum amount of reduction. A microworld is "a computer simulation of a complex dynamic system used to study problem solving and decision making" (Brehmer 1990). The aim is here to study general cognitive processes in dynamic situations. Observations made in work situations have brought laboratory researchers to study more complex and dynamic activities than before. Microworlds have thus this advantage to constitute an intermediary between real-life situations and laboratory experiments.

Whatever the kind of simulation, it makes reductions towards reality. As far as microworlds are concerned, the validity criteria is the identity of psychological mechanisms subjects develop in the simulation and in some real-life activity. The aim of microworlds is indeed to reproduce general characteristics of a complex and dynamic task, with a 'cover-story' for relating the experiment to the general experience of the subject.

3 The Question of Time

Process control situations involve specific cognitive requirements : operators must construct a mental model of the situation allowing them to anticipate, and estimate temporal intervals.We think that this mental model is strongly linked to the temporal structure of the environment. This structure can be described as a "Temporal Reference System" (Javaux, Grosjean and Van Daele 1991) Representing this kind of temporal structure may help operators to regulate actions in time.

Reproducing the temporal structure of a dynamic environment in a microworld experiment allows to handle this structure by creating, for example, different complexity levels. We can then compare subjects' performance according to different temporal structures. Indeed, the main interests of simulation for studying temporal management is that it allows to test the separate effect of different variables or their interactions, and to handle the temporality of the process (Eyrolle, Mariné and Mailles 1996).

4 Describing a Temporal Structure

Temporality of the controlled process can be expressed by a succession of different stages with various durations. For example, anaesthesia situation can be broken down into different stages (or units) characterized by temporal relations : the pre-anaesthesia unit, the induction unit, the anaesthesia regulation unit, and the post-surgical unit (Nyssen and Javaux 1996). These units can be again decomposed in different subtasks with temporal relations. The sequential order of these macro-units is strict ; however, the order of subtasks can be more fuzzy and stage durations depend on different variables that are not all *a priori* predictable.

In some situations, like in a fibreglass factory (Grosjean 1994), operators have to face multiple subprocesses evolving in parallel. Here again, each subprocess is composed of a serie of different elements. Thus, from a formal point of view, the process temporality can be described by complex order relations between time intervals, as those developped by Allen (1984) : *Before*, *Equal*, *Meets*, *Overlaps*, *During*, *Starts* and *Finishes*. These relations allow the description of a complex temporal structure based on intervals. Each element of a task can be qualified by a begining, an end and some temporal extent (Javaux 1996).

Synchronization of actions constraints can be expressed by the fact that actions must take place at very specific moments in order to respect temporal constraints of the process (duration of transformation stages, order of actions, delays…). The temporal constraints are not only inherent to the process itself but are also related to the more global environment. For controlling the process, operators must synchronize their actions to the temporal structure of the environment.

5 The Microworld "Epure"

"Epure" has been created in order to study temporal management. It allows the experimenter to design different temporal structures of a dynamic process. These structures can be more or less complex depending on durations of stages and number of parallel subprocesses. So, we can study adjustment of actions to distinct objects in simultaneous evolution. These objects are transformed across different successive stages characterized by a temporal extent. The designed microworld is based on a water purification process. The process is composed of four successive stages (floculation, decantation, filtration, ozonation) that take place in 4 different tanks on line. Except for filtration, each stage is separated by a "transit" tank. The figure 1 represents one treatment line.

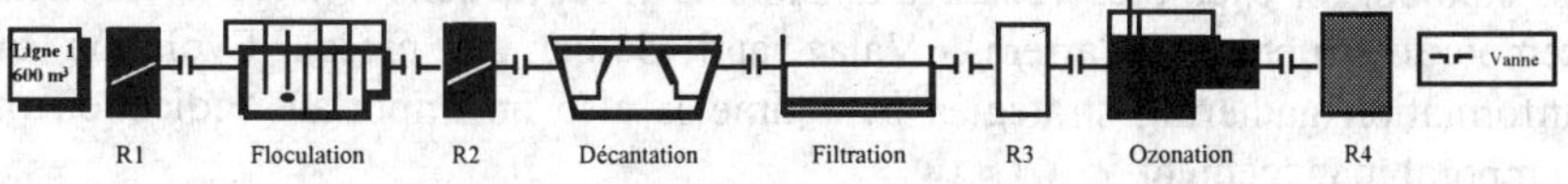

Figure 1 : representation of a treatment line

The realization of each stage requires two operations : the filling of the tank, and the treatment of water. The circulation of water is done by subject's opening and closing floodgates situated between each tank. 14 successive actions are required to set the 11 operations for purifying one mass of water. 4

different masses of water can be treated in one line, and the subject can carry up to 5 parallel lines. Producing the more possible water in a limited time requires the crossing of the purification procedures of each mass of water. The 14 actions relative to "mass A" must be inserted in the 14 actions relative to "mass B" and so on…

Besides this succession constraint, actions must take place at very specific moments, according to operation durations. Durations depend on two variables : the operation nature (decantation is longer than filtration that is longer than floculation and ozonation) and the line capacity. Each line can carry a particular volume of water in the tanks. The variation of operation durations requires the coordination of the different procedures for one mass of water. This coordination must be done, not only within one line but also between all the other lines.

Information about the state of a specific stage is available when the subject activate the window relative to a particular tank. The software records all actions done regarding the process : activation of windows and floodgates with each occurrence date. This recording permit then to study when the subject needed a particular information, and the strategies used to control the system.

6 Conclusion

The main interests of "Epure" is that it allows to compare subjects according to performance and strategies. Measurement of performance in microworlds is a difficulty sometimes discussed in litterature (see e.g. Howie and Vicente 1998), because most of the time there is no 'one right way' to perform the task. In our case, this problem does not arise because "Epure" is relatively simple and the procedure for purifying a mass of water is fixed. The micro-world presented here is not really of the same kind as usual. It is an environment designed to study temporal relations exclusively, there is no causal relations between variables, which are of course present in real process control situations. Performance is here measured by reference to an optimal procedure exclusively tied to the *moment* actions are taken. Temporal management strategies can also be studied: for example, we saw that subjects group actions relative to identical temporal structures (Carreras, Valax and Cellier, in press). Evolution of information gathering strategies over time is also an important indication of temporal management.

This kind of research permit the isolation of a particular activity, in our case, temporal management. It allows an experimental control that cannot be realized in real-work settings. Nevertheless, reductions taken towards reality involve generalization difficulties. We think that microworld simulations must take place in a kind of "alternative" research process. It can be carried out ponctualy for answering specific questions which emerged from real work situations, or/and, like in our case, results can be used as a guide-line for observing real-work situations. Indeed, we plan to observe an anaesthesia situation in order to validate the importance of representing temporal structure for regulate actions in time. The articulation of these different research methods can lead, to our view, to fruitful reflexions and strong results.

7 References

Allen, J. A. (1984). Towards a general theory of action and time. *A.I. Journal,* 23 (2), 123-154.

Brehmer, B. (1990). Towards a taxonomy of microworlds. *Proceedings of the First MOHAWC Workshop, Liège, May 15-16.*

Carreras, O., Valax, M. F. & Cellier, J. M. (in press). Réduction de la complexité temporelle dans le contrôle d'un micro-monde dynamique. *Le Travail Humain.*

Eyrolle, H., Mariné, C. & Mailles, S. (1996). La simulation des environnements dynamiques : intérêts et limites. In J. M. Cellier, V. De Keyser, & C. Valot (Eds.): *La gestion du temps dans les environnements dynamiques*, pp. 103-121. Paris: PUF.

Grosjean, V. (1994). Temporal strategies in multiple parallel processes control. *Paper presented at the IPA Symposium, Time and the dynamic control of behavior, Liège, Belgium.*

Howie, D. E. & Vicente, K. J. (1998). Measures of operator performance in complex, dynamic microworlds: advancing the state of the art. *Ergonomics*, 41 (4), 485-500.

Javaux, D. (1996). La formalisation des tâches temporelles. In J. M. Cellier, V. De Keyser & C. Valot (Eds.): *La gestion du temps dans les environnements dynamiques*, pp. 122-158. Paris: PUF.

Javaux, D., Grosjean, V. & Van Daele, A. (1991). Temporal Reference Systems: a particularly heuristic extension to the classical notion of time. *Proceedings of the Fourth MOHAWC Workshop.*

Nyssen, A. S. & Javaux, D. (1996). Analysis of synchronization constraints and associated errors in collective work environments. *Ergonomics,* 39 (10), 1249-1264.

Training Simulators in Anesthesia : Towards a Hierarchy of Learning Situations

Anne-Sophie Nyssen, PhD

University of Liege - Faculty of Psychology and Educational Sciences
Work Psychology Department - bd. du Rectorat, 5 - B32
B - 4000 Liege (Belgium)
E-mail : asnyssen@ulg.ac.be

1. Introduction

The use of very realistic, "high-fidelity" simulators for training is becoming more and more widespread in industrial environments. Civil and military aviation are not the only areas that benefit from this technology. The Navy, the Army, and the areas of space exploration and nuclear energy have also developed simulators. They are generally called upon in high-risk sectors where it is not possible to resort to training by direct execution of the task. This impossibility can be due to deontological reasons (security and safety), economic reasons (the material cost and the cost of errors), or technical reasons (the very weak probability of certain incidents).
In medicine, especially in anaesthesia, simulation is an emerging field. Its major purpose is to rehearse management of both frequently occurring and rare events during anaesthesia. Although common sense makes us believe that simulators can be useful in training and teaching, hard evidence to support this assumption is not available. Will the use of simulators be better than the current methods of training? Do the perceived advantages of such systems necessarily translate to more effective human performance? How do the simulators preserve the syntax and the semantics of naturalistic situations? This is the core of a debate on *ecological validity* , crucial for methodological, safety and economical reasons. Eyrolles, Marine & Mailles (1996) have distinguished different types of simulators with respect to the characteristics of the reference situation and to the control task requirements - active versus passive, static versus dynamic, interactive versus non interactive, realistic versus partial. But this typology has not been related to empirical data on operators performance and training efficacy. Several authors have emphasized the major role played by the instructor in the training process (Nyssen & De Keyser,1998). In this current, Samurçay and Rogalski (1998) have described the simulation situation as a

mediation process between the reference naturalistic situations and the training process. Following these authors, simulators are always a simplification of naturalistic situations. But the instructor may, by an adequate training, compensate possible side effects of this simplification; moreover he can use them in the training process. Since simulators may become the standard of training in many high risk environments, their role in training must be better understood. The basic question is less of ecological validity than of *didactical use* of simulated situtions. In this line of though, we compare and assess the efficacy of two types of simulators : screen only simulator and full scale simulator as a training tool in anesthesia.

2. Subjects and Method

Twenty anaesthetist trainees with more or less experience (10 x 4 or 5 years of training and 10 x 2 or 3 years of training) have been confronted to two plausible anaesthesia incidents on a screen only simulator and on a full scale simulator. The control scenario consisted of a pre-scripted simulation of malignant hyperthermia (MH) while the test scenario was an anaphylactic shock (AS). The experimental setting was inspired by Chopra et al. (1994), and the performance of the trainees has been assessed by experienced instructors on the base of a list of criteria (see Nyssen 1997).

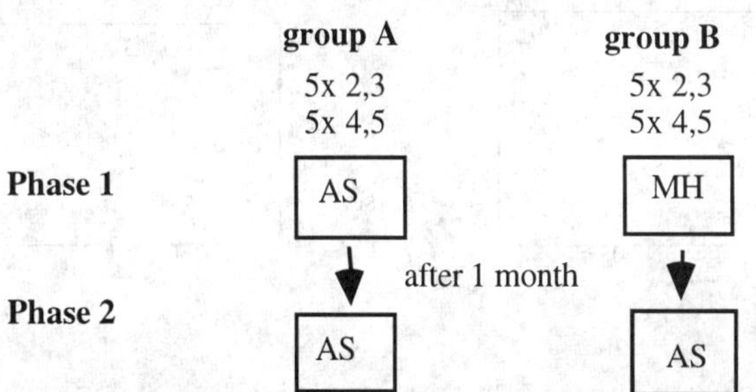

AS= Anaphylactic shock ; MH = malignant hyperthermia

Fig. 1. Experimental Design

The screen simulator (ASC) used in the study is entirely decoupled of naturalistic anesthetic situations. There are no real equipment, no physical patient, no team, no hospital, no history, no social stakes, no conflicting goals. Only causal relations between the drugs and the physiological variables of the patient are displayed on the screen; monitorings are simulated. The order of the actions and events is maintained but not the real durations of the scenario.

The full scale simulator system used is part of the CASE Series is presented in the form of a convential operating room equipped with all the necessary equipment for an operation but where the patient has been replaced by a dummy best representing the clinical reality. The dummy can be intubated and can regurgitate; it is equipped with lungs able to "consume" oxygen (O_2) and produce carbon dioxide (CO_2). It can be hooked up to a classical respirator or ventilate spontaneously. The anesthetist can listen to the cardiac and lung sounds and take the pulse in the neck; he can inject drugs to which the dummy responds in real time. Other variables can be monitored according to the desired degree of sophistication.

3. Results

Twenty anesthetists (10x2,3rd year of expererience / 10x4,5rd year of experience) were trained on the screen simulators and twenty anesthetists (10x2,3rd year of expererience / 10x4,5rd year of experience) were trained on the full scale simulator. The performance scores were signifcantly better in group A during phase 2 on both simulators. The comparaison of the group showed that the performance scores were signifcantly better in group A during phase 2 on both simulators.

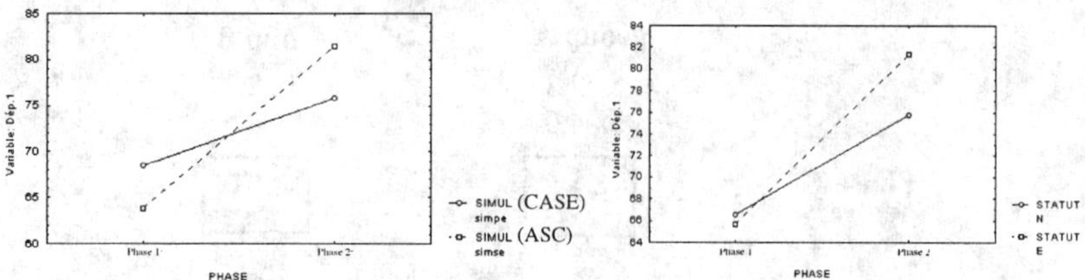

Fig. 2 - The ASC simulator does not lead to a better performance of the subjects. But in the second phase, the diagnosis performance is better, if not significantly, than in the CASE simulator.

Fig. 3- The experts have a slightly - but not significantly - worse performance than the novices on the ASC, in the first phase of the experiment. But they improve fastly in the second phase.

We also observed, even if it is not significant, that the performance scores were better in the more experienced group on both simulators during phase 2 and that the evolution of the performance scores between phase 1 and 2 was very different on the two simulators.

3. Discussion "*Towards a hierarchy of simulators*"

From these first results, we can say that each type of simulators can contribute to the improvement of performance and we can imagine in the training a structure of learning situations, using less and more complex training tools. Each tool must be used with a specific goal which must take into account the propreties of the tool and must be derived from an in-depth understanding of the problem situations that must be solved in the context. In each learning tool, the instructor plays a major role by choosing the adequate learning situation in regard to the trainee experience and by orienting the instruction process. From the analysis of our data, we can say that :

- the screen simulator focuses on causal relations existing in the patient's variables and can be used to learn basic recognition schema of limited patterns informations. However, there is a stress on diagnosis; There is a lack of potential anticipation, and no attention is paid to the recovery process so important in reality. These aspects, as well as team management, can be trained on the full scale simulator.
- the full-scale simulator can be used for learning objectives that concern coordination with the team and dynamic management of the problem situation. There are numerous contextual variables and interferences capable of distracting the attention of the subjects; But the hospital, the history, the social stakes and conflicting goals are still missing.
- the natural situation underlines the inclusion of the operation in a socio-historic context, within a team and in a health-care institution which possess their own constraints, which are very difficult to simulate.

Simulators are components of a larger training process that has to be continuously fullfilled by the field - by observations, collection of errors, safety meetings, questionnaires - which stress the main difficulties in the expertise structuring. The tight connection between field data and simulator training insures that no critical aspects of the structuring of the anesthetist expertise remain in the shadow, and it facilitates the transfer of knowledge from one side to the other one. The table below summarized the results that emerged from our data analyzes.

	Training contents	Advantages	Limits
Screen only simulator *ASC*	basic intellectual skills recall of concepts & rules problem diagnosis	prior knowlegde acquisition training package economy	weak ecological validity unique solution to complex problems "response's computer" incorrect temporal syntax

Full scale simulator *CASE*	basic tasks & procedures dynamic integration (anticipation, planning) management tasks practice on rare events	high ecological validity (but still limited) favors explicit learning and performance elicitation : - metacomponents - high order process control of training	simplified social relationships simplified temporal syntax overrepresentation of rare events high cost
Reality	task context : complexity social aspects synchronization constraints	increases the domain expertise favors implicit learning provides data collection to feed-up the training process	risk poor guidance and poor feedback improves performance but no clear explicit learning

References

Chopra, V., Gesink, B.J., De Jong, J., Bovill, J.G.., Spierdijk, J. & Brand, R.P. (1994). Does training on anesthesia simulator lead to improvements in performance? *British Journal of Psychology,* 73. 293-297.

Eyrolle, H., Mariné, C. & Mailles, S. (1996). La simulation des environnements dynamiques: intérêts et limites. In J.M. Cellier, V. De Keyser & C. Valot (Eds), *La gestion du temps dans)les environnements dynamiques.* Col. Le Travail Humain. Paris : Presses Universitaires de France.

Gaba, D., Fish, K., & Horward, S. (1994). *Crisis management in anesthesiology.* New York : Churchill Livingstone.

NYSSEN, A.S. (1997).Vers une nouvelle approche de l'erreur humaine dans les systèmes complexes : exploration des mécanismes de production de l'erreur en anesthésie. Thèse de doctorat, Université de Liège, Liège.

Nyssen AS, & De Keyser V. (1998). Improving Training in Problem Solving Skills : Analysis of Anesthetist's Performance in Simulated Problem Situations. *Le travail humain*, 61,4, 387-401.

Samurçay, R. & Rogalski, J. (1998). Exploration didactique des situations de simulation. *Le travail humain*, 61, 4, 1998, 387-401.

A New Integrated Support Environment for Requirements Analysis of User Interface Development

Y. Tokuda*, T. Murakami*, Y. Tanaka*, E.S. Lee**
* HI Laboratory, Products CS Dept., CS Center, Sony Corp., Japan
** School of Electrical and Computer Engineering,
Sung Kyun Kwan University (SKKU), Korea

1 Introduction

In developing highly user-satisfiable software systems of User Interface (UI), Requirements Analysis (RA) is one of the most important and difficult processes (Faulk 1996). Most traditional RA based on individual analyst's experiential methods and insufficient tool support has a strict limitation for helping the analysts in the process. So it is strongly expected to develop methodologies and tools for supporting the RA process efficiently and entirely.

In this paper, we suggest a model of RA process in UI design and propose a support system named USE, User-friendly Support Environment for UI design, to give a systematic and integrated support.

2 Design Concepts and Policies

2.1 Modeling the RA Process through Real-world Task Analysis

We design a model of the RA process through real-world task analysis, especially by analyzing the actual activities of our present UI development groups, and their new requirements for the future. The RA process model roughly consists of the following 8 stages: (1) problem description, (2) user identification, (3) goal analysis, (4) task analysis, (5) UI function definition, (6) UI function parts extraction, (7) situation organization, (8) GUI image sketching, and (9) image attachment on the storyboard. These stages capture the essential activities to make up the requirements analysis clearly and systematically.

2.2 Easy File Management per Project based on the Model

We support to manage product files per project based on the model. Some related files in a project are connected with each other based on analytical relationships between files, e.g. a user goal file is connected with user description files and goal hierarchy files, respectively.

USE also provides its users with "Makigami" style multi windows display function. We regard a sequence of files related to a project as a "Makigami", which means a kind of rolled paper in Japanese and can unfold only focused parts of whole paper, and think of managing them entirely. In opening or closing any files, USE rearranges all files opened that time in consideration of their analytical relationships. Concretely, files of different type are arranged horizontally, and ones of the same type are arranged vertically in that type.

The above file management promotes re-engineering among project because analysts can use existing results at various grain from parts of a file to whole project. For example, in developing a derivational system, the analyst will be able to reuse the results of its original system by adapting parts of them.

2.3 Integrated, Consistent User-friendly User Interface

USE provides its users with an integrated user-friendly UI based on the "Makigami" style as shown in the below figure. USE also provides a consistent UI in overall analysis tasks by means of some similar hierarchical expressions. The expressions are based on a hierarchical structure which is represented as a indented list like the Windows Explorer and whose component nodes are represented as items of that list. Each item consists of item icon, item label, and operation icon (optional). The item icon shows a type of item as well as its iterativeness. The operation icon specifies a relationship among sub-nodes of the item, e.g. sequential and selective relationships.

3 System Structure of USE

Based on the above design concepts and policies, we construct USE as an integrated environment including the following supporting tools:(1) problem description editor, (2) user goal analyzer, (3) goal/task analyzer, (4) UI function definition editor, (5) situation organizer, (6) GUI sketch editor, and (7) storyboard editor. USE provides its users with these tools through an integrated UI on a project window, which reflects the "Makigami" style and in whose work area, multiple tools can be usable. All tools except for the problem description and the GUI sketch editors support their analysis tasks consistently by means of the above hierarchical notations, as shown in Figure 1. As a results, their users, especially Windows PC users, feel familiarity because those tools offer them operational manners like the Explorer.

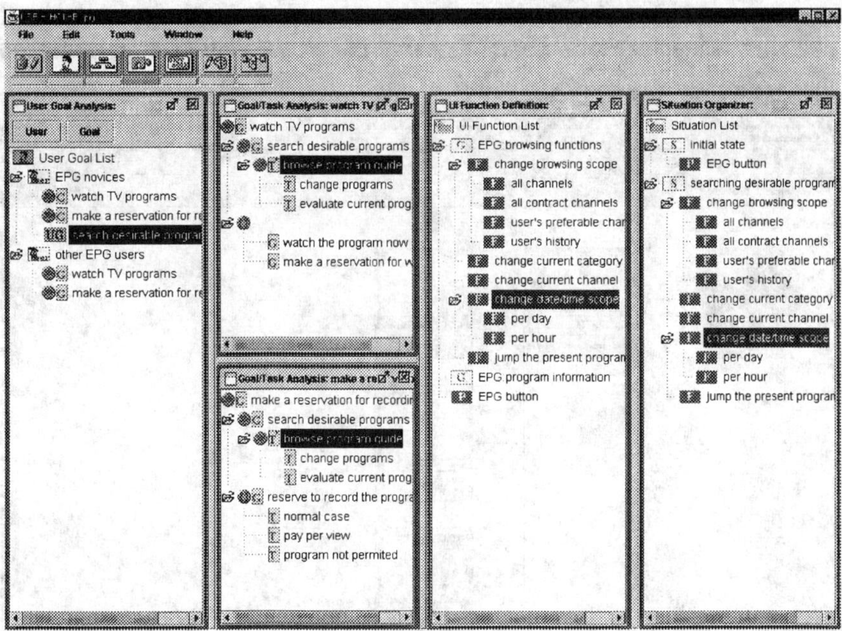

Figure 1. A screen shot of USE system

Problem description editor is a template-based text editor for identifying user requirements and constraints. User goal analyzer supports to identify target user groups and to extract their goals from the problem description. Goal/task analyzer is for decomposing goals/tasks and supports to define the relationships such as sequential, selective and iterative. UI function definition editor supports to extract UI functions from user's aspect to execute the tasks and goals, and a set of functional parts for each function. Situation organizer constructs a set of functions for a specific situation to give an intuitive image to the next stage, GUI sketch editor. GUI sketch editor is a drawing tool for sketching GUI images with some primitive objects including button, checkbox and so on. Storyboard editor supports to make some storyboards, which are sequences of the output of GUI sketch Editor or other drawing tools along with user's possible operations. It also provides slide show simulations in order to verify the storyboards. With the aim of providing a consistent UI and making the story building task efficient, these storyboard are represented as a variation of the above hierarchical notations whose links between nodes are directed links and which has two specialized branches, back and goto, as shown in Figure 2.

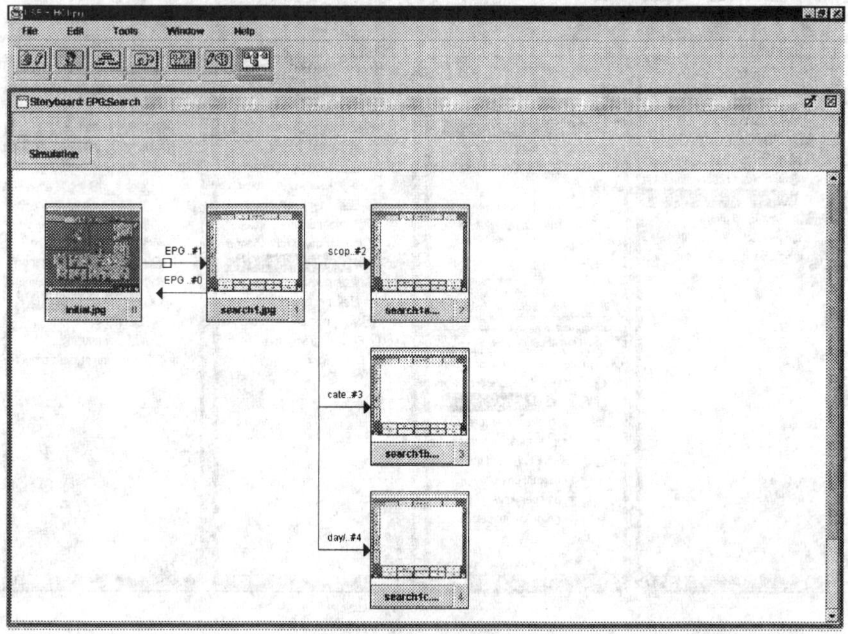

Figure 2. An example of storyboard editor's hierarchical notation

4 Implementation and Evaluation

We have implemented the prototype of USE using JDK 1.2 on Windows PCs. We have actually applied USE to requirements analysis tasks for different applications such as electronic program guides (EPG) and information retrieval on WWW.

We show a partial analysis example about the EPG. The USE window shown in Figure 1 contains four types of tools: user goal analyzer, goal/task analyzer, UI function definition editor, and situation organizer, for supporting each analysis tasks. In the user goal analyzer, two type of user groups, "EPG novices" and "other EPG users", are identified, and their typical user goals, e.g. "make a reservation for recording", with explanatory comments are also specified. Each user goal is decomposed into sub-goals and tasks in the next goal/task analyzer. For example, the goal "make a reservation for recording" is decomposed into two sub-goals "search desirable programs" and "reserve to record the programs", and the relationship between the sub-goals is expressed with sequential notation, ">>". In the following UI function definition editor, UI functions, e.g. "change browsing scope", and UI parts, e.g. "EPG button" are extracted and defined in detail. In the situation organizer, various situations in use, e.g. "searching desirable programs", are defined in terms of a set of UI functions and UI parts, and those situations are visualized using GUI sketch editor or the other existing drawing tools such as paint brush and Visio.

Figure 2 shows the storyboard editor, in which various types of scenarios are developed by using the produced GUI images. The example is an early partial flow which starts at "initial state" image, and transfers to "searching desirable programs" one.

We evaluated the effectiveness of the proposed concepts and system, through the actually application examples, so it was the first step for the future full evaluation.

For each analysis tasks based on the RA process model, the USE can support its users as follows: the users (analysts) can identify and decompose user goals per assumed user group, effectively extract essential functions from analysed goal/task hierarchies, and easily construct GUI images and stories through extracting situations where a set of UI functions can be provided.

The "Makigami" style multi window display function can efficiently support user's management of many tools and files related complicatedly. The users can construct their work environment suitable for various contexts such as analytic situation, user preference and so on.

The consistent UI using similar hierarchical expressions can provide the users with familiarity and efficiency because of operational manners like the Windows Explorer.

5 Conclusions

In this paper, we built the RA process model and proposed a support system USE to support the RA process systematically, totally and efficiently. We also applied it to practical domains and evaluated its usefulness.

6 Reference

Faulk, S. (1996). Software Requirements: A Tutorial. In Dorfman, M. & Thayer, R. H. (Eds.): *Software Engineering*, 82–103.

A Method for Extracting Requirements that a User Interface Prototype Contains

Alon Ravid and Daniel M. Berry
Faculty of Computer Science, Technion—Israel Institute of Technology, Haifa 32000, Israel, dberry@csg.uwaterloo.ca

1 Introduction

User interface (UI) prototyping (UIPing) (Connell and Shafer 1989, 1995) is a commonly employed requirements elicitation and validation technique, used to determine the functionality, UI, data structure, and other characteristics of a system. User requirements are explored through experimental development, demonstration, refinement, and iteration. The UI prototype (UIP) is built during the requirements analysis and specification phases of a software project. The goal of UIPing is to discover user requirements through early implementation of the UI and the functionality behind it. A UIP permits users to relate to something tangible and to obtain a requirements definition that is as complete as possible, so that the requirements can be validated by the user on the basis of realistic examples. The products of this process are various documents such as a software requirements specification (SRS), an occupation analysis document (OAD), and the prototype itself. (An OAD is essentially a draft version of a tutorial and user's manual for the system to be built.) The process of creating a UIP involves some difficulties. Once the prototype is developed and agreed upon, how is the information that it contains captured and represented in the other analysis documents?

2 The Project that Exposed the Problem

In recent years, Ravid, the first author, was the software system engineer for a highly complicated simulator for generating infrared (IR) scenes, called the Target Scene Generator (TSG). Because of the many difficulties faced during the early stages of the system specification, The team decided to rapidly develop a throwaway UIP (Andriole 1994). This UIP was to improve the communication with the customer and to help complete requirements specification within a reasonable amount of time. Software engineers, human engineering people, and users group representatives were involved in the development of the prototype. This approach turned out to be successful. Some major misunderstandings with the customer and many contradicting requirements were discovered. As is recommended by many including Fred Brooks (1995), the prototype was thrown away, and the development of the production version was started from scratch.

3 Problem Description

The process of creating the TSG prototype involved some difficulties, which are common to other prototyping-oriented projects as well. Given an informal problem description, it is not obvious how to systematically and efficiently construct a prototype. It is hard to distinguish between requirements and design details. After completing the prototype, a method is needed to integrate it to other models of the system. A portion of the information implicit in the prototype is left implicit and is not known until the prototype is used to answer questions. A method is needed to capture the prototype's semantics, and state them formally, in order that they will provide a suitable and traceable base for further development and testing. This problem was discovered unexpectedly while the team was preparing the project to be reviewed for ISO 9000.3 compliance by the Israeli Standards Institute. During a preceding internal review, Ravid was asked to present the project and the software development documents produced up to the day the review was conducted. Ravid presented the system specification documents, the SRS, the OAD, and the prototype. The prototype aroused additional questions about the system. Obviously, the prototype was used to answer these questions. After Ravid finished answering these questions, one of the reviewers asked "Where is all this information written?" Part of the information appeared in the SRS, part was expressed indirectly by other statements in the SRS, part was written in the OAD, and part was not written anywhere even though it was known, understood, and agreed upon by all the people involved in the project by virtue of their having worked together to produce the prototype. This undocumented information included indispensable knowledge about the system, knowledge which seemed to be essential for new programmers joining the group and for maintenance personnel who will have to support the system in the near and far future.

4 Key Question

We ask two key questions about UIPs.

1. What does a UIP say and what does it not say?

2. What is the right way to formalize and present requirements which are specified and embodied in a UIP?

We have concluded that there is no way to extract this information from a previously written prototype in the absence of knowledge of how it was developed. This is because there are aspects of the prototype's behavior that are artifacts of the prototyping process, often rapid, that are not intended as specified behavior. Just examining a prototype does not allow distinguishing intended from non-intended behavior. The problem is the same as that of reverse engineering from code. Therefore, it is necessary to have carried out prototyping explicitly as part of a fully traced requirements elicitation process in which it is decided and documented ahead of time what behavioral aspects, usually in the user-interface, are being modeled in the prototype. The tracing links allow easy access to the documentation of this decision so that in the future, when one is following the trace links to track down an answer to a requirements issue, one sees the explicit decision and knows whether to consult the prototype or another document in the requirements specification suite.

5 The Proposed Solution

We decided to use existing methods to model requirements, because we believe there are more than enough. The modeling technique should be chosen based on the characteristics of the system being modeled. Most importantly, the problem addressed by this research does not result from the lack of modeling methods, but rather from the fact that existing methods fail to identify the kinds of information that a UIP contains and thus do not provide a way to capture and present that information. Instead, the solution presented here consists of steps to be taken in addition to existing UIPing and requirements modeling methods, to be applied *before* and *during* UIPing and requirements modeling, to ensure that later, developers are able to answer the questions of what the UIP specifies and what it does not.

There is no advice here on what to prototype and what not to prototype. What is prototyped and what is not are the specifier's decisions. This decision is made based on his or her experience in specifying other systems in the same application domains; it will be based on what is already well understood and what is not. There is advice here on what to do once the decision is made so that the developers and maintainers know what of the prototype is intended and where to look, in the prototype or in other documents, for answers about their questions. The advice concerns documenting this choice and creating tracing links among the documents that help search for answers to these questions.

The basic idea of the proposed approach is to perform exploratory requirements prototyping in a systematic fashion that is based on tailoring a prototype construction process to the system under development, on identifying, before UIPing begins, the kinds of requirements information that the UIP will contain and will not contain, and on choosing modeling techniques that properly represent this information. Tailoring a prototype construction process is done in six recurrent steps, recurrent in that as one is following the steps in sequence, one can go back to any previous step to redo it based on new information.

1. Define the system's operational environment and its interfaces to other systems.
2. Identify to which application domains the system belongs.
3. Characterize the principal properties and the main features of an application belonging to these domains.
4. Identify which of these properties are applicable to the system under development.
5. Decide which of the identified properties requirements will be prototyped.
6. Prototype the system's interfaces and chosen properties.

The concept of application domains is an important one. All programs in a domain share common data, attributes, and operations. Examples of domains are simulation systems, information systems, data acquisition systems, and UI-intensive systems. An application can find itself in several domains, depending on the functions it must provide. The point about a domain is that from past experience, much is known about the data, attributes, and operations that are needed by any application in the domain. Often, there are even libraries of data definitions, procedures, and functions that can be used by simple inclusion to simplify programming the application. For example, there are many GUI building libraries available to be used to build any UI-intensive application.

The kinds of requirements information that a UIP contains can be classified into the following types,

1. the functionality and behavior of the application, that is, its reactive nature, including constraints placed on this behavior and operational logic,
2. the application's data model, data dictionary, and data processing capabilities,
3. the application's taxonomy along with a dictionary for it,
4. a partial specification of the interfaces to other systems, and
5. general knowledge about the system and intent.

Prototyping of the system interfaces and chosen properties is conducted an iterative manner as recommended in many publications about prototype-oriented software development. Developers and users can go over this list of properties jointly in a systematic fashion. They can discuss them by means of interviews, modeling, implementation, demonstration, refinement, and validation. The requirements information can be grouped and organized according to the kinds of information that the developers want to discover, which is also the kind of information the UIP will contain. The developers and customers can address all related issues and, when they come to an agreement, state the requirements formally. Appropriate models for stating the agreed upon requirements formally should be chosen based on the application domain, on the nature of the application to be developed, and on many other factors. The number and types of these models and the level of detail can be chosen after deciding if the prototype will be thrown away. All prototyped aspects have to be modeled. The chosen method should have the capability of representing the aspects listed above, in order to produce a good requirements model and avoid losing valuable information about software requirements.

6 Conclusions

The presented approach attempts to deal with the every-day prototyping reality, the approach to prototype construction, the distinction between the information the prototype contains and the modeling method, and the use of the prototype as a requirements elicitation aid and as a requirements model itself.

To validate the approach, we have carried out a case study applying the solution approach to the development of the TSG system. Space limitations preclude giving any details of the case study. The reader is referred to the first author's thesis for the missing details (Ravid 1999). In that document, the reader will find an adaptation of the general approach to the specific UIPing technique used to learn the requirements for the TSG, a detailed description of the case study, an evaluation of the effectiveness of the solution, and lessons learned.

7 References

Andriole, S.J. (1994). Fast Cheap Requirements: Prototype or Else!. *IEEE Software*, 14(2), 85–87.

Brooks, F.P., Jr. (1995). *The Mythical Man Month*, Second Edition. Reading: Addison Wesley.

Connell, J.L. & Shafer, L.B. (1989). *Structured Rapid Prototyping*. Englewood-Cliffs: Prentice Hall, Yourdon Press.

Connell, J.L. & Shafer, L.B. (1995). *Object-Oriented Rapid Prototyping*. Englewood-Cliffs: Prentice Hall, Yourdon Press.

Ravid, A. (1999). A Method for Extracting and Stating Software Requirements that a User Interface Prototype Contains. M.Sc. Thesis, Faculty of Computer Science, Technion, Haifa, Israel, available at ftp://www.cs.technion.ac.il/pub/misc/dberry/Thesis.doc .

Documentation of Prototypes in Terms of Use Scenarios: Nice to Have or Indispensable?

Wolfgang Dzida and Regine Freitag
GMD German National Research Center of Information Technology

Introduction

A prototype enables the designer to demonstrate an understanding of requirements in terms of a designer's language. The usability designer is used to express requirements in terms of "look and feel" attributes in compliance with style guides. These attributes are in response to user requirements, which are expressed in terms of the user's language describing tasks, objectives, user performance, etc. Two types of scenarios which we call *context* and *use scenario* may help catch the requirements from the user in a way that is comprehensible to the user. (For the role of scenarios in requirements engineering see the paper of Wolfgang Dzida in this volume.) A prototype is intended to express these requirements in other words. Both languages, scenarios and prototypes, can be used to produce a double representation of requirements. However, double representation requires double effort and may be questioned by managers. This paper argues in favour of double representation for two reasons: 1) according to ISO 8402 a requirements definition is an "expression of the needs and/or their translation into stated requirements for the characteristics of an entity to enable its realization . . ." (par. 2.3); and 2) double representation has often proven its worth in creative processes which the analysis and design process should be.

Bridging the Communication Gap

Winograd and Flores (1986) pointed out that we create our world (e.g., our systems) through language. Until recently, the designer's modeling language dominated the requirements definition process. In traditional systems analysis there was a lack of awareness of the user's problem situation. It may be necessary, however, to enter this situation in terms of the user's language, thereby considering a different representation of the design problem as the designer usually does. Domain representation and design representation, both provide a more comprehensive view of the problem, with the user's view focusing tasks, intended results, and complaints and with the designer's view focusing a design proposal for the user's problem setting. The user's view is complementary to the designer's view and serves as a rationale for the designer's solutions. Scenarios turned out to be much closer to the user's mother tongue than any professional problem statement language. To bridge the communication gap between user and designer we contrasted the user's context scenarios with the designer's initial idea (use concept) of a system to come. In context scenarios we described in a narrative way the main user taks, problems with task conduct in the current context and the user's visions of overcoming the deficiencies in view of a new technology. "Success in design involves understanding the context into which the designed artefact will be introduced" (Newman and Lamming, 1995, p. xii). User-validated scenarios served as a basis for deriving requirements in terms of user performance. Confronted with this information the designer was challenged to revise or refine the initial design idea. In response to the context scenario the designer aimed at expressing the requirements in terms of a design language, which is the repertoire of user-interface attributes. The bridge between user and designer was established by defining the requirements in two languages according to the previously mentioned standard (ISO 8402): user needs and corresponding system attributes.

The Virtues of Double Representation

Klix (1979) investigated the effects of double representation in creative achievements of physics. He observed that physicists developed individual but very different representations of the same problem, e.g., one representation in terms of a kinetic equation (Newton) and another one in high imagery terms (Leibniz). "Pictures of physical reality" are linked to "complementary logic concepts (p. 6). The lesson learned from this observation is "that this kind of double representation of information in human memory can be a source of intellectual creativity . . . The transposition of one kind of representation into the other makes the structural identification possible, reveals possible transformations which could not be disclosed from the other one" (p.6).

In interdisciplinary cooperation double representation of a problem has often proved quite effective in engendering synthetic solutions. Each partner represents an understanding of the problem by means of own concepts, words, graphics, and the like, thereby emphasizing or neglecting aspects of the problem but also uncovering hidden features, thereby favouring both partners in forming a shared understanding.

Analysis and design can also be taken as an interdisciplinary cooperation. Cooperating partners such as user and designer may benefit from double representations when they help clarify hidden misunderstandings. But user and designer may also be surprised by the creative transformation of requirements in both directions, user and system requirements. A seemingly double representation may then turn out as being complementary. "By making some additional assumptions it is possible to infer one representation from the other. Even if this is the case, so that one representation implicitly contains the information in the other, it is still the case that one representation may be most useful for one purpose and the other for another" (Monk, 1998, p.111). Hence, it is still possible that a prototype tells the designer more than thousand words of a corresponding scenario. Nevertheless, the purpose of double representation is primarily seen in stimulating creativity.

Double Representation in Prototyping Projects

In prototyping projects we attempted to induce such creative transformation processes by double representation of requirements. We employed another type of scenarios which we call use scenarios (user language) to represent prototypes (designer language) in a way that is comprehensible to both the user and the designer.

In response to the previously described context scenario (and the derived requirements) we asked the designer to develop a prototype for some of the key tasks. Prototyping was taken as an opportunity to demonstrate the designer's initial understanding of the requirements. The first prototype provided a more realistic impression of the system to come, particularly regarding its interactive features. A representative user was then asked to conduct the key tasks at the prototype interface and describe the task performance in terms of use scenarios. The analyst helped the user in editing the scenarios for each key task. Writing down the scenario made it possible to reflect on and reason about user performance (Monk, 1998). The content of a use scenario mirrors the system behaviour in terms of user performance. (For the symbiosis of scenarios and prototypes see the paper of Regine Freitag in this volume.) All partners involved in the prototyping process achieved a more elaborated understanding of the requirements and the design proposal, thus evaluating the prototyped user performance but also creating a number of new requirements. Noteworthy, evaluation was focusing the efficiency of user performance, with the "look and feel" attributes being considered secondarily.

Documenting Prototypes by Use Scenarios

We documented user performance in two ways: use scenarios and use cases, with the latter being a very concise form of a use scenario. Each version of a prototyped key task was documented by a separate use scenario. In iterative prototyping versions of use scenarios need to be produced. To minimize the documentation effort, new versions of a use scenario were documented in a more concise use-case description, especially for those parts of user performance which remained unchanged. Fragments of typical use scenarios were only inserted in the use case structure at points which reflect the revised parts of the prototype. Any improvement of a prototype goes along with an improvement of the user scenario.

The major problem with this approach is to keep consistency among requirements specification, versions of scenarios and prototypes (see also Weidenhaupt et al, 1998, p.39). We prefered a table-structured representation of the use scenario to establish links between each segment of the use scenario (a specific cell in the table) and the requirement (a corresponding cell in another column).

A use scenario provides a representation of an effective user performance, thus making it more likely that ineffective dialogue steps cannot be failed to notice. In other words, the final use scenario is a success story about the prototype. It is validated by the user and accompanied by a list of usability requirements. Hence, the virtue of use scenarios in the prototyping process is its communicativeness for all parties involved. Even managers can be satisfied with this kind of documentation, since the final use scenario represents the agreed upon understanding in the design team, including the users. Use scenarios turned out to be indispensable in evolving a consensus among user and designer and for validating requirements as well.

Use scenarios also serve as product documents of prototypes. Monk (1998) recommends to include other documents with a scenario, e.g., photocopies of real work objects, screen dumps, commented design decisions, flow charts, and so on. The primary purpose of amending the scenario is to represent the scenario in ways that can be communicated easily to the different members (roles) of the design team (McGraw and Harbison, 1997). For instance, scenarios can serve as a valuable source for any kind of user documentation.

Conclusion

A user being confronted with a prototype may be overwhelmed with design details. Use scenarios can help the user to restructure this information in more familiar terms which is the user's task performance, thereby producing a representation complementary to the system behaviour. Double representation of a prototype in terms of use scenarios did not turn out as waste of effort but as a means to improve an understanding of requirements and design.

References

ISO 8402 (1994). Quality management and quality assurance – Vocabulary.

Klix, F. Ed. (1979). *Human and artificial intelligence*. Amsterdam: North-Holland.

McGraw, K.L. and Harbison, K. (1997). *User-Centered Requirements: The Scenario-Based Engineering Process*. Mahwah, N.J.: Lawrence Erlbaum.

Newman, W.M. and Lamming, M.G. (1995). *Interactive System Design.*. Wokingham: Addison-Wesley.

Weidenhaupt, K., Pohl, K., Jarke, M., and Haumer, P. (1998). Scenarios in system development: Current practice. *IEEE Software*, March/April, 34-45.

Winograd, T. and Flores, F. (1986). *Understanding Computers and Cognition – A New Foundation for Design.* Norwood, N.J.: Ablex.

Monk, A. (1998). Lightweight Techniques to Encourage Innovative User Interface Design. In: Wood, L.E. (ed.): *User Interface Design. Bridging the Gap from User Requirements to Design*. Boca Raton: CRC Press.

Design Patterns: Bridges between Application Domain and Software Labs

Peter W. Fach
RWG Stuttgart, Germany

1 Introduction

The role of frameworks as a method for modern software development is continually increasing in importance. Essentially, design patterns are metaphors which describe behavior and structure of a single framework or the cooperation of several ones. Beside these features, which are highly estimated by software engineers, design patterns offer unique means to help application experts participate in the development process - a methodological step which definitely leads to more usable systems. These features as well as their potential to structure user interfaces will be discussed with the help of two types of design patterns which are used in the application system provided by the RWG for about 350 cooperative banks in southwestern Germany.

1.1 The Tools-Material-Pattern

A very well-known design pattern which describes the interplay between objects of the application domain (e.g. account) and objects of the interaction domain (e.g. editor) refers to the Tools-Material-Metaphor. On the one hand this metaphor, i.e. that one works on materials by using tools, illustrates how two software objects refer to one another. For instance, an object 'account-editor' is especially tailored to work on an object 'account' so that users can pay in or pay out money, whilst the object 'account' is designed with the sole aim to reflect its counterpart in the application domain in the best possible way. Thus, all interaction is handled by tools (the 'MVC-paradigm'; e.g. Reenskaug 1996). On the other hand, this metaphor also serves the purpose of structuring the user interface. This means that users are offered tools which they consciously and selectively apply to materials represented as icons on the user interface. By this particular quality, users are enabled to precisely name occurring problems in terms of tools and/or materials used in their application domain, and software designers, in turn, are enabled to locate and classify problems or new requirements by referring to the same terminology and entities. Thus, users and designers use the same terminology and refer to the same technical entities.

1.2 The Role-Pattern

In a universal bank, customers often appear in diverse roles: as an investor, borrower, guarantor, or as a customer of other institutions such as investment or assurance companies. In simple terms, the notion underlying the Role-Pattern is to map the structure found in the application domain into the technical solution. In doing so, one gains many technical advantages, for instance, that the above mentioned roles can be implemented independently from one another thus enabling frequent alterations in response to legal changes, new sales strategies, etc. (Bäumer, Gryczan, Knoll, Lilienthal, Riehle and Züllighoven 1997). Obviously, this pattern, too, makes users and designers speak the same language and share the same entities. Note, however, that unlike the Tools-Material-Pattern, this pattern has its roots in the application domain and carries terminology and artifacts into the software labs.

2 Discussion

Both design patterns are completely integrated into the user interface of the application system provided by the RWG. For instance, users apply a particular search tool which works on the material ‘customer register’ in order to find a particular customer, or they use similar tools to find accounts or loan records. In any case, users are trained to be aware of the fact that the user interface is made up of tools and materials. The Role-Pattern is integrated into the user interface in a similar way, i.e. by means of a particular tool, which displays the diverse components belonging to the material ‘person’ (see Figure 1). The upper window of this tool displays personal data provided by all roles, whilst the lower left-hand part lists all the roles that a person consists of. Finally, in the lower right-hand corner several extensions of a selected role are displayed.

These two examples should illustrate as well to what extent design patterns can contribute to the field of requirements engineering. For example, the development process established at the RWG does not rely on ‘traditional’ customer requirement specifications, but on so-called ‘system visions’, documents which describe the to-be-built system on the basis of tools and materials (Bürkle, Gryczan and Züllighoven 1995). During the collection and specification of requirements these documents are continually reviewed by application experts, thus initiating an author-critic cycle even in the very early stages of development. In these phases, especially when extending or revising an already existing application system, it is extremely fruitful to work with application experts who are familiar with the Tools-Material-Pattern. Due to the fact that the tools of the application system (e.g. editors, browsers, or the special-purpose-tool for role objects, see Figure 1) are part of their everyday work, it is often sufficient to achieve a common understanding of the domain-specific materials for fixing the requirements (e.g. attributes of an account such as 'balance' or its services such as 'pay-in' or 'pay-out'). How such a particular material then is 'worked' (i.e. that aspect of a software system which often is referred to as the dialogue structure) can in many cases be described and even implemented by reusing already existing tools (Fach 1997). It should be obvious that the Role-Pattern supports revision or extension of the material 'person' in a comparably efficient manner.

In any case, to make the development process more accessible for application experts and the application domain more transparent to software designers, an interesting and promising approach is to look for further patterns which have the same potential as the presented ones. The objective of this contribution is to provide a basis for collecting, examining, and discussing further design patterns of which a great number can be found in modern software development.

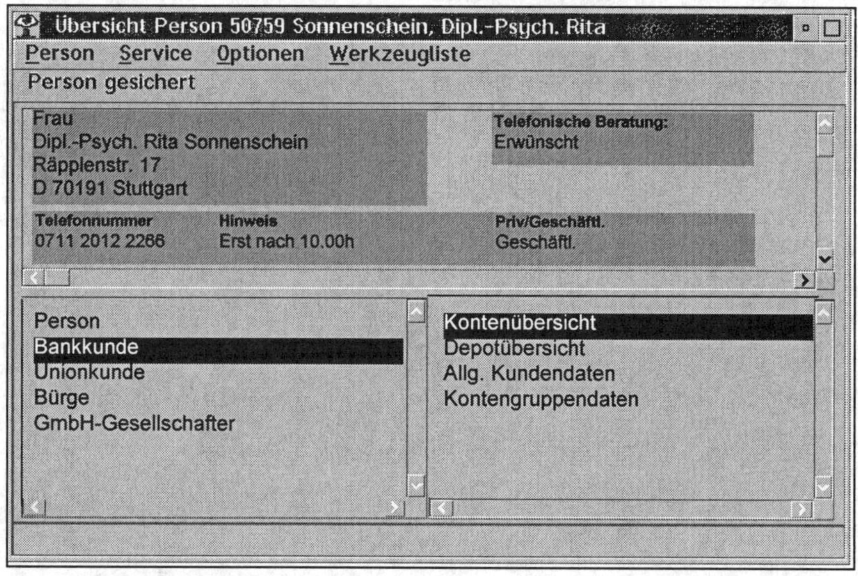

Figure 1: Tool which Displays all Components of a Particular Person. The Material 'Person' is Implemented by Applying the Role-Pattern.

3 References

Bäumer, D., Gryczan, G., Knoll, R., Lilienthal, C., Riehle, D. & Züllighoven, H. (1997). Framework Development for Large Systems. *Communications of the ACM*, 40 (10), 52 - 95.

Bürkle, U., Gryczan, G. & Züllighoven, H. (1995). Object-Oriented System Development in a Banking Project: Methodology, Experience, and Conclusions. *Human Computer Interaction*, 10, (4), 293-336.

Fach, P. (1997). Prototypen im Rahmen der Werkzeug-Automat-Material-Metapher. In G.Szwillus (Eds.): PB'97 Prototypen für Benutzungsschnittstellen. *Notizen zu Interaktiven Systemen*, 19, 12-18.

Reenskaug, T. (1996). *Working with Objects*. New York: Prentice-Hall.

Specification of the User Interface: State of the Art in Real Projects

Friedrich Strauss
sd&m GmbH, Munich, Germany
strauss@sdm.de

1 Introduction

Most specification methods concentrate on describing the functional requirements of a system. Especially object oriented notations like UML don't support an explicit specification of the user interface and its coupling to the application. We first state some requirements on (UI-)specifications for commercial projects: The user should understand the specification, the specification should contain also (non-) functional requirements on the UI, and the specification should be sufficiently detailed to design and implement the system.

We present then an example of a user interface specification representing the notations used in different projects within sd&m. The main element of these specifications is the combination of an end-user documentation of the UI with connections to the data model and the functions called by the user interface. The kind of user interface specification we use has to be refined and to be integrated with specification notations like UML. This description of how real projects collect UI-requirements and document them should help the workshop to consider the real ugly world. The proposed notation for specifications can be a basis for further discussion in the workshop.

2 Software development at sd&m

Software design & management (sd&m) is a German company with 650 employees. sd&m develops individual information systems for the back office of large companies like DaimlerChrysler, BMW, T-Mobil, NUR-Touristic, Dresdner Bank, Thyssen. The typical kind of information system we build handles customer orders with special consideration of peculiarities of the line of business and the customer organization itself. We mainly carry out projects between 5 and 50 person years.

2.1 The software development cycle

Normally our projects follow the waterfall model. The first step is a requirements analysis which yields a non-technical specification of the system describing the functions of the system and its user interface. The system design yields a system architecture, a modularization and a design oriented object model or entity relationship model. Programming, test, integration, and roll-out are the final phases.

To handle large projects, we try to develop an information system in several steps to minimize risks, to concentrate on the most important functions, and to be able to learn from the first steps. Hence in the big view, we follow some kind of spiral model though within each project, we strictly design and develop the software as described in the requirements specification.

2.2 Working with the Customer

Developing individual information systems for different lines of businesses requires much flexibility. Most companies have their own style of how to write a specification. Still we usually have to refine the notation for the requirements specification and the design specification.

In our view the requirements specification has to be accomplished in collaboration with the customer. We prefer to write the requirements in mixed teams consisting of software developers of sd&m and the customer and some future users of the system.

3 UI Specification: Requirements from the practice

High quality requirements specifications are essential to build individual information systems for our customers. The functions of the information systems we build usually have to consider a heap of special cases and a usable user interface is crucial because the users have to handle hundreds of orders each day. Most user interfaces are form based. The requirements specification documents have to fulfill different needs:

- *Understandable by non-IT-people*
  Only the prospective users have the knowledge how to handle the information in the system. Therefore the users must understand the specification documents in order to review them.
- *Detailed enough to design the system*
  The requirements specification is the basis to build the system. Anything which is not completely described will not be build or will delay the project because software developers have to ask somebody how to implement a function.

- *Documentation oriented quality assurance*
  Specification tools without sufficient reporting facilities are worthless: Reviewers work with documents on paper, reviewing something in a case tool is tedious and commenting is difficult in comparison to reviewing a document on paper.
- *Specification of (non)-functional requirements on the user interface*
  Examples of non-functional are grouping of information in a dialog to ease recognition, efficient support of standard tasks (ie. through shortcuts).

Especially the two requirements *understandable* and *detailed enough* are partially contradicting. Therefore a specification contains specific elements for these needs: Use Cases and literal descriptions of functions are user oriented; structured descriptions of functions (sometimes with pseudo-code to describe the flow of control), object models etc. are oriented towards the software developer. The software developer have to keep these different parts of the specification consistent. Still the user has to get an understanding of the specification to check whether all requirements are met.

It is difficult to explicitly capture and document the user interface requirements. Designing and evaluating a user interface seems to be the most effective method. The early design of the user interface has several advantages. First, user interfaces help the users to understand the system specification: users can complete the object models or entity relationship models with the help of a user interface prototype, because they easily detect missing attributes in a user interface. Second, this approach avoids the problem of transforming an 'abstract,' possibly incomplete and inconsistent set of user interface requirements into a valid design. Especially non-functional requirements on the user interface are difficult to describe without a user interface design. For example think of dialogue shortcuts for tasks which are often correlated.

The final user interface prototype needs a documentation in order to share the information with the programmers and to support reviews. Common specification methods help little on the UI specification

Object oriented analysis methods like OOT or notations like UML concentrate on describing the functional requirements of a system and they primarily supply documents for the software developer: event sequence diagrams, class association diagrams or object models. With Use Cases it is possible to describe the tasks of the user and how the system should support them. Use Cases also provide a user oriented description and a good basis for the construction of the user interface, but they are not a full substitute for a user interface specification as described above.

Formal specification techniques are popular within the scientific community. However, they have two main disadvantages: First, only software engineers can understand such notations, a user has no chance to understand a notation like dialogue nets, UAN, etc. Second, they are either too complicated to use, to inflexible and to time consuming. Third they do not really shorten the coding (i.e. through automatic conversion of specifications into code) as research on UIMS and UIDEs (Myers 95) did not yield satisfactory results yet.

4 An Example of a user interface specification

Here we illustrate the elements of the user interface specification as they are often used at sd&m. The context is a client information system for a building society (in German: 'Bausparkasse') with more than a million clients. The example handles a Use Case where a contract for an existing or a new client is entered.

The user interface specification is de facto a documentation of the prototype and describes the attributes and the actions for each window (see figure 1). The descriptions of the attributes are organized similar to the hierarchical organization of the widgets in the window. The description contains the type of the widget (unless it is an edit-text), the usage (input/output, special restrictions), and the connection to some database attribute:

Surrounding_Widget (type of widget)

- **Widget_label** (tab-no., widget-type) attribute-name_in_database
  description

In addition to the description of the static part of the window, the dynamic part has to be described. For complex dialogues, we prefer rule based descriptions:

- **Action-Name (event on widget)**
  description of the action

Rule based descriptions are most effective if many simultaneously possible actions exist within a window, as usually is the case in graphical user interfaces. Sometimes we distinguish different states for a window like *empty*, *editing* or *consistent*. Actions may be restricted to special states and they may change the state. This approach has been inspired by (Strauss 1995), a UIMS approach with a rule based dialog controller.

We usually complete these descriptions with diagrams for the flow of control through multiple windows (finite automata). For simple dialogs (i.e., for character user interfaces) we use only a finite automata with actions and states, where we associate a screen with each state (Denert and Siedersleben 1991).

5 Summary

In our view prototyping of the user interface is an important task within the specification. It supports the communication with users of the software and it should be a documented part of the specification. Object oriented analysis notations and tools have to be enhanced: They should allow integration of user interface prototypes and support some kind of documentation of the user interface. The proposed notation can be a starting point for such a user interface documentation.

WINDOW-ELEMENTS:

Registration (group box)

– **BSC-No.** (1) AT_VBSVT_BSC_Number
this field can contain a customer-no. or a full BSC-no. (customer-no. with a postfixed contract-no.)

– **Type of client** (3, Poptext) AT_PPERS_Person_type
individual (default), couple or company

ACTIONS

•**Read customer (on exit of BSC-No.)**
if the user enters a customer-no. without a contract-no. then the system reads the customer from the database (new contract for existing customer).

•**Input of a BSC-No. (on exit of BSC-No.)**
if the user enters a BSV-No. then no customer is read (new customer).

•**Adjustment to type_of_client (on exit of type_of_client)**
if type_of_client is set to individual, then disable:
on register Client: group-box 2. client, name client-group, ...

Figure1: Example of a dialog specification (shortened) for a new-client contract dialog.

6 References

Denert, E., Siedersleben, J. (1991). *Software-Engineering*. Berlin: Springer.

Strauss, F. (1995). A situation based dialog model for complex direct manipulation interfaces. In Hasan, H, Nicastri, C. (Eds.): OZCHI '95, pp. 249-255, Australia: EMS.

Myers, B. A. (1995). User Interface Software Tools. In ACM Transactions on Computer-Human Interaction., vol. 2, no. 1, March, 1995. pp. 64-103.

Cooperative Development: Underspecification in External Representations and Software Usability

Eamonn O'Neill
Human-Computer Interaction Laboratory, Department of Computer Science
Queen Mary and Westfield College, University of London,
London E1 4NS, UK
eamonn@dcs.qmw.ac.uk

1 Shared Understandings and Cooperative Development

Given that a user is a primary source of task knowledge and requirements, the developer must act as a catalyst for the synthesis of requirements from the user's situated task knowledge. In order to derive an understanding of the requirements for the system which arise from the situation, the developer must first derive an understanding of the situation itself. While simply giving developers access to users, as a passive on-line source of information, may well go some way towards facilitating this process, substantive improvement may require more direct and active user cooperation with developers.

At the heart of cooperative, or participatory, development is the assumption that active user participation in software development should facilitate the integration of users' requirements and task knowledge into the software design (see Schuler and Namioka, 1993). The sharing of users' requirements and task knowledge demands a shared vocabulary and shared representations between users and developers. The user cannot be expected to provide these and so they must be introduced and used with guidance from the developer to produce a directed dialogue between developer and user. The result of this process is shared understandings, or common ground (Clark, 1996), between user and developer.

The work reported in small part here adopted a systems development approach of active user-developer cooperation throughout the development lifecycle

(O'Neill, Johnson and Johnson, 1999). Direct user-developer collaboration facilitated rapid, iterative generation, validation and representation of participants' understandings of current and envisioned work situations, requirements and envisioned system designs. Patterns of four distinct contributions were identified in the joint user-developer development activities (O'Neill, 1996). These were contributions to work analysis, requirements analysis, work design and software design.

Various representations were used to model the objects of these development activities. Users' work situations were represented graphically as task models. System designs were represented as paper and, later, software prototypes. Contributions to work analysis, work design and software design could be assimilated directly into one of the forms of external, shared representation which were used in the cooperative development sessions: current task models, envisioned task models and prototypes. However, for the development activity of requirements analysis in the studied projects, there was no explicit external shared model; that is, users and developers did not work together on a shared, external representation of requirements. In one project, a natural language list of requirements was drawn up by the developers early in the project on the basis of interviews with the users. There was some initial discussion with users on validating the list and it was subsequently used simply as a checklist by developers. In the other project, a quite substantial text based requirements specification was produced but its circulation and use were confined to developers.

2 Representations and Cooperative Development Activities

In both projects, users and developers did cooperate in discussing and defining requirements but this work was performed not explicitly in relation to an external shared representation of the requirements but in relation to the external shared representations which were available: those of the current work situation, of the envisioned work situation and of the software system. Cooperative work with each of these external shared representations raised requirements. Previously identified requirements were indirectly embodied in envisioned task models and prototypes in the form of design proposals to meet these requirements. In addition, participants explicitly annotated requirements on the task models, recording such a contribution on a task model chart next to the task, object or role to which the requirement related. In this way, the task model provided a structure for the requirements.

In many cases the available external shared representations did not lend themselves to the assimilation or appending of contributions made, constraining what could effectively be represented (cf. Palmer, 1978). Frequent examples of this occurred when software design activities, with software prototypes as the external shared representation, included contributions to requirements analysis. Whilst such contributions could readily be assimilated into the participants' internal models or understandings, and might influence future versions of the software prototypes, they were not, indeed often could not be, assimilated into the prototype in use. At best such contributions might be assimilated into a subsidiary external shared representation, such as public lists or notes, or into a participant's personal notes. However, in practice it seems that participants often committed such contributions only to their internal representations of the requirements.

Task models and prototypes failed fully to specify their represented designs and associated requirements. Design specification gaps in the representations were of two kinds. First, there were areas where those constructing the representation explicitly agreed that there was a gap which they could not fill. In one of the projects, for example, a separate developer was producing technically advanced software to support a particular task at the same time that the envisioned task model was being produced. Those constructing the envisioned task model did not know how this new technology would work and so could not model in detail the tasks which relied upon it.

Ironically, the second reason for design specification gaps in external representations was the very strength of common ground built within the teams constructing the task models. Common ground established within the development team consisted in participants' internal models *and* shared external models. Thus, part of this common ground (i.e. the internal models) became unavailable when software implementation was not performed by the same people, revealing gaps in the explicit external representations. Also, common ground established during construction of the external representations could allow a relatively simple symbol in the external representation to carry a lot of meaning which was not conveyed outside the cooperative development situation.

Specification gaps bridged by common ground within the group creating the external representation were more insidious than explicitly recognised gaps in that the existence of the former kind of gaps often did not become apparent until software implementation was attempted, again leaving the implementors to attempt quick and dirty design.

Design specification gaps due to the limitations of paper prototypes are widely reported (e.g. Rudd, Stern and Isensee, 1996). Paper prototypes poorly model the dynamic aspects of software. This resulted, in both projects, in navigational features of the software - how to move from one task to another - being underspecified at the paper prototyping stage. Hence, it was necessary for the implementors to perform detailed software design. Specification gaps in the paper prototypes led to implementors' making poorly supported software design decisions which often led to lower level usability issues, while specification gaps in the task models of the envisioned work led to implementors' making poorly supported task design decisions which often led to higher level usability issues.

As noted above, the nature of the external representations, specifically the lack of an explicit and distinct shared external representation of requirements, enforced on the participants even stronger reliance on internal representations to bridge gaps in the external specification of requirements and on shared understandings of complex meanings attached to quite simple annotations.

Thus, the task models and prototypes effectively facilitated the cooperative identification of requirements but not their public specification and capture. This led directly to problems with detailed design and implementation of the software. Requirements recorded only in the participants' internal models frequently failed to become embodied in subsequent editions of the task and software designs. This was greatly exacerbated when implementation was performed by developers who had not been involved in the cooperative analysis and design activities. The common ground amongst the participants in these activities was distributed across both the external, shared representations and their respective internal models. Thus, the implementors, in receiving the external representations, were being given only part of the design and requirements specifications.

This underspecification in the external shared representations forced the implementors to make ungrounded detailed design decisions. Usability evaluations conducted on early versions of the implemented systems found that poor usability issues attributable to these ungrounded design decisions accounted for 27.6% of all identified usability problems in one of the systems and 74.1% in the other.

3 Conclusion

Graphical task models and paper and software prototypes were used as representational supports for cooperative development activities. They provided effective support as product of and basis for the activities of work analysis, work design and software design. They did not, however, provide effective support for the activity of requirements analysis.

A lesson for practice is that basing a meeting around a form of model which supports the declared development activity of the meeting, for example a task model when the development meeting is called to analyse the users' work situation, may inhibit the recording of potentially significant contributions which are made in the meeting. It may be more effective to focus a meeting on the relevant form of model but also to have permanently visible, and accessible for changes, other forms of representation and structuring. For example, when prototyping it might be productive also to have available modifiable task models and requirements specifications in which to assimilate contributions to work situation analysis or design and requirements analysis.

A key question remains: what forms of representation may provide effective support for cooperative requirements analysis? Such external shared representations must both present the recorded specification of previously determined requirements and facilitate the identification and public recording of new requirements. Crucially, they must be comprehensible and modifiable by both professional developers and users.

A potential partial solution to underspecification is to involve the implementors as participants in the earlier development activities, so making them partners in the personal common ground which is constructed therein. This, however, is not an approach which could be followed in many software development situations, especially very large projects with multiple development teams each with several members. The most effective and efficient means of resolving specification and implementation gaps due to gaps in common ground is likely to remain rapid iteration around the design, evaluate, redesign cycle.

4 References

Clark, H.H. (1996). *Using language*. Cambridge: Cambridge University Press.

O'Neill, E.J. (1996). Task model support for cooperative analysis. *Proceedings of CHI'96 Conference on Human Factors in Computing Systems, Conference Companion*, pp. 259-60. New York: ACM Press.

O'Neill, E., Johnson, P. & Johnson, H. (1999). Representations and user-developer interaction in cooperative analysis and design. *Human Computer Interaction*, 14 (1).

Palmer, S.E. (1978). Fundamental aspects of cognitive representation. In Rosch, E. & Lloyd, B.B. (Eds.): *Cognition and categorization*, pp. 259-303. Hillsdale, N.J.: Lawrence Erlbaum.

Rudd, J., Stern, K. & Isensee, S. (1996). Low vs high-fidelity prototyping debate. *Interactions*, January 1996, pp. 76-85.

Schuler, D. & Namioka, A. (1993). *Participatory design: principles and practices*. Hillsdale, N.J.: Lawrence Erlbaum.

Effects of the Internet marketplace on the usability practice

Masako Itoh
HIT (Human Interface Assessment and Design) Center
NTT Advanced Technology Corporation,
12-1 Ekimae-honcho, Kawasaki-ku, Kawasaki-shi, Kanagawa, 210-0007, Japan

Introduction

Traditionally, usability practice meant usability testing. Testing was almost at the end of the product development process, and the purpose of testing was to find problems in using the products for practical tasks. Though usability problems were found, there was no established route to give feedback to the development teams. Generally speaking, usability was not considered as a quality of the product but as a time and cost consuming constraint. Many companies were concerned more about increasing productivity and reducing cost than in improving usability. They strictly managed qualities such as reliability and safety but they did not consider the optimum interaction between user's goal-oriented behavior and products in real situations.

Recently, the situation has changed as the electronic marketplace developed on the Internet. With the rapid increase in the number of on-line users, growing numbers of companies have set up their Web presence in order to market their products and services. These companies have become serious about usability.

1 Reasons for the change

There are several reasons why usability awareness is increasing.

1. Web sites offer "friction-free capitalism(Gates, Myhrvold, Sheridan & Rinearson, 1995)" where buyers and sellers are directly connected. It is easy for users to write and send their opinions and requirements to sellers and providers.
2. As Rao, Salam & DosSantos (1998) say, users and consumers on the Internet are likely to be better educated, more demanding, and have greater control of the using and purchasing process from initiation to selection. They want to enjoy the site and efficiently find information they need. If they do not experience comfortable interaction with the site, they are reluctant to access

it again.

3. The Internet is an efficient channel for advertising. If a Web site is attractive and popular, it is a promising place to advertise.

4. Corporate Web sites have come to be a 'testbed' for new technologies. Services based on a leading-edge technology are introduced to the site where users' goal-oriented and needs-oriented activities are carried out. Technical engineers and business managers witness and realize the gap between their expectation and users' real behavior.

5. Another situation that accelerates the change is development of intranets. The intranet was introduced to the workplace so that workers can share information across divisions and increase productivity of their intellectual work. However, workers often complain about inconsistency of GUI design, inconvenient search operation, poor navigation, and so forth. Through this intranet experience, IT departments of companies have come to understand that usability of their system is an important quality of their services.

2 What is required for the new usability practice?

As companies have become serious about usability, usability practice has also changed. Usability practice has gone beyond traditional usability testing to encompass activities in the product development process and other corporate requirements for human-centered development (Norman, 1998). The nature of usability testing has changed to keep pace with the broader scope of practice.

2-1 Changes in Usability Testing

From finding problems to identifying needs.

Usability testing becomes a means for collecting and identifying needs in the real use situation. Kannan, Chang, & Whinston (1998) say that the Internet is an efficient channel for marketing research because it provides a faster, cheaper, and more precise collection of marketing information. But marketing research, via both Internet and conventional means, focuses on collecting consumers' needs that they consciously notice and remember. On the other hand, usability testing focuses on the process of interaction between the interface, the user's goal-oriented and needs-oriented behavior, and the environments around them. Usability engineers analyze the interactive process and often find users' behavioral patterns that they are not aware of and can not articulate. Usability engineers must identify needs as well as find problems with the products.

Business managers as a main audiences

Business managers, instead of development engineers, become a more important audience of usability reports. They are concerned with the following:

- How well does the information on the Web site and its interface support users' activity?
- Do users easily understand what to do on the site? Does it look and feel right? If not, how should it be improved?
- What kind of behaviors or expectations do users have when they are on the Internet?
- How do users use multiple media differently in order to get information? What media-mix plans should be developed for future products/services?
- How users like or dislike the leading-edge technologies which are introduced to the Web site?

Usability engineers are required to contribute not only to improving present products but also to planning and designing new products/services. They visit customers, and make observations, and carry out interviews. They plan and design new systems such as a distance learning system and a network operation system, which support people's real life activities. After the system is completed they execute a field study in which people use the service for months. The report includes users' needs, characteristics, contexts where the users are situated, problems of the present system, and sometimes new designs for the system. The audience consists of people from business departments, marketing divisions, designing divisions and product development teams.

Data

Secondary data (data that has been previously gathered for other projects but might be relevant to the problem at hand) is important in conducting the new usability practice. Usability engineers need user information that a business department or a marketing division has in order to plan usability testing effectively. Organizations having the relevant database gain an advantage in effectively testing and diagnosing problems in order to create new design solutions.

2-2 Changes in the product development process

As an organization realizes that usability is an important quality of their products, managers and practitioners of the company change activities in the product development process.

Previously, conventional product planning was based on a new technology, marketing analysis, and market study. Now usability champions advocate observing and analyzing users' activities in real context of use for product planning. They need models which show physical, cognitive, emotional, social, cultural, and organizational variables which influence users' behavior.

They are stimulating people of their divisions to learn research methods and theories of human science and social science through working with the scientists in a usability test and clarifying requirements for a new design. This inquiry process is gradually introduced ahead of designing prototypes.

In addition, long-term monitoring of use of the system is increasing. To systematically collect users' access and input, an appropriate mechanism for collecting necessary data must be designed into the product. Unfortunately, as interface designers and system engineers tend to concentrate on making the product itself, it often happens that certain kinds of data can not be collected after the product is implemented. This is why usability engineers should join in the earlier stages of product development process so that he or she can give a list of functions for long-term monitoring as well as short-term evaluations before the coding starts.

2-3 Design resource management

Design resources are beginning to be managed through an intranet. Human-centered design requires multi-disciplinary scientific knowledge, internal style guides, product knowledge, previous design solutions, national and international standards, marketing information, and so forth. It is hard for an individual division to collect and manage all kinds of information. Cross-functional groups must collaboratively construct the database on the intranet so that designers, technical writers, and usability engineers can easily access the database, acquire the new information they need, and provide appropriate feedback. This environment would promote shared knowledge and better communication among divisions.

3 Towards an integrated usability practice

The above-mentioned effects of Internet and intranet upon usability practice have happened within the business market. Other relevant movements, that is, accessibility and the ISO/FDIS-13407 process have also become salient. Accessibility is the design concept of promoting a high degree of usability for people with disabilities (Vanderheiden, 1997). Particularly with the rapid increase in the number of the Internet users and the quantity of information on the Internet, concerns about accessibility are increasing in Japanese society. The ISO/FDIS-13407 process is "human-centered design processes for interactive systems" in which the key concept is usability (quality in use). Usability champion in companies are learning the ISO/FDIS-13407 process (Bevan & Curson, 1997) and practical methods of human-centered design.

The business, accessibility, and ISO/FDIS-13407 movements are ripe for integration in terms of human-centered and universal design. Companies are gradually changing from technology to human-centered development. A remaining issue is the practice of academics. More and more collaborative research involving engineering, human science, social science, and design is required. Traditionally Japanese academic circles tended to separate themselves from the real world. Academics must change this culture and attain a more integrated usability practice in order to contribute to designing high-usability artifacts for the people and the environment around them.

4 REFERENCES

Bevan, N. & Curson, I. (1997). Planning and implementing user centered design using ISO13407. *In Proceedings of the Sixth IFIP Conference on Human-Computer Interaction*, Sydney, Australia, July 1997.

Gates, B., Myhrvold, N., Sheridan, & Rinearson, P. (1995). *The Road Ahead.* Viking Penguin, New York.

Kannan, P.K., Chang, Ai-Mei, & Whinston, A.B. (1998). Marketing information on the I-Way. *Communications of the ACM*, 4, 35-43.

Norman, D.A. (1998). *The Invisible Computer.* The MIT Press. pp.185-201.

Rao H.R., Salam, A.F., & DosSantos, B. (1998). Marketing and the internet. *Communications of the ACM*, 4, 32-34.

Vanderheiden, G.C. (1997). Nomadicity, disability access, and the every-citizen interface. In toward an every-citizen interface to the Nation's information infrastructure steering committee, Computer science and telecommunications board, Commission on physical sciences, mathematics, and applications, & National Research Council. *More than Screen Deep: Toward an Every-Citizen Interface to the Nation's Information Infrastructure.* National Academy Press.

Method to Build Good Usability: Task Analysis and User Interface Design Using Operation Flowcharts

Haruhiko Urokohara
Managing Director, Chief Designer of User Interface Design Dept., NOVAS Inc.

1 Current Problems

1.1 Introduction

The method of user interface design introduced here is a design technique actively utilizing operation flowcharts. This technique attempts to determine usability logically and visually by describing all the elements forming usability on operation flowcharts. By taking examples from 15 development project cases performed to date, the concept of flowcharts, the methods of describing them and the effects of introduction of the technique will be reported.

1.2 Current problems associated with examination of usability

Problem 1: Naturally, designing a user interface is to consider a design from the standpoint of end users by thoroughly understanding the operation requirements of the customer. In examining user-centered usability, it is desirable that designers should speculate what users may consider or desire as much as possible, but there are no methods established to date in which designers can experience users' positions, which is one of the factors that makes reflection of users' desire for usability difficult.

Problem 2: To solve more fundamental problems of usability or examine more complicated usability in detail, the conventional method for description relying on screen transition diagrams and flowcharts is still insufficient. This is because technical ready-made ideas and limitations often penetrate into prototype screen images and flowcharts produced in the early stage of development.

2 Application To User Interface Design

2.1 Various operation flowcharts

We have been drawing unique flowcharts for the following 15 systems.

(1992) ¥ Semiconductor production apparatus, ¥ ATM control system

(1993) ¥ High-quality car audio system

(1994) ¥ TV conference system, ¥ Airline ticket reservation system

(1995) ¥ Network management system, ¥ Semiconductor production control system

(1996) ¥ Production data generation system, ¥ Information collection system

(1997) ¥ 24-hour unattended store, ¥ Mobile information terminal, ¥ POS system

(1998) ¥ Examination sample transportation system, ¥ Medical diagnostic system

(1999) ¥ Digital still camera

3 Missions Of Operation Flowcharts

3.1 Types of operation flowcharts

Operation flowcharts can be categorized by the four missions {(1) grasp of the overall system, (2) standards for evaluating usability, (3) idea source for screen configuration, and (4) verification of consistency in operation rules} and the three types (A: Screen transition diagram, B: Operation flowchart and C: Thought flowchart). In one development, flowcharts are used to judge screen configurations, while they are utilized as evaluation standards of usability in another development. Although roles played by flowcharts may vary from one development to another, an overall flowchart for the purpose of grasping the outlines of the entire operation has been able to be drawn for any development project.

3.2 Visualization of usability

Usability and its user friendliness are invisible. Therefore, considering that it is an important step to have the invisible usability visualized as much as possible, I have tried out several visualization methods. First of all, I tried to describe the entire current operation as finely as possible. This is different from assembling story boards by utilizing prepared screen images beforehand, instead, various operation elements including phenomena took place and users' thought are to be drawn along the time axis in this process. (Figure 1) To perform this task, it is necessary to ask users very closely and put down every information obtained from observation of users in their routine activities along the time axis. A chart obtained in the first instance is some unorganized scattering of tasks jotted on a sheet. What is shown on the sheet is just a scribbled description which is never called an operation flow nor classification of tasks. But this messy description itself is what truly represents the users' actual operation.

3.3 To understand users' operation

It may not be too much to say that understanding of users' operation through observation of their routine activities is the most important process in designing a user interface. Actual user operations are often affected by manners of operation or customary practices, and therefore, it is important to understand what judgment criteria users have in conducting their routine or emergency operations. However, most of the system engineers are not aware of the importance of understanding users' operation.

3.4 To draw flowcharts

Detailed drawing rules are now under analysis. Only the concept of rules are discussed here.The rule is to establish a firm time axis and to draw the importance and frequency of operations in bold lines.(describe operation procedures along the time axis faithfully) Figure 2 is an operation flowchart of a 24-hour unattended store drawn by using the rules. As for this flowchart, following two flowcharts were united. One is "operation flowcharts" to understand the entire flow of work and the other is "thought flowcharts" to understand customers' thought. And the purchasing operation block (background is gray) in this flowchart is drawn as a separate flowchart. This block is accompanied by detailed flowcharts prepared on separate sheets.

3.5 To judge logically and visually

Operation flowcharts which are easier to understand visually can be drawn when an operation system is organized logically. It is also understood from a number of experiments that an operation flowchart is much easier to understand visually, when it is classified precisely with one of the axis as time (sequence of operation opportunity) and the other as business operation factors unique to the system.Furthermore, it is possible to judge effectiveness of operation procedures, arrangement of displayed information and arrangement of buttons, when a simple operation flowchart is drawn according to fixed drawing rules. Fixed drawing rules refer to the drawing of flowcharts that distinguish the importance and frequency of operations based on business and operational information, while simple drawing refers to the work to simplify lines drawn on a flowchart.

4 Practice By Engineers

4.1 Possibility of easy introduction of this technique

There was a case in which this technique was utilized by a team consisting of only a company's designing engineers which produced and sold a exclusive-use POS system for gas stations. This company had quite a high level of interests in user interface design, but never had participation of a user interface designer into development due to various reasons inhibiting such arrangements. Therefore, this technique was introduced to this company and its development team, securing the work hours they can afford for added tasks, utilized the technique for clarification of specifications by drawing operation flowcharts and active discussions on screen configurations.

4.2 GUI study for the exclusive-use POS system

Situations: Prior to the commencement of consultation with us, they were in a transition phase from using there own hardware and OS to the Windows NT environment and their plan to convert CUI base screen displays to GUI was in progress. However, as they did not have judgment criteria for GUI, their plan was stopped after completion of designing of the primary study screen configurations in the paper model stage of the fundamental specification examination phase, because various opinions were raised by directors, the project leader and project team members. After consulting, the company's engineers decided to introduce two items out of the four missions mentioned earlier, to the GUI, namely (1) grasp of the overall system and (2) standards for evaluating usability, and their development work ceased in mid-stage was resumed in consequence.(Figure 3)

4.3 The engineers' impressions of introduction of the technique

Engineers in general who are engaged in designing of a user interface have very strong interests in usability. However, what they have learned so far in their experience is mostly centered on programming techniques for accurate performance of functions, in other words they have never had chances to learn user-centered methods. What they recognized and appreciated among things they experienced in the consultation is (1) to define relationship between primary and secondary things, (2) to have an overview of the operation and (3) the fact that it can be done by themselves.

5 Future Potential

Drawing of operation flowcharts is an effective tool to clarify required or fundamental specifications. They also give various labor- and cost-saving effects such as reduction of design work hours and lowering a program modification rate with respect to usability. Of course, the largest advantage is that they provide a means to explain the effects of a user interface concisely to designing engineers, which leads to successful development of commercial products with good usability.To improve usability, it is desirable that a user interface designer participates in a project from the stage to examine required and fundamental specifications so that he/she can utilize his/her logical thinking methods. The current method to draw flowcharts uses quite a primitive drawing method, because there is no specialized tool available. I am trying to develop a specialized tool for this technique or an add-in tool to an existing application software.

6 References

Haruhiko Urokohara (1996), Operation Flow Charts to straighten up tasks and specifications for programming (Human interface News and Report).

Yasuyuki Kikuchi, Toshiki Yamaoka, Haruhiko Urokohara (1995), GUI Design Guidebook.

Ben Shuneiderman (1992), Designing the User Interface 2nd edition .

Kageyu Noro (1990), Illustrated Ergonomics .

QUIS: applying a new walkthrough method to a product design process

Suma Suzuki, Ayako Kaneko, Kazunori Ohmura
Sony Corporation

1 Introduction

Practical usability evaluation techniques for product development should be available early in the development process as well as time- and cost-effective. Inspection methods such as cognitive walkthroughs and heuristic evaluation have been developed to satisfy these requirements. However, both methods are less practical in that heuristic evaluation needs usability specialists (Nielsen 1992) and cognitive walkthroughs are difficult for novice evaluators (Wharton et al. 1992).

Being usable for novice evaluators is another requirement for practical methods to be widely employed for product development. Therefore, we developed and introduced QUIS (Quick Usability Inspection System) into our company, a new inspection method designed for product engineers without usability expertise, based on the cognitive walkthrough (Suzuki et al. 1995). Despite its effectiveness verified by experiments, however, QUIS has been employed in only a few products. In this paper we first outline QUIS and our activities of promoting QUIS and then discuss issues we should consider in introducing evaluation tools into a product design process.

2 Outline of QUIS

The basic procedure of QUIS is identical with the cognitive walkthrough. Inspectors evaluate an interface following task scenarios, focusing on whether users can perform all operations for a task without difficulty. Because QUIS has been developed especially for novice evaluators such as software engineers, it has the following features in comparison with the cognitive walkthrough.

1) Easier way of describing user's goals

In evaluating an interface, it is very important to consider not only how but also why users perform certain operations. Cognitive walkthroughs require inspectors to suppose user's goal structure for every task independent of the flow of operations, which is difficult and confusing even for usability specialists. In QUIS inspectors have only to describe a goal for each operation (Figure 1).

2) Improved rating of severity

In cognitive walkthroughs inspectors rate severity of identified problems into five levels by percentage of users who would experience the problems. However, the severity is not determined only by the percentage of users, and the five-level rating is too detailed. So QUIS's rating has two axes: the percentage of users (frequency) and an impact of the problem on the task (seriousness), each with three levels (Figure 2).

3) Reduced number of questions

In QUIS 7 questions cover the 16 ones used in the cognitive walkthrough. We also added 6 questions about recovery from errors, which the cognitive walkthrough does not provide (Figure 3).

...

S2 Describe the correct operation procedure and the goals of each step.

No.	Operation	Goal
1	*Press PLAY*	*Specifying the track.*
2	*Slide EDIT to the right.*	*Selecting the erase function.*
3	*Press ENTER/REPEAT.*	*Erasing.*
4	*Press ENTER/REPEAT again.*	*Confirming that the track is to be erased.*
5	*Press STOP.*	*Ending the erase operation.*

...

Figure 1 An example of user's goal in QUIS form.

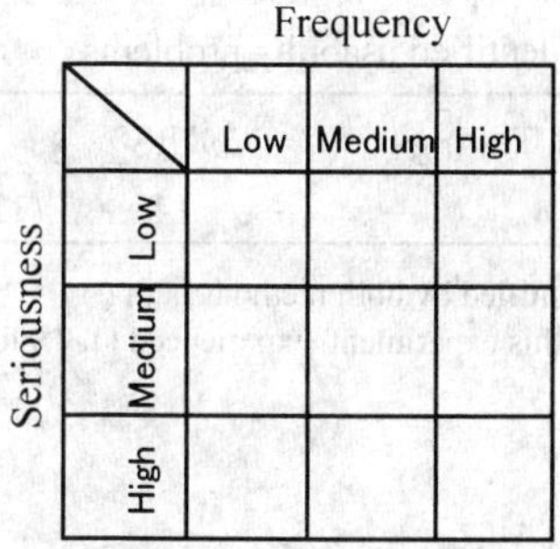

Figure 2 Rating matrix of problem severity. *Frequency* means the percentage of users and *seriousness* means an impact of the problem on the task.

...

Is there a possibility that the response will not cause the user to realize that he/she performed the incorrect operation?

Yes ☐ No ☐

Is there a possibility that the user will not understand what he/she needs to do in order to return to the correct procedure?

Yes ☐ No ☐

...

Figure 3 Excerpt from questions about recovery from errors.

We conducted two comparative experiments of QUIS with cognitive walkthroughs and usability testing and the results showed the effectiveness of QUIS (Suzuki et al. 1995). Compared with cognitive walkthroughs, QUIS found more problems and took less time to identify them (Table 1). Moreover, even novice inspectors found problems equivalent to those found in usability testing (Table 2).

Table 1 Evaluation time and identified usability problems.

	Cognitive Walkthrough	QUIS	
Task 1			
Evaluation time (min.)	75	40	
Identified problems	7	9	(6)
Task 2			
Evaluation time (min.)	60	40	
Identified problems	5	9	(5)

The number of problems identified by both methods is in ( ).

Table 2 Identified usability problems.

Usability Testing	QUIS	
18	14	(11)

The number of problems identified by both methods is in ( ).
Inspectors who participated this experiment experienced QUIS for the first time.

3 Promotion of QUIS

We introduced QUIS into Sony's product development process through the following promotions.

- Publicizing QUIS at events, in booklets, and on the intranet
- Consulting with engineers who want to use QUIS
- Opening a half-day training course of QUIS

Although 185 engineers have been registered to use QUIS, it has been applied to only 14 products. Some interviews with those who have not used QUIS yet show that they wish to use it but they have little time for usability evaluation. These results imply that there are some potential needs of usability evaluation tools but QUIS is not the best solution.

About 60 engineers have taken the training course. They say that although QUIS enabled them to find usability problems that they would have missed, they got tired in evaluating tasks of long sequence of operations because they had to answer the same question set for every operation.

4 Discussion

Our purpose of developing QUIS was to invent a usability evaluation tool for product developers. So QUIS was required to be: (1) usable for engineers without usability expertise, and (2) more efficient than existing methods such as cognitive walkthroughs and usability testing. While the results of our experiments showed that QUIS satisfied these requirements, QUIS has scarcely been accepted in Sony as a practical tool. What prevents it from acceptance?

The main factor is the redundancy of QUIS, which is an unexpected effect of our attempt to provide a reliable tool for novice evaluators. Certainly QUIS is more efficient than cognitive walkthroughs and usability testing. But product developers expect a much quicker method.

Further, the more fundamental cause is that usability evaluation has not been considered indispensable in Sony's product development process. So even if engineers want to do usability evaluation, their manager are unwilling to allow it. Our activities to promote QUIS lacked consideration for this problem. We realized that educating engineers, especially their managers, on the importance of usability evaluation beforehand was necessary to successfully introduce a usability evaluation tool, no matter how effective it was.

At the same time, however, we found that QUIS was a powerful tool to learn the viewpoints of usability evaluation: that is, how users interact with products and what prevents them from smooth interaction. Using QUIS, inspectors can simultaneously learn the viewpoints while answering each question. The engineers who finished the QUIS training course say that they can remember these points in designing user interfaces. Consequently QUIS is usable for training user interface designers.

Through introducing QUIS, we have learned that having engineers understand the importance of usability evaluation is one of our urgent tasks to incorporate usability evaluation into Sony's product development process. Now we are developing a curriculum of usability studies for the engineers and managers.

5 References

Nielsen, J. (1992). Finding usability problems through heuristic evaluation. *Proceedings ACM CHI'92 conference* (Monterey, CA, May 3-7, 1992): 373-380.

Wharton, C., Bradford, J., Jefflies, R., and Franzke, M. (1992). Applying cognitive walkthroughs to more complex user interfaces: Experiencies, issues, and recomendations. *Proceedings ACM CHI'92 conference* (Monterey, CA, May 3-7, 1992): 381-388.

Suzuki, S., Kaneko, A., and Ohmura, K. (1995). User interface evaluation system QUIS. *Proceedings of the 11th symposium on human interface* (Kyoto, Japan, October 18-20, 1995): 747-752 (in Japanese).

A Comparative Study of sHEM (structured heuristic evaluation method)

Masaaki Kurosu 1), Masamori Sugizaki 2), Sachiyo Matsuura 2)
1) Faculty of Information, Shizuoka University
2) Humanonics Research Group, Yamaha Motors Ltd.

1. Introduction

Among the various inspection methods, the heuristic evaluation method (HEM) (Nielsen and Molich 1990) and the cognitive walkthrough method (Polson et al. 1992) are famous and frequently used. These two methods are contrasting in that the former inspects widely but shallowly and the latter inspects narrowly but deeply. But because of the higher flexibility and the higher productivity of the heuristic evaluation method, it is reported that this method is used more frequently than the cognitive walkthrough method (Nielsen 1995). He claims the heuristic evaluation method to be the best of the inspection methods.
This is why we adopted this method as the basis for developing a new inspection method. What we focused on was the number of usability guidelines used in the heuristic evaluation method. In the original heuristic evaluation method, Nielsen proposed 10 guideline items. All the items are important for the user interface design, but when we think of the larger sets of guidelines which are used in the interface design, it should be noted that what he picked up was just a part of the whole set and/or an abstracted superset of the whole guideline items. Considering the mental process of the usability evaluator when s/he is inspecting the product by the heuristic evaluation method, not the whole set of the usability guidelines but just 10 guideline items are activated in his/her working memory. Because of the small number of the guideline items, there is a possibility that some aspects of the usability may not be actively inspected.
In order to improve the productivity of the heuristic evaluation, it is necessary to increase the number of guidelines presented to the evaluator. But if we simply give him/her a large set of the guideline items at the same time, s/he might be confused because of the capacity of the working memory and we cannot expect the better performance. What we proposed was that the whole guideline items

should be divided into subsets with the reasonable number of items and secondly the whole evaluation session should be divided into several subsessions. If we assign a different subset of the guideline to a different subsession, we can attain both of the goals of increasing the number of guidelines used in the evaluation session and of increasing the productivity of detecting the usability problems. This idea of structuring the evaluation session is the basic principle of the structured heuristic evaluation (sHEM) (Kurosu et al. 1997).

In order to confirm our hypothesis that the sHEM should have a better productivity compared to the original HEM, an experimental validation should be done (Experiment 1). Besides the comparison with the HEM, we performed the comparison with the cognitive walkthrough method and the user testing (Experiment 2).

Our ultimate goal was to develop the usability evaluation method with highest cost/performance. Usually, the performance of the evaluation method was measured only by the productivity or the number of problems detected by the method. But we thought that the cost of the method should also be considered. The cost of the usability evaluation method can be measured by the length of the evaluation process. And, of course, the shorter the better. For this purpose, we adopted the cost/performance measure named as PPPH or the number of Problems detected Per Person and Hours. This measure can be calculated by dividing the number of problems by the number of usability evaluators and the total length of the evaluation in hours.

2. Experiment 1

In this experiment, the performance of the HEM, the sHEM and the revised version of sHEM were compared. We then compared the cost/performance of the methods using the measure of PPPH. (Matsuura et al. 1997, Sugizaki et al. 1997, Kurosu et al. 1998a)

2.1 Method

Subjects. 15 employees of Yamaha motors participated in the experiment as the evaluators.

Experimental conditions. Following the lecture on the guidelines, there was 3 hours evaluation session. The evaluation session was divided into 6 subsessions of 30 minutes. In terms of the HEM, each of the 6 subsessions was the same, i.e. the evaluators inspected the targeted device by following 10 guidelines. For the sHEM and the revised-sHEM group, 6 subsessions were assigned different subsets of the usability guidelines. The revised-sHEM composed of larger set of guidelines compared to the sHEM.

Targeted device. Portable mini-disc (MD) was the targeted device in the experiment.

Procedure. In the evaluation session, the evaluators were requested to write down the usability problem on the Post-It and at the end of each subsession they were asked to paste the Post-Its on a A4 paper to deliver it to the experimenter.

2.2 Results

The number of problems detected in each of 6 subsessions are shown in Figure 1 for three different methods. It is evident that the number of problems detected

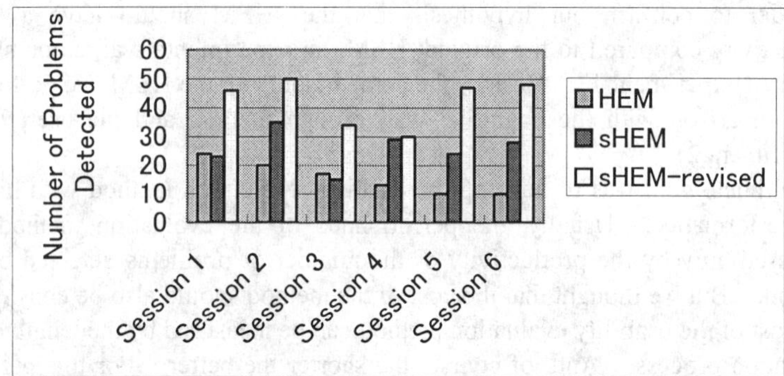

Figure 1. Number of problems detected by the HEM, the sHEM and the revised-sHEM

by the HEM is decreasing exponentially. This might have been caused by the saturation on the part of the evaluator by focusing on the same 10 guidelines all the time. On the contrary, the performance of the sHEM and the revised sHEM is rather constant. The result of the T-test for the HEM and the sHEM was $t=2.7183$, $p<0.05$ and for the HEM and the revised-sHEM was $t=6.8914$, $p<0.005$. This result shows that the shift of focus of attention in these two methods worked positively for maintaining the productivity.

The results of this experiment is summarized in Table 1. The total number of problems detected in 3 hours session were 94 with 11 duplications for the HEM, 153 with 25 duplications for the sHEM and 254 with 61 duplications for the revised-sHEM. Thus the revised-sHEM showed 2.3 times of productivity compared to the HEM.

Table 1. Results of the experiment for the HEM, the sHEM and the revised-sHEM

	Problems detected	Importance Rating	Evaluator	Time to evaluate (hour)	PPPH
HEM	83	2.49	5	3	5.53
sHEM	128	2.30	5	3	8.53
revised-sHEM	193	2.61	5	3	12.87

PPPH as the measure of the cost/performance is also shown in this table. PPPH means the number of problems that can be detected by one evaluator for one hour of evaluation.

3. Experiment 2

In this experiment, the sHEM was compared with the cognitive walkthrough method (CW) and the user testing (UT). The purpose was to confirm the high cost/performance of sHEM compared to other usability evaluation techniques than the HEM. (Matsuura et al. 1998, Sugizaki et al. 1998)

3.1 Method

Subjects. 5 employees of Yamaha motors, 3 of them were male and 2 of them were female, participated in the cognitive walkthrough evaluation as the evaluators.

Experimental conditions. Both of the cognitive walkthrough method and the user testing trace the user's behavior to find out the usability problems. There were five benchmark tasks.

Targeted device. A personal digital assistant (PDA) was used as the target device.

Procedure. In the cognitive walkthrough session, the evaluators wrote down the possible usability problems onto the check sheet. In the user testing, the subject simply performed the tasks. The think-aloud method was used.

3.2 Results

Total number of usability problems found for 5 different tasks by the cognitive walkthrough method were 38 including 12 duplications for the novice evaluators, and 60 including 19 duplications for the expert evaluators. The usability problems found by the user testing was 24. But because there are differences in the number of evaluators and the length of the evaluation, the comparison should be done in terms of the PPPH.

Table 2. Comparison of the user testing, the cognitive walkthrough and the sHEM.

	Problems detected	Number of evaluator	Time to evaluate (hour)	Time to analyze (hour)	PPPH
UT	24	1	1	10	2.18
CW(novice)	26	5	3	2	1.04
CW(expert)	41	3	1	2	4.56
sHEM(novice)	130	5	3	0	8.67
sHEM(expert)	131	3	1.5	0	29.11

The result is shown in table 2. The PPPH we obtained shows that the sHEM is more efficient than the cognitive walkthrough method about 8.3 times for the novice evaluator and 6.4 times for the expert evaluator. Compared to the user testing, the sHEM was found to be 4.0 to 13.4 times better.

4. Discussion

For the purpose of comparing the cost/performance of different methods, the measure of PPPH was proposed. This measure means the number of usability problems detected by one evaluator in one hour. For the novice evaluators, the PPPH value of the sHEM is 1.54 to 2.33 times larger than the HEM, and 8.34 times larger than the cognitive walkthrough method. And for the expert evaluators, the value of the sHEM is 6.38 times larger than the cognitive walkthrough method. Thus the performance of the sHEM was found to be the best among the methods used in this study.

5. References

Kurosu, M. (1996) The structure of the usability concept, 11th Symposium on Human Interface 351-356 (In Japanese)

Kurosu, M., Matsuura, S., and Sugizaki, M. (1997) Categorical inspection method: Structured heuristic evaluation (sHEM), IEEE System Man and Cybernetics'97, 2613-2618

Kurosu, M., Sugizaki, M., and Matsuura, S. (1997) Usability Evaluation Technique with High Efficiency 1. Structured Heuristic Evaluation (sHEM), 2. Experimental validation of sHEM, 3. Experimental validation of modified sHEM 13th Symposium on Human Interface, 481-488, 489-494 (In Japanese)

Kurosu, M., Sugizaki, M., and Matsuura, S. (1998a) Structured heuristic evaluation (sHEM), Usability Professional's Association 7th Annual Conference Proceedings, 3-5

Kurosu, M., Sugizaki, M., and Matsuura, S. (1998b) Experimental Analysis of the Cognitive Walkthrough Method, 14th Symposium on Human Interface, 37-42 (In Japanese)

Matsuura, S., Sugizaki, M., and Kurosu, M. (1998) Experimental Approach to Cognitive Walkthrough Method 1, 2 14th Symposium on Human Interface, 29-32, 33-36 (In Japanese)

Nielsen, J. (1993) Usability Engineering, Wiley

Nielsen, J. and Mack, R.L. (1994) Usability Inspection Methods, Wiley

Nielsen, J. and Molich, R. (1990) Heuristic evaluation of user interfaces, Proc. ACM SIGCHI '90 Conf. 249-256

Polson, P.G., Lewis, C., Rieman, J., and Wharton, C. (1992) Cognitive walkthrough: a method for theory based evaluation of user interfaces, Int. J. Man-machine Studies, 36, 741-773

3D Visualization of Plant Parameters and its Evaluation

Tatsuro Sano, Yoshio Nakatani, Akira Sugimoto
IESL, MITSUBISHI Electric Corporation, 8-1-1, 661-8661, Japan

1. Introduction

Recently, high-quality 3D graphics are relatively inexpensive, so there are many attempts to visualize information of very large systems by using 3D graphics. Although a 3D graphical user interface (3D-GUI) system usually costs more to develop than a 2D graphical user interface (2D-GUI) system. To solve this problem, some programming libraries for constructing 3D-GUIs have been developed (Strauss and Carey 1992, Wernecke 1994). However, for developers to use these libraries to construct 3D-GUIs, they need to understand the many kinds of Application Programming Interfaces (API) of these 3D graphic libraries. It requires much work for developers to write many lines of source code required for 3D-GUIs.

We have previously proposed a 2D-GUI construction method („the GUI Constructing & Customizing Method (GCM)“) to help programmers develop 2D-GUIs very easily. This method has been used in actual fields, such as plant supervisory control systems (Kitamura and Sugimoto 1995). In this paper, to help the programmers develop 3D-GUIs very easily, we propose an effective method for constructing 3D-GUIs semi-automatically by applying GCM -- we call this 3D-GUI construction method „the 3D Generating & Customizing Method (3D-GCM)". In addition, we show the advantage of 3D-GCM through human experiments comparing 2D-GUIs with 3D-GUIs, both of which are generated with our method.

2. 3D Generating & Customizing Method

2.1 The Application of GCM in Constructing a 3D-GUI

In the original GCM, there are two main steps to generate a 2D-GUI. First, a default 2D-GUI is automatically generated from the data and functions which are included in the application program (AP). Second, the default 2D-GUI is customized interactively on line, and the final 2D-GUI is constructed.

Our proposed 3D-GCM follows the same model as the original GCM. The 3D-GCM also takes two steps to construct a 3D-GUI. First, a default 3D-GUI is generated from an existing 2D-GUI. Second, the default 3D-GUI is customized interactively on line, and the final 3D-GUI is generated. In this method, the relations between the 3D-GUI and the AP data in display and control are the same as the relations between 2D-GUI and the AP data. Therefore, the developer does not have to define the relations between the 3D-GUI and the AP data. In addition, because the relations between the 3D-GUI and the AP data are also kept while the 3D-GUI is being customized, the developer can customize the 3D-GUI and look at its practical behavior by executing the AP. The developer can make changes to 3D-GUIs, including replacing 3D-GUIs, adding 3D-GUIs, changing 3D-GUIs' colors and shapes, and setting action links. In this method, AP developers can generate 3D-GUIs by using existing 2D-GUIs and can even construct 3D-GUIs without knowledge of 3D-GUI programming. In addition, this method allows end-users to customize 3D-GUIs flexibly while the AP is being executed.

2.1 Implementation

We earlier developed a C++ class library called GhostHouse for the GCM, we then expanded this library to design the 3D-GCCM. In the GhostHouse library, we introduced two classes, one called a Model-Ghost and the other called a View-Ghost (Fig.1). These classes are designed to enable flexible customization while keeping the relations between the GUI parts and the AP data. In this library, the model-Ghost manages AP data and the View-Ghost manages the AP GUI parts. These Ghosts ensure that interactions between GUI parts and data run smoothly. Because the references between View-Ghost and Model-Ghost are implemented by a link object that has a normalized communication protocol, these references can be kept even when a 3D-GUI part is replaced. Because of this function, it is not necessary to redefine the relations between data and 3D-GUI parts when 3D-GUIs are generated automatically or 3D-GUI parts are changed, which means that 3D-GUIs can be generated easily.

2.2 Application Example

We constructed a water pipe network operation support system (a kind of simulator system) using a 3D-GCM. The water pipe network operation support system is used for deciding configuration of a water pipe network and for operation training of a water pipe network. Figure 2 shows an existing water pipe network operation support system which GhostHouse constructed. A window in Figure 2 shows a network structure.

We applied a 3D-GCM to this system. an example of a water pipe network operation support system that has been constructed automatically with a 3-D GUI. Figure 3 shows an example of a automatically generated 3D-GUI with a 3D-GCM. The shapes of the automatically generated 3D-GUI nodes are default shapes that were decided beforehand. Their X-Y locations in the 2D plane are the same as the X-Y locations in the original 2D-GUI and their Z-axis values also show values which their original 2D-GUIs express.

Figure 4 shows an example of the final design of a customized 3D-GUI. In it, we use thickness of arc to show waterflow, and added texture to the map in order to correspond to topography locations. By this simple customization, a developer can construct a real system very easily.

3 Evaluation of a 3D-GUI Constructed with a 3D-GCM

Our proposed 3D-GUI construction method is especially effective in displaying values of information placed in a 2D plane to a Z-axis. A major characteristic of displaying values to a Z-axis is that it is easy to understand a entire condition of a system at a glance. And it is natural that values of information are displayed on Z-axis keeping their placement in a 2D plane.

On the other hand, there may be fundamental questions: Whether are these 3D-GUIs which are constructed by our method really more effective for users than existing 2D-GUIs for users or not? And if they are effective, then for what amount of information.

In order to determine the effectiveness of 3D-GUIs constructed using a 3D-GCM, we conducted a comparative experiment. This experiment also was used to develop a guideline for when to use 3D-GUIs. The experiment consisted of a 2-D GUI system and a 3-D GUI system doing similar tasks that were generated by a 3-D GCM from 2-D GUIs. We then compared the time it took the human subjects to finish these tasks and do questionnaires.

For this experiment we chose the water pipe network operation support system described in section 2.2(Application Example). This example was used because a water pipe network has a fixed layout in a 2D plane(landscape), and we judged it was suitable for estimating the effectiveness of a 3D-GCM which was especially effective in displaying values of information placed in a 2D plane on Z-axis.

3.1 A Water Pipe Network Operation Support System

A water pipe network consists of pumps, valves, and water supply tanks, which expressed by nodes, and of water pipes which are represented by arcs. The water pipe network operation support system calculates the water pressure of nodes and water flow of arcs, and so on.

3.2 Experiment

We executed three tasks as follows:

Task 1: Finding the node that shows maximum water pressure and the node that shows minimum water pressure.

Task 2: Finding three consecutive nodes that have a lower water pressure than the nodes surrounding them.

The numbers of nodes in the water pipe network used in Task 1 are 5, 10, 20, 50, 100, and 500. The number ofnodes in the water pipe networkused in Task 2 are 10, 20, 50, 100, and 500. For both the 2D-GUI and 3D-GUI tests, subjects could do similar mouse operations -- parallel translation, zooming, and attribute setting. In addition, in the 3D-GUI test, subjects could also operate rotations with the mouse. Twelve subjects were used in all, all of whom were skilled in operating a mouse.

We assumed that the average task execution times for the 2D-GUI and 3D-GUI takes were equal and did a non-ordered sign test with the significance level of 0.05. Table 1 shows the results. In Table 1, the circle shows the GUI system which is significantly shorter in average task execution time; the X mark shows the GUI which is significantly longer; and the dashed mark shows the others. Graph 1 shows the average task execution times.

3.3 Discussion

As for task execution time, the 3D-GUI system was significantly shorter when the number of nodes exceeded 100. We got a similar result from questionnaires. This is because when the number of nodes exceeded 100, the 2D-GUI could not display all of the data at one time. In other words, it is generally appropriate to use a 3D-GUI system when all the data can not be displayed at one time with a 2D-GUI and on a scale in which we can recognize the parameter values. Conversely, it is not necessary or appropriate to use a 3D-GUI system when all values can be displayed at one time using a 2D-GUI.

4 Summary & Future Work

In this paper, we have proposed a effective method for constructing 3D-GUIs semi-automatically. We have also conducted experiments in which human subjects rated 2D-GUIs against 3D-GUIs both of which were generated by our method. These experiments show the effectiveness of 3D-GUIs. However, in our current version of the system, the automatically created 3D-GUIs are still simple and domain-free parts, and the users are required to customize the generated parts to some extent. We will need to expand our method further in order to be implemented to generate domain-specific and more sophisticated parts automatically.

Reference

Misayo Kitamura, and Akira Sugimoto. (1995). GhostHouse: A Class Library for Generating Customizable Graphical User Interface. Trans. on Information Processing Society Japan, Vol.36, No.4, pp.944-958.

Paul S. Strauss, and Rikk Carey. (1992). An object-oriented 3D graphics toolkit. Proceeding of SIGGRAPH'92, pp.341-349, ACM Press.

Josie Wernecke. (1994). The Inventor Mentor. Addison Wesley.

K. M. Fairchild, S. E. Poltrock, and G. W. Furnas. (1988). SemNet: Three-dimensional graphic representation of large knowledge bases. Cognitive Science and its Applications for Human-Computer Interaction, pp.201-233, Lawrence Erlbaum Associates.

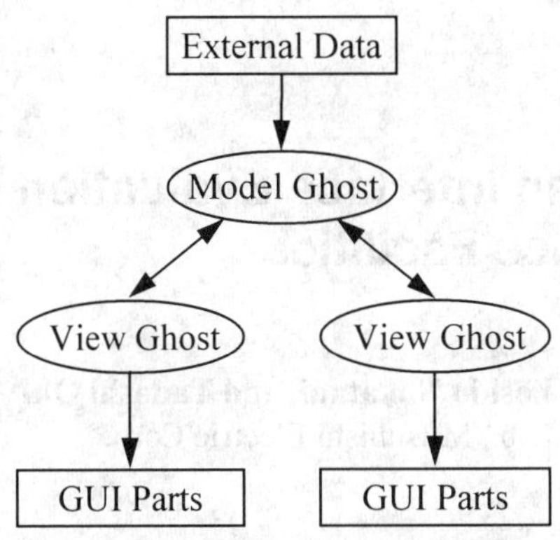

Fig. 1 The frame-work of GhostHouse

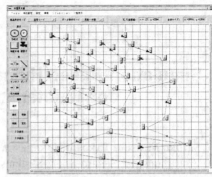

Fig.2 Screen example of a 2D-GUI

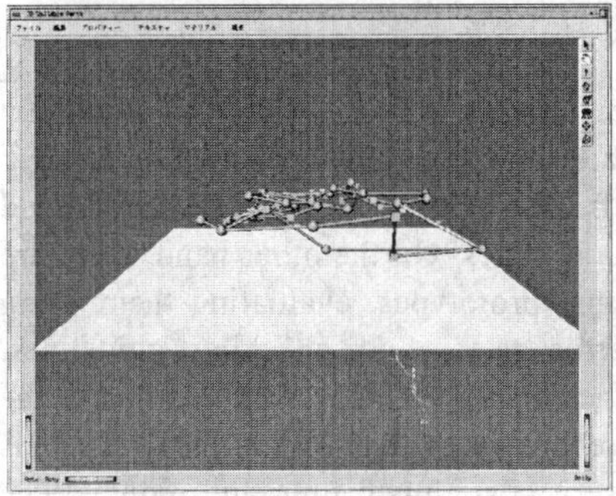

Fig. 3 Screen example of a default 3D-GUI

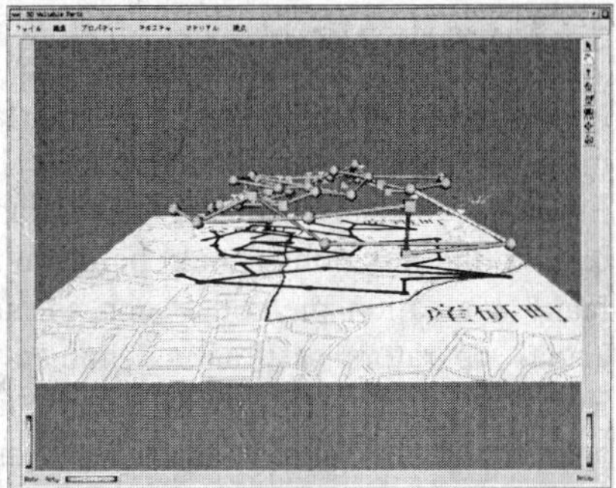

Fig.4 Screen example of a customized 3D-GUI

Table 1 Sign test results

		number of nodes					
		5	10	20	50	100	500
Task1	2DGUI	○	○	—	—	×	×
	3DGUI	×	×	—	—	○	○
Task2	2DGUI		○	—	—	×	×
	3DGUI		×	—	—	○	○

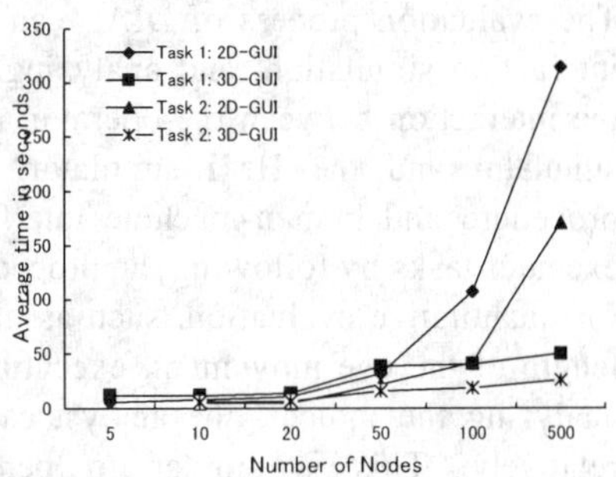

Graph 1 Examination results

Simulation-Based Human Interface Evaluation for Maintenance Facilities

Satoko Matsuo, Takashi Nakagawa, Yoshio Nakatani, and Tadashi Ohi
Industrial Electronics & Systems Lab., Mitsubishi Electric Corp.

1 Introduction

In designing a human-machine interface (HMI), it is important to iteratively make prototypes of the product, evaluate, and give feedback on the design. Especially, because nuclear power plants are required to have HMIs of good quality, the iterations should be made many times. On the other hand, there are many evaluation methods, such as making prototypes, evaluating them using actual operators, estimating human error rates, and checking guidelines. Because each of them deals with particular aspects of HMI problems, the analyst should combine them for comprehensive evaluation and analysis. It is difficult to iterate the evaluation because it costs much time and money. The authors proposed a new method to evaluate and analyze the HMI design of plant equipment easily and rapidly. They have implemented the system as DIAS[1], an HMI evaluation method by using computer simulation.

The evaluation process of DIAS can be divided into three parts: data editing, interaction simulation, and analyzing. The simulation part of DIAS simulates the interaction between the operator and the plant equipment with the operator simulator and the HMI simulator. These simulators require an operating procedure and human-machine interface design data. The operator simulator executes tasks by following the procedure. The analyzing part calculates indices for quantitative evaluation, such as moving course and length of the operator, length of the eye movement, execution time, and human error probability. By analyzing the indices, the analyst can compare the HMIs of several designs relatively. This is similar to benchmark testing to evaluate computer performance of given tasks.

---

[1] DIAS : Dynamic Interaction Analysis Support system

2 DIAS

2.1 Distribution Simulation System

The distribution simulation system consists of two simulators, the Operator Simulator and the Human-Machine Interface Simulator. It simulates the interaction between the operator and the HMI (Fig. 1). The data editors are subsystems for the distribution simulation system and edit the operating procedure data and the HMI design data.

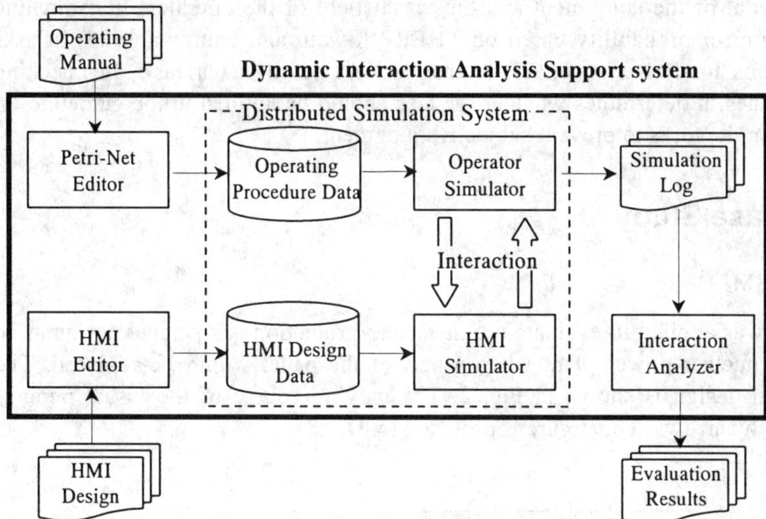

Fig. 1 Configuration of DIAS

Operator Simulator. The operator simulator (Nakagawa et al. 1996) simulates the operator execute tasks based on the procedure, taking into account the cognitive and behavioral features such as visual aspects (focal view and peripheral view), human cognitive model (Reason 1990), operational speed, and walking speed. The operating procedure data is represented by the petri-net model, which is edited by the petri-net editor. By using the petri-net model, the analyst can perceive the structure of the procedure and the dynamic process of the state transition comprehensively.

HMI Simulator. The HMI simulator constructs a virtual control room, and the virtual operator in the operator simulator works there. It provides the layout of the equipment, the moving course and the visual field of the operator, and other information. The HMI data is edited by the HMI editor. The analyst can easily make and modify the design data.

2.2 Interaction Analyzer

The distributed simulation system outputs a simulation log, in which shows the sequential recording of the executed task, standing position, visual point, time, and so on. The interaction analyzer calculates the moving course and length of the operator, the moving course and length of the observing point, execution time, and other variables. Some evaluation methods, such as checking guidelines or estimating human error probability, require the analyst's judgment depending on the interaction situation. In such cases, the interaction analyzer supports the judgment in certain ways. It shows detail of the situation, such as the layout of the equipment and the visual field of the operator. In estimating human error probability based on THERP (Swain and Guttman 1983), it asks questions to derive a suitable human error rate for each task. In checking guidelines, it determines whether the task should be applied to the guideline. In addition, it works to prevent subjective judgment.

3 Case Study

3.1 HMI Designs

DIAS was applied to evaluate a transformer protection relay panel design at an actual nuclear power plant when a part of the panel would be replaced. The original design is shown in Fig. 2. The moving course of the visual point is shown by arrows. This figure is provided by DIAS.

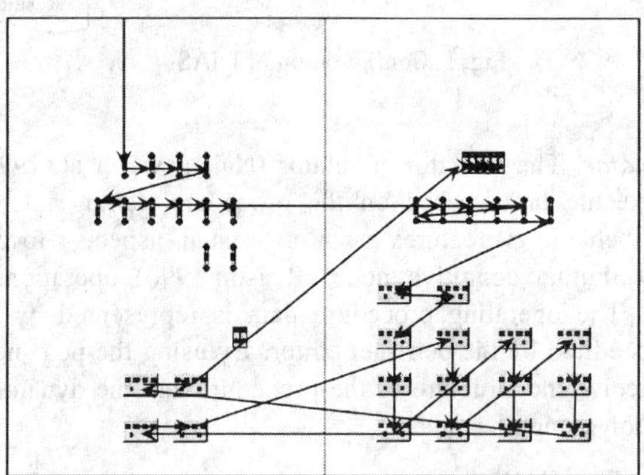

Fig. 2 Original design of transformer protection relay panel

The interaction analyzer counted the human error probability for each task by applying THERP. It also obtained the execution time, the moving length of the operator, and the length of the eye movement by analyzing the simulation results.

3.2 Results

In analyzing the results by DIAS, it became clear that the increment of human error probability in the original design was mainly caused by two factors: the lowness of the equipment's position and the mismatch between procedure and configuration of the equipment. After considering these observations, the authors proposed the new design shown in Fig. 3, which is improved to overcome these limitations. DIAS was also applied to the new design. In the original design, the switches seemed to be laid out without any explicit design concept. On the other hand, in the new design they are classified into various groups, which are laid out separately. Moreover, the movement of the operator's view point is generally downward and rightward.

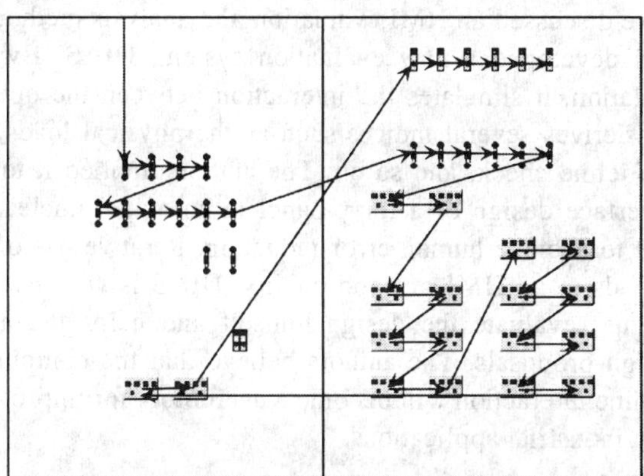

Fig. 3 Modified design of transformer protection relay panel

As a result of re-evaluation (Fig. 4), the new design is expected to reduce human error probability by about 30%. Because of the improvement of the lowness of the equipment's position, the length of the movement (up and down) will also be improved. Fig. 4 also shows that the configuration appropriate to the procedure reduces the length of movement (back and forth) and the length of eye movement.

DIAS enables the design process (evaluation, analysis, feedback of the results to the design) to be carried out easily and rapidly.

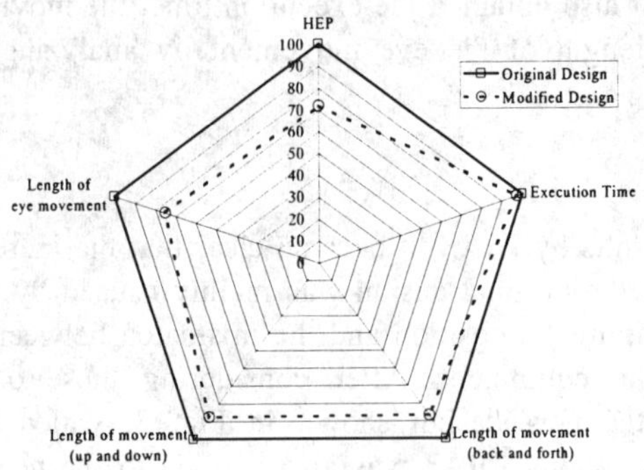

Fig. 4 Comparison of the evaluation results (Original Design : 100)

4 Conclusions

In this paper, we discussed an HMI evaluation and analysis method in industrial application and developed a new evaluation system, DIAS. By applying the computer simulation, it simulates the interaction between the operator and the equipment and derives several indices such as the physical loads, human error probability, guideline check, and so on. The authors applied it to evaluate and analyze the interface design of a relay panel in an actual nuclear power plant and modified it to achieve human error reduction. Iterative use of DIAS in the design phase leads to an HMI of good quality. DIAS is very useful because it helps the designer evaluate the design himself and helps the analyst design alternative design proposals. The authors believe that the computer simulation of human-machine interaction will become a useful tool for improvements in the human factor in industrial applications.

5 References

Nakagawa, T., Nakatani, Y., Yoshikawa, H., Takahashi, M., Hasegawa, A., and Furuta, T. (1996). Simulation Based Interface Evaluation of Man-Machine Interface Design in Power Plants, *Cognitive Systems Engineering in Process Control (CSEPC 96)*.

Reason, J. (1990). *Human Error.* Cambridge University Press.

Swain, A. D. and Guttman, H.E. (1983). *Handbook of Human Reliability Analysis with Emphasis on Nuclear Power Plant Applications, NUREG/CR-1278*

Sensory Evaluation Method for Determining Portable Information Terminal Requirements for Nursing Care Applications

Naotsune Hosono Oki Electric Ind.Co.,Ltd

Hiromitsu Inoue Chiba College of Health Science

Yutaka Tomita Ph.D. Keio University

1. INTRODUCTION

It is often stated that we are living in the Information Age, an age in which complex information must be quickly and easily accessed. As we consider the 8th International HCI Conference theme "Creating New Relationships", we explore the emergence of microprocessors and how new and sophisticated microprocessors allow for smaller and more powerful machines. Those new machines have the potential to be used by anyone irrespective of age, gender, location, nationality, or time considerations. This paper explores a sensory evaluation method using an experimental Portable Information Terminal (PIT) platform (Knaster, B. 1994) to meet hospital nursing care requirements. We explore the various aspects of multimedia access to evaluate the users need to employ the PIT in a ubiquitous manner. Because of generally scarce resources designers must tradeoff various elements of these basic requirements. However, in the initial stages of development, designers will have little or no direct experience in developing a specific machine to meet the exacting requirements of nursing care professionals. Therefore, this paper proposes that the sensory evaluation can guide perplexed designers in the initial development stage (Nielsen, J. 1993, Norman, D, A. 1988).

2. SENSORY EVALUATION

2.1 PIT Nursing System

The major users of medical support systems are doctors, nurses, therapists and patients. This paper discusses a case study of a PIT vertical application and focuses on the nurse/therapist users. The medical support system is a centralized server system managing the patients database and numerous peripherals. The peripherals are personal computers (PCs) located in doctor's offices, and nurse stations. PCs in the nurse station are connected to a Dock Station which facilitates up and downloading data between the PIT and patients database and recharges the PIT power supply (Fig.1).

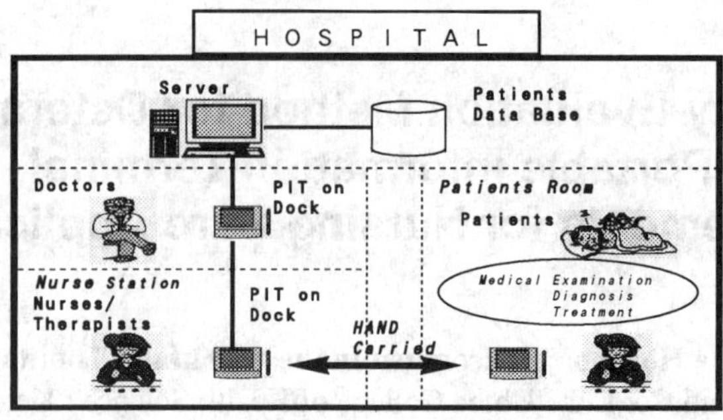

Fig.1 Nursing Care System

2.2 Users Analysis and Subjects

PIT users can be segmented into horizontal and vertical users. Mobile professionals are recognized as horizontal users. Whereas vertical users, such as hospital nurses/therapists, follow well established hospital policies and procedures so that they can take urgent, extreme and instant action with regard to determining and addressing patients critical needs. They cannot afford to focus on manipulating complex PIT functions when they must concentrate all of their immediate attention on their patients. Given the initial lack of a specific requirement balance, PIT interface designers are required to measure and evaluate the value of a user's initial comfort level while reminding themselves that users gradually and unconsciously evolve from a novice into an experts.

2.3 Required Elements

The major usability testing items are reference PIT hardware and software features. Required elements of the PIT device include: a compact size, portability, a high processing speed, low power consumption, ability to handle various communication needs, be inexpensive, et cetra (Schneiderman, R. 1994). Additional functions are derived from the standpoint of users specific needs. They are screen sizes, types of displays, methods of communications, battery life expectancy, extendibility of input/output ports, keyboard layouts, volume, weight, and price. The ten major components with parameters that need to be considered when designing for nurses/therapists in a Medical Care vertical application are: Screen Size (SS), Display Type (DT), Communication (CC), Battery Life (BL), PCMCIA card slots (PC), Keyboard (KB), Volume (VM), Weight (WT), Price (PR), Robustness (RB).

2.4 Extract Samples

In the nurses/therapists interviews the questionnaires were modified to express specific technical questions such as Volume (VM) as preferred styles (Watch, Pocket/Necklace or Shoulder hung) rather than specific technical volumetric terms that may have been less well understood. Battery Life (BL) was fixed to five hours and Communications (CC) were not considered an issue as information was transferred to the central database via the Dock Station located at the nurses station where the PIT battery was recharged. Back-lights were required to extract the patient medical history when the devices were tested in patients rooms where the environment could be quite extreme, usually dark and quiet. PCMCIA card slots (PC) were converted from the number of card slots to the size of data media needed for each patient. The Price (PR) was not an issue as the hospital provided the PIT to the nurses/therapists as part of the system. Robustness (RB) was specifically added to the design equation to address the PIT striking an object or falling to the floor as the nurses/therapists concentrated on addressing the immediate needs of the patients. Note that the Weight (WT) is proportional to the Volume (VM). For the preparation of this sensory evaluation, seven samples (Table 1) were selected from 972 combinations of user derived functions determined through a heuristic sorting algorithm.

Table 1 PIT Samples

Samples	Screen Size (SS) [inch]	Data size (PC) [Pages] (MB)	Keyboard (KB)	Volume/Weight (VM) / (WT)	Robustness (RB)
T	2	100 (2MB)	without	Watch (80g)	insignificant
D	5	400 (8MB)	without	Pocket/Necklace (350g)	insignificant
E	5	400 (8MB)	with	Pocket/Necklace (350g)	insignificant
F	5	400 (8MB)	without	Pocket/Necklace (350g)	significant
G	5	400 (8MB)	with	Pocket/Necklace (350g)	significant
H	8	800 (16MB)	without	Shoulder hung (850g)	insignificant
J	8	800 (16MB)	with	Shoulder hung (850g)	significant

2.5 Evaluation Procedure

The required measurement and evaluation can be accomplished by utilizing Marble Method and Correspondence Analysis (CA). The subjects (panels), professional nurses/therapists were given 25 marbles (tokens) to distribute as votes among the seven samples (Inoue, H. 1993). They were permitted to vote with up to ten marbles for any one sample among the seven. The subjects were shown three major mockup models (2-inch, 5-inch and 8-inch screen sizes) before voting for their reference device. The advantage of this method is that it is possible to extract indecisive requirements from the users rather than to allow only one marble per subject. Before the start of the evaluation, subjects were interviewed to determine personal information (Face Information) which could potentially be referenced into the evaluation. The Face Information included name, gender, age, familiarity with a word processor, personal computer, E-mail/Internet, personal digital assistants (PDA), and cellular phone. They were also asked how they preferred to carry the PIT from the Nurse Station to the patients rooms.

2.6 Data Processing

The collected data was processed and analyzed using a statistical algorithm; CA and its SPSS package (SPSS. 1998). The Advantage of the CA is that it can easily forecast the relationships between subjects and samples since they are plotted in the same plane. Additionally, multiple points relationships are plotted by relational not by absolute relationships which usually result from Principal Component Analysis (PCA).

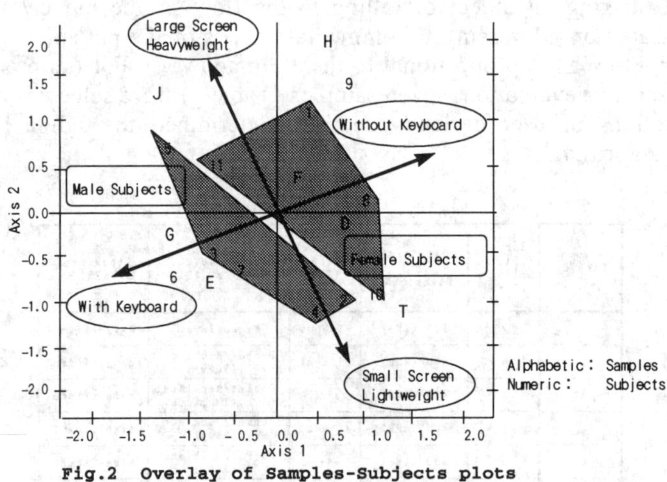

Fig.2 Overlay of Samples-Subjects plots in Correspondence Analysis

3. RESULTS AND CONSIDERATIONS

A previous evaluation of the PIT samples utilized people outside the health care environment; students and business people, and was structured so that the device was to be used either in their office or outside the office (N. Hosono 1997). The results were categorized into three groups: portability, battery life and wireless functions. In this latest evaluation study, seven different featured PIT models were selected and eleven nurses and therapists test subjects participated in the survey. Marble Method was employed by giving each test subject 25 marbles to assign to the PIT samples they most wanted to use. The preferences among the seven samples by each test subject were plotted by CA (Fig.2). The Eigenvalue shows that the two components (Axes) are found to account for 75.9% of the test subject's preferences. The first two Eigenvalues were comparatively large and the third solution drops rather suddenly. This suggests that the first two solutions were main components for the data set. They are: Screen Size

(SS) and Keyboard (KB) followed by Robustness (RB). Screen Size (SS) is quite proportional to Volume (VM) and Weight (WT). Therefore the most favorable parameters in components are: 5-inch in Screen Size (SS), Display Type (DT) with back-lighting, Robustness (RB) and adequate room for patient data relating to the PCMCIA card slots(PC). The male subjects preferred with Keyboard (KB) whereas females in their 20's did not, although almost all subjects said in their Face Information that they used word processors and personal computers. No distinctive feature could be found regarding the users of E-mail/Internet. Only one subject used a PDA and this did not affect the results. Those subjects who do not use cellular phones preferred light-weight samples and Robustness (RB) was not always considered as a requirement.

4. CONCLUSION

This paper proposes a way to develop a specific machine to meet exacting requirements in the initial stages of development when designers have little or no direct experience or reference products to draw upon in developing. This study focused on a nursing care professional application using of Portable Information Terminals. The combination of Marble Method and Correspondence Analysis was applied. Seven samples were comprised from ten major components and 972 parameter combinations. Two major components (Axes) which accounts 75.9% in Eigenvalue were found using a questionnaire survey. They are: 5-inch in screen size, the display type with back-lighting, with robustness and adequate room for patient data. Additionally, it was found that the male subjects preferred with a keyboard whereas females did not. Through this study, the required PIT features for the nursing care professional application were initially defined. A follow-up study must be performed to determine any specialized doctors and patients feature requirements.

5. REFERENCES

Hosono, N. (1997) Usability Study on PIT Platform. *Proceeding of the 13th Triennial IEA'97* Vol. 5, 98-100.

Hosono, N. (1997) Usability Study on PIT Platform. *Proceeding of the 5th WWDU'97 Tokyo* 19-20.

Inoue, H. (1993) Item Selection and its Importance. *23rd Nikkagiren Symposium.*

Knaster, B. (1994) Presenting Magic Cap. *Addison-Wesley.*

Nielsen, J. (1993) Usability Engineering. *Academic Press, Inc.*

Norman, D, A. (1988) The Psychology of Everyday things. *Addison-Wesley.*

Schneiderman, R. (1994) Wireless Personal Communications. *IEEE Press.*

SPSS. (1998) Statistical Package for Social Science for Windows 8.0J Categories. *SPSS Japan.*

Quality Control and the Usability Evaluation

Haruo Imai, Akiko Irie, Hiroko Mayuzumi
Quality Management Headquarters, Canon Inc.

1 Introduction

In recent years, user group of Canon products has widened, due to diversification of our sales to peripheral products of computers, such as printers, digital cameras, etc. Accordingly, we have experienced sharp increase in inquiries and complaints from customers on the installation and operation of such products. Not only to reduce the increasing cost involved in handling the calls from customers, but to protect the brand image, it has become an urgent matter to set up a system for developing products from the customer's viewpoint; the products that would "satisfy the customer and are easy to use".

Quality Management Headquarters has, therefore, started 'Activity to Improve "Quality for Customers"' to set up a new quality assurance system including the requirements of ISO 13407. In this system, product planning is examined from the customer's viewpoint and compatibility is evaluated during the product development process. The methods used in this system include market research, breakdown of inquiries, comparison with competitor products, etc. Among various methods, we regard usability evaluation as the most important factor and have adopted Canon's own practical evaluation methods.

2 Practical Usability Evaluation Methods

Various approaches have been studied as usability evaluation methods [Nielsen 1993; Rubin 1994; Nielsen and Mack 1994; Bias and Mayhew 1994; Kaiho and Harada 1993], but to make it useful at the product development stage, we value the following three points.

1) The degree of usability is measurable in the comparable scale with other quality elements such as reliability, safety, performance, etc., and that in an easily understandable manner.

2) With the limited and tight development schedule, it is possible to perform effectively in a short period.

3) It shall be able to be performed without highly technical knowledge, but with certain level of training.

In order to solve these points, we are studying to develop our own original usability evaluation methods that evaluators and designers will find easy to become familiar with, by incorporating the concept of Canon's quality evaluation to the popular usability evaluation methods.

2-1 Quantitative Grasping of Problems Found by User Testing

User testing is one of the popular methods for finding out the usability problems most objectively, but carrying out that method even more efficiently, we have originated a method of grasping those usability problems quantitatively.

The main characteristics of this method is that, when analyzing the usability problems, analysis method used in quality evaluation is applied. By monitoring the participants during the test, we can study to what degree the usability problems affect the users. Following the predetermined standard, problems are classified into four ranks depending on its seriousness, and the points are given respectively. Also, depending on the frequency of its appearance, we would be able to work out the degree of impact on the user, and calculate these as 'Usability Defect Index'. (Fig.1)

Usability Defect Index

Usability Defect Index (Total) = Σ (Usability Defect Index of each problem)

Usability Defect Index (Each problem) = Σ (Ranking of the problem × Frequency of its appearance)

Ranking of the problem

AA:Has possibility to destroy the product itself or data, or may be hazardous to user.

A: Unable to continue the operation, which may result in inquiry to service center.

B: User feels uncertain of the operation (not as serious as rank A).

C: Do not give major affect to the operation (user's comments, unconscious mistake in touch operation, etc.,).

Fig.1 Usability Defect Index

In considering the acceptability of the product's usability, the developers shall take in mind the following two points:

- The product do not give any impact to the user in the most serious rank.
- The total of the Usability Defect Index do not exceed certain standard.

Using this method, we find it possible to set standards for usability problems, similar to those for other quality problems. Making it possible to get the result of analysis in shorter period, we have enabled to feed back the information to the product, leading to effective use in actual product development. (Fig. 2)

Problems	Initial Test				After 1st Modification			
	Rank		Frequency	Defect Index	Rank		Frequency	Defect Index
Problem_01	**A**	**6**	**100%**	**6.0**	A	6	29%	1.7
Problem_02	**A**	**6**	**43%**	**2.6**	-	0	0%	0.0
Problem_03	**B**	**3**	**71%**	**2.1**	**B**	**3**	**71%**	**2.1**
Problem_04	**B**	**3**	**57%**	**1.7**	**B**	**3**	**57%**	**1.7**
Problem_05	C	1	57%	0.6	-	0	0%	0.0
Problem_06	C	1	57%	[illegible]	[illegible]	[illegible]	0%	[illegible]
...								
					C	1	14%	
Problem_13	C	1	14%	0.1	-	0	0%	0.0
Total				**13.9**				**5.7**

Fig.2 Example: User Testing

2-2 Application of Inspection Methods

The efficiency of user testing is limited by some issues, such as having to recruit participants, arranging for prototype products, and other aspects. For efficient evaluation, we applied the inspection methods and are carrying out user testing in conjunction with it.

Inspection methods also have some problems from the point of objectivity such as being unable to know what degree the usability problems affect the users and the frequency of its appearance, so we can't feed back the evaluation result smoothly. But when we applied sHEM [Kurosu *et al.* 1997] guidelines which systematically summarize evaluations items, we were able to detect many problems and feed back the evaluation result smoothly, so we are now adopting sHEM guidelines into our quality evaluation.

When applied to Canon products, however, there appeared to be some polarization in the categories of the detected problems, so we are studying to develop our own original guideline. For example, in the evaluation using sHEM guidelines had been applied to the software having 9 main operating screens and 41 sub-screens, 99 problems were detected by three evaluators. Reviewing the classification, these problems applied only to 12 out of 26 items, that are categorized into 3 groups; operability, cognitivity, and pleasantness. Among the problems, those that applied to "plainness of cognitivity" were by far the greatest in number, so there was quite a polarization in the distribution of applicable category items. (Fig.3) This is just an example, but since the trend is particularly striking, when applied to our software products, a further subdivision of the categories is now being studied.

Also, evaluation time with sHEM is set at about 30 minutes for each category, but when usability is considered as one item of quality evaluation, it is desirable that evaluation is carried out until the detected problems become more or less saturated. In the above example, as the total actual time to detect the problems and rank them by seriousness took 10 hours/person, we are studying to set up our own original standards on evaluation times to agree with program sizes and number of screens, etc.

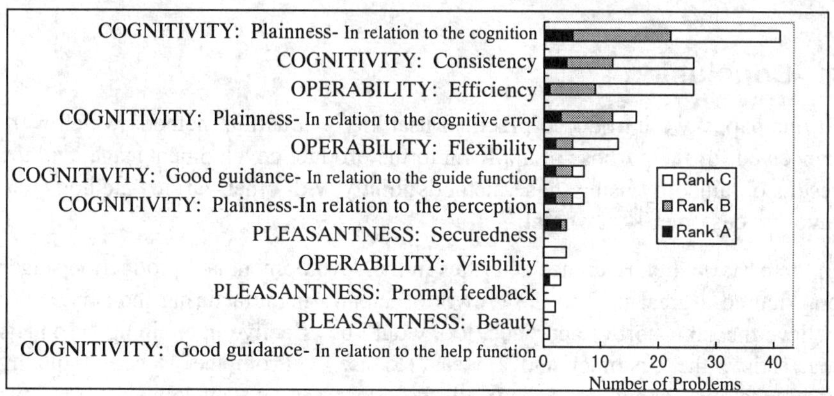

Fig.3 Example: Inspection Method applying sHEM Guidelines

3 Results

The following case given of a printer is an example of the results obtained by carrying out the usability evaluations we introduced herewith jointly with other various methods which involves the customer's viewpoint.

Up to now, most of the problems concerning Bubble Jet Printers were inquiries from customers who couldn't connect the printers to their personal computer. Accordingly, we first analyzed the inquiries from customers and determined the cause of the problems and then, based on results of Customer Satisfaction surveys, analysis of competitors' products, etc., made improvements in our operation manual. Next, by repeatedly carrying out the usability evaluations before shipment, we added more improvements and finally completed a manual that was easy for anyone to understand. Thus, we succeeded in greatly reducing the number of the most frequent inquiries on set up (particularly on installation of printer drivers). (Fig. 4)

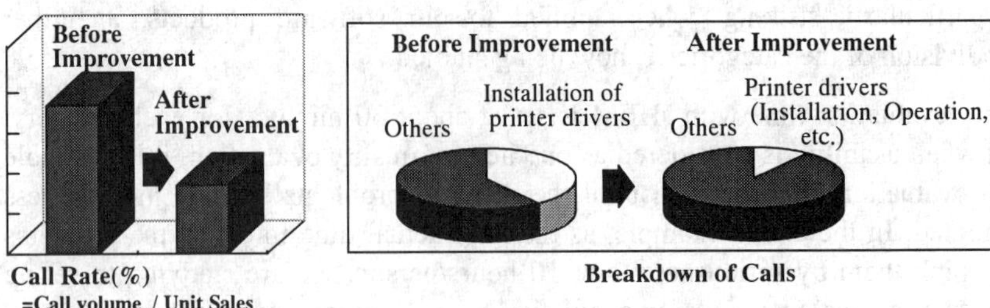

Fig.4 Results (Calls from Customers)

4 Conclusion

In this paper, we introduced practical usability evaluation methods which were conceived for the purpose of applying to the product development stage, and the results obtained by using those methods jointly with other various methods that involves customer's viewpoint.

In the past, quality assurance activities in the development and production stages contributed a great deal to the growth of Japan's manufacturing industry. We believe the same sort of approach, too, would be effective in attaining "products that satisfy the customer and are easy to use". From the aspect of quality assurance, we intend to review all the processes of development from the customer's viewpoint.

5 References

Nielsen, J. (1993). *Usability Engineering.* Boston: Academic Press.

Rubin, J. (1994). *Handbook of Usability Testing.* New York: John Wiley & Sons.

Nielsen, J. & Mack, R.L. (Eds.) (1994). *Usability Inspection Methods.* New York: John Wiley & Sons.

Bias, R.G. & Mayhew, D.J. (Eds.) (1994). *Cost-Justifying Usability.* Boston: Academic Press.

Kaiho, H. & Harada, E. (Eds.) (1993). *Purotokoru Bunseki Nyumon. (Introduction to Protocol Analysis).* Tokyo: Shinyou-sha

Kurosu, M. & Sugizaki, M. & Matsuura, S. (1997). Usability Evaluation Technique with High Efficiency -1, -2, -3. *Proceedings of the 13th Symposium on Human Interface* (HIS '97,Osaka,Japan,October 21-23,1997), 481-500.

Evaluation of Taste for Colours on Motorcycle

Masamori Sugizaki
Human Technology Dep. Advanced Technology Research Div.,
Yamaha Motor Co., Ltd.
2500 Shingai, Iwata, Shizuoka 438-8501,Japan

In a motorcycle catering much for individuals' different taste, the functional components play another important role as a design element. This study focuses on the colours of a motorcycle body, especially the colours of the fuel tank and fenders and shows the relationship between these colours and the individual's taste. For the motorcycle employed for our test, silver was preferred with a high evaluation. For the other colours, the individual's taste differed a great deal according to the location where they are used or how they are used. It is, therefore, very important to achieve a firm grasp of the users' taste in colours.

1. Introduction

Colour is given very great weight to the design of a motorcycle. This is especially the case with a product like a motorcycle that caters much for the individual's different taste, and the image of a particular motorcycle greatly varies according to its colours, of which the selection and arrangement in turn have a great effect on its sales.

We made an experiment to investigate any effect that colours, which are a very complicated yet very important element for products, might have upon people's taste for motorcycles.

2. Method of Investigation

2.1. Evaluators

Forty-one (41) Japanese men and women were selected at random as evaluators for this investigation from among motorcycle owners in and around Tokyo. Their age ranged from 22 to 55.

2.2. Motorcycles for evaluation

For the purpose of this experiment, eight (8) panels were made in computer graphics for an on-off road dual purpose motorcycle capable of a ride on irregular terrain (Fig. 1) with particular colours for the fuel tank and fenders.

Fig1 On-off road dual purpose motorcycle

The combinations of colours as shown in Table 1 were used

Table 1 Motorcycle specification for evaluation

Motorcycle	Fuel tank	Front fender	Rear fender
A	Silver	Silver	Silver
B	Red	White	White
C	Dark blue	White	White
D	Silver	White	White
E	Red	Red	Red
F	Dark blue	Dark blue	Dark blue
G	Black	White	White
H	Black	Black	Black

2.3. Evaluation method

Two (2) each were picked out of these eight (8) panels and evaluated as to which was preferred by paired comparison.

2.4. Analysis

The results of this paired comparison were employed for the following analysis.

(1) The evaluators were divided into groups by the cluster analysis method according to their taste for colours.

(2) The taste of each group for the motorcycle was ranked by the paired comparison method.

(3) For each group, the degree of the effect produced on this ranking by the colours was found, and based on that degree, a formula for estimating a favorable rank was obtained.

3. Result

3.1. Grouping of the evaluators

As a result of the cluster analysis, two groups of evaluators, one of 24 (Group 1) and the other of 17 (Group 2) were formed.

3.2. Ranking of taste for motorcycles
All evaluators and the two groups formed in the above 3.1 were ranked by the paired comparison method in the level of taste for the evaluated motorcycles (Figs. 2, 3,and4).

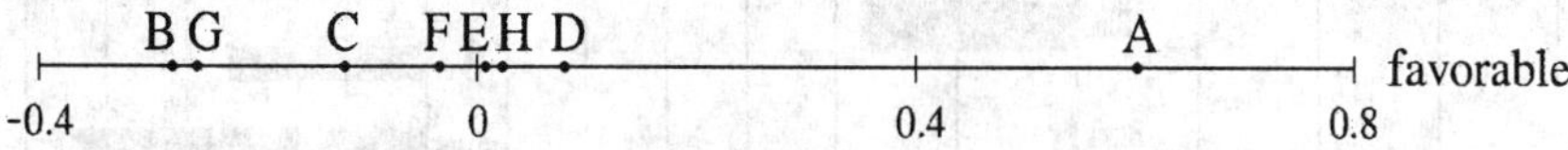

Fig.2 Scale of main effects (All evaluators)

All analysis of data of evaluators indicated that silver was particularly liked for the fuel tank amd fenders.

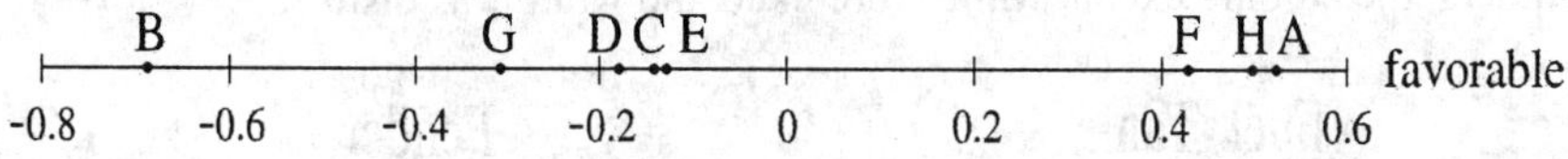

Fig.3 Scale of main effects (Group 1)

For Group 1, the combination of similar colours was liked both on the fuel tank and on the fenders.

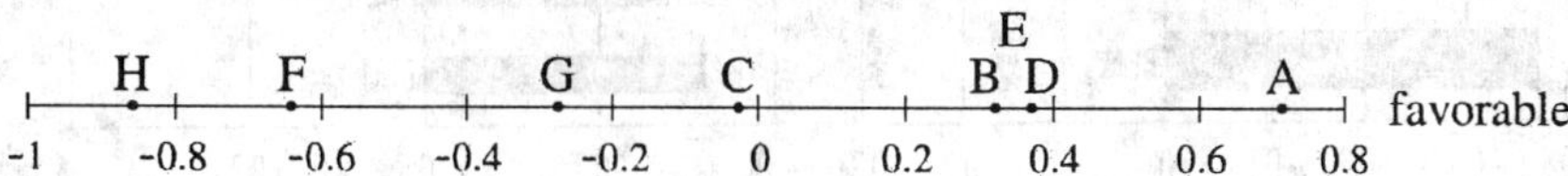

Fig.4 Scale of main effects (Group 2)

For Group 2, the combination of different colours was liked both on the fuel tank and on the fenders.

3.3. Degree of effect of each colour and rank estimation formula
With the value representing the average degree of favorite in each motorcycle as an outside criterion and with the items consisting of the fuel tank and fenders and the category of their colours, a calculation was made of the magnitude of the effect of each colour on the outside criterion by employing the quantification theory type I (Figs 5,6).

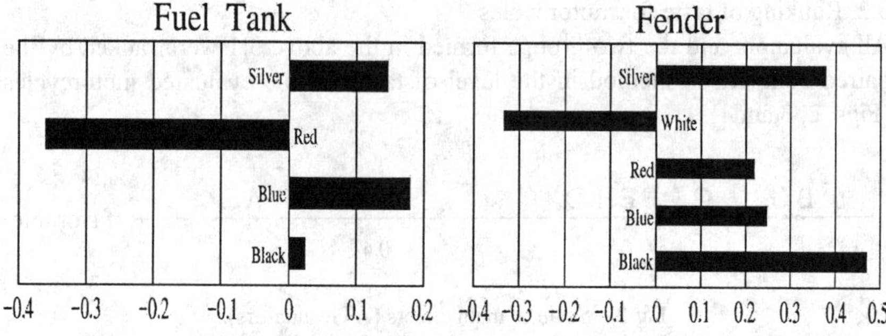

Fig.5 Effect of colour (Group 1)

For the fuel tank, dark blue and silver were liked and red was disliked. For the fenders, the colours except white were liked and white was disliked.

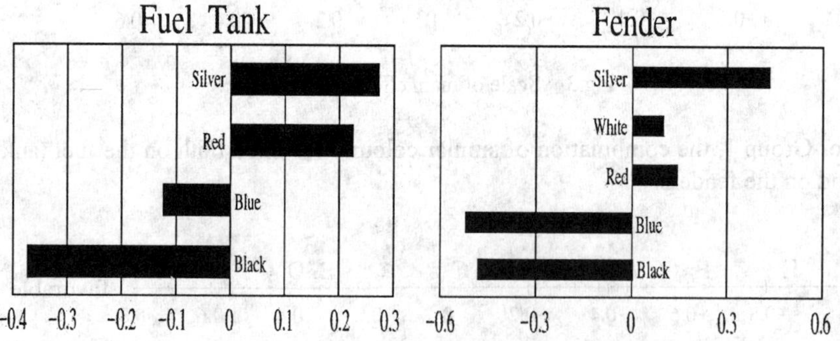

Fig.6 Effect of colour (Group 2)

For the fuel tank, silver and red were liked and black and dark blue were disliked. For the fenders, silver was liked and dark blue and black were disliked. With this motorcycle, every evaluator liked silver both on the fuel tank and on the fenders.

This result indicates that many evaluators felt that the silver as it was used on this motorcycle matched its image. Thus, people would accept this motorcycle on the market if its fuel tank and fenders are both coloured silver. On the other hand, if its fuel tank is coloured black or red and the fenders white, dark blue, or black, people would turn away.

Here is an example of an estimation formula for favorite ranking(Y), based on the data from Group 1.

Estimation formula:

$$Y=0.14915*\delta(\text{Fuel Tank :Silver})-0.3565*\delta(\text{F Tank :Red})+0.18324*\delta(\text{F Tank :Dark blue})+0.02415*\delta(\text{F Tank :Black})+0.37926*\delta(\text{Fender : Silver})-0.331*\delta(\text{Fender : White})+0.22017*\delta(\text{Fender : Red})+0.24858*\delta(\text{Fender : Dark blue})+0.47585*\delta(\text{Fender : Black})$$

For $\delta(A{:}B)=1$ where A=B, and $\delta(A{:}B)=0$ where $A\neq B$

3.4. Characteristic of colour evaluation

Our experiment showed that silver was liked by most of the evaluators. Silver and white may not look much different, yet they indicate a great difference if measured on the scale of taste. This may be partly because white gives a monotonous impression, whereas silver looks refined and elegant, providing the motorcycle with a sort of refinement leading to the user's satisfaction. In contract, red, dark blue and black should be treated with care as it can be evaluated in very many different ways.

4. Conclusion

In our experiment we used a on-off road dual purpose motorcycle to investigate what effect might be brought about on the individual taste by colours on its fuel tank and fenders.

The result showed that any group of evaluators liked silver both on the fuel tank and on the fenders. The group who liked similar colors on the fuel tank and on the fenders very much disliked red and liked dark blue on the fuel tank, whereas on the fenders they disliked white and very much liked black.

On the other hand, the group who liked different colours on the fuel tank and on the fenders were found to like red either on the fuel tank or on the fenders and to dislike black and dark blue very much. In this way, which colour to use on a particular motorcycle has a great influence upon the individual taste and consequently its sales. It is, therefore, very important to acquire a grasp of the taste of the targeted users.

5. Reference

Research Committee of Sensory Evaluation, Union of Japanese Scientists and Engineers. *Sensory Evaluation Handbook* (Japanese)

Proposal and Evaluation of User Interface for Personal Digital Assistants

Nobuhiko MASUI*, Tetsuo OKAZAKI and Yoshinobu TONOMURA***
*NTT Cyber Solutions Laboratories, **NTT Software Corporation
1-1 Hikarinooka Yokosuka-Shi Kanagawa 239-0847 Japan

1 INTRODUCTION

Personal Digital Assistants (PDA) are becoming essential private and business tools but their displays remain rather small. Moreover, their operating systems are limited. Obviously, they should allow the user to efficiently view a large amount of information within a limited time. An information display method that efficiently displays a large amount of information in a small display area and excels in visual recognition and ease of operation is thus essential. In addition, new PDA applications could be developed if a new information display method were developed that could efficiently display complex information. We propose and evaluate such a method for the user interface of PDA's. Our assumptions are an A5 size color display with about 640*480 pixels.
Chapter 2 introduces the new method and a realization technique. Chapter 3 describes the experiment used to clarify the performance of this method, and experiment results are discussed. Chapter 4 provides examples of suitable applications for this method.

2 INFORMATION DISPLAY METHOD

There are three methods for displaying a large amount of information: the scroll method, the multi-window method, and the overlay method. The scroll method demands frequent scrolling to refer to all information. The multi-window method occludes information lying underneath a window. Therefore, window switching is needed to view all information.

The overlay method tries to visually combine information on two or more separate layers and is often used in CAD and mapping systems. The overlay method can display a large amount of information in a small display area simultaneously. However, the existing overlay method suffers from 3 main problems: information occlusion, insufficient identification, and weak linkage between layers or objects.

We have extended the overlay method to create a new display method suitable for efficiently displaying complex information in a small display area. The method alters the visual attributes of the information on each layer as is described below.

The features of this method are as follows.

(1) To eliminate occlusion, the layers are made partially transparent.

(2) To display different types of information efficiently, layer order is carefully managed according to the contents to maximize readability. Layer display can be activated either manually or automatically and layer visual attributes can be specified to suit the task at hand. For example, specific information can be emphatically displayed by switching layers. According to the emphasized layer, layer order and visual attributes can be altered automatically.

(3) To clearly establish the importance or relation of information, the method alters the visual attributes of each layer. For example, layer order can be indicated by increasing the size of information or the density of the layers. Relationships can be expressed by using similar colors or similar density.

(4) Each object (including text blocks) on each layer is assigned a point, e.g. its center point, to permit direct access to the layers. Clicking or pointing on the screen selects the nearest center point and thus the layer on which the point is located. Layer/object selection brings the layer to the top or triggers the display of related information.

In this method, information is controlled at three levels of administration: all information, each layer, and each object. Main control information is as shown in Table 1.

Table 1. Main control information

administration unit	control information
all information	- to allocate information on each layer - the number of displayed layer - displayed layer order - layer switching
each layer	- visual attributes
each object	- displayed position - presence of related information

3 EVALUATION

3.1 The first experiment

Method. It is inevitable that information will be overlapped using this method. Thus, it is important to understand the relation between the overlapping of information and visual recognition. It is necessary to clarify what visual attribute or combination of visual attributes is effective.

To use small screens efficiently, one technique placed transparent or semi-transparent user interface elements over objects (Bier 1993), (Kamba 1996). The impact of transparency level using semi-transparent windows and menus has been evaluated (Harrison 1995), (Harrison 1996). The first experiment is evaluated by the suitable visual attribute viewpoint.

The basic evaluation experiment presented the subjects with two or more overlapped texts on the above PDA's display. The instruction was to find a specific word in the specified text. Two parameters were varied in this experiment. One was visual attribute, and the other was the number of transparent layers. We measured both the correct answer rate and the time required to retrieve the word (including confirmation).

Condition. 24 subjects were instructed to locate words within text displayed on one of several layers (2 to 5). The total number of trials was 60. The position of the target layer was randomly set. Headlines of newspapers that were not too long and didn't cause bias in the content were displayed. The size, color, and density of each layer were varied. The setting of visual attributes in each layer is shown in Table 2. Experiment details are shown in Table 3. An example of the test material with three layers is shown in Figure 1.

Table 2. Setting of visual attributes

layer		1	2	3	4	5
size(%)		100	95	90	85	80
density(%)		100	90	80	70	60
color	R	255	255	0	255	255
	G	255	255	255	0	0
	B	255	0	255	255	0
		white	yellow	cyan	magenta	red

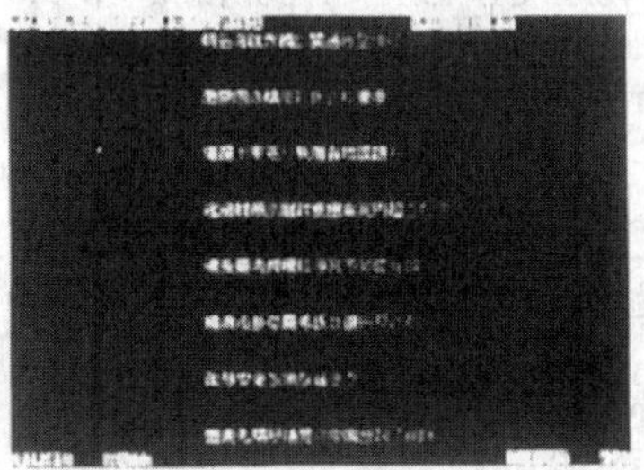

Figure 1. An example of material

Table 3. Experiment conditions

Item	Condition
equipment	color display A5 size PDA with 640*480 pixels
input device	pen
displayed information	headlines of newspapers
text attributes	size: 12 point font: MS Gothic
the number of displayed layers	2 to 5
the number of text lines	8 per layer
position of text line	display on same start point
visual attributes	size, color, density and these combination
system requirement	PDA with special stand on table subjects were seated
subjects	24 women from twenties to forties
instructions	prior to actual testing, instructions were given by an experimenter find a word quickly and correctly possible do over again possible to skip (give up) pen operation and practice trials for experimental tasks were executed

Results and discussion. Results of the average correct answer rate and the average execution time are shown Figure 2. Text overlapping increases with the number of layers; this decreases the correct answer rate and increases the execution time. The decrease in the correct answer rate was small when the size and color of each layer were changed at the same time. With this setting, the correct answer rate was 80% or more even when there were five layers. The increase in the execution time was small when the size of each layer or the size and the other visual attributes of each layer were changed. For example, when the size and color were changed, the execution time with five layers was 14.0 seconds while the execution time with two layers was 9.2 seconds. It is thought that this decrease in visual recognition speed is acceptable considering that the amount of displayed text is 2.5 times larger. The decrease in visual recognition can be suppressed by changing text size on each layer for the severest display condition, which texts were displayed overlapping in the same place.

A comparison was made against the scroll method. As a result, the execution time increases in proportion to the number of layers in the scroll method, and increases in a geometrical progression in this method. This method becomes advantageous when the number of displayed layers aren't too abundant (about five).

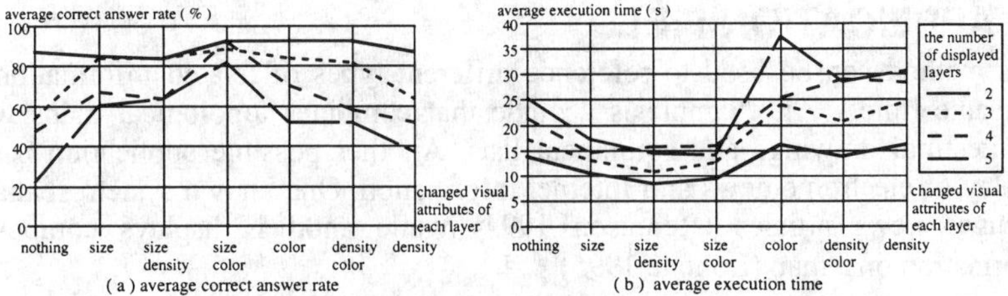

Figure 2. First experiment results

3.2 The second experiment

Method. Accessing all of the information on the screen requires the user to switch layers. Thus, it is important to understand this action. Supplemental experiments were performed to this end. The second experiment is basically same as the first experiment to the exclusion of layer switching. We measured the skip (give up) rate and the number of times required to retrieve the word in addition to the correct answer rate and the time required to retrieve the word.

Condition. Another 8 subjects were instructed to locate words within text displayed on one of several layers (2 to 5). The total number of trials was 60. Layer switching was done manually. The other experiment details were the same as in the first experiment.

Results and discussion. Results of both experiments are compared in Figure 3. Compared to the first experiment, the number of execution increases but the skip rate decreases. For example, when the size and density were different for all five layers, the number of execution with the second experiment was 7.4 times but only 3.1 times in the first experiment. The skip rates were 5.0 % and 30.0 % for the second and first experiments, respectively. It is thought that layer switching is effective in the point of decreasing the skip rate considering that the skip rate is 0.17 times large while the number of execution is 2.4 times large. Obviously, layer switching becomes more advantageous as the number of layer increases. The correct answer rate increases. And the increase rate in the execution time is not so large considering to the increase rate in the number of execution.

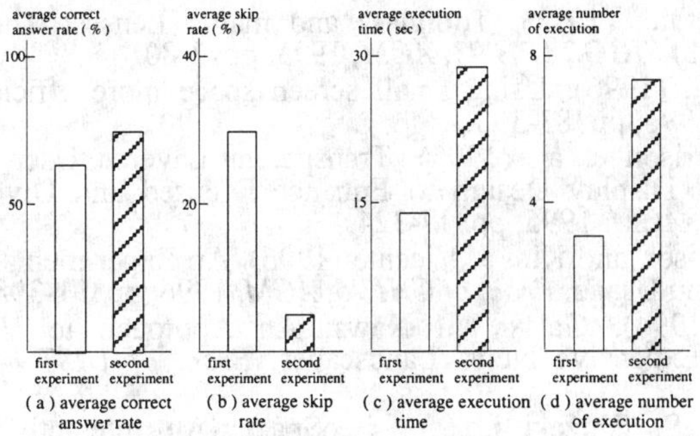

Figure 3. Examples of both experiment results
(the size and density are different for all five layers)

4 APPLICATION FIELD

This method can be used to reference different types of related information at the same time. An example is a guide that combines topological maps with architectural drawings and equipment lists. Another possible application is the display of electronic news and Internet information. One known system spatially arranges news articles (Rennison 1994) while another displays compound information on a map (Lokuge 1995).

Examples of cable ducting (underground facility) combined with above-ground equipment management are shown in Figure 4. Specific information is emphatically displayed by pushing the button. In addition, layer order and visual attributes of layers are automatically decided. This method is very effective in combination with GIS software. By this application, the overlapping information is made easy to compare and recognize more, and the information on height and depth can be expressed. A more detailed examination of benefit and convenience on the proposed method is advanced.

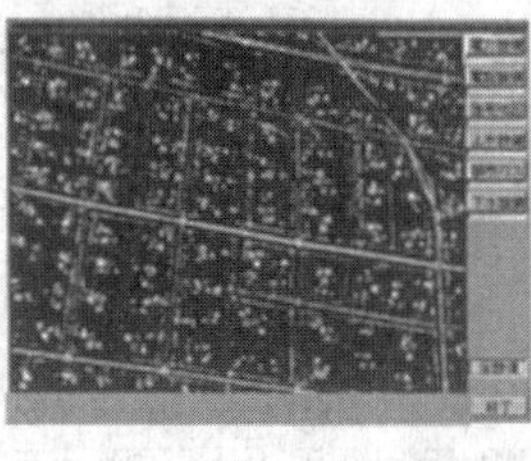

(a) normal

(b) telecommunication equipment is emphatically displayed

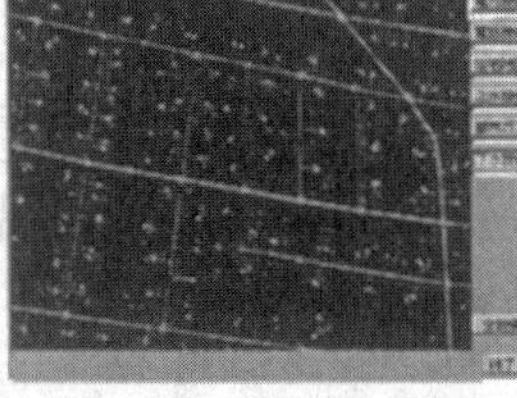

(c) drains are emphatically displayed

Figure 4. Examples of application

5 CONCLUSION

To make PDA easier to use, this paper proposed the information display method by which a large amount of information can be displayed in a small display area. Tests confirmed that, for text information, text size is important discrimination features. The example of application was illustrated.

6 REFERENCES

Eric A. Bier et al. (1993). Toolglass and magic Lenses: The See-Through Interface. *Proc. of SIGGRAPH'93, ACM*, 1993, pp73-80.

T. Kamba et al. (1996). Using small screen space more efficiently. *Proc. of CHI'96, ACM*, 1996, pp383-390.

Beverly L. Harrison et al. (1995). Transparent Layered User Interfaces: An Evaluation of a Display Design to Enhance Focused and Divided Attention. *Proc. of CHI'95, ACM*, 1995, pp317-324.

Beverly L. Harrison and Kim J. Vicente (1996). An Experimental Evaluation of Transparent Menu Usage. *Proc. of CHI'96, ACM*, 1996, pp391-398.

E. Rennison (1994). Galaxy of News: An Approach to Visualizing and Understanding Expansive News Landscapes. *Proc. of UIST'94, ACM*, 1994, pp3-12.

I. Lokuge and S. Ishizaki (1995). GeoSpace: An Interactive Visualization System for Exploring Complex Information Spaces. *Proc. of CHI'95, ACM*, 1995, pp409-414.

Software Tool for the Use of Human Factors Engineering Guidelines to Conduct Control Room Evaluations

John M. O'Hara and William S. Brown
Brookhaven National Laboratory, Department of Advanced Technology
Brookhaven National Laboratory, Upton, New York 11973
E-mail : ohara@bnl.gov, brown@bn.gov

1 Introduction

User interfaces for complex systems employ technologies that can have significant implications for human performance and system safety. To achieve safe and efficient system operation, human system interfaces (HSIs) must be evaluated to ensure that operator performance and reliability are appropriately supported.

One of the primary methods used to evaluate HSIs is the use of human factors engineering (HFE) guidelines. The *Human System Interface Design Review Guideline* (O'Hara, Brown, Stubler, Wachtel, and Persensky, 1996) was developed to support such evaluations. The objective of the HSI evaluation is to verify that the HSI components (e.g., alarms, displays, and controls) will support crew tasks and are designed according to accepted HFE principles.

A method was developed to establish guidance with high internal and external validity (O'Hara, Brown, and Nasta, 1996). Internal validity is the degree to which the individual guidelines are based on an auditable technical basis. The technical basis is the information upon which the guideline is established and justified. External validity is the degree to which the guidelines are subjected to independent peer review. Peer review is a good method of screening guidelines for conformance to accepted HFE practices and for evaluating them in the context of the operational experience of HSIs in real systems. Using the method, HFE guidelines were developed to address both advanced and conventional HSIs. The general contents of the individual sections of the guidelines are described below.

A section on *Information Display* addresses visual displays in top-down fashion, with guidance on display formats (such as mimic displays and trend graphs), display format elements (such as labels, icons, symbols, text, and coding), data quality and update rate, and display devices (such as CRTs). A section on *User-System Interaction* addresses the modes of interaction between user and HSI. Topics include dialogue formats (such as command language and direct manipulation), navigation, display controls, entering information, system messages, and prompts. This section also contains guidelines on data integrity and the prevention of inadvertent change or deletion of data, minimization of data loss due to computer failure, and protection of data, such as setpoints, from unauthorized access. A section on *Process Control and Input Devices* addresses information entry, dialogue types, display control, information manipulation, system response time, and display-control integration. A section on *Alarms* addresses the selection of alarm conditions, choice of setpoints, alarm processing, alarm availability (such as filtering and suppression of alarms), unique aspects of the display of alarm information (such as organization, coding, and alarm message content), and alarm controls. A section on *Analysis and Decision Aids* addresses aids provided to personnel for situation analysis and decision making and their functional requirements, such as explanation and simulation facilities. A section on *Inter-Personnel Communication* addresses speech and computer-mediated communication between plant personnel, e.g., preparing, addressing, transmitting and receiving messages. The final section on *Workplace Design* addresses general workplace design considerations, including the design of consoles, workstations, instrument panels, furnishings, general control room arrangement, and environmental factors (e.g., temperature, ventilation, illumination, and noise).

A total of over 1500 guidelines were developed. The logistics of dealing with so many guidelines can be difficult, and in any specific design evaluation only a small number of individual guidelines are used. Thus, a software tool was developed to provide support for using the guidelines to conduct an evaluation.

2 DRG Software Development

The following functions were deemed highly desirable based on a consideration of how the document will be used: search, hyper-links for easy retrieval of related or supplemental information and navigation, and browsing. Usability objectives called for minimizing learning time and user workload, and providing meaningful help and error feedback.

A prototype interactive *Design Review Guideline* (DRG) was implemented based on these requirements and then evaluated, both in terms of the technical merit of the HFE guidelines and the software design. The evaluation was conducted in three phases: Development Test, User Test, and Peer-Review Workshop. The Development Test provided a preliminary evaluation of the DRG and an opportunity to correct interface problems before subsequent testing. The User Test was a field test of the DRG in advanced control rooms by independent users. Finally, a peer-review workshop was conducted to address the validity and technical basis of the guidelines.

The results supported the validity of the HFE guidelines. The primary source documents were judged an appropriate technical basis. The results indicated that the DRG was rated highly on indices of usability (such as visual clarity, consistency, explicitness, ease of use, ease of learning, low memory load). Some difficulties were encountered, mainly concerning input devices, reporting and help functions. These were addressed in the final design implementation.

3 DRG Implementation and Description

The DRG contains the complete set of HFE guidelines in a hierarchical structure. The HFE guidelines are stored in a master database file composed of several primary fields: guideline number and title, guideline statement, additional information (providing explanatory information, examples, and supporting documentation), and source (link to technical basis on which the guideline was developed). The software makes it easy to edit the guidelines and incorporate new guidelines as they become available, and provides a variety of aids for using the guidance to review HSIs; some of these features and functions are described below.

3.1 Creating and Modifying Evaluation Files

The first step in using the software is to tailor the guidance to the requirements of the specific review by using the table of contents displayed by the DRG to identify the required sections. Sections can be excluded by first clicking on the section titles in the "Select" window and then clicking the *Exclude* button. When the *Include* button is clicked, the section is included back in the guidelines and can be used for the evaluation. To assist the user in determining the appropriateness of the guidelines within a section, the *Browse* button displays the titles and text of the individual guidelines associated with the section currently highlighted.

It may be necessary to use a given section of guidelines many times during a single evaluation. This can be accomplished by highlighting the title of the subsection containing that particular guidance, pressing the *Copies* button, and entering the number of copies required. When the selection process is completed, the file contains only the guidelines needed for the review. During an evaluation, the user may wish to include a section that was excluded from the file. This can be accomplished using the *Modify* function. At that point the process is the same as described above for including and excluding sections.

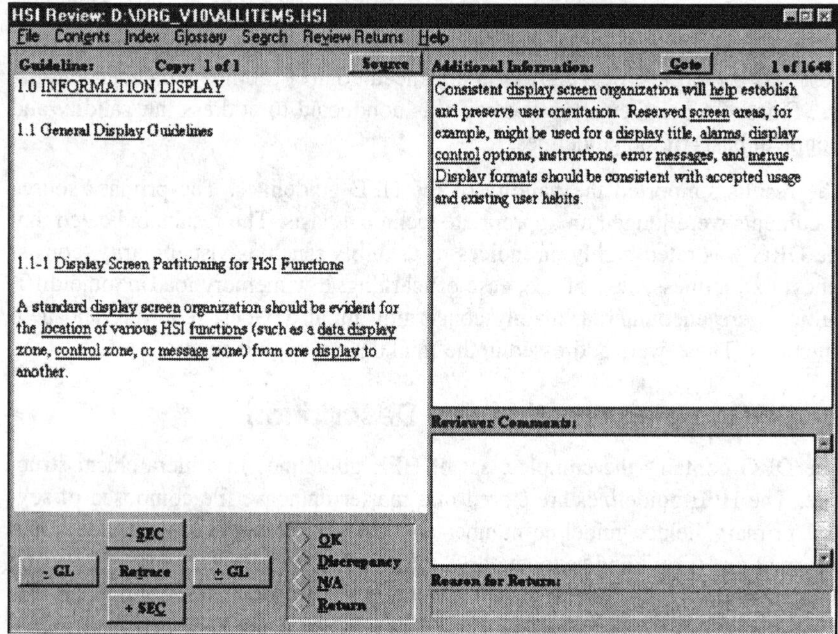

Figure 1. Evaluation screen.

3.2 Evaluating Human-System Interfaces

When the evaluation file is opened, the DRG displays the sections and associated guidelines for that evaluation (see Figure 1). The evaluation buttons, (*OK*, *Discrepancy*, *N/A*, and *Return*) record how well the HSI being reviewed conforms to the guideline. The *Return* button indicates that a guideline applies, but more information is needed to assess conformance (in other words, the guideline will be returned to at a later time). Guidelines to be returned to are collected and reviewed by using the *Review Returns* function on the menu bar of the "HSI Review" window. Comments can be added using the "Reviewer Comments" area. These comments are stored with the guideline and can be printed in an evaluation report summary. Underlined words have hypertext links to glossary definitions, tables, and figures. In addition to obtaining word definitions through hypertext links, an on-line glossary is available by selecting *Glossary* from the menu bar.

3.3 Navigating and Searching the Guidelines

Movement between entire sections can be made in two ways: using the *Contents* function to move to the section directly, or by using the +/– section buttons. The *+SEC* and *-SEC* buttons can be used to move from section to section. Within a section, the *+GL* and *-GL* buttons display the next or previous guideline. The

Retrace button retraces the user's path through the guidelines. Two navigation functions, *Search* and *Index*, enable the user to find specific guidance of interest. *Search* can be applied to any combination of guideline names, guideline text, additional information fields, and comment fields for any term specified. *Index* displays instances of the selected term in context. Clicking on a line of text brings the reviewer to the guideline in which the text string appears.

3.4 On-Line Help

A *Help* feature is available within every window of the DRG providing on-line assistance for performing a desired function. The help was developed in standard windows help format.

3.5 Preparing Reports

The *Report* function lists the number of guidelines currently evaluated as *OK*, *Discrepancy*, *Not Applicable*, *Incomplete* (i.e., not yet evaluated), or marked *Return* for later review. Any or all of these evaluation categories can be included in the report. The numbers and titles of the guidelines, along with any comments entered by the user, can be viewed. The report can be printed, or saved as a generic text file for importing into word processing software.

4 Conclusion

The DRG was developed to support users in evaluating HSIs using technically valid HFE guidelines. The software has been used for a couple of years. We have received very positive feedback from users, and several organizations have asked to use the software to present their own guidance for different purposes. We are currently conducting an evaluation of the guidance, including the software, in order to identify problems and obtain recommendations for new features and functions. These modifications will be made in the near future when the next version of the software will be released.

5 References

O'Hara, J., Brown, W., Stubler, W., Wachtel. J. & Persensky, J. (1996). *Human-system interface design review guideline* (NUREG-0700, Rev. 1). Washington, D.C.: U.S. Nuclear Regulatory Commission.

O'Hara, J., Brown, W. & Nasta, K. (1996). *Development of NUREG-0700, Revision 1* (BNL Technical Report L-1317-2-12/96). Upton, New York: Brookhaven National Laboratory.

Embedding HCI guideline input to iterative user interface development

C. Stephanidis, D. Grammenos and D. Akoumianakis
Institute of Computer Science, Foundation for Research and Technology Hellas
Science and Technology Park of Crete, GR-71110 Heraklion, Crete, Greece
Tel: +30-81-391741, Fax: +30-81-391740, E-mail: cs@ics.forth.gr

1 Introduction

For many years, guidelines have constituted an inexpensive and widely used tool for communicating human factors knowledge for the creation of more usable and effective user interfaces. Although the value and importance of guidelines is indisputable, so is the apparent ineffectiveness and lack of user-friendliness of their present incarnation; i.e. paper-based manuals. The paper medium imposes a number of limitations (see, Grammenos et al., 1999a); in addition to impediments related to the medium of propagation, the content of guidelines imposes further limitations, as: (a) it often comes in a context-independent form and requires intensive interpretation; and (b) sometimes the language and style used is inadequate or unfamiliar to designers and developers. The above limitations in combination with the emerging need for interactive tools to provide more effective and efficient human factors input in early design activities, have created a compelling need for a new generation of tools, namely tools for working with guidelines.

2 SHERLOCK: A Guideline Management System

2.1 Overview

Sherlock is a Guideline Management System which aims to support the design team, during the early phases of prototypical development, as well as, the usability analyst, when inspecting prototype versions, to assess compliance against an agreed set of rules, or a standard. There is some common ground shared between Sherlock and other available tools for working with guidelines, but there are also distinctive differences. Sherlock provides the framework and tools for automatically evaluating user interfaces, depositing and managing guidelines, depositing experience on fixing usability problems, and, finally, for documenting background knowledge on guidelines. Thus, Sherlock can act as: (a) an automatic inspection system; (b) a component for ensuring consistency across different corporate developments; (c) a companion tool to usability experts, acting both as critiquing tool and as an interactive reminder and data base of guideline-related information; (d) an on-line design aid for user interface designers; and (e) a knowledge repository for frequently encountered usability problems, as well as their respective solutions. Sherlock follows a client / server model. Both the server and the client run under MS Windows '95. The client module is an add-in to the Visual Basic 5.0 Integrated Development Environment (IDE). More details about the development of Sherlock can be found in (Grammenos et al., 1999a).

2.2 Advantages of the Sherlock Guideline Management System

(Grammenos et al., 1999b) presents a list of shortcomings of the existing tools for working with guidelines which impede their wider use and adoption by practitioners. Below, we will account how Sherlock overcomes these shortcomings:

2.2.1 *Context specificity and guidelines customisation*

Problem: Tools for working with guidelines cannot cope with context parameters and customisation issues
Solution: Sherlock provides facilities for activating / deactivating rules or clusters of rules, depending on the current context of the evaluation (Figure 1). Additionally, whenever it is feasible, rules can be automatically deactivated if they do not fit in the current context (e.g., a cluster of rules for form-based interfaces may not be executed for an interface that is not a form).

2.2.2 *Loose coupling/integration with user interface development systems*

Problem: Tools for working with guidelines have traditionally not been integrated within popular user interface development systems.
Solution: Sherlock is integrated in the Visual Basic 5.0 IDE, a very popular user interface development environment, so that recommendations can be directly propagated to the actual implementation.

2.2.3 Extensibility, maintenance and versioning of guideline knowledge

Problem: Currently available tools offer no computer-aided support for: (a) maintaining the guideline rule base (e.g., identifying competing guidelines, conflicting recommendations, automatic updates); (b) extending its scope with new rules; (c) versioning of guideline reference manuals, so as to depict specific requirements of particular design cases.
Solution: (a) Sherlock provides mechanisms for identifying conflicting or duplicate guidelines and offers the appropriate help and tools to resolve the conflict; (b) The rule base is easily extensible and open to modification or customi-

sation. In fact, Sherlock does not contain any evaluation rules per se; it just provides a common framework under which guidelines from various sources, instantiated as evaluation rules, can be integrated. The type, scope and content of the rules, as well as the implementation details (e.g., programming language, structure) of the respective assessment routines are only restricted by the imagination and capabilities of their developer; and (c) The rule base may contain more than one versions / implementations of the same rule or clusters of rules. Then, the user can select which of them will be active through the respective customisation and conflict resolution mechanisms.

2.2.4 Design augmentation is beyond guideline access

Problem: There is no way to capitalise on, and reuse past experience, explore alternatives before committing to a particular design option, document problems and design deficiencies, so that they can be referred to by future activities.
Solution: Sherlock keeps track of, and documents, the usability problems detected, as well as, the solutions that were given to these problems. These data are then provided to the users as background information whenever a relevant usability problem is detected.

2.2.5 Corporate support

Problem: Another important shortcoming of existing tools, is their lack for supporting corporate practices.
Solution: As mentioned before, Sherlock' s rule base is open to extensions and modifications. Organisation may easily develop and add new rules to the rule base and even select and customise a subset of already existing ones.

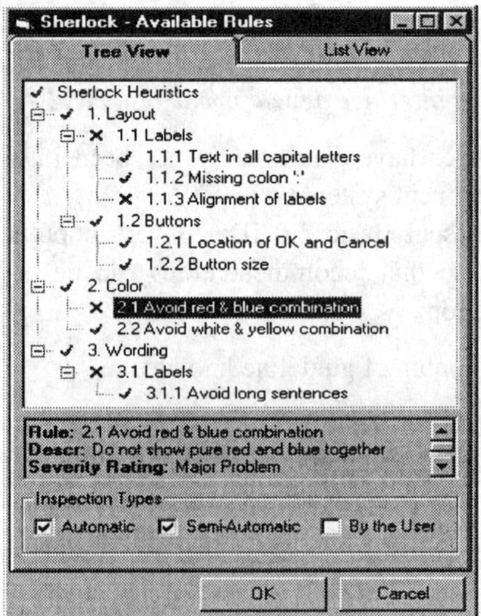

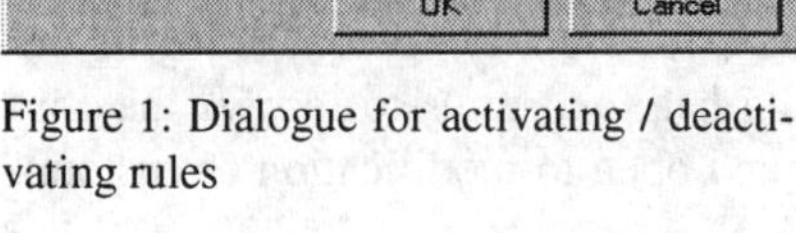

Figure 1: Dialogue for activating / deactivating rules

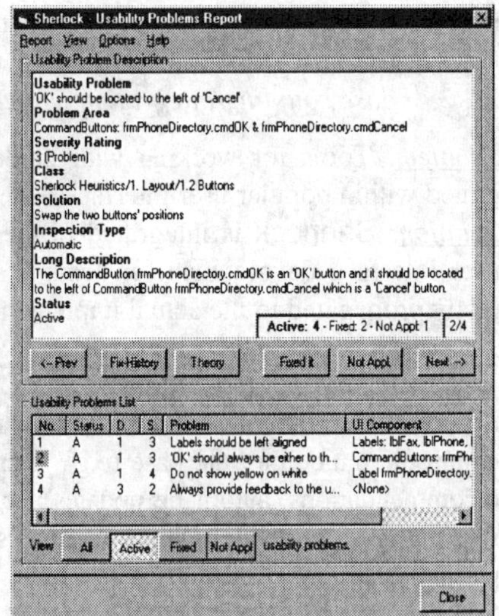

Figure 2: Reporting usability problems

2.2.6 *Reporting*

Problem: Reporting design defects and alternative solutions is another important issue that needs to be supported by tools for working with guidelines, if they are to provide an effective and efficient medium for integrating human factors knowledge into software design and management.
Solution: Sherlock, has a comprehensive report module (Figure 2) which provides analytic description of each rule violation detected, offers background information about the guideline (such as theory and examples) as well as information about how the same problem was solved in the past. Additionally, this module offers facilities for grouping and classifying usability problems to assist the developer to organise the task at hand.

2.3 Supporting the user-centred design process through Sherlock

One of the most important aspects of user-centred design is the emphasis given on the iteration between design and early evaluation of design solutions. Sherlock can be easily introduced in this iterative procedure, since it can provide fast and easy evaluation of user interface prototypes while they are still designed. This way, evaluation becomes an integrated part of design instead of being a separate phase. According to (ISO, 1997) the user-centred design process consists of four basic and interdependent activities that are iterating until the predefined objectives are satisfied: (a) understand and specify the context of use; (b) specify the user and organisational requirements; (c) produce design solutions; and (d) evaluate design against requirements. In order for Sherlock to be successfully introduced in the process, the following actions have to take place during each of the aforementioned activities:

a) *Understand and specify the context of use:* During this phase guideline collections and manuals that are relevant to the current context of use have to be collected and selected.
b) *Specify the user and organisational requirements:* The selected guidelines should be translated into concrete evaluation rules. Rules that are not included in the current version of the Sherlock' s rule base will have to be converted to computer programs (in the form of ActiveX DLLs). Additionally, existing rules may need to be modified. In case new rules conflict with existing ones, conflicts will have to be resolved.
c) *Produce design solutions:* Instead of the traditional approach, where a design solution was evaluated only after being completed, Sherlock offers the opportunity of evaluating a user interface while it is created. Thus, before a final prototype is ready for expert or user evaluation many usability problems will be detected and eliminated. Additionally, areas of interest, where further evaluation is needed (e.g., trade-offs between competing guidelines) will be identified and will be given as input to the next phase.

d) *Evaluate design against requirements:* During this phase Sherlock can act as an electronic guidelines handbook or as a "reminder" tool for assisting experts in identifying potential usability problems that can not be automatically evaluated or as a guide for identifying areas where user testing should focus.

3 Conclusion

As is the case with all available design tools, Sherlock is not a panacea for the usability analyst. The limits of automated user interface evaluation (Farenc et al., 1996) as well as those of expert evaluation (Desurvire, 1994) have already been investigated and are well documented. These limitations are often used as arguments for condemning automated user interface evaluation. But, the fact that this type of evaluation cannot do everything should not diminish the value of the technique. Sherlock is intended to be a 'usability aspirin'; i.e., a small-scale, useful and friendly to use tool that allows user interface designers to easily and quickly get rid of 'usability headaches'. In other words it allows them to spot and correct tedious, frequently repeated usability problems, so that their focus of attention can be shifted towards larger and more important usability issues.

4 References

Cohen, A., Crow, D., Dilli, I., Gorny, P., Hoffman, H.-J., Ianella, R., Ogawa, K., Reiterer, H., Ueno, K. & Vanderdonckt, J. (1995). Tools for Working with Guidelines. *SIGCHI Bulletin*, 27 (2), 30–32.

Desurvire, H.W. (1994). Faster, Cheaper!! Are Usability Inspection Methods as Effective as Empirical Testing? In Nielsen, J. & Mack R.L. (Eds.): *Usability Inspection Methods*, pp. 173-202. New York: John Wiley & Sons Inc.

Farenc, C., Liberati, V. & Barthet M.-F. (1996). Automatic Ergonomic Evaluation: What are the limits? In Vanderdonckt, J. (Ed), *Proc. of the 2nd International Workshop on Computer-Aided Design of User Interfaces* (CADUI'96, Namur, June 5-7, 1996), pp. 159-170, Namur: Presses Universitaires de Namur.

Gorny, P. (1995). EXPOSE: An HCI-Counselling tool for User Interface Design. *Proceedings of INTERACT '95*, pp. 297-304, London: Chapman & Hall.

Grammenos, D., Akoumianakis, D. & Stephanidis, C. (1999a). Integrated Support for Working with Guidelines: The Sherlock Guideline Management System. *Interacting with Computers,* 11 (4), to appear.

Grammenos, D., Akoumianakis, D. & Stephanidis, C. (1999b). Support for Iterative User Interface Prototyping: The Sherlock Guideline management System. *Proceedings of the IFIP Working Conference on Engineering for Human-Computer Interaction* (EHCI '98, Crete, Greece), to appear.

ISO/DIS 13407 (1997). *Human-centred design processes for interactive systems*. Geneva, Switzerland:.International Organisation for Standardisation.

Completing Human Factor Guidelines by Interactive Examples

Hartmut Wandke, Jens Hüttner
Humboldt-Universität zu Berlin, Mathematisch-Naturwissenschaftliche Fak. II
Institut für Psychologie, Kognitive Ergonomie / Ingenieurpsychologie
Oranienburger Str. 18, D-10178 Berlin, Germany
Phone: ++49 30 285 165 245 /244 (secr.) - Fax: ++49 30 282 40 46
E-mail: hartmut.wandke@psychologie.hu-berlin.de

1 Introduction

The demands currently being made on modern, ergonomical software as a work tool cannot be sufficiently handled by the usual sources in software programming, i.e. by the intuition and common sense of the software designer. Software determines just how much of a cooperative and communicative partner the computer's "behavior" will be. Clearly this "soft-tool" for the work place must therefore be made to fit human possibilities and needs. "Job design of computer-aided work is thus to a great extent software design, i.e. the design of the user-interface component of the software that is relevant for the job" (Hacker, 1989). However, the designers of computer-aided work tend to be software designers specialized in computer programming and not specialists trained in work science or work psychology. Software-ergonomic knowledge could therefore form a very necessary and basic supplement for designers of user-interfaces, who due to their specialization have a very different perspective on the problems of software development.

2 Software Designers' Software-Ergonomic Knowledge

In a number of questionnaire surveys over the past 8 years we have ascertained the software-ergonomic knowledge of designers in software companies (e.g., Hüttner & Wandke, 1993). The significance of user-friendliness was recognized by all the questioned software designers; 50% of those asked estimated that at least 30% of the total product development was dedicated to the user-interface. However, they barely had any basic knowledge of it, and relied for the most part on their intuition and personal experience when programming. Selected extracts from the results of these surveys are given below.

- Questionnaire survey 1992

  82 people from 50 companies took part in the survey. As their participation was as result of replying to adverts, a certain amount of "positive selection" must be assumed (interest for aspects of user-friendliness). Which guidelines, norms and standards are known about?

Norms, standards, guidelines from companies	known %	used %
DIN 66 234	22	11
ISO 92 41	1	0
CUA / SAA from IBM	54	31
Apple Guidelines	15	7
none explicit	10	26

Table 1. What System designer know and use

How well know are software-ergonomic terms?

Only 6 from 50 software-ergonomic terms were classified as "known" or "well-known" by 75% of those questioned, and 66 % of the terms had to be classified as "unknown".

- Interview survey CeBIT 1996

  At the worlds largest computer fair 49 exhibitors from the "Medium Sized Software Companies" section were questioned with a standard interview at their stands. To summarize, we established amongst other things that the software could not be used without the help of bulky manuals (25% had between 600 and 5000 pages), 38 % of those questioned had **very seldom** contact to their the clients, and only 25 % had during training courses they had attended heard even a reference to user-interface design. When developing software 76% allowed themselves to be lead by their intuition and their own experience. All the same, 68% admitted that they were **not** guided by specialist literature, and 29% **not** by norms or standards.

- Questionnaire survey 1997/98

  In this large scale survey 1500 software companies in German speaking countries were contacted by mail. 404 participants could be included in the analysis. Generally speaking software-ergonomic knowledge had increased, though there was a lot of catching up to be done. Here are two results from the investigation.

Rank	
1	Putting oneself in the position of the future user
2	Intuition and experience
3	Aesthetic sense and feeling
4	Testing on the user

5	Company guidelines
6	Norms and standards

Table 2. What guides the designers when developing software and how much?

German and international norms, standards and guidelines	Yes = 32,8 %
Manufacturer guidelines, recommendations, handbooks	Yes = 57,7 %
Specialist literature	Yes = 33,9 %
Internet sources (newsgroups, FAQ´s, mailing lists)	Yes = 12,6 %

Table 3. Which software-ergonomic information sources were known to the designers?

The results of the different surveys demonstrate how important is the task of providing and communicating software-ergonomic knowledge. With the **inra** System we are providing a possible way of gradually doing this. The **inra** System consists of a hypertext system for Windows with interactive examples and an accompanying reference book (Hüttner et al., 1995). Both parts complement each other with regard to both their focus of content and their medium.

3 The inra System and its Interactive Examples

One part of **inra** is a collection of guidelines in hypertext format. These guidelines are supplemented by the following components:

Interactive examples consisting of mini programs and scenarios, Annotated screen shots taken from the interactive examples explaining selected guidelines, Glossary, List of key words, List of publications, Theoretical and empirical background of guidelines (mostly printed text), Methods for testing usability (printed text only).

inra is not a tool, but a hypertext-based information system. The core of **inra** is made up of mini programs and scenarios which allow the user to gain experience with interfaces that either comply or violate several human factors guidelines in dialogue design and information presentation. Of the great many guidelines that are implicitly addressed in both compliance and violation cases 49 are explicitly explained in another **inra** component showing annotated screen shots. The user can click one of the arrows (coloured on the screen) and in another window be given information about the actual design problems. In the following is an example of **inra** giving information about "grouping".

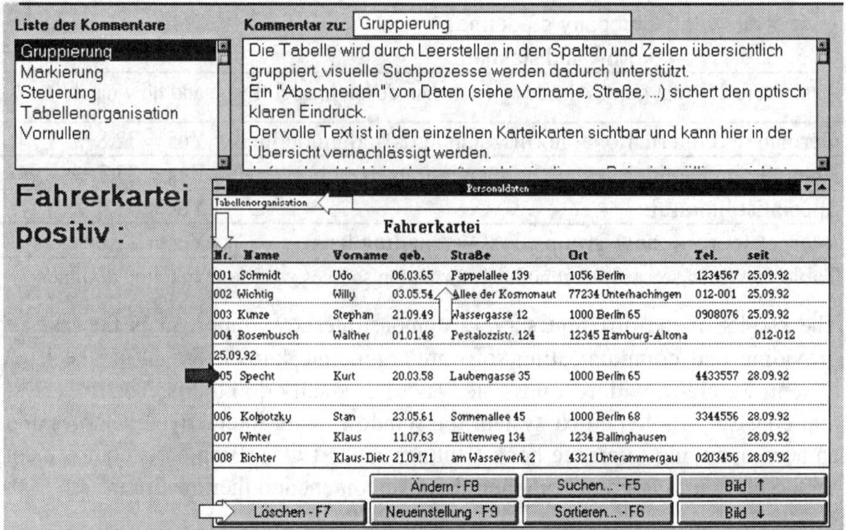

Figure 1. Example page of the **inra**-system

The use of these interactive examples was the subject of an empirical study with software designers.

4 Conclusions

According to our study, most of the identified design problems regarded the middle levels of the human-computer-interaction (syntax, semantic). Problems on the lower levels (information presentation) were rarely identified. Clearly it would be particularly important to supplement these problems with explicit explanations and complementary text.

A number of conclusions for the use of interactive examples in knowledge transference can be drawn from the presented results. Briefly, learning from interactive examples in dialog ...

1. ...is an effective form of knowledge transfer. This is demonstrated by the large number of design problems that were identified by the subjects themselves. Generalization and transfer of the example ideas took place.

2. ...is not enough on its own. It is meaningful to have various different explicatory supplements, such as explicit commentaries and additional texts, because, for example, in the empirical study some of the explicitly "hidden" guideline violations were not found by the subjects.

3. ... enables violations and compliances to be identified with equal frequency. The designers did not just criticize, they also praised or said how things could be improved upon.
4. ... leads to the identification of design problems independent of the order of negative and positive examples. User-controlled learning is possible as examples can be chosen at will. The user does not have to follow rigid programs.
5. ... is interesting and motivating. In an interview 10 out of the 12 pairs said they had gained ideas and stimulation that they could apply in their day to day work.

5 References

Hacker, W., Raum, H., Rentzsch, M. & Völker, K. (1989). *Bildschirmarbeit - Arbeitswissenschaftliche Empfehlungen*. Berlin: Die Wirtschaft.

Hüttner, J., Wandke, H. & A. Rätz (1995). *Benutzerfreundliche Software. Psychologisches Wissen für die ergonomische Schnittstellengestaltung.* Berlin: BMP-Verlag.

Hüttner, J. & Wandke, H. (1993). What do system designers know about software ergonomics and how to improve their knowledge? In Luczak, H., Cakir, A. & Cakir, G. (Eds.) *Work With Display Units 92* (p. 304-308). Amsterdam: Elsevier.

A Petri Nets Method for Specification and Automatic Generation of User Interface

Faouzi Moussa*, Meriem Riahi*, Mohamed Moalla*, Christophe Kolski**
* Faculté des Sciences de Tunis - Département des Sciences de l'Informatique - Campus Universitaire 1060 Le Belvédère – Tunis, Tunisia
email: faouzi.moussa@ensi.rnu.tn
** LAMIH - URA CNRS 1775 Le Mont Houy, BP 311
59304 Valenciennes Cedex, France – email: kolski@univ-valenciennes.fr

1 Introduction

In process control, many automated applications with high degree of safety require the human operator's permanent presence. The operators have to control the system's evolution and to react to unexpected events. Graphical user interfaces inform the operators about the process evolution and assist them in their complex task of problem solving. Several researchers are interested in the improvement of the quality of such interfaces (e.g., Rasmussen 1986; Gilmore et al. 1989; Kolski 1997). Nowadays, the tendency is to provide formal techniques for interface specification. So that, it becomes easier to verify the consistency of the specifications before the interface generation. This is particularly crucial when the interfaces are used in critical domains such as process control or transportation systems. Many works have been carried out within this framework and different approaches have been proposed (Moussa 1992; Bodart et al. 1995; Mahfoudhi 1997; Palanque and Bastide 1997). But, most of them presents slackness concerning for example system analysis, interface specification or automatic generation of the interface. Thus, we try to palliate them by proposing a global approach which covers the different aspects of the user interface design. This approach benefits from Petri Nets formalism and contributes to the "Tools for working with guidelines" domain.

In this paper, we present, firstly, the proposed approach for the design of user interfaces dedicated to process control. Then, we focus on using guidelines and Petri Nets model for the User Interface (UI) specification.

2 Proposed Approach for User Interface Design

We are especially interested in the identification of the user requirements from the Human-Machine System (HMS) analysis. The objective is to take into account these user requirements in a UI design methodology integrating automatic generation based on ergonomic guidelines (Riahi et al. 1998). This approach is made up of six steps. In the **first step**, a preliminary and necessary analysis of the process and its command system is carried out in the first step. This analysis provides a document containing the data of the process and its different technical and functional constraints.

The **second step** consists in analyzing the whole MMS in order to identify the operator's tasks, the command system and the process features. A combination of different methods and techniques are proposed for that. Indeed, to identify the appropriate graphical displays for the process control, we consider that a first hierarchical decomposition of the HMS is needed. The well-known SADT method is here proposed. This decomposition allows us to identify the appropriate elementary sub-systems to be studied. Then, we propose a malfunctioning analysis in order to identify for each sub-system, the possible failures, invoking their reasons and their effects on the system. For that goal, two complementary methods can be used: FMEA method (Failure Modes and Effect Analysis) and FTA method (Fault Tree Analysis) (Fadier, 1990). Afterwards, we identify for each sub-system and for each functional context (normal and abnormal), the associated operator's tasks. The analysis of these tasks provides later, the deduction of the user requirements. Thus, we have to model the HMS dynamics describing the system's evolution inherent to the different events. Petri Nets, which can express aspects of concurrency and parallelism are used. The places of Petri Nets represent the different sub-system states. The transitions express the different possible evolutions of the system's state. The Petri Nets proposed here are the Interpreted Petri Nets (Moalla, 1985). They introduce the notion of events and conditions as well as the notion of actions. So, we associate to each transition, a passing condition, a triggering event and an action expressing the execution of an operator's task. One can notice that any modification of the system's state will imply a changing in the graphical displays. This change can affect either the object's parameters (e.g., colors, shapes) or the display contents (e.g., appearance or disappearance of some graphical objects), thus allowing the deduction of the user requirements.

The **third step** concerns the deduction of the user requirements. Controlling the process, the operators need to understand the changes of the process state. This information will be transmitted to them through different displays of the interface (e.g., messages, numeric values, graphic symbols, alarms). Thus, we identify for each state, the appropriate informational variables and associate them to the corresponding place in the Petri Net modelling the HMS dynamics. Likewise, for performing their tasks, the operators need to command some variables of the system in order to correct abnormal situations. For that, the interface will propose a set of control objects to command the process. So, we associate the necessary command variables to the appropriate transitions expressing the human tasks. The set of these command and informational variables constitute the user requirements.

Once, the user requirements identified, the **fourth step** consists in specifying the UI. It is necessary to identify the adequate graphical displays, to decide on their presentation modes and to describe the human-machine dialogue. The principle adopted for these aims is developed below.

In the **fifth step**, according with Palanque and Bastide (1997), it is possible to benefit from the formal technique used (Petri Nets), to verify the UI specifications. Many researches are yet required for that. The **last step** is dedicated to the automatic UI generation based on the validated UI specifications. First promising results have been obtained thanks to the experience of the Ergo-Conceptor system (Moussa 1992; Moussa et al. 1999). Of course, a more advanced research is required.

3 User Interface Specification

The transition from process analysis, task analysis, or HMS analysis to the UI specification is considered as difficult, and is often highly empirical. Few formal approaches are described in the literature. To identify the necessary graphical displays for process control, specific knowledge in Human-Computer Interaction and Software Ergonomics is especially required. In this stage, we propose to benefit from the researches carried out since the eighties, by some researchers concerning knowledge-based approaches for automatic evaluation and design of UI used in process control. In fact, there are many ergonomic rules available in the literature. For instance, Gilmore and al. (1989), inspired by their works with the U.S. Nuclear Regulatory Commission (USNRC), have written a handbook of guidelines for user-computer interface design in process control; these guidelines were organized into four categories, namely, (1) video displays, (2) control and input devices, (3) control/display integration, (4) workplace layout and environmental factors. Each guideline was presented in a structured format. Hundreds of guidelines adapted to this particular field (process control) were listed in the handbook. Other important contributions in this field exist (see for instance O'Hara and Furtado's papers in this special session).

Several model-based tools for the interface development are described in the literature (Vanderdonckt 1996). The Ergo-Conceptor system, one of these tools, is oriented towards process control interactive applications. So, following the principle of Ergo-Conceptor, our objective is to decide automatically on the appropriate displays to associate to each sub-system. For that, it will takes into consideration, on the first hand, the characteristics of each sub-system (e.g., the list of its functioning states, the user requirements associated to each state) and on the other hand, specific formalized guidelines stored in its knowledge bases.

Once the different graphical displays are identified, we have to specify the presentation and the dialogue associated to each of them. Interface objects are identified according to the user requirements deduced in step 2 of our approach (figure 1). Their presentations will, also, be chosen according to a knowledge based system, dealing with guidelines, in order to ensure the better ergonomic quality. At this level of abstraction, guidelines are more frequent and it is easier to formalize them (Moussa et al., 1999). Then, we have to model the dialogue component of the interface. For that, we consider that each graphical object presents a control structure. This control structure describes the evolution of the object's state according to the system evolution and to the operator's actions. The Interpreted Petri Nets are, again, proposed for modeling the behavior of each graphical object. The HMS dialogue is, therefore, completely described by the different Petri Nets used for modeling the HMS dynamics and the control structures of the different UI objects (figure 1).

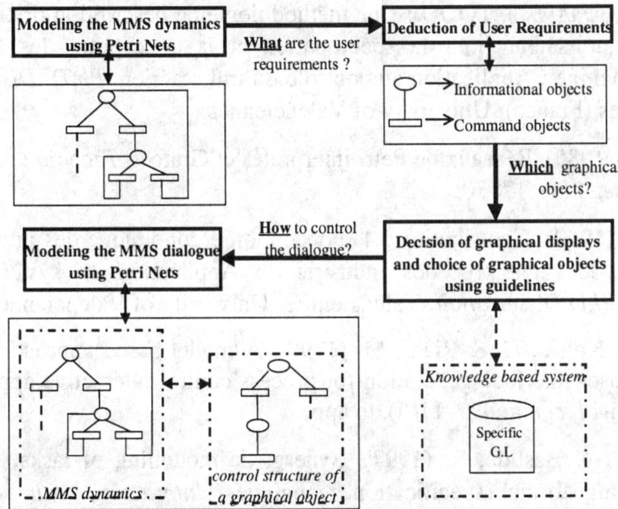

Figure 1: Principle of specifications' generation using Petri Nets and guidelines.

4 Conclusion

In order to study the efficiency of our approach, we are implementing the Petri net model in a C++ environment. Simultaneously, we are implementing a rule based system using a first inference engine and the object oriented paradigm. This system deals with guidelines concerning UI specifications. The validation concerns a simplified example of nuclear power plant. First results are planed to be published in near future.

5 References

Bodart, F., Hennebert, A.-M., Leheureux, J.-M., Provot, I., Vanderdonckt, J. & Zucchinetti, G. (1995). Key activities for a development Methodology of Interactive Applications. In Benyon, D. & Palanque, Ph. (Eds): *Critical Issues in User Interface Systems Engineering*, pp. 109-134. Berlin: Springer-Verlag.

Fadier, E. (1990). Fiabilité Humaine : Méthodes d'analyse et domaine d'application. In Leplat, J. & de Terssac, G. (Eds.): *Les facteurs humains de la fiabilité dans les systèmes complexes*. Marseille: Editions Octarès.

Gilmore, W.E., Gertman, D.I. & Blackman, H.S. (1989). *User Computer Interface in process Control: A Human factors Engineering Handbook.* New York: Academic Press.

Kolski, C. (1997). *Interfaces Homme-Machine : Application aux systèmes Industriels complexes.* Paris: Editions Hermès.

Mahfoudhi, A. (1997). TOOD : Une méthodologie de description orientée objet des tâches utilisateur pour la spécification et la conception des Interfaces Homme-Machine : Application au contrôle du trafic aérien. *Ph.D. Dissertation.* Valenciennes (France): University of Valenciennes.

Moalla, M. (1985). Réseaux de Petri interprétés et Grafcet. *Technique et Science Informatique,* 4 (1), 17–30.

Moussa, F. (1992). Contribution à la conception ergonomique des interfaces de supervision dans les procédés Industriels : Application au système Ergo-Conceptor. *PhD Dissertation*, Valenciennes: University of Valenciennes.

Moussa, F., Kolski, C. & Riahi, M. (1999). A model based approach to semi-automated user interface generation for process control interactive applications. *Interacting with computers*, 11(5), to appear.

Palanque, P. & Bastide, R. (1997). Synergistic modelling of tasks, users and systems using formal specification techniques. *Interacting with computers,* 9(12), 129–153.

Rasmussen, J. (1986). *Information processing and human-machine interaction, an approach to cognitive engineering*. Amsterdam: Elsevier Science Publishing.

Riahi, M., Moussa, F., Moalla, M. & Kolski C. (1998). Vers une spécification formelle des Interfaces Homme-Machine basée sur l'utilisation des Réseaux de Pétri. *Proceedings ERGO IA' 98* (Biarritz, November 4-6, 1998), pp. 196-205.

Vanderdonckt, J. (Ed.) (1996). *Computer-Aided Design of User Interfaces, Proc. of the 2nd Int. Workshop on Computer-Aided Design of User Interfaces* (CADUI'96, Namur, June 5-7, 1996). Namur: Presses Universitaires de Namur.

An Approach to Improve Design and Usability of User Interfaces for Supervision Systems by using Human Factors

Elizabeth Furtado
Department of computer sciences - CCT -University of Fortaleza
Rua Washington Soares, 1321 - Bairro Edson Queiroz
Fortaleza - CE - 60455770 – Brasil; Email : Elizabet@feq.unifor.br

1 Introduction

Design and implementation of interactive systems is a complex work due to the use of a large variety of models and extensive notations (Montoy 1995; Larman 1998). Computer-based tools for generating interfaces automatically allow to reduce this complexity. Using these tools the divers models used during the design of the system are deduced from the specification of the domain classes and the tasks performed by the user - user tasks. Moreover, the automatic generation of interfaces from conceptual specifications allows to keep the consistency among different models and to reduce the system development time. Our research is focused on developing a method to build automatically user interfaces from conceptual specifications of supervision systems, more specifically, of their decision support systems. These later systems determine the assistance (such as alarms, diagnosis, and plan of actions and prediction or recovery procedures), which help the operator in recovering a problem when a failure occurs. In our work, the interfaces are able both to adapt themselves to the operator's needs during a problem solution and to respect some ergonomic principles. Adaptability improves handling and learning of the interface. Ergonomic principles are guidelines about the ergonomic aspects of the interfaces (such as to shown only the necessary information, to let the user to control the system dialog). Taking into account ergonomic principles in the interface design allows the designer to determine the best way which information must be given to the user during his interaction with the system. The article gives an overview of adaptive user interfaces in the supervision domain and presents the method, called MACIA, which is necessary to follow for generating interfaces.

2 Adaptive User Interfaces for Supervision Systems

Interfaces play a vital role in improving the communication between the operator and the supervision system. The interface has to adapt itself to the operator's needs in order to determine the assistance he should receive from the supervision system during a problem solution. In order to construct interfaces that adapt themselves to human preferences, we have studied some human cognitive models that try to understand the human's behavior and can help the designer in identifying individual differences. The problem resolution model described in (Hoc 1996) describes the phases that an operator takes to solve a problem when he is controlling an industrial process, namely identification of an abnormal situation; evaluation of this situation; take of decision and execution of an action. The phases followed by an operator depends on his experience level. An expert operator can identify an abnormal situation, then execute an action directly. However, a novice operator verifies at first all the assistance that are given to him, and so he takes a decision. It is important to point out that an operator can be an expert in the treatment of a failure, but a novice in the treatment of another one. According to (Ujita 1992), an operator performs many actions in order to treat a failure. For example, in order to decrease the water level, it is necessary to open the valve X and to decrease the pressure of the turbine Y. Each action corresponds to a different cycle, which is composed of the phases of operator's cognitive process. We have observed some problems with the usability of the interfaces in several works on the supervision domain. The problems are partly because ergonomic principles have been rarely taken into account. Using the information system developed by (Harmut 1999), the user can make some self-experiences, in order to know when an interface violates several guidelines. The usability problems are also because the adaptation process consists in changing the interfaces so frequently that the operator becomes himself confusing (Montoy 1995). Therefore, the adaptations should be constrained to the most important user tasks. In order to describe the user tasks efficiently and the relevant information, the designer should choose the appropriated methods aiming at system analysis.

3 Description of MACIA

MACIA describes a process where the interface specifications are automatically generated from a task model made by the designer. The three principal phases of MACIA, defined in (Furtado 1997) are: task modeling, conceptual interface specification and real interface generation. A conceptual interface notion represents the information that is associated to an interface, with no representation of how they appear on the screen. The "real interface" deals with what appears on the screen. MACIA aims at decreasing the interface design complexity and increasing the interface usability. To decrease the interface design complexity, the method integrates human factors, that are expressed through both the initial task model and the ergonomic rules. The initial task model, which is based on operator's problem resolution model, is used like a guide for helping the designer in modeling the operator tasks. The ergonomic rules, which are deduced from ergonomic recommendations, are used for generating the interfaces in two stages. First, the specification of the conceptual interfaces is made by the application of ergonomic rules over the modeled tasks. These interfaces are represented in the Conceptual Interface Model (MIC). Finally, the real interfaces are generated from the MIC structure. To increase the interface usability, MACIA determines which assistance should be shown to the operator and how it should be shown on the screen according to his experience level. Follow we describe the initial task model and the ergonomic rules, which we have elaborated.

3.1 The Initial Task Model

Definition and structure of the initial task model. The initial task model is used to describe the initial tasks that must be performed by the operator. The initial tasks concern both the supervision of an industrial process and the decision-making tasks, which correspond to the operator's problem resolution model. The initial task model is a skeleton, which is used like a guide to help the designer to describe the process control tasks for a supervision application. In order to define a specific task model, the designer should both add other tasks to the initial model and initialize or modify some task characteristics (such as the execution conditions, the data). The initial tasks of decision making are the following: *Identifying alarm* and *Problem solving*. This later is composed of subtasks, which are responsible for determining and showing the assistance to the operator. To define these tasks, we have associated each operator's cognitive phase to one or more tasks. For example, when the operator is in the cognitive phase of evaluating a situation, he can receive as assistance a diagnosis and/or a plan of actions. Basically, the functioning of the initial task model is the following. If the operator does not receive any alarm, then he should supervise the process. When a failure occurs, it means that the supervision task execution is interrupted and the operator should solve the problem provoked by such failure. At this moment, he can make a diagnostic of the problem and its correction and obtain some assistance according to his experience level.

Utilization of the task model in the adaptation process. The mechanism to determine the operator's experience level when he is treating a failure is based on the model defined by (Ujita 1992) and implemented through of the initial task model. This model allows identifying the different cycles which the operator passes when he is treating a failure. A cycle is initialized when the *Identifying alarm* task is triggered and it is composed of the tasks of problem solving. A cycle is finished when the operator validates the correction task. In order to analyze the operator's behavior, all these cycles are recorded in his interaction log. In addition, the operator's preferences are also recorded concerning an assistance of a cycle (such as if he has asked the assistance by himself or rejected it). The analysis of the interaction log of an operator is used to determine his experience level. As it is difficult to know all the characteristics of an operator concerned to his experience level, we have chosen to categorize the operators, which have some characteristics in common in stereotype. The different features that describe a stereotype are described in details in (Furtado 1996). Considering that the operator's experience level can be different to each occurrence of a failure, his stereotype is determined each time the *Identifying alarm* task occurs.

3.2 Ergonomic Rules

Ergonomic rules for conceptual interface specification. The ergonomic rules used to specify the conceptual interfaces were empirically obtained by integrating expertise about ergonomic in the mapping of the task model to the MIC. This integration was facilitated because a task model is generated from the initial task model, which makes it possible to determine partly the way, which the operator should follow to perform a task. The MIC is a hierarchical model made of two components: workspace and interaction object. A workspace is a logical screen used by the user to perform different tasks. An interaction object represents the interaction aspects of the tasks, such as a data handled by the user during a running task. This object is a component of the corresponding workspace associated to this task.

The tasks that will be associated to a workspace depend on their characteristics and the task associations are defined by the ergonomic rules. For instance, we have elaborated a rule, which defines that a workspace should be associated to each parallel task. A parallel task interrupts the execution of another parallel task. This rule aims to show precisely to the user the moment of interruption. This rule is based on the following recommendation concerning the supervision domain: *any window shouldn't cover one window that shows some kind of alarm* (Kolski 1993). According to the initial model, the task that shows some kind of alarm (called *identifying alarm* task) is a parallel task. A task can also create one or several interaction objects in its corresponding workspace associated. To define the rules of creation, we have considered recommendations like (Bastien 1993): *to give the opportunity to the user to either perform a task or cancel it.* The deduced rule has been the following: *If a task must be only performed when the user validates it, then two interaction objects must be created.* The first one allows the activation of this task and the second one is used to cancel the task that is running. It is also important to point out that the number of interaction objects in a workspace and the data type of the tasks are also taken into account. To define the number of interaction objects in a workspace, we have considered that: *it is necessary to respect the memory capacity limit of a person* and *to show the information which is meaningful in the task.* The process of workspace definition finished, rules of optimization are applied over the MIC structure to join two workspaces. It allows, for instance, limiting the steps that the user should go through to run a task. After the optimization phase, the designer can make some modifications and the real interface can be generated.

Ergonomic rules for real interface generation. The real interface can be automatically obtained from the MIC by using both ergonomic rules and interactive objects from a toolkit. The ergonomic rules used here aim at choosing the real interface aspects, such as the denomination of names and colors, the position of information in the window, the interaction styles, the interactive objects to represent the interaction objects on the screen and so on. For instance, a list box is the most appropriated interactive object, when the following recommendation is applied: *it is to be avoided that an user inputs a string, in giving him the opportunity to select this string from a list* (Bodart and Vanderdonckt 1993).

4 Conclusion

The method described here can be considered suitable for designing the interfaces, because it doesn't require an important effort from the designer: he has only to learn the task model notation to obtain the first specifications of the interface; he doesn't care about both ergonomic principles and individual differences in the interface design, and finally he is guided to identify the most important operator's tasks. In order to improve the usability of the user interfaces, the adaptation process only addresses the assistance tasks based on the operator's cognitive model and takes into account the operator's preferences concerning the received assistance to update his experience level.

5 References

Bastien, J.M.C. & Scapin D.L.. (1993). *Ergonomic criteria for the evaluation of user interfaces*. INRIA Research report No.156, Rocquencourt: INRIA.

Berthomé-Montoy, A. (1995). *Une approche descriptive de l'auto-adaptativité des interfaces homme-machine*. Ph.D. Dissertation, Lyon: Université Claude Bernard.

Bodart, F. & Vanderdonckt, J. (1993). Expressing Guidelines into an Ergonomical Style-Guide for highly Interactive Applications. In S. Ashlund, K. Mullet, A. Henderson, EL Hollnagel, T. White (Eds.): Adjunct Proceedings of InterCHI'93, pp. 35-36.

Furtado, E. (1996). Conception, Réalisation et Evaluation d'interfaces à partir des Spécifications Conceptuelles. *Revue en Sciences cognitives*. France.

Furtado, E. (1997). Mise en oeuvre d'une méthode de conception d'interfaces adaptatives pour des systèmes de supervision à partir des Spécifications Conceptuelles. Ph.D.

Harmut, W. & Huttner J. (1999). Completing Human Factor Guidelines by Interactive Examples, *Proc. of 8th Int. Conf. on Human-Computer Interaction* (HCI International'99)

Hoc, J.M. (1996). *Supervision et contrôle de processus, La cognition en situation dynamique*. Grenoble: Presses universitaires de Grenoble.

Kolski, C. (1993). Ingénierie des interfaces homme-machine. Editions Hermès.

Larman C. (1998). Applying UML and Patterns. An Introduction to Object-oriented Analysis and Design. Prentice-Hall.

Ujita, H. 1992. Human characteristics of plant operation and man-machine interface. *Reliability engineering and system safety*, 38, 119-124.

Automated Generation of an On-Line Guidelines Repository

Jean Vanderdonckt
Université catholique de Louvain, Institut d'Administration et de Gestion
Place des Doyens, 1 - B-1348 Louvain-la-Neuve, Belgium
Phone: +32-(0)10-47 85 25 - Fax: +32-(0)10-47 83 24
E-mail: vanderdonckt@qant.ucl.ac.be, vanderdoncktj@acm.org
URL: http://www.qant.ucl.ac.be/membres/jv/jv.html

1 Introduction

Involving User Interface (UI) guidelines is one possible way to improve usability of interactive applications either at design/programming time or at evaluation time. At design time, they can provide designers with some assistance by helping them to orient design options to obtain UIs which are more usable, more adapted, tailored to contextual needs (i.e., user population needs, physical environment constraints, and task needs). Guidelines can also serve as requirements to be achieved by developers at programming time. At evaluation time, a new UI or a previously designed one can be submitted to guidelines checking to guarantee a minimal threshold of usability.

Guidelines today exist for a wide spectrum of interactive applications ranging from general business oriented applications, such as (Smith and Mosier 1986) and (Scapin 1986) to specific UIs in control rooms, such as (O'Hara *et al.*1994), from traditional applications (MIL-STD-1472D 1989) to World-Wide-Web-based applications. Guidelines are gathered into five basic types of ergonomic sources depending on their domain of human activity, software and hardware platform, corporate environment: design rules, set of guidelines, style guides, standards, and ergonomic algorithms (Vanderdonckt, 1999). A *style guide* is a set of guidelines and/or functional or non-functional specifications aiming at consistency for a family of distinct UIs. This family can be based on an operating system (such as Windows'95), on a software editor (such as Borland's products), by a particular physical environment (such as IBM CUA), by a domain of human activity (such as medicine) or by a corporate (in-house style guide). Several corporate environments have committed themselves to the deployment of their own style guide, called *custom style guide* in contrast to *general style guides*, for several purposes:

- to avoid catastrophically unusable UIs,
- to ensure a better consistency across the family of interactive applications they are developing, using, disseminating,
- to communicate a unique and recognizable UI look & feel that reflects the organization philosophy, products and services,
- to inform the decision of design options decided by designers at design time,
- to provide assistance to evaluators of UIs at evaluation time,
- to reduce costs at both design and evaluation time by letting people to concentrate on major potential issues rather than loosing resources on insignificant details,
- to capture the progressive experience (e.g., pitfalls, errors, trial, good choices, usable windows) that an organization can gain in designing interactive applications not only to avoid reinventing the wheel but moreover to build new interactive applications by relying on top of past experience;
- to widely disseminate this experience throughout the organization.

2 Paper-based versus on-line style guides

Style guides are essentially presented in two forms: paper-based style guides (e.g., document, internal report, technical report, handbook, reference manual) or on-line style guides (e.g., downloadable document, HTML conversion of documents, pieces of software, on-line reference manual). Examples of the second form are called "tools for working with guidelines". Advantages, depicted by a "👍" sign, and shortcomings of paper-based style guides, depicted by a "👎" sign, are reproduced in the first column of Table 1, whereas advantages and shortcomings of on-line style guides are reproduced in the second column.

Paper-based style guide	**On-line style guide**
👍 Easiness	👎 Difficulty
👍 Accessibility	👎 Limited accessibility
👍 Legibility	👎 Reduced legibility
👍 Timeliness	👎 Inconvenience
👍 Hardware/software independence	👎 Hardware/software dependence
👎 Important size	👍 Size independence
👎 Rigidity	👍 Flexibility
👎 Fixed level of detail	👍 Varying level of detail
👎 Static format	👍 Dynamic format
👎 Limited reusability	👍 Increased reusability

Table 1. Advantages and shortcomings of paper-based versus on-line style guides.

As observed in Table 1, the characteristics which are considered as advantages for a paper-based style guide are evaluated as shortcomings for an on-line style guide, and vice versa. Both versions are consequently seen as complementary to each other. To overcome major shortcomings of producing an on-line style guide, we developed SIERRA (Système Interactif pour l'ERgonomie de Réalisation des Applications interactives – Interactive System for Software Ergonomics of Interactive Applications), a tool for working with guidelines that automatically produces an on-line style guide from a paper-based one.

3 Automatic Generation of an On-line Style Guide

A simple comparative analysis can establish that the following set of functions is minimal for a tool for working with guidelines:

- the reading and understanding of the complete contents of a guideline;
- sequential access to a guideline, a section or a division of guidelines;
- random access to a guideline, a section or a division of guidelines;
- the reading and understanding of related concepts (e.g., ergonomic criteria);
- the link from one guideline to another related, to other associated concepts;
- the searching of guidelines by title or any information found in its contents;
- the searching of guidelines by reference, by design ergonomic criteria, by evaluation ergonomic criteria, by linguistic level);
- the comparison of two selected guidelines;
- random access to illustrations and examples of a guideline;
- annotation, printing and copy-pasting a guideline;
- gathering of guidelines in a group for report and evaluation purposes,...

The idea of automatic generation of an on-line style guide consists in automatically generating a hypermedia providing the above functions from the paper-based style guide. This document should be structured into taxonomical sections ending up with guidelines, each guideline being itself decomposed into the attributes of a general guideline model described in (Vanderdonckt 1995). To make this generation as general as possible, the document is decomposed into information units in the hypermedia. Each *information unit* is characterized by a title (e.g., « Introduction », « Good example »), a title level (e.g., « Chapter 1 », « Section 1.1 », « Sub-section 1.1.1 », « Guideline 1.1.1•1 »), and an informational content which is structured or not according to the general guideline model:

1. if the contents are structured, i.e. if the information unit consists in a single guideline, then the model definition determines the information type ;

2. if the contents are not structured, i.e. if the information unit is anything else than a guideline, then its information type should be specified in heterogeneous, text, table, list, image, sound, video, macro, application, environment.

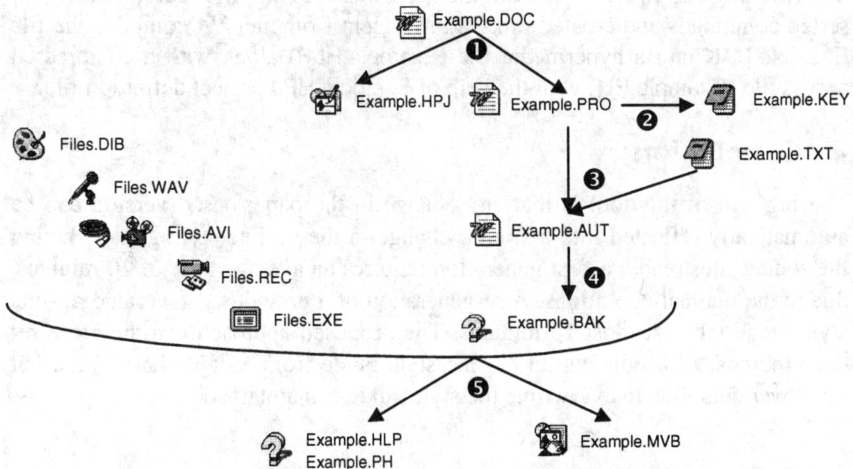

Figure 1. Process for automated generation of an on-line style guide.

Figure 1 depicts the generation process followed to automatically generate a on-line style guide from a paper-based document (Example.DOC) written in Microsoft Word. In this word processor, several WordBasic macros are executed:

- the automatic processing macro ❶ generates a first file decomposing each structured unit into sections and leaving each unstructured section as a whole with standard browsing links (Example.PRO) as well as a project definition file (Example.HPJ). This file, editable with MS Project Editor, contains the table of contents, its structure and the type of all information units;
- the keywords creation macro ❷ sorts in the Example.KEY file the alphabetical list of all words the document is containing with their occurrences in information units. Some specific significant words related to guidelines topics can be automatically added and other insignificant words can be automatically removed with additive and subtractive filters; the resulting file can be edited with any ASCII editor;
- the intra-link macro ❸ automatically generates Example.AUT with links in the hypermedia from a file containing words (e.g. [Smith & Mosier 1986], interaction style) to their definition or reference in the hypermedia (e.g., to the bibliographical reference of Smith & Mosier's document, to a definition of « interaction style »);
- the extra-link macro ❹ inserts in Example.BAK corresponding commands to access external files containing resources containing images (Files.DIB), sounds (Files.WAV), video sequences (Files.AVI), recorded macros (Files. REC), executable applications (Files.EXE), and applications that demonstrate behaviors in other environments (Files.EXE).

The resulting file can then be submitted to a manual edition of automatically inserted commands and created links ❺. MS Help Compiler V4 compiles the file Example.BAK into a hypermedia file (Example.HLP) along with a compressed access file (Example.PH) with the help of Example.HPJ project definition file.

4 Conclusion

The big win of this tool is that any change in the paper-based version can be automatically reflected into a similar change in the on-line style guide. Having the source files ready, a first generation requires an average time of 20 minutes, due to the manual operations. A re-generation of a previously generated on-line style guide takes at most 10 minutes. The proposed approach can therefore reduce the cost of producing an on-line style guide from a paper-based one, but moreover cuts time in every time the style guide is maintained.

5 References

McFarland, A. & Dayton, T. (1995). Design Guide for Multiplatform Graphical User Interfaces, Issue 3. Bellcore Document Number LP-R13. Piscataway (New Jersey): Bellcore.

Military Standard: Human Engineering Design Criteria for Military Systems, Equipment, and Facilities. (1989). Document MIL-STD-1472D, Notice 3, U.S. Department of Defense.

O'Hara, J.M., Brown, W.S., Baker, C.C., Welch, D.L., Granda, T.M. & Vingelis, P.J. (1994). *Advanced Human-System Interface Design Review Guideline*, Evaluation Procedures and Guidelines for Human Factors Engineering Reviews. Document NUREG/CR-5908 - BNL-NUREG-52333, Vol. 2. U.S. Nuclear Regulatory Commission, Upton: Brookhaven National Laboratory/Carlow International Inc.

Scapin, D.L. (1986). *Guide ergonomique de conception des interfaces homme-ordinateur*. Research Report N°77. Rocquencourt: INRIA.

Smith, S.L. & Mosier, J.N. (1986). *Design guidelines for the user interface software*. Technical Report ESD-TR-86-278 (NTIS No. AD A177198). Hanscom Air Force Base (Massachusetts): U.S.A.F. Electronic Systems Division.

Vanderdonckt, J. (1995). Accessing Guidelines Information with SIERRA. In Nordbyn, K., Helmersen, P.H., Gilmore, D.J. & Arnesen, S.A. (Eds.): *Proc. of the 5th IFIP TC13 Conf. on Human-Computer Interaction* (INTERACT'95, Lillehammer, June 25-29, 1995), pp. 311-316. London: Chapman & Hall.

Vanderdonckt, J. (1999). Development Milestones towards a Tool for Working with Guidelines. *Interacting with Computers*, 11 (4), to appear.

Mental effort and evaluation of user-interfaces: a questionnaire approach

Albert G. Arnold
Faculty of Technology, Policy and Management
Delft University of Technology
De Vries van Heystplantsoen 2
2628 RZ Delft
Tel: +31 15 2783752
Fax: +31 15 2782950
a.g.arnold@wtm.tudelft.nl

1 Introduction

'A newspaper published an article about a Dutch university. This university provided to its students access to INTERNET and other applications. From the article it becomes clear that the quality of the students' products did not really increase, but they achieved their goals with less effort. Therefore it was concluded that all these electronic gadgets did not make a difference.'

In our view this is a wrong conclusion! The mere fact that products with the same quality were produced with less user effort can be seen as a real benefit. From a work psychological perspective the performance of users should only be considered in combination with the effort users invest (Arnold, 1998). This is important because users are capable of achieving high levels of performance even with a user-unfriendly system if they are motivated to a high degree. However, the costs (in terms of user effort expenditure) of achieving these high levels of performance in a sub-optimal software environment might be quite high.

More specific, computer systems should accommodate the user's cognitive, perceptual, and motoric processes in such a way that the user's work activities are supported or facilitated (Roe, 1984; 1988). Herewith, the concept of *action facilitation* is introduced. Action facilitation may be operationalised as follows: 'An interactive computer system is said to facilitate user actions if its system characteristics enable the user to maintain or increase his/her task performance (effectiveness) under condition of a decrease in mental or physical effort' (Arnold, 1998). The concept of action facilitation is a part of a broader framework developed by Roe (1984; 1988) and is based on the action theoretical work of Hacker (1985; 1986) and Rasmussen (1986). One of the action facilitatioin research activities was the development of an interface evaluation approach (Zijlstra, 1993; Arnold, 1998). Within this approach the value of user's mental effort expenditure as a usability evaluation criterion has been proven.

2 Effort measurement

User effort is the amount of energy a user has to activate to meet the perceived task demands. The amount of energy one has actually to spend depends on the load of the task (mentally or physically), his/her coping strategy, the available knowledge and experience, and the psycho-physiological state of the worker. Basically, there are two ways to measure mental effort. First, one can apply the heart rate registration method. Three measurements can be inferred from the heart rate registration: the mean interbeat interval, heart rate variability related to blood pressure regulation, and heart rate variability related to temperature regulation. Heart rate variability is a good physiological indicator of mental effort expenditure (Zijlstra, 1993) The application of his kind of measurement should be left in the hands of experts. Especially, the pre-processing and statistical analysis of the data is rather complex and time consuming. Only in very specific circumstances the heart rate registration approach is indicated, e.g. in cases of systems with high health and safety risks.

Second, one can administer questionnaires. In particular, two questionnaires have successfully been applied in the earlier mentioned interface design and evaluation approach (Arnold, 1998): the Task Load indeX (TLX) and the Subjective Mental Effort Questionnaire (SMEQ). The TLX is a well known rating procedure developed by NASA (NASA Ames Research Centre, 1986). The TLX provides a broad view of the taskload experienced by users. Much research has been done with the TLX with positive results. A more focussed and much simpler questionnaire is the SMEQ (Zijlstra, 1993). This questionnaire is purely related to mental effort expenditure. The SMEQ is not so popular as the TLX and therefore more information on this subjective mental effort measurement tool is presented in this contribution.

The application of questionnaires to measure the workload or the mental effort expenditure is indicated in the evaluation of interfaces used in contexts, which do not contain safety and health critical factors, for example office environments. The use of these questionnaires provides valuable user information on a cost-effective basis.

3 Description of the SMEQ

The SMEQ is a paper-and-pencil scale (Zijlstra, 1993). It consists of a short instruction for the subjects (in which the goal and use of the SMEQ are described) and the rating scale itself. The SMEQ answer sheet uses one page of A4 size containing the question ('Indicate the amount of effort you invested in the task you just performed') and an answering device. The answering device is a vertical line of 150 millimetres length start with 0 ('no effort invested'). Along the line nine anchor points are printed to help individuals to estimate their own perceived effort. The subjects are invited to indicate their experienced effort by putting a mark on the vertical line. The scores on the SMEQ are given by the distance from the beginning of the scale (zero point) up to the heart of the subject's cross in millimetres. The time necessary to fill in the questionnaire by respondents comes to less then a minute.

Significant positive correlations have been found between the SMEQ scores and user performance measured in net task time, and also significant negative correlations between the SMEQ and the SUMI. It can be argued that a higher performance goes together with a higher expenditure of effort by the users. And when users have to expend more effort with a certain system they very likely dislike this system more. These outcomes have been obtained in controlled experiments conducted in the context of the ESPRIT project 5429 'Measuring Usability of Systems in Context (MUSiC)'[1].

The SMEQ can be used when a stable and working prototype of a system is available. In general, the SMEQ is administered to users when they have finished a particular task with an interactive system, which is subject to evaluation. In order to gather ecological valid results (i.e. results which are also valid in the users' daily context), it is recommended to develop a number standardised real-life tasks, to simulate important aspects of the user context, and to invite users belonging to the user target group. The following evaluation designs can be considered:

---

[1] In this project a number usability tools and measurement methods have been developed: context of use analysis method (Macleod *et al*., (1993), the formal analysis method (Bösser & Melchior, 1990), the performance measurement method (NPL, 1995), the cognitive measurement method (Houwing *et al*., 1993), and the user satisfaction measurement method (Porteous & Kirakowsi, 1992). MUSiC was a three-year project and started in November 1989.

- Comparison between packages with same functionality (e.g. in the case of purchasing).
- Comparison between different versions (e.g. in the case of interface development).
- Comparison between various tasks covering different functions of a package (e.g. in the case of a one-shot usability test).

4 Conclusions

The SMEQ is easy to administer and is a good indicator of mental effort expenditure together with performance achieved. The SMEQ can be used in usability laboratory settings as well as real life office situations. And therefore the SMEQ is a proper evaluation tool to use in a software development process and to estimate the users' effort preferably in combination with user performance. As soon as a working and stable prototype is available the SMEQ can be used.

It should be noted that subjective mental effort measurement is not the same as objective mental effort. In particular, in cases of user enjoyment the estimation of the own effort expenditure might be too low. When users are feeling good and are enjoying their interaction with the system they do not fully experience their effort expenditure. In these kinds of situations they are reporting less effort expenditure then objectively measured.

5 References

Arnold, A.G. (1998) *Action Facilitation and Interface Evaluation. A work psychological approach to the development of usable software*. Delft: Delft University Press.

Arnold, A.G. (1998) *Action Facilitation and Interface Evaluation: A work psychological approach to the development of usable software*. Delft: Delft University Press.

Bösser, T. & E.M. Melchior (1990) Cognitive modeling, Rapid Prototyping and User Centered Design with the SANE toolkit. Commission of the European Community (Ed.), *ESPRIT '90 Conference Proceedings*. Dordrecht: Kluwer, pp. 589-606.

Hacker, W. (1985) Activity: A Fruitful Concept in Industrial Psychology. In: M. Frese & J. Sabini (Eds.) Goal Directed Behavior: The Concept of Action in Psychology. Hillsdale, New Jersey: Lawrence Erlbaum Associates, Publishers.

Hacker, W. (1986) *Allgemeine Arbeits- und Ingenieurpsychologie, Psychische Struktur und Regulation von Arbeitstätigkeiten*. Bern: Verlag Hans Huber.

Houwing, E.M., M. Wiethoff & A.G. Arnold (1993) *Introduction to Cognitive Workload Measurement*. Delft: Laboratory for Work and Interaction Technology, Delft University of Technology.

Macleod, M., C. Thomas, A. Dillon, J. Maissel, R. Rengger, M. Maguire, M. Sweeney, R. Corcoran & N. Bevan (1993) *Usability Context Analysis: A Practical Guide, V3.1*. National Physical Laboratory DICT, Teddington, UK..

NASA Ames Research Centre, Human Performance Group (1986) *Collecting NASA Workload Ratings: A Paper-and-Pencil Package*. Moffe Field, CA:NASA Ames Research Centre.

NPL Usability Services (1995) *Performance Measurement Handbook*. Teddington, Middlesex, UK: National Physical Laboratory.

Porteous, M.A. & J. Kirakowsi (1992) *SUMI - Subjective Usability Measurement Inventory*. Cork: Department of Applied Psychology, University College Cork.

Rasmussen, J. (1986) *Information Processing and Human-Computer Interaction*. Amsterdam: North-Holland.

Roe, R.A. (1984) Taakstructuur en software-ontwerp. In: G.C. van der Veer & E. Lammers (Eds.) *Programmatuur naar menselijke maat*. Amsterdam: Stichting Informatica Congressen.

Roe, R.A. (1988) Acting Systems Design - an Action Theoretical Approach to the Design of Man-Computer Systems. In: V. de Keyser, T. Qvale, B. wilpert & S.A. Ruiz Quintanilla (Eds.) *The Meaning of Work and Technical Oprions*. Chichester: John Wiley & Sons.

Zijlstra, F.R.H. (1993) *Efficiency in Work Behaviour. A Design Approach for Modern Tools*. Delft: Delft Universitaire Pers.

User Satisfaction Measurement Methodologies: Extending the User Satisfaction Questionnaire

Gerard Hollemans
Philips Research N.V.[1]

1 Introduction

As usability is no longer a relatively unimportant add-on to consumer electronics products, attention is paid to the ease-of-use during the development of these products. As a consequence, testing for usability is becoming more and more important. Unlike for software, however, tools to assess the usability for consumer electronics products as perceived by the user are rare. A need was felt to construct a measurement instrument for subjective usability, i.e. usability as perceived by the user for the consumer electronics products specifically.

Since the measurement instrument is to be used in the development process in industry, the data delivered by the measurement instrument have to be formative, reusable and comparable over different usability test situations at low cost and high speed. A questionnaire was chosen as type of instrument. It was named the User Satisfaction Questionnaire (USQ).

In this paper the theoretical framework underlying the USQ is described and problematic aspects of the methodological model underlying the development of the USQ are discussed and possible solutions are suggested. Finally, a technique used for the administration the USQ is presented.

2 The theoretical framework underlying the USQ

A theoretical framework for the USQ is needed to guide the construction, selection and modification of items and the interpretation of the data resulting

---

[1] Part of the work described here was done at IPO, Center for Research on User-System Interaction that is part of the Eindhoven University of Technology.

from the use of the questionnaire. The framework used for the development of the USQ consists of **a set of usability related system qualities**, such as "Self descriptiveness" and "Learnability", **linked to a set of aspects of the user-system interaction**, such as "Feedback" and "Error recovery" (see table 1). In a small scale field survey HCI experts linked the system qualities to the aspects of the user-system interaction, thus defining the usability related system qualities and identifying the relevant combinations of system quality and interaction aspect (De Ruyter and Hollemans 1997). This methodology can also be used to facilitate the process of coming to shared definitions of usability dimensions in groups of people working together in HCI (related) projects.

Table 1: Part of the theoretical framework of the USQ

System qualities	Aspects of user-system interaction		
	Feedback	Error recovery	…
Self descriptiveness	•		…
Learnability	•		…
Efficiency	•	•	…
…	…	…	…

The provisional definition of satisfaction that goes along with the framework is: "Satisfaction is the degree to which the confidence of the user that (the interaction with) the product has qualities the user desires gives rise to a positive affection towards the product".

For every combination of system quality and interaction aspect that the expert judged to influence the usability of a product at least two items were gathered. Some items were adapted from existing questionnaires, but most items were phrased from scratch.

The next step was to filter the inadequate items, i.e. items that users are unable to respond to properly, from the pool of items that was constructed. In a series of experiments items that appeared not suitable for the USQ were removed, replaced, or modified (Hollemans and De Ruyter 1998). Currently a set of well-tested items is available that have proven to work under various circumstances and for various types of users. Validation and standardisation of the USQ remain to be done.

3 The methodology applied to construct the USQ

The theoretical framework that guides the construction of the USQ and the interpretation of the data have been described along with the steps that have been taken in the development of the USQ. However, the methodology used to construct the USQ has remained implicit.

3.1 Classical Testing Theory

Classical Testing Theory (CTT) has been applied to construct the USQ. In CTT the response to an item is modelled as consisting of a 'true' score plus some 'error'. The true score is considered to be the 'real' opinion of the respondent. The aim in developing a questionnaire is to maximise the size of the 'true' component. The ultimately reliable questionnaire measures the intended concept without error.

To assess the reliability of a questionnaire Cronbach's alpha is computed:

$$\alpha = \frac{n}{(n-1)}\left[1-\frac{\sum_{i=1}^{n}\mathrm{var}(X_i)}{\mathrm{var}(\sum_{i=1}^{n}X_i)}\right],$$

where *var(x)* is the variance of *x* and *n* is the number of items in the questionnaire. Cronbach's alpha can be considered as the ratio of the variance of the true score to the variance of the total score, i.e. the sum of all the observed scores (proof of this is beyond the scope of this paper and can be found in Traub 1994). As Cronbach's alpha is a generalisation of the test-retest reliability coefficient, it is necessary that the items of a questionnaire can be considered replications. In other words, a questionnaire is considered to be series of equivalent tests of a concept.

Since Cronbach's alpha takes all the separate variances of the items into account, removing one item from a questionnaire changes its reliability. The calculated reliability only holds for a questionnaire *in form in which it was tested.* Because mutual influences of items are not accounted for in the total score on a questionnaire, the established validity of a questionnaire too holds only for a questionnaire *in form in which it was tested.* It is impossible to predict what happens to the reliability and validity of a questionnaire if items are added or left out. This makes questionnaires developed using CTT inflexible, only to be used 'as is'. In practice, however, questionnaires developed with CTT are adapted often due to the time and other constraints present in the usability test. A solution to the inflexibility is clearly needed.

3.2 Item Response Theory

An alternative to CTT is Item Response Theory (IRT). In IRT the focus is not on the development of a questionnaire, but rather on the development of items. Furthermore, instead of a one to one relation between the real opinion of the user and the response, a likelihood relation is modelled:

$$\log\left(\frac{P_{nijk}}{P_{nijk-1}}\right) = U_n - L_j - D_i - S_k.$$

Consider a user that rates a video cassette recorder (VCR) 4 out of 5 on a certain item *x*. The IRT model describes that the log of the probability that this user *j* rates this VCR *n* 4 (=*k*) out of 5 divided by the probability that this user *j* rates this VCR *n* 3 (=*k-1*) out of 5 can be described by the usability U of this VCR *n* minus the leniency L (actually, the severity) of this user *j* minus the difficulty D of obtaining a high score on item *i* minus the difficulty S of stepping from a rating of 3 (*k-1*) to a rating of 4 (*k*).

There are many advantages to IRT compared to CTT. As items are developed independently from each other, IRT allows for adapting a questionnaire, adding and removing items as necessary. The choice of the items should be based on theory, as it is done in the construction phase of a questionnaire. Due to the form of the model, items are also developed independent of the sample of users used to estimate the parameters. What is more, the model gives information about the users. Finally, IRT allows for the construction of questionnaires that cover more than one concept. All of this makes it possible to build a item bank with a large series of established items from which for usability test a series of suitable items can be selected. Other advantages of IRT can be found in Wright and Stone (1979) and Linacre (1993).

The model given here is not the only one possible. Within the limits of the maximum number of parameters that can be estimated for a given sample size, virtually any model can be fitted to the data at hand. Currently included factors can be left out and other factors can be included if necessary.

4 Administration of the USQ

The administration of the USQ is, of course, directly dependent upon the format of the items. All items in the set of the USQ have the form of a statement to which the users can respond by indicating their level of agreement.

The most commonly used technique to obtain responses is to present the user with a five point rating scale at which the users are asked to mark their response. It can be frequently observed, however, that users are undecided about the point on the scale they want to mark. The hesitation may have two causes: The user is not sure what to respond, or the available response alternatives do not match the intended response well enough.

In those cases where the available response alternatives do not match the intended response the user is forced to choose a less than optimal response alternative. This implies that the observed response unintentionally deviates from the response the user intended to give.

To prevent this in the administration of the USQ, a slider (see also ITU-R BT.500-7 Rev. 1995) is presented to the users. Users can mark their response on the slider exactly at the place they want and are no longer forced to select a response from a limited set of discrete responses. The distance between the left side of the scale and the mark that the user made on the scale is considered to be the quantification of the response. Currently a clustering algorithm is being developed to counteract the noise in the data that is introduced by quantifying responses at a higher resolution than is used by the respondent.

5 Discussion

The results so far in the development of the USQ are promising. A series of tested items that have proven to work under various circumstances is available. If the remainder of the development of the USQ is cast in the framework of IRT, the questionnaire becomes a flexible tool that can be adapted to the needs of the researcher to assess users' satisfaction. Eventually the scope of consumer electronics products may be broadened to encompass other domains in the item bank as well.

The users are also equipped with flexible means to contribute to the satisfaction measurement. By letting the user mark their opinion on a slider, a continuous rating scale, the researcher allows them maximal freedom, thus enhancing the validity of the test results.

6 References

De Ruyter, B.E.R., & Hollemans, G. (1998). *Towards a User Satisfaction Questionnaire for Consumer Electronics: Theoretical Basis* [NL-TN 406/97]. Eindhoven: Nat.Lab. Philips Electronics N.V.

Hollemans, G., & De Ruyter, B.E.R. (1998) *Selection of Items for the User Satisfaction Questionnaire.* Eindhoven University of Technology: IPO-report 1185.

ITU-R Recommendation BT.500-7 (Revised) (1995). *Methodology for the subjective assessment of the quality of television pictures.*

Linacre, J. M. (1993). *Many-Facet Rasch Measurement.* Chicago: MESA Press.

Traub, R.E. (1994). *Reliability for the Social Sciences.* Thousand Oaks: Sage.

Wright, B.D., Stone, M.H. (1979). *Best Test Design.* Chicago: MESA Press.

WorkLAB – An interactive interface evaluation approach

Robert Baggen* and Rainer Wieland+
* TÜV Informationstechnik GmbH, Essen
+ Bergische Universität – Gesamthochschule Wuppertal, Projektgruppe MenBIT

1 The WorkLAB approach

The user interface evaluation approach presented in the following paper was inspired by the concept of ambulatory assessment proposed by Fahrenberg & Myrtek (1996). In their studies, subjects carried portable devices for continuous physiological recordings and recurring assessments of several subjective states in a natural setting. In software evaluation however, physiological recordings are seldom applicable since most indicators correlate only weakly with the details of a design. Here, the focus is not on the effect of working conditions on the user, but on the conditions themselves as imposed by the design of a user interface.

Another starting point for the development of the approach was the observation, that computers with graphical user interfaces control dialog windows, menus and direct manipulations etc. and so literally "know" something about the work that is done with them. Nevertheless, logfiles as classical method for interface evaluation are considered difficult to process and lack important information about the subjective side of interface usage.

In the WorkLAB approach, we thus combined logfile recording with the recurring assessment of subjective judgements on dialog quality to a method we called context-related inquiry (Figure 1). Whenever the logfile tool finds a dialog situation worth an assessment, the users are prompted to rate the dialog quality on some scales which were developed according to the interface standards DIN EN ISO 9241 Part 10 and 11. Besides the automated triggering of ratings, the users are encouraged to give dialog quality ratings in any moment they feel it is necessary.

The technical setup for the WorkLAB approach mainly consists of a normal computer workplace which is equipped with the logfile tool. The tool records the logfile and sends trigger signals to PROTEUS, a psychophysiological measurement system. PROTEUS offers the opportunity to put auditive questions with a soundcard and to collect answers via a special keyboard. It was developed by the MenBIT group at the University of Wuppertal and the ELK GmbH, Krefeld, Germany (Baggen, Fürth, Wieland-Eckelmann & Lieser, 1996; Wieland-Eckelmann, Baggen, Saßmannshausen, Schwarz, Schmitz, Ademmer & Rose, 1996).

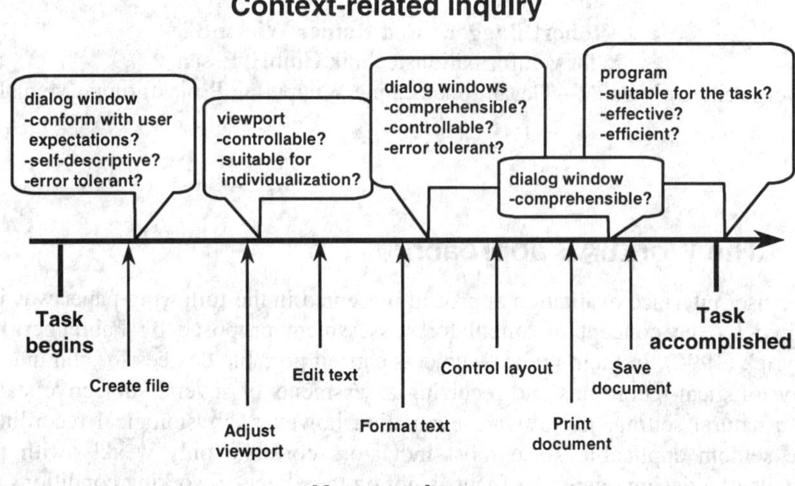

Figure 1: Context-related inquiry.

Since the inquiries can be taylored to almost any design level within a user interface, subjective ratings on almost any aspect of dialog quality are available. On the task level, user support and the concepts of effectiveness and effiency in task accomplishment can be evaluated. On the overall system level, ratings on the suitability for individualization as well learnability can be optained. Finally, on the dialog level most of the concepts from DIN EN ISO 9241 Part 10 as well as a wealth of information about the behavioural side of the user interaction are available.

As the last component in the WorkLAB approach, a display tool was developed for immediate processing of the data. The tool can be used by the evaluator to screen the results and to provide reference information for later discussion with the users.

2 Evaluation of a text processor

The WorkLAB approach was used in an evaluation of a text processor to check if context-related inquiry is a feasible way to test user interfaces and to develop a stock of reliable inquiry items. 24 professional secretaries and students participated in the study. They worked for four hours on four tasks that were representative for their daily use of the software.

On the global level, we found that the program was rated weakly positive by the participants. This result could be confirmed with ratings optained from the ISONORM questionnaire by Prümper & Anft (1993). But since there was no comparison with other software packages so far, this information was of limited value for design purposes.

On the task level, we optained significant differences amoung the four tasks only for the item "was the required amount of work reasonable?". It turned out that table editing was the task with the best score, whereas work with templates got the worst (Figure 2). Nevertheless, this information was also found relatively useless for redesign since no clues showed up about the cause of the problems.

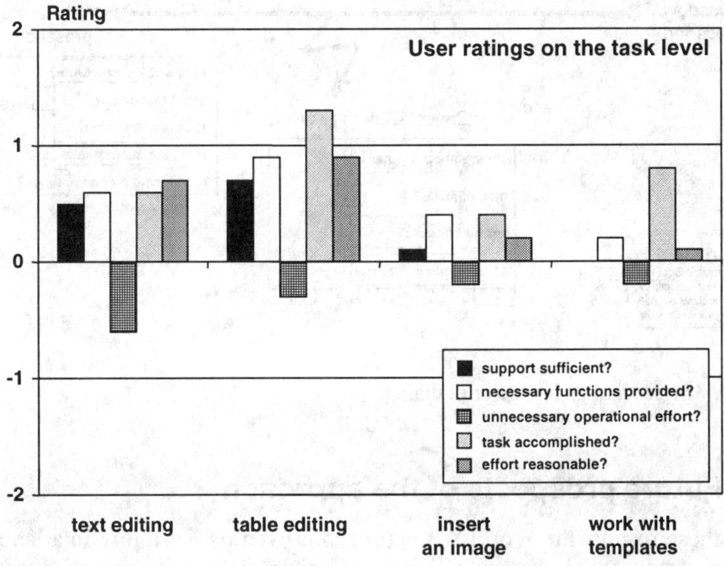

Figure 2: User ratings on the task level (-2=fully disagree..+2=fully agree).

Finally, on the dialog level we were able to evaluate some twenty dialog windows with at least 20 ratings each. Besides several others, we evaluated the print dialog of the word processor (Figure 3). Four different inquiries with up to six questions each were applied. These were presented at different times during the dialog: On opening of the window, after some interaction with the window, when the window was closed with "OK" and when it was interrupted with "Cancel".

Several problems could be identified for the print dialog. So the participants rated, that in their view there was no way to interrupt an ongoing dialog although it is simple to cancel the dialog with the appropriate button. The participants also complained, that errors resulting in a bad print were not shown. Modern text processors provide either a WYSIWYG-functionality or a preview that might be better linked to the print dialog.

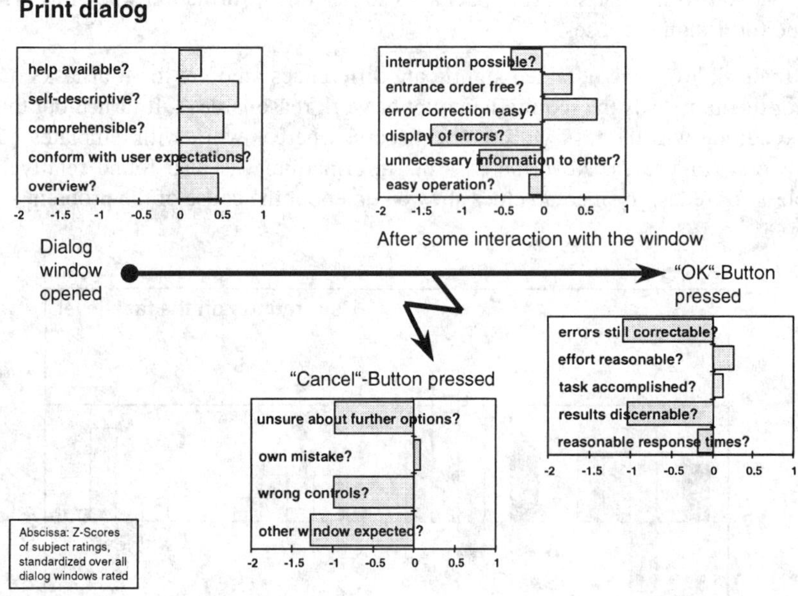

Figure 3: User ratings for the print dialog.

3 Future prospects of the approach

With these results, the WorkLAB approach proved its feasibility in a laboratory setting. As expected, the evaluation on the dialog level provided the most valuable results which could be immediately translated in design recommandations. Thus, it seems possible to support detailed software design with the use of such an adapted questionnaire method. Criteria for good, acceptable and poor design solutions can be raised easily. But of course, a complete coverage of all functionality of complex applications will still require some more development.

Requiring a complicated technical setup, the WorkLAB approach seems difficult to apply in projects with a tight schedule. But since the most time consuming preparation is the programming of the user interaction detection and recording, the easiest way to support the WorkLAB approach will be to integrate special methods into the class libraries of software development tools.

Given further improvements of the method, the WorkLAB approach provides interesting prospects for software evaluation in some future. Applied by users themselves, it seems to be a quite natural approach for the self-evaluation of interfaces because usability problems can be reported immediately from within the very dialog context. Since all information transfer can be automated, it can be useful for web pages or software which is distributed via the internet. WorkLAB software plugged into a web browser could provide the designer of web pages with much more information about his product than the simple concept of hit rates known today.

4 References

Baggen, R., Fürth, K., Wieland-Eckelmann, R. & Lieser, U. (1996) *PROTEUS audio - Das auditiv-physiologische Meßsystem. Programmier- und Referenzhandbuch, Bedienungsanleitung*. Göttingen: Hogrefe.

Fahrenberg, J. & Myrtek, M. (1996) (Eds.) *Ambulatory assessment. Computer-Assisted Psychological and Psychophysiological Methods.* Göttingen: Hogrefe & Huber.

Prümper, J. & Anft, M. (1993) Die Evaluation von Software auf der Grundlage des Entwurfs zur internationalen Ergonomie-Norm ISO 9241, Teil 10 als Beitrag zur partizipativen Systemgestaltung – ein Fallbeispiel. In K.H. Rödiger (Hrsg.): *Software-Ergonomie '93 – Von der Benutzungsoberfläche zur Arbeitsgestaltung* (S. 145-156). Suttgart: Teubner.

Wieland-Eckelmann, R., Baggen, R., Saßmannshausen, A., Schwarz, R., Schmitz, U., Ademmer, C. & Rose, M. (1996) *Gestaltung beanspruchungsoptimaler Bildschirmarbeit.* Schriftenreihe der Bundesanstalt für Arbeitsmedizin, Fb 12.002. Bremerhaven: Wirtschaftsverlag NW.

5 Adress for correspondance

Dipl.-Psych. Robert Baggen

TÜV Informationstechnik GmbH
Informationsgestaltung
Am Technologiepark 1
45307 Essen, Germany

Phone: ++49 201 8999 435
Fax: ++49 201 8999 444
Email: R.Baggen@tuvit.de

IsoMetrics: An usability inventory supporting summative and formative evaluation of software systems

Günther Gediga & Kai-Christoph Hamborg
Fachbereich Psychologie, Universität Osnabrück, Germany

1. Introduction

The IsoMetrics usability inventory provides an user-oriented, summative as well as formative approach in software evaluation on the basis of ISO 9241 Part 10. The summative version of IsoMetrics ($IsoMetrics^S$) gathers valid information about differences in the usability of software systems. Moreover the formative version of IsoMetrics ($IsoMetrics^L$) is a tool for supporting the identification of software weaknesses.

2. Design of the Inventory

The actual version of IsoMetrics comprises 75 items operationalizing the seven design principles of ISO 9241 Part 10. The statement of each item has to be assessed on a five point rating scale starting from 1 („predominantly disagree") to 5 („predominantly agree"). A further category („no opinion") is offered to reduce arbitrary answers. $IsoMetrics^L$ consists of the same items as $IsoMetrics^S$ and uses the same rating procedure. Additionally, each user is asked to give a second rating, based upon the following request:

"Please rate the importance of the above item in terms of supporting your general impression of the software."

This rating ranges from 1 („unimportant") to 5 („important"), and a further „no opinion" category may also be selected. Thus, each item is supplied with a weighting index.

To evoke information about malfunctions and weak points of the system under study, the question:

"Can you give a concrete example where you can (not) agree with the above statement?"

is posed. This gives users the opportunity to report problems with the software, which they attribute to the actual usability item. An example of one item of IsoMetrics^{L} is presented in Figure 1.

		Predominantly disagree		So - so		Predominantly agree	No opinion
S.6	When menu items are not available in certain situations, this fact is visually communicated to me.	1	2	3	4	5	
		Unimportant		So - so		Important	No opinion
Please rate the importance of the above item in terms of supporting your general impression of the software?		1	2	3	4	5	
Can you give a concrete example where you cannot agree with the above statement?							
... plenty more space ...							

Figure 1: Example item of IsoMetrics^{L}

This design provides information that can be used within an iterative software development. In summary, these are

- Scores of the usability dimension to measure the progress of development,
- Concrete information about malfunctions and their user-perceived attributes,
- Mean weight of any user-perceived attribute given a class of system malfunctions.

3. Reliability of IsoMetrics

The reliability of IsoMetrics was examined in two studies. The first study based on the evaluation of a R/3 based database system (Release 2.0; SAP AG) with the beta version of IsoMetrics. 111 subjects who had frequently used the application participated in this study. The second study concentrated on the replication of the results of the reliability analysis of the Beta-Version. A sample of 229 employees of the University of Osnabrück used IsoMetrics^{S} to rate the software they were frequently working with. The systems with more than 10 mentions were analysed, namely, LATEX (a text processing system for mathematicians), Pica (a system used by the university library), Word (Microsoft Word), and WinWord (Microsoft Word for Windows).

To justify that the reliability of usability scales are the same given different software systems, we tested the difference of Z-scores of the reliability. The results of this procedure show that the statistical quality of the scales is at least satisfactory and the replication in terms of reliability successful (Table 1).

	Software system				Construction
Scale	LATEX	Pica	WinWord	Word	R/3
Suitability for the task	.81	.32	*.33*	.69	.84
	(.85)	(.56)	(.53)	(.75)	
Self descriptiveness	.82	.83	.82	.80	.78
Controllability	.80	.79	.79	.76	.69
Conformity with user expectations	.41	.10	*.32*	*.44*	.71
	(.75)	(.47)	(.70)	(.75)	
Error tolerance	.49	.71	.63	.56	.75
Suitability for individualisation	.86	.81	.78	.76	.86
Suitability for learning	.14	.57	*.32*	*.37*	.84
	(.50)	(.80)	(.54)	(.54)	

Table 1: Reliability analysis within 5 different software systems

There were only a few significant differences due to three items in contrast to the construction set which showed a negative item-total correlation. These three items were eliminated from the scales for further analysis. Table 1 shows in brackets that the elimination of these items sufficed to adjust the reliability of the scales.

4. Validation of IsoMetrics

In order to validate the IsoMetrics inventory, we compared the scale means of the five different software systems analysed in the reliability studies (see above). Table 2 shows the summary results of the comparisons.

The scales "suitability for the task" and "conformity with user expectations" show no differences. This was to be expected, since the users had gauged those systems, which they were used to during their daily work. Even though a system reaction might not be plausible to a user, it still is an expected one – thus eliminating the differences between user expectation and system reaction. Therefore, the conformity with user expectation is given.

Scale	F(4, 230)	Tail-prob.	Significant Scheffé contrasts
Suitability for the task	2.31	n.s.	--
Self descriptiveness	12.34	<0.1%	{WinWord} > {Pica, Word, R/3} > {LATEX}
Controllability	7.98	<0.1%	{WinWord} > {Pica, Word, R/3, LATEX}
Conformity with user expectations	0.87	n.s	--
Error tolerance	3.78	<1%	--
Suitability for individualisation	11.55	<0.1%	{WinWord} > {Word}> {Pica, R/3}
Suitability for learning	4.46	<1%	{WinWord, R/3} > {Word}

Table 2: Comparison of 5 different software systems

Given the other dimensions of usability, we should expect some marked differences between the systems. As a replication we expect the GUI interface of WinWord to be rated better than the interfaces of the other systems. The DOS-based environment of the LATEX software should exhibit deficiencies in terms of „self – descriptiveness“, since its creators did not consider „suitability for learning“ a priority but the aspect of „customisability“ (adaptation of the system to the working environment by experts and not by users) and „tailorability“ (expert based adaptation of the software to user's needs on the workstation level). For the same reason it can be expected that R/3 shows comparatively high ratings in terms of „suitability for learning" and low ratings in terms of „suitability for individualisation“.

Although we can´t discuss the results in detail in this place, we find that the results validate the summative version of IsoMetrics: There are no differences (in terms of no medium effect size) in the scales where we do not expect them, and the differences we found are quite plausible.

5. Application of IsoMetrics

Gould (1988) considers continuous evaluation and modification within the software development process as one prerequisite to design usable systems. IsoMetrics supports this approach by getting quantitative and qualitative informations about deficiencies of the system from the users point of view and thereby permits its improvement.

IsoMetrics has been applied in several software development projects in which the inventory has proven its practicability. Given ten evaluators IsoMetrics^L evokes about one hundred remarks (the validity of this outcomes has been reported elsewhere, Willumeit et al., 1996). They are prioritized due to their rated importance and their frequency (Gediga et al., in press; Willumeit et al, 1996). The content analysis of the remarks results in weak points of the evaluated software providing the input to a usability review. In such a review users, software-engineers and human factors specialists develop remedies for the systems weak points and discuss its (re)design. The acceptance of this procedure could be proved on either side by the developers as well as the users.

A version of IsoMetrics which is applicable in group settings is discussed in Hamborg et al. (1999).

6. References

Gediga, G., Hamborg K.-C. & Düntsch, I. (in print). The IsoMetrics Usability Inventory: An operationalisation of ISO 9241-10. *Behaviour and Information Technology*.

Gould, J.D. (1998). How to Design Usable Systems. In M. Helander (ed.): *Handbook of Human-Computer-Interaction.* Amsterdam: Elsevier Science Publishers, pp. 757-789.

Hamborg, K.-C., Gediga, G., Döhl, M., Janssen, P. & Ollermann, F. (1999) Softwareevaluation in Gruppen oder Einzelevaluation: Sehen zwei Augen mehr als vier? In U. Arend & K. Pitschke (Hrsg.): *Software-Ergonomie `99. Design von Informationswelten.* Stuttgart: Teubner, pp. 97-109.

Willumeit, H., Gediga, G. & Hamborg, K.-C. (1996). IsoMetrics^L: Ein Verfahren zur formativen Evaluation von Software nach ISO9241/10. *Ergonomie und Informatik*, 27, pp. 5-12.

Tradeoffs in the Design of the IBM Computer Usability Satisfaction Questionnaires

James R. Lewis
International Business Machines Corp.

1 Introduction

Psychometrics is a well-developed field in psychology, and usability researchers began to use psychometric methods to develop and evaluate questionnaires to assess usability a little over ten years ago (Sweeney & Dillon, 1987). The goal of psychometrics is to establish the quality of psychological measures (Nunnally, 1978). Is a measure reliable (consistent)? Given a reliable measure, is it valid (measures the intended attribute)? Finally, is the measure appropriately sensitive to experimental manipulations? Here is a brief review of some basic elements of standard psychometric practice.

2 Brief Review of Psychometric Practice

Reliability goals. In psychometrics, reliability is quantified consistency, typically estimated using coefficient alpha (Nunnally, 1978). Coefficient alpha can range from 0 (no reliability) to 1 (perfect reliability). Measures of individual aptitude (such as IQ tests or college entrance exams) should have a minimum reliability of .90 (preferably a reliability of .95). For other research or evaluation, measurement reliability should be at least .70 (Landauer, 1988).

Validity goals. Validity is the measurement of the extent to which a questionnaire measures what it claims to measure. Researchers commonly use the Pearson correlation coefficient to assess criterion-related validity (the relationship between the measure of interest and a different concurrent or predictive measure). Moderate correlations (with absolute values as small as .30 to .40) are often large enough to justify the use of psychometric instruments (Nunnally, 1978).

Sensitivity goals. A questionnaire that is reliable and valid should also be sensitive – capable of detecting appropriate differences. Statistically significant differences in the magnitudes of questionnaire scores for different systems or other usability-related manipulations provide evidence for sensitivity.

Goals of factor analysis. Factor analysis is a statistical procedure that examines the correlations among variables to discover clusters of related variables (Nunnally, 1978). Because summated (Likert) scales are more reliable than single-item scales (Nunnally, 1978) and it is easier to present and interpret a smaller number of scores, it is common to conduct a factor analysis to determine if there is a statistical basis for the formation of summative scales.

3 Tradeoffs Considered in the Development of the IBM Questionnaires

Number of scale steps. The more scale steps in a questionnaire the better, but with rapidly diminishing returns (Nunnally, 1978). As the number of scale steps increases from 2 to 20, there is an initial rapid increase in reliability, but it tends to level off at about 7 steps. After 11 steps there is little gain in reliability from increasing the number of steps. The number of steps is most important for single-item assessments, but is usually less important when summing scores over a number of items. Attitude scales tend to be highly reliable because the items tend to correlate rather highly with one another. Reliability, then, usually is not a problem in the construction of summated attitude scales.

This turned out to be true in the case of the IBM questionnaires (Lewis, 1995). Coefficient alpha exceeded .89 for all instruments using 7-point scales. Coefficient alpha for a questionnaire using 5-point scales ranged from .64 to .93 and averaged .80. A related analysis using the same data (Lewis, 1993) showed that the mean difference of the 7-point scales correlated more strongly than the mean difference of the 5-point scales with the observed significance levels of *t*-tests. For these reasons, we currently use 7-point rather than 5-point scales.

Calculating scale scores. From psychometric theory (Nunnally, 1978), scale reliability is a function of the interrelatedness of scale items, the number of scale steps per item, and the number of items in a scale. If a participant chooses not to answer an item, the effect would be to slightly reduce the reliability of the scale in that instance. In most cases, the remaining items should offer a reasonable estimate of the appropriate scale score. From a practical standpoint, averaging the answered items to obtain the scale score enhances the flexibility of use of the questionnaire, because if an item is not appropriate in a specific context and users choose not to answer it, the questionnaire is still useful. Also, users who do not answer every item can stay in the sample. Finally, averaging items to obtain scale scores does not affect the statistical properties of the scores, and standardizes the range of scale scores, making them easier to interpret and compare. For example, with items based on 7-point scales, all the summative scales would also have scores that range from 1 to 7. For these reasons, we average the responses given by a participant across the items for each scale.

Unidimensional or multidimensional instrument. The developer of a questionnaire can have the goal of creating a unidimensional or multidimensional instrument (McIver & Carmines, 1981). A unidimensional instrument will typically require fewer items, so it will take less time to administer and provides a straightforward measurement because it has no subscales. A multidimensional instrument, because it measures several subscales related to the higher-level, overall scale, typically requires more items. For example, the System Usability Scale (Brooke, 199?), a unidimensional instrument, contains ten items. The PSSUQ, a multidimensional instrument that provides measurements for three subscales as well as the overall measurement, contains 19 items.

I actually can't claim that we set out to create a multidimensional instrument when we put together the first version of the PSSUQ. A group of usability evaluators selected the items on the basis of their comprehensive content regarding hypothesized constituents of usability. However, we have found its subscales to be informative and useful. For our purposes, this benefit clearly outweighs its slightly longer administration time relative to shorter instruments.

Control of potential response bias or consistency in item alignment. It is a common practice in questionnaire development to vary the tone of items so that, typically, half of the items elicit agreement and the other half elicit disagreement. The purpose of this is to control potential response bias. An alternative approach, less commonly used, is to align the items consistently.

Probably the most common criticism I've seen of the IBM questionnaires is that they do not use the standard control for potential response bias. Our rationale in consistently aligning the items was to make it as easy as possible for participants to complete the questionnaire. With consistent item alignment, the proper way to mark responses on the scales is clearer and requires less interpretive effort on the part of the participant. Even if this results in some response bias, typical use of the IBM questionnaires is to compare systems or experimental conditions. In this context of use, any systematic response bias will cancel out across comparisons.

I have seen the caution expressed that a frustrated or lazy participant will simply choose one end point or the other and mark all items the same way. With all items aligned in the same way, this could lead to the erroneous conclusion that the participant held a strong belief (either positive or negative) regarding the usability of the system. With items constructed in the standard way, such a set of responses would indicate a neutral opinion. Although this characteristic of the standard approach is appealing, I have seen no evidence of such participant behavior, at least not in the hundreds of PSSUQs that I have personally scored. I am sure it is a valid concern in other areas of psychology – especially some areas of clinical or counseling psychology, where the emphasis is on the individual rather than group comparisons. It is possible that constructing a usability assessment questionnaire in the standard way could lead to more item-marking errors on the part of sincere participants than the approach of consistently aligning items (although I know of no research in this area).

To norm or not to norm. When a questionnaire has norms, data exists that allows researchers to interpret individual and average scores as greater or smaller than the expected norm scores. In some contexts (field studies, standard single-system usability studies), this can be a tremendous advantage. In other contexts (multiple-system comparative usability studies, other types of experiments), it might provide no particular advantage.

When I performed the psychometric qualification of the CSUQ, I acquired a fair amount of data suitable for norms. I never published the norms because they were considered IBM Confidential. Those norms are now about 10 years out of date, and I no longer use them. The only instruments I know of that appear to have useful norms are those created by Kirakowski and his colleagues (Kirakowski & Corbett, 1993; Kirakowski & Dillon, 1988). Researchers should be cautious in the use of such norms, however, because differences between the contexts in which the norms were gathered and the use of the instrument could be misleading. Norms are of clear value in many situations, but it is important not to overgeneralize their applicability in usability evaluation.

4 Advantages of Using Psychometrically Qualified Instruments

Despite any controversies regarding decisions made in the development of such questionnaires, standardized satisfaction measurements (whichever questionnaire you choose to use) offer many advantages to the usability practitioner (Nunnally, 1978). Specifically, standardized measurements (even without norms) provide objectivity, replicability, quantification, economy, communication, and scientific generalization. Standardization also permits practitioners to use powerful methods of mathematics and statistics to better understand their results (Nunnally, 1978). The level of measurement of an instrument (ratio, interval, ordinal) does not limit permissible arithmetic operations or related statistical operations, but does limit the permissible interpretations of the results of these operations (Harris, 1985). Measurements using Likert scales are ordinal. Suppose you compare two products with the

PSSUQ, and Product A receives a score of 2.0 versus Product B's score of 4.0. Given a significant comparison, you could say that Product A had more satisfying usability characteristics than Product B (an ordinal claim), but you could not say that Product A was twice as satisfying as B (a ratio claim).

In conclusion, psychometrically qualified, standardized questionnaires can be valuable additions to practitioners' repertoire of usability evaluation techniques.

5 References

Brooke, J. (199?). SUS – A quick and dirty usability scale. Unpublished paper.

Harris, R. J. (1985). *A primer of multivariate statistics*. Orlando, FL: Academic Press.

Kirakowski, J., & Corbett, M. (1993). SUMI: The software usability measurement inventory. *British Journal of Educational Technology*, *24*, 210-212.

Kirakowski, J., & Dillon, A. (1988). *The computer user satisfaction inventory (CUSI): Manual and scoring key*. Cork, Ireland: Human Factors Research Group, University College of Cork.

Landauer, T. K. (1988). Research methods in human-computer interaction. In M. Helander (Ed.), *Handbook of Human-Computer Interaction* (pp. 905-928). New York, NY: Elsevier.

Lewis, J. R. (1993). Multipoint scales: Mean and median differences and observed significance levels. *International Journal of Human-Computer Interaction*, 5, 383-392.

Lewis, J. R. (1995). IBM computer usability satisfaction questionnaires: Psychometric evaluation and instructions for use. *International Journal of Human-Computer Interaction*, 7, 57-78.

McIver, J. P., & Carmines, E. G. (1981). *Unidimensional scaling*. Sage University Paper Series on Quantitative Applications in the Social Sciences, series no. 07-024, Beverly Hills, CA: Sage Publications.

Nunnally, J. C. (1978). *Psychometric Theory*. New York, NY: McGraw-Hill.

Sweeney, M., & Dillon A. (1987). Methodologies employed in the psychological evaluation of HCI. In *Proceedings of Human-Computer Interaction -- INTERACT '87* (pp. 367-373).

Test IT: ISONORM 9241/10

Jochen Prümper
FHTW - Fachhochschule für Technik und Wirtschaft, D-10313 Berlin
j.pruemper@fhtw-berlin.de

1 Introduction

„ISO 9241: Ergonomic requirements for office work with visual display terminals (VDTs), Part 10: Dialogue Principles“ (1995) is an official international standard and describes seven general ergonomic principles, which are independent of any specific dialogue technique; i.e. they are presented without reference to situations of use, applications, environments, or technology. The seven principles are as follows:

Table 1: The Dialog Principles of ISO 9241/10 (ISO 9241-10 1995, p. 5ff.)

Dialog Principle	Description
• Suitability for the task	A dialog is suitable for a task when it supports the user in the effective and efficient completion of the task.
• Self-descriptiveness	A dialog is self-descriptive when each dialog step is immediately comprehensible through feedback from the system or is explained to the user on request.
• Controllability	A dialog is controllable when the user is able to initiate and control the direction and pace of the interaction until the point at which the goal has been met.
• Conformity with user expectations	A dialog conforms with user expectations when it is consistent and corresponds to the user characteristics, such as task knowledge, education, experience, and to commonly accepted conventions.
• Error tolerance	A dialog is error tolerant if despite evident errors in input, the intended result may be achieved with either no or minimal corrective action by the user.
• Suitability for individualization	A dialog is capable of individualization when the interface software can be modified to suit the task needs, individual preferences, and skills of the user.
• Suitability for learning	A dialog is suitable for learning when it supports and guides the user in learning to use the system.

In order to analyse whether a software-system meets the dialog principles of ISO 9241-10, those principles must be characterized through a valid and reliable evaluation instrument. In this paper the software evaluation instrument ISONORM 9241/10 is presented; some reliability and validity results are discussed.

2 The questionnaire „ISONORM 9241/10"

The questionnaire „ISONORM 9241/10" was designed as an evaluation instrument that is economical to use. Therefore, each of the seven principles was operationalized by five items only. The questionnaire has a seven-tier, bi-polar question format. The answers range from „- - -" to „+ + +" (coded: 1-7). Filling out the questionnaire takes approximately 10 minutes. Figure 1 shows a sample item referring to the principle „suitability for the task".

Figure 1: Sample item from ISONORM 9241/10

The software ...	---	--	-	-/+	+	++	+++	*The software ...*
requires unnecessary inputs.	❍	❍	❍	❍	❍	❍	❍	does not require unnecessary inputs.

3 Reliability

In the following the results with regard to Cronbach's alpha and the re-test reliability are presented.

3.1 Subjects

1265 users have up to now used ISONORM 9241/10 in the evaluation of software. The average age of the subjects was 34.5 years, 51.8% were female, 48.2% male. The users evaluated 178 different software programs. Their general computer experience was on average 77 months and their experience with the evaluated software 25 months. The question „How well do you know the evaluated software?", was answered on a seven-point scale ranging from „very bad" (1) to „very good" (7) on average with 5.1.

3.2 Cronbach's alpha

As can be seen from table 1, the values for Cronbach's alpha with regard to the seven scales of ISONORM 9241/10 are satisfactory (scale means range between 4.4 and 5.3; standard deviations between 1.0 and 1.6).

3.3 Re-Test Reliability

To determine the re-test reliability, 49 users from the sample were asked at two times of measurement (on average after a period of 6.7 months) to evaluate the software they most frequently used. Total re-test reliability amounted to r = .77 (p < .001, N = 49). Taking into consideration the possible factors which could reduce the re-test reliability (e.g. expertise changing over time might have an influence on how the software is evaluated), the ISONORM 9241/10 evaluations can be assessed as being stable over time (see tab. 2).

Table 2: Reliability of ISONORM 9241/10
(⊙ N between 1208 and 1251; ⊠ N= 49, * p < .001)

Principle	**alpha**⊙	**Re-Test**⊠
• Suitability for the task	.81	.67*
• Self-descriptiveness	.86	.62*
• Controllability	.84	.64*
• Conformity with user expectations	.84	.60*
• Error tolerance	.87	.68*
• Suitability for individualization	.89	.63*
• Suitability for learning	.83	.59*

4 Validity

A first validity study was conducted by Prümper (1993). In this study it was shown that the user-friendliness of systems with a GUI was judged to be significantly better than those systems without a GUI, across all seven principles of ISO 9241-10. A second study includes correlating ISONORM 9241/10 with other software-evaluation instruments. For this purpose, two user-oriented questionnaires and one expert-evaluation inventory were used. The results will be presented in the following.

4.1 The User-oriented Questionnaires

The first questionnaire is a german translation of the „*QUIS*“ (long form; Shneiderman 1987) by Kinder (1991) and the second the „*BBD*“ from Spinas (1987). 31 users were requested to evaluate their software by means of ISONORM 9241/10 and the two questionnaires mentioned above. The users' general computer experience was on average 58 months, the experience with the evaluated software 18 months on average. The question „How well do you know the evaluated software?“ was answered on a seven-point scale ranging from „very bad“ (1) to „very good“ (7) on average with 5.6. As can be seen from table 3, ISONORM 9241/10 significantly correlates with the other two user-oriented questionnaires.

Table 3: Validity of ISONORM 9241/10 (N = 31 users, * p < .001)

	QUIS	BBD
ISONORM 9241/10	.73*	.71*

4.2 The Expert-Evaluation Inventory

EVADIS II is a comprehensive evaluation system to be used by experts in the field of software ergonomics (Reiterer & Oppermann 1993). For the validity study 13 different software systems were evaluated by a specialist in software ergonomics together with an experienced user of the software. Each evaluation session lasted approximately three hours. A subsample (N=383) of the full sample was used for this analysis. Each software was evaluated by 29.5 users on average, $N_{min.} = 11$, $N_{max.} = 79$. The correlation between EVADIS II and the mean ISONORM 9241/10 judgements was $r = .59$ ($p < .01$, N = 13 software programs).

This shows that the user-oriented questionnaire generates results similar to the outcomes of the expert-evaluation system, but with much less effort in time and money.

5 Discussion

In this paper the software evaluation instrument ISONORM 9241/10 was introduced, and some reliability and validity results were presented.

The reliability (Cronbach's alpha and re-test reliability) as well as the validity yielded satisfactory results. A further advantage of ISONORM 9241/10 is that it was constructed on the basis of an international standard. Therefore, it is likely to become internationally accepted. The questionnaire ISONORM 9241/10 is currently available in German (Prümper & Anft 1997), Dutch (Prümper & Anft 1998) and English (Prümper & Anft 1999). Further translations are in the offing.

6 References

ISO 9241-10 (1995). *Ergonomic requirements for office work with visual display terminals (VDTs) - Part 10: Dialogue principles.* Brussels: CEN - European Committee for Standardization.

Kinder, A. (1991). *Testen und Bewerten von Software durch Benutzer*. Gießen: Justus-Liebig-Universität Gießen (unpublished master thesis).

Prümper, J. (1993). Software-Evaluation based upon ISO 9241 Part 10. In T. Grechenig & M. Tscheligi (Eds.) *Human Computer Interaction* (p. 255-265). Berlin: Springer.

Prümper, J. (1997). Der Benutzungsfragebogen ISONORM 9241/10: Ergebnisse zur Reliabilität und Validität. In: R. Liskowsky, B.M. Velichkovsky & W. Wünschmann (Eds.), *Software-Ergonomie '97 – Usability Engineering: Integration von Mensch-Computer-Interaktion und Software-Entwicklung* (p. 253-262). Stuttgart: Teubner.

Prümper, J. & Anft, M. (1997). ISONORM 9241/10 - Beurteilung von Software auf Grundlage der Internationalen Ergonomie-Norm ISO 9241/10 (Questionnaire). In: Döbele-Martin, C., Martin, G., Richenhagen, G. & Prümper, J. (1997). *Ergonomie-Prüfer – Handlungshilfen zur ergonomischen Gestaltung der Bildschirmarbeit* (p. 103-113). Oberhausen: TBS.

Prümper, J. & Anft, M. (1998). *ISONORM 9241/10 - Beoordeling van software op basis van de Internationale Ergonomie-Norm ISO 9241/10* (unpublished questionnaire).

Prümper, J. & Anft, M. (1999). *ISONORM 9241/10 - Evaluation of Software based upon International Standard for Ergonomic Design ISO 9241/10* (unpublished questionnaire).

Reiterer, H. & Oppermann, R. (1993). Evaluation of user interfaces: EVADIS II - a comprehensive evaluation approach. *Behaviour & Information Technology*, 3, 137-148.

Shneiderman, B. (1987). *Designing the user interface*. Reading MA: Addison-Wesley.

Spinas, P. (1987). *Arbeitspsychologische Aspekte der Benutzerfreundlichkeit von Bildschirmsystemen*. Zürich: ADAG.

Smelcer, J.B. (1995). User errors in database query composition. *International Journal of Human-Computer Studies*, 42, 353-381.

Zapf, D., Brodbeck, F.C., Frese, M., Peters, H. & Prümper, J. (1992). Errors in working with office computers: a first validation of a taxonomy for observed errors in a field setting. *International Journal of Human-Computer Interaction*, 4 (4), 311-339.

Task Modelling for Database Interface Development

Tony Griffiths, Norman W. Paton, Carole A. Goble, Adrian J. West
Department of Computer Science,
University of Manchester, Oxford Road,
Manchester M13 9PL, United Kingdom
E-mail: {griffitt, norm, carole, ajw}@cs.man.ac.uk

1 Introduction

Although task modelling serves a pivotal role in the model-based paradigm for developing user interfaces, task models are often simplistic in their view of the modelled domain, and fail to address many of the fundamental processing requirements of real world applications.

The Teallach project is developing a model-based user interface development environment (MB-UIDE) which is primarily concerned with the development of user interfaces to applications built upon object databases, and as such its task modelling component pays particular attention to the information processing and linkage requirements of complex database applications. This paper describes the Teallach approach to modelling the tasks necessary to interact with an object database, and how these tasks can be used to reflect more accurately the often complex information flows between a user interface and its underlying application.

2 The Role of Task Modelling in MB-UIDEs

2.1 Task Model Scope and Semantics

There appear to be two general schools of thought about what a task model is and where it fits into the model-based development life-cycle. One school characterises task modelling as an early, goal oriented, design activity that is subsequently refined in the process of producing an executable prototype (Markopoulos et al. 1992), and the other school, which describes tasks in terms of "user actions including both motor steps and mental steps when performing tasks" (Szekely et al. 1996), thus focussing on a much lower level dialogue between human and computer.

The Teallach task model (TM) favours the latter of these two schools, assuming that a high-level task analysis has already been completed, and is hence outside its scope. The description of the tasks captured by the TM need not, however, descend to the very low-level dialogue favoured by the latter school, since low-level dialogue can be specified through the Teallach presentation model.

2.2 Model-based Development Life-Cycle

MB-UIDEs typically support a single fixed method for developing their component models, frequently stipulating that their task model must be constructed before any other model. This results in an inflexible interface development life-cycle, imposing a methodology that not all developers will be comfortable with.

Teallach attempts to circumvent these problems by minimising constraints on the order in which its three models (task, domain and presentation) can be constructed. As a result of this flexible method, Teallach does not impose restrictions upon the order in which links between related model components can be defined, thus providing a designer with the ability to either create the models in isolation from each other, or to incrementally build a design using the facilities provided by the three models.

3 Linking to the Underlying Application

Typically, a MB-UIDE's only interface with its underlying application is through its domain model (often represented in terms of an object model). Teallach's domain model (DM) is a rich representation of an object database application. This provides the TM with a representation of the underlying application meta data (i.e., its object model), and also a means of manipulating and querying the application data (through the standardised ODMG query language and programming language bindings) (Cattell et al. 1997). The tight binding between the TM and the DM's high-level view of the underlying application provide the UI designer with easy access to the services provided by both the object database and the underlying application. This provides the designer with a highly integrated view of the application domain in terms of its data and available functionality.

4 The Teallach Task Model

Many MB-UIDEs have concentrated on capturing the declarative semantics of user tasks through both task and domain-centric methods, yet have failed to provide comprehensive support for data flow between user and application, and internal data flow and state information. The following sections discuss how the TM represents these concepts through the use of a simple example centred around the task of connecting to a library database.

To successfully connect, the user must specify whether they are a librarian or a borrower, and provide their user name and password. Incorrect data will result in repetition of the task. From even this simple example, it can be seen that a task model should provide basic requirements for declaration of state information, information flow, and a linkage mechanism to the underlying application. The TM for this example and its associated DM are shown in figure 1. The TM shares its basic structure with the task models of many MB-UIDEs in that it is a task sub-task hierarchy organised by temporal relations, with its leaf nodes representing interaction or action tasks.

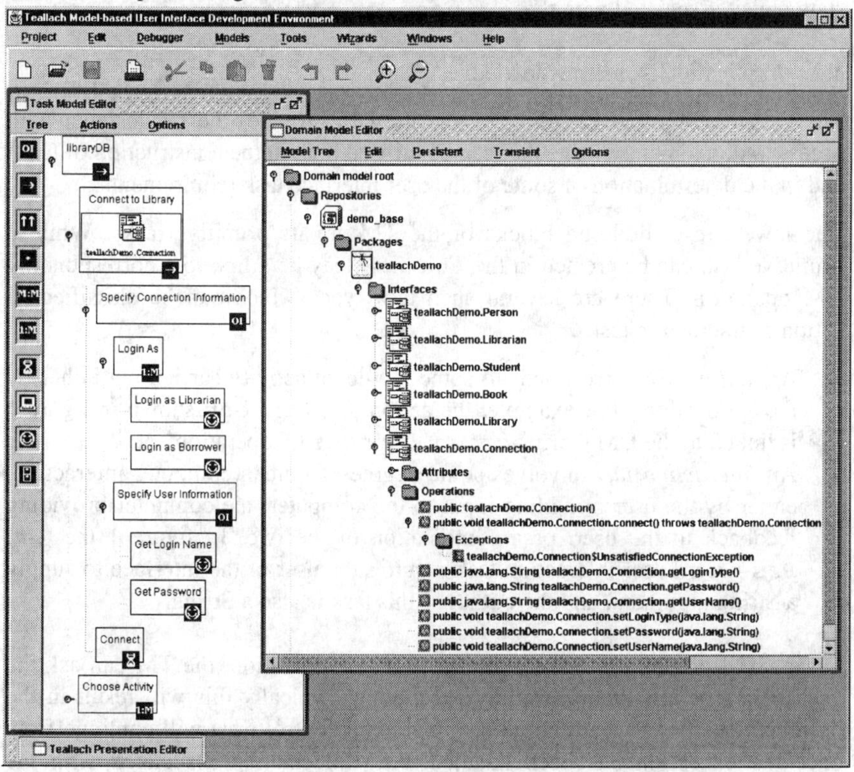

Figure 1: An Example Teallach Task and Domain Model

4.1 State Information and Information Parameters

The TM allows the declaration of local state, and the association of this state with a task. State information can be associated with any non-primitive task, with the associated task providing its scope. A state variable has type information corresponding to a DM class definition. Figure 1 shows that a state object of type `Connection` has been declared as state local to the `Connect to Library` task.

In the TM, information flow is specified by utilising a task's input and output parameters. In general, the output of a task may be linked to either one of the inputs of another task, or to an attribute of some state object. For example, the output String of the `Get Login Name` user interaction task is linked to the input parameter `setUserName(String)` operation of the `Connection` state object associated with the `Connect to Library` task.

4.2 Application Domain Knowledge

The tight binding between the TM and the DM provides the UI designer with access to domain knowledge that can be utilised both in the construction of tasks and in the determination of some of the user interface task requirements.

The lowest level building blocks of the TM are its primitive tasks. While a primitive task can be created in the TM, there may be a link to a corresponding DM operation. There are several such task types which can be classified as action or interaction tasks.

- An *action task* corresponds to some low-level activity carried out either by the application. For example, the action task `Connect` shown in figure 1 is linked to the DM `Connection.connect()` operation.
- An *interaction task* involves some degree of human-computer interaction, either by the user providing input to the computer, the computer providing feedback to the user, or a combination of the two. In figure 1 the `Get Password` user interaction task requires the user of the interface to supply a String argument, and the output of this task is also a String.

Once a TM primitive task been linked to a DM operation, the TM can ask the DM if the operation can raise any exceptions. Typically this will result in the designer specifying where in the task hierarchy control flow will continue from once the exception has been handled. Any information associated with the exception will be passed along the control flow. For example, the `Connect` task in figure 1 can be modelled to handle the exception thrown by the linked `Connection.connect()` DM operation. The designer can specify that if an exception is raised, then the TM should return to the `Connect to Library` task, as shown in figure 2.

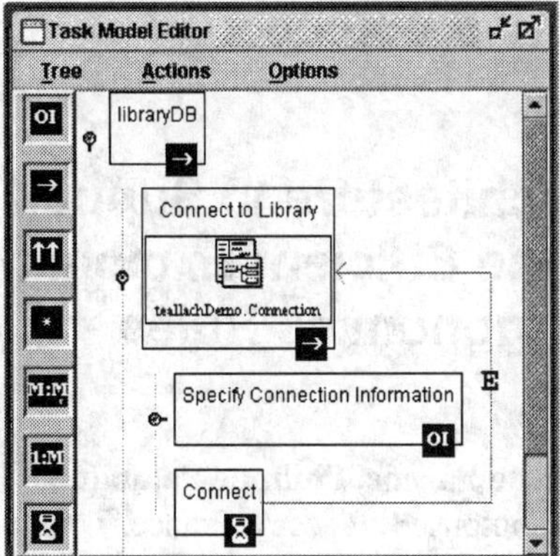

Figure 2: Declaring Exceptions in the Teallach Task Model

5 Conclusions

This paper has presented an overview of the Teallach task model, and has described how, using model-based techniques, it can capture the often complex information processing requirements and human-computer interaction necessitated by user interfaces to object database applications. Through its tightly coupled domain and task models, Teallach can easily and transparently access domain knowledge and functionality provided by the underlying application. This allows Teallach to model more accurately the information processing requirements of an application user interface.

Acknowledgements: This work is funded by the UK EPSRC. We also thank our partners on the Teallach project for their contributions to the development of the overall Teallach system. They are Peter Barclay, Richard Cooper, Phil Gray, Jessie Kennedy, Jo McKirdy, and Michael Smyth.

6 References

Cattell, R. et al. (1997). *The Object Database Standard: 2.0.* Morgan Kaufmann.

Markopoulos, P., Pycock, J., Wilson, S. and Johnson, P. (1992). Adept - A task based design environment. In *Proceedings of the 25th Hawaii International Conference on System Sciences*, IEEE Computer Society Press, pp. 587-596.

Szekely, P., Sukaviriya, P., Castells, P., Muhtkumarasamy, J., Salcher, E. (1996). Declarative Interface Models For User Interface Construction Tools: The MASTERMIND Approach. In Wierse, et al. (Eds.): *Engineering For Human-Computer Interaction. IEEE Visualization '95 Workshop*. Springer.

An Extensible Architecture to Support the Structuring and the Efficient Exploitation of Ergonomics Rules

Christelle Farenc, Philippe Palanque
L.I.H.S., Université Toulouse I, Place A. France, Toulouse, France.
Tel: +33 - 05 61.63.35.88, E-mail:{farenc, palanque}@univ-tlse1.fr

1 Introduction

Incorporating Human Factors in the development life cycle of interactive system has been the corner stone of human computer interaction research for many years. One possible way for such an incorporation is an explicit use of Human Factors expressed in terms of ergonomic rules. These ergonomic rules (also called guidelines, human-computer interaction principles, design rules or maxims) correspond to the explanation of the ergonomic knowledge dedicated to interactive systems.

An ergonomic rule can be considered as a principle that has to be taken into account for the building or the evaluation of user interfaces (UI) in order to respect cognitive capabilities of users. The scope of these principles is usually not general and vary according to the "context of use" that can be as different as tasks, user models, user environment (organizational aspects), etc.

Ergonomic rules are supposed to help developers to build UI respecting human factor principles. Unfortunately, studies carried out with designers show that guidelines are difficult to apply at design time (Smith 1988), (De Souza 1990). The difficulties encountered by designers is mainly due to the structuring of the guidelines and the way they are formulated. Indeed, most of the developers consider a UI in terms of input/output whatever the development methodology used, and thus need to have recommendations structured according to the components used for building the UI. On the opposite, ergonomic rules are expressed in terms of ergonomic criteria, cognitive principles, etc. This discrepancy between ergonomic knowledge expression and developers' needs is at the basis of difficulties found by developers in order to embed this knowledge in

UI. This is due to the fact that, at the origin, ergonomic rules were dedicated to people with skill knowledge in cognitive science and ergonomics. At present time these rules are widely available and developers want to use them.

This paper presents a method for structuring ergonomic rules in order to make them usable by developers. Indeed, in order for an artifact to be usable, it must be designed according to the way users use it (Norman 1988), thus this structuring organizes ergonomic rules according to developer's use. The next section of the paper presents the method for structuring ergonomic rules. This method is then applied to build a full evaluation method called ERGOVAL. Section 3 presents a discussion about the possibility of using ERGOVAL for automatic evaluation of user interfaces. The interests and the scope of this automatic evaluation is discussed. Due to lack of place, the example of the evaluation method and the discussion can not appear in this article. However, the example is available on request by email.

2 Putting Ergonomic Rules at Developers' Disposal

The aim of this section is two fold: first to propose a set of rules for structuring ergonomic rules in order to make them usable by developers with limited knowledge in ergonomics or cognitive science and then to present an evaluation method ERGOVAL structured according to the previous rules.

2.1 Rules for Structuring Ergonomic Rules

The first rule. In order to provide information to developers in a way they can handle it, it is mandatory to structure it according to developers' knowledge. The first rule is thus to produce a structure of the knowledge of the developers. The main part of developers' knowledge that can be embedded in ergonomic rules concerns the interaction objects (IO), thus structuring developers' knowledge consists in organizing IO.

The second rule. This rule is to use the canvas produced by the first rule in order to structure ergonomic rules. This only corresponds to rephrase and instantiate rules according to the IO structure produce in the previous phase.

The third rule. This rule must be considered at a different level form the previous ones. Its aim is not to organize or structure information but at the opposite to structure the evaluation process in order to provide results that will be usable directly by developers. This means that the recommendations have to concern IO and not any more usability criteria. This rule can be applied directly by interpreting the previous structuring.

These three rules for structuring and using the ergonomic rules can be applied to any set of ergonomic rules. However as stated in rule one, these rules heavily rely on interface presentation and thus the products of these rules cannot be considered as general. For example the same rules cannot be applied for user interfaces with direct manipulation or WIMP dialogue styles.

2.2 The ERGOVAL Method

We have applied this method for structuring ergonomic rules for WIMP interfaces and built an evaluation method called ERGOVAL. This method is dedicated to the evaluation of Graphical User Interface developed in the Windows environment. The method has been designed according to the rules presented in section 2.1.

The first rule. We first have to build a representation of the UI in terms of the interaction objects of the norm CUA that we call the decomposition of graphical objects. The inclusion mechanisms is at the basis of this decomposition thus all the objects are linked according to the relation composed of. Objects may also entertain other kind of relationships such as: activation, chaining, aggregation, positioning and semantics. for instance. A subset of the structure of all the IO of the norm CUA is presented in Figure 1.

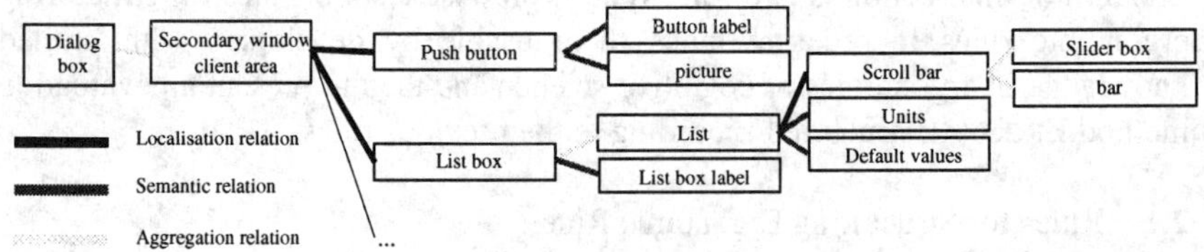

Figure 1 : Extract of the structural decomposition

The main principle of building this structure is only to take into account all the interaction objects even though there is no ergonomic rule that can be applied to them. Indeed, this is out the scope of this stage. It is important to notice that part of the interface will be represented within objects by attributes. For instance, the title of a menu option has an attribute *opening* which value is "..." if a dialogue box will be opened when the option is selected.

The second rule. This stage aims at refining ergonomic rules according to the IO structure built at the previous stage. Thus each ergonomic rule is associated to a set of IO. As input we have gathered ergonomic recommendations coming from different research work that can be found in the literature: (Bastien 1991), (Vanderdonckt 1994).

These recommendations were selected according to two main criteria:

- a good level of accuracy i.e. these recommendations are refined enough and are thus close to IO,

- a good covering of the various elements involved in ergonomic expertise, namely: the diversity of objects involved : lexical, syntactic, pragmatic, semantics, levels and ergonomic design principles. As for the pragmatic level, only guidelines that do not require in-depth analysis of the task were incorporated. However, this is only due to the fact that automation of the evaluation is considered. Otherwise, such rules would have been introduced.

All the ergonomic rules have been reformulated according to the IO structure. This reformulation process consists in :

- *finding all the IO* of the decomposition that are concerned by the ergonomic rules,
- *verifying that ergonomic recommendations are not redundant.* This verification is very important as ergonomic recommendations come from different guidelines.
- *verifying that ergonomic recommendations are not conflicting.* Conflicts only occur when high level rules are considered. When refinement is done (for example by clarifying context) conflicts are removed.

The third rule. As this rule concerns the evaluation process, we have decided (in the ERGOVAL method), to define another knowledge structuring. All IO concerned by the same set of recommendations are grouped together into classes of objects: ***the typology of graphic objects.***

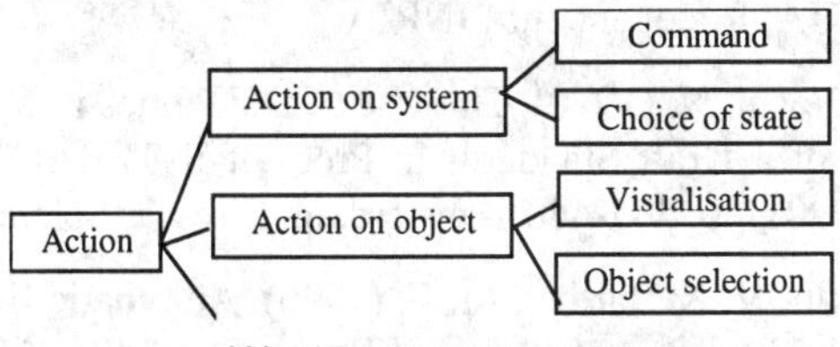

Figure 2: An extract of the typology.

Optimization problem. This typology has been done by grouping (in classes) all the objects that are interested in the same set of rules. The resulting typology consists in several levels of abstraction, the graphical objects being the leaves of this typology. Figure 2 presents an extract of this typology. The entire typology can be found in (Farenc 1997). The links presented in Figure 2 are *types of* links which have hierarchical properties, i.e. each type inherits attributes from the parent type and associated rules. In this way, the *action on system* type inherits attributes from the *action* type and the *command* type inherits attributes from the *action on system* type.

An object may belong to several types. For instance, an object icon-button (called *graphic key* in the CUA terminology) belongs both to the *command* and *choice of state* types and therefore inherits attributes from those types.

By structuring objects and rules it this typology allows:

- to reduce the design deviation search time - all rules associated to types to which the object under evaluation does not belong are not explored;
- the completeness and cohesion of the base to be improved by matching types to rules;
- at implementation level the maintainability of the rule base to be increased, particularly by reducing the number of rules to be implemented - rules are only implemented once, at the highest possible level of abstraction of the typology, and are then inherited by lower level graphical objects.

As far as the development of the rule base is concerned, it is assisted by the typology which offers two entry points - graphical objects and ergonomic recommendations.

3 References

Bastien, C. & Scapin, D.L. (1991). Critères ergonomiques pour l'évaluation des interfaces utilisateurs: définitions, commentaires, justifications et exemples. Research report INRIA, Rocquencourt: INRIA.

De Souza, F. & Bevan, N. (1990). The Use of Guidelines in Menu Interface Design: Evaluation of a Draft Standard. In Proc. of INTERACT'90 (Cambridge, August 27-31, 1990), pp. 435-440, Amsterdam: Elsevier Science Publishers.

Farenc, C., Liberati, V. & Barthet, M.-F. (1996). Automatic Ergonomic Evaluation: What are the Limits? In *Proc. of 2nd International Workshop on Computer-Aided Design of User Interfaces* (CADUI'96, Namur, June 5-7, 1996), pp. 159-170. Namur: Presses Universitaires de Namur.

Farenc, C. (1997). L'évaluation Ergonomique des Interfaces Homme-Machine, Ph.D. thesis. Toulouse: Université Toulouse I.

Norman, D. (1988). *The psychology of everyday things*. Harper and Collins.

Smith, S.L. (1988). Standards Versus Guidelines for Designing User Interface Software. In Helander, M. (Ed.): *Handbook of Human-Computer Interaction*, pp. 877-889. Amsterdam: North-Holland.

Vanderdonckt, J. (1994). *Guide ergonomique des interfaces homme-machine*. Namur: Presses Universitaires de Namur. ISBN 2-87037-189-6.

Assisting Designers in Developing Interactive Business Oriented Applications

Jean Vanderdonckt
Université catholique de Louvain, Institut d'Administration et de Gestion
Place des Doyens, 1 - B-1348 Louvain-la-Neuve, Belgium
Phone: +32-(0)10-47 85 25 - Fax: +32-(0)10-47 83 24
E-mail: vanderdonckt@qant.ucl.ac.be, vanderdoncktj@acm.org
URL: http://www.qant.ucl.ac.be/membres/jv/jv.html

1 Introduction

Up to some recent years, many software tools, techniques and methods have been developed to systematically produce parts or whole of a user interface for a target interactive application. For example, software tools that automatically generate a working user interface have been demonstrated feasible and operational in the domain of interactive business oriented applications. These days, there is a shift of focus (Vanderdonckt 1996) from automated generation of user interfaces (where the process is blindly driven without human intervention) to computer-aided design of user interface (where the task analyst, the designer, the developer, the evaluator are more active in the production process by deciding options, by driving the process, and by being assisted by software tools and guided by methods in this process).

This paper will primarily focus on assisting designers in a particular task involved in the global development life cycle: the presentation design. A visual, argumented, manipulable and computer-aided technique is presented to assist designers in cooperating with the system to build a first sketch of the presentation. This technique has been implemented in the SEGUIA tool (Système Expert Générant une « User Interface » de manière Assistée).

2 The SEGUIA Presentation Generator

From a **descriptive** viewpoint, SEGUIA consists in a software that automatically produces the presentation of a Windows-based user interface of a business oriented application (e.g., data management, database systems, office automation). This user interface is composed of a main application window with child windows and dialog boxes. The main application window comes with a menu bar and pull-down menus and the other compound objects are filled with traditional interaction objects (e.g., edit box, list box, combination box). Since the produced user interface is workable, it can be shown, executed and tested by anyone. SEGUIA is the software that supports the presentation design in the TRIDENT methodology to develop highly interactive business oriented applications (Bodart *et al.* 1995).

From a **technical** viewpoint, it is an expert system built on top of the AION/DS expert system shell marketed by Trinzic Software (Trinzic 1993). This expert system shell initiates an inference engine capable of processing rules contained in a knowledge base in a forward or backward chaining approach. Here, the forward chaining is exploited to process ergonomic rules, sometimes called guidelines, (Smith and Mosier 1986) to progressively produce a final user interface.

From a **methodological** viewpoint, SEGUIA follows a model-based approach (Puerta, 1997; Puerta, 1999) to derive a presentation from two models: a data model augmented with meta-information (practically, an entity-relationship model detailed with additional properties for attributes) and an activity-chaining graph model (practically, a model which specifies how the data flow is regulated among semantic functions belonging to the semantic core of an application).

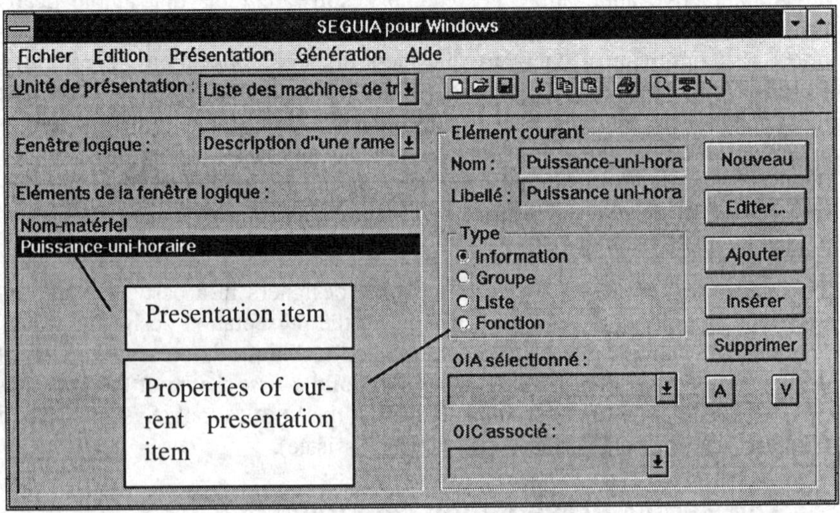

Figure 1. The SEGUIA environment showing properties of a presentation item.

From a **procedural** viewpoint, the attributes of entities and relationships, along with their meta-information (fig. 1), are first considered to select appropriate Abstract Interaction Objects (AIOs). Since AION/DS allows the dynamic creation of objects, each selected AIO is mapped onto a physical Concrete Interaction Object (CIO) of the Windows environment. Then, the activity chaining graph and other properties are analyzed to derive the positioning, the size and the arrangement of individual CIOs into a compound CIO (e.g., a dialog box or a child window). Two placement algorithms have been developed for this purpose. The processing of each algorithm produces the physical value of previously created CIO, such as location, height and length, color, font, selection properties, default values and so on (fig. 2).

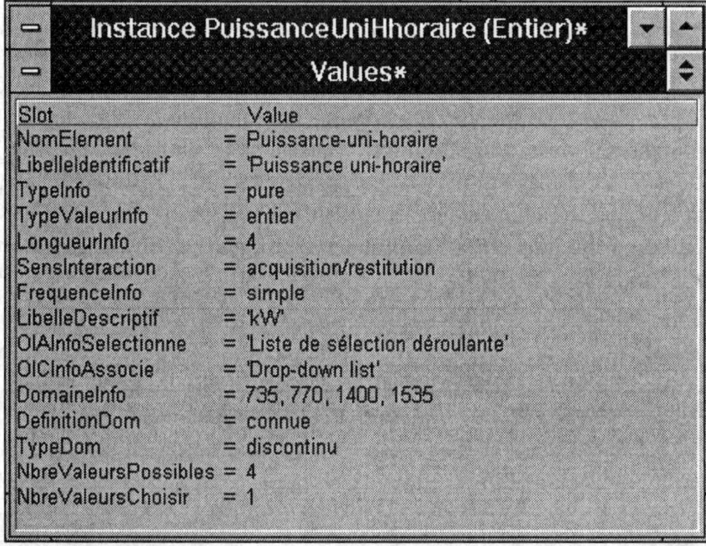

Figure 2. Generated properties for a presentation item with AIO and CIO.

Finally, this dynamically created user interface can be edited in a direct manipulation graphical editor to edit any property of generated CIOs. After that, the final results are converted and stored as Windows resource files to be used in a Windows-based development environment (e.g., Borland Delphi or Microsoft Basic V5.0).

3 Automatic versus Computer-Aided Design

The generation process can be executed in two ways:

1. in an automated way: the designer is supposed to provide the complete information required by the starting models (i.e., the entity-relationship model

and the activity-chaining graph) and to launch the generation within SEGUIA. This process is therefore considered as *single phased* (the designer launches the generation by selecting the appropriate menu item in SEGUIA) and *blind* (the designers does not see any intermediate results, only the final presentation can be finally seen);

2. in a computer-aided way: the designer is supposed to progressively input and refine the information required by the starting models so that the software can take advantage of information as it is provided by the designer. The software then interacts with the designer to progressively produces an intermediate solution for both selection of AIOs and positioning of CIOs. This process is therefore considered (fig. 3) as *three phased* (an automatic generation of a starting proposal from existing information, a phase of progressive refinement where the designer is presented with several options to decide from, and a phase considering the task context by exploiting domain properties), *visual* (the current working presentation is generated on the fly and presented during generation to the designer), and *justified* (each design option is accompanied with some argumentation explaining what the advantages and the inconveniences are on the usability). This last justification is based on the guidelines that just served to generate the ongoing presentation (Vanderdonckt 1999).

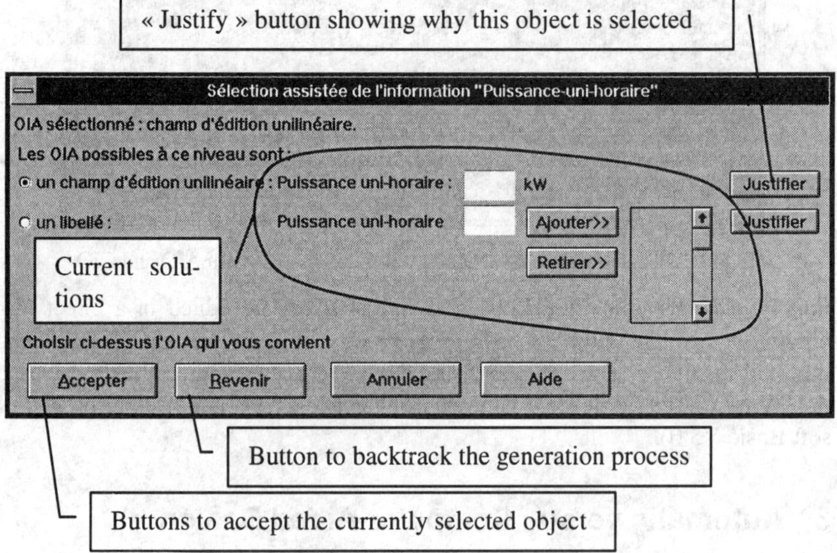

Figure 3. Computer-aided selection of an Abstract Interaction Object.

4 Conclusion

The big win of SEGUIA is that any change in the underlying models can be automatically reflected into a similar change in the generated presentation. Having the underlying models fully specified, an automated generation requires an average time of 15 minutes, mostly due to the manual operations. A regeneration of a previously generated presentation takes at most 10 minutes. The proposed approach can therefore reduce the cost of producing a first presentation, but also cuts the time every time the models are evolving. Of course, the final presentation should be interpreted as a first sketch to be further adapted to user's needs. If time and resources are insufficient, this first proposal is a workable user interface that can be incorporated in a final Windows application. It seems that designers are appreciating automated generation and computer-aided design of such a presentation unequally. When resources are tight, the "blind" process is launched; when time constraints allow some flexibility or when the designer wants to be more involved in the process, the computer-aided process is preferred. This conclusion seems to confirm that fully automated generation tend to be replaced by a designer-controllable process, which is more desirable.

5 References

Bodart, F., Hennebert, A.-M., Leheureux, J.-M., Provot, I., Vanderdonckt, J. & Zucchinetti, G. (1995). Key activities for a development Methodology of Interactive Applications. In Benyon, D. & Palanque, Ph. (Eds): *Critical Issues in User Interface Systems Engineering*, pp. 109-134. Berlin: Springer-Verlag.

Smith, S.L. & Mosier, J.N. (1986). *Design guidelines for the user interface software*. Technical Report ESD-TR-86-278 (NTIS No. AD A177198). Hanscom Air Force Base (Massachusetts): U.S.A.F. Electronic Systems Division.

Puerta, A.R. (July 1997). Model-Based Interface Development Environments. *IEEE Software*, ?(?), ?-?.

Puerta, A.R., *??*. In Proc. of ACM Conference on Human Aspects in Computing Systems (CHI'99, ??), New York: ACM Press, to appear.

Trinzic Corp. (1993). *AION Development System (ADS) V6.04*, General Reference. Palo Alto: Trinzic Corporation.

Vanderdonckt, J. (Ed.) (1996). *Computer-Aided Design of User Interfaces, Proc. of the 2nd Int. Workshop on Computer-Aided Design of User Interfaces* (CADUI'96, Namur, June 5-7, 1996). Namur: Presses Universitaires de Namur.

Vanderdonckt, J. (1999). Development Milestones towards a Tool for Working with Guidelines. *Interacting with Computers*, 11 (4), to appear.

Human-Centered Model-Based Interface Development

Angel R. Puerta
RedWhale Software and Stanford University
192 Walter Hays Drive
Palo Alto, CA 94303 USA
puerta@redwhale.com - http://www.redwhale.com

1 Introduction

Traditionally, model-based interface development environments have been designed with the goal of supporting an *automated process.* This process generates a working interface from a declarative interface model and with minimal user interaction (Vanderdonckt and Bodart 1993; Johnson, Wilson et al. 1994; Puerta and Eriksson 1994). In doing so, these environments enforce an engineering-oriented approach on their users (generally user interface designers). It is somewhat paradoxical that being user-centered design methodologies so well-accepted as sound approaches to user interface design, the tools in the model-based approach that support user-centered interface design are not *themselves* user-centered in nature. Instead, these tools force designers to follow a task flow that is not natural to them and that does not correspond to their normal design flow.

Over the past two years, our group has been developing a new-generation model-based environment called MOBI-D (Model-Based Interface Designer) (Puerta 1997) that has as one of its central goals to support a *human-centered* design process. In this paper, we enumerate the components and principal features of MOBI-D and present a number of design principles that we believe essential in order to enable a human-centered task flow in model-based interface development environments. We then describe how each of the tools in MOBI-D attempts to follow such design principles. The result is a suite of tools that supports designers in their regular activities without forcing significant changes in how those designers complete their tasks.

2 MOBI-D

The Model-Based Interface Designer (Puerta 1997) is an integrated environment for the design of user interfaces based on declarative interface models. An *interface model* is a knowledge representation of all the relevant aspects of a user interface. Interface models are made up of component submodels. These submodels typically represent collections of interface elements. An interface model in MOBI-D has components for user tasks, domain objects, user types, presentation elements, and dialog elements. A comprehensive description of the MOBI-D interface models has been reported elsewhere (Puerta 1996).

MOBI-D supports designers throughout a user-centered process for interface design. This process includes phases for the elicitation of user-task models, the definition and refinement of task-domain-user model components, and the design of layouts and dialogs. Individual tools within MOBI-D are normally aimed at one of the phases. At all times, the actions of designers are recorded into an interface model, thus the complete design process can be viewed as the evolution of an interface model, component by component. The following tools are the most important ones within the MOBI-D framework.

- The user-task elicitation tool (U-TEL) (Tam, Maulsby et al. 1998). This tool allows interface designers, to create an informal outline of the user task that the target interface will support. The outline gives an organizational overview of a user-task model. It is normally accompanied by lists identifying the main user types and domain objects that the interface will support. Users of U-TEL can begin construction of the outline directly in the outline editor or by extracting from text relevant information.
- A set of interface model editors (Puerta and Maulsby 1997). MOBI-D includes a specific editing tool for each of the components in an interface model, as enumerated above. The editors allow the complete specification and visualization of all the elements and attributes of a component. Normally, the outlines created in U-TEL are refined into user-task, domain, and user model components using these editors.
- The task-interface model mapper (TIMM) (Puerta and Eisenstein 1999). This tools uses an knowledge base of interface design guidelines to assist designers in selecting appropriate widgets for each object and datum to be accessed or displayed on the interface.
- A model-based interface builder (MOBILE) (Puerta, Cheng et al. 1999). This tool allows designers to define the layout and window navigation schema for the interface. It differs from conventional interface builders in that all design activities in the tool are linked to the existing user-task

model. In this manner, designers are always aware of how each widget in the interface relates to specific subtasks in the user-task model.

We have produced two versions of a research prototype of MOBI-D and used them to build a number of user interfaces in various domains. Additionally, we are conducting new research on dialog management and reasoning for interfaces designed with MOBI-D.

3 Design Principles for Human-Centered Model-Based Tools

We have been studying the area of model-based interface development for a number of years and have built various model-based interface development tools (Puerta and Eriksson 1994). These tools had the goal of automating the interface design process and were not human-centered in principle. As a result, they had limited success and acceptance. Our experience in building and using those tools has led us to a series of principles that we believe to be essential in order to construct a model-based environment that is human-centered. These principles are as follows:

- *Hiding formalisms*. Model-based tools should support formal methods for modeling tasks on the background. They should not require designers to become experts on the formalisms used or to directly use the terminology associated with such formalisms. Both of those requirements would deviate the designer from its main concern: producing an user interface. Tools that impose formalisms on their users are likely to become more of burden to the designer than a supporting instrument.
- *Informality*. Formal methods supported by model-based tools should allow for reasonable deviation from the established procedures. Obviously, each formal method will have a minimum set of procedural requirements imposed on designers. However, beyond those minimum parameters, designers should be able to be *informal* in their definition and manipulation of user tasks and all other elements in an interface model.
- *Interactivity*. Model-based tools should support an interactive process with designers. They should not attempt to automate any of the creative elements of user interface design. The tools, however, should automate mechanical, time-consuming tasks that are well understood. In this regard, they should mimic commercially-available programming environments.
- *Decision support*. Even though model-based tools should not automate creative design tasks, they should offer support to the designers for those tasks. Such support should exploit the formalisms and knowledge

representations associated with the tools. A common example is the use of knowledge bases of interface design guidelines to assist designers in the selection of appropriate widgets.

- *Phase integration.* It is essential that sets of task-based tools are integrated in such a manner that they support designers from the initial phase of elicitation of user tasks to the final phases of layout and dialog design. Without such integration, the appeal of model-based tools to designers is limited because the end product of using the tools is not a completed user interface.
- *Design rationale.* Throughout every step of the user interface design cycle, model-based tools should allow for the annotation and specification of design decisions. When applicable, these design rationale statements should be linked to the underlying formalism in a meaningful way.
- *Avoiding complexity.* Model-based tools should not impose on designers levels of complexity that turn the use of the tools into a burden rather than a benefit. For example, if a given model-based tool requires elaborate and detailed user-task models for even simple applications, then designers are unlikely to use such a tool because they will consider such activities beyond their natural goal of building an interface.

4 Design Principles and MOBI-D

In developing MOBI-D, we have attempted to follow the principles outlined above. For each tool within the environment, there is a subset of those principles that is relevant. All tools in MOBI-D hide the formalism of the interface model from designers. Each tool is aimed at a step of a user-centered interface design process. Even the model editors, which are the closest access to interface models that designers have, are built to support that process not to support the direct editing of interface models. As a by-product of this implementation feature, the tools allow deviations from the underlying formalism which increases the flexibility offered to designers.

As a main feature of MOBI-D, no automatic generation of interfaces is attempted. All the creative elements of interface design are left up to the designers themselves. The environment, however, does offer extensive decision support throughout, especially for widget selection and user-task elicitation. Furthermore, the tools are integrated in such a manner as to allow the output of one tool to be used on the next step of design by another one of the tools, or, to concurrently have the output of one tool be reflected immediately on the view of another one of the tools. Additionally, the tools allow for the most part the annotation of design decisions via comments. These comments serve as a design rationale and introduce yet another level of flexibility for users.

In sum, we believe that it is essential for model-based systems to conform to a human-centered interaction style and that the principles outlined above, when followed, lead to systems that are more likely to be human-centered in essence.

5 Acknowledgements

I thank Hung-Yut Chen, Eric Cheng, Jacob Eisenstein, James J. Kim, Kjetil Larsen, David Maulsby, Justin Min Dat Nguyen, Tunhow Ou, David Selinger, and Chung-Man Tam for their work in the design and development of MOBI-D.

6 References

Johnson, P., S. Wilson, et al. (1994). Scenarios, Task Analysis, and the ADEPT Design Environment. Scenario Based Design. J. Carrol, Addison-Wesley.

Puerta, A., E. Cheng, et al. (1999). MOBILE: User-Centered Interface Building. CHI99, Pittsburgh, ACM Press.

Puerta, A. and J. Eisenstein (1999). Towards a General Computational Framework for Model-Based Interface Development Systems. IUI99: International Conference on Intelligent User Interfaces, Los Angeles, ACM Press.

Puerta, A. and H. Eriksson (1994). Model-Based Automated Generation of User Interfaces. AAAI'94, AAAI Press.

Puerta, A. and D. Maulsby (1997). Management of Interface Design Knowledge with MOBI-D. IUI97: 1997 International Conference on Intelligent User Interfaces.

Puerta, A. R. (1996). The MECANO Project: Comprehensive and Integrated Support for Model-Based Interface Development. CADUI96: Computer-Aided Design of User Interfaces, Namur, Belgium.

Puerta, A. R. (1997). A Model-Based Interface Development Environment. IEEE Software. **14:** 40-47.

Tam, R. C.-M., D. Maulsby, et al. (1998). U-TEL: A Tool for Eliciting User Task Models from Domain Experts. IUI98: 1998 International Conference on Intelligent User Interfaces, San Francisco, CA, ACM Press.

Vanderdonckt, J. M. and F. Bodart (1993). Encapsulating Knowledge for Intelligent Automatic Interaction Objects Selection. InterCHI'93, ACM Press.

Testing the Usability of Visual Languages: A Web-Based Methodology

Mauro Mosconi – Marco Porta
Dipartimento di Informatica e Sistemistica – Università di Pavia
Via Ferrata, 1 – 27100 – Pavia – Italy
mauro@vision.unipv.it – porta@vision.unipv.it

1 The Objective: Testing the Expressiveness of Visual Control Structures

The purpose of this paper is to illustrate the methodology we developed in order to start a comparative usability study for different implementations of visual control flow constructs.

Since loops in data-flow visual languages (Hils 1992) may be difficult to understand (Green and Petre 1996), we decided to test the usability of the solutions we devised for *VIPERS*, a data-flow visual language developed at the University of Pavia (Mosconi and Porta 1998). To get useful indications, we opted for a comparative analysis, also referring to the well-known data-flow language *LabView* (Vose 1986), where iterative constructs are implemented according to a totally different philosophy.

Among the possible evaluation methods (Preece 1993), we elicited *observational evaluation*, which involves observing or monitoring users' behavior while they are using an interface, and *survey evaluation*, which means seeking users' subjective opinions. To collect data about what users do when they interact with the test interface, employment of *direct observation* was avoided. In fact, if users are constantly aware that their performance is being monitored, their behavior may be strongly influenced (*Hawthorne effect*). Instead, we used *software logging* to record the dialog between user and system. In particular, our methodology is based on the use of the *log files* of a web server, as will be illustrated. Moreover, we elicited *questionnaire* forms to support the survey evaluation.

2 The Testing Methodology

2.1 Selecting the Test Context

Every usability evaluation is meaningful within a precise context, including the practice level of testers, the types of task being undertaken and the environment in which the test is carried out.

For our first experiments, we decided to work with high school students (17 to 19 years old) with little skill in textual programming and no experience at all in visual programming. We set proper mathematical applications as test tasks. Even though we were aware that many other application domains would be more suitable for a data-flow approach, we opted for problems close to their school experience.

As far as the programming environment was concerned, our aim was to make the interaction independent of the computer platforms used (by carrying out tests in an heterogeneous environment –our lab– with both PCs and Mac and UNIX machines). This last consideration also influenced our choice to focus on the program understanding process rather than on program construction.

We stress the fact that we did not want to compare the usability of the whole VIPERS and LabView environments, but only to observe how efficiently these two languages visually express control constructs (loops, in particular).

2.2 Planning the Tests

We planned two sessions, each one with twelve users. Altogether, twelve users tackled a set of three problems through VIPERS and twelve the same set through LabView.

Each user had to examine, in sequence, three visual programs displayed on the computer screen and translate them into correspondent textual programs (in pseudo-Pascal, since they all knew this language). We considered the number of right solutions as a first indicator of the comprehensibility of the languages in the loop implementation. Time for the test was set at one and a half hours.

2.3 Implementing the Tests: the Technical Approach

The idea making the creation of our tests (and generally any tests that can exploit this set up) particularly economical is the use of a web server and its log files. Each problem is presented to the user in the form of a web page: during the test, the user interacts solely and exclusively with a web browser, independently of the platform used. In the web page (see Figure 1) the problem is visualized as an image, or rather, as an *image map*. Image maps are graphics

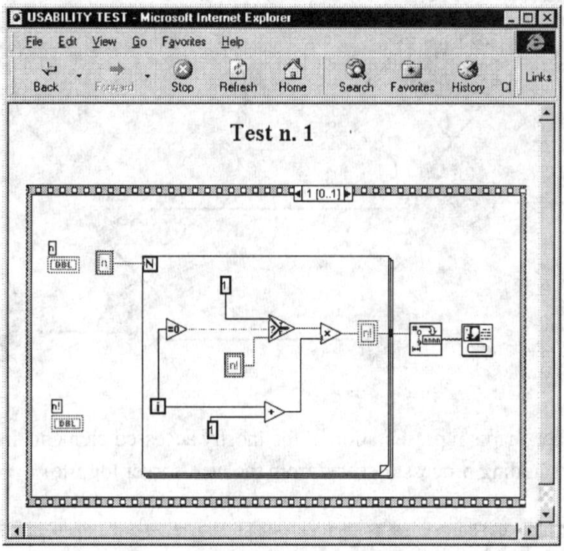

Figure 1: a (LabView) visual program to be examined is shown within a web browser

where certain regions are mapped to URLs. By clicking on different regions, different resources can be accessed from the same graphic.

In our case, we associated each graphic symbol with a page illustrating its meaning. The log file of the web server, by registering the actions of the users, was able to reveal how many times (and in what order) each user had clicked on a symbol of the program to ask for the help pages, and how much time the user stopped to read the explanations before returning to the main page using the BACK key.

Figure 2 points out, using "fingers marks", those parts of one of the test programs that were the most studied by an average of testers. For a complete report of the results obtained see (Ghittori 1998): here, we concentrate on illustrating the methodology used.

One type of result we believe worth showing is that given in Figure 3, where the path followed by a tester in examining a program was reconstructed on the basis of the log file. Our impression is that this scan path (which can easily be obtained automatically) may prove to be a very interesting tool (as analyzed by cognitive psychologists) with which to verify old hypotheses regarding the mental representation of visual programs and the cognitive processes involved in programming, in these new environments.

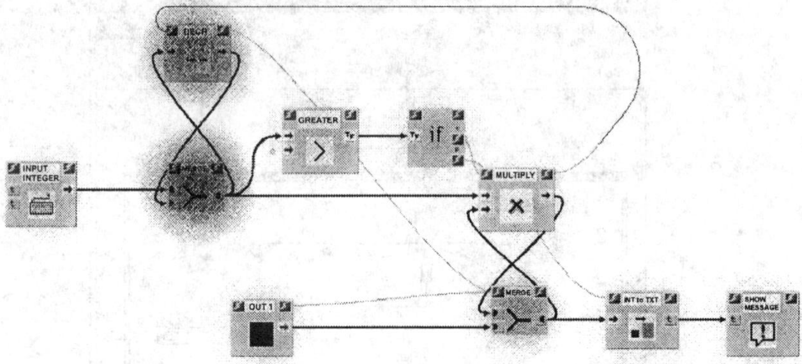

Figure 2: a finger-marks representation of the mostly accessed elements in a (VIPERS) program understanding process (derived from the web-server log file)

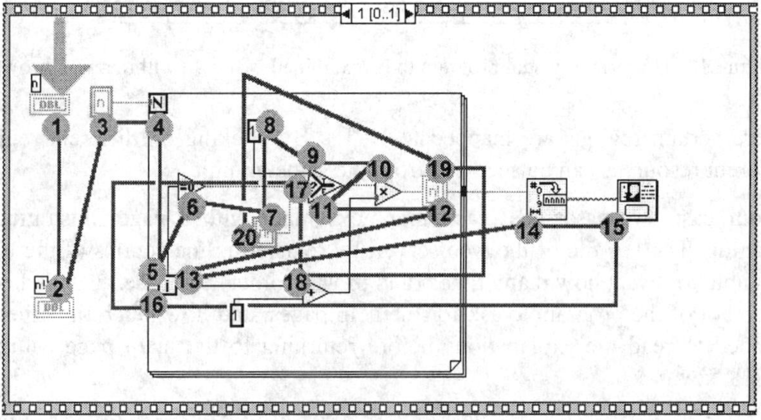

Figure 3: reconstruction of the mental scan-path of the user inspecting a visual program

3 Considerations About the Testing Methodology

We believe we have perfected a methodology for the usability analysis which is effective, supplies a wealth of feedback, and is moreover fast and economical. In particular, we would like to point out that:

- the method proposed is non-invasive (the user does not know he/she is being monitored and is therefore less subject to external conditioning).
- Data collection is objective because it is given by the log file path.

- This method can be used on different hardware platforms at the same time and allows more than one user to be monitored at the same time.

- The tests can be carried out without having to remove the tester from his/her work.

- The processing of the data gathered is made easier by numerous elaboration tools for the web log file information. Moreover, setting the test does not require particularly extensive knowledge of informatics.

- The hypermedial means used make it easier to prepare the material needed to instruct the testers (in our case, these are the pages explaining how the visual language "blocks" function). In other words, the problem is solved of how to organize the presentation of information to the user by allowing him/her to directly access only the information needed (which varies according to the ability of the user and the type of problem).

- By registering the position of the click on the image it is also possible to automatically obtain a graphic representation of the "mental path" taken by the subject analyzing a visual program.

- In general, it makes sense to state that this methodology can be applied to a variety of different research environments where interaction between users and complex images is being studied.

References

Ghittori, E. (1998). *Usabilità dei linguaggi visuali dataflow: il problema dei costrutti di controllo*. Master's Thesis, University of Pavia.

Green, T. R. G., Petre, M. (1996). Usability Analysis of Visual Programming Environments: A 'Cognitive Dimension' Framework. *Journal of Visual Languages and Computing*, 7(2), 131-174.

Hils, D. D. (1992). Visual Languages and Computing Survey: Data Flow Visual Programming Languages. *Journal of Visual Languages and Computing*, vol. 3, 69-101.

Mosconi, M., Porta, M. (1998). Designing new Programming Constructs in a Data Flow VL. *Proceedings of the 14th IEEE International Conference on Visual Languages* (VL'98, 1-4 September 1998, Nova Scotia, Canada).

Preece, J. (1993). *A Guide to Usability. Human Factors in Computing*. Addison Wesley.

Vose, G. M. (1986). LabView: Laboratory Virtual Instrument Engineering Workbench. *BYTE*, vol. 11, n. 9, 82-84.

THE USER ACTION FRAMEWORK: A THEORY-BASED FOUNDATION FOR INSPECTION AND CLASSIFICATION OF USABILITY PROBLEMS

H. Rex Hartson[1], Terence S. Andre[2], Robert C. Williges[2], and Linda van Rens[1]

Department of Computer Science[1]
Department of Industrial and Systems Engineering[2]
Virginia Tech, Blacksburg, VA 24061 USA

1 Motivation

Because of growing awareness of the importance of usability, organizations are expending ever-increasing resources for "doing usability;" enviable usability laboratories are built, developers are trained in usability methods and considerable resources are devoted to conducting usability evaluations. However, most organizations experience limited returns on their usability investment. Methods for usability inspection, usability data classification, and usability data management are ad hoc at best; lack of a unifying framework for usability tools limits the effectiveness of usability development activities. There is a clear need for more focused usability inspection methods and higher quality descriptions and classification in usability problem reports.

In this paper we report on a unifying framework, called the User Action Framework (UAF), we have developed as a foundation for structured methods and integrated tools to get the most value from usability data by way of analysis, classification, storage and retrieval, reporting, and redesign. The UAF is a theory-based structure for organizing usability concepts, issues, design features, usability problems, and design guidelines that has evolved as a common core for a multiplicity of usability methods and tools for usability practitioners. The UAF is manifest within a usability support tool by mapping the content and structure into an expression specific to the purpose of the tool. We describe the UAF concept and one such tool, the Usability Problem Inspector (UPI) tool.

2 Integrating Framework

Previous work on the Usability Problem Taxonomy by Keenan (in press) and the Usability Problem Classifier by van Rens (1997) provided useful methods for classifying usability problems by type. However these classification schemes needed an improved structure before reliable tools could be based on them. The solution came in the form of an adaptation and extension of Norman's theory of action (Norman, 1986), resulting in what we call the Interaction Cycle, which provides a high-level structure for organizing the contents of the Usability Problem Classifier. The UAF is a tool-independent knowledge base of usability concepts and issues organized under the relevant parts of the Interaction Cycle, as shown in Figure 1. Although there are other usability inspection methods, for example, based on Norman's model, this underlying knowledge base of usability concepts and issues distinguishes the UAF and its derivative tools from other such approaches.

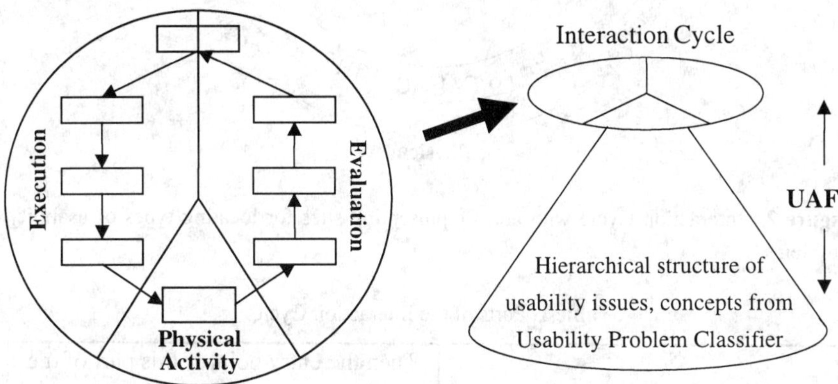

Figure 1. Combining an adaptation of Norman's seven stages of action (on the left) with the hierarchical structure of the Usability Problem Classifier to form the User Action Framework.

3 The Interaction Cycle

Our research makes a significant contribution by operationalizing Norman's seven-stage theory of action into the organizing structure of a usability concept framework and an evaluation tool of practical utility. The Interaction Cycle shown in full detail in Figure 2, adapted from Norman (1986), is the core of the UAF and provides high level organization and entry points to the underlying structure for classifying details. Finding the correct entry point for a usability issue, concept, guideline, or problem is based on determining the part of the Interaction Cycle where the user is affected.

The Interaction Cycle gives a picture of how interaction happens, expressed in terms of effects on users doing tasks and with a focus on user actions (cognitive and physical). As in Norman's model, the top part of the cycle is for cognitive actions, with the remaining sector for physical actions. Table 1 illustrates how usability issues are related to a specific part of the Interaction Cycle.

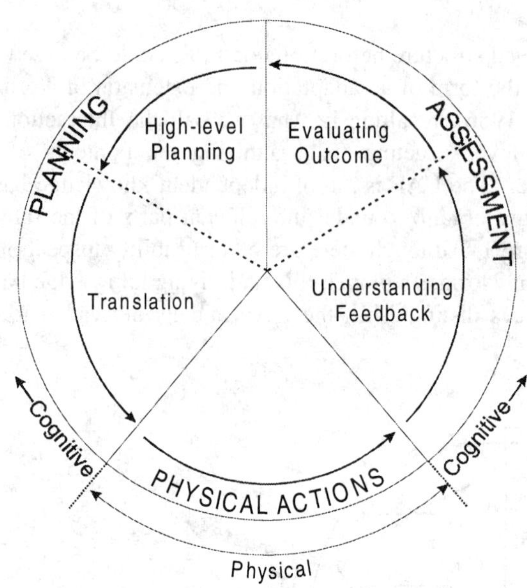

Figure 2. Interaction Cycle with areas representing sites for locating types of usability problems.

Table 1. Parts of the Interaction Cycle.

If the issue is about:	Then the entry point is this part of the Interaction Cycle
Establishing goals, tasks, and/or intentions	High-level Planning (up to Translation)
Translating intentions into plans for physical actions	Translation part of Planning
Making physical input actions	Physical Actions
Perceiving, understanding and evaluating outcome based on feedback in system response	Assessment

Where other methods simply identify problems, the UAF classifies and contextualizes design problems in terms of user's interaction-based behavior. This allows us to understand the significance of design problems in the totality of the cognitive and physical aspects of task performance. For example, using a heuristic or guideline, one might decide a given usability problem was a "consistency problem." In contrast, because the deeper classification structure of the UAF aids description, we can formulate a much more complete identification of the problem as "a problem where lack of consistency in a data format led to a failure in cognitive affordance for avoiding errors in a data field during a form-filling interaction." By placing the effect of the usability problem more precisely within the user interaction cycle, a more complete usability problem report supports a more specific focus for redesign solutions.

4 Mapping to an Inspection Tool

The UAF has become a common conceptual framework for our usability work. We apply the framework by mapping it into specific usability support tools such as the Usability Problem Classifier and the UPI. The meaning and structure are retained, but the concepts are *expressed* in a way that is tailored to the purpose of that tool. For the UPI, the expression is in terms of the types of problems to look for in a usability inspection. The UPI offers several benefits to practitioners. First, the UPI is based on the supporting infrastructure of detailed usability concepts in the UAF. Therefore, a capability to document problems found (i.e., complete description in terms of problem type and subtype, including the effect on the user within the interaction process) is built into the inspection tool, allowing the evaluator to focus on finding problems. Second, because of entry through the Interaction Cycle, evaluators and developers have a greater chance of understanding the description of multiple task-thread problems that are usually encountered in user interaction design. The UPI does this by proactively directing the inspection process to follow-up on potential related problems. For example, by finding an Assessment problem with the content of the feedback part of a system response, the UPI reminds the evaluator of a possible Planning problem if the prompt part of the system response also does not easily direct the user to the next, correct intention. Finally, the UPI allows the evaluator to consider significant physical usability issues (e.g., Fitts' law, object design, disability accommodation) in addition to the traditional cognitive problems

5 Usability Evaluation Study Using the UPI Tool

To test the concepts of the UPI, and the Interaction Cycle and UAF it is based on, we responded to an opportunity to evaluate a commercial message management service developed by the CrossMedia Networks Corporation. As the goal of the evaluation was rapid turn-around of results, we focused on comparing the UPI to the traditional heuristic method developed by Nielsen and Molich (1990).

Each expert evaluator produced a list of problems with respect to the evaluation technique used. Figure 3 shows the general distribution of different problem types identified by each group. As shown, ten of these problems were identified by both methods. All six evaluators rated severity of the 40 unique problem types on a scale from 1 to 4. We examined severity by splitting problems into most severe (median of 3.0 or higher) and least severe (median of 2.50 or lower). The Heuristic method isolated primarily the least severe problems; whereas the UPI method found significantly more severe problems (χ^2=7.36, df=2, p <.05).

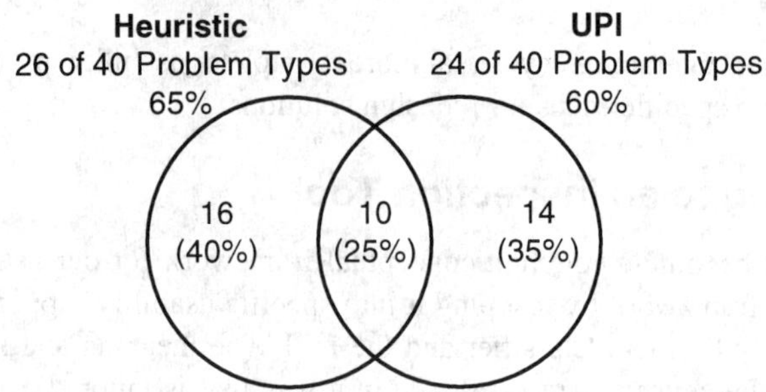

Figure 3. Comparison of number of unique problem types identified in Heuristic and UPI methods.

6 Conclusion

The UAF provides a model-based framework for examining usability issues in the context of user action. From the UAF, we are able to map to various usability tools, including the UPI for prediction of usability problems. We expect that the UAF, and associated mappings, will provide usability professionals with comprehensive tools to conduct more efficient and effective usability evaluations through an easily understood framework, a complete way to understand problems, and built-in links to possible design solutions.

7 Acknowledgments

This work has benefited from comments and suggestions provided by José Castillo, Deborah Hix, Steve Belz, and Faith McCreary. This research was supported by the Virginia Center for Innovative Technology and the CrossMedia Networks Corporation. MyInBox is a registered trademark of the CrossMedia Networks Corporation.

8 References

Keenan, S. L., Hartson, H. R., Kafura, D. G., & Schulman, R. S. (in press). The Usability Problem Taxonomy: A framework for classification and analysis. Empirical Software Engineering.

Nielsen, J., & Molich, R. (1990). Heuristic evaluation of user interfaces. In CHI '90 Conference Proceedings (pp. 249-256). New York: ACM Press.

Norman, D. A. (1986). Cognitive engineering. In D. A. Norman & S. W. Draper (Eds.), User centered system design: New perspectives on human-computer interaction (pp. 31-61). Hillsdale, NJ: Lawrence Erlbaum Associates.

van Rens, L. S. (1997). Usability problem classifier. Unpublished master's thesis, Virginia Polytechnic Institute and State University, Blacksburg, VA.

The Evaluator Effect during First-Time Use of the Cognitive Walkthrough Technique†

Morten Hertzum
Centre for Human-Machine Interaction
Risø National Laboratory, Denmark

Niels Ebbe Jacobsen
Department of Psychology
University of Copenhagen, Denmark

1 Introduction

Practising system developers without a human factors background need robust, easy-to-use usability evaluation methods. The cognitive walkthrough (CW) technique (Lewis et al. 1990, Wharton et al. 1994) has been devised to provide such a method and is particularly suited to evaluate designs before testing with users becomes feasible and as a supplement to user testing.

While several studies have evaluated how well CW predicts the problems encountered in thinking-aloud studies (e.g. John and Mashyna 1997, Lewis et al. 1990), only Lewis et al. have assessed to what extent different evaluators obtain the same results when evaluating the same interface. Data from Lewis et al. suggests that the variability in performance among evaluators using CW is much lower than that of evaluators using heuristic evaluation or thinking-aloud studies (Jacobsen et al. 1998, Nielsen 1994). One reason for this seemingly higher robustness of CW might be that it is a quite structured process. CW has however evolved considerably since the study of Lewis et al. Moreover, their data was limited in sample size and applicability to actual CW evaluators.

To inform practitioners and methods developers about the robustness of CW this paper investigates to what extent novice evaluators who perform a CW of the same tasks detect the same problems in the evaluated interface. While acknowledging the importance of choosing the right tasks in a CW, we have decided to focus on the actual walkthrough process.

---

† This work was supported by grants from the Danish National Research Foundation and the Danish Research Councils. We wish to thank the evaluators for their time and effort.

2 The CW Technique

Our study is based on the version of CW described in Wharton et al. (1994). CW consists of a preparation phase and an execution phase. In the preparation phase the evaluator describes a typical user, chooses the tasks to be evaluated, and constructs a correct action sequence for each task. In the execution phase the evaluator asks four questions for each action in the action sequences: (1) Will the user try to achieve the right effect? (2) Will the user notice that the correct action is available? (3) Will the user associate the correct action with the effect trying to be achieved? (4) If the correct action is performed, will the user see that progress is being made toward solution of the task? With the description of the user in mind the evaluator decides whether each question leads to a success or failure story. In case of a failure story a usability problem has been detected.

3 Method

Eleven graduate students in computer science evaluated a prototype of a Web-based system against set tasks. Half of the evaluators had design experience from industry, but they had no prior knowledge of the system to be evaluated. The evaluated system, called HCILIB, was a prototype of a Web-based library giving access to a collection of scientific articles on human-computer interaction. HCILIB (Perstrup et al. 1997) integrates Boolean search with a scatter-gather inspired technique to display a browsable structure of the collection. Boolean searches can be expressed as conventional Boolean queries (using ANDs and ORs) or by means of a Venn diagram metaphor. The Venn diagram metaphor relieves the user from direct interaction with logical expressions. Instead, query terms are entered into two search boxes, A and B, and the search results are automatically sorted into three disjunctive sets corresponding to A–B, $A \cap B$, and B–A.

The experiment was embedded in a grade-giving assignment where the students were asked to construct action sequences and do a cognitive walkthrough of three set tasks. Just before the assignment was handed out the evaluators received two hours of instructions in CW based on a lecture on the practitioner's guide to CW (Wharton et al. 1994). The instructions also offered the evaluators some hands-on experience followed by instant feedback. The evaluators documented their cognitive walkthroughs in a problem list describing each detected problem. As a rough estimate each evaluator spent 2-3 hours completing his/her CW. Based on the problem lists from the 11 evaluators the two authors independently constructed a master list of unique problem tokens. Combining these master lists we had an inter rater reliability of 80%; disagreements were resolved through discussion and a consensus was reached.

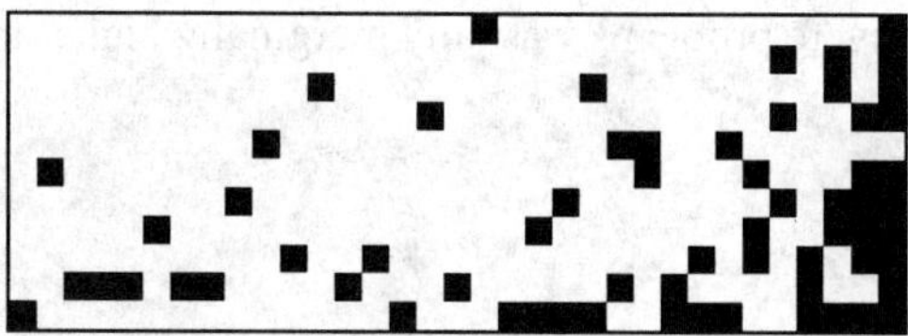

Figure 1. Matrix showing who found which problems. Each row represents an evaluator, each column a problem, and each black square that the evaluator detected the problem.

4 Results

The eleven evaluators reported a total of 74 problem instances from their CWs. These problem instances made up 33 unique problem tokens (in the following just termed problems). As much as 58% of the problems were detected by only a single evaluator, and no single problem was detected by all evaluators, see Figure 1. A single evaluator found on average 18% of the 33 known problems.

We were curious to know how groups of evaluators performed compared to single evaluators. Figure 2 shows the average number of problems that would be found by aggregating the sets of problems found by different groups of evaluators. For each group a given problem was considered found if it was found by at least one of the evaluators in the group. The results suggest a great deal of misses – or false alarms – in the performance of single evaluators. An analysis of problem severity could not explain this evaluator effect, as the

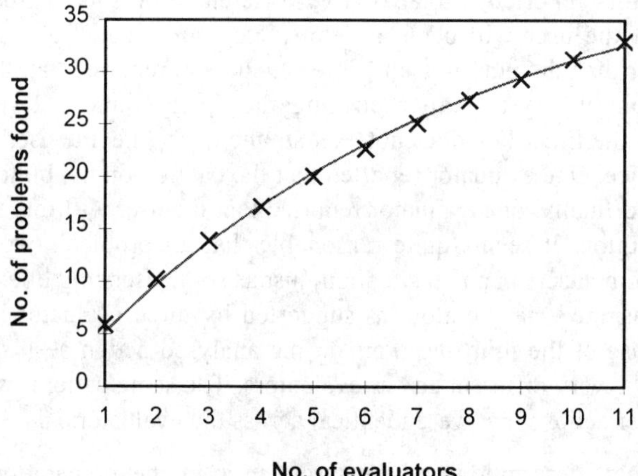

Figure 2. The number of problems detected shown as a function of the number of evaluators. The data points are the aggregated values from the experiment. The curve plots an estimate given by the formula $f(k) = n(1 - (1 - p)^k)$, for $n = 43$ and $p = 0.121$.

detection rate for severe problems was only marginally higher than for the entire set of problems.

5 Discussion

As for other usability evaluation methods it is crucial to CW that the walkthrough leads to a reliable problem list. Studies on heuristic evaluation and usability studies have found substantial individual differences in the evaluators' performance (Jacobsen et al. 1998, Nielsen 1994). This suggests that our results are partly attributable to usability evaluation in general, rather than solely to CW. We believe, however, that the CW procedure falls short of providing the evaluators with a feel for the users and thus becomes inaccurate for two reasons: (1) Anchoring, i.e. despite the evaluator's efforts the walkthrough will end up evaluating the system against a user who is much too similar to the evaluator to be representative of the actual users. (2) Stereotyping, i.e. the walkthrough will end up reflecting a user that is much too homogeneous to accommodate the diversity of the actual users of the system evaluated.

We investigated the anchoring and stereotyping hypotheses by looking closer on how the evaluators answered the four questions on identical actions. Though the evaluators constructed their action sequences from the same three tasks only 4 out of an average of 15 actions were identical across all evaluators. One of these actions is to execute a query by activating a Query button. In evaluating this action three evaluators reported success stories on all four questions, while eight evaluators reported a total of five different problems: Three evaluators reported that the user will click a Venn pictogram situated above the Query button, rather than the button itself. Three evaluators reported that there is weak feedback from the system after clicking the Query button. Two evaluators reported that the Enter key does not execute the query, i.e. the user has to use a pointing device. One evaluator reported that the caption on the button should be changed. And finally, one evaluator reported that the user will forget to activate the Query button. It seems quite reasonable that all problems would actually happen for some users in a real situation, just as some users might experience no troubles using the Query button, as suggested by three evaluators. Though all evaluators' use of the four questions on the analysed action seems reasonable, the outcome is very different across evaluators. The same pattern was found for the three other actions that were identical across the evaluators.

The evaluators' descriptions of the target user in the preparation phase are similar in content, and they generally provide a broad description of a large, homogeneous group of users. The descriptions are in many respects similar to the descriptions of users given as examples by Wharton et al. (1994). Despite the formal description of the user, or perhaps because of the generality of these descriptions, the evaluators might not fully realise the heterogeneity of the user group or their walkthrough might be anchored to their own experience with the system. Each of the four questions drives the evaluator to think of the user's behaviour in a certain situation. When the fictive user description becomes too fuzzy or lacks details to judge the user's behaviour, the evaluator unintentionally substitutes the description with a particular user much like herself/himself. Thus, evaluators tend to produce success stories if they imagine themselves having no troubles using the feature in question, and they report problems when they imagine themselves having troubles in the particular situation. In this sense a single evaluator using CW resembles an evaluator performing a thinking-aloud study with one user, namely himself/herself.

Wharton et al. (1994) state that CWs can be performed by individual evaluators as well as by groups of co-operating evaluators. For inexperienced CW evaluators our study strongly indicates that several evaluators are necessary to achieve a performance that is acceptable for practical use of the CW technique. Additional studies are required to learn how more experienced evaluators perform and to study more closely *why* we see these individual differences.

6 References

Jacobsen, N. E., Hertzum, M. & John, B. E. (1998). The evaluator effect in usability studies: problem detection and severity judgments. In *Proceedings of the Human Factors and Ergonomics Society 42nd Annual Meeting* (Chicago, October 5-9, 1998), pp. 1336-1340. Santa Monica: HFES.

John, B. E. & Mashyna, M. M. (1997). Evaluating a multimedia authoring tool. *Journal of the American Society for Information Science*, 48(11), 1004-1022.

Lewis, C., Polson, P., Wharton, C. & Rieman, J. (1990). Testing a walkthrough methodology for theory-based design of walk-up-and-use interfaces. In *Proceedings of the ACM CHI'90 Conference* (Seattle, April 1990), pp. 235-242. New York: ACM Press.

Nielsen, J. (1994). Heuristic evaluation. In Nielsen, J. & Mack, R. L. (Eds.): *Usability Inspection Methods*, pp. 25-62. New York: John Wiley.

Perstrup, K., Frøkjær, E., Konstantinovitz, M., Konstantinovitz, T., Sørensen, F. S. & Varming, J. (1997). A World Wide Web-based HCI-library designed for interaction studies. In *Third ERCIM User Interfaces for All Workshop* (Obernai, France, November 1997).

Wharton, C., Rieman, J., Lewis, C. & Polson, P. (1994). The cognitive walkthrough method: a practitioner's guide. In Nielsen, J. & Mack, R. L. (Eds.): *Usability Inspection Methods*, pp. 105-140. New York: John Wiley.

A Process for Appraising Commercial Usability Evaluation Methods

Ronan Fitzpatrick
Department of Mathematics,
Statistics and Computer Science,
Dublin Institute of Technology,
Kevin Street, Dublin 8, Ireland.
Tel: +353 (1) 4024835,
Fax: +353 (1) 4024994
Email: rfitzpatrick@maths.kst.dit.ie

Alan Dix
School of Computing,
Staffordshire University,
Beaconside, Stafford ST18 ODG,
United Kingdom.
Tel: +44 1785 353428,
Fax: +44 1785 353431
Email: A.J.Dix@soc.staffs.ac.uk

1. Introduction

To support software usability evaluation, commercial usability evaluation methods have been developed. These methods are practical implementations of generic methods and incorporate good professional practice and industry standards. Many of these methods predate ISO 9000-3 (1997), ISO/DIS 9241-11 (1995) and the EU Directive on display screen equipment (Council Directive (90/270/EEC 1990).

Strategic managers and IS professionals need to know that new systems (being developed, purchased or re-engineered legacy systems) comply with these standards and legislation. Developers wishing to comply with evaluation of designs against user requirements (ISO 13407 1997) must show their "*process and rationale for the selection of methods and measures used*". These professionals need usability evaluation support to help them to select and justify commercial usability evaluation methods appropriate to their projects.

The commercial methods for usability evaluation that exist do not necessarily evaluate similar usability attributes. SUMI measures user attitudes to an existing interface while MUSE (a structured Method for USability Engineering) is concerned with validating usability requirements analysis and usability specification. More commercial usability evaluation methods are evolving (Lin *et al.* 1997) and this gives rise to a number of need-to-know issues which include:

- What commercial usability methods are available and how they are used.
- Which usability evaluation strategies the commercial methods apply to (i.e. Virtual Engineering, Soft Modeling, Hard Review and Real World).
- If the commercial method is standalone or is part of a wider methodology.
- When and where in the system life cycle a commercial method is of benefit.
- What usability characteristics are evaluated by individual methods.
- The relative merits of methods that evaluate similar usability characteristics.
- The reliability of the results produced by these commercial methods.
- Which combination of methods is most appropriate to achieving the highest usability for a specific project.

3. The process defined

This process for appraising commercial usability evaluation methods involves two stages and uses a specially designed Method Appraisal Grid as shown in figure 1.

Usability evaluation method															Usability considerations																		Sheet ___ of ___
	Methodology				Strategic Application				Life cycle processes						Context of use				Usability measures	Attributes of a usable software product													
															%	%	%	%		%	%	%	%	%	%	%	%	%	%	%	%	%	Product Weighting ___
Commercial method name	Philosophy	Model	Generic Method	Tool	1 Virtual Engineering	2 Soft Modelling	3 Hard Review	4 Real World	Analysis	Design	Build	Buy	Operation	Maintenance	User	Goals/Tasks	Equipment	Environment		Suitability	Installability	Functionality	Adaptability	Ease-of-use	Learnability	Interoperability	Reliability	Safety	Security	Correctness	Efficiency	Overall	
					HR			RW											Effectiveness														Method Reliability
					3	4													Efficiency														
																			Satisfaction														
					1	2																											
					VE			SM																									Score
									Rating ___																								
					HR			RW											Effectiveness														Method Reliability
					3	4													Efficiency														
																			Satisfaction														
					1	2																											
					VE			SM																									Score
									Rating ___																								

Figure 1 - The Method Appraisal Grid (MAG).

Stage one is a comprehensive review of the method and the recording of its features and usefulness to the usability evaluation process. It is headed Usability evaluation method, and includes space for each method to be named and for recording its Methodology, Strategic Application and support for Life cycle processes. Stage two, headed Usability considerations, indicates *"what"* the method evaluates and includes Context of use, Usability measures and the Attributes of a usable software product. (Bevan and Macleod 1994: ISO/DIS 9241-11 1995: Fitzpatrick and Higgins 1998).

A weighting and rating matrix is used to calculate a score for each commercial method being appraised. Elements are weighted to suit the evaluating organisation's specific requirements and rated if these requirements are met. Finally, there is space to record a proven reliability measure for each method (as published by its owner).

3.1 Methodology

The Methodology category records the motivating philosophy for the method, and shows at a glance if it is a model, a standalone method or part of an integrated suite. The divisions, *Philosophy, Model, Generic Method* and *Tool*, are based on the view of methodologies given by Avison and Fitzgerald (1988). This category records if the method being appraised is supported by a computerised tool.

3.2 Strategic Application

The Strategic Application of the commercial method is based on four strategies appropriate to the stage in the life cycle when usability evaluation takes place. They also take account of the resources that are available and the need to use multiple methods during the evaluation process (Nielsen 1993) These strategies are *Real World, Virtual Engineering, Soft Modeling and Hard Review* (Fitzpatrick 1997)

3.3 Life cycle processes

This category records which life cycle processes the method can be applied to. It identifies "*where*" in the life cycle the evaluation method under review can be used. Support is included for those who wish to evaluate product they buy. The category sub-divisions are based on the primary processes outlined in ISO/IEC 12207 (1995) and are *Analysis, Design, Build, Buy, Operation* and *Maintenance*.

3.4 Context of use

When evaluating usability it is necessary to consider the user's profile, the tasks to be completed, the equipment to be used and the user's working environment (Bevan and Macleod 1994). This category identifies if a method being appraised addresses these items under headings *User, Goals/Tasks, Equipment and Environment.*

3.5 Usability measures

The Usability measures are those recommended in ISO/DIS 9241-11 (1995) and are *Effectiveness, Efficiency* and *Satisfaction.* Empty rows are included on the Grid for tailoring the appraisal to suit the appraiser's special usability measures - like *Usage.*

3.6 Attributes of a usable software product

The Attributes of a usable software product are the external software quality factors identified by Fitzpatrick and Higgins (1998). Each attribute is reviewed to establish if the commercial method addresses that attribute. During user-satisfaction evaluation it is also necessary to establish overall product satisfaction (Kirakowski 1995). So, an "overall" attribute is added to MAG for appraisal. The attributes are, *Suitability, Installability, Functionality, Adaptability, Ease-of-use, Learnability, Interoperability, Reliability, Safety, Security, Correctness, Efficiency* and *Overall* .

3.7 Weighting

Context of use and the Attributes of a usable software product are both scored. Weighting factors are used to indicate the importance of the elements of the context of use and the attributes of a usable software product. Organisations will have different usability needs, so, before appraising a commercial method, the evaluating organisation must decide the relative importance of the various elements and then allocate a percent weighting to each, such that the total is 100%. Weighting appears once on MAG as the same values must apply to all methods being appraised.

3.8 Rating

To rate a method each component of Context of use and each Attribute of a usable software product is subjectively grading on a scale of 0 to 100. This indicates the extent to which the different components of contexts of use and attributes are, in the opinion of the appraiser, satisfied by the method under review. The weighting and rating are then used to calculate a score for the method being appraised.

3.9 Scoring

A method is scored by multiplying each weighting by its rating and adding all the quotients. This gives a score that can be used for method comparison.

3.10 Method Reliability

System professionals need to be confident that the results yielded by the chosen commercial method are reliable (Kirakowski 1995). So, the final record on the Method Appraisal Grid is Method Reliability, which is the method owner's reliability rating or a independent verification of that rating.

4. Using the Method Appraisal Grid

- The primary use of MAG is to appraise commercial evaluation methods.
- When an evaluating team must devise its own or customise an existing commercial method, MAG provides focus for categories that must be addressed.
- Usability evaluation is essential to product quality control but is only useful if it

assess the appropriate attributes of the software product. MAG supports the correct choice of usability methods thus ensuring the highest quality software. So, to show compliance with a quality development process like the CMM, organisations can use MAG to demonstrate their quality practice.

- Developers can cite MAG as "*The process and rationale for the selection of [usability] methods and measures used*" in order to demonstrate compliance with ISO 13407 (Table A5) for evaluation of designs against user requirements.

5. References

Avison, D. and Fitzgerald, G. (1988) *Information systems development: methodologies*, techniques and tools, Oxford, Blackwell Scientific Publications.

Bevan, N. and Macleod, M. (1994) Usability measurement in context, *Behaviour and Information Technology*, Basingstoke, Taylor & Francis, (1 & 2)

Council Directive (90/270/EEC) (1990) Minimum safety and health requirements for work with display screen equipment", *Official journal of the EU*, p. L 156/14-18

Fitzpatrick, R. (1997). An investigation and analysis of current methods for measuring software usability, MSc dissertation, SOC, Staffordshire University, UK

Fitzpatrick, R. and Higgins, C. (1998). Usable software and its attributes: A synthesis of software quality, European Community law and HCI, In: *People and Computers XIII. Proceedings of HCI'98 Conference*, p 3-21

ISO 9000-3 (1997) *International Standard. Quality management and quality assurance standards - part 3:Guidelines for the application of ISO 9001:1994 to the development, supply, installation and maintenance of computer software*,

ISO/DIS 9241-11 (1995) *International Standard. Ergonomic requirements for office work with visual display terminals (VDTs). Part 11:Guidance on usability*,

ISO/IEC 12207 (1995) *International Standard. Information technology - Software life cycle processes*, International Organisation for Standardisation, Genève.

ISO/DIS 13407 (1997) *International Standard - Human-centered design processes for interactive systems*, International Organisation for Standardisation, Genève

Kirakowski, J. (1995) "The use of questionnaire methods for usability assessment", *Unpublished readings*, Human Factors Research Group, University College Cork.

Lin, Han X., Choong, Yee-Yin and Salvendy, Gavriel (1997) A proposed index of usability: a method for comparing the relative usability of different software. behaviour and Information Technology, 16(4/5) p267-278

Nielsen, J. (1993) *Usability engineering*, Academic Press Limited, London, UK

From Usability to Actability

Stefan Cronholm[1] , Pär J. Ågerfalk[2,4], Göran Goldkuhl [3,4]

[1] Dept. of Computer and Information Science
Linköping University, SE-581 83 Linköping, Sweden
stecr@ida.liu.se

[2] Dept. of Informatics (ESA)
Örebro University, SE-701 82 Örebro, Sweden
pak@esa.oru.se

[3] Dept. of Informatics, Jönköping International Business School
P.O. 1026, SE-551 11 Jönköping, Sweden
ggo@ida.liu.se

[4] Centre for Studies on Humans, Technology and Organization (CMTO)
Linköping University, SE-581 83 Linköping, Sweden

1 Introduction

To be able to design an information system (IS) that supports users performing business actions, it is necessary to understand both the users and their needs. One way to achieve a better understanding of the users' needs is to use information systems development (ISD) methods.

In the traditional ISD field several methods have been proposed. According to Lif (1998) these ISD methods offer little or no support for usability factors. The ISD methods traditionally focus on how to structure data and how to describe business flows or business processes. They seldom deal with how users should interact with an IS in a business, or how interfaces should be designed in order to permit, promote and facilitate users' business actions.

This paper discusses how theories from the Human-Computer-Interaction (HCI) and ISD fields can be combined in order to achieve better information systems.

More specifically, one aim of this research is to create a reconciliation of the HCI perspectives of usability with the language action (LA) perspective into what we call *actability*. The paper discusses advantages and limitations found in both the LA perspective and in prevalent HCI theories. This theoretical discussion will act as a base for developing a method based on actability.

2 Theoretical Analysis Framework

In order to highlight how theories from the HCI field and the LA perspective can be combined, we use a simple and general framework of IS usage (Shackel, 1984), which we believe covers most ISD use-situations. The framework consists of four components – user, task, tool, and environment (see Figure 1). The point of the framework is that none of the components can be considered in isolation from the others.

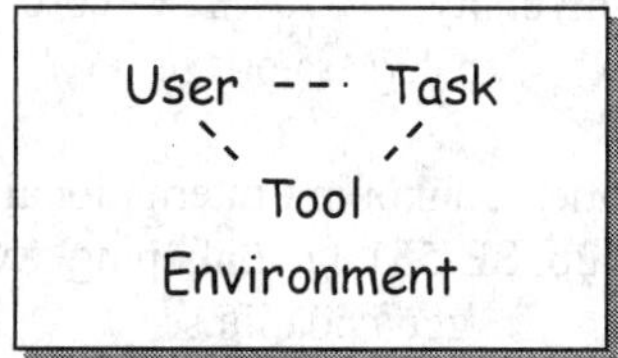

Figure 1. The four components of an IS use-situation (Shackel, 1984)

In the general framework there are relationships between user and task, task and tool, user and tool, as well as between those three and the environment. Our plan is to examine these relationships from each perspective in order to identify strengths and deficiencies.

3 Action and Information Systems

From an action perspective, information systems are viewed as communication systems, as distinct from strict representational views of information. A representational view of information means that designers try to create an 'image' of the reality in order to have the analysed piece of reality properly represented in the systems database. This strict representational view can be challenged, which an action perspective certainly does (e.g. Goldkuhl & Lyytinen, 1982; Winograd & Flores, 1986). In the LA perspective, information systems are not considered as "containers of facts" or "instruments for information transmission" (Ågerfalk & Goldkuhl, 1998). The LA perspective emphasises what users do while communicating through an IS (*ibid.*). Information systems are systems for business action, and business action is the means by which business relations are created. The aim of an IS is to support, facilitate and enable business actions.

In the LA perspective, the notion of information systems can be defined in the following way (Ågerfalk & Goldkuhl, 1998): an IS consists of 1) an action potential (a repertoire of actions and vocabulary); 2) a record of earlier actions and other prerequisites; and 3) actions performed interactively by the user and the system and/or automatically by the system.

Designing an IS means suggesting and establishing an action potential. An action potential both enables and delimits actions. It entails a repertoire of actions and a related vocabulary. The vocabulary consists of concepts related to the business language. An IS must also offer a record of actions performed. Information about these performed actions can normally be found in the IS database.

It is obvious that the LA perspective focuses on users performing actions (tasks). The meaning and purpose of acting is emphasised. As we can see, the LA perspective also discusses the tool needed for performing tasks. Furthermore, the LA perspective also discusses the performance of actions within a social context (environment). To date, LA approaches do not include sufficient descriptions of the relationships between the user and the tool, even though Ågerfalk & Goldkuhl (1998) have made some preliminary contributions.

4 Usability and Information Systems

In the HCI literature there are several definitions of the concept 'usability'. One definition is: "Usability is the result of relevance, efficiency, attitude and learnability" (Löwgren, 1993). Another similar definition states: "Usability, a key concept in HCI, is concerned with making systems easy to learn and easy to use" (Preece *et al*, 1994). HCI-research is mainly focused on the interface between a human and a computer. Both human and computer aspects are considered. A popular research area in HCI covers different interaction styles and forms (e.g. Preece *et al*, 1994; Sims, 1994).

In a use-situation (consisting of a user, a task and a tool) the usability perspective aims to cover all these three components. It is well known that all these three components have to be studied equally. As mentioned in section 2, they also have to be studied in a context/environment. However, we argue that there are differences between the usability perspective contributions to each of these three components, and in the relationships between them.

In the HCI literature it is clear that the major contributions to the concept of usability are to the relationship between the user and the tool components (in other words to the Human-Computer relationship), but little has been written about the relationship between the tool and the task component, or about the relationship between the user and the task component.

Our understanding is also that usability is considered mainly from an individual perspective. Human-Computer-Interaction has traditionally focused the dyad of one user using one computer system (Löwgren, 1995). This means that the traditional usability perspective misses the surrounding social context. There is, however, an emerging perspective on usability that consider factors such as the social organisation of work and how computers can be used to support it (*ibid.*).

Traditional ISD methods suffer from limitations in their treatment of cognitive and human factors as well as in their recommendations for the analysis of different interaction styles. Within the HCI field, on the other hand, these aspects are discussed frequently (e.g. Norman, 1988). Nielsen (1993) also discusses the importance of taking into account differences in the experiences of different users. These topics are not stressed in the ISD field and have so far been completely left out in the LA perspective. Hence, we think that it would be fruitful to combine the usability perspective with that of LA, since the latter primarily focuses on acts (tasks).

5 Actability and Information Systems

Our approach is to combine theories from different fields in order to achieve better information systems. The analysis indicates that the weakness of the LA perspective lies in the relationship between the user and the tool. This relationship is particularly focused on the usability perspective, which offers it good support. Our analysis also indicates that the strength of the LA perspective lies in the relationship between the user and the task, and between the task and the tool. The usability perspective does not offer the same support for these relationships.

When designing IS interaction, usability is important to consider. However, we believe that the common notion of usability is too narrow, since it is often perceived as dealing only with how to design user interfaces. When designing communication through an IS, the question of *how* to interact is, of course, important. Equally important, however, is *what* to communicate and *why*. Moreover, all three aspects must be considered in a *context* where the communication is taking place – that is, a social context that is never static and fully predictable.

We propose the concept 'actability', which is based on theories from the LA perspective and of usability, to assist discussion about the use of information systems in business processes. An information system's actability is its ability to perform actions, and to permit, promote and facilitate users to perform their actions both through the system and based on messages from the system, in some business context. The 'degree' of actability possessed by a certain IS is always related to the particular business context. The business context includes actors' pre-knowledge and skills relating both to the IS and the business task to be performed. Therefore, IS actability is not a static property of an IS, but depends on the social structures surrounding it.

Please note that the issue is not whether usability should be considered part of actability, and actability an extension of usability, or vice versa. The issue is to make information systems more actable and thus more usable.

6 References

Goldkuhl, G. & Lyytinen, K. (1982). *A language action view of information systems*. SYSLAB report no 14, SYSLAB, University of Stockholm. Sweden

Lif, M. (1998). *Adding Usability*. (Unpublished PhD thesis). Faculty of Science and Technology. Uppsala University. Sweden.

Löwgren, J. (1993). *Human-computer interaction – What every system developer should know.* Lund: Studentlitteratur.

Löwgren, J. (1995). *Perspectives on Usability*. Department of Computer and Information Science, Linköping University. Sweden.

Nielsen, J. (1993). *Usability Engineering*. Academic Press, San Diego, CA.

Norman, D. A. (1988). *The Psychology of Everyday Things*. Basic Books. New York.

Preece, J., Rogers, Y., Sharp, H., Benyon, D., Holland, S & Carey, T. (1994). *Human-Computer Interaction*. Addison & Wesley.

Shackel, B. (1984). The Concept of Usability. In Bennet, J., Case, D., Sandelin, J. & Smith, M. (Eds.): *Visual Display Terminals: Usability Issues and Health Concerns*. Englewood Cliffs NJ: Prentice Hall.

Sims, O. (1994). *Business Objects – delivering co-operative objects for client-server*, McGraw-Hill, Berkshire, England.

Winograd, T. & Flores, F. (1986). *Understanding Computers and Cognition*. New Jersey: Ablex Publishing Corporation.

Ågerfalk, P. J., Goldkuhl, G. (1998). Elicitation and Analysis of Actability Requirements. In Fowler & Dawson (Eds): *Proceedings of 3rd Australian Conference on Requirements Engineering*. Geelong, Vic, Australia. October 26-27, 1998, pp 14-28. Deakin University.

Consequences of Computer Breakdowns on Time Usage

Arzu ÇÖLTEKIN[1], Matti VARTIAINEN[2], I. Murat KOÇ[3]
[1]Institute of Photogrammetry & Remote Sensing, [2]Laboratory of Work Psychology and Leadership, Helsinki University of Technology, 02015 HUT, [3]Avaruuskatu 4 B 20, 02210 Espoo. E-mail: Arzu.Coltekin@hut.fi, Matti.Vartiainen@hut.fi, Mkoc@netscape.net

1 Problem-Setting

A well-known problem among computer users is unexpected and annoying disturbances and breakdowns of computers. They do not only cause stress and strain, but also a lot of work-time waste. In this study, we investigate time-waste in computer use. Two different operating systems (Microsoft WinNT and Linux) are also compared and their effect on the problem is analysed.

1.1 Background

The technological progress has brought computers and computerised systems to everybody's life. Information and communication technologies have penetrated all the fields of leisure and working time with different applications and tools.

All of us agree on the benefits of computers, but we must also pay attention to the complications and problems they bring along these benefits. For instance, new physical health problems on back, hands and eyes are much discussed in the field of occupational health. Some diagnosis and treatment methods are being standardised for computer-related problems. One more important issue is their effect on work processes, how interruptions and breakdowns influence on work? Computers do have crashes, updates, network breaks, memory shortages, viruses, format mismatches, network congestion, new incoming technologies to learn and compete with and many more small obstacles.

1.2 Research questions

The stated challenges and time waste as their consequence lead to decreased productivity and motivation, which are important to analyze. Our research questions are:

1. *How much* time is wasted when computers break down for some reason and employees are waiting or fixing them to get their job done?
2. Are there differences in breakdowns and time-waste between two different operating systems (MS Win NT and Linux)?

2 Data and Method

2.1 Sample

A technical-scientific team (n=17) at Helsinki University of Technology (Institute of Photogrammetry and Remote Sensing) was chosen as the test group (Table 1). There were four females and 13 males. All were research scientists with engineering background except one secretary. The team had a computer expert particularly to solve computer-related problems. All used computers over half of their working time. Their work was very much dependent on computers. The employee may be helpless to produce anything without the computer, the data may be needed immediately and the server where the back-up files are may be down at the moment. All the participants had two operating systems installed in their PCs: Linux and MS Windows NT. Seven on them used both systems, five only Linux and five NT. To test the breakdowns and time-waste while using different operating systems, Linux- and Windows-groups were compared with each other.

2.2 Method and Procedure

The data was collected by *self-observation questionnaire* during 15 working days from each subject. The following questions were asked: the length of working day, hours working with computer, type of the problem, time used to fix the problem, time lost because of the problem, and operating system used.

In the end of the experiment, *interviews* were made. Interview questions dealt, for instance, with following topics: effects of problems on motivation, the source of problem, and evaluations of the questionnaire forms. Since one of the co-writers was working at the institute, observations and self-observations were also made.

The subjects were given all the 15 questionnaire forms at the beginning of the test period. In fact, the test days distributed along a longer period of two months. One of the writers visited all the subjects often to obtain the commitment and to collect the answers. The results were calculated using MS-Excel, and expressed in several graphs and percents comparatively. The interviews were written down and interpreted qualitatively with the results of the questionnaire.

3 RESULTS

3.1 Breakdowns and wasted time

Two hundred fifty-four working days were observed, and there were problems in 81 days (Table 1). Sixty-three percents of the 2111 working hours were worked with computers. Over four percents of the total working time were lost because of breakdowns. On an average, every scientist lost over half an hour every day because of breakdowns. Ninety active working hours, that is eleven working days were lost while fixing the problems among 17 scientists.

Table 1. The obtained results.

	NWD	DWP	WH	CT	TSF	TL	Linux	MS	OS usage	Profes-sion	age	Sex
Subj. 1	15	9	121.5	88	4.07	7.57	3	6	both	DI	25	F
2	15	2	145.2	114	24	0.5	1	1	both	DI	27	M
3	15	10	131.8	81.5	9.7	13.2	0	10	both	DI	29	F
4	15	6	104.5	94.5	6.25	6.25	1	5	both	cmp.sc.	29	M
5	15	5	118	77	2.34	3.84	3	2	both	DI	30	M
6	15	5	119	62	6.5	7.71	1	4	both	Li.Tech.	32	M
7	15	3	120	68	3	5	0	3	both	DI	49	M
8	15	3	117.05	71	2	3.5	3	0	linux	DI	29	M
9	15	1	125	104.5	3.5	3.5	1	0	linux	DI	32	M
10	15	5	122	89	2.15	2.5	5	0	linux	Li.Tech.	34	M
11	15	0	118	62	0	0	0	0	linux	DI	36	M
12	15	1	130	112	0.5	0.5	0	1	linux	Dr.	42	M
13	15	7	125	57.5	10.5	11.17	0	7	ms	DI	27	M
14	15	5	109.5	53.5	3.66	4.84	1	4	ms	DI	29	M
15	15	7	105	45	7	6.5	0	7	ms	Secret.	38	F
16	13	7	115	80	2.41	8.25	0	7	ms	DI	44	F
17	15	5	184	80	7	5	0	5	ms	Prof.Dr.	47	M
Total	253	81	2110.55	1339.5	94.58	89.83	19	62				

The abbriviations in the table: *NWD:* Number of working days, *DWP:* Number of 'days with problems', *WH:* total working hours, *CT:* Total computing time in hours, *TSF:* Total "time spent for fixing" the problem in hours, *TL:* Total time (hours) lost, *Linux:* The number of 'days with problems' with Linux operating system, *MS:* The number of 'days with problems' with MS Windows NT operating system, *OS Usage:* The

Operating System usage. In addition, *Profession*, *Age* and *Sex* were documented. Also in table: Subj.: Subject, ms: Microsoft, DI: Master of Science, Li.Tech.: Licentiate of Technology (a higher degree than MSc), Secret.: Secretary, F: Female, M: Male

About seven percents of the 'computing time' were lost and used for fixing the computer problems (Figure 1). Time used for fixing the problem is more than what the subjects expressed as the 'lost' time.

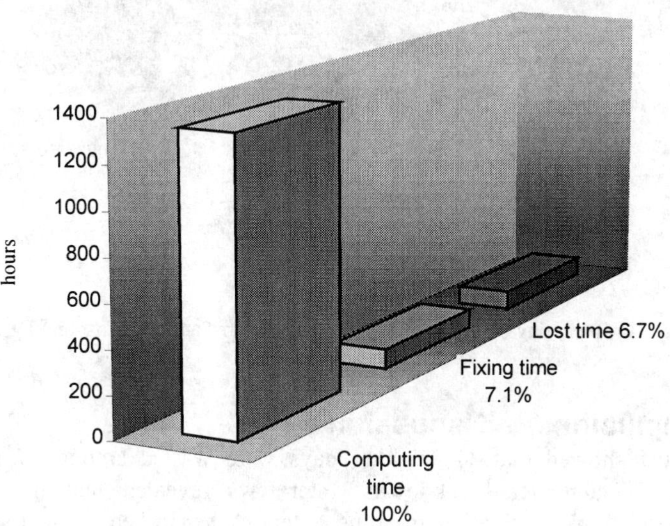

Figure 1. Time lost and used for fixing the problems from the total computing time.

3.2 The effect of operating systems

In all, MS users reported 62 and Linux users 19 problem days. Ten of the subjects used both systems. Because we do not know the exact amount of time for each operating system in these cases, we compared only those subjects who used either Linux or MS system. The data shows that MS users had more problem days, spent approximately 67 % more time for fixing problems, and consequently lost more working time (Figure 2).

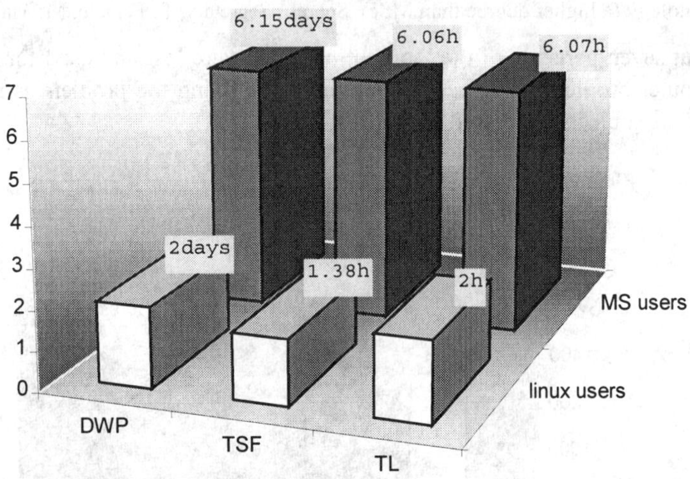

Figure 2. Problem days, time spent for fixing problems and time lost among MS and Linux users.

4 Conclusions & Discussion

The study showed that 11 working days were wasted among 17 scientists because of computer breakdowns. Interviews revealed that this lowered employees' work motivation to some extent. When asked about the biggest problem source, the number one was: "Most of the time, it is the user". It is a bit surprising that subjects did not blame hardware or software and usability problems. Users' tactics to solve problems are based on their own know-how. Our study group was a team of highly educated specialists. This has an effect on their action strategies. It would be important to repeat the study among different employee groups. If the study would be carried out in another type of organisation, which is less technology-oriented, but still works computer-dependably, more problems are likely to be reported.

There were differences in breakdowns and time-waste of different operating systems in favour of Linux operating system. It seems that operating systems have big influence on problems. The amount of subjects is, however, far too small to make any final conclusions.

Why extending ERP software with multi-user interfaces?

Roelof J. van den Berg
Ilse M. Breedvelt-Schouten
Baan Development B.V.

1 Introduction

In the past years many companies have invested in integrated ICT-support of their business processes through implementation of software systems for Enterprise Resource Planning (ERP). These ERP systems provide integrated support of processes related to e.g. manufacturing, warehousing and distribution, finance and accounting, human resources and customer relationship management.

ERP implementations often fall within a bigger scheme of increasing the potential of the organization as a knowledge processing entity. Market demands force organizations to restructure their operations with a "design for concurrency" approach in mind. The integrated support which ERP provides helps to structure the corporate flow of information. It makes the operations more transparant throughout the company, increasing awareness about interdependencies and thus triggering opportunities for more efficient ways of working.

At the same time the objective of boosting the knowledge processing capacity exposes room for improvement in current ERP systems. This is also reflected in the current global installed base of ERP systems - just over 3 million users - versus the potential one, i.e. the number of people doing work that could be supported with ERP functionality. This is more than 200 million people. Naturally, the difference stems from a complex of reasons - e.g. the relatively young age of ERP systems and their relative high price - but an important one among these is the current systems' lack of support of groupwork.

2 Towards ERP-support of group processes

An increasing number of organisations realise they have to invest in means for better internal communication, co-ordination and co-operation. Although ERP systems contribute to integrated support of processes, they are still underperforming in the areas of integrated support of groupwork in these processes. ERP systems are still very much based on a tayloristic view on the world. Their current architectures reflect the assumption that the complete set of work in the organization can be split up in atomic units of work (tasks, handled by an individual) which **do not interact during execution**.

As an integrated system ERP software of course supports dependencies between tasks, but only the sequential relationships: output of one task is input for another. For instance in warehousing the results of a quality check on inbound goods are needed before their actual storage can be processed. But interactions between tasks during task execution are not supported. Whether or not this is done makes the difference between actually supporting groupwork and only storing its results. Currently an ERP system only does the latter. An ERP system that would really **support** multi-user activities would have dramatically more appeal to a significant part of the world. This can be illustrated with the support for the Master Production Schedule, a central planning document in the MRP-II philosophy, in current ERP systems

Many systems allow the MPS (a table with planned production dates and expected sales) to be put in and used in further lower level planning. But they do not support the **creation** of the MPS. This can be a process of tough negotiations between representatives from the sales department and production. The first generally try to boost production to prevent it will limit their sales, the second generally try to limit production to prevent problems on the shop floor e.g. necessity to work overtime and long lead-times. Before an actual control action, the realization of a specific MPS, is taken a variety of arguments can go back and forth and a range of data sources will be used to support these. If ERP systems would support this unpredicatable process of negotation too with CSCW style functionality, effectively providing a virtual planning room, it would significantly increase their power and appeal.

3 Example: sales order processing

The scenario in this section relates to sales order processing. The most common way of executing this process for ERP users is order processing during telephone contact with the customer. Some ERP systems also have provisions which allow the customer to place an order via the internet (McKie 1998). In this last case there is no direct contact between the sales clerk and the customer. The example below is based on the situation of web-sales with real-time interaction between sales clerk and the customer, who is already known in the organization of the supplier. Their communication takes place via an alternative channel.

A customer visits the web page of a sales organisation and enters his name and/or customer ID in order to get into contact with a sales representative. The system identifies the customer and links him/her on-line to the appropriate sales representative. Once the connection is established, both the customer and the sales representative can communicate about the shared object via a chosen communication channel, which is in this case a 'chat' program. The shared object is a draft (for a sales order). It is not known beforehand whether this draft can be converted into a quotation or an order, since this depends on the wish of the customer.

The goal of the sales representative is to identify the needs of the customer and make a sound deal on price and delivery dates and finally be able to convert a draft into a quotation or an order. The goal of the customer is to get the correct products just in time at the right place, and for a good price.

The sales representative has access to the appropriate data for selling products, while the customer can only see or enter data that is provided by the sales organisation/or sales person. In fact, the sales representative's task of entering the draft is the basis of the co-operation with the customer. During the interaction, the sales representative and the customer are totally free in determining the order of entering the data, although only one value can be entered at a time. The customer's wishes are the basis for the actual working order, while the sales representative has all the 'power' on the task's functionality.

When the customer and sales representative agree on the draft, the sales representative can trigger the conversion of the draft into a quotation or an order. There is also a possibility for the sales representative to save the draft temporarily for later conversion. See Figure 1.

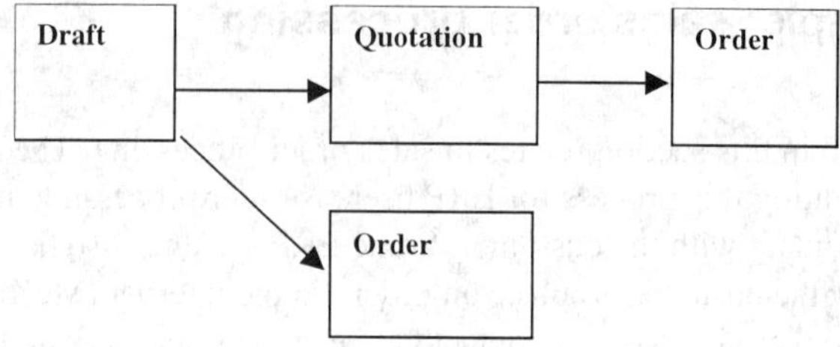

Figure 1: The available conversions for the objects Draft and Quotation.

Besides the 'standard' situation of having the draft as basis of the co-operative interaction, the possibility exists to interact about a quotation or an order as shared object. In these cases, the sales representative opens these objects. The sales representative only can edit the quotation attributes and convert the quotation into an order. The order, however, cannot be edited by the sales representative. The sales representative can share the quotation and the order with the customer.

4 Definition of multi-user tasks: cornerstone of effective distributed enterprise systems

More explicit support for tasks such as the one in the previous section may seem to represent merely marginal enhancements of the current ERP systems, especially on an individual basis. Yet, it is important to realize that support of multi-user tasks goes beyond the level of "nice to have" on the ERP-requirements list. The "death of distance" in organizational work has dramatic repercussions for the way dependencies between tasks, especially group tasks, have to be specified and supported in systems, in order to make them effective (Fielding et al. 1998). In a group task execution in a traditional same place/same time setting the interactions are completely transparent for everyone involved. Everybody is witness to the complete process, which accordingly can be tuned informally. It suffices to define (and support) only the output in advance. But when group tasks can be distributed in space and time even beyond the limits of the own organization a number of group task aspects immediately become much less obvious to those involved (Stiemerling et al. 1998), e.g.:

- identities, roles and authorizations of contributors;

- triggers to contributions and follow-up actions;

- timing and deadlines for contributions;

- deliverables of the task and checks for completeness.

It is essential that systems that support group work in distributed (virtual) environments are equipped to give explicit attention to such matters. An approach to define multi-user tasks in a much richer way than has been prevalent so far thus becomes a cornerstone of effective distributed enterprise software. In this light the two subsequent papers in this session will respectively concentrate on the modeling of multi-user tasks and the architecture to generate multi-user interfaces respectively.

5 Conclusion

Current ERP systems are powerful to integrate support for a wide range of business processes, but they do so on the assumption of single-user tasks only. ERP considers group tasks to be executed as "one man", it ignores the sophistication of the group process and focuses on the group output. Yet, against the background of the "death of distance" and the according level of freedom in distribution of business processes it becomes essential for organizations to use enterprise systems that do not treat multi-user tasks as a black box. Those ERP vendors who will meet the requirements for more sophisticated support of group work will be rewarded with the interest of a multitude, larger than their current customer base.

6 References

Fielding, Roy T. et al., "Web Based Development of Complex Information Products", Communications of the ACM, Vol. 41, No. 8, pg. 84-92

McKie, Stewart, "ERP meets Web E-Commerce", DBMS, Vol. 11, No. 8, July 1998, pg. 38-45

Stiemerling, Oliver and Cremers, Armin B., "The Use of Cooperation Scenarios in the Design and Evaluation of a CSCW System", IEEE Transactions on Software Engineering, Vol. 24, No. 12, pg. 1171-1181

Modelling Multi-Users Tasks

F.Paternò, G.Ballardin, C.Mancini
CNUCE, C.N.R.

1 Introduction

Task models are the meeting point about multiple views of an interactive software application because their full development often require the involvement of many expertises: the designer, the expert of the application domain, the end user, the software developer and so on. The need for this possible multiple contributions is derived also for their possible multiple use:

- *To improve understanding of the application domain*, in this case designers develop task models corresponding to the existing application and the modelling work can stimulate the request of many clarifications whose need may not be perceived just reading textual descriptions or documents;
- *To support effective design*, in this case designers develop an envisioned task model of a new application incorporating requirements received (for example to use some specific technology or to overcome some limitations for end user that occur in current applications) and the various information contained in the new task model can support the design and implementation of a concrete user interface;
- *To perform usability evaluation*, to compare how users interact with an application and the related task model (Lecerof & Paternò 1999) can be useful to understand limitations of the current design, whose improvement may imply changes in the corresponding task model.

Despite of the increasing recognition of the importance of task models in the HCI area there is a lack of engineered tools supporting their use. This limitation is a consequence of the lack of structured methods and precise notations for task models. In this paper we introduce and discuss how we are solving such a problem by developing an automatic environment for task models specified in the ConcurTaskTrees notation (Paternò 1999).

2 Modelling Multi-Users Applications

An important aspect to consider is that with the increasing availability of network applications in all type of environments there is a natural tendency to develop applications where multiple users can remotely cooperate to reach common goals. It is thus important to be able to develop task models where it is easy to express relationships among activities performed by multiple users.

Moreover, even if the idea of using task models to support design and development of user interfaces for single user interfaces has been considered in various research proposals we can note that less attention has been paid to the case of multi-users applications. While even interesting research environment supporting development for multi-users applications (Roseman & Greenberg 1996) have solved problems related on how to support flexible internal mechanisms but they are rather difficult to use. We can find modelling approaches that consider multi-users application such as UML but they give rather poor support to the design of the user interface and their description of tasks has some limitations.

We use the ConcurTaskTrees notation to represent task models. It is a graphical notation where tasks are logically structured in a hierarchy and with a wide variety of operators that allow designers to describe flexible, dynamic and interactive environments. It uses different icons to indicate how the task performance is allocated (to the user or the application, or an interaction of both).

The notation has been designed so as to be intuitively understandable even by people who have not background in notations and to be powerful enough to express a rich set of possibilities. In this respect it has been an improvement with respect to previous practice because some of the previous approaches either considered only sequential tasks (Card et al. 1983) or tended to specify a lot of details that are not particularly important in designing task models.

We have also developed a set of criteria to support the concrete design and development of user interfaces taking into consideration various aspects of the task model and the information contained in it. For example, the temporal relationships among tasks are a source of various useful indications. They allow designers to identify the tasks that are logically available for performance at any time. This can be useful to identify when activate the related interaction techniques thus contributing to define the dialogue part of the user interface. There are various levels of consistency that can be considered: a complete consistency can mean that only the interaction techniques associated with logically enabled tasks are perceivable and reacting. A weaker level of consistency is that there may be interaction techniques associated with tasks disabled that can be perceivable (but not reactive) to allow the user to understand their relationships with those reactive. A further possibility is to allow some interaction techniques associated with not enabled tasks to be reactive. This could allow users to change how to use the application. Even if this last option can easily generate the possibility to allow the user to make errors, actions not useful for the current tasks.

We have also developed a method supporting the possibility to obtain user interfaces consistent with the information contained in the task model. The first problem we have addressed is to identify the tasks that can be supported by the same presentation unit. To this end we have developed a tool that is able to take as input a ConcurTaskTrees specification and then to find the enabled task sets which are sets containing tasks that are enabled during the same period of time. Tasks belonging to the same set should be supported by the interaction and presentation techniques provided by the same presentation unit because they are logically enabled at the same time. Even if, as we have seen, this is not mandatory and designers in some cases may want to follow other rules.

We have then developed some further rules to identify the possible presentation units and also the specific interaction techniques associated with each task depending on its type and the type and cardinality of the objects that it manipulates.

The description of cooperative activities is currently done in ConcurTaskTrees by developing a task model for each user role involved and then developing a similar task model dedicated to what we call cooperative tasks, tasks that in order to be performed require activities by multiple users. In this part, the cooperative part we indicate the temporal relationships among such cooperative tasks and we decompose them until we reach tasks that are performed by only one user. These single user tasks are indicated in the cooperative part and then, if they are further refined, this decomposition is described in the task model of the corresponding role. This type of solution has various advantages: it is modular so that if designers are interested in the activities of a specific user they can consider only the related part as well as they have the possibility to focus on the cooperative aspects.

3 Tool Support for Task Modelling

We have developed an approach (Paternò & Mancini 1999) supported by a tool (available at http://giove.cnuce.cnr.it/EL_TM.html) that allows designers in the initial phase when they have to develop their task model from scratch. We took into account that usually a lot of information is available in informal material, such as scenarios or use cases descriptions. Usually this information contains many indications of what tasks should be supported and the objects that they have to manipulate. Thus our tool supports their selection and their placement in a logical framework from where further development of the task model would be easier especially with respect to when designers have to start from scratch.

The ConcurTaskTrees notation allows designers to specify task models hierarchically structured with the possibility to indicate temporal relationships among tasks, objects that they manipulate, how their performance is allocated and so on. Besides, it is possible to specify cooperative applications where different roles are involved, for this purpose there is a part dedicated to the cooperative aspect that, using the same notation, allow designers to indicate relationships among activities performed by different users.

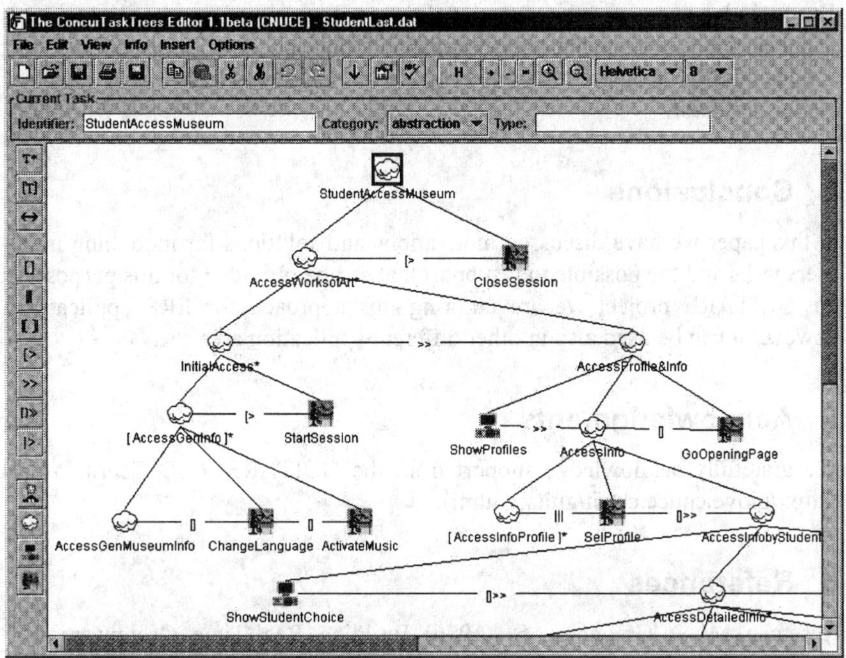

Figure 1: The editor of ConcurTaskTrees task models.

The editor (see Figure 1) supporting the development of ConcurTaskTrees task models (available at http://giove.cnuce.cnr.it/ctte.html) currently provides mainly three types of functionalities (we plan to introduce further features):

- *to allow designers to input the information requested by the notation*: tasks and their attributes (name, category, type, related objects, …), temporal relationships among tasks and other information;

- *to help designers in improving the layout of their specification*: supporting automatic lining up, movement of levels or subtrees of the task model, cut and paste of pieces of task model, folding and unfolding of subtrees, possibility to add tasks as brothers or as child of the current task and providing statistics concerning the task model (number of tasks by category, number of temporal relationships by operator, number of levels, ...);
- *to check correctness of the specification*, for example it is possible to automatically detect whether temporal operators among tasks at the same level are missing or whether there is a task with only one child.

The tool also supports other functionalities such as inserting other task models in the current one or saving images containing the task model or part of it so that they can easily included in reports or documents. Besides it also includes a task model simulator.

4 Conclusions

In this paper we have discussed motivations and solutions for modelling multi-users tasks and the possible tool support that can be provided for this purpose. In the GUITARE project we are tailoring this approach for ERP applications, however it can be used also in other different application areas.

5 Acknowledgments

We gratefully acknowledge support from the GUITARE R&D Esprit Project (http://giove.cnuce.cnr.it/guitare.html).

6 References

Roseman M. & Greenberg S. (1996) Building Real-Time Groupware with GroupKit, A Gropware Toolkit; *ACM Transaction on Computer-Human Interaction*, Vol 3 N°1 March 1996; 66-106.

Card, S., Moran, T., Newell A. (1983) *The Psychology of Human-Computer Interaction*, Lawrence Erlbaum, Hillsdale, N.J., 1983

Lecerof A., Paternò F. (1988) Automatic Support for Usability Evaluation, *IEEE Transactions on Software Engineering*, October 1998, pp.863-888.

Paternò, F., *Model-Based Design and Evaluation of Interactive Applica*tion. Springer Verlag, 1999.

Paternò F., Mancini C. (1999) Developing Task Models from Informal Scenarios, *Late-Breaking Results ACM CHI'99*.

Architecture for Multi-User Interfaces

Ilse Breedvelt-Schouten
Baan Labs, Ede, Netherlands

1 Introduction

Below an approach is presented that uses task models as the basis for development of user interfaces. This approach was chosen for the future user interface development and runtime environment for ERP systems at Baan. Since communication via computer systems is and will remain an important part of daily life, we need to incorporate Computer Supported Cooperative Work (CSCW) features into this environment. This paper discusses the architecture for the task modeling environment and the implications of incorporating CSCW characteristics of this architecture.

2 Generating User Interfaces for a Single User

A user interface of ERP systems consists of both interaction semantics and a graphical presentation, which are independent elements. For instance changes in button appearance are independent from the interaction structure of a dialog. Next to the user interface, there is also application logic captured in Business Objects. In the ERP application domain, there is a strong requirement that the application can be customized to specific business and user needs. Therefore, the presentation and the interaction semantics must be separately customizable.

Baan chose the task modeling approach (ConcurTaskTree notation) of CNUCE as a basis for the development of a user interface and runtime environment in order to fulfil the requirements of separation of presentation, interaction semantics and application logic (see Figure 1). At runtime the UI and the Business Objects are coupled to each other via the Task Interpreter which connects actions of the user to actions of the application logic and vice versa. The Task Interpreter needs a Task Model and a Form as input for the runtime application instantiation. The Form is used to create the graphical presentation and the Task Model is used to compose the interaction semantics. In the *Task Model Editor*, the developer designs a task model that specifies the interaction semantics between the user and the application at the level of attributes and methods of Business Objects. The leaf nodes of a Task manipulate an attribute or a trigger of a method. Based on the task model, pre-defined user interface elements are available in the *Form Editor*, where the developer can design the presentation and layout of the user interface. The interaction semantics of the user interface do not need to be created in this Form Editor, since the task model contains this information. In the future we plan to use presentation and layout rules to automatically create a user interface in the Form Editor from the Task Model.

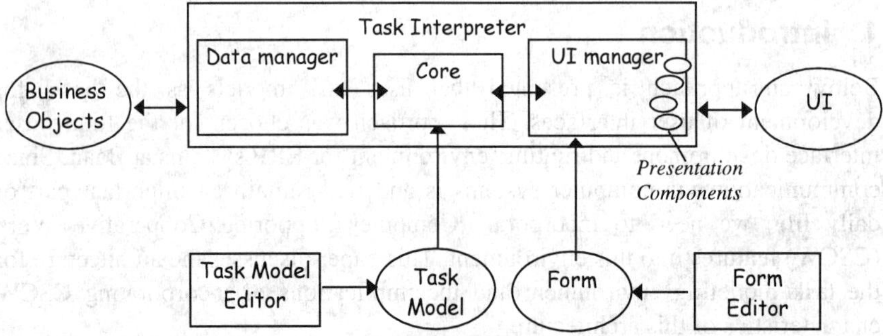

Figure 1: The Architecture of the Task Modeling Environment

The Task Interpreter contains three parts: the Core, the UI manager and the Data Manager. The Core holds the runtime task model. The Data Manager handles the interaction with the Business Objects, while the UI manager handles the interaction between the user and the user interface elements with help of abstract Presentation Components defined by the form. These components are the abstract descriptions of the UI elements and have a direct relation to nodes in the task model.

3 Multi-user Requirements for ERP

Current ERP systems are typically designed for single users who interact with a business system in order to perform one predefined activity. However, in reality often more people are involved in performing one task. Since Baan is working on Task Modeling for user interface development and runtime environment, the need exists to look ahead and make the environment suitable for multiple users. This section describes a set of ERP-specific Computer Supported Cooperative Work requirements for the Task Modeling architecture.

Synchronous interactions must be supported. In ERP systems there is already an asynchronous facility for multiple users who are cooperating: Workflow. To make ERP systems really multi user enabled, you also need to cover synchronous interactions.

Users must be able to communicate 'informally' about a 'shared object'. A shared object in an ERP environment can be anything from Sales Order to Purchase Order to Warehouse Inventory. The users act on these objects via a task, which can be shared with other users. In order to communicate about these shared objects the user interface must support the ability to communicate easily and 'informally'.

Specific UI support is needed to help the users in communicating about a shared object and to be aware of each other's actions. When two users are working remotely on the same attributes of an object, they need to be aware of the actions of the other users. Since both users do not have the same screen in front of them, we need to use WYSIWISY (What You See Is What I Show You), also known as relaxed WYSIWIS (What You See Is What I See). In order to still see where the other user has the focus on, specific UI support is needed (e.g., semantic Telepointers, highlighting of activated text fields).

The communication about the shared objects should take place via a separate communication channel, depending on the functionality of the user's client software. This channel can be anything from telephone to videoconference. The task model should not fix the communication channel to be used. It should be left to the user to choose a channel, depending on his possibilities.

Only two users should be involved in a cooperative session. In the current ERP system, communication via the system is very limited. A maximum of two users involved in one cooperative task is already a large improvement. Therefore, we do not need to consider more than two users. This however, does not limit the amount of cooperative tasks one user can be involved in. E.g., a Sales Representative can cooperate with a Customer, but when a certain question arises about the Inventory, the Sales Representative can contact a Warehouse person and have a separate cooperative task with him.

Once a cooperative session is started, the identification of the other users must be provided. In ERP systems, a lot of information is legally sensitive and often limited to certain roles. Next to the fact that the system will recognize the users, it is also necessary and important that each user is aware of the identity of the other cooperation .

There must be a strict separation of public and private elements. It is important for users of cooperative tasks to be able to limit or expand the public data that might be shared with the other. The granularity of the information that can be made public or private can differ, but the minimum must be based on the level of one attribute of a Business Object or one method trigger, because the smallest unit in task modeling is also on that level.

In case of data sharing and data hiding, flexibility is desirable to change access rights for data in order to transform public data into private data and vice versa. It is very difficult to know in advance what information may be shared or must be kept private for every cooperative session. Of course, there is information limited to certain roles, but in non-routine situations people might decide that someone else is allowed to see more or less than 'normal'. Therefore, it should be possible to change the access rights for data during a cooperative session in a flexible manner when necessary.

It is important to note that only 'senders' of a shared object can determine what is public and private. The 'receiver' is dependent on what a sender of the information is 'willing' to give. Thus it would be nice if senders should be able to over-rule access rights and make public what can be shared. A sender can set the following settings on attribute and method trigger level for a receiver:

- not shared with receiver
- shared
  - output only: receiver can only see an attribute or method trigger
  - editable: receiver can edit an attribute or trigger a method

4 Multi-user Interface Architecture

The Task Modeling environment must be made suitable for multiple users cooperating on shared objects. In continuation of the separation of interaction semantics, presentation and application logic, the cooperation features should be captured in a separate layer in the Task Modeling architecture. Based on the thesis of Brinley Smith[1995] we identify the need of three models to support cooperation in the architecture: Access Control, Policy Nodes, and Roles.

Access Control. In the Access Control Editor one can indicate whether a task may be shared and if so, how it is shared (editable or output only). These sharing properties have direct consequences on the user interface presentation of the different users. If, for example, one user has access to all the information of a task, and the other user can access only a part of this information, the user interfaces are not similar. However, both users are interacting on this shared task. The Access Control properties need to be coupled to the Presentation Components in the UI manager of the Task Interpreter.

Policy Nodes. Next to presentation issues, also the interaction with the application contains cooperative aspects. The question is what to do when multiple people are triggering one method? In such a case, the input of the users must be interpreted before a method in the application logic can be called. An example is a situation where multiple users can vote for one certain type of pizza to be ordered. Depending on what the 'majority' chooses, the correct type of pizza will be ordered. For this, simple rules in the "If <trigger> then <response>"style can be used. Triggers can either be selective (based on user role) or consensus (based on (relative) number of users selecting it). For this type of functionality, *policy nodes* are coupled to the task nodes representing the trigger. When this task is triggered, the policy node determines what information is passed to the application logic.

Roles. Modeling the roles has a direct relation with the Access Control and the Policy Nodes. For the Access Control, the role indicates what part of the information is available. The relation with the Policy Nodes is that the roles of the users involved in a cooperative session are used in the interpretation of triggers before the method in the application is called. In the Role Modeller the developer indicates what roles are involved in the cooperative tasks: names, types of users, etc.

Architecture. The cooperative features need to be incorporated in the architectural design of the Task Modeling environment. The architecture needs to be extended with Access Control, Policy Nodes and Roles. These aspects have a direct relation with the Presentation Components, which are based on the task model leaf nodes. At development time the task models should be used to indicate what the access rights and input interpretations are. At runtime, this information must be used as filter between the UI manager and the Presentation Components.

5 References

Gareth Brinley Smith (1995). *A Shared Object Layer to Support Cooperative User Interfaces*, Computing Department, Lancaster University, UK.

Richard Michael Bentley (1994), *Supporting Multi-User Interface Development for Cooperative Systems*, Department of Computing, Lancaster University, England.

Paternò F., Mancini C., Meniconi S. (1997), *ConcurTaskTrees: A Diagrammatic Notation for Specifying Task Models,* Proceedings Interact '97, Chapman&Hall, pp.362-369

TOWARD USABLE BROWSE HIERARCHIES FOR THE WEB

Kirsten Risden
Microsoft Research

1 Introduction

The World Wide Web presents both new challenges and opportunities for conducting Human Factors work. Browse hierarchies used to classify web content present an interesting case in point. On the one hand, the size of the domain to be classified and its general-purpose intent (i.e., it is intended for all users and retrieval of all types of information) make traditional techniques such as card sorts too unwieldy. On the other hand, the nature of the Web makes efficient data collection from a relatively large number of people a possibility. The purpose of this initial, small-scale study was to begin to investigate usability methods that have the potential to scale to the range of users and tasks and at the same time take advantage of the data collection possibilities that exist for the web (e.g., server log data.)

A coherent, learnable category structure is a central goal for browse hierarchies such as those on Yahoo, Excite, msn.com and other Internet portals. Such a structure will allow users to efficiently find the information they need and to become more and more proficient in using the hierarchy over time. We know from cognitive psychology (Rosch 1975) that coherent, learnable category structures have high within-category similarity and high between category discriminability. For abstract categories, such as those found in browse hierarchies for the web, we also know that linguistic cues that highlight relevant features of the categories can be important (Horton and Markman 1980). Categories whose members go together in a loose way, have high overlap with other categories, or are represented with general labels that do not highlight reasons for category membership should be difficult for people to use.

Beyond the difficulties of size and the general-purpose nature of browse hierarchies in the creation of usable browse hierarchies, the fact that once a browse hierarchy is released to the Web content continues to be added makes maintaining a good user experience challenging. Changes to the makeup and size of categories mean that the category structure can continue to grow and evolve new categories and new category structure. Accommodating changes may have unanticipated effects on the user's ability to find information. Taking the hypothetical example in Table 1 below, evolving from an "Arts" to an "Arts & Humanities" category may lead to confusion between this new category and a "Culture & Society" category within the same browse hierarchy. This is presumably because increases in the generality of a category demand more encompassing labels, which in turn, allow for greater overlap between category content. So in addition to doing usability evaluation during the creation of a browse hierarchy, there is a need to do usability "check ups" on and ongoing basis.

ARTS	ARTS & HUMANITIES	CULTURE & SOCIETY
Art History	**Architecture**	Architecture
Artists	Art History	Art History
Design Arts	Artists	Culture
Museums	**Culture**	Death and Dying
Theory	Design Arts	Fashion
Visual Arts	**Humanities**	Food and Drink
	Museums	Gender
	Musical Arts	Religion
	Performing Arts	Holidays
	Photography	Mythology
	Theory	
	Visual Arts	

Table 1. Illustration of how changes within one category (Arts) may lead to greater similarity and user confusion across categories (Arts & Humanities versus Culture and Society). Added categories are bolded.

Clearly, it would be nice to have a way to continually monitor how effectively and efficiently users can locate information as well as the source of any difficulties they are experiencing. Users should be able to make a direct path through the hierarchy to the content they want easily identifying which categories lead to the information and differentiating them from categories that will not lead to the desired information. Traversals between major categories of the browse hierarchy on the same information retrieval task would be evidence of confusion and an indication of usability problems in areas where traversals are common. Such data may be obtained through server logs under certain circumstances. The goal of the following study was to determine the potential usefulness of tracking traversal patterns through a browse hierarchy as a way to monitor confusion and determine its source.

2. Study design and methodology

5 participants were asked to complete 35 information retrieval tasks using an experimental browse hierarchy. The hierarchy was presented to subjects within a software tool that displays categories in a simple hierarchical format and records user paths. This software set the tasks in a context in which user interface problems would not interfere with the specific interest in the usability of the categories. The 35 information retrieval tasks were based on popular activities on the Web, the participants did the tasks in different orders, and they were allowed to "back up" in the hierarchy if they thought they needed to use a different area.

Of particular interest were the top-level categories explored on a given task. This was, in part, to simplify analysis and, in part, because top level categories pose challenges to finding information in a browse hierarchy. Top-level categories tend to be more general. As a result it is often quite difficult for users to determine which top-level category a particular topic is likely to be in. If the methodology I am proposing is useful, it should be sensitive to this difficulty and should expose the primary sources of user problems. The top-level categories of the browse hierarchy used in this study are shown in Table 2 below.

Business & Finance
Computers & Internet
Entertainment & Media
Health & Fitness
Home & Family
Interests & Lifestyles
People & Communities
News & Information
Reference & Education
Sports & Recreation
Travel & Leisure

Table 2. Top level categories of the browse hierarchy used in this study.

3 Results

Exploration of multiple top-level categories on the same task was assumed to indicate a lack of certainty or confusion regarding where the information would be located. Each top-level category explored on a given task was scored as "confused" with each of the others explored on that task. For example, if the task was to look for information about bike riding and the user looked in Health & Fitness first, Interests & Lifestyles second and finally settled on Sports & Recreation, then each of these categories would be scored as being confused with one another on this trial. Confusability matrices were constructed by tallying the number of times each pair of categories was confused across tasks and subjects.

Analysis of these data was organized around three questions. 1) How prevalent is confusion between top-level categories in the browse hierarchy? 2) What is the structure of this confusion? 3) What is the source of the problem?

The average frequency with which top-level categories were confused with one another was 18.55 across the 35 trials. 44% of that occurred during the first ten tasks for an average frequency of 8.18. The average frequency with which top-level categories were confused on the last ten tasks was 3.82. This indicated that there was a substantial problem with differentiating the categories from one another; one that diminished but was nonetheless present even after 25 trials.

To determine the structure of confusion, a network representation was created. (See Chi 1983 and Chen 1997 for other examples of using networks to understand concept relationship in complex information sets.) Categories that were confused were linked together in the network[1]. This network is shown in Figure 1. The overall pattern reveals that the vast majority of user confusion involved the Interests & Lifestyles and News & Information categories. A separate network constructed for just the last 10 tasks (not shown here) that users performed showed that Interests & Lifestyles continued to cause confusion even after a substantial number of trials.

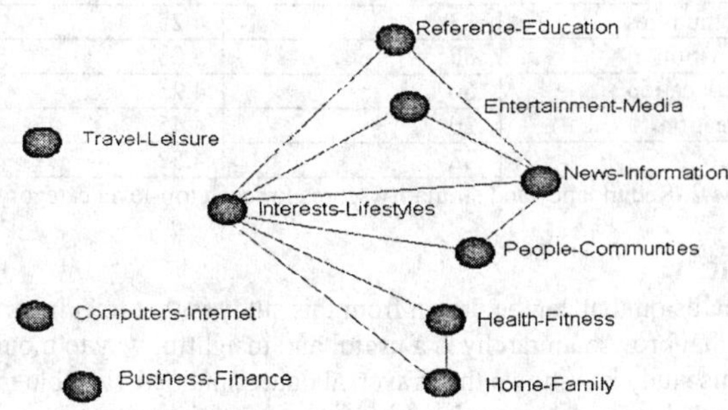

Figure 1. Network representation of category confusion. Categories that were frequently confused with one another are linked.

Examination of the content of these and other categories in the browse hierarchy showed substantial redundancy. Many sub categories were "members" of two or more of the top-level categories. The proportion of redundant sub categories ranged from a high of .89 to a low of .00 as shown in Table 3. Follow up analyses showed a strong correlation between the proportion of redundant sub categories and the frequency with which a top-level category was confused with other categories ($\underline{r}$ (9) = .76, $\underline{p} < .05$). This finding suggests that a high level of redundancy makes it very difficult for users to learn to differentiate one category from another.

Overly general labels may also fail to provide linguistic cues that highlight differences between categories. For example, most of the categories could be viewed as "interests" or "information". The use of these highly general words in Interests & Lifestyles and News & Information may make it difficult for users to distinguish between these and other categories. In other words, the more similar a label is to other labels the more frequently it will be confused with other categories during information retrieval tasks. To determine whether this is the case, a separate set of subjects was asked to rate each pair of labels on a 7 point Likert scale. Average similarity scores are provided in Table 3. Higher numbers indicated greater similarity. The correlation between similarity ratings and frequency of confusion was strong and significant ($\underline{r}$ (9) = .71, $\underline{p} < .05$) indicating that categories with more similar labels were more likely to be confused with other categories in the set.

---

[1] For clarity of presentation, only those category pairs confused four or more times are linked. This accounts for 56% of the total confusion data and clearly illustrates the major patterns in the data.

Top level category	Proportion redundant sub categories	Average similarity to other labels.
Business & Finance	.00	4.35
Computers & Internet	.14	4.12
Entertainment & Media	.20	4.30
Health & Fitness	.00	4.35
Home & Family	.80	4.75
Interests & Lifestyles	.89	5.15
People & Communities	.83	4.25
News & Information	.80	5.08
Reference & Education	.60	4.25
Sports & Recreation	.20	4.45
Travel & Leisure	.33	4.57

Table 3. Redundancy and similarity scores for each top-level category.

4 Discussion

The major conclusion that can be drawn from this study is that tracking traversal patterns through a browse hierarchy is a useful and insightful way to monitor user experience. This study has shown that traversal data contained valuable information about usability problems in one browse hierarchy. More importantly, analysis of the traversal data revealed the source of confusion, and permitted diagnosis of why it occurred. Specifically, the users in this study experienced a significant amount of confusion in using a browse hierarchy. The source of confusion was pin pointed to two major categories by mapping the data into a network representation that makes inter relations among categories explicit. Finally, the patterns observed in the network representation were validated against measures of learnable category structures. This demonstrated that those patterns were indeed rooted in users' psychological experience of the browse hierarchy and provided both an explanation for why confusion was occurring where it did and what to do to alleviate it.

The next step is to generalize the methods and approach used in this small scale study to data from large numbers of people carrying out their own information retrieval tasks in real world settings. Tools for automatically collecting and analyzing such data will need to be developed to make such work tractable. However, the opportunity to continually monitor user experience in a dynamic and changing software environment is likely to be worth the effort.

5 References

Chen, C. (1997). Structuring and visualizing the WWW by generalized similarity analysis. In *Proceedings of the 8th ACM Conference on Hypertext*, (Southampton, U.K., April). Pp. 177-186.

Chi, M.T.K. & Koeske, R.D. (1983). Network representation of a child's dinosaur knowledge. *Developmental Psychology*, 19, 29-39.

Markman, E. M. (1980). Developmental differences in the acquisition of basic and superordinate categories. *Child Development*, 51, 708-719.

Rosch, E. H. (1975). Cognitive representations of semantic categories. *Journal of Experimental Psychology: General*, 104, 192-233.

Research Methods for Next Generation HCI

Mary Czerwinski
Microsoft Research

1 A Formal Research Approach

Human-computer interaction research benefits from the scientific method as much as any other field and it should be theory-driven to the greatest degree possible in order to make significant advancements. The leading researchers and their corresponding important contributions have all followed this path, and the quality of their work is a testimony to the success of this approach (e.g., Card, Moran & Newell 1983; Guiard 1987; Shneiderman 1998; Tullis 1997). As Landauer (1997) pointed out, there are at least 3 goals for research in the field of HCI. First, research is required in order to understand what system to build and what features and functions to design for which users and tasks. Secondly, we need to perform research to develop better scientific theories, models and principles about user behavior during computing tasks. Finally, the goal of obtaining high quality guidelines or standards for user interface design that can be handed off to designers and developers requires formal research. But how do we go about performing these badly needed areas of study? According to Landauer, research in HCI differs to varying degrees from the scientific processes and methods used in such areas as psychology. Specifically with regard to formal lab studies, significance testing and control, the ideal cannot always be attained in HCI research. In addition, the design process does not always lend itself to systematically testing one variable at a time. This perhaps should not be so surprising, since the discipline of HCI uses mostly an engineering process model, and pulls from such varied areas of expertise as Computer Science, AI, Linguistics, and Psychology, among others (Preece 1994). In what follows I will briefly attempt to formalize a scientific approach to research in HCI that can be practically used in the design of user interfaces now and in the future.

It is extremely important that we follow a formal scientific process in HCI in order to move our domain expertise forward and communicate broadly amongst practitioners. Of course, there are practical demands on the HCI practitioner that directly conflict with performing research in this manner. Product team members want study results yesterday, the prototype is not fully featured enough, or our competitors' products have too different a set of features and functions to be directly comparable in a rigorous experimental setting. Still, the practitioner can always isolate a question or prediction about a planned user study that will lend itself to an empirical answer if enough creative energy is applied to the problem. Once that question has been identified, the following set of methods can be put into play.

- *Identify the Question*--Make it a deep enough one in every study so as to move your team/product/field of interest forward
- *Literature Review*—Use resources such as the HCI Bibliography (http://www.hcibib.org) for a quick first pass to see what is out there in our field's journals, books and conference proceedings. Then read the two or three key papers to get a good overview of where the field thinks the important research issues lie. It may actually be that your question has already been addressed and you can communicate this knowledge back to the team!
- *Design the Study*—This is the most challenging task and requires creativity to apply paradigms and design solutions from other fields that might assist in answering your problem at hand. Often this is where we can help our teams or field the most—by contributing a new method for asking a question that hasn't been used to date. These new paradigms can be reapplied to answer similar questions and hence check reliability, validity and generalizability of the original results.
- *Determine Measurement Method*—Task Time, Error Time, Number of Tasks Completed Successfully, etc. are all valid measures to use in studies. However, creating new measures that are meaningful to your team is a completely valid solution and should be encouraged.
- *Determine the Number of Participants Needed to Answer Question*— Run fully counterbalanced, within subjects or Latin Square designs to boost statistical power since running more subjects is usually prohibitively expensive.
- *Isolate Variable(s) of Interest*—Ensuring that you have controlled for order effects and task/question/item effects is key to teasing apart the contribution of each variable and their interactions.
- *Perform Careful, Unbiased Observations*—Fully instrumenting your system, so that potentially every keystroke is time stamped and every path

navigated is identifiable, is the best defense here. It is often the case that in retrospect an unidentified confound or question is brought into focus as you interpret your data. In addition, check to make sure that your random number generating programs are really working, that your time stamping is fine-grained enough for the question at hand, and that you take care notes synchronized to your video equipment in case you need to go back to the tapes for more data. To the greatest degree possible, remain blind to the condition any subject is in when recording data.

- *Analyze and Interpret the Data*—If you do not have a statistics background, be sure to enlist expertise so that you know that you are running the proper statistical tests without violating any of that test's assumptions. It may only be necessary to look at measures of central tendency in your data (e.g., averages, medians or modes of your data distributions), for practical purposes. Determine confidence intervals around your measures of central tendency so that you can properly indicate graphically which variables or conditions were significantly different from the others.

- *Communicate your Findings*—Product teams will want your results from your study as soon as you finish running the last participant. Go ahead and satisfy them by sending out a quick write up (1 page is usually sufficient) of the study's findings and any key usability issues requiring immediate redesign. Then, take your time to carefully examine the data you have collected, so that you can tell a meaningful story. Give yourself time to digest the data—don't be surprised if the story you develop explain the findings changes as you think about it more and more deeply. This is normal, and is most likely an indication that you are learning something very important about design. When you are fairly sure of your findings, write up a formal report of the study. The exact instructions and materials given to subjects should always be provided as an appendix. The report will provide an important historical snapshot for others interested in these particular HCI questions.

2 Example Cases

I will go through a few cases wherein we have used the above approach, and learned something significant in the process. These projects include research topics' ranging from multi-modal input to adaptive user interfaces and provide a nice overview of the wide variety of user interface research projects going on in advanced user interface design today.

Understanding how to measure cognitive load, selective attention, memory and situational awareness have become critical to the HCI professional. As a first example, psychophysical studies in our group have shown that extremely subtle auditory cues (often no more than 50-200 msec. in duration) can provide a wealth of semantic meaning to computing events associated with large information spaces. However, the key is to utilize these cues as an extra, complementary channel of information that does not interfere with the current task at hand (Wickens 1984). In order to ensure that the auditory display is indeed accomplishing the goal of a high meaning to disruption ratio, dual task studies are being performed in which the user is interrupted by an auditory cue while engrossed in a primary visual task. A familiarity with this dual task paradigm, and how to analyze data from it, has gone a long way toward helping us to formalize the key questions involved in peripheral, auditory displays. In addition, we're learning a lot about auditory cueing, as well as the cost of interruption based on how busy the user is.

Another example comes from our work with 3D information visualization. In this research area a working knowledge of effective ways of combining andutilizing 3D depth cues has been critical. In addition, a solid background in spatial memory and the updating of spatial orientation in cognition has allowed us to make interesting predictions and interpretations on our findings in virtual environments. Some of our studies have been heavily influenced by psychologists studying these issues, both in the physical environment--the "real world" (Franklin & Tversky 1990) and electronically (Hightower et al. 1998). Interesting parallels between the real world, desktop VR (virtual reality), and immersive VR have not only helped us build extremely effective designs, they have also led to new theories of perception and spatial cognition in VR. Understanding the guidelines and principles from years of basic research is necessary for professionals working in the domain of user modeling or the design of novel input devices (Card et al. 1983), (Tullis 1997), (Kabbash et al. 1994). In our work on two-handed input plus speech and touch-sensing devices, for example, we have adopted the paradigms, theories and models of those in the literature to develop formal experiments looking at the costs or benefits of using these new techniques. Now that these methods have been developed, it has been easy to think of ways in which to swap out certain variables and replace them with others as we explore the design space. In the area of user modeling, we can easily model the effectiveness of visual designs based on models of visual search and attention, without running studies if necessary.

3 Conclusion

In this article I attempted to outline a solid research methodology for an HCI professional. It is my hope that I have hinted at what it might take to contribute to a large body of knowledge that can be shared within our community as we advance toward a robust theory of human-computer interaction.

4 References

Card, S., Moran, T.P., & Newell, A. (1983). *The psychology of human-computer interaction.* Hillsdale, New Jersey: Lawrence Erlbaum.

Franklin, N. and Tversky, B. (1990). Searching imagined environments, *Journal of Experimental Psychology: General, 1*, 63-76.

Guiard, Y. (1987). Asymmetric division of labor in human skilled bimanual action: The kinematic chain as a model, *Journal of Motor Behavior, 4*, p. 486-517.

Hightower, R.R., Ring, L.T., Helfman, J.I., Bederson, B.B., & Hollan, J.D. (1998). Graphical multiscale web histories: A study of PadPrints. *In Hypertext '98: The Proceedings of the 9 th Conference on Hypertext and Hypermedia*, Pittsburgh, PA: ACM, 58-65.

Kabbash, P., Buxton, W., & Sellen, A. (1994). Two-handed input in a compound task. In the Proceedings *of CHI '94: Human Factors in Computing Systems*, Boston, MA: ACM, 417-423.

Landauer, T.K. (1997). Behavioral research methods in human-computer interaction. Chapter in Helander, M.G., Landauer, T.K. & Prabhu, P.V. (Eds.), *Handbook of Human-Computer Interaction*, 2 nd Edition, Amsterdam: Elsevier, pp. 203-228.

Shneiderman, B. (1998*). Designing the User Interface: Strategies for Effective Human-Computer Interaction*, 3 rd Edition, Reading, MA: Addison-Wesley.
Tullis, T.S. (1997). Screen design. Chapter in Helander, M.G., Landauer, T.K. & Prabhu, P.V. (Eds.), *Handbook of Human-Computer Interaction*, 2 nd Edition, Amsterdam: Elsevier, pp. 503-532.

Wickens, C.D. (1984). *Engineering psychology and human performance.* Glenville, IL: Scott, Foresman.

Remote Usability Testing through the Internet

Bo Chen[1]**, Mark Mitsock**[1]**, Jose Coronado**[1] **and Gavriel Salvendy**[1]
[1] Hyperion Solutions Corporation, Stamford, CT, U.S.A.
[2] School of Industrial Engineering, Purdue University, West Lafayette, IN, U.S.A.

1. Introduction and background

Conventional usability testing takes place in a usability lab equipped with audio-visual equipment and scanning software. The lab environment allows the experimenter to scan the screen to record the user's interaction, and combined with a think-aloud protocol, to record the users real-time comments and facial expressions. Logging software can gather performance data. The results are high quality data for analysis (Nielsen 1994).

Remote usability testing through the Internet has the potential to allow the experimenter to gather data of equally high quality without the test subjects having to be physically present at the usability lab. With remote testing, the experimenter at the usability lab uses Internet conferencing software to initiate a collaboration session and to share a software prototype on the usability lab's computer with a subject at a remote site. Because the prototype is running on the usability lab computer, the remote subject's interactions with the software are scanned by the lab's equipment and performance data is logged by the logging software. The remote subjects' real-time comments are recorded by the lab's audio equipment. Optionally facial expression can be recorded if the remote subject has video equipment at their site. Lab-quality data can thus be obtained without incurring prohibitive costs in travel time and expense (Hartson *et al.* 1996).

Because many of our clients are multi-national companies with locations all around the globe, we find the potential for gathering high quality usability data over the Internet very attractive. We wanted to determine whether being at a remote location would have an impact, positive or negative, on the performance of usability test subjects or on their satisfaction ratings of the software. We also wanted to discover the strengths and weaknesses of the remote testing environment as compared to the in-person testing environment. We conducted an 'in-person' usability test on a real software prototype using a group of locally recruited subjects, and then repeated the same test with a group of remote subjects via the Internet.

The concept of remote usability testing through the Internet combines two major research areas, CSCW or "Computer-Supported Cooperative Work", and usability testing. CSCW focuses on how people work together using computer technology. Typical applications for CSCW include email, awareness and notification systems, video conferencing, chat systems, multi-player games, and real-time shared applications, such as collaborative writing or drawing. Remote usability testing is a natural application for CSCW. It can be thought of as a collaboration over the network between the experimenter and the usability test subject to evaluate the ease of use of a software product (Hammontree *et al.* 1994).

Many methods of gathering usability data remotely are available, include remote questionnaire/survey, remote-control evaluation, semi-instrumented remote evaluation, instrumented remote evaluation, and video conferencing as an extension of the usability laboratory. The last one, video conferencing, uses video teleconferencing over the network as a mechanism to transfer video data in real time. It comes the closest to the effect of laboratory testing (Hartson *et al.* 1996). We decided that the critical components that Internet video conferencing software could provide were real-time observation of the users interaction with the software and their verbal comments. Observation of facial expressions, although helpful, was not considered critical. There are several internet-based software tools that provide these features. We chose Microsoft NetMeeting™ due to its low cost and easy configuration. As a multimedia collaboration tool, NetMeeting provides capabilities of transferring real-time video and audio data, sending text message, sharing applications, and transferring files (http://www.microsoft.com/netmeeting/).

2. Experimental design

The session structure was based on a between-subjects design. The independent variable (testing method) had two conditions. It compared a regular in-person usability lab evaluation (lab condition – 8 participants) and a remote usability evaluation (remote condition – 8 participants) using Microsoft NetMeeting as the conferencing software. The in-person subjects came to our usability lab in Stamford, CT. The remote subjects participated from their workplace in Foster City, CA using the company intranet and the conferencing software. Performance time and satisfaction ratings data collection was automated.

Success rate data was collected through observation. For both conditions (in-person and remote), the software prototype being evaluated ran on a machine in the usability lab in Stamford CT. The screen was scanned, and an audio-visual system (cameras, microphones, and speakers) facilitated the collection of subjective data such as the participant comments during the session and the debriefing. Automated logging software gathered performance data. Remote subjects were put on speakerphone so that their comments were recorded by the lab's audio equipment. For the in-person subjects, facial expressions were captured on the usability lab video equipment.

For the in-person participants, one observer was with them in the testing room, and the other observer was in the control room and received a scan-converted view of the computer used by the subject. The participant received the task sheet from the administrator, and was asked to use the think aloud protocol. For the remote participants, the session protocol was slightly different. The session instructions and the task sheet were sent via e-mail. The participants were required to pre-install NetMeeting. The observer initiated the session by calling the participant at their workplace in Foster City, CA. A collaboration session with the conferencing software was then initiated over the Internet. All remote subjects used speakerphone or headsets at their location, so they had both hands free to operate the computer.

The session structure was the same for both in-person and remote users. After greeting and introduction, the participants were asked to fill an on-line pre-session questionnaire. Before starting the tasks, subjects performed a network speed test. Two data points were collected for this speed test. Then four tasks were given one at a time. In addition to the data collection software, the observers used a data collection sheet with specific points. After the tasks were completed, users answered an on-line post session questionnaire and a short debriefing interview was conducted. The session took 25 to 30 minutes.

3. Results and discussion

3.1. Performance comparison

Network speed plays an important role in time-related measurements on the network. The mean time for speed test for in-person subjects was 0.72 seconds with a standard deviation of 0.2973. The mean time for speed test for remote subjects was 1.13 seconds with a standard deviation of 0.7053. T-test doesn't show a significant difference due to the small sample size (t (7, 7) = 1.50, p = .15). We take network delay into account when analyzing the performance data.

All users, both in-person and remote, completed tasks 1 and 3. All in-person users and seven out of eight remote users completed task 2. Task 4 was a more complex task. Four remote subjects completed task 4, whereas only one in-person subject completed it. Only the data for successfully completed tasks was analyzed. The mean completion time comparison is shown in Figure 1.

Figure 1 shows an interesting result. Even without factoring in the network delay, the remote subjects mean completion time is shorter than the in-person participants in all four tasks. The difference in the performance time between the two conditions may be accounted for by individual differences of the users in each group. The user profiles indicated that all eight of the in-person subjects were recruited from the company's documentation group. Of the remote subjects, three were from the documentation group, four from the hotline group and one from product management.

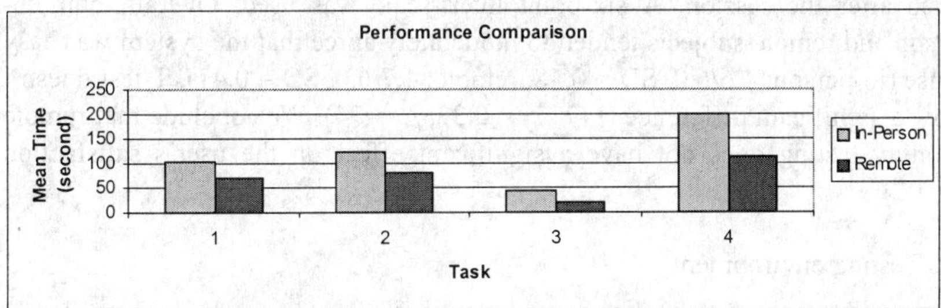

Figure 1. Performance comparison between in-person and remote users.

Figure 2 shows the performance time comparison for remote versus in-person subject grouped by user types.

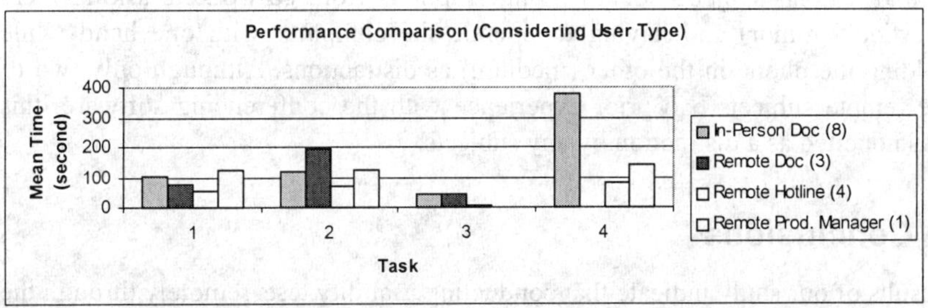

Figure 2. Performance comparison considering user types.

Because all users finished tasks 1 and 3, we could compare the performance time both for in-person versus remote subjects and for different user types for these two tasks. The performance times of the members of the documentation group do not show significant difference between in-person and remote subjects. The hotline subjects had a faster mean performance time than either in-person or remote documentation subjects. Because only one out of three remote documentation subjects finished task 2, there are not enough data points to perform a reliable comparison of their performance time with the in-person subjects. Subjects from hotline (remote) show a faster performance time in task 2 than the in-person documentation subjects. These results indicate that the users with similar profile had similar performance regardless of the testing condition. Skilled users (hotline users) performed consistently better even though they were remote. Due to the small sample size, we consider that further testing is necessary to confirm the validity of these conclusions.

3.2. Satisfaction ratings

Ten questions about the user's satisfaction with the prototype software were asked after the session. A six-point likert-scale was used. Overall, both in-person and remote subjects tended to moderately agree that the system was easy to use (in-person: 4.6/6.0, SD = 0.48; remote: 4.7/6.0, SD = 0.41). T-test doesn't show a significant difference ($t\ (7, 7) = 0.33$, $p = .74$). We conclude that remote usability testing does not have a significant effect on the user's satisfaction level.

3.3. Testing environment

In the post session questionnaire we asked subjects to identify distractions in the test environment and to rank them as minor, medium, or major. Remote users mentioned the time delay caused by working on the network and the need to adjust their screen resolution to match the lab computer's resolution as distraction factors. Both were ranked as minor. In a pilot study previous to the formal testing, subjects identified interruptions from co-workers asking them questions (minor) and having to operate the computer with one hand while holding the phone in the other (medium) as distractions. Although only two of the remote subjects had prior experience with the conferencing software, this was not cited as a distraction by any subjects.

4. Conclusions

Results of our study indicate that conducting usability tests remotely through the Internet does not have a significant impact, positive or negative on performance and satisfaction as compared to in-person testing. Some minor distractions were present in the remote environment, but we judged that they had no impact on the quality of the data. However, due to the small sample size, further testing is needed to support the conclusion.

References

Hammontree, M., Weiler, P., & Nayak, N. (1994). Remote usability testing. *Interaction*, July 1994.

Hartson, H., Castillo, J., Kelso, J. & Neale, W. (1996). Remote evaluation: the network as an extension of the usability laboratory. *CHI 96 Conference Proceedings*, pp. 228-235.

Nielsen, J. (1994). *Usability Engineering*. Boston, MA: AP Professional.

Software Tools for Collection and Analysis of Observational Data

Lucas Noldus[1], Aurelia Kwint[1], Wibo ten Hove[1] and Ruud Derix[2]
[1] Noldus Information Technology b.v., Wageningen, The Netherlands
[2] Noldus Information Technology GmbH, Freiburg, Germany

1. Observing human-computer interaction

User-centered design of computer systems involves a variety of methods and techniques to gather data about the quality of a design, prototype or product in relation to the requirements and desires of the user. Which method is used at any given moment depends on the phase of the development process. Interviews, questionnaires and focus group discussions are commonly used to gather user requirements or views on new design concepts. As soon as a prototype is available, walk-through sessions allow users to comment on details of the user interface and system functions relevant to the tasks to be carried out. Later on, user testing takes place in order to evaluate how well the system can be used to perform actual tasks. Provided they are well-designed and realistic, user trials show how users will respond to the future system and provide invaluable evidence concerning problems users may encounter. User tests also allow a quantitative measurement of productivity issues.

User testing involves the measurement of all relevant aspects of the interaction between the user and the system. Interaction involves the behavior of two interactants, so both user behavior and system behavior must be evaluated. Relevant aspects of the user's behavior include the use of input devices (keyboard, mouse, etc.), manipulation of objects (e.g. documentation materials), body posture, facial expression, gaze direction, verbal comments (e.g. thinking aloud, talking to colleague). System behavior is usually limited to screen display (menu, dialog or message being activated), auditory cues and triggering of peripheral apparatus.

Measuring human-computer interaction often results in a wealth of data to be analyzed. To increase the efficiency of user testing as a component of the design process, there is an ongoing interest to automate data collection and analysis. New tools are becoming available for automatic logging of user operation (Harel 1999) and eye tracking (Teiwes *et al.* 1999). In this presentation, we will focus on observation, either in a laboratory or in a work setting. The design of usability labs is the topic of another presentation in this session (Johnson and Connolly 1999), while new methods for remote usability testing are addressed by Chen *et al.* (1999). We will present two software tools for collection and analysis of observational data: The Observer® and Usabilityware™. More information can be found at http://www.noldus.com.

2. The Observer

The Observer is an integrated system for the collection, analysis, presentation and management of observational data (Noldus *et al.* 1999). It allows the investigator to record activities, postures, movements, positions, facial expressions, social interactions, verbal comments or any other observable aspect of human behavior. Direct keyboard data entry eliminates time-consuming and error-prone data transcription and makes analysis reports available instantly.

2.1. Data collection

The Observer distinguishes between two different types of data entry, *coding* and *annotation*, which can be used simultaneously. With coding, key presses (or mouse clicks) are used to log events and the time at which they occur, in accordance with standard observational methods (Martin and Bateson 1993; Lehner 1996). Each code is time-stamped and validated against a user-defined set of categories. Any data-entry errors can be corrected on-line or after a session has been completed. With annotation, text typed by the observer is time-stamped and stored in an observational data file without constraints as to what is entered.

Observation is either *direct*, where the investigator is present during the task, or *indirect*, where the task is viewed by some other means, usually through use of a video recorder. During direct or 'live' observation, data are entered directly into a PC or handheld computer (e.g. Psion Workabout). The latter are especially useful when the investigator has to be mobile during observations, e.g. when recording activities in a work setting. Indirect observation implies that the test is first recorded on videotape (or stored on disk as a digital video file) after which data are collected during video playback. The Observer® Video-Pro supports this mode of operation by interfacing with a wide range of VCRs, audio and time code equipment, scan converters and multimedia boards. It reads time codes directly from analog or digital video tape or media file (AVI, QuickTime, MPEG-1, MPEG-2, DV, etc.), which allows accurate event timing at the playback speed of the investigator's choice. To prevent look-away errors, the system presents the video image in a window on the computer screen. Furthermore, the playback of the video recording is controlled via the PC, which facilitates event searching, editing and production of highlight tapes, CDs or DVDs.

2.2. Data analysis

Once data collection has been completed, one can explore the observational data in a variety of ways:

- *Time-event table*: a chronological listing of all recorded events, sorted in columns by subject and class of behavior. It visualizes the change of states and the occurrences of events in the course of time.
- *Time-event plot*: a graph in which observational data are plotted against time. This presents a quick overview of what happened and when.
- *Elementary statistics*: descriptive statistics on the frequency and duration of events or states. Statistics can be calculated for subjects, individual behaviors, modifiers, and combinations of those. One can split or lump categories, using nesting levels to create complex queries, define (iterative) time windows, average data across observations, etc. In a usability evaluation, this analysis will report task completion times, how many errors were made, what proportion of the time was spent on navigation or reading help text, etc.
- *Lag sequential analysis*: this examines how often certain events are preceded or followed by other events. For instance, in a usability evaluation, The Observer will compute how often a certain operation is followed by an error.
- *Reliability analysis*: this measures the level of agreement between pairs of observers, with a user-defined level of tolerance. This technique is often used when training new observers.
- *Export functions*: for additional calculations and inferential analysis, one can export the summary tables to spreadsheets, databases or statistics packages.

3. Usabilityware

Usabilityware was designed specifically for the usability industry, in collaboration with Usability Systems Inc. (Johnson and Connolly, 1999). It is an event logging software application that was designed to increase the speed and accuracy of usability projects. The program is built around the kernel that also forms the core of The Observer Video-Pro, but the functionality is quite different (Table 1).

Table 1. Feature comparison between Usabilityware and The Observer Video-Pro.

Program feature	Usabilityware™ 1.0	The Observer® Video-Pro 4.0
Program architecture	One integrated program	Modular design, with Project Manager, Configuration Designer, Event Recorder and 5 analysis modules
User interface terminology	Adjusted to usability practitioners	Generic terminology for observational research
Hardware installation wizard	Yes	No
Project setup wizard	Yes	No
Sampling method(s)	Focal	Focal, ad lib, scan, one-zero
Number of actors (focal subjects)	1	1 or more (up to 120)
Grouping of events	'Tasks' and 'Categories'	Up to 16 user-defined classes of behavioral elements
Add keyboard codes during event recording	Yes	No
Event summary	Yes ('Usability Hit List' with duration of tasks, comments, suggestions and priority)	Yes (no statistics included)
Video play list	Yes	Yes
Time-event table	No	Yes
Time-event plot	No	Yes
Reliability analysis	No	Yes
Elementary statistics	Duration of tasks (integrated in 'Usability Hit List')	Duration, frequency, latency of (combinations of) actors, behaviors and modifiers
Lag sequential analysis	No	Yes
Report generator	Comprehensive	Basic

While The Observer is a generic package for observational research, the user interface and terminology of Usabilityware have been tailored to the usability practitioner. For instance, where The Observer uses data records with *actor*, *behavior* and *modifiers*, records in Usabilityware have *task* and *category* as standard fields. Usabilityware makes use of 'Quick Key Codes' which are unique codes for each Task and Category within a project. So instead of having to type 'Task One - the participant installed the software', one can just hit the '1' key on the keyboard (or whatever key one assigns to Task One), and Usabilityware does the rest. This allows one to concentrate on observing, not typing. One can also add new key codes during a logging session.

Observed events can be summarized in 'Usability Hit Lists', where comments can be added and observations can be sorted by task or category. This minimizes the route from the test lab back to the design or documentation department. Along with allowing one to create highlight tapes, Usabilityware allows one to work with digital video files (AVI, MPEG, QuickTime, etc.), which can be analyzed right on screen. From there, one can also create 'Edit Decision Lists' which can be exported to a digital video editing package giving one the ability to efficiently create highlight videos.

4. References

Chen, B., Mitsock, M., Coronado, J. & Salvendy, G. (1999). Remote usability testing through the Internet. *This volume.*

Harel, A. (1999). Automatic operation logging and usability validation. *This volume.*

Johnson, R. & Connolly, E. (1999). Mobile and stationary usability labs: technical issues, trends and perspectives. *This volume.*

Lehner, P.H. (1996). *Handbook of Ethological Methods*. Second Edition. Cambridge: Cambridge University Press.

Martin, P. & Bateson, P. (1993). *Measuring Behaviour: An Introductory Guide*. Second Edition. Cambridge: Cambridge University Press.

Noldus, L.P.J.J., Trienes, R.J.H., Hendriksen, A.H.M., Jansen, H. & Jansen, R.G. (1999). The Observer® Video-Pro: an integrated system for collection, management, analysis and presentation of time-structured data from live observations, video tapes and digital media files. *Submitted for publication.*

Teiwes, W., Bachofer, M., Edwards, G., Marshall, S., Schmidt, E. & Teiwes, W. (1999). The use of eye tracking for human-computer interaction research and usability testing. *This volume.*

The Use of Eye Tracking for Human-Computer Interaction Research and Usability Testing

Winfried Teiwes[1], Michael Bachofer[2], Greg Edwards[3], Sandra Marshall[4], Eberhard Schmidt[5] and Wulf Teiwes[6]
[1] SensoMotoric Instruments GmbH, Teltow/Berlin, Germany
[2] Fachhochschule für Druck und Medien, Stuttgart, Germany
[3] Center for Study of Language and Information, Stanford University, U.S.A.
[4] Cognitive Ergonomics Research Facility, San Diego State University, U.S.A.
[5] SensoMotoric Instruments Inc., Needham/Boston, MA, U.S.A.
[6] Visual Interaction GmbH, Berlin, Germany

1. Introduction

Many research projects have been conducted in the past using point of regard measurement to analyze human-computer interfaces to improve usability. A variety of eye tracking methods have been developed with the aim of improving accuracy, ease of use and non-intrusive measurement. Recently developed video-based techniques for measuring eye movements allow contact-free recording with moderate head movements while the subject is looking at displays such as video or computer monitors (http://www.smi.de).

Accuracy of these techniques is subject to ongoing improvement and validation (Teiwes *et al.* 1997). Ease of use of these devices for human-computer interaction research and usability testing requires further enhancements such as operator-free handling, semi-automatic setup and integrated calibration, as well as semi-automated analysis and its synchronization to changes of the visual scene. This paper presents particular examples of usability testing to highlight these needs as well as approaches for automated analysis and interpretation.

This development of automated tools for eye movement interpretation may improve the design of visual user interfaces as well as the use eye movement for computer interaction.

2. Internet banner research

A study of different banner designs on internet web pages was performed by Bachofer (1997), in which eye movements of several subjects were recorded non-obtrusively while browsing through an internet magazine under realistic conditions. Forty-one different banners in the internet magazine were tested and analyzed eye movements were correlated to the memorization from the subjects evaluated by recall and recognition tests in order to assist in validation of web page designs and to get deeper insight in users' behavior and their decision making process while using the Web. Furthermore, the banner's effect on recognizing consumer brand images using different forms, layouts, animations, positioning on the Web page and the influence of integrated audio components (i.e. midi jingles) were evaluated.

The study has shown that average view time of banners increases with banner size, animated text banners have only a low recall value and audio jingles improve brand awareness. Furthermore, users lose interest in banner frames (position fixed) and concentrate on the scrolled text frame.

Experiences from this study show a significant value of this type of application, however, an automated visual context related gaze path analysis needs to be implemented for an efficient broader use.

3. Eye tracking in ergonomics

The Cognitive Ergonomics Research Facility of San Diego State University is investigating eye movements while using a decision support system based on two computer displays presenting different graphical and numerical information. Several types of analysis have been developed to compare the overall use of different visual presentations, calculations of percentage of use of various areas on the display (*Figure 1*), as well as scan patterns over defined screen regions while searching specific information.

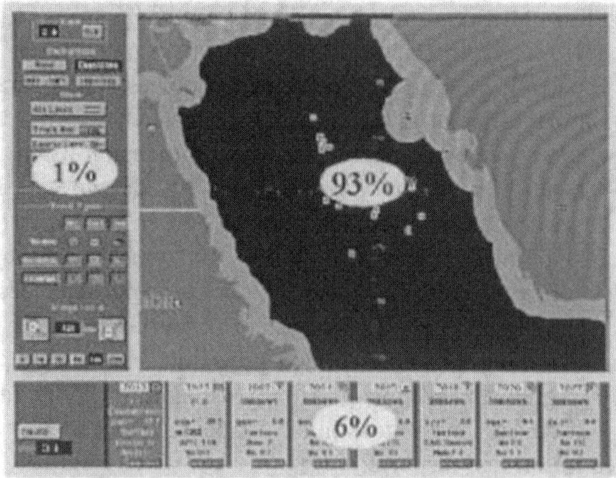

Figure 1. Calculation of percentage of various areas of display (Marshall, 1997).

4. Eye movement interpretation

Increasingly the integration of point of regard measurements and pupil size changes together with the associated content of the visual scene and other multimodal interactions such as keyboard, mouse and voice events is playing a major role. Furthermore, in order to implement semi-automatic processes for analyzing interactions from several subjects, interpretation of eye movement data into relevant information and effective methods of presentation need to be identified and implemented in order to derive to a useful, productive and affordable solution.

Edwards (1998) implemented an Eye Interpretation Engine to recognize patterns of eye movements caused by typical human behavior. It differentiates three behaviors: Knowledgeable Movement, Searching and Prolonged Searching by analyzing eye movement patterns, i.e. saccades, fixations, significant fixations, blinks and revisits.

The implementation of this analysis online while performing tests in applications mentioned above would allow a more sophisticated context-free or context-specific analysis.

5. Multimodal input and advanced user interfaces

The interpretation of eye movements to obtain a higher level of information will not only allow an improvement of user interfaces, but possibly also enable the use of eye tracking in human computer interaction as an additional multi-modal input device. First attempts are already using gaze fixations in handicapped applications, but mostly in standard graphical environments not designed for eye control. Especially new mobile computing devices (i.e. wearable PCs and PICs) require more efficient and if possible hands-free input devices such as voice control. Eye movements could add a new dimension to the interaction if integrated as a further multimodal input device. However, the graphical interface needs to be designed for this purpose and automatic interpretation of eye movements to differentiate its multiple functions, such as scanning an image, gaze stabilization, reading, pointing, clicking and other desires needs to be provided. The latter becomes of further importance since eye tracking technology is becoming available at high precision, low weight and low cost and therefore viable to be integrated as an additional input device for professional and consumer applications in binocular head mounted displays or monocular head-up micro displays.

6. References

Bachofer, M. (1997). *Wie wirkt Werbung im Web*. Hamburg: Gruner & Jahr.

Edwards, G. (1998). A tool for creating eye-aware applications that adapt to changes in user behavior. *Proceedings of ASSETS 98* (Marina del Rey, California).

Marshall, S.P., Morrison, J.G., Allred, L.E., Gillikin, S., McAllister, J.A. (1997). Eye tracking in tactical decision-making environments: implementation and analysis. *Proceedings of the 1997 Command and Control Research and Technology Symposium*, pp. 347-355. Washington, DC: National Defense University, ACTIS.

Teiwes, W., Merfeld, D.M., Young, L.R. & Clarke, A.H. (1997). Comparison of the scleral search coil and video-oculography techniques for three-dimensional eye movement measurement. In Fetter, M., Haslwanter, T., Misslisch, H. & Tweed, D. (Eds.): *Three-Dimensional Kinematics of Eye, Head and Limb Movements*, pp. 429-443. Harwood Academic Publishers.

Mobile and Stationary Usability Labs: Technical Issues, Trends and Perspectives

Reed Johnson and Ed Connolly
Usability Systems Inc., Alpharetta, GA, U.S.A.

1. Introduction

When business executives are asked "What do you consider to be your most valuable asset and why?", the answers vary between money, time and people. Time and money are certainly important and nothing would get done without people. The correct answer, however, is people because people make decisions on how all of the corporate assets are to be used. Companies strive for better results. The way to have better results is for people to make better decisions. People make better decisions when they have better information.

2. Perspectives

The Usability Engineering Industry is about people and obtaining information. It is interesting to see this field grow and realize the tremendous impact it could have; where so many people can benefit and find themselves in a win-win situation. An amazing aspect of usability engineering is that **any** organization, company, educational institution or government agency can have happy customers and workers if they choose to. The biggest obstacle preventing this transformation is lack of information.

Two companies may be customer-oriented and both may have an ongoing usability engineering program, yet their customer satisfaction results may differ substantially. One company recognizes the benefits of usability engineering better than the other.

The difference is a matter of **Perspective**; a matter of one's vantagepoint, one's approach, one's viewpoint, one's outlook. And it is not just a single person's perspective but a few select people that have the power to make or break the process. The ripple effect of their decisions can be tremendously important to the health and wellbeing of the entire company.

There are three key teams that can have very important perspectives on the topic: Upper Management, Developers and Usability Practitioners.

The most influential perspective is that of **Upper Management**. A personal, visible, executive endorsement, referred to as "buy-in", at this level will legitimize a usability program. Without buy-in at this level, a usability program is going to have problems being accepted. Getting buy-in from the chief executive level or from a Very Important Top Officer (VITO) must be the highest priority of any usability group. There is an emerging trend that indicates top management is becoming more aware of and desirous of the benefits of usability. They see and hear about usability in visits to suppliers, in conversations with peers and in published testimonials. Those who solicit top management's buy-in must speak management's language. Too often that approach is not taken by usability's leading proponents, the Practitioners.

The perspective of the **Usability Practitioner** originates from experience and academic background. An additional plus is that the practitioner is a user advocate. Past history indicates, however, that without the executive level buy-in, practitioners are fighting an uphill battle. Some have been very successful but for most it seems to be a weary, never ending effort. The key here is that the 'language' used by practitioners to promote usability is not the same as the 'language' used by those who write the checks and release the funds. In other words the benefits conveyed are not hitting the target. A usability lab has to be perceived as more than something 'nice to have'. Usability program budget cuts and program shutdowns indicate that corrective measures are in order.

Historically, the perspective of the **Developers** has been that change is neither desired nor sought after. There is, however, a growing population of developers that have been enlightened and have joined the movement to adopt a more user-centered design mentality. Even though executive level buy-in is key to the overall usability effort, the developers can delay or derail a usability program's momentum.

3. Trends

In every organization there is a 'moment of profound insight' for Upper Management, Developers and Practitioners. A moment when the light bulb goes on and a loud "Aha, now I get it!" occurs. Is there a common denominator that causes that moment?

The answer is Yes. It is the corporate reality check. It is when a 'usability lab system' is used to showcase real users struggling with a product that may have already absorbed large amounts of resources. Reality says that the end is not in sight. A usability lab system produces an amalgamation of perspectives, and provides a communication breakthrough. It is the prime source for obtaining user information from which better decisions can be made - thus better results.

However, not all 'labs' are usability labs. And not all 'usability labs' are 'usability lab systems'. There are major strategic differences here and it directly affects the perspective taken by Upper Management and Developers. The perceived corporate benefits of an integrated 'usability lab system' are far superior to the low cost camcorder 'lab' and the home-grown 'usability lab' alternatives. A growing trend is to incorporate the physical lab as a marketing tool as well as a development tool.

Here's why. Upper Management looks for information that gives them a better idea of what the future has in store. If the information shows them how to reduce future risks and better allocate resources then they are very receptive. Money is released for investment in tools that add value to the company's efforts to grow earnings, improve productivity, evidence customer centeredness, gain a competitive advantage, improve product quality, demonstrate technological acumen and build shareholder value.

The question of whether to have a stationary lab system or a portable lab system is a common one. The trend now is that a need exists for both within a company. It makes sense to have a stationary system if you have a centralized internal customer base and you can use the usability program as a showcase-marketing tool for prospective external customers. There will always be times, however, when going on site with a portable lab system is the faster, more effective way to gather user input. Waiting for users to fit you into their schedule also impedes the progress of a product to market.

There is also a trend for academic institutions to pick the portable lab as the system of choice. It enables campus-wide usage in various schools and administrative departments and the applications are numerous.

4. Technical issues

In a classic 1994 analysis of the software industry's record of success and failure for software projects, The Standish Group reports that the major reasons for software project failures, cost overruns and delays are due to lack of user input. The report also points out that the major reasons for projects being successful, that is on time and on budget, was the ability to obtain user input and management's involvement. This report has helped practitioners provide very convincing data for establishing a usability program complete with the right tools.

When designing and building an integrated 'usability lab system' you will want to choose a firm that is experienced in this industry. The firm should have experience in designing and manufacturing based on durability, functionality, reliability, acceptability and, of course, usability. It is important to recognize that practitioners themselves will spend many hours observing users and in performing certain tasks during these observations. A usable system becomes a combination of art and engineering that is usability tested on a number of levels.

The labs in general are controlled environments and use video cameras remotely controlled from an ergonomically designed, patented console (http://www.usabilitysystems.com). Audio systems capture the user, observer and others as necessary. The media of choice is analog videotape or digital videotape. The most cost-effective storage medium is videotape. Highlight tapes, which are those selected clips of an evaluation or project, are generated via logging software such as Usabilityware (Noldus *et al.* 1999), played back on a computer and then available for distribution on CD, DVD, intranet, internet and videotape. The future of digital video and the Internet will continue to have a positive impact on future usability engineering products and services.

How do you cost-justify an integrated 'usability lab system'? There are many ways to demonstrate the value of a system to the company's bottom line. Cost savings, cost avoidance, increased productivity as a result of having a lab are some of the popular approaches. In a comprehensive book entitled *Cost Justifying Usability* (Bias and Mayhew 1994) there are numerous real life examples of how to obtain 'numbers'. Another way to approach cost justification is to investigate the key business drivers and see what the constrictions are that usability can identify and correct.

Clients are reluctant to disclose the savings that their usability programs have produced. This is easy to accept once you realize that the savings are always quite significant and details invade proprietary information. Here are some that indicate the scope of what can be achieved. A company saved $300,000 in training costs. A single project in one company paid for the lab and then produced savings on 40 other projects in a year. A procedural change in one company produced savings of over $400,000. A franchise saved $4 million with a self-install project.

To summarize, the most important step is to obtain top management's buy-in to a usability engineering program. This can be achieved by presenting your business plan as a solution to closing the gap between reality and results needed. No product, process or policy will be successful without real user input throughout its life cycle. Your user profile will change as product usage changes so user evaluation must be done early and often. The most effective way to obtain user input, improve internal communications, present a customer-oriented image and obtain corporate-wide buy-in is to utilize integrated usability lab systems - both stationary and portable.

5. References

Bias, R. & Mayhew, D. (1994). *Cost Justifying Usability.* San Diego: Academic Press.

Nielsen, J. (1993). *Usability Engineering.* Boston, MA: AP Professional.

Noldus, L., Kwint, A., ten Hove, W. & Derix, R. (1999). Software tools for collection and analysis of observational data. *This volume.*

Parinello, A. (1994). *Selling to VITO.* Holbrook, MA: Adams Media Corporation.

Callahan, C. & Nemec, J. (1999). The CEO's information technology challenge. *Strategy & Business*, 1Q1999 (14), 78-89. New York: Booz, Allen & Hamilton.

Heiman, S., Sanchez, D. & Tuleja, T. (1998). *The New Strategic Selling.* New York: Wm. Morrow & Co.

The Standish Group International, Inc. (1994). *CHAOS.* Dennis, MA.

Automatic Operation Logging and Usability Validation

Avi Harel
Ergolight Ltd., Haifa, Israel

1. Introduction

Landauer (1995) argues that the average software program has 40 design flaws that impair employees' ability to use it. Usability validation is the process of confirming that the user is able to work with the product as designed. A common method of usability validation is to ask the user to provide feedback, either in free form or by filling in questionnaires. This method provides very low yield, since users do not report problems that they cannot repeat and because users tend to attribute unexpected product behavior to their own limitations rather than to deficiencies in the product design.

Another method of usability validation that provides higher yield is by testing. Usability testing is the process of identifying the reasons for the users' failure to accomplish tasks within reasonable time limits (Rubin 1994). Common practices of usability testing target mainly the product learnability, namely, identifying design flaws that prevent new users from using the product at all. In order to expand the testing to target experienced users as well, one needs to test for user errors. An overview of user errors is provided by Reason (1990).

2. Automatic operation logging

Automatic operation logging (also know as 'UI event tracking') is a set of techniques for recording the user actions in order to raise the product quality. Automatic operation logging is used for obtaining 'operation profiles' and for usability testing. For usability testing, it enables identification of user errors by presenting the sequence of user actions that are overlooked using regular testing practices.

Hilbert and Redmiles (1998) provide an overview of research and tools developed for operation tracking. Recently, new methods have been proposed for using operation tracking for identifying user errors. According to these methods, usability is validated against a designer's model, expressed as usage expectations (Hilbert and Redmiles 1998) or task models (Lecerof and Paterno 1998). Hilbert and Redmiles (1998) propose using their method for automatic error detection over the Internet.

Implicitly, these methods assume that the user intention can be automatically interpreted from the log of user actions. Such assumptions should be challenged, whether the test subject is experienced or not. Novices are not able to transform their intentions to actions, because they do not know the operational procedures yet. Experienced users do know the procedures but make errors that should be identified during the testing. Theoretically, the drawback of these methods is that the user errors are regarded as deviations from the designer's model, instead of from the user's model. Suppose, for example, that the test subject unintentionally activates the wrong control. When using a designer's model, this action may be out of context and the user error might be overlooked. Interpreting the user intention from the user actions (e.g. Lecerof and Paterno 1998) is useful solely for ideal error-free operation. However, when testing for user errors, such methods become useless.

A better approach might be that the user intention should be elicited independently of the log of user actions. To identify unintentional user actions, one should synchronize the log of user actions with a log of user intentions, generated using common methods such as thinking aloud, and recorded using an observer logger such as The Observer® or Usabilityware™ (Noldus *et al.* 1999).

3. Understanding user errors

A user error can be defined as a situation in which the user action is either not in accordance with the user intention or out of context. Accordingly, the two main types of user errors are:

- *Psychomotoric errors* - user actions that do not match the user's intention.
- *Context errors* - user actions that do not match the product's operating context.

The identification of an instance of user difficulty typically consists of the following steps:

1. Identify the situations in which the user experiences operational difficulties.
2. Identify the intended task or step that the user failed to accomplish.
3. Analyze the reasons for the user's difficulty.
4. Facilitate the resumption of the operation in spite of the difficulties encountered.

This procedure applies to all kinds of user difficulties. In particular, it applies to user errors. The main difference between the various kinds of user difficulties is in the third step - analysis of the reasons for the user's difficulty. To analyze user errors, one should record the user actions, the context and the user intention. Next, the log of user actions should be matched to the record of user intention and to the record of product context. For the matching, each record should have a time stamp and the recording should be synchronized. Operation logging is commonly considered as the recording of the user's actions. In the ErgoLight™ software, this notion is extended to also refer to the recording of the product context. Examples of how ErgoLight tools identify user errors can be found in http://www.ergolight-sw.com/www/Examples.html.

4. Instrumentation

Modern operating systems support automatic capturing of the user actions. For example, Microsoft Windows provides hooking APIs that enable tracking of GUI components (controls and menu items) provided that they are 'standard', which means that the GUI was designed using proper builders. To enable operation tracking, a developer should be able to identify GUI components that are not standard, to add code for hooking those non-standard components and to verify that the code hooks properly. ErgoLight provides tools for identification and verification of both standard and non-standard controls and examples of code for hooking non-standard components.

Context buttons, such as check boxes and radio buttons, may impose restrictions on the applicability of software features. For example, for regular printing, the **Print to file** check box in a **Print** dialog box should always be cleared. A common user error is trying to work in the wrong context. For example, trying to print when the **Print to file** check box is selected. This error is typical to preliminary versions of new Windows products, as developers often forget to reset the check box when the user activates the standard **Print** toolbar button. Windows APIs allow reading the state of context buttons. ErgoLight tracking tools use these APIs to capture context buttons automatically.

Context buttons provide explicit context of standard format. Often however, a feature is constrained by compound context, specified implicitly by the application code. For example, when a software flag is set in response to an internal error to protect a document against accidental changes. ErgoLight provides means for hooking the state of implicit context and examples of code for the hooking.

5. Integration in laboratory testing

Common laboratory testing practices involve inspection by a test operator or an observer. Video recording is helpful for reviewing the inspections off-line. In usability labs, identification of the user's intention is achieved easily, by asking the users to think aloud. An observation data logger, such as The Observer or Usabilityware, may be used to record the user's intentions. A scan converter may be used to record the system response to the user's actions. Analysis of the reasons for the user's difficulties typically relies on the test operator, who should learn the details of the product operation before the testing begins. In addition, the test operator may facilitate the user either by directing him/her to the right actions or by applying workaround procedures. Rubin (1994) provides a comprehensive overview of applying various usability lab techniques.

'Traditional' usability testing, as described above, is probably the best method available whenever the main marketing concern is to increase the product learnability. However, empirical studies show that during usability inspections, inspectors often detect only a small portion of the existing usability problems (Desurvire 1994; Nielsen 1992). Apparently, testing solely by inspection is not adequate for two very important types of software products:

- **Performance critical systems**, for which the marketing goal is to improve the user performance. User errors degrade user performance in two ways: by introducing incorrect data into the product database and by slowing down the product operation. Research on the rate of slowing down as a result of the user's errors has yet to be conducted. Initial results are that the degradation in operation speed may be as high as 50%;
- **Safety critical systems**, for which the main marketing concern is to prevent accidents caused by user errors. There are indications that for each actual accident there are about 100 situations of 'almost accident'.

For products of these types, the main testing goal is to understand the user errors rather than to increase the product learnability.

6. Operation tracking in remote testing

Remote testing is commonly considered an extension of laboratory testing, where the user is observed and monitored remotely (Castillo *et al.* 1997; Hartson *et al.* 1996). The motivation for performing this kind of remote testing is to save the expenses involved in bringing the users to the lab. Chen *et al.* (1999) provide data showing that when using video conferencing software, this kind of remote testing might be almost as good as real laboratory testing.

Another motivation for remote testing is the need to collect data from many anonymous users. A typical example is when a vendor makes a beta version of a new software product available for downloading from a Web site. In this example, the user typically does not communicate with the developer, except for getting technical support. Kaasgaard *et al.* (1999) present a method of remote testing wherein user intentions are collected in 'use case signatures'. User intention is obtained by prompting the users to fill in their intention in case of an operational difficulty. This method applies to situations where the main marketing concern is the product learnability. However, because it ignores the user operation, it is not applicable for testing performance critical and safety critical systems.

ErgoLight tools for remote testing are designed to capture all the data required to understand user errors remotely. The user actions and the operating context are recorded continuously. The user intention is elicited only when the user experiences an operational difficulty. User errors are identified by matching the actions to the intention and to the operating context, as in usability labs.

7. Conclusion

Understanding user errors is crucial for increasing the usability of performance critical and safety critical systems. Traditional usability testing methods - video recording, thinking aloud and observations - provide data about the user intention. To understand the user errors, one should also capture the sequence of user actions and the operating context.

By integration with observation loggers such as The Observer or Usabilityware, automatic operation logging enables identification of user errors that are practically impossible to identify using traditional testing tools alone.

8. References

Chen, B., Mitsock, M., Coronado, J. & Salvendy, G. (1999). Remote usability testing through the Internet. *This volume.*

Desurvir, H.W. (1994). Faster, cheaper!! Are usability inspection methods as effective as empirical testing? In Nielsen, J. & Mack, R.L. (Eds.): *Usability Inspection Methods*, pp. 173-202. Ney York: Wiley.

Hartson, H., Castillo, J., Kelso, J. & Neale, W. (1996). Remote evaluation: the network as an extension of the usability laboratory. *CHI 96 Conference Proceedings*, pp. 228-235.

Hilbert, D.M. & Redmiles, D.F. (1998). Extracting usability information from user interface events. *Department of Information and Computer Science, University of California, Irvine, 29 Oct. 1998.*

Kaasgaard, K., Myhlendorph, T., Snitker, T. & Sorensen, H.E. (1999). Remote usability testing of a Web site information architecture: "testing for a dollar a day". *Submitted*.

Landauer, T. (1995). *The Trouble with Computers*. Cambridge: MIT Press.

Lecerof, A. & Paterno, F. (1998). Automatic support for usability evaluation. *IEEE Transactions on Software Engineering*, 24/10.

Nielsen, J. (1992). Finding usability problems through heuristic evaluation. *CHI 92 Conference Proceedings*, pp. 373-380.

Noldus, L., Kwint, A., ten Hove, W. & Derix, R. (1999). Software tools for collection and analysis of observational data. *This volume*.

Reason, J.T. (1990). *Human Error*. Cambridge: Cambridge University Press.

Rubin, J. (1994). *Handbook of Usability Testing*. New York: Wiley.

PART 5

DESIGN AND DEVELOPMENT

User Interface Design in the Post-PC Era

Michael F. Mohageg, Ph.D.
Sun Microsystems, Inc.

1 Introduction

Post-PC products are computer-enhanced, electronic devices dedicated to a very restricted set of tasks. Typically, these devices are also characterized by what they are not: Unlike personal computers (PCs) they are not generic computing platforms, tend not to have standard PC input and output devices (mouse, keyboard, CRT), and are intended for a broader user population than PCs. The most common post-PC devices are "Information Appliances," such as TV set-top boxes (STB), Personal Digital Assistant (PDA), Internet-enabled screen phones, and pagers. The concept of information appliances is borrowed from the traditional notion of an appliance. It is a device that performs only a few tasks, but does them well, efficiently, and with little hardship or confusion for the user. For instance, refrigerators are bought solely for the purpose of keeping items either cool or frozen; it does little else other than blend aesthetically with the kitchen. A dishwasher washes dishes. A microwave oven has the task of heating food. Information appliances apply this notion of a dedicated device to computing technology, with the purpose of creating small, easy-to-use, low cost devices that perform only a few tasks. I will use information appliances as the running example for this paper.

2 Why Should Post-PC User Interfaces be Different?

Information appliances (post-PC UIs) need user interface solutions different from those of a PC for several reasons. The two most important are: 1) the target users, and 2) the nature of information appliances.

2.1 The Target User

Information appliances are intended for a wider user population than PC users. These users may have no computer experience, so human interface metaphors and models borrowed from desktop environments may not be appropriate. Popup menus, scroll bars, drag and drop, or the computer desktop may be quite unfamiliar. Users of post-PC devices may feel uncomfortable dealing with anything they consider too "high-tech" and tend to be unwilling to learn complex interaction models. On the other hand, they are familiar with appliances such as push-button phones, microwave oven control panels, and TV remote controls.

2.2 The Nature of Information Appliances

Information appliances such as STBs or smart pagers are inherently built and used for very different reasons than a PC. The strength of the devices lies in their specialization for particular tasks. For example, a TV is inherently an entertainment device and a cellular phone is a communication device.

Products can be designed to provide an environment that perfectly matches the requirements of the device. For instance, small, monochrome LCDs with a touch panel may be more appropriate for a portable device. A screen-phone may not use a physical keyboard at all, and instead rely mostly on gesture-based input with a stylus. A PDA user may not need a "desktop" metaphor with files and folders to manage his PDA "objects."

While designers of information appliances are freed from the trappings of the PC, there are still considerable design challenges, due largely to technical limitations of these devices. Compared to PCs, information appliances have less memory (both storage and run-time), smaller displays, potentially less powerful processors, and different input and output devices. Especially in the case of input devices, information appliances tend to provide lower bandwidth mechanisms, which limit the richness of users' inputs. These factors place constraints on the user interface that do not normally exist on a desktop computer.

3 Design Principles

This paper describes some of the design principles used in creating a user interface for information appliances. The three principles included here are not an exhaustive list; rather, they are used to provide a flavor of the approach to designing post-PC UIs. Also, these philosophies are not rules or prescriptions for design. Instead, they represent a design bias that is quite different from that of traditional PCs and workstations. The presentation will include more of the design principles.

3.1 Identify a Target Domain

One of the keys to providing a post-PC user interface is to identify and define the target domain for the device. A "domain" is defined as the environment, the applications, and how the device is likely to be used. The device is intended to be dedicated to a set of very specific and related tasks in a defined setting. Both hardware and software are optimized for this application domain.

There are three general domains for information appliances:

- Entertainment - e.g., STB
- Information access and communication - e.g., screen phone, PDA, pager
- Assistants - e.g., a trouble shooting electronic handbook for an assembly line supervisor

Each of these domains has particular characteristics, which have considerable design implications for the user interface (see Table 1). For instance, devices in the information access and communication domain are typified by short, but intense bursts of interaction; these devices tend to be portable. Users of information access and communication devices may be more interested in efficiency than an "engaging" experience. Conversely, devices in the entertainment domain are characterized by longer and more relaxed usage sessions; entertainment devices tend to be stationary. Users of entertainment devices are more interested in having fun and enjoying their interaction with the content as opposed to being efficient. While these characterizations are broad, the basic theme is accurate. The characteristics of each domain heavily bias design decisions for the user interface.

Table 1. Samples of characteristics and design implications for each target domain.

DOMAIN	**CHARACTERISTICS**	**SAMPLE DESIGN IMPLICATIONS**
Entertainment	1. Long interactions (> 30 min.) 2. Less structured interaction (vs. PC) 3. Not very "directed" tasks 4. More relaxed interaction 5. Various levels of concentration	1. Content is critical so devote significant I/O bandwidth to content. 2. Pleasant experience preferred to efficient one
Information Access & Communication	1. Short interactions (< 10 min.) 2. Structured interaction (vs. Ent't) 3. Usually "directed" tasks 4. Various levels of concentration	1. Ease of learning and long term use are critical. 2. Efficiency can be a key feature of the UI.
Assistant Devices	Similar to Information Access & Communication	Similar to Info. Access & Communication

3.2 Dedicated Devices Mean Dedicated Human Interfaces

Post-PC products are targeted in very focused fashion to the domain in which they will be used. Given this inherent specialization, a single solution does not fulfill the user interface needs of such varied products. User interfaces must be developed that best meet the needs of each device. Therefore, the user interface must cleanly integrate with the applications, I/O, and other system components. This tight focus and high level of integration work to the user's advantage by designing pieces to work well together. For example, information appliances use I/O methods optimized for particular uses. A PDA might use a stylus to allow handwriting recognition and touch input. A screen phone may have a keyboard for text input, while a cellular phone may use the number pad. Therefore, the same E-mail application will have a different user interface depending on the device. It's important to ensure that the characteristics of each device are appropriately handled in the UI.

3.3 Simplify

Simplifying user interfaces has been a prime objective for designers. However, despite repeated attempts, we've been largely unsuccessful in keeping PCs and workstations as simple as possible. There are many reasons for this increasing complexity in products, such as the computer industry business model ("next version should have more features for less cost") and continual obsolescence (Moore's Law). I will avoid a discussion of these issues and instead focus on two methods of simplification that are key to post-PC UIs.

Functionality vs. simplicity. A long-standing truth in the industry is that the more functionality is included in a product, the more complex it becomes. In designing an information appliance, it is critical to reduce functionality to the most essential and most frequently needed functions. Information appliances offer a unique opportunity to justifiably eliminate unneeded functionality because the devices will be targeted to a less technically savvy audience, support a limited set of tasks, and will have dedicated user interfaces. A good way to consider this trade-off in the interface design is to think of the 80/20 rule. For each application or feature set, it's helpful to identify of the 20% of functions that will meet 80% of the users' task needs. Those are the functions to support in the product and optimize the design around. The remaining 80% of functions are included based on other criteria (user tasks, utility, necessity, competitive edge, price, etc.), but the design is not dictated by providing the same access to these functions as for the critical 20%.

Gratuitous elimination of functions is not useful either. Removing too much functionality can lead to a product that is either too limited to support users' tasks or simply underwhelming. Therefore, it's important to provide enough functionality to have a compelling and useful product without complicating the user experience. I refer to that "sweet spot" as the Functionality Threshold: It is just the right collection and number of features to strike the balance between functionality and simplicity.

Choice vs. Simplicity. As with functionality, a trade-off exists between simplicity and how much choice is available to a user. Choice is not an inherently negative construct. Choice is useful when it is appropriate, but post-PC UIs can benefit from reduced choice. This approach is contrary to traditional design practices in the industry, where the norm is to provide users with choices on everything from their desktop color scheme to the "tool bars" they want in the UI. Given the reduced functionality, targeted user population, and focused nature of these devices, reasonable defaults can be provided and non-essential options removed completely. An example of reasonable choice reduction is found in the TV set-top box Picture-in-Picture (PIP) window. Most Internet STBs allow users to watch TV in a PIP window while performing standard internet functions. However, this PIP window is limited to being accessed and put away; that's it! Users do not have the "choice" to move the PIP window, select its size, or change any of its characteristics. This approach, while limited, meets the needs of the overwhelming majority of users. While some users may find the lack of choice limiting, the design bias is towards simplicity and supporting the most common scenarios. Clearly, there are functions and features for which users should have choices. However, limiting choice to the essential few can reduce complexity.

4 Discussion

The principles presented above provide an approach to the design of post-PC UIs. But, aside from the user interface design issues, there are a number of long term usage questions that must be addressed if post-PC and information appliances are successful. Some of those questions are:

- Will post-PC devices lead to a proliferation of wildly different user interfaces? And if so, is this not a threat to users' transfer of learning?

- Seems there is a natural tendency to reduce the number of "gadgets" one carries around. Will this tendency lead to combining multiple devices into one, thereby defeating the purpose of a simple, streamlined device?

- Will the use of multiple devices lead to a management problem? That is, how can a user keep all his information synchronized across many devices? How will a user be able to keep his information synchronized with those of other users?

I shall discuss these questions during the panel presentation. While there are significant challenges in this new post-PC era, the promise of making computing technology more accessible and available to all consumers is very exciting and worthwhile.

5 Acknowledgments

Many thanks to Annette Wagner for her help in honing the concepts in this paper.

Vehicle-Navigation User-Interface Design: Lessons for Consumer Devices

Aaron Marcus, John Armitage, Volker Frank, and Edward Guttman
Aaron Marcus and Associates, Inc. (AM+A), 1144 65th Street, Suite F,
Emeryville, CA 94608-1053 USA;Tel: 510-601-0994x19, Fax: 510-547-6125
Email: Aaron@AmandA.com, Web: http://www.AMandA.com

1 Introduction

Vehicles with computers, global positioning satellite (GPS) systems, and Internet access can provide drivers and passengers with information about their location, trips, and sites of interest. This paper discusses lessons learned from an early (1989-92) project in user-interface and information-visualization design (UI+IVD) that solved constraints relevant to today's consumer devices with "baby faces" (Marcus 1998). They must provide easy access to functions and data that are easy for users to comprehend, remember, and use. High-quality UI design improves the likelihood users will be more productive and satisfied.

2 Project Description

During 1988-94, Motorola developed an intelligent-vehicle navigation system as part of the Advanced Driver and Vehicle Advisory Navigation Concept (ADVANCE) project (Tucker 1994, Marcus 1999). The authors' firm (AM+A) designed precise, interactive prototypes with detailed interaction and appearance, based on requirements documents prepared by other development team members. These prototypes enabled the development team to eliminate major errors in metaphors, mental models, and navigation and to show management, selected key customers, and prospective users in focus groups, even in testing sessions, in order to gain buy-in and gather information.

The hardware display used a Sharp five-inch liquid crystal display (LCD) with touch-screen surface. At the time, the LCD was state-of-the-art and exceeded the visual quality of current commercial navigation displays. Nevertheless, the display had significant limitations: 16 colors, limited font design, low resolution (one- quarter of SVGA, or 320 x 240 pixels). In addition, the non-square pixels appeared in a staggered brick pattern, not in a right-angled row-and-column layout, which meant horizontal and vertical one-pixel lines differed in width.

3 User-Interface Components

Primary UI components of the final prototype included the following:

Metaphors: Based on user research, the product had to appear as an extension of existing dashboard controls rather than use typical desktop metaphors (windows, mice, etc). The basic references were maps, with roads, scales, north arrows, etc., and trips, with planners, destinations, route preferences, etc. Also: typewriter and telephone keypads, dashboard control panels, and (rarely) computer pop-up control panels, all with labels and buttons.

Mental model: One significant challenge was the initial screen contents upon turning on the device. AM+A believed the design proposed by other team groups, which had two screens of nine items each, was too complicated. AM+A felt the initial screen should have 7±2 items to denote and connote a simple product. Motorola eventually reconfigured the mental model more simply.

Navigation: Following upon the developers' desire to emphasize "simple" dashboard controls rather than "complex" and daunting computer navigation, the screen elements changed, with few exceptions, by complete panes. This navigation choice differs from typical desktop personal computer interaction with detailed internal hypertext-like navigation in which any part of the display might be interactive and show small-scale pop-up controls, or widgets.

Appearance: The product had to seem simple and appealing, but all typography, colors, and layout also served complex functions, e.g., legibility and readability of small map symbols under varying light conditions, from bright daylight to night-time. A simple grid for menu title-bars, primary navigation buttons, lists, and other screen elements accommodated most displays. In part because AM+A felt the default fonts were inadequate, especially for maps, Motorola used a Helvetica sans-serif font. A set of 16 colors served semantic references economically but necessarily repetitively (see Figures below). Four colors in the main title bars acted as visual cues to represent the main functional modules of the product. Other colors represented the levels of road capacity, road traffic, warning messages for impending maneuvers, and geographic sites.

Interaction: Touch interaction usually located and selected items, generally with a small number of targets in a screen. The user might have long fingernails or be wearing gloves; consequently, most primary targets (*e.g.,* regular menu buttons) were large in area, often four percent of the display area (essentially in a 5x5 matrix). Interaction techniques included selecting by using single-line Scroll Down and Scroll Up buttons at the bottom of the screen (to preserve usable screen width), instead of a more typical, computer-like scroll bar at the right side. Experiments showed controlling a scrollbar at arms-length seemed more difficult than taps on screen buttons. A telephone keypad was used for future telephony functions and current alphabetic selection of menu-list items.

Unique items: Users could edit short text for usernames, addresses, destinations, etc., in a Querty keyboard. Because the keys were small, an additional location cursor assisted in identifying the appropriate target area: green crosshairs along the entire row and column of the selected key. Another cursor was the "off-center zoom-box" used to locate and select map targets. This control showed a 100 x 80-pixel horizontal rectangle outline with a translucent interior and a cross-hair cursor located in the upper left. The off-center cursor enabled the user to place a finger in the lower right of the zoom box to move the cursor and still see the desired target. The user could turn off these optional acoustic messages globally in preference settings. Note: warning messages (visual, verbal, and tonal) of impending maneuvers were timed according to the complexity of the maneuver (single or multiple maneuvers in rapid succession) and the speed of the vehicle (as determined by sensors).

The following figures show two screen examples.

Figure 1: The opening screen enables users to understand the basic content, the essential metaphors and the mental model. The design challenge was to reduce the number of choices so users would not be intimidated but could still obtain useful information quickly and easily. Color codes established on this first screen are used throughout the product as reminders and guides.

Figure 2: Route-guidance information is displayed by arrow pictograms, maps, and text, which account for the three cognitive abilities/preferences of users, as determined from focus group interviews. In each of the three displays, red, yellow, and green indicate initial, interim, and immediate-decision stages of route navigation during maneuvers. Optional audio cues (five tones) and spoken instructions (in English and other languages) present redundant information.

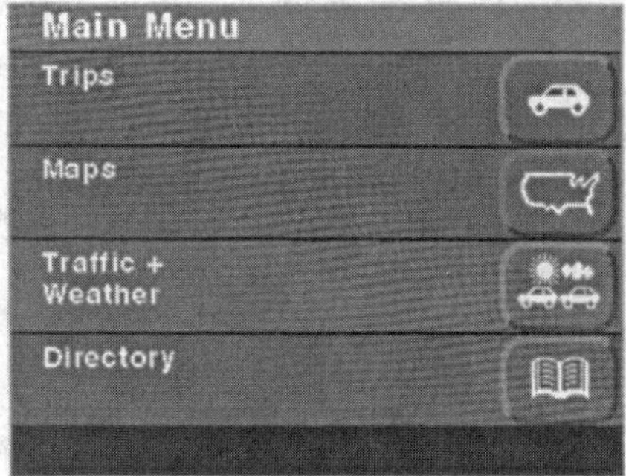

Figure 1

Figure 2

4 Lessons Learned

The project did not reach production while the author's firm was involved in UI design. Nevertheless, the development team learned important lessons that seem transferable to other contexts of UI+IVD, especially for consumer products:

- The value of multi-disciplinary teams in UI development as well as the need for good communication among team members from different professional disciplines. In particular, interactive design-space exploration tools were useful to enable clients to understand what variations were possible and to appreciate the professional skill of designers in synthesizing solutions.

- The necessity of setting design goals for both ease of comprehension and speed of access. Over the last twenty years one author (Marcus) has noted that technophiles (from either engineering or marketing departments) often seek to enable users to move ever faster whether or not their comprehension keeps pace. Designers must strive to balance these orthogonal forces in the design-space.

- The importance of accounting for different cognitive preferences for absorbing information. When designers are targeting "wide-band audiences" with multiple demographic characteristics of age, gender, education-level, cultural background, etc., even if the users are united in their immediate tasks, it seems important to account for diversity in the fundamental components of the UI.

- The impact of user testing on the design process on the cost of user testing on the business model. Motorola was exemplary in its devotion to user testing. In the end, the cost of these efforts may have defeated the business case for the product development. Again, over the last twenty years, the author has noted some clients hesitate to convene focus groups and test usability during development. The design community must gather valid, precise, accurate, clear statistics of success cases and make these available to make the case for good design and to clarify what professional practice requires.

5 Conclusion

Motorola decided not to complete the product and was obliged to place documents in the public domain (Marcus 1999). By the mid- to late-nineties, commercial vehicle navigation systems appeared in European, Japanese, and American automobiles, but few achieved the level of graphic quality of the Motorola prototype. A comparison of some systems appears in (Paul 1996).

Although Websites now offer detailed, up-to-the-minute traffic and weather information and are potentially displayable on monitors in the vehicle or on portable personal communicators, they have the limitations as driver-oriented displays. They are too verbose, and the graphic images are not simplified enough for rapid comprehension. The organization of contents seems oriented to a user browsing at work or home, not the driver. The colors and fonts are not legible and readable for viewing in the typical ambient light or viewing-distance conditions of vehicles. The interactions are oriented to detailed desktop targets, not the simplified dashboard layouts. Internet-based displays still seem in need of design for the specific target market and conditions of the driver en route.

The approach to the UI development process, the specific UI design results, and the lessons learned seem to have value for designers of current and next-generation traveler information systems and consumer devices in general. Note: This paper is adapted from (Marcus 1999).

6 References

Clarke, Cathy, and Lee Swearingen. “Motorola Smart Car,” Chapter 2.3, in *Macromedia Director Design Guide*. Hayden Books, Indianapolis, IN, 1994.

Marcus, Aaron, *Graphic Design for Electronic Documents and User Interf*aces, Addison-Wesley, Reading, 1992.

Marcus, Aaron, "Principles of Effective Visual Communication for Graphical User Interface Design," *Readings in Human-Computer Interaction*, 2nd Ed., ed. Baecker *et al.*, Morgan Kauffman, Palo Alto, 1995, pp. 425-441.

Marcus, Aaron. “Baby Faces: User-Interface Design for Small Displays.” Panel Description, *Conference Summary*, CHI-98, National Conference of ACM/SIGCHI, 18-23 April 1998, Los Angeles, CA, pp. 96-97.

Marcus, Aaron, "Metaphor Design in User Interfaces," *The Journal of Computer Documentation*, ACM/SIGDOC, Vol. 22, No. 2, May 1998, pp. 43-57.

Marcus, Aaron, “Finding the Right Way: A Case Study about Designing the User Interface a Motorola Vehicle-Navigation System,” in Bergman, Eric, ed., *Beyond the Desktop,* Morgan Kaufman, Palo Alto, 1999 (in press).

Paul, Rik. "Lost? Or Found? Finding our Way through the High-Tech Terrain of Onboard Navigation Systems." *Motor Trend*, December 1996, 6 pp.

Tucker, Fred. "Why Run the Race for Better Transportation." *Washington Times*, 23 May 1994, p. 17.

The Design of Microsoft Windows CE

Sarah Zuberec
Microsoft Corporation

Introduction

Windows CE user interfaces have faced a variety of design challenges over the last five years. While some may argue that this UI is nothing more than the traditional desktop Microsoft Windows miniaturized to fit a smaller display, UI research has been carried out on each product to help guide the design process. As the research continues, the understanding of users, their tasks and environments evolves. In order to support tasks in various environments, new form factors and input/output methods have been developed. Each product was designed to be a desktop companion. To that end, consistency with the desktop versions of the applications and operating system has played a large role in defining the user interface for each product. The challenge facing Windows CE designers is to extend the traditional desktop Windows metaphors and designs to the broad spectrum of CE devices.

Users, Tasks and Environments

Users of the Handheld PC (H/PC), Palm-sized PC (P/PC) and Auto PC (A/PC) are loosely defined as mobile professionals. These are people who spend a set amount of time away from their work desk, who have an intermediate understanding of Microsoft Windows, use Microsoft Word or Microsoft Excel on a daily basis and who maintain both an address book and calendar to manage their personal information.

When considering the characteristics of the mobile professional and tasks that needed to be supported, designers for Windows CE realised that a number of different form factors and feature sets would be needed. While not all the user tasks, scenarios and features are outlined in this paper, here are a few that were considered when developing various products.

- A lawyer spends a lot of time in meetings away from her desk. Although the meetings take place in her office building, she is required to take notes and write up summaries for future reference. She frequently references other documents during these meetings. One meeting usually triggers another, so the lawyer needs access to her schedule at all times.

- A real estate agent spends a majority of the day in his car driving from property to property. He uses his cellular phone to set up meeting times and locations with clients. His car becomes his "second office".

- Walking from building to building, a university student needs to find out where her next class is. In class she quickly writes down her homework assignment. A new tutor provides both office hours and contact information that the student also records.

In the first scenario, it is clear that a keyboard for touch-typing and an application for note taking are necessary. While the lawyer doesn't leave her building, she is still mobile and requires the ability to view and modify her personal information. The H/PC was designed to support the creation of document outlines and access to transferred documents from the desktop. In this case, desktop integration is maintained through the application design of Pocket Word and the data transfer of calendar information between the PC and the device.

In the second scenario, the real estate agent uses his cellular phone to contact clients to set up appointments. Managing names, phone numbers and addresses as well as finding directions are tasks supported by the Auto PC. Using voice recognition and an integrated cell phone, the real estate agent can place a phone call without taking his eyes off the road or his hands off the wheel. The directions to the next property can be spoken to him from the direction application so he doesn't need to rely on a printed map. Contact information has been transferred from his PC to his A/PC via his H/PC.

In the third scenario, the student needs to access her information quickly. The Palm-sized PC was designed to support both quick look up and entry of information. Hardware buttons were incorporated to support one handed look up of information. Handwriting recognition and voice recording were implemented to support quick data entry. The product needed to be small and easy to carry. Desktop integration is achieved through data sharing of contacts, notes and appointments with related desktop applications.

As mentioned earlier, these are by no means all the tasks and scenarios that are supported by Windows CE products. At a base level however, all products remain a desktop companion. This does not mean that all Windows CE applications will look and feel exactly like their desktop counterparts.

Ultimately the characteristic to define a companion product will be its ability to share information through some kind of connectivity.

Products and Influencers

Windows CE is an operating system designed to run on computers that are much smaller than traditional desktop PCs and laptop PCs. The first version of Windows CE was created to run on H/PC 1.0, which shipped in 1996. While this paper will mention the H/PC, P/PC and A/PC, Windows CE also runs TV related products, internet related products and vertical applications.

As a quick overview, the H/PC is a clamshell shaped device with both a physical keyboard and stylus used to interact with the touch screen. The Palm-sized PC is a smaller touch screen device with an on screen keyboard and handwriting recognition. Both products support the management of personal information such as names and addresses, appointments and tasks. They also allow the creation and editing of Microsoft Pocket Word and Microsoft Pocket Excel documents. Each product also supports features unique to the platform. As an example, the H/PC supports Pocket Powerpoint Viewer for portable presentations and Pocket Access for database creation. The P/PC supports the recognition of Rich Ink. The user can ink on the screen and the feature will transform the ink into printed text. All data can be kept up to date on both the device and the PC using a serial cable connection and a synchronisation process. The Auto PC is an in-vehicle multimedia information, control and entertainment system that uses a small display, hardware buttons and speech recognition. It supports vehicle navigation, digital and analogue entertainment and wireless communications.

Both usability lab testing and field research played an important role in helping the design team better understand the users, their task and the design space. In order to maintain a strong affiliation with the desktop, the design team was tasked with creating an interface solution that resembled Windows 95. Although the design direction was set, fundamental interaction questions still remained for each product. H/PC 1.0: How large do on-screen targets need to be to be accurately activated with either a finger or a stylus? Can an auto-save model support and satisfy users' data management needs? Can single tap activation be a successful method of interaction? H/PC 2.0: Which font could satisfy screen real estate constraints and still be easy to read? P/PC 1.0: What data entry method (on screen keyboard or handwriting recogniser) would support both quick entry and maintain a high level of accuracy? A/PC 1.0: How should common tasks like phone number look up be supported in an auto environment?

While results of the above questions are not reported here, it is easy to understand that Windows 95 did not have the answers. Leveraging the desktop promoted the "walk up and recognise" scenario however it was clear that Windows 95 was optimised for a large screen display. Suddenly something that was very familiar and had been shown to satisfy an array of user profiles, became a huge design challenge.

The Issue of Consistency

How much Windows is too much Windows? It is difficult to argue the fact that a large number of people use Windows on a daily basis. These people have invested time in learning and optimising their use of the product. They have also invested data in formats supported by the system. Companies have invested time training people how to use Windows. Any new system introduced into the workplace would increase training costs. Any new system introduced into the consumer space is another system to learn. It would be foolish to ignore these facts. These points map almost exactly to the advantages of consistency outlined by Nielsen (1989) *in Coordinating User Interfaces for Consistency*. If considered literally, it would appear that all Windows CE products should match Windows desktop counterparts identically. Windows CE designers have begun to reaslise that the device form factors and the tasks that the user needs to accomplish should dictate the degree of desktop behaviour retained.

If Windows CE products take into consideration consistency with desktop Windows, users can gain a lot. Users know what they are getting; they have a general idea of how it works. Unfortunately not all devices support the same functionality because not all devices serve the same purposes. Form factors are different. Methods of input and output are different. User expectations can be different. In the case of the first iterations of the Palm-sized PC, interface consistency was not enough to ensure success. Although the product looked like both its handheld and desktop counterparts, the design increased the number of steps it took to accomplish a task. If quick data look up was a key scenario of the product, then it wasn't being satisfied with the multiple screen taps necessary to acquire information. So while familiarity and functionality were satisfied, usability was not. In this case, ease of learning conflicted with ease of use. While the "walk up and use" scenario was satisfied, long term use was significantly hampered.

Unpracticed or inexperienced designers may search for the recipe of good design to ensure product success. What steps can be followed to produce a usable, aesthetically pleasing, successful products? When considering this, it would appear that consistency alone could provide the obvious solutions to the hardest problems. Take something that users understand, use frequently and successfully and copy it. However successful designers understand that successful products support users and their work. As Grudin (1989) states, "interface objects must be designed and placed in accordance to users' tasks". So while it would appear that the desktop design should be the way to go, time will only tell if portable products become more like desktops, desktops become more like portable products or the two diverge completely.

Conclusion

Consistency of cross platform interfaces has become a hot topic. When surveying the design space of an array of existing products, it can been seen that the space for Windows CE is quite huge: from four colour grey scale to full colour, through an array of screen sizes and resolutions, with interaction styles directed by stylus, mouse and voice, on platforms for the hand, lap, car and living room. New platforms and new environments fuel many of the interface developments. One thing remains clear: to date, Windows CE interfaces have always contained a flavour of desktop Windows. Due to the familiarity of the desktop software, it would appear to make sense to leverage the familiar designs in new emerging platforms. However, the user tasks and device designs can be diverse. There is no reason to think that in the future, the tasks and designs will become any less diverse. Specialised devices will continuously be built to suit user needs and environments. However it would be shortsighted to conclude that the same interface would suit all tasks in all potential environments regardless of how recognizable the interface. There are no clear rules to follow when and when not to be consistent. On one hand it would appear that a common interface would satisfy all user needs. On the other hand it would be impossible to design an interface to satisfy this goal. To create successful Windows CE products, the device cannot ignore desktop Windows but Windows cannot be held ultimate over design realities. Therein lies the design challenge facing Windows CE designers.

References

Grudin, J. (1989). The Case Against User Interface Consistency. *Communications of the ACM, 32, 1164-1173.*

Nielsen, J. (Ed.). (1989). Coordinating User Interfaces for Consistency. Academic Press, Inc. San Diego, 1989.

Inattentive use: A new Design Concept

Georg Stroem
L. M. Ericsson A/S

1 Inattentive use may be the rule rather than the exception

Inattentive use of electronic equipment may be the rule rather than the exception. One common example is writing with a word processing program: The user concentrates on what he is writing, not on how to operate the word processor.

Often, the same user operates a multitude of user interfaces. Instead of learning the specifics of each, he may browse an unfamiliar user interface as an automatic process, until he recognises what he is looking for.

If the user interface is not designed for inattentive use, the user will often make mistakes or be annoyed when forced to focus his attention on the equipment.

2 Methods used investigating inattentive use

I have used introspection (James 1890) and observations of users for identifying the most critical limitations of the inattentive user. I have then used experimental results from the cognitive psychology to provide more precise guidelines for design of electronic equipment for inattentive use.

3 Results

The problem of inattentive use was identified by Miyahta and Norman (1986). However, they used a computer (UNIX) inspired model leaving out one of the most essential aspects of inattentive activities: *Even when the user does not pay attention, he is to some extent aware of what is happening, and he is aware of what he is doing.*

The user experiences, that each object of thought is surrounded by a fringe, he is aware of without paying attention to (James 1890). The fringe may draw his attention or influence an association from one object of thought to another (James 1890). It consists of a complex and continuously shifting pattern of perceptions, processes and at least three types of inattentive activities.

An automatic process where each action is triggered by the result of the former, or the actions overlap and are parts of one smooth flow of movements. The process may be highly complex as, for instance, driving a car while thinking about something else.

Shifting attention consists of the user focusing his attention on the equipment only for brief periods of time.

Dissociation (Schumacher 1995) or reduced conception (Stroem 1996) is a state of mind where the user is capable of sensing the equipment and of acting but not capable of thinking about his actions.

3.1 Perception while inattentive

The inattentive user cannot react on conjunctions of unrelated elements (Treisman et al. 1977). The user of a web site tends to react to single key words, but not to combinations of two or three words, for instance one word preceded by a modifier (observation).

In addition, specific targets tend to pop out from the background and overshadow other items (Kahneman and Treisman 1984). The user of a web site tends to select links, where the title includes a word, which the user have many and strong associations to, even if that word have little or no relation to the topic he is searching for (observation). The usability can then be improved significantly by replacing such words with slightly more neutral terms (observation).

3.2 Drawing the attention

The user interface shall draw the attention of the user when, and only when, he must pay attention to a decision. Different color, slope, granularity and in particular movements are effective at drawing the attention. These characteristics are to some extent discriminated even before the signals leave the eye (Bruce and Green 1990), such that it requires an effort *not* to pay attention to them. No wonder, most users hate dancing figures on top of web pages (observation).

3.3 Decisions as part of an automatic process

Unless the user makes a special effort he will follow the most used automatic process. He will read and follow a misleading text displayed by the equipment, even when he knows, how the equipment should be operated (observation).

The automatic process in itself will consist of the most frequent flow of actions, not the one that from a logical point of view is consistent (observation). Even though the operation of a user interface is logically consistent, the user may still experience it as inconsistent (observation).

Finally, the user will make errors while carrying out an automatic process, for instance confuse two similar looking objects or two similar steps in different automatic processes (Reason 1979, observation).

3.4 Decisions with shifting attention

The most comfortable length of each thought appears to be between 10 and 15 seconds (Blumenthal 1977). If the user on average is interrupted every 20 seconds, the user interface shall be designed, so the user can decide upon a proper decision after going through a minimum of thoughts, ideally only one or two.

Otherwise, the user will never reach a decision, because he has to start all over after each interruption, or, more likely, he will become impatient and at last just do something (observation).

3.5 Decisions while dissociated

During dissociation or reduced conception things may appear new and extremely bright as if seen for the first time (Stroem 1996). However, the user cannot consider the consequences of his actions or of inaction (Schumacher 1995, Stroem 1996). If the operation of the equipment may lead to irreversible damage, it shall be impossible to operate it, unless the user occasionally pays attention.

3.6 Physical actions as an automatic process

Effortless physical actions are always controlled as automatic processes. They require that the need for visual guidance and feedback are minimized (Stroem 1996), and that the risk of damages to the user's body are minimized (Pascarelli and Kella 1993). When the user does not pay attention, he is less sensitive to any beginning pains and discomforts.

3.7 Applications in design

I have applied the concept of inattentive use in a number of designs.

Kitchen timer used when preparing food. The characteristics of inattentive use could explain the problems experienced by users of existing types, and it could be used as basis of an improved design (Stroem 1996).

Communication equipment: The characteristics of inattentive use could be used as a basis for designing more flexible and consistent communication equipment and for developing new features for such equipment (Stroem 1996).

Web pages: The characteristics of inattentive use were effective for interpreting usability tests and for identifying the most effective improvements of the design (observation).

4 Discussion

The characteristics and limitations of inattentive use makes it possible to analyse and explaining the actions of the user, in particular when the user appears to be irrational and not thinking about his actions. One example is when the user repeats the same error one time after the other, even though he knows, how the equipment should be operated (as when he repeatedly reads and follows a misleading text on the equipment).

Changes to the user interface can then more precisely address the causes of errors made by the users. One example is the web pages where keywords with strong associations had to be replaced with words with a similar meaning but less strong associations. Without an understanding of the inattentive process the conclusion may have been, that the terms were difficult to understand for the user, when in fact they were too strong, too easy to understand.

Computer inspired models of the user (as described in Miyahta and Norman 1986, Mayhew 1992) are not precise enough to be used for similar analysis of the user's actions.

5 Conclusion

Inattentive actions and perception are probably a bigger area of study than attentive actions and perception. What has been done so far is only a first exploration of it. The results are promising, and the work is necessary.

As described in section 1, inattentive use of electronic equipment may be the rule rather than the exception. In addition, even when the user focuses his attention on a decision, his perceptions and associations are still influenced by processes outside his volitional control.

When designing electronic equipment, it is therefore essential that we take the capabilities and limitations of the inattentive user into account.

6 References

Blumenthal, A. L. (1977). *The process of cognition.* New Jersey: Prentice Hall.

Bruce, Vicki and Patrick R. Green (1990*) Visual perception.* UK: Lawrence Erlbaum Associates.

Card, S. K. (1981). The Model Human Processor: A Model for Making Engineering Calculations of Human Performance. *Proceedings of the Human Factors Society 25th Annual Meeting,* 301-305.

James, W. (1890). The principles of Psychology.

Kahneman, D. & Treisman A. (1984). Changing views of attention and automaticity. In Parasuraman R. & Davies D. R. (Eds.): *Varieties of attention.* Florida: Academic Press.

Mayhew, D. J. (1992*). Principles and guidelines in software user interface design.* New Jersey: Prentice Hall.

Miyata, Y. & Norman D. A. (1986). Psychological Issues in Support of Multiple Activities. In Norman D. A. & Draper, S. W. (Eds.): *User Centered System Design.* New Jersey: Lawrence Erlbaum Associates.

Pascarelli, , E. F. and Kella, J.. Soft-Tissue Injuries Related to Use of the Computer Keyboard. *Journal of Occupational Medicine,* 35(5), 522-532.

Reason, J. (1979). Actions not as planned: The prize of automatization. In Underwood, G. & Stevens, R. (Eds.): *Aspects of Consciousness, volume 1 Psychological Issues.* London: Academic Press.

Schumacher, J. F. (1995). *The corruption of reality A Unified Theory of Religion, Hypnosis and Psychopathology.* USA: Prometheus Books.

Stroem, G. (1996). *Inattentive use of Electronic Equipment.* Copenhagen: Unpublished Ph. D. thesis (copies may be ordered through www.georg.dk).

Treisman, A. M., Sykes M. & Gelade G. (1977). Selective attention and stimulus integration. In Dornic, S. (Ed.): *Attention and performance VI, Proc. 6th Int. Symposium on Attention and Performance.* New Jersey: Lawrence Erlbaum Associates.

200,000,000,000 Call Records—Now What Do We Do?

Randall S. Hansen and R. Kevin Stone
AT&T Labs, Middletown, New Jersey USA

1 Keeping Track of 250 Million Calls per Day

Representing and providing ready access to large amounts of complex data has always been a difficult design issue for those who design the data structures and their representations in computing systems. Nowhere is this issue more apparent than in the telecommunications industry, where each call generates its own trace in the network and in legacy systems in which storage space is always a consideration of paramount importance. Each of these calls results in the generation of a Call Detail Record (CDR) containing information about the call. The AT&T network processes over 250 million calls—and their associated CDRs—on an average business day. The CDR information is used not only for customer billing, but also for network planning, engineering, marketing, and for security investigations. CDRs are the key to providing information about customer network usage. Because of their business value, AT&T stores each CDR for two years in a single database. In the near future, this database will contain approximately 200 billion call records.

2 Recording Applications Manager

Recording Applications Manager (RAM) is a system for monitoring and managing the call detail recording process (for a more complete description of RAM and the user-centered design method that produced it, see Somberg, 1999). RAM provides the user with an integrated view of the various data sources and systems involved in call detail collection, management, and use. It provides a consistent way of interacting with those data sources and systems, and it places common functions on a platform that can be used by all data sources and systems. A significant design challenge we encountered when designing RAM was to provide users with a means of requesting data from AT&T's huge and complex CDR database.

3 The Trouble with Call Detail Records

Unfortunately, the information for each call is recorded and stored as a single hexadecimal string of data based on rules that, as with most legacy systems, have grown to fit their ever-expanding environment. Thus, a CDR is stored using any one of hundreds of data structures (assigned based on characteristics of the call, such as service type), each structure having as many as fifty fields of information. Many fields are common across many structures, but many are not, and there are over 500 fields that may be found in one or more structures. Each of these structures can in turn have any number of sets of addendum rules (called modules) modifying its contents and the overall information layout of the call record. Additionally, different sources of data (that is to say, different network element types, or different vendors) and different data formats are kept in (and supplied by) different systems.

The complexity problem is exacerbated by the fact that CDR information elements are not defined from the user's perspective. The names of fields that appear in the CDR data structures are frequently not the names that users employ for the information contained in the field. The same information can appear in fields with different labels, depending on the CDR's data structure. A single piece of information can appear spread across multiple fields, or as only a part of a single field. This requires users to know what types of call records they are looking for and be able to locate the relevant information in each of the CDR structures being inspected. These problems are illustrated in the following query statement:

(STRUCTURE_CODE = 1 AND ORIGINATING_NPA = 123 AND ORIGINATING_NUMBER = 4567890) OR

(STRUCTURE_CODE = 2 AND TERMINATING_NPA = 123 AND TERMINATING_NUMBER = 4567890) OR

(STRUCTURE_CODE = 3 AND TANDEM_DIGITS_DIALED_1 = 1234567 AND TANDEM_DIGITS_DIALED_2 = 8901234)

In this example, the user is looking for calls that were billed to a particular telephone number (1234567890) or credit card number (12345678901234) belonging to some individual. The first problem the user encounters is that there is no single field corresponding to "billing number" in which the numbers of interest might be found. The user must search in fields with different labels, depending on the type of call (different call types result in different structures).

If the call record uses Structure 1 (i.e., the call is billed to the originating number), then the information is found in ORIGINATING_NPA and ORIGINATING_NUMBER (a single piece of information spread across two fields). If the call record uses Structure 2 (i.e., the charges are reversed, and the call is billed to the terminating number), then the information is found in TERMINATING_NPA and TERMINATING_NUMBER (again, a single piece of information spread across two fields). If the call record uses Structure 3 (i.e., the call is billed to a credit card), then the information is found in TANDEM_DIGITS_DIALED_1 and TANDEM_DIGITS_DIALED_2 (once again, a single piece of information spread across two fields, and these fields have very strange names).

The above example illustrates another problem: the syntax required to express the query is fairly complicated, given that humans aren't particularly good at the type of formal reasoning that is required for the expression of SQL-like queries (Greene, Devlin, Cannata & Gomez, 1990; Wason & Johnson-Laird, 1972). The complexity of call record formats, however, along with the number of different fields and formats that can occur, seem to leave no alternative other than an SQL-like form of query expression.

Together, the complexity of call detail recording and the use of a complex syntax to express database queries mean that querying the CDR database can be a fairly laborious process that requires a great deal of training and skill (and time) in order to accurately retrieve the necessary information. Without a more user-oriented query method, the people who must use CDR information face an imposing environment with a prodigious learning curve.

4 A User-Oriented Query Method

To address the problem of call detail records just described, we created an environment in RAM that provides users with a powerful, easy-to-use interface for call data retrieval. We designed a data access method that allows users to access data without requiring an understanding of the structural minutiae of CDRs. We also provided a guided interface that allows the user to create very complex data queries without knowledge of SQL operators and syntax.

Central to these concepts and central to our design is the formation of a user-oriented vocabulary. The construction of the vocabulary entailed a process of user interviews and a systematic review of historical logs where we were able to determine the most frequently referenced CDR fields. From this information, and in concert with the users, we were able to create a powerful translation device that makes use of common, user-oriented vocabulary terms, hiding the complexity of call detail records and the complexity of SQL query construction from the user. The translation device uses complex rules that map a particular piece of user-oriented call information to a field or fields in the call record. The mapping scheme takes into account the structure code and the values of other fields within the call record in locating the user-oriented call information. The result is that when users need to search for calls associated with a particular telephone number or credit card number belonging to some individual (returning to the previous example), they can refer to the information using common language terms (e.g., "billing number" instead of "tandem digits dialed 1") that are representative of the information sought, and can avoid the use of SQL operators and syntax. This query method allows users to simply associate values of interest with the information of interest, as shown in Figure 1.

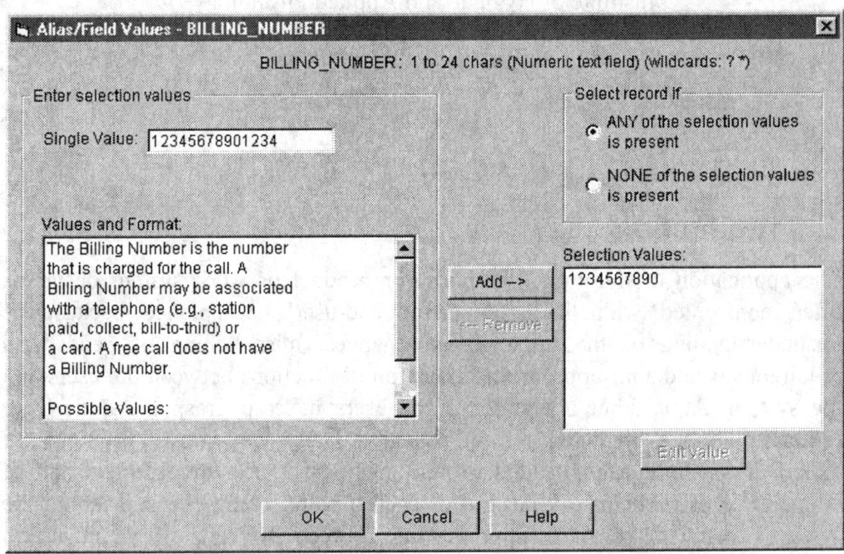

Figure 1. Associating values with user-oriented call information.

5 References

Greene, S. L., Devlin, S. J., Cannata, P. E. & Gomez, L. M. (1990). No IFs, ANDs, or ORs: a study of database querying. *International Journal of Man-Machine Studies*, 32, 303-326.

Somberg, B. L. (1999). A case study of end-to-end human factors. *Ergonomics in Design*, in press.

Wason, P. C. & Johnson-Laird, P. N. (1972). *Psychology of Reasoning*. Cambridge, MA, USA: Harvard University Press.

Who is the Designer?
The B-VOR Process of Participatory Design

Jürgen Held and Helmut Krueger
Swiss Federal Institute of Technology
Institute of Hygiene and Applied Physiology
Clausiusstrasse 25 - 8092 Zurich - Switzerland
held@iha.bepr.ethz.ch

1 Introduction

The application of traditional methods for product or work system design is often confronted with the problem of end-user's acceptance because of misunderstanding of the implemented changes, mismatching of needs and requirements and a misappropriate allocation of functions between the users and the system. At last but not at least: the users have to grasp the underlying rationality in the predicted system changes. Those well-known problems of expert driven approaches leaded to new methods of user involvement and of cooperative or participatory design to reach a better comprehension about the system's problems and to improve the problem solving with the user's knowledge about their work processes.

But those methods are also confronted with the well-known problem mentioned already above: the end-users, in this case the traditional designers or ergonomic consultants has to understand the rationality behind the new methods for to adapt them to his own context of work as a basis for success, acceptance and further development in their work, i.e. their design processes.

Therefore it is the aim of this study to develop a general model of a participatory design strategy, which shows the underlying principles and the interactions between designers and users. The name of this model is „B-VOR", with the two meanings of the German abbreviation: Beteiligungsorientierte Vorgehensweise (Engl. Participatory Method) and the German word „Bevor" (Engl.: before) to point out the importance of a certain process of mutual comprehension between designer and users before problem solving starts.

2 Method

The conception of the B-VOR model follows three steps:
1. Analysis of co-operative and participatory design methods in work systems
2. Structuring of principles, characteristics and related user reactions-actions
3. Model of designer-user processes, their roles and the related project phases

3 Results

3.1 Analysis of participatory design methods

Experiences of certain participatory design projects (Noro and Imada 1991, Held 1998) shows a structure of four principles: simplicity, confrontation, game and overview. In contrary to complex methods of analysing or measurements, participatory tools are often simple, robust and have a direct alignment to the participants to relieve their approach to the process of co-operation. Therefore inhibitions, as they are usual, when people of different social and professional context shall work together, can relieve and gave those tools the name of „Ice-breaking" devices (Noro and Imada 1991, Held 1999).

One further difficulty in designer-user interactions is the lack of problem awareness, especially on the user's side. Used to do their work in the usual surrounding over a long period of time, they get in a certain way indifferent toward their familiar work situation. To improve this, participatory tools tries to change their point of view by the confrontation with unusual perspectives of the usual. (figure 1).

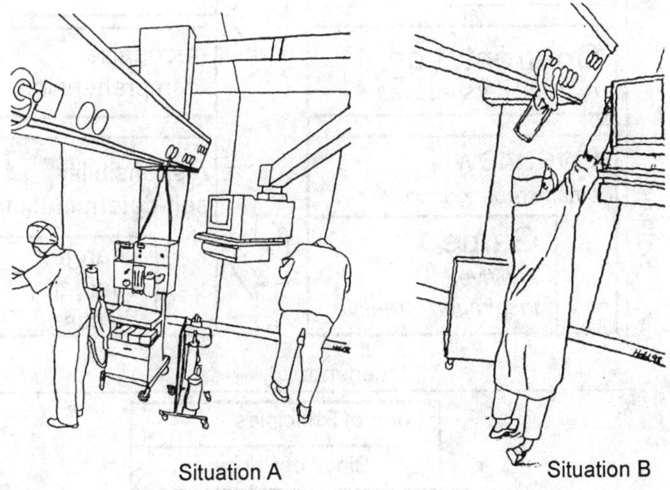

Figure 1: Photo-Confrontation: Users saw themselves for the first time in those situations

Users knowledge, imagination and anticipation is a valuable source of design information but often difficult to elicit. During the work most of the user's attention is bounded to the work process. In designer-users discussions beside the work, the important context of the situation and the concerned activities are missing. A game of the work procedures, playing at the workplace and with the support of improvise equipment and prototypes can have a positive impact for problem recognition, knowledge explication and the users imagination of possible improvements (Held 1998).

To take into account of participatory system design in small concrete steps of change has the effect, that the users can overview and anticipate future changes. They profit from stepwise and small success („small wins", „creating change" Noro and Imada 1991) and can take responsibility for the design process as well as for the future implementation and further development of changes.

3.2 Structuring of principles, characteristic, and related user reactions-actions

With regard to the conception of the B-VOR model the four principles can structured in sections, described by characteristics and related to the reactions or actions of the concerned users. The main idea is to divide the design process in the two sections of comprehension building and problem solving (figure 2).

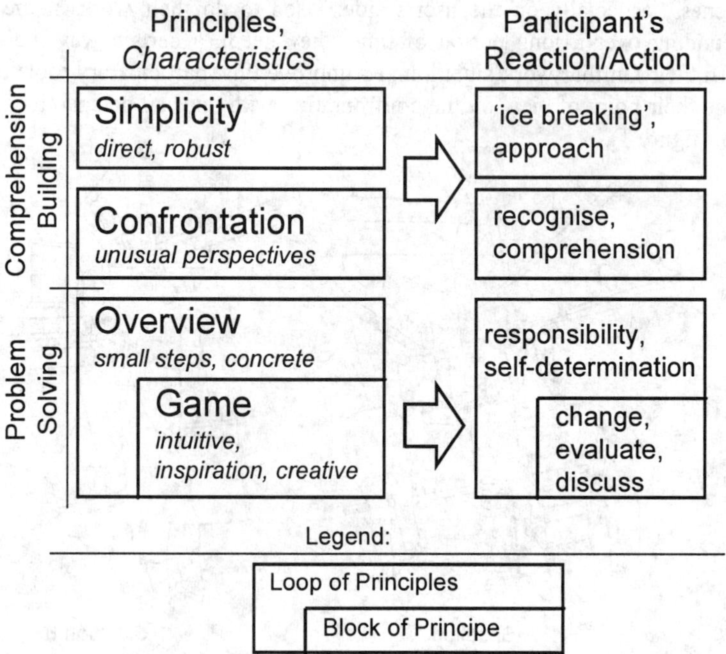

Figure 2: Principles and participant's reaction/action in a participatory design process

3.3 Model of designer-user processes their roles and the related project phases

The first section (comprehension building) of the B-VOR model, include the principles of simplicity and confrontation. The designer starts with observation and interviews, he tries to recognise, interpret and understand the user's work processes. In the following phase he reflects his interpretation to the users side. Now vice versa the users are in the position to recognise, to interpret and to understand. This can pass through several times until a consensus of both interpretation is reached (figure 3).

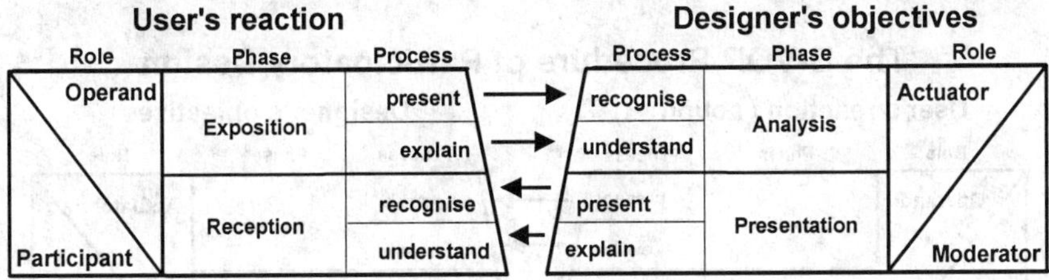

Figure 3: The change of processes in the section of comprehension building

The further problem solving and production and tests of design solutions has to be done under the requirements of the principles: overview and game. Consecutively the procedure is an iterative cycle, i.e. a trial and error process of the stepwise changes. The users became active participants, the designer role is to moderate the process of change and further more and more to facilitate changes until the consensus of acceptance. (figure 4).

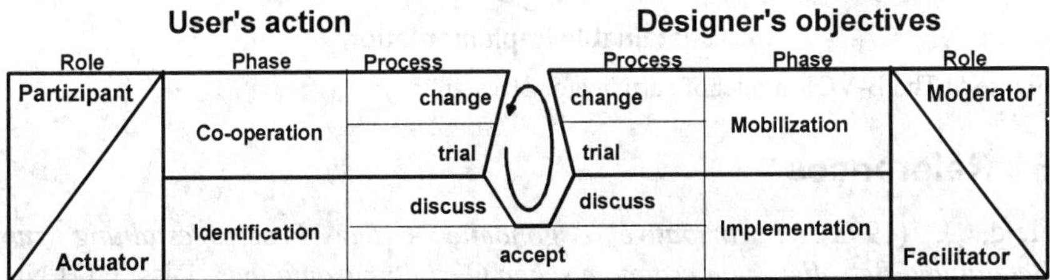

Figure 4: The cycle of problem solving

4 Conclusion and Discussion

Since the first systems approaches in project management derives in the 60th, the field of design has been flooded with design methods, procedure flow charts or models for systems design. Beside some authors (Huerlimann 1981) collected systematically over 3000 methods for problem solving and efforts for international standards (DIS/ISO 13407:1998) document reason and ratio to involve the end-users in an early stage of the design process. Overall it seems, that problem solving is no longer the problem in design, but problem definition -

that might be the problem and causes further problems over the entire process! Than the question is, if the end-user involvement in the early stage of problem solving is as reasonable as it seems to be, and leads to the need of a better comprehension between designer and users before the definition of the problem. The latter is naturally the start of a project, but has to be seen as a section of interactions-recognition and interpretations-understanding. It is the objective of the B-VOR model to communicate this importance of mutual interpretation presenting. But who is at least the designer in participatory design? It is as before the designer, as a designer of the co-operative process (Figure 5)!

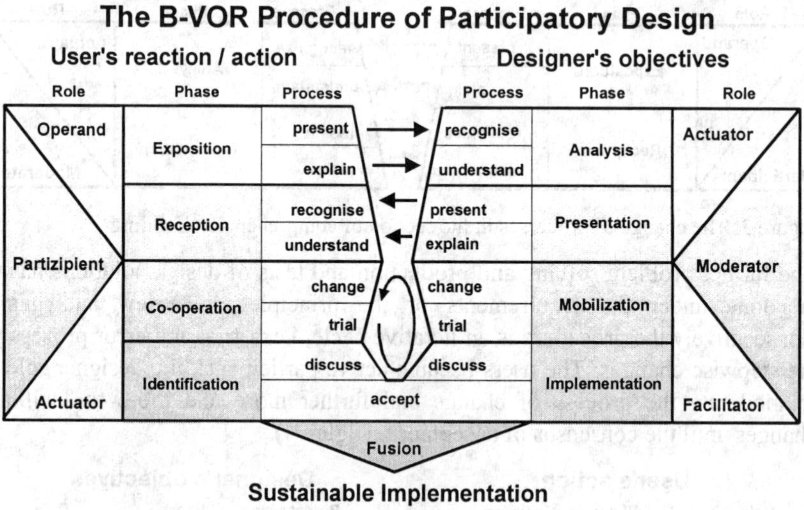

Figure 5: The B-VOR model of participatory Design

5 References

Held, J. (1998). *Partizipative Ergonomie - Die Prozessgestaltung zur Beteiligung Betroffener an ergonomischen Gestaltungsaufgaben.* Diss. ETH Nr. 12825. Swiss Federal Institute of Technology.

Held, J. and Krueger, H. (1999). The Ice-breaking VALAMO - A Tool for Participatory Processes. In *Proc. 8th Int. Conference on Human-Computer Interaction* (Munich, Germany, August 22-27, 1999), in print.

Huerlimann, W. (1981). *Methodenkatalog: Ein systematisches Inventar von ueber 3000 Problemlösungsmethoden.* Bern: Lang.

Noro, K. & Imada, A. (1991). *Participatory Ergonomics.* London, New York, Philadelphia: Taylor & Francis.

ISO/DIS 13407 (1998) *Human-centred design for interactive systems.* Draft International Standard. ISO/TC 159/SC4 1997-08-21.

Legacy Interface Migration: A Task-Centered Approach

Lanyan Kong (1), Eleni Stroulia (1) and Bruce Matichuk (2)
(1) 615 General Services Building, Computing Science Department, University of Alberta, Edmonton, AB, T6G 2H1, Canada {lanyan,stroulia}@cs.ualberta.ca
(2) CEL Corporation, 9637-45 Avenue, Edmonton, AB, T6E 5Z8, Canada
bmatichuk@celcorp.com

1 Introduction

Legacy systems constitute repositories of valuable corporate knowledge collected over a long time. However their unintuitive text-based interfaces make them difficult to learn and to use, and as a result the problem of migrating legacy interfaces to GUIs has emerged as one among the most interesting and challenging ones in the software industry (Moore 1996). The approach that we have been investigating in the CelLEST project to the problem of legacy interface migration involves understanding how the system is currently being used, that is how its users interact with it. In this approach, the interaction between the user and the system is recorded, while the user accomplishes her task. Then the recorded interaction trace is abstracted to a specification of the information exchanged between the system and the user. Finally, this specification gives rise to a specific task-oriented GUI design. This approach localizes the reverse-engineering process to these aspects of the legacy system relevant to a specific task, instead of having to deal with the whole body of the system code (Merlo et. al., 1995). Furthermore, it produces an understanding of the information and the logic of the organization's current processes as they have actually evolved through the system's lifetime. Finally, it enables the development of intuitive, user-friendly graphical interfaces tailored to simplifying the interaction of the user with the system in service of a specific task (Wilson & Johnson 1995, Sanz & Gomez 1995).

2 The Framework

Our work relies on the CelWare toolkit, developed by CEL corporation. CelWare enables (a) the modeling of the legacy system text interface in terms of its screens and the transitions among them, (b) the recording of the interaction between the user and the system, and (c) the control of the legacy system from a front-end graphical interface.

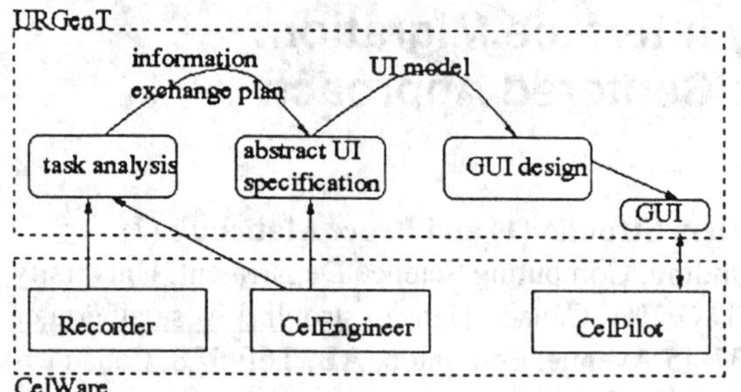

Figure 1:

URGenT and CelWare:

An illustration of the interdependencies among the CelWare toolkit components and **URGenT**.

CelEngineer is a repository of meta-data, modeling the legacy system. In this model, an interface is viewed as a collection of uniquely identifiable *Screens*, each of which allows a set of possible *Actions* to transition from this *screen* to other *screens*. Each *action* consists of a sequence of *ActionItems*, i.e., primitive interactions between the user and the *screen,* such as function keys, and data entry. The **Recorder** is an enhanced terminal emulator that records the system-user interaction in terms of the **CelEngineer** model. Each recorded *session* consists of a sequence of the visited *screens* and the *actions* performed on them by the user in order to achieve a specific task. **CelPilot** is the runtime component of **CelWare**, which handles the communication between a desktop GUI application and the host legacy system. Once a system has been mapped in **CelEngineer, CelPilot** can be used by the front-end GUI to translate the user's actions on the GUI to functionally equivalent actions applied to the legacy system's original interface, and thus control the system to perform the desired task.

URGenT (User interface ReGENeration Tool) interacts with the legacy system through the **CelWare** toolkit. It uses as input a small set of system-user interaction traces, recorded by the **Recorder**. The screens visited by the user in these traces are recognized in terms of the **CelEngineer** meta-data and annotated by their unique ID. **URGenT**'s task is to parse the recorded traces into an abstract screen navigation and information exchange plan that the user carries out to accomplish her task. Then, based on this abstract task, it specifies a user interface that can properly convey all types of information that need to be exchanged between the user and the system in service of the task at hand. Finally, given a profile of the users that perform the task, it generates a front-end GUI appropriate for users' requirements, based on standard interface design guidelines. At this point, the graphical interface generated by **URGenT** is integrated with **CelPilot,** so that the user's interaction with the GUI is translated to an equivalent interaction with the underlying legacy interface.

3 The URGenT Process

URGenT adopts an "information exchange" model for describing the user's task. The user's interaction with the system allows her to provide (obtain) some pieces(s) of information to (from) the system, and she accomplishes that by a sequence of elementary information-exchange actions on the system screens. There are three types of such elementary actions. Data-entry operations are *tell* actions: the user provides information to the system. Alternatively the user may obtain information from the system, which she indicates by highlighting the screen area where the interesting information appears. If the information in question always appears in the same area, then this is an *ask-standard* action; if the screen is dynamic and the information may appear in different positions in the screen, then this is an *ask-select* action. These three types of actions imply the need for different graphical objects in the front end GUI.

URGenT classifies the different pieces of information exchanged, i.e., "asked" and "told" by the user, as *System Constants, User Variables, Task Constants,* and *Problem Variables*. *System Constants* are constant strings that must be entered in the host system screens every time they are visited, independent of the user's task when visiting these screens. *User Variables* are data items associated with the user who performs the task, such as her login and password. *Task Constants* are constant strings that need to be entered on a screen when it is visited in service of a specific task. *Problem Variables* are data items that flow through the screens of a task session; they are either original user input or intermediate system output used as input in subsequent screens.

The **URGenT** process for reverse-engineering the system-user information-exchange plan consists of three steps: Task Analysis, Abstract User Interface Specification and Graphical User Interface Generation. These steps will be explained in terms of a hypothetical task in an Insurance Information System. Consider a situation where the insurance company computerized its claims department separately from their customer's database, and therefore owns two separate subsystems, i.e., *subsystem1* and *subsystem2*, containing their customer and claims information. Suppose further that in *subsystem2*, the user must enter the customer's claim number in order to retrieve the data relevant to generating a report on the customer's accident, but she only knows the name of the customer. So she has to first search for the claim number in the *subsystem1* by entering the customer name, before she can go in *subsystem2* to retrieve the report data.

Task Analysis

Task analysis occurs in three phases: the first examines a single trace of the task, the second phase examines multiple traces of the same task performed by the same user, and the third one examines multiple traces of the same task performed by different users.

Table 2. Part of a Recorded Trace and its Analysis

Screen Name	Data items	1st phase Single trace	2nd phase Multiple traces	3rd phase Different users
Signon	"lanyan"@T "t65j"@E	Tell var2 Tell var3	Task-specific Task-specific	User variable User variable
Name search	"scott"@E	Tell var6	Problem-spec.	Problem variable
#Claim retrieval	mouse(3,12)to(3,17)	Ask var7	Task-specific	Ask standard
Menu	"7889"@T "a1"@E	Tell var7 Tell var8	Problem-spec. Task-specific	Problem variable Task Constant

The first two columns of Table 2 describe a part of a recorded session, in terms of screens visited and data entered and obtained, for the above reporting task. The variables in quote mark are data items appearing on a screen or entered by the user, while the variables beginning with "@" are special keystrokes, e.g., "@9" is function key F9. "MouseTrack (x1,y1)to(x2,y2)" means that the user highlights an area with starting and ending coordinates (x1,y1) and (x2,y2) correspondingly. The last three columns on the right of the table show the results of the three phases of the task analysis process. At the end of this phase, **URGenT** has identified two user-specific variables (the user's login and password), a task constant ("a1" is the string entered in "Menu" to get the desired information about the claim), and two problem variables (the customer's name and the number of his claim). It has also identified that the claim number can be obtained form a standard position in the "#Claim retrieval" screen.

Abstract User Interface Specification

After the analysis of a task, **URGenT** specifies an abstract GUI, which requires the user to input her data items (user variables and user input problem variables) only once, and buffers them appropriately to deliver them to all screens that use them. This GUI also retrieves from the legacy interface the "asked data items" and feeds them to the appropriate screens. So in the above example, the user will enter her personal data and the customer name only. This interaction process is drastically simpler than the original one, where the user enters her personal data twice (to the two subsystems), writes down the claim number on paper and then enters this same number in three different screens to get different elements of the final report. At this point, for all data items to be manipulated, depending on their type, **URGenT** identifies a class of graphical interaction objects appropriate for the data-entry action at hand (Lewis & Rieman 1993). For example, for a *tell* action manipulating a date, appropriate graphical objects might be a calendar, a combination of three scrolling lists for year, month and day selection, a simple text entry box, etc.

Graphical User Interface Generation

The final step in the **URGenT** process is the actual generation of a GUI for implementing the user's task. Having identified a class of graphical objects appropriate for each data item, **URGenT** proceeds to develop a dynamic HTML GUI with graphical objects appropriate for the users for which this GUI is intended. At this point, since we are dealing with report tasks, which are fairly small, this problem is simple in that only a single screen is needed with a few objects placed in a simple sequence according to the order of their use in the process. More complex tasks will require more elaborate approaches to the object layout problem.

4 Summary and Discussion

In this paper, we discussed a process for the migration of a legacy system's interface to a GUI. This process is based on the understanding of the user's information-processing task, as extracted from traces of the user's interaction with the system, and results in the design of task- and user- specific GUIs. We are currently working to extend this method from deterministic tasks, for which all traces of the same task consist of the same sequences of screens and actions, to more complex, non-deterministic tasks.

Acknowledgements: This work has benefited from a lot of lively and in-depth discussions with the rest of the CelLest group, Paul Sorenson, Mohammad ElRamly and Roland Penner. It was supported by NSERC CRD 215451-98.

5 References

Lewis C. & Rieman, J. (1993) Task-Centered User Interface Design, http://www.acm.org/~perlman/uidesign.html.

Merlo E., Gagné P.Y., Girard J.F., Kontogiannis K., Hendren L.J., Panangaden P. & De Mori R. (1995) Reverse engineering and reengineering of user interfaces, IEEE Software, 12(1), 64-73.

Moore M. (1996) Representation Issues for Reengineering Interactive Systems, ACM Computing Surveys, 28(4), 199-es.

Sanz, M. & Gomez E.J. (1995) Task Model for Graphical User Interface Development, Grupo de Bioingenieriay Telemedicina, Universidad Politecnica de Madrid, Technical Report gbt-hf-95-1.

Vanniamparampil A., Shneiderman B., Plaisant C. & Rose A. (1995) User interface reengineering: A diagnostic approach, University of Maryland, Department of Computer Science, Technical Report CS-TR-767.

Wilson S., & Johnson P. (1995) Empowering Users in a Task-Based Approach to Design, Proceedings of DIS'95, Symposium on Designing Interactive Systems, Ann Arbor, Michigan, August 23-25, pp.25-31, ACM Press.

FormGen: A Generator for Adaptive Forms Based on EasyGUI

Alfons Brandl, Gerwin Klein
Institut für Informatik, Technische Universität München

As a common task in modern applications users need to enter, edit or browse large, complex, maybe even recursive data structures. Many users prefer form based user interfaces. The form fillin interface style (Szekely 1998) is in widespread use e.g. for database queries, e-commerce orders etc. and exploits the user's familiarity with the paper equivalent. Many use cases are cumbersome on actual paper forms (e.g. "fill in lines 7-9 only if you answered yes to question 6, proceed to 10 if you answered no") but can be handled quite elegantly with electronic forms by displaying relevant sections only - depending on the users former input. Those forms are called adaptive (Frank and Szekely 1998) or dynamic (Girgensohn, Zimmermann et al.1995).

Coding adaptive forms by hand or implementing them with a layout based development tool can grow rather expensive for large and more complex data structures. This paper focuses on how the development of adaptive form based user interfaces can be supported by the automatic code generation tool *FormGen*.

Apparently there are three points to consider when developing a form based UI:

- the application domain, i.e. the data structure to be edited,
- the dynamic behavior and
- the layout of each form to be displayed.

In the case of form fillins, layout and dynamics of the user interface can be quite easily deduced from the logical structure of the application domain. Given the form paradigm, the data type determines to a very large degree what to present to the user at a given time. It also determines in which order data needs to be entered and which parts can be entered independently of one another.

Using these relationships FormGen generates a complete set of forms from a formal description of the input data type and in this way enables the development of adaptive forms without any coding in a programming language.

In FormGen, like in many programming languages, data types can be described using tuples (records), variants, lists and basic predefined types. FormGen offers the developer a context free grammar notation for defining abstract data types. The following example illustrates this notation by defining an abstract data structure for search queries in an internet search engine:

Example:

```
Query           ::= CompositeQuery | Text | DateRange
CompositeQuery  ::= Query* Operation
Operation       ::= and | or
Text            ::= String:searchText
DateRange       ::= Date:from Date:to
Date            ::= int:day int:month int:year
```

In this example a query is either a composite query containing a list of sub-queries connected with the "and" or the "or" operation, a piece of text, or a range of dates. Note that this grammar is ambiguous as well as recursive - both cases are handled by FormGen.

As shown in figure 1, FormGen actually consists of two generators: *classgen* and *formgen*. classgen generates a set of Java classes representing the data type described by the specification, formgen generates Java classes that implement the actual form based user interface. The modular architecture of the generator enables the use of classgen not only for the generation of user interfaces, but also in other contexts such as generating attributed abstract syntax tree representations in programming language parsers, intermediate representations in compilers or any other recursive data type expressed by a context free grammar.

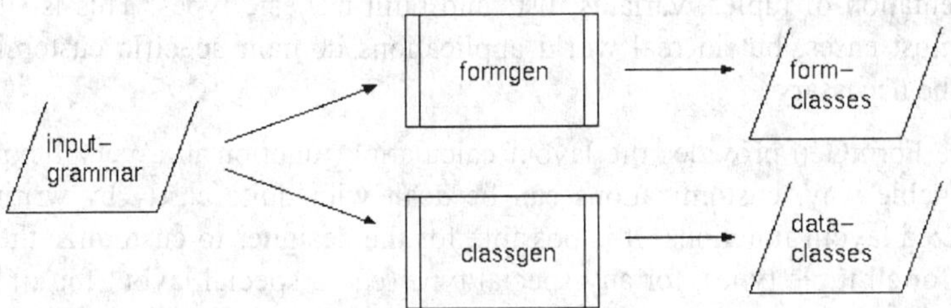

Figure 1: FormGen architecture

Our tool for generating form components is surprisingly simple, accomplishing its task in one pass. This is possible, because the generated classes are based on the most recent version of the Java UI framework *EasyGUI* (Klein 1997). The EasyGUI framework differs from other toolkits such as Java AWT in declaring the layout of an user interface per state rather than per state transition. This feature of EasyGUI allows the presentation aspects to be distinct from the dynamic aspects of the user interface, simplifying the form generator's work enormously.

The underlying concepts of EasyGUI stem from the field of compiler construction and were originally applied in the scope of user interfaces in the *BOSS* system (Schreiber 1994, Schreiber 1997). EasyGUI has been applied in different graduate and undergraduate student courses at the Technische Universität München since 1995. These courses were arranged by the chair for formal languages, compiler and software construction of professor Eickel.

Internally the user edits and navigates an abstract, attributed data tree e.g. adding new nodes or changing the content of existing nodes. FormGen regards the presentation part of the user interface as an attribute of the tree. The content of this attribute is a relatively abstract set (in comparison to e.g. Java AWT) of interaction objects: covering basics like buttons, and composite interaction objects like boxes which order their child objects horizontally or vertically in a TEX-like manner. EasyGUI provides direct support for this kind of layout.

The value of each layout attribute instance is calculated by an arbitrarily complex function. For example this function could display the whole data set entered by the user during the whole session, marking the current editing position in some way, or it could just display the current position in the tree together with the currently available actions the user may want to perform (such as changing the content of the node or adding a new sub node).

FormGen provides a standard layout calculating function that handles the presentation of tuples, variants, lists and built-in basic types. This is sufficient for most cases, but in real world applications domain specific customizations will be necessary.

Since FormGen provides the layout calculating function in a very flexible and pluggable way, customizations can be dealt with quite easily by writing customized layout functions. It is possible for the designer to customize the layout e.g. for all tuple types, for any special type (e.g. a special layout for all "`and`"-queries) or even for a concrete instance of a tree node (e.g. the root query could be given another layout than the sub-queries).

The screenshots of some generated forms in figure 2 show the correlation between the layout of the form and the corresponding abstract tree.

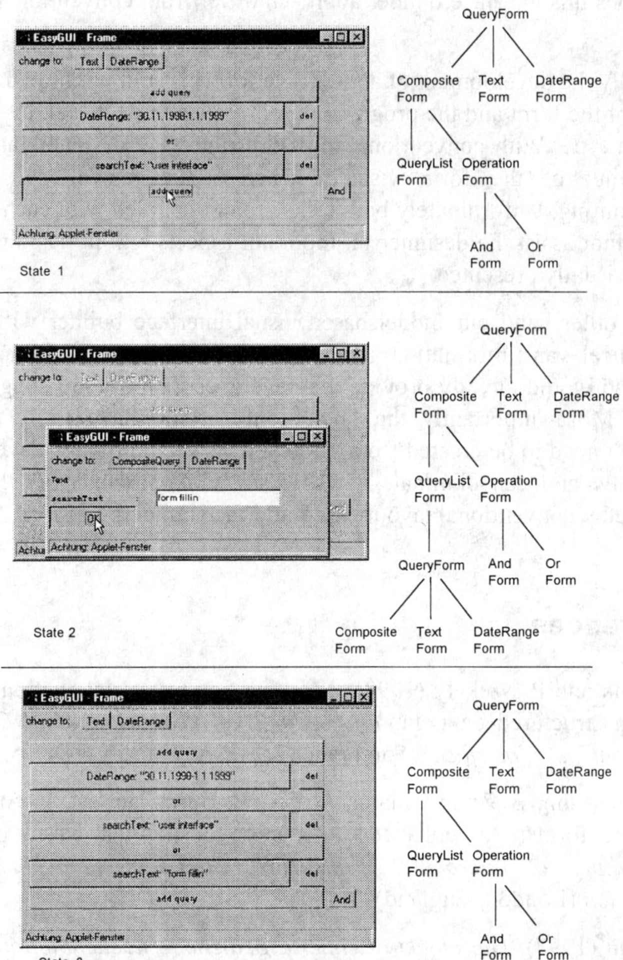

Figure 2: Screenshots of generated forms

Even if the application domain and therefore the input grammar has changed, the customized layout functions can be reused because of the applied object oriented generation concept.

In the current version the developer implements the customized layout functions in Java. Our future work includes a wysiwyg interface builder which will allow to customize the form layout without having to write any code in a programming language.

Why does this interface builder approach differ from conventional layout based tools?

Generally the development of appropriate forms is split up into the design of the layout of the form and the programming of logical aspects e.g. by implementing the data type. With conventional tools user interface designers are able to edit the layout of the form visually whereas the logical part is done by programming. Unfortunately both tasks are not coupled with each other and the tool cannot assist the designer in important aspects e.g. in deciding whether all data is visually presented.

On the other hand our model based visual interface builder will be equipped with all relevant information about the logical structure of the application domain and would already provide the developer with a good suggestion for the layout. Most importantly the layout only would have to be adjusted and wouldn't need to be created from scratch. This should reduce development time drastically and can eliminate a whole set of common errors that cannot be checked by conventional layout based interface builders.

References

M. Frank and P. Szekely (1998). Adaptive Forms: An interaction paradigm for entering structured data. In: *Proceedings of the International Conference on Intelligent User Interfaces* (San Francisco, USA, January 6-8), 153-160.

A. Girgensohn, B. Zimmermann, A. Lee, B. Burns, and M. E. Atwood (1995). Dynamic forms: An enhanced interaction abstraction based on forms. In: *Proceedings of Interact'95, Fifth IFIP Conference on Human-Computer Interaction* (London, England), 362-367.

G. Klein (1997). *Die objekt-orientierte Bedienoberflächenbibliothek EasyGUI.* Fortgeschrittenenpraktikum, Technische Universität München.

S. Schreiber (1994). Specification and Generation of User Interfaces with the BOSS-System. In *Human computer Interaction, Selected Papers EWHCI'94 Conference*, Springer LNCS 876.

S.Schreiber (1997). *Spezifikationstechniken und Generierungstechniken für graphische Benutzungsoberflächen.* Dissertation, Technische Universität München.

Exploiting Knowledge in Large Industrial Companies: A Combined Approach to Information Retrieval from Legacy Databases

Martin Atkinson*, Ole Martin Winnem**
*Joint Research Center of the European Commission, Ispra, Italy.
**Sintef Telecom and Informatics, Trondhiem, Norway

1 Introduction

Corporate experience of large industrial companies exists in many forms, most commonly, it is stored in databases that are located at different sites. These databases are often designed and constructed for specific industrial activities or domains, such as: design, engineering, manufacturing, marketing, operations and maintenance. There is often a strong correlation between the information stored in these databases, especially with respect to particular activities or product lines. However collaboration between different domain experts is necessary in order to extract pertinent information concerning a particular product or activity.

The objective of the Esprit Project Noemie was to automate as much as possible the access of information across industrial legacy databases in order to aid corporate decision making processes. This implied capturing the knowledge of domain experts as well as the domain specific constraints of users, whilst at the same time combining available software technologies for distributed data access, data correlation and knowledge retrieval.

The results from this project provide a methodology for domain modelling and problem solicitation, a software architecture supporting different data search, inference and correlation modules and a user interface supporting the methodology and tool configuration. These have been proved through configuration into two user applications.

2 Knowledge Representation

The developed methodology was based on mixture of Common KADS (Wielinga et al 1994) knowledge engineering Components of Expertise (Steels,

1990) and Inreca (Althoff et al 1995) approaches. This provides for the identification and structuring of the elements of a general domain model.

A general domain model is a core component of a Noemie application. The role of this model is to provide a means for the component software modules and the user to interact effectively in the partial or complete resolution of user supplied problems. Therefore this model serves as a repository for both system related configuration information, general domain knowledge and an experience base of use cases. The conceptualization of the model being a network of Entities linked by Relations forming Entity-Relation-Entity or Entity-Relation-Value assertions.

The general domain knowledge encapsulates information on the particular target domain, concerning background information to problems for which a user will make a request. The Entity-Relation-Entity structure of the model allows the construction of a semantic network representing the users problem space. Moreover, entities of the domain model can be sub-classed to represent: the kind of problems that the domain model supports (Tasks) incorporating the goals and sub-goals that need to be achieved to complete the task, how the kernel should respond to such problems (Actions) combining the methods to be invoked from the comprising modules, experience of how the kernel has previously resolved problems (Cases) and the knowledge that kernel needs to activate tasks (Domain Objects). Incorporated into the domain objects is also information concerning the users profile, the preferences for the user interface display and previous session interaction history.

The encapsulation of Cases within the domain model is an important aspect of this representation system; supporting the problem solving activity of a user through the exploitation of previous or preprogrammed solution strategies. New cases or problem situations can be formed in terms of domain objects. Goals can be selected, then new cases can be formed based on the their success or achievement.

3 General Architecture

A general software architecture decomposes a Noemie application into three principle modules that group principal software components. The objective of this architecture being to provide a common base for integrating access to all pertinent information, from both a software system and user perspective, whilst allowing different search and analysis modules to be "plugged" into the system. In essence, the architecture integrates: the structure and nature of available historical data, a general domain model capable of capturing any given problem relating to a users decision, and selectivity of data search or analysis tools. The diagram in the figure 1 shows the composition of the modules, together with the interfaces between components.

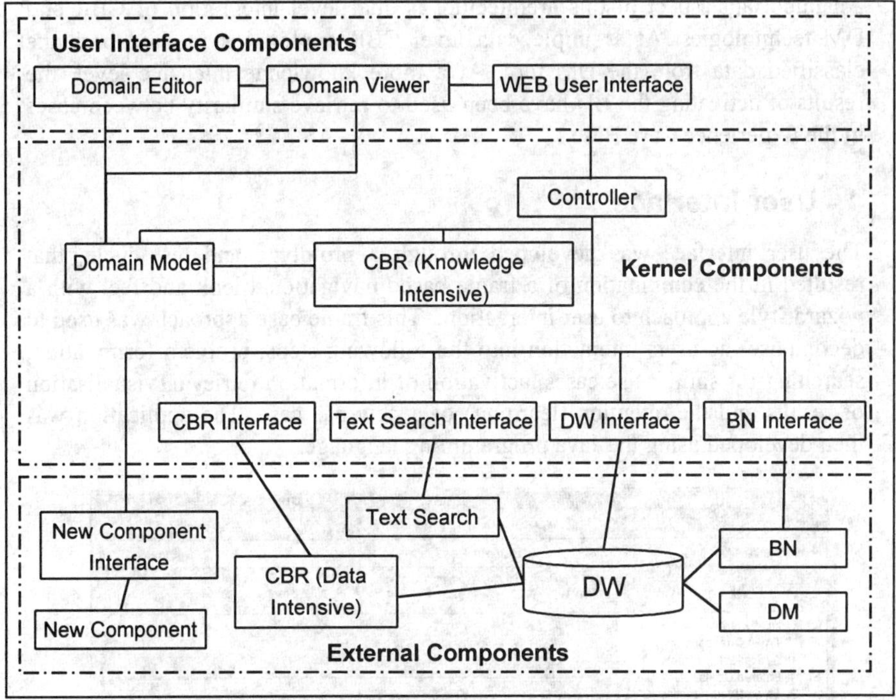

Figure 1: Noemie module architecture and component interfaces.

The User Interface consists of two principle modules: a Web User Interface that allows interrogation and activation of problem resolution, and a Domain Model Editor that is an expert user Domain Model management tool.

The role of the Kernel module is twofold: to manage the process of information retrieval based on requests from the user interface hence mediating data handling between the plugged in modules, and passing results back to the user interface. It therefore provides a bus oriented architecture of components that integrate the domain model with a knowledge intensive Case Based Reasoning (CBR) component, as well as providing the possibility to have a configurable analytical tool set. Pre-integrated into the kernel are dedicated interfaces that allow interrogation or activation of: the Data Warehouse (DW), Text search engine, external CBR (data oriented) (AcknoSoft 1998), causal data mining (through a belief network – BN) (Aamodt et al 1999) and symbolically clustered data mining (DM) (Diday 1995).

The External Component module contains all information processing tools that are used by a particular application. It includes by default the DW that federates the legacy databases.

An important aspect of this architecture is dual level integration of CBR and DM technologies. At a simple data level CBR exploits as its case base the classified data from the DM tool. At a more knowledge intensive level, the results of activating the BN have been used to retrieve similarity between cases in the Experience Case Base.

4 User Interface

The user interface was developed through a prototype and test cycle, that resulted in the combination of a frame based navigational look and feel with a wizard style approach to user interaction.. This frame base approach was used to decompose the users interaction into the following steps: problem formulation, searching for similar use cases, activation of information retrieval, visualisation of results and the retention (learning) of a new use case. The application was then developed using the Java programming language.

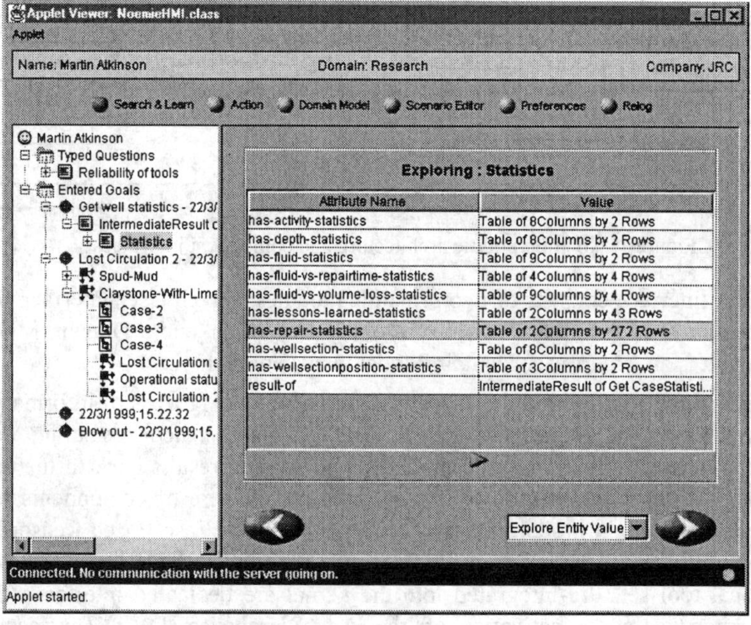

Figure 2: Screenshot of WEB user interface: showing the combined approach to navigation and wizard assisted interaction.

A permanently visible navigation frame (left frame) shows both the current stage of the interaction with the system as well as significant domain model Objects, Tasks and Cases, that can be selected and expanded allowing depthwise interrogation of the domain model. Next to this frame is the wizard screen that prompts the user for information concerning the current step of interaction.

5 Conclusion

The application of the methodology and customisation of the Noemie Software has been made for two pilot applications in the oil industry. These have demonstrated significant changes to the way of preparing, handling and presenting information to the problem holder. The same information structure is used to explain the user situation similarity with concrete experiences in a dynamically evolving experience base; to present domain information to the user; and to control the system performance.

The result of the project is an application configurable system where the user is supported in solving problem by using others experiences. Either by using general knowledge (domain model or task hierarchy) to find solutions, or by using concrete experiences that contains the solution method.

6 References

Aamodt (1994) Explanation-Driven Case-Based Reasoning, Topics in case-based reasoning. Springer Verlag, 1994:274-288.

Aamodt et al (1999) Learning Retrieval Knowledge from Data, Proceedings of IJCAI-99 Workshop on Automating the Construction of Case Based Reasoners, 1999.

AcknoSoft (1998) The Kate Toolbox for Reasoning from Cases, 1998.

Althoff et al (1995) Case-Based Reasoning for Decision Support and Diagnostic Problem Solving: The INRECA Approach. Proceedings of the 3rd workshop of the German special interest group on CBR at the 3rd German expert system conference, 1995.

Diday (1995). From Data to Knowledge: Probabilist Objects for a Symbolic Data Analysis, DIMACS Series in Discrete Mathematics and Theoretical Computer Science Volume 19, 1995.

Steels (1990). The Components of Expertise, AI Magazine 11(2), 1990:30-49.

Wielinga et al (1994), Expertise model definition document. ESPRIT Project P-5248 /KADS-II/M2/UvA/026/5.0, University of Amsterdam, Free University of Brussels and Netherlands Energy Research Center ECN, June 1994.

DISTRIBUTED EXPERT SYSTEM FOR INTERACTIVE REASONING AND EVALUATION

Celestine A. Ntuen and Eui H. Park
The Institute for Human-Machine Studies, College of Engineering
North Carolina A&T State University, Greensboro, NC 27411 USA

1 INTRODUCTION

The forms in which information is presented to the decision maker (DM) has been a research topic of interest in Human Factors and Cognitive Science arena (see, e.g. Freedy & Johnson, 1982; MacGregor & Slovic, 1986; Wickens & Kramer, 1985, Jones, Wickens, & Deutsch, 1990). Significant part of these studies deal with how DMs make their decisions with computer graphics or intelligent decision aids (Benbasat & Dexter, 1985; Desanctis, 1984; Dickinson, Desanctis & McBride, 1986). The basic measures used are decision accuracy, decision time, and the ability to predict effect of uncertainties (Zachary, 1986). Other studies have attempted to measure cognitive effort and mental workload (Payne, Bettman, & Johnson, 1990; Todd & Benbasat, 1994), and meta-analysis of decision trade-off protocol (Beach & Mitchel, 1978; Johnson, Payne, & Bettman, 1988). Most of these studies use business and economic data, and are often simplified with respect to data display rendering (dimensionality effect: 2D, versus 3D, etc.).

In the Human-Computer Interaction (HCI) domain, one of the objectives of designing display and visualization aids is to enhance the human cognitive usability of the system. In general, a cognitive model is a dominant criterion in information display and visualization tools. This paper presents an experimental study conducted to assess the effects of information display and visualization on cognitive reasoning skills of the human operator. The domain application is a simulated, scaled-down version of tactical decision-making (Ntuen, Chi & Park; Ntuen, Mountjoy & Yarbrough, 1998).

2 COGNITIVE REASONING

Cognitive reasoning was selected for the study because reasoning accuracy is known to generally correlate to the degree of correspondence between the decision task and decision maker's level of experience (Hammond, Hamm, Grassia, & Pearson, 1987).

The human cognitive reasoning studied can be classified into three broad areas. These are analogical genre, process tracing, and constructive elaboration. In analogical reasoning, the decision-maker uses familiar objects and instances as a reference point for a rough-cut inference. In process tracing, the decision-maker uses cognitive maps of object-attribute-value (OAV) triplet to infer outcomes based on characteristics from a map cluster. In constructive elaboration, the decision-maker applies mean-ends analysis to compare the end state of the problem with the most immediate partial solutions.

All the reasoning processes above can be, and have been enhanced by the use of information display and visualization tools (Johnson, Payne & Bettman, 1988). In the tactical decision making, the decision-maker must acquire and model information in a short time. An example of a decision domain with these spatio-temporal characteristics is military maneuver in which two opposing forces try to predict the movement and the behaviors of each other. The display of the commander's mental strategy can be used to enhance the predictive outcomes of the opponent's behavior in response to a give strategy.

3 METHOD

3.1 Participants

Nineteen students (12 graduates and 7undergraduates) volunteers from North Carolina A&T State University were used. The average age of the participants was 23.6 years. There were 8 females and 11 males. All the undergraduate students were from ROTC detachment unit. All subjects accepted for the study had 20/20 vision.

3.2 Apparatus

A battle event testbed (BET) display was developed using Rapid™ and EXCEL™ software on an IBM compatible PC with color graphics board monitor and a mouse. The color monitor has a resolution of 1152 x 864 pixels. Figure 1 shows an example screen capture of the BET scenario.

3.3 Cognitive Reasoning Tasks

Two general tasks associated with information visualization were tested. These are **information extraction** and **decision tasks**, respectively. Information extraction tasks are dependent on the human cognitive and display design factors. These include:

Realism of display cues. Realism is used here to mean the user's perception of display objects and their closeness to physical objects in the real world.
Correlation of mental models. This is a measure of how similar display representation of physical objects matches what the user already knows about the objects.

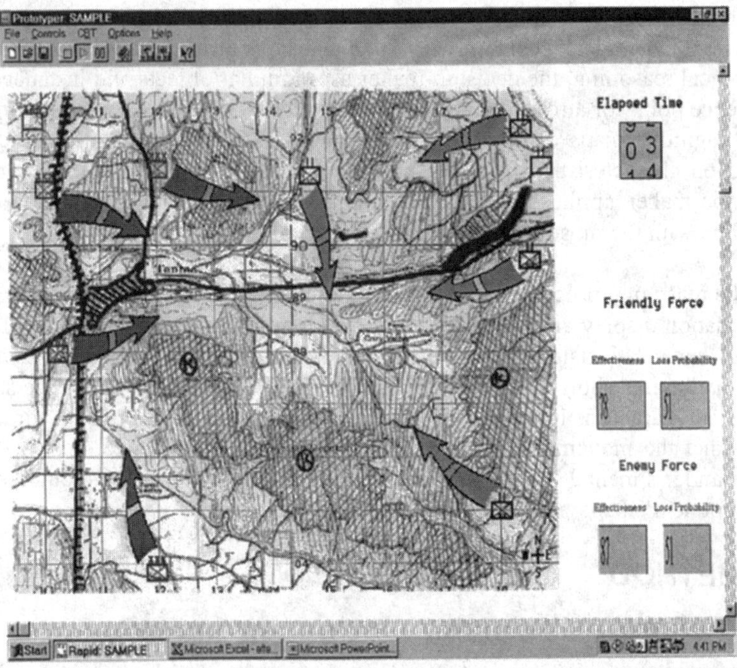

Figure 1. Testbed Display

The cognitive decision tasks studied were feature detection and recognition of decision attributes. Feature detection deals with the decision-maker's ability to detect changes in the problem space as decision variable changes. Recognition refers to the ability to infer "satisficing" solution from a group of many solutions.

We hypothesize that realism is a highly subjective measure based on the user's experience of the context. Mental model correlation and reminding cues are both a measure of 'cognitive fit' of the display elements.

3.4 The Experiment

The major decision tasks used in this experiment are as follows:

- **Feature Detection:**
  (1) Determining the spatial position of enemy troops relative to the friendly troops.

(2) Discriminating the enemy and friendly (forces) symbols in a highly cluttered display environment.
(3) Using the avenue of approach (AA) arrows to estimate direction of troop movements.
(4) Using the size of AA arrows to estimate troop strength.

- **Recognition Tasks:**
  (1) Ability to recognize geographical direction of troop movement using the AA arrows.
  (2) Ability to recognize the enemy and friendly troop engagement in a randomly selected battle envelop.
  (3) Ability to recognize troop loss based on the disappearance of unit symbols on the display scenario.

The experiment measured "error" as the dependent variable. Error was chosen because errors are often used as critic mechanisms in design and as a measure of cognitive misfits in mental tasks (Benbasart & Dexter, 1985). The errors associated with the decision tasks are shown in Table 1.

Table 1. Decision Errors in Display Tasks

Decision Task	Decision Error
Feature Detection Tasks: • Spatial location of troops • Symbol discrimination in a cluttered scene • Estimating direction of troop movement	Perceptual (distal) error Feature detection error Geometric error (angular and distance)
Recognition Tasks: • Recognize troop movements • Recognize troop engagement • Recognize troop loss	Inertial error Complexity error Attention (focus) error

Various display scenarios in the BET display were generated to conform to each of the tasks in Table 1. Although task response times were recorded, such analysis is irrelevant here since time is insensitive to design critique. The subjects were exposed to each display condition for ten seconds; during this time, the subjects were expected to provide answers to each of the task conditions.

4 RESULTS AND CONCLUSION

The relationships between the decision tasks and information extraction tasks were examined with correlation coefficients. The results are shown in Table 2.

Table 2. Pearson Correlation Coefficients between Decision and Feature Extraction Tasks.

Decision Tasks	Feature Extraction Tasks		
	Realism	Mental Model	Reminders
Spatial location of troops	0.410	0.779	-0.375
Symbol discrimination in a clutter	-0.633	-0.174	-0.233
Direction of troop estimation	0.728	0.580	0.639
Estimating direction of troop strength	-0.275	-0.105	-0.126
Recognize troop movements	0.233	0.314	0.514
Recognize troop engagement	0.490	0.536	0.733
Recognize troop loss	0.664	0.620	-0.422

The significant levels of correlation from Table 2 are as follows:

- The ability to discriminate enemy and friendly troop symbols in a clutter environment was negatively correlated with the realism of the display elements. The present of clutter has little or no effect on both the subject's mental model and as a decision reminder.
- The ability of the subjects to estimate the direction of the troops was significant in terms of display realism, mental model, and attention focus to past events (reminders).
- Both the ability to locate troop locations on the display and the ability to recognize the tactical engagement of enemy and friendly forces was "some what" correlated equally.
- The subjects' conceptual inferences about goal states were more predictable when their mental models were portrayed as similar (analog) previously held knowledge. This result is reflected in the mental model correlation coefficients with: (a) spatial tasks ($r = 0.779$, $p = 0.034$); (b) direction (or mental rotation) task ($r = 0.580$, $p = 0.107$); (c) troop engagement (or strategy task) ($r = 0.536$, $p = 0.01$); (d) risk assessment (troop loss estimate task ($r = 0.620$, $p = 0.109$).
- In process tracing tasks, the subjects tend to use the display to diagnose their potential correlation of display realism with the tasks. The major concern by the subjects was how the design of the display (process) improves human ability to execute cognitive tasks. For example, the correlation coefficient of –0.633 ($p = 0.071$) with symbol discrimination in a cluttered environment reflected the subject's opinion that displays should be able to filter out unwanted information clutters before presenting the symbols for them.
- It was noted that subjects apply constructive elaboration strategy when they seek help or reminders from the display agents. Most perceptual estimation tasks employ this strategy. For example, in Table 2, reminders were highly

correlated with estimating troop direction ($r = 0.639$, $p = 0.097$), recognizing troop movement ($r = 0.514$, $p = 0.004$), and recognizing engagement formulation ($r - 0.733$, $p = 0.024$).

5 References

Beach, L. R. & Mitchell, T. R. (1978). A contingency model for the selection of decision strategies. *Academy of Management Review*, 3, 439-449.

Benbasat, I, & Dexter, A. S. (1985). An experimental evaluation of graphical and color-enhanced information presentation. *Management Science*, 31, 134-1364.

Desantis, G. (1984). Computer graphics as decision aids: direction for research. *Decision Sciences*, 5, 463-487.

Dicson, G. W., Desanctis, G., & McBride, D. J. (1986). Understanding the effectiveness of computer graphics of decision support; A cumulative experiment approach. *Communication of ACM*, 29, 40—47.

Freedy, A. & Johnson, E. M. (1982). Human factors issues in computer management of information for decision making. *IEEE Transactions on Systems, Man and Cybernetics*, SMC 12, 437-438.

Johnson, E., Payne, J., & Bettman, J. (1988). Informational displays and performance reverals. *Organizational Behavior & Human Decision Processes*, 42, 1-21.

Jones, P. M., Wickens, C. D., & Deutsch, S.J. (1990). The display of multivariate information: An experimental study of an information integration task. *Human Performance*, 3, 1-17.

MacGregor, D. & Slovic, P. (1986). Graphic representation of judgemental information. *Human-Computer Interaction*, 2, 179-2000.

Ntuen, C. A., Chi, C, & Park, E.H. (1999). An interactive display decision support system for visualization of battle events. *Proc. 8th Annual Industrial Engineering Research Conference*, Phoenix, Arizona (to appear).

Ntuen, C. A., Mountjoy, D. N., & Yarborough, L. (1998). A collective asset display for commander's decision making. *Proc. 1998 IEEE Conference on Systems, Man, and Cybernetics*, San Diego, CA, (CDROM).

Payne, J., Bettman, J., Johnson, E., J., & Coupey, E. (1990). Understanding contingent choice: A computer simulation approach. *IEEE Transaction on Systems, Man, and Cybernetics*, 20, 296-309.

Wickens, C. D. & Kramer, A. (1985). Engineering psychology. *Annual Review of Psychology*, 36. 307-348.

Zachary, W. (1986). A cognitively based functional taxonomy of decision support system techniques. *Human-Computer Interaction*, 2, 25-63.

Role Concept in Software Development

Sandra Frings, Dr. Anette Weisbecker
Fraunhofer Institut fuer Arbeitswirtschaft und Organisation

1 Introduction

In software-development the use of new technologies on one hand opens new perspectives for software applications and the support of business processes. On the other hand it implies a modification of the requirements toward software engineering. Trying out and using new technologies calls for new activities in software development and requires new or different qualities of an employee.

Having these facts in mind the project PROMPT (Organisational composition / formation and methods for software development processes) was funded by the German Department for Education (BMBF) in which a role concept was developed (Weisbecker and Groh 1998).

2 The role concept

In the role concept a role is defined by the necessary experience, knowledge and abilities which are needed to fulfil the tasks and activities asked for in the specific role. The actual person filling the role is called the role bearer. One person can fill one role, or even more roles. But the tasks of one role can also be spread among more persons depending on the size and complexity of the project or the available resources.

This concept describes the interplay between all roles in software engineering and states a method for defining, detecting and integrating roles into the whole software development cycle. Consequently a company specific approach in the development of a role concept is to be followed. This approach includes the integration of roles into the software development process model and the management aspects of software projects (e.g. setting up project teams and human project resource estimation). Furthermore the approach includes the integration of roles into the qualification concept for software engineering. The determination of the qualificational demand and the derivation of educational measures is supported by this role concept.

There are several reasons why a company should introduce a role concept for software engineering (Frings and Weisbecker 1998a). The clearly stated definition of roles evokes a transparent knowledge of tasks and responsibilities. The support in project management is called forth by a suitable team grouping and resource planing. Future educational measures can finally be planned by determining the need for education in specific areas.

The first step of introducing a role concept includes the analysis of the existing processes already established in software development. Since those processes are described by activities - among other things - the corresponding roles can be determined.

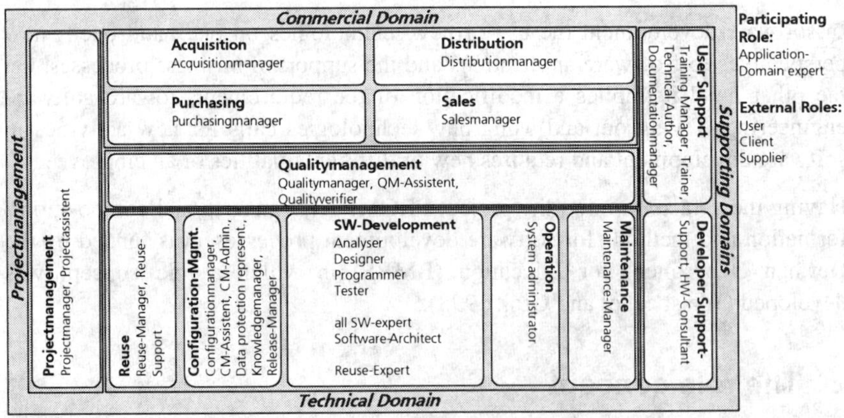

Figure 1: Process model

Once all processes have been covered and a set of roles has been identified (in the introduction phase) a structure is created to divide all roles into different areas. The phase depended roles (e.g. the role “designer”) refer to the roles of a specific software development phase like the analysis or the design phase. The phase independent roles (e.g. the role “quality manager”) do not belong to one specific phase but have activities located in more than one phase. A third area, the domain specific roles relate to a single scope of duties like the area of data bases in which the role “data base expert” would be situated.

A different view of how roles can be categorised is described by the kind of execution of the role tasks. We have distinguished between three kinds: the management and leading roles cover a whole area or process of the software development cycle which means the role is responsible for all process activities (e.g. the role "configuration manager"). The executing roles are responsible for the actual execution of the main activities of the process (e.g. the role "analysts") and the consulting roles have the task to support other roles in different areas, e.g. the role "expert in software-ergonomics".

When identifying roles this predefined structure makes sure that for each area in software development all necessary roles are considered.

A consistent role description was defined to guarantee a definite mutual delimitation of the roles. First a role is described by the tasks, activities and expected results as well as the necessary responsibilities. Furthermore a role description includes the needed qualification requirements, which are divided into technical, methodical, social and media competencies which the role bearer has to fulfil to a certain degree .The relationship between roles also has to be stated to provide for a transparency of possible conflicts or discrepancies between certain roles at the time of project planing and role assignation.

Using this role description and adding a certain peculiarity (qualification degree) to a qualification requirement defined by one of the four competencies a role profile can be set up. This should be done considering certain project or company characteristics (Frings and Weisbecker 1998b).

An employee profile can be created by estimating the employees qualification degree of the individual competencies according to the role profile. Once such a current employee profile is assessed it can be compared to the profile of the role to be selected. This way the actual educational need can be identified for this employee considering a certain project. According to the severity of the educational demand certain measures can be selected to guarantee for fulfilling a role by an employee in the best possible way.

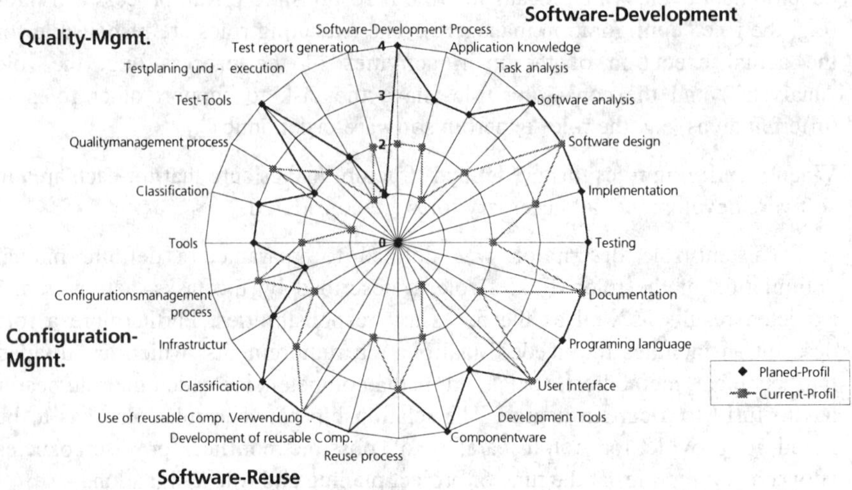

Figure 2: Qualification Profile

Since new technologies ask for a dynamic and flexible way of modification and advancement of the role concept a method was developed to help identify new roles, to describe and integrated them into the existing process model and role concept.

To support the process of project management this method also proposes how to assign roles to a project, how to assign employees to a role and how to integrate roles into a project team without being part of the team for the whole project duration. The recommendations rest upon the main project characteristics identified at a previous stage. Therefore project types were declared for which different approaches are proposed to assign roles to projects.

3 Outlook

As has been stated the role concept is made up of different approaches and methods to support the project management, to increase the transparency of activities and responsibilities, and to better plan the educational measures needed for a best possible accomplishment of software project. However the concept does not consist of a computer system.

In the process of setting up the requirements for a system to support the role concept so called “skill management systems” have been analysed. Those systems have proven to be too extensive on the one hand since they include too many aspects not needed for the concept but on the other hand do not cover the most important aspect – the definition and assignment of roles in and to projects. By now a prototype of a role administration system has been implemented.

Concluding my contribution for the HCI '99 will cover the potentials for, the description of and the experiences gained in the process of introducing a role concept into part of a company.

4 References

Frings, S.; Weisbecker, A. (1998). Zur Qualifizierung der Mitarbeiter ein Rollenkonzept einsetzen. In *Computerwoche* (No. 9 vom 27. Februar 1998), pp. 117-120.

Frings, S.; Weisbecker, A. (1998) Für jeden die passende Rolle. In: *it-Management,* July 1998.

Weisbecker, A.; Groh, G. (Eds.) (1998). Organisationsgestaltung und Methoden für menschengerechte Software-Entwicklungsprozesse - PROMPT. Stuttgart: Fraunhofer IRB Verlag.

Acquiring Tasks: A Better Way Than Asking?

Martin Cierjacks
University of Trier

1 Introduction

For supplying software tools a software engineer has to know what workers do on their jobs. The necessary data are derived from organisational specifications of jobs (e.g. job orders, specifications of results or work flow), checklists, scenarios, prototypes, interviews, and modelling (Partsch,1991). As tasks of people are not necessarily congruent to their instructions (Rautenberger, Spinas, Strohm, Ulich, & Waeber, 1994; Hacker, 1998; Hacker, Großmann, & Teske-El Kodwa, 1991), the organisational specifications may not be sufficient for a proper task modelling.

There is a broad consensus among software engineers that it is a basic need to do user-centred engineering (e.g. Helander, 1988; Ziegler & Ilg, 1993; Jacobsen, Christerson, Jonsson & Övergaard, 1992). Only - as the understanding of the models used in progress requires a certain amount of specific knowledge - the user's inclusion is reduced to being asked. The usual procedure in task modelling is as follows: first the analyser interviews the task-owner about his job and he himself does the modelling afterwards. It would be preferable to make the task-owner model the job by himself, which mostly is prohibited by the lack of modelling know-how of the owner. As software engineers have obtained many skills on how to model tasks and most users have not, a tool for the user's inclusion in the process should guarantee the same quality of the model, with no respect to the modellers previous modelling capacities.

2 How to acquire subjective task models and how to validate them

The process of software engineering can be seen as a series of translations from reality to the actual computer program (Schneider, 1986; see figure 1). As many transformation steps seem to be necessary for programming a proper computer tool, the probability for errors increases with every translation step in the process. A way to solve this problem can be a decreasing number of transformations.

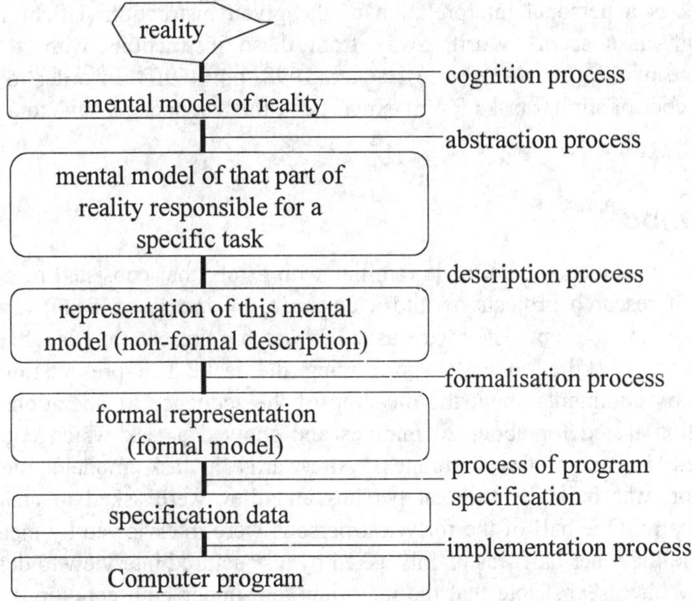

Figure 1: Transformation process from reality to computer program (Schneider, 1986)

Fewer transformations steps can be achieved by getting a formal description directly from the task owner. The description and the formalisation process are merged into one transformation. Thus the engineering has saved one potential error source. An instrument to support a direct formalisation from a user's view on specific tasks has to meet with two demands: The software engineer is more willing to use a formal tool, if either he is an expert in this special instrument (Finney, 1996), or the tool does not need specific abilities (Saiedian, 1996). The task owner needs a simple and understandable paper-and-pencil modelling tool, because most of the task owners may not be used to modelling business processes by means of software.

To meet these demands, IVA (**I**nstrument zur **V**organgs **A**nalyse; Cierjacks, Antoni, Resch & Mangold, 1996) is an instrument to formalise the task owner's model of his own task and the surrounding business process. It uses a combination of a structured interview and a certain structure-laying technique (Scheele & Groeben, 1988). It assumes that tasks are a composition of seven different elements. Each element is symbolised by a special card. The task owner is guided by the interview to label the cards and to form a "picture" of the task.

The experimental validation of this kind instrument is extremely difficult, as each task is a personal interpretation of the given instruction (Ulich, 1994). Resulting, tasks seem to drift away from these instructions with growing experience of the task owner (see Hacker, 1998, pp. 369ff). In order to get a large number of similar tasks for an experiment, the test persons have to acquire a new task.

3 Method

A number of students were made familiar with a job, that consisted of getting data about research projects of into a database, by means of a software-tool which was new to them. They were asked to watch a presentation using several examples of a skilled operator performing the task. The presentation was enriched by comments about the meaning of the sequence of operations. The presentation lasted for about 20 minutes and showed a task which could be performed in about three minutes. Afterwards a task model/a detailed description was build by the test persons and they were asked to answer a questionnaire. One half of the forty test persons were interviewed by means of IVA while the other half was plainly asked by a structured interview to describe what they have seen. Note that the interview tool under both conditions is the same, in the IVA situation additionally the structure laying technique is being used. One half of the subjects were graduate students of information sciences (experts). This group was chosen to represent persons with expert modelling skills. The group without trained modelling know-how was recruited from under-graduate students of psychology (generalists). The two dimensions IVA vs. interview and expert vs. generalist were combined to four conditions with ten test persons in each cell. Three different interviewers examined at least three persons from each condition. All descriptions and task models were compared with a generic task model.

4 Results

Every subject was able to describe the task in detail/make a model and only few syntax errors were made by the group using IVA. The IVA group built models that were more complete than the group describing the task (missing elements: IVA mean 5.15, SD 4.79; Interview mean 6.29, SD 5.35). Even though IVA was a new modelling technique for the test persons, there were no significant differences in confidence in the task descriptions. Also there was no significant difference in the ratings about understanding and the appropriateness of the modelling tools neither between IVA and Interview nor between experts and generalists. The only exception to this tendency is the statement "I could try the task now on my own" which was significantly less agreed by generalists under the IVA condition. This could be an argument for the growing problem awareness through the modelling process.

The task could be modelled on different levels of description. The test persons were free to choose the most suitable level. The experts tended to describe their task from an abstract level, while more generalists outlined the details in their models. When using IVA the differences in abstraction decreased and both groups tried to combine the details with the underlying structure.

Most of the significant results were found in comparing the different interviewer's results. The number of cards used in the models, the mistakes in syntax and semantic, and the shape of the IVA models was compared. No significant difference could be found between the interviewers under the IVA condition. In contrast there was a significant difference in shape, abstraction level, and completeness under the interview condition.

5 Discussion

The results show, that there is a better way than plainly asking. Leaving data sampling and modelling to the task owner while supporting him through a suitable modelling technique leads to obtain a decreasing interviewer effect and to more completeness of data. It helps to a more error-free model of reality and extracts a more standardised level of description. No differences were found in the confidence in the modelling. The most important finding is, that a modelling technique in addition to the interview does help to decrease interviewer effects.

To put it all in a nutshell, to make task owners formalise task descriptions takes no effect on the confidence in the correctness of the model, but it saves time and decreases the influence of the interviewer.

6 References

Cierjacks, M., Antoni, C., Resch, D. & Mangold, R. (1995). *Instrument zur Vorgangsanalyse (IVA): Handanweisung.* Unveröffentlichter Forschungsbericht: Mannheim.

Finney, K. (1996). Mathematical Notation in formal Specification: To difficult for the Masses? *IEEE Transactions on Software Engineering*, vol. 9, no. 6, pp. 733-744.

Hacker, W. (1998). *Allgemeine Arbeitspsychologie: psychische Regulation von Arbeitstätigkeit*. 1. Aufl. Bern: Huber.

Hacker, W., Großmann, N. & Teske-El Kodwa, S. (1991). Knowledge elicitation: A comparison of models and methods. In H.-J. Bullinger (ed.), *Human aspects in computing*, pp. 861-865. Amsterdam: Elsevier.

Helander, M. (1988). *Handbook of Human-Computer Interaction*. Amsterdam: Elsevier Science Publishers B.V.

Jacobsen, I., Christerson, M.; Jonsson, P. & Övergaard, G. (1992). *Object-Oriented Software Engeneering*. Workingham: Addison Wesley.

Partsch, H. (1991) *Requirements Engineering*. München; Wien: Oldenbourg.

Rautenberger, M., Spinas, P. Strohm, O. Ulich, E. & Waeber, D. (1994). *Benutzerorientierte Software- Entwicklung*. Stuttgart: B.G. Teubner.

Saiedian, H. (1996). An Invitation to Formal Methods. *IEEE Computer*, pp.16-30.

Scheele, B. & Groeben, N. (1988). *Dialog- Konsens- Methoden zur Rekonstruktion Subjektiver Theorien*. Tübingen: A. Francke Verlag GmbH.

Schneider, H.J. (1986). Formale Gestaltungsaspekte in der Systementwicklung. In: *Handbuch der Modernen Datenverarbeitung*, Heft 130, S. 3-17.

Ulich, E. (1994). *Arbeitspsychologie* (3. Aufl.) Stuttgart: Schäffler Poeschel.

Ziegler, J. & Ilg, R. (1993) *Benutzergerechte Softwaregestaltung*. München; Wien: R. Oldenbourg Verlag.

Tasks and Situations: Considerations for Models and Design Principles in Human Computer Interaction

PeterJohnson
Department of Computer Science, Queen Mary & Westfield College
University of London, London E1 4NS
pete@dcs.qmw.ac.uk

1 Introduction

How do we design computer systems so that they are useful and efficient artefacts that improve the quality and productivity of our lives? What explanations can we offer as to why some systems achieve this and others fail? How can we predict a'priori which systems and system features are going to improve our lives and which ones will not? If we knew the answers to these questions then we would be in a better position to design and evaluate computer systems, and we would be developing an explanatory understanding of human computer interaction. I use the words explanatory understanding to differentiate such an account of human computer interaction from other, alternative, forms of account that seek purely to describe or to assess human computer interaction and interactive system designs, without giving any explanatory, causal account of how such interaction might have come about. In recent writings in human computer interaction and software engineering much attention has been given to ethnomethodological studies of human activity as either inputs to design or the analysis of peoples' use of design in the work place or elsewhere. I will argue that, while such accounts and approaches are useful, they are inadequate since they fail to provide any explanatory causal accounts as to how or why designs fail or succeed. Hence, they offer little in the way of guidance or principles for design. It is not the use of ethnomethodological approaches that I am dissatisfied with, far from it, I have long argued for the forms of observation, data collection and analysis that such approaches bring. It is the failure to recognise that a description is not an understanding and that causal explanations are needed to arrive at an understanding of what counts for good HCI. Further problems arise in HCI because of the over emphasis on assessment in evaluation where all that is asked is, „is A better than B" or „ is A good for C". Such questions are regularly asked and answered using various degrees of empirical rigour. It is clear that for some purposes, such questions are valid and the accuracy to which they can be answered will, to some degree, depend upon the rigour with which the empirical assessment was undertaken. However, by failing to provide an underlying explanatory causal account of why „A is better than B" or „A is good for C", HCI is failing to provide the basis for reasoning and reflection about the fundamental questions as to why some computer systems improve our lives and others do not. Empirical and evaluative approaches form an important methodological part of HCI and play a significant role in our search for causal explanations. It seems that, in the design and evaluation of HCI we must understand what causes some interactions to be seen as good, productive, and useful, and what causes others to fail. Thus, we need an explanatory and causal understanding of HCI.

One aspect of seeking such a causal and explanatory understanding of HCI is that we need to consider the people, computers, tasks, situations and contexts in which the interactions occur, not as individual components but as they come together in HCI. Partitioning HCI into separate components is going to fail to give a full account since the very nature of the problem to be understood is how they interact. Each component constrains and influences the interaction of the other components. Therefore, we need to be able to construct models of interaction that identify and provide a basis for reasoning about the influences and constraints these components place upon each other, not just as individual phenomena, models or theories. This paper provides an example of an explanatory causal account of HCI and shows how models and principles for HCI might be developed from such an approach.

2 Developing a Theoretical Basis for Modelling

In previous research (Johnson & Johnson, 1991, Johnson, Johnson & Wilson 1995, Hamilton, Johnson & Johnson 1998, Markopoulos et al 1998) we have developed a theory of Task Knowledge Structures (TKS) for the purposes of understanding and explaining the structure of human task activity so as to provide designers of computer systems with a basis for understanding how to design systems to support people in carrying out both existing and new tasks. From this applied theory of tasks we developed a model based approach to user interaction design (ADEPT) and demonstrated how the TKS based models of human activity could be used to inform and facilitate design. More recently, following from the theoretical work we have been able to formulate four principles of design that have a theoretical and explanatory basis, provide predictive guidance to designers and that have been empirically shown to produce improved usability. From TKS theory we have developed four principles for design, again in summary form these principles are:

Four Principles from TKS

Taxonomic Structure: Objects that are the same or similar will be conceptually grouped together, and actions on the same or similar objects will be carried out together.
Procedural Dependency: Actions which are causally related to each other through a task goal structure will be conceptually grouped together.
Conformance: User interfaces which conform to the users conceptual grouping will be easier to use.
Transformation: transforming a conceptual grouping to accommodate changes in the level or structure of concepts is cognitively expensive.

We do not propose to expand upon these principles here since our objective in this paper is to go beyond these and begin to develop theoretical models and design principles for supporting collaborative work. However, it is worth stating that these four principles have direct implications for the design of all classes of systems.

Some design implications arising from TKS principles:

Categorise objects based on object taxonomy and reflect this category structure in the user interface through perceptual properties of the display.
Actions should be categorised according to their procedural, sequential causal relations and this should be reflected in the user interface display through the temporal and perceptual properties of the user interface design.
Users can and do follow the user interface structure to build their own task knowledge structures provided that the user interface supports a task structure.
Where the user interface does not support a task structure the user will impose their own task structure but this will take longer to achieve and their performance with the user interface will be slower than if the interface reflected a task structure.

The particular scope and form of the original TKS theory is towards individual users carrying out tasks by themselves. While it accommodated a role perspective on work it did not consider how people work together to perform group tasks or how individual and group tasks interact. To extend this theoretical basis we must consider the nature of group tasks, the nature of collaboration and the interaction between group tasks and individual tasks. Moreover, we must approach this from the motivation and for the purpose of informing the design and evaluation of computer supported collaborative work.

3 Extending TKS Theory and Principles to Collaborative Group Work.

Much attention has been given to supporting collaboration, through assumptions about sharing data/information, reciprocal views of participants and creating awareness of each group member. Similarly, attempts have been made to base the design around quasi-metaphors such as 'rooms' and 'locales' in which group and individual activity is intended to occur. However, these intuitions and assumptions do not provide any insight into the nature of collaboration or group work. Consequently, computer systems designed to support collaborative work often fail to support any particular form of collaborative group work and contribute little to HCI theory.

In developing an HCI account of collaborative work that can contribute to the design and evaluation of collaborative systems we must be able to understand the structure of group work and how such human activity can be affected by computer technologies. To accommodate group work in a collaborative context, extensions to TKS are required both to the underlying theory and to the principles.

In an attempt to model collaborative group work for HCI design, we have proposed the following group task modelling framework:

Group goals; purpose of group activity, responsibilities of group, focus of common ground.
Individual goals; purpose of individual activity, responsibilities of individuals, individuals' contributions to common ground.
Group objects; artefacts available to the group, properties, relations, states, content of common ground.
Individual objects; artefacts available to the individual, properties, relations, states, excluded from common ground.
Group processes; actions available to the group, procedural relations, affects content of common ground.
Individual processes; actions available to the individual, procedural relations, effects, excluded from common ground.

In considering the nature of collaborative group work, three important activities need to be understood. These activities can themselves be thought of as tasks in so far as they require resources and are purposeful. The three forms of activity that are synthesised in collaborative group work are group tasks, individual tasks and the additional tasks required to achieve collaboration.

For example, consider a team of doctors and nurses engaged in performing a neurosurgical operation. Together, the group work at a common goal such as removing a tumor from the patient's brain. Each member of the team will carry out various acts, some in coordination with other members of the team and others on their own. Some of these acts will be part of the team activity and some not. For instance, the consultant surgeon may be accessing and removing the tumor, the senior registrar may have prepared the patient and be observing the consultant, and subsequently may be involved in making good the surgery once the consultant has removed the tumor. The anesthetist may be involved in monitoring the patient's body state and life support system while the surgery is in progress. The nurses may be involved in keeping everything that is needed during the operation 'to hand' and in ensuring that everything in the operating theatre that comes into contact with the patient is kept sterile at all times.

At times, the surgeon and the anesthetist may be working completely independently on tasks that form part of the group activity but do not *per se* require any overt collaboration. At other times, the surgeon may require the anaesthetist to alter the body state of the patient (e.g. the oxygen level) in concert with a particular part of her removing the tumor. At other times, the anesthetist may be filling in a crossword puzzle while monitoring the displays and alarms.

The point is that there are various forms of group and individual task structures involved and in addition there are tasks of collaboration, as for example how the surgeon signals to the anesthetist that she wants him to alter the oxygen level. The amount of effort expended and resources required to do this may vary. The collaboration task is an additional task and requires effort and resources above and beyond those of the individual and group tasks. The collaboration aims to establish common ground and imposes costs on the participants. The costs of these tasks (the group tasks, the individual tasks and the collaboration tasks) are all incurred in performing the activity and must be borne.

Thus, we can begin to evaluate collaborative work systems in terms of the resource costs. We can think of these as both group task resources and group-computer interaction resources. Group task resources are what knowledge, both procedural and declarative, the individual and the group as a whole need to carry out the tasks in which they are engaged. The group-computer interaction resources are the above plus the resources the computer system needs in order to support the group activity. Additionally, the costs of establishing and maintaining common ground are resources expended to support. Hence, we have a basis for designing and assessing collaborative systems in terms of resource costs.

To account for collaborative work, the theoretical basis of TKS needs to be extended to consider group, individual and collaboration tasks. In addition we must also extend the HCI principles derived from TKS theory to reflect that, (a) collaborative tasks require common ground, (b) common ground is derived from taxonomic and procedural knowledge of individuals in the group and from the context in which the collaboration occurs, and (c) resources are expended in establishing common ground amongst the group and in performing transformations of individual knowledge and information to shared knowledge and vice versa.

4 References

Hamilton, F., Johnson, P. and Johnson, H. (1998). Task related principles for user interface design. Proceedings of the Schärding workshop on Task Analysis, June 1998.

Johnson, H. and Johnson, P. (1991). Task knowledge structures: psychological basis and integration into system design. Acta Psychologica 78, 3-26.

Johnson, P., Johnson, H. and Wilson, S. (1995). Rapid prototyping of user interfaces driven by task models. In J.M. Carroll (ed) Scenario-based design: envisioning work and technology in system development, New York, John Wiley and Sons, 209-246.

Markopoulos, P., Johnson, P. and Rowson, J. (1998). Formal aspects of task based design. Proceedings of DSV-IS'97, Springer-Wien.

Developing Scenario-Based Requirements and Testing them for Minimum Quality

Wolfgang Dzida
GMD German National Research Center for Information Technology

Introduction

Scenarios have increasingly been used as a source of domain knowledge enabling analysts and designers to deepen their understanding of the users'/customer's requirements (Weidenhaupt et al., 1998). This paper is aimed at concisely describing the process of scenario acquisition, the extraction of requirements from scenarios, the validation of both scenarios and requirements and a test for compliance with an international usability standard (ISO 9241-10), thereby achieving validated minimum requirements as early as possible in the design process.

The focus of this paper is an analysis of non-functional and usability requirements elicited from the context of product use. Our approach can be seen as an amendment to an otherwise accomplished functional analysis, for instance, a use-case analysis. The knowledge to be captured for requirements analysis is represented in two types of scenarios, context and use scenarios, the acquisition of which being conducted iteratively in a two-steps process.

Context scenario is defined as a narrative and episodical description of the user's tasks conducted in his current context. The product to come is not yet considered in this type of scenario, however, its envisaged usage is described as far as the prospective user can give an idea of it. Drawing on concepts of Carroll (1994, 1995), we want to define *use scenario* as a narrative and task-related description of the actual use of a product (or prototype). The distinction between both kinds of scenarios goes along with an acquisition of knowledge in two steps, first the domain knowledge, then the knowledge of how to use a system. The elicited knowledge serves as a basis for deriving usability requirements, first the context related then the use related ones. The analyst thus specifies requirements, which can be easily validated by the user, since each requirement is linked to a statement in the scenario. Usability standards, in particular ISO 9241-10, serve as an additional source to transform the elicited context and use requirements in terms of minimum usability requirements. The analyst thus tests the requirements for compliance with usability standards.

Theoretical and methodical reasons have been published arguing that the analysis of domain-specific requirements and usability design (i.e. prototyping) are inevitably intertwined (Dzida and Freitag, 1998). For this reason, we regard the development of requirements as a constructive activity relying on user-validated scenarios, prototypes and usability standards.

Scenario Acquisition

The *context scenario* is edited by the analyst after having interviewed the domain expert or a representative user of the product to come. Guidance is provided to the analyst how to capture the knowledge and document it in terms of the user's language; also, how to structure a scenario according to the structure of tasks. Questions are derived from a list of attributes typically characterizing the context of an office product (see ISO 9241-11, 1998). Example scenario: "Jim works in a service company. Sometimes, his task performance is interrupted by phone calls, which may require him to start a dialogue with the system on a new task in order to answer the call. But Jim wants to continue the previous task after interruption." This fragment scenario clearly indicates the dependency of the user's activity from the context.

After having validated the final context scenario by asking the domain expert (or user) for confirmation, a prototype needs to be developed challenging the designer to demonstrate what has been understood so far about the problem domain and its transformation into a new operative form.

In view of prototypes, dialogue characteristics are analyzed for specifying dialogue requirements in more detail. (For the symbiosis of scenarios and prototypes see the paper of Regine Freitag in this volume.) Prototypes enable the users to demonstrate the efficient conduct of typical tasks and comment their interactive characteristics, such as steps of interactive performance, intended and achieved results, features of dialogue and information presentation. For validating the (required) interactive characteristics of the prototypes the actual use of the product is described in terms of a *use scenario*, thereby documenting the efficient conduct of a required task, which serves as a rationale for the specified dialogue attributes. (For the documentation of prototypes in terms of scenarios see the paper of Dzida and Freitag in this volume.)

Extraction of Requirements from Scenarios

The extraction of requirements from a *context scenario* is an intermediary step between scenario documentation and prototyping. Requirements are derived in two steps. First, the implied task and organizational requirements are reflected with regard to ergonomic quality criteria of human task design (see ISO 9241-2). Second, dialogue requirements are derived from both the scenario and the implied task design requirements, thereby considering that dialogue and task design need to be analyzed conjointly. The extraction of dialogue requirements is aimed at constructing an efficient task performance at the user interface as far as the context of use is concerned. For example: From the above mentioned fragment scenario the analyst may derive an implied need, which has to be considered without any question: "Service always implies that phone calls must be answered immediately." Based on this requirement a user performance requirement can be derived: "Can the user interrupt the system dialogue in order to start a new task before resuming the previous one without loosing data?".

While extracting dialogue requirements the analyst encounters a number of questions as regards the details of the context and the problem domain, which need to be clarified by the domain expert. Clarification can be taken as a first step of validating the scenario content, but can also induce new ideas about technology-supported task and organization. For example, the interviewed user clarifies the problem of feeling disturbed by phone calls while working highly concentrated on a complicated case. Jim envisions some technical support allowing him to escape occasionally from interruptions which cause stress and error-prone situations.

A more complete set of task performance requirements needs to be extracted from the *use scenario*. This scenario is documented in response to a prototype. To validate the use scenario it is passed to users who used and assessed the prototype. Further details of interactive task performance are then acquired, thus enabling the analyst to extract further dialogue requirements. To ensure that the extraction of requirements results in a constructive process, each requirement derived from the scenarios is accompanied by a dialogue attribute, which is recommended as a design proposal. For example, the need for switching between tasks can be satisfied by a technical interruption feature (e.g. multi-windowing dialogue) that allows to resume the interrupted task when necessary.

Noteworthy, the user's requirement for keeping control on phone calls which may disturbe him while being concentrated on a complicated task can be further satisfied by recommending an additional, but non-functional support feature: call diversion. This proposal is to design the user's work context according to the requirements derived from the user's context of system use.

The definition of a requirement in terms of both a required task performance and an adequate design proposal falls in line with ISO 8402. A requirement according to this quality assurance standard is an "expression of the needs and/or their translation into stated requirements for the characteristics of an entity to enable its realization . . ." (par. 2.3).

Testing Requirements for Minimum Quality

Testing dialogue requirements for the minimum of quality requires compliance with usability standards, especially ISO 9241-10, which enables the analyst to specify the user's task performance requirements according to design principles. Most of the Part 10 standard requirements are stated in terms of user performance. The analyst must understand the context of use (see context scenario) and the actual performance of tasks (see use scenario) as perfect as possible. In view of these data the standard requirements can be interpreted, so as to specify the required minimum performance. For example, from the principle of "controllability" (ISO 9241-10, par. 3.3) the analyst can derive a minimum usability test criterion concerning interrupting and resuming task performance after having investigated the contextual conditions of user's task performance.

With the aid of further usability standards (ISO 9241, Parts 13-17)), the analyst can suggest user interface attributes serving as minimum design proposals to amend the list of performance requirements. Of course, the designer may feel free to create a design solution above the recommended minimum level. Nevertheless, all partners involved in the early analysis and design stage can be satisfied with an early test of the requirements for compliance with the required minimum quality of human-computer interaction represented by the standard.

Conclusion

This paper wants to show that opposite to many quality assurance standards (e.g. ISO 9000-3 [1991], ISO 13407 [1999]) the validation of usability requirements can be done prior to verification and very early in the development process. Scenarios are an effective means for supporting this strategy. They allow the domain expert (user) to really participate in the development of requirements and their validation, and they enable the analyst to better understand the problem domain.

References

Carroll, J.M. (1994). Making *use* a design representation. *Communications of the ACM*, 37/12, 29-35.

Carroll, J.M. (Ed.) (1995). *Scenario-Based Design: Envisioning Work and Technology in System Development*. New York: John Wiley.

Dzida, W. and Freitag, R. (1998). Making use of scenarios for validating analysis and design. *IEEE Transactions on Software Engineering*, vol. 24, no 12, 1182-1196.

ISO 8402 (1994). Quality management and quality assurance – Vocabulary.

ISO 9000-3 (1991). Quality management and quality assurance standards. Part 3: Guidelines for the application of ISO 9001 to the development, supply and maintenance of software.

ISO 9241-2 (1995). Ergonomic requirements for office work with visual display terminals (VDTs) - Part 10: Guidance on task requirements.

ISO 9241-10 (1995). Ergonomic requirements for office work with visual display terminals (VDTs) - Part 10: Dialogue principles.

ISO 9241-11 (1998). Ergonomic requirements for office work with visual display terminals (VDTs) - Part 11: Guidance on usability.

ISO 13407 (1999). Human-centred design processes for interactive systems.

Weidenhaupt, K., Pohl, K., Jarke, M., and Haumer, P. (1998). Scenarios in system development: Current practice. *IEEE Software*, March/April, 34-45.

The Symbiosis of Scenarios and Prototypes in Usability Requirements Engineering

Regine Freitag
GMD German National Research Center for Information Technology

Introduction

The major purpose of requirements engineering is to achieve a common understanding between the user and the designer of an intended software product. Scenarios describe the user's view of performing tasks with a software product in terms of the user's language (Carroll 1994, 1995). Prototypes represent the designer's ideas initially realized as an arrangement of functions at the envisaged user interface. Both kind of representations, scenarios and prototypes, form a symbiosis in the design process (Weidenhaupt et al., 1998). Scenarios serve as a catalyst when designing and validating prototypes. Prototypes are in response to scenarios and serve as stimuli to users for describing more elaborated scenarios. This paper illustrates how the use of scenarios can guide the prototyping process by directing the designer's focus on task and context-related issues when arguing about technical features concerning the user interface. The role of two types of scenarios is considered: context and use scenarios. (For the usage of scenarios in usability requirements engineering see also the paper of Wolfgang Dzida in this volume).

The Role of Prototyping in Usability Engineering

Prototyping has been accepted as a method to involve the user early in the development process of a software product. A prototype is intended to give the prospective user a chance to argue about the fitness of tangible design proposals to the required results of user task performance. In usability engineering prototyping serves two purposes: first, prototyping as a process is challenging the designer to demonstrate an elaborated understanding of the user requirements by transforming them into product attributes and second, the prototype as a design proposal is enabling the user to check whether a consensus about the suggested attributes can be achieved. Furthermore, the user is encouraged to reflect his requirements in view of the potential of a technology and concrete technical alternatives. This kind of prototyping is also refered to as exploratory.

The idea of *exploratory* prototyping is to demonstrate and examine alternative design options (Budde et al., 1992) in order to come to an initial design proposal. Discussing various ways of performing a task in view of a prototype reveals trade-offs and misunderstandings but leads also to more concrete, well-founded requirements on the user's side (Ryan and Doubleday, 1997). Alternative design options can be submitted to claims analysis (Carroll and

Rosson, 1992) in order to assess the pros and cons of critical design proposals and document their design rationales. These elaborated requirements have to be considered in redesigning the prototype. The process of iteratively developing and evaluating a design proposal by reformulating or adding requirements and adapting the prototype is called *evolutionary* prototyping. Prototyping as an engineering paradigm supports the requirements definition process and "eases communication problems between developers and clients" (Bischofberger, Pomberger, 1992, p. 19). But installing a prototyping-oriented design process alone does not guarantee that misunderstandings among the members of the prototyping team vanish. Context and use scenarios can help guide and validate the prototyping process.

Context Scenario: Focusing Task Performance in Context

If a new system is to be developed it is vital for the potential users and the designers to start a dialectic process [(Dahlbom and Mathiassen, 1995), (Dzida and Freitag, 1998)] in the course of which user requirements are developed. The dialectic process is initiated by contrasting a design idea (developer's intention) with the task performance in its current context (user's view) including the envisaged understanding prospective users create of the system to come. This process is supported by formulating and discussing context scenarios in the user's language. In a context scenario the user's key tasks and his work situation are put down in a narrative way including the expected improvement of the current task situation.

The dialectic process helps uncover mismatches between the user's and the analyst's understanding of the system. It is likely that the designer wants to demonstrate a meanwhile more elaborated understanding in form of a prototype. But prototyping should not start until the final context scenario has been validated by the user. The final context scenario causes the designer to revise the initial concept or create an initial use scenario for a main task which has been achieved consensus upon. In the initial concept the designer describes an understanding of the user's task in view of system attributes. On the basis of this concept further questions on the side of the designer can be resolved before prototyping starts. Therefore the context scenario(s) and the initial concept help structure the prototyping activity (Weidenhaupt, 1998).

Another reason for not starting with prototyping too early is the risk of importing a technical bias into the requirements elicitation process. This is also the case when there is already a system in use which is to be replaced by a different one. Users are inclined to argue in terms of the familiar system (insisting on seemingly comfortable dialogue parts). The use of context scenarios allows the user and the designer to step back and think about envisaged task and work conditions instead of a habitual system use, perhaps even opening one's mind for radically different solutions.

On the other side context scenarios should not be misunderstood as a justification for mirroring the task flow described in the scenario into the dialogue with the system. The task descriptions contained in a context scenario should not be taken as the "one best way" of pursuing a task but only as one possible way of performance. Context scenarios offer a way to combine all the information which is collected and listed separately in traditional context analysis. Therefore, context scenarios shall encourage the designer to consider the entire work situation when planning the prototype.

Use Scenario: Validating the Prototype

The most salient feature of scenarios is that they include requirements in terms of expected results of user task performance (user requirements), whilst a prototype provides a tangible representation of these results in terms of system attributes (system requirements). A correspondence between user and system requirements must be achieved during the development process. There may be alternative system requirements for a specific user requirement. The aim of usability requirements engineering, however, is to figure out which of the alternative system attributes fits well. This can best be achieved by presenting a prototype, which enables the user to assess alternative design proposals and confirm that the expected effects or results have been achieved. The requirement elicitation process ends up in the definition of usability requirements, with usability being defined as providing an efficient and satisfying solution to the user's requirements (ISO 9241-11). (For the documentation of requirements see the paper of Dzida and Freitag in this volume).

If the prototype is considered as being sufficiently elaborated it has to be validated. It does not suffice to merely present the prototype to the user. The user must have the opportunity to relate the prototype to his tasks and to imagine the use of the prototype in his context. This is when use scenarios come into play. A use scenario describes the user's perspective while performing a task with the prototype. The prototype need not support real performance of a task but has to give a realistic feel about the presentation of the offered functions and the interaction with these functions.

Use scenarios should be described after the prototype is constructed. As with the context scenarios, use scenarios are edited by the analyst but written in terms of the user's language. Again, the user is asked for describing an envisaged use of the system in view of the prototype. With the help of use scenarios the user can validate if the prototype is a useful technical realization of his requirements. Use scenarios help focus on those attributes of the prototype, which support the user in performing the task at hand, thereby guiding the attention on task and context-related features instead of evaluating arbitrary presentation details as often been practiced by heuristic evaluation.

Use scenarios reveal further requirements in terms of user and task performance and provide a rationale to revise the prototype. Once again it is not intended to prescibe a specific task performance. A use scenario is nothing but an example task execution which helps evaluate the prototype's fitness in fulfilling user performance requirements. This process leads to a validated design proposal which is more than just an evaluated user interface and can thus serve as a consensual basis for the final requirements definition.

Summary

The construction of an exploratory prototype urges the designer to transform user requirements formulated in a task-dependent way into software and dialogue functions. Context scenarios help focusing on the task and context-related issues when it comes to prototyping. On the other hand, the evaluation of a prototype urges the user to reflect attributes of task performance in view of technical features. Use scenarios serve as a means to concentrate on the task and context-related issues when evaluating the prototype. Use scenarios as well as prototypes thus help both the designer and the users to create synthetic effects regarding user performance and product design.

References

Bischofberger, W., Pomberger, G. (1992). *Prototyping-Oriented Software Development. Concepts and Tools*. Berlin: Springer.

Budde, R., Kautz, K., Kuhlenkamp, K., Züllighoven, H. (1992). *Prototyping: An Approach to Evolutionary System Development*. Springer, Berlin.

Carroll, J.M. and Rosson, M.B. (1992). Getting around the task-artifact cycle: How to make claims and design by scenario. *ACM Transactions of Information Systems*, 10, 181-212.

Carroll, J.M. (1994). Making *use* a design representation. *Communications of the ACM*, 37/12, 29-35.

Carroll, J.M. Ed. (1995). *Scenario-Based Design: Envisioning Work and Technology in System Development*. John Wiley, New York.

Dahlbom, B. and Mathiassen, L. (1995). *Computers in Context*. NCC Blackwell, Oxford.

Dzida, W. and Freitag, R. (1998). Making use of scenarios for validating analysis and design. *IEEE Transactions on Software Engineering*, 24(12), 1182-1196.

ISO 9241-11 (1998). Ergonomic requirements for office work with visual display terminals (VDTs) - Part 11: Guidance on usability.

Ryan, M. and Doubleday, A., (1997). 'Throw Away' Prototyping for Requirements Capture. In: Szwillus, G. (ed.): *PB'97: Prototypen für Benutzungsschnittstellen*, Tagungsband, Paderborn, 27-34

Weidenhaupt, K., Pohl, K., Jarke, M., and Haumer, P. (1998). Scenarios in system development: Current practice. *IEEE Software*, March/April, 34-45.

Application-Oriented Software Development for Supporting Cooperative Work

Heinz Züllighoven, Guido Gryczan, Anita Krabbel, Ingrid Wetzel
Software Engineering Group, Computer Science Dept., University of Hamburg
[zuellighoven, gryczan, krabbel, wetzel]@informatik.uni-hamburg.de

1 What is Application-Orientation?

Customer relations have become one of the central concerns of companies both in the manufacturing and the service sector. Software engineering thus should rethink its aims and methodologies. IT usage and software development have to be seen as means to an end which means providing professional users with useful and usable software so that the can offer adequate services to their customers. This way of looking at IT and software development is what we call application-orientation.

Application-orientation means that analysis, design and construction of software is firmly based on the tasks and the way of dealing with them in everyday work situations. Understanding the tasks at hand and the concepts behind them is the main challenge for software developers. Application software, therefore, should reflect and represent the core concepts and the familiar objects and means of work of the application area.

2 The Tools & Materials Approach

Building application-oriented software is a task that has to combine usage quality with work organisation. The ultimate goal of our so-called Tools & Materials approach, is to adequately support users in their tasks. But THE user as such is a fiction. Also, tasks are of very different nature. Complexity increases as users have to fulfil their tasks in different organisational settings.

This means that software developers have to cope with a design task in its original sense, creating artifacts with a suitable functionality and an adequate way of handling. Our guideline for this design task is what we call a leitmotif.

A *leitmotif* makes the way of looking at software explicit. It helps developers and users to understand and design a software system on a general level.

A leitmotif can be made tangible with the help of a set of design metaphors which solidify the general guidelines in a pictorial way.

A *design metaphor* describes a component of a software system by means of an artifact from the users' 'every-day' world. With the help of these metaphors it is possible for software developers and users to relate design and implementation components to familiar implements and terms, so that all parties share a common background and have a basis for communication.

We have chosen Tools and Materials as our predominant design metaphors, since we can frequently describe a work situation by saying how tools and materials are used to achieve a goal. Materials are those objects of the application domain that are worked upon and then become part of the work results. Tools reflect the organisation of work. They objectify our experience of working with materials in a similar way in different situations. Two additional design metaphors are automaton and desktop. An automaton realizes a task completely so that user-interaction is reduced to input data. The user has little means to change the implemented flow of control. The desktop captures the notion of space. Design metaphors mentioned are almost always used in the context of a desktop. Design metaphors have an application-oriented interpretation ("I use the cashier-tool to deposit an amount of money in my account") and a technical interpretation ("The material class *account* must have a method *deposit* that takes an *amount of money* as parameter"). They play an important role when partitioning frameworks (Bäumer et al. 1997).

3 Cooperative Work and Workplace Types

Originally, we have build support for the individual workplace mainly in the domains of software engineering environments and in the banking sector. But we soon had to face the task of integrating components that could support cooperative work. Over the last two years we have extended our approach into this area. The original approach can be characterised by task-orientation, involvement of users, and a reification of relevant issues as components of the software system. With the shift of focus towards cooperative work, the new issues are: (a) intertwining of functionality and cooperation within tasks (Grudin 1994) (b) a distinction between explicit and implicit ways of cooperation and their support and, (c) a reification of means and media of cooperation and of the cooperation process itself.

We do not start by asking, how people cooperate but we focus on the tasks and the characteristics of the relevant work situation. We try to understand what type of work we are faced with, who does what with whom and, most importantly, why.

A basis for supporting various types of cooperation is to identify different workplace types. Taking, for example, normal office-type of work, we have identified (1) the well-equipped workplace for expert users suited for flexible, situated tasks, the (2) function workplace for expert users with repetitious tasks, and the (3) back-office workplace for a small set of routine tasks for low qualified users.

Each type of workplace is equipped with the adequate items (i.e. tools, materials, automata). This categorisation of types of workplaces is by no means exhaustive. We emphasize the underlying principle: Workplace types are related to the work organisation in an application domain.

4 Cooperation Types

Based on workplace types and the cooperative work within an application domain, we have set up an explicit model of cooperation which consists of actors, objects and means of work and explicit means and media of cooperation. The central idea is to extend the different workplace types by components which can support cooperation but which fit within the overall leitmotif and the metaphors for applications in this domain. So, the design issue is fitting different workplace types to the cooperation model of the application domain.

Evaluating different projects, we have identified similarities among various ways of cooperation and introduce cooperation types. A cooperation type abstracts from concrete ways of cooperation. It characterizes work division, used cooperation and coordination mechanisms and tasks. For cooperation types we distinguish *cooperation media* from *cooperation means*. Cooperation means are objects used for structuring and coordinating the cooperation while cooperation media are objects supporting the exchange of information or material. Our interpretation of coordination is based on (Malone and Crowston 1994).

In the following we sketch three examples of cooperation types: implicit cooperation through a common archive, explicit cooperation by PO boxes and explicit coordination with dockets.

4.1 Cooperation with Archives

- *Situation/Example*: A typical situation for using archives is a small group of specialized advisors in a bank (e.g. for company credits) who are in charge of a group of customer companies. They share the same customer records.
- *Cooperation Model*: An advisor team provides continuous services to a selected group of customers. The team member share folders and files holding the customer records. A common archive storing these folders can be accessed from each work place. Usually, a team member will retrieve only part of the customer records (e.g. a credit folder) for an individual task. While working with the archive each workplace receives awareness information of who works on the same material in parallel or who has worked on it in the past. There are conventions on how to solve conflicting accesses. These conventions might be enforced by the system or they are organizationally respected by the users supported by appropriate awareness information (Mark et al. 1997).
- *Cooperation Means*: Common archive with folders and files. Awareness information.
- *Cooperation Media*: Browser for accessing the archive.

4.2 Cooperation with Archives

- *Situation/Example*: PO Boxes are used in situations where, e.g., credit officers or customer advisors send credit application forms to their back-office. These forms have to be distributed among the people working in that back-office in order to mail routine letters and perform standard checks.

- *Cooperation Model*: The management of a unit has to distribute simple tasks among a group of persons. In principle, every task can be handled by each person of the team and it can be completed by a single person. So we need an easy way of forwarding the tasks together with dossiers and documents. In addition, it must be easy to find a substitution, if a person is not on duty. PO boxes for each team member are connected to a general group mailbox. The group mailbox is linked to an email service. Other groups and units send their documents and forms to this general mailbox. One person is in charge for distributing the incoming mail to the different PO boxes. The PO boxes with their name tags indicate which work places are available and additionally (by showing their content) who is actually there or whose work assignment is unbalanced. There is no implemented access mechanism. Each person involved knows by convention who has the right to put something into a box and who is allowed to empty a box.
- *Cooperation Means*: General group mailbox, individual PO boxes, distributed documents like dossiers and forms.
- *Cooperation Media*: Email system, browser tool for accessing mail boxes.

4.3 Cooperation with Dockets

- *Situation/Example*: Dockets are used when a complex task like granting credits is accomplished in several steps. These sub-tasks are handled according to well-known rules and with a clear work division e.g. between back office, department head, controller and account manager.
- *Cooperation Model*: A small fixed set of people works on complex tasks exchanging information and documents. The task is subdivided into well-known steps and is being performed by known rules including given sequences and responsibilities. Every actual work step has to be documented. But the concrete work situation frequently calls for flexibility in the selection of individual work steps and the overall working sequence. Dockets (i.e. to-do lists) can be fixed to folders or dossiers which are passed between the work places involved. These dockets are used as process pattern signalizing the usual work sequence and the persons or workplaces involved. Additionally, they indicate the status of the process. Each receiver of a document or folder with an attached docket has control over the next steps by possibly changing the process pattern. The docket works together with a mail system which delivers the attached material according to the status of the docket. While a complex task is processed each person involved can request its current state and location from a task monitor.
- *Cooperation Means*: Dockets attached to folders or documents, task monitor with status information.
- *Cooperation Media*: Mail system.

5 A Flexible Cooperation Support

Once we have identified a set of workplace types and cooperation types, we have a basis for coping with different cooperative work situations in an application domain. Because if we start to combine workplace types with cooperation types, we realize that in general almost all combinations are feasible. As figure 1 shows, these combinations only differ to their degree of usefulness.

	Cooperation with Archives	Cooperation with PO boxes	Cooperation with Dockets
Expert workplace	**	**	**
Function workplace	*	-	**
Back-office workplace	*	**	**

Figure 1: Suitability of cooperation types for workplace types
[(**) very useful (*) useful (-) just working]

What does figure 1 mean? We found out that cooperation types and workplace types can almost freely be intertwined. This means in a concrete application domain, that we can provide substantial support for cooperative work for a wide range of working situations by recombining a few workplace types with a small set of cooperation means and media. The combination we actually choose, depends on what work analysis and job design (or business process reengineering) tells us.

6 References

Bäumer et al. (1997) D. Bäumer, G. Gryczan, R. Knoll, C. Lilienthal, D. Riehle, H. Züllighoven: Framework Development for Large Systems. CACM, October 1997, Vol. 40, No 10, pp. 52 - 59.

Grudin (1994). J. Grudin: Groupware and Social Dynamics: Eight Challenges for Developers. CACM, Vol 37, No 1, 1994, pp. 92-105.

Malone and Crowston (1994) T.W. Malone, K. Crowston: The interdisciplinary study of coordination. ACM Computing Surveys, Vol.26, No.1, pp.87-119

Mark et al. (1997). G. Mark, L. Fuchs, M. Sohlenkamp: Supporting Groupware conventions through Contextual Awareness, In W. Prinz, T. Rodden, J. Hughes, and K. Schmidt (eds.), Proceedings of ECSCW'97, Sept. 7-11, Lancaster, England, Kluwer Academic Publishers, Dordrecht, 1997, pp. 253-268.

Schmidt and Bannon (1992) K. Schmidt, L. Bannon: Taking CSCW Seriously: Supporting Articulation Work. In: Computer Supported Cooperative Work: An International Journal, 1 (1992) 1, pp. 1–33.

Internationalization and Localization of the Web Sites

Nuray Aykin
AT&T Labs

1 Introduction

With the explosion of the Internet, the need for multilingual Web sites becomes more than a luxury for the companies who wish to reach worldwide customers. Although the majority of the Internet users speak English and over 80% of the Internet content is in English, the statistics show that this is changing at a fast pace. The Asia Pacific countries have the fastest growth of Internet infrastructure and usage. Today, there are 5.6 million Internet users in Japan, 1.3 million in Taiwan, and 1.2 million in Australia. The Internet users in Europe show a similar trend, with 4.5 million in Germany, 4 million in UK, and 1.5 million in France. By year 2002, it is expected that there will be 300 million Internet users, and 50% of them will be non-English speakers. Another fact is that the internationalization and localization is not just a country issue. Even within the United States, multilingual and cultural issues exist. People in the U.S. speak different languages, and belong to diverse ethnic groups. For example, the percent of people speaking languages other than English is 78% in Miami, 49% in Los Angeles, 45% in San Francisco, and 42% in New York. Based on these statistics, we expect that the need for multilingual and multicultural sites is increasing dramatically. For example, The Los Angeles Times is putting up a Spanish page Web site that translates daily news into Spanish (Woods, 1998).

From just about any country, you can reach anywhere and view any Web site. When you publish a Web site on the Internet, you are already reaching global users. Anyone in the world can access your Web site, can look at what you offer, and can engage himself/herself with the world that has no boundaries. Your site can be a great source for branding, electronic commerce, or market analyzer. Accessing to a wider audience in their own language and cultural content provides great opportunities. But we cannot reach our customers if we do not address their needs, culture, language, and preferences. We also need to ensure that the content is appropriate for the targeted cultures, and not offensive.

A fully localized Web site is one that is not just translated, but contains local content, respects cultural and language preferences, and considers technical capabilities of the locale that it is designed for. The users should feel that they are at home, not visiting a site that alienates them due to the language and the content of the site.

2 Internationalization and Localization Design Guidelines

If you are planning to have a Web site designed for several locales, it is important to identify what needs to be internationalized and what needs to be localized. Your company logo, company information, product information, technical background and some other general topics can fall into the "internationalized" category. This means you need to generate one version of the content, and have it translated to the desired languages. The localization here can be just the translation. However, some of the content needs to be locale specific, such as calendar of events, locale specific formats (e.g. date, time, and address formats), contact information, locale specific products, and some locale specific events that attract visitors to the Web site. In this case the locale specific contents must be generated and translated into the desired languages. The following guidelines cover design issues related to internationalization and localization (Aykin, 1998).

2.1 Guidelines on Internationalization

- Devise common strategies that can accommodate the many ways in which global users work: The life and work styles of people vary around the world. Their interests can be different from country to country. However, if you are planning to have an internationalized Web site it is possible to design a Web site that could be suitable to people around the world.

- Avoid culture-specific examples: If the Web site is designed for international use and not addressed to a particular locale, then there should be no culture-specific information and examples on the Web site.

- Avoid ethnic stereotypes: Stereotyping of any sorts should be avoided since it can be offensive in certain cultures.

- Avoid religious references: Unless the site is designed for religious purpose, there should be no references to the religion.

- Avoid showing body parts: The body parts can have different meanings in different cultures. The best way to ensure that the Web site does not offend any culture is to not show any body parts as part of the graphical images.
- Avoid showing "flesh": Some religions do not allow showing body "flesh". If the Web site is likely to be visited by people in these religions, it is important not to show any "flesh" on the Web site, regardless of age and gender.
- Beware of gender-specific roles in other cultures: In some cultures, males and females have specific roles and ranking in the society. It would be best not to refer to gender-specific roles as part of the Web site content.
- Internationalize your graphics: The graphical images increase the aesthetics of the Web site. However, if the right images and icons are not used, the Web site may become offensive to certain cultures, and/or the intended meaning of the images may get lost.
- Use appropriate colors that are suitable for international use: It is important to choose the right colors for the Web site that can reflect the images of different cultures. Each culture has its own preferences. For example, pastel colors are suitable for Asian countries, and bright colors are suitable for Latin American countries.
- Allow flexibility for formatting: This applies to address, fax/telephone number, numbers, date and time formatting. The name and address formatting includes identifying the number of address lines, names, suffixes, prefixes, addressing formats, zip/postal codes, labeling of fields, and character support. To create a common address format use 2 lines for street entry, use appropriate labeling (such as zip/postal code, use 8-digit postal code, and allow wrapping on the text fields. For common phone/fax formatting, provide at least 15-digit field size to accommodate the world standard. For number representation use "." for thousands separator, and "," for decimal separator. For date formatting use either the ISO standard (yyyy/mm/dd) or commonly accepted dd/mm/yyyy format, with month shown as text rather than numeric.
- Register with international search engines and international registration services such as www. GlobalPromote.com, international versions of Yahoo!, Alta Vista, Excite, Lycos, Infoseek, etc., and local countries directory services.

2.2 Guidelines on Localization

- Identify the areas that change when going from one country to another: In designing Web sites for different locales, the first step should be to define the content that is internationalized and common across all Web sites, and the content that will be locale specific.

- Understand the culture of the target market: It is necessary to understand the culture of the target market, i.e. the patterns of thought, behavior, and preferences. This is very crucial in determining the acceptability of a Web site content within a particular locale. What do the behavior of the user look like? What are the users' expectations and preferences? It is important to gain a better understanding of users, and to propose user models that best describe how users think about their work.

- Create guidelines and design specifications to ensure consistency across locale contents: To maintain the image of the Web site across locales, it is important to create guidelines and design specifications for the Web designers. Auditing of locale-specific Web sites becomes very crucial since the company image that is conveyed to the different locales should stay the same.

- Make sure the content is culturally appropriate: Consulting with the in-country employees or performing in-country usability testing can help define the culturally appropriate content for targeted locales.

- Consider translation of text and its impact on design: Translation of text into other languages is one of the major aspects of internationalization. It involves localizing the contents of the text for the target locale, translation of text, changes in the reading/writing directions (uni- and bi-directional text), and accommodating expansion during translation. Expansion of text can range from 10% to 200% depending on the language. Usage of grammar and controlled English, as well as limitations in acronyms, jargon, and abbreviations become important.

- Choose a translation vendor with which you can form a long-term relationship to handle all your translation needs: Working with one vendor, and forming a long-term relationship with them make the vendor feel as part of a team.

- Choose appropriate character set and fonts: Different languages require different character sets. The 8-bit character set is necessary for most languages. The U.S. 7-bit character is sufficient only for English, Hawaiian, and Swahili languages. The issue on character sets includes number of bytes per character, screen width per character, differing number of bytes per character, special characters, capitalization rules, printable and white space characters, and universal character sets including Unicode, UTF-8, etc.

- Use appropriate colors for the locale: Choosing colors for a particular locale can be quite a task. You should seek locale expertise on colors that are appropriate and aesthetically pleasing for that culture.
- Use appropriate icons and graphics for the locale: Designing icons and graphical images that are appropriate and meaningful in the target locale requires understanding the target culture. It is important that the graphical images are meaningful and are not offensive.
- Do not assume that the world uses credit cards (consider local content such as bank direct debit, credit cards, checks, etc.)
- Consider legal limitations for the targeted countries: It is important to understand the impact of laws and regulations related to Internet content and use. For example, Germany prohibits product comparisons, and Sweden prohibits sweepstakes, etc.)
- Use appropriate formats for numeric data, date/time, monetary, and address representation. This requires knowledge of the preferences, locale standards and people's daily usage.
- Involve country people, lawyers, consultants to evaluate your site: It is crucial to have your Web site evaluated for its content, layout and format in order to ensure it is culturally appropriate,

3 Conclusion

In order to have a truly global presence on the Internet, we need to understand the issues surrounding internationalization and localization. It is important to make human factors practitioners aware of how to approach these issues, and how to solve design problems to make the Web sites global. Creating internationalized and localized Web sites is a new challenge for human factors practitioners. It adds a new value to the adage "design for the user." It will make the product easier to extend to different locales, improve quality and ease of use, and provide a competitive edge in the global market.

4 References

Aykin, N. (1998) Internationalization and Localization of the Web Sites, Workshop presented at the Asia-Pacific Computer Human Interaction Conference (APCHI), July ,1998.

Woods, J. Dynamic Database-Driven Multilingual Web Solutions. Multilingual Web Sites Conference. February 1998. San Diego: CA.

Trends in Future Web Designs

Mary Czerwinski
Microsoft Research

1 New Web Browsers and Visualization Techniques

As computing bandwidth increases and the graphics capabilities of consumer hardware become more sophisticated, the trend on the World Wide Web is toward leveraging these technical shifts. In this paper, I will describe a sampling of the new browser designs that have been emerging with greater frequency over the last several months. While many of these browsers and visualization techniques were designed for quite specific tasks, in this article I am interested in browsers that were specifically designed to help organize and make sense of large collections of Web pages, files, or documents available on the Web. Target tasks for this analysis included:

- Finding a Web document you know exists, and that you have visited before (*targeted revisitation*)
- Finding a Web document you know exists, but that you've never seen before (*targeted search*)
- Finding a Web document and most of the pages related to it on a particular *topic (comprehensive browsing*)
- Finding a Web document on a topic that is "close enough" to the subject at hand *(satisficing during browsing*)

The dimensions we are using for evaluation came from three different aspects of research in our lab:

- Our on-going usability analysis of Web browsers (e.g, Czerwinski & Larson 1997).
- Literature review of important browser usage characteristics, (e.g., Abrams & Baecker 1997; Tauscher & Greenberg 1997)
- Cross-product comparison and historical tracking

Targeted revisitation is the ability to go back to a page you have previously visited. This is the ability for which the History and the Favorites features in most Web browsers were designed. PadPrints (Hightower et al. 1998) has been shown empirically to do a superior job of providing the user with an easy way to return to a previously visited page. PadPrints works by creating a visual tree of all the web pages that the user has previously visited. As the user moves around a web site the visualization keeps growing to show the overall structure of the site, and more importantly, the users' present location and navigation pattern within the site. At any time, the user can select any previously visited page on the tree and revisit that page. Recently, Jul (1998) has extended the zoomable user interface idea by indicating "critical zones" visually when the user is zoomed out so far as to effectively be in a "desert fog".

Empirical studies (Robertson et al. 1998) have also shown that a 3D user interface solution, the Data Mountain, is a more efficient way of navigating to previously viewed Web pages compared to a 1D folder hierarchy alternative, such as Microsoft's Internet Explorer v. 4.0. The Data Mountain's primary strength is in targeted revisitation tasks, although it does effectively support casual browsing as well. The main benefits of the Data Mountain include its "lightweight" policy with regard to grouping (there is no policy but users group their web pages anyway) and its use of 3D depth cues using only 2D interaction techniques. Subsequent research efforts added "implicit query" highlighting to the browser. Using standard similarity metrics, web pages that are similar in content to the currently selected web page are highlighted with a simple, green surrounding halo. The user can quickly glance to see which clusters of web pages are mostly strongly related to the current page, and either open those pages, or store the currently selected page near that cluster. Empirical studies show that users can find their targeted web page more quickly still if implicit query highlighting is turned on when the web pages are stored (Czerwinski et al. 1999).

For *targeted search* tasks, getting great overview information and maintaining a global/local viewpoint is critical. So is knowing what categories you have already examined during a search task. We chose to examine browsers like InXight's Hyperbolic Browser (Lamping et al. 1995). The hyperbolic browser is a wheel of information with browsing beginning at the center node. When a spoke of the wheel is selected from the first ring around the center node, the wheel's shape warps leaving the first ring smaller but still visible, while allocating more screen space to subcategories and nearest neighbors of the selected item. In user studies, we compared the hyperbolic browser to a traditional hierarchical tree control user interface. Our studies demonstrated that the hyperbolic browser was effective for finding search targets during browsing, due primarily to its effective use of overviews and related item expansions via animation. Users frequently found target items in local spaces that they may not have found if they were required to physically navigate up and down a tree. This type of browser does depend on a very careful layout of the physical space, and an alternative, alphabetical layout would not have provided the affordances necessary for serendipitously finding information as we observed in our studies. One serious usability issue we observed was that items on the outside rim of the wheel were grouped strongly together, with users often assuming that they all belonged to the same category. Careful use of alternative perceptual coding for semantic categories could alleviate this usability problem in future versions of the browser.

For *satisficing during browsing*, e.g., looking for some non-specific information in a topic area, Perspecta's_ Smart Content Viewer [http://www.perspecta.com], which was predicated by earlier work (Rennison 1994) holds some promise. The Perspecta browser has a fly-through navigation system. While many users will have difficulty using the navigation controls of such a browser, it is generally regarded as "cool". Visible on the screen are the top-level items spread evenly across the display. As the user flies towards a node (either via holding a mouse button down or double clicking on a node), the node begins to appear much larger. Concurrently, the items subordinate to that node now appear closer to the user, although their size is smaller (i.e. they are more distant from the user) in relation to the parent node. The advantage to this system is that it is quite easy to fly towards a node, quickly see previews of what's there and then fly elsewhere if the desired information doesn't appear to be in that category. On the negative side, users do get lost in hyperspace with this browser, and have a difficult time remembering where they have already traversed, as well as flying back out to a global point of view.

Comprehensive browsing necessitates an effective global/local perspective, getting great overviews and previews before navigating to categories, having semantically meaningful category labels, as well as the ability to visualize related clusters of information in an intuitive manner. Few to no novel browsers have sufficiently managed this rather ambitious task to date, although our group at Microsoft, as well as many others, is working hard on this problem. So far, we have concentrated our efforts on 3D information visualization techniques that rely on 3D cues and the use of lightweight, subject-defined layouts in 3D. In recent work, we have redesigned the Windows browser as a 3D hallway. Tasks can live on the walls, the floor or the front and back walls of the hallway—the user decides what categorizing scheme to assign to the various spatial locations in the hallways and arranges her information accordingly. Simple navigational metaphors, essentially forwards and backwards, allow users to interact in 3D using primarily 2D techniques. Our first two studies have already shown that users prefer this 3D visualization to their current desktop or browser. Future research will focus on better methods for scaling this visualization up so that literally hundreds of web pages and documents can be easily viewed, compared and selected.

2 Conclusion

It is our hope that by highlighting a set of novel web designs and focusing on what has been observed to work effectively or not in laboratory usability studies, we have identified trends that will prove useful to designers working in this area. We are currently working on methods and paradigms for more adequately evaluating novel browsers (e.g., Larson & Czerwinski 1997), and these techniques will be communicated in future publications, once proven valid and reliable.

3 References

Abrams, D. & Baecker, R. (1997). How People Use WWW Bookmarks *Proceedings of ACM CHI 97 Conference on Human Factors in Computing Systems,* pp. 341-342.

Chen, C. & Czerwinski, M.P. (1997). Spatial ability and visual navigation: An empirical study. In the *New Review for Hypertext and Multimedia*, 3, 69-89.

Czerwinski, M.P., Dumais, S.T., Robertson, G.G., van Dantzich, M. & Tiernan, S. (1999). Visualizing implicit queries for information management and retrieval. In the Proceedings of *CHI '99: Human Factors in Computing Systems*, Pittsburgh, PA: ACM.

Czerwinski, M.P. & Larson, K. (1997). The New Web Browsers: They're Cool but Are They Useful? In H. Thimbleby, B. O'Conaill and P. Thomas (Eds), *People and Computers XII: Proceedings of HCI'97*, Springer Verlag, Berlin.

Gershon, N., Card, S.K. & Eick, S.G. (1997). Information visualization. In the Proceedings *of CHI '97: Human Factors in Computing Systems*, Atlanta, GA: ACM.

Hightower, R.R., Ring, L.T., Helfman, J.I., Bederson, B.B., & Hollan, J.D. (1998). Graphical multiscale web histories: A study of PadPrints. *In Hypertext '98: The Proceedings of the 9 th Conference on Hypertext and Hypermedia*, Pittsburgh, PA: ACM, 58-65.

Jul, S. & Furnas, G.W. (1998). Critical zones in desert fog: Aids to multiscale navigation. In *Proceedings of UIST '98: Symposium on User Interface Software & Technology*, November, San Francisco, CA: ACM.

Lamping, J., Rao, R. & Pirolli, P. (1995). A focus+context technique based on hyperbolic geometry for visualizing large hierarchies. In the Proceedings *of CHI '95: Human Factors in Computing Systems*, Denver, CO: ACM, 401-408.

Larson, K. & Czerwinski, M.P. (1998). Web design: The implications of structure, memory & scent for information retrieval, In the proceedings of the *Association for Computing Machinery's (ACM) CHI '98 Conference,* April, Los Angeles, CA, pp. 25-32.

Rennison, E. (1994). Galaxy of news: An approach to visualizing and understanding expansive news landscapes. In *Proceedings of ACM UIST '94 Symposium on User Interface Software & Technology*, Marina del Ray, CA: ACM, 3-12

Robertson, G., Czerwinski, M., Larson, K., Robbins, D., Thiel, D. & van Dantzich, M. (1998). Data Mountain: Using spatial memory for document management, In *Proceedings of UIST '98: Symposium on User Interface Software & Technology*, November, San Francisco, CA: ACM.

Tauscher, L. & Greenberg, S. (1997). Revisitation patterns in World Wide Web navigation, *Proceedings of ACM CHI 97 Conference on Human Factors in Computing Systems*, pp. 399-406.

User-Interface Development: Lessons for the Future from Two Past Projects

Aaron Marcus, John Armitage, Volker Frank, and Edward Guttman
Aaron Marcus and Associates, Inc., 1144 65th Street, Suite F, Emeryville, CA 94608-1053 USA;Tel:+1-510-601-0994x19, Fax:+1-510-547-6125
Email: Aaron@AmandA.com, Web: http://www.AMandA.com

1 Introduction

In 1998-99, the authors' firm, Aaron Marcus and Associates, Inc. (AM+A), designed and implemented the user interface (UI) for Eye-to-Mind™ CD-ROM-based multimedia products of Cogito Learning Media (CLM), which summarize knowledge of a discipline. AM+A also designed and implemented the UI of a prototype for Unigraphics Solutions, Inc.'s (USI) Computer-Assisted Self-Teaching ™ (CAST) Web-based online help to serve as a guide for a major redesign of the company's online training for its computer-aided design and manufacturing software. Based on two different technologies, audiences, and products, AM+A has derived recommendations for user-interface development relevant to CD-ROM- and Web-based education and training.

Developing a user interface for a system enabling access, search, retrieval, and rapid decision making about competing elements of knowledge typically has these steps: planning, research, analysis, design, implementation, evaluation, documentation, and training. Development is cyclical and may be partially repetitive. For example, evaluation may be carried out prior to, during, or after the design step. For specific users (defined by their demographics, experience, education, and roles in organizations of work or play) and their tasks, user interfaces must provide these components: *Metaphors* are essential concepts conveyed visually through words, images, and acoustic or tactile cues. *Mental Models* are organizations of data, functions, tasks, roles, and people in groups at work or play. *Navigation* is the movement through mental models afforded by windows, menus, dialogue areas, control panels, etc. *Interaction* is the means by which users input changes, and the system supplies feedback. *Appearance* consists of verbal, visual, acoustic, and tactile perceptual characteristics.

2 Cogito Learning Media

CLM's objective for the Eye-to-Mind Series was to develop inexpensive, informal media-rich "knowledge summaries" of college-level content. Each product needed visual impact to compete in the consumer marketplace while also being technically informatiave. This series is an "attractor" for knowledge, which likely will be found on the Web if many topics in distance learning compete for a student's attention. Images, animation, video, sound, a glossary, and a bibliography supplement explanatory text. Teachers can complement course lectures with attractive, engaging, well-designed content. Students find useful summaries that introduce a subject and/or confirm their understanding.

AM+A designed a UI-framework for stories by expert authors edited by CLM and proposed several non-standard metaphors, then decided on an innovative, minimal approach to a table of contents and a set of tools for discovering the content (see Figure 1). One design challenge was balancing the right amount of mystery and exploration with contents that some users might expect to be passively. In the final design, a simple visual, iconic reminder of the table of contents sections (without labels, but with roll-over captions) appears in every screen reminding the viewer of the content organization and current location.

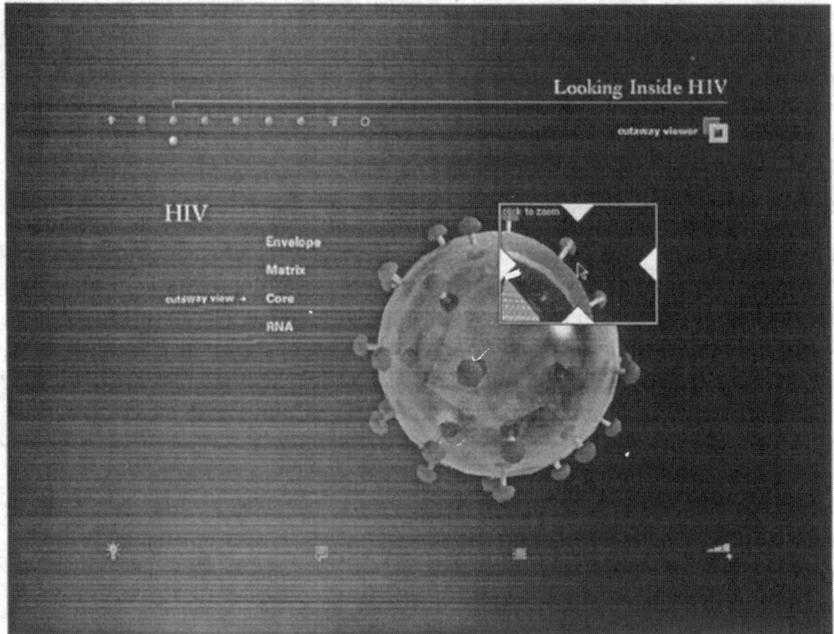

Figure 1

The emphasis in these products is on visual variety and engaging content, which is not in itself extensive or highly structured. The layout is simple; the colors and typography, too, are spare, but distinctive and memorable, distinguishing the products from others in the field. The primary design objective was to capture the attention and excitement of the viewer. Based on reviews by teachers, the objective seems to have achieved successThe designers began by exploring metaphors, general appearance styles, and navigation systems that matched the product's functional and marketing requirements. The desire to present an executive summary of a topic while allowing the user to explore in greater depth persisted. This concept drove the metaphor, appearance, and navigational design process.

Early in the design process, AM+A considered metaphors such as these: science education/entertainment nostalgia , e.g. using a "Mr. Wizard"-like subject guide; traditional audio/visual equipment, e.g., film and slide projectors; postmodern gadgetry, e. g., using knobs, switches, dials and view screens to interact with the information; theatrical or game-show presentations with curtains, announcers, and commercial breaks; and a minimal/elegant program-structure driven system, in which the metaphor, mental model and navigation are all joined in a single visual element.

The minimal/elegant approach was selected primarily because it highlights the large-scale beautifully-rendered animations, complex static or animated technical illustrations, video segments, and compelling typography that make up the product's content. Additionally, this design approach will adapt easily and inexpensively to new content in future products.

The navigation system presents the user an overall linear summary of the topic while always allowing the user to access detailed information. Accessing this detailed information does not cause a user to lose the context in the linear summary from which an exploration embarked. The user also has constant access to previews of other sections of the product and simple online help.

In an effort to make this product stand out among the competition and enrich the user's experience, certain unique interactions were also included. Tools were developed that allow the user to zoom-in on, pan across, or see "photographs" of the imagery to reveal detailed content. The tools give context-sensitive visual hints to promote an exploratory style of learning. Such specialized content-dependent interaction tools can be developed for each new title in the Eye-to-Mind series.

3 Unigraphics Solutions

A very different set of circumstances obtains for the online content available to USI's CAST™ prototype (see Figure 2). As with typical computer-based training, the content has a highly evolved organization of courses and procedures. Courses are step-by-step means to accomplish a goal within a task. Procedures are step-by-step means to achieve a specific goal that might be used within one or more tasks. Essential design tasks were (1) to improve the mental model and navigation and (2) to design a novel appearance and interaction for a user interface whose metaphorical references and basic mental model were well-established for well-understood user communities and their tasks.

Among specific design tasks were the following: Use the heritage of the existing print-based content. Account for very different user classes, e.g., novices, experts, and occasional users. Do not overwhelm users who are new to Unigraphics. Do not under-assist power users who use Unigraphics specific functions frequently (e.g., drafting) but who need to use other functions within applications, or even new applications, occasionally. Achieve the marketing goal of competing favorably with the visual appeal of another competitive

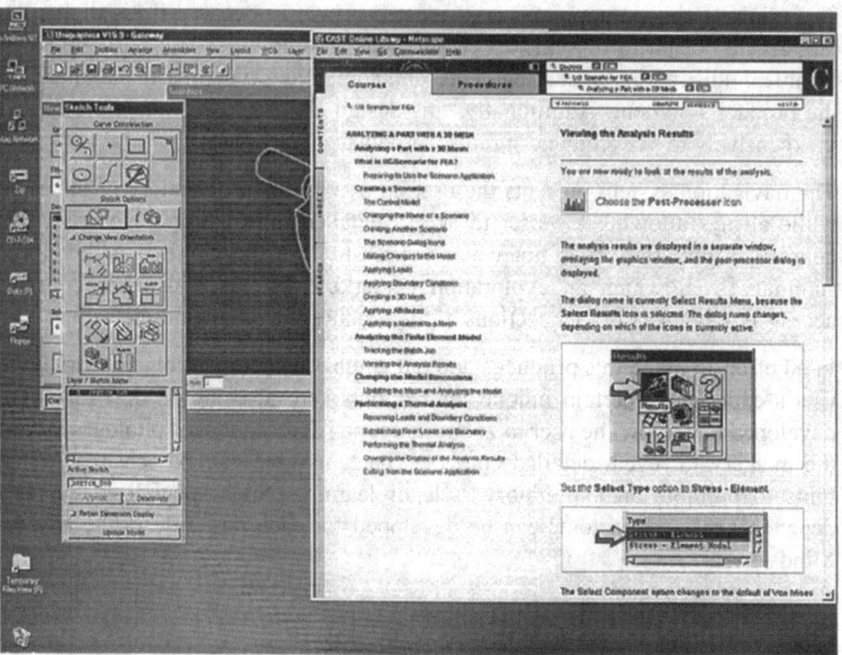

Figures 2

product, CAD Potential ™. USI's competitive advantage is that the tutorial content comes from the software developer, and would be perceived to be better, but at the time, the competing product looked like a better product.

In the previous versions of the product, the content organization was ambiguous, users did not have access to both kinds of content (courses and procedures) at all times, and it was not easy for the user to get from one to the other kind of content. The revised design begins with new items at the very beginning. The initial page, or home page, features specially designed imagery that establishes a product identity and brand identity. The redesigned Courses screen shows several design innovations. Left-edge tabs provide easy access to viewing tools. Bands of different backgrounds make the text more readable. Bullets appear, or highlighting appears when items are located but not selected. In addition, the design does not retain all of the labels of the current hierarchy. The design de-emphasizes or omits "Lessons" and "Topics," because their meaning to the user is ambiguous. Instead, the design emphasizes only "Courses" and "Procedures." The "Navigation Bar" shows the navigation path. New navigation controls, Abridged, Previous, Next, and Complete, enable the user quickly to select the preferred view. Text and figure layout has improved visual consistency. A new Search control, appears both in the Course and in Procedures windows. A new function allows multiple retrievals, from which the user may select one to show in greater detail. The Collapse command enables the user to remove the search and retrieval list. The existing product required three windows to accomplish a task. The new design combines them into one with expanding side panes. The legibility and readability of type and labels is improved, together with a more consistent layout. Navigation orients users to groups of content (search, index, contents) and content hierarchy (courses or procedures).

4 Conclusion

The two products show design approaches that are fundamental for successful UIs of Web-based education and training: Content summaries must present engaging imagery and sound to attract users to specific knowledge. Mental models must avoid overwhelming users with content and enable them to navigate simply and quickly.

Cross-cultural Usability Engineering: development and state of the art

Pia Honold
Siemens AG, Corporate Technology - User Interface Design, D-81730 Munich
pia.honold@mchp.siemens.de

1 Introduction

In recent years, the interest in the influence of culture on human-computer interaction has grown significantly within the HCI community. Global markets are in need of culture-fair products and practitioners are in need of a common knowledge base about Cross-cultural Usability Engineering. As different research communities have been dealing with Cross-cultural Usability Engineering, this field is today marked by a wide heterogeneity, as far as aims of research, methods and terms are concerned.

The aim of this article is to show developmental lines in the field of cross-cultural usability engineering. Therefore, publications concerning human-machine-interaction and culture were reviewed and clustered according to contents. Three main phases of Cross-cultural Usability Engineering emerged (see Table 1)[1].

Using this as a base, the question for a new structuring of Cross-cultural Usability Engineering into more homogeneous fields of research is raised. In the following, the three main phases will be described in detail.

---

[1] Reference list and detailed clustering can be obtained from the author

Table 1: Clustering of publications between 1975 an1998 according to contents

CATEGORY	N
Phase 1 (1975-88): Classical ergonomic research is applied to non-Western countries	
Antropomethry	4
Population stereotypes	3
Field studies for ergonomic design in developmental countries	4
Communicating ergonomic or technical knowledge to culturally diverse users	3
Comparison of design solutions between cultural diverse users	2
Design of text entry device for pictographical language	1
Others	3
Phase 2 (1990-95): Practical solutions of UI-design for non-Western markets become a necessity	
Guidelines for software internationalization	11
Theoretical thoughts about culture fair design	8
Description of design solutions for internationalization	3
Comparison of design solution between cultural diverse user groups	2
Usability testing	2
Phase 3 (1996-98) The need for a theoretical foundation of cross-cultural usability engineering is recognized	
Theoretical thoughts about culture fair design	10
Comparison of design solution between cultural diverse user groups	10
Culture models	7
Description of design solutions for internationalization	7
GroupWare for international users	5
Culture specific software solutions	4
Guidelines for managing the process of internationalization	3
Ethnomethodology	1
Usability testing	1
Others	3

2 1975-1988: Classical ergonomic research is applied to non-Western countries

In the early 1970s, the ergonomic community became interested in the influence of national and cultural factors. The growing industrialization of developmental countries showed that cultural differences can lead to serious practical difficulties. As economy and research grew to be more global, the universal validity of *ergonomic principles* became questionable (Chapanis, 1975b). Therefore, research involving people from non-Western cultures was demanded. In spite of this, for the next 20 years there will be only a little response in the human factors community (Kaplan, 1995).

The content-analysis of the papers showed that the broadening of the research field remained solely topological. Theories and methods of classical ergonomics were merely imposed on other countries and cultures. There was no discussion about whether not only ergonomic principles, but although *theories and methods* were adequate for the new topic of culture.

Most probably, this is the reason why the vision of "cultural ergonomics" did not find many followers.

3 1990-95: Practical solutions of UI-design for non-Western markets become a necessity

In 1990, an interest in "user interface design" and "culture" suddenly came up in the community of computer science. As markets grew outside the USA, practical economical needs for "localized" user interfaces rose. This is reflected by a huge amount of guidelines for software internationalization (see Tab. 1, phase 2, first line). Case studies of practical design solutions were given, but little experimental comparison was conducted. The computer science community does not refer to earlier publications. Up to 1995, difficulties with highly visible cultural differences (character sets, formats etc.) hade been solved. Nevertheless, there are some exceptions (Fernandes, 1995, Hoft, 1995, Nielsen, 1990, del Galdo & Nielsen, 1996), which try to build up practical cultural models which could help to identify cultural influences which lie less visible "beyond the interface" (Bannon & Bødker, 1991).

4 1996-98: The need for a theoretical foundation of cross-cultural usability engineering is recognized

Since 1995, there have been attempts to deal with Cross-cultural Usability Engineering in a less pragmatic and more scientific way. This is reflected by the growth of comparisons between culturally diverse user groups and publications which connect cross-cultural psychology with usability engineering theories and models (Gobbin, 1998; Waldegg-Bourges & Scrivener, 1998; Stathis & Sergot, 1998). Nevertheless, they do so in a quite heterogeneous way.

5 Conclusion

Cross-cultural Usability Engineering shows a huge diversity regarding research topics, methods, terms and theories. Now is time to build up common ground, with common theories and terminology. Therefore it seems to be useful to structure the heterogeneous field into more homogeneous sub-groups which share common objectives and to bring out their distinctive viewpoint.

One way of structuring is a diversion, which follows the process of product planning: cross-cultural user requirements, cross-cultural UI-Design and the management of the process of internationalization (see figure 1).

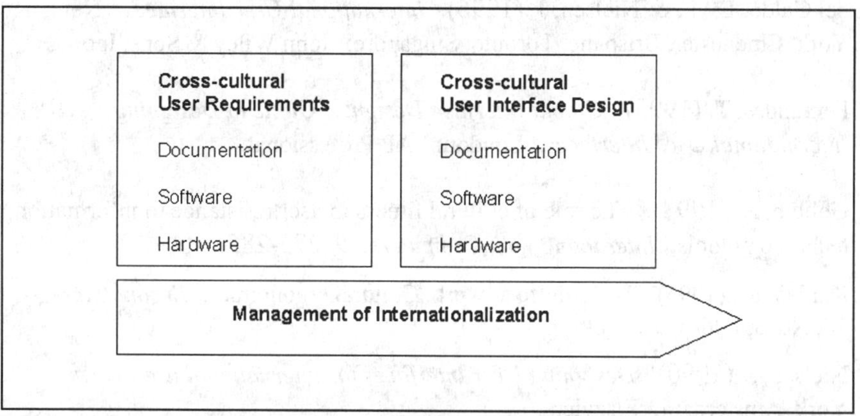

Figure 1: Possibility of structuring the field of Cross-cultural Usability Engineering.

Firsttly, in these different fields, the impact of culture on the use of technologies (and vice versa) have to be studied in a *more scientific* way.

Secondly, the mutual exchange between the different research communities and practicioners has to be enhanced. A good starting point is for example the newsgroup CHI-INTERCULTURAL@ACM.ORG.

Thirdly, findings of cultural research have to be incorporated more deeply in the process of product development. To take a step forward this goal, the User Interfaces Design department at Siemens has established a Usability Lab in the USA and China.

6 References:

Bannon, L.J., & Bødker, S. (1991). Beyond the interface: Encountering artifacts in use . In J.M. Carroll (Ed.), *Designing Interaction: Psychology at the Human-ComputerInteraction.* (pp. 227-253). Cambridge: Cambridge University Press.

Bourges-Waldegg, P., & Scrivener, S.A. (1998). Meaning, the central issue in cross-cultural HCI design. *Interacting with Computers*, *9,* 287-309.

Chapanis, A. (1975a). *Ethnic variables in Human Factors Engineering.* Baltimore: The John Hopkins University Press.

Chapanis, A. (1975b). Cosmopolitanism: A new area in the evolution of Human Factors Engineering. In A. Chapanis (Ed.), *Ethnic variables in Human Factors Engineering.* (pp. 1-9). Baltimore: The John Hopkins University Press.

del Galdo, E.M., & Nielsen, J. (1996). *International User Interfaces.* New York, Chichester, Brisbane, Toronto, Singapore: John Wiley & Sons, Inc.

Fernandes, T. (1995). *Global Interface Design. A Guide to Designing International User Interfaces.* London: AP Professional.

Gobbin, R. (1998). The role of cultural fitness in user resistance to information technology tools. *Interacting with Computers*, *9,* 275-285.

Kaplan, M. (1995). The culture at work: Cultural ergonomics. *Ergonomics*, 38, No. 3, 606-615.

Nielsen, J. (1990). *Designing User interfaces for international use.* New York, Amsterdam: Elsevier.

Stathis, K., & Sergot, M. (1998). An abstract framework for globalising interactive systems. *Interacting with Computers*, *9,* 401-416.

Communicating Entities: a Semiotic-Based Methodology for Interface Design

Osvaldo Luiz de Oliveira, Maria Cecília Calani Braranauskas
São Francisco University – USF, State University of Campinas - UNICAMP

1 Introduction

An important idea underlying this paper is that computer systems can be understood as media, i.e., they are technological engines that mediate our communication and as such they are used to transmit information, knowledge, feelings and requests, to allow conversation, entertainment etc.. In this sense, computers should not be compared to tools, as it is often the case, but to books, theatre, movies, telephone, newspapers and so on.. We are not referring to communication-based applications like electronic mail, video-conference, or applications classified as Groupware and CSCW (Computer Supported Cooperative Work), for which the media role becomes evident. We refer to any type of software.

Essentially being media, computer systems are used to transmit signs among human beings. The concept of sign is the very heart of the previous statement. A sign is something that stands for another thing, for somebody, under certain aspects (Peirce 1974). Any mark, physical movement, symbol, token etc., used to indicate and convey thoughts, information and commands constitute a sign.

The study of the signs and the way they work in the production of meanings is denominated Semiotics. Santaella (1996) argues that Semiotics proposes to view the world as a language. She doesn't only refer to the spoken or written verbal languages, but to all other types of language: deaf-and-dumb language, dance, cookery, fashion, rituals of primitive tribes, music, sculptures, scenography, hieroglyph, dreams, wind etc.. Semiotics investigates all possible languages as a phenomenon of meaning-making.

In this paper we used Semiotics to support a methodology development for interface design. In section 2 we discuss semiotic principles underlying the interface design and we present the framework for the methodology. In the section 3 we conclude.

2 The methodology for interface design: principles and framework

Winograd (1996) defines software not only as a device for users interaction with the computer but also as a place in which the user lives. When an architect designs a home or an office building, a structure is being specified. More significantly, though, the patterns of life for its inhabitants are being shaped. *People are thought of as inhabitants rather than as users of buildings. ... we approach software users as inhabitants, focusing on how they live in the spaces that designers create* (Winograd 1996, p. xvii). Thus, the interface should be understood as being composed of entities with communicative capacity, including human beings (users).

We conceptualise interface as a space to be inhabited by human and non-human **entities** (buttons, windows, game heroes etc.). People are thought of as inhabitants rather than as users of software. Designers are the builders of the space to be inhabited. Under this perspective, we are interested in studying the computer-mediated interaction among entities of the interface, one or more of which are human beings. Experimental results motivate and support this perspective (Oliveira and Baranauskas 1998), while recent literature shares part of it with us (Winograd 1996; Rheinfrank and Evenson 1996; Laurel 1993). The understanding of interface design under our perspective involves considering each entity as a sign with capacity of reciprocal communication by means of languages (sign systems).

Each interface entity has a **semiosis capacity** within the interface world, i. e., it has mechanisms to perceive and to interpret the interface world. Humans in the real world have a semiosis capacity given by their capacity to perceive the world, through their senses, and their capacity to interpret it. The designer chooses a semiosis capacity for an entity within a continuum of semiosis capacities relative to the interface world. This continuum varies from the "null" semiosis capacity (an entity does not perceive nor interpret the interface world) to the "total" semiosis (the entity perceives everything and interprets everything in the interface world), going through intermediary points equivalent to semiosis capacities typical of human beings in the real world. For example, the designer could give an entity of the interface (a button, a puppet or perhaps something as a wall) a "null" semiosis capacity (as ornamental interface entities), anthropomorphic semiosis capacity (as a pilot in a Grand Prix game) or semiosis capacities higher than the human semiosis capacity (as a super-hero who can see through a wall).

Each entity also has a **sign emission possibility** within the interface world. This sign emission possibility is revealed through a complex of relationships of three types: (1) an entity A communicates something to an entity B through its form, colour, texture, voice tone, noise produced, a smell in the air, a way of acting, a way of moving, its personality etc.. Thus, each entity can work itself as a sign; (2) an entity A can emit sentences to an entity B according to some understood language among them; (3) two entities A and B interact (physically, chemically, socially etc.) and the result of this interaction communicates something to a third entity C. For example, an entity "button" interact with the entity "rolling bar" sliding on it. The button position relative to the scroll bar can communicate to a third entity the quantity of scroll in a document of typical desktop interface.

The interface design should emerge from **conviviality** among the entities that inhabit the interface, i. e., from the local language established continually by the relationship established from the conviviality among them. The local language corresponds to the dialect resulting from the confront between the private languages "spoken" by human entities in the presence of the interface and the private languages "spoken" by the non-human entities. Through the acquisition of new habits, the entities assimilate the environment language up to a certain point. Also, they contribute to the consolidation of it. Human entities make this naturally. It is a designer's task to accommodate the non-human interface entities to make them assimilate and contribute to the establishment of the local language.

In short, the semiotic principles underlying our concept of interface design are:

1. the interface should be understood as a group of entities that communicate, one or more of the which are human beings;
2. every entity in the interface possesses a semiosis capacity;
3. every entity in the interface has a possibility of sign emission;
4. the language used by the entities of the interface emerges from the conviviality among them.

Our methodology framework presupposes design as an iterative process in which interface prototypes are continually developed, used and evaluated (fig. 1). An initial prototype should be developed establishing the initial entities, its respective semiosis capacities and its possibility of sign generation. In each design cycle the prototype should be put into practice by users and the private language of each interface entity should be evaluated with reference to its self-consistency and with reference to the consistency with other languages "spoken" by the other interface entities. The designer must redesign the prototype seeking to eliminate the possible inconsistencies. The designer can make this acting upon the semiosis capacities and the possibility of sign emission of the entities. Thus the language of each entity tends to adapt to the languages practiced by other entities of the interface world. Modifications in the semiosis capacity of an entity can regulate the way it is capable of perceiving and interpreting the interface world. Modifications in the sign emission possibility of an entity regulate the way it is capable of emitting signs for the interface world. The semiotic theory of Hjelmslev (1968) supports this formalism by allowing to describe, to analyze and to compare different semiotic systems (languages).

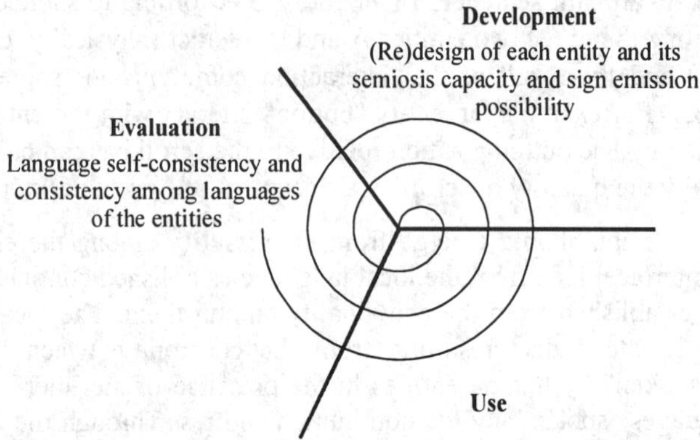

Figure 1: Interface Design as an iterative process

We are applying this methodology to the development of the TC - Theatre in the Computer: a software for children to write and to act in plays (Oliveira and Baranauskas 1999). The child can inhabit the TC interface world in different ways: as an author, a director, an actor or a spectator of plays. Each one of these ways is a function of the semiosis capacity and sign emission possibility of the entity inhabited by the child and creates different educational possibilities for him.

3 Discussion and Conclusion

The development of cognitive theories on Human-Computer Interaction brought us a vision of the computer as a cognitive tool enabling people to enlarge their understanding, memorisation and decision-making capacities. Semiotic approaches to software design allow to consider not only the immediate aspects of Human-Computer Interaction but, also, the underlying aspects of the cultural and social context in which the interaction happens. The semiotic approaches facilitate an interpersonal, social, cultural perspective, focusing on the expression and interpretation of the elements in the software interface.

Our proposal moves the understanding of the interface from "the thing in between" (Laurel and Mountford 1990; Nadin 1988) or "the thing to be read" (Andersen 1997), of many characterisations of the interface, to "the thing to be inhabited". Considering that the languages of the interface entities tend to accommodate, in a continuous process, our proposal understands the interface design as something provisory, something ready to be used, re-evaluated and re-done, making explicit the evolutionary character of the interface, as it happens with all the things in the Universe and inside the human mind. By considering the user (or users) as entities that inhabit the interface, our proposal includes the phenomena of the new interfaces for GroupWare, CSCW and virtual reality, also considering the phenomena underlying the traditional interfaces.

4 References

Andersen, P. B. (1997). *A Theory of Computer Semiotics.* Updated ed. of 1990. New York: Cambridge University Press.

Hjelmslev, L. (1968). *Prolégomènes a Une Théorie du Langage.* Paris: Les Éditions de Minuit.

Laurel, B. (1993). *Computers as Teatre.* Reading: Addison-Wesley Publishing Company.

Laurel, B., Mountford, J. (1990). Introduction. In Laurel, B. K. (ed.): *The art of Human-Computer Interface Design.* Reading: Addison-Wesley Publishing Company.

Nadin, M. (1988). Interface Design. *Semiotica,* 69 (3/4), 269-302.

Oliveira, O. L., Baranauskas, M. C. C. (1998). *Interface Understood as Communicating Entities – a Semiotic perspective.* Technical Report IC-98-41. December, 1998. Campinas: Available in http://www.dcc.unicamp.br.

Oliveira, O. L., Baranauskas, M. C. C. (1999). The Theatre Though the Computer: a virtual space to be inhabited . In: *Virtuality in Education – What are the future educational contexts?, Proc. of the CAL'99* (CAL'99, London, UK, March 29-31, 1999), forthcoming. London: Elsevier.

Peirce, C. S. (1974). *Collected Papers of Charles Sanders Peirce.* Charles Hartshorme e Paul Weiss (ed.). Vol. II: Elements of Logic. Thrird Printing. Cambridge: Harvard University Press.

Rheinfrank, J., Evenson, S. (1996). Design Languages. In Winograd, T. (ed.): *Bringing Design to Software.* New York: Addison-Wesley Publishing Company.

Santaella, L. (1996). *O que é Semiótica.* 12.ed.. São Paulo: Editora Brasiliense.

Winograd, T. (1996). Introduction. In Winograd, T. (ed.): *Bringing Design to Software.* New York: Addison-Wesley Publishing Company.

Discovering Latent Relationships among Ideas : A Methodology for Facilitating New Idea Creation

Yosuke Kinoe[A], Hirohiko Mori[B]

[A] IBM Japan, Ltd., 1623-14, Shimotsuruma, Yamato-shi, Kanagawa, Japan.
[B] Musashi Institute of Technology, 1-28-1, Tamazutsumi, Setagaya, Tokyo, Japan.

1 Introduction

Augmentation of creativity has recently become an important aspect of the design of computer-supported work (e.g. Edmonds, Fischer, et.al, 1995). The objective of this paper is to propose a new methodology for facilitating the processes whereby humans and computer systems develop new ideas collaboratively. This methodology is based on Distributed Genetic Programming (GP) (Koza, 1992), an approach to program induction based on the mechanics of natural selection and natural genetics. We developed an experimental system with the aim of providing an intelligent environment that stimulates analysts to create new ideas by applying the methodology (Kinoe, Mori, Sugita & Hayashi, 1997). This paper also introduces a practical study of idea development in which the methodology is used to discuss basic issues involved in supporting a creative process whereby humans and computer systems can create new ideas collaboratively.

2 Methodology and Experimental System

In this methodology, the process of formulating new ideas can be divided into two different stages, analytical and creative (Kinoe, Mori & Hayashi, 1993).

2.1 Analytical stage

Analysts (i.e. users) prepare a starting set of ideas that may include, for example, hundreds of raw ideas gathered during a brainstorming session, observational findings, and ideas that have previously used for solving problems in other domains.

In the analytical stage, analysts systematically characterize a set of existing ideas according to a formalized procedure by using the system. This process is based on a formalized analysis procedure originally developed for verbal protocol analysis (Kinoe, Mori & Hayashi, 1993). Analysts establish analysis viewpoints and define an analytical model that consists of a set of attributes for characterizing the ideas, categorized according to the analysis viewpoints. Each of the ideas is characterized with a combination of attributes selected from each analysis viewpoint, on the basis of the analytical model. In accordance with the results of analysts' characterization of ideas, the system classifies ideas according to their attributes and constructs an initial map of existing ideas by applying Hayashi's Quantification Method III (for example, see Figure 1).

2.2 Creative Stage

The second stage, the creative stage, is more dynamic. Analysts try to formulate new ideas by discovering latent relationships among existing ideas. The system provides an intelligent support environment based on the results of the analytical stage. The process can be described as an iteration of the following steps, which include collaboration between analysts and the system.

At first, analysts choose a central idea by clicking on a map, then they select similar or related ideas, and reorganize the map by re-grouping those related ideas. In this step, a traditional methodology for creating new ideas by using cards, such as the KJ method (Kawakita, 1967), can be applied. Next, on the basis of the analysts' choice, the system suggests several candidate ideas that may be related to those focused on by the analysts. By applying the program induction mechanism of GP, the system provides various ways of estimating the closeness between ideas, and generates various types of candidates that include some outside of the analysts' initial focus. The analysts then evaluate these candidates; that is, they adopt some of them and reject the others. While the analysts are evaluating the candidates, they try to find latent relationships between them and their central ideas. The system thus stimulates new ideas by expanding the analysts' perspectives on the central ideas, especially when they adopt suggestions outside the area they originally focused on. On the other hand, rejection of candidates can greatly help analysts to clarify their intentions. Finally, by using an inductive learning mechanism of GP, the system adapts to the analysts' intentions according to their evaluations, so that it can provide better suggestions during the next iteration.

3 Case Study

We applied this methodology to a practical study of idea development. Two hundred ideas had been gathered through a series of brainstorming sessions on the question, "What kinds of activities are effective for creating attractive, human-centered software products?" Analysts input a list of the ideas and characterized all the ideas with a set of attributes from eleven analysis viewpoints, on the basis of an analytical model that contains one hundred and four attributes. Figure 1 shows an idea map based on the results of the analytical stage. Analysts tried to create new ideas by using the system's intelligent support environment. In this case study, we carefully analyzed the analysts' processes of idea development using the system, and found several fundamental issues related to supporting the development of new ideas.

3.1 Types of System's Suggestions

From the analysts' viewpoint, the system's suggestions could be classified into four types (Figure 1): (1) *equivalent ideas* to the central ideas, (2) different but *analogous ideas* within the analysts' focus, which stimulate analysts to consider the central ideas from a different viewpoint, (3) *stimulating ideas* outside of the analysts' focus, which help analysts to expand their perspectives on the central ideas, and (4) *confusing ideas* outside of the analysts' focus. According to our findings, one of the most significant advantages using the learning mechanism of GP to provide intelligent support in the system, is that it provides analysts with second and third types of suggestions.

3.2 Types of Analyst's Situations in Creative Stage

The results of the study suggested that analysts' understanding of the central ideas was not consistent throughout their idea development processes, but changed dynamically. Analysts sometimes actively expanded and contracted their perspectives on the ideas as their ideas developed. We found at least five different kinds of situation in processes whereby analysts and the system developed new ideas collaboratively.

In the first situation, (a) analysts tried to understand the central ideas and *constructed an initial criterion* for classifying the existing ideas. They mainly gathered "equivalent ideas" to the central ideas. After analysts had established a stable criterion, (b) they *efficiently classified* the systems' suggestions according to their well-defined criterion. However, their classification work shortly became stereotyped. In order to break through such mannerism, (c) analysts often *explored alternative ideas* that might have latent similarities with the central ideas. Especially by discovering latent similarities between the central ideas and the suggested "analogous ideas" within the area they were focusing on, analysts acquired different aspects on understanding of the central ideas. When they discovered the latent similarities, they were able to create new ideas. Furthermore, (d) analysts *actively explored latent relationships* between the central ideas and the suggested "stimulating ideas" outside of the area they had

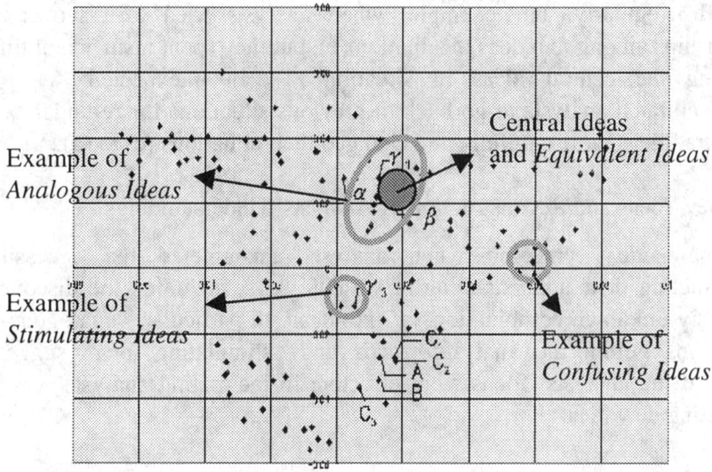

Figure 1. An Idea Map: Central Ideas and *System's Suggestions*

focused on. Discovering new combinations of ideas that were originally considered unrelated stimulated the creation of new ideas. Finally, (e) analysts *reached a more higher level of understanding* of the central issue by establishing a new criterion. In this way, they deepened their understanding and proceeded spirally to another next stage of iteration.

3.3 Relationship between Analyst's Situations and Expected Suggestions

Analysts developed new perspectives on the central issue by actively expanding and contracting their perspectives through a cycle of situations (a) to (e). The type of suggestion analysts hoped to receive from the system often changed according to their own situations. In this case study, *"Equivalent idea"* suggestions by the system were helpful for analysts' initial understanding and classification of ideas in situations (a) and (b), but they were not useful in situation (d). On the other hand, *"stimulating idea"* suggestions were extremely effective for analysts to discover new combinations of ideas that were originally considered unrelated in situation (d), although they were usually rejected in situations (a) and (b). "Stimulating idea" suggestions sometimes harmful for analysts' initial understanding and classification of ideas in situations (a) and (b). The result suggested that the relationship between analysts' situations and expected suggestions should be considered as an important issue for improving the design of an intelligent support environment of our system.

Analysts often changed their situations when they received unexpected type of system's suggestions. Analysts' situations usually changed in alphabetical order. For example, an interesting "*stimulating idea*" facilitated analysts to change their situations from (c) to (d). On the other hand, analysts' situations sometimes changed irregularly. For example, when analysts tried to discover latent relationships among the ideas in situation (c), but they spent insufficient time for analyzing the central ideas in situation (a), an unexpected *"equivalent idea"* facilitated analysts to go back to previous situations for re-understanding the central ideas more carefully and changed their situations (c) to (a).

3.4 New Ideas and Reconstruction of Analysts' Understandings

Most new ideas emerged when analysts encountered the necessity of reconstructing their understanding in various ways, including the discovery of previously unknown combinations of the ideas. In particular, "analogous idea" suggestions within the analysts' focus and "stimulating idea" suggestions outside of the analysts' focus were effective in facilitating analysts' processes for creating new ideas.

4 Conclusion

In this paper, we proposed a new methodology for facilitating the processes whereby humans and computer systems develop new ideas collaboratively. We developed an experimental system based on the methodology. To improve our intelligent support environment, it is essential that we focus on enhancements of our inductive learning mechanism, such as GP/CBR integration and creation of psychological framework of analysts' dynamic processes in the creative stage in our future studies.

5 References

Edmonds, E., Fischer, G., Mountford, S.J., Nake, F., Riecken, D. and Spence, R. (1995). Creativity Interacting with Computers. *ACM, Proceedings of CHI'95*. Addison Wesley.

Koza, J. R. (1992). Genetic Programming: On the Programming of Computers by Means of Natural Selection. MIT Press.

Kinoe, Y., Mori, H., Sugita, N., and Hayashi, Y. (1997). Intelligent Support for Discovering Latent Relationships among Ideas: A Methodology Based on Genetic Programming. *Proceedings of the 13th Congress of the International Ergonomics Association '97, Vol.5, Tampere, Finland.* pp.193-195.

Kinoe, Y., Mori, H. and Hayashi, Y. (1993). Integrating Analytical and Creative Processes for User Interface Re-Design. In G. Salvendy, M.J. Smith (Eds.): *Human-Computer Interaction.* pp.163-168. Elsevier.

Kawakita, J. (1967). Hassouhou. Tokyo: Chuohkohron. (*In Japanese*).

From Focus Group to Functional Specification – A Linguistic Approach to the Transfer of Knowledge

Claus Knapheide
Siemens AG
Corporate Technology
User Inter Face Design
81730 München

User related software development has recently been estimated as a key factor for the success of electronical products in the market. If users are related with the process from the very beginning, the software will meet the customers' expectations and can be used with a high degree of security and joy of use. Software will be streamlined and inexpensive when only those requirements are covered that proved to be really necessary under the aspect of usage.

For these reasons, the software certification criteria do not only cover the process of inspection (ISO 9241, Part 10) and Usability Testing (ISO 9241, Part 11), but also a direct cooperation with users in the beginning of the process where the context of use and task analysis (ISO 13407) are in question. For this purpose, a discourse called focus group discussion has been established (cf. Greenbaum 1988, Stewart & Shamdasani 1990).

Group discussion originates from market research, where it was used as an economical version of deep interview with many participants. Still there is only little methodological knowledge on focus group discussion, since the technical disciplines that take part in the process hardly have theoretical access to the discourse as such, while psychological research tends to stress emotional, affective, or physiological aspects of behaviour (cf. Ehlich 1997). As a result, focus groups are still being used more or less accidental and as a non-systematic method to reveal any kind of knowledge.

A linguistic approach, however, looks into the verbal actions of participants and tries to relate these with mental categories such as knowledge. When designing the user software for a very complex medical equipment made by Siemens, we held more than 80 hours of focus group discussions and established them as a strictly methodological technique in the overall process of requirement engineering.

The depth of knowledge, the way how users treat their experience during the discussion, the exactitude of information and how wide it was spread convinced everybody of the necessity to give requirements as they emerge during focus groups the reliability they deserve. This happened as the moderator forced the participants to always relate their requirements (or any other statement) through personal experience (*Erfahrungswissen*, Rehbein 1977), their knowledge of the situation (*situatives Wissen*, ibd.) and the topic of the discussion (*Diskurswissen*, ibd.).

From the character of linguistic material used by the participants, their expressions and wording, it is possible to qualify the discourse as a description (*Beschreibung, Schilderung*), report (*Bericht*), or narration (*Erzählen*, cf. Ehlich 1983, Redder 1990, Rehbein 1982, Rehbein 1989). Narration (used as an architerm like *Erzählen*, cf. Ehlich 1983, 139) is only possible if the speaker is motivated to provide others with or generate new knowledge.

In opposite to the literal kind of narration (story or *Erzählung*) where it is exactly the divergency between the listener's (daily) experience and the story content that creates the point of interest, narration in its general sense allows for the verbalisation of knowledge that meets the listener's expectations and is therefore worth being told. In this case, the result is not a story but an illustration, a report, a description, and it contains pictures (*Bilder*) or estimations (*Einschätzung*; see *Wissenstypen*, Ehlich & Rehbein 1977, Ehlich 1991).

Each of these has a specific domain of validity and generalizability; their practical relevance varies and so does their degree of specification. This means that the verbalised knowledge shows its quality with respect to these categories.

What about alternatives to focus group discussions, e.g. interviews? Completely different from narration are lists (cf. Ehlich 183, 146 f.), for example questionnaires. Narration is always complete and relevant, and there is a perfect interactive apparatus to see to relevance and completeness. For interviews, the responsibility of completeness and relevance lies in the hands of the interviewer. The interviewee has little possibilities to qualify the questions, and he is not meant to do so. This means that interviews can never be user driven enough to guarantee user conformance to models of functionality or user related interface concepts. We only use interviews to record biographical and job related data in order to relate these with the results of our focus group discussions.

A focus group discussion gathers 5 to 7 authentic users of a former software / product, or people who represent the target market for a totally new product. We try to bring together people to form a socially homogenous and an inhomogenous group - with respect to their experience. Mostly software development and marketing staff are present in the same room, though separately seated. These can only jump into the discussion via the moderator. Of course a moderator is present, as well as somebody to assure technical support.

The most important interaction is that the users produce a cooperative discourse and try to get shared knowledge on the topic (*Verständigungshandeln*, Rehbein 1977). This discourse has to be cooperative and it has to be topic driven such that the requirements, their context, and the tasks related can properly be understood by the internal auditorium. The moderator's task is to assure the necessary broadness and depth of discussion, and the integration of all the participants. He has to keep the game running.

With this, the customer can be sure that the participants all narrate (not speculate), that they all relate their reports (not conjectures - *Mutmaßungen*) and descriptions (not lies or inventions) on each other and do not miss each other's points. He can be sure that the participants combine their estimations to create a valid picture on the topic, but do not persuade each other into irrelevant ideas.

Requirements that have been qualified during the process of focus group discussions can more or less directly be brought into the requirement specifications or even functional specifications. During our project we followed the rule that would not allow any function be put into the product without having a positive feedback by users during focus group discussions. Everything, of course, that had been asked for by users found its place in one of the software versions. The later is only true for statements that had been made by users during focus group discussions, not for statements given by single users on the phone, via the help desk, nor for user statements that came to the team in an indirect communicative process (by sales representatives as they refer to users etc.).

If statements can easily be formulated and written down, we do so. Everything that had been 'published' on the wall can, after participant feedback, be judged as common knowledge of the group.

It has to be stated that focus group moderators should know the linguistic means that allow for the qualification of user's statements in order to recognize the patterns used during the discussions. Knowledge has to be established as leading category in order to understand how valuable users can participate in the product development process.

References

Ehlich, K. (1983). Alltägliches Erzählen. In Sanders, W. & Wegenast, K. (Eds.) (1983) Erzählen für Kinder - Erzählen von Gott. Stuttgart: Kohlhammer, 128 - 150

Ehlich, K. (1984). Sprechhandlungsanalyse. In: Haft & Kordes (Eds.) (1984). Enzyklopädie Erziehungswissenschaft Band 2. Stuttgart: Klett-Cotta, 525 - 538

Ehlich, K. (1991). Funktional-pragmatische Diskursanalyse. Ziele und Verfahren. In: Flader, D. (Ed.) (1991) Verbale Interaktion. Stuttgart: Metzler, 127 – 143

Ehlich, K. (1997) Vorurteile, Vor-Urteile, Wissenstypen, mentale und diskursive Strukturen. München: Ludwig-Maximilians-Universität, mimeo

Ehlich, K. & Rehbein, J. (1972) Erwarten. In: Wunderlich, D. (Ed.) (1972). Linguistische Pragmatik. Wiesbaden: Athenaion, 99 - 115

Ehlich, K. & Rehbein, J. (1977). Wissen, kommunikatives Handeln und die Schule. In: Goeppert, H. (Ed.) (1977). Sprachverhalten im Unterricht. München: Fink, 36 - 113

Ehlich, K. & Rehbein, J. (1978). On Effective Reasoning. In: Nickel, G. (Ed.) (1978) Pragmalinguistics. Stuttgart: Hochschul-Verlag, 67 - 92

Greenbaum, T.L. (1988). The practical Handbook and Guide to Focus Group Research, Lexington MA: D.C. Heath and Company

Redder, A. (1990). Grammatiktheorie und sprachliches Handeln: >>denn<< und >>da<<. Tübingen: Niemeyer

Rehbein, J. (1977). Komplexes Handeln. Stuttgart: Metzler

Rehbein, J. (1982). Biografisches Erzählen. In: Lämmert, E. (Ed.) (1982). Erzählforschung: Ein Symposion. Stuttgart: Metzler, 51 - 73

Rehbein, J. (1984). Beschreiben, Berichten und Erzählen. In: Ehlich, K. (Ed.) (1984). Erzählen in der Schule. Tübingen: Narr, 66 - 124

Rehbein, J. (1989). Biographiefragmente. Nicht-erzählende, rekonstruktive Diskursformen in der Hochschulkommunikation. In: Kokemohr, R. & Marotzki, W. (Eds.) (1989). Studentenbiographien. Frankfurt: Lang, 163 - 253

Stewart, D.W. & Shamdasani, P. N. (1990). Focus Groups: Theory and Practice. Applied Social Research Methods Series, Vol. 20, SAGE Publications: Newbury Park

Designing Interactions through Meaning

Manuel Imaz
Napier University, Edinburgh, imaz@accessnet.es

1 Introduction

The introduction of patterns in software design means making aware some unconscious schemas used in the design activity. The use of patterns currently extends from programming to analysis including process definition. The central discovery with patterns is that most of our mental activity is a recurring one only changing the degree of granularity of phenomena to which patterns are applied. What we propose in this paper is to extend the usual schema of patterns in order to be able to capture some general cognitive processes known as *conceptual integration* or *blends*.

Encompassing cognitive processes to which they apply, patterns are defined at different levels of granularity and they may be combined in nested structures to form -as Christopher Alexander proposed- a language of its own. But design patterns, in general, are based on already defined software constructs such as classes, objects or other.

In order for a pattern to be able to capture conceptual integration, it has to be defined as a network to mimic the network of conceptual integration the pattern will represent. What we are proposing in this paper is to generalise such patterns to include other type of mechanisms, which could be used to design new constructs. This special activity of design -not usually reflected in designing artifacts- is *conceptual integration* in the form of *blends*

2 Cognitive Background

Cognitive semantics or experientialism offers us some interesting tools, which can be used to analyse the design activity. These concepts are mental space, projection, blend and stories. A *mental space* is a medium for conceptualisation and thought. Any fixed or ongoing state of affairs as we conceptualise it is represented by a mental space. Examples include our immediate reality, fictional situations, past or future situations, abstract domains and so on (Lakoff 1987). Another notion associated to mental spaces is that of *cognitive mapping* or *projection*. It has been argued that "a variety of constructions involving analogy, metaphor, and hedges set up multispace configurations with source, target, generic, and blended spaces that project onto each other in several directions" (Fauconnier 1994).

Blending – integrating partial structures from different domains – is another special case of imaginative projection or mapping. Blending receives a partial structure from two or more input spaces, producing a new space that has emergent structure of its own. All spaces involved in a blend (input, generic and the blend) determines a *conceptual integration network*. The importance of blend in design is becoming more and more clear as blends are the central elements for producing innovation.

3 Patterns for Conceptual Integration

Pattern Name: *RefiningTheMetaphor*

Context: Metaphors are good at suggesting a general orientation, but not good at accurately encoding precise semantics (Nardi and Zarmer, 1993). This is a quite acceptable position, considering that a metaphor is a very general and vaguely defined mapping between two different domains: one is the well known and the other the less known and requiring further investigation. The question is that, in the design process, we are looking for new and helpful ideas that can contribute to our final product. What we have to do is transform the original metaphor in a refined conceptual framework that could be implemented as new constructs (signs or visual formalisms). The metaphor is part of the raw material needed, the new constructs are the final products.

Problem: Up to date, we have not been given a conceptual tool nor guidelines about how to apply a given metaphor in a disciplined way. Once a new metaphor is conceived, it is just applied in an ad-hoc way using most handicraft methods.

Forces: We are considering, on one hand, a mainly creative process based on a new metaphor, which has to be explored and developed. On the other, this creative process has the usual characteristics of handcrafting and non-predictable results.

Solution: We must differentiate the input conceptual spaces. In this case we have four input spaces:

1) Real Workplace Desktop, 2) Graphical Representations, 3) Computer Commands and 4) Manipulative Operations.

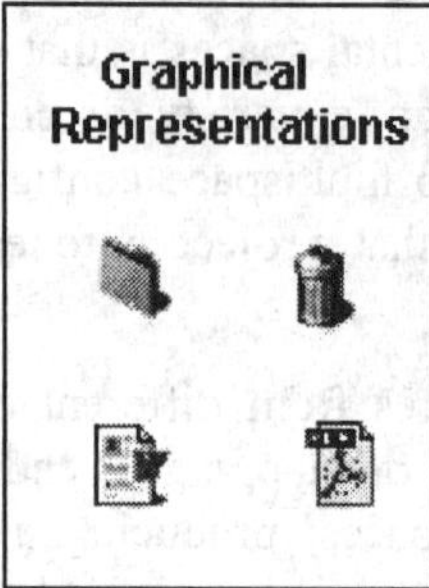

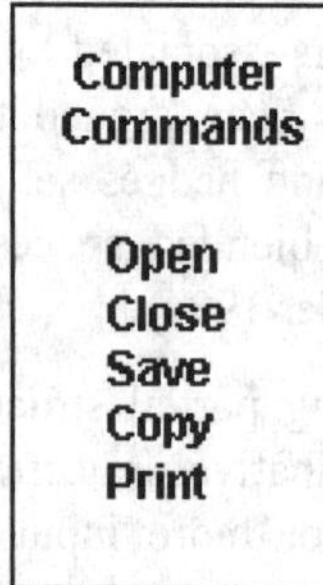

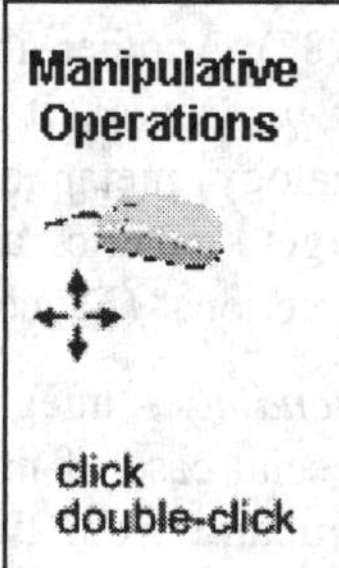

Figure 1

We are here proposing a more complex structure for patterns whereby we establish partial links between them, such as proposed for input spaces in modelling a blend. In order for us to follow the proposed schema, each input space will be considered as a pattern on its own.

Real Workplace Desktop

Each of the input spaces has a different organising frame. The frame of the real workplace desktop, for example, has elements like folders, documents, and trashcans. In such a frame, we organise our documents into folders, so we have to *open* a folder to get a document; we *write* documents and then we *save* them and so on.

Graphical Representations

The frame of graphics has forms, colours, shades and effects in order to represent elements from the real world. Sometimes, these both spaces are considered the main input spaces of the metaphor.

Computer Commands

The frame of commands with orders like open, close, copy or print. These are the usual commands used in previous versions of operating systems.

Manipulative Operations

The frame of manipulations involves clicks, double-clicks, drags and so on. The actions are derived from the use of devices such as mouses or joysticks.

Emerging Space: Blend

The next step is framing the blend. As we are in presence of a four-sided network -because we are projecting some topology from each of the input spaces- we are going to have a blend-specific emergent frame. In the real desktop frame, there are actions such as open a folder, get out a document or put in a document. In the frame of graphical representations (signs), we have to design some graphical transformations to indicate changing states of the sign. In the case of computer commands frame, we make reference to a classical command set, and in the blend-specific emergent frame we *click a folder icon in order to select it* without an equivalent in the real desktop frame.

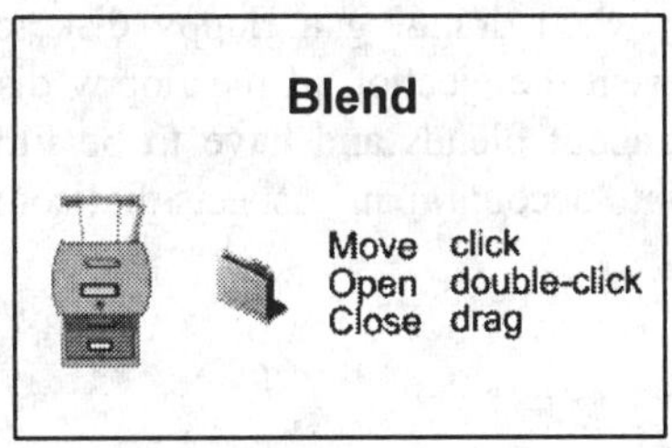

Figure 2

The emergent constructs -the signs or visual formalisms- have the form of an icon to facilitate the task of recognising what element of the real desktop it is referring to. The sign has some topology incorporated from the real desktop input (folders are open to take out a document, folders are included into others, documents are placed into folders). The first design solution for the sign may be closer to the real desktop equivalent: open a folder and move a document into it. But the activity of using such signs might evolve in a totally new solution as when we put a document *on* a folder in order to place the former *into* the last. The conceptual integration network is constructed dynamically over a period of time and signs may evolve gradually and autonomously.

Another aspect to be taken into account is the way we handle signs. We have included a fourth input space of manipulative operations based on a mouse, a joystick or another device. This input space participates in the blend in order to determine the way we manipulate signs either direct or indirectly.

Some Salient Characteristics of Blends: Optimality principles

In any conceptual integration network there is evidence of some principles, which are satisfied to a certain degree: *Integration*, *Web*, *Unpacking*, *Topology* and *Good reasons*. Here we will focus on integration and topology. In order to satisfy integration, "a blend must constitute a tightly integrated scene that can be manipulated as a unit. More generally, every space in the network should have integration" (Fauconnier and Turner 1998). Regarding topology, it is applied when each "input space and any element in that space projected into the blend, it is optimal for the relations of the elements in the blend to match the relations in its counterpart".

It has been shown how the use of the trashcan in MacOS is a problematic solution. In terms of these principles, we can say that it is a failure to satisfy both Integration and Topology. For example, one ejects the floppy disk to keep it rather than discard it, or the dragging operation of one icon to another results in the first icon to be contained in the second, but not so in this. If we have applied the topology principle, we would have designed an icon to represent the floppy disk driver and, when dragging a floppy disk to that icon, it would transform itself to represent the ejection of the floppy disk. All these are only minimal considerations about blends and have to be further developed in an extended paper to take into account many aspects that could not be presented in this paper.

4 Conclusion

The application of conceptual-integration patterns as a network of input spaces and a blend space would allow follow the avatars of different concepts that will combine in new design constructs. The proposed schema follows the model applied to explain how meaning develops from a given sentence or paragraph applied by cognitive semantics scholars. The new types of patterns proposed will be able to connect partial structure from different input spaces and to give the design basis for all blends obtained from the original spaces. These gradual transformations also reflect the transition from informal to formal description of design elements through intermediate -semiformal- stages.

5 References

Fauconnier, G (1994). *Mental Spaces. Aspects of Meaning Construction in Natural Language.* Cambridge University Press.

Fauconnier, G & Turner, M (1998). Principles of Conceptual Integration. In Koenig, J-P. (Ed.): *Discourse and Cognition: Bridging the Gap*, pp. 269-283. CSLI Publications. Standford. California.

Lakoff, G. (1987). *Women, Fire and Dangerous Things: What Categories Reveal About the Mind* . Chicago University Press.

Nardi, B & Zarmer, C. (1993). Beyond Models and metaphors: Visual formalisms in user interface design. *Journal of Visual Languages and Computing,* March.

Turner, M. (1996). *The Literary Mind.* Oxford University Press.

Towards Complex Object Oriented Analysis and Design

Ludwik Kuźniarz
Maciej Piasecki
Computer Science Department
Wrocław University of Technology

1. Introduction

The *Object Oriented Analysis and Design Methodologies* (OOADMs) have been developed for many years and with the emergence of Unified Modelling Language (UML) they are now becoming a matured standard for computer systems development. Advanced tools supporting integrated analysis, designing and development of computer systems are appearing in parallel. Most of them allow easy integration with *Graphical User Interface* (GUI) and impose *object oriented* (OO) construction of *User Interface* (UI). However, in OOADMs the problem of proper construction of UI is often limited to technical and presentational issues. In system development OOADMs focus on modelling of system structure and behaviour but do not propose any systematic way of describing users' preferences for UI.

The main goal of our work was to construct a complex OO method for system development integrating OOADM and UI development processes as two parallel but interleaving and mutually depended activities. To achieve this the sequence of development phases (together with their goals and notations) is introduced.

2. Traditional Approach to UI Design

There are many different methods of UI design. However, most of them cover only some parts of the whole *UI development process* (UIDP) and are incompatible with OOADM. In particular they are often based on early prototyping and do not have explicit links with notations and results of OOADMs' subsequent steps of development. In the literature a lot of attention is paid to *require-*

ments acquisition - the first phase of UIDP – where the future users and context of use are described (Newman and Lamming 1995). Many guidelines are also formulated for the last phase concerning the problems of choice of proper GUI standard, application of GUI tools of interaction, layout of the screen, presentation details etc., e.g. (Galitz 1996). However, there is a large gap between the first and the last phase, described only as a "*conceptual design*" which on the other hand is very important to the final shape of UI, because it determines the future semantics of UI i.e. conceptual structures communicated by UI.

3. Object Oriented Approach to UI Design

The starting point for the work presented in the paper was an attempt to use OO diagrams to model the conceptual structure of UI and to do this in correlation

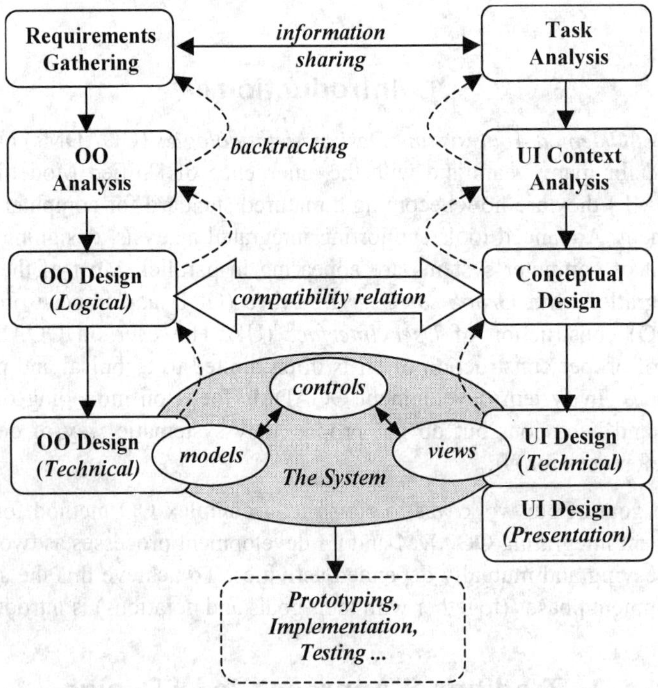

Fig. 1 Scheme of Complex Object Oriented Analysis and Design Method

with OO analysis and design of the system.

The work was inspired by the definition of UI given by L. Barfield (Barfield 1993): "It is composed of those parts of a system that are designed to be apparent to and manipulable by the user and those models and impression that are built up in the mind of the user in response to interacting with these features."

As well as the emphasis given in the book to the role played by the *user model* (or *user mental model* (Newman and Lamming 1995) i.e. the model of the system possessed by the user) in the interaction between the man and computer. Moreover, the interaction with the system and the presented feedback are identified as the most important factors influencing the construction of user model. It means that UI must 'explain' the construction of the system to the user in such a way that the user can understand it and build the user model of the system. To achieve this UI does not have to present all details of the system internal construction, but it has to support the user in building a proper picture of the system.

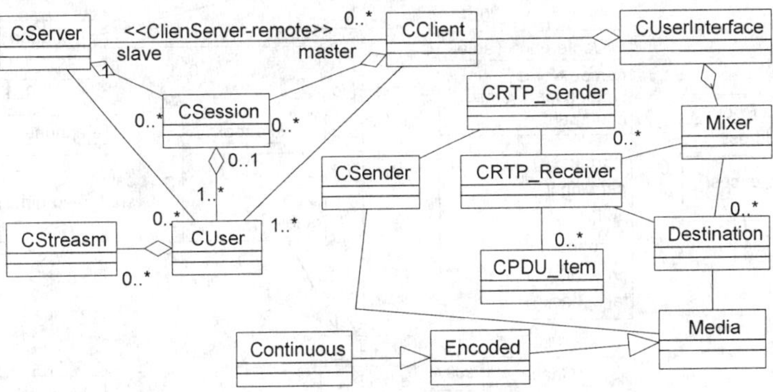

Fig. 2 The system model created during OOAD

Finally, L. Barfield noticed that the basic function of UI is to allow the user to fulfil his tasks using the system and introduced the four levels of UI description: *tasks*, *semantics*, *syntax* and *presentation levels*.

According to this the proposed Complex OOAD (COOAD) UIDP is split into four phases presented on the Fig. 1.

The Task Analysis phase, well described in the literature, e.g. (Hackos and Redish 1998) is parallel to requirements acquisition. *Use cases diagrams* (Rumbaugh J.,et al. 1999) can be adjusted to describe the results of this phase.

The main goal of the second phase – *UI Context Analysis* is not only to identify and describe users classes and context of use, as it is proposed in literature (e.g. ISO 9241), but also to include identification of possible metaphors and users concepts. The metaphors can be described in natural language as well as using simple OO diagrams to capture schemes, structures, patterns of behaviour, which can further be used as the base for some parts of UI. The identified user concepts (and the vocabulary used by the user to express them) should be preserved in further UI design.

The main goal of the third phase - *Conceptual Design* - is to define the conceptual structure of UI i.e. to define what kinds of user models users should build as an effect of the interaction with the system. These intended user models are formally described by *user views*. The user view is an OO model describing the structure of intended user model. There is at least one user view for each class of

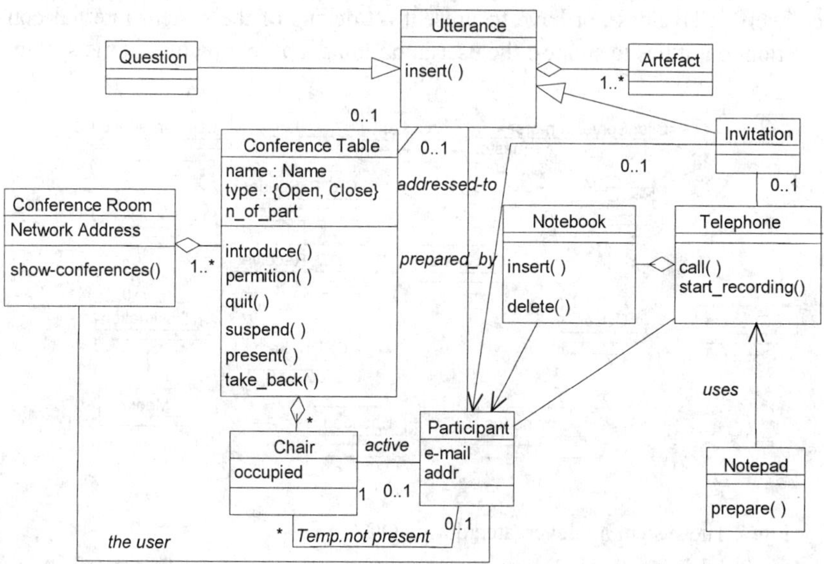

Fig. 3 A user view created during conceptual design phase

users. Each user view establishes a kind of *perspective* on the system and must be *compatible* with the model of the system created during analysis. The corresponding phase in the system development process is logical design (abstracting from technical details). The starting point for construction of user views is the set of metaphors identified in the second phase. The names of objects and relations come also from concepts collected there. The whole process is a combination of *analysis* done on the information collected earlier and *synthesis* from the structures of the model of the system (simplification) created in parallel.

The next phase – *UI Design* (including presentation design) starts with a choice of GUI standard. Then the user views are transformed into the OO model of UI. The overall architecture of UI is based on the MVC (*models, views, controls*) architecture originally proposed for Smalltalk. The system design model plays the role of MVC *models*, the models created from user views become MVC *views* and the compatibility relation between the system model and user views becomes the OO model of MVC *controls*.

4. Simple Example

As an example we consider the construction of a teleconferencing system. The system should allow many users to communicate across a computer network using many different media. Users can establish many discussion groups. A particular user can participate in many groups in the same time.

The scheme of the system model created by the independent designer is presented on the Fig. 2. The vocabulary used there is the vocabulary of the system analyst and the diagram represents his view of the system. The user view of the system is presented on the Fig. 3. It shows the different perspective of the system created on the base of simple metaphors: conference room, conference table with a spatial order of participants (chairs) etc.

5. Conclusions

COOAD is an attempt toward unified method of the system development process based on OO notation. It is not quite matured at the moment. However, it was included into a course delivered to software engineering students and applied to some practical cases realised as student projects. Interesting observation was that designers tried to take different user perspectives during construction of user views. Limitation of the method is assumption of 'a static user' - the natural process of development of the user from 'novice' to 'expert' is ignored at present.

6. References

Barfield L. *The User Interface Concepts & Design.* Addison-Wesley 1993.

Galitz W.O. *Essential Guide to User Interface Design.* Wiley Comp. Pub. 1996.

Hackos J., Redish J. *User and Task Analysis for Interface Design*, Wiley Comp. Pub. 1998.

Draft International Standard ISO DIS 9241-11 Ergonomic requirements for office work with visual display terminals (VDTs):- Part 11 Guidance on Usability.

Newman W., Lamming M. *Interactive System Design.* Addison-Wesley 1995.

Rumbaugh J.,et al *The Unified Modeling Language Reference Manual*, Addison Wesley, 1999

INTELLIGENT OBJECTS IN HUMAN-COMPUTER INTERACTION

Celestine A. Ntuen and Kanwal Hanspal
The Institute for Human-Machine Studies, College of Engineering
North Carolina A&T State University, Greensboro, NC 27411 USA

1 INTRODUCTION

Objects are traditionally known as artifacts used to describe both mental and physical entities. However, objects are usually described by attribute – value relationships. Such description makes objects and their technical tools such as Object-Oriented Programming (OOP) limited in application to situations in which knowledge of task are to be shared between and among several agents. Obviously, the single object inheritance metaphor is no more adequate.

The history of OOPs (Danford & Tomlinson, 1988; Coud & Yourdon, 1991) suggests that it emerged as a knowledge representation metaphor whose primary goals were to support property inheritance in a hierarchy of information space. Recent observations and critics of OOPs (Trajkovic, Gievska, & Dacev, 1995; Pen & Carrico, 1993) note that OOP paradigms do not cover a global set of object knowledge: for example, while objects in the abstraction hierarchy space can inherit properties of the parent in the upper echelon, they do not have the "intelligence" to share their domain task knowledge with other objects, especially when collaboration and execution of conjunctive tasks are desired. In other words, in agent-based and collaborative problem solving systems and interfaces, O-A-V (object-attribute-value) triplet in OOP representation falls short of its acclaimed performance (Mahfoudhi, Abed & Augu, 1995).

2 INTELLIGENT OBJECTS IN HCI DESIGN

2.1 Theoretical Background

Evolving software systems developed from agent-based paradigms require, among other things, shared knowledge in an object space. There are recent interests in the HCI community to develop more intelligent user interface (IUI) using object properties in order to achieve this knowledge sharing goal (Gorlen, 1990; Vlissides and Linton, 1990).

In our approach, we view an intelligent object (IO) as an artifact to support human activity. The design constructs are based on representation of task contexts: task characteristics, division of work, language of interaction, and task activities. Traykovic, Gievaska, and Davcev (1995) describe these concepts in terms of action processor and semantic knowledge.

In our context of application: an IO has the following properties:

(a) Knowledge of tasks, which represents the task characteristics to be performed.

(b) Knowledge of the user, which represents the level of interaction desired.

(c) Command knowledge, which represents the arbiter of actions to be performed on and by objects.

(d) Control knowledge, which represents the operational control of tasks by psychomotor activities (e.g., direct manipulation of widgets) and internal or virtual control by one object manipulating the behavior of other objects in time and space.

(e) Collaborative knowledge, which allows objects to share task knowledge and learns through property inheritance as well as by adaptation to new roles.

2.2 Development of IO Representation

Our concept of IO began during knowledge acquisition phase for decision support display design project (Ntuen & Hanspal, 1998). An open questionnaire form with O-A-V properties were initially used as a knowledge acquisition tool. The subjects used in the experiment were retired military strategists in military planning and operation. The subjects indicated that the O-A-V structure failed to capture the knowledge of the user's expectations. Rather, because of our physical concepts about objects, designers tend to emphasize the prosthesis view of objects, which often lacks the internal, structure processor characteristics. Through the subject's comments, we added further attributes for IOs. These are:

(i) Functionality. This attempts to define what the object does to the task (internal abstraction attribute at the system level), and what the object does to the user (the physical abstraction at the human-system interaction level). An IO, when functioning, is engaged in exchanging knowledge about tasks, concepts of next events, translation of human actions, and discerning when operations are to take place.

(ii) Interactivity. This defines the physical structure of interface and interaction language. For example, how human will use the objects, and the level of dimensionality of interaction. Interaction knowledge can be represented as widgets, buttons, or use of speech interface.

(iii) Conceptualization. This represents the (mental) internal structure used to define what operations are allocated to the objects and which are for humans. An object is conceptually defined to be useful at two levels:

operations and properties. Operations are typically executed by the user through direct manipulation (Sheridenman, 1982).

3 DESIGN REPRESENTION AND IMPLEMENTION

3.1 Design Representation

The HCI objects were developed using a collection of OOPs based design tools: RapidApp's Developer Magic tools, including cvd, cvstatic, cvbuild, Delta C++, and Smart Build (Developer Magic, 1995). RapidApp can generate C++ codes with interface classes based on the IRIS ViewKit toolkit. The codes produced by RapidAPP were automatically integrated into the Indigo Magic Desktop environment.

The general pseudocode structure of the IO representation is given in two phases. Phase I gives the design structure defined by:
/* Design Structure*/

```
Object:  :      Send_Message [Message_type];
         :      goal_object (.);
         :              Behavior_type: Share _message;
         :              Interaction_object:
                        user_task [object];
                        object _class (class_list)
                End_Interaction
         :      Function:
                        call_object [get_object_name];
                        fetch_data [get_data_source];
                        file [get filename * action_type];
                Return (Function);
                End_Function

         /*     Experiment Level structure */
Goal_object:    :
                Goal_instant: = Message_type;
                Inheritance:  = [Object_list*behavior-type];
                Object   : =        Query [goal_instant];
                Return (object)
End_goal
```

3.2 Design Implementation

The IO concept is used to design widgets, windows, and menus (WWM) for information display interface (IDI). Figure 1 shows an example IDI for high level concepts of information abstraction. Because of the IO knowledge structures the WWM objects, icons, and radio buttons are in the active mode, behave as display agents, and each with the capability of querying information from database and communicating with another object to display particular information based on the user's intention. The IDI objects are highly task driven in the lower level of abstractions, but knowledge-driven at higher echelon of object-to-object relationships.

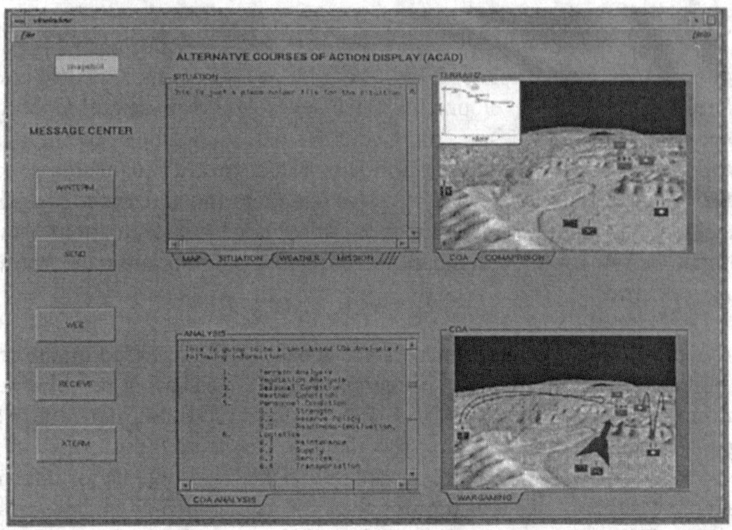

Figure1. Sample Object-Based Display

The application is developed for a UNIX environment using SGI INDY desktop tools. As can be seen in Figure 1, the message center has different widgets for direct manipulation. On the right hand side, we have four small multiview windows with task buttons attached to them. The user can use the window or buttons to explore into the embedded knowledge levels designed into each object category. Messages and information are parsed from and between windows because of the definition of shared knowledge. An example code for this shared knowledge property is accomplished as defined in Exhibit-1.

```
Deck2TabbedDeck::Deck2TabbedDeck (const char *name, Widget parent):
                        VkkTabbedDeck (name, parent)
{
        // Load any class-default resources for this object
setDefaultResources (baseWidget (), _defaultDeck2TabbedDeckResources  );

_mediaviewer2 = new MediaViewer ("mediaviewer2", _vkdeck->baseWidget ()  );
_mediaviewer2->sgiw ();
_neduavuewer2->vuew/fuke ("/usr/people/kanwal/ANALYSIS");
```

```
registerChild ( _mediaviewer2, "mediaviewer2");
//--- Add application code here:
```

Exhibit-1: Shared knowledge and inheritance code.

3 CONCLUSION

Object-attribute-value triplets have provided a substantial contribution to software designs and developments. However, the representation of design knowledge is based mostly on hierarchical organization of knowledge. Typically, the knowledge organization is oriented towards sequential execution of tasks. In agent-based software systems, collaboration and knowledge sharing are the major emphasis. The current knowledge representation using the traditional O-A-V triplet is deficient in addressing collaboration and knowledge sharing properties.

We have developed widgets, windows and menus (WWM) based on the concept of intelligent objects designed to encourage knowledge sharing between humans and abstract objects during task execution. With the IDI prototype, the user can reorganize the object's behavior by altering the spatial positions of icons, buttons, and other widgets. The user can define database objects for information management support, visualization of the objects for display and visualization support, and decision support objects for rule-based task execution. Thus IOs are adaptable, reusable, and provides transparent low-to-high end granular levels of interaction with the user.

5 References

Coud, P. and Yourdon, E. (1990). Object Oriented Analysis (2nd Edition). *Englewood Cliffs*, New Jersey: Prentice Hall.

Danforth, S. and Tomlinson, C. (1988). Type theories and object-oriented programming. *ACM Computing Surveys*, 123-128.

Developer Magic (1995). RapidAPP User's Guide. Silicon Graphics, Inc.

Gorlen, K. E. (1990). Data Abstraction and object-oriented programming in C^{++}. New York: *John Wiley & Sons*.

Mahfoudhi, A., Abed, M., and Augue, J-C (1995). An object oriented methodology for man-machine systems analysis and design. *In Symbiosis of Human Artifact* (Y. Anzai, K. Ogwawa & H. Mori, Eds.). New York, Elsevier Science, 965-970.

Ntuen, C. A. and Hanspal, K. (1998). Design and implementation of graphical user interface for courses of action display. Report #HMSEL0-CIE/98. The Institute for Human-Machine Studies, North Carolina A&T State University, Greensboro, NC.

Penz, F., & Carrico, L. (1993). Objects feeling agents in a multiview object space. *In Human Computer Interaction* (T. Grechenig & M. Tscheligi, Eds.). *Vienna Conference*, VCHCI '93, Fin de Siécle, Vienna, September 20-22, (24-62). New York: Springer-Verlag).

Trajkovic, V., Gievska, S., and Davcew, D. (1995), Object-oriented multimedia user interface. *In symbiosis of Human Artifact (Y. Anzai, K Ogawa, & H. Mori*, Eds.). New York, Elsevier Science, 231-236.

The role of external memory in a complex task: Effects of device and memory restrictions on program generation

Simon P. Davies

Department of Psychology, University of Hull,
Cottingham Road, HULL, HU7 6RX, UK

1 Introduction

One significant finding in programming research is that code is not generated in a linear fashion - that is, in a strict first-to-last order (Davies, 1991). Typically, programmers make many deviations from linear development, leaving gaps in the emerging program to be filled in later. Hence, the final text order of the program rarely corresponds to its generative order. Green et al (1987) proposed a model to account for this finding. Their parsing/gnisrap model describes the process by which a skeletal plan is instantiated in a programming notation. This model introduces a working memory component into the analysis of coding behaviour that forces the model to use an external medium (eg the VDU screen) when program fragments are completed or when working memory is overloaded. Hence, programs are not simply built up internally and then output to an external media with a generative order that reflects the final text order of the program. Rather, programmers will frequently need to refer back to generated code in order to recreate the original plan structure. The parsing element of the model describes this process, while gnisrap describes the generative process.

Given the cognitive costs that are involved in continually evaluating and modifying generated code, we require an explanation as to why skilled programmers rely so extensively on external rather than internal memory sources. Two experiments are reported here which attempt to address this issue directly. The first experiment considers the role of working memory in the determination of strategy for novice and expert programmers. The second experiment looks at the effects upon certain error forms of restricting the kinds of manipulations programmers can make within an environment.

2 Method

In the first experiment participants carried out an articulatory suppression task while engaged in a program generation activity. This experiment addresses a number of specific hypotheses. Firstly, if working memory limitations cause programmers to make use of an external medium, as suggested by Green et al, then the act of loading working memory through a concurrent task should give rise to an increase in nonlinearities. Given the effort required to use an external medium, in terms of the number of times a programmer must engage in the parsing/gnisrap cycle, one would expert experienced programmers to rely more extensively upon internal sources. Additional support for this hypothesis also arises from studies that suggest a strong link between expertise and working memory availability. However, the results of Davies (1991) give rise to an opposing hypothesis. This work suggests that skilled programmers make less use of internal sources than do novices and tend to rely more extensively upon using an external medium to record partial code fragments as they are generated.

The second experiment considers the role of working memory from a different perspective. Here interest is directed towards the way in which restricting the use of an external medium affects performance. In terms of the above analysis, if programmers are not able to correct already generated code at later stages in the coding process, then this should have some effect upon their performance. In this experiment, subjects created a program using a full-screen editor that provided no opportunity for the revision of existing text. The use of such an editor clearly places a significant load upon a subjects working memory capacity since they will be required to internally generate as much of the program as possible before externalising it. By placing emphasis upon the use of working memory it should be possible to induce error prone behaviour that parallels that which is evident when working memory is loaded in other ways, for instance via articulatory suppression.

Subjects: Twenty subjects participated in these experiments. One group of ten subjects were professional programmers. All the subjects in this group used Pascal extensively and all had substantial training in its use. Members of this group were classified as experts. A second group consisted of 2nd year undergraduate students all of whom had been formally instructed in Pascal syntax and language use during the first year of their course. This group were classified as novices.

Procedure and Design: Subjects were asked to carry out a simple articulatory suppression task which involved repeating a string of five randomly presented digits. At the same time, subjects were requested to generate a simple Pascal program that could read a series of input values, calculate a running total, output an average value and stop given a specific terminating condition. This specification was derived from Johnson and Soloway (1985) and was chosen because it has formed the basis of many empirical studies. Three independent raters were asked to analyse all the resulting program transcripts for errors (using the classification described above). The experiment was a two-factor design, with the following independent variables: 1. Suppression/No suppression and 2. Level of expertise (Novice/Expert). The dependent was the number and type of errors remaining in the final program.

Results (Experiment 1): Analysis revealed a main effect of expertise ($F1,36 = 9.37$, $p<0.01$) and suppression ($F1,36 = 4.54$, $p<0.05$) and an interaction between these two factors ($F1,36 = 15.89$, $p<0.01$). A number of post-hoc comparisons were carried out using the Newman-Keules test with an adopted significance level of $p<0.01$. This indicated a significant difference in error rates in the both experimental conditions when comparing the novice and expert groups. In addition, a significant difference between error rates across conditions was evident for the novice group.

Results (Experiment 2): These data were analysed using a two-way analysis of variance with the following factors; Environment (restricted/unrestricted) and Level of expertise (Novice/Expert) This analysis revealed a main effect of Environment ($F1,36 = 5.74$, $p<0.05$), and of Level of expertise ($F1,36 = 4.21$, $p<0.05$) and an interaction between these two factors ($F1,36 = 9.76$, $p<0.01$). Post-hoc comparisons revealed a significant difference between the number of errors produced by novices and experts in condition 1.

Error classification analysis: In the case of experts, there is a fairly even distribution of error types across the two experimental conditions. Indeed, further statistical analysis revealed no significant differences between error types both within and between conditions (t-tests). In the case of the novice group, the distribution of error types is less straightforward. In the non-suppression condition, novices produced a significantly greater number of plan errors in comparison to the other categories (t-test). Moreover, the only significant difference between the novice and experts groups in this condition was the number of plan errors produced by the novice group (t-test). In the second condition, the distribution of errors across classification types for expert subjects was again fairly even. No significant differences between any of the error classifications were evident. For the novice group, significantly more control-flow and interaction errors were evident in comparison to the other two error classifications (t-test).

4 Discussion

The results of the first experiment demonstrate that expert performance in programming tasks is not significantly affected by articulatory suppression. Hence, for experts the number of errors produced in the two conditions is similar. Moreover, it appears that strategy is similarly unaffected. Conversely, the novice group produced significantly more errors in the suppression condition compared to the non-suppression condition. In addition, the nature of the coding strategy that they adopt is also affected. In particular, novice programmers revert from a linear generation strategy characterised by the prevalence of within-plan jumps, to a strategy more characteristic of experts. That is, to a strategy reflecting a significant increase in non-linearities. Further support for this view is evident in the error data. In the non-suppression condition, novice subjects are clearly more error prone than experts. This finding is not unexpected. However, in the suppression condition, the error rate for the expert group changes little from this base line whereas the novice error rate more than doubles. This may indicate that when working memory is loaded novices must externalise information and that this constitutes a strategy that they find unnatural, thus leading to an increased error rate.

A more detailed analysis of these errors revealed a change in the nature of errors for novice subjects between the two experimental conditions. In the non-suppression condition, the novice group make a greater number of plan errors, suggesting knowledge-based difficulties. Conversely, in the suppression condition a greater proportion of control-flow and interaction errors are evident. In terms of the present analysis, the preponderance of control-flow and interaction errors may reflect problems keeping track of the interdependences between elements in the emerging program. When working memory availability is reduced it appears that novices experience some difficulty with these interdependences. Unlike experts, it seems that novices cannot use the external display as an aid to memory to its full extent.

An alternative explanation for these findings is that experts simply have an extended working memory capacity. Such an account would presumably have no difficulty predicting the results of the experiment reported in this paper. In order to assess the cogency of this alternative explanation, the second experiment reported in this paper adopts a different approach for exploring the relationship between working memory and the development of programming skill. Whereas the first experiment attempted to reduce the subjects' available working memory capacity, the second experiment was designed to encourage subjects to rely upon working memory. Hence, if experts have an extended working memory capacity they should demonstrate performance similar to that displayed in the first experiment. Moreover, if the extended capacity notion is correct, then experts should out perform novices even in the situation where the task environment is restricted as in this second experiment.

These experiments demonstrate that the relationship between skill development in programming and working memory is not as predicted. It appears that experts rely significantly upon external sources to record code fragments as these are generated and then return later, in terms of the temporal sequence of program generation, to further elaborate these fragments. It has been suggested that a major determinant of expertise in programming may be related to the adoption or the development of strategies that facilitate the efficient use of external sources. The externalisation of information clearly has a high cost in terms of the reparsing or recomprehension of generated code that is implied. Hence, it might seem counterintuitive to suggest that problem solvers will tend to rely upon this kind of strategy rather than upon a strategy that involves the more extensive use of working memory. However, this explanation is consonant with existing work which has implicated display-based recognition skills in theoretical analyses of complex problem solving (Larkin, 1989). The contribution of these analyses has been important, but they have neglected to consider the relationship between display use and expertise and the consequent effect that this may have upon the nature of problem solving strategies.

5 References

Davies, S. P. (1991). The role of notation and knowledge representation in the determination of programming strategy: A framework for integrating models of programming behaviour. Cognitive Science, 15, 547 – 572.

Green, T. R. G. (1991). Describing information artifacts with cognitive dimensions and structure maps. In D. Diaper and N. Hammond (Eds.), People and Computers 6, Cambridge University Press.

Green, T. R. G., Bellamy, R. K. E. and Parker, J. M. (1987). Parsing and gnisrap: a model of device use, Proc. INTERACT'87, H. J. Bullinger and B. Shackel (Eds.), Elsevier Science Publishers B. V., North-Holland.

Larkin, J. H. (1989). Display-based problem solving. In D. Klahr and K. Kotovsky, (Eds.), Complex Information Processing; The impact of Herbert A. Simon.

Pennington, N. (1987). Stimulus structures and mental representation in expert comprehension of computer programs. Cognitive psychology, 19, 295 - 341.

Expertise in computer programming: Exploring commonalities between code comprehension and generation activities

Simon P. Davies
Department of Psychology
University of Hull, Cottingham Road,
Hull, HU6 7RX, UK

1 Introduction

This paper reports an experiment on the comprehension of computer programs viewed through a limited access window by novice and expert programmers. The intention of the study was to explore the role played by the external environment in mediating the cognitive processes involved in computer programming.

Previous work by the author Davies (1993) using a text editor which only allowed movement between adjacent lines in a program, hence enforcing a strict linear generation strategy, showed that this device restriction can systematically affect error patterns and solution times in a code generation activity. Moreover, this effect was more marked for experts than for novices, suggesting potential disruption to 'normal' problem solving strategies.

This finding may be explained in terms of display-based competence (Larkin, 1989). Models of display-based competence suggest that part of the process of becoming an expert involves developing strategies which lead to extensive externalisation behaviour (Davies, 1996; Green et al, 1987) where information is output to some external source, rather than rely upon working memory. In contrast, novices appear to externalise little and their comprehension, whilst poorer than the expert group overall, is not similarly affected by restrictions to the device used to create the program measured in terms of their overall error patterns.

One issue that arose from this work was concerned with the question of whether program generation is underpined by similar cognitive process as comprehension activities. This question is important since there have been suggestions that there may be an asymmetry between generation and comprehension activities (Davies, 1993) while others think the processes are the same (Detienne, 1990). In this paper an experiment is presented which considers the effects on program comprehension of constraining a programmers ability to use certain forms of strategy by presenting the program via a restricted text editor. We consider in this context the comprehension of programs presented in this way by novices and experts. In order to assess comprehension, participants viewed the stimuli for a brief period and were then asked questions intended to tap various forms of programming knowledge. We are also interested in whether this device restriction affects the amount and nature of material that is externalised. In order to evaluate this, the participants were allowed to make notes and these were subsequently analysed in terms of their quantity and nature.

2 Method

2.1 Participants

Twenty participants were recruited for the experiment reported in this paper. One group of ten subjects were experienced programmers who were either professional programmers are teachers of advanced Pascal programming courses (Mean length of programming experience, 5.4 years). The remaining ten participants were all first year undergraduate students on computer science/engineering courses. All of this novice group had some (no more than 6 months) programming experience.

2.2 Design and procedure

There were three experimental conditions:

1. View code through one line window in fixed, linear order

2. View code through one line window in subject determined order

3. View whole program

All of these conditions involved a fixed total program presentation time of 3 minutes. Participants viewed the same program in all three conditions, but the order of presentation of the different conditions was fully counterbalanced. The experiment was a mixed design with one between subjects factor, namely expertise and one within subject factor, the programming environment that was used. Hence, there were two independent variables in this study - Expertise which has two levels (Novice/Expert) and Environment which has three levels (one line-linear, one line-subject controlled and whole program presented). The dependent variable was performance in response to various comprehension questions derived from Pennington's classification scheme (Ramalingam and Wiedenbeck, 1997; Pennington, 1987). These questions fell into five categories following Pennington's original classification scheme: function, sequence, data-flow, control-flow and state. In all, the participants were presented with 3 questions derived from each category (A total of 15 questions overall). Note that different questions were posed in relation to the three different presentation modes, and the order was fully randomised. A secondary dependent variable was the amount of notes generated at different levels of abstraction. All participants were given a large sheet of paper upon which to record their notes and all activities were video taped. If experts tend to externalise then restricting the environments will affect normal strategies and experts will tend to externalise more in this condition. In contrast, we have argued that novices may not be affected by restrictions to the environment in the same way, since they tend to rely upon working memory to develop code.

3 Results

The results of this study were analysed using a mixed model ANOVA. Analysis revealed a main effect of Expertise ($F_{1,18}=18.43$, $P<0.01$), a main effect of Environment ($F_{2,18}=9.32$, $P<0.01$) and a significant interaction between these factors ($F_{2,18}=10.65$, $P<0.01$). A post hoc analysis of the comprehension questions showed that novices answered all question types with similar levels of accuracy, regardless of the environment used to present the program. In contrast, the expert group performed poorly on the control-flow and sequence questions, but only in the restricted environments. In the non-restricted environment, the expert group answered all comprehension category questions more accurately than the novice group.

The average time devoted to producing notes at different levels of abstraction (defined here as low, medium and high) was also assessed. The experimenter and two other independent raters studied the notes produced by each participant. Preliminary analysis suggested that these notes could be classified into three categories; notes of a high level of abstraction which consisted simply of code sketches (typically these reproduced elements of the actual stimuli), notes of intermediate levels of abstraction which were code-like, but not reproductions of the stimuli (typically, these were in a form of pseudo-code), and finally, notes of a high-level of abstraction, consisting predominantly of diagrammatic forms of notion (for instance, data-flow charts or structure diagrams). The initial classification segmented the written protocol of each subject into one of the three categories just outlined. The raters classified all notes and other external representations into the three categories. There was a high level of inter-rater correlation ($W = 0.82$, $p<0.01$). This classification was then retrospectively applied to the video protocol to determine how long each participant spent generating notes in the three categories. An attempt was made to normalise these data by dividing the time spend producing each category of note with the total time spent producing notes for all three categories. These data revealed that the expert group externalised more low level notes than the novices ($F_{1,18}=10.43$, $P<0.01$), that the experts externalised more in the restricted conditions than the unrestricted condition $F_{2,18}=12.32$, $P<0.01$), and that the novices were not affected by the environment in terms of the type of notes ($F_{2,18}=12.76$, $P<0.01$) or the quantity of notes that were externalised ($F_{2,18}=5.34$, $P<0.05$).

4 Conclusions

The findings of the study reported here suggest firstly that different restrictions to the task environment tend to affect the performance of experts more than novices, but only under certain conditions. Secondly, that the pattern of errors in response to different categories of comprehension question is affected systematically by restrictions to the task environment, but again only in the case of expert programmers. Finally, that the amount of information that is externalised is similarly affected by restrictions to the task environment and modified by expertise. These findings suggest systematic differences in the strategies used by novice and expert programmers and support the view that experts attempt to externalise as much information as possible, whereas novices tend to rely upon working memory. When either of these strategies is disrupted then the behaviour and performance of the retrospective groups is significantly affected. These findings mirror those of previous work by the author who showed that the generation of programs under conditions that were similar to those reported in the present study gave rise to almost identical patterns of performance (Davies, 1996). This may in turn provide credence for a display-based view of expert problem solving strategies. In conclusion it is argued that comprehension and generation strategies may well be underpinned by similar cognitive processes.

5 References

Davies, S. P. (1993). Models and Theories of Programming Strategy. International Journal of Man-Machine Studies, 39, (2), 237-267

Davies, S. P. (1996). Display-Based Problem Solving Strategies in Computer Programming. W. D. Gray and D. A. Boehm-Davis (Eds.), Empirical Studies of Programmers: Sixth Workshop. Ablex, New Jersey.

Detienne, F. (1990). Expert programming knowledge: A schema-based approach. In J.-M. Hoc, T. R. G. Green, R Samurcay and D. J. Gilmore (Eds.), Psychology of Programming, Academic Press, London.

Green, T. R. G., Bellamy, R. K. E. and Parker, J. M. (1987). Parsing and gnisrap: a model of device use, Proc. INTERACT'87, H. J. Bullinger and B. Shackel (Eds.), Elsevier Science Publishers B. V., North-Holland.

Larkin, J. (1989). Display-based problem solving. In D. Klahr and K. Kotovsky (Eds.), Complex information processing: A tribute to Herbert A. Simon. Lawrence Erlbaum, Hillsdale, NJ.

Pennington, N. (1987). Stimulus structures and mental representations in expert comprehension of computer programs. Cognitive Psychology, 19, 295-341.

Ramalingam, V., and Wiedenbeck, S. (1997). An empirical Study of Novice Program Comprehension in the Imperative and Object-Oriented Styles. Empirical studies of programmers: Seventh workshop, Ablex, Norwood, NJ.

Active vs. Passive Systems for Automatic Program Diagnosis

Haider Ali Ramadhan
Department of Computer Science, Sultan Qaboos University
PO Box 36 Code 123 Muscat, Sultanate of Oman
haider@squ.edu.om Fax: 968-513145

1 Introduction

Automatic program diagnosis and debugging systems can be classified by their primary means of program analysis. The most distinctive split is between those systems that are unable to analyze partial code segments as they are provided by the user and must wait until the entire solution code is completed before attempting any diagnosis, and those that are capable of analyzing partial solutions. The former perform *passive analysis* while the later perform *active analysis*.

Systems using passive analysis do not trace the intentions of the user or his design decisions while being developed and require him to explicitly request the automatic debugging of his code segments. These systems localize errors in the user programs either by looking for surface structural forms (plans) or by accounting for differences between forms and actual code segments. Examples of these systems include Bridge (Bonar 1992) and Proust (Johnson 1990). Generally speaking, these systems rely on some sort of pre-stored requirements for a complete solution. It is worth noting here that systems which rely on pre-stored requirements for a successful solution cannot solve the problems themselves, and hence cannot reason about the solutions and designs provided by the users.

On the other hand, systems using active analysis perform automatic debugging by implementing model-tracing (Anderson 1990). Through this approach, these systems subdivide tasks into smaller steps that must be solved one at a time. The user's design decisions are traced as he develops the solution. On each step taken by the user, these systems check to see if the user is following a design path known to be correct or buggy. Buggy paths are pruned as soon as they are detected by giving the user intelligent feedback and allowing him to try again. Example of such systems include The Lisp Tutor (Anderson 1990) and GIL (Resier 1992). These systems tend to be quite directive. However, through rich interaction and flexible immediate feedback, these systems detect very specific bugs and misconceptions.

Several advantages can be outlined of the active approach to automatic program debugging. First, very specific errors can be diagnosed and immediate feedback can be given in proper context of the error, hence the users can be explicitly guided in the process of acquiring problem-solving skills. Second, systems using this approach are capable of analyzing partial solutions as they are provided by the user and therefore have access to all intermediate states, hence they work with more information than systems that are capable of only analyzing complete solutions. This in turn provides these systems with the capability of reasoning about the programming process itself and thus generate very specific explanations and advice. Third, the impact of multiple bugs on the diagnosis process is minimized. Many of the post-event based systems such as Proust and Talus, have to deal with disentangling multiple bugs which require them to generate all possible alternative treatments of these bugs and pick the best from among them. Systems using active analysis simply prevent the user from making multiple bugs and explain each bug immediately, therefore, the code never contains more than one bug at a time. In addition to cognitive justifications, see (Anderson 1990), this approach greatly simplifies the engineering and the implementation of automatic program diagnosis.

Despite these advantages, this approach to automatic program diagnosis tends to be very directive. In addition, this approach ignores the issue of providing the user with some chance to detect and correct bugs on his own, since the user is not allowed to go wrong. However, these disadvantages can be overcome by supporting a more flexible style of user interaction while still retaining close ties to model-tracing. This was successfully accomplished by the DISCOVER (Ramadhan 1992, 1998) system through (1) supporting an ability to give delayed feedback by increasing the grain size of automatic diagnosis to a complete program statement, not just a single word or symbol, and (2) allowing the user to do limited backtracking by giving him some chance to delete previously entered code and restart, while in full interaction with the system.

2 An Overview of the Discover System

DISCOVER is a discovery programming environment which integrates software visualization and immediacy features with model-tracing based diagnosis. The figure below shows the user interface of the system and how the user interacts while attempting to solve the *Ending Value Averaging Problem.* The interface appears to a user as a collection of seven windows. The four windows on the left side of the interface, namely the *Memory Space,* the *Input Space,* the *Output Space* and the *Algorithm Space,* represent the components of the underlying programming machine. However, DISCOVER presents it as a visible machine in its real-time action, and not as a static, textbook-like picture of language execution and machine behavior. Through this dynamic, visible machine, novices (1) can observe how program statements are executed in an animated way, (2) can see hidden and internal changes in some conceptual parts of the underlying computer, such as the memory space, and (3) can relate problem solving with the properties of the machine they are interacting with. In the figure, the user has failed to accumulate the numbers read by the program for the averaging purpose. The system detects this misconception and considers this step as a deviation from the solution path, and hence decides to interfere by guiding the user toward the expected step.

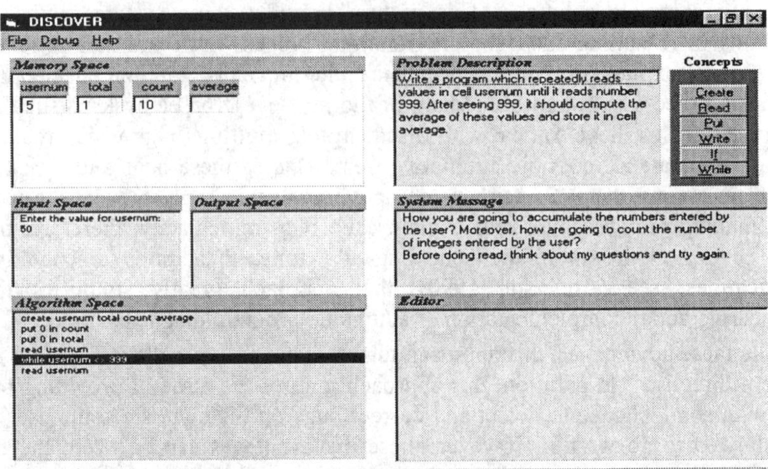

The interface is designed to expose the users to whatever is being manipulated and experienced, and hence brings them closer to the language and the machine. By doing that, the system brings the affordance into users knowledge of perception and action. The graphical model of the notional machine, through visualization and immediacy features, is expected to help the users perceive and manipulate the dynamic behavior of the program and its unfolding computation with less effort and accelerate the debugging task. By allowing the users to have such graphical view of program behvaior within an integrated and coherent image of the programming machine, the system can reduce the mental overload on the novice users, hence more emphasis can be put on program understanding, debugging and problem solving. In short, interactions in DISCOVER among the program, the language and the machine are designed to produce an environment which makes it effortless for users to examine a program, figure out its effects and connections, and to relate problem solving with the properties of the machine they are interacting with.

The programming language of the system is a simple pseudo-code based and algorithm like language. At present, the language has no provision for functions, procedures, recursion and complicated data structures such as records, arrays and lists, thus focusing users attention on basic programming concepts and simplifying the learning process. Programming concepts supported include CREATE, PUT, READ, WRITE, WHILE-END-WHILE and IF-ISTRUE-ISFALSE. The selections in the *Concepts* menu, shown in the top right-most position, contain the beginning of phrases. Each phrase corresponds to one programming concept.

3 Program Diagnosis

DISCOVER analyzes the surface code of the completed concept (partial solution code) without much specific knowledge about the problem to be solved or about how to design and construct an algorithm (i.e. DISCOVER cannot solve the problem itself). The above figure shows the interaction between the system and the user while attempting to solve the *Ending Value Averaging Problem.* In this example, the user has failed to accumulate the numbers read by the program for the averaging purpose. The system detects this misconception and considers this step as a deviation from the solution path, and hence decides to interfere by guiding the user toward the expected step. For more detail regarding the diagnosis process, see (Ramadhan 1998).

Much like and Bridge, DISCOVER relies on a pre-stored reference solution (the ideal student model) for a given problem and applies various heuristics and pattern matching techniques to match the solution code provided by the novice with the reference solution to spot errors. Unlike this systems, however, DISCOVER is capable of interactively analyzing partial solution code. By doing that, DISCOVER explicitly guides the novice in the process of putting together programming concepts to solve a given problem. As stated before, DISCOVER implements a more flexible style of model-tracing (e.g. than the Lisp Tutor and GIL). This is accomplished by increasing the grain size of diagnosis to a complete program statement.

4 Evaluation

To get some insights into the superiority of active approach to diagnosis over passive approach or vice versa, we have developed a new version of DISCOVER. In this version, the system supports both active as well passive modes of automatic diagnosis, hence the user is given more control over when feedback is presented. In passive mode, the feedback is delayed until the user finishes typing the solution code and explicitly requesting the diagnosis of the code. However, the diagnosis of the user solution is carried out in the same way: the system scans the code in a top-down, left-to-right manner, stops at the first error encountered, and ignores the rest of the solution. This version is similar in functionality to the latest version of the Lisp Tutor (Anderson 1990).

At present we are in the process of completing an empirical evaluation addressing the hypotheses that active analysis, through its ability to diagnose partial solution steps and provide immediate feedback on errors, helps users more in problem solving than passive approach. Twenty users, mostly novices, were divided randomly into two groups: Active and Passive. The Active group had to solve programming problems using the model-tracing based environment of DISCOVER, with immediate feedback supported, while the Passive group had to solve the same problems using the passive environment of DISCOVER, with feedback being delayed until the user explicitly requests it. In analyzing the results, only two measures were used: total logical errors made and total times spent while solving problems. So far, only five users in each group have completed the evaluation.

Results that have been compiled so far showed that the passive environment did not prove to be as effective as the active one. Overall, the Passive group made about 46% more errors to complete their programming problems than the Active group. Regarding the time spent on problem solving, the Passive group took about 63% longer to complete their problems than the Active group. These results may suggest that the effectiveness of the Active approach to intelligent diagnosis is achieved by maintaining strong control over the progress of the user during problem solving. Broadly speaking, these partial results seem to be consistent with an earlier evaluation conducted to test the *Student Controlled* version of the Lisp Tutor (Anderson 1990).

5 References

Anderson, J. (1990). Cognitive Modeling and Intelligent Tutoring. *Artificial Intelligence and Learning Environments,* Clancy and Soloway (Eds.), MIT/Elsevier.

Bonar, G. (1992). Intelligent Tutoring with Intermediate Representations. *Proceedings of the Second Conference on Intelligent Tutoring Systems (ITS-92),* Canada.

Johnson, W. (1990). Understanding and Debugging Novice Programs. *Artificial Intelligence and Learning Environments,* Clancey and Soloway (Eds.), MIT/Elsevier.

Ramadhan, H. (1992). Intelligent vs. Unintelligent Programming Systems for Novices. *Proceedings of the IEEE 15th International Conference on Computer Applications and Systems, USA.*

Ramadhan, H. (1998). Model tracing based approach to intelligent program diagnosis. *SQU Journal of Science & Technology, Vol. 2.*

Reiser, B. (1992). Making Process Visible: Scaffolding Learning with Reasoning-Congruent Representations. *Proceedings of the 2nd Conference on Intelligent Tutoring Systems (ITS '92),* Montreal.

Improving the Engineering of Immediate Feedback for Model-Tracing Based Program Diagnosis

Haider Ali Ramadhan
Department of Computer Science, Sultan Qaboos University
PO Box 36 Code 123 Muscat, Sultanate of Oman
haider@squ.edu.om Fax: 968-513145

1 Introduction

Normally, novices are required to produce programs for tasks they can perform by hand. In other words, the goals are familiar to them and they have procedures at their disposal so that these goals can be reached. These familiar tasks become difficult because there is one new condition that must be satisfied: programs have to be understandable by a computer through a definite means of communications, namely the programming language. Programmers thus have to restrict themselves to the use of operations the computer can perform. To be able to use these operations and express them in a program, novices must elaborate representations of these operations; such representations may not be necessary in usual and real-world problem-solving situations. In real-world situations the operations can be elaborated at execution time, where immediate concrete feedback is provided after the execution of each operation. It is this feedback that makes it easier for novices to solve the same tasks in real-world environments. To provide novices with a simulated and real-world like environment when learning computer programming, a program diagnosis system needs to make the immediate concrete feedback available to novices.

Empirical evaluations studying how novices understand and debug their programs showed that novice programmers cannot deal with multiple errors at the same time and that they tend to focus entirely on one error at a time. When faced with multiple error messages, novices ignore all but first error message and use a depth-first debugging approach concentrating on only one error at a time. These results also showed that when novices get error messages, they tend to repair only the first error and ignore the remaining error messages. Hence, it was suggested that novices should not be allowed to make multiple errors and that immediate feedback should be provided whenever an error is discovered (Reiser 1992, Anderson 1990). In fact, our own experience in teaching introductory computer programming courses has reached a similar conclusion.

2 The Nature of immediate Feedback

The principle of immediate feedback is derived from theoretical and practical as well as technical considerations. For a full theoretical account of immediate feedback and of the application of the ACT* model of skill acquisition and its learning assumptions in general, see (Anderson 1990). Practically, there are two reasons for preferring immediate feedback over delayed feedback. First, if the user is allowed to continue on an erroneous path, the error can get compounded by additional errors and the user can spend a lot of time and become quite frustrated trying to recover. A second reason for providing immediate feedback is that very specific misconceptions can be explained in their very immediate context.

Technically, immediate feedback simplifies the implementation of model-tracing based program diagnosis. Systems performing program diagnosis using this approach monitor novices actions as they move along the solution path during problem solving, automatically analyze partial solution steps for errors and misconceptions, and offer guidance whenever the users deviate from a correct solution path. This is achieved by using a set of problem solving rules that allow these systems to generate possible steps that a user might take while solving a given problem.

Program diagnosis systems like Bridge (Bonar 1992) and Proust (Johnson 1990) have to deal with disentangling multiple bugs which require them to generate all possible alternative treatments of these bugs and pick the best treatment from among them. In some cases the problem of dealing with multiple bugs can easily lead to a combinatorial explosion. The simplest solution to this problem is to prevent the user from making multiple bugs and diagnose each bug immediately, as it occurs, and thus minimizing the impact on the diagnosis process. The advantage is very clear: the problem of detecting multiple bugs and then deciding which bug to diagnose first does not even arise. The main task in this case is to concentrate on deciding which strategy to use with the user upon detecting a bug.

Despite these justifications and advantages of immediate feedback, the way it was implemented in the Lisp Tutor has been proven to be quite controversial. The user's knowledge of Lisp is represented in the Lisp tutor at about the finest grain size that has some functional meaning in Lisp. In other words, the system models the performance of the user at the level of individual Lisp symbols. This approach tends to be very directive and greatly restricts the user during problem solving. It also makes it very difficult for the system to diagnose individual bugs as they are created because a single symbol does not generate enough context for good diagnosis and tutoring. In addition, this approach completely ignores the issue of providing the user with some chance for self detection and correction of errors (debugging), since the user is never allowed to really go wrong. When learning to program, it is very important for the user, especially a novice, to get some chance to discover some of his own flaws and misconceptions without any intervention from the system.

3 Need for flexible Interactions

The principal features of the interaction style embodied in model-tracing systems mentioned earlier can be summarized as follows. First, these systems insist that the novice stay on a correct solution path and immediately flag errors. They react to every symbol the novice types and provide immediate feedback as soon as the novice deviates from the solution path. Second, these systems do not allow the novice to backtrack and delete previously entered code. Third, these systems use a menu-based dialogue to track planning decisions and behaviors when they fail to trace these behaviors nonintrusively. Finally, these systems force the novice to enter the code in a left-to-right, top-down manner. This implies that the next piece of code or the next step on a solution path is decided by the system and not by the novice. Occasionally though, the user is given some freedom in dealing with arguments whose ordering is not important, or even with functions which have the same underlying functionality, such as *cons*, *append* and *list*.

While each of these features has pedagogical justification and close ties to the underlying cognitive modeling, there is no reason why some of these features, especially the first two, cannot be improved to support a more flexible style of tutorial interaction while preserving a close relationship to the model-tracing approach. Some amount of self-detection and correction of errors may lead to a clearer understanding of the problem and a better explanation of the programming process by the novice and certainly is something that users using the Lisp Tutor have said they wanted (Anderson 1990). Providing immediate feedback upon every single Lisp symbol is also extremely undesirable and restricting in situations where not enough context has been established for the novice to understand why his solution is wrong.

3.1 An Improved Interaction Style

To compromise the issues related to immediate feedback, we have developed a prototype system for program diagnosis using model-tracing approach. The system, called DISCOVER (Ramadhan 1992), supports a more flexible style of tutorial interaction that is based on improving the first two features of the Lisp Tutor's interaction style mentioned above while retaining very close ties to model-tracing and immediate feedback.. This is achieved by increasing the grain size of automatic tutoring and by providing novices with some opportunity for self-correction of errors. The principal features of DISCOVER's interaction style can be summarized as follows. First, the system reacts to every complete programming statement and expression, not to a single symbol, and provides immediate feedback as soon as the novice wanders of the correct solution path. Second, the system supports limited backtracking by allowing novices to delete previously entered code (e.g. parts of the statement currently being completed). Third, the system supports an explicit planning mechanism to trace the intentions and high-level goals of the novices. Novices externalize their planning decisions by choosing form a menu of programming concepts rather than through a dialogue. Finally, the system requires the novice to enter the code in a top-down manner. By increasing the grain size of tutoring to a complete statement and expression, DISCOVER provides novices some opportunity for self-correction and also a larger context for instruction. Since the grain size of tutoring is confined to a single symbol, the Lisp Tutor finds it difficult to explain why a novice's action is wrong at the point which the misconception is first manifested because there is not enough context.

To consider an example, compare a novice who provides '(append (list x)y)' where 'cons x y)' is better. It would become easier to explain the choice after the complete statement has been provided rather than after '(append' has been entered. In the case of DISCOVER, this problem does not arise. If the novice provides the following statement in DISCOVER's algorithmic language, for example, 'READ 5 IN num' where 'PUT 5 IN num' is more appropriate, the system explains the choice after the complete statement has been typed in rather than immediately after 'READ' has been selected. This allows DISCOVER to generate more appropriate explanations and advice that can derive mapping, generalization and coordination that exist between similar programming concepts. For example, in the case of 'READ' instead of 'PUT', DISCOVER informs the novice that it would be better in normal cases where getting an input from the user is not required to use the 'PUT' concept for assigning values to cells. This explanation would not become possible to generate if DISCOVER could not wait to see whether the novice indeed wanted to read 5 and not some other values in cell 'num'.

DISCOVER also supports limited backtracking by allowing novices to delete previously entered parameters and operators of the statement currently being completed. Unfortunately, at present the backtracking is confined to the current statement only. The novice can also cancel the selection of a concept and select a new one that represents best his next goal. For example, if the novice selects the 'READ' concept where 'WHILE' is expected and realizes after completing the selected concept, but before submitting it, that he made an error, he can backspace over the statement. The system would ignore the selection without considering it a deviation from a solution path. This gives the novice some opportunity for self-correction. In fact, there are cases in which the novice may be confused about what goals and plans are appropriate in the current situation and would realize only if he is given a little more time to self-correct. This is not possible with the classical version of the Lisp Tutor.

4 References

Anderson, J. (1990). Cognitive Modeling and Intelligent Tutoring. *Artificial Intelligence and Learning Environments,* Clancy and Soloway (Eds.), MIT/Elsevier.

Bonar, G. (1992). Intelligent Tutoring with Intermediate Representations. *Proceedings of the Second Conference on Intelligent Tutoring Systems (ITS-92),* Canada.

Johnson, W. (1990). Understanding and Debugging Novice Programs. *Artificial Intelligence and Learning Environments,* Clancey and Soloway (Eds.), MIT/Elsevier.

Reiser, B. (1992). Making Process Visible: Scaffolding Learning with Reasoning-Congruent Representations. *Proceedings of the 2nd Conference on Intelligent Tutoring Systems (ITS '92),* Montreal.

Ramadhan, H. (1992). Intelligent vs. Unintelligent Programming Systems for Novices. Pro*ceedings of the IEEE 15th* International Conference on Computer Applications and *Systems (COMPSAC'92),* USA.

HandsOn: Dynamic Interface Presentations by Example*

Pablo Castells
E.T.S.I. Informática
Universidad Autónoma de Madrid
pablo.castells@ii.uam.es

Pedro Szekely
Information Sciences Institute
University of Southern California
szekely@isi.edu

1 Introduction

Visual tools for GUI development have greatly contributed to alleviate the effort involved in interface construction (NeXT 1990). Visual builders save time, require very little knowledge from the developer, and help improve the quality of displays. However, the tools we know today are confined to the construction of the static portion of presentations and provide very little or no support for the dynamic aspects of interface displays. The main reason for this is the lack of abstraction of the visual languages these tools provide, which on the one hand favors their ease of use, but on the other makes it very hard to specify procedural information.

Our research aims at extending the expressive power of existing visual tools for the construction of a significant range of dynamic displays while retaining the ease of use of direct manipulation. We do so by a) using the model-based approach (Wiecha et al. 1990, Szekely et al. 1993, Szekely et al. 1996) for the internal representation of the constructed displays, with models that support dynamic presentation functionalities, and b) developing an extended visual language that incorporates abstractions and Programming By Example (PBE) techniques (Cypher 1993) for the interactive specification of dynamic presentations by manipulating interface presentation objects in a visual tool. Tighly integrated in both the visual language and the presentation model,

* This work was partially supported by the *Plan Nacional de Investigación, Spain, Project Number TIC96-0723-C02-02*

HandsOn provides graphic design tools like guides and dynamic grids to build well-structured and visually appealing designs.

2 Related Work

Several systems in the field of Programming By Example (PBE) have shown that it is possible to overcome the limitations of interface builders by including inference capabilities and domain knowledge to make it possible to build abstractions by manipulating concrete objects (Cypher 1993). However the results achieved to date tend to lack the reliability required for a wide implantation in GUI technology. We believe one of the key aspects for a PBE system to be workable is to avoid hiding relevant information to the user, which otherwise results in impredictability and lack of control. The difficulty resides in finding the appropriate form to convey this information.

On the other end, high-level systems like UIMSs and model-based tools support sophisticated interface features (Wiecha et al. 1990, Szekely et al. 1993, Szekely et al. 1996) but they are hard to use as they require learning a particular specification language and understanding non-trivial abstract concepts. While some of these tools have been complemented with graphical editors, the interaction with the developer tends to be based on menus, property-sheets, and the like.

Graphic design has become an essential part of the development of GUI products (Williams 1994). Yet few if any interface construction tools support it, or they do in a very limited way. The layout facilities provided by interface builders are similar to those provided by drawing editors where groups of elements can be left-aligned, right-aligned, etc., but they do not support the way graphic designers do page design using guides and grids.

3 System Overview

HandsOn takes its presentation model from previous work in Mastermind (Castells et al. 1997). The model language has been conceived to be easily amenable to interactive specification and bridges the gap between the declarative descriptions obtained from a graphical tool and the procedural information needed to execute the interfaces described in the model (figure 1).

The HandsOn development environment integrates a graphical presentation editor and an application data builder (see figure 2). The designer constructs presentation components in the editor by selecting graphical primitives and widgets from a catalog of predefined components. Presentations are built by working on a representation that is very similar to the final resulting interface.

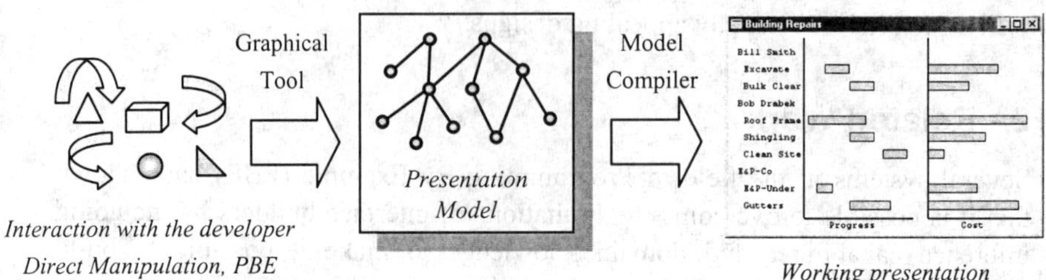

Figure 1. Model construction and interface generation

Graphic designers often work using reference lines and grids to organize page layouts. The HandsOn editor provides guides and grids for global screen space organization, not merely as passive visual aids, but in the form of design objects that interact actively with other objects of the display, playing a central role in defining the dynamic behavior of presentations. These reference lines are not present in the final interface.

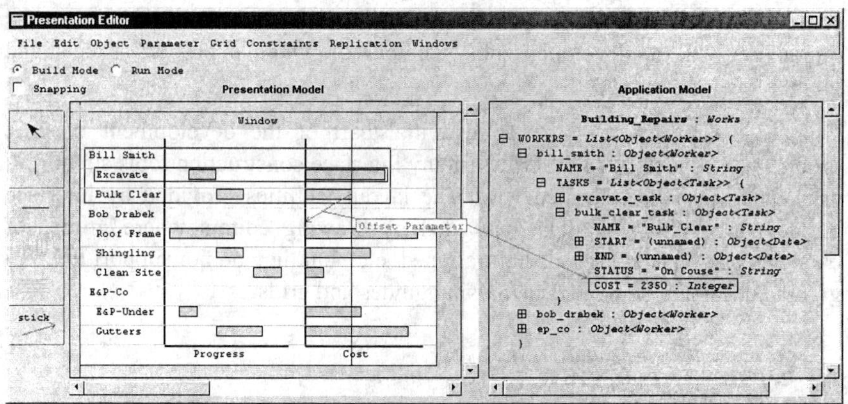

Figure 2. The HandsOn environment

The application data builder allows the designer to browse the classes defined in the application, create objects, and connect data to graphical components of the presentation, building generic presentations that display application values. This way the designer can deal with abstract presentation structures in terms of concrete display objects without introducing new unintuitive and awkward visual notations. When the resulting interface display is generated, example values give rise to placeholders that store real application data at runtime.

The application model includes standard types, sequences, object classes, and specific example objects created by the developer. Our application model requires a programming language that supports dynamic object creation and method invocation. We have used Amulet (Myers 1996) in our prototype implementation, but a more general platform like CORBA would meet our requirements as well, and would impose less restrictions on the application programmer.

4 Presentation Design by Example

The designer can define visually one-directional constraints on presentation properties by manipulating guides and by drag-and-drop within or across the presentation and the application models. By altering constrained values, the designer can modify the characteristics of the constraints demonstrationally.

The designer can also create collections of variable amounts of components associated to an application data sequence by constructing an example of data sequence and dragging one of its values onto a graphic component. Upon confirmation, the system creates one copy of the component per value contained in the data set, associating values to replicas. To modify the characteristics of the replicas only the first object needs to be manipulated. A similar mechanism can be found in Peridot (Cypher 1993).

HandsOn has facilities for easily specifying the layout of replicated components with respect to one or two grids. The designer attaches the primary replica to one or more grid lines and the system infers the positioning of the rest of the objects across the grid(s). HandsOn includes rules for the adjustment of the layout strategy (e.g. do line wrapping, avoid overlapping, etc.) by manipulating individual replicas.

In order to give the designer the appropriate level of control, the editor provides at all times complete feedback of the designer's operations as well as operations performed by the system and their effects. The information is presented in-place by introducing when necessary virtual manipulable objects (e.g. labeled rectangles for parameters, arrows for constraints) consistently integrated in the environment (i.e. they can be selected, moved around, etc.), avoiding as much as possible the use of property sheets and other extraneous elements that force the designer out from the design being constructed.

5 Conclusions

HandsOn is an attempt to combine the expressive power of a model-based system with the ease of use of a programming by example tool. Demonstrational techniques benefit from the model-based paradigm because the model provides an explicit declarative representation of the interface the tool can reason about. Balance between expressive power and ease of use is achieved by a) keeping limited the amount and complexity of model abstraction the developer has to deal with and b) mapping underlying model abstractions to visible presentation objects, reducing the mental transformation effort required from the developer between different representations of the interface.

Our work focuses on the visual part of interface design and does not currently address aspects related to the dialog with the end-user. Our plans for the near future include the extension of our work to a more comprehensive environment that incorporates user task modeling (Contreras 1997).

6 References

Castells, P., Szekely, P. & Salcher, E. (1997). Declarative Models of Presentation. *Proceedings International Conference on Intelligent Interfaces* (IUI'96, Orlando (Florida), USA, January 1997).

Contreras, J. & Saiz F. (1996). A Framework for the Automatic Generation of Software Tutoring. In Vanderdonckt, J. (Ed.): *Computer-Aided Design of User Interfaces*. Presses Universitaires de Namur.

Cypher, A. (ed.) (1993). *Watch What I Do Programming by Demonstration*. The MIT Press.

Myers, B.A. et al. (1996). *The Amulet 2.0 Reference Manual*. Carnegie Mellon University Tech. Report.

NeXT, Inc. (1990). *NeXT Interface Builder*. Palo Alto.

Szekely, P., Luo, P. & Neches, R. (1993). Beyond Interface Builders: Model-Based Interface Tools. *Proceedings International Conference on Computer-Human Interaction* (INTERCHI'93, Amsterdam, the Netherlands, April 1993).

Szekely, P., Sukaviriya, P., Castells, P., Muthukumarasamy P. & Salcher, E. (1996). Declarative Interface Models for User Interface Construction: The Mastermind Approach. In Bass, L. & Unger, C. (Eds.): *Engineering for Human-Computer Interaction*. Chapman & Hall.

Wiecha, A., Bennett, W., Boies, S., Gould, J. & Greene, S. (1990). ITS: A Tool For Rapidly Developing Interactive Applications. *ACM Transactions on Information Systems*, 8 (3), 204-236.

Williams, R. (1994). *The Non-Designer Design Book*. Berkeley: Peachpit Press Inc.

AUDIOTEST: Utilising Audio to Communicate Information in Program Debugging

Dimitrios I Rigas[1], Mark A R Kirby[2] and Daniel O'Connel[2]
[1]Department of Computer Science, University of Hull, Hull Hu6 7RX, UK
e-mail: D.Rigas@dcs.hull.ac.uk, Tel: (+44) (0) 1482 465038
[2]School of Computing, University of Huddersfield, Huddersfield, HD1 3DH,UK
e-mail: M.A.R.Kirby@hud.ac.uk, Tel: (+44) (0) 1484 472938

1 Utilising Audio to Communicate Information

This paper describes three experiments in which audio, non-speech sound and particularly structured musical stimuli, was utilised to communicate information in interfaces for program debugging. In this paper, the term audio or non-speech sound refers to structured musical elements such as pitch, rhythm, melody, tunes, harmonic sequences, and timbre. The experiments investigated pitch, tunes and timbre produced by multiple timbre synthesisers, sound cards or computational equivalents. The goal of these initial experiments was to obtain a substantiated view of the potential capabilities and limitations of using sound to communicate information in program debugging activities.

Auditory feedback can be generally divided into synthesised speech, environmental sounds (or auditory icons) and structured musical sounds (or earcons). Musical sound is a rich medium containing numerous structures introduced by musicians over many years of human evolution. Also, given that we live in an age where multimedia systems are fully capable of producing musical sounds relatively easily and effortlessly, the use of structured musical stimuli in interfaces is currently at a relatively low level. The auditory channel, as a whole, has been neglected in the development of user-interfaces, possibly because there is very little known about how humans understand and process auditory stimuli. It is not intuitively obvious how to use musical structures in interface design. Current user interfaces focus heavily on visual interaction. The consequence of this is that user interfaces have become more and more visually crowded as the user's needs to interact with the computer increase. This creates considerable difficulties for blind users. Sound has already been utilised successfully to communicate graphical information to blind users (Rigas and Alty 1997). There are also other examples of user interfaces that accommodate special needs (Edwards 1995). Other experiments in the literature suggest that the auditory channel provides alternative ways of conveying information to general user interfaces (Brewster 1994).

Sound has also been used to communicate the execution of programs (Rigas and Alty 1998) and the contents of complex databases (Rigas et al. 1997). Other work includes the CAITLIN tool. The tool assists novice programmers in debugging activities by communicating statements and structures (e.g., loops, selection) within programs (Vickers and Alty 1996). It is also believed that sound could be used as an additional communication metaphor to communicate compilation errors and abstract location of those errors in object-oriented programs. This paper examines this hypothesis by identifying some common musical structures and experimenting with these structures in a variety of circumstances with users.

2 AUDIOTEST: Communicating Program Information

Three experiments were performed using a prototype tool, called AUDIOTEST. This tool was developed in C++ under UNIX and communicates C++ programming errors with audio files. The purpose of this tool was to automate experiments in auditory information processing for program debugging activities. Structured laboratory experiments, post-experimental interviews and questionnaires were used for this empirical investigation. All of the subjects participated in the experiments described below were postgraduate students in Computer Science courses in the age range of 25 to 35. They all had knowledge of C++ and program debugging activities.

2.1 Communicating Types of Compilation Errors

This first experiment aimed to test the possibility of using audio samples to communicate compilation errors. Six audio samples were designed to communicate six types of errors. These were *omission, class, function, operator, arithmetic,* and *fatal* errors. Various tunes were chosen with an exception to the fatal error, which was communicated by a short sequence of random notes using an organ. A 'Happy Birthday' tune played by piano (electronically, see section 3) communicated omission, 'Teletubbies' (BBC, UK, 1997) for class, 'Egyptian Magician' (Blanco Negro, Spain, 1994) for function, 'My Desire' (VC: Recordings, UK) for operator, and 'Gym Tonic' (Yellow Records, France) for arithmetic errors. All audio samples were between 6 and 8 seconds and 16 to 25 notes. Seven subjects participated in this experiment by using the AUDIOTEST tool (one subject at a time). The tool trained subjects by presenting each sound sample and the associated type of error communicated five times. Subjects were then presented with samples in a random order and they were requested to recall the error that the auditory stimuli communicated.

Successful recall rates of compilation errors were 71% for the operator error, 100% for the arithmetic and class errors, 85.7% for the function error, 42.8% for the fatal error, and 92.8% for the omission error. Recall performance for each subject was 84.6% for S1 and S3 (S denotes subject), 92.3% for S2, 100% for S4, 92.3% for S5 and S7, and 61.5% for S6. Statistical significance was observed in five out of the six types of errors communicated (operator χ^2=2.57, critical value 1.64, $p< 0.10$, arithmetic χ^2=21, critical value 10.38, $p<0.0005$, function χ^2=7.14, critical value 6.64, $p<0.005$, omission χ^2=10.28, critical value 6.64, $p<0.005$, class χ^2=14, critical value 10.83, $p<0.0005$). Fatal error failed to reach statistical significance (χ^2= 0.28, critical value 1.64, $p<0.10$). The low performance of the fatal error indicates that the short sequence of random notes was not so memorable when it is compared with the scores of the other errors, which were constructed using sequences of tunes. Subjects reported memory problems in identifying the fatal error.

2.2 Communicating Abstract Location of Errors within a Program

A second experiment was performed in an attempt to communicate *abstract locations* of errors within a program by using a tune with a number of variations. Six subjects were requested to consider the tune as a communication metaphor that conveys information about the compilation and execution of a program. Initially, they were presented with the original tune three times. During the experiment, variations (pitch was changed at particular parts) of the original tune were presented and subjects were requested to identify the altered parts of the tune. This experiment examines if these altered parts of the tune can communicate the *abstract location* of errors encountered within a program during compilation or execution. A tune sequence of 37 seconds was used. Six subjects (from the ones who took part in the previous experiment) were tested using the AUDIOTEST tool. Five altered versions of the tune communicated 126 locations of compilation errors within various programs.

Individual subject performance, in terms of subjects identifying abstract locations of the communicated errors, was 33.3% for S1 (S1 denotes subject 1), 66.6% for S2, 95.2% for S3, 76.1% for S4, 57.1% for S5, and 52.3% for S6. Overall, a statistical significance was reached (χ^2=8.12, critical value 6.64, $p<0.005$). In post-experimental interviews, subjects reported that they could easily identify abstract locations of errors when there was a large pitch difference (more than an octave) at various parts within the tune. They also remarked that some confusion occurred in understanding whether the tone changed naturally or artificially (i.e., to communicate an error). For program debugging activities, however, the type as well as the location of errors needs to be communicated.

2.3 Communicating Type and Abstract Location of Errors within a Program

A third experiment was also performed in which the same tune was used, as in section 2.2, but this time different musical instruments were introduced to play small parts of the tune in an attempt to communicate *type* and *location* of errors within a program. Omission was communicated using a piano, class using a electric guitar, function using a harpsichord, operator using a wooden glockenspiel, arithmetic using a celeste, and fatal using an organ. Subjects were explained that the tune can be regarded as a metaphor of a program and the changes of instruments within the tune can be regarded as errors of the program. In this manner, each musical instrument communicated a particular type of error. The position at which a new instrument was introduced within the tune communicated the abstract location of the error in the program.

The results of this experiment, for each individual subject, were 91.6% for S1 (S denotes subject), 75% for S2, 83.3% for S3, 66.6% for S4, 91.6% for S5, and 80% for S6. The overall average recognition rate across all subjects was 81.3%. A statistical significance was reached for the overall result (χ^2=23.4, critical value 10.38, p<0.0005). In a detailed analysis of the results, it was observed that some subjects confused instruments deriving from the same families. This re-enforces previous findings in timbre selection (Brewster 1994, Rigas 1996).

3 Discussion and Conclusions

This work indicates that there is a prima facie case for utilising sound. Subjects, with no special musical knowledge or ability, were in a position to extract information and meaningfully relate compilation errors and abstract locations to samples of auditory stimuli (pitch, tunes and timbre). This could potentially prove to be a useful interface mechanism. For example, software engineers could potentially take advantage by an audio-visual feedback in software testing. Users with special needs, such as the visually impaired, could also benefit. Auditory user interfaces could offer a promising alternative for blind computer users who currently have considerable difficulty in using computer systems with their emphasis on visual interfaces could also involve speech and environmental sounds.

Results indicate that random auditory-musical designs are less memorable than tunes, as indicated in the low recall rate of the fatal error (see section 2.1). A combination of musical parameters and structures also appears to be useful, as indicated in the use of a tune and various musical instruments to communicate more information (see section 2.2). The use of a tune as a communication metaphor was found to be useful. Users identify pitch changes, within a tune, better when it is across octaves than within an octave.

Intensive memory and concentration demands upon subjects were also observed. There were instances (see section 2) where the auditory stimuli failed to offer the same degree of perceptual detail, as one would have expected by the visual medium. This has also been observed in other experiments (Rigas 1996). A combination of audio and visual feedback could alleviate some of these difficulties. Following this initial investigation, larger experiments are currently being performed using visual stimuli (e.g., graphics, colours, animation) and various forms of auditory stimuli (e.g., structured musical stimuli, speech and environmental sounds) as communication metaphors either on its own or in synergy.

4 References

Brewster, S. A. (1994). Providing a Structured Method for Integrating Non-Speech Audio into Human-Computer Interfaces. PhD Thesis, University of York, England, UK.

Edwards, A. D. N. (ed.) (1995). Extra-Ordinary Human-Computer Interaction. Cambridge: University Press.

Rigas, D., Alty, J. L. & Long, F. W. (1997). Can Music Support Interfaces to Complex Databases? In EUROMICRO-97, New Frontiers of Information Technology, pp. 78-84. IEEE Computer Society Press.

Rigas, D. & Alty J. L. (1997). The Use of Music in a Graphical Interface for the Visually Impaired. In Howard, S. Hammond, J. & Lindgaard K. (Eds.): *Human-Computer Interaction, INTERACT'97: Proc. of the 5th IFP Conference on Human-Computer Interaction*, pp. 228-235. Sydney, Australia: Chapman and Hall.

Rigas, D. & Alty, J. L. (1998). Using Sound to Communicate Program Execution. In EUROMICRO-98, Engineering Systems and Software for the Next Decade, pp. 625-632, IEEE Computer Society Press.

Rigas, D. (1996). Guidelines for Auditory Interface Design: An empirical Investigation. PhD Thesis. Loughborough University, England, UK.

Vickers, P. & Alty, J. L. (1996). CAITLIN: A Musical Program Auralisation Tool to Assist Novice Programmers with Debugging. In Frysinger, S. & Kramer, G. (eds.), Proc. of ICAD'96, pp. 17-23.

A Flow-chart Based Learning System for Computer Programming

Kuo-Hung Huang,[1] Kuohua Wang,[2] S. Y. Chiu[2]
[1] Hung Kuang Institute of Technology, Taiwan; [2] National Changhua University of Education, Taiwan

1 Introduction

After years of action research on computer programming instruction, we found that the novice programmers encountered the difficulties of converting their plans into codes when solving problems. Although they know how to solve the problems, most of the beginners are not able to use the programming language to express their plans. It seems that there is a gap between plans and code in computer language. We decide to develop a computer-based scaffolding system to help the students to cross the gap.

2 Design Strategies

Many research results showed, using flow charts to teach computer programming would help the novices to concentrate on the program design without knowing the syntactic details of the language (Dalbey and Linn 1985; Scanlan and Clark 1988). Novices can easily use flow charts to express their naive plans. However, the task of implementing their plans in computer language will confront the novice programmers with an enormous gap (Bonar and Liffick 1990). Our flow-chart based learning system is aimed to bridge the gap. Combining the flow charts and graphic interface tools, we design a visual learning system to enhance the beginners' learning by improving the communication qualities and reducing the cognitive loads (Baroth and Hartsough 1994).

We choose C language as the target language of our learning system. Based on the outcome of several empirical studies (Eisenberg 1995; McCalla 1992), we adopt the following design strategies:

2.1 Object-Oriented Design

Since the functions and features of this learning system are upgrading frequently, we decide to adopt the object-oriented techniques to attain the flexibility on system development. Object-oriented technology is used in the phase of system design and implementation. Icons, flow charts, variables, user interface, and code translation system are designed as objects.

This system is developed in C++ language. We use Borland C++ Builder to develop the system. Because of the RAD feature, the BCB compiler makes our graphic user interface design more efficient.

2.2 Present Powerful Ideas Only

Because the anticipated users of this system are novice programmers, we decide to implement some important features of the computer language only. For example, loop, selection, variables, and function/procedure, are the powerful ideas of the procedural languages. We exclude some important features of C language such as pointers, recursion, and struct/union.

2.3 Software Visualization

Software visualization has had a major impact on software development and education(Brown 1987; Mulholland and Eisenstadt 1998). Students can learn by experimenting and exploring this system when the immediate feedback is provided. We design debugger-like functions to help students visualize the execution of their programs. Through step-by-step execution, students can observe the changing values of the variables.

2.4 Code Translation

The goal of this system is to help students learn how to program in C language. We encourage students to use flow charts implementing their plans. Then, this system will translate the flow charts into codes in C language. By mapping the flow charts and the codes, students can gradually acquire the syntactic and semantic knowledge of C language.

3 System Overview

Flow-chart based learning system is a windows-platform application with user-friendly interface. This system consists of three components(see figure 1):

1. Flow chart designing tool. A user-friendly graphic tool for students to design flow charts. By point-and-click, students can easily create their flow charts and variables.

2. Flow chart execution tool. A system provides one-step and step-by-step execution. While running, the values of the variables declared in the programs will be dynamically displayed. Students can use this tool to test their plans and flow charts.

3. Code translator. A system converts the flow charts into the computer programs in C language. It provides the user with an opportunity to compare and map the flow chats and computer codes.

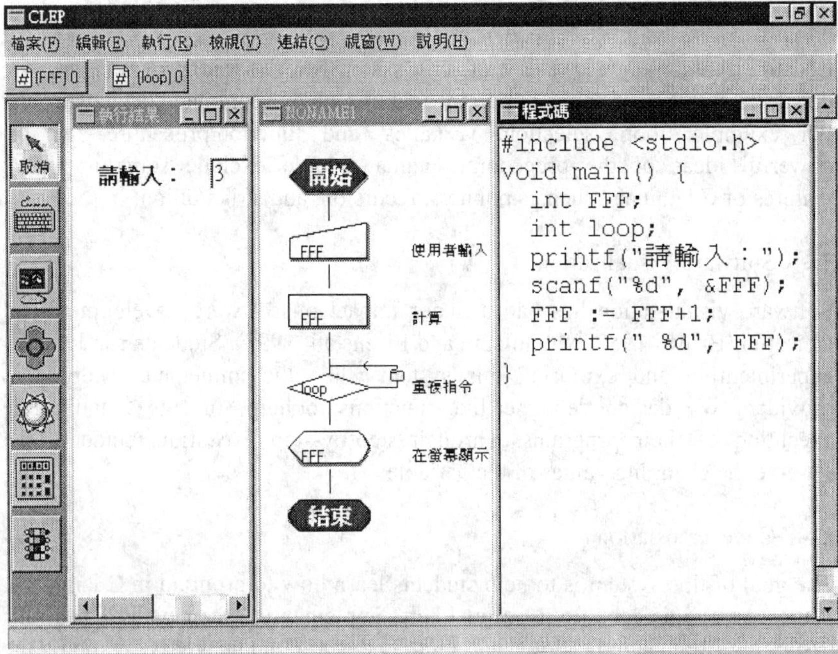

Figure 1. The Flow-chart based Learning System

According to their informal plans, the novices can use mouse to produce flow charts simply by clicking on icons and entering values for variables. The users can then run the flow charts to see if the results are correct. By tracing and mapping the codes, generated by this system, to their plans, the novices will gradually learn how to implement their plans in computer language. Eventually, the novices can develope the abilities of self-explanation.

4 Learning with the System

While the learning system has been developing, we conduct case studies to gather information for further improvement. Using qualitative research method, we intensively observe five novice programmers when they use this learning system to solve problems. These novices are computer-major students in a college. They have taken a course of the introduction to programming last semester and four of them failed the course. At first, we ask them to solve problems individually by writing programs in C language. Then, we teach them how to use this system to solve problems. Their operations on the computer screens are recorded with VCR. Finally, they are given a different set of problems and asked to write programs.

By examining their codes, observing the videotapes, and interviewing with the students, we have found several facts:

1. For those students who were unconfident in their computer programming abilities or intimidated by computer programming, this system has helped them change their views on computer programming.
2. Some students tended to solve problems by assembling segments of computer codes in their memory. This system resteered their views toward learning programming. They became aware that they could learn how to program by exploring and manipulating, not merely memorizing the computer programs.
3. Some students tended to quit when encountering difficult problems even with this system. The excuse is that for them it is not worthwhile to 'think hard'.

5 Discussion

Although we are still not sure how much this system will help the novices learn to program, we can see the changes the system has brought up. At least the students reacted positively toward this learning system. When using this learning system in the class, we found that the learning atmosphere of the classroom changed dramatically.

We used this system in a course of computer programming. This class of students, major in computer science, had variety of backgrounds. Some of them had years of experience of coding in BASIC, Cobol, or Clipper. But most of them are not familiar with programming. For the beginners, we encouraged them to elaborate on the solution and produced the flow chart with this system. After testing the flow chart to verify the correctness, they then started to comprehend the generated codes. During the process of code comprehension, they would solicit help from the teacher or the peers whenever they could not master some of the codes.

Equipped with this system, the learning environment enables students with different experiences and abilities help one another to explore actively, and teacher no longer dominates the class and changes his role into a counselor by giving advice and guidance when students need help. For future improvement, we will expand this learning system to include more tools of diversity. Students with different experiences and backgrounds can access to the more appropriate tool for their learning styles and abilities.

6 References

Baroth, E. & Hartsough, C. (1994). Visual Programming in the Real World. In Burnett, Margaret, Goldberg, Adele, & Lewis, Ted (Eds.): *Visual Object-Oriented Programming*, pp. 21-42. Greenwich: Manning.

Bonar, J., & Liffick, B. (1990). Programming Languages for Novices. In Chang, S. (Ed.): *Principles of Visual Programming Systems,* pp. 326-366. Englewood Cliffs, NJ: Prentice-Hall.

Brown, M. H. (1987). *Algorithm Animation.* Cambridge: The MIT Press.

Dalbey, J. & Linn, M. C. (1986). Cognitive Consequences of Programming: Augmentations to BASIC Instruction. *Journal of Educational Computing Research*, 2(1), 75-93.

Eisenberg, M. (1995). Creating Software Applications for Children: Some Thoughts About Design. In diSessa, A. A., Hoyles, C., Noss, R. & Edwards, L. D. (Eds.): *Computers & Exploratory Learning,* pp. 175-196. Berlin:Springer.

McCalla, G. I.(1992). The Search for Adaptability, Flexibility, and Individualization: Approaches to Curriculum in Intelligent Tutoring Systems. In Jones, M. & Winne, P. (Eds.): *Adaptive Learning Environments*, pp. 91-121. Berlin:Springer-Verlag.

Mulholland, P. & Eisenstadt, M. (1997). Using Software to Teach Computer Programming: Past, Present and Future. In Stasko, J., Domingue, J., Brown, M. H. & Price, B. A. (Eds.): Software Visualization, pp. 399-408. Cambridge: The MIT Press.

Scanlan, D. & Clark, L. (1988). An Empirical Investigation of Flowchart Preference. *Journal of Computers in Mathematics & Science Teaching*, 8(2), 56-64.

Automatic Construction of Intelligent Diagrammatic Environments

Bernd Meyer[1], Hubert Zweckstetter[1], Luis Mandel[2], Zoltan Gassmann[1]
[1]Univ. of Munich, Dept. of Computer Science, Oettingenstr. 67, 80538 Munich
[2]FAST e.V., Arabellastr. 17, 81925 Munich, Germany
contact email: bernd.meyer@acm.org

1 Introduction

Graphical user interfaces have become an integral part of almost every modern application type and it can be claimed that they are among the driving forces that have made the computer accessible to non-expert users. However, comparing the use of graphics in existent user interfaces with that in non-computer-based work, the inadequacy of standard GUIs for complex visual communication is revealed: Most GUIs are still WIMP interfaces centered around such simple interaction devices like icons, buttons, menus or image maps. On the contrary, in non-computer-based work rich and highly structured graphical notations prevail. There are diagrammatic languages in almost every technical discipline, for example circuit diagrams, architectural floor plans or chemical formulas, and modern software engineering is embracing all kinds of diagrammatic specification methods. Likewise, non-technical fields use their own well-established diagrammatic systems, for example choreography notation. But even in such application domains where diagrammatic notations are a natural element of discourse and where their meaning is well-understood, they are still rarely utilized to their full extent in the human-computer interface. It is hardly ever possible for the user to communicate with a computer just by sketching an annotated diagram like two human experts would do in their discourse. This poor integration of richly structured visual communication into graphical user interfaces is mainly due to a lack of method and tool support for the implementation of the required interfaces. While countless specification mechanisms and interface builders for WIMP interfaces exist, there are almost no tools and only few specification formalisms for building intelligent graphical interfaces for such richly structured diagrammatic notations.

This work is supported by DFG Grant Wi-841 and Bayerische Forschungsstiftung.

If we want to unleash the full power of diagram languages, we have to turn to a syntax and semantics-based high-level approach for the implementation of diagrammatic front-ends.

In the following we present Recopla, a meta-environment for the automatic construction of diagram environments from declarative high-level specifications.

From a specification of appearance, syntax and semantics of a diagram language Recopla generates a specialized interface for the specified diagram type. These automatically constructed front-ends are capable of checking the correctness of a given diagram and can interpret the defined diagram types and translate them into a representation suitable for processing by the back-end application. The generated front-ends are self-contained Java applications that can flexibly communicate with arbitrary back-end applications via dedicated bidirectional Internet communication channels. Due to Recopla's non-monotonic incremental interpretation capabilities, these interfaces do not limit the user to syntax-directed interaction but allow natural interaction with the diagram. In contrast to virtually all other systems, Recopla's interpretation component incrementally maintains a consistent interpretation even under object deletion and other arbitrary modifications of the diagram.

2 Building Diagrammatic Interfaces from Declarative Specifications with Recopla

Our meta-environment for the automatic construction of intelligent diagrammatic interfaces from declarative formal specifications consists of a syntax-based high-level interpretation component (Meyer and Zweckstetter 1998) on top of the generic editor-builder Recopla (Gassmann 1998).

The syntax and semantics specification of the diagrammatic language is achieved by using attributed grammars in the form of conditional relational grammars (CRG, Meyer and Zweckstetter 1998). CRGs are a multidimensional grammar formalism that inherits from both relational grammars and constraint multiset grammars (Marriott, Meyer and Wittenburg 1998). In addition to specifying the syntax and semantics (i.e. the interpretation) of diagrammatic notations, conditional relational grammars also provide mechanisms to specify interpretation-based modal interaction, so that immediate semantic feedback can be given to the user without requiring the intervention of the back-end.

The appearance of the basic graphical entities, i.e. the visual vocabulary (such as icons, line types, etc.), that may be used in an editor instance is specified interactively on the meta-level by drawing prototype entities with Recopla's object-oriented graphics editor. The syntax, i.e. the permissible ways to arrange these graphical entities spatially and topologically into valid diagrams, is defined with a CRG. The interpretation and translation of diagrams is specified with the same formalism by using attributed productions and/or by attaching actions to productions. As a toy example consider the following production which is taken from a simplified grammar for state transition diagrams:

```
Transition --> Arrow, Label ;    % where
  exists(State_1, State_2),      % such that
     starts_at(State_1, Arrow),
     ends_at(State_2, Arrow),
     close_to(Label, Arrow) ;    % and set
  Transition.condition = Label.string,
  Transition.from = State_1.name,
  Transition.to = State_2.name.
```

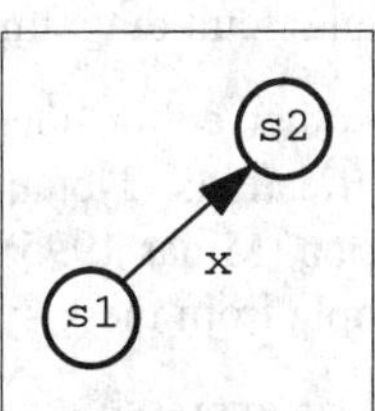

CRGs allow the usage of fully context-sensitive productions and of arbitrary user-defined predicates that are evaluated as conditions on the objects' attributes. A CRG is essentially processed in bottom-up fashion by simultaneously rewriting a set of graphical objects and a set of spatial relations between these objects.

In interactive systems it is highly desirable to support immediate semantic feedback from the interpretation without restricting the admissible user-interactions. For systems working on the basis of a formal syntax definition this is usually a difficult problem since they have to address the issue of incremental parsing and, more importantly, of non-monotonic incremental interpretation if they want to support unrestricted interaction.

In Recopla's interpretation component this is solved by viewing parsing as logical deduction: The grammar execution can essentially be understood as a bottom-up derivation of definite clauses with set-valued attributes. This perspective allows us to employ deduction techniques developed in the area of Truth Maintenance Systems (Doyle 1979, 1981; DeKleer 1985, 1989). In this way a consistent interpretation can incrementally be maintained when the diagram is changing without that a complete re-interpretation has to be performed from scratch.

Despite the rigorously specified syntax this schema frees the user from the restraints of syntax-directed editing. If required, immediate feedback during editing as well as modal interaction situations can be specified in a flexible manner using one of two mechanisms: (a) In a declarative manner by grouping the productions according to editing modes and activating/deactivating groups depending on the current context, (b) in a procedural manner by attaching arbitrary user-defined actions to the productions.

Since a vast number of diagrammatic notations used in practice essentially have a (hierarchical) graph as their underlying backbone structure, Recopla is designed to offer special support for the definition of and interaction with graph structures. However, the system also supports editing and interpretation of more general, non-graph-like diagrammatic notations.

3 Evaluation and Future Plans

Recopla was successfully used to build editors for a number of standard notations, among these animated Petri nets, object-oriented class diagrams, timing diagrams, digital circuit diagrams, extended flow charts and even interactive board games.

Few other editor construction toolkits exist that support declarative high-level specifications. Notable exceptions are Penguins (Chok and Marriott 1998), DiaGen (Minas 1995) and GenEd (Haarslev 1998a). The features distinguishing Recopla from these systems are:

1. an expressive and fully context-sensitive specification formalism,
2. the capability to handle non-monotonic interaction with a fully incremental interpretation method that supports instantaneous semantic feedback,
3. support for flexible specification of modal interaction,
4. built-in support for hierarchical graph structures,
5. generation of portable, monolithic Java-based stand-alone editors.

While all of the above mentioned systems support some of these features, none of them supports all. Penguins additionally performs a form of error correction, a capability not yet offered by Recopla's interpretation component. While the formalism underlying GenEd could in principle support error correction, the system does not yet utilize these capabilities (Haarslev 1998b).

Current development plans for future versions of Recopla include automatic layout mechanisms, an animation specification language such as Villon (Meyer 1997), and the tighter integration into web-based settings by generating signed applets instead of applications.

	Portable Stand-alone Editors	Interaction Specification	full Context-Sensitivity	incremental	non-monotonic	Error Correction	Graph Support
Recopla	+	+	+	+	+		+
Penguins			+	+	+	+	
DiaGen		+		+			
GenEd			+	+	+	O	

4 References

Chok, S.S., Marriott, K. (1998). Constructing User Interfaces for Pen-based Computers. In *Formalizing Reasoning with Visual and Diagrammatic Representations* (FRVDR '98: AAAI Fall Symposium, Orlando, USA, October 23-25, 1998), pp. 67-77. Menlo Park: AAAI Press.

De Kleer, J. (1985). An Assumption-based Truth Maintenance System. *Artificial Intelligence*, 26(1): 127-162.

De Kleer, J. (1989). A Comparison of Assumption-Based Truth Maintenance and Constraint Satisfaction. In Proc. *International Joint Conference on Artificial Intelligence* (IJCAI 1989, Detroit, USA), pp. 290-296.

Doyle, J. (1979). A Truth Maintenance System. *Artificial Intelligence*, 12(3):231-272.

Doyle, J. (1981). A Truth Maintenance System. In Webber, B., Nilson, N. (Eds.): *Readings in Artificial Intelligence*, pp. 496-516. Palo Alto: Tioga.

Gassmann, Z. (1998). *A Graphic Meta-Editor for the Generation of Syntax-Oriented Graph Editors*. Diplomarbeit, University of Munich.

Haarslev, V. (1998a). A Fully Formalized Theory for Describing Visual Notations. In Marriott, K., Meyer, B. (Eds.): *Visual Language Theory*, pp. 261-292. New York: Springer Verlag.

Haarslev, V. (1998b). *Personal Communication*. September 1998.

Marriott, K., Meyer, B., Wittenburg, K. (1998). A Survey of Visual Language Specification and Recognition. In Marriott, K., Meyer, B. (Eds.): *Visual Language Theory*, pp. 5-85. New York: Springer Verlag.

Meyer, B., Zweckstetter, H. (1998). Interpretation of Visual Notations in the Recopla Editor Generator. In *Formalizing Reasoning with Visual and Diagrammatic Representations* (FRVDR '98: AAAI Fall Symposium, Orlando, USA, October 23-25, 1998), pp. 107. Menlo Park: AAAI Press.

Meyer, B. (1997). Formalization of Visual Mathematical Notations. In *Reasoning with Diagrammatic Representations* (DR-II: AAAI Fall Symposium, Cambridge, USA, November 8-10, 1997), pp. 58-68. Menlo Park: AAAI Press.

Minas, M. (1995). DiaGen: A Generator for Diagram Editors Providing Direct Manipulation and Execution of Diagrams. In Proc. *IEEE Symposium on Visual Languages* (VL '95, Darmstadt, Germany, September 5-9), pp. 203-210. Los Alamitos: IEEE Press.

Integrating Perspectives for UI Design

Chris Stary
University of Linz, Department of Business Information Systems,
Communications Engineering, Freistädterstr. 315, A-4040 Linz

1 Introduction

Since human-computer interaction does not occur in purely technical, organizational or social systems, but socio-technical systems (being part of work systems), design of user interfaces as well as their evaluation have to reflect these dimensions. This paper emphasizes the issue of mutual integration of perspectives, namely migrating design elements that have been isolated for several development purposes, such as semantic data descriptions, and have to be related to interaction elements. We first reveal the understanding of perspectives in general before we deal with user interface design and detail the perspectives involved in human-computer interaction. We proceed with the issue of integration, discussing the roles of concepts, procedures, communication and notation in the course of development. We then give a case from a project where technically-driven development has led to organizational deficiencies, as well as cognitive and social problems.

Perspectives relate to the understanding of the environment by an individual. It means 'to look through' either as a standpoint or as a process, namely to express a way observations are interpreted. It may also denote some kind of filter. In this case, the concept operates as a selection of features when referring to an observation or phenomenon (Nygaard et al., 1989). Perspective as a selection is a commonly used concept in user interface development, as can be easily demonstrated through the upraise of object-oriented systems. Here, filtering of information is performed for several purposes, such as hiding, coupling, generalizing, and aggregating.

For the integration of isolated elements we have to be aware that the selection of particular properties of an object or phenomenon influences the ''choice of operative cognitions'' (Nygaard et al., 1989, p. 382), i.e. the way we think about it and represent the object or the phenomenon, e.g., through specifying dialog models. These cognitions are relevant for the interpretation (also considered an interpretation of perspective) of the selected properties, and finally lead to the definition of standpoints (again, an interpretation of perspective). The latter influences ''the choice of position with respect to the situation of phenomenon in question'' (ibid.), thus, leading to schools of thoughts or paradigms.

Long (1986) urged developers to take into account cognitive aspects, task requirements, organization and design of work in the course of development of usable systems. From then on, the context of interactive technology has been emphasized several times for development activities, e.g., Johnson (1992). However, in-depth integration has not been achieved yet, due to a variety of problems, e.g., Rodriguez et al. (1997), that stem from coupling user interface specifications with software design processes, requirements, and methods. The consequences are disrupted procedures, misunderstandings among developers along the development phases, and a significant loss of context from analysis to implementation. As a result, interfaces might be inconsistent with the requirements, incomplete with respect to task accomplishment, and/or inaccurate in their presentation of data and functions, as well as with respect to navigation. Hence, any integration of perspectives has to deal with (i) the methodology to follow for development, (ii) the representation of development knowledge, and with (iii) the language used for representation and communication.

2 A Case

The case study reveals deficiencies that are caused by technology-driven developments, neglecting the organizational, cognitive and social context of interactive work. The study has been performed on behalf of the Austrian Ministry of Public Economics and Transportation (under contract GZ 527.023/-V/A/61/95) in the field of computer support for supply chain management. Recent developments in that field are characterized through:

- Workplaces involved in transportation are more and more industrially organized. This fact is reflected through the intensity of work, such as route planning becoming part of the actual transport operation and supply chains (Sydow, 1992).
- The use of information and communication technologies (ICTs) in the field of transportation and logistics has undergone significant changes, such as networking previously de-coupled business processes along transportation chains. Networking supports (i) globalization and expansion across different types of businesses, and (ii) the diffusion of ICT applications into previously isolated operations. For instance, traditional transport providers offer commissioning, information processing, packing, finishing, quality control, recycling, and factoring of goods, in addition to transportation. These fundamental changes in work contents require novel qualifications of the employees in this business (Alleweldt et al., 1990).
- In addition, employees have to be empowered with social abilities, such as customer-oriented behavior (Nitzinger, 1984).

As a consequence, workplaces along logistic processes heavily rely on context-sensitive and user-conform computer support. According to Danckwerts (1991) current workplaces in the field of logistics can be categorized as follows: Dispatching, Transport, Handling. These activities are the major ones performed in supply chain management. Dispatching comprises the entire range of planning, scheduling, and organizing the transport of goods. In particular, it focuses on commissioning of goods and carriers, planning and scheduling of routes, managing carriers, optimization, monitoring, and the communication with all organizational units and workers involved. Information system support is considered to be crucial for dispatching. This statement holds for both, namely for automated decision or planning support, and direct advice for workers, in case of manual dispatching.

Handling and transporting goods do not only concern (re)loading and packing with respect to carriers and storage anymore, but also steering and control activities. This enrichment of work requires a higher level of skills. In particular, handling of goods comprises the acquisition of the raw data that are further processed in the course of dispatching and transportation.

The evaluation of the technology delivered for supply chain management in the respective project has been performed with the EU-CON technique (Stary et al., 1998) investigating a SME in the field of waste management. Due to space limits we can only give the results for one type of workplace, namely dispatching: The combined ***dispatching/customer service workplace*** has been affected through a novel supply chain management software in a mediate way: Although the new software is now in operation for dispatching, the data for customer service might not be available in time and in a consistent way. Required updates are not performed in time, and the accessibility of data is not given, as required for completing customer service tasks. As a consequence, customer service has to contact several co-workers to receive the required data, otherwise the integrity of data cannot be ensured.

Invoice management, the second main task to be performed at the customer service work place, is hindered the same way like the customer service itself. In case, data sets that are required for invoice processing, are missing, the dispatcher has (i) to switch between software applications, and (ii) to contact other workers, in order to proceed with a now complete set of invoice data. Overall, the organization of work has been affected through the software in a way, that

(i) required inputs for work are not provided accurately - they might not be in time, correct, or complete,
(ii) task-related contacts among workers have increased due to technological deficiencies,

(iii) the provided feedback loops do not enable immediate quality checks.

From the perspective of customer service and dispatching several consequences can be drawn: Services cannot be delivered in time; Customer satisfaction will be influenced by the delays in providing services; The social setting will be affected due to the increase of task-related contacts.

3 Conclusions

Technology-driven development might affect those workplaces that are not directly concerned with the introduction of computer support. However, these implications might lead to major perturbations of the entire set of business processes, as the interviewed dispatcher puts it: 'Even when everything works [from the technological perspective], business does not proceed as smoothly as it did before.'

The enrichment or enlargement of work as well as the methodologies used for software development require rethinking, in terms of considering organizational, social, and cognitive aspects of computer-supported work. Participatory design, being a communicational and methodological issue, as well as proper skill development of employees seem to influence usability and finally, customer satisfaction. Continuous organizational development as well as human resource management provide the proper context of human-centered technology development. An integration of perspectives has to be performed at the methodological, representational and notational level. It requires:

- **Transparency of processes have to be given prior to software development.**
  In case process models are missing in the course of technical and organizational developments, the interfaces between the activities and components are not very likely to work as smoothly as supposed. For instance, in the field of gas logistics, gas stations have to be provided with gas according to round trips instead of single orders. Hence, the logistics process is not initiated by orders as in other fields, but rather by dispatching activities.
- **Participation of end users is mandatory for software development.**
  Workers possess the competitive knowledge that helps in producing adequate support. End user participation is a way to keep workers involved from the beginning of a project until user acceptance can be provided. Front line users do not only adjust knowledge in the phases of analysis and design, but also provide early feedback in the course of implementation or prototyping. For instance, in the area of frozen good storage special modes and devices have to be developed for data input due to the particular clothing of workers. In order to capture that knowledge (leading to particular

requirements for interaction) workers should become part of the development process.

- **Qualification and training for workers are required for the skilled use of ICTs.**
  Due to the complexity stemming from integrating previously isolated processes and activities workers require thorough knowledge not only about the models represented in the computer systems, but also how to use them via technical interfaces and organizational entry points. For instance, job enlargement due to the reduction of manual work in the field of storage management is reflected through complex human-computer interaction which requires skill management by human resource manager instead of try-and-error-procedures by end users.

References

Alleweldt, K.; Heger-Danckwerts, U.: Logistik und Gütertransport - Chancen für qualifizierte Arbeit?, in: Die Mitbestimmung, 37 (6/7), 1990, pp. 417-420.

Danckwerts, D.: Logistik und Arbeit im Gütertransportwesen. Rahmenbedingungen, Verlaufsformen und soziale Folgen der Rationalisierung, Westdeutscher Verlag, Opladen 1991.

Johnson, P.: Human-Computer Interaction, McGraw Hill, London, 1992.

Long, J.B.: Designing for Usability, in: People and Computers: Designing for Usability, eds: Harrison, M.D.; Monk, A.F., pp. 3-23, Cambridge University Press, 1986.

Nitzinger, U.: Verbesserung der Zusammenarbeit an Logistikschnittstellen zwischen Verlader und Spediteur, in: ZfB-Ergänzungsheft 2/84, p. 131-149.

Nygaard, K.; Sorgaard, P.: The Perspective Concept in Informatics, in: Computers and Democracy, eds: Bjerknes, G.; Ehn, P.; Kyng, M., Avebury, Aldershot, 1989.

Rodriguez, F.G.; Scapin, D.L.: Editing MAD* Task Descriptions for Specifying User Interfaces, at Both the Semantic and Presentation Level, in: Proceedings DSVIS'97, pp. 215-225.

Stary, Ch.; Riesenecker-Caba, Th.: WP 3600 - Final Report, Umbrella Project LLZ (in German), Austrian Ministry of Economic Affairs, 1997.

Stary, Ch.; Riesenecker-Caba, Th.; Flecker, J.: The EU State of the Art - Implementing the EU-Directive on Man-Machine Communication, in: Behavior and Information Technology, Vol. 17, 1998.

Sydow, J.: Strategische Netzwerke und Personalmanagement, in: Hans-Böckler-Stiftung (Hrsg.), Transport und Logistik. Neue Unternehmenskonzepte und Mitbestimmung, Düsseldorf 1992, p. 31-36.

THE TASK OF INTEGRATING PERSPECTIVES: LESSONS LEARNT FROM EVALUATION

Alexandra Totter
University of Linz, Department of Business Information Systems, Communications Engineering, Freistädterstr. 315, A-4040 Linz

1 Introduction

As interactive systems are socio-technical systems, they integrate technical components, namely hard- and software, with a social component, namely the human user. The latter is assumed to be embedded into an organizational setting, such as the work environment (Stary, 1996). Given this characteristic feature the design as well as the evaluation of interactive systems have to take into account several perspectives, according to Totter et al. (1997), and Frese et al. (1988): (i) The *technical perspective,* focusing on interaction modalities, media and styles as well as the (combinations of) software and hardware used in the accomplishment of tasks; (ii) The *organizational perspective,* concentrating on the global and individual organization of work tasks, (iii) The *cognitive perspective,* focusing on the individual needs and performance of the end-user.

These perspectives have also been addressed in the field of usability engineering. Usability has been referred to as the quality of a product in use and is defined in the ISO 9241-11 standard (ISO, 1997) as follows: „Usability of a product is the extent to which the product can be used by ***specified users*** to achieve ***specified goals*** with effectiveness, efficiency, and satisfaction in a ***specified context of use***." This explanation identifies usability as a complex, multidimensional concept, requiring the migration of cognitive components (specified users), the organization of the environment (specified as a set of goals in a specified context of use), such as the workplace, technical features (the product), and their intertwining (interaction).

Hence, for each technique proposed for evaluation it has to be questioned in how far it takes into account different perspectives and their mutual tuning at a user interface. In this paper, a comparative study in that respect with a sample of selected evaluation techniques is presented: Firstly, it has to be analyzed how detailed the procedures guide their users along the process of evaluation, in order to identify categories of elements under evaluation. Secondly, it has to be identified the extent to which each perspective (cognitive, organizational and technical) and interrelationships are considered for evaluation.

The next section will briefly present the evaluation techniques under investigation. In the subsequent section the analysis is presented. The results are given by listing supported and unsupported perspectives. The list enables for each of the investigated techniques to propose improvements to become a multi-perspective technique measuring the usability of interactive systems.

2 The Techniques Under Investigation

The selection of the techniques has been performed according to their objective, namely measuring usability, and their availability. The selected techniques for usability evaluation were:

Cognitive Walkthrough (Lewis & Wharton, 1997): This instrument focuses on ***evaluating a design for ease of learning***. It attempts to provide a detailed, step by step evaluation of the user's interaction with an interface in the process of carrying out a specific task. The method is performed by analysts and reflects the analysts´ judgments. They identify problems by tracing the likely mental processes of a hypothetical user.

EU-Con (Stary et al., 1998): This technique has been developed to implement the ***EU-directive 90/270/EEC*** (EU, 1990) on man-machine communication. The procedure consists of four phases: preparations, execution, tuning, and rework. The evaluation process is supported by a questionnaire to be filled in by users, a guide for evaluation and a handbook for evaluation and engineering.

ISO 9241 evaluator (Oppermann et al., 1997): This approach is an example of a expert-based evaluation method developed for ***conformance testing with the ISO 9241 standard part 10-17***. The primary target of the evaluation with the ISO 9241 evaluator is the user interface of a software system.

Heuristic Evaluation (Nielsen, 1994): This instrument is a usability engineering method for locating usability problems in user interface design in a way that it can be performed as part of an iterative design process. Heuristic evaluation involves the participation of a small group of usability experts, who ***examine the interface and judge its compliance with recognized usability principles*** (so called usability heuristics).

MUSiC performance measurement method (MacLeod et al., 1997): This technique ***measures usability in terms of task performance***, i.e. the achievement of frequent and critical task goals, by particular users, in a context simulating the work environment. This approach provides both measures and design feedback.

3 Comparative Analysis

According to its concept, the evaluation of usability has to be placed in the given context of use addressing user characteristics, jobs and task elements, hard/software, and the organizational, technical and physical environment. In order to gather information to the extent to which the techniques enable different perspectives for evaluation, each of the techniques has been detailed to the procedure evaluators have to perform to successfully apply the respective technique. Since interactive systems are socio-technical systems usability problems do not necessarily arise at the level of functionality of a software system, but rather at the user's or organization's level, however, due to technical deficiencies or disturbances in the social setting. The analysis of the procedure to be followed for applying each of the selected evaluation techniques has led to the identification of (i) the activities to be performed in a certain sequence, (ii) the tools that have been developed to support the particular evaluation technique, and (iii) the persons (roles) that have to perform the evaluation. These data provide insight into the level of support that techniques offer for each of the perspectives.

Cognitive Walkthrough requires analysts to think about the mental processes of users. It focuses on the cognitive perspective with some limitations: Since end users are not involved, the validity of the analysts′ assumptions about the mental models of the end users are not compared to the end users' mental models. Furthermore, the technique does not offer tools that support the analysts in how to perform the evaluation. The focus on the organizational perspective is also limited, since the designer is responsible for determining a correct sequence of actions, no matter whether there actually exists more than one (best) way of performing the task. The purpose of this technique is to suggest to the designer where the design is likely to fail and why. Thus, it also implements the technical perspective of evaluation.

EU-CON: Since all users of an organization (in effect, all users in the working environment) have to perform the evaluation and they can identify the tasks considered for the evaluation by themselves, this technique initially focuses on the organizational perspective. One disadvantage of the approach can be seen in the level of detail in task description that may vary from end user to end user. In particular, the technique does not provide a particular method / tool to describe the tasks in detail. It is up to the users to decide how and to what level of detail they describe their task. Through the participation of end users, and questions collecting information about end users, the cognitive perspective is considered, too. The discussion between end users and designers to improve the interface and the support through the handbook for engineering (which provides suggestions for improving the user interface to compensate for specific categories of problems) finally lead to the inclusion of the technical perspective.

ISO 9241 Evaluator: This expert- and guideline-oriented technique evaluates the user interface focusing mainly on the technical perspective. The detected problems are related to software usability and hardly to the quality of work or user characteristics. It does neither take into account the cognitive nor the organizational perspectives in the course of evaluating interactive systems.

Heuristic Evaluation: This task- and user-independent technique directs the attention mainly to the characteristics of the interface. Through checking the interface against heuristics it focuses on the technical perspective whilst evaluating interactive systems. Although in the debriefing session the outcome of the evaluation is discussed and suggestions for improving the interface can be gathered, no tools have been developed to support the aforementioned processes or the re-design.

MUSiC Performance Measurement Method: The specification and identification of the context of use provide adequate support for capturing the organizational perspective. The development of user profiles and the participation of end users in the course of the evaluation takes into account the cognitive perspective. The technical perspective is addressed through performance data, such as task effectiveness, time for task accomplishment for each task, snag, search and help times, etc. No direct support for improving the interface is given.

4 Conclusions

This paper has introduced a method and a case to compare several techniques for evaluating the usability of interactive systems. Although usability is a property of the overall interactive system, the results of the analysis show clearly, that the focus of attention of the existing evaluation techniques is usually on specific elements within the overall system, and not on all three of the perspectives or a combination of them that characterize it (i.e., the cognitive, organizational and technical perspective).

Most of the techniques have deficiencies in capturing the organizational perspective - the global and individual organization of work tasks. In particular, few integrated tools exist for the detailed description of the task and determination of the organizational settings, although several approaches for task analysis exist.

Within the preparation phase of most of the evaluation techniques, user characteristics should be collected. Unfortunately, information to systematically support this activity in a transparent way is lacking. Additionally, little is said about techniques to determine representative users or how to adapt techniques stemming from behavioral and social science. Further work is necessary to overcome the identified deficiencies. For example, techniques should be enriched to integrate information about the task and the organizational settings:

Techniques and concepts stemming from occupational psychology could be considered to achieve this goal. Techniques to reflect differences in mental models between designers and end users could enable the migration of the cognitive perspective into the evaluation process.

In order to bridge the gap between design and evaluation, isolated usability problems have to be considered in the context of the design knowledge. Only through the development of underlying theories dealing with human-computer interaction, presenting clear analytical definitions of usability principles, multi-perspective usability evaluation techniques can be developed. They enable placing usability in the natural context of use consisting of the users, their jobs and tasks, their hard- and software, and the organizational, technical and physical environment.

References

EU (1990): Directive 90/270/EWG on Man-Machine Communication, in: Directive 29.5.1990 about Safety and Health of VDU-Work, Official EC-Newsletter, 33, L156, Minimal Standards, Par. 3, p. 18, 21.6.

Frese, M.; Brodbeck, F.C. (1988): Computers in Office and Administration: Psychological Knowledge in Practice (in German); Springer: Berlin, Heidelberg.

International Standardization Organization DIS 9241-11 (1997): Ergonomic requirements for office work with visual display terminals (VDTs), part 11: Guidance on usability.

Lewis, C.; Wharton, C. (19997): Cognitive Walkthroughs, in: Helander M.; Landauer, T.K.; Prabhu, P. (eds.): Handbook of Human-Computer Interaction (2nd ed.); Elsevier Science B.V.; pp. 717-732.

MacLeod, M.; Bowden, R.; Bevan, N.; Curson, I. (1997): The MUSiC performance measurement method, in: Behaviour & Information Technology, 1997, Vol. 14, No. 4/5, pp. 279-293.

Nielsen, J. (1994): Heuristic Evaluation, in: Nielsen, J.; Mack, R.L. (eds.): Usability Inspection Methods, John Wiley & Sons, Inc.: New York, pp. 25-62.

Opperman, R.; Reiterer, H. (1997): Software evaluation using the 9241 evaluator, in: Behaviour & Information Technology, 1997, Vol. 14, No. 4/5, pp. 232-245.

Stary, Ch. (1996): Interactive Systems: Software Engineering and Software Ergonomics (2nd ed., in German); Vieweg: Braunschweig/Wiesbaden.

Stary, Ch.; Riesenecker-Caba, T.; Flecker, J. (1998): Implementing the directive for VDU work - the EU-state of the art, in: Behaviour & Information Technology, 1998, Vol. 17, No. 2, pp. 65-81.

Totter A.; Stary Ch. (1997): Re-Thinking Function Allocation in Terms of Task Conformance, in: Proceedings of the 1st International Conference on Allocation of Functions, ALLFN'97: Revisiting The Allocation of Functions Issue: New Perspectives, Oct. 1st-3rd 1997, Vol. 1, pp. 135-142.

THE ROLE OF TASK DRIVEN DESIGN FOR THE INTEGRATION OF PERSPECTIVES

Carmela Serfaty Architect DPLG
ADEST Workspace architecture 61 Rue de la Garenne 92310 Sèvres France
Tel : 33 1 45 34 74 07 Fax : 01.45.07.84.76 e-mail :serfaty@club-internet.fr

1 Project overview

Projects in architecture feature an interdisciplinary effort between geographically spanned groups of participants involved in the process of architectural design, such as :

I. Client based groups and ergonomists that define, alone or together with the architect, the project's requirements

II. Engineering firms, quantity- surveyors, architects, interior designers, decorators, computerised- design groups, secretarial pools, documentation centres, contractors and other partners that collaborate in preparing joint design proposals.

How do we approach the client and the project?

2 End user and task analysis : The participatory and global design

Technological shifts have changed working structures and techniques dramatically making it impossible to approach the problem with the prevalent linear logical method. At the same time architectural traditional techniques of individual or group space allocations as well as mainly task driven design are proving insufficient. We more and more experience the necessity to apply the global complex approach methodology, tools and techniques. These tools include knowledge drawn from human sciences such as sociology, psychology and business administration but also from mathematical models, statistics as well as computer sciences. These allow both new observational methods and new design techniques more participatory, global and interactive in order to understand workflow, company structure, living conditions and environments of the working world.

In the process of design we start by

1. task analysis

2. observation of working positions and overflow

3. participatory discussions

4. programming

5. designing workspace

The more we analyse working models the more we tend to envision offices as places for computer augmented co-operative work in co-operative environments in which space sharing, meeting and transaction are the most critical features together with mobile, wireless and networked technology.

These features go along with new construction techniques using new materials such as up-graded glass to allow more light, blur the boundaries between exterior and interior settings and atmospheric conditions. Master multifunctional use of space etc,.

New construction techniques must be looked for in order to allow mobile technology to function as glass permits voice of freeset telephones to be tracked out the building; creating a major security problem. On the other side, concrete slabs between floors stops voice transmission. We must look for “softer”, “intelligent” materials to comply with new requirements of virtual and mobile technologies. Office buildings and environments have to be completely rethought in terms of construction techniques as well as in terms of integration in physical, social and urban environments.

3 Telecommunicated, computer-aided and mediated design.

Design partners should participate in a collaborative computer-mediated design process augmented with advanced computer aided design technologies. New technologies help emerging innovative design environments. These environments are ones in which the aggregation, sharing and understanding of intellectual resources and expertise from a diversity of peoples and disciplines act as a major factor in a collaborative design process aiming at a joint design proposal. On the other hand, aided design technologies also allow creating very subtle working environments that augment creativity and productivity.

Computerised technology in the form of programs such as Autocad, Photoshop, scanning tools that permit to scan materials and photos of architectural environments permits more skills and allow a better mastering and mediating of the project.

Virtual reality computerised techniques open new perspectives of studying a project, making alternative offers and showing projects to clients in an interactive process of designing with the client or with other partners.

4 " The studio of the future " : Integrating perspectives

Unfortunately the actual state of the art does not allow all these applications as many firms and contractors are not computer-networked and are do not give enough thought to new technological innovations, leaving , thus, the process of design at a more traditional state and holding up emerging innovative design environments. Yet the necessity of looking for new techniques to analyse the working world as well as the possibilities offered by new technologies for underlying representations of this world and of working space have revolutionised both communication between design and building partners, representation of architectural projects and possibilities offered to comply with requirements of creating more productive and creative working spaces by using human computer interfaces.

5 Office planning projects in Paris

We will show and highlight these points by architectural projects of office planning in Paris :

1 VARIOUS

The office as a meeting place- The most critical space is allocated to informal meeting of workers from all grades

2 AKZO NOBEL

Space sharing and multipurpose space

The "nomadic office" – spaces and furniture can be interchanged to serve multipurpose and multifunctional working techniques.

3 PROBTP

Creating work- enhancing settings on computers

4 COPROSA

The flexible office – the design less office. Technology must be implemented to allow frequent changes of structure.

6 Alternative design alternative offices alternative environments

The process of design has become a comprehensive and complex process on the boundary between human sciences approach, human observation methodology and computerised sciences. It is the interface that allows us to both understand and create more valuable environments for the working world. One that blurs the boundaries between living and working environments as well as those between the material and immaterial office environment distinguished by free set telephone technology, "passage" between materials such as tables or walls and implementations such as computerised table- screens or wall- screens and why not electronic beam screens instead of walls? These technologies seem to be most adapted to understand and create environments in which technology is ubiquitous and embedded. These subtle environments are the true interface between man and computerised technologies and remain in pace with global environmental issues. They seem most adapted for corporations searching to breakthrough technological shifts, competitiveness needs and environmental issues into the new era of knowledge.

7 References

Morin, E. *Appréhender et comprendre la complexité et la pensée complexe.* Colloque La complexité et la pensée complexe, des atouts pour les organisations françaises. Paris 1998

Egea, H. *Le livre de l'auto-management ou l'entreprise adaptative du XXI siècle.* Editions Systems 1996

Streitz, N. A. & all. *Cooperative buildings.* Proceedings of CoBuild'98. Berlin : Springer 98

How Software Engineers deal with Task Models

Peter Forbrig

University of Rostock, Department of Computer Science
Albert-Einstein-Str. 21, 18051 Rostock Germany

1 Model-Based Development of Interactive Software

Up to now, there have been different points of view on how to develop interactive software. On one hand, the community of traditional software engineers insisted and part of it still insists in a more or less complete and more or less formal specification of the functionality of an application in the course of design that has finally to be implemented. After finishing these steps the specification of user interfaces is performed, thus leading to a more or less complete integration of the user interface into the application. As an example have a look at the papers concerning UML. On the other hand the community of software ergonomists mostly pursues a completely orthogonal strategy: The user interface is designed and specified before the functionality of an application is going to be specified. As a consequence, the desired functionality of a software system has to be derived from more or less complex interaction feature and modality specifications.

We are going to show the mutual relationships involved in model-based development based on end-user tasks (section 2). We then proceed with discussing the role of the existing and envisioned models (section 3). We will finally give a strategy for the implementation (section 4). Section 5 concludes the paper summarising the achievements and identifying topics for further research.

2 When and How to Focus on a Model of the Dialog

In this section we are going to develop an understanding on how many steps have to be performed before a dialog or application model can be specified in the context of end-user tasks. Interactive system development that takes into account end-user tasks has to comprise some representation of these tasks. For instance, in the approach by [Stary et al. 97] interactive software development is based on the methodological integration of a task model, a user model, a problem-domain model (data model), and an interaction model. The design process is a loose order of specification activities and mutual adaptation procedures in and between different models. This approach still enables designer to start either with the interface specification or the data modelling activities, based on a task model.

Now the question is: Does a task model (and the related models) reflect the existing tasks (and interaction modalities) or does it have to reflect envisioned tasks, as claimed by Johnson et al. (1996)?

3 Existing and Envisioned Models

Although it is widely accepted to distinguish between existing and envisioned task models, the same constraint might not hold for the other models. The task, object, and user model have very much interrelations. They are not independent dimensions of a system. Each model depends on the other two, as the following examples demonstrate:

i. A task depends on the user who has to perform it and the objects (tools) which are available.

ii. A multimedia application cannot be built without tools (media players).

iii. The characterisation of the skills of a user depends on the tasks she or he has to perform and the objects (tools) she or he has to use. An expert in data base development has not to be an expert in compiler construction.

Once we assume that the development of interactive systems might be based on envisioned task models, the consequences of this assumption have to be elaborated. Due to the close interrelationship of the models each modification of a model has to be reflected by the other once. As a consequence, we do not only have to distinguish between an existing and envisioned task model, but also between an existing and envisioned user and object model (see also Forbrig et al (1997), and figure 2).

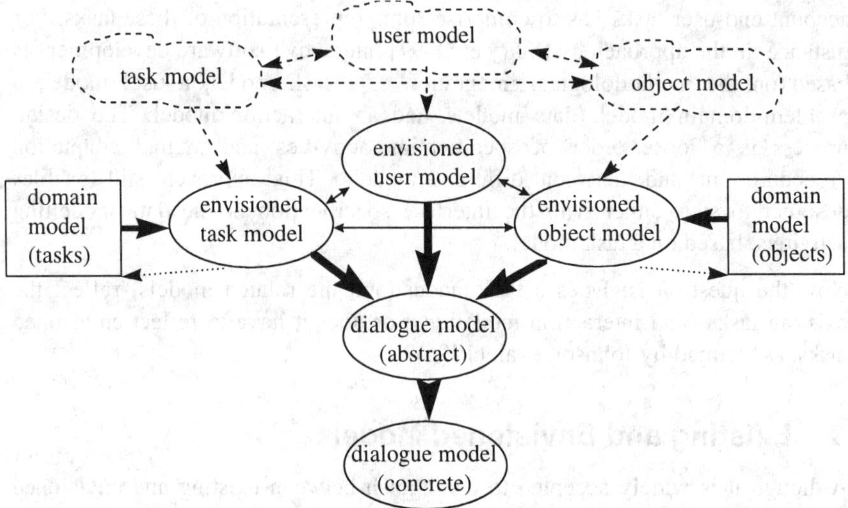

Figure 1: Relationships between existing and envisioned models.

In addition to the different kinds of each category of models different terms according to the previous discussion are used. The role of the previous problem domain model is now played by the object model. The problem domain model is now defined in a broader sense. It contains tasks and objects. Originally it has been defined for objects exclusively. However, it makes sense to incorporate into the development of envisioned task models and envisioned object models. This could significantly increase the reuse of parts of existing models.

4 Models to Support Programming

In the following we discuss the relationship between design and implementation. Since the gap between the community of software ergonomic people and software engineers should be bridged, the treatment of the models has to reflect support to the development process itself. That means that the development of the user interface as well as the development of the application core should be supported. We will at first have a look at the task models.

4.1 The importance of tasks

A task model for the software developer is needed for proper understanding of the application domain at hand. A task is described by its goal, a list of roles performing the task, a list of artefacts which were manipulated, and a list of tools supporting the task (**Task = (Goal, Role(s), Artefact(s), Tool(s))**). As software engineers we are also interested in some hints how the software can be developed by a programmer. For the development of user interfaces the importance of the envisioned task model has been already mentioned, e.g. [Johnson et al. 96]. When we take a deeper look at the tasks of a programmer in it becomes evident that the structure of the programming task is highly related to the existing task model. This situation occurs, since the implementation has to be performed for those tasks that have to be automated. Subtrees of the existing task model are equivalent to subtrees of the task model of the programmer. The objects of the task model provide their names to classes in the programming model. It contains the information of the task, which have to be implemented by functions of the system.

4.2 The importance of objects

We have already mentioned the object-oriented metaphor. Our approach is a unified strategy of structured and object-oriented ideas. Focussing our first analysis on task descriptions does not imply to ignore the importance of defining objects: There are special relationships between artefacts, parts of artefacts and tools, which are specified within the object model. These relationships can be extracted from the task model but they can also guide the development of the task model, in case they already exist.

4.3 Relations between models

Figure 2 tries to visualise the relations between the suggested models during software development.

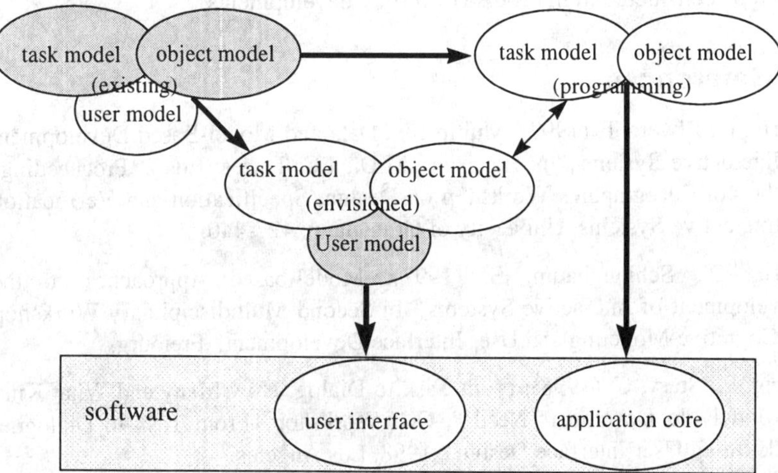

Figure 2: Relations between all suggested models.

The task model and the object model for programming are mainly influenced by the models of the existing work situation. The influence of the envisioned models is very limited. It is almost completely restricted to the rough structure and the envisioned user model which has much more influence to programming than the existing one.

5 Conclusions

In this paper we have revealed the origins and different applications of model-based development approaches with respect to task-oriented user interface and general software development. The gap between function-oriented and user interface-oriented approaches has been bridged for object-oriented development techniques. The crucial issues in this context that have been addressed were:

1. What type of models are required, and how do these models have to be related mutually to enable seamless software development.
2. How can the programming task be supported through those models that have been defined in the preceding design phase.

A framework of task models, object models and user models was introduced where each of the models has an existing, an envisioned and a programming dimension. This multiple dimensions require tools enabling the manipulation of all these models along the different phases of software development. These tools should also enable to propagate changes from one model to the other models. In order to achieve this goal algorithms checking consistency and completeness have to be considered in the course of further developments.

6 References

Forbrig, P.; Elwert, T. (1997): Multimedia Data and Model-Based Development of Interactive Systems,' in: Harrison , M.D.; J.C. Torres (Eds.), Proceeedings of the 4th Eurographics Workshop on Design, Specification and Verification of Interactive Systems, University of Granada, p. 427-440.

Forbrig, P.; Schlungbaum, E. (1997): Model-based Approaches to the Development of Interactive Systems," In Second Multidisciplinary Workshop on Cognitive Modeling and User Interface Development, Freiburg.

Forbrig, P.; Stary, C. (1998): From Task to Dialog: How Many and What Kind of Models do Developers Need ?, CHI Workshop "From Task to Dialogue: Task-Based User Interface Design", 1998, Los Angeles.

S. Wilson, P. Johnson (1996): Bridging the Generation Gap: From Work Tasks to User Interface Design, in Proc. of CADUI'96, Namur, Belgium, 5-7 June 96, pp. 77-94.

Bringing In the Social Perspective: User Centred Design

Jan Gulliksen
Department for Human Computer Interaction, Uppsala university, Lägerhyddsvägen 18, SE-75237 Uppsala, Sweden. and Center for user-oriented IT-design (CID), Royal Institute of Technology, Stockholm, Sweden.

1 Introduction

The social setting of end users does not only influence their task performance and attitudes towards computer systems but also their willingness to contribute their knowledge actively to the design of artefacts. Standish Group [Dennis, 1995] analysed IT-development projects in US showing that 31,1 % of the projects were aborted without having accomplished their goals, 52,7 % were completed with serious delays and at much higher costs (in general the costs increased with 189 %), and 16,2 % were completed within the budget and time frame. The most important factors for success in the projects that where completed within the timeframe were; user participation, management support, a clear requirement specification and good planning.

Successful software development requires *usable* systems. ISO 9241 defines Usability as; *the extent to which a product can be used by specified users to achieve specified goals with effectiveness, efficiency and satisfaction, in a specified context of use* [ISO 9241].

To design systems that are usable in a specific use situation, active involvement of representatives of the user population is essential, but no guarantee. Therefore user centred design is a necessity for designing usable systems. Our view on *user centred design (UCD)* of usable systems is based on ISO draft international standard on human centred design process for interactive systems [ISO 13407] and Gould & Lewis principles for designing for usability [Gould & Lewis, 1985], that is:

- *Work controlled development*. Early focus on users and tasks. The designer must understand the users, their cognitive behaviour, attitudes and the characteristics of the work. Appropriate allocation of function between the user and the system is also important to prevent unnecessary control and preemption.

- *Active user participation* throughout the project, in analysis, design, development and evaluation. This requires a careful user selection process emphasising the skills of typical users, both:
  - Work domain experts (continuously through the development project)
  - Actual end-users (for evaluation of various design results)
- *Early prototyping* to evaluate and develop design solutions.
- *Continuous iteration* of design solutions. A cyclic process of design, evaluation and redesign should be repeated as often as possible. The evaluation process should mean empirical measurement in which experiments are performed with prototypes with which real users perform real tasks with the purpose of observing, monitoring and analysing the users' reactions and attitudes.
- *Multidisciplinary design teams*. Include a usability designer in the process.
- *Integrated design*. Continuous developments of the system, the work activity, help, education, organisation, etc. in the development work.

2 Common Problems with UCD

So, how come so many projects fail to reach their goals within the defined limit of time and budget? Several problems have been observed when having user participation in a development project [Gulliksen & Lantz, 1998] and we have classified them as:

- *Attitude problem.* Many system developers regard computer system development as an artistic occupation with an expressive task, or a task of breaking new technical limitations, rather than, as preferable, service in a work context.
- *Communication problems*. Trying to understand and interpret the worlds of the various roles involved in the development work, indicate a problem of communication. Power relations, group processes, communication languages, lacking time, ability and interest influence.
- *Methods and tools problems.* Although accessible to the public, methods and tools do not always support UCD and the design process.
- *Lacking time.* Iterations, as one of the fundamentals of UCD often tend to delay the project in the construction phase. Conversely, problems of

different types are generated by the fact that many large system development projects tend to last for several years.

- *Organisational problems*. Management seldom supports the allocation of sufficient time and resources for UCD. Adverse managerial influence, various conflicting power relations and the lack of minimal organisational support for usability-related work can present severe obstacles to this process.
- *Participants support*. There must be support for UCD both at the managerial level and at the user level. Time and resources need to be allocated to a project for the users to participate as much as necessary.
- *Competence problems*. The participants seldom have the knowledge, skill, special abilities or even interest in UCD. HCI knowledge is still difficult to apply in practical system development.
- *External aspects* Various unexpected incidents can disturb UCD work, e.g. processes of change, political or strategically important decisions. Differing interests can be represented, and conflicts may occur that requires external control.

Why do we face all of these different problems related to user participation and the usability problems with the resulting systems?

3 Overcoming Obstacles to UCD

From several studies of UCD in practice we can observe difficulties in adopting UCD throughout the entire lifecycle. Usability related activities are usually adopted late in the process, evaluating the product right before it reaches the market, mostly without possibilities of correcting the problems.

3.1 Iterations and Lacking Development Time

The difficulty adopting an iterative approach is an obstacle to UCD. One iteration must involve 1) a proper analysis of the user task and context, 2) a prototype design phase, and 3) a documented usability evaluation of the design prototype that produce evaluation results that need to be addressed in the following process. Merely claiming that usability aspects are considered, without performing documented evaluations with users, can not be regarded as iterative design.

When development projects are delayed, usability evaluations late in the process are very easy to exclude. Project management can therefore vote against iterative UCD for the fright of loosing control over the development times. Therefore, UCD must be adopted early in the process to be beneficiary and to prevent increasing development times later on.

3.2 Guidelines for User Participation and Selection

Despite extensive participation we see a lack of real communication and understanding between computer professionals and user representatives. The user population can be known or unknown, available or not, and different approaches needs to be adopted depending on the user population. The important thing, however, is that the users that participates are skilled, dedicated and willing to contribute to changes. Depending on the user population various ways of selecting and using the user population in the development projects can be adopted.

By providing guidelines on user participation we address several of the problems related to communication, skills and attitudes in a UCD project. These domain specific guidelines describe whom, where, when, how and why users should participate, thereby providing practical knowledge useful for project management and development work. The guidelines give practical advice on the processes in which the users are involved. Like for example that it is important to address all aspects from the users and communicate the decisions made based on the user aspects back to the users to keep the users confidence.

3.3 Communication Problems

How do we avoid having a HCI expert standing in one corner screaming? We propose the "Usability Designer" to promote UCD in system development projects. A usability designer should have extensive knowledge on human cognitive characteristics, abilities for aesthetic design, possibilities of understanding the work domain, some development tool knowledge and finally the social competence required to be a communication link between users and developers. Tool knowledge is essential to receive confidence with the developers. Understanding the work domain can only be achieved through participation in the preceding work modelling sessions. So far, in practice, this role has been shouldered by GUI-programmers, which unfortunately has lead to several of the above mentioned problems.

3.4 Prototyping for a More Efficient Development Process

Typically in in-house development projects, up to a year of analysis work precedes the actual user interface design process. We pursue early prototyping as a means for efficient user participation and promote better iterative design of usable systems. So far, according to our observations, early user interface prototyping can make the user representatives feel contributing and make several of the other modelling steps faster and more efficient.

3.5 Holistic Integrated Design

Changing the technology evidently cause changes for the work activities, the organisational structures and the human beings with their skills and expertise [Leavitt, 1958]. However, this observation has not lead to any efforts to meet these challenges when developing the technology. We are currently striving for integrated design by defining methods for organisational change and learning simultaneously when developing computer support for a work situation. This requires an interdisciplinary research setting.

4 Conclusions

Successfully adopted UCD has, when adopted, been regarded as an absolute necessity to arrive at usable systems. However, few projects really adopt UCD throughout the entire development process. Traditional in-house development projects have a well defined user population, participating as a part of their work. With the recent increase in network services for the public, the user population has become all the more difficult to specify and access. We need to develop new and efficient ways of capturing user requirements, performing user centred prototyping and usability evaluations for these new technologies. As researchers we here have the responsibility to educate software developers on UCD and enhance the possibilities for efficient user participation to be able to increase the maturity and awareness of these questions.

5 References

Standish Group, (1995) *The Chaos Report.* Dennis MA, USA, Available: http://www.standishgroup.com/chaos.html

International Organisation of Standardisation (1998) ISO IS 9241 *Ergonomic requirements for office work with visual display terminals (VDTs) Part 11 – Guidance on usability*. (International standard)

International Organisation of Standardisation (1998) ISO/DIS 13407 *Human centred design process of interactive systems.* (Draft international standard)

Gould, J.D., & Lewis, C. (1985). Designing for Usability: Key Principles and What Designers Think. *Comm. of the ACM, Vol. 38, No. 3,* pp. 300-311.

Gulliksen, J. & Lantz, A. (1998) User orientation – does it work or is it just an aggravation? *I^3-magazine No. 2,* March 1998, The European Network for Intelligent Information Interfaces.

Leavitt, H.J. (1958) *Managerial Psychology.* University of Chicago Press, Ltd. London.

Prosody and User Acceptance of TTS

Kevin H. Richardson
AT&T, Rm D2-2C34, 200 Laurel Avenue, Middletown NJ; USA 07748

1 Introduction

With the advent of automatic call handling, AT&T, and more specifically, the User Interface community, has devoted a great deal of time and resources to the development, modification, coordination and storage of recorded human speech. One solution to all these problems lies in the development of Text-To-Speech (TTS) technology. TTS holds the promise of not only eliminating the need for recorded voice announcements but also of providing a common AT&T "sound". The potential benefits are obvious. However, the traditional drawbacks of TTS technology are also obvious: It doesn't sound as natural as recorded human speech (and is occasionally difficult to understand) and users don't like it.

One of the keys to "naturalness" and user acceptance of TTS-generated announcements lies in its ability to adequately model the prosodic, or stylistic, elements of human speech (e.g., stress patterns, intonation, rate and rhythm). The psychological literature has demonstrated quite clearly that prosodic information plays an important role in both how natural a phrase sounds as well as how easy it is to recognize, understand and remember (Diehl, Souther, & Convis, 1980; Martin, 1972; Miller, 1981). Current research by AT&T Labs has resulted in much more understandable and natural sounding "next generation" TTS algorithms (see Beutnagel, Conkie, Schroeter, Stylianou & Syrdal, 1998). In order to successfully implement TTS in place of recorded announcements, two studies were conducted to understand the prosodic differences between traditional TTS, AT&T's "NextGen" TTS, and recorded human speech.

2 Customer Rating Study

The first study determined how successful researchers at AT&T Labs have been in bridging the "prosody gap" between Text-To-Speech and human speech. Thirty-nine participants between the ages of 16 and 70 with no reported speech or hearing disorders were recruited from a local shopping mall to take part in the study. Participants were instructed that their task would be to *"listen to some different voices that might be used for an automated telephone service and tell me what you think of the* ***voice*** *(not the automated service)."* Participants were also told that they would be making a test phone call to each of three different automated phone services and that after each call they would be required to answer a short survey regarding their overall impressions of the recorded voices.

Using a 3 (Voice: Old TTS, NextGen TTS and Natural) X 3 (Task: Card Call A, Dialing Instructions, and Card Call B) within-subject design, participants completed each of the three phone service tasks in each of the following three sets of stimuli for a total of 9 trials:

1. "Old" TTS – formant synthesized speech of the type typically encountered in interactive voice response telephony systems.
2. "NextGen" TTS – concatenative synthesized speech, derived from natural speech as recorded by a professional voice talent (DM), currently under development at AT&T Labs, and
3. Natural Speech – recorded by the same professional voice talent (DM) used to generate the NextGen TTS.

After each task, participants rated the "voice" they heard on five scales:

a) Overall Quality,

b) Listening Effort,

c) Intelligibility,

d) Naturalness, and

e) Pleasantness.

Results of participants' ratings on the 4 point Mean Opinion Scale (MOS) in response to the question, *"Overall, how would you rate the voice you just heard?"* are shown in Figure 1. Rating score differences between all sets of stimuli were significant, suggesting that participants preferred the NextGen TTS over the Old TTS but still prefer naturally recorded speech when forced to make a comparison.

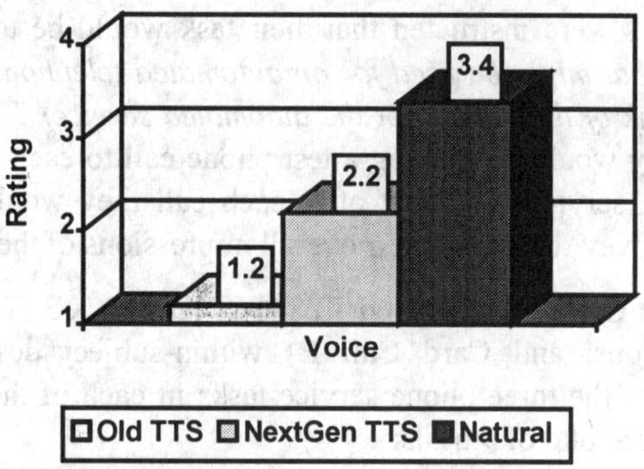

Figure 1. Overall voice quality ratings: (1) Poor, (2) Fair, (3) Good, and (4) Excellent.

Results of participants' ratings on the 5 point scales of Ease of Listening, Intelligibility, Naturalness and Pleasantness collapsed across scale and task are shown in Figure 2. Rating score differences between all sets of stimuli were significant. Participants clearly perceive the NextGen TTS as being significantly easier to listen to and more intelligible, natural and pleasant than the Old TTS, though significantly less so than naturally recorded speech.

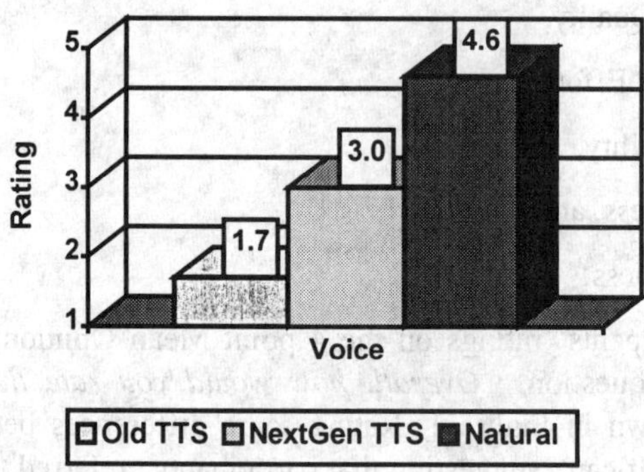

Figure 2. Five point Ease of Listening, Intelligibility, Naturalness & Pleasantness scores collapsed across scale and task.

3 Acoustic Analysis Study

Given that each set of stimuli delivered the same linguistic information and the well-delineated differences in participant ratings, an acoustic analysis of certain prosodic elements was conducted in order to uncover links between prosodic differences and caller preference. The second study examined three acoustic parameters:

1. Overall Speaking Rate – measured in words per second,
2. Inter-Phrase Pause Length – pauses at phrase boundaries (e.g., commas),
3. Inter-Sentence Pause Length – pauses between sentences.

The ten announcements most frequently heard by the participants in Study 1 were examined. The stimuli varied in length between 2 and 33 words and were composed of intonational phrases (set off by commas) and/or multiple sentences.

Results of the speaking rate analysis showed that the naturally recorded speech was produced at a slower rate (2.6 words per second) than the NextGen TTS (2.8 words per second) and the Old TTS (2.9 words per second). The differences between stimulus sets, while in the expected direction, were not significant.

Results of the inter-phrase pause length analysis revealed large, significant differences between the stimuli, as shown in Figure 3.. The mean pause length between phrases was significantly longer for the naturally recorded speech (260 msec) than for the NextGen TTS (180 msec) or the Old TTS (100 msec).

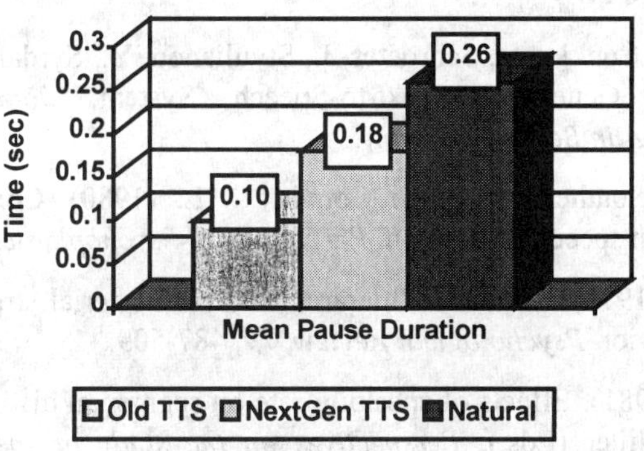

Figure 3. Between-phrase pause duration.

Results of the inter-sentence pause length analysis also revealed large differences (naturally recorded speech - 550 msec, NextGen TTS - 300 msec, Old TTS - 210 msec). It should be noted, however, that due to the small sample size (N=8) and the method used to generate the NextGen TTS multi-sentence announcements, it was not possible to reliably determine significance.

4 Discussion

Although test subjects still prefer announcements recorded from natural speech when forced to choose, the results of Studies 1 and 2 suggest that AT&T's new text-to-speech has clearly advanced beyond the traditional TTS, currently found in many telephone services. While these results suggest that work remains to be done, it is unclear that the goal of TTS needs to be announcements indistinguishable from natural speech. One can argue that user acceptance of TTS is a function, not only of its fidelity to natural speech, but also of the types of advanced and personalized services that such technology affords. With this in mind, the question facing telephony service developers becomes, not, "Do we have a TTS technology that is indistinguishable from natural speech?", but, "Is our TTS natural-sounding ***enough***, given the types of features and services we are able to provide, that callers will accept it.". This is not to say that the practice of employing recorded natural speech can be abandoned. There are certainly telephone services for which either naturally recorded speech or AT&T's NextGen TTS would be most appropriate (e.g., 800CALLATT vs. an Email Reader service). Callers do, after all, prefer naturally recorded speech, as the results of Study 1 demonstrate. The trick, will be to determine which one to choose.

5 References

Beutnagel, M., Conkie, A., Schroeter, J., Styulianou, Y., Syrdal, A. (1999). The AT&T Next Generation Text-to-Speech System. *Joint Meeting of ASA/EAA/DAGA in Berlin, Germany.*

Diehl, R. L., Souther, A. F., & Convis, C. L. (1980). Conditions on rate normalization in speech perception. *Perception & Psychophysics, 27*, 435-443.

Martin, J. G. (1972). Rhythmic (hierarchical) versus serial structure in speech and other behavior. *Psychological Review*, *79*, 487-509.

Miller, J. L. (1981). Effects of speaking rate on segmental distinctions. In P. D. Eimas & J. Miller (Eds.), *Perspectives on the study of speech* (p. 39-74). Hillsdale, NJ: Erlbaum.

Benefits of Internationalizing Software and Successful On-Demand Multilingual Web Publishing

Tiziana Perinotti
TGP Consulting USA Tiziana Perinotti
TGP Consulting, founder, 365 West Charleston Road,
Palo Alto, CALIFORNIA 94036 USA
Developer of the award winning Silicon Valley Localization Forum at:
http://www.TGPConsulting.com
mailto:info@TGPConsulting.com
voice: 650.494.7485, fax: 650.493.6848

When you translate a document, the end result is a passive object. The translated document doesn't perform any function.

On the contrary, when you translate a SW program, the resulting object is active because it can process the data (text and numerals) in another language.

Therefore, SW Internationalization and Localization always imply two fundamental processes:

- the conversion of the application's UI (all the components displayed on the screen);
- the conversion of the program's logic so that it can handle the special requirements of a target writing system.

Internationalization (I18N) is the process of enabling a SW product so that it can support all the linguistic features of different locales. In other words, I18N is the set of processes, procedures, tools and technologies that facilitate the localization (L10N) process.

Then, the Localization of an Internationalized SW product becomes the specific adaptation to a locale or multiple locales/writing systems without the requirement to re-design the original US product.

An Internationalized SW application is an application whose UI is still in English but with built-in support for all the languages/writing systems for the countries where the application will be sold.

I18N built-in support has to do with many SW Development issues that no translation process can resolve. Some of the most important Engineering issues are: support character sets for all the locales where the product will be sold; support double-byte for languages such as Japanese, Chinese, Korean; sorting and case conversions; extract all of the text strings and other UI's objects' position and size from the code and put them into resource files so that they can be translated; design menus, dialogs, screens, and status bars that can accommodate the translated text. For a complete list of engineering issues when developing Internationalized SW, go to: <http://www.tgpconsulting.com/articles/check.html>.

There are specific guidelines, techniques and procedures that allow product teams, Testers and SW Developers, to avoid the typical mistakes when writing code for an Int'l audience and develop SW which is Internationalized from the very beginning. There are also tools that will facilitate the developer and tester's tasks. You will find lots of information about these tools at the Silicon Valley Localization Forum website at: <http://www.tgpconsulting.com/tools/index.htm>.

Companies will avoid costly SW retrofitting when they implement a sound Internationalization and Localization strategy for their product/service line.

Why is it so important to focus on Internationalization first, before doing any translation?

Because of very compelling reasons:

1. Time to Market: when the English base product is enabled to support all the locales, the specific Localization process only requires the translation of the User's Interface and the overall Translation cycle is much shorter (one development cycle instead of multiple development cycles, each one for a different locale). Localized products can be released right after the US version ships. And you can release your Int'l English version everywhere in the world where a fully translated product is not necessary.

2. Maintenance: from a product evolution, it's less expensive and easier to maintain one source code set instead of multiple source code sets, each one for a different locale.

3. Revenues: the goal is to reduce the gap between the date you ship the US product and the date you release the localized versions to generate as much revenues as possible. If the gap becomes too big, you will loose the window of

opportunity to obtain the extra sales spike which is typical of any successful product lunch, and you will loose revenues for the time period you don't have localized products available.

The revenue loss can be substantial and unfortunately there are still many companies, from startups to large corporations, that are unclear on the benefits of Internationalizing their products as part of their SW Development process, thus loosing a lot of revenues and paying for huge opportunity costs.

Important concepts to be successful global web publishers include.

It's a paradigm shift. It's like real time TV, constant change/update. Content creation is no longer static, and information doesn't last long. Cost/profitability model is different than traditional Localization of SW applications.

You must have a super partnership with your vendor. You need to rely on a localization vendor completely integrated into your web development cycle and with the right skills set and attitude.

You must have top-notch Project Management. One web week is equivalent to at least two traditional SW development weeks. You need experienced project leaders, excellent communicators with years of simultaneous multilingual SW products' management.

What's your goal? Do you have a worldwide plan for product/services, distribution, marketing, launch, budget and revenue?

You must have technology and tools Based on your goal, you need the appropriate web tools for your lingua QA, Translation memory, Glossary, SW/Content Version Control, multilingual website maintenance and upgrades and archiving.

Think in terms of Locales. Have a plan for Local content production, meaning "in-the-country" resources, content-development plan and strategy.

Do you have a quality control mechanism in place? Do you have a clear process established, do you have a company-wide strategy, do the teams know their roles and responsibilities? If you don't, your chances of success will be really limited.

NOTE: For more information on how to develop and test Internationalized and Localized SW, go to the Silicon Valley Localization Forum at: http://www.TGPConsulting.com This site's objective is to help companies increase their revenues through cost effective SW Solutions for the Global Market.

Index Author

Index Subject